Essentials of Flow Assurance Solids in Oil and Gas Operations

Essentials of Flow Assurance Solids in Oil and Gas Operations

Understanding Fundamentals, Characterization, Prediction, Environmental Safety, and Management

Abdullah Hussein

ELSEVIER

Gulf Professional Publishing
An imprint of Elsevier

Gulf Professional Publishing is an imprint of Elsevier
50 Hampshire Street, 5th Floor, Cambridge, MA 02139, United States
The Boulevard, Langford Lane, Kidlington, Oxford, OX5 1GB, United Kingdom

ISBN: 978-0-323-99118-6

For information on all Gulf Professional publications
visit our website at https://www.elsevier.com/books-and-journals

Publisher: Charlotte Cockle
Acquisitions Editor: Katie Hammon
Editorial Project Manager: Rupinder K. Heron
Production Project Manager: Prem Kumar Kaliamoorthi
Cover Designer: Matthew Limbert

Typeset by STRAIVE, India

This book is dedicated to my parents and family in Egypt,
and to my wife and family in Canada.

Contents

CHAPTER

Oil and Gas Production Operations and Production Fluids

1

1.1 Introduction

Oil and gas were formed over millions of years from the dead bodies of plankton (tiny marine organisms). These prehistoric dead plankton and their naturally engineered product have dominated and shaped modern times into "the Great Oil Age," as McKenzie-Brown, Jaremko, and Finch titled their book on the subject [1].

Petroleum shapes our modern times in two main ways: first by serving as the main source of energy, and secondly by being the raw material for thousands of products that play critical roles in modern civilization.

Fossil fuels still constitute the major global source of energy, despite the recent advances in renewable sources of energy. Fossil fuels represent 85% of global primary energy consumption [2,3], with oil remaining as the most used source in the energy mix (33%–34%), as shown in Fig. 1.1.

The global consumption of and demand for hydrocarbons (oil and gas) outweigh that of other sources of energy such as coal, nuclear, and renewable energies, due to the fast-growing population and newly emerging technologies. The global energy demand is projected to more than double by 2050 due to the growth in population and economies [4]. The demand for gas is growing faster than the demand for oil, especially in developing countries [5,6].

Oil demand and price depend on global political, industrial, social, and environmental factors, such as wars, pandemics, climate change, rising or falling economies, and emerging industries. For example, in 2020 with the outbreak of the COVID-19 pandemic, primary energy consumption fell by 4.5%, the largest decline since 1945. The drop in energy consumption was driven mainly by oil, which contributed almost three-quarters of the net decline, although natural gas and coal also saw significant declines. The oil price (Dated Brent) averaged $41.84/bbl in 2020—the lowest since 2004 [7].

Besides being a main source of energy, petroleum also is the raw material for the petrochemical industry. Petroleum feedstock is used in petrochemical plants and turned into plastic to make essential products used in our everyday lives. Most refineries convert just 5%–20% of incoming oil into petrochemicals [8]. More than 6000 everyday products get their start from oil, including electronics, textiles, sporting goods, health and beauty products, medical supplies, and many household products [9]. All these products are shaping modern civilization—thus giving up on petroleum production is an issue with dire consequences.

Petroleum can be extracted either from conventional or unconventional resources. *Conventional resources* are discrete accumulations or pools of oil or gas, where the rock formations hosting these pools usually have high porosity and permeability and are found below impermeable rock formations. It is more

Essentials of Flow Assurance Solids in Oil and Gas Operations. https://doi.org/10.1016/B978-0-323-99118-6.00012-5

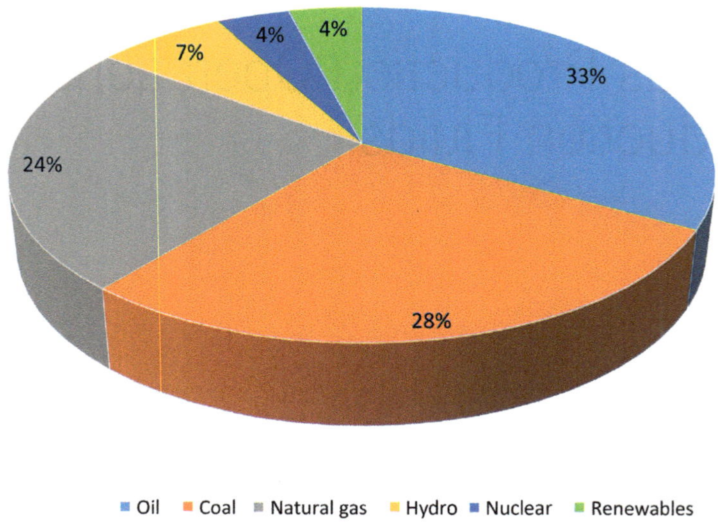

FIG. 1.1

Global primary energy consumption by fuel type.

Data from BP Energy, 2018, BP Energy Outlook 2018 Edition. https://www.bp.com/content/dam/bp/business-sites/en/global/corporate/pdfs/energy-economics/energy-outlook/bp-energy-outlook-2018.pdf.

straightforward and less expensive to extract petroleum hydrocarbons from conventional sources. *Unconventional resources*, on the other hand, are oil- or gas-bearing units where the permeability and porosity are so low that the resource cannot be extracted economically through a vertical wellbore and instead a horizontal wellbore is required, followed by multistage hydraulic fracturing to achieve economic production. Examples of unconventional oil resources include shale oil, extra heavy oil, tar sands, and bitumen [10]. Fig. 1.2 illustrates the difference between the conventional and unconventional resources.

The global petroleum resource potential is massive; conventional and unconventional hydrocarbon resources, at a ratio of about 2:8, amount to about 5×10^{12} t in total. With a current recovery percentage of conventional hydrocarbon resources of only 25%, and that of unconventional hydrocarbon resources being nominal, the oil industry life is still projected to extend for more than 150 years [11].

The global reserves seem to be increasing after taking in account the unconventional reserves that have been recently disclosed and adding them up to fullfil the energy demand. Fig. 1.3 indicates the countries with the highest proven reserves.

Due to its finite nature, the world's oil production will reach a maximum value when approximately half of the existing resources have been extracted, which is called the peak oil theory or the Hubbert peak theory, in which oil production takes a symmetrical bell shape, as depicted in Fig. 1.4. The theory predictions were partially correct; however, new discoveries of unconventional resources have caused this theory to fade. Peak oil does not necessarily mean running out of oil; it simply means that the yield of extraction, in economic and energy terms, gradually declines to the point that it is no longer logical to invest the huge amounts of financial resources needed to keep production increasing [12].

Moreover, according to the abiotic oil theory, peak oil theory may not be correct if the rate of oil formation is the same as the extraction rate.

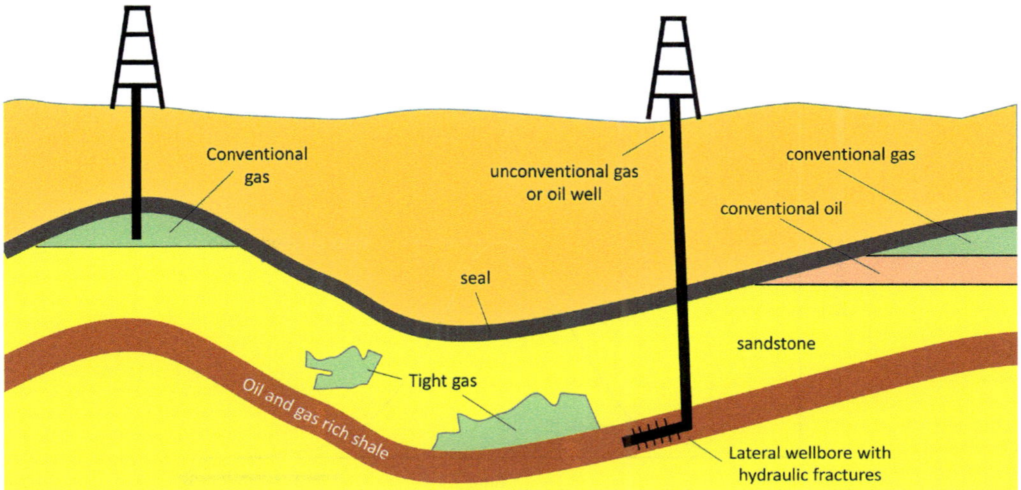

FIG. 1.2

Conventional vs unconventional petroleum resources.

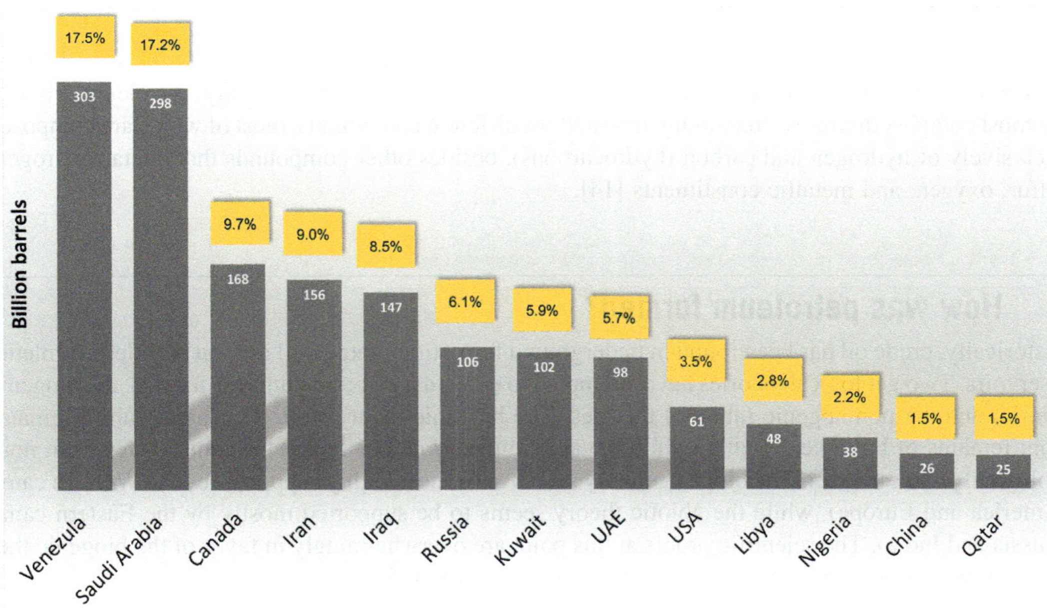

FIG. 1.3

The total proven oil reserves (in billion barrels) by country and their percentage of the global reserves. The numbers includes gas condensate, natural gas liquids (NGL), and liquid oil.

Data from BP, BP statistical review of world energy, in: Statistical Review of World Energy, vol. 68, 2019, pp. 1–62.

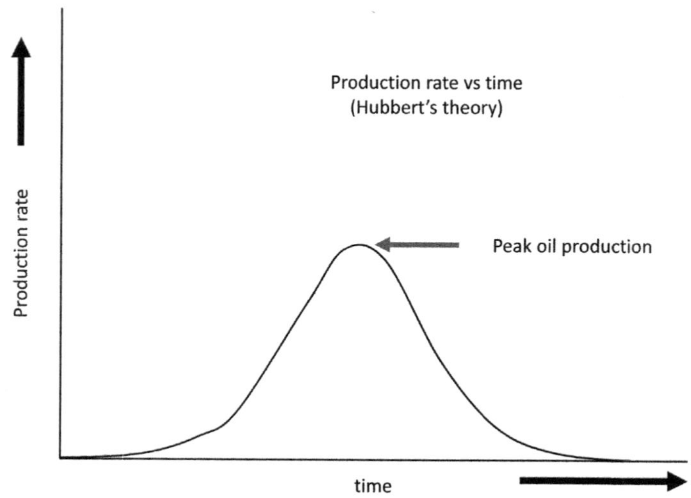

FIG. 1.4
Peak oil theory.

1.2 What is petroleum?

Petroleum can be defined as a gas or oil mixture of hydrocarbons extracted from the rocks of the earth by drilling down into a reservoir rock and piping it to the surface [13]. Chemically, petroleum is one of the most complex mixtures, containing thousands of different compounds, most of which are composed exclusively of hydrogen and carbon (hydrocarbons), besides other compounds that contain nitrogen, sulfur, oxygen, and metallic constituents [14].

1.3 How was petroleum formed?

Historically, crude oil has been found in underground formations entrapped in what is called petroleum reservoirs. Two geological theories have attempted to explain how petroleum was formed: the biogenic (biotic) and the nonbiogenic (abiotic) theories. The biogenic theory states that petroleum originates from remains of biological matter, while the abiotic theory claims that petroleum derives from non-biological processes [12]. The biogenic theory seems to be most highly supported by the Western camp (America and Europe), while the abiotic theory seems to be supported mostly by the Eastern camp (Russia and India). The scientific proofs at this point are overwhelmingly in favor of the biogenic theory, but there is nothing to prevent both theories being true [15].

1.3.1 The biogenic theory of formation of petroleum

A biogenetic origin of petroleum suggests that petroleum is a product of a long-term transformation of dead biological species subjected to high temperature and pressure. Fig. 1.5 illustrates the biogenic theory of the formation of petroleum. According to this theory, petroleum was formed millions of years

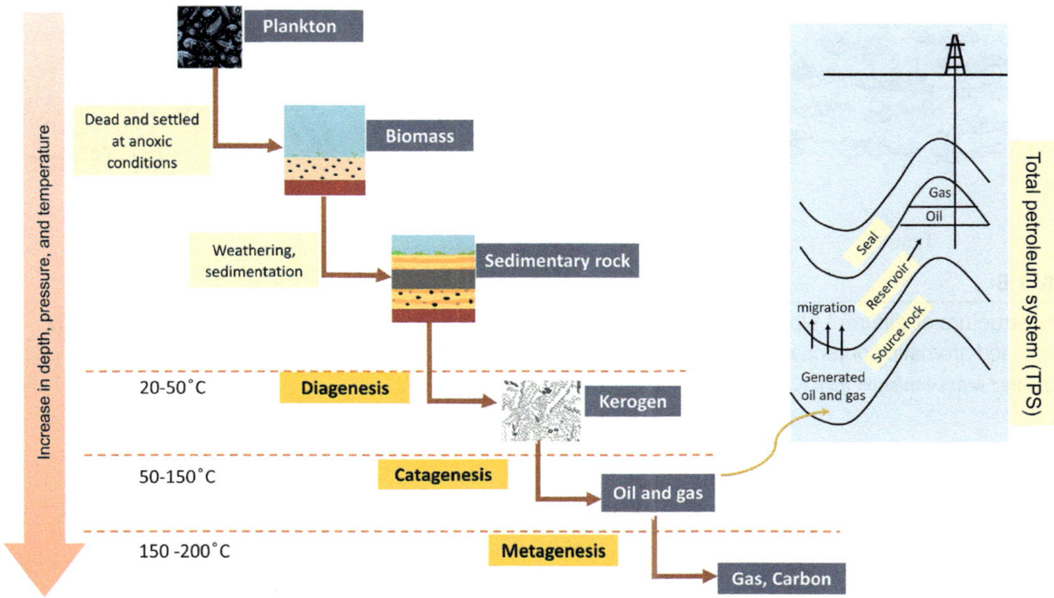

FIG. 1.5

Biogenic theory of formation of petroleum.

ago, in a marine environment [16]. In this theory, dead plankton, algae, and other microscopic species settled down in the stagnant and anoxic (oxygen free) conditions of the ocean floor, which favored the accumulation and preservation of these biomasses and prevented their decomposition. As a result, a thick layer of biomass or organic mush was formed [17]. Such large biomasses were buried with the mud, sand, debris, and plant and animal remains that have been carried in streams and rivers, flowing down to the ocean over millions of years, to spread and accumulate on the ocean floor/sea bed, forming what is called sedimentary rock [16]. Hence, these sedimentary rocks contain organic matter, and a sedimentary rock with sufficient organic material is known as shale [17,18].

The organic material accumulated in the sedimentary rocks then underwent multiphase transformations, as illustrated in Fig. 1.5, finally leading to the formation of fluid hydrocarbons. These transformation processes include

- **Diagenesis**: Diagenesis is the sum total of physical, chemical, and biological processes that occur in recently deposited sediments under mild conditions of temperature (20–50°C) and pressure that lead to the formation of kerogen [19]. Different types of kerogen are formed, depending on the contributing organisms: Type I forms mainly from algae, Type II from a mixture of algae and land plants, and Type III primarily from land plants. Fig. 1.6 illustrates the differences between the three kerogen types [20].

- **Catagenesis**: Catagenesis is the thermal cracking of the kerogen, which leads to the generation of most hydrocarbons, i.e., oil and gas [19]. Catagenesis generally involves heating in the range of 50°C to 150°C. The conditions of catagenesis determine the product, such that higher temperature and pressure lead to more complete "cracking" of the kerogen, progressively producing lighter and

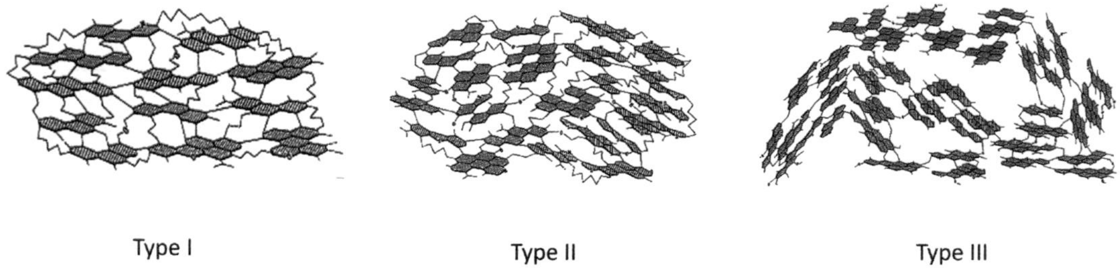

FIG. 1.6

The structure of different types of kerogens. Type I: low aromatic content and long aliphatic chains; Type II: increased aromatic content, medium aliphatic chains; Type III: high aromatic content and short aliphatic chains.

Reprinted with permission from F. Behar, M. Vandenbroucke, Chemical modelling of kerogens, Org. Geochem. 11(1) (1987) 15–24, https://doi.org/10.1016/0146-6380(87)90047-7. Copyright (1987), Elsevier.

smaller hydrocarbons. Generally, Type I and Type II kerogens are predisposed to generate oil (oil is produced by Type I kerogen, waxy oil is produced by type II kerogen), and gas is produced by Type III kerogen [21,22]. The processes of kerogen alteration are known as maturation [21].

- **Metagenesis**: Metagenesis is reached only at great depths, and it is the last stage of maturation and conversion of organic matter into hydrocarbons or into carbon [21]. Metagenesis occurs at temperatures from 150°C to 200°C. At the end of metagenesis, methane, or dry gas, is evolved along with nonhydrocarbon gases such as CO_2, N_2, and H_2S, as oil molecules are cracked into smaller gas molecules [19].

After formation, petroleum migrates from the rocks where it was formed (known as the source rocks) to adjacent porous rocks (known as the reservoir rocks), where it accumulates in large, economically attractive amounts. The structures in which oil and gas accumulate are called traps, and the whole system (source rock, reservoir rock, cap rock, overburden rocks, migration path) is called the petroleum system, or the total petroleum system (TPS).

After this outline of the biogenic theory of petroleum formation, a question emerges: were these accumulated volumes of plankton and tiny organisms sufficient to supply the enormous amounts of oil and gas that have been discovered?

The answer is yes: the organic matter produced by photosynthesis in the oceans is estimated to be sufficient to produce **11** million metric tons of hydrocarbon precursors annually [23]. A very small amount of this organic material preserved in sedimentary rocks each year, through geologic time, would supply all of the known oil and gas fields, plus many undiscovered giant fields [24]. The total amount of organic matter dispersed in the sedimentary rocks of the earth has been estimated to be about 2700 trillion metric tons; of this amount **50** trillion metric tons are dispersed petroleum hydrocarbons, of which **0.5** trillion metric tons exist in petroleum reservoirs [25].

1.3.2 Abiogenic theory of formation of petroleum

The abiogenic theory suggested early on that the hydrocarbons are of primordial origins, are as old as the earth itself as they arrived on earth from primitive solar nebula, and were trapped deep in the earth when the earth was formed long before the appearance of any form of life. This consequently expands the source of natural gases from crustal organic materials into primordial hydrocarbons from deep earth. The high pressure in deep earth favors the stability of methane and other hydrocarbons. Under proper geological conditions, the primitive abiogenic hydrocarbons can form natural gas reservoirs. These deep hydrocarbons are subsequently released by the mantle and migrate toward the surface, utilizing weaknesses in the crust such as plate boundaries, faults, and sites of meteorite impacts [22]. This theory was supported by iconic scientists such as Dimitri Mendeleev and Fred Hoyle.

Other mechanisms have been recently suggested to explain abiogenic petroleum formation: methane polymerization, the reaction of minerals with water at high temperature and pressure in the mantle (Eqs. 1.1, 1.2, 1.3), the conversion of carbon monoxide or carbon dioxide into hydrocarbons (Fischer-Tropsch synthesis) (Eq. 1.4), thermal metamorphism of minerals (Eq. 1.5), and serpentinization (Eqs. 1.6, 1.7, 1.8) [22,26,27].

The main mechanisms in this theory are

$$3FeO + H_2O \rightarrow Fe_3O_4 + H_2 \tag{1.1}$$

$$4H_2 + CaCO_3 \rightarrow CH_4 + CaO + 2H_2O \tag{1.2}$$

$$nCH_4 + nFe_3O_4 + nH_2O \rightarrow C_2H_6 + Fe_2O_3 + HCO_3^- + H^+ \tag{1.3}$$

$$nCO + (2n+1)H_2 \rightarrow C_nH_{2n+2} + nH_2O \tag{1.4}$$

$$FeO + CaCO_3 + H_2O \rightarrow Fe_3O_4 + CH_4 + CaO \tag{1.5}$$

$$Mg_{1.8}Fe_{0.2}SiO_4 + 1.37H_2O \rightarrow 0.5Mg_3Si_2O_5(OH)_4 + 0.3Mg(OH)_2 + 0.067Fe_3O_4 + 0.067H_2 \tag{1.6}$$

$$(2n+1)H_2 + nCO \rightarrow C_nH_{(2n+2)} + nH_2O \tag{1.7}$$

$$4H_2 + CO_2 \rightarrow CH_4 + 2H_2O \tag{1.8}$$

The main requirements for these mechanisms and theories are [28]:

- Adequately high pressure and temperature
- Donors/sources of carbon and hydrogen
- Thermodynamically favorable reaction environment

Evidence for the abiogenic petroleum formation theory includes [26,28–30]:

- The existence of methane on other planets of the solar system, meteors, moons, and comets.
- The fact that hydrocarbons are being evolved from the inner parts of the earth is evident from the presence of mud volcanoes.
- Flames seen during earthquakes.
- Abundant hydrocarbons in carbonaceous chondrites. Also the presence of bitumen and hydrocarbons in native diamonds, carbonado, and kimberlites could be taken into a consideration as evidence confirming the abyssal petroleum origin.
- The most convincing evidence for the inorganic theory is considered to be the occurrence of commercial quantities of oil in crystalline and metamorphic basement rock, contrary to the biogenic theory, which states that petroleum is formed in sedimentary rocks.

The potential resources containing this type of abiogenic petroleum are estimated to be much larger than biogenic petroleum on earth, and they can be found anywhere, provided that the required formation conditions are met [26,29].

1.4 Life cycle of oil and gas fields and stages of development

The oil and gas sector comprises three main parts with different activities: upstream (exploration and production), midstream (transportation, storage, and processing), and downstream (refinery, petrochemical, and distribution). Fig. 1.7 illustrates these three main parts and their activities.

There are five phases of the life cycle of oil and gas fields: exploration, appraisal, development, production, and abandonment. These stages are summarized in Fig. 1.8.

Exploration aims to identify and locate potentially viable oil and gas sources through geological surveys and drilling exploration wells to identify areas of potential interest. In the appraisal stage, the explored viable oil/gas sources and the field description are examined in more detail, infrastructure may be developed to access sites, and the next steps are planned. The development stage involves reviewing the documents (contracts, permits), outlining a conceptual development plan, preparing the design for the production wells; at the end of this stage, the first oil and gas will be produced. In production the oil and gas will be produced in significant and economically profitable amounts. When the oil and gas production is no longer cost-effective, the wells are plugged and abandoned and production facilities are removed; this is the last stage of the oil and gas field's life cycle [31,32].

The development and production planning are based on the expected production profile, which will determine the size of the facilities required to treat and dispose of the fluids and the number and phasing of wells to be drilled. A typical production profile is made up of three phases [33], as illustrated in Fig. 1.9:

1. Build-up period: during this period production wells are progressively brought on flow.
2. Plateau period: a constant production rate is maintained.
3. Decline period: all producers show a declining of production rates.

The main upstream operations during the field development of the production system involve, but are not limited to, exploration, drilling, completion, production, workover, and abandonment. These operations are discussed briefly in the next sections.

1.4.1 Exploration

This is the early stage in which companies try to locate and identify potential oil and natural gas resources through geological surveys.

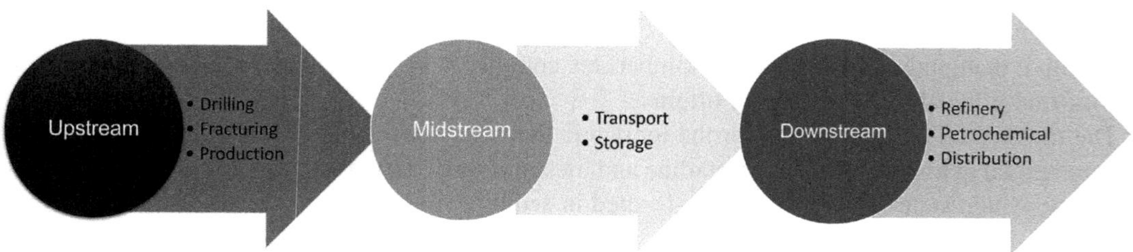

FIG. 1.7

Activities in oil and gas industry sectors.

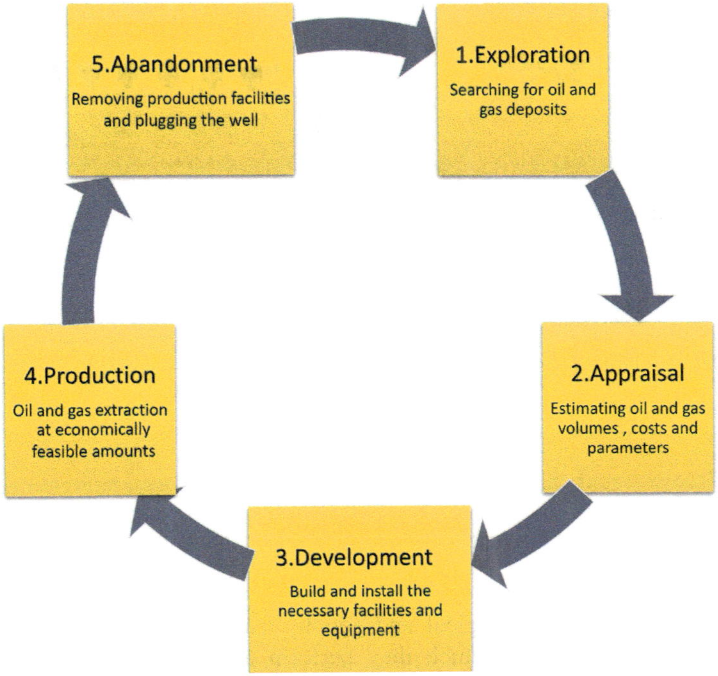

FIG. 1.8

Life cycle of an oilfield.

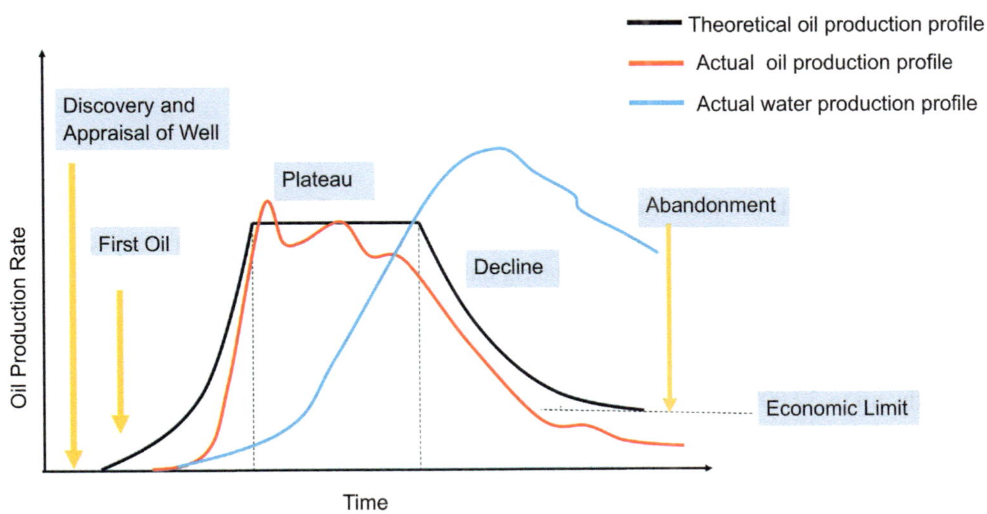

FIG. 1.9

Oil production profile.

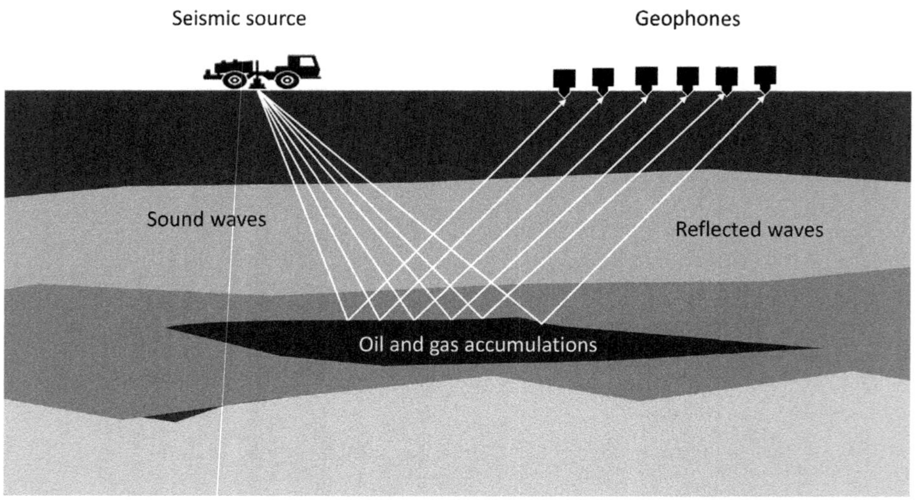

FIG. 1.10

Onshore seismic surveys.

After reviewing the existing geological and geophysical data, companies/governments pursue licenses and acquisition of land on which they perform technical surveys, i.e., seismic surveys (Fig. 1.10) and exploratory drilling, to identify and locate potential oil and gas deposits.

The government can seek investment for its own exploration or grant access for firms to explore, through direct negotiation or bidding processes. International companies may explore alone, or two or more companies may form joint ventures to explore together, with one company being appointed the operator. After identifying potential oil/gas sources, the companies will develop work plans for the next phases [32].

1.4.2 Drilling

After the oil/gas deposits are identified and confirmed, a drilling rig (Fig. 1.11) bores a hole through geological formations until it reaches the petroleum reservoir where oil and gas are trapped. The size of the borehole differs from well to well [34]. The rig daily rate will vary according to the rig type, water depth, distance from shore, and drilling depth. For onshore, it will be >\$100,000/day, and for deep-water offshore Gulf of Mexico, it can be very high—up to \$600,000 to \$800,000/day. The number of days will be a function of depth. For a usual depth up to 20,000 ft, we can assume 70 to 80 days, and for deeper depths up to 32,000 ft, a maximum of 150 days (values are from 2010) [35].

1.4.3 Completion

Well completion is defined as a single operation involving the installation of production casing and equipment in order to bring the well into production from one or more zones [36].

FIG. 1.11

Offshore drilling rig.

Photo credit: GeoffreyWhiteway; source: https://freerangestock.com/photos/65814/oil-and-gas-drilling-platform-with-iceberg.html, public domain.

1.4.4 Production

The production process starts when significant amounts of reservoir fluids are produced. Oil production usually takes place in different phases, namely primary, secondary, and tertiary recovery. The produced fluids will undergo further treatments, primarily including phase separation, chemical/physical treatment, reuse, or disposal. Finally the separated oil or gas is stored in tanks and is ready for sale.

Fig. 1.12 summarizes the different oil recovery processes [37].

1.4.4.1 Primary recovery

Primary recovery refers to a process in which the hydrocarbons in the reservoir trap are forced to the surface by the natural energy contained in the trap [38]. Such driving energy may be derived from liquid expansion and evolution of dissolved gases from the oil as reservoir pressure is lowered during production, expansion of free gas or a gas cap, influx of natural water, gravity, or combinations of these effects [39].

With time, the reservoir pressure will decline and hydrocarbon production rates become uneconomical. In this case pumps such as a rod pump, an electrical submersible pump, or a gas-lift installation can be used to reduce the bottomhole pressure or increase the differential pressure to increase hydrocarbon production; in this case, it is known as artificial lift.

In both types of primary recovery (natural flow or artificial lift), nothing is added to the reservoir to increase or maintain the reservoir energy or to sweep the oil toward the well [39].

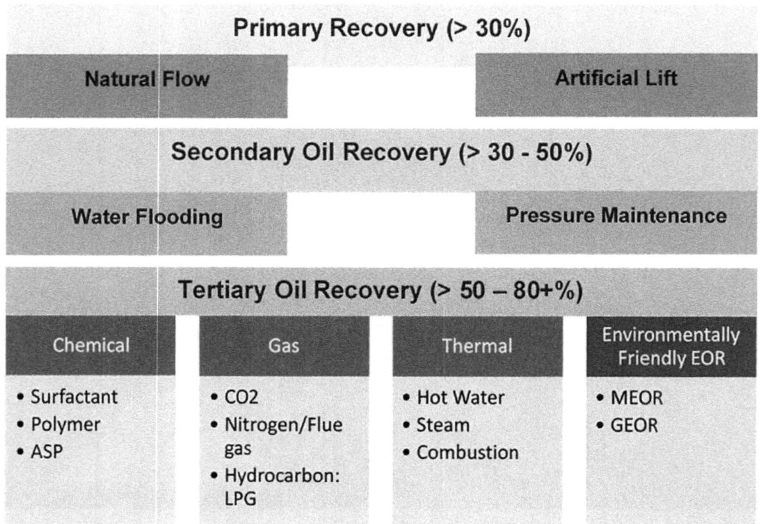

FIG. 1.12

Classification of oil recovery processes.

Source B. Haq, The role of microbial products in green enhanced oil recovery: acetone and butanone, Polymers 13(12) (2021). https://doi.org/10.3390/polym13121946.

1.4.4.2 Secondary recovery

With continuous hydrocarbon production, the reservoir pressure decreases, until at some point the reservoir pressure is insufficient to force the crude oil to the surface [39]. Then the secondary recovery methods are used to maintain the depleted reservoir pressure and improve hydrocarbon sweeping by injecting an external fluid, such as water or gas, into the reservoir through injection wells, which have fluid communication with production wells.

The gas (such as CO_2, N_2, or produced gas) is injected above the oil, whereas water is injected below the oil, forcing it upwards, commonly known as a waterflooding operation, using water sources such as sea water [38].

1.4.4.3 Enhanced oil recovery

Enhanced, also known as tertiary, oil recovery is used to restore the depleted pressure of the reservoir, displace hydrocarbons toward production wells, or alter reservoir fluids to improve hydrocarbon flow.

Tertiary oil recovery is usually applied before the secondary oil recovery techniques are no longer enough to sustain production, and when the reservoir has poor properties, or the crude oil has low API and is highly viscous [39].

Different methods are used in enhanced oil recovery, including chemical flooding (alkaline, polymer, surfactant), miscible displacement (CO_2, N_2, or hydrocarbon injection), thermal recovery (steamflood or in situ combustion), or microbial methods [40].

1.4.4.4 Crude oil and gas processing

The fluids produced from the well contain a mixture of oil, condensate, hydrocarbon gas, brine water, solids (sand, fines, scale, corrosion products, etc.) and other gases (N_2, CO_2, H_2S). This mixture is processed so that the fluid meets pipeline and production equipment specifications for fluid transportation, improving the quality, metering, and storage of the produced hydrocarbons, and for optimal operations at the refinery facilities.

The main operations for crude oil processing include phase separation and desalting, besides other operations to meet the required specifications of bottom sediment and water (BS&W), salt content, vapor pressure, and others, before the products are sent to their transport and refinery terminals. These operations are completed depending on the location of the field and the type of the facility used. In onshore fields, production wells and processing facilities are usually in close vicinity. A typical crude oil processing facility is illustrated in Fig. 1.13. In offshore, the production platform usually partially processes the produced fluids before sending them to another onshore processing facility to finish their processing. However, offshore vessels such as Floating Production Storage and Offloading vessels (FPSOs) are able to both process and store produced hydrocarbons before offloading periodically to shuttle tankers or transmitting processed petroleum via pipelines.

Natural gas processing involves separation from liquids, water removal (gas dehydration), removal of CO_2 and H_2S (gas sweetening), compression, and natural gas liquids separation. A typical gas processing plant is illustrated in Fig. 1.14.

1.4.4.5 Produced water treatment

Along with the oil and gas phases, significant amounts of water are coproduced from a producing well. Such waters are produced carrying significant amounts of dissolved and dispersed contaminants. This produced water can be reused (for produced water reinjection (PWR), for irrigation), or disposed (into the ocean or underground through injection wells). Thus it is very important to treat produced waters to meet the regulatory requirements. Examples of the produced water treatments are deoiling, suspended solids removal, desalination, and contaminants removal. A produced water treatment plant will vary depending on the technology used. A flowchart of the produced water treatment processes is illustrated in Fig. 1.15.

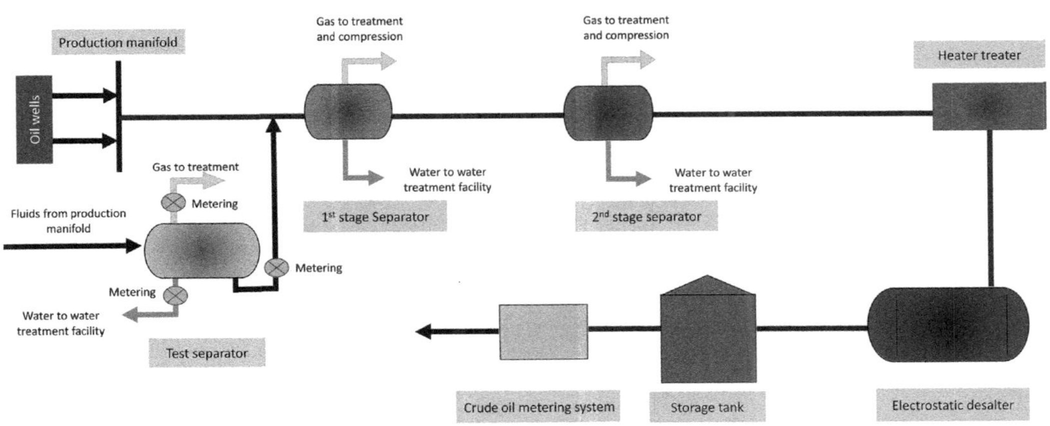

FIG. 1.13

Typical crude oil processing plant.

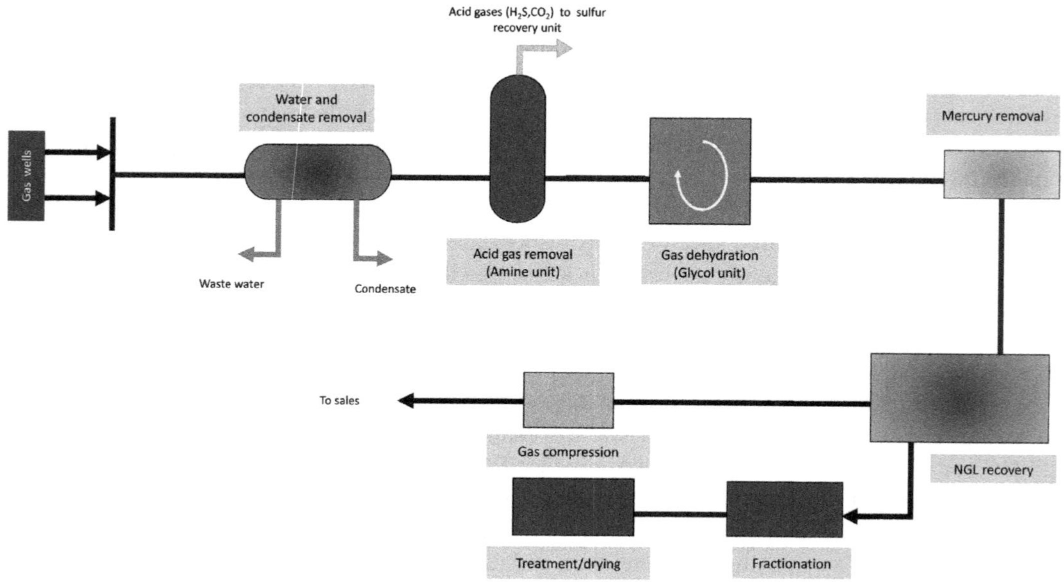

FIG. 1.14

Typical gas processing plant.

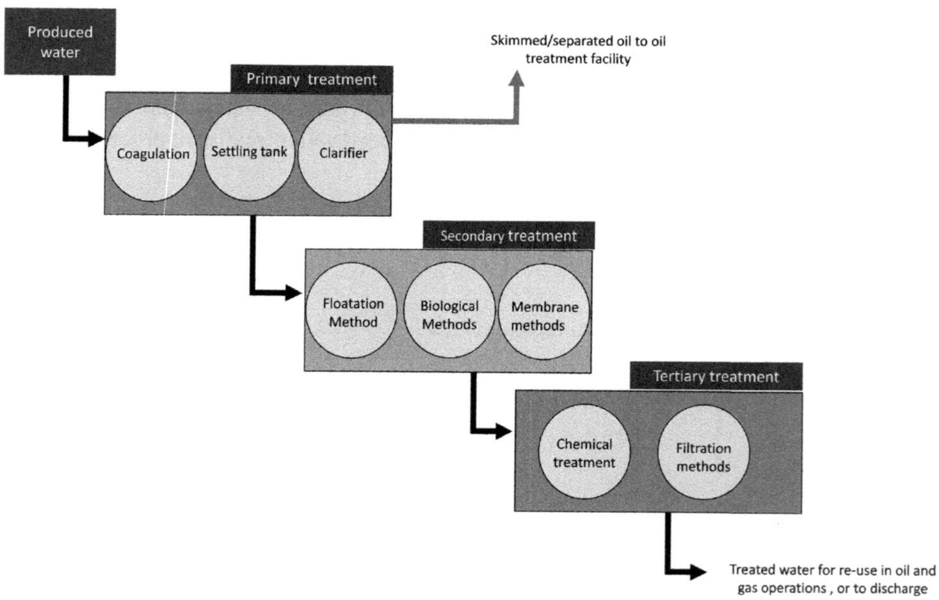

FIG. 1.15

A flowchart of produced water treatment techniques and processes.

1.4.5 Workover/recompletion

Well workover refers to oil and gas well intervention involving invasive techniques, such as wireline, coiled tubing, or snubbing. It is a process of pulling and replacing a well completion to repair an existing production well for the purpose of restoring, prolonging, or enhancing the production of hydrocarbons [41].

1.4.6 Eventual abandonment

Once the oil and gas wells show uneconomical reserves/fluids production, they are abandoned as part of decommissioning activities and plugged with cement for operational and environmental reasons [42].

With regard to technical knowledge, these operations can be classified into two major disciplines that cover the whole production technology [43]:

- Production engineering:

This covers the fluids flow, reservoir dynamics, piping and equipment design, installation, operation monitoring. and fault diagnosis.

- Production chemistry:

This basically covers the different fluid interactions with each other, with the formation rocks, and with the production facilities, and the problems associated with these interactions. It also involves optimizing produced fluids processing, treatment, storage, and disposal.

Necessarily, the production team usually comprises professionals with different specialties in engineering, physics, geology, chemistry, and biology, all integrated to achieve the main objectives of any oil company, which are producing with maximum revenue at extended periods with minimum costs, and operating safely with fewer incidents and impacts on the environment.

1.5 The production system

After exploring oil and gas accumulations in a petroleum reservoir and drilling a conduit into that reservoir, petroleum production can start where economical volumes of hydrocarbons are delivered from the reservoir to the surface facilities, where the fluids are treated and stored before being sent off to refineries or sale. This whole system through which the fluids are transported is known as the production system. The *production system* is a composite term describing the entire production process and it includes the following principal components, as shown in Fig. 1.16 [5,43]:

(1) Reservoir—contains highly compressible fluids at an elevated temperature and pressure.
(2) Wellbore—comprising the production interval, the sump, and the fluids in the wellbore.
(3) Production conduit—comprising the tubing and the tubing components.
(4) Wellhead, Christmas tree, and flow lines.
(5) Treatment facilities, with different facilities for each type of fluid (gas, oil, and water).

For onshore fields, the pipeline tieback is normally in the form of a pipeline, from wellheads to headers to the process facility, and then the export terminal [5].

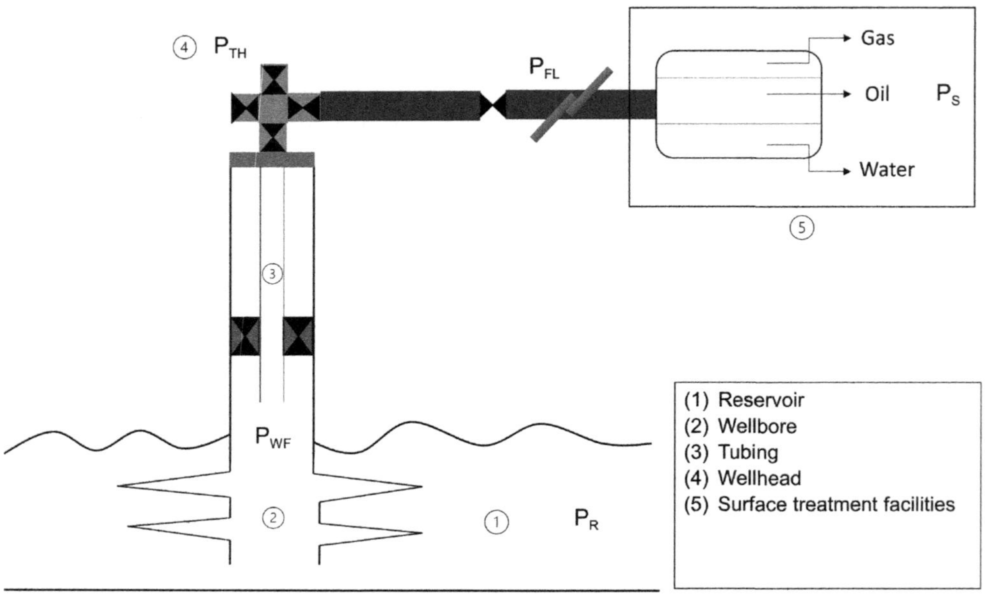

FIG. 1.16

The main components of a production system.

In offshore/deepwater environments, subsea tiebacks connecting all deepwater wellhead manifolds to the process facility are now recognized as one of the cheapest ways to develop. The deepwater environment can be categorized based on water depth tiers into deep (500–2000m), very deep (2000–3000m), and ultra deepwater (> 3000m) [5].

1.6 Production system parameters

1.6.1 Pressure

Pressure is considered the main driving force in petroleum production operations. Fluids are pushed upwards by the driving force of reservoir pressure, and when it depletes, operators attempt to maintain the reservoir pressure by some means of pumps or fluids injection, as mentioned in the oil recovery section. The pressure of fluids within the pores of a reservoir is usually hydrostatic pressure, or the pressure exerted by a column of water from the formation's depth to sea level.

Reservoir pressure affects the system design and the various production operations. System design, metallurgy, production equipment, and recovery operations all depend on the reservoir pressure in addition to the produced fluid properties [44,45].

Reservoir pressure is directly related to the wellhead pressure, which affects the pipeline operating pressure.

During production the pressure drops along the system, as illustrated in Fig. 1.17. Pressure drop is one of the main factors that leads to the formation of the different flow assurance solids, such as mineral scales,

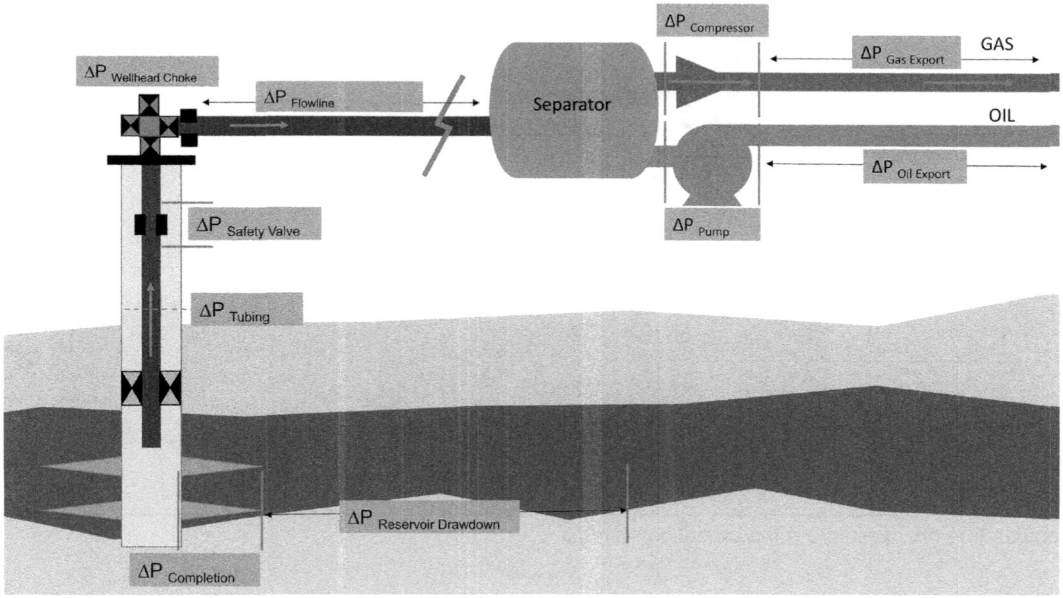

FIG. 1.17

Pressure changes across the production system.

paraffins, and asphaltenes, which cause further pressure drop and interrupt production operations. So, pressure loss or pressure drop is undesirable, and keeping pressure under control to run production within the optimum rates and preventing flow assurance issues is the best oilfield practice. Some solid deposits can form at high pressure, such as gas hydrates, which usually form at high pressure and low temperatures.

1.6.2 Temperature

Like pressure, temperature is another driving force of fluids flow. An increase in temperature increases the pressure, changes the rheology of the flowing fluids, and changes the reservoir rock properties as well [46]. System temperature affects system design, material selection, equipment, and the main operations.

As fluids are produced from the reservoir, their temperature drops and their viscosity increases, which affects their flow, causing pressure drop. Temperature drop across the production system is illustrated in Fig. 1.18.

Temperature drop decreases fluid viscosity, causing flowability issues. Besides, temperature reduction causes major flow assurance issues like paraffin gelation and deposition, gas hydrates deposition, mineral scales formation, and asphaltenes aggregation, in addition to its significant impact on fluids separation and demulsification.

1.6.3 pH

pH is a measure of acidity or basicity of a solution. Mathematically, $pH = \log_{10} (1/[H^+])$, where the brackets [ ] represents the concentration in moles/liter. The pH scale ranges from 0 to 14; pH values below 7 are acidic and above 7 are basic, whereas at pH 7 the solution is neutral. The pH of a solution depends on

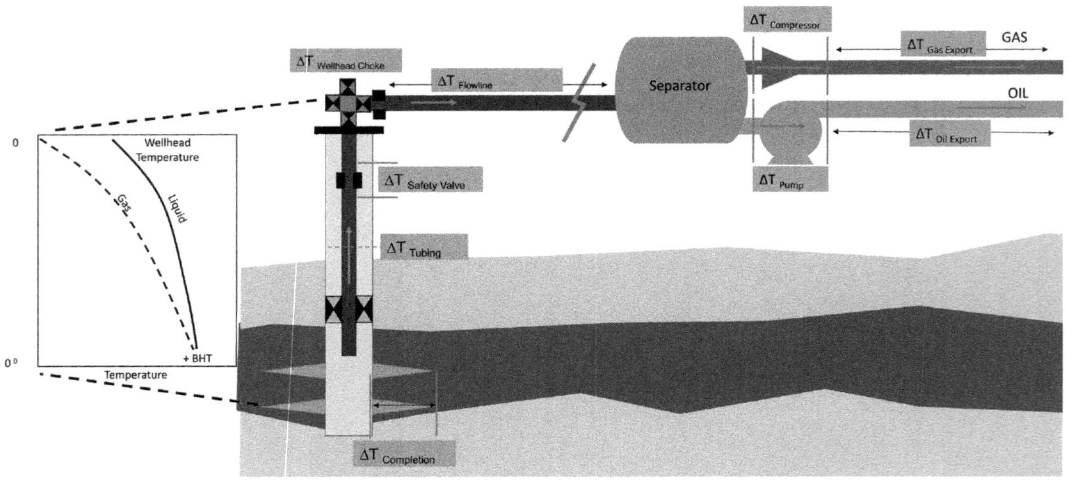

FIG. 1.18

Temperature changes across the prodution system.

its composition. In crude oil, the presence of naphthenic acids and dissolved gases like H_2S and CO_2 will have a big impact on the crude pH, while CO_2 and H_2S will have the bigger impact on produced water.

When pressure drops during production, the dissolved gases are expelled from the solution, which lead to changes in the fluid pH. Temperature changes also cause changes in the fluid pH, as the pH of the solution decreases when temperature increases.

Changes of the fluid pH also occur when fluids are introduced to the production system, like drilling fluids, fracturing fluids, acid stimulation and cleaning jobs, scale inhibitors, corrosion inhibitors, biocides, oxygen scavengers, H_2S scavengers, and other treating chemicals.

Generally speaking, acidic pH has deleterious effects on the integrity of the production system, as low pH causes corrosion problems and the severity of the problem increases with decreasing the pH. Acidic solutions also cause destabilization of asphaltenes, leading to their precipitation. On the other hand, high pH was reported to cause deposition of mineral scales like carbonates, hydroxides, and sulfides. Naphthenate soaps were reported to form at a pH above 6.

The pH of the produced fluids can be measured in the lab using electrochemical pH probes or by colorimetric methods. These lab measurements are not representative of the downhole reservoir fluids due to the changes the sample encounters during sampling and analysis like depressurization, loss of gases, and precipitations. Therefore more precise measurements of the fluid pH can be conducted using a downhole pH probe, which is affordable from service companies. Another approach to obtain a precise pH value is by calculating the pH using scale prediction, corrosion prediction, or other flow assurance prediction software.

1.6.4 Flow rate

Flow rate is the amount of substance that passes per unit of time; it is a measure of the speed of fluids flow through a conduit or vessel. Volumetric flow rate is the volume of fluid that passes per unit time, while the mass flow rate is mass of fluid passes per unit time.

A simple correlation of volumetric flow rate is given by Eq. (1.9):

$$Q = v \cdot A \tag{1.9}$$

where Q is the volumetric flow rate (m^3/s), v is the flow velocity m/s, and A is the cross-sectional area (m^2). Note the difference between flow rate Q (upper case) and volumetric flux q (lower case), which is volumetric flow rate of a fluid per unit area, with units of volume/(time × area) or $m^3/(m^2\ s)$, that is, $m\,s^{-1}$.

In multiphase flow, only a fraction of pipe cross-section is occupied by one phase, because at least two phases flow simultaneously. Thus the preceding formula gives imaginary values for the individual phases, frequently denoted as superficial velocities [47]. For oil, gas, and water mixtures, the multiphase flow rate is given by Eq. (1.10):

$$Q = A\left(\propto v_g + \beta v_w + x v_o\right) \tag{1.10}$$

where α, β, and x are the gas void fraction, water fraction, and oil fraction, respectively; v_g, v_w, and v_o are the instantaneous velocities of gas, water, and oil.

The oil production rate Q_o is measured in stock tank barrels per day (STB/day), while gas production rate (Q_g) is measured in standard cubic feet per day (SCF/day) or million standard cubic feet per day (MMSCFD).

The flow rate is related to the productivity through the productivity index (J), which is given by Eq. (1.11):

$$J = \frac{Q_o}{P_R - P_{Wf}} \tag{1.11}$$

where J (STB/day/psi), Q_o is the liquid production rate (STB/day), P_R is the reservoir pressure (psi), and P_{wf} is the wellbore (bottomhole) flowing pressure (psi).

Flow metering is essential during production operations. Poor accuracy of multiphase flow measurements can have a huge effect on [48]:

- Model prediction, history matching, and future of reservoir
- Control of flow patterns
- Phase separation
- Chemical injection
- Emulsion layer
- Corrosion rate

Since subsea instrumentation is extremely expensive and cumbersome, monitoring of the multiphase flow is often reduced to the top of the vertical riser-pipeline and following pipelines, which is located above sea [48].

A producing well flow rate is the total well production rate; well flow rate is measured during well tests and production logging, which are conducted regularly to verify the performance of an oil well.

Different types of flow meters are used for topside flow rate measurements, including coriolis flow meters, ultrasonic flow meters, thermal flow meters, turbine flow meters, differential flow meters, positive displacement flow meters, and vortex flow meters.

1.6.5 System design

The system design must consider a variety of factors to account for the different stages of production and field development. Some of these factors are

- Project size and proven hydrocarbon volumes
- Project life cycle and future considerations
- Reservoir performance
- Produced fluid volumes, chemical and physical properties
- Sand, fines and solids production, concentration, and distribution
- Geotechnical survey data
- Safety and environmental considerations.

The size of the project, the proven hydrocarbons, and project life cycle are the main investment drives for governments and production companies. They also determine future field plans, like typing with other fields or using it as a backup for other developments.

Reservoir conditions determine system design aspects: for example, how long hydrocarbons will be produced naturally and whether artificial lift or other recovery methods will be needed depend on reservoir conditions. Reservoirs with very high pressure and high temperature require special metallurgy. If too low, the reservoir temperature reduces fluid flowability and requires some investment in pumps.

Fluid volumes, composition, and physical properties determine design aspects. Fluids with high H_2S, CO_2, and naphthenic require acids corrosion resistant alloys (CRAs). Paraffinic crudes in a low-temperature environment or producing from a low-temperature reservoir require heating equipment and thermal insulation. Also, sand and fines control measures must be considered during system design.

Onshore production is usually easier than offshore, which requires complex installations with high expenses, and offshore production from shallow water is much easier than production from deep and ultra-deepwater. Offshore production requires geotechnical survey data that can affect both pipeline mechanical design and operations. Deepwater development faces operational challenges like depth, complexity of lithologies, complexity of reservoir, and long-distance pipeline tieback to beach [5]. Rigs/platforms in use worldwide vary and include jackup and concrete platforms, SPARS, drillships, TLPs (tension leg platforms), FPSOs (floating, drilling, production, and storage systems), and semi-submersibles [5], as illustrated in Fig. 1.19.

1.7 Production fluids

The fluids produced from a reservoir are basically hydrocarbons (gas, condensate, oil) and water. The main goal of production operations is to produce significant economical amounts of hydrocarbons. At the beginning, hydrocarbons are usually accompanied by relatively trivial amounts of water, which increases with time until it overcomes hydrocarbon production.

1.7.1 Hydrocarbons

A vast variety of hydrocarbon compounds are produced from petroleum reservoirs in the form of crude oil, condensate, and natural gas.

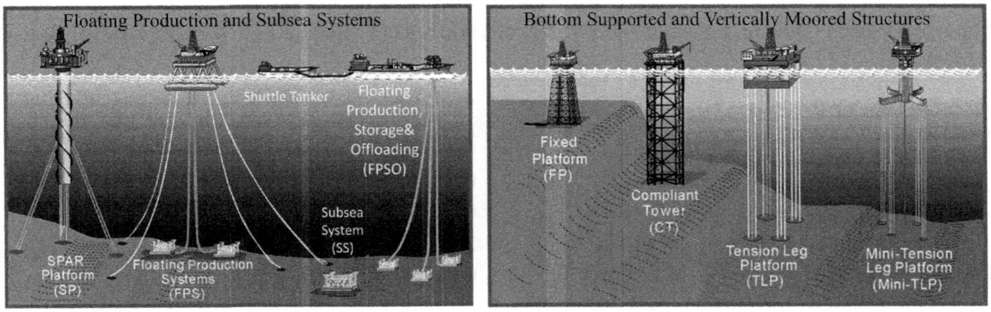

FIG. 1.19

The different deepwater development systems.

Photo courtesy Bureau of Safety and Environmental Enforcement (BSEE), public domain.

1.7.1.1 Natural gas

Natural gas is the gaseous mixture that occurs alone or associated with crude oil reservoirs, which is predominantly methane, along with other combustible hydrocarbon compounds as well as nonhydrocarbon compounds (CO_2, Ar, N_2) [49,50].

Hydrocarbon gases are generated during the successive stages of kerogen evolution. Three successive stages of gas generation take place during the evolution of sediments: biogenic generation during diagenesis, and thermally during catagenesis and metagenesis [19]. Other origins of gases include: atmospheric air (entrapping O_2, N_2, Ar), volcanic and geothermal gases (producing CO_2, H_2S, N_2, CH_4, He, Ar), and radioactivity (by disintegration of uranium and thorium producing He and Ar) [19].

The gas occurs either alone or with accumulations of crude oil, and in that case forms a gas cap (gas phase on top of oil phase in a trap), and when the reservoir pressure is high enough the natural gas may be dissolved in the crude oil and then released upon release of pressure by drilling and production [39].

1.7.1.1.1 Natural gas composition and properties

There is no single composition of the components that might be representative of a so-called *typical* natural gas. Natural gas from different wells varies widely in composition and analysis, and the proportion of nonhydrocarbon constituents can vary over a very wide range [39,51,52]. Table 1.1 shows the different compositions of various gas sources.

Some notes on the gas composition:

- Natural gas consists of mostly methane (C_1), but it contains other hydrocarbons, principally ethane (C_2), propane (C_3), butanes (C_4), and pentanes C_5.
- Raw natural gas also contains water vapor, hydrogen sulfide (H_2S), carbon dioxide, nitrogen, helium, argon, and other impurities, such as mercury.
- Condensate (liquid hydrocarbons dissolved in the gas) is separated when the gas reaches surface or shallow subsurface positions.
- Some of these constituents are considered impurities and are required to be removed due to their detrimental effect on the gas quality or the integrity of the production system. These include: water, hydrogen, hydrogen sulfide, oxygen, radon (norm), arsenic, helium, mercury, BTEX (benzene; toluene; ethylbenzene; and *o*-, *m*-, and *p*-xylenes).

Table 1.1 Typical gas composition from different sources (a) Ref. [51], (b) Ref. [53], (c) Ref. [54], (d) Ref. [55].

Component	Canada[a] (mole %)	Egypt (mole %)	Nigeria[b] (volume %)	Kuwait[c] (mole %)	China[d] (volume %)
He	0	0.2	0	0	0
N_2	3.2	0.58	0.61	0.5	0.04
CO_2	1.7	0.001	2.59	12.0	3.02
H_2S	3.3	0.013	0.001	4.0	0.0264
C_1	77.1	77.0	78.81	65.0	95.6
C_2	6.6	12.0	10.46	10.0	0.60
C_3	3.1	6.1	4.62	5.0	0.08
C_4	2	1.7	1.76	2.5	0.03
C_{5+}	3	2.401	0.89	–	0.04

Natural gas is virtually colorless, tasteless, and odorless in its pure state. For leak detection purposes, an odorant (mercaptan) is added that can be smelled in concentrations as low as 1%. Some of the physical properties of natural gas are illustrated in Table 1.2 [51]:

The state of ideal gases is defined by the equation of state, and the simplest form is represented by the gas laws in Eq. (1.12) [56]:

$$PV = nRT \tag{1.12}$$

where P is the pressure, kPa; V is the volume, L; T is the temperature, K; R is the universal gas constant ($R = 8.3142$ J/K mole), and n is the number of moles. Eq. (1.12) represents the ideal gas behavior; in the case of real gases, this equation can be rewritten as:

$$PV = ZnRT \tag{1.13}$$

where Z is known as the deviation factor or the compressibility factor. It is defined as the ratio of the volume of a real gas to the volume of a perfect one at identical temperature and pressure. The most common approach to the determination of deviation factors is based on the theorem of corresponding states. This principle states that real gas mixtures, like natural gases, behave similarly if their

Table 1.2 Properties of natural gas.

Properties	Value
Relative molar mass	17–20
Relative density, 15°C	0.72–0.81
Boiling point, °C	−162
Autoignition temperature, °C	540–560
Vapor falmability limits, volume %	5–15
Lower calorific value, MJ/kg	38–50

Reprinted with permission from S. Mokhatab, W.A. Poe, J.G. Speight, Handbook of Natural Gas Transmission and Processing, Elsevier B.V., Amsterdam, Netherlands, 2006. Copyright (2006), Elsevier.

pseudo-reduced parameters are identical. The most widely accepted Z-factor correlation was given by Standing-Katz [47].

The formations volume factor for gases (B_g) is widely used in the oil and gas industry and expresses the volume taken up by 1 m^3 (at standard state conditions, P^0 and V^o, T^o) at a pressure p and a temperature T, defined by Eq. (1.14) [56]:

$$B_g = \frac{V}{V^o} = \frac{P^o Z T}{P Z^o T^o} \tag{1.14}$$

where the standard conditions P^0 and T^o are generally defined in Europe as $P^0 = 1.00$ bar, $T^o = 273.15$ K, and in the United States as $P^0 = 14.7$ psia (1.01325 bar), $T^o = 60.0$°F (15°C), and B_g has units of volume/volume and can be represented in a variety of units (e.g., rcf/scf, rbbl/scf, and rbbl/Mscf).

The gas density is defined by Eq. (1.15) [56]:

$$\rho_g = \frac{\rho_g^o}{B_g} \tag{1.15}$$

Another way to represent gas density is given by Eq. (1.16):

$$\rho_g = \frac{PM}{ZRT} \tag{1.16}$$

The atmospheric density of the gas (ρ_g) is often related to the atmospheric density of air (ρ_a) at the same standard temperature, hence the "specific gravity" γ_g of the gas is defined by [56]:

$$\gamma_g = \frac{\rho_g}{\rho_a} \tag{1.17}$$

Another way to represent gas specific gravity is according to Eq. (1.18):

$$\gamma_g = \frac{M_g}{M_{\text{air}}} \tag{1.18}$$

where M_g is the molecular weight of the gas molecules (for methane = 16 g/mol) and M_{air} is the molecular weight of air (28.97 g/mol). For a gas mixture, the molecular weight is calculated as the summation of the mole fraction of each component times its molecular weight.

Based on the gas law, the isothermal gas compressibility is given by Eq. (1.19) [56]:

$$c_g = \frac{1}{V}\left(\frac{\partial V}{\partial P}\right)_T = -\frac{1}{B_g}\left(\frac{\partial B_g}{\partial P}\right)_T = \frac{1}{P} - \frac{1}{Z}\left(\frac{\partial Z}{\partial P}\right) \tag{1.19}$$

$$\text{For ideal gas } Z = 1, \text{so } \frac{\partial Z}{\partial P} = 0.$$

Therefore the gas compressibility is only a function of pressure and is inversely proportional to the pressure.

After extraction, natural gas is usually treated for removal of acid gases, moisture dew-point adjustment, and odorization. After treatment, it is sold within prescribed limits of pressure, calorific value, and possibly *Wobbe index* [51]. Natural gas is the cleanest burning fossil fuel, which makes it a highly desirable fuel for many applications; it is nontoxic and creates no hazard when inhaled in limited quantities. It is lighter than air and will dissipate rapidly if it escapes into the atmosphere.

Table 1.3 Classification of natural gas [39,52,57,58].

Classification	Description
Based on the occurrence of gas	*Nonassociated* gas: natural gas is found in reservoirs in which there is no, or minimal amounts of, crude oil or gas condensate. Usually it is richer in methane and leaner in higher molecular weight hydrocarbons
	Associated natural gas (dissolved natural gas) occurs either as free gas or as gas in solution in the crude oil. Usually it is leaner in methane and richer in high molecular weight hydrocarbons
	Residue gas is natural gas from which the higher-molecular-weight hydrocarbons have been extracted
	Casing head gas is derived from crude oil but is separated at the separation facility at the wellhead
Based on its methane and heavy hydrocarbons content	*Lean (dry)* gas is gas in which methane is the major constituent, with a few or no liquefiable liquid hydrocarbons
	Rich (wet) gas contains considerable amounts of higher-molecular-weight and liquefiable hydrocarbons such as ethane (C2), propane (C3), butane (C4), and hydrocarbons as high as octane (C8)
Based on its sulfur content	A sweet gas contains negligible amounts of H_2S (less than 4 ppmv of H_2S)
	A sour gas has unacceptable quantities of H_2S (higher than 4 ppmv)
	Acid gas is natural gas with significant concentrations of both CO_2 and H_2S
Based on its carbon and hydrogen isotopes	Biogases are characterized by $\delta^{13}C_1 \leq -55.0‰$
	Thermogenic methane has $\delta^{13}C_1 \geq 45$ to $-50‰$
	Oil-type gases are characterized by $\delta^{13}C_2 \leq -29.0‰$, $-55.0\% \leq \delta^{13}C_1 \leq -30.0‰$
	Coal-type gases are characterized by $\delta^{13}C_2 \geq 27.5‰$ and $-43.0‰ \leq \delta^{13}C_1 \leq -10.0‰$

1.7.1.1.2 Classification of natural gas

Natural gas can be classified in many ways, based on its occurrence, hydrocarbon composition, sulfur content, and isotopes. Classification and identification based on occurrence helps in optimizing the production parameters. Classifying based on hydrocarbon and sulfur content can help with processing using the proper facilities, and identification using isotopes content provides information about the source, generation, migration, and alteration of hydrocarbons. Table 1.3 summarizes the classification of natural gas.

1.7.1.2 Condensate

Gas condensates are hydrocarbons in which conditions of temperature and pressure have resulted in the condensation of heavier hydrocarbons from the reservoir gas [5]. The liquid normally drops out of the gas when the pressure falls below the dew-point pressure in the reservoir. Further liquid condensation will occur on the surface [59].

Starobinets et al. [60] distinguished between the primary and secondary condensates. The primary condensate forms in a compressed gas due to dissolution of liquid hydrocarbons from bituminous rocks. The secondary condensate forms when liquid hydrocarbons from the oil accumulation underneath the gas cap dissolve in the gas.

1.7.1.2.1 Condensate composition and properties

The condensate is composed of a mixture of gasoline and heavier oil fractions [61]. Typical condensate composition is predominantly pentane (C5) or heavier hydrocarbon liquids (up to C8), with relatively small amounts of lighter fractions C1, C2, C3, and C4. Depending upon the source of the condensate, benzene, toluene, xylene isomers, and ethyl benzene may also be present. This is also known as natural gasoline C5 plus and pentanes plus, and is a liquid at normal temperatures and pressure [50,51,62]. The API gravity of condensates is on the order of 50°–70°, equivalent to an SG of 0.70–0.78 [5].

1.7.1.3 Crude oil

Crude oil is regarded as the current most-used fuel worldwide. Total world oil production in 2019 averaged over 80 million barrels per day [63].

1.7.1.3.1 Crude oil composition

Crude oil is a naturally occurring, multicomponent mixture, consisting predominantly of hydrogen and carbon as well as a small percentage of organic-derived sulfur, nitrogen, oxygen, and organometallic compounds.

Elemental analysis of petroleum shows that the major constituents are carbon and hydrogen, with smaller amounts of sulfur, nitrogen, oxygen, and trace elements such as vanadium, nickel, iron, copper, and others [64]. The basic elemental composition of crude oil is given in Table 1.4, and this typically exhibits little variation from source to source.

Details of the elemental analysis and methods of measurement are discussed in the literature by Nadkarni [65].

Hydrocarbons in crude oil are principally grouped into aliphatic hydrocarbons (*n*-alkanes, isoalkanes, and cycloalkanes), aromatics/polycyclic aromatic hydrocarbons (PAHs) and their alkylated derivatives, heteroatomic organic compounds (S, N, O compounds), and organometallic compounds [66]. Fig. 1.20 summarizes these compounds.

A more comprehensive method of characterizing crude oil is by determining the concentration of four major fractions, namely saturates, aromatics, resins, and asphaltenes. This composition is called the SARA composition or SARA analysis of crude oil, with the first letter of each group taken to build this acronym [19]. The first three fractions (saturates, aromatics, and resins) are also known as maltenes in the literature [67]. Table 1.5 shows the properties of these fractions.

These four fractions and their properties are very useful and are commonly used in studying flow assurance and production chemistry issues. Fig. 1.21 illustrates the SARA analysis.

Table 1.4 The elemental composition of crude oil.

Element	Typical composition (wt%)
C	83.0–87.0
H	10.0–14.0
S	0.05–6.0
N	0.1–2.0
O	0.05–1.5
Metals	< 0.1

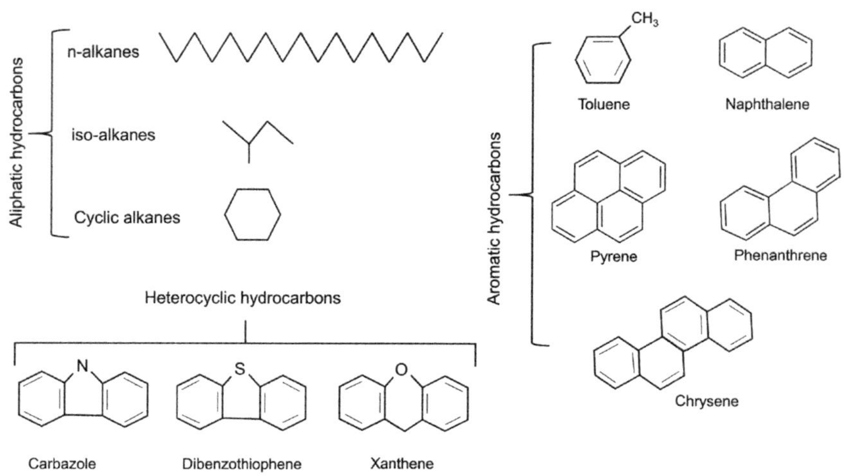

FIG. 1.20

The different compounds in crude oil.

The concentrations of the various crude oil constituents are not independent parameters. The gross composition of crude oil is described by Eq. (1.20):

$$\text{Saturated} + \textbf{Aromatics} + \textbf{Resins} + \textbf{Asphaltenes} = 1 \tag{1.20}$$

- Saturates

Saturates are nonpolar hydrocarbons without double bonds, including straight chain, branched alkanes (paraffins), and saturated alkanes with ring structure, or cycloalkanes (naphthenes) [64,68].

Table 1.5 The properties of SARA fractions of crude oil and examples of light and heavy crude [68–70].

Name of the fraction	Solubility in alkanes	Polarity	Main elements	Examples	
				North Sea, wt % (density 0.796 g/mL)	Alaska crude, wt% (density 0.997 g/mL)
Saturates	Completely soluble	Nonpolar	C, H	79.8	23
Aromatics	Completely soluble	Nonpolar	C, H	16.5	22
Resins	Soluble in light fractions (insoluble in liquid propane)	Polar	C, H, N, S	3.6	35
Asphaltenes	Insoluble	Polar	C, H, N, S, O, Fe, Ni, V	0.1	18

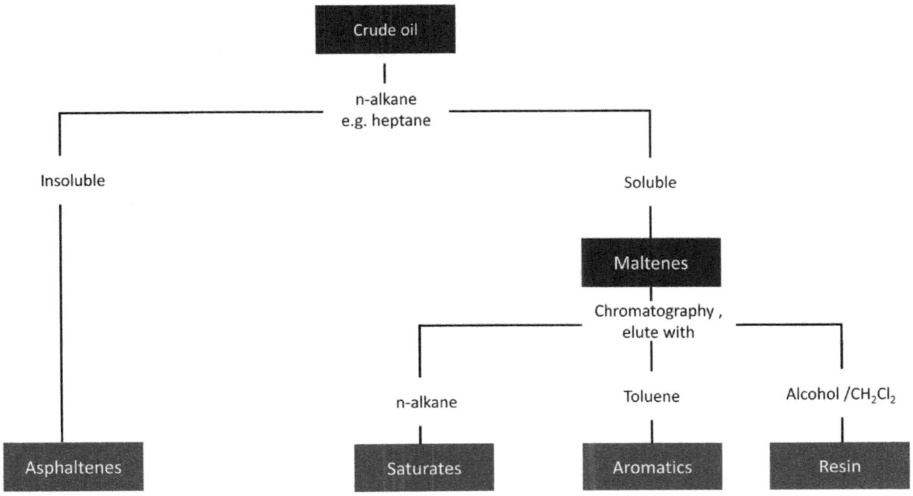

FIG. 1.21

SARA analysis of crude oil.

The possible origin of *n*-alkanes in petroleum consists of three main sources: n-alkanes produced by living organisms, e.g., cuticular waxes, retained in kerogen and transferred to petroleum; straight chain aliphatic acids (fatty acids) formed in hydrolysis of triglycerides (fats); and esters containing straight chain acidic and alcoholic moieties. Heavy n-alkanes are formed in kerogen in decarboxylation reactions of fatty acids and esters, while the isoalkanes are attributed to the olefins produced in the thermo-cracking reactions, which are catalytically transformed in the presence of acidic mineral clays into complex mixtures of isoalkanes [71].

Saturates are the main source of the flow assurance paraffin deposit problems. Paraffin deposition is discussed in detail in Chapter 7.

- Aromatics

Aromatic compounds include at least one benzene ring. The more volatile monoaromatic (single-ring) compounds found in crude oil are often referred to as BTEX (or benzene, toluene, ethylbenzene, and xylene) [72]. Those with more than one ring are commonly referred to as polycyclic aromatic hydrocarbons (PAHs), including aromatic hydrocarbons with two rings, naphthalenes, the three-ring phenanthrenes, dibenzothiophenes, and fluorenes, and also the four-ring chrysenes [73].

- Resin

Resins are a heavier fraction than aromatics and saturates. They consist of large polar molecules, often containing heteroatoms such as nitrogen, oxygen, and sulfur. The resin constituents contain a variety of functional groups including thiophene, benzothiophene, and dibenzothiophene systems, hydrogen-bonded hydroxyl groups, pyrrole (and indole) N-H functions, ester functions, acid functions, carbonyl (ketone or quinone) functions, and sulfur-oxygen functions [74]. Resins are structurally similar to asphaltenes but have a lower molecular weight [75], and they usually contain a higher percentage of saturated aliphatic and naphthenic structures in their molecules (the H/C ratio of resins is between 1.2 and 1.7 times higher than that of asphaltenes; the H/C ratio of asphaltenes is between 0.9 and 1.2) [76].

Naphthenic acids are also usually classified as part of the resins fraction [76]. They are believed to be incorporated into crude oil either by acidic compounds in the source rocks, acidic compounds formed during biodegradation of hydrocarbons, or acids produced by bacteria itself [77].

The resin fraction is soluble in light alkanes such as pentane and heptane, but insoluble in liquid propane [76].

Resins are known for their adsorption abilities, especially for asphaltenes [78]. This is why they are called "resins," as they remain in solution after asphaltenes have been removed and adsorb on active surface materials [79]. Resin adsorption and interaction with asphaltenes play a significant role in crude oil asphaltene stability.

- Asphaltenes

Asphaltenes are defined by their solubility behavior; asphaltenes are soluble in aromatic solvents and insoluble in alkane solvents. Asphaltenes are generally considered to consist of condensed aromatic nuclei that carry alkyl and alicyclic systems with heteroatoms (nitrogen, oxygen, sulfur, metals) scattered throughout in various locations. Asphaltene molecules can have carbon numbers over 40, and molecular weights from 500 to 10,000 have been cited in the literature [79,80].

Asphaltenes were assumed to occur within the oil rather than as a breakdown of kerogen during the maturation process [80]. However, some indicators of maturation have been detected. Asphaltenes are currently recognized as dispersed, chemically altered fragments of kerogen, which migrated out of the source rock of the oil during oil catagenesis [81].

Asphaltenes have been widely studied in the literature due to their confusing characteristics, and due to their deleterious effects on production. They exist in the reservoir as nanoaggregates that are stabilized by resins adsorbed on their surface, and during production they can form viscoelastic films at the O/W interfaces, causing emulsion problems or forming microaggregates that finally deposit in the production system. Asphaltene deposition problems are discussed in detail in Chapter 8.

1.7.1.3.2 Physical properties of crude oil

The production and behavior of a crude oil is shaped not only by its chemical composition but also by its physical properties.

- **Density, specific gravity, API gravity**

The densities of crude oils commonly range from 0.7 to 0.99 g/cm^3; most oils will float on freshwater (1.00 g/cm^3) and on seawater (1.028 g/cm^3) [82,83].

The specific gravity is the ratio of weight of equal volumes of a substance to that of water. API gravity is usually used to reflect the density of petroleum in the United States. The API correlation is given by Eq. (1.21):

$$\text{API}^{\circ} = \frac{141.5}{\text{Specific gravity at } 60^{\circ}\text{F}} - 131.5 \quad (1.21)$$

Table 1.6 Crude oil types based on their density and API.

Crude oil	Density (kg/m^3)	°API
Light	<870	>31
Medium	870–920	22–31
Heavy	920–1000	10–22
Extra-heavy	>1000	<10

The number of API degrees decreases as the density increases. Crude oils are classified as light, medium, heavy, and bitumen based on their density or their API gravity degree. Table 1.6 shows the classification of crude oils based on their API°.

Density, specific gravity, and API are usually obtained by the use of a hydrometer guided by standard methods ASTM D287 or ASTM D1298. New emerging methods using digital density meters like the Stabinger Viscometer density meter can also be used according to standard method ASTM D4052.

- **Flash point**

Flash point can be defined as the lowest temperature at which the liquid gives off enough vapor to form a flammable mixture with air [84]. Flash point is important from a safety standpoint in transporting crude oils, and also during workover and maintenance it is important to know the crude oil and the flash point of the chemicals used.

Flash points are measured by "open cup" and "closed cup" methods. The open-cup method is conducted where the sample vessel is exposed to the air; this method is required by some regulations. Closed cup is conducted when the sample vessel is enclosed, to represent actual conditions (such as a fuel tank). More details about these tests are found in ASTM standard methods E-502, D-56, D-92, D-93, D-1310, D-3278, and D-3828 [84]. Heavier fuels and crude oils are best measured by ASTM D7094. Smaller sample volumes of about 1–2 mL can be measured using ASTM D6450 [85].

- **Pour point**

Pour point is the lowest temperature at which crude oil will pour as a flow when it is cooled down under test static conditions, with no stirring. Pour point plays a critical role in pipeline operations and crude oil flow properties, especially with paraffinic oils, during shutdowns and startups. Pour point can also be critical for the correct operation of lubricating and fuel systems [86]. ASTM standards D97, D5853, and D6749 can be used to measure the pour point of crude oil.

- **Viscosity**

Viscosity is defined as the resistance to flow of a liquid: the lower the viscosity, the more readily a liquid flows.

The viscosity of oil is largely determined by its composition. The greater the percentage of light components such as saturates and the lower the amount of asphaltenes, the lower the viscosity [73]. Viscosity of crude oil also varies with pressure, temperature, and the quantities of dissolved gas in the solution [87].

The effect of pressure is confusing, depending on fluids saturation. Generally, in the regions below the bubble-point, viscosity decreases with increasing pressure due to the thinning effect of gas going into solution, whereas in the region above the bubble-point, the viscosity increases with increasing pressure as liquids are compressed. For a crude oil that is undersaturated at the initial reservoir pressure, the viscosity will slightly decrease as the reservoir pressure declines, and the lowest viscosity will occur at the saturation pressure. At pressures below the bubble-point, evolution of gas from solution will increase the density and viscosity of the crude oil as the reservoir pressure is decreased further [87,88].

Temperature also affects viscosity, with a lower temperature resulting in a higher viscosity. The variations with temperature are commonly large. Oil that flows readily at 40°C can become a slow-moving, viscous mass at 10°C [73].

Two types of viscosity are commonly used: dynamic viscosity μ and kinematic viscosity ν. Dynamic viscosity is related to kinematic viscosity by the relation $\eta = \rho\nu$ where ρ is the density of the fluid. The unit of dynamic viscosity μ is centipoise, while that of the kinematic viscosity is usually the centistoke [89].

The standard methods used to measure viscosity of crude petroleum and petroleum products include: ASTM D88; ASTM D341; ASTM D445; ASTM D2161; ASTM D2270, and ASTM D7042.

- **Reid vapor pressure**

Reid vapor pressure (RVP) is a measure of volatility and is defined as the pressure at which a hydrocarbon liquid will begin to flash to vapor under specific conditions (100°F, 37.8°C). It is of importance for production, storage, handling, and refinery applications.

The standard method for measuring RVP is ASTM D323, and ASTM D6377 is the standard method for measuring vapor pressure of crude oil.

- **Sulfur content**

Elemental sulfur can be present in the reservoir rock, hence it dissolves and reacts in the crude oil to produce various sulfur compounds and/or hydrogen sulfide, which may react further with certain components of the oil [39].

Sulfur is the third most abundant atomic constituent of crude oils following carbon and hydrogen. It is present in medium as well as in heavy fractions of crude oil [19]. Usually crude oils with sulfur content $<0.5\%$ are called sweet crude oils, while crude oils with sulfur content $>0.5\%$ are called sour crude oils [90]. Crude oils are also classified as light, medium, and heavy sulfur oil, depending on their sulfur content [90].

Sulfur content was used by Moldowan et al. [91] as an indicator of oil source origin, as oils of marine origin have $>0.5\%$ sulfur and those originating from nonmarine origin have $<0.5\%$ sulfur.

- **Nickel/vanadium ratio**

A lot of metals exist in crude oils from different sources. These metals exist in petroleum in one of two forms: as an organometallic porphyrin complex related to the original organic material or complexed with asphaltenes and resins. Among all the metals, Ni and V have the utmost importance, as they are used in identifying the origin and environment of deposition of petroleum as well as regional geochemical prospection [92,93].

Barwise [94] reported that low V/Ni ratios (<0.5) are expected for petroleum or crude oil derived from marine organic matter with high to moderate sulfur content, while crude oil or petroleum from lacustrine and terrestrial organic matter has V/Ni ratios ranging from 1 to 10. The NI/V ratio is also related to the maturation and the age of crude oil. In addition, it is used to identify areas of alteration due to the hydrocarbon seepage [95]. The Ni/V ratio is also used in identification of the crude oil source of spills in environmental studies. Ni, V, and Fe are related to aggregation and deposition of asphaltenes in crude oil.

1.7.1.3.3 Classification of crude oils

Crude oil can be generally classified based on crude oil quality and composition parameters like API, sulfur, origin, etc. The Ref. [61] comprehensively discusses different classes of crude oils.

- **Classification according to density, specific gravity, and API**

Density, specific gravity, and API are actually used to a large extent in expressing the quality of crude oils. Crudes are classified into light, medium, and bitumen. Table 1.6 summarizes these types.

Trading of crude oil is conventionally based on prices relative to WTI (West Texas Intermediate API 39.6°) and Brent (Brent Blend 38.6°). Both of these crude oils are light and sweet [75].

- **Classification according to sulfur content**

Crude oils are designated sweet if the sulfur content is less than 0.5% and otherwise they are called sour crude oils. A more specific classification of crude oils based on sulfur content is given by Chilingar et al. [61]: low, sulfur oils <0.5%; medium, sulfur oils 0.5%–2.0%, high, sulfur oils >2%.

- **Classification according to composition**

In American practice, crude oils have long been roughly classified as (a) paraffin base, (b) naphthenic or asphaltic base, and (c) mixed base if they contain both paraffin and asphalt. Tissot and Welte [19], classified crude oils based on the ratios of different compositions into six classes of crude oils, namely: paraffinic, naphthenic, paraffinic-naphthenic, aromatic intermediate, aromatic naphthenic, and aromatic asphaltic crude oils.

The Watson characterization factor K_w is used to characterize crude oils into their hydrocarbon composition based on physical properties like boiling point and specific gravity. It is given by Eq. (1.22):

$$K_w = \frac{\frac{1}{8} {T_b}^{1/3}}{\mathrm{SG}\left(15^\circ\mathrm{C}\right)} \tag{1.22}$$

where T_b is the boiling point in $K°$, and SG is the specific gravity at 15°C. Depending on the value of the Watson characterization factor, crude oils are classified as paraffinic ($K_w = 11$–12.9), naphthenic ($K_w = 10$–11), or aromatic ($K_w < 10$) [96]; this system is also referred to as the UOP system.

The US Bureau of Mines Correlation Index (BMCI) or (CI) is useful for characterization of crude oil fractions. CI is defined in terms of mean average boiling point (T_b) (°R) and specific gravity (SG) at 60°F, as shown in Eq. (1.23):

$$\mathrm{CI} = \frac{87552}{T_b} + 473.7\mathrm{SG} - 456.8 \tag{1.23}$$

Crude oils can be classified as paraffinic if $\mathrm{CI} < 29.8$, naphthenic if $\mathrm{CI} < 57.0$, and aromatic if $\mathrm{CI} > 75.0$.

- **Geochemical classification of crude oil**

The geochemical classification of crude oils includes the following types [61]:

- Type 1. Catagenetically altered crude oils (lower crude oil density, lower content of resins plus asphalthenes, lower sometimes to zero, of metal-porphyrin complexes; and high in the content of gasoline fraction).
- Type 2. Hypergenetically altered crude oils (usually heavier, lower API gravity, higher in resins and asphaltenes contents, and lower in the gasoline boiling point fraction).
- Type 3. Crude oils that underwent various alterations during migration (high paraffinic, less naphthenic content).

1.7.1.4 Hydrocarbon analysis methods and techniques

Different standard methods and techniques are used to measure the quality of crude oil. In addition, during production it is critical to monitor the crude oil composition and properties and use the data to track and relate it to the changes happening in the production systems, so that production can be optimized and potential production problems can be avoided. Different standard methods for testing crude oil, gas, and condensate can be used, including ASTM, IP, and UOP. Table 1.7 summarizes common hydrocarbon testing methods.

Table 1.7 Crude oil analysis methods and techniques.

Standard	Test	Technique
ASTM 4006	Bottom sediment and water	Distillation
ASTM 4007		Centrifuge
ASTM D4377	Water in crude oil	Potentiometric Karl Fischer titration
ASTM D4928		Coulometric Karl Fischer titration
ASTM D4807	Sediment in crude oil	Membrane filtration
ASTM D473		Extraction method
ASTM D6470	Salt in crude oil	Potentiometric method
ASTM D3230		Electrical method
UOP46	Wax content	Gravimetric method
ASTM D 3279	Asphaltenes in crude oil	Gravimetric method
ASTM D6560		
IP 143		
ASTM D445	Kinematic viscosity of transparent and opaque liquids	Capillary viscometer
ASTM D2007	SARA analysis	Chromatographic method
ASTM D7900	Crude oil light components	Gas chromatography
ASTM D1945	Natural gas composition analysis	Gas chromatography

1.7.2 Water

Oil and gas fields produce and handle massive amounts of water on a daily basis. The ratio of produced water to oil equivalents (WOR) or the ratio of water to gas (WGR) produced from a well varies widely from essentially zero to more than 50, where the WGR usually is higher than the WOR, and the WOR usually increases in the older fields [24,97]. The average global water and oil production ratio is 3:1 (approximately 300 million barrels of produced water per day) [98], but this ratio is different from well to well and can be more than 20 in some parts of the world [99]. In the United States the WOR is 7:1, with estimations of 15 to 20 billion barrels generated each year [100]. On the other hand, when the reservoir pressure depletes, significant amounts of water are injected into the reservoir to maintain its pressure and enhance oil production. Besides produced water and injection water, water is also used for different purposes in oil fields, such as washing (desalters, tubing cleaning, workover maintenance), as general solvent for acid jobs, for fracturing jobs, and scale squeeze. Also potable and utility water is used in the oilfields, camps, and accommodations.

Produced water (also referred to as brine, saltwater, or formation water) is water that comes out of the well with the crude oil during crude oil production. Produced water usually contains soluble and nonsoluble gas or oil organics, suspended solids, dissolved salts, and various chemicals used in the production process.

1.7.2.1 Origins of produced water

The origin of water found in oil and gas reservoirs is attributed to the incidents that took place during fossil fuels accumulation and formation.

When meteoric waters (in the form of rain) hit the face of the earth, they cause weathering and erosion of rocks, soil, and organic matter. These waters then form channels and flow until they find their way to lakes or to the sea, where they lose their suspended solids, causing them to settle down at the bottom of the sea or the lake, and they also react with their waters, causing precipitations. With further sedimentation and water evaporation, sediment compaction increases, squeezing water out of the pores and leaving some trapped, which causes changes in the composition of the interstitial water [24]. Due to sedimentation and compaction, now the water in any pore is not the same as that which was there when the particles that surround it were deposited [24]. Water can also flow from the same hydrocarbon zone due to hydrocarbon production, and can flow from other hydrocarbon zones due to hydrocarbon migration [101].

Mechanisms that cause the oilfield waters to differ in composition from water originally deposited with the sediments include ion exchange, infiltrating waters, sediment leaching, mineral formation, sulfate reduction, and ultrafiltration through clay-shale layers [24].

Another origin of the produced water is seawater being injected into the producing formation to maintain the depleting pressure; it reacts with the reservoir fluids and rocks and becomes part of the reservoir fluids, and then is produced again as produced water when the injected water sweeps and breaks through the formation.

Some of the terminology used to describe water based on origin is summarized in Table 1.8.

1.7.2.2 Chemical composition of produced water

The physical and chemical properties of produced water vary considerably depending on the geographic location of the field, the geologic formation from where the water was produced, geochemistry of the producing formation, lifetime of a reservoir, and the type of hydrocarbon product being produced [97,100,108].

Table 1.8 Water terms and classifications.

Term	Definition
Meteoric water	Water from recent atmospheric precipitation mainly as rain and snow, which provides surface runoff of rivers, streams, and creeks [102]
Interstitial water	Water contained in the small pores of spaces between the minute grains or units of rock. Interstitial waters are: (1) syngenetic (formed at the same time as the enclosing rocks); or (2) epigenetic (originated by subsequent infiltration into rocks) [24]
Connate water	Water trapped in the pores of a rock during formation of the rock [103]. Therefore connate water probably should be considered to be an interstitial water of syngenetic origin [24]. Connate water is also called fossil water
Diagenetic water	Those waters that have changed chemically and physically, both before, during, and after sediment consolidation
Formation water	Water that naturally occurs in the rocks and is present in them immediately before drilling [104]
Juvenile water	Water that is in primary magma or derived from primary magma [105]
Aquifer	A geologic unit that is highly permeable and can store and transmit a significant amount of groundwater [106]
Aquitard	Also known as a confining bed, is a much less permeable geologic unit; water flows very slowly in aquitards [106,107]
Chemically bound water	Thin layer of water surrounding mineral surfaces associated with rock strata. This water is associated with mineral changes like anhydrite to gypsum [107]
Water dissolved in hydrocarbon phase	Water that is miscible in hydrocarbon phase; its concentration, its salt content, and the salt content in the hydrocarbon phase depend on pressure and temperature, with evaporation and condensation being the main processes controlling it

Due to the various activities involved during production, produced water from oil production includes formation water, enhanced oil recovery fluids (seawater, freshwater, steam miscible gas, or solvents) and the various production chemicals used in enhancing the production. Produced water from gas production includes formation water, condensed water (gas miscible water vapor that condenses into a liquid state upon production), and the various production chemicals [100].

In general, major compositions of produced water include but are not limited to [100,109]:

- Dissolved minerals (often expressed as salinity, conductivity, or total dissolved solids (TDS)).
- Dissolved and dispersed oil, also called oil and grease (various organic compounds associated with hydrocarbons in the formation).
- Dissolved gases (H_2S, CO_2, etc.).
- Treatment chemicals, which are inorganic and organic compounds introduced as chemical additives to improve drilling and production operations.
- Produced solids including reservoir rocks, formed mineral scales, and corrosion products.
- Naturally occurring radioactive material.
- Bacteria and microorganisms.

Tables 1.9 and 1.10 summarize the range of composition of produced water from oil and gas fields and also the seawater composition. These data are quoted from the literature [110–115]. It is worth noting

Table 1.9 Oilfield Produced water composition.

Parameter	Values
Density (kgIm^3)	1014–1140
Surface tension (dynes/cm)	43–78
pH	4.3–10
Total oil (mg/L)	2–565
COD (mg/L)	1220
Total organic carbon (TOC) (mg/L)	0–1500
Total suspended solids (TSS) (mg/L)	1.2–1000
BTEX (mg/L)	0.39–35
Total polars (mg/L)	9.7–600
Volatile fatty acids (VFA) (mg/L)	2–4900
Higher acids (mg/L)	<1–63
Phenols (mg/L)	0.009–23
Chloride (mg/L)	80–200,000
Sulfate (mg/L)	<2–1650
Sulfide (mg/L)	10
Ammoniacal nitrogen (mg/L)	10–300
Bicarbonate (mg/L)	77–3990
Metals (mg/L)	
Boron (B)	5–95
Lithium (Li)	3–50
Sodium (Na)	132–97,000
Calcium (Ca)	13–25,800
Magnesium (Mg)	8–6000
Potassium (K)	24–4300
Strontium (Sr)	0.02–1000
Barium (Ba)	1.3–650
Cadmium (Cd)	0.005–0.2
Aluminum (Al)	310–410
Copper (Cu)	<0.002–1.5
Chromium (Cr)	0.02–1.1
Mercury (Hg)	<0.001–0.002
Arsenic (As)	<0.005–0.3
Lead (Pd)	0.002–8.8
Zinc (Zn)	0.01–35
Iron (Fe)	<0.1–100
Manganese (Mn)	<0.004–175
Beryllium (Be)	<0.001–0.004
Titanium	0.01–0.7
Silver (Ag)	<0.001–0.15

Reprinted by permission from P.J.C. Tibbetts, I.T. Buchanan, L.J. Gawel, R. Large, A comprehensive determination of produced water composition, in: J.P. Ray, F.R. Engelhardt (Eds.), Produced Water. Environmental Science Research, vol. 46, Springer, Boston, MA, 1992. https://doi.org/10.1007/978-1-4615-2902-6_9. Copyright (1992), Springer Nature.

Table 1.10 Constituents (mg/L) in natural gas produced waters.

Parameter	Minimum value	Maximum value	Parameter	Minimum value	Maximum value
pH[a]	4.4	7.0	Iron[a]	ND	1100
pH[b]	3.1	6.47	Iron[b]	39	680
Conductivity[a] (μmhos/cm)	4200	180,000	Lead[b]	<0.2	10.2
Conductivity[b] (μmhos/cm)	136,000	586,000	Lithium[b]	18.6	235
Alkalinity[b]	0	285	Magnesium[a]	0.9	4300
TDS[a]	2600	310,000	Magnesium[b]	1300	3900
TDS[b]	139,00	360,000	Manganese[a]	0.045	6.5
TSS[a]	14	800	Manganese[b]	3.59	65
TSS[b]	8	5484	Nickel[a]	ND	0.02
BOD[a]	75	2870	Nickel[b]	<0.08	9.2
COD[a]	2600	120,000	Potassium[b]	149	3870
Aluminum[a]	ND	0.4	Silver[b]	0.047	7
Aluminum[b]	<0.50	83	Sodium[a]	520	45,000
Arsenic[a]	0.004	1	Sodium[b]	37,500	120,000
Arsenic[b]	<0.005	1.51	Strontium[a]	–	6200
Barium[a]	ND	26	Sulfate[a]	<0.1	47
Barium[b]	9.65	1740	Sulfate[b]	ND	19
Boron[a]	ND	56	Tin[a]	ND	1.1
Bromide[b]	150	1149	Zinc[a]	ND	0.022
Cadmium[a]	ND	0.015	Zinc[b]	<0.02	5
Cadmium[b]	<0.02	1.21	TOC[a]	67	38,000
Calcium[a]	ND	25,000	Surfactants[b]	0.08	1200
Calcium[b]	9400	51,300	Benzene[a]	1.8	6.9
Chloride[a]	1400	190,000	Benzene[c]	<0.010	10.3
Chloride[b]	81,500	167,448	Toluene[a]	0.857	3.37
Chromium[a]	ND	0.03	Toluene[c]	<0.010	18
Copper[a]	ND	0.02	Oil/grease[a]	6	60
Copper[b]	<0.02	5	Oil/grease[b]	2.3	38.8

[a]*Ref. [110].*
[b]*Ref. [111].*
[c]*Ref. [112].*
Reprinted by permission from Johnson, B.M., Kanagy, L.E., Rodgers, J.H. et al., Chemical, physical, and risk characterization of natural gas storage produced waters, Water Air Soil Pollut. 191 (2008) 33–54, https://doi.org/10.1007/s11270-007-9605-8. Copyright (2008), Springer Nature.

that these values represent the range of mean concentrations and higher concentrations are expected in some fields.

- **Dissolved minerals**

The amounts of dissolved constituents found in oilfield waters range from less than 100 mg/L to more than 400,000 mg/L [108,116] (the highest TDS reported was 463,000 mg/L, as mentioned by Otten and Mercier [117]). This salinity distribution is dependent upon several factors, including hydraulic gradients, depth of occurrence, distance from outcrops, mobility of the dissolved chemical elements, soluble material in the associated rocks, ion exchange reactions, and clay membrane filtration mechanisms [24].

Some general characteristics of the mineral constituents of produced waters are:

- Generally the concentrations of calcium, sodium, and chloride increase with increasing salinity, while the sulfate concentrations decrease [118].
- The Ca/Na ratio usually increases with depth and age of the associated rocks, while the Mg/Na ratio decreases [24].
- Generally the Ca/Mg ratio increases with increasing salinity while the ratio Na/(Ca + Mg) decreases irregularly with increasing salinity and depth [119].
- Ca, Ba, Sr, Zn, Pb, Fe are considered high potential scaling species. Changes in their concentration may indicate scale formation (if the geological interactions in the reservoir were excluded).
- High salinity waters are always associated with high corrosion in production facilities.
- pH of produced water is affected by the water composition and strongly affected by dissolved gases such as CO_2 and H_2S. Many techniques have been developed to measure pH, for downhole (downhole probe) and topsides applications. However, a good practice of obtaining fairly precise pH is by calculating it using scale prediction software.
- According to its mineral composition, the produced water can be classified into four basic categories and subcategorized into more classes, according to Table 1.11 [120]:

This classification was first proposed by Sulin [121], and he determined that hydrocarbon accumulations are most commonly related to water types, in this order: chloride-calcium > bicarbonate-sodium > chloride-magnesium > sulfate-sodium. Most oilfield waters are of the chloride-calcium type.

Table 1.11 Classification of produced waters.

Type of water	Concentration ratio expressed by milligram equivalent percentage		
	Na^+/Cl^-	Na^+-Cl^-/SO_4^{2-}	Cl^--Na^+/Mg^{2+}
Calcium chloride	<1	<0	>1
Magnesium chloride	<1	<0	<1
Sodium bicarbonate	>1	>1	<0
Sodium sulfate	>1	<1	<0

Reprinted with permission from W. Renpu, Basis of well completion engineering, in: Advanced Well Completion Engineering, Elsevier, 2011, pp. 1–74. https://doi.org/10.1016/b978-0-12-385868-9.00001-4. Copyright (2011) Elsevier.

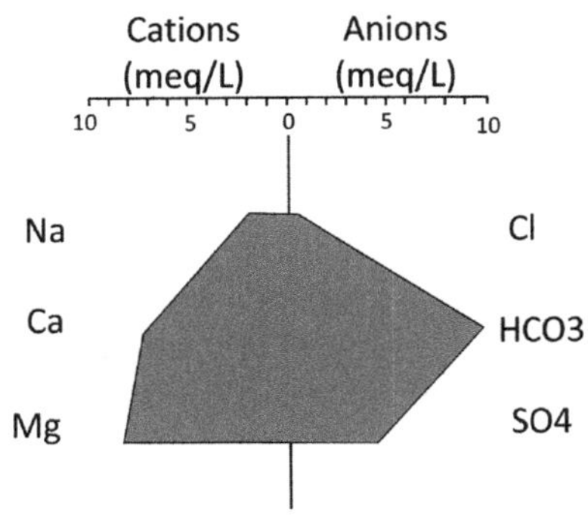

FIG. 1.22

Stiff diagram (butterfly diagram).

Source Wikipedia https://en.wikipedia.org/wiki/File:Example_Stiff_Diagram.jpg (public domain).

Other classifications of produced water are Palmer's, Chebotarev's, and Schoeller's classifications, which are also based on inorganic or mineral analysis of the produced water.

- The analytical data of mineral content can be represented by graphical plots, which aid in rapid identification of the water type, provide a quick way to spot differences between different waters, highlight important points about the analysis that might be missed by simply reading a report, and are also useful in pattern and trend comparisons. These include methods like the Tickell diagram, Reistle system, Piper diagram, and the most commonly used Stiff diagram (Fig. 1.22) as it was adopted by API. A logarithmic or linear plot of meq/L of each radical is plotted [24,122].

Cations are usually measured using techniques like ICP-OES, ICP-MS, and AA.

- **Dissolved and dispersed oil**

In produced water, most of the hydrocarbons are in dispersed form as small droplets of organics suspend in produced water, including PAHs and alkyl phenols. Dissolved oil is defined as water-soluble organics in produced water, including BTEX, phenols, aliphatic hydrocarbons, carboxylic acid, and low molecular weight aromatic compounds [98,123,124]. Produced waters from gas production have higher content of low molecular-weight aromatic hydrocarbons, such as benzene, toluene, ethylbenzene, and xylene (BTEX), than those from oil operations; hence they are relatively more toxic than produced waters from oil production. Studies indicate that the produced waters discharged from gas/condensate platforms are about 10 times more toxic than the produced waters discharged from oil platforms [125].

The amounts of dissolved and suspended oil present in produced water (prior to treatment) depend on the following factors [109]:

- Oil composition
- pH, salinity, TDS (total dissolved solids), temperature
- Oil/water ratio
- Type and quantity of oilfield chemicals
- Type and quantity of various stability compounds (waxes, asphaltenes, fine solids)

Dissolved and dispersed hydrocarbons are usually measured using GC, GC-MS techniques.

- **Dissolved gases**

The major gases in produced water are CO_2, O_2, and H_2S, which commonly cause production problems like corrosion and scaling. The origin of these gases can be due to bacterial activity or chemical dissolution of reservoir rocks, or thermochemical reaction between fluids in the formation. Other hydrocarbon gases that can be found dissolved in the produced water include methane, with measurable amounts of ethane, propane, and butane.

- **Treating chemicals**

Treatment chemicals are added to enhance the production or to control flow assurance problems. Some of these chemicals are toxic and considered environmentally hazardous, which requires extra treatments and costs. Operating companies and service companies are seeking for more environmentally friendly chemicals. A list of these chemicals is seen in Table 1.12 [126,127].

- **Production solids**

Production solids are a wide range of materials including formation solids or formation rocks from the unconsolidated reservoirs, corrosion and scale products, bacteria, waxes, and asphaltenes. These solids are the main study of this book and are discussed in detail, type by type, in the coming chapters.

Table 1.12 Some of the production chemicals used in oil and gas operations.

Function	Chemical class	Targeted phase and primary solubility
Corrosion inhibitor	Imidazolines	Water/oil
	Amines	Water/oil
	Quaternary ammonium salts	Water/oil
Oxygen scavenger	Ammonium bisulfite	Water
Biocide	Quaternary amine salt	Water
	Amine acetate	Water
	Glutaraldehyde	Water
Hydrate inhibitor	Methanol	Water
	Ethylene glycol	Water
Gas dehydration	Triethylene glycol	Water
Scale inhibitor	Phosphate ester	Water
	Poly carboxylates	Water
	Poly vinyl sulfonate	Water
	Phosphonates	Water
Demulsifier	Oxyalkylated resins	Oil
	Polyglycol esters	Oil
	Sulfonates	Oil
Solid removal	Polyamine	Water
	Quaternary polyamine	Water

- **Radioactive materials**

The geological formations contain plenty of radioactive materials, including uranium and its decay, thorium and its decay, radium and its decay, potassium 40 and its decay, lead 210 and its decay, and polonium 210 and its decay.

The two radium isotopes present in produced water and barite scale are 226Ra (half-life = 1600 years) and 228Ra (half-life =5.8 years). These two isotopes are produced by radioactive decay of uranium and thorium present in rocks of the oil-producing formations. The concentration of dissolved radium is therefore influenced by the abundance of uranium and thorium in reservoir rock and by the accessibility of water to the sites containing uranium and thorium [128].

- **Bacteria and organisms**

Various types of bacteria are found in the produced waters including sulfate-reducing bacteria, iron bacteria, acid-producing bacteria, nitrate-reducing bacteria, and others. These bacteria can thrive when sufficient nutrients exist in the reservoir at the optimum pressure, temperature, and aerobic and anaerobic conditions.

1.7.2.3 Physical properties of produced water

- **Resistivity**

The resistivity of formation water is a measure of the resistance offered by the water to electrical current. It can be measured directly or calculated [24]. Formation water resistivity, R_{wg}, is used in electric-log interpretation; the value is adjusted to formation temperature.

- **Surface tension (interfacial tension)**

Surface tension (interfacial tension or IFT) is a measure of the attractive force acting at a boundary between two phases. Produced waters have surface tension between 43 and 78 dyn/cm. This value is dependent on the type of production chemicals injected in the system and their concentration.

- **Viscosity**

The viscosity of formation water, η_w, is a function of [129]:

- Pressure, brine viscosity increases with increasing pressure.
- Temperature, brine viscosity decreases with increasing temperature.
- Dissolved solids, brine viscosity increases with increasing dissolved salts.

- **Density and specific gravity**

The density of formation water is a function of:

- Pressure
- Temperature
- Dissolved constituents

Density of produced water is determined in many ways. Pycnometer and the Stabinger method are among the common ones/standard methods.

An approach to calculating water density is to first calculate the density of formation water at standard conditions with McCain's correlation [130,131]:

$$\rho_w = 62.368 + 0.438603 \times S + 1.60074 \times 10^{-3} \times S^2 \tag{1.24}$$

where density is in lbm/ft^3, and S is salinity in weight percent. Then, density at reservoir conditions is calculated by dividing the density by the brine formation volume factor at the reservoir temperature and pressure of interest.

The specific gravity of formation water γ_w can be estimated, if the TDS is known, with:

$$\gamma_w = 1 + C_{sd} \times 0.695 \times 10^{-6} \tag{1.25}$$

where C_{sd} = concentration of dissolved solids (also known as TDS), mg/L.

Importance of produced water composition and properties:

- The chemical composition of formation waters can be used as an indicator of the presence of petroleum as outlined by Sulin's method [121].
- Produced water composition can indicate the origin of the produced water and the various interactions (diagenic, catagenic) during hydrocarbon formation [132].
- The chemical composition analysis of formation water can help us determine the source of water as a fingerprint for a specific production formation and also determine the possible water production rate.
- This is important in the drilling design for drilling-in target using gas base fluid.
- Reservoir sensitivity and degree of working fluid damage can be evaluated using formation water property data.
- Oil and gas saturations can be determined, and oil and gas reservoirs can be identified using logging data of formation water resistivity, *Rw*.
- Compatibility and suitability of production and completion fluids that are introduced to the reservoir all depend on the produced water composition, i.e., drilling and completion, perforating, killing, fracturing, and acidizing fluids; all of these can be identified and optimized using formation water analysis data.
- Production chemistry problems like corrosion, scaling, and microbiology can be predicted and controlled by knowing the produced water composition.
- To forecast the drive mechanisms originated by water, either water drive or water injection can be used.
- Knowing the produced water composition is a critical point to build and design wastewater treatment strategy.

1.7.2.4 Water analysis methods and techniques

Different methods and techniques are adopted to determine produced water composition and properties. Some of the main parameters and methods are summarized in Table 1.13.

Various other methods have also been applied, such as those in the standard methods for the examination of water and wastewater, which is a comprehensive reference for testing water properties, EPA methods, NACE methods, and others.

Water cations can be measured by various techniques such as ICP-OES, ICP-MS, ion chromatograph, atomic absorption, and X-ray fluorescence. Anions can be measured by different methods such as ion chromatograph, titrimetric methods, spectroscopic methods, and others.

Table 1.13 Methods and techniques used in water analysis.

Parameter	Method	Technique
pH	ASTM D1293	Glass electrode or pH electrode
Conductivity and resistivity	ASTM D 1125	Conductivity cell
TDS and TSS	ASTM D5907	Gravimetric methods
TSS	USGS I-3765-85 EPA Method 160.2 NACE TM0173	Gravimetric methods
Metals	EPA Method 200.7	ICP-OES
Chloride in water	ASTM D512	Titrimetric methods, ion selective electrode
Sulfate in water	ASTM D516 ASTM D4130	Turbidimetric method
Alkalinity	Standard method 2320 B	Titration method

1.7.3 Phase behavior of petroleum fluids

1.7.3.1 Phase behavior of hydrocarbon systems

A phase is generally defined as any homogeneous part of a system that is physically distinct and separated from other parts of the system by definitive boundaries [133]. Oil and gas fields encounter different phases like solids, liquids, and gases. The phase behavior studies the various interactions of these phases. The experimental or mathematical determinations of the conditions under which these phases occur are conveniently expressed in different types of diagrams, commonly called phase diagrams [133]. Equilibrium phase diagrams offer convenient representations of the ranges of temperature, pressure, and composition within which various combinations of phases coexist [134]. Fig. 1.23 illustrates the phase diagram of a single component.

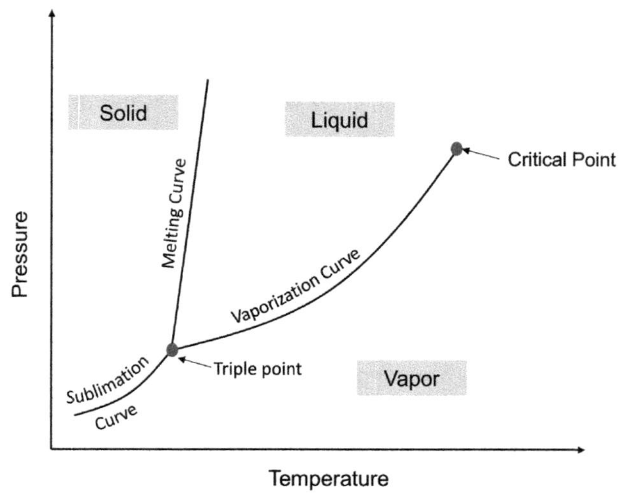

FIG. 1.23

Typical phase diagram of a single component.

At temperatures below the triple point and pressures below the sublimation curve the component forms a vapor phase, while a solid phase is formed at pressures above the sublimation curve. The sublimation curve represents solid-vapor equilibrium, where solid and vapor can coexist. Similarly, the melting curve represents the solid-liquid equilibrium. Liquid and vapor can coexist along the vaporization or vapor-pressure curve. At pressures above the vapor pressure, a liquid forms, while at pressures lower than the vapor pressure, a vapor forms. The vapor-pressure curve ends at the critical point, which is the maximum temperature at which two phases can exist. At temperatures above the critical temperature, a single phase forms over the entire range of pressures [134]. The point of intersection of all three curves represents the substance's triple point—the temperature and pressure at which all three phases are in equilibrium.

In oil and gas fields, reservoir fluid studies involve studying liquid phases such as crude oil and condensate, and gas phases such as natural gas. During hydrocarbons production, the continuous changes in the operating conditions (pressure, temperature) from the reservoir to the surface lead to various interactions between these phases and multiple phases transitions.

To understand that, think of the fact that crude oil/gas has to flow through different domains: reservoir (a porous medium), wellbore (near vertical), and surface facilities and flow lines (near horizontal). During the flow through these domains, the phases change continuously. For example, undersaturated oil will enter a wellbore in the liquid phase, evolving to two-phase flow in the production tubing when the pressure reduces below the bubble point. Saturated oil will evolve to two-phase flow in the reservoir formation upon production due to pressure drawdown. Natural gas will enter a wellbore in the gas phase. Gas condensate evolves from natural gas in the reservoir formation, condensing out the heavy constituents due to pressure drawdown in the near-wellbore region. Superimposed on these simple hydrocarbon flows is the flow of produced water [75,135].

It is a common practice in oil and gas to represent the phase behavior or reservoir fluids in a PVT diagram. The phase behavior of oil and gas is generally treated independently of the water phase, considering that the water effect is minor, unless water-hydrocarbon solid structures, known as hydrates, are formed [136]. A typical phase diagram is illustrated in Fig. 1.24 [137].

Although a reservoir fluid may be composed of many thousands of compounds, the phase behavior fundamentals can be explained by examining the behavior of pure and simple multicomponent mixtures. The behavior of all real reservoir fluids basically follows the same principle, but to facilitate the application of the technology in the industry, reservoir fluids have been classified into various groups such as dry gas, wet gas, gas condensate, volatile oil, and black oil [136]. Table 1.14 summarizes these various groups of reservoir fluids [138].

Generally, a system in thermodynamic equilibrium possesses a defined relationship between the state variables. These are united in the so-called *equation of state* or EOS, which states that for a given amount of substance contained in a system, the temperature, volume, and pressure are not independent quantities; they are connected by a relationship of the general form:

$$F(P,V,T)=0 \tag{1.26}$$

An equation of state (EOS) is an analytical expression relating pressure to the volume and temperature. The expression is used to describe the volumetric behavior, the vapor/liquid equilibria (VLE), and the thermal properties of fluids. Numerous EOSs have been proposed since van der Waals introduced his expression in 1873, and they have been successfully used to calculate reservoir fluid properties [139]. The equations of state are divided into two main groups: cubic and noncubic. Noncubic equations can

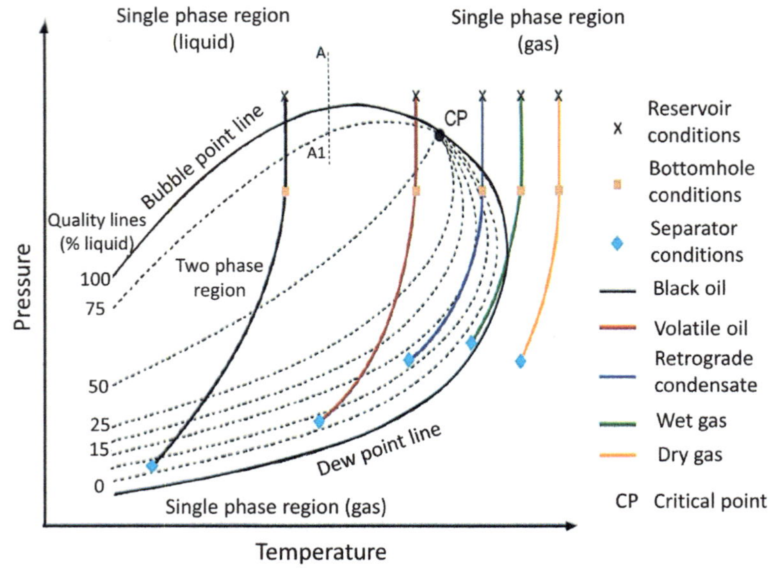

FIG. 1.24

Phase diagram of reservoir hydrocarbons.

Reprinted with permission from J. Bellarby, Tubing well performance, heat transfer and sizing in: J. Bellarby (Ed.), Well Completion Desin, Developments in Petroleum Science, vol. 56, Elsevier, 2009, pp. 247–302, https://doi.org/10.1016/S0376-7361(08)00205-7. *Copyright (2009) Elsevier.*

Table 1.14 Reservoir fluids types [138].

Reservoir fluid type	Properties
Black oils	Liquid oil including large, heavy, and nonvolatile molecules. Black oil reservoirs usually have temperatures below the critical temperature of a hydrocarbon mixture. Point A in Fig. 1.24 represents unsaturated black oil, while A1 shows saturated black oil, where oil is at full capacity of dissolved gas and where a decrease of system pressure will release gas to form the gas phase
Volatile oils	Volatile oils contain more C2-C6 compounds and fewer heavy molecules than black oil. Reservoir conditions are very close to the critical temperature and pressure. Any small reduction in pressure can cause the release of a large amount of gas
Condensate	Retrograde gas is a fluid that is gas at reservoir pressure and temperature, whereas large quantities of liquid are formed by retrograde condensation when pressure and temperature drop. Retrograde gases are also called retrograde gas condensates, gas condensates, or condensates
Wet gases	Wet gas contains less methane (<85%), and more ethane and higher hydrocarbons; it exists in a pure gas phase in the reservoir, and with pressure drop it becomes a two-phase mixture (liquid/gas) in a flowline from the well tube to the separator at the topside platform, where liquid condensation takes place
Dry gases	Dry gas is primarily methane. It exists as gas under all conditions of pressure and temperature encountered during the production phases from reservoir conditions involving transport and process conditions, and no hydrocarbon liquids are formed. Dry gas reservoirs have temperatures above the cricondentherm

better describe the volumetric behavior of pure substances, but may not be suitable for complex hydrocarbon mixtures [140].

Commonly used EOSs in reservoir engineering calculations include the Redlich and Kwong two-parameter cubic equation of state (RK EOS), Soave-Redlich-Kwong (SRK EOS), Peng-Robinson (PR EOS), Esmaeilzadeh-Roshanfekr (ER EOS), Schmidt-Wenzel (SW EOS), Patel-Teja (PT EOS), and others.

Phase behavior plays a vital role in many petroleum applications, such as:

- Enhanced oil recovery
- Compositional simulation
- Geochemical behavior
- Wellbore stability
- Geothermal energy
- Multiphase flow in wellbores and pipes
- Surface facilities

1.7.3.2 Phase behavior of water-hydrocarbon system

Phase equilibrium in water hydrocarbon systems is critical for comprehensive understanding of the reservoir and produced fluids. That can be attributed to modification of the hydrocarbon composition, especially when reservoir pressure becomes very low; this modification can be due to:

- Solubility of water in hydrocarbons
- Solubility of hydrocarbons in water
- Effect of water on saturation pressure
- Portioning of specific compounds between the different phases

Thus this can affect the general phase equilibrium of the system.

Water-hydrocarbon equilibrium is critical in the calculations of mineral scale tendencies. It is also critical for calculating and predicting the gas hydrates formation. Further details of these are discussed in Chapter 12.

Cubic equations of state can describe such equilibrium after some modification, knowing that cubic EOSs are not suitable for electrolyte solution calculations.

Water-hydrocarbon phase equilibria are also studied for their importance in major operations, e.g., drilling, its importance in multiphase flow, system design, and other production operations challenges.

1.8 Summary

Petroleum forms a major underpinning of modern life, constituting the main energy source and a raw material for other industrial sectors. Two main theories of petroleum formation exist: the biogenic and abiogenic theories. These theories have been presented here, along with a general overview of the petroleum production system and the major production operations. In addition, the main operating parameters have been outlined. The physical properties and chemical phase behavior of the production fluids, namely gas, condensate, oil, and water, were addressed.

References

[1] P. McKenzie-Brown, G. Jaremko, D. Finch, The Great Oil Age, first ed., Brush Education, Alberta, Canada, 1993.

[2] BP Energy, BP Energy Outlook, 2018, 2018 Edition https://www.bp.com/content/dam/bp/business-sites/en/global/corporate/pdfs/energy-economics/energy-outlook/bp-energy-outlook-2018.pdf.

[3] BP Energy, BP statistical review of world energy, in: Statistical Review of World Energy, vol. 68, BP plc, 2019, pp. 1–62.

[4] J. Yoon, Y. Sun, J.A. Rogers, Flexible solar cells made of nanowires/microwires, in: Semiconductor Nanomaterials for Flexible Technologies, Elsevier, 2010, https://doi.org/10.1016/c2009-0-20347-3.

[5] B. Oyeneyin, Integrated sand management for effective hydrocarbon flow assurance, in: Developments in Petroleum Science, vol. 63, Elsevier, 2015, pp. 1–9, ISBN: 978-0-444-62637-0, https://doi.org/10.1016/B978-0-444-62637 0.00001-4.

[6] IEA, Data and Statistics, Retrieved from: https://www.iea.org/data-and-statistics. (Last visited November 2019).

[7] BP Energy, Statistical Review of World Energy 2021, seventieth ed., BP plc, 2021.

[8] A.H. Tullo, Why the future of oil is in chemicals, not fuels, Chem. Eng. News 97 (8) (2019) 26–29.

[9] The Canadian Association of Petroleum Producers (CAPP), Uses for Oil, 2019, Retrieved from: https://www.capp.ca/oil/uses-for-oil/. (Last visit April 2020).

[10] A. Satter, G.M. Iqbal, Conventional and Unconventional Petroleum Reserves—definitions and World Outlook, Reservoir Engineering, Elsevier, 2016, pp. 427–438. https://doi.org/10.1016/b978-0-12-800219-3.00023-1.

[11] Z. Caineng, Y. Zhi, Z. Guosheng, H. Lianhua, Z. Rukai, T. Shizhen, Y. Xuanjun, et al., Conventional and unconventional petroleum 'orderly accumulation': concept and practical significance, Pet. Explor. Dev. 41 (1) (2014) 14–30, https://doi.org/10.1016/S1876-3804(14)60002-1.

[12] M. Höök, U. Bardi, L. Feng, X. Pang, Development of oil formation theories and their importance for peak oil, Mar. Pet. Geol. 27 (9) (2010) 1995–2004, https://doi.org/10.1016/j.marpetgeo.2010.06.005.

[13] American Petroleum Institute, Glossary of Oilfield Production Terminology (GOT) (Definitions and Abbreviations), American Petroleum Institute, Washington, DC, 1998.

[14] D. Tiab, E.C. Donaldson, Introduction to petroleum geology, Petrophysics (2012) 27–83, https://doi.org/10.1016/b978-0-12-383848-3.00002-5.

[15] R. Robinson, The origins of petroleum, Nature 212 (1966) 1291–1295.

[16] American Petroleum Institute, Introduction to Oil and Gas Production, fifth ed., University of Texas at Austin Petroleum, 1996.

[17] R. Smith, Crude: The Incredible Journey of Oil, Australian Broadcasting Corporation (ABC) (Australia), 2007.

[18] S. Boggs, Principles of Sedimentology and Stratigraphy, fourth ed., Pearson Prentice Hall, Upper Saddle River, NJ, 2006, p. 662.

[19] B.P. Tissot, D.H. Welte, Petroleum Formation and Occurrence: A New Approach to Oil and Gas Exploration, second ed., Springer-Verlag, Berlin, Heidelberg, 1984.

[20] F. Behar, M. Vandenbroucke, Chemical modelling of kerogens, Org. Geochem. 11 (1) (1987) 15–24, https://doi.org/10.1016/0146-6380(87)90047-7.

[21] C. Boyer, J. Kieschnick, R. Suarez-Rivera, R.E. Lewis, G. Waters, Producing gas from its source, Oilfield Rev. 18 (3) (2006) 36–49.

[22] M.A. Sephton, R.M. Hazen, On the origins of deep hydrocarbons, Rev. Mineral. Geochem. 75 (2013) 449–465, https://doi.org/10.2138/rmg.2013.75.14.

[23] G.A. Riley, The carbon metabolism and photosynthetic efficiency of the earth as a whole, J. Am. Sci. 32 (1944) 134.

[24] A.G. Collins, Geochemistry of Oilfield Waters, Elsevier Scient. Publ., 1975.
[25] J.M. Hunt, How gas and oil form and migrate, World Oil 163 (1968) 140–150.
[26] X.B. Wang, Z.Y. Ouyang, S.G. Zhuo, M.F. Zhang, G.D. Zheng, Y.L. Wang, Serpentinization, abiogenic organic compounds, and deep life, Sci. China Earth Sci. 57 (5) (2014) 878–887, https://doi.org/10.1007/s11430-014-4821-8.
[27] J.L. Charlou, J.P. Donval, P. Jean-Baptiste, D. Levaché, Y. Fouquet, J.P. Foucher, P. Cochonat, Abiogenic petroleum generated by serpentinization of oceanic mantellic rocks, in: AAPG Research Conference, Calgary, Canada, 2005, 2005.
[28] K. Vladimir, Abiogenic deep origin of hydrocarbons and oil and gas deposits formation, in: Hydrocarbon, InTech, 2013, https://doi.org/10.5772/51549.
[29] S. Abbas, The non-organic theory of the genesis of petroleum, Curr. Sci. 71 (9) (1996) 677–684.
[30] V.G. Kutcherov, V.A. Krayushkin, Deep-seated abiogenic origin of petroleum: from geological assessment to physical theory, Rev. Geophys. 48 (1) (2010), https://doi.org/10.1029/2008RG000270.
[31] T. Paulauskiene, Petroleum extraction engineering, in: Recent Insights in Petroleum Science and Engineering, InTech, 2018, https://doi.org/10.5772/intechopen.70360.
[32] E. Darko, Short guide summarising the oil and gas industry lifecycle for a non-technical audience, in: Economic and Private Sector Professional Evidence and Applied Knowledge Services Helpdesk Request, No. October, EPS-PEAKS, United Kingdom, 2014, pp. 5–15. https://assets.publishing.service.gov.uk/.../Short_guide_summarising_the_oil_and_gas.
[33] F. Robelius, Giant Oil Fields—The Highway to Oil. Giant Oil Fields and Their Importance for Future Oil Production, Acta Universitatis Upsaliensis. Digital Comprehensive Summaries of Uppsala Dissertations from the Faculty of Science and Technology. 168 pp. Uppsala, 2007, ISBN: 978-91-554-6823-1.
[34] J. Hanania, K. Stenhouse, J. Donev, Energy Education—Oil Well, 2017, Available from: https://energyeducation.ca/encyclopedia/Oil_well. (Accessed 2 April 2020).
[35] L. Amado, Field case evaluations, 2013, in: L. Amado (Ed.), Reservoir Exploration and Appraisal, Elsevier, 2013, pp. 1–157, https://doi.org/10.1016/C2009-0-30581-4.
[36] D. Perrin, Well Completion and Servicing: Oil and Gas Field Development Techniques, 1999. Publications de l'Institut français du pétrole.
[37] B. Haq, The role of microbial products in green enhanced oil recovery: acetone and butanone, Polymers 13 (12) (2021), https://doi.org/10.3390/polym13121946.
[38] J.G. Speight, Enhanced recovery methods for heavy oil and tar sands, in: Enhanced Recovery Methods for Heavy Oil and Tar Sands, Gulf, 2013, https://doi.org/10.1016/C2013-0-15525-0.
[39] J.G. Speight, Production, in: Handbook of Offshore Oil and Gas Operations, Elsevier, 2015, pp. 149–189 (Chapter 6).
[40] A. Mandal, Chemical flood enhanced oil recovery: a review, Int. J. Oil Gas Coal Technol. 9 (3) (2015) 241–264, https://doi.org/10.1504/IJOGCT.2015.069001.
[41] B. Guo, X. Liu, X. Tan, Well Workover in Petroleum Production Engineering, second ed., Elsevier, Amsterdam, 2017.
[42] J.M. Davison, M. Salehabadi, S. De Gennaro, D. Wilkinson, H. Hogg, C. Hunter, P. Schutjens, Plugging and abandonment of oil and gas wells: a geomechanics perspective, in: In 51st US Rock Mechanics/Geomechanics Symposium 2017, vol. 3, American Rock Mechanics Association (ARMA), 2017, pp. 1693–1701.
[43] Heriot-Watt University, Production Technology 1a, Department of Petroleum Engineering—Heriot-Watt University, Edinburgh, UK, 1999.
[44] B. Guo, S. Song, A. Ghalambor, T.R. Lin, General design information, in: Offshore Pipelines, Elsevier, Amsterdam, 2016, pp. 13–20, https://doi.org/10.1016/B978-0-12-397949-0.00002-9.
[45] A. Satter, G.M. Iqbal, Reservoir fluid properties, in: Reservoir Engineering, Elsevier, 2016, pp. 81–105, https://doi.org/10.1016/b978-0-12-800219-3.00004-8.

[46] S.K. Sanyal, S.S. Marsden, H.J. Ramey, Effect of Temperature on Petrophysical Properties of Reservoir Rocks, Society of Petroleum Engineers, 1974, January 1, https://doi.org/10.2118/4898-MS.
[47] G. Takacs, A review of production engineering fundamentals, in: Sucker-Rod Pumping Handbook, Elsevier, 2015, pp. 13–56 (Chapter 2).
[48] L.S. Hansen, S. Pedersen, P. Durdevic, Multi-phase flow metering in offshore oil and gas transportation pipelines: trends and perspectives, Sensors (Switzerland) (2019), https://doi.org/10.3390/s19092184.
[49] J.G. Speight, Natural Gas: A Basic Handbook, GPC Books, Gulf Publishing Company, Houston, TX, 2007.
[50] J.G. Speight, The Chemistry and Technology of Crude Oil, fifth ed., CRC Press, Taylor & Francis Group, Boca Raton, FL, 2014.
[51] S. Mokhatab, W.A. Poe, J.G. Speight, Handbook of Natural Gas Transmission and Processing, Elsevier B.V, Amsterdam, Netherlands, 2006.
[52] A.J. Kidnay, W.R. Parrish, Fundamentals of Natural Gas Processing, CRC Press, Boca Raton, FL, 2006, p. 9.
[53] N. Anosike, A. El-Suleiman, P. Pilidis, Associated gas utilization using gas turbine engine, performance implication—Nigerian case study, Energy Power Eng. 08 (03) (2016) 137–145, https://doi.org/10.4236/epe.2016.83012.
[54] Y. Alqaheem, A simulation study for the treatment of Kuwait sour gas by membranes, Heliyon 7 (1) (2021), https://doi.org/10.1016/j.heliyon.2021.e05953.
[55] X. Hu, The physical properties of reservoir fluids, in: X.T. Hu, et al. (Eds.), Physics of Petroleum Reservoirs, Springer Mineralogy, 2017, pp. 165–324, https://doi.org/10.1007/978-3-662-53284-3_3.
[56] Z.E. Heinemann, W. Brigitte, Texbook 2. Reservoir Fluids, PHDG Textook Series, University of Leoben, 2004. 2004. 155 p.
[57] L. Pei, W. Gang, C. Zhu, Y. Liu, W. He, Y. Dong, B. Xiang, Carbon isotope and origin of the hydrocarbon gases in the Junggar Basin, China, J. Nat. Gas Geosci. 3 (5) (2018) 253–261, https://doi.org/10.1016/j.jnggs.2018.11.002.
[58] M.A. Thomas, Chemical and isotopic characteristics of methane in groundwater of Ohio, 2016, in: U.S. Geological Survey Scientific Investigations Report 2018–5097, 2018, https://doi.org/10.3133/sir20185097. 42 p.
[59] H.X. Vo, Composition Variation During Flow of Gas Condensate Wells, 2010. A Report Submitted to the Department of Energy Resources Engineering of Stanford University.
[60] I.S. Starobinets, L.S. Kondratov, G.S. Fedorova, Role of physical fields in formation of hydrocarbon and geochemical concentration fields in sedimentary rocks, in: Modeling of Geochemical Field of Oil and Gas Deposits, Nedra Press, Moscow, 1986, pp. 47–54.
[61] G.V. Chilingar, L. Buryakovsky, N.A. Eremenko, M.V. Gorfunkel, Geology and geochemistry of oil and gas, Dev. Pet. Sci. 52 (2005) 390, https://doi.org/10.1017/CBO9781107415324.004.
[62] D.P. Nolan, Physical properties of hydrocarbons and petrochemicals, in: Handbook of Fire and Explosion Protection Engineering Principles, 2014, pp. 55–77, https://doi.org/10.1016/b978-0-323-31301-8.00004-0.
[63] EIA, Production of Crude Oil, US Energy Information Administration, 2019.
[64] J.G. Speight, The Chemistry and Technology of Petroleum, third ed., Marcel Dekker, New York, 1999.
[65] Elemental Analysis of Crude Oils, in: K. Nadkarni (Ed.), Elemental Analysis of Fossil Fuels and Related Materials, ASTM International, West Conshohocken, PA, 2014, pp. 457–508.
[66] Mathpro, An Introduction to Petroleum Refining and the Production of Ultra Low Sulfur Gasoline and Diesel Fuel, MathPro, Inc, Bethesda, MD, 2011, pp. 1–35.
[67] D. Lesueur, The colloidal structure of bitumen: consequences on the rheology and on the mechanisms of bitumen modification, Adv. Colloid Interf. Sci. (2009), https://doi.org/10.1016/j.cis.2008.08.011.
[68] M.Z. Hasanvand, M. Montazer, M. Salehzadeh, M. Amiri, M. Fathinasab, A literature review of asphaltene entity, precipitation, and deposition, introducing recently models of deposition in the well column, J. Oil Gas Petrochemical Sci. 1 (2) (2018) 83–89, https://doi.org/10.30881/jogps.00016.

[69] A. Hinkle, E.J. Shin, M.W. Liberatore, A.M. Herring, M. Batzle, Correlating the chemical and physical properties of set of heavy oils from around the world, Fuel 87 (2008) 3065–3070.
[70] N. Aske, Characterisation of Crude Oil Components, Asphaltene Aggregation and Emulsion Stability by Means of Near Infrared Spectroscopy and Multivariate Analysis (PhD thesis), Norwegian University of Science and Technology Trondheim, 2002.
[71] Y.V. Kissin, Catagenesis and composition of petroleum: origin of n-alkanes and isoalkanes in petroleum crudes, Geochim. Cosmochim. Acta 51 (9) (1987) 2445–2457, https://doi.org/10.1016/0016-7037(87)90296-1.
[72] J.G. Speight, The Chemistry and Technology of Petroleum, Marcel Dekker, New York, 1991.
[73] National Academies of Sciences, Engineering, and Medicine, Spills of Diluted Bitumen From Pipelines: A Comparative Study of Environmental Fate, Effects, and Response, National Academies Press, 2016. https://doi.org/10.17226/21834.
[74] J.G. Speight, Petroleum asphaltenes—part 1: asphaltenes, resins and the structure of petroleum, Oil Gas Sci. Technol. 59 (5) (2004) 467–477, https://doi.org/10.2516/ogst:2004032.
[75] J.S. Gudmundsson, Asphaltene, in: Flow Assurance Solids in Oil and Gas Production, CRC Press, 2017, pp. 45–83, https://doi.org/10.1201/9781315185118-3.
[76] A. Demirbas, O. Taylan, Removing of resins from crude oils, Pet. Sci. Technol. 34 (8) (2016) 771–777, https://doi.org/10.1080/10916466.2016.1163397.
[77] J.G. Speight, Occurrence and formation of crude oil and natural gas, in: Subsea and Deepwater Oil and Gas Science and Technology, Elsevier, 2015, pp. 1–43, https://doi.org/10.1016/b978-1-85617-558-6.00001-5.
[78] O. León, E. Contreras, E. Rogel, G. Dambakli, S. Acevedo, L. Carbognani, J. Espidel, Adsorption of native resins on asphaltene particles: a correlation between adsorption and activity, Langmuir 18 (13) (2002) 5106–5112, https://doi.org/10.1021/la011394q.
[79] M. Bobra, A study of water-in-oil emulsification, Int. Oil Spill Conf. Proc. 1991 (1) (1991) 483–488. Retrieved from https://www.bsee.gov/sites/bsee.gov/files/osrr-oil-spill-response-research//120az.pdf.
[80] J.G. Speight, Chemical and physical studies of petroleum asphaltenes, Dev. Pet. Sci. 40 (PA) (1994) 7–65, https://doi.org/10.1016/S0376-7361(09)70249-3.
[81] L. Goual, Petroleum asphaltenes, in: Crude Oil Emulsions—Composition Stability and Characterization, InTech, 2012, https://doi.org/10.5772/35875.
[82] B. Hollebone, The oil properties data appendix, in: M. Fingas (Ed.), Handbook of Oil Spill Science and Technology, John Wiley and Sons, NY, 2015, pp. 577–681. 2015. B.P. Hollebone, Appendix A. The oil properties data appendix, in: Handbook of Oil Spill Science and Technology, John Wiley & Sons, 2015, pp. 575–681, https://doi.org/10.1002/9781118989982.app1.
[83] R.G. Santos, W. Loh, A.C. Bannwart, O.V. Trevisan, An overview of heavy oil properties and its recovery and transportation methods, Braz. J. Chem. Eng. (2014), https://doi.org/10.1590/0104-6632.20140313s00001853.
[84] A. Kayode Coker, Process safety and pressure-relieving devices, in: Ludwig's Applied Process Design for Chemical and Petrochemical Plants, fourth ed., vol. 2, Gulf Professional, 2010, pp. 575–770, https://doi.org/10.1016/C2009-0-27075-9.
[85] B. Hollebone, Oil Physical Properties: Measurement and Correlation in Oil Spill Science and Technology, second ed., Gulf Professional, 2017, pp. 185–207, https://doi.org/10.1016/B978-0-12-809413-6.00003-5.
[86] ASTM D6749-02(2018), Standard Test Method for Pour Point of Petroleum Products (Automatic Air Pressure Method), ASTM International, West Conshohocken, PA, 2018, www.astm.org, https://doi.org/10.1520/D6749-02R18.
[87] V.P. Dimri, R.P. Srivastava, N. Vedanti, Reservoir geophysics. Some basic concepts, in: Handbook of Geophysical Exploration: Seismic Exploration, vol. 41, Elsevier B.V., 2012, pp. 89–118, https://doi.org/10.1016/B978-0-08-045158-9.00005-1.

[88] F. David Martin, R.M. Colpitts, Reservoir engineering A2, in: W.C. Lyons (Ed.), Standard Handbook of Petroleum and Natural Gas Engineering, Gulf Professional Publishing, Houston, 1996, pp. 1–362, https://doi.org/10.1016/B978-088415643-7/50009-1 (Chapter 5).
[89] J.R. Fanchi, Fluid Properties in Shared Earth Modeling, Gulf Professional Publishing, 2002, pp. 87–107.
[90] S. Wlazlowski, B. Hagströmer, M. Giulietti, Causality in crude oil prices, Appl. Econ. 43 (24) (2011) 3337–3347, https://doi.org/10.1080/00036841003636250.
[91] J.M. Moldowan, W.K. Seifert, E.J. Gallegos, Relationship between petroleum composition and depositional environment of petroleum source rocks, AAPG Bull. 69 (1985) 1255–1268.
[92] J. Elirich, A.V. Hirner, H. Stark, Distribution of trace elements in crude oils from southern Germany, Chem. Geol. 48 (1985) 313–323.
[93] A.F. Oluwole, O.I. Asubiojo, J.I. Nwachukwu, J.A. Ojo, A. Ogunsola, R. Filby, S. Fitzgerald, C. Grimmy, Neutron activation analysis of Nigerian crude oil, J. Radioanal. Nucl. Chem. 168 (1993) 145–152.
[94] A.J.G. Barwise, Role of nickel and vanadium in petroleum classification, Energy Fuel 4 (1990) 647–652.
[95] M.A. Rasheed, P.L. Srinivasa Rao, A. Patil, H. Patel, S. Kala, A. Boruah, S.Z. Hasan, P.H. Rao, T. Harinarayana, Identification of hydrocarbon microseepage using trace metal indicators in petroliferous region of South Cambay Basin, Gujarat, India, Int. J. Petroleum Petrochemical Eng. 1 (1) (2015) 12.
[96] K.M. Watson, E.F. Nelson, G.B. Murphy, Characterization of petroleum fractions, Ind. Eng. Chem. 27 (12) (1935) 1460–1464.
[97] J. Neff, K. Lee, E.M. DeBlois, Produced water: overview of composition, fates, and effects, in: Produced Water, Springer, New York, 2011, pp. 3–54, https://doi.org/10.1007/978-1-4614-0046-2_1.
[98] Y. Liang, Y. Ning, L. Liao, B. Yuan, Special focus on produced water in oil and gas fields, in: Formation Damage During Improved Oil Recovery, Elsevier, 2018, pp. 515–586, https://doi.org/10.1016/b978-0-12-813782-6.00014-2.
[99] A. Larson, Produced Water: Oil and Gas Terminology Glossary, FACT SHEET, WSEC-2017-FS-013 IWWC, Water Environment Federation, 2018.
[100] C.E. Clark, J.A. Veil, Produced water volumes and management practices in the United States, in: Further Explorations in the Energy-Water Nexus, Nova Science Publishers, 2012, pp. 49–105.
[101] J.A. Veil, M.G. Puder, D. Elcock, R.J. Redweik Jr., A White Paper Describing Produced Water From Production of Crude Oil, Natural Gas, and Coal Bed Methane, Argonne National Laboratory (ANL), Lemont, IL, 2004.
[102] R.M. Flores, Co-Produced Water Management and Environmental Impacts in Coal and Coalbed Gas, Elsevier, 2014.
[103] Oilfield Glossary, Connate Water, Schlumberger Limited, 2019. Retrieved from: https://glossary.oilfield.slb.com/en/terms/c/connate_water. (Last visited December 2019).
[104] L.C. Case, Origin and current usage of the term "connate water", Bull. Am. Assoc. Pet. Geol. 39 (1955) 1879–1882.
[105] D.E. White, Magmatic, connate, and metamorphic waters, Geol. Soc. Am. Bull. 68 (1957) 1659–1682.
[106] S. Ge, S.M. Gorelick, Hydrology, floods and droughts: groundwater and surface water, in: Encyclopedia of Atmospheric Sciences, second ed., Elsevier, 2015, pp. 209–216, https://doi.org/10.1016/B978-0-12-382225-3.00171-7.
[107] P.J. Webb, Enhanced Scale Management Through the Application of Inorganic Geochemistry and Statistics, SPE, 2004.
[108] K. Guerra, K. Dahm, S. Dundorf, Oil and Gas Produced Water Management and Beneficial Use in the Western United States, U.S. Department of the Interior Bureau of Reclamation, Managing Water in the West, no. 157, 2011, p. 129. doi:3180.
[109] B.R. Hansen, S.R. Davies, Review of potential technologies for the removal of dissolved components from produced water, Chem. Eng. Res. Des. 72 (1994) 176–188.

[110] J.P. Fillo, J.M. Evans, Characterization and management of produced waters from underground natural gas storage reservoirs, in: American Gas Association Operation Section Proceedings, 1990, pp. 448–459.
[111] United States Environmental Protection Agency (USEPA), EPA Office of Compliance Sector Notebook Project: Profile of the Oil and Gas Extraction Industry, 2000. EPA/310-R-99-006.
[112] M.C. Shepherd, F.L. Shore, S.K. Mertens, J.S. Gibson, Characterization of produced waters from natural gas production and storage operations: regulatory analysis of a complex matrix, in: J.P. Ray, F.R. Engelhardt (Eds.), Produced Water: Technological/Environmental Issues and Solutions, Plenum, New York, 1992, pp. 163–174.
[113] A. Fakhru'l-Razi, A. Pendashteh, L.C. Abdullah, D.R.A. Biak, S.S. Madaeni, Z.Z. Abidin, Review of technologies for oil and gas produced water treatment, J. Hazard. Mater. 170 (2009) 530–551.
[114] P.J.C. Tibbetts, I.T. Buchanan, L.J. Gawel, R. Large, A comprehensive determination of produced water composition, in: J.P. Ray, F.R. Engelhardt (Eds.), Produced Water. Environmental Science Research, vol. 46, Springer, Boston, MA, 1992, https://doi.org/10.1007/978-1-4615-2902-6_9.
[115] B.M. Johnson, L.E. Kanagy, J.H. Rodgers, et al., Chemical, physical, and risk characterization of natural gas storage produced waters, Water Air Soil Pollut. 191 (2008) 33–54, https://doi.org/10.1007/s11270-007-9605-8.
[116] K. Benko, J. Drewes, Produced water in the western United States: geographical distribution, occurrence, and composition, Environ. Eng. Sci. 25 (2) (2008) 239–246.
[117] J.K. Otten, T. Mercier, Produced Water Brine and Stream Salinity, US Geological Survey (USGS), 1995. http://water.usgs.gov/orh/nrwww/Otten.pdf. (Accessed October 2019).
[118] R.J. Dingman, E.E. Angino, Chemical composition of selected Kansas brines as a aid to interpreting change in water chemistry with depth, Chem. Geol. 4 (3) (1969) 25–339.
[119] P.A. Dickey, Increasing concentration of subsurface brine with depth, Chem. Geol. 4 (1969) 361–370.
[120] W. Renpu, Basis of well completion engineering, in: Advanced Well Completion Engineering, Elsevier, 2011, pp. 1–74, https://doi.org/10.1016/b978-0-12-385868-9.00001-4.
[121] V.A. Sulin, Waters of Petroleum Formations in the System of Nature Waters, Gostoptekhizdat, Moscow, 1946. 96 pp. (in Russian).
[122] C.C. Patton, J.M Campbell, Applied Water Technology, Campbell Petroleum Series, 1986.
[123] M.T. Stephenson, A survey of produced water studies, in: J.P. Ray, F.R. Engelhardt (Eds.), Produced Water: Technological/Environmental Issues and Solutions, Plenum Publishing Corp, New York, 1992, pp. 1–12.
[124] P. Ekins, R. Vanner, J. Firebrace, Zero emissions of oil in water from offshore oil and gas installations: economic and environmental implications, J. Clean. Prod. 15 (2007) 1302–1315.
[125] R.P.W.M. Jacobs, R.O.H. Grant, J. Kwant, J.M. Marqueine, E. Mentzer, The composition of produced water from shell operated oil and gas production in the North Sea, in: J.P. Ray, F.R. Englehart (Eds.), Produced Water, Plenum Press, New York, 1992.
[126] S. Johnsen, T. Roe Utvik, E. Garland, B. Vde Vals, J. Campbell, Environmental fate and effects of contaminants in produced water, in: Seventh SPE International Conference on Health. Safety and Environment in Oil and Gas Exploration and Production, Richardson, TX, 2004, p. 9.
[127] R.C. Ayers, M. Parker, Produced Water Waste Management Technical Report, Canadian Association of Petroleum Producers (CAPP), 2001.
[128] U.S Geological Survey, Naturally Occurring Radioactive Materials (NORM) in Produced Water and Oil-Field Equipment—An Issue for the Energy Industry. USGS Fact Sheet FS–142–99, 1999.
[129] J.W. Amyx, C.M. Bass Jr., R.L. Whiting, Petroleum Reservoir Engineering, McGraw-Hill Book, New York City, 1960.
[130] W.D. McCain Jr., Reservoir-fluid property correlations-state of the art (includes associated papers 23583 and 23594), SPE Reserv. Eng. 6 (2) (1991) 266–272. SPE-18571-PA https://doi.org/10.2118/18571-PA.
[131] W.D. McCain Jr., The Properties of Petroleum Fluids, second ed., PennWell Books, Tulsa, Oklahoma, 1990.

[132] I. Demir, Formation Water Chemistry and Modeling of Fluid-Rock Interaction for Improved Oil Recovery in Aux Vases and Cypress Formations, Illinois Basin, 1995. Illinois Petroleum 148.

[133] T. Ahmed, Fundamentals of hydrocarbon phase behavior, in: Equations of State and PVT Analysis, Elsevier, 2016, pp. 1–69, https://doi.org/10.1016/b978-0-12-801570-4.00001-5.

[134] F.M. Orr Jr., K. Jessen, U. Stanford, in: J.R. Fanchi (Ed.), Phase Diagrams in Volume I—General Engineering, Society of Petroleum Engineers, 2007, pp. 371–396 (Chapter 8).

[135] W.C. Lyons, G.J. Plisga, M.D. Lorenz, Standard Handbook of Petroleum and Natural Gas Engineering, Elsevier, 2015, pp. 1–1822, https://doi.org/10.1016/C2010-0-64695-5.

[136] A. Danesh, Phase behaviour fundamentals, in: PVT and Phase Behaviour of Petroleum Reservoir Fluids, Elsevier, 1998, pp. 1–31.

[137] J. Bellarby, Tubing Well Performance, Heat Transfer and Sizing, in: J. Bellarby (Ed.), Well Completion Desin, Developments in Petroleum Science, vol. 56, Elsevier, 2009, pp. 247–302.

[138] Y. Bai, Q. Bai, Subsea pipelines and risers, in: Subsea Pipelines and Risers, Elsevier, 2005, https://doi.org/10.1016/B978-0-08-044566-3.X5000-3.

[139] I. Ashour, N. Al-Rawahi, A. Fatemi, G. Vakili-Nezh, Applications of equations of state in the oil and gas industry, in: Thermodynamics—Kinetics of Dynamic Systems, InTech, 2011, https://doi.org/10.5772/23668.

[140] A. Firoozabadi, Thermodynamics of Hydrocarbon Reservoirs, McGraw-Hill, 1989.

CHAPTER

Flow Assurance

2

2.1 Introduction

The main motivation for the development of oil and gas fields is to produce at a stable and maximized pace from the reservoir to the sales terminals, with the fewest obstacles and the lowest operating costs.

Various types of obstacles are frequently encountered during production operations, as illustrated in Fig. 2.1. These obstacles can be production chemistry related, such as emulsions, corrosion, solid deposits, microbial activity, and reservoir souring, or they can be fluid hydrodynamics related, such as friction loss, intermittent flow, slugging, and metering issues. Basically, flow assurance is that management strategy that incorporates diverse disciplines of science and engineering to ensure a smooth flow of the produced fluids with the fewest obstacles.

The term flow assurance is composed of two parts: *flow* and *assurance*. Accordingly, the discussion in this chapter begins with understanding the concept of flow, and then it is wrapped up with the concept of the assurance of flow, and its components.

2.2 The concept of fluid flow

Fluid flow is defined as the motion of a fluid subjected to unbalanced forces or stresses; the motion continues as long as the unbalanced forces are applied. A comprehensive way of describing flow is as a concept that describes how fluids behave and how they interact with their surrounding environment: for example, water moving through a channel or pipe, or over a surface. Flow can be either steady or unsteady. If the properties of the fluid (velocity, pressure, density, etc.) at any point in the flow field are independent of time, then the flow is steady; otherwise, it is unsteady. The unsteady flow is time dependent and is also called transient flow [1].

Flow can also be either laminar or turbulent. Laminar flows are smoother, as the fluid flows in parallel layers with no disruption between the layers, while turbulent flows are more chaotic and are characterized by recirculation, swirling and reverse currents, called eddy currents, or eddies. Fig. 2.2 illustrates laminar and turbulent flow.

Another way to define laminar and turbulent flow is through the fluid's Reynolds number. The Reynolds number is the ratio of inertia forces to viscous forces and is a convenient parameter for predicting if a flow condition will be laminar or turbulent [2,3]. It is defined according to Eq. (2.1):

$$\mathrm{Re}_D = \frac{\rho V D}{\eta} \tag{2.1}$$

Essentials of Flow Assurance Solids in Oil and Gas Operations. https://doi.org/10.1016/B978-0-323-99118-6.00015-0

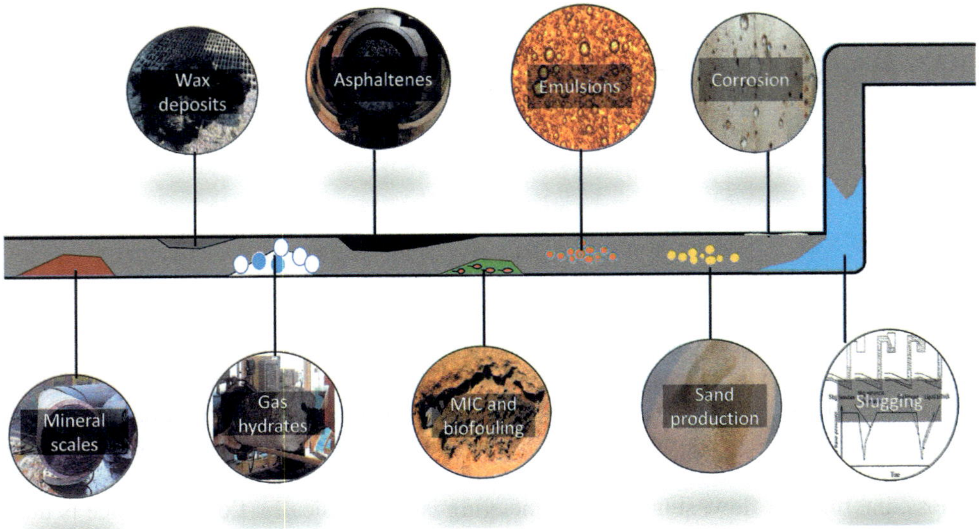

FIG. 2.1

Flow assurance problems in oil and gas fields.

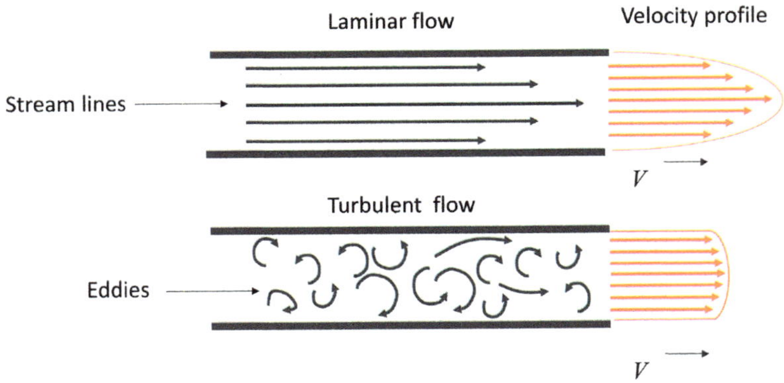

FIG. 2.2

Laminar vs turbulent flow.

where Re_D is Reynolds number, dimensionless; D is pipe diameter, ft; V is fluid velocity, ft/s; ρ is fluid density, lb/ft^3; and η is fluid viscosity, lb/ft s.

- if $Re < 2000 \rightarrow$ Laminar
- if $2000 < Re < 3000 \rightarrow$ Transient
- if $3000 < Re \rightarrow$ Turbulent

The question that arises now is what actually makes fluids flow? Or what is the driving force behind the flow of fluids?

And the answer is energy....

Thermal, chemical, electrical, or gravitational forms of energy can be the driving force of fluid flow. The applied energy form must lead to enough pressure difference (also known as the pressure gradient) between the starting point and the terminal point so that fluids can flow.

In a petroleum system, the fluids in the reservoir are naturally compressed under very high temperatures, and they store a considerable amount of energy converted to high reservoir pressure, which is the main driving force for pushing the fluids upwards in the production tubing. Pumps are also used to keep fluids flowing; they operate by converting the mechanical shaft work to flow energy in the form of increasing the fluid pressure.

So in the interest of simplicity, we can say that pressure (or pressure gradient) is the conventional driving force of fluid flow. In other words, a pressure drop in the production system means weakening its driving force; therefore the main task of the production operator is to maintain a minimum pressure drop.

2.3 Pressure drop

Pressure drop (ΔP) is a decrease in fluid pressure during flow between two points. At the start in a petroleum production system, the production fluids are flowing upwards by the natural reservoir pressure resulting from the compressibility of the reservoir fluids. During production, the system encounters different types of pressure drops, including the following [4]:

- Pressure drop due to the loss of the natural pumping power of the reservoir due to the continuous discharging of the reservoir fluids.
- Pressure drop due to fluid flow (flow through porous media, perforations, restrictions, increases in elevation, friction, turbulence, shaft work, etc.).

As the production continues, the pressure keeps decreasing and pressure drops keep increasing until reaching a point that significantly diminishes the production rates. Therefore it is critical to keep pressure drop as low as possible to keep the production rates steady, as planned.

Pressure drop is also called the head loss. The head loss can be expressed according to Eq. (2.2):

$$h_{\text{loss}} = \sum h_{\text{major\_losses}} + \sum h_{\text{minor\_losses}} \tag{2.2}$$

where

h_{loss} is total head loss in the pipe or duct system.
$h_{\text{major\_losses}}$ is the major loss due to friction in the pipe or duct system.
$h_{\text{minor\_losses}}$ is the minor loss due to the components in the system, like system design, valves, bends, composition of fluids flowing through the pipelines, presence of solids within the fluids, etc.

The terms minor and major do not really reflect the implication of the names, as a minor loss sometimes can be significantly higher than a major loss.

The mechanical energy balance equation expresses the conservation of the sum of pressure, kinetic and potential energies, the net heat transfer q, the work done by the system ω, and the frictional energy e_f, which makes the phenomena of pressure loss and the factors affecting it more understandable [4]:

$$\frac{\Delta P}{\rho} = g\Delta Z + \frac{\Delta V^2}{2} + e_f + \omega - q \tag{2.3}$$

where P is pressure (psi), ρ is fluid density (lb/ft^3), g is acceleration due to gravity (32.2 ft/s^2), Z is the vertical height (ft), V is fluid velocity (ft/s), e_f is irreversible energy dissipation between two points (ft^2/s^2), ω is work done per unit mass of liquid, and for most practical applications, there is no work done by or on the fluid and the kinetic energy correction factor is assumed to be 1; q is the net amount of heat transferred into the system (mechanical energy/time) and by assuming that very little heat transferred into the system, then Eq. (2.3) can be rewritten as:

$$\frac{\Delta P}{\rho} = g\Delta Z + \frac{\Delta V^2}{2} + e_f \tag{2.4}$$

Eq. (2.4) represents Bernoulli's equation, adding the friction loss (e_f) to it. Here, to represent the pressure drop, Eq. (2.4) will be:

$$\Delta P = \underbrace{\rho g \Delta Z}_{\Delta P \text{ gravity}} + \underbrace{\rho \frac{\Delta V^2}{2}}_{\Delta P \text{ acceleration}} + \underbrace{\rho e_f}_{\Delta P \text{ friction}} \tag{2.5}$$

Eq. (2.5) represents a general correlation to understand pressure loss in a production system. It is clear from this equation that three main pressure drop modules are responsible for the total pressure drop in a system: pressure drop due to gravity, acceleration, and friction.

2.4 Factors affecting pressure drop

In oil and gas production systems, the pressure drop is experienced as the fluid moves from the reservoir sandface to the surface facilities. This pressure drop is a function of the mechanical configuration of the wellbore and surface facilities, the properties of the fluids, and the producing rate [5].

The total pressure drop from the reservoir pressure to the surface is illustrated in Fig. 2.3. The pressure drop at different points or nodes in the production system is known as the nodal analysis. If the separator represents the end of the production system, the total pressure drop in the system is the difference between the average reservoir pressure and the separator pressure, represented by $\Delta P_T = P_R - P_S$.

The main factors and contributors in pressure drop in oil and gas production systems are discussed in the following sections.

2.4.1 Wellbore pressure drop

Production systems experience significant pressure drop in the wellbore, as up to 80% of the total pressure loss in a flowing well can occur in lifting the reservoir fluid to the surface. The pressure drop experienced within the reservoir (due to fluid flow in porous media) and the fluid movement from the reservoir sandface to the surface are functions of the configuration of the wellbore (the different

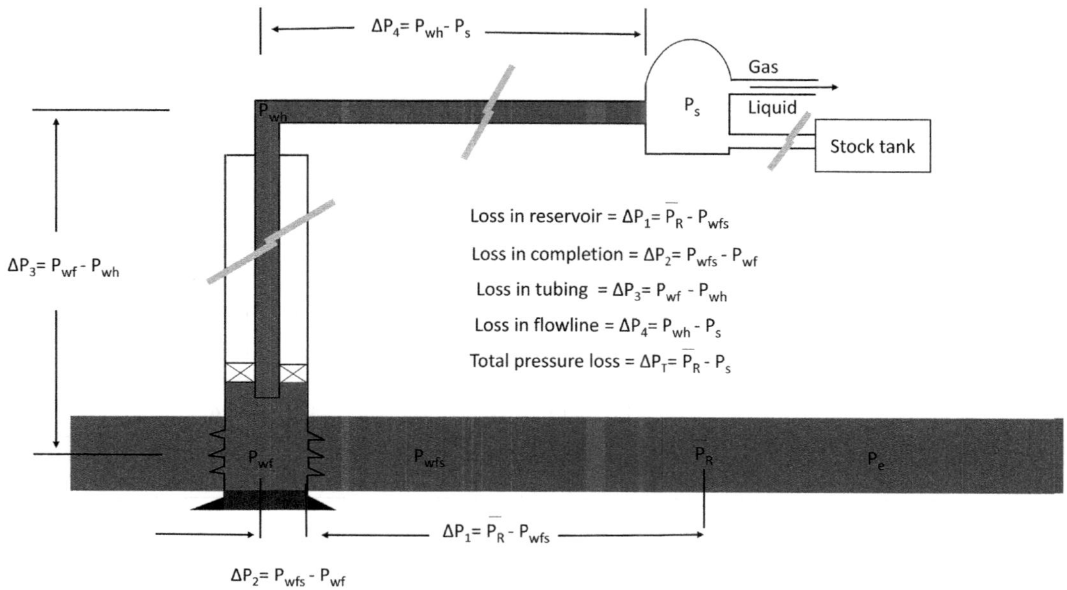

FIG. 2.3

Pressure losses in a production system.

installation that restricts the flow, e.g., nipples, sliding sleeves, or subsurface flow-control devices), the properties of the fluids (viscosity and density), and the producing rate [5].

2.4.2 System design and installations

Throughout their journey into the production system, fluids pass through different installations that cause pressure drops, such as pipe entrances and exits, valves, fittings, elbows, bends, tees, orifices, and others. The term e_f in Eq. (2.5) represents the loss of mechanical energy due to friction and includes losses due to the flow through lengths of pipe installations. The loss at these installations is known as the minor loss, or the local loss [4]. For each frictional device, a loss term is used, as per Eq. (2.6):

$$e_f = K_f \frac{V^2}{2} \tag{2.6}$$

where V is fluid velocity (ft/s), and K_f is the *resistance coefficient*, or sometimes called the *excess head loss* (or *loss coefficient*) due to the pipe or pipe fitting, and its value is different for every valve and installation. Table 2.1 represents the different values of K_f for the different installations.

Table 2.1 The value of K_f of some pipeline installations.

Installation	Description	K_f
Gate valve	1″	0.18
	4″	0.14
	12″–16″	0.1
Ball valve	1″	0.07
	4″	0.05
	12″–16″	0.04
Standard elbow-90°	1″	0.69
	12″–16″	0.39
Standard Tee thru-flo	1″	0.46
	12″–16″	0.26

Reprinted with permission from E.S. Menon, P.S. Menon, Pressure loss through piping systems, in: Working Guide to Pumps and Pumping Stations, Elsevier, 2010, pp. 69–111, doi:10.1016/b978-1-85617-828-0.00004-4. Copyright (2010), Elsevier.

2.4.3 Friction

The calculation of friction loss in pipes and ducts depends on whether the flow is laminar or turbulent [6]. The fluid normally flows in the pipelines in layers, as indicated in Fig. 2.4: a first is the viscous layer, which is the layer that is touching the pipe wall at which the flow is laminar, followed by the buffer layer, and finally the buffer layer is followed by the turbulent region.

Friction forces are related to fluid velocity. As fluid velocity increases, until it reaches a critical point at which the fluid flow changes from laminar flow to turbulent flow, instead of using the energy to pump and push the fluids in the line, it will be wasted in vortex currents perpendicular to the flow direction, known as eddies. These currents are accompanied by increasing fluid resistance and friction forces, which lead to energy loss in the form of pressure drop.

The pressure loss due to friction in a given length and diameter of pipe at a certain liquid flow velocity can be calculated by redefining the K_f in Eq. (2.6) ($K_f = f\ L/D$), which gives us the Darcy-Weisbach equation that fully represents the terms e_f in Eq. (2.5), to be represented by Eq. (2.7) [4]:

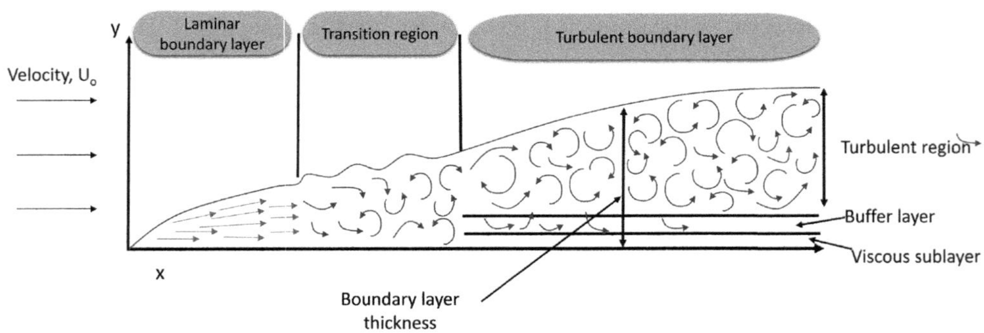

FIG. 2.4

Laminar vs turbulent flow with boundary layers.

$$e_f = f\frac{LV^2}{2D} \tag{2.7}$$

where V is the fluid velocity (ft/s), f is the friction factor (dimensionless), L is the flow path length (ft), and D is the flow path diameter (ft).

From this equation, it is obvious that pressure loss increases as the pipe length increases and diameter decreases.

The friction factor f (also known as the Darcy friction factor) in Eq. (2.7) depends on flow type. For laminar flow, f depends only on the Reynolds number and it can be given by Eq. (2.8):

$$f = \frac{64}{\text{Re}} \tag{2.8}$$

This means the only fluid property that influences friction loss is the viscosity (because the density cancels out in Eq. 2.1) [4].

However, for turbulent flow the friction factor cannot be calculated by Eq. (2.8), and it depends on the internal roughness of the pipe, the inside diameter of the pipe, and the Reynolds number. In practice, the friction factor in the case of turbulent flow can only be obtained by using the Moody diagram, with its values ranging from 0.008 to 0.10, as shown in the Moody diagram in Fig. 2.5 [2–4]. The Moody

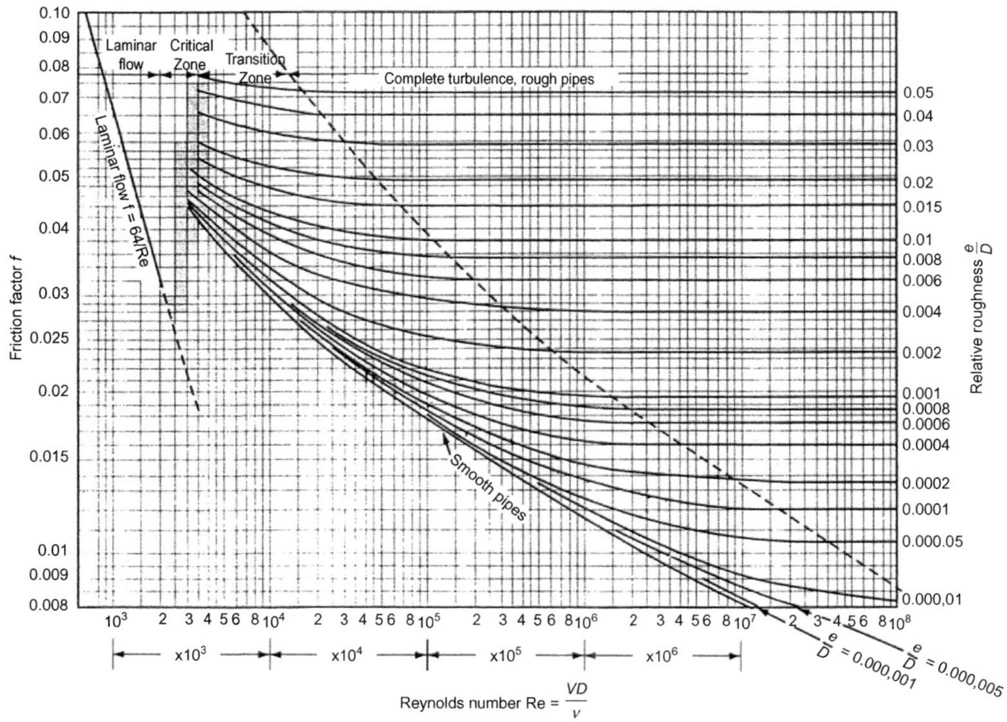

FIG. 2.5

Moody diagram.

Reprinted with permission from E.S. Menon, P.S. Menon, Pressure loss through piping systems, in: Working Guide to Pumps and Pumping Stations, Elsevier, 2010, pp. 69–111, doi:10.1016/b978-1-85617-828-0.00004-4. Copyright (2010), Elsevier.

diagram is a graph in nondimensional form that relates the Darcy friction factor, Reynolds number, and the relative roughness.

The Darcy friction factor is also referred to as the Moody friction factor, since it can be read off the Moody diagram. The term Fanning friction factor is another way of addressing the friction factor: it is simply one-fourth the value of the Darcy (or Moody) friction factor, as shown in Eq. (2.9):

$$\text{Fanning friction factor} = \text{Darcy friction factor}/4 \tag{2.9}$$

2.4.4 Surface roughness

As the flow becomes turbulent (at Re > 4000), the friction factor depends basically on both the value of Re as well as the relative roughness of the pipe (*e*/*D*). However, with increasing Re above a certain value (> 10 million), the friction factor becomes a function of the relative roughness alone, and that region is designated in the Moody diagram as complete turbulence in rough pipes. The region between the critical and turbulent (the dashed line) zones is known as the transition zone (Fig. 2.5) [3,6]. The relative roughness is given by Eq. (2.10).

$$\text{Relative roughness} = \text{Absolute roughness}/\text{inside diameter of pipe} = e/D \tag{2.10}$$

For new steel pipe, the absolute roughness of $e = 0.0018$ in. may be used. Table 2.2 lists the absolute roughness of some pipe materials.

2.4.5 Fluid properties

The flowing fluid volumes and properties affect pressure drop in many different ways. Produced fluid volumes and flow rates affect the pressure drop. Fluid properties like viscosity and density are among the major factors that affect the pressure drop. The more viscous the liquids are, the more friction they encounter in the pipe. Less dense (more volatile) and compressible fluids flow with the least pressure drop [7].

In gas systems, the difference in properties between gas and liquid makes pressure drop a complex problem. High gas rates sufficient to remove liquid are required to reach stable production rates. If gas rates are too low, pressure drop increases due to liquid accumulation, increasing back pressure on the formation and, with liquid accumulation in the wellbore, this might ultimately kill the well.

Furthermore, in oil systems, fluid mixing and formation of intermediate phases like emulsions and foams affect fluid flow and consequently pressure drop. Stable emulsions increase the flow viscosity, resulting in greater pressure drop [8]. Pressure drop due to emulsion was found to be greater than pressure drop due to water or oil alone [9].

Table 2.2 Surface roughness of some piping materials.

Pipe material	Pipe roughness (in.)	Pipe roughness (mm)
Cast iron	0.0102	0.26
Galvanized iron	0.0059	0.15
Commercial steel	0.0018	0.045

2.4.6 Temperature

Temperature generally affects fluid properties (viscosity, density, compressibility). High temperature decreases fluid viscosity and density and improves fluid flow, thus reducing fluid pressure drop. On the other hand, high temperature affects fluid compressibility, which may lead to changes in fluid velocity that finally affect pressure drop [7].

2.4.7 Gravity forces

The effect of gravitational forces is demonstrated in Eq. (2.5). The weight of the height column of fluids will be important in the vertical part of the well, long flowlines with elevations, and risers, due to the gravity issues [7]. The effect of gravity on fluid flow and pressure drop was thoroughly addressed by MacGillicray [10] and Ohta et al. [11].

2.4.8 Fluid flow regimes

Most oil wells ultimately produce both oil and gas at the beginning and often also produce water at certain stages of the well life cycle. Consequently, multiphase flow is common in oil wells [12]. Multiphase flow in a pipeline is usually studied by the flow regime and corresponding flow rate. Flow regimes describe the flow structure and phase distributions. Flow regimes can affect pressure drop and cause troubled production. Slug flow is a common example of a troublesome flow regime that causes significant pressure drop in the production system.

2.4.9 Solid particle transport and deposition

Solid particles can originate from the reservoir rocks, or precipitate due to changes in thermodynamic equilibrium of produced fluids. Sand, clay, paraffin, asphaltenes, gas hydrates, and mineral scales are examples of the different solids that can exist in a system. These solids can stay suspended in the produced fluids and transport with them, or can deposit on the internal pipeline walls. In either case, such solids affect pressure drop in different ways:

(a) Solids transported with the fluids cause changes in the fluid rheology and flow regimes, leading to pressure drop.
(b) Solids cause changes in surface roughness, which affects frictional pressure drop.
(c) Solids deposition causes changes in pipelines and equipment internal dimensions, causing flow restrictions and pressure drop.

These impacts of solids on pressure drop are addressed in detail in Chapter 3, which covers problems associated with solids formation and deposition.

2.5 The flow assurance concept

Now, in studying the flow assurance concept, it plainly means assuring a steady flow of fluids with the least number of issues that hinder the flow or cause pressure drop, as discussed earlier. These issues or challenges evidently stem from two main origins, which represent the main themes of flow assurance:

fluid dynamics origins and fluid production chemistry, or simply the fluid physics and fluid chemistry. Fig. 2.6 illustrates the concept of flow assurance and its two pillars: fluid dynamics and production chemistry.

Flow assurance in the general sense is a risk analysis and management process. Flow assurance engineers use science and engineering disciplines to analyze and assess process risks, and then design the risk control methods. Flow assurance studies require understanding both steady-state and transient multiphase flows, as well as fluid properties, fluid composition, pressure volume temperature (PVT) characteristics, and other fluid phase behaviors from the reservoir to the topsides process equipment throughout the life of the field, to prevent challenging conditions [13].

Flow assurance is a relatively new term in the oil and gas industry. It was first used by Petrobras in the early 1990s in Portuguese, as *garantia do escoamento* meaning literally "guarantee of flow," or flow assurance. Flow assurance was originally developed to maintain fluid flow in deepwater operations, which encountered solids issues such as wax and gas hydrates deposition problems, where the main concerns are extreme temperature, pressure, very long tiebacks of pipelines with complex system installations, and difficulty in accessing the system. A historical review of the development of the flow assurance strategies for gas hydrates is given by Sloan [14]. The main flow assurance strategy back then was initially the "avoidance" of gas hydrates formation by operating outside their formation envelope [7]. Moving forward, the flow assurance strategy covered more fluid flow obstacles and adopted a "preventing the problem" strategy (using chemical and engineering means) rather than "avoiding" them. Further development in flow assurance occurred, as a discipline of science in the petroleum industry, by addressing the possible production problems and laying them out in a risk assessment matrix and providing possible solutions, hence providing "management" of the production problems. Nowadays, with the aid of the digital revolution, Big Data, and the incorporation of different realms of science and

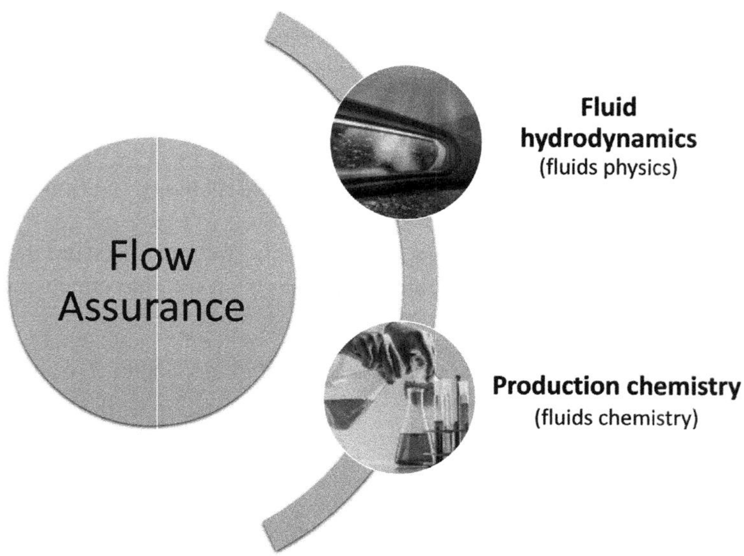

FIG. 2.6

Basic pillars of flow assurance.

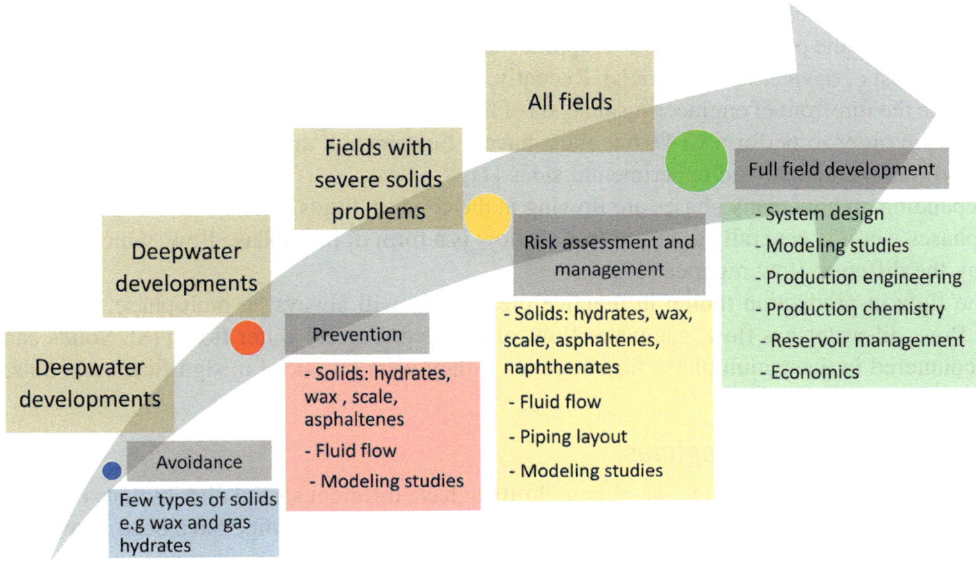

FIG. 2.7

Development of the flow assurance concept.

engineering, flow assurance extensively contributes to the development of oil and gas fields from the starting point of the commissioning, to system design, startup, production, shutdown, maintenance, up until field abandonment. Fig. 2.7 represents the development of the flow assurance concept over time.

In the coming sections, the two pillars of flow assurance and their topics are addressed, and then the flow assurance as a management strategy is discussed.

2.6 Fluid dynamics

Fluid dynamics is a branch of fluid mechanics that studies the movement of liquids and gases. The study of the movements of gases and their interactions with moving bodies is usually called aerodynamics, whereas the study of liquids is called hydrodynamics.

In the fundamental study of fluid dynamics three main approaches, namely, theoretical, experimental, and computational, are encountered [15].

The central mathematical description for all theoretical fluid dynamics models is given by the Navier-Stokes equations, which describe the motion of viscous fluid domains [15]. The main axioms of fluid dynamics are directed by governing equations that are based on the three physics laws of conservation [16]:

1. Conservation of Mass: Continuity Equation.
2. Conservation of Momentum: Momentum Equation of Newton's Second Law.
3. Conservation of Energy: First Law of Thermodynamics or Energy Equation.

In fluid dynamics, these three laws are expressed using the Reynolds transport theorem.

In addition to the previous theoretical approaches is the experimental part, which is a bit expensive (due to the cost of devices and materials). Recently, computational fluid dynamics (CFD) emerged and surfaced in the forefront of engineering analysis of fluid flow, due to the decreasing costs of computing. However, in order to perform CFD, one must have a fundamental understanding of fluid flow itself, from both the theoretical and experimental sides [1].

Depending on how many phases are flowing in the conduit, fluids can flow in one phase or in multiple phases, yet it is generally assumed that all flow is a form of multiphase flow, which is characterized by the different flow regimes [17].

The flow in production tubing in practice presumably will always be multiphase, including oil-water flow, oil-water-gas flow, gas-water flow, and gas-condensate-water flow [18]. Solids can also be encountered in these multiphase flow regimes if they were produced in significant amounts.

2.6.1 Multiphase flow regimes

A special feature of multiphase flows is their ability to have different spatial distributions of the phases in a pipe. These different flow patterns and arrangements of fluid in a pipe have been known as flow regimes, which in turn can be altered by the direction of flow relative to gravitational acceleration [19]. Knowledge of the flow regime determines the selection of the appropriate model for pressure gradient and liquid holdup [20].

Since there are three phases, possible flows can be solid-liquid, solid-gas, liquid-liquid, liquid-gas (liquid-gas can be viewed as the extreme case of liquid-liquid where the density ratio is extremely large) and solid-liquid-gas flow; if water and oil are well mixed, they can form emulsion as an intermediate phase; when gas and liquid are well mixed, they can form foam, which can significantly affect the flow across the production system [17].

Materials in multiphase flow can flow in the same direction, known as cocurrent flow, or they can flow in the opposite direction, known as countercurrent flow. In general, cocurrent is more common [17].

In cocurrent flow, two liquids can have three main categories: vertical, horizontal, and sloped (or near vertical and near horizontal). There is no exact meaning for the terms "near vertical" or "near horizontal" and there is no consensus on the limiting angles. Moreover, the vertical flow may be with gravity or against gravity; while the counter gravity flow is more common industrially, the vertical flow with gravity is also common in oil and gas operations, especially in the fluids injection wells during secondary and tertiary recovery operations [17].

2.6.1.1 Gas-liquid and liquid-liquid flow regimes

In vertical pipes and for a constant liquid flow rate (assuming pressure is above the bubble point in oil/gas systems) the pipe is filled in the bottom with liquid due to the gravity forces, and then the gas phase tends to be distributed in the liquid oil as small bubbles at low gas flow rates, which is called the "bubble flow," then the gas will start to generate when the pressure falls below the bubble point. Increasing gas flow rate causes agglomeration of bubbles into larger slugs and plugs, and the flow in this case is known as "slug flow." Further increasing gas flow rate causes further gas expansion and liberation, resulting in slug expansion and coalescence and separation of the phases into annular patterns or "annulur flow," wherein liquid concentrates at the channel walls and gas flows in the central core for vertical ducts. At extremely high velocities of the central gas column, shear at the gas-oil interface can lead to oil dispersion in the gas in the form of a "mist." This "mist flow" pattern will occur at very high flow velocities in the tubing and for systems with a high gas-oil ratio (GOR) [19,21]. Fig. 2.8 illustrates the gas-liquid flow regimes in vertical pipes.

The case of liquid-liquid flow in production wells may occur in low GOR wells that produce water. The flow regimes for liquid-liquid systems may not follow the same behavior as the gas-liquid system. If oil and water enter the wellbore from the reservoir and flow up the tubing to the surface, the physical distribution of the phases will depend upon their relative volumetric properties, i.e., one phase will be continuous and the other dispersed [21]. For example:

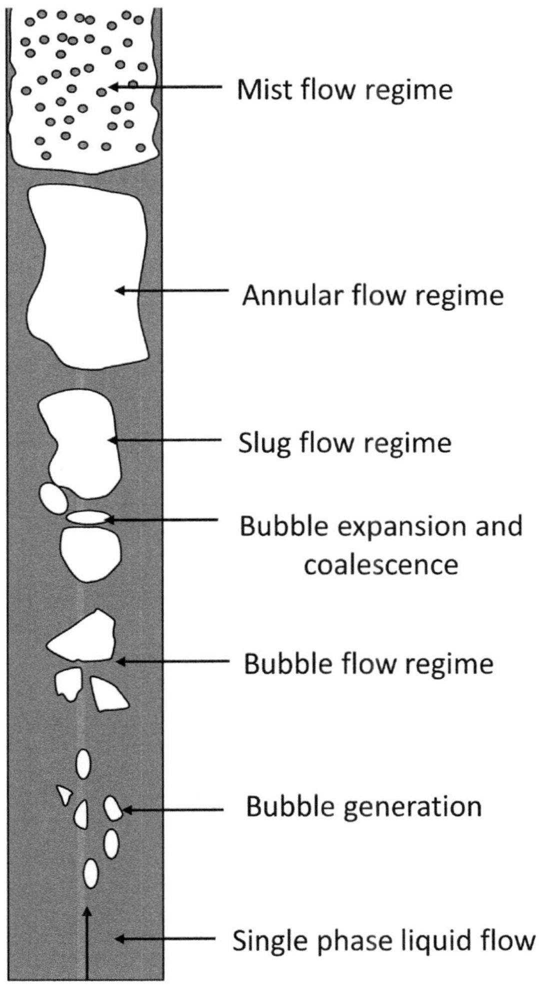

FIG. 2.8

Gas-liquid flow regimes in vertical pipes.

(a) In a high WOR well (say, 90%), the oil would be dispersed in the continuous water phase.
(b) At a low WOR, the oil would be in a continuous phase.

For horizontal ducts, gravitational force (buoyancy) tends to drain the liquid annulus toward the bottom of the channel so the bubbles and slugs tend to flow in the upper part of the pipe; further expanding the gas slugs results in stratified flow and increasing the velocity will lead to stratified wavy flow. This downward segregation of the liquid phase can be overcome by kinetic forces at high flow rates, causing stratified flows to revert to annular flows. At high gas flow rates, more of the liquid tends to be entrained as dispersed drops; in the limit one obtains completely dispersed mist flow [19]. Fig. 2.9 illustrates the gas-liquid flow regimes in horizontal pipes.

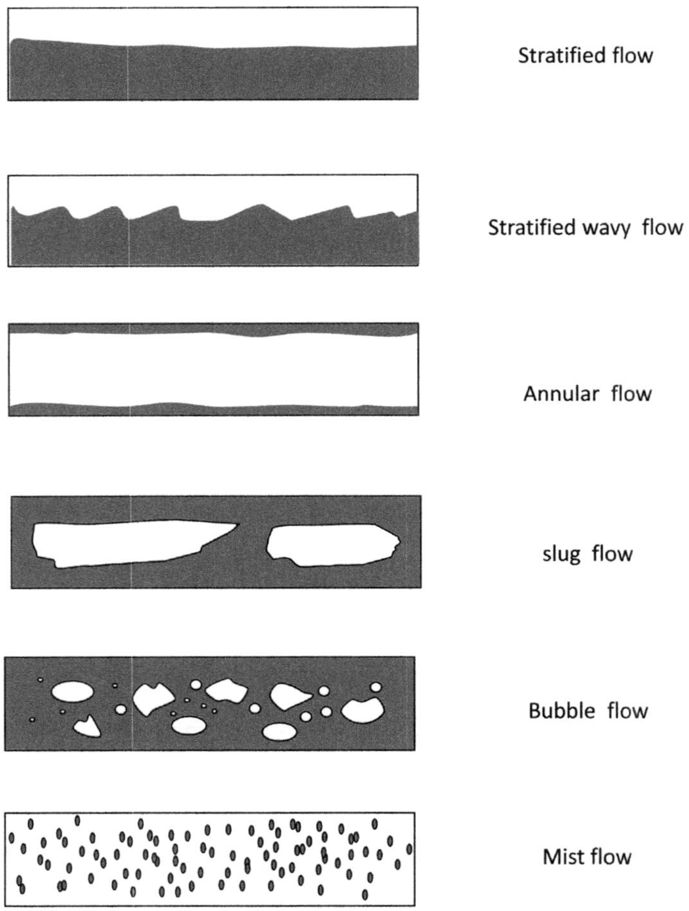

FIG. 2.9

Gas-liquid flow regimes in horizontal pipes.

2.6.1.2 Solid-fluid flow regimes

Solids can be transported with liquids, which is called hydraulic transport or slurry flow, or they can be transported with gases, which is called pneumatic conveying or pneumatic transport.

The conveying or transport of solids by a fluid in a pipe can involve a wide range of flow conditions and phase distributions, depending on the density, size, shape, and concentration of the solid particles and the density, viscosity, and velocity of the fluid [22].

Consider vertical upflow of a fluid (gas or liquid) with solid particles. At low flow rates (small gas/liquid velocity), the fluid phase percolates between stationary particles; this is termed flow through a fixed bed. With the fluid velocity increasing beyond a minimum, the particles will move and the flow regime is referred to as mixed fluidized bed. Additional increase of the fluid velocity will move all the particles and this is called a full fluidized bed. In the case of liquid, a further increase in velocity will create a slug flow (slug shapes or domes are almost empty of solid particles). In the case of gas, additional increase creates "tunnels" that are almost empty of solid particles. Any additional increase in the fluid velocity causes large turbulence and the ordinary domes are replaced by churn type flow or large bubbles that are almost empty of solid particles. Further increase of the fluid flow increases the empty spots to the whole flow. In that case, the sparse solid particles are dispersed all over, which is called a dispersed flow regime [17]. Fig. 2.10 summarizes solid-fluid flow regimes in vertical pipes.

In a horizontal pipe in the case of solid-fluid transportation, five flow regimes can be discerned depending on the properties of the conveyed solids, the conveying fluids, and the characteristics of the pipe: homogeneous flow, heterogeneous flow, intermediate flow regime, saltation, and capsule flow or moving bed [23].

The homogeneous regime is when the particles are fine and light (when all the particle sizes are on the order of tens of microns or less) and their settling velocity is much less than the turbulent mixing velocity, and the mean velocity of the flow is high enough to keep them uniformly suspended throughout the pipe cross-section.

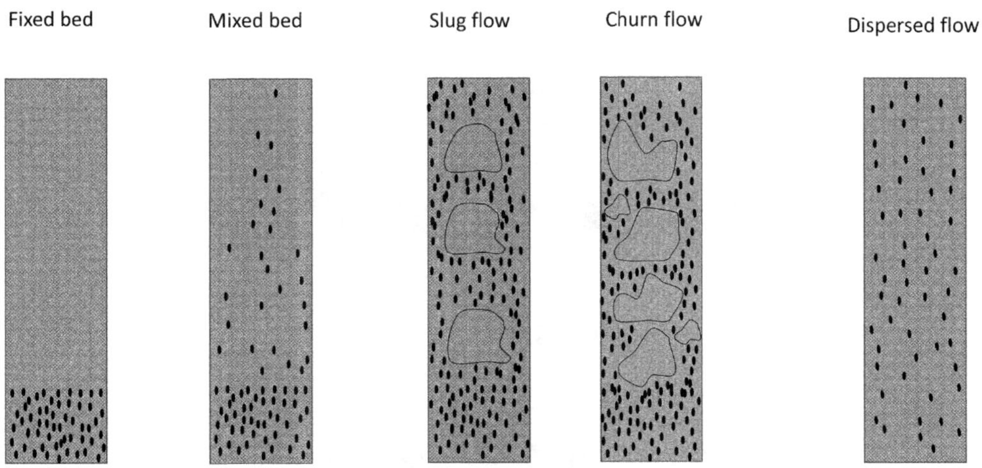

FIG. 2.10

Flow regimes of solid-fluid flow in vertical pipes.

When somewhat larger particles are present, vertical gradients will occur in the concentration and the regime is termed heterogeneous; moreover, the larger particles will tend to sediment faster and so a vertical size gradient will also occur. The intermediate flow regime occurs when conditions of the homogeneous and heterogeneous flow exist simultaneously. The limit of the heterogeneous flow regime occurs when the particles form a packed bed in the bottom of the pipe. When a packed bed develops, the flow regime is known as a saltation flow. In a saltation flow, solids join the bed and proceed in discontinuous jumps, slides, or rolls. When the flow velocity is slow, the turbulence is not enough to carry the solids, and then the solids will settle and flow at the bottom of the pipe as a sliding bed. This is considered a type of stratified flow. Capsule flow occurs when solids are packed into cylindrical capsules and transported in series [24–26]. Fig. 2.11 summarizes solid-fluid flow regimes in horizontal pipes.

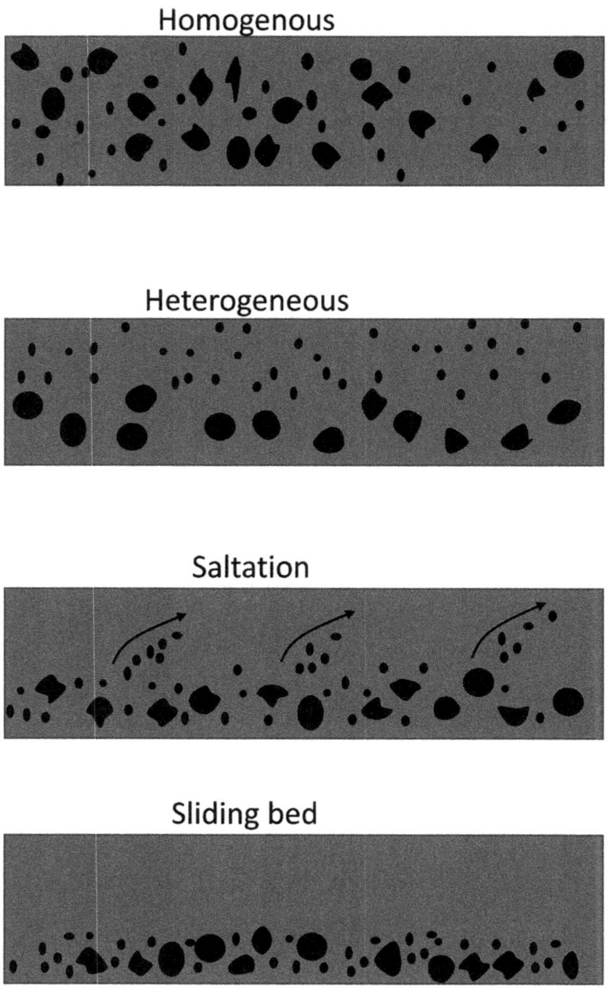

FIG. 2.11

Flow regimes of solid-liquid flow in horizontal pipes.

The effect of different flow regimes on the flow and pressure drop is well discussed in the literature. In the oilfield, it is very important to know the flow regime. Operators prefer continuous flow, not intermittent flow. Slug flow is intermittent and is definitely not desired in oil and gas operations due to its detrimental impacts on the integrity of components in processing/receiving facilities.

Slugs can cause damage to facilities, instabilities in liquid control systems, separator flooding, increased corrosion, starving compressors, slug-induced oscillations, and high backpressure, which lead to shutdowns and production loss. There are different forms of slugs, such as hydrodynamic slugs, terrain-induced slugs, turn-up slugs, and pigging slugs. Terrain-induced slugging normally occurs when liquid is trapped in the pipeline at low spots. Riser slugging is a type of terrain-induced slugging that occurs at the riser base and normally depends on the flow rate [27,28]. A severe slugging mechanism is depicted in Fig. 2.12 [29].

Therefore slugging must be detected and prevented in production systems. Slug detection can be achieved using a flow regimes map, modeling and simulation, or some devices in the system to collect slug-relevant signals. Slugging can be prevented by improving the process design, optimizing the process operations and parameters (e.g., feedback control of a riser/well topside choke valve and feedback control of artificial gas-lifting), and using equipment like slug catchers or slug compression systems that can be installed in the oilfield where medium-to-long slugs are expected [30].

Flow patterns are commonly represented in flow regime maps. The flow regime map was proposed by Hewitt and Roberts in 1969, and it represents a fairly wide range of experimental data for upward two-phase

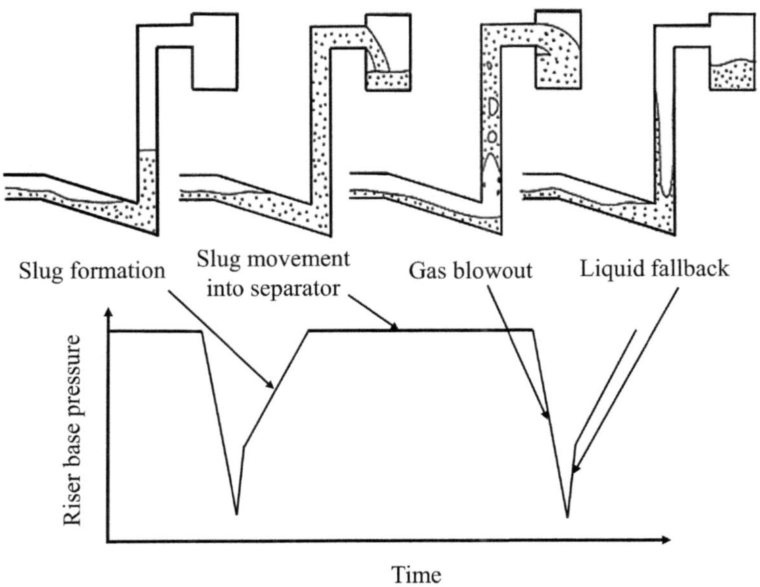

FIG. 2.12

Description of severe slugging.

Reprinted with permission from P. Han, L. Guo, Numerical simulation of terrain-induced severe slugging coupled by hydrodynamic slugs in a pipeline-riser system, Int. J. Heat Fluid Flow 56 (2015) 355–366, doi:10.1016/j.ijheatfluidflow.2015.10.005. Copyright (2015), Elsevier.

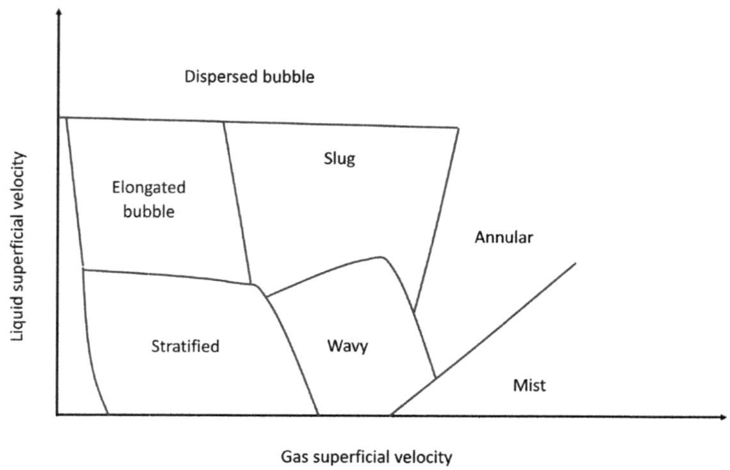

FIG. 2.13

Flow regime map.

flow [31]. Various variables have been used in such maps, including superficial velocity, mass flux, and Froude number. There is no universal flow regime map for a flow in all lines; every pipe flow case should be studied and flow regimes map should be elucidated. Fig. 2.13 gives an example of a flow regime map.

2.6.2 Computational fluid dynamics

Computational fluid dynamics (CFD) is a branch of fluid dynamics that mathematically models fluid behavior using digital computers to produce quantitative predictions of fluid-flow phenomena based on the conservation laws that govern fluid motion (conservation of mass, momentum, and energy). CFD has increased considerably in importance and in accuracy; however, its predictions are never completely exact. Because many potential sources of error may be involved, one must be very careful when interpreting the results produced by CFD techniques [32].

In the oil and gas industry, CFD has principally been used to model fluid flow in surface pipelines, production tubing, and different production equipment [33]. Moreover, recently CFD applications were extended to studying optimization of bit performance, convective heat transfer between casings and annular fluids, flow in porous media, simulation of multiphase flow for designing and sizing flowlines and separation equipment, cement displacement, corrosion and erosion prediction, optimization of drilling hydraulics and hole cleaning, and wellbore stability analysis [34]. These software packages can also include different chemical reactions, fluid structures mixing, and interaction modules used for deposit simulation and prediction. A number of commercial CFD packages are available for use in different oil and gas field applications. A few of these packages and their features are discussed in the following text.

- **ANSYS**: this software is a product of ANSYS Inc. ANSYS oil and gas simulation software enables designing, verifying, and delivering more reliable equipment for drilling, production, transport, processing, and refining at lower cost. In addition it has a module that studies flow assurance problems, including equipment structural integrity, slugging, and sand transport and it can be used for solid deposits formation prediction.

- **Aspen HYSYS**: this software is a product of AspenTech. It has been widely used to model many facets of the oil and gas production fields, including separation systems, environmental control systems, gas dehydration, H_2S and CO_2 removal, and more. It is the tool of choice to determine the heat and material balance, separation performance, and regulatory compliance. Among other key performance criteria, the flow assurance module in Aspen HYSYS includes predicting hydrate, wax, asphaltene, and mineral scale formation.
- **PVTsim**: this is a product of Calsep. It is a versatile equation of state (EOS) modeling software that allows the user to simulate fluid properties and experimental PVT data. PVTsim provides detailed information about simulating and predicting gas hydrates, wax, asphaltene, and mineral scales formation.
- **PIPESIM**: the PIPESIM simulator is a product of Schlumberger. It offers comprehensive steady-state flow assurance workflows for front-end system design and production operations. The flow assurance capabilities of the simulator enable engineers to ensure safe and effective fluid transport—from sizing of facilities, pipelines, and lift systems, to ensuring effective liquids and solids management, to well and pipeline integrity. It also gives information about solids formation and deposition.
- **FlowManager**: this is a product of TechnipFMC plc. It is a dynamic multiphase flow simulator for online surveillance, optimization, and prediction of subsea oil and gas production systems. The simulator covers all operational scenarios (e.g., production, shut-in, restart) for a complete subsea field layout (reservoir, wells, pipelines, subsea process equipment). It also provides information about solids precipitation and deposition.
- **LedaFlow**: this is a new dynamic multiphase flow simulator. It is the product of a decade of collaboration between Total, ConocoPhillips, and SINTEF, and is commercialized by Kongsberg Oil & Gas Technologies. It is used by flow assurance, process, production, and reservoir engineers through the project life cycle to address gas/liquid, thermal, solid, and integrity challenges.

2.7 Production chemistry

Production chemistry is essentially concerned with the interactions of the produced fluids with each other, and with their surroundings (reservoir rocks, pipes, or equipment) under the system operating parameters (temperature, pressure, flow rate, pH, etc.). These interactions can lead to serious issues that cause costly maintenance, downtime, and production loss. Kelland [35] categorized these production issues into four main types:

- Fouling problems due to the precipitation and deposition of solids in the production system. Examples are mineral scales, paraffin, asphaltenes, gas hydrates, and biofouling.
- Fluid physical properties problems. Examples are emulsions, foams, and viscous fluids.
- Facilities integrity problems due to the deterioration or failure of the system structural materials when they interact with the flowing fluids or fall under heavy stress. These issues are covered under the corrosion problem.

- Environmental problems due to the effect of the produced fluids on the environment when they are finally discharged or disposed of, or when they accidentally leak into the environment due to failed operations. Examples are the produced water discharge and oil spills.

The main problems of production chemistry are discussed comprehensively in the following sections.

2.7.1 Solid deposits

Solid deposits formation is one of the major flow assurance issues in oil and gas operations. Mitigating solids deposition problems is not an easy task for a production chemist, due to their diversity, their rapid formation rate, and impact even at low concentrations.

Different types of solids are encountered in oil and gas operations, which can be classified into inorganic solids, including mineral scales, sulfur, sand, and fines or organic solids such as asphaltenes, paraffin waxes, gas hydrates, diamondoids, and naphthenates.

With a quick look into the production system, we can realize that usually these flow assurance solids are derived from the produced fluids themselves when their equilibrium is disturbed during production operations. To understand that phenomena, imagine production fluids at thermodynamic equilibrium at the reservoir high temperature and pressure for millions of years. Once production started and temperature and pressure changed, this equilibrium was disturbed; therefore the fluids tend to retain their equilibrium through changes in the fluid phases, which finally leads to the occurrence of various flow assurance solids like mineral scales, asphaltenes, waxes, and hydrates. Speaking of flow assurance solids formation, Fig. 2.14 illustrates the formation of different flow assurance solids as a function of system pressure/temperature.

As seen in Fig. 2.14, the production journey of reservoir fluids starts at reservoir pressure (P) and temperature (T) and ends at the production facility. On their way, these fluids traverse a pressure and

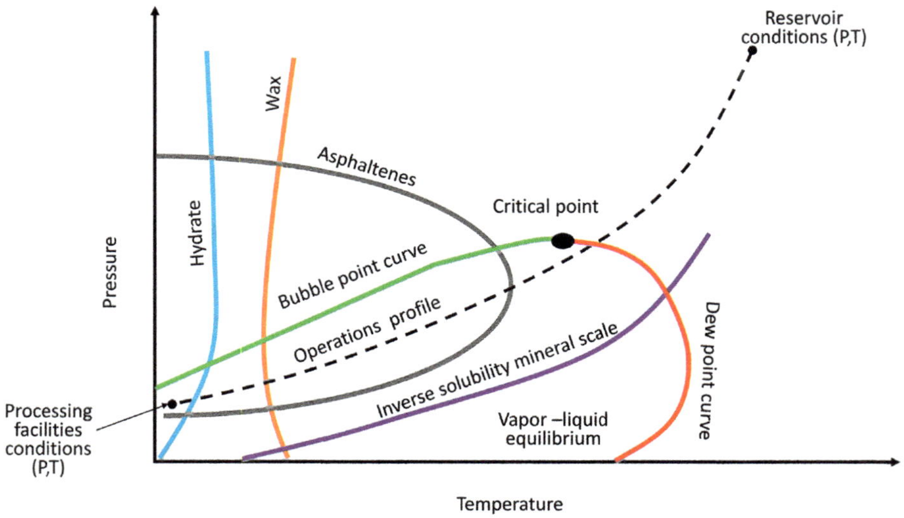

FIG. 2.14

Pressure-temperature curves of flow assurance solids.

temperature operating profile *(dotted black line)* that may cross the gas-liquid equilibrium line *(solid red-green line)*, which may cause slugging. In addition, the operating profile may cross one or more organic or inorganic solid equilibrium zones, which will result in precipitation that may deposit as solids to impede fluid flow. Pressure and temperature decline affects various types of mineral scales, i.e., sulfates and sulfides, whereas reverse solubility scales like calcite precipitate at high temperatures. Gas hydrates usually form at high pressures and low temperatures, while asphaltenes are mostly affected and precipitated due to pressure reduction, whereas pressure and temperature decline commonly cause wax deposition. The critical point defines the temperature and pressure on the bubble point-dew point line where the properties of gas and liquid are identical [36].

Besides natural thermodynamic pressure, temperature equilibrium disturbance, changes of fluid pH, mixing between incompatible fluids, fluid hydrodynamics, and microbial activity are other factors of solid deposits formation as indicated in Table 2.3. Fig. 2.15 shows full blockage by mineral scales and Fig. 2.16 shows mixed organic and inorganic deposits during pigging operations.

The solids types, characterization, monitoring, prediction, and management are the main topics of this book and every type is investigated thoroughly in the coming chapters.

Table 2.3 Summary of the major oil and gas field solid deposits types and their main contributing factors.

Deposit type	Contributing factors
Mineral scales	– Incompatible waters mixing – Pressure – Temperature – pH – Hydrodynamics
Gas hydrates	– Pressure – Temperature – Surfactants – Hydrodynamics
Wax deposits	– Temperature – Pressure – Hydrodynamics
Asphaltenes	– Mixing incompatible crudes – Pressure – Temperature – pH shift, acidizing – Streaming potential – Corrosion – Hydrodynamics
Naphthenates	– Mixing incompatible fluids – pH – Pressure – Temperature
Biofouling	– Bacteria activity – pH – Temperature – Nutrition availability

FIG. 2.15

Mineral scales blockage in production system

FIG. 2.16

Mixed organic and inorganic deposits during pigging operations.

2.7.2 Emulsions

2.7.2.1 Emulsification mechanism

Water is a major component of petroleum reservoirs. During oil production, water is mixed vigorously with oil at the choking spots in the production system (perforations, valves, and other equipment). As a result, water gets dispersed in oil and, with the aid of some surface active materials (like asphaltenes, paraffins, scales, and fines), forms a stable colloidal phase called an emulsion.

An emulsion is thus a special kind of colloidal dispersion in which a liquid is dispersed in a continuous liquid phase of different composition. The dispersed phase is sometimes referred to as the internal phase, and the continuous phase as the external phase [37].

Schubert and Armbruster [38] concluded that there are three basic reasons to form emulsions, as illustrated in Fig. 2.17:

- Interaction between two immiscible fluids, oil and water.
- Existence of emulsifying agents that can adjust themselves in the O/W interface to stabilize emulsions. Emulsifying agents can be a component of the crude oil such as asphaltenes, resins, and waxes, or they can be formed during production such as inorganic scales and fines, or they can be added to the system such as production chemicals with a surfactant nature.
- Diffusion of one liquid into another due to turbulent flow or mixing energy. Mechanical agitation is the basic cause of emulsion formation. While mixing is very low in the reservoir, it is higher at chokes, wellhead valves, etc.; extreme mixing conditions are experienced at the pressure gradient and this forms new W/O interfaces.

The factors affecting emulsion formation and stability are [39]:

- Heavy polar materials in the crude oil.
- Fine solids including organics (asphaltenes, waxes) and inorganics (clays, scales, corrosion products).
- Temperature.
- Droplet size and droplet size distribution.
- pH of the brine.
- Brine composition.

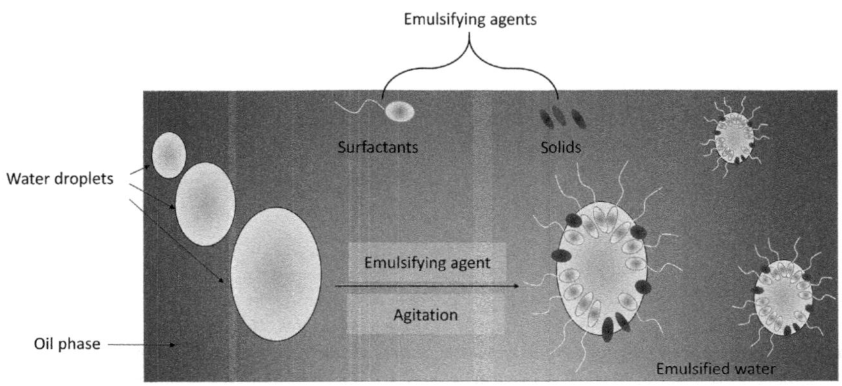

FIG. 2.17

Emulsion formation and stabilization.

2.7.2.2 Types of emulsions

Emulsions can be classified according to the continuous or dispersed phase into three types [40]: *Water-in-oil* (W/O) emulsion (ordinary emulsion where water drops are diffused in the oil constant phase), *oil-in-water* (O/W) emulsion (reverse emulsion where oil is dispersed in water), and this is in contrast to the previous type of emulsion. These two main types are the common types of emulsions in the petroleum industry. The third type is the multifaceted emulsion, or complex multiple emulsion, including double emulsions, i.e., *water-in-oil-in-water* (W/O/W) emulsion or *oil-in-water-in-oil* (O/W/O) emulsion; triple emulsions, i.e., W/O/W/O, or O/WO/W, and quadruple emulsions W/O/W/O/W. W/O and O/W emulsions are the most studied in oil and gas operations. Fig. 2.18 shows the different types of emulsions [41,42].

Operators and chemists care particularly about the kinetic stability of emulsions, which can also be classified into loose emulsions that last for only a few minutes, medium emulsions lasting for tens of minutes, and tight emulsions lasting for hours or days [39]. An example of a very tight emulsion is shown in Fig. 2.19.

Another way of classifying emulsions is according to their droplet size and stability, where macroemulsions are 1–100 μm in size and less stable, microemulsions are 20–500 nm in size and thermodynamically stable, and finally nanoemulsions are 10–100 nm in size and are thermodynamically unstable [43].

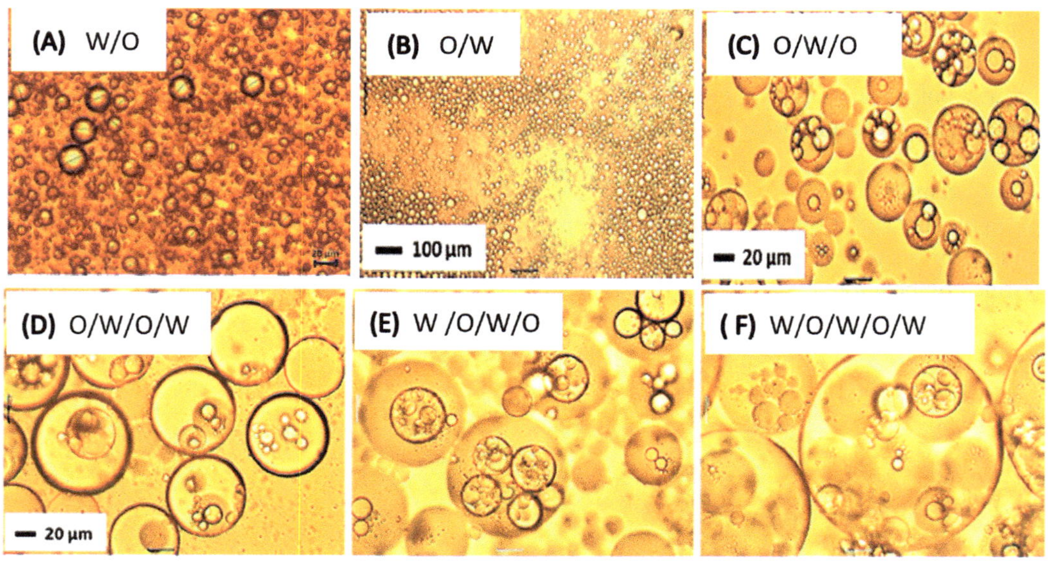

FIG. 2.18

The different types of emulsions.

Reprinted with permission from (A) G.R. Borges, G.B. Farias, T.M. Braz, L.M. Santos, M.J. Amaral, M. Fortuny, ... A.F. Santos, Use of near infrared for evaluation of droplet size distribution and water content in water-in-crude oil emulsions in pressurized pipeline, Fuel 147 (2015) 43–52, doi:10.1016/j.fuel.2015.01.053. Copyright (2015), Elsevier. (B–F) K.L. Thompson, C.J. Mable, J.A. Lane, M.J. Derry, L.A. Fielding, S.P. Armes, Preparation of pickering double emulsions using block copolymer worms, Langmuir 31(14) (2015) 4137–4144, doi:10.1021/acs.langmuir.5b00741. Copyright (2015), American Chemical Society.

FIG. 2.19

An example of a very tight emulsion.

Failure to manage emulsion formation and oil/water phase separation leads to significant challenges during production operations, including increasing fluid pumping costs and increased pressure drop, which in the worst cases can lead to down times in addition to high chemical treatment costs [44].

2.7.2.3 Emulsion control methods

Crude-oil emulsions are not thermodynamically stable and, given adequate time, will separate into oil and brine phases, in contrast to some well-known domestic examples of emulsions, e.g., mayonnaise, which will remain stable for years [44]. The process of destabilizing and breaking the emulsion into its water and oil components is conventionally known as "demulsification."

The destabilization/demulsification of water in oil (w/o) emulsions occurs in stages [44], as depicted in Fig. 2.20:

- Flocculation stage (droplets clump together forming clusters).
- Coalescence (droplets fuse together forming larger size drop).
- Creaming (for oil) or sedimentation (for water).
- Phase separation.

Demulsification of W/O emulsions in crude oil is usually covered under oil dehydration, while the treatment of O/W emulsions is usually studied during the produced water treatment or deoiling.

- **Oil demulsification (oil dehydration)**

Oil demulsification can be achieved in many ways. In field practices, a combination of methods is usually used to achieve the targeted oil production and water phase separation [35,39,45]. The common oil demulsification methods are listed in Table 2.4.

- **Oil in water treatment (deoiling)**

A wide variety of methods can be used these days to separate O/W emulsions. Some of these methods are listed in Table 2.5.

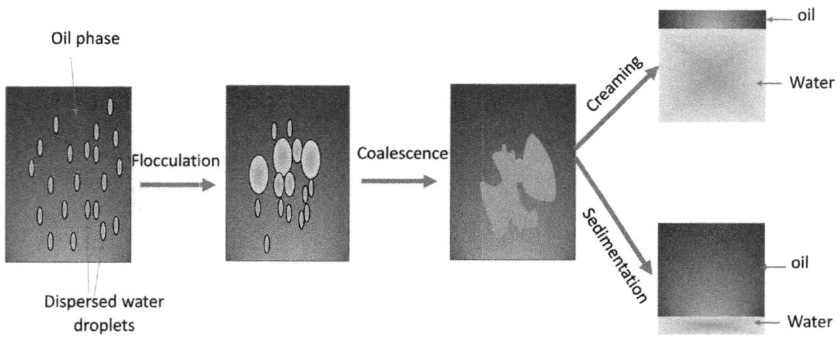

FIG. 2.20

The demulsification process.

Table 2.4 The different oil demulsification techniques.

Method	Description
Operational methods	Controlling the variables that cause emulsion formation, like solids, fluid mixing, shear, and production rates
Thermal methods	Heater treaters and heat exchangers are used to heat the fluids to reduce their viscosity and increase water coalescing and settling
Mechanical methods	Mechanical means are used to disrupt the physical barriers between oil and water and the density variation between them leads to separation. Examples are free-water knockout drum, two- and three-phase separators, desalters, gun barrels, and settling tanks
Electrical methods	The application of electric fields for separation of oil and water. Desalters are a common example
Chemical methods	The use of chemicals called demulsifiers to break the emulsions by neutralizing the stabilizing effect of emulsifying agents at the W/O interface. Demulsifiers rupture or weaken the rigid film, and enhance water droplet coalescence and separation

Table 2.5 Reverse emulsion treatment (deoiling).

Methods	Description
Mechanical methods	Examples: gravity settling tanks and ponds, skimming tanks, plate separators, dissolved air floatation (DAF), induced air floatation (IAF), hydrocyclones, centrifuges, and coalescers
Filtration methods	Use of filters to remove oil droplets
Chemical methods	Examples: polyelectrolytes, inorganic salts, demulsifiers, latex chemicals

2.7.3 Sludge

Petroleum sludge is usually a mixture of various quantities of waste oil, wastewater, sand, fines, and mineral matter [46]. It is considered a stable, tight, and highly viscous form of emulsion [47]. Fig. 2.21 illustrates sludge collected from a separator bottom.

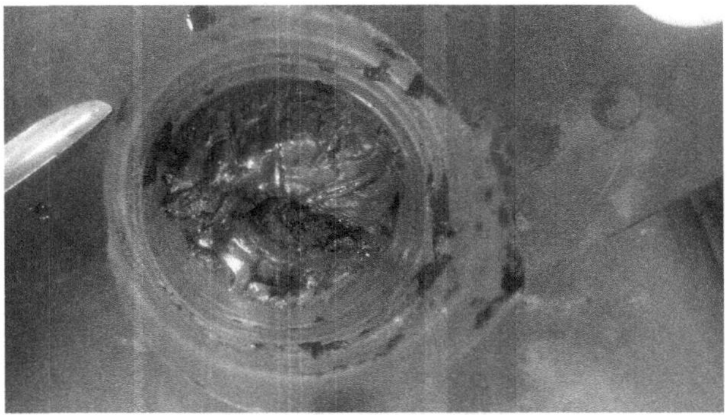

FIG. 2.21

An example of sludge sample from separator bottom.

Oily sludge is considered hazardous waste according to the Environmental Protection Act and Hazardous Wastes Handling Rules; thus, unless totally remediated, petroleum sludge cannot be disposed of as landfill [46].

Petroleum sludge is formed by fluids cooling below the cloud point, evaporation of light ends, mixing with incompatible materials, and the introduction of water to form stable thick emulsions [48]. Sludge usually accumulates in crude oil tanks, refinery product tanks, desalters, and elsewhere during oil production and processing. Sludges with $<40\%$ recoverable oil are considered to be low oil content sludge [46].

Sludge can be treated for oil recovery or can be disposed of due to its hazardous nature. Various technologies for oil recovery and redemption of crude sludge include chemical treatments, distillation processes, cracking, hydrotreating, irradiation, centrifugation, solvent treatment, and bioremediation [46,48].

Sludge disposal methods are burial (only allowed where no leaching of chemicals can occur), road surfacing (either via direct application or by mixing in asphalt), solidification to prevent leaching, and land farming in contained areas where toxic materials evaporate and/or degrade naturally [47].

2.7.4 Petroleum foams

A foam is a gas-liquid colloidal state where gas is dispersed in liquid, similar to an emulsion, whereas an emulsion is a liquid dispersed in another different liquid. In petroleum foams, the liquid is the continuous phase and the gas is the dispersed phase.

Foam formation requires the presence of liquid, gas, and a foam stabilizing agent (surfactant). They are generated anywhere from the wellhead to the refinery at points where gas release and related pressure drops occur in processing equipment, pipework, chokes, control valves, and in separator inlet devices where poor design causes inefficient separation (Fig. 2.22) [49].

Crude oil foam creates operational problems, including the following [50]:

- Poor level control that can lead to platform shutdowns.
- Liquid carryover in the gas outlet that can lead to flooding of downstream scrubbers and compressors.

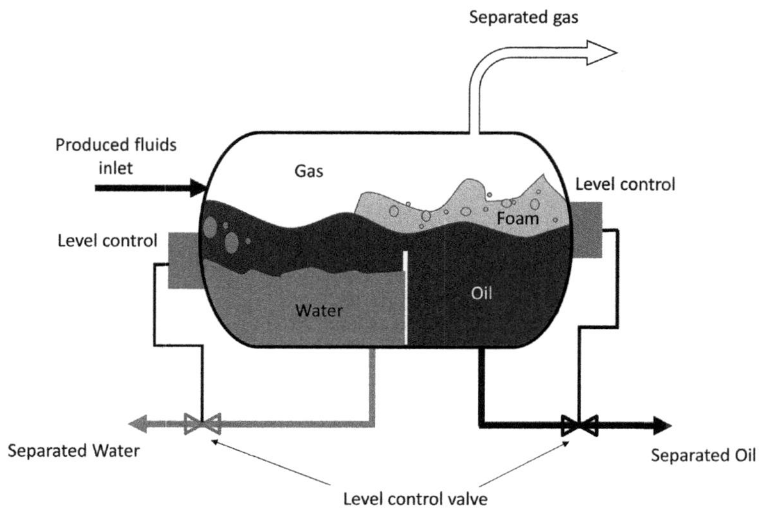

FIG. 2.22

Foam in separator and its effect on level control.

- Gas carry-under in the liquid outlet that can lead to increased compression requirements.

Control of foams is done either by chemical treatment using what is branded as antifoam or defoamer chemicals; an example is silicon or fluorosilicon chemicals. Another way to control foams is by properly sizing and adjusting the separator components.

2.7.5 Corrosion

Corrosion can be defined as the destructive attack of a metal by a chemical or electrochemical reaction with its environment, which eventually leads to the formation of one or more of the metal ores (i.e., oxides, sulfides, carbonates). The total annual cost of corrosion in the oil and gas production industry is estimated to be US$1.372 billion, broken down into $589 million annually in surface pipeline and facility costs, $463 million in downhole tubing expenses, and another $320 million in capital expenditures related to corrosion [51,52]. Globally, corrosion incurs a total cost of $US 2.5 trillion dollars. This figure represents roughly 3.4% of the global gross domestic product (GDP) [53].

2.7.5.1 Corrosion mechanisms

The main driving force for corrosion is that the metal ores are at a lower energy state than the metal itself and consequently the metals have a natural tendency to return to their lower energy state, or "corrode" [54].

For corrosion to happen, the corrosion cell components must exist; these include

- Electrodes (anode and cathode).
- Ionic current pathway.
- Electronic path.

When these components are combined, the corrosion process starts by dissolving the metal and converting it into metal ions in what is called the anodic reaction (oxidation reaction); then the electrons generated from this anodic reaction transfer through the electronic pathway to reach the cathode, where the cathodic reaction (or reduction reaction) takes place. The corrosion process is illustrated in Fig. 2.23.

The typical anodic and cathodic reactions are listed in Eqs. (2.11)–(2.14):

$$Fe \rightarrow Fe^{+2} + 2e^{-}\,(\text{anodic}) \tag{2.11}$$

$$O_2 + 4e^{-} + 4H^{+} \rightarrow 2H_2O\,(\text{cathodic}) \tag{2.12}$$

$$2H^{+} + 2e^{-} \rightarrow 2H_2\,(\text{cathodic}) \tag{2.13}$$

$$M^{3+} + e^{-} \rightarrow M^{2+}, M = \text{metal ion}\,(\text{cathodic}) \tag{2.14}$$

Generally speaking, the corrosion process typically depends on various factors, including the following:

- Metal type.
- Electrolyte composition and properties (pH, dissolved salts, dissolved gases, etc.).
- Operating conditions (temperature, pressure, flow rates, etc.).

In oil and gas fields, there are many contributors to the corrosion process, including the following [55]:

- Dissolved gases (H_2S, CO_2, O_2).
- Dissolved salts (produced water salinity can be up to 400,000 ppm).
- Microorganisms.
- Solids, scale formation and accumulation, and sand production.
- Detrimental operating practices (reckless pigging, inhibition, operating out of safe operations window).

2.7.5.2 Forms of corrosion

Corrosion is a complex problem that often takes place by different mechanisms, depending on the contributing factors in the surrounding environment, which can result in numerous forms of corrosion. Understanding the corrosion mechanism and identifying the corrosion form is key in corrosion mitigation and management plans. Many forms of corrosion have been studied in different industries, and some of these forms are illustrated in Fig. 2.24 [56–66].

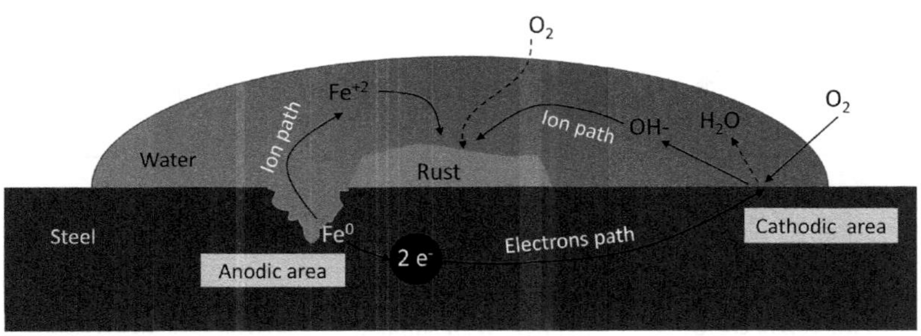

FIG. 2.23

The corrosion process.

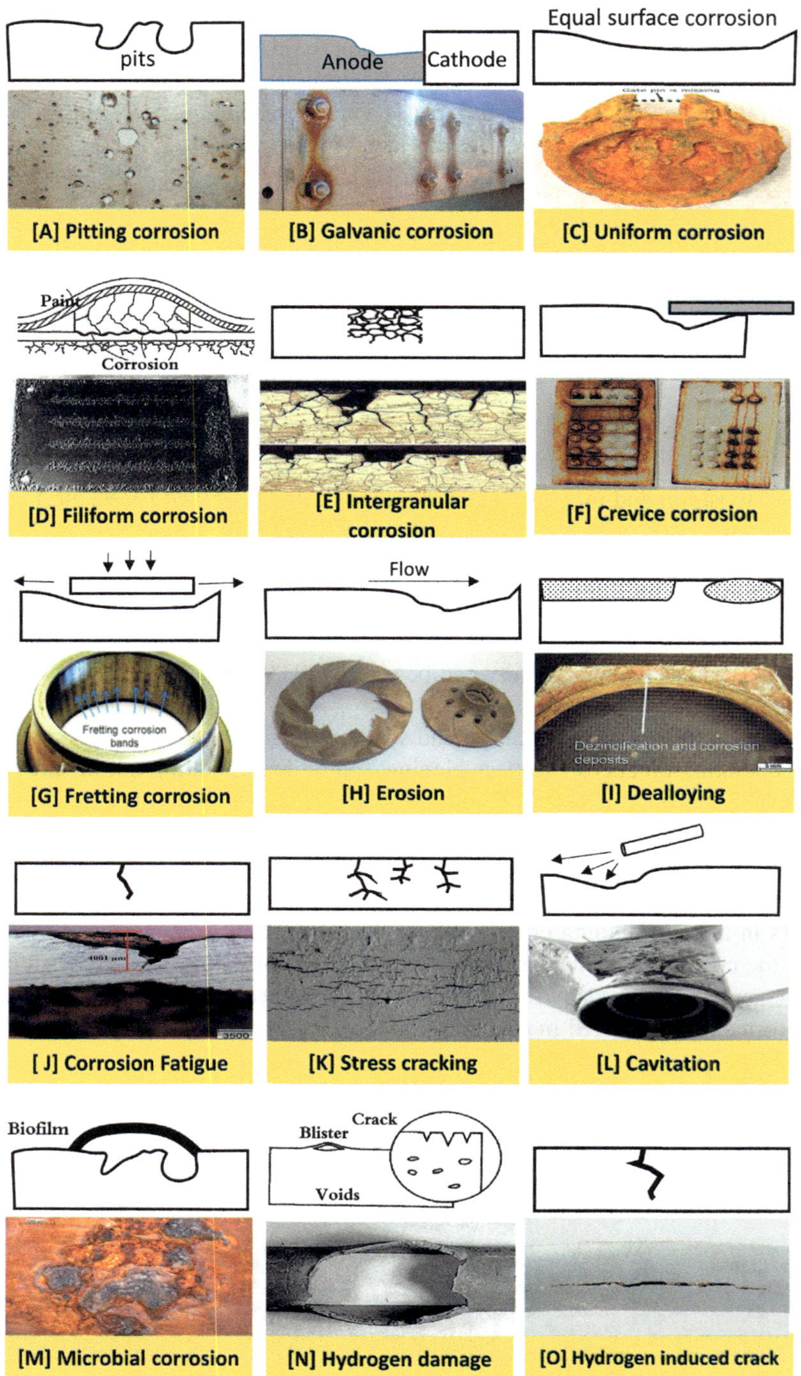
pits
Anode
Cathode
Equal surface corrosion
[A] Pitting corrosion
[B] Galvanic corrosion
[C] Uniform corrosion
Paint
Corrosion
[D] Filiform corrosion
[E] Intergranular corrosion
[F] Crevice corrosion
Flow
Fretting corrosion bands
Dezincification and corrosion deposits
[G] Fretting corrosion
[H] Erosion
[I] Dealloying
[J] Corrosion Fatigue
[K] Stress cracking
[L] Cavitation
Biofilm
Blister
Crack
Voids
[M] Microbial corrosion
[N] Hydrogen damage
[O] Hydrogen induced crack

FIG. 2.24

Forms of corrosion.

Sources: (A) F.J. Cárcel-Carrasco, M. Pascual-Guillamón, L.S. García, F.S. Vicente, M.A. Pérez-Puig, Pitting corrosion in AISI 304 rolled stainless steel welding at different deformation levels, Appl. Sci. 9(16) (2019), https://doi.org/10.3390/app9163265. (B) Photo credit D3j4vu https://commons.wikimedia.org/wiki/File:Stainless-steel-mild-steel.jpg license CC BY 3.0. (C) W.S. Tait, Controlling corrosion of chemical processing equipment, in: Handbook of Environmental Degradation of Materials, third ed., Elsevier, 2018, pp. 583–600, https://doi.org/10.1016/B978-0-323-52472-8.00028-9. Copyright (2018) with permission from Elsevier. (D) C. Vargel, Types of corrosion on aluminium, in: Corrosion of Aluminium, Elsevier, 2004, pp. 113–146, https://doi.org/10.1016/b978-008044495-6/50012-4. Copyright (2004), with permission from Elsevier, (E) D.L. Engelberg, Intergranular corrosion, in: Shreir's Corrosion, Elsevier, 2010, pp. 810–827, https://doi.org/10.1016/B978-044452787-5.00032-9. Copyright (2010) with permission from Elsevier. (F) Photo courtesy KTA-Tator, Inc https://kta.com/, reprinted with permission. (G) J. Lai, K. Stadler, Investigation on the mechanisms of white etching crack (WEC) formation in rolling contact fatigue and identification of a root cause for bearing premature failure, Wear 364–365 (October) (2016), 244–56, https://doi.org/10.1016/j.wear.2016.08.001. Copyright (2016), with permission from Elsevier. (H) R.J.K. Wood, 6.09—Erosion/corrosion, in: Comprehensive Structural Integrity, vol. 6, Elsevier, 2007, pp. 395–427, https://doi.org/10.1016/B978-008043749-1/00315-2. Copyright (2003) with permission from Elsevier, (I) G.A. Pantazopoulos, A.I. Toulfatzis, Failure analysis of a machinable brass connector in a boiler unit installation, Case Stud. Eng. Fail. Anal. 1(1) (2013) 18–23, https://doi.org/10.1016/j.csefa.2012.11.002. Copyright (2013), with permission from Elsevier. (J) Photo courtesy US Corrosion Services, https://uscorrosion.com/, reprinted with permission. (K) O.I. Zvirko, S.F. Savula, V.M. Tsependa, G. Gabetta, H.M. Nykyforchyn, Stress corrosion cracking of gas pipeline steels of different strength, in: Procedia Structural Integrity, vol. 2, Elsevier B.V., 2016, pp. 509–516, https://doi.org/10.1016/j.prostr.2016.06.066. Copyright (2016), with permission from Elsevier. (L) W.S. Tait, Corrosion prevention and control of chemical processing equipment, in: Handbook of Environmental Degradation of Materials, second ed., Elsevier, 2012, pp. 863–86, https://doi.org/10.1016/B978-1-4377-3455-3.00028-6. Copyright (2012), with permission from Elsevier. (M) E. Huttunen-Saarivirta, M. Honkanen, T. Lepistö, V.T. Kuokkala, L. Koivisto, C.G. Berg, Microbiologically influenced corrosion (MIC) in stainless steel heat exchanger, Appl. Surf. Sci., 258(17) (2012) 6512–6526, https://doi.org/10.1016/j.apsusc.2012.03.068. Copyright (2012), with permission from Elsevier. (N) M.B. Djukic, V. Sijacki Zeravcic, G.M. Bakic, A. Sedmak, B. Rajicic, Hydrogen damage of steels: a case study and hydrogen embrittlement model, Eng. Fail. Anal. 58(December) (2015), pp. 485–98, https://doi.org/10.1016/j.engfailanal.2015.05.017. Copyright (2015), with permission from Elsevier. (O) Photo by CEphoto, Uwe Aranas, https://commons.wikimedia.org/wiki/File:Steel-with-Hydrogen-Induced-Cracks-01.jpg, CC BY-SA 4.0.

2.7.5.3 Corrosion control

The corrosion control starts with studying the system components, fluids, and operating conditions, followed by proper material selection, and then developing and implementing a corrosion control (prevention, mitigation) program and rigorously following up and monitoring for corrosion signs during operations. Depending on the validity of the applied corrosion control plan, changes can be done to improve it or to switch to a more effective plan. Table 2.6 summarizes the common corrosion control methods.

2.7.6 Oilfield microbiology

Oil and gas were originally formed in the ancient sedimentary rocks due to induced microbial activity. However, when microbes are encountered during production operations, they become the culprit in a number of flow assurance problems, including microbiologically influenced corrosion (MIC), H_2S production, reservoir souring, mineral scale formation, biofouling, and degrading of drilling mud and production chemicals. Fig. 2.25 illustrates oil and gas field problems associated with microorganisms.

A vast variety of microorganisms can exist in oil and gas fields; these include bacteria, archaea, fungi, algae, and protozoa. These types can exist individually, or coexist independently, or live

Table 2.6 Corrosion mitigation methods.

Mitigation method	Description
Improving of structures	– Avoid sharp corners – Avoid dead legs – Avoid dissimilar metals – Avoid mechanical stress – Avoid surface imperfections
Material selection	– Corrosion resistant alloys – Compatible with produced fluids – Optimum mechanical properties – Economical – Use of coatings
Optimize operating conditions	– pH – Temperature – Pressure – Flow rate – Remove dissolved and acid gases: O_2, CO_2, H_2S
Chemical methods	– Anodic corrosion inhibitors – Cathodic corrosion inhibitors – Adsorption (mixed) corrosion inhibitors
Potential change	– Anodic protection – Cathodic protection

syntrophically depending on each other. Among these microorganisms, bacteria are considered the most common and most studied species in oil and gas fields. Examples of bacteria in oil and gas fields are sulfate-reducing bacteria (SRB), iron bacteria, and acid-producing bacteria. The different kinds of microorganisms and their impact on production are discussed in Chapter 10, Biofouling.

If any microbial activity is detected during production operations, quick action should be taken to control these activities due to their quick and severely impacting nature. Fig. 2.26 illustrates a severe case of MIC corrosion in an elbow. These control actions are based on the use of biocide chemicals, which are used to kill or stop the reproduction of these organisms. Another way of controlling microorganisms is by the use of routine pigging and ultraviolet radiation.

On the other hand, bacteria can be used in the reservoirs to enhance production in a technology known as microbiologically enhanced oil recovery (MEOR). In that technology, bacteria are added to the petroleum reservoirs to produce acids, gases, and biosurfactants that can increase reservoir pressure, reduce oil viscosity, increase formation permeability, and create more flow channels for the oil to flow.

2.7.7 Reservoir souring

Reservoir souring is a process whereby a previously sweet reservoir (i.e., one containing little or no H_2S) starts to produce sour fluids (i.e., H_2S) due to the activities of sulfate-reducing bacteria or other geochemical reactions [67].

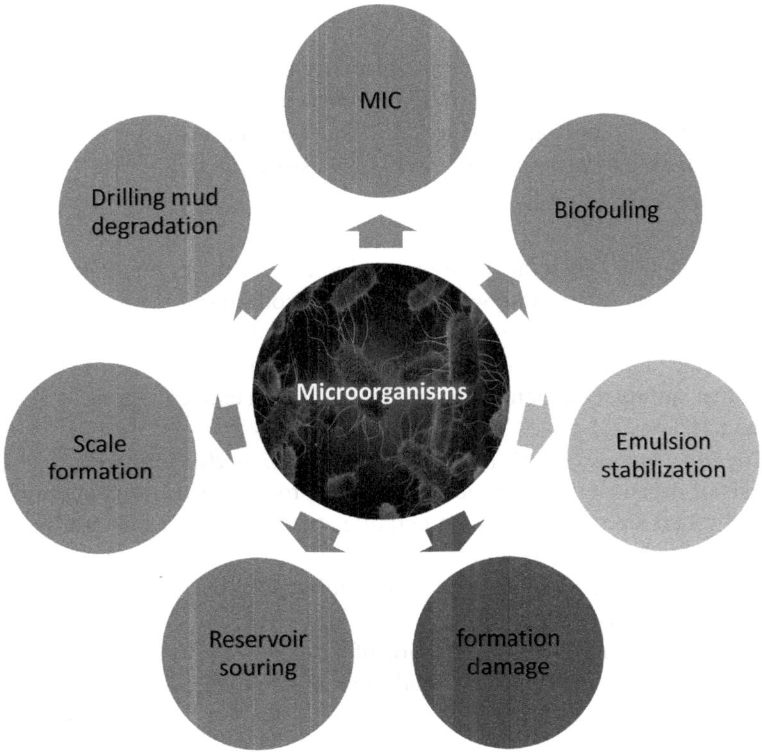

FIG. 2.25

Problems associated with microbial activity.

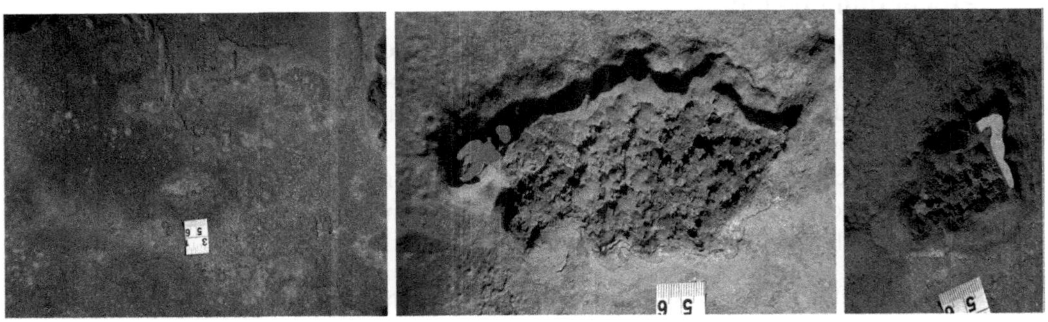

FIG. 2.26

Severe MIC case in an elbow.

Generally speaking, reservoir souring can occur due to the following:

- *Microbial activity*: sulfate-reducing bacteria and archaea are known for reducing sulfate (breathing sulfate) into sulfide (H_2S) under many different environmental conditions at the expense of a range of electron donors and carbon sources (Eq. 2.15). Sulfate is basically introduced into the reservoir during seawater flooding to maintain the reservoir pressure [68,69].

$$SO_4^{--} \xrightarrow{SRB,\ \text{Optimum Nutrition, Optimum Conditions}} H_2S \tag{2.15}$$

- *Thermal sulfate reduction*: Thermal sulfate reduction chemically is the direct reduction of sulfate (sulfate, derived from seawater, pore water, or from dissolution of solid calcium sulfate, mainly gypsum and anhydrite in the reservoir rocks) by hydrocarbons in order to produce hydrogen sulfide at temperatures in the range of 250–325°C and modest pressures [70,71]. This thermal redox reaction or thermochemical sulfate reduction (TSR) can result in high concentrations of H_2S in the reservoir fluid (more than 10%) [72].
- *Thermolysis and/or aquathermolysis of organosulfur compounds*: Thermolysis and/or aquathermolysis of organosulfur compounds can also produce H_2S. Aquathermolysis reactions between oil and steam are dominant at temperatures below 240°C, while thermolysis (i.e., in the absence of water) dominates at higher temperatures [73,74].
- *Mineral solubility*: reductive dissolution of sulfide minerals like pyrite, sphalerite, and galena is another source of reservoir souring [72].
- *Degradation of corrosion inhibitors*: degradation of the common corrosion inhibitors that are usually added to the fracking acidizing fluids can lead to production of H_2S [75].

Controlling reservoir souring is usually done by removing the sulfate from the injected sea water, or killing and minimizing the microbial activity by the use of biocides. Another method is the use of H_2S scavengers, which react chemically with hydrogen sulfide and prevent its deleterious impacts on the production.

2.7.8 Production chemicals

From the beginning of exploring oil and gas fields until the end of operations and abandoning the field, a large number of chemicals are used on a daily basis to support production. They are usually used (either alone or in addition to other physical and operational methods) to mitigate production problems and to improve the process performance.

Some of the typical production treating chemicals and their uses are listed in Table 2.7.

Before field application of any of these chemicals, the system should be studied from different perspectives and the following questions must be answered:

- Is this chemical effective at the current operating parameters?
- Is it used in the right solvent and in the right physical form?
- Will this chemical disperse and spread homogeneouly in the production system?
- To what extent is this chemical efficiency sensitive to the applied dose, as sometimes chemical pumps do not precisely deliver the designed dose?
- Is this chemical compatible with the produced fluids and the other applied production chemicals? Also, is it compatible with the structural materials along the production system?

Table 2.7 Common production chemicals used during oil and gas production.

Problem	Chemicals
Emulsions	Demulsifiers: oxyalkylated resins, polyglycol esters, alkyl aryl sulfonates
Foam	Defoamers: silicones, polyglycol esters
Corrosion	Corrosion inhibitors: imidazolines, amines, quaternary ammonium salts, oxygen scavengers: bisulfite
Bacteria	Biocides: quanternary ammonium salts, gluteraldehyde, tetrakis (hydroxymethyl)phosphonium sulfate (THPS), bronopol
H_2S	H_2S scavengers: triazines
Mineral scales	Scale inhibitors: phosphates, phosphate esters, phosphonates, polycarboxylates, polysulfonates, Polyphosphino carboxylic acid Scale dissolvers: polyaminocarboxylates, mineral and organic acids
Asphaltenes	Asphaltene inhibitors: polyester or polyamide, alkylalcohols, alkylamins, and alkylphenols Asphaltene dispersants: alkylaryl sulfonic acids, phosphoric esters, and phosphonocarboxylic acids
Wax	Wax inhibitors and PPD: polyethylene vinyl acetate, maleic anhydrides, acrylonitriles Wax dispersants: alkyl sulfonates, alkyl aryl sulfonates, fatty amine ethoxylates
Gas hydrates	Thermodynamic hydrate inhibitors: methanol, glycol, salts Low dosage hydrate inhibitors: poly vinylcaprolactams, poly vinylpyrrolidone
Drag reducers	Ultrahigh molecular weight polyethylene, copolymers of α-olefins, aluminum carboxylate

- What are the side effects of this chemical on the production system, in case of regular doses and in case of overdosing? Could it trigger more production chemistry problems? Or could it work synergistically with other production chemicals?
- Will it cause downstream problems?
- Are there any injection problems—viscosity, cloud point, foaming?
- Is this chemical feasible economically?

Production chemicals are applied after rigorous lab tests, followed by field trials to determine their feasibility in the field; finally, after field application, an optimization of the chemical dose is achieved.

The global market for oil and gas treating chemicals has seen many advances in the last two decades due to the increase in oil demand and production. For the coming years, the market for oil and gas water treatment chemicals is expected to grow due to the expansion of oil refineries around the world, and shifting focus toward the usage of green chemicals, while scale and corrosion inhibitors are expected to dominate the market [76].

2.7.9 Complexity of production chemistry problems

From the preceding discussion, one can imagine how complex production chemistry problems are! This is attributed to the different phases encountered, the different interactions between these phases, and the continuous changes in volumes and operating conditions during production.

By looking at production chemistry at the microlevel, we can see that certain interfaces are a good stage for these chemical and physical interactions. For example, the oil/water interface is where emulsion problems show up, the oil/gas interface is where foam problems start, the water/gas or gas/oil/water interfaces are where gas hydrate problems start, the solids/pipe wall interface is where deposition problems start, the biofilm/pipe wall interface is where microbiology problems worsen, and the water/pipe wall or oil/pipe wall interface is where corrosion problems start.

These problems can enhance and support each other like a chain reaction. For example, corrosion of the pipes causes mineral scaling when the corrosion products accumulate on the pipe wall; at the same time, mineral scales cause underdeposit corrosion (UDC), and can also create a good environment for bacterial blooming, not to mention that the bacterial activity causes corrosion, mineral scaling (known as biomineralization), and biofouling. Mineral scales, wax deposits and asphaltenes are known to stabilize petroleum emulsions. As we stipulated, it is like a chain reaction, and once one starts, others undoubtedly can show up.

Unfortunately, it is not only limited to the chain reaction of a problem causing another problem, but also solving a problem can trigger another problem. For example, solving an emulsion problem by separating oil/water can trigger corrosion and scaling problems; mixing oils (one hot and the other cold) to overcome paraffinic oil viscosity problems can trigger asphaltenes and emulsion problems; mixing different waters to enhance production triggers mineral scales; the use of specialty chemicals can stabilize emulsions; the methanol used in treating gas hydrates causes corrosion and solid deposition problems; acids used in removing mineral scales cause corrosion and destabilize asphaltenes; paraffinic dissolvers also destabilize asphaltenes, and the list goes on.

These complexities and interferences in production chemistry are illustrated in Fig. 2.27.

Therefore for smooth and economical production operations, the production chemist has to be always vigilant and proactive in handling production chemistry problems. Continuous monitoring of process parameters, fluid characterization, fluid modeling, and evaluation of the applied mitigation to problems are key steps in controlling production chemistry problems. These are also the key steps in the flow assurance management strategies.

Production chemistry issues management can be thought of as the healthcare system of the production system. Illnesses of the physical body are not too different from troubled production systems. Viral/bacterial infection, thrombosis, hypertension, hypotension, plaque, and blood clots are health issues that are similar to microbiological issues, fluid thickening, pressure buildup, pressure drop, or solids deposition in a production system. In the same way, a production chemist looks a lot like a physician trying to diagnose and treat illnesses. Analogous to monitoring body vitals like temperature and blood pressure for any symptoms, production system vital signs such as unusual pressure and temperature changes are also indicative of flow assurance problems. In cases of physical illness, samples are pulled and bloodwork is done for further investigation; likewise in a production system samples are pulled and further analysis and modeling are implemented for deeper understanding of the root cause of the problem. Sometimes blood tests are not enough, so scans like X-rays, CAT scans, MRI, or ultrasound are needed; fortuitously, the same scans can be done on different parts of the production system to detect damage in the pipelines. After the investigation is done, medications are prescribed, whereas in a production system, the treating chemicals are usually injected at the proper dose.

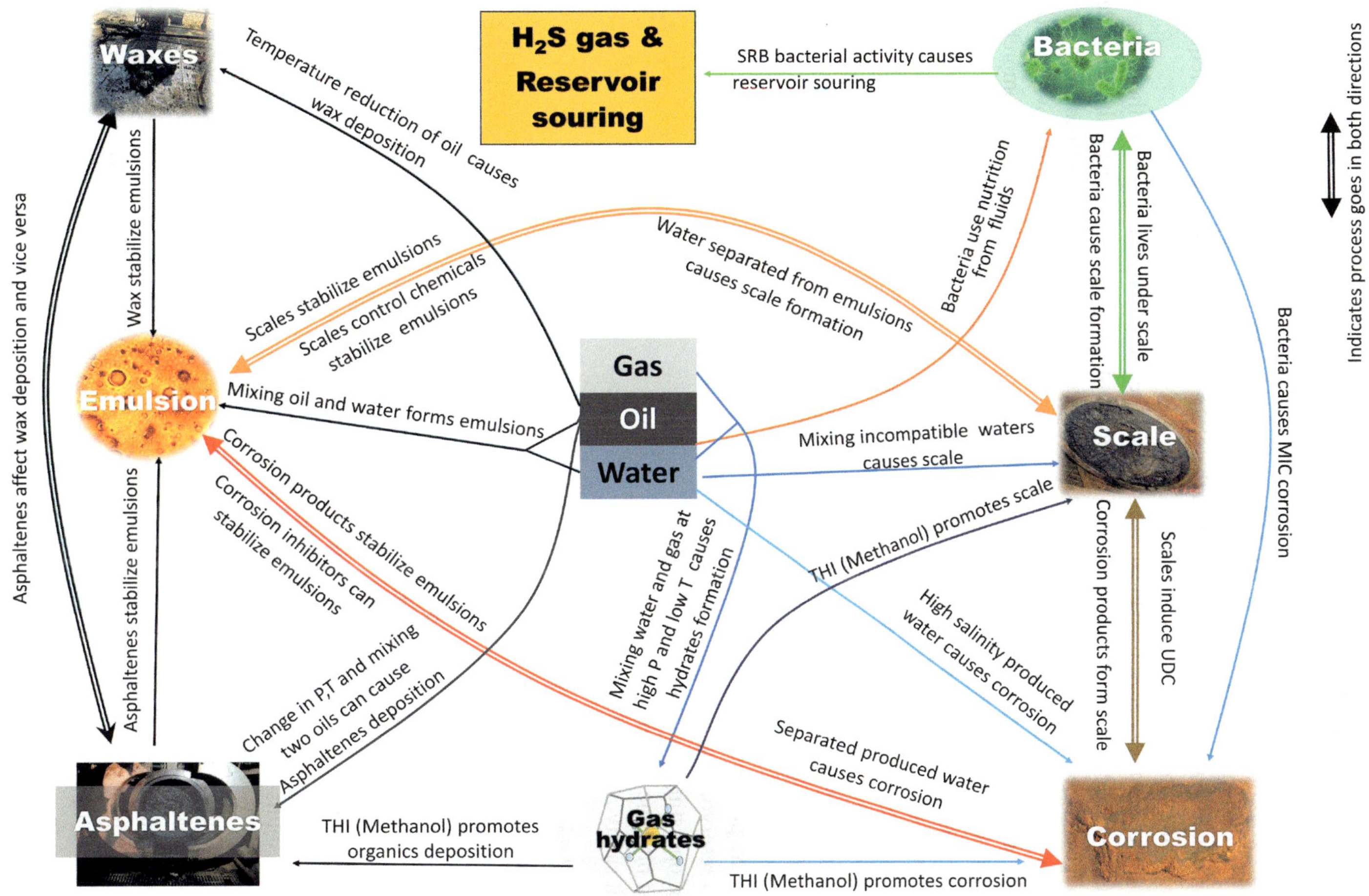

FIG. 2.27

The complexity of production chemistry.

2.8 Flow assurance strategy

A generic flow assurance strategy generally encompasses data collection, modeling, and mitigation or management. The bulk of the flow assurance analysis should be done prior to or in the earlier front end engineering and design (FEED), where the system design, specifications, and operations are planned. The requirements for each project are different and therefore project-specific strategies are required for flow assurance problems [12].

Flow assurance strategy depends on the phase at which the project is running, i.e., in the exploratory phase, when no samples are available; with an estimated fluid composition, and a preliminary idea of what the production system may look like, one can generate a basic scenario that defines the main risks expected in the life of the field. Also, the experience from analogous fields can be used with caution in this step. In the next phase, the development phase where samples are available, after comprehensive and proper fluid characterization, the basic scenario can be corrected and further detailed by a combination of a thorough experimental campaign and modeling. This information will be critical during concept selection. Finally, once the field starts its production, the FA assessment can be improved to understand the reasons for production impairment that were not identified during the development phase.

The flow assurance should be changed or verified during the progress from one phase to another during field development and when specific changes occur, including

- Change in the produced fluid composition, i.e., change in GOR, water cut, etc.
- Change in reservoir producing formation.
- Change in operating conditions.
- Change in pipeline routing.
- Change in safety and environmental aspects.

A flowchart of flow assurance assessment strategy is illustrated in Fig. 2.28.

2.8.1 Sampling

The main goal of any sampling procedure is to collect a fluid sample that is representative of process bulk fluids, or of the reservoir fluids in the case of petroleum production systems. The validity of the next steps in the flow assurance strategy strongly relies on the accuracy of fluid sampling and analysis. It is highly important that the sample remains representative during handling and storage, until all required measurements have been completed. Therefore the sample contamination, phase changes, and precipitation of solids due to changes in temperature and pressure should be minimized and avoided as much as possible. An ideal sample should be collected contamination-free from the reservoir at the reservoir temperature and pressure and transported to the laboratory maintaining the same temperature and pressure. If that is not possible, a more realistic goal is to reduce the potential for phase changes through pressure and potentially temperature compensation. The samples also should represent a steady-state flow to represent the mean properties of the reservoir fluids. Usually sample cylinders are used for this purpose. Fig. 2.29 illustrates a sample cylinder.

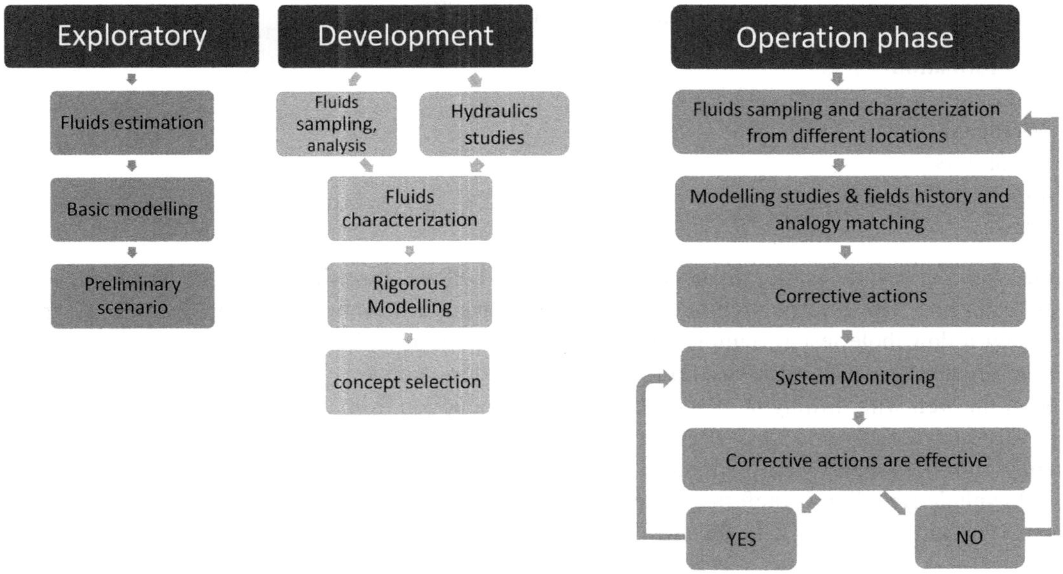

FIG. 2.28

Flow assurance flow chart.

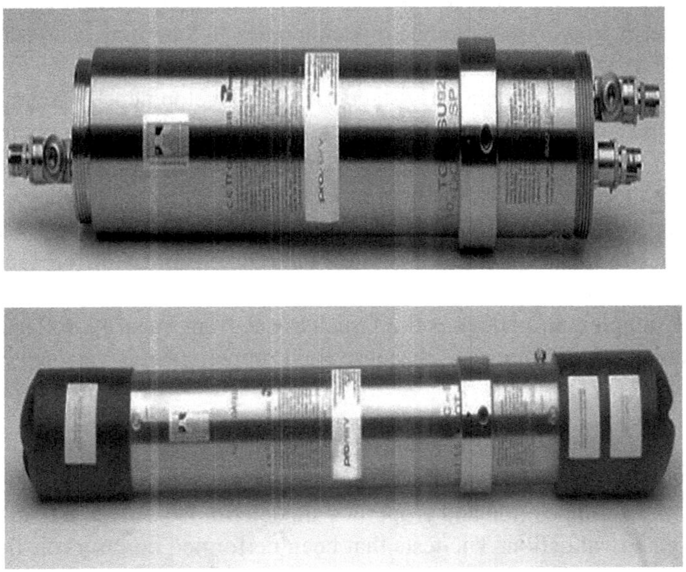

FIG. 2.29

Sample cylinders.

Photo courtesy Proserv UK Ltd. (https://www.proserv.com/) reprinted with permission.

Samples are usually collected from a wide range of locations such as:

- The formation
- Wellbores
- Wellhead
- Separators
- Pipelines
- Tanks

American Petroleum Institute Standard Procedure API-PR-44 furnishes the procedure for reservoir fluid sampling. Downhole sampling is implemented using a downhole sampler fixed on a wireline or using a downhole tester. Samples from multiple depths or aerial positions should be considered to identify and quantify variations. Some interferences during downhole sampling include contamination of the well with drilling mud, gas liberation, or production from different zones. Fluid properties can vary both aerially and vertically even in well-connected reservoirs. Samples taken with openhole formation testers represent a point value in the reservoir.

Downhole sampling is preferred; recently different manufacturers have been supplying downhole sampling equipment. Fig. 2.30 illustrates a downhole sampling tools. Wellhead sampling is usually performed for routine analysis. When downhole sampling is not available, loss of gases and sample cooling are expected. For better representative hydrocarbon samples, this should be limited to single phase hydrocarbons, i.e., dry gas, low GOR oils, or high pressure/high temperature.

Separator sampling is usually done during well testing. Two main methods are recommended for sampling from a separator: filling an evacuated container, and filling a piston-type container. Samples from a well test represent an average over the producing interval. Sampling from tanks is covered under standard method ASTM **D 4057**.

Water samples may be scarce in the early stages of field development. Water salinity is critical for many operations; in the case of low water content at the beginning, it can be obtained from the resistivity logs. It is better for water samples to be collected from downhole, or upstream, of the problem under investigation, and kept pressurized and temperature controlled until analyzed. Water sampling must be done in proper sampling bottles and the samples must be well preserved using the proper method. Fig. 2.31 illustrates subsea sampling, and Fig. 2.32 illustrates different topside manual and automated sampling systems.

In the absence of samples, an analogous fluid, such as one from a nearby well in production, may be used. This always entails significant risks, because fluid properties may vary widely, even within the same reservoir.

2.8.2 Analysis

In the analysis step, fluid composition and properties are determined. A first step in the analysis will be sample conditioning and validation. The tests that been performed on reservoir fluids include:

PVT analysis to determine reservoir fluid behavior and properties to simulate the interactions in the reservoir and at the surface during production. PVT analysis includes determining bubble point, dew point, oil formation volume factor, gas formation volume factor, and gas-to-oil ratio (GOR). These tests are conducted using the differential liberation test and flash liberation tests in PVT cells; at the end this will give a phase diagram of the fluids at the specified temperature and pressure.

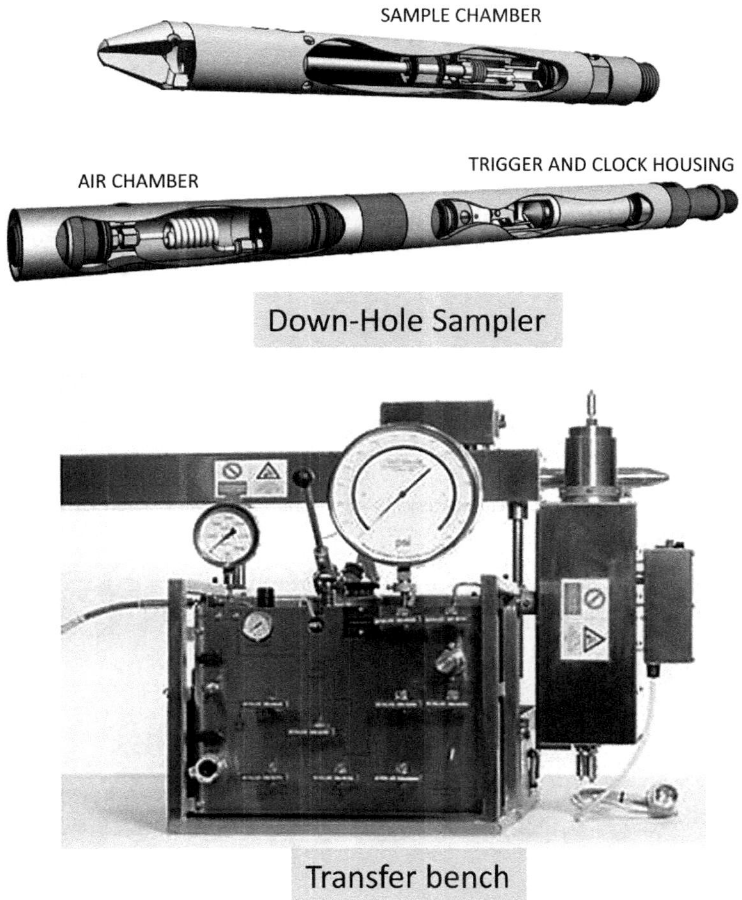

FIG. 2.30

Downhole sampling tools.

Photo courtesy Proserv UK Ltd. (https://www.proserv.com/) reprinted with permission.

For wax, the following are measured on a dead oil: the normal paraffin distribution, using high temperature gas chromatography (HTGC), wax appearance temperature, viscosity, and pour point. If these parameters indicate potential wax deposition, elevated viscosity, or gelling problems, a more thorough analysis program is needed, including cold finger, flow loop, and measurements made under live oil conditions and chemical evaluation.

For asphaltenes, dead oil characterization data, including SARA (Saturate Aromatic Resin Asphaltene) and paraffinic solvent (typically *n*-pentane or *n*-heptane) titration endpoint, are used as screens for fluid stability, along with asphaltenes onset precipitation or asphaltenes onset pressure (AOP).

For hydrates gas, composition from a standard PVT or validation study and water composition are used in a thermodynamic model to generate the expected hydrate formation boundary.

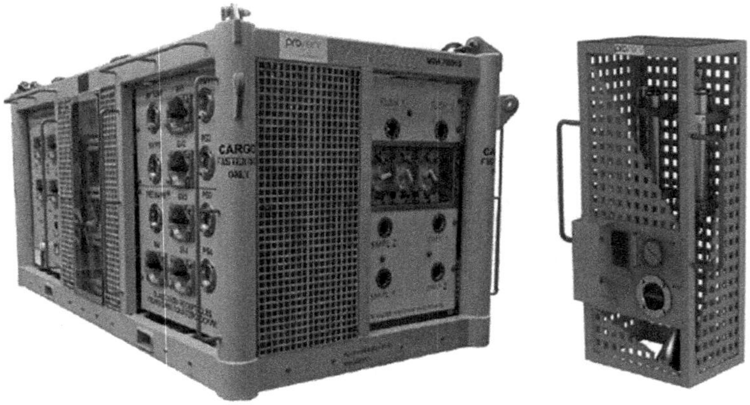

FIG. 2.31

Subsea sampling system.

Photo courtesy Proserv UK Ltd. (https://www.proserv.com/) reprinted with permission.

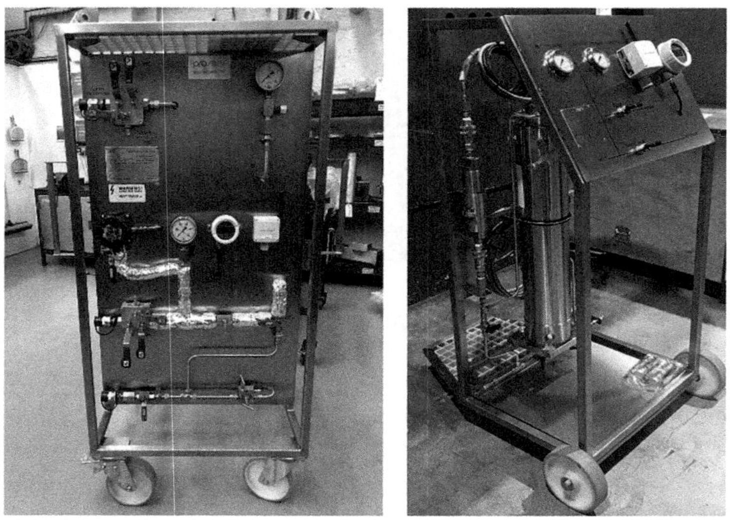

FIG. 2.32

Different topside sampling systems.

Photo courtesy Proserv UK Ltd. (https://www.proserv.com/) reprinted with permission.

For naphthenates, FT-IR, total acid number (TAN), and naphthenic acid identification method (using spectroscopic or mass spectroscopic methods) are usually used to measure them.

Water composition is determined using ICP technique (ICP-OES or ICP-MS), which is used to determine the cation content of the water; anions are determined using either titration, spectroscopic, ion chromatography, or potentiometric titration methods; also water pH, resistivity, TDS, and TSS are of major importance.

2.8.3 Modeling

The measured physical and chemical properties of the fluids are then used in the modeling packages to simulate the phases of production. System design, multiphase flow, solids formation, and system integrity issues are modeled using a wide range of commercial simulation packages like those mentioned in Section 2.6.2.

Hydrates formation curves, asphaltene deposition envelope, and paraffin deposition envelope are the primary results of thermodynamic simulations, which are critical to optimize system design and operations, especially in long tiebacks and subsea installations. Scaling tendency is also produced in the case of availability of produced water data or using analog fields. Corrosion and other system integrity issues are also being simulated. Solids prediction/simulation and the available academic and commercial modeling packages are covered in Chapter 12, "Flow assurance solids prediction and modeling." It is critical to assure the use of the proper modeling tools at the preliminary stages of the project, since their data will be used extensively in optimizing the system design and operations, especially in the startup and until the system reaches the steady state.

At the preliminary stage, the main concept includes modeling the steady state and the transient state operations to optimize and maintain them. The steady state involves but is not limited to modeling the system design, flow rates and working pressures, flowline sizing, flow regimes, system temperature and insulation, use of heat, phase separation, monitoring devices, and other system aspects. Transient state includes simulating the startup, shutdown, and other process interruptions and unsteady events. The outcomes from thermodynamic solids prediction should be combined with the system multiphase flow modeling to maintain these main operations. For fields with severe hydrate or paraffin risks, startup and shutdown (planned and unplanned) must be studied and modeled thoroughly due to their damaging and costly effects.

During the operations or production phases, more fluid samples will be available, more water will be available, the system will be more conditioned with the fluids, system operating parameters will be well maintained, and the effect of third-party processes (indirectly affecting processes) will be experienced. So it is convenient to characterize the fluids and perform modeling to assure nothing is changed from the main concept that was stipulated in the preliminary step. Furthermore, field observations and experienced problems can be combined with the modeling to be more relevant and representative of the final concept. For example, some modeling tools may overestimate or underestimate specific solids problem, so can be corrected with the field experience and reflected in a final robust concept.

2.8.4 Management strategy

After modeling and simulation, the proper concept is selected with the optimum management and control strategy being applied. The concept and the applied flow assurance management strategy depend on the expected problem, the number of problems, environmental issues, and on the economic feasibility of the treatment strategy.

For solids management, the flow assurance strategies can be operational by optimizing system design, applying insulation, applying heat, and optimizing flow rates and operating conditions in solids free zones. Other operational methods can include capital projects like water softening, sulfate removal, and cold flow. Chemical methods include designing and optimizing chemical treatments, i.e., injection points location and doses. The chemicals used must be compatible with each other and the produced fluids, and not cause any side problems. Nonchemical methods may include using electromagnetic, acoustic, coating, smart installation, or mechanical methods.

Preventative measures must be in place from the start of a project's design phase, and must include all necessary information about the system.

Remediation methods should also be ready for any upsets in the system and exaggerated conditions. For deposits remediation, common methods are depressurization, use of heat, chemical cleaning, pigging, gauging, coil tubing, and others.

2.8.5 Monitoring and improvement

After applying the flow assurance management strategy, system monitoring must be implemented to ensure the feasibility of the applied strategy. The easiest method of monitoring is checking and observing the system parameters for major changes. Changes in system pressure and temperature are strongly indicative of solids and other flow assurance issues. Routine analysis of the produced fluids is conventionally used to monitor changes and flow assurance issues; in addition, total suspended solids testing is strongly indicative of solids deposition problems. Furthermore, physical sensors that are based on acoustic, electric, electromagnetic, radiographic, or tomographic methods have also been successfully used as feasible tools in monitoring system vulnerability to solids formation and assessing the applied mitigation methods. These sensors are examined in Chapter 13, Monitoring of flow assurance solids in oil and gas fields.

2.9 Flow assurance case studies

In this section we review some of the flow assurance cases in brief, with our comments at the end to uncover the lessons learned from them.

CASE STUDY 1

Drilling exploratory wells within marginal oil fields in Vietnam indicated that the maximum total daily production rate from the wells could potentially range to about 20,000 barrels of oil per day (BOPD). Unfortunately, numerous challenges exist in the field and are considered stumbling blocks for production, as described in Nguyen et al. [77].

The main operations were to establish a pipeline system that is technically and economically affordable to transport crude oil from the marginal field of Ca Ngu Vang (CNV) to the CPP-3 platform, Bach Ho oil field. The transportation diagram is described in Fig. 2.33.

The oil and gas well stream fluids from CNV WHP (wellhead platform) will be transported to the CPP-3 platform via an insulated production pipeline for processing and onward transporting to an FSO tanker (oil) or to Dinh Co gas terminal. The length of the pipeline CNV WHP to CPP-3 is 25 km. Both steady-state and transient analysis were considered to ensure optimizing the pipeline mechanical, process, and thermal design, and to utilize all the available laboratory measurements to determine the operational strategies to alleviate wax deposition and/or gel formation during all steady-state and transient events.

The production pipeline routes from CNV WHP to CPP-3 consist of two vertical risers, and one horizontal pipeline. The water depth along the pipeline varies from 55.5 m at the CNV WHP to 46.0 m at the CPP-3 platform.

Fluids were characterized to assess the possible risks, and hydraulics were modeled in steady state using PIPESIM and PIPEPHASE, and in transient state using OLGA software. Because the CNV fluid wax appearance temperature (WAT) is 64°C, the primary flow assurance risk associated with steady-state production of this field is wax deposition. Passive insulation was determined to be the most economical method to avoid wax deposition. Therefore it was necessary to insulate

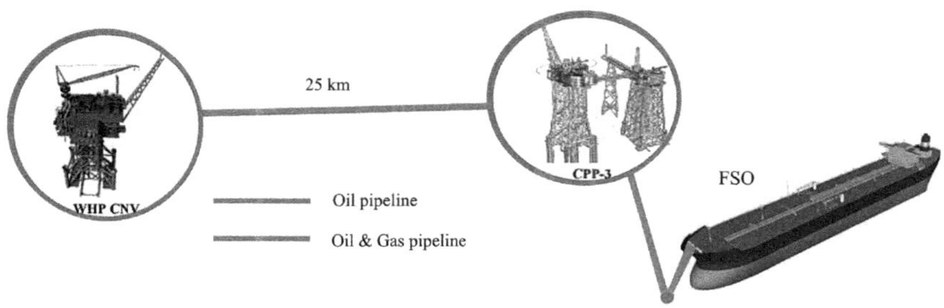

FIG. 2.33

Layout of the field under study.

the subsea tieback with PU foam insulation with nominal thermal transmittance or U-value (U-factor) of 0.2, 0.3, and 0.4 BTU/ft^2 h °F. To select the U-value of the insulation required, the steady-state thermal-hydraulic calculation of the flowline CNV WHP-CPP-3 was performed by PIPESIME. The simulation result indicates that an insulation layer with U-value of 0.2 BTU/ft^2 h °F is ideal for maintaining the temperature of the transported product above the WAT at a flow rate of 20,000 STB/d; the expense is still an important factor that should be considered. Based on experience, for the sake of safety and cost, the U-value of 0.241 BTU/ft^2 h °F was empirically chosen to insulate the pipeline system CNV-CPP-3.

The initial start-up was simulated to understand and optimize the start-up conditions, especially after replacing the hydrostatic testing fluids. For shutdowns, wax gelation and deposition are highly possible; therefore the restart conditions (restart) pressure were simulated to evaluate the optimum. In addition, the holdup in the pipeline was at the steady state before the shutdown. Here, the holdup of the transported fluid is distributed evenly along the flowline, and there is no observation of slugs. However, the longer the shutdown time, the higher the temperature drop. Therefore the formation of the holdup is inevitable. With a shutdown time from 6h to 6days or more, liquid holdup calculated results proved the formation of slugs along the pipeline, which could subsequently affect the transportation. The temperature of the flowline decreases dramatically with the increase of shutdown time, and thus increases the restart pressure. If the system stops working for more than 20h, temperature varies from about 70°C to below 30°C; however, it is compulsory to keep the temperature of the pipeline system from more than 64°C (gel point) to more than 36°C (wax point) for the pipeline CNV-CPP-3. Therefore the shutdown time, in any case, should not exceed this threshold in order to remove the risks of relevant pressure and temperature when restarting the transport system.

CASE STUDY 2

An oil producer in an Egyptian field encountered complex solids deposition issues. The well started production with 13,000 BPD before decreasing down to 3000 BPD. Due to the very low water cut at the beginning, no water samples were collected to assess the mineral scales tendency. With increasing water production, the well encountered mineral scale formation. Hence, a scale management plan was initiated by characterizing the collected scale samples, collecting fluids samples, and performing scale tendency predictions. Analyzing the scale sample showed that the composition of the scale sample was a mixture of hydrocarbon deposits, (Fe, Zn, Pb) sulfides 35%, and sand 45%. However, the scaling tendency calculations showed the potential formation of only barite and anhydrite scales and no scale tendency for any type of sulfides. Following this, the plan was made to inject scale inhibitor using a downhole injection line; the scale inhibitor was mainly targeting sulfate scale, since it is harder to remove and faster forming than the sulfides, which were already detected in the scale sample, plus the fact that the sulfate scale inhibitor can still inhibit sulfide scales. The mitigation plan seemed to work fine at the beginning. However, as the water production increased, there was a slight increase in Zn and Pb concentrations. That consequently led to formation of severe (Zn, Pb) sulfide scales, which were detected in the scale retrieved from well gauging. The severity of the sulfide scales increased, which led to significant production drops due to severe fouling in the production choke valve. Analysis of the scale samples collected from the production choke and the tubing gauging showed that the major composition was hydrocarbons 10%, (Zn, Pb)S 72%, FeS 5%, (Ba, Sr)SO_4 3%, and sand 10%. Therefore this necessitated a change in the mitigation plan. Hence, more water samples were collected and the scale prediction was performed; suspended solids tests (TSS) were performed at the wellhead where the test filter was later examined, using the environmental scanning electron microscope to identify the scale type and coverage on its surface. The scale tendency calculations in both software packages again showed only sulfate scale potential and no sulfide scale tendency, which required another batch of sampling with more cautious sampling and sample preservation procedures. After collecting the second batch of samples and performing scaling tendency in the two different software applications, surprisingly sulfide scale showed a small tendency in one of the software packages in addition to sulfate scales, while the other software only showed sulfate scales and no sulfide scales tendency. Moreover, the TSS results showed the presence of (Zn, Pb)S and $BaSO_4$ scales. Combining these results, the plan was to inject a mixture of two different scale inhibitors (70% SI for sulfate scale +30% SI for sulfide scales). The scale inhibitors were tested for their compatibility and their mixture was tested for compatibility with the production fluids, and compatibility with the injection lines. After field application, the produced fluids were analyzed to monitor the changes in the cations analysis, and TSS testing was routinely performed. The results showed a decline in the received scale on the TSS filter, which was confirmed with the well gauging jobs and the improved well production rate.

Now, what are the lessons learned from these case studies?

Case study 1 indicated the use of flow assurance methods to design and optimize the production system. It is a holistic application of the flow assurance phenomena. Fluids composition and operating parameters were used in a flow assurance software (for steady state and for transient state) to determine the optimum parameters of pipeline sizes, capacity, and insulation, besides other possible flow assurance obstacles including slugging and wax deposition. The fluid physicochemical properties and system parameters were the basic inputs, while the flow assurance software was the main tool used that defines the importance of modeling.

Case study 2, however, dives deeper into production chemistry and monitoring. As mentioned in case study 2, proper fluids sampling and analysis are key in flow assurance investigations. Samples have to be properly preserved and analyzed to get the expected results from modeling. Furthermore, not all modeling tools are accurate; the two models used gave different results: one of them was correct while the other was not representative. Therefore it is optimal and highly recommended to use more than one modeling software package. Also using the actual deposit sample analysis is representative of the problem and usually used to develope treatment plans. Moreover, monitoring is a crucial factor in the flow assurance strategy. Monitoring the production parameters and monitoring the produced fluids using TSS testing gave a good indication of the deposition rates and the deposit types. Most importantly, with after-treatment monitoring if sufficiently applied, the production choke blockages could have been avoided.

2.10 Summary

Flow assurance is a fundamental discipline of science in oil and gas operations that has been gaining momentum in the last few decades. The term, which was originated in the 1990s, basically means the methods and measures used to assure the smooth and safe delivery of the produced fluids from the reservoir to the final terminals. A wide variety of obstacles challenge the operators on a daily basis, including multiphase flow issues, solids formation, phase separation, and system integrity issues.

Generally speaking, the fluid flow is highly affected by the pressure drop in the system, which in turn is impacted by many factors, among them system design, friction, temperature, surface roughness, fluid properties, and flow regimes.

Flow assurance entails two major branches: fluid hydrodynamics and production chemistry. The former investigates the multiphase flow issues, i.e., flow regimes, slugging, liquid holdup, etc., and the latter investigates fluid phase changes and interactions, i.e., emulsions, foaming, sludges, mineral scales, gas hydrates, paraffins, asphaltenes, naphthenates, microbiological activities, and corrosion issues.

Recently, the assurance phenomena was extended to cover a bigger footprint in the science and engineering realm in the oil and gas industry, upgrading from just developing prevention methods for solids like gas hydrates, to be involved in the earliest steps of field development and commissioning up to the final stages of the field abandonment and decommissioning.

Nonetheless, a major strategy that production engineers and chemists conventionally apply in their work to solve flow assurance issues involves sampling the fluids, modeling, optimizing the system or applying mitigation strategies, monitoring the system, and finally adjusting and optimizing the system based on the outcomes of the management strategy.

References

[1] J.M. McDonough, Lectures in Elementary Fluid Dynamics: Physics, Mathematics and Applications, fourth ed., Departments of Mechanical Engineering and Mathematics University of Kentucky, 2009.

[2] G.E. Miller, Biomedical transport processes, in: Introduction to Biomedical Engineering, Elsevier, 2011, pp. 937–993, https://doi.org/10.1016/B978-0-12-374979-6.00014-9.

[3] E.S. Menon, P.S. Menon, Pressure loss through piping systems, in: Working Guide to Pumps and Pumping Stations, Elsevier, 2010, pp. 69–111, https://doi.org/10.1016/b978-1-85617-828-0.00004-4.

[4] A. Kayode Coker, Fluid flow, in: Ludwig's Applied Process Design for Chemical and Petrochemical Plants, fourth ed., vol. 1, Gulf Publishing Company, 2007, pp. 133–302.

[5] M.L. Wiggins, Inflow and outflow performance, in: Volume IV—Production Operations Engineering, Society of Petroleum Engineers, 2006.

[6] J. Paul Tullis, Hydraulics of pipe systems, in: F. Kreith, S.A. Berger, et al. (Eds.), Fluid Mechanics. Mechanical Engineering Handbook, 1999, CRC Press LLC, 1999.

[7] T.B. Irmann-Jacobsen, Flow Assurance—A System Perspective, FMC Subsea Technologies, 2013.

[8] S. Sham Dol, L. Jit Sen, The effect of flow-induced oil-water emulsions on pressure drop, international, Int. J. Theor. Appl. Mech. 2 (2017) 1–6. http://www.iaras.org/iaras/journals/ijtam.

[9] A. Mukhaimer, A. Al-Sarkhi, M. El Nakla, W.H. Ahmed, L. Al-Hadhrami, Pressure drop and flow pattern of oil-water flow for low viscosity oils: role of mixture viscosity, Int. J. Multiphase Flow 73 (July) (2015) 90–96, https://doi.org/10.1016/j.ijmultiphaseflow.2015.02.018.

[10] R. MacGillicray, Gravity and Gas Density Effects on Annular Flow Average Film Thickness and Frictional Pressure Drop, 2004 (Water, September) http://www.collectionscanada.gc.ca/obj/s4/f2/dsk3/SSU/TC-SSU-09222004160524.pdf.

[11] H. Ohta, R. Kataoka, S. Morioka, Y. Shinmoto, Effect of gravity on the pressure drop in gas-liquid two-phase flow, microgravity research and applications in physical sciences and biotechnology, in: O. Minster, B. Schürmann (Eds.), Proceedings of the First International Symposium Held 10–15 September, 2000 in Sorrento, Italy, European Space Agency, ESA SP-454, 2000–2001, p. 83.

[12] Y. Bai, Q. Bai, Subsea corrosion and scale, in: Subsea Engineering Handbook, Elsevier, 2019, pp. 455–487, https://doi.org/10.1016/b978-0-12-812622-6.00017-8.

[13] P. Kondapi, J. Creek, Y. Doreen Chin, R. Moe, Technologies optimize flow assurance, in: The American Oil & Gas Reporter, National Publishers Group, 2019. https://www.aogr.com/magazine/editors-choice/technologies-optimize-flow-assurance. (Accessed October 2021).

[14] E.D. Sloan, Hydrocarbon hydrate flow assurance history as a guide to a conceptual model, Molecules (2021), https://doi.org/10.3390/molecules26154476.

[15] K.N. Ghia, J. Yang, G.A. Osswald, U. Ghia, Analysis of experimental and computational results for incipient separation and dynamic stall, in: Experimental Heat Transfer, Fluid Mechanics and Thermodynamics 1993, Elsevier, 1993.

[16] F.M. White, Fluid Mechanics, McGraw-Hill, New York, 2003.

[17] G. Bar-Meir, Basics of Fluid Mechanics, Last modified: Version 0.3.4.0 17 March 2013 www.potto.org/downloads.php.

[18] J.S. Gudmundsson, Flow Assurance Solids in Oil and Gas Production. Flow Assurance Solids in Oil and Gas Production, CRC Press, 2017, https://doi.org/10.1201/9781315185118.

[19] J.C. Chen, Multiphase flow, in: F. Kreith, S.A. Berger, et al. (Eds.), Fluid Mechanics. Mechanical Engineering Handbook, CRC Press LLC, 1999. 1999.

[20] W.C. Lyons, G.J. Plisga, M.D. Lorenz, Standard Handbook of Petroleum and Natural Gas Engineering, Elsevier, 2015, pp. 1–1822, https://doi.org/10.1016/C2010-0-64695-5.

[21] Heriot-Watt University, Production Technology 1a, Department of Petroleum Engineering, Heriot-Watt University, Edinburgh, UK, 1999.

[22] R. Darby, R. Darby, R.P. Chhabra, Chemical Engineering Fluid Mechanics, second ed. revised and expanded, CRC Press, 2017, pp. 1–559, https://doi.org/10.1201/9781315274492.

[23] S.K. Sastry, C.A. Zuritz, A review of particle behaviour in tube flow: applications to aseptic processing, J. Food Process Eng. 10 (1987) 27–52.

[24] C.E. Brennen, Fundamentals of Multiphase Flow, vol. 9780521848046, Cambridge University Press, 2013, https://doi.org/10.1017/CBO9780511807169.

[25] I. Zandi, Advances in Solid–Liquid Flow in Pipes and Its Application, Elsevier, 1971, https://doi.org/10.1016/c2013-0-02346-8.

[26] A. Legrand, M. Renuy, E. Goudaliez, L. Fillaudeau, S. Bourneaud, et al., Characterization of solid-liquid suspensions (real, large and non-spherical particles in non Newtonian carrier fluid) flowing in horizontal and vertical pipes. 3, in: International Symposium on Two-Phase Flow modelling and experimentation, September 2004, Pisa, Italy, 2004. ffhal-02334241f.

[27] S. Mokhatab, W.A. Poe, G.J. Speight, Handbook of Natural Gas Transmission and Processing, Gulf Professional Publishing, USA, 2006.

[28] R.H. Nemoto, R.N. Br, J.L. Baliño, R.L. Tanaka, C.A. Godinho, A case study in flow assurance of a pipeline-riser system using OLGA, in: Proceedings of ENCIT, 2010. Retrieved from: http://abcm.org.br/anais/encit/2010/PDF/ENC10-0333.pdf.

[29] P. Han, L. Guo, Numerical simulation of terrain-induced severe slugging coupled by hydrodynamic slugs in a pipeline-riser system, Int. J. Heat Fluid Flow 56 (2015) 355–366, https://doi.org/10.1016/j.ijheatfluidflow.2015.10.005.

[30] S. Pedersen, P. Durdevic, Z. Yang, Review of slug detection, modeling and control techniques for offshore oil & gas production processes, in: IFAC-PapersOnLine, vol. 28, 2015, pp. 89–96, https://doi.org/10.1016/j.ifacol.2015.08.015.

[31] A. Faghri, Y. Zhang, Two-phase flow and heat transfer, in: Transport Phenomena in Multiphase Systems, Academic Press Elsevier, 2006.

[32] H.H. Hu, Computational Fluid Dynamics, in Fluid Mechanics, fifth ed., Elsevier Publishing, 2012, pp. 421–472.

[33] M.T. Byrne, M.A. Jimenez, E.A. Rojas, E. Castillo, Computational Fluid Dynamics for Reservoir and Well Fluid Flow Performance Modelling, Society of Petroleum Engineers, 2011, January 1, https://doi.org/10.2118/144130-MS.

[34] A.E. Osgouei, G. Altun, A review of application of computational fluid dynamics in prediction of oil and gas wells, in: Conference Proceedings, 19th International Petroleum and Natural Gas Congress and Exhibition of Turkey, May 2013, 2013. cp-380-00043.

[35] M.A. Kelland, Production Chemicals for the Oil and Gas Industry, CRC Press, 2014, https://doi.org/10.1201/b16648.

[36] SLB, Flow Assurance, Schlumberger, 2019. Retrieved from: https://www.glossary.oilfield.slb.com/en/Terms/f/flow_assurance.aspx. (Accessed November 2019).

[37] L.L. Schramm, Petroleum emulsions: basic principles, in: Emulsions: Fundamentals and Applications in the Petroleum Industry, American Chemical Society, 1992, pp. 1–49.

[38] H. Schubert, H. Armbruster, Principles of formation and stability of emulsions, Chem. Ing. Tech. 61 (9) (1989) 701–711.

[39] S. Kokal, Crude Oil Emulsions: Everything You Wanted to Know but Were Afraid to Ask, Society of Petroleum Engineers, 2008, January 1.

[40] B.A. Khan, N. Akhtar, H.M.S. Khan, K. Waseem, T. Mahmood, A. Rasul, H. Khan, et al., Basics of pharmaceutical emulsions: a review, Afr. J. Pharm. Pharmacol 5 (25) (2011) 2715–2725.

[41] G.R. Borges, G.B. Farias, T.M. Braz, L.M. Santos, M.J. Amaral, M. Fortuny, A.F. Santos, Use of near infrared for evaluation of droplet size distribution and water content in water-in-crude oil emulsions in pressurized pipeline, Fuel 147 (2015) 43–52, https://doi.org/10.1016/j.fuel.2015.01.053.

[42] K.L. Thompson, C.J. Mable, J.A. Lane, M.J. Derry, L.A. Fielding, S.P. Armes, Preparation of pickering double emulsions using block copolymer worms, Langmuir 31 (14) (2015) 4137–4144, https://doi.org/10.1021/acs.langmuir.5b00741.

[43] A. Gupta, H.B. Eral, T.A. Hatton, P.S. Doyle, Nanoemulsions: formation, properties and applications, Soft Matter 12 (2016) 2826–2841.

[44] B. Oyeneyin, Introduction to the hydrocarbon composite production system, in: Developments in Petroleum Science, 63, Elsevier, 2015, pp. 11–128, https://doi.org/10.1016/B978-0-444-62637-0.00002-6.

[45] M.A. Saad, M. Kamil, N.H. Abdurahman, R. Mohd Yunus, O.I. Awad, An overview of recent advances in state-of-the-art techniques in the demulsification of crude oil emulsions, Processes 7 (7) (2019), https://doi.org/10.3390/pr7070470. MDPI AG: 470.

[46] B. Islam, Petroleum sludge, its treatment and disposal: a review, Int. J. Chem. Sci. 13 (4) (2012) 1584–1602.

[47] R. Hollamby, G. Nuis, Process Chemistry, Shell International Exploration and Production B.V. Research and Technical Services, Rijswijk, 2000.

[48] O.A. Johnson, A. Chioma Affam, Petroleum sludge treatment and disposal: a review, Environ. Eng. Res. 24 (2) (2019) 191–201, https://doi.org/10.4491/EER.2018.134.

[49] W. Georgie, Foaming in Separators: Handling and Operation, 2013. SPE Webinar.

[50] R.W. Chin, et al., Chemical defoamer reduction with new internals in the Mars TLP separators, in: SPE Annual Technical Conference and Exhibition, 3–6 October 1999, Houston, Texas, 1999.

[51] T.E. Perez, Corrosion in the oil and gas industry: an increasing challenge for materials, JOM 65 (8) (2013) 1033–1042, https://doi.org/10.1007/s11837-013-0675-3.

[52] M. Simons, Report of Offshore Technology Conference (OTC) Presentation, NACE International Oil and Gas Production, 2008.

[53] G. Koch, J. Varney, N. Thompson, O. Moghissi, M. Gould, J. Payer, The International Measures of Prevention, Application and Economics of Corrosion Technology (IMPACT), NACE International, 2016.

[54] C.C. Patton, Applied Water Technology, Campbell Petroleum Series, first ed., 1987. October 1987.

[55] The Canadian Association of Petroleum Producers (CAPP), Best Management Practice Mitigation of Internal Corrosion in Carbon Steel Water Pipeline Systems, 2018.

[56] F.J. Cárcel-Carrasco, M. Pascual-Guillamón, L.S. García, F.S. Vicente, M.A. Pérez-Puig, Pitting corrosion in AISI 304 rolled stainless steel welding at different deformation levels, Appl. Sci. 9 (16) (2019), https://doi.org/10.3390/app9163265.
[57] W.S. Tait, Controlling corrosion of chemical processing equipment, in: Handbook of Environmental Degradation of Materials, third ed., Elsevier, 2018, pp. 583–600, https://doi.org/10.1016/B978-0-323-52472-8.00028-9.
[58] C. Vargel, Types of corrosion on aluminium, in: Corrosion of Aluminium, Elsevier, 2004, pp. 113–146, https://doi.org/10.1016/b978-008044495-6/50012-4.
[59] D.L. Engelberg, Intergranular corrosion, in: Shreir's Corrosion, Elsevier, 2010, pp. 810–827, https://doi.org/10.1016/B978-044452787-5.00032-9.
[60] J. Lai, K. Stadler, Investigation on the mechanisms of white etching crack (WEC) formation in rolling contact fatigue and identification of a root cause for bearing premature failure, Wear 364–365 (October) (2016) 244–256, https://doi.org/10.1016/j.wear.2016.08.001.
[61] R.J.K. Wood, 6.09—Erosion/corrosion, in: Comprehensive Structural Integrity, vol. 6, Elsevier, 2007, pp. 395–427, https://doi.org/10.1016/B978-008043749-1/00315-2.
[62] G.A. Pantazopoulos, A.I. Toulfatzis, Failure analysis of a machinable brass connector in a boiler unit installation, Case Stud. Eng. Fail. Anal. 1 (1) (2013) 18–23, https://doi.org/10.1016/j.csefa.2012.11.002.
[63] O.I. Zvirko, S.F. Savula, V.M. Tsependa, G. Gabetta, H.M. Nykyforchyn, Stress corrosion cracking of gas pipeline steels of different strength, in: Procedia Structural Integrity, 2, Elsevier B.V., 2016, pp. 509–516, https://doi.org/10.1016/j.prostr.2016.06.066.
[64] W.S. Tait, Corrosion prevention and control of chemical processing equipment, in: Handbook of Environmental Degradation of Materials, second ed., Elsevier, 2012, pp. 863–886, https://doi.org/10.1016/B978-1-4377-3455-3.00028-6.
[65] E. Huttunen-Saarivirta, M. Honkanen, T. Lepistö, V.T. Kuokkala, L. Koivisto, C.G. Berg, Microbiologically influenced corrosion (MIC) in stainless steel heat exchanger, Appl. Surf. Sci. 258 (17) (2012) 6512–6526,- https://doi.org/10.1016/j.apsusc.2012.03.068.
[66] M.B. Djukic, V. Sijacki Zeravcic, G.M. Bakic, A. Sedmak, B. Rajicic, Hydrogen damage of steels: a case study and hydrogen embrittlement model, Eng. Fail. Anal. 58 (December) (2015) 485–498, https://doi.org/10.1016/j.engfailanal.2015.05.017.
[67] R.J. Johnson, B.D. Folwell, A. Wirekoh, M. Frenzel, T.L. Skovhus, Reservoir souring—latest developments for application and mitigation, J. Biotechnol. 256 (August) (2017) 57–67, https://doi.org/10.1016/j.jbiotec.2017.04.003.
[68] Y. Holubnyak, J.M. Bremer, J.A. Hamling, B.L. Huffman, B. Mibeck, R.J. Klapperich, S.A. Smith, J.A. Sorensen, J.A. Harju, Understanding the souring at Bakken oil reservoirs, in: SPE, 2011, p. 141434.
[69] Z.I. Khatib, J.R. Salanitro, Reservoir souring: analysis of surveys and experience in sour waterfloods, in: Society of Petroleum Engineers. SPE-38795, 1997.
[70] W.G. Toland, Oxidation of organic compounds with aqueous sulfate, J. Am. Chem. Soc. 82 (1960) 1911.
[71] Y. Kiyosu, H.R. Krouse, The role of organic acid in the abiogenic reduction of sulfate and the sulfur isotope effect, Geochem. J. 24 (1990) 21–27.
[72] M. Basafa, K. Hawboldt, Reservoir souring: sulfur chemistry in offshore oil and gas reservoir fluids, J. Pet. Explor. Prod. Technol. (2019), https://doi.org/10.1007/s13202-018-0528-2.
[73] G.G. Hoffmann, T. David, Aquathrmolysis of Organic Compounds in the Presence of H2S and Sulfate, 2018. https://web.anl.gov/PCS/acsfuel/preprint%20archive/Files/42_2_SAN%20 FRANCISCO_04-97_0772.pdf.
[74] N. Marcano, S. Larter, L. Snowdon, B. Bennett, An overview of the origin, pathways and controls of H2S production during thermal recovery operations of heavy and extra-heavy oil, in: Integration, GeoConvention, 2013.

[75] J.J. Marrugo-Hernandez, R. Prinsloo, J. Fischer, R.A. Marriott, Downhole chemical degradation of corrosion inhibitors commonly used in shale gas fracturing and stimulation, J. Nat. Gas Sci. Eng. 72 (December) (2019), https://doi.org/10.1016/j.jngse.2019.103041.

[76] Global Oil & Gas Water Treatment Chemicals Market 2019–2024—Scale and Corrosion Inhibitors to Dominate the Market, Retrieved from https://www.globenewswire.com/news-release/2019/07/02/1877168/0/en/Global-Oil-Gas-Water-Treatment-Chemicals-Market-2019-2024-Scale-and-Corrosion-Inhibitors-to-Dominate-the-Market.html. (Accessed November 2019).

[77] V.T. Nguyen, T.H.Y. Nguyen, T.T. Trinh, C.S. Tong, D.H. Le, Flow assurance study on offshore marginal oil field development in Vietnam: the case of Ca Ngu Vang oil field, Inz. Miner. 1 (1) (2020) 15–22, https://doi.org/10.29227/IM-2020-02-04.

CHAPTER

3 Problems Associated With Flow Assurance Solids in Production

3.1 Introduction

Solids deposition is one of the major flow assurance issues in the oil and gas industry. Precipitation and deposition of different solids is detrimental to the production system in many ways. Loss of production, decreases in heat transfer efficiency, reduction of pumping efficiency, increases in fuel consumption, long downtimes for maintenance, and additional expenses for treatment are a few examples of the damaging effects of solids deposition. In extreme cases, solids deposition can result in the abandonment of a pipeline or production facility, as in the reported case of UK Lasmo when it abandoned and decommissioned its platform in November 1994 due to a recurrence of wax blockage.

The next few sections examine the details of some of the challenges posed by solids deposition for a production system, in order to give the reader an in-depth understanding of how they actually affect the production operations, which helps in raising awareness of their risks and developing mitigation plans.

3.2 Where do deposits form in a production system?

Field observations have shown that solid deposits can form almost anywhere in the production system, in reservoir formations, perforations, along production tubing and downhole equipment, in the wellhead, flowlines, topside installations and equipment, processing facilities, and storage tanks. However, there are still some locations in the production system that are more vulnerable to solids deposition than others. Locations where incompatible fluids commingle or where system parameters like temperature and pressure significantly change, driving the system thermodynamically toward solids precipitation, are vulnerable to solids deposition. Furthermore, fluid hydrodynamics play a considerable role in determining if the solid precipitates will stick to surfaces and form a deposit, or be transported with the flowing fluids, and where they will finally be deposited. Fig. 3.1 represents the locations having higher potential of solids formation. These involve but are not limited to:

- Pressure drop locations: in the reservoir, perforation, along production tubing, valves, separators, and other system installations.
- Temperature drop: temperature drop in the formation, at the perforation, along the production tubing, on the topside facilities and flowlines.

Essentials of Flow Assurance Solids in Oil and Gas Operations. https://doi.org/10.1016/B978-0-323-99118-6.00006-X

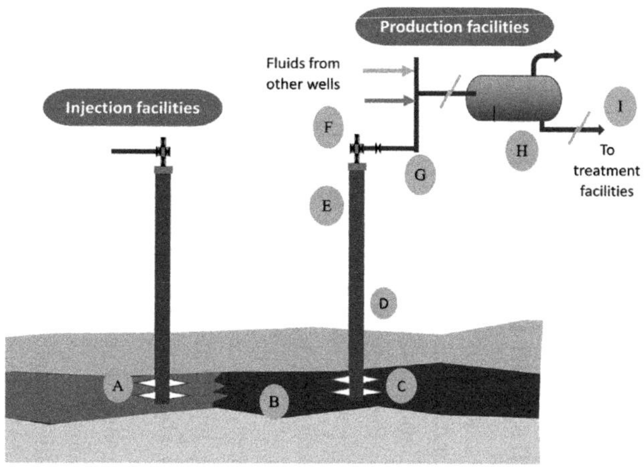

	Location	Contributing Factors
A	Downhole injection well	Temperature increase , pressure change Mixing of injection water with reservoir water Geochemical interactions between injected water and the reservoir rocks
B	Reservoir	Mixing of injection water and reservoir water Geochemical interactions between injection water and reservoir rocks Pressure change through a porous media Joule Thomson cooling/heating
C	Perforations and downhole production well	Mixing of injection water and produced water Mixing between different production zones (commingled production) Pressure drop through perforations Joule-Thomson cooling
D	Production tubing and downhole equipment (gas lift mandrels, ESP,	Incompatible fluids mixing Pressure fluctuations Temperature drop
E	Top part of production tubing	Pressure drop Temperature drop
F	Wellhead and production choke	Pressure drop Temperature drop
G	Production header	Mixing of incompatible fluids from different wells streams Pressure and temperature changes
H	separators	Pressure drop Low velocity flow , high retention time, suspended solids accumulation, Mixing fluids from different streams
I	Processing facilities &downstream Flowlines	Pressure changes, temperature changes , low velocity flow , mixing incompatible streams

FIG. 3.1

The various locations where solids can form in a production system.

– Mixing zones between injection water/gas and formation fluids, mixing between fluids in the topsides.

These locations are as indicated in Fig. 3.1.

In addition to their ability to accumulate anywhere in the production system, solid deposits are operational challenges due to their peculiarities, which include the following:

- Solid deposits are quick forming and quick impacting. For example, in Miller Field in the North Sea, a well production rate declined from 30,000 BPD to zero BPD in just 24 h after having been plugged with barium sulfate deposits [1]. Similar situations were reported with paraffin wax deposits during shutdowns (even short-term shutdowns) if the temperature fell below the wax appearance temperature, which required costly restart procedures.
- They are severely impacting even at very small concentrations. For example, a case study reported that a few milligrams per liter of naphthenate deposits over 6 weeks had led to accumulation of 30 tons of deposits, costing US$6 M to be removed [2]. Moreover, small concentrations of the solids can stabilize emulsions and interfere with phase separation operations.
- They are diverse and perpetually changing. Solids usually do not deposit in a pure form, but rather as a composite of different compounds, and often involve organic and inorganic types mixed together. In this wide range of compounds, although they may share some common properties, their stability, solubility behavior, and the factors affecting them are quite different, which complicates the problem. Adding to this, the changing operating parameters and the changing water/oil/gas volumes relentlessly change the chemical equilibrium that controls these solid formations and the amounts of solids that can form, and this exacerbates the problem.
- With regard to the previous point, the mitigating strategies depend on the solid type and volume, which in the case of a composite of deposits will necessarily require complex mitigation programs. Moreover, the mitigation plans must be designed and implemented before any significant changes take place in the system, which time delay could result in changes in the solids' nature and quantities, which would render the mitigation plans ineffective or in vain.

Therefore, in fighting solids deposition, quick and proactive actions are always necessary to avert their stealthy and damaging impacts.

3.3 The cost of solids formation and deposition

Solids deposition causes tremendous problems in oil and gas production systems, including formation damages, tubular blockages, equipment impairment, other production chemistry issues, extra fuel and power consumption, extra expenses for treatment, in addition to the safety and environmentally related issues. The overall results of such problems are loss of production, additional expenses, and reduction in revenues. The problems that are caused by solids deposition are categorized and summarized in Fig. 3.2.

The worldwide loss of production due to scale deposits is estimated to be 28%. During the period of 1999–2003, BP estimates that about 18% of their well losses were attributed to scale deposition [3]. The additional expenses due to deposits are the maintenance and mitigation expenses, i.e., workover, recompletion, reperforation and other well intervention operations, chemical expenses, maintenance

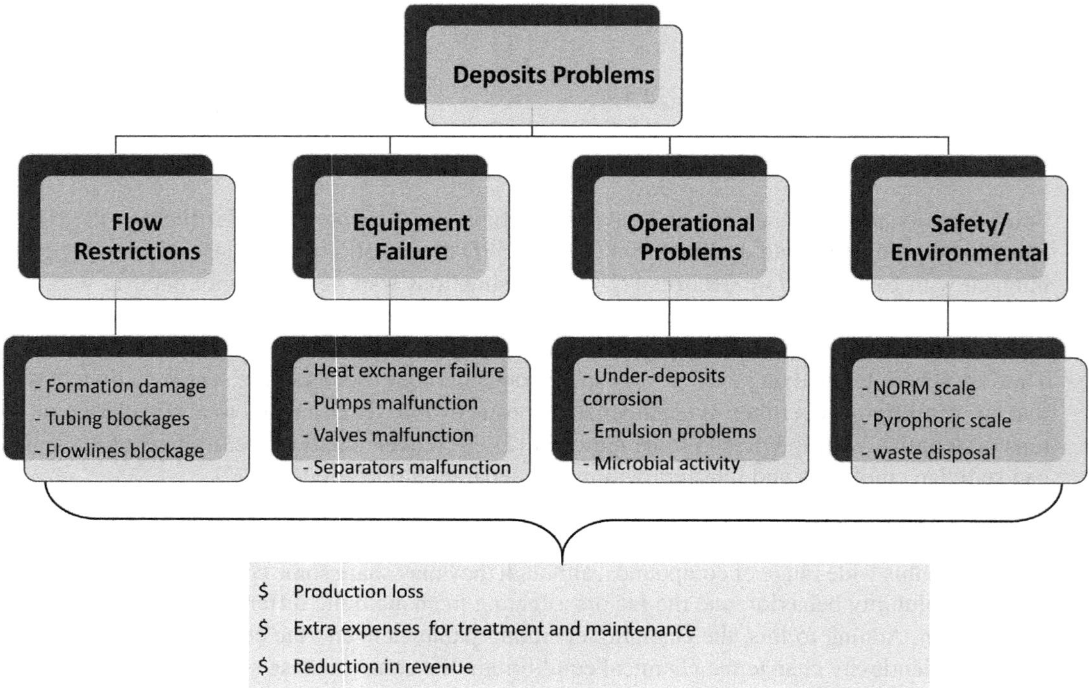

FIG. 3.2

The problems associated with flow assurance solid deposits in a production system.

equipment rental, production equipment spare parts, capital mitigation project expenses (like sulfate reduction plant and injection water softening), and labor. Other extra expenses come from the equipment overconsumption of fuel due to diminished performance. Furthermore, the problem entails safety and environmental issues like NORMs, pyrophoric deposits, waste disposal, and treating the produced water from the spent injected chemicals, which of course all come at a large cost for remediation.

Reports show that the mineral scale problems lead to an estimated cost of US$1.5–2 billion per year [4,5]. Worldwide costs associated with heat exchanger fouling alone have been estimated at US$26 billion/year, or 0.2% of the gross national product of the industrialized countries [6]. The cost of scaling and fouling worldwide, based on 1984, is given in Table 3.1 [7].

There are few reported studies detailing the costs of the different types of solid deposits in oil and gas fields. However, multiple reported case studies can give us a good indication of how costly solids deposition problems can be. The direct cost of removing scale from one well can be as high as $2.5 M and the cost of deferred production can be even higher [8]. Sometimes the cost can be scary and persistent, such as what happened in Miller Field when the well was fully blocked by barium sulfate deposits in 24h and production went down from 30,000 BPD to zero BPD, which in terms of money is a loss of US$384,000/day according to the oil prices at that time (US$12.8/barrel as per 1998 average price) [1]. The annual scale damage in the oil and gas fields in different regions is estimated to be $100 MM in the Middle East, $400 MM in the North Sea, $600 MM in North and South America, $140 MM in Russia [3]. Fouling in the crude preheat train (PHT) leads to an enormous cost penalty (on the order of $1 billion per annum in the United States alone) [9]. Van Nostrand et al. [10] estimated that process-side fouling costs

Table 3.1 Estimations of fouling costs in some countries.

Country	Fouling costs US $ million	GNP[a] (1984) US $ million	Fouling costs % of GNP
United States (1982)	3860–7000 8000–10,000	3,634,000	0.12–0.22 0.28–0.35
Japan	3062	1,225,000	0.25
West Germany	1533	613,000	0.25
United Kingdom (1978)	700–930	285,000	0.2–0.33
Australia	260	173,000	0.15
New Zealand	35	23,000	0.15
Total industrial world	26,850	13,429,000	0.20

[a]GNP, *gross national product.*

Reproduced with permission from S.N. Kazi, Particulate matter: interfacial properties, fouling, and its mitigation, in: Mineral Scales and Deposits: Scientific and Technological Approaches, Elsevier, 2015, pp. 141–168, https://doi.org/10.1016/B978-0-444-63228-9.00007-3. Copyright (2015), Elsevier, license: 5274240739476.

the United States refineries US$1.36 billion per year, and US$861 MM of that in crude PHT alone. Adjusting for inflation to 2014, this is equivalent to US$3.6 billion and US$2.26 billion, respectively. Wax deposition in production tubing and transport pipelines drastically reduces operating efficiency and its removal requires significant additional operating costs. The direct cost of removing paraffin blockage from a pipeline 40,000 ft long with diameter 6 in. can be in six figures and reach up to $1,000,000. The removal of a wax deposit plug or gas hydrates plug using conventional methods can take days to weeks of downtime, costing millions of dollar in production loss. Creek [11] estimated the cost of methanol for complete hydrate inhibition in an oil field producing 50,000 bbl/day water may reach 50% of the operating revenues, at the current prices of oil and methanol. The US oil and gas disciplines suffer from severe costs, reaching approximately $1.372 billion per year due to biofouling according to the National Association of Corrosion Engineers [12]. It is also well accepted that the major problem concerning biofilm formation on heat transfer systems is the resultant costly reduction in heat exchanger performance. Ibrahim [13] demonstrated that the estimated costs for biofilm formation in heat exchangers are 7.5% of the total maintenance cost. It is reported that annual costs due to heat exchanger fouling in the United Kingdom are £900 million; in the United States the cost is $2.5 million annually [14]. Table 3.2 summarizes the cost of solids deposition from different oil and gas field case studies.

In the next few sections, the different types of solids deposition that affect production operations are thoroughly discussed.

3.4 Flow restrictions

The stereotypical impact of solids deposition in a production system is restricting the flow of the produced fluids. The problem usually begins with a thin film of deposits that quickly grows into thick layers that internally coat the production tubulars and equipment, causing a reduction of their internal

Table 3.2 The cost of solid deposits in some oil and gas fields, MM = millions [1,11,15–20].

Field/country	Cost US$	Cost explained	Deposit type
Miller—North Sea—United Kingdom	384,000/day (1998 oil price)	Production loss	$BaSO_4$
Forties—North Sea—United Kingdom	3.8 MM	Treatment	Mineral scales
Tatham Offshore, Texas and Lasmo, United Kingdom	100 MM	Treatment, production loss, and field was abandoned	Wax and asphaltene
Duri Field—Indonesia	3.6 MM	Built revenue after scale removal	$CaCO_3$
Veslefrikk field—Norway	1.1 billion	Treatment and production loss	Mineral scales
Egyptian field	3 MM	Price of treating chemicals only (production loss not included)	Mineral scale, wax, and asphaltenes
Nigerian fields	6 MM	Removal (production loss not included)	Naphthenate
United States	4.5–5 MM	Wax control	Wax deposits
Gulf of Mexico GOM	25.12 MM	Asphaltenes remediation, not including production deferral	Asphaltenes
Ras Shukair Field—Egypt	>1 MM	Treatment	Biocorrosion and biofouling

dimensions and resulting in fluid flow restrictions, which are reflected as pressure drop and decreased productivity. The thicker the deposit layer, the higher the effect on the fluid flows, and in severe cases full blockage can be encountered, leading to a complete shutdown of production.

3.4.1 Formation damage

Solid deposits can form in the reservoir, clogging the small fluids flow channels in the reservoir rocks or the pore throat, causing what is known as formation damage. Formation damage is a generic term that refers to the impairment of the permeability of petroleum bearing formations by various adverse processes [21]. The consequences of formation damage are reduction of the oil and gas productivity of reservoirs and uneconomical operation.

Physicochemical, chemical, hydrodynamic, biologic, and mechanical processes often lead to the mobilization, generation, migration, and deposition of fine particles, which, in turn, cause formation damage in petroleum-bearing formations [21,22]. Organic deposition is also quite common; perhaps the most common formation damage problem reported in the mature oil-producing regions of the world is organic deposits forming both in and around the wellbore [23]. Also, plugging the pores of the formation by solids results in injectivity decline over time [24].

These solids can deposit on the surface of the grain or the pore throat or pore bridging between the deposited particles and the grains, and then it can form an internal or external cake that fully plugs the pore throats [25,26]. The different particulate processes that lead to the damage are schematically depicted in Fig. 3.3.

In the case of reservoir damage, the most commonly used measure of formation damage in a well is the skin factor, S. The skin factor is a dimensionless pressure drop caused by a flow restriction

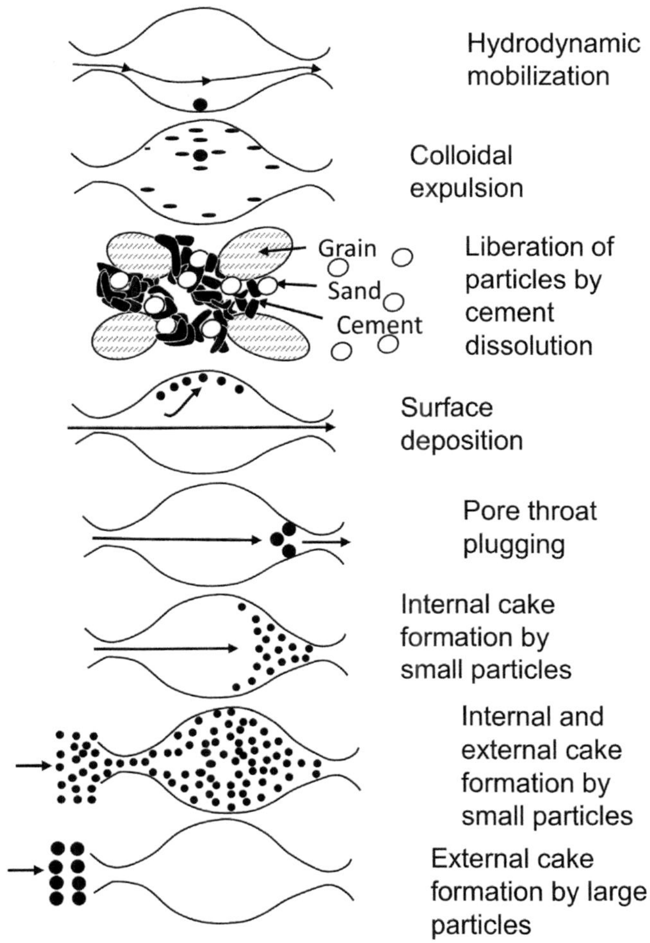

FIG. 3.3

Formation damage mechanisms.

Reproduced with permission from F. Civan, Reservoir Formation Damage: Fundamentals, Modeling, Assessment, and Mitigation, third ed., Elsevier, 2015, https://doi.org/10.1016/C2014-0-01087-8. Copyright (2015), Elsevier.

(permeability reduction) in the near-wellbore region. If the skin factor is known, the pressure drops across the damaged zones can be estimated by Eq. (3.1) [27]:

$$\Delta P_{\text{skin}} = \left[\frac{141.2\,\mu\beta}{K}\right] \cdot \frac{q}{h}\,(\text{s}) \tag{3.1}$$

where μ is the fluid viscosity (CP), β is the fluid formation volume factor (rb/stb for oil, ft^3/scf for gas), h is the net reservoir thickness (ft), K is the reservoir permeability in darcy (d), q is the well rate (STB/d for oil, MMSCFD for gas).

3.4.2 Production tubular flow restrictions and blockages

Solids deposition is a common problem in production tubulars, i.e., tubing, risers, and flowlines in oil and gas fields. Fig. 3.4 shows some severe cases of flowline blockages by scale deposits.

As mentioned earlier, a common result of solids formation in production tubulars is pressure drop. Solids formation affects pressure drop in the tubulars due to many mechanisms, including the following:

FIG. 3.4

Scale deposit blockages in pipelines from Egyptian oilfield.

(a) Particles transportation: solids transportation with the fluids, causing changes in the fluid properties and flow regimes, consequently causing pressure drop.
(b) Particle interactions: the transported solid particles can have different types of interactions that cause pressure drop, most commonly: particle-particle interactions (collisions), particle-fluids interactions, and particle-pipe wall interactions.
(c) Changes in surface roughness: When the particles deposit on the pipe wall, they cause changes in the surface roughness. In turbulent flow, which is the most encountered flow industrially, the friction loss primarily depends on the pipeline surface roughness.
(d) Changes in conduit internal dimensions: solid deposit layers normally coat the internal surfaces of conduits and equipment, leading to changes in their internal dimensions, which causes significant pressure drop.

A brief overview of the different mechanisms by which solids affect pressure drop will be given.

3.4.2.1 Pressure drop due to solids transportation and interactions

Not all the precipitated solids have the opportunity to stick to the pipe wall and accumulate in layers, but some of them can be conveyed with the flowing stream circulating in the production system until they reach deadends or final terminals. Such transportation of solids with the fluids poses extra pressure drop on the system, given by Eq. (3.2) [28]:

$$\Delta P_t = \Delta P_f + \Delta P_s \tag{3.2}$$

where ΔP_t is the total pressure drop, ΔP_f is the pressure drop due to fluids flow, and ΔP_s is the pressure drop due to solids transported in the fluids.

The effect of transported solids on pressure drop increases with increasing concentration of the transported particles, which in petroleum production systems is given by a high rate of solids precipitation, or high rates of sand and fines production from unconsolidated reservoirs. About 70% of the total world's hydrocarbons are located in poorly consolidated reservoirs [29,30]. Field observations reported in the literature indicate volumetric concentration of sand in oil pipe systems varies from 1% to 40% [29–31]. Pressure drop due to sand production in horizontal and vertical wells was demonstrated by Almedeij et al. [32], and Olufemi and Darlington [33].

Various types of interactions take place with particles suspended in a flowing fluid, including particle-particle interactions, particle-fluid interactions, and particle-pipe wall interactions. These interactions can change the rheology and other physical properties of the fluids, thus causing pressure drop. Particle-particle physical interactions include [34]:

- Interparticle attraction promotes the formation of flocs and aggregates. This phenomenon occurs mostly in fine particle suspensions.
- Hydrodynamic interactions give rise to viscous dissipation in the liquid.
- Particle-particle contact brings into play frictional interactions.

The fluid-solids suspension viscosity and yield stress increase with solids concentration and particle size at the different pseudo-shear rates. The increase in suspension viscosity with solids concentration is attributed to the increased frequency of particle-particle interactions, whereas the increase of viscosity with particle size is attributed to the increased inertial effects [35].

The fluid-particle interaction force can be divided into five main contributors [36]:

- Buoyancy force.
- Local fluid acceleration force, which is parallel to the mean acceleration of the fluid.
- Drag force, which acts in the direction of the fluid-particle slip velocity.
- Virtual mass force, which is parallel to the relative acceleration between the phases.
- Lift force, which is normal to the slip velocity.

Two types of friction losses are encountered when solids are transported with fluids in pipelines: mechanical friction and viscous friction. The mechanical friction is created through the contact between the solid particles and the pipe wall and it can be permanent or random. The viscous friction is created due to the existence of solid particles in the near wall layer of the carrying liquid as addition to the fluid viscous friction losses. Therefore the properties of the carrying liquid near the wall layer are changed [37,38].

Generally speaking, the pressure loss due to solid-fluid transport depends on:

- Solids concentration: the pressure drop increases with increasing solid particle concentration [39,40].
- Solid particle size: the smaller particle size will have lower pressure drop at lower velocities and higher pressure drop at higher velocities, compared to larger sized particles [39,40].
- Flow pattern and flow rate: the different flow regimes have significant effects on pressure drop. Flow regimes of solids transported in fluids can be classified into settling and nonsettling regimes [38,41], or more effectively can be classified into homogeneous, heterogeneous, and bedload [32], and each one has an effect on pressure drop.

In the case of a homogeneous flow regime, the particles are uniformly distributed throughout the pipe cross-section, which is the case with small or fine particle sizes, and the pressure loss in this case can be calculated on the basis of the rheological behavior of this suspension [32,42], which can be generally calculated by the Darcy and Weisbach equation (Eq. 2.5) in Chapter 2, Flow Assurance.

In the case of a heterogeneous flow regime, the particles are not uniformly distributed throughout the pipe cross-section, representing the transport of coarser particles, and the pressure drop may be calculated by the modified (extended) Durand equation (Eq. 3.3), which can also be used to determine pressure drop for any flow regime [42,43]:

$$\Delta P = \left[\frac{1}{83m}\left[\frac{gD}{v^2}(\rho_s-\rho_f)\rho_f\sqrt{C_D}\right]^{\frac{1.5}{m^3}} C_o+1\right] f\rho_f 2v^2\frac{L}{D} \tag{3.3}$$

where ΔP is the pressure drop due to coarse particles, f is the friction factor, v is fluid velocity (m/s), L is pipe length (m), C_o is the delivered total concentration (vol%), ρ_f is the fluid density (kg/m^3), ρ_s is the solids density (kg/m^3), C_D is the drag coefficient, D is the pipe diameter (m), m is a correlation factor that depends on the particle diameter, and g is the gravity constant (9.8 m/s^2).

For a more generic formula of the Durand equation that covers polydispersed mixtures of solids-liquid, which covers homogeneous and heterogeneous cases, the following equation (Eq. 3.4) presented by Weber [44] can be used:

$$\Delta P = \left[\frac{1}{83m}\left[\frac{gD}{v_m^2}(\rho_s-\rho_{fs})\rho_f\sqrt{C_D}\right]^{\frac{1.5}{m^3}} (1-C_f)C_o+1\right] f\rho_{fs} 2v_m^2\frac{L}{D} \tag{3.4}$$

where C_f is a homogeneous part of particle size, and ρ_{fs} is the density of enriched fluid.

The bedload transport regime occurs when the gravitational forces of sediment particles become larger than the forces produced by turbulence. Here, sediment particles start moving very close to the pipe invert by rolling, sliding, or sometimes by jumping a few grain diameters.

3.4.2.2 Pressure drop due to changes in surface roughness

Friction loss is another means of interaction between flowing fluids and the pipe walls, and it represents energy consumption or pressure drop. Pressure drop due to friction is examined in Chapter 2, Flow Assurance. Pressure drop due to friction in turbulent flows like those encountered in oil and gas systems predominantly depends on surface roughness. Hence, another way solid deposits are affecting the pressure drop in a pipe by changing the surface roughness occurs when they are deposited on the internal surface of that pipe.

The structure of the roughness is also an important factor. For example, the friction factor for uniformly distributed roughness will be different from a structured deposit such as transverse ripples. In addition, the friction factor in liquid pipelines is not exactly the same as in gas pipelines [45].

Many correlations have been used to describe the surface roughness of a pipeline, including [46,47], but a more convenient one is the correlation developed by Haaland [48]:

$$\frac{1}{\sqrt{f}} = -\frac{1.8}{n}\log\left[\left(\frac{6.9}{\mathrm{Re}}\right)^{n} + \left(\frac{k}{3.75d}\right)^{1.11n}\right] \tag{3.5}$$

where f is the friction factor, Re is Reynolds number, d is the pipe diameter, n is an exponent equal to 1 for liquids and 3 for gases, the ratio k/d is the relative roughness, where k stands for sand-grain roughness. It is not the measured roughness of a pipe wall, but the roughness equivalent to the diameter of sand grains glued to a pipe wall that gives the same pressure drop [45].

This correlation is a practical correlation over a wide range of roughness and Reynolds number. One main problem with this correlation and the previously mentioned correlations is that they all are based on pipe wall surfaces with uniform roughness, which is not the case with rippled deposits. Therefore sometimes it is more practical to use the rectangular transverse roughness correlation by Baumann and Rehme [49] to model rippled deposits in pipelines.

3.4.2.3 Pressure drop due to changes in conduit dimensions

When the solids reach their resting velocity, they deposit on the pipe wall, and thereafter they gradually grow from thin films to thick layers, resulting in reduction in the internal diameter of the pipe, and that consequently causes significant pressure drop. Sometimes this deposition process is uniform, in which the solids accumulate evenly on the internal surfaces of the pipe, whereas at other times the solids accumulate unevenly or bumpily on the pipe surfaces, causing throttling of the flow through the formation of a "Venturi effect" or "Vena contracta" similar shapes that can chock the flow in no time. Above all, further growing of deposit layers can plug the whole internal volume of the conduits and fully block the fluid flow. Figs. 3.5 and 3.6 illustrate the distribution of deposits inside a pipeline.

The pressure loss in the case of uniform deposit layers will be different from nonuniform scale layers. In the case of a uniform deposit layer inside the pipeline, the pressure drop can be calculated from the Poiseuille formula [50]:

$$\Delta P = \frac{8\eta LQ}{\pi(r-t)^{4}} \tag{3.6}$$

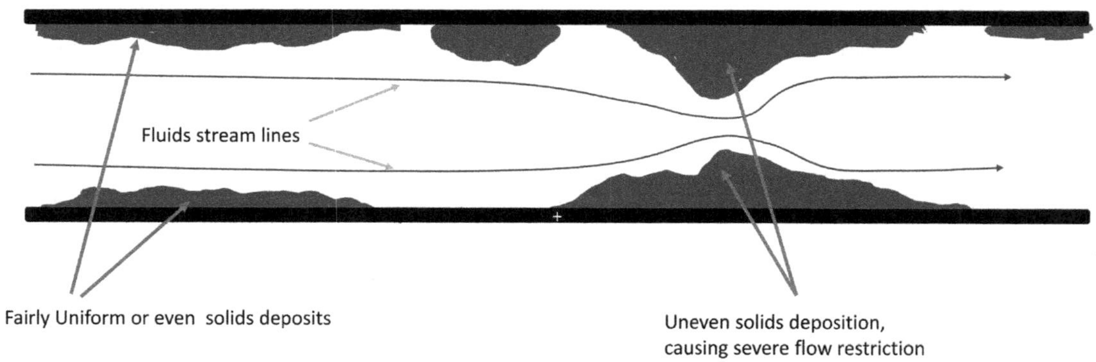

FIG. 3.5

The formation of even and uneven solid deposit layers and their effects on the fluid flow.

where ΔP is the pressure drop, L is the length of the pipe, Q is the volumetric flow rate, r is the pipe radius, t is the thickness of the deposits layer (assumed to be uniform over L), and η is a coefficient of viscosity of the fluid.

The reduction of the pipeline diameter in both cases (even and uneven) is significant; however, it is more severe in the case of uneven deposits formation. To realize the difference, consider a 12″ pipeline,

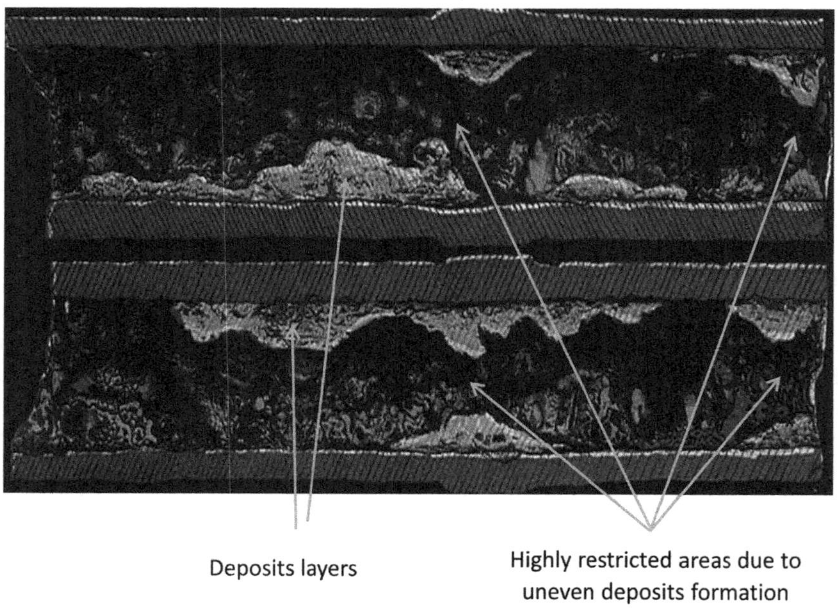

FIG. 3.6

Cross-section of scaled pipeline.

Reproduced with permission from Annual Report of the Chief Engineer of the Water Department of the City of Philadelphia, J. Franklin Inst. 71(5) (1861), 349–350, https://doi.org/10.1016/0016-0032(61)90410-0, Elsevier.

where the inside diameter is reduced by 15 mm by smooth even deposits; here the throughput loss at a given pressure will be over 10% and to compensate for the loss and to restore the original throughput, level pressure has to be increased by ~30%. However, in the case of uneven deposits, the resulting turbulent flow reduces the throughput by ~35% and pressure has to be increased by 140% to overcome such losses.

In heat exchangers, pressure loss is considered more critical than the loss in heat transfer due to fouling. The deposited layer roughens the surface, diminishes the inner diameter, and raises the outer dimension of the tubes, causing an increase in pressure drop. Pressure drop inside a tube of a heat exchanger under fouled and clean states can be correlated according to Eq. (3.7) [51]:

$$\frac{\Delta p_f}{\Delta P_c} = \frac{f_f}{f_c}\left(\frac{d_f}{d_c}\right)\left(\frac{q_f}{q_c}\right)^2 \tag{3.7}$$

where f is the friction factor, d is the tube diameter, q is the flow rate, and the subscript f stands for fouled tube and c stands for clean tube.

If the mass flow rates under clean and fouled conditions are considered to be the same, the previous equation becomes [51]:

$$\frac{\Delta p_f}{\Delta P_c} = \frac{f_f}{f_c}\left(\frac{d_f}{d_c}\right)^5 \tag{3.8}$$

The thickness t_f of the deposit layer can be obtained from:

$$t_f = 0.5 d_c \left[1 - \exp\left(-\frac{2k_f R_f}{d_c}\right)\right] \tag{3.9}$$

where d_c is the fouled tube diameter and R_f is the fouling factor.

3.5 Equipment impairment and failure

Equipment failure due to solids deposition is very common in oil and gas fields. Field operators always encounter valves, pumps, separators, flowmeters, and other downhole and topside equipment that requires maintenance or replacement due to solids build-up that degrades their performance. A few examples of equipment failure due to solids deposition are presented in the following sections.

3.5.1 Heat transfer equipment impairment and failure

One of the main problems caused by deposits accumulation on surfaces in many industries is the drastic reduction in heat transfer rates in equipment such as boilers, steam generators, evaporators, exchangers, engine jackets, cooling towers, etc. The thermal conductivity of most of the deposits is considerably less than that of steel, and the consequence of such deposits, even extremely thin ones, is an insulating effect [52].

For example, in a heat exchanger, the rate of heat transfer is expressed by Eq. (3.10) [53]:

$$Q = UA\Delta T_m \tag{3.10}$$

where Q is the rate of heat transfer (Btu/h), U is the heat transfer coefficient (Btu/h/ft^2 F) that represents the thermal conductance of the heat exchanger (the higher the value of U, the more easily heat is transferred from one fluid to the other), A is the heat transfer surface area (ft^2), ΔT_m = log mean temperature difference between fluids (°F).

In order to efficiently transfer a given number of Btus through a heat exchange surface in a given period of time, it is necessary to maintain a certain temperature differential. If scale forms, this differential temperature has to be higher in order to compensate for the insulating effect of the deposits layers due to the lower thermal conductivity of deposits compared to steel [52]. Various types of solids that have been reported to deposit in heat transfer equipment are: calcite (the most vulnerable type due to the direct effect of temperature on its solubility), siderite, silicates, calcium sulfates, sulfur, and iron sulfides. Asphaltenes, oxidative polymerization, and coke formation arising from components in the oil are other examples of solids that can deposit in heat transfer equipment [54]. In addition to these types, biofouling is common in some heat transfer equipment like heat exchangers [55]. Table 3.3 shows the thermal conductivity of metals and those of some solid deposits that are found in oil and gas fields.

As the thickness of the deposit layers increases with time, the heat loss increases, which diminishes the heat transfer efficiency of the equipment. In addition, the change in the deposit layer morphology has significant impact on the heat transfer process.

Under clean conditions, the overall heat transfer coefficient, U_c, is given by the sum of the individual thermal resistances [59]:

$$\frac{1}{U_c} = \frac{D_o}{D_i}\frac{1}{\alpha_i} + R_w + \frac{1}{\alpha_o} \tag{3.11}$$

where α_i and α_o are, respectively, the inside and outside tube convective heat transfer coefficient and R_w is the thermal resistance of the metal wall (x_m/λ_m), D_i is the inner diameter, and D_o is the outer diameter. When deposits form on both sides of the tube (inside and outside), the heat transfer is impaired, and that is defined by adding what is known as the respective heat transfer resistances [53,60].

Thus, fouling in heat exchangers is traditionally expressed using the concept of a fouling resistance or the fouling factor R_f (or fouling resistance), which represents the resistance to the flow of heat due to the buildup of a fouling layer on the tube surfaces of the heat exchanger. The fouling factor is not

Table 3.3 Thermal conductivity of metals and foulants [56–58].

Material	Thermal conductivity (W/mK)
Copper	386
Stainless steel	25
Mild steel	50
Calcium carbonate	2.9
Iron oxide scale	2.2–2.9
Calcium sulfate	0.74–2.3
Wax	0.24
Asphalt	0.75–1.26
Biofilm	0.6

usually measured directly, but must be determined from the degradation of the overall heat transfer coefficient [57]:

$$R_f = \frac{1}{U_f} - \frac{1}{U_c} \tag{3.12}$$

where R_f is the fouling factor or unit thermal resistance of deposit (m^2K/W), U_f is the overall heat transfer coefficient of the heat exchanger after fouling (W/m^2K), and U_c is the overall heat transfer coefficient of a clean heat exchanger (W/m^2K). Another generalized way of expressing the fouling resistance is by using the resistance to heat flow across a solid surface, which is given as [53]:

$$R = \frac{x}{\lambda} \tag{3.13}$$

where x is the solid surface thickness, λ is the thermal conductivity of that solid material, and for metal pipe wall $R_w = x_m/\lambda_m$, for deposits $R_f = X_f/\lambda_f$.

The total resistance to heat flow will be the sum of the individual resistances, which is given by Eq. (3.14):

$$R_T = \left[\frac{x_o}{\lambda_o}\right] + \left[\frac{x_i}{\lambda_i}\right] + \left[\frac{x_m}{\lambda_m}\right] + \frac{1}{\alpha_o} + \frac{1}{\alpha_i} \tag{3.14}$$

where the subscripts *o* and *i* represent outer and inner.

Another way to express R_T is by Eq. (3.15):

$$R_T = r_1 + r_2 + r_3 + r_4 + r_5 \tag{3.15}$$

where R_T = total heat flow resistance.

r_1 = heat flow resistance of the process-side film.
r_2 = heat flow resistance of the deposit layer (the process-side fouling).
r_3 = heat flow resistance of the exchanger tube wall.
r_4 = heat flow resistance of the deposit layer (the water-side fouling).
r_5 = heat flow resistance of the water-side film.

In other words: tube-side fluid, tube-side scale, tube metal, shell-side scale, and shell-side fluid. A schematic illustration of the fouled heat exchanger tube is depicted in Fig. 3.7.

A good example to picture the severity of heat exchanger fouling due to solids deposition is that only 0.1 in. of $CaCO_3$ on a heat transfer surface was found to cause a heat transfer loss of 40% [61]. An old study found that boiler water treatment inadequacies amounted for 26% of the failures in the boiler tubes [52]. Pilavachi and Isdale [62] concluded over the European Community as a whole that the cost of heat exchanger fouling was on the order of 10×10^9 ECU and of this total 20%–30% was due to the cost of additional energy [53]. Fig. 3.8 shows a heavily fouled heat exchanger.

In addition to reducing the heat transfer efficiency, increasing fuel and power consumption is a problem. As a rule of thumb, one millimeter of scale build-up in boiler tubes can increase fuel consumption by 2% [63]. Scale formation in heating elements in a power plant increased the power consumption by approximately 4% to 12% [64]. Some numbers related to energy loss due to scale are listed in Table 3.4.

Similarly, black powder accumulation in gas systems reduces compressor efficiency and increases fuel consumption.

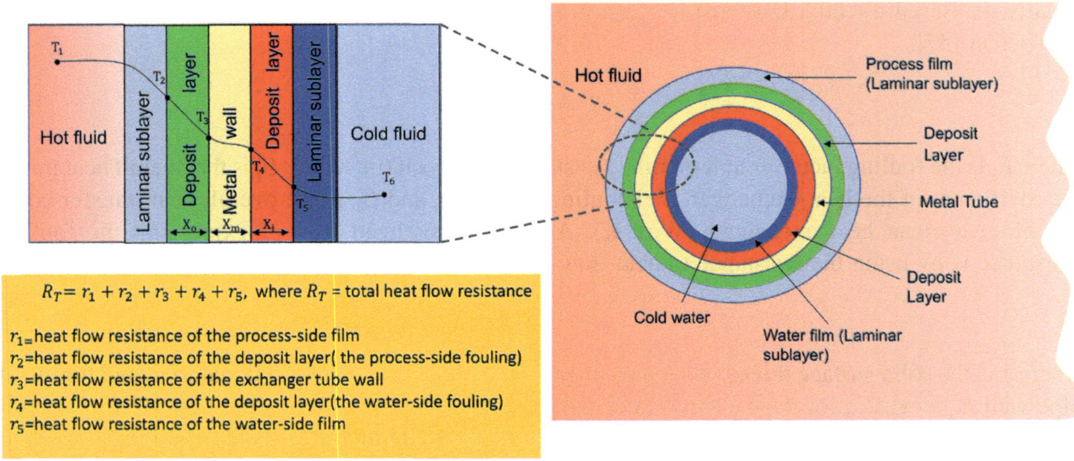

FIG. 3.7

Schematic of the total heat flow resistance in heat exchanger tube.

FIG. 3.8

Heavily fouled heat exchanger.

Courtesy H.A.H. Ibrahim, in: V.N. Katsikis (Ed.), Fouling in Heat Exchangers in MATLAB—A Fundamental Tool for Scientific Computing and Engineering Applications, vol. 3, 2012, https://doi.org/10.5772/46462.

Table 3.4 The effect of scale thickness on fuel loss.

	Energy loss %		
	Scale type		
Scale thickness (inches)	Normal	High iron	Iron plus silica
1/64	1.0	1.6	3.5
1/32	2.0	3.0	7.0
3/64	3.0	4.7	–
1/16	3.9	6.2	–

Data from National Institute of Standards and Technology, Handbook 115, (1975) Supplement 1. Normal scale is formed under low pressure, while the high iron and iron plus silica scale composition results from high-pressure service conditions.

3.5.2 Pumps impairment and failure

In pumps, the deposit can accumulate in the different compartments, i.e., impeller, casing, suction strainer, or bearings, causing serious reduction in the effective diameters inside these compartments, increase in friction losses, and general loss in performance and efficiency. For example, for a centrifugal pump to operate, the liquids enter the eye of the impeller under pressure; this pressure is called the net positive suction head (NPSH) and that deals with the suction side of the pump. There are two main NPSH values: one is the available NPSH (NPSHa) and this is calculated by the system the pump is assembled in, and the other is the minimum required NPSH (NPSHr) and this is specified by the manufacturer. For proper operation of the pump, the NPSHa must be greater than the NPSHr; if not, the pump will not only produce little or no flow, but will cavitate and may quickly self-destruct. The problem is that NPSH is not a fixed or constant value, and can be affected by friction losses in the piping, which are affected by the flow rate and the degree of fouling [65]. Fouling results in a drop in NPSHa and, consequently, in the margin between the NPSHa and NPSHr. The point where this margin reduces to a minimum positive value is specified as the maximum flow rate, Q_{max}. The fouling of the suction header causes a steeper drop in the NPSHa curve and this reduces the Q_{max} flow rate [66]. Fig. 3.9 shows the effect of fouling in the suction header on the NPSHa.

Another way of affecting the pumps occurs when the deposits accumulate in pipelines and different locations in the system that are related to the pump design and specifications, which exerts extra resistance on the pump (affects NPSH values) and leads to lower performance or early failure of the pump.

Crude oil pumps (COPs) have been reported to be impaired and to fail from wax deposits and waxy crudes. Also, sulfate scales (Ca, Ba, Sr-sulfates) and carbonates (Ca, Fe-carbonates) have been observed to impair COP. The composition of deposits collected from different crude oil pumps in Egyptian fields are illustrated in Fig. 3.10.

Fouling of downhole pumps like electrical submersible pumps (ESPs) is very common. ESP fouling was reported in many incidents with different deposits types: calcium sulfates [67], barium sulfate, calcium carbonate, and iron sulfide scales [68], asphaltenes, paraffin, and hydrates [69,70]. Fig. 3.11 shows an ESP fouled with mixed deposits (organic and inorganic composite deposits).

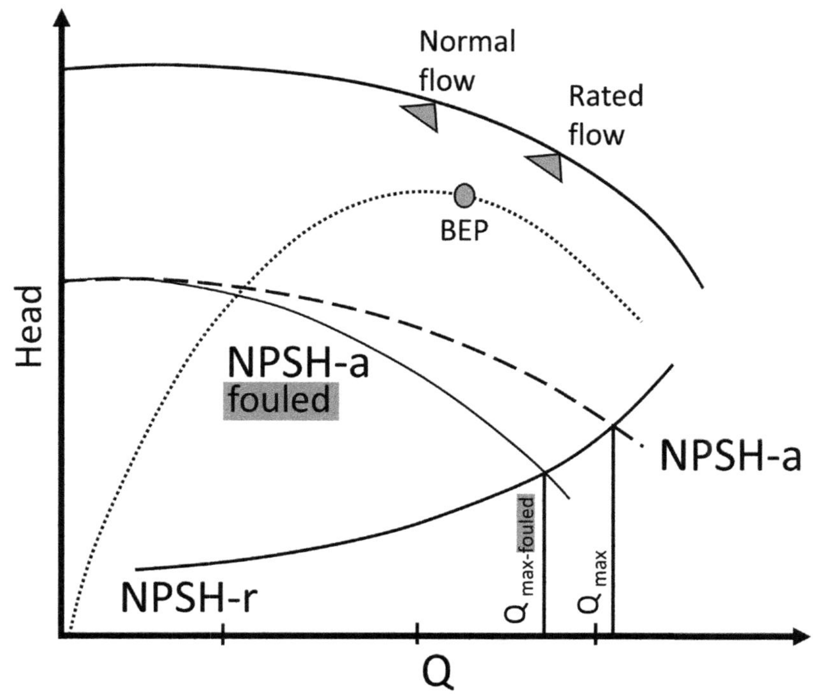

FIG. 3.9

Effect of fouling on NPSH-a and flow rate.

Reproduced with permission from G. Paresh, M. Octo, Practical centrifugal pumps design, operation and maintenance, British Library Cataloguing in Publication Data Elsevier, 2005, pp. 1–260, https://doi.org/10.1016/B978-0-7506-6273-4.X5000-4. Copyright (2005) Elsevier.

Fouling of gas compressors is another example of fouling in pumping equipment. Compressor fouling is probably the most common cause of performance deterioration [71]. Solids contamination of the compressor impeller blades changes their aerodynamic shape, roughens their surfaces, and reduces the gas flow, thus reducing the flow coefficient and finally causing a drop in the compressor overall efficiency [71–73]. Fig. 3.12 illustrates the change in compressor efficiency due to fouling. Compressor foulants include mineral scales, hydrocarbons, glycol, and oil and grease from the compressor lubricating system.

Different types of deposits can form in compressors, including inorganic scales such as carbonate, sulfide, chloride scales, and organic deposits. Fig. 3.13 illustrates the composition of deposits collected from different compressors in Egyptian fields. The carbonate and sulfide scales can be attributed to corrosion as corrosion products, while the chloride scales are attributed to water carryover.

3.5.3 Separation equipment impairment and failure

In addition to the heat transfer equipment impairment, deposits can impair other production equipment like separators, slug catchers, strainers, and amine and glycol units.

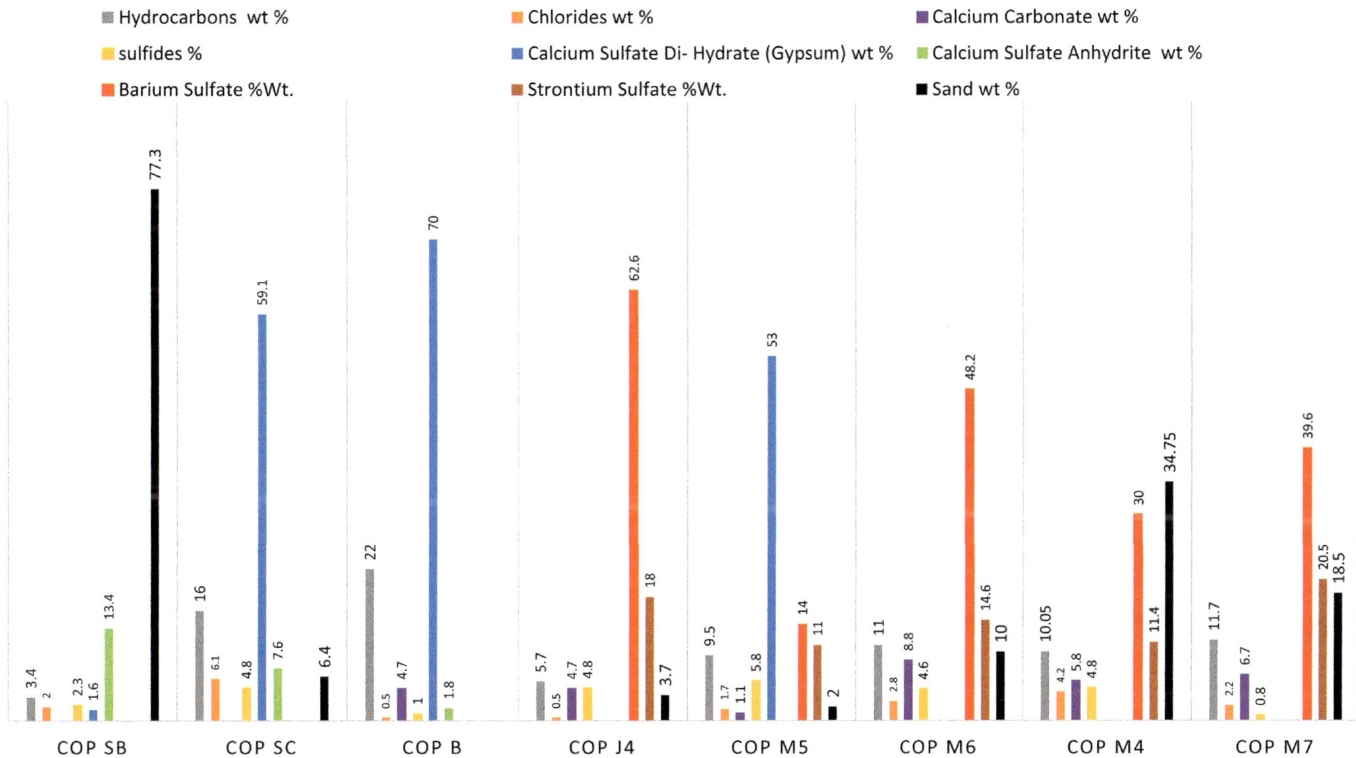

FIG. 3.10

Composition of deposits from different pumps.

FIG. 3.11

Composite deposits in ESP from Egyptian field.

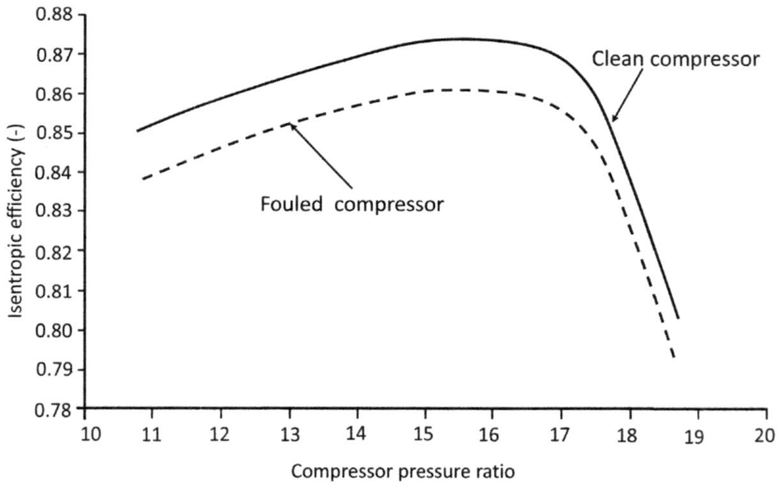

FIG. 3.12

Effect of compressor fouling on the compressor efficiency characteristic for a given compressor nondimensional speed.

Reproduced with permission from A.M.Y. Razak, Gas turbine performance deterioration, in: Industrial Gas Turbines, Elsevier, 2007, pp. 241–261, https://doi.org/10.1533/9781845693404.1.241. Copyright (2007), Elsevier.

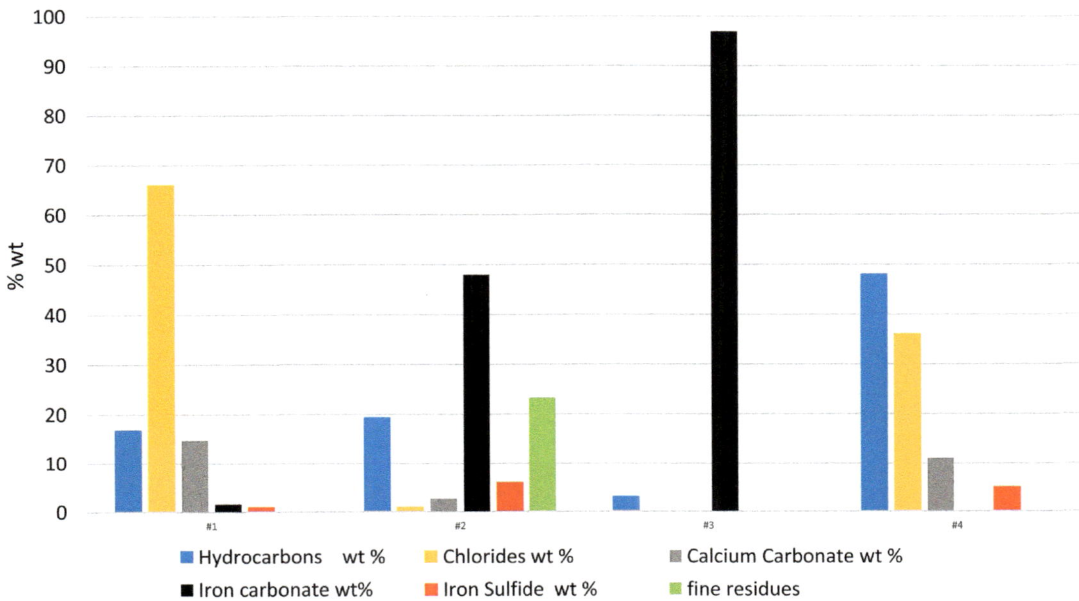

FIG. 3.13

Composition of deposits from different compressors.

In separators, solids deposition leads to plugging of separator bridles, hydrocyclone liners, and level control valves, solid loading in overboard water, poor oil-water interface, loss of retention time in separators, and plugging of reject lines of the produced water treating vessels. Fig. 3.14 illustrates a sludge sample from a separator, and Fig. 3.15 shows the composition of deposit samples collected from different separators from Egyptian fields.

Fouling of mist eliminators in slug catchers pushes them to their limit and they become inefficient. As a result, downstream facilities may be exposed to high liquid and solid entrainment in the gas outlet stream [74,75]. Fig. 3.16 illustrates deposit samples from a slug catcher.

Corrosion and mineral scaling are common problems in amine and glycol units. In addition, charred materials and polymerized materials have been observed in these systems during regeneration processes at very high temperatures.

Black powder accumulation in gas systems impairs filter systems and increases their maintenance frequency and cartridge replacement.

3.5.4 Flowmeter impairment and failure

Deposition and fouling in flowmeters can occur in different forms as uniform or intermittent layers in the mechanical internal parts and also on the transducer [76]. These deposits cause changes in the surface roughness, effective diameter, and attenuation or alteration of the received signal intensity, and that can significantly degrade the flowmeter accuracy.

Jordan et al. [77] reported that multiphase flowmeters on North Sea wells have shown a decline in performance due to barium sulfate scale.

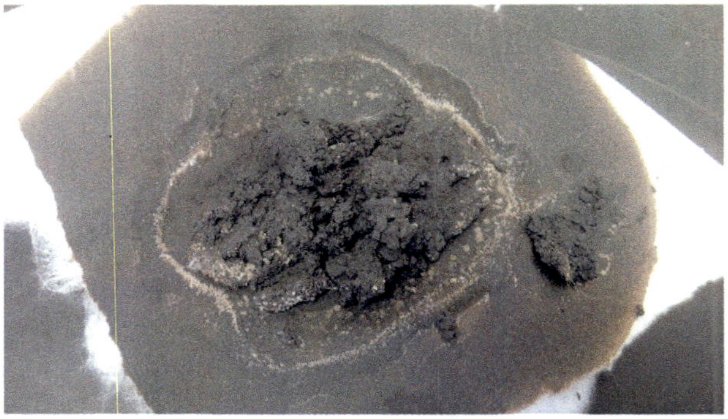

FIG. 3.14

Sludge sample collected from separator.

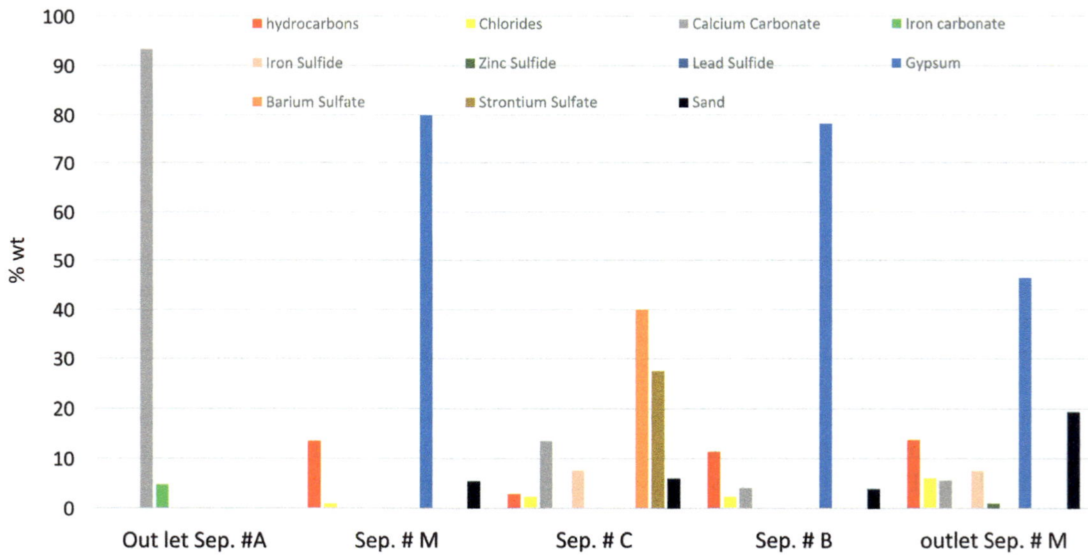

FIG. 3.15

Composition of deposit samples from different separators from Egyptian fields.

Solid particles and fines are also known to interfere with the metering system. Black powder (mixture of iron compounds) is known to interfere and affect the metering devices in gas metering pipelines and stations.

3.5.5 Valve impairment and failure

Downhole and topsides valves are prone to scaling and fouling with mineral sales and organic deposits, which finally leads to loss and their failure. Fig. 3.17 illustrates full blockage of a valve. Pressure drop within the valves is considered the major reason for precipitation and deposition of the solids.

Fig. 3.18 shows production choke failure due to zinc and lead sulfides, even though scale inhibitors were injected in the downhole.

FIG. 3.16

Deposit samples from slug catcher.

FIG. 3.17

Valve blockage due to calcium sulfate deposits.

Mineral scale deposition is widely recognized as one of the most common causes of failure in gas lift valve check valves. When the latch becomes coated in scale, intervention operations to replace the failed valve can become very challenging. All types of deposits are expected to occur in the gas lift mandrels: mineral scales, wax, asphaltenes, sand, proppants, and cement debris; also, viscous fluids can cause the same issue [78,79]. Gas lift valves that are stuck open or stuck closed due to mineral scale and other deposits are very common, which leads to improper operation of the gas lift system, and usually the problem ends up with changing the gas lift mandrel. These gas lift valves are very sensitive to solid deposits, as even very thin layers of deposits can cause their impairment, as indicated in Fig. 3.19, because very thin layers of scale can cause mandrel malfunctioning.

FIG. 3.18

ZnS deposits plugging the holes of a production choke (left), and choke after cleaning (right).

FIG. 3.19

Thin scale layers in gas lift mandrel; these very thin layers caused failure of the gas lift valves.

Subsurface safety valves (SSSVs) are a critical component installed in both onshore and offshore operations to provide emergency closure of the production conduit in the event of a catastrophe, such as a well blowout. Mineral deposits can build up in the inner wall and spring cavities of the SSSV and cause serious technical and integrity issues; i.e., scales can cause the flapper to become blocked in an open or closed state, resulting in the shut-in of the well [80,81].

Other valves that can be impaired and fail due to deposits include the downhole safety valves (DHSVs), inflow control valves (ICVs), inflow control devices (ICDs), choke valves [82].

Payne [83] reported observations of calcium carbonate scale in the safety valves at the North Sea Murchison oil field. The reduced bore of a downhole safety valve (DHSV), according to Payne, will give a pressure drop that increases the calcium carbonate saturation ratio, and also creates turbulence and shear forces that enhance the precipitation reaction kinetics. Statoil's internal DHSV failure frequency statistics also show a correlation between oil fields with scaling potential and failure frequency [82], as illustrated in Fig. 3.20. This indicates that in fact scale may be a frequent reason for DHSV failure [82].

3.5.6 Sand control equipment impairment and failure

Deposits can form in filters, strainers, downhole screens, gravel pack systems, and other sand control equipment.

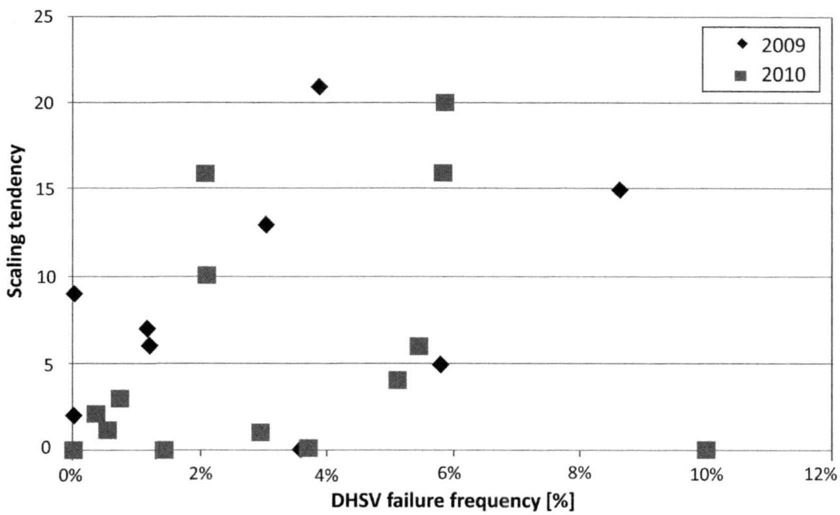

FIG. 3.20

Scaling tendency plotted versus DHSV failure frequency for Statoil operated oil and gas fields on the Norwegian continental shelf. The scaling tendency index has been derived from the number of scale control-related well interventions performed per field.

Reproduced from L.O. Jøsang, M.N. Psarrou, T. Østvold, T. Tjomsland, K. Ramstad, B. Marie Hustad, Scale under turbulent flow, in: Tekna 23rd International Oil Field Chemistry Symposium, 18–21 March 2012, Geilo, Norway, 2012, with permission from the author, L.O. Josang.

Accumulated fines, mineral scales, paraffin, and asphaltenes are examples of deposits that can plug gravel packs [84]. Scale plugging of sand control equipment was reported in horizontal wells in Gulf of Mexico and North Sea fields [85].

Besides affecting the production equipment, solids also affect metering, monitoring, and analyzing devices in the system. They interfere with their signals, giving false readings that render them inaccurate. Fine solids and black powder are known to interfere with inline inspection (ILI) smart pigs. Thick solid layers can interfere with nondestructive testing (NDT) devices.

3.6 Production chemistry problems induced by solids deposition

Solids deposits not only impair production equipment and restrict fluids flow, but they also extend their detrimental effects and interfere with production operations to cause larger flow assurance-related problems, like emulsion and corrosion. These production chemistry problems that are generated by solids deposition can become very complex and concerning, since the conventional treatments in this case can turn inefficient or be totally in vain if the problem root cause (which might be the interference from solids formation) is not professionally studied and treated beforehand.

3.6.1 Emulsion stabilization by solids

Fine solids generally stabilize an emulsion mechanically. Emulsions that are stabilized by solids are called pickering emulsions [86].

Solid particles stabilize emulsions in different ways, depending on different factors:

- Solid particles must be able to diffuse to the oil/water interface as they form rigid films that can sterically inhibit the coalescence of emulsion droplets.
- Solid particles at the interface may be electrically charged, which may also enhance the stability of the emulsion [87].
- Solid particles also stabilize emulsions based on their wettability. The wettability of solids is widely determined using the contact angle. The contact angle at the oil-water-solid three-phase contact line is measured through the aqueous phase, as shown in Fig. 3.21. When the water contact angle, θ, is less than 90 degrees, the solid is preferentially water-wet. Similarly, when the water contact angle is greater than 90 degrees, the solid is preferentially oil-wet. Contact angles close to 90 degrees result in an intermediately wetted solid (biwettable) that generally leads to the tightest emulsions [88,89]. If the solid remains entirely in the oil or water phase, it will not be an emulsion stabilizer. For the solid to act as an emulsion stabilizer, it must be present at the interface and must be wetted by both the oil and water phases. In general, oil-wet solids stabilize a water-in-oil emulsion. Similarly, water-wet solids stabilize a water-continuous or an oil-in-water emulsion. Examples of oil-wet solids are asphaltenes and waxes, and iron sulfides are more oil wet than the other mineral scales. Examples of water-wet solids are inorganic scales ($CaCO_3$, $CaSO_4$), clays, sand, and corrosion products. The

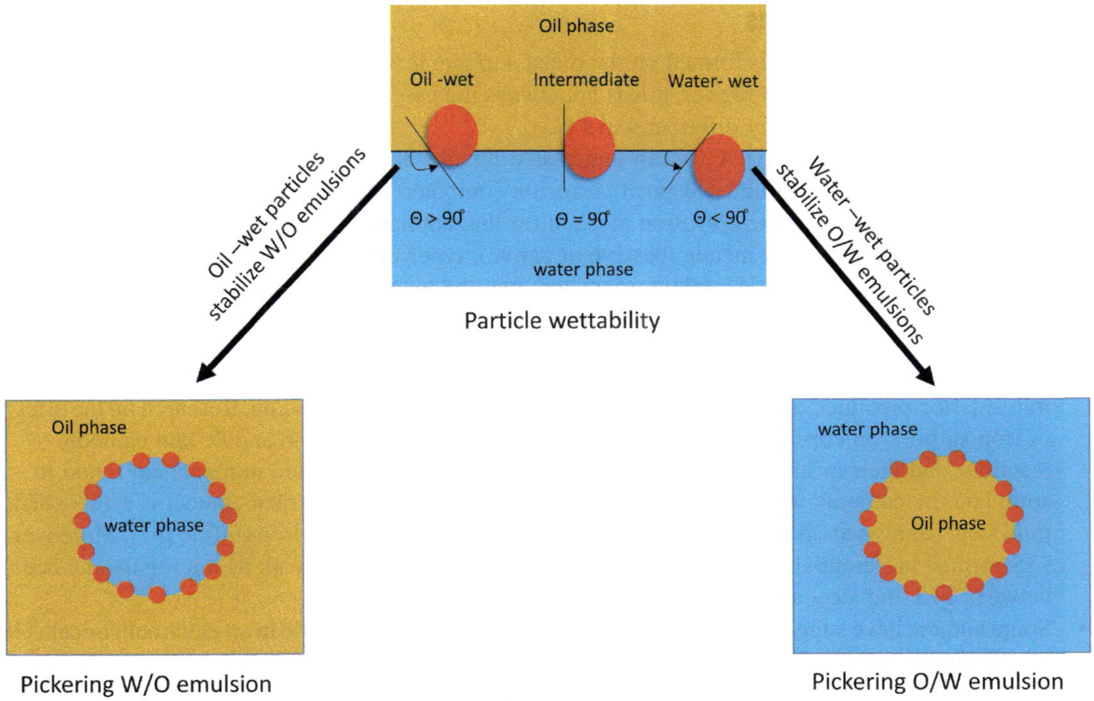

FIG. 3.21

The particles wettability and emulsion stabilization by solids.

effectiveness of these particles in stabilizing emulsions is strongly dependent on several factors, such as interparticle interactions, particle size, and wettability of the material [90].

- Surfactant-like molecules such as asphaltenes and resins (particularly the naphthenic acids) use the advantage of being polar compounds (have hydrophilic heads) with aliphatic side chains (hydrophobic tails), so they can adjust themselves in the oil/water interface to reduce the interfacial tension (IFT) and, consequently, enhance the droplet dispersion and stabilize either water/oil emulsions or water/oil emulsions [91]. Asphaltenes can adsorb onto oil/water interfaces, creating a viscoelastic network that stabilizes the emulsions [92].

Accordingly, sometimes emulsion problems require absolute attention and in-depth investigation to determine the root cause of stabilizing the emulsions; otherwise the conventional treatments using typical demulsifiers will be in vain. Refs. [93, 94] reported examples of tight emulsions that, after proper studies to detemine the root cause by suspended solids analysis using the ESEM technique, were found to be induced and stabilized by mineral scales, and after applying an optimum dose of scale inhibitor, the problem was successfully mitigated.

3.6.2 Corrosion problems

Usually when a deposit layer is formed on the metal surface, it works as a protective layer from successive corrosion processes. However, this is not always the case, as deposits can actually exaggerate the corrosion problem, in many different ways [95].

Under-deposit corrosion (UDC) is an aggressive attack that is associated with solid deposits. Fig. 3.22 illustrates a case of under-deposit corrosion combined with biocorrosion.

Generally, there is no single corrosion mechanism that applies to all cases of corrosion occurring under deposits, since different metals (passive or active), types of deposits (reactive, inert), and corrosive species (O_2, CO_2, H_2S, H^+, microorganisms) may be involved [96]. A few examples of UDC mechanisms include:

- Some deposits, such as iron sulfides, are semiconductive deposit types in oil and gas production and transmission pipelines, and that property may lead to exacerbating corrosion. Research on the effect of iron sulfides on corrosion of mild steel has shown their influence on anodic and cathodic reactions and their influence on accelerating corrosion [95]. Iron sulfide minerals are found to stimulate the cathodic reaction (cathodic polarization), and once electrical contact is established, mild steel behaves as an anode and electron transfer occurs through the iron sulfide [97]. Moreover, FeS formed preferential cathodic sites with a low overpotential that work as microgalvanic cells between FeS and steel and the selective dissolution of steel [98].
- Some authors have suggested that siderite scale can function as a cathode in an electrochemical cell (to the anodic steel surface), and propagate metal dissolution [99,100].
- Some of the UDC can be attributed to the differential chemistry between the under deposit (dominant internal anode) and bare metal (dominant external cathode) and here the corrosion is more like crevice corrosion; or both the anode and cathode are underneath the deposit and only separated by the deposit irregularities, and in this case the corrosion is more like pitting. Also, an anode and cathode can have no separation between them under the deposit and in this case it

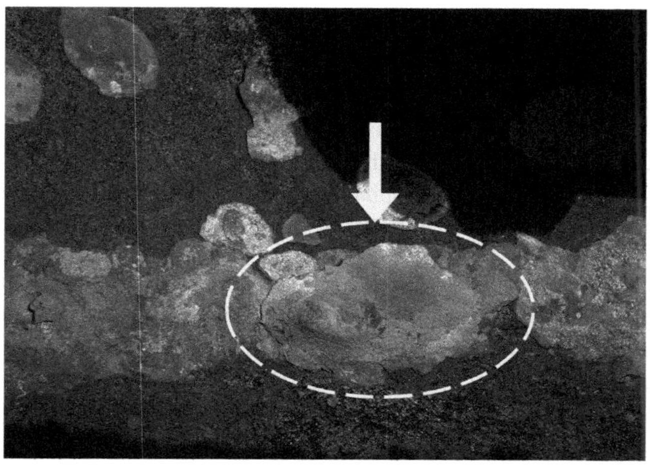

FIG. 3.22

A case of under-deposit corrosion associated with microbial activity.

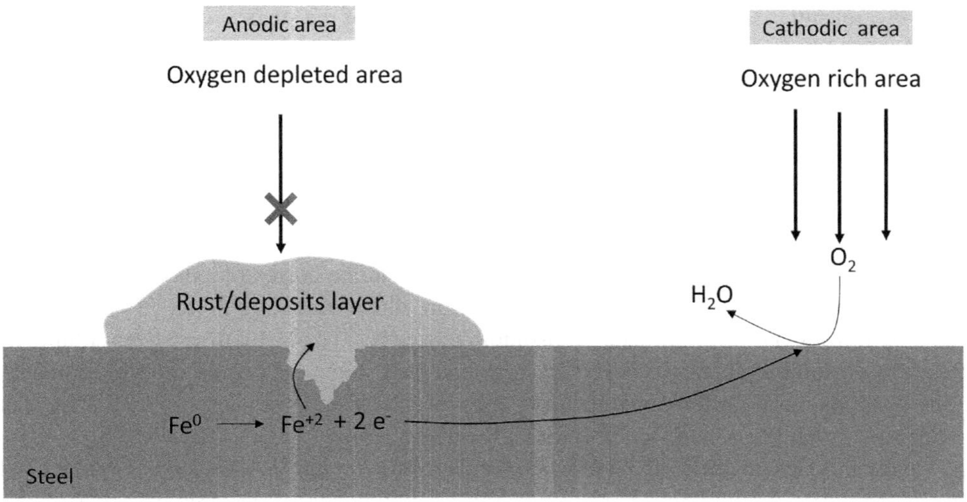

FIG. 3.23

Under-deposit corrosion due to differential aeration cell.

behaves more like uniform corrosion [95,96]. The deposits causing the UDC could be inorganic, organic, biofilms, or a mix of all of them.

- Deposit layer on a metal surface can create a differential aeration cell, when they form irregularly on the metal surface, restricting oxygen access to the sites covered with deposits, making them anodes to those bare metal sites that are exposed to oxygen, which work as cathodes. This increases the corrosion rates in anodic or deposit covered areas. This phenomena is illustrated in Fig. 3.23.

Aside from UDC, flowing solid particles at high speed also cause erosion corrosion, depending on the particle properties, flow rate, and location (elbow). They can also erode the protective film of the corrosion inhibitors, or erode the internal coating [96].

Another way of affecting corrosion is by decreasing inhibition efficiency by reducing the available concentration of corrosion inhibitor, due to its adsorption onto the solids, which typically have high surface area [96].

Additionally, methanol, which is a common thermodynamic hydrates inhibitor commonly used in gas operations, was reported to cause multiple corrosion problems in production systems. For instance, methanol dissolves the corrosion inhibitor protective film, increases the ingressed dissolved oxygen content, and increases the risk of sulfide stress cracking (SSC) and stress orientated hydrogen-induced cracking (SOHIC) in sour gas systems [101].

3.6.3 Microbial activity

Another problem associated with deposits is the bacterial activity. In biofouled systems, bacteria thrive under a biofilm of a matrix of an extracellular polymeric substance that they produce. Microorganisms can colonize metal surfaces within organic deposits and under other solids in aqueous environments (especially under corrosion deposits). Yebra et al. [102] demonstrated that the biofouling process comprises several steps, starting from the initial conditioning of the surface via organic and inorganic molecules to the colonization of microorganisms (microbial biofouling), and eventually the establishment and growth of macroorganisms (microbial biofouling). Treating bacteria under deposits or under biofilm is difficult, because once they are established, they are difficult to remove by chemical treatment alone, and they usually require a combination of chemicals or a combination of mechanical (pigging) and chemical treatments to be able to breach the deposit layer, and then the biofilm layer, to reach the targeted microorganisms [103].

Furthermore, some microbiologically influenced corrosion mechanisms are in fact attributed to the formation of deposits by these bacteria, which tend to accelerate corrosion rates. A good example is corrosion by iron and manganese oxidizing bacteria, in which oxidizing Fe^o to Fe^{2+} is not seen, but rather soluble ferrous (Fe^{+2}) is oxidized to insoluble ferric (Fe^{+3}) and MnOB oxidizes soluble manganous (Mn^{+2}) to insoluble manganic (Mn^{+4}), which can lead to the deposition of oxides on solid surfaces; these solids form differential aeration cells accelerating corrosion, and also increase the metal corrosion potential, which finally leads to localized pitting [104].

Besides the production chemistry problems, solids formation interferes with some major field operations. For example, gas hydrates deposits are considered the major flow assurance issue in deepwater drilling operations.

3.7 Safety and environmental problems

Solids deposition encompasses safety and environmental impacts. Some of these are:

- One of the major safety hazards associated with deposits formation is naturally occurring radioactive materials (NORMs), which include the radioactive element radium (Ra). NORMs are common hazards associated with barium strontium sulfate deposits. The two radium isotopes present in produced water and barite scale are ^{226}Ra (half-life =1600 years) and ^{228}Ra (half-life =5.8 years). Oil-field scale from the United States was reported to have several thousand pCi/g, although very small quantities of scale have been reported with as much as 400,000 pCi/g of radium. For comparison, most natural soils and rocks contain approximately 0.5–5 pCi/g of total radium. A uranium ore sample containing 1 wt% uranium has approximately 3300 pCi/g of ^{226}Ra. Most of the radium in older oil-field scale is ^{226}Ra, because the shorter lived ^{228}Ra decays with a half-life of 5.8 years [105]. NORMs are hazardous materials that, depending on the exposure time and dose, can lead to severe side effects in the human body in the short and long term.
- Deposits formation increases CO_2 gas emissions. Recent studies have shown that heat exchanger fouling may be responsible for 1%–2.5% of global CO_2 emissions. CO_2 emissions will result in environmental penalties. Considering that there are currently 750 refineries worldwide, the total amount of emitted CO_2 due to fouling is approximately 88 million tons per annum [106]. The increase in temperature due to greenhouse gases can also trigger the dissolution of gas hydrates in the arctic and permafrost regions that might exaggerate the global warming issue.
- Another type of hazard is that iron sulfide deposits are pyrophoric. A pyrophoric material reacts with air, or with moisture in air. Typical reactions that occur are oxidation and hydrolysis, and the heat generated by the reactions may ignite the material. Equipment can be scaled with iron sulfide deposits. After the equipment is emptied, preparatory to cleaning and during ventilation, iron

sulfide deposits will dry out and react with oxygen in the air, generating heat and spontaneously igniting. The equipment should be purged with gas containing low (e.g., 5%) levels of oxygen and kept wet. This approach keeps the pyrophoric deposits wet until the atmosphere is noncombustible and the deposits are either oxidized or removed [107].

- The chemicals that are used in removing and mitigating solid deposits were reported to bear some hazards due to their toxicity, biodegradability, and other effects on the ecosystem and the environment in general. These chemicals are usually partitioned and end up in the produced water, which must be treated before discharged into the environment or the ocean.
- Failure of safety valves and full blockage of flowlines are catastrophic during operations that might cause oil spills and other fluids discharge into the environment as oil. Besides, flowlines blockages and solids carryover along the production system can lead to serious incidents, such as operator injuries, and even death; a few examples include:
    - In Feb. 2014 a Hiland gas pipeline exploded about 6 miles south of Tioga, North Dakota due to gas hydrates blockage [108].
    - Lives were lost due to solids blockages in an incident of gas hydrates blockage in a buried sour gas pipeline, when the operators depressurized the pipeline from one side, which led to partially dissociating the hydrate plug and rupturing and erupting the pipeline from the ground, which struck the operator and killed him before arriving at the hospital [109].
    - A Siberian incident occurred in February 2000, when a pipefitter was attempting to remove a hydrate plug by heating an exposed pipeline with a torch. The gas pressure from a dissociated midhydrate plug rose rapidly, perhaps being confined by the plug ends. The pipeline exploded and, in the resulting fire, one man died, and four others were badly injured.
    - Recent investigations showed that the Deepwater Horizon disaster might have occurred due to sudden formation of gas hydrates, causing the worst oil spill catastrophe as well as the loss of many operators' lives.

3.8 Summary

Solids depositions in oil and gas operations always carry a very high cost. Solid deposits are known to be formed anywhere in the production system and to have rapid impacts, even at low concentrations. Several production interruptions are inevitable in oil and gas fields infested with solid deposits, with major consequences, basically including production loss, extra maintenance and mitigation expenses, and reduction in revenue. These implications can be categorized and discussed in terms of flow restrictions, equipment failure, production chemistry-related problems, and environmental-associated issues. Flow restrictions principally cause permeability reduction of the reservoir formation, known as formation damage, changing the internal dimensions of production tubulars or fully blocking fluids flow. Besides, equipment failure, e.g., heat exchangers, pump, separators, etc., is another common problem with solids formation. These equipment problems may also be associated with other problems: for instance, fouling in heat transfer equipment leads to reducing their heat transfer efficiency, increasing fuel consumption, and increasing gas emissions. Furthermore, solids formation induces other production chemistry issues like emulsions, corrosion, and bacteria-related problems. Last but not least, solids deposition entails multiple environmental challenges, including the hazardous NORMS, pyrophoric materials, operator safety, and climate change problems. These issues were discussed to give the readers an intensified glimpse of how devastating solid deposits can be in a production system.

References

[1] M. Brown, Full scale attack, BP Technol. Mag. 30 (1998) 30–32.

[2] C. Igwebueze, L. Oduola, O. Smith, P. Vijn, A.G. Shepherd, Calcium naphthenate solid deposit identification and control in offshore Nigerian fields, in: Proceedings—SPE International Symposium on Oilfield Chemistry, vol. 1, 2013, pp. 156–168, https://doi.org/10.2118/164055-ms.

[3] W. Wayne, M. Frenier, Formation, Removal and Inhibition of Inorganic Scale in the Oilfield Environment, Society of Petroleum Engineers, 2009.

[4] W.W. Frenier, M. Ziauddin, A multifaceted approach for controlling complex deposits in oil and gas production, in: SPE 132707 Paper Presented at the SPE Annual Technical Conference and Exhibition, Florence, Tuscany, Italy, 20–22 September 2010, 2010.

[5] W. Shih, Formation and Control of Calcium Sulfate Dihydrate (Gypsum) Crystallization on RO Membranes and Surrogate Polymeric Surfaces (Ph.D. dissertation), University of California, Los Angeles, 2007.

[6] H. Muller-Steinhagen, Handbook Heat Exchanger Fouling. Mitigation and Cleaning Technologies, Publico Publications, Essen, Germany, 2000.

[7] S.N. Kazi, Particulate matter: interfacial properties, fouling, and its mitigation, in: Mineral Scales and Deposits: Scientific and Technological Approaches, Elsevier, 2015, pp. 141–168, https://doi.org/10.1016/B978-0-444-63228-9.00007-3.

[8] H. Wigg, Establishing the true cost of downhole scale control, in: Presented at the "Solving Oilfield Scaling" Conference, Aberdeen, UK, 1995.

[9] G.F. Hewitt, F. Coletti, Crude Oil Fouling: Deposit Characterization, Measurements, and Modeling, Elsevier, 2014, https://doi.org/10.1016/C2013-0-18845-9.

[10] W.L.J. Van Nostrand, S.H. Leach, J.L. Haluska, Economic penalties associated with the fouling of refinery heat transfer equipment, in: E.F.C. Somerscales, J.G. Knudsen (Eds.), Fouling of Heat Transfer Equipment, August 1979, Troy, New York, Hemisphere, Washington, 1981, pp. 619–643.

[11] J.L. Creek, Efficient hydrate plug prevention, Energy Fuel 26 (2012) 4112–4116.

[12] C. Campbell, Advances in testing and monitoring of biocides in oil and gashyb, in: Trends in Oil and Gas Corrosion Research and Technologies: Production and Transmission, Elsevier, 2017, pp. 489–511, https://doi.org/10.1016/B978-0-08-101105-8.00021-8.

[13] H.A.H. Ibrahim, in: V.N. Katsikis (Ed.), Fouling in Heat Exchangers in MATLAB—A Fundamental Tool for Scientific Computing and Engineering Applications, vol. 3, 2012, https://doi.org/10.5772/46462.

[14] T.S. Rao, Biofouling in industrial water systems, in: Mineral Scales and Deposits: Scientific and Technological Approaches, Elsevier, 2015, pp. 123–140, https://doi.org/10.1016/B978-0-444-63228-9.00006-1.

[15] Schlumberger, Scale Control, 2013, Available from: http://www.slb.com/services/completions/stimulation/sandstone/scale_control.aspx. (Accessed 13 December 2013).

[16] T. Tjomsland, M.N. Grotle, O. Vikane, Scale Control Strategy and Economical Consequences of Scale at Veslefrikk, Society of Petroleum Engineers, 2001, January 1, https://doi.org/10.2118/68308-MS.

[17] L. Oduola, C.U. Igwebueze, O. Smith, P. Vijn, A.G. Shepherd, Calcium Naphthenate Solid Deposit Identification and Control in Offshore Nigerian Fields, Society of Petroleum Engineers, 2013, April 8, https://doi.org/10.2118/164055-MS.

[18] C.A. Bilderback, L.A. McDougall, Complete paraffin control in petroleum production, J. Pet. Technol. 21 (1969) 1151–1156.

[19] D. Gonzalez, F. Gonzalez, M. Pietrobon, M. Haghshenas, M. Shurn, A. Mees, C. Stewart, C. Ogugbue, Strategies to monitor and mitigate asphaltene issues in the production system of a Gulf of Mexico Deepwater subsea development, in: Proceedings of the Annual Offshore Technology Conference, Offshore Technology Conference, vol. 3, 2016, pp. 2679–2694, https://doi.org/10.4043/27123-ms.

[20] S.M. El-Raghy, H.M.A. El-Leil, H.H. Ghazal, Microbial induced corrosion of subsea pipeline in the Gulf of Suez, in: Paper Presented at the Middle East Oil Show and Conference, Bahrain, March, 1997, https://doi.org/10.2118/37791-MS.

[21] F. Civan, Reservoir Formation Damage: Fundamentals, Modeling, Assessment, and Mitigation, third ed., Elsevier, 2015, https://doi.org/10.1016/C2014-0-01087-8.

[22] J. Moghadasi, M. Jamialahmadi, H. Müller-Steinhagen, A. Sharif, Formation Damage Due to Scale Formation in Porous Media Resulting From Water Injection, Society of Petroleum Engineers, 2004, January 1, https://doi.org/10.2118/86524-MS.

[23] M.E. Newberry, K.M. Barker, Formation damage prevention through the control of paraffin and asphaltene deposition, in: Presented at the SPE Production Operations Symposium, Oklahoma City, Oklahoma, 10-12 March 1985. SPE-13796-MS, 1985, https://doi.org/10.2118/13796-MS.

[24] S.J. Houston, B.W.D. Yardley, P.C. Smalley, I. Collins, Precipitation and dissolution of minerals Durina Waterfloodina of a North Sea oil field, in: SPE Eighth International Symposium on Oilfield Scale 2006, vol. 2006, 2006, pp. 300–308, https://doi.org/10.2523/100603-ms.

[25] J.H. Barkman, D.H. Davidson, Measuring water quality and predicting well impairment, JPT 24 (1972) 865–873.

[26] C. Roque, C. Chauveteau, M.E.A. Renard, Mechanism of formation damage by retention of particles suspended in injection water, in: Paper SPE 30110 Presented at the 1995 European Formation Damage Conference, den Haague, The Netherlands, 1995.

[27] A.U. Chaudhry, Gas Well Testing Handbook, Elsevier, 2003, https://doi.org/10.1016/B978-0-7506-7705-9.X5000-X.

[28] R. Darby, R.P. Chhabra, Chemical Engineering Fluids Mechanics, second ed., Marcel Dekker, USA, 2001.

[29] A. Nouri, H. Vaziri, H. Belhaj, R. Islam, Effect of volumetric failure on sand production in oil-wellbore, in: SPE Paper 84448 Presented at the SPE Asia Pacific Oil and Gas Conference and Exhibition Jakarta, Indonesia, April 15–17, 2003.

[30] A. Nouri, H. Vaziri, H. Belhaj, R. Islam, Comprehensive transient modeling of sand production in horizontal wellbores, in: SPE Paper 84500 Presented at the SPE Annual Technical Conference and Exhibition Denver, Colorado, October 5–8, 2003.

[31] N. Morita, P.A. Boyd, Typical sand production problems: case studies and strategies for sand control, in: SPE Paper 22739 Presented at the 66th Annual Technical Conference and Exhibition of the Society of Petroleum Engineers Dallas, Texas, October 6–9, 1991.

[32] J.H. Almedeij, M.K. Algharaib, Influence of sand production on pressure drawdown in horizontal wells: theoretical evidence, J. Pet. Sci. Eng. 47 (3–4) (2005) 137–145, https://doi.org/10.1016/j.petrol.2005.03.005.

[33] A. Olufemi, E. Darlington, Effect of sand production on pressure drop for vertical wells in gas storage reservoir, in: Society of Petroleum Engineers—Nigeria Annual International Conference and Exhibition 2010, NAICE, vol. 1, 2010, pp. 448–452, https://doi.org/10.2118/140610-ms.

[34] D.C.-H. Cheng, Viscosity-concentration equations and flow curves for suspensions, Chem. Ind. 17 (1980) 403–406.

[35] N. Mangesana, R.S. Chikuku, A.N. Mainza, I. Govender, A.P. van der Westhuizen, M. Narashima, The effect of particle sizes and solids concentration on the rheology of silica sand based suspensions, J. South. Afr. Inst. Min. Metall. 108 (2008) 237–243. ISSN:2225-6253.

[36] R. Jamshidi, L. Mazzei, CFD modeling of fluidized beds, in: Reference Module in Chemistry, Molecular Sciences and Chemical Engineering, 2018, Elsevier, 2018.

[37] H.H. Al-Kayiem, T.M. Huong, J.A. Khan, Evaluation of pressure losses due to solids-in-liquid pipe flow by CFD simulation, ARPN J. Eng. Appl. Sci. 11 (16) (2016) 9908–9916.

[38] V. Matousek, Pressure drops and flow patterns in sand-mixture pipes, Exp. Thermal Fluid Sci. 26 (6–7) (2002) 693–702, https://doi.org/10.1016/S0894-1777(02)00176-0.

[39] D.R. Kaushal, K. Sato, T. Toyota, K. Funatsu, Y. Tomita, Effect of particle size distribution on pressure drop and concentration profile in pipeline flow of highly concentrated slurry, Int. J. Multiphase Flow 31 (2005) 809–823.

[40] T. Nabil, I. El-Sawaf, K. El-Nahhas, Computational fluid dynamics simulation of the solid-liquid slurry flow in a pipeline, in: Seventeenth International Water Technology Conference, IWTC 17, IWTC, 2013.

[41] K. El-Nahhas, N.G. El-Hak, M.A. Rayan, I. El-Sawaf, Effect of particle size distribution on the hydraulic transport of settling slurries, in: Thirteenth International Water Technology Conference, IWTC13, Hurghada, Egypt, 2009, pp. 463–474.

[42] J. Faitli, Pressure loss calculation model for well-graded solid-liquid pipe flows on the basis of systematic pilot plant investigations, in: Intellectual Service for Oil and Gas Industry: Analysis, Solution, Perspectives, Co-Proceedings of Ufa State Petroleum Technical University and University of Miskolc, Ufa, 2000.

[43] A.I. Abd Al Aziz, H.I. Mohamed, A study of the factors affecting transporting solid–liquid suspension through pipelines, Open J. Fluid Dyn. 3 (3) (2013) 152–162, https://doi.org/10.4236/ojfd.2013.33020.

[44] M. Weber, Improved durand–equation for multiple application, in: International Symposium on Slurry Flows, 1986, pp. 61–66.

[45] J.S. Gudmundsson, Asphaltene, in: Flow Assurance Solids in Oil and Gas Production, CRC Press, 2017, pp. 45–83, https://doi.org/10.1201/9781315185118-3.

[46] J. Nikuradse, Stromnungsgesetze in Rauhen Rohren, Forschungsheft 361, Vol. B, VDI Verlang, Berlin, 1933 (translated in NACA Technical Memorandum Nr. 1292, 1950).

[47] C.F. Colebrook, C.M. White, Experiments with fluid-friction in roughened pipes, Proc. R. Soc. Lond. 161 (1937) 367–381.

[48] S. Haaland, Simple and explicit formulas for the friction factor in turbulent pipe flow, J. Fluid Mech. 105 (1983) 89–90.

[49] W. Baumann, K. Rehme, Friction correlations for rectangular roughness, Int. J. Heat Mass Transf. 18 (1975) 1189–1197.

[50] G.P.P. Gunarathne, Measurement and monitoring techniques for scale deposits in petroleum pipelines, in: Conference Record—IEEE Instrumentation and Measurement Technology Conference, vol. 2, Institute of Electrical and Electronics Engineers, 1997, pp. 841–847, https://doi.org/10.1109/IMTC.1997.610200.

[51] S.N. Kazi, Fouling and fouling mitigation on heat exchanger surfaces, in: J. Mitrovic (Ed.), Heat Exchangers—Basics Design Applications, IntechOpen, 2012 March 9, https://doi.org/10.5772/32990. Available from: https://www.intechopen.com/books/heat-exchangers-basics-design-applications/heat-exchanger-fouling-and-its-mitigation.

[52] J.C. Cowan, D.J. Weintritt, Water Foemed Scale Deposits, Gulf Publishing Company, TX, USA, 1976.

[53] T.R. Bott, Fouling of Heat Exchangers A Volume in Chemical Engineering Monographs, 1995.

[54] S.K. Ogbonnaya, O.O. Ajayi, Fouling phenomenon and its effect on heat exchanger: a review, in: Frontiers in Heat and Mass Transfer 9, Global Digital Central, 2017, https://doi.org/10.5098/hmt.9.31.

[55] K.D. Demadis, Combating heat exchanger fouling and corrosion phenomena in process waters, in: Compact Heat Exchangers and Enhancement Technology for the Process Industries, Begell House, 2003, pp. 483–490.

[56] J. Carvill, Thermodynamics and heat transfer, in: J. Carvill (Ed.), Mechanical Engineer's Data Handbook, Elsevier, 1993, https://doi.org/10.1016/B978-0-08-051135-1.50008-X (Chapter 3).

[57] M.M. Awad, Fouling of heat transfer surfaces, Chem. Eng. Technol. 10 (1) (2011) 113–125.

[58] A.C.J. de Korte, H.J.H. Brouwers, Calculation of thermal conductivity of gypsum plasterboards at ambient and elevated temperature, Fire Mater. (2009), https://doi.org/10.1002/fam.1009.

[59] F. Coletti, B.D. Crittenden, A.J. Haslam, G.F. Hewitt, G. Jackson, G. Jimenez-Serratos, S. Macchietto, et al., Modeling of fouling from molecular to plant scale, in: Crude Oil Fouling: Deposit Characterization, Measurements, and Modeling, Elsevier, 2014, pp. 179–320, https://doi.org/10.1016/B978-0-12-801256-7.00005-1.

[60] G.F. Hewitt, G.L. Shires, T.R. Bott, Process Heat Transfer, CRC Press, London, 1994.

[61] J.M. Donohue, C.V. Sarno, Pollution abatement pressures influence cooling water conditioning, Mater. Prot. Perform. 10 (12) (1971) 19–21.

[62] P.A. Pilavachi, J.D. Isdale, European community R&D strategy in the field of heat exchanger fouling: projects, Heat Recovery Syst. CHP 13 (2) (1993) 133–138, https://doi.org/10.1016/0890-4332(93)90032-Q.

[63] Natural Resources Canada (NRCAN), Increasing the Energy Efficiency of Boiler and Heater Installations, Natural Resources Canada (NRCAN), 2017 (Accessed 16 September 2020, Modified 17 February 2016).

[64] L. Lethea, Impact of water hardness on energy consumption of geyser heating elements, Water SA 43 (4) (2017) 614–625, https://doi.org/10.4314/wsa.v43i4.09.

[65] J. Colt, J.E. Huguenin, Design and operating guide for aquaculture seawater systems, Aquaculture 217 (1–4) (2003) 683–684, https://doi.org/10.1016/s0044-8486(02)00500-8.

[66] G. Paresh, M. Octo, Practical centrifugal pumps design, operation and maintenance, British Library Cataloguing in Publication Data, Elsevier, 2005, pp. 1–260, https://doi.org/10.1016/B978-0-7506-6273-4.X5000-4.

[67] S. Ren, H. Xu, M. Li, Y. Wang, F. Shen, The application of using EDTA to dissolve calcium sulphate deposits for plugged ESP wells in Rumaila oilfield, in: IOP Conference Series: Earth and Environmental Science, vol. 170, Institute of Physics Publishing, 2018, https://doi.org/10.1088/1755-1315/170/2/022081.

[68] R.W. Cramer, J.L. Bearden, Development and Application of a Downhole Chemical Injection Pump for Use in ESP Applications, Society of Petroleum Engineers, 1985, January 1, https://doi.org/10.2118/14403-MS.

[69] K.U. Stanghelle, Evaluation of artificial lift methods on the Gyda field, Mycotoxin Res. 6 (2009) 100, https://doi.org/10.1007/BF03192151.

[70] D. Denney, Asphaltene Deposition in Electrical-Submersible-Pump Applications, Society of Petroleum Engineers, 2000, https://doi.org/10.2118/0599-0035-JPT.

[71] A.M.Y. Razak, Gas turbine performance deterioration, in: Industrial Gas Turbines, Elsevier, 2007, pp. 241–261, https://doi.org/10.1533/9781845693404.1.241.

[72] H. Saadawi, Oil Field Experience With Gas Turbine-Driven Compressor Stations, American Society of Mechanical Engineers, 1982.

[73] C.B. Meher-Homji, Gas turbine axial compressor fouling: a unified treatment of its effect, detection and control, Int. J. Turbo Jet Engines 9 (4) (1992) 99–111.

[74] B. Dumas, Tracerco Assists a Large LNG Facility Experiencing Problems Within Their Slug Catcher, second ed., vol. 6, Insight, 2019.

[75] S. Aoun, Slug Catcher Debottleneck Protects Downstream Processing (TIA), Digitalrefining, 2020.

[76] J.G. Drenthen, M. Vermeulen, M. Kurth, H. den Hollander, The detection of corrosion and fouling and the operational influence on ultrasonic flow meters using reflecting paths, in: 29th International North Sea Flow Measurement Workshop, 2011.

[77] M.M. Jordan, K. Sjursaether, M.C. Edgerton, R. Bruce, Inhibition of lead and zinc sulphide scale deposits formed during production from high temperature oil and condensate reservoirs, in: SPE—Asia Pacific Oil and Gas Conference, Society of Petroleum Engineers (SPE), 2000, pp. 599–610, https://doi.org/10.2118/64427-ms.

[78] S. Murchie, J. Gjertsen, M. Grødem, M.S. Sætrevik, E. Osaland, P.L. Sørensen, S.K. Hansen, Electric line deployed orientated suction cleanout device enables gas lift valve retrieval from debris blocked side pocket mandrel in a highly deviated well, in: SPE/ICoTA Coiled Tubing and Well Intervention Conference and Exhibition, 2018, https://doi.org/10.2118/189952-ms.

[79] R.J. Tailby, C.B. Amor, A. Mc Donough, Scale removal from the recesses of side-pocket mandrels, in: SPE/ICoTA Coiled Tubing Roundtable, 1999, https://doi.org/10.2118/54477-ms.

[80] T.V.J. Charpentier, A. Neville, S. Baraka-Lokmane, C. Hurtevent, J.-R. Ordonez-Varela, F.M. Nielsen, Ø. Bache, Evaluation of Anti-Fouling Surfaces for Prevention of Mineral Scaling in Sub-Surface Safety Valves, Society of Petroleum Engineers, 2014, May 14, https://doi.org/10.2118/169750-MS.

[81] D. Kumar, J.C. Welch, Z. Xu, Reduction in Scale Build-Up From Sub-Surface Safety Valve Using Hydrophobic Material Coating, Society of Petroleum Engineers, 2013, September 30, https://doi.org/10.2118/166218-MS.

[82] L.O. Jøsang, M.N. Psarrou, T. Østvold, T. Tjomsland, K. Ramstad, B. Marie Hustad, Scale under turbulent flow, in: Tekna 23rd International Oil Field Chemistry Symposium, 18–21 March 2012, Geilo, Norway, 2012.

[83] G.E. Payne, A history of downhole scale inhibition by squeeze treatments on the murchison platform, in: SPE 16539, Offshore Europe, Aberdeen, United Kingdom, 1987.

[84] S. Ali, R. Dickerson, C. Bennett, P. Bixenman, M. Parlar, C. Price-Smith, S. Cooper, et al., High-productivity horizontal gravel packs, Oilfield Rev. 13 (2) (2001) 52–73.

[85] S. Schulien, J. Ovsthus, L.A. Hestenes, B. Neigart, T.A. Nistad, Scale Formation and Treatment in and Around Sand Control Screens, Society of Petroleum Engineers, 1997, January 1, https://doi.org/10.2118/37305-MS.

[86] Y. Yang, Z. Fang, X. Chen, W. Zhang, Y. Xie, Y. Chen, Z. Liu, W. Yuan, An overview of pickering emulsions: solid-particle materials, classification, morphology, and applications, Front. Pharmacol. (2017), https://doi.org/10.3389/fphar.2017.00287.

[87] D.E. Tambe, M.M. Sharma, Factors controlling the stability of colloid-stabilized emulsions: I. An experimental investigation, J. Colloid Interface Sci. 157 (1) (1993) 244–253.

[88] A.I.A. Mohamed, I.A. Hussein, A.S. Sultan, G.A. Al-Muntasheri, Use of organoclay as a stabilizer for water-in-oil emulsions under high-temperature high-salinity conditions, J. Pet. Sci. Eng. 160 (January) (2018) 302–312, https://doi.org/10.1016/j.petrol.2017.10.077.

[89] T. Tadros, Encyclopedia of Colloid and Interface Science, Springer, Berlin, Heidelberg, 2013, https://doi.org/10.1007/978-3-642-20665-8.

[90] C. Dalmazzone, C. Noik, P. Glenat, H.M. Dang, Development of a methodology for the optimization of dehydration of extra heavy-oil emulsions, SPE J. 15 (2010) 726.

[91] F. Goodarzi, S. Zendehboudi, A comprehensive review on emulsions and emulsion stability in chemical and energy industries, Can. J. Chem. Eng. 97 (1) (2019) 281–309, https://doi.org/10.1002/cjce.23336.

[92] K. Akbarzadeh, A. Hammami, A. Kharrat, D. Zhang, S. Allenson, J. Creek, S. Kabir, et al., Asphaltenes—problematic but rich in potential, Oilfield Rev. 19 (2007) 22–43.

[93] M.M. Jordan, J. Buckman, C.J. Johnston, Direct Monitoring of the Performance of Scale Control Programs Across the Produced Water Life Cycle via Suspended Solids Analysis, Society of Petroleum Engineers, 2005, January 1, https://doi.org/10.2118/93892-MS.

[94] M.M. Jordan, C.J. Johnston, M. Robb, Evaluation Methods for Suspended Solids and Produced Water as an Aid in Determining Effectiveness of Scale Control Both Downhole and Topside, Society of Petroleum Engineers, 2005, January 1, https://doi.org/10.2118/92663-MS.

[95] B. Brown, J. Moloney, Under-deposit corrosion, in: Trends in Oil and Gas Corrosion Research and Technologies: Production and Transmission, Elsevier, 2017, pp. 363–383, https://doi.org/10.1016/B978-0-08-101105-8.00015-2.

[96] J.R. Vera, D. Daniels, M.H. Achour, Under deposit corrosion (UDC) in the oil and gas industry: a review of mechanisms, testing and mitigation, in: NACE—International Corrosion Conference Series, vol. 4, 2012, pp. 3028–3040.

[97] B.J. Little, R.I. Ray, R.K. Pope, Bioactive environments: corrosion, in: Encyclopedia of Materials: Science and Technology, second ed., Elsevier, Amsterdam, 2001.

[98] P. Bai, Y. Liang, S. Zheng, C. Chen, Effect of amorphous FeS semiconductor on the corrosion behavior of pipe steel in H2S-containing environments, Ind. Eng. Chem. Res. 55 (41) (2016) 10932–10940, https://doi.org/10.1021/acs.iecr.6b03000.

[99] J. Han, B.N. Brown, S. Nesic, Investigation of the galvanic mechanism for localized carbon dioxide corrosion propagation using the artificial pit technique, Corrosion 66 (2010) 1–12.

[100] R.M. Fernández-Domene, R. Leiva-garcía, J. Andrews, R. Akid, Galvanic corrosion following local breakdown of a scale formed on X-65 in CO2 saturated solutions, in: NACE Corrosion 2015, Paper No. 5921, 2015.

[101] N.G. Park, L. Morello, J.E. Wong, S.A. Maksoud, The effect of oxygenated methanol on corrosion of carbon steel in sour wet gas environments, in: Paper Presented at the CORROSION 2007, Nashville, Tennessee, March 2007, NACE International, 2007, January 1.

[102] D.M. Yebra, S. Kiil, J.K. Dam, Antifouling technology-past, present and future steps towards efficient and environmentally friendly antifouling coatings, Prog. Org. Coat. 50 (2004) 75.

[103] R.B. Eckert, Residual life predictions—extending service life, in: Trends in Oil and Gas Corrosion Research and Technologies Production and Transmission, Woodhead Publishing Series in Energy 2017, Elsevier, 2017, pp. 765–786.

[104] B.J. Little, J. Hinks, D.J. Blackwood, Microbially influenced corrosion: towards an interdisciplinary perspective on mechanisms, Int. Biodeterior. Biodegrad. 154 (October) (2020), https://doi.org/10.1016/j.ibiod.2020.105062.

[105] USGS, Naturally occurring radioactive materials (NORM) in produced water and oil-field equipment—an issue for the energy industry, in: USGS Fact Sheet FS–142–99 September 1999, 1999.

[106] H. Müller-Steinhagen, M.R. Malayeiri, A.P. Watkinson, Heat exchanger fouling: environmental impacts, Heat Transfer Eng. 30 (2009) 773–776, https://doi.org/10.1080/01457630902744119.

[107] D.P. Nolan, Control of ignition sources, in: Handbook of Fire and Explosion Protection Engineering Principles for Oil, Gas, Chemical and Related Facilities, 2011, Elsevier, 2011, pp. 145–155.

[108] J. Holdman, Natural gas explosion reported near Tioga, Bakken News (2014, February 11). bismarcktribune.com. Retrieved 4 October 2016.

[109] Sloan, Hydrates Engineering, Monograph 21, Society of Petroleum Engineers, Richardson, TX, 2000.

CHAPTER

Principles of Flow Assurance Solids Formation Mechanisms

4

4.1 Introduction

Solids formation is a global production operations problem in various industries. In the oil and gas industry, various types of organic and inorganic solids form in production systems, causing major flow assurance challenges. Understanding the mechanisms by which these solids form helps the operators and researchers to develop effective solutions to avoid expensive maintenance costs and reduce production loss and downtime. While these deposits differ in the details of how they actually form, they share some basic steps; these include supersaturation, nucleation, crystal growth, adhesion, and aging. These main steps basically depend on thermodynamic, kinetic, and hydrodynamic factors.

4.2 What is a solid deposit?

The simple and operational answer to this question is that it is a "troublemaker," based on the list of deleterious impacts solid deposits have on production systems, as discussed in Chapter 3, Problems Associated With Flow Assurance Solids in Production.

A more comprehensive definition is given by Hoang [1], which states that a solid deposit is a solid layer attached to equipment surfaces and piping systems through a multistep process conventionally called fouling, scaling, or encrustation, in which some materials, originally dissolved in fluids, are precipitated and deposited on the surfaces when their thermodynamic equilibrium is disturbed under certain conditions. Scale, deposits, sludges, sediments, precipitates, foulants, etc., are all different terms that are conventionally used to describe solid deposits in oil and gas operations.

Moreover, scaling, fouling, incrustation, sedimentation, and sludging are different ways of introducing the solid deposits formation process, although fouling is considered a general industrial term that describes solids accumulation on surfaces. Cowan and Weintritt [2] described these different terms in their encyclopedic book. *Scaling* and *scale* are terms that are conventionally used in connection with deposits associated with water environments, and they usually denote hard, adherent deposits of inorganic mineral constituents. *Incrustation* is a term describing some marine boiler deposits. *Fouling* is a general term; however, sometimes it is used to describe deposits on a heat exchanger surface. While *microbiological fouling* may be used when deposits contain substantial amounts of biologically generated materials, *marine fouling* describes growth of marine organisms on conduits, on ship hulls, and in water intakes. *Sediment* or *sludge* is used to describe loose deposits in lines and vessels. A *gyp* is used

Essentials of Flow Assurance Solids in Oil and Gas Operations. https://doi.org/10.1016/B978-0-323-99118-6.00017-4

to describe all types of scale deposits in the oil industry. A *stone* refers to deposits in the brewing and dairy trade. A *calculus* or *stone* describes some mineral deposits in humans and animals. The organic deposits are termed ARPD, the abbreviation for asphaltene, resins, and paraffin deposits [3].

4.3 Types of solid deposits in oil and gas fields

There are many types of solid deposits encountered in oil and gas fields. They are generally classified as organic and inorganic deposits, or mineral deposits. Fig. 4.1 illustrates the different organic and inorganic deposits in oil and gas fields.

A large variety of mineral deposits are formed in oil and gas fields, with sulfates and carbonates of group II metals (alkaline earth metals), iron oxides, sulfides and carbonate, zinc, and lead sulfides, and sodium and calcium chlorides being the most common types.

Organic deposits include asphaltenes, paraffin wax, resins, diamondoids, organometallics, mercaptans, and their combinations [4]. Naphthenate deposits are considered a newcomer to the organic deposits list.

In addition to these types, biofouling is yet another type of organic deposit, not originating from oil or gas hydrocarbons but being composed of organic matter formed by biological activity [5]. Moreover, biofouling involves mineral scale formation in a process known as biomineralization through different processes of biocorrosion and minerals precipitation. That makes biofouling a mixed type of organic and inorganic deposits.

Additionally, there are the pseudo-deposits, which originate from the incompatibility between the different oilfield chemicals, which are polymeric or organic in nature. The deposits can form as single type or most commonly as a mixture of two or more different types of solids combined in one layer or a multilayered deposit. Fig. 4.2 illustrates multilayered scale deposits collected from 18″ pipeline.

FIG. 4.1

Types of flow assurance solid deposits in oil and gas fields.

FIG. 4.2

Multilayered scale deposit comprised main composition of $BaSO_4$, $SrSO_4$, with minor $CaCO_3$, (Fe, Zn, Pb)S, besides organic components.

Gudmundsson [6] designated the solid deposits in oil and gas fields as the big five and drew analogies between these solids and other substances in real life. For example, the mineral scale deposits are saltlike deposits, hydrates are icelike deposits, asphaltenes are tarlike solids, waxes are candlelike solids, and naphthenates are soaplike solids.

4.4 Precipitation vs deposition

Before we start the discussion on solid deposition mechanisms, it is important to understand the difference between two main terms that can be confusing sometimes: precipitation and deposition.

Precipitation is basically the formation of a solid phase from a bulk liquid phase, principally as a function of thermodynamic variables (i.e., pressure, temperature, pH, and composition). All of the solid deposits indicated previously basically originate from the produced fluids due to disturbance in the thermodynamic equilibrium of the produced fluids that was established in the reservoir formation over millions of years, and solids precipitate as a consequence of the system readjusting to a new equilibrium.

Deposition, on the other hand, is the formation and growth of a solid layer on the surface of a pipe or equipment. Deposition depends on fluid hydrodynamics, such as flow rate, flow regime, and particle properties. Solids deposit on surfaces by different mechanisms, including diffusion, dispersion, inertial, and gravitational forces.

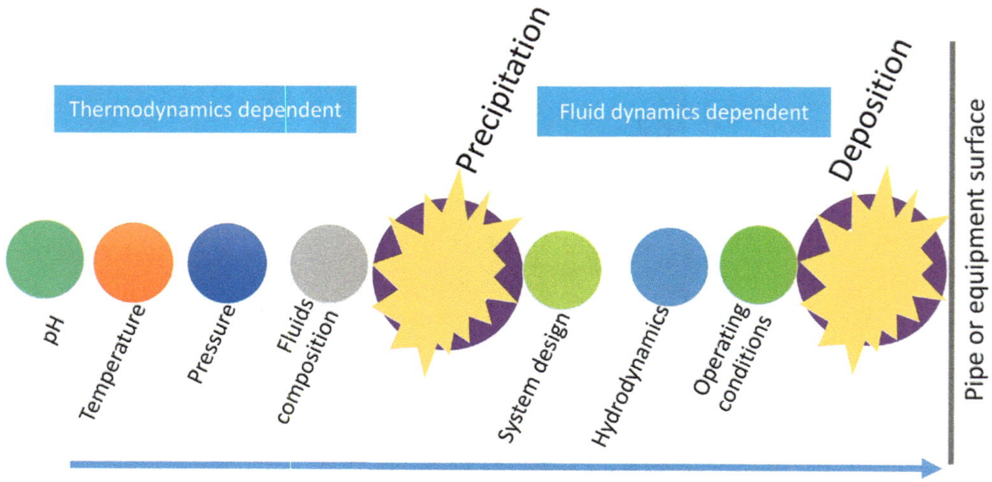

FIG. 4.3

Precipitation vs deposition.

It is clear that precipitation is a prerequisite for deposition and solid deposits formation can be generalized as precipitation followed by deposition steps. Fig. 4.3 illustrates the difference between precipitation and deposition.

Now, the question is, does every precipitation lead to a deposition? The answer is no, not necessarily. In fact, there are many scenarios for precipitation and deposition:

- If the fluid thermodynamics is not in favor of precipitation, then no solids will form originally, and the system should be clean of deposits.
- If the fluid thermodynamics is in favor of precipitation but the fluid hydrodynamics is not in favor of deposition (no deposition regime), then the precipitated solids can be transported with the fluids along the production system; part of them may redissolve, but otherwise the solids can reach low rate areas or dead ends where they can settle and accumulate, like the bottoms of tanks and separators.
- If the thermodynamics is in favor of precipitation and the hydrodynamics are in favor of deposition, then solid deposits will form.

From experience in oil and gas fields, there is no such field that is free of solid deposits.

4.5 Mechanism of formation of solid deposits

Now, let's see how solid deposits actually form. A quick search of the literature shows that solid deposition mechanisms follow five basic steps, as illustrated in Fig. 4.4:

- Supersaturation
- Nucleation
- Crystal growth
- Adhesion
- Aging

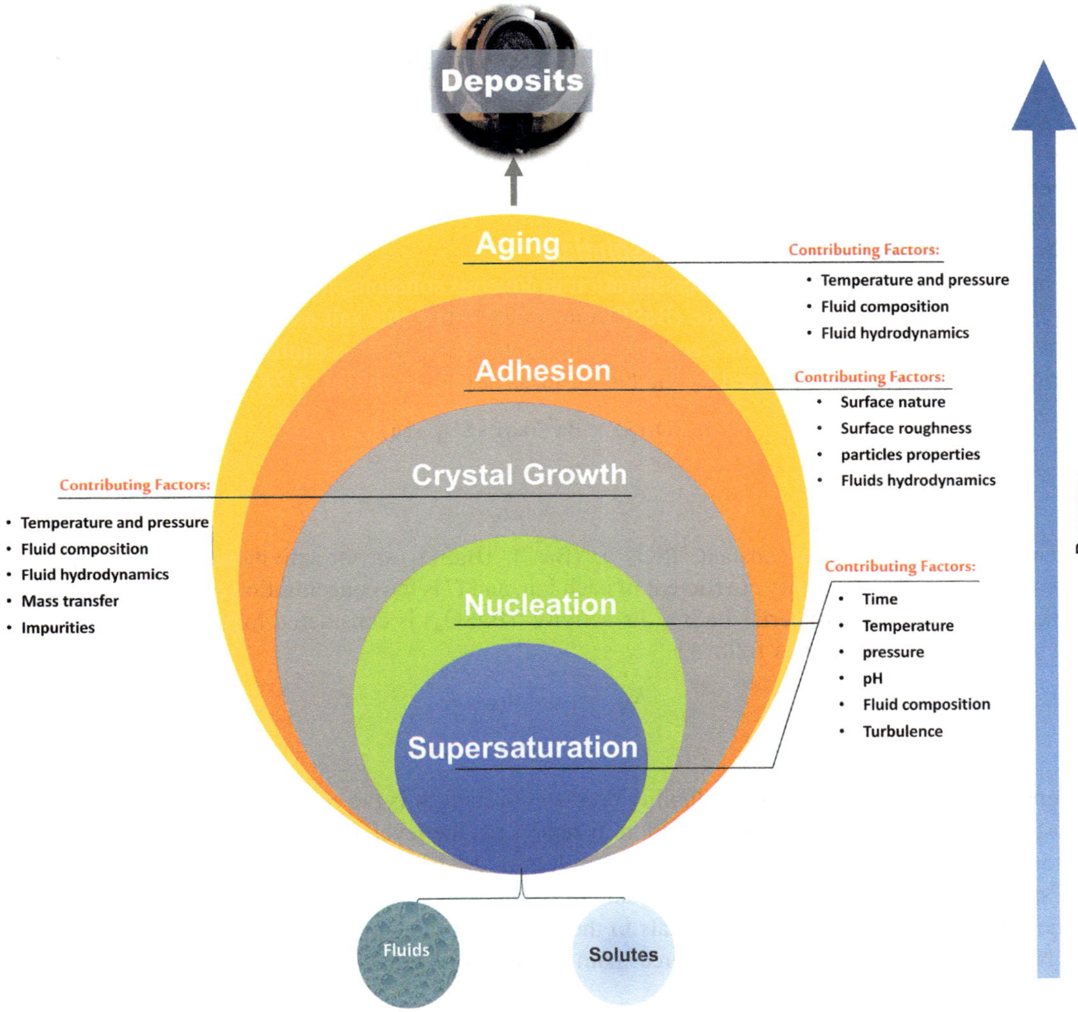

FIG. 4.4

The general mechanism of solids deposits formation.

These steps apply to all kinds of deposits (organic and inorganic) with some differences in the details. For example, paraffin is studied using the approach of crystallization classic theory, and more particularly with the scale formation approach [7]. Another example is the asphaltene deposition mechanism, which suggests that the nanoaggregates grow into microaggregates, and then they form primary particles; then later they age and harden to form the asphaltene deposits. This mechanism has similarities to the recent approach of some mineral scales formation, such as $BaSO_4$ in which nanoparticles grow to microparticles, which then age to form hard deposits [8,9].

The basic steps involved in solid deposit formation are discussed in some detail in the following sections.

4.5.1 Step (1) supersaturation

The first step in the solid formation mechanism is supersaturation. Supersaturation acts as the driving force for precipitation.

Supersaturation is defined as a state of thermodynamic instability when the dissolved materials (solutes) in the solution exceed their solubility limit and hence the solution becomes prone to be relieved of supersaturation through precipitation [10].

4.5.1.1 Supersaturation in aqueous solutions

To understand the phenomena of supersaturation in aqueous solutions, let's take an example of a sparingly soluble salt like barium sulfate ($BaSO_4$) in water. When this salt exists in water, it dissociates partially into its ions until it reaches an equilibrium state (Eq. 4.1). This equilibrium reaction and equilibrium constant can be expressed by the law of mass action, as in (Eq. 4.2):

$$BaSO_4(S) \rightleftharpoons Ba^{2+}(aq) + SO_4^{2-}(aq) \tag{4.1}$$

$$K_{eq} = \frac{a_{(Ba^{2+})}\, a_{(SO_4^{2-})}}{a_{(BaSO_4)}} \tag{4.2}$$

where K_{eq} is the equilibrium constant, ${}^{a}(SO_4^{2-})$, ${}^{a}(Ba^{2+})$, ${}^{a}(BaSO_4)$ are the activities of these species, and where γ represents the activity coefficient of each ion, and C is the concentration of each ion in molar. In the case of the solid material $BaSO_4$, its activity ${}^{a}(BaSO_4)$ is considered to be the unit; therefore Eq. (4.2) can be expressed as follows:

$$K_{sp} = a_{(Ba^{2+})}\, a_{(SO_4^{2-})} \tag{4.3}$$

$$K_{sp} = \text{IAP} \tag{4.4}$$

where K_{sp} is a constant called the solubility product constant. It is the equilibrium constant for a solid substance dissolving in an aqueous solution. It represents the level at which a solute dissolves in solution. The higher the K_{sp} value, the higher the solubility of the material. The K_{sp} value is constant for the same material at constant temperature and pressure. The righthand side of Eq. (4.4) is called the ionic activity product (IAP), and it equals to the product of the activities of the ionic species.

Now, for any solution, there are three main situations of Eq. (4.4) that govern the precipitation process:

1. $K_{sp} > \text{IAP}$, which means the solution is undersaturated, as the ionic product of the ions is less than the solubility limit. In this situation no precipitation occurs.
2. $K_{sp} = \text{IAP}$, which means the solution is saturated and $BaSO_4$ reached its maximum solubility limit. This situation implies an equilibrium state where the $BaSO_4$ is at equilibrium with its ions in the solution.
3. $K_{sp} < \text{IAP}$, which means the solution is supersaturated, as $BaSO_4$ has exceeded its solubility limit. In this situation $BaSO_4$ solids precipitate and separate from the solution.

Thus the salt will precipitate when the product of molar concentration of its ions (or the ion activity) exceeds the solubility product value.

In Fig. 4.5 the solid line in the figure represents the normal solubility curve for a solute in a solvent, or the point of equilibrium state between the solid and the solution of the salt (saturation state). The area

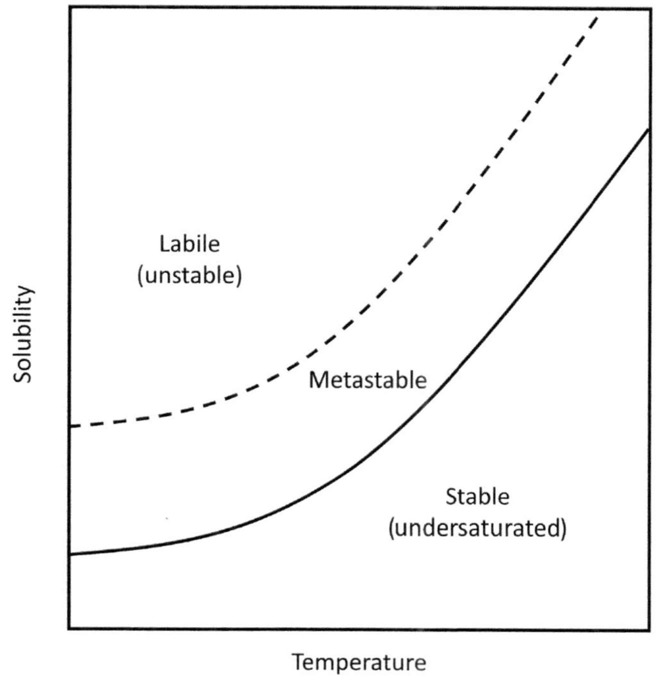

FIG. 4.5

Solubility diagram, showing the stable, metastable, and labile zones for the precipiation of a solid from solution.

above the upper broken line represents the conditions at which spontaneous crystallization can occur. Three zones are represented in the diagram: stable, metastable, and labile [2,11].

- The stable region (undersaturated): crystallization will not occur no matter how much salt is added, as it will dissolve.
- The metastable region, which is thermodynamically unstable but may be stabilized kinetically: in this region spontaneous crystallization is not likely to occur, but it still can be destabilized by changes in the system conditions.
- The unstable, or labile region: spontaneous crystallization is probable.

Eq. (4.4) can also be rewritten as follows:

$$SR = \frac{IAP}{K_{sp}} \tag{4.5}$$

where SR is the saturation ratio, and similarly this will give us three different scenarios:

- SR <1 represents the undersaturated solution
- SR $=1$ represents the saturated solution
- SR >1 represents the supersaturated solution

Supersaturation can be expressed in many different ways:

$$\Omega = \Delta C = (Cs - Ce) \tag{4.6}$$

where Ω and ΔC are the *supersaturation degree*, Cs is the solute concentration at supersaturation, Ce is the concentration at equilibrium, and also there is the *supersaturation ratio* (S), and the *relative supersaturation* (σ) where:

$$S = (Cs/Ce) \tag{4.7}$$

$$\sigma = (Cs - Ce)/Ce \tag{4.8}$$

The question here is what is the driving force for a salt to originally become supersaturated and precipitate?

The answer from thermodynamics is the chemical potential. The difference between the chemical potentials of the solute in the supersaturated (unstable state) and saturated (stable state) solutions is the driving force for supersaturation. The chemical potential, μ (J/mol), can be expressed as:

$$\Delta\mu = \mu_2 - \mu_1 \tag{4.9}$$

where $\Delta\mu$ is the chemical potential difference, μ_2 is the chemical potential of the solute in the supersaturated state, and μ_1 is the chemical potential of the solute at equilibrium in the solution. Chemical potential is expressed in the units of J/kg, or J/mol. The chemical potential can be expressed by the following equation:

$$\mu = \mu_o + RT \ln C \tag{4.10}$$

where μ_o is the standard chemical potential of the salt at STP, R is the gas constant (8.314 J/kmol K), T is the temperature (K), and C is the concentration. Therefore,

$$\Delta\mu = RT \ln \frac{C_s}{C_e} = RT \ln S \tag{4.11}$$

Knowing the number of moles of the solute n, this equation can be expressed in terms of Gibbs free energy ΔG:

$$\Delta G = nRT \ln \frac{C_s}{C_e} = nRT \ln S \tag{4.12}$$

This means that the solute will have less free energy when it precipitates from the solution than it does when it stays in the supersaturated solution.

Now, how does a precipitate actually start in a solution?

As it seems from Eq. (4.1), when both ions Ba, SO_4 come together in the solution and their concentration exceeds the solubility limit, they form a precipitate. That's generally right, but it is not the full picture, as it is not a single-step process; in fact, it has to go through an intermediate step. First, it has to form an ion pair as an intermediate state, and then the ion pair dissociates/transform into a precipitate [12,13], as shown in Eq. (4.13):

$$Ba^{2+}(aq) + SO_4^{2-}(aq) \rightleftharpoons (Ba\,SO_4)^{o}(aq) \rightleftharpoons BaSO_4(S) \tag{4.13}$$

where $(Ba\,SO_4)^{o}(aq)$ is the ion pair of barium sulfate in aqueous solution, which has to dissociate into the solid barium sulfate.

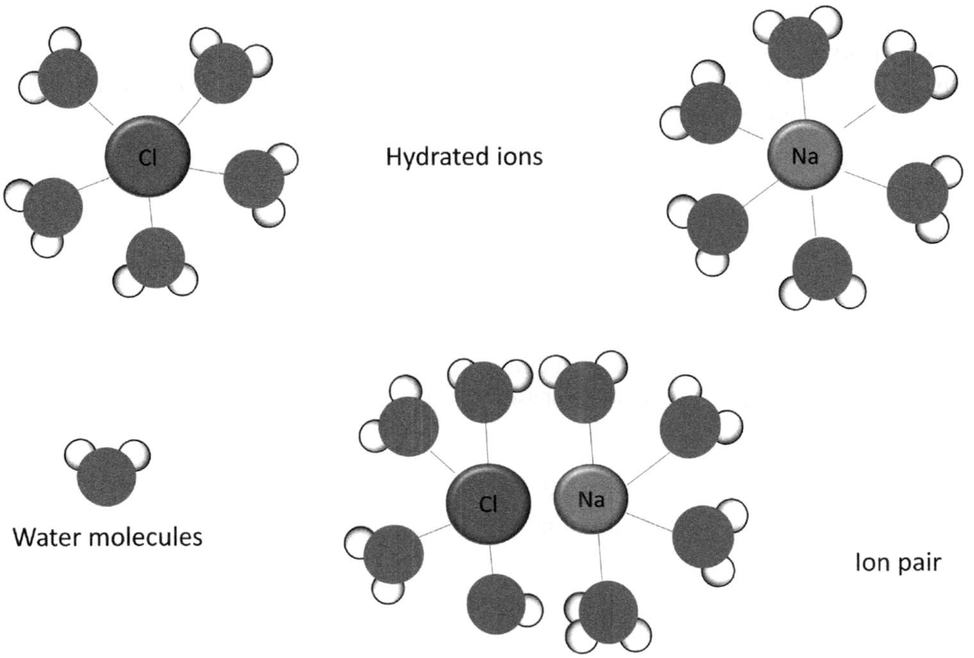

FIG. 4.6

Ion pairs in aqueous solution

A salt in water results in the association between the oppositely charged ions, each associated with a hydration sphere (hydration shells) of water molecules (Fig 4.6). Depending on the size of the ion, its charge, the pH, and other conditions, hydration shells with a size between 2 and 21 water molecules can exist [14].

The extent of formation of ion pairs in dilute electrolyte solutions is much lower than it is in concentrated solutions [1]. Ion pairs are rather weak complexes, as each of the ions retains its hydration sphere; however, they affect the distribution of the ion species in the solution, and therefore affect the solubility and precipitation of salts in the solution [15]. Thus they are used extensively during scale prediction calculations, because they affect the available free ions for the precipitation reaction.

For example, the measured K_{sp} for calcium sulfate is 4.93×10^{-5} at 25°C. The solubility of $CaSO_4$ should be 7.02×10^{-3} M if the only equilibrium involved were as follows:

$$Ca^{2+}(aq) + SO_4^{2-}(aq) \rightleftharpoons CaSO_4(S) \tag{4.14}$$

In fact, the experimentally measured solubility of calcium sulfate at 25°C is 1.6×10^{-2} M, almost twice the value predicted from its K_{sp}. The reason for the discrepancy is that the concentration of ion pairs in a saturated $CaSO_4$ solution is almost as high as the concentration of the hydrated ions [16].

4.5.1.2 Supersaturation in molecular solutions

In the case of organic solids, supersaturation will be studied in the molecular solutions. Generally speaking, in a molecular solution, the partitioning of component i between the liquid and solid phases is represented according to the following reaction [17]:

$$i(S) \leftrightarrow i(L) \tag{4.15}$$

The equilibrium constant ($K_{i(T)}$) is given by:

$$K_{i(T)} = \frac{\gamma^L{}_i\,(x,T)x^L{}_i}{\gamma_i^S x_i^S} \tag{4.16}$$

where the superscripts L and S denote the liquid and solid phases, respectively. γ is the activity coefficient, calculated by an adequate thermodynamic model. x is the component mole fraction in a phase. For pure components, such as pure solid phase, $\gamma = 1.0$ and $x = 1.0$. Then, $\gamma_i^S\ x_i^S = 1.0$ and that equation becomes [17]:

$$K_{i(T)} = \gamma^L{}_i\,(x,T)x^L{}_i \tag{4.17}$$

The equilibrium constant is defined according to:

$$K_{i(T)} = \exp\left[-\frac{\Delta H_{fi}}{R}\left(\frac{1}{T}-\frac{1}{T_m}\right)\right] \tag{4.18}$$

where ΔH_{fi} denotes the heat of fusion (J/mol). T and T_m denote the absolute solution temperature and the melting point temperature of the solid (K), respectively. R is the gas constant (J/mol/K).

Also, a general way of expressing saturation in molecular solutions is by defining the saturation ratio (S), which is given by [17]:

$$S = x_i^L / \left(x_i^L\right)_S \tag{4.19}$$

where x_i^L is the mole fraction of the dissolved organic in oil, and $(x_i^L)_s$ is the organic solubility at saturation conditions and it is predicted using a thermodynamic model, such as those given by Chung [18]. Like the aqueous solutions, we have three situations:

- $S < 1$ for undersaturated solution.
- $S = 1$ for saturated solution.
- $S > 1$ for supersaturated solution.

In the case of paraffin wax, the crystallization of solid paraffin is actually an example of liquid-solid phase equilibrium. It is explained in terms of established principles of thermodynamics of solution; the solution of higher molecular weight hydrocarbons (HMWHs) in low molecular weight hydrocarbons that act as a solvent. Generally, high molecular weight solids precipitate whenever the carrying capacity of the fluid solvent decreases. Temperature is the main parameter that controls the fluid solvent saturation [19,20].

In the case of paraffins, the value of $(x_i^L)_s$ from Eq. (4.19) is denoted as x_p^L and can be estimated by applying the following real solution model of Eq. (4.20) [18]:

$$x_P^L = x_P^S \exp\left[-\frac{\Delta H_P}{R}\left(\frac{1}{T}-\frac{1}{T_{PM}}\right) - \frac{V_P}{RT}(\delta_m - \delta_P)^2 - \ln\left(\frac{V_P}{V_m}\right) - 1 + \frac{V_P}{V_m}\right] \tag{4.20}$$

where x_P^L denotes the dissolved paraffin mole fraction in oil, x_P^S is the paraffin precipitate mole fraction in oil, ΔH_P is the latent heat of fusion of the paraffin (J/mol), R is the gas constant (J/mol/K), T is temperature (K), T_{PM} is the melting point or fusion temperature for paraffin (K), V_P and V_m are the molar volumes of paraffin and oil (L/mol), and δ_P and δ_m are the solubility parameters of paraffin and oil ($J^{0.5}/mol^{0.5}/L^{0.5}$).

The solubility parameter here is the Hildebrand solubility parameter, and it describes the degree of solubility for nonpolar or slightly polar substances without hydrogen bonding. It also describes the amount of energy required to evaporate one-unit volume of a liquid [21]; the solubility parameter can be expressed according to Eq. (4.21):

$$\delta = \left(\frac{\Delta H_{\text{vap}} - RT}{v}\right)^{1/2} \tag{4.21}$$

where ΔH_{vap} is the heat of vaporization (J/mol), the RT term is the gas constant and absolute temperature, v is the molar volume.

Generally the solubility of petroleum waxes can be expressed empirically, thermodynamically, or graphically. The empirical correlations are generally based on the boiling point of the solvent, the solubility parameters, the wax melting point, and/or the molecular weight of the solvent, while the thermodynamic relations are derived by equating the fugacities of the wax in the solid and liquid phases [22], and the graphical methods are described by the solubility curves [23].

The solubility of asphaltenes can explained by two models: the real solution model and the colloidal model. The real-solution theory assumes that asphaltene is a solid that is dissolved in crude oil, and it defines the value of $(x_i^L)_s$ from Eq. (4.17), which is denoted x_a^L and can be explained by Eq. (4.22) [18]:

$$x_a^L = \exp\left[-\frac{\Delta H_a}{R}\left(\frac{1}{T} - \frac{1}{T_{af}}\right) - \frac{V_{aL}}{RT}(\delta_{mL} - \delta_{aL})^2 - \ln\left(\frac{V_{aL}}{V_m}\right) - 1 + \frac{V_{aL}}{V_m}\right] \tag{4.22}$$

where x_a^L is the mole fraction of asphaltenes in solution, ΔH_a is the latent heat of fusion of the asphaltenes, R is the gas constant, T is temperature, T_{af} is the melting point or fusion temperature for asphaltenes, V_{aL} and V_m are the molar volumes of asphaltenes and oil, and δ_{aL} and δ_{mL} are the solubility parameters of asphaltenes and oil.

The colloidal theory assumes that asphaltenes are suspended as a colloid in crude oil and stabilized in the crude oil by the resins adsorbed on their surface, and therefore can better describe the asphaltene peptization/flocculation phenomena. According to this approach, asphaltenes tend to self-associate at a specific capacity known as the critical micellization concentration (CMC), or the critical nanoaggregate concentration (CNAC), after which further flocculation and agglomeration of asphaltene particles lead to their precipitation and hardening [24]. Crude oil asphaltene instability can also be described using the colloidal instability index (CII), which relates the instability and precipitation of asphaltenes to the crude oil composition, which is elucidated from the SARA analysis. The CII is expressed by the following correlation [25]:

$$CII = \frac{(\text{saturates} + \text{asphaltenes})}{(\text{resins} + \text{aromatics})} \tag{4.23}$$

Oils with a CII of below 0.7 are considered stable (no asphaltene precipitation) while those with a CII of above 0.9 are considered very unstable (asphaltene supersaturation and precipitation); a CII of 0.7–0.9 is indicative of an oil with moderate instability [25].

Unlike paraffins and asphaltenes, naphthenate deposition occurs when the naphthenate ion interacts with the metals at the oil/water interface, forming soap emulsion/solids, a mechanism quite similar to scale precipitation. The naphthenic acids (RCOOH) are partitioned between the hydrocarbon (HC) and the aqueous phases (aq), as described by Eq. (4.24):

$$\mathrm{RCOOH_{HC}} \rightleftharpoons \mathrm{RCOOH_{aq}} \tag{4.24}$$

The partitioning coefficient is defined by:

$$K_{\mathrm{part}} = \frac{[\mathrm{RCOOH}]_{\mathrm{HC}}}{[\mathrm{RCOOH}]_{\mathrm{aq}}} \tag{4.25}$$

The naphthenate (R-COO_2) concentration increases at the oil-water interface by dissociation of the naphthenic acid according to Eq. (4.26), and reaches supersaturation and precipitate according to Eq. (4.27), forming the metal soap:

$$\mathrm{RCOOH} + \mathrm{OH^-} \rightleftharpoons \mathrm{RCOO^-} + \mathrm{H_2O} \tag{4.26}$$

$$\mathrm{RCOO^-} + \mathrm{M^{z+}} \rightleftharpoons \mathrm{RCOOM(S)} \tag{4.27}$$

Like the inorganic scales, the K_{sp} of the metal naphthenates can be estimated from Eq. (4.28):

$$K_{sp} = [\mathrm{RCOO^-}][\mathrm{M^+}] \tag{4.28}$$

The naphthenate supersaturation is triggered by an increase in naphthenate, metal ion concentration, pH, pressure drop, or a change in temperature.

In gas hydrate crystallization, one molecule (G) of the dissolved gas and n_w water molecules of the solution form one building unit ($G\ n_w H_2O$) of the hydrate crystal, according to the equilibrium in Eq. (4.29) [26]:

$$G + n_w\mathrm{H_2O} \rightleftharpoons G \cdot n_w\mathrm{H_2O} \tag{4.29}$$

The solubility of gases in the aqueous phase also plays a significant role in the crystallization process. The portioning and solubility of the gases in liquids is defined by Henry's law. The solubility of an organic gas A in water is described by Eq. (4.30):

$$A(\mathrm{g}) \rightleftharpoons A(\mathrm{aq}) \tag{4.30}$$

According to Henry's law, in Eq. (4.31), the solubility is related to the gas concentration:

$$K_H = \frac{\gamma_A m_A}{p_A} \tag{4.31}$$

where K_H (mol kg^{-1} atm^{-1}) is the Henry's law constant, m_A is the molality of A (moles of A per kilogram of liquid water), and p_A (atm) is the gas-phase partial pressure of A. The quantity γ_A is the activity coefficient of A in the aqueous phase and in an infinitely dilute aqueous solution has a value of unity.

4.5.1.3 Factors affecting supersaturation

Supersaturation is the primary cause and the driving force of solids precipitation and crystallization. However, supersaturation occurs only under proper conditions, and these conditions are called the secondary cause of the deposition.

Both undersaturated and supersaturated solutions can precipitate solids-forming materials when the proper conditions or the secondary cause is attained for precipitation. A water can be undersaturated with respect to a scale-forming substance, but can be supersaturated if conditions such as changes in pH, pressure, or temperature decrease the solubility of the material. Precipitation can also occur when a solution is supersaturated and conditions are altered, thus changing the solubility of the scale-forming product. As another example, high paraffin or asphaltene content in the crude oil does not necessarily stipulate their precipitation. For example, deposition of asphaltenes is more likely to happen in low asphaltene crudes (<0.5%, like Hessi Masoud/Algeria), or from low paraffinic crudes rather than from heavy asphaltenic crudes. The same applies for some paraffinic crudes, as they were found to flow with no wax deposition problems. These species rather precipitate when significant change in thermodynamic conditions takes place. In this case, the actual supersaturation point of paraffins is when the temperature falls below the wax appearance temperature (WAT), while asphaltenes start to precipitate when pressure falls below asphaltenes onset pressure (AOP).

- **Temperature**

The solubility of the solid deposits is affected by the temperature. Basically there are two different types of materials behavior when temperature is applied to their solutions: normal solubility and inverse solubility behavior. Cooling a normal solubility salt solution results in supersaturation, while heating an inverse solubility salt solution results in supersaturation. To understand this behavior, think of a solid material i_s dissociated into liquid state i_l, which may be expressed by one of the following:

$$i(\mathrm{S}) + \mathrm{heat} \leftrightarrow i(\mathrm{L}) \tag{4.32}$$

$$i(\mathrm{S}) \leftrightarrow i(\mathrm{L}) + \mathrm{heat} \tag{4.33}$$

The first case (Eq. 4.32) represents the endothermic dissolution, which means the solid material needs heat to dissolve; thus by adding heat to the system, and according to Le Chaterlier's principle, the reaction proceeds toward the righthand side, dissociating more of that material. This case is referred to as the normal solubility materials, in which their solubility increases with increasing temperature. Examples are halites, barium and sulfate, iron sulfides, paraffin waxes, and gas hydrates.

The second case (Eq. 4.33) is what is known as exothermic dissolution, as the solid material gives off heat when it dissolves, so in this case if you add heat to that system, according to Le Chaterlier's principle, the reaction will shift to the lefthand side to absorb the added heat to retain its equilibrium, which means precipitation or crystallization of solids. This case is called inverse solubility or retrograde solubility materials, in which their solubility decreases by increasing temperature. Examples of these materials are calcium carbonate, calcium sulfate, calcium phosphate, naphthenate soaps; asphaltenes in some cases also precipitate with increasing temperature.

In case of solid mineral scales, the dissolution is endothermic ($+\Delta H_{\mathrm{dissolution}}$) or exothermic ($-\Delta H_{\mathrm{dissolution}}$) depending on the relative magnitude of the heat of hydration ($\Delta H_{\mathrm{solvent}}$) and the lattice energy (U_L) of the crystal, according to Eq. (4.34):

$$\Delta H_{\mathrm{diss}} = U_L - \Delta H_{\mathrm{solvent}} \tag{4.34}$$

Another way the temperature affects the supersaturation is that, when heat is applied, it can lead to evaporation of the solvent and increase the concentration of the dissolved species to the supersaturation level (IAP $> K_{sp}$); hence precipitation occurs, which is considered a primary cause of depositing halite scale.

Temperature also affects the solubility constant K_{sp} value, so different values of K_{sp} for the same material are expected at different temperatures and constant pressure [27].

The change in temperature is the main cause of deposition of paraffin waxes. Paraffin waxes start to precipitate when the temperature of the crude oil drops below WAT. Temperature also plays a critical role in hydrate deposits formation, since gas hydrates usually form at low temperature, reaching the freezing point of the water, and at high pressures. The effect of temperature on asphaltene supersaturation and precipitation is not fully understood. Conflicting results of the effect of temperature on the solubility and precipitation of asphaltenes have been reported. However, most of the fouling in heat exchangers in refineries is because of asphaltenes deposition. Also, temperature was found to increase the precipitation of naphthenate soaps.

- **Pressure**

Pressure affects many types of solid deposits. The effect of pressure is significant in systems that encounter gas addition, gas loss, or volume changes. Generally, pressure drop increases precipitation of calcium carbonate, calcium sulfate, and sulfides. Pressure drop is one of the main drives for asphaltenes deposition, as asphaltenes precipitate when pressure falls below AOP. Naphthenate soaps deposit due to pH changes that are triggered by pressure drops. Additionally, pressure drop though orifices leads to the Joule Tomson cooling effect, which consequently precipitates paraffin waxes.

On the other hand, pressure increase causes hydrate precipitation when temperature drops low enough to form hydrates.

- **pH**

Change in pH triggers supersaturation in many compounds, because their solubility is pH dependent. Among them are calcium and iron carbonate, iron, zinc, and lead sulfides, iron oxides, and silicates. In addition, some organic deposits are affected and depend on pH changes such as naphthenates and asphaltenes.

- **Change in concentration**

Supersaturation occurs when the concentration of the solute exceeds its solubility limit in the solution.

The increase in solute concentration takes place by one of the following:

- Mixing of solutions (incompatible solutions).
- Solvent evaporation.
- In situ change in concentration due to bacterial action; i.e., production of H_2S.

Mixing between two or more different waters is inevitable, and it is always the main cause of mineral scale precipitation and deposition in oil and fields. Mixing causes supersaturation and precipitation; then the fluid temperature, pressure, hydrodynamics, and residence time determine where the precipitates deposit in the system. Mixing between SO_4^{2-} rich sea water and Ba^{2+} rich produced waters lead to the formation of hard $BaSO_4$ scale. Mixing between brines that are highly concentrated with Na and Ca ions and crude oil that is rich with naphthenic acids leads to the formation of naphthenate deposits at the water/oil interface, especially in the separators. Mixing between different crudes is one of the main reasons for asphaltenes deposition, as one of the crudes strips off the resins that are stabilizing the asphaltenes in the other crude oil. Another example of mixing that causes asphaltenes deposition is mixing between the CO_2 gas during miscible gas injection and crude oil.

The second cause of change in concentration is solvent evaporation. Evaporative loss of liquid water is generally the result of gas breakout from undersaturated condensate and oil wells, as well as the expansion of gas in gas wells. This increase in water vapor can leave behind insufficient liquid water to maintain halite solubility in the coproduced brine phase. Halite self-scaling is found with both high-temperature and low-temperature wells [e.g., with 125°F and 350°F bottomhole temperature (BHT) gas/gas condensate wells]. Water evaporation at a reduced pressure is the main mechanism for halite deposition in the near wellbore formation and the wellbore of a gas producer [28].

The third cause of change in concentration and precipitation is bacterial action. Sulfate reducing bacteria (SRB) uses the sulfate in the injected water or in the connate water to produce H_2S gas, which when reacting with iron, zinc, or lead leads to the formation of sulfide deposits. It is also proven that a byproduct of the bacterial action of SRB is increasing the carbonate species concentration, which leads to formation of carbonate scales. Iron bacteria oxidizes ferrous ions into ferric ions, which precipitate as iron oxides and iron hydroxides. Methanogens can contribute to microbially influenced corrosion, leading to formation of iron carbonate deposits [29].

- **Dissolved salts**

The dissolved salts in a solution affect the solubility of some salts in water. This effect takes place in two distinct ways: first, the effect of the ionic strength on the solid solubility, and second, the effect of specific ions on the solid solubility.

Increasing the dissolved salts increases the ionic strength and that decreases the ionic activity of the ions in the solution, which increases the salt solubility. The ionic strength can be expressed according to Eq. (4.35):

$$I_s = 1/2\sum_i Z_i^2 C_i \tag{4.35}$$

where Z is the charge of the ion, and C is the concentration. The ionic strength is related to the activity coefficient according to the Debye-Hückel rule:

$$log\gamma_z = \frac{-0.51\,Z^2\sqrt{I_s}}{1+(\alpha\sqrt{I_s}/305)} \tag{4.36}$$

where γ_z is the activity coefficient of the dissolved ions, α is the ion size, in picometers (pm). Increasing ionic strength means a net reduction in attractive forces between the ions, so there is a lesser tendency for them to come together, meaning that their solubility is increased [30].This phenomenon is called the diverse ion effect. The diverse ion effect or the increase in ionic strength is known to increase the solubility of barite, celestite, calcite, and gypsum.

On the other hand, when the solution has an ion that is common with the sparingly soluble salt, that will decrease the solubility of the sparingly soluble salt. This phenomenon is known as the common ion effect, like adding carbonate ions to a solution of calcium carbonate.

Apart from this ionic strength effect, some specific ions can affect the precipitation of some salts: for example, Mg^{2+}, Ni^{2+}, Co^{2+}, Fe^{3+}, Zn^{2+}, and Cu^{2+} ions were found to inhibit the precipitation of calcium carbonate (aragonite form) [31].

$CaCl_2$ salt was found to have a greater effect on asphaltene aggregation compared to $NaCl_2$ salt [32]. Shojaati et al. [33] concluded that the instability of asphaltenes decreased according to the following order: crude oil with no brine > $MgCl_2$ > $CaCl_2$ > $MgSO_4$ > Na_2SO_4.

- **Time**

Time is the kinetic term of the supersaturation and precipitation reactions, and it is particularly known as the "induction time." The induction time corresponds to the moment at which the phase transformation is accomplished, or in other words it is the time required before a supersaturated solution will form a precipitate and it is a function of the solution supersaturation degree [2,17]. Higher induction time means slower reaction, and shorter induction time means faster precipitation reaction. Variations in induction time can result in periodic supersaturation, nucleation, precipitation, and eventual scale formation. Liesegang rings (Fig. 4.7) are examples of periodic supersaturation and precipitation. Such rhythmic precipitation usually results in "banded" scale deposits [2].

The inhibitory effect of some metal ions on calcium carbonate deposition is in fact attributed to the fact that these impurities can result in a prolonged induction period, leading to slower precipitation [31]. The same effect is (prolonged induction time) expected from the different scale inhibitors added to water to inhibit precipitation and deposition of mineral scale deposits.

According to Mullin [34] the mean time of critical nucleus formation (t_{ind}) is inversely proportional to nucleation rate (J_n).

$$t_{ind} = \frac{1}{J_n} \tag{4.37}$$

Or, more consistently,

$$t_{ind} = \frac{n_o}{J_n} \tag{4.38}$$

where n_o is nuclei population density.

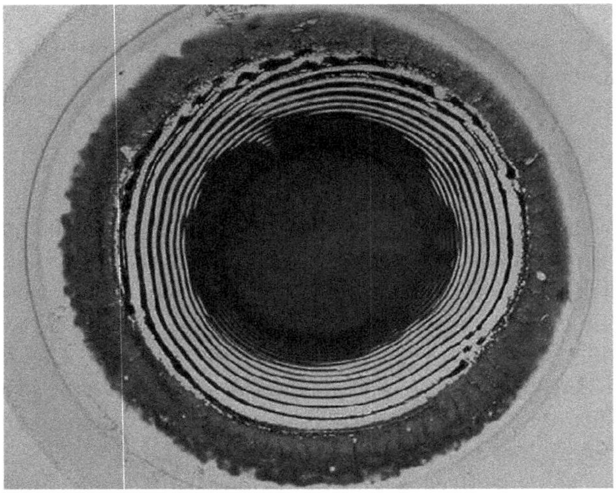

FIG. 4.7

Liesegang rings in silver-chromate precipitate.

Photo credit András Büki, source, https://commons.wikimedia.org/wiki/File:Liesegang_rings.jpg.

After discussing all these factors, or the secondary causes of precipitation, it worth mentioning that the degree of supersaturation in the bulk solution does not always indicate what can occur in localized areas. Supersaturation can be highly localized, as in a thin film or boundary layer next to a heated surface, at the interface of a gas-liquid contact such as a bubble, in a microscopic crevice, or under previously deposited solids [2].

4.5.2 Step (2) nucleation

While supersaturation is the driving force in precipitation and deposition, nucleation represents the initial stage of precipitation.

By definition, the term "nucleation" is generally used to describe the onset of the formation of a new phase from a parent phase [35]. The formation of liquid droplets from a vapor phase or from another liquid, and the formation of crystalline particles from a liquid phase or from another solid are examples of nucleation. In solid-liquid systems, nucleation describes the process in which the smallest stable aggregates of a crystalline phase are formed in a crystallizing system [1].

During nucleation a series of atomic or molecular processes take place by which the atoms or molecules of a reactant phase rearrange into a cluster or aggregate of the product phase, known as nucleus or precritical nuclei. Precritical nuclei are in dynamic equilibrium with the solution (can dissolve or reaggregate). When an aggregate grows to some critical size known as the critical nucleus, nucleation of a solid-state particle occurs. These larger nuclei (known as supercritical nuclei) do not readily dissolve and they have the ability to grow irreversibly to a macroscopically larger size [36,37]. Once formed, seed crystals subsequently grow by ions/molecules adsorbing onto imperfections on the crystal surfaces—extending the crystal size [38].

Nucleation can be induced under some conditions by seeding, agitation, mechanical shock, friction, and extreme pressures in solutions: examples are dust particles, vibrations, by rubbing, tapping, or scratching a rod in a supersaturated solution [2].

4.5.2.1 Types of nucleation

There are two types of nucleation mechanisms: primary nucleation and secondary nucleation. Primary nucleation occurs in systems that do not contain crystals of the same crystallizing material, and it subdivides into two different types: homogeneous and heterogeneous nucleation. If a solution contains no foreign particles or crystals of its own type, a nucleus can only be formed by homogeneous nucleation. Heterogeneous nucleation takes place if foreign surfaces (particles) are present in the system, as a nuclei can form on those surfaces at less supersaturation compared to the case of homogeneous nucleation. On the other hand, if nuclei are generated in the vicinity of crystals of the same material and act as attrition agents or seed crystals, this is referred to as secondary nucleation. The different types of nucleation mechanisms are illustrated in Fig. 4.8.

Heterogeneous nucleation is more common than homogeneous nucleation in industrial water handling systems, as it requires less energy and therefore it occurs at lower supersaturations, delaying the homogeneous nucleation until the available heterogeneous nuclei are used up [2]. Sanz et al. [39] investigated homogeneous ice nucleation by molecular dynamics (MD) simulations. The results indicated that we have to wait an outrageous 10^{83} s ($\sim 3 \times 10^{76}$ years) to witness a single ice nucleus in 1 m^3 water at -20°C. Knott et al. [40] indicated that homogeneous nucleation of gas hydrates occurs at extremely low rates, i.e., on orders of 10^{-111} (nuclei/cm^3 s). This means that the homogeneous nucleation of gas hydrates is unrealistic in reality [40], due to the technical challenge of removing all

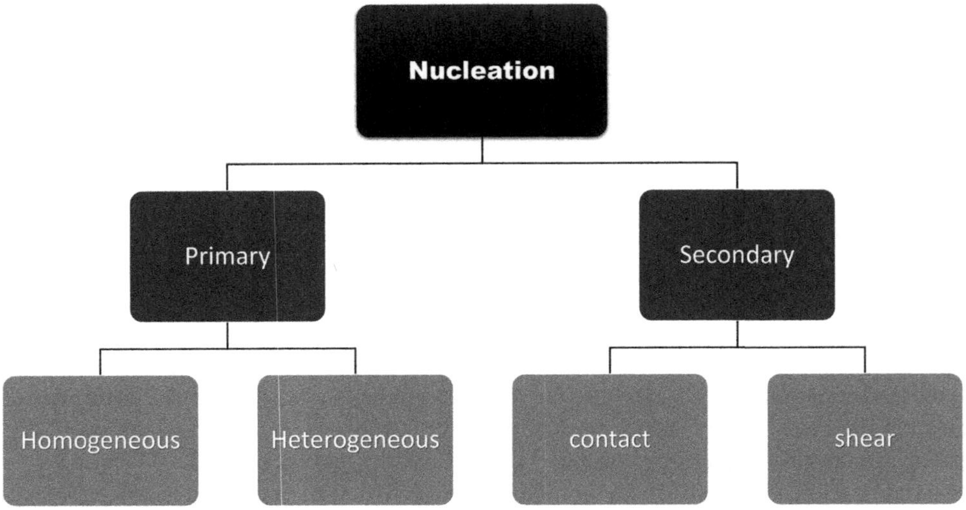

FIG. 4.8

Scheme showing the types of nucleation.

impurities in the system. In fact, it is doubtful whether homogeneous nucleation even exists, since it is virtually impossible to keep a system free of foreign nucleating sites.

4.5.2.2 Nucleation theories

How exactly a stable crystal nucleus is formed within a homogeneous fluid is not known with certainty; however, two main theories have been used to explain the nucleation: classical and nonclassical approaches (also called two-step nucleation theories). Fig. 4.9 illustrates the difference between the two theories [41].

Classical nucleation theory (CNT) is based on the notion of critical nuclei serving as transition states between supersaturated solutions and growing particles. According to classical nucleation theory (CNT), the ions undergo stochastic collisions in supersaturated solutions, which lead to the formation of thermodynamically unstable precritical nuclei. Upon reaching a critical size, the probability of which depends on the level of supersaturation, the transition to postcritical nuclei that can grow without limit, given a sufficient supply of ions or molecules, is facilitated. Importantly, within CNT, precritical, critical, and postcritical nuclei are considered to be characterized by a solid-liquid interfacial surface, which is assumed to correspond to that of the macroscopic interface. CNT is useful for predicting and explaining nucleation phenomena, since it is currently the only quantitative framework at hand [41,42].

The nonclassical approach (also called prenucleation cluster, or PNC, pathway, or the two-step pathway) proceeds through two steps: the formation of the intermediate stage and critical nucleus, before the crystal is formed. According to this approach, ions or molecules form thermodynamically stable and highly dynamic PNCs. PNCs thus do not have an interfacial surface and must be considered as solutes. Once a certain ion activity product (IAP) has been reached, the clusters can change their structural form, leading to a significant decrease of the dynamics and rendering them phase-separated

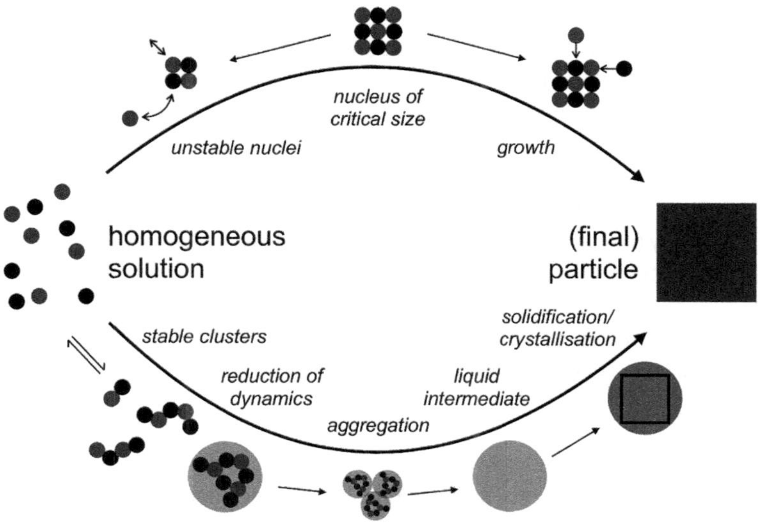

FIG. 4.9

Nucleation mechanisms according to CNT (top) and PNC (bottom).

Source D. Gebauer, How can additives control the early stages of mineralisation?, Minerals (2018), https://doi.org/10.3390/min8050179.

nanodroplets. Driven by the reduction of interfacial surface area, the nanodroplets undergo aggregation and/or coalescence, yielding larger liquid intermediate phases. These dehydrate and solidify toward amorphous intermediates, which eventually transform into crystals [41].

The key difference between the CNT and PNC approach is that the fundamental precursors are thermodynamically stable and form independently of the level of supersaturation. The PNCs are basically solutes and do not have a phase interface; the solute PNCs become phase-separated upon crossing a specific IAP known as the liquid-liquid binodal limit, due to a significant decrease of cluster dynamics and not based on the size of the species forming [41]. A thorough review of the classical and nonclassical nucleation theories is given by Karthika et al. [43].

4.5.2.3 Homogeneous nucleation

According to the classical theory of nucleation, the nucleation process arises from a sequence of bimolecular additions according to the scheme in Eq. (4.39):

$$\begin{gathered} A + A \rightleftarrows A_2 \\ A_2 + A \rightleftarrows A_3 \\ \cdot \quad \cdot \quad \cdot \\ \cdot \quad \cdot \quad \cdot \\ A_{n-1} + A \leftrightarrows A_{n\,(\text{critical cluster})} \end{gathered} \tag{4.39}$$

where A_2, A_3 are clusters and A_n is a critical cluster or critical nucleus; further molecular additions to the critical cluster would result in nucleation and subsequent growth of the nucleus. According to this theory, the chemical species (in a local region of high supersaturation) coagulate and specifically are oriented into a fixed lattice, whereas other embryos or subnuclei that are unstable redissolve. Nuclei that grow above a certain critical size will continue to grow to well-formed crystals.

The structure of the so-called critical nucleus is not known for certainty, and it is too small to be directly observed. It could be a miniature crystal, nearly perfect in form. On the other hand, it could be a rather diffuse body with molecules or solvated ions in a state not too different from that in the bulk fluid, with no clearly defined surface or interfaces [44].

Nucleation is a dynamic process in which surface excess free energy (positive and unfavorable) and volume excess free energy (negative and favorable) battle against each other when embryos nucleate. The former is for solute molecules to become part of the surface of the clustering subcritical sized embryos, while the latter is for solute molecules to integrate into the bulk or interior of the embryonic nuclei [45].

The change in overall free energy of formation is the sum of the surface free energy (ΔG_s) and the volume free energy (ΔG_v). This is represented by Eq. (4.40):

$$\Delta G = \Delta G_s + \Delta G_v \tag{4.40}$$

where ΔG is normally measured in $\text{kJ}\,\text{mol}^{-1}$.

The surface free energy represents surface formation (interface surface formation between the crystal/nuclei and the liquid) and it is the excess free energy between the particle surface and interior. Assuming that the clusters or nuclei are spherical with raduis r, the surface free energy can be expressed as follows:

$$\Delta G_s = 4\pi r^2 \gamma_{\text{i}} \tag{4.41}$$

where γ_{i} is the interfacial tension.

The volume free energy ΔG_v represents the volume change; it is the difference in energy between the crystal/nuclei with very big volume ($r = \infty$) and the dissolved material in the solution and it has a negative value in supersaturated solutions. For spherical nuclei/clusters, it can be expressed by Eq. (4.42):

$$\Delta G_v = \frac{4}{3}\pi r^3 \Delta G^* \tag{4.42}$$

where ΔG^* is the change in free energy per volume unit, and it is expressed by Eq. (4.43):

$$\Delta G^* = \frac{KT}{v} \ln S \tag{4.43}$$

where K is the Boltzmann constant, v is the molar volume, and S is the supersaturation ratio.

Adding all terms in one equation, we get:

$$\Delta G = 4\pi r^2 \gamma_{\text{i}} + \frac{4}{3}\pi r^3 \Delta G^* \tag{4.44}$$

Nucleation is a thermally activated process with an energy barrier, as indicated in Fig. 4.10. The change in overall free energy of formation with respect to the nucleus size passes through a maximum value. The critical size results in decrease in free energy of the particle and it will participate in the nucleation process. The critical size can be calculated by minimizing the free energy function with respect to the radius, or when $d\Delta G/dr = 0$, thus

$$r_c = \frac{-2\gamma_{\text{i}}}{\Delta G^*} \tag{4.45}$$

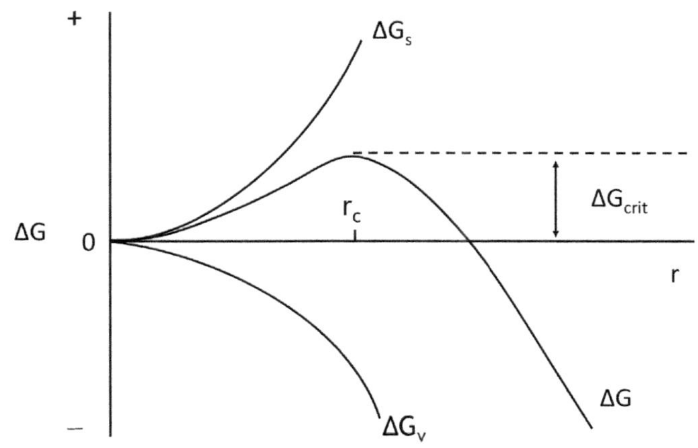

FIG. 4.10

Free energy diagram of nucleation process, showing ΔG_v, ΔG_s, and critical nucleus size.

where ΔG* is a negative value, or in other words:

$$r_c = \frac{2\gamma_i v}{KT \ \ln S} \tag{4.46}$$

Eq. (4.46) indicates that the radius of the critical nucleus decreases by increasing the supersaturation ratio, or in other words it means that nucleation will increase by increasing the supersaturation ratio.

The critical free energy can be calculated from:

$$\Delta G_{\text{crit}} = \frac{16\pi\gamma_i^{\,3}}{3(\Delta G^*)^2} = \frac{4\pi r_c^2 \gamma_i}{3} \tag{4.47}$$

The newly created crystalline lattice structure can redissolve or grow further, depending on its size, where the critical nucleus size (radius) represents the minimum stable nucleus size. Therefore the new crystal will continue to grow if its size is bigger than the critical size, while it will dissolve if it is below the critical size.

There is currently no consensus among researchers on the average size of critical nuclei. They have been claimed to be from as small as 10 molecules [46] to close to 1000 [47] or even several million molecules [48].

4.5.2.4 Heterogeneous nucleation

As mentioned previously, homogeneous nucleation is thought to occur quite rarely in real-life situations since the presence of foreign particles is difficult to avoid in the supersaturated solution. Generally, heterogeneous nucleation takes place more quickly since the foreign particles act as a catalyst for the crystal growth, thus eliminating the necessity of creating a new surface and the incipient surface energy requirements. The heteronucleation will thus occur at lower energy requirements, as indicated in Fig. 4.11 [44]. In addition, heteronucleation requires lower supersaturation than the homogeneous nucleation; this was demonstrated by Walton [49] when he showed that heterogeneous nucleation can occur in the metastable region, while the homogeneous nucleation occurs in the labile region [44].

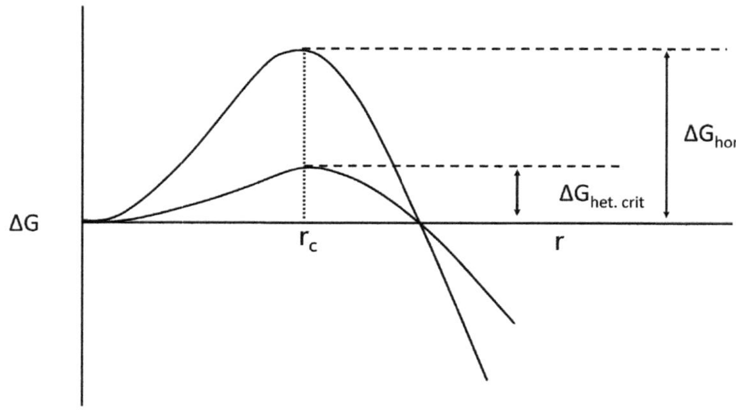

FIG. 4.11

Difference in free energy between homogenous and heterogeneous nucleation, which indicates the lower energy required for heterogeneous nucleation compared to the homogenous nucleation.

The decrease in free energy depends on a factor (Ø) having a value less than 1 and depends on the contact (or wetting) angle between the crystalline deposits and the foreign solid surface. The relationship is formulated in Eq. (4.48) [1]:

$$\Delta G_{\text{het. crit}} = \text{Ø}\Delta G_{\text{hom. crit}} \tag{4.48}$$

where Ø is a factor that has the value ($0 < \text{Ø} < 1$) related to the contact angles, θ, between the substrate and the heteroembryo forming on the substrate. It can be calculated from Eq. (4.49) [1]:

$$\text{Ø} = \frac{(2+\cos\theta)(1-\cos\theta)^2}{4} \tag{4.49}$$

where θ is the contact or wetting angle, which can be calculated, based on Fig. 4.12, by Eq. (4.50) [11]:

$$\cos\theta = \frac{\gamma_{sl} - \gamma_{cs}}{\gamma_{cl}} \tag{4.50}$$

where γ_{sl} is the interfacial tension between the substrate and liquid, γ_{cs} is the interfacial tension between the cluster and substrate, and γ_{cl} is the interfacial tension between the cluster and liquid.

There are three cases of the contact angle, as indicated in Fig. 4.12: if $\theta = 180$ then $\text{Ø} = 1$ and $\Delta G_{\text{het. crit}} = \Delta G_{\text{hom. crit}}$ and that is the case of incoherent nucleus (incompatible crystal and substrate); when $0 < \theta < 180$ then $\text{Ø} < 1$ and $\Delta G_{\text{het. crit}} < \Delta G_{\text{hom. crit}}$ and that is the case of partial or intermediate coherence; the third case occurs when $\theta = 0$ then $\text{Ø} = 0$ and $\Delta G_{\text{het. crit}} = 0$ and that is the case of complete coherence between the crystal and the substrate, and that means the substrate is the same material being crystallized, representing the secondary nucleation [11].

Foreign substrates (surfaces, or solids) used during the heterogeneous nucleation include the impurities in the walls of a pipe or vessel, and when the walls are corroded they offer more nucleating sites; corrosion products, welds, and other internal stress points in metals are ideal, and even scratches or cold working of surfaces, oxide films, fingerprints, and microscopic crevices and cracks are suitable [2].

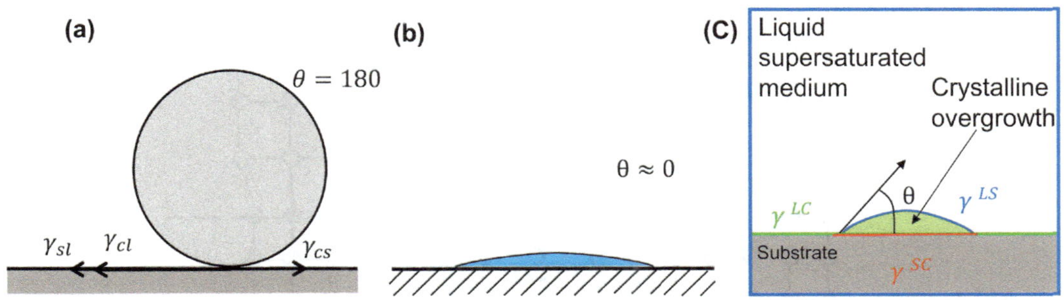

FIG. 4.12

The scenarios of nucleation of crystalline solid *C* on a substrate *S*, in liquid medium *l*: (a) incoherent, (b) fully coherent, (c) intermediate.

Reprinted with permission from P.G. Koutsoukos, Water chemistry and its role in industrial water systems, in: Mineral Scales and Deposits, Elsevier, 2015, pp. 25–45, https://doi.org/10.1016/B978-0-444-63228-9.00002-4. Copyright (2015), Elsevier.

Suspended solids affect the deposition in two main ways: first, they can adhere to the surfaces, accumulate and build up amorphous deposit layers that are held together by physical forces (e.g., electrostatic or agglomeration); second, they work as nucleating sites (substrate) by lowering the energy barrier for the nucleation of solids and also to facilitate further the growth of the dissolved salts supplying the appropriate template.

Suspended solids work best as nucleating sites or substrate for the overgrowth or embryos when they exhibit geometrical compatibility of their crystal lattices. Parameters like spacing, structure, and relative orientation of the lattices define the kinetics and energetics of solid/solid interfaces [1]. When the atomic dimensions of one or more commonly occurring faces of each are similar, this is commonly called "epitaxy" and the growing layer is called the "epilayer" [2]. A quantitative measure for the lattice compatibility between substrates and overgrown phases has been termed lattice misfit by van der Merwe [50]. Assuming a rectangular symmetry for both the substrate and the overgrowth, the lattice misfit may be defined as [1]:

$$\delta_i = \frac{\alpha_o - \alpha_s}{\alpha_s} \tag{4.51}$$

where the subscripts s and o refer to the lattice constants of the substrate and of the overgrowth, respectively. Eq. (4.51) is used to describe the misfit for thin epilayers (one or two layers thick) where the interatomic distances at equilibrium may differ by 5%; however, for thick epilayers, this misfit is described by Eq. (4.52) [1,51]:

$$\delta_i = \frac{2(\alpha_{si} - \alpha_{oi})}{\alpha_{si} + \alpha_{oi}} \tag{4.52}$$

Two basic scenarios exist between the substrate and the overgrowth: (1) complete coherence or matched lattice where substrate and overgrowth are compatible with similar crystal parameters, and (2) complete incoherence, where they are not compatible. It is usually impossible to achieve coherence between a substrate and an overgrowth when their crystal structures are entirely different. In most

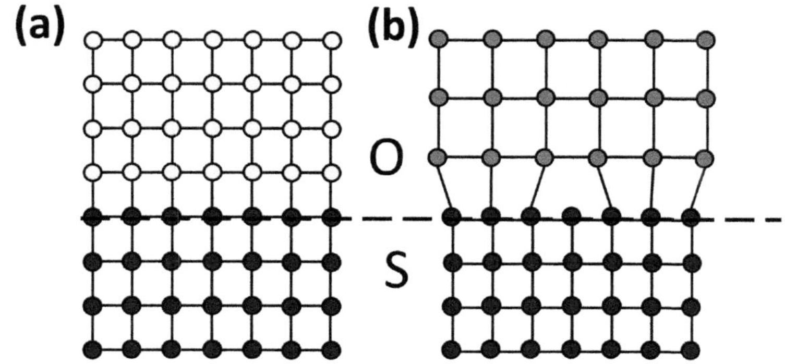

FIG. 4.13

The matching between growing crystal and substrate. (A) Completely coherent interface (lattice matched); (B) partially coherent interface (relaxed heteroepitaxy. With misfit dislocation) O, overgrowth; S, substrate.

Reprinted with permission from P.G. Koutsoukos, Water chemistry and its role in industrial water systems, in: Mineral Scales and Deposits, Elsevier, 2015, pp. 25–45, https://doi.org/10.1016/B978-0-444-63228-9.00002-4. Copyright (2015), Elsevier.

cases, however, partial coherence is encountered, known as relaxed (with misfit defect) or strained epitaxy growth, as illustrated in Fig. 4.13.

4.5.2.5 Rate of nucleation

The rate of nucleation, J_n, is the number of nuclei formed per unit time per unit volume. Following the primary homogeneous nucleation mechanism, the rate of nucleation can be expressed in the form of the Arrhenius reaction velocity equation (Eq. 4.53) [52,53]:

$$J_n = A\exp\left(\frac{-\Delta G_{\text{crit}}}{KT}\right) \tag{4.53}$$

where A is the preexponential factor and has a theoretical value of 1030 nuclei/cm^{3}s, k is the Boltzmann constant, ΔG is the free energy change between a small particle of solute and solute in solution, and T is the temperature. By defining the ΔG_{crit}, the nucleation rate can now be obtained from Eq. (4.54):

$$J_n = A\exp\left(\frac{16\pi\gamma_i{}^3 v^2}{3K^3T^3(\ \text{In}S)^2}\right) \tag{4.54}$$

From this equation, it is clear that the increasing supersaturated solution, interfacial tension, and temperature can increase nucleation rate [44].

4.5.2.6 Induction time

The rate of nucleation can be determined by observing induction time. Therefore the induction time represents the kinetics of the nucleation process, and it is defined as the amount of time elapsed between the creation of supersaturation and the point at which crystals are first detected [1,44]. The induction time is divided into three periods. At the beginning, the solution requires a time to achieve a quasisteady-state distribution of molecules. This period is called a certain relaxation time, t_r. Then time is required for the formation of a stable nucleus, t_n, and then for the nucleus to grow to a detectable size, t_g [34,54]. So, the induction time, t_{ind}, may be written as:

$$t_{ind} = t_r + t_n + t_g \tag{4.55}$$

The relaxation time depends to a great extent on the system viscosity and diffusivity. The nucleation time depends on the supersaturation, which affects the size of the critical nucleus. The growth time depends on the size at which nuclei are detectable and the growth rate applicable to this early stage of development [34].

In low supersaturation solutions as in Fig. 4.14, at point A, the initial supersaturation is created when time is zero and after a certain time, at point B', the solution forms critical-sized nuclei in which crystals cannot be visually detected. This point is called the critical nuclei time (t_n). At point B, a certain induction time t_{ind} elapses and crystals are first detected. However, at point B, no significant changes in the solution can be detected until point C, sometimes referred to as the end of the latent period, t_{lp}, followed by rapid decay of supersaturation (D). Crystal growth is predominant during the decay of the supersaturation region. Growth will slow down near equilibrium concentration at point E. At very high supersaturation, the induction time and latent period can be extremely short and are virtually indistinguishable [34,44].

The induction time is usually expressed as the inverse of the rate of nucleation ($t = 1/J_n$). In the case of true homogeneous nucleation, and based on classical nucleation relationship, the induction time can be given by Eq. (4.56):

$$\tau_{ind} = A \exp\left(\frac{\Delta G}{kT}\right) \tag{4.56}$$

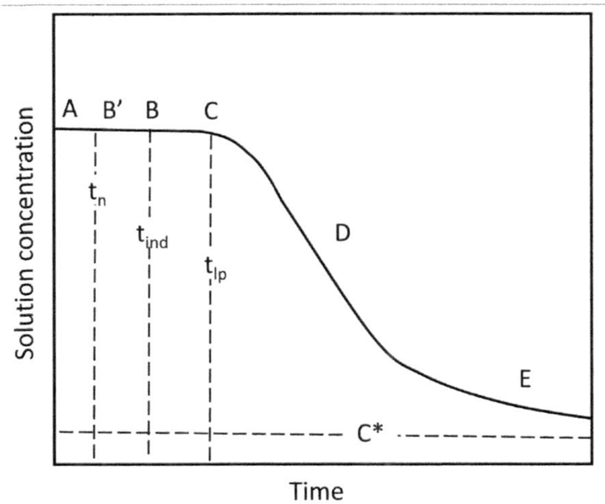

FIG. 4.14

Desupersaturation curve, with different time events.

Reprinted with permission from J.W. Mullin, Crystallization, Butterworth-Heinemann, Oxford, 2001. Copyright (2001), Elsevier.

and the induction time in the case of heterogeneous nucleation will be given by Eq. (4.57):

$$t_{ind} = \exp\left(\frac{16\pi \varnothing \gamma_i^{\,3} v^2}{3K^3T^3\,(\ln S)^2}\right) \tag{4.57}$$

4.5.2.7 Secondary nucleation

Secondary nucleation is a nucleation process resulting from the presence of seed crystals in the solution. The size of seed crystals influences secondary nucleation. For example, large seed crystals generate more secondary nuclei in an agitation system than do small seed crystals, because of their greater contact probabilities and collision energies.

Secondary nucleation can happen due to many factors, including [44,55,56]:

1. Apparent secondary nucleation when the nuclei are brought into the system along with inoculating crystals, and that type includes seeding with crystal dust (dust breeding), polycrystalline breeding, and macroabrasion.
2. True secondary nucleation where new nuclei are formed by interaction between the crystal and solution. The type of true secondary nucleation is shear nucleation caused by fluid shear. This type can be divided into three groups: formation of nuclei from the solid phase, i.e., from a seed crystal; formation of nuclei from a dissolved substance in solution; and formation of nuclei from a transition phase at the crystal surface.
3. Contact nucleation, which is probably most predominant in any stirred crystallizer. Nucleation is caused by crystal-crystal collision, crystal-stirrer collision, or crystal-crystallizer wall collision.

The model form used to describe the kinetics of secondary nucleation [57] is:

$$J_n = K_N\,(C - C_s)^i \tag{4.58}$$

where $(C - C_s)$ is the supersaturation concentration difference, and K_N and i are empirically determined coefficients.

The induction time for secondary nucleation can be determined from Eq. (4.59):

$$\ln t_{ind} = \ln\frac{n_o}{K_N} - i\,\ln(C - C_s) \tag{4.59}$$

The slope of the plot of $ln\ t_{ind}$ versus $\ln (C - C_s)$ gives the value of i. The value of i is associated with the nature of nucleation. Secondary nucleation corresponds to a low value of i in contrast to high values of homogeneous or heterogeneous primary nucleation [1].

4.5.2.8 Gas hydrate nucleation

Hydrate nucleation is considered a complicated process due to the variable phases and driving forces involved, including subcooling, variation in chemical potential, change in fugacity, or simply the overpressure, supersaturation, or the total molar Gibbs free energy change. Generally speaking, subcooling (ΔT) is often adopted to represent the driving force of hydrate formation [45].

Hydrate nucleation often occurs at a gas-water interface not in the bulk phase (due to the low solubility of gas in water and the low concentration of water in the vapor phase), where a sufficient concentration gradient for either phase exists at the interface, and the phases most probably contain impurities [45]. However, more recent MD simulations by Jacobson et al. [58] indicate that there is no observable preference of location for nucleation to occur and it could be either on the interface or in the bulk solution. In addition, foreign surfaces such as a pipe wall or container wall may be involved in the process. Thus most if not all hydrate nucleation in real systems is expected to be heterogeneous [39,45].

Hydrate nucleation and formation is widely described using classical nucleation theory (CNT), which postulates a growing spherical solid nucleus in a supercooled liquid phase [59]. Recently, a number of nonclassical approaches have emerged. Common gas nucleation theories include labile cluster hypothesis, the local structuring nucleation hypothesis, interfacial nucleation hypothesis, blob mechanism, and others. These mechanisms are reviewed by Ke et al. [45] and are discussed in further detail in Chapter 6, Gas Hydrates and Diamondoids. Gas hydrates nucleation studies are classified into: (1) nucleation at constant temperature and degree of subcooling where induction time and nucleation rate are studied, and (2) nucleation during constant cooling experiments, which study the subcooling level at which nucleation occur [45].

Again, like the mineral scales, the energy barrier ΔG_{crit} has to be passed to form the critical nuclei with critical radius r_c, which can be calculated from Eqs. (4.47), (4.48). Continuously increasing the supersaturation level/subcooling level eventually causes spontaneous nucleation when the free energy barrier is reduced to be sufficiently small [60].

A basic relationship between critical size and temperature (supercooling level, ΔT) [34] is

$$r_c \, \alpha \, \frac{1}{\Delta T} \tag{4.60}$$

and the energy barrier is also related to ΔT:

$$\Delta G_{crit} \, \alpha \, \frac{1}{\Delta T^2} \tag{4.61}$$

These equations indicate that the size of critical nuclei and the energy barrier would both decrease with decreasing temperature (i.e., enhanced supercooling). Thus the probability of nucleation should increase with increasing supercooling inside the hydrate region.

The nucleation rate of gas hydrates is usually expressed the same way as the mineral scales. However, the induction time calculations are controversial and apparatus-dependent, due to the stochastic nature of the hydrate nucleation process, and the different parameters that affect the induction time, e.g., impurities, driving forces, contact areas, etc. [45]. It is extremely difficult or maybe unrealistic to observe in-situ nucleation with existing experimental technologies [45].

4.5.2.9 Wax nucleation

Temperature reduction is also another driving force for paraffin wax nucleation. As temperature reduces, the wax molecules solubility limit is approached, and the kinetic energy of the paraffin molecules is reduced; consequently, the motion of the wax molecules is hindered, and the molecules tend to be closer to each other. As this process continues, the wax molecules becoming tangled, forming clusters that grow larger and become stable upon reaching a certain critical size, while those clusters for which the critical size is not attained will redissolve [61]. As the temperature of the solution is lowered to its WAT, the paraffin molecules begin to nucleate until the clusters reach a critical size and become stable crystals [62]. Some techniques were able to detect wax crystals at the microscopic level (at the nucleation stage) [61].

4.5.2.10 Asphaltenes nucleation

Generally, asphaltenes come out of solution when the difference of solubility parameters between the asphaltene macromolecular assemblies (nanoaggregates) and the medium reaches a certain threshold [63]. However, the formation of a new phase requires the appearance of small particles or nuclei that overcome an energy barrier and some degree of supersaturation is necessary for nucleation and precipitation to begin [64]. Nucleation is expected to occur simultaneously for a partially soluble homogeneous material, but not necessarily for a polydisperse material like asphaltenes. Asphaltenes are thought to exist in the oil as a polydisperse distribution of solubilities, molecular weights, and other properties [65,66]. Hence, it follows that molecular aggregates having different properties could have different stability thresholds with different degrees of supersaturation at a given condition, and therefore come out of solution at different times. The most insoluble asphaltenes (molecules or nanoaggregates) might nucleate first, followed by the less insoluble material. It is also possible that initially soluble nanoaggregates undergo a molecular rearrangement to form a mixture of soluble and insoluble nanoaggregates and molecules. Hence, nucleation may continue over time, which means that only some of the ultimately unstable asphaltenes will be available at first and the remainder will appear over time [67].

4.5.2.11 Naphthenate nucleation

In the case of naphthenates, the rate of nucleation depends on various parameters, such as particle radius, solid-solution interfacial tension, molecular volume, temperature, and supersaturation. Supersaturation is a function of the concentration of metal ion (ionic strength) and dissociated naphthenic acid [68]. Presence of suspended solids, corrosion and scale particles affect naphthenates nucleation, as they work as nucleating sites. Shear and turbulence effects accelerate the nucleation of calcium naphthenates, since naphthenates are formed by molecular diffusion of acid from the bulk to the interface. The growth rate is also influenced by supersaturation level, temperature, particle size as well as shear/turbulence, and cation/anion ratio [69].

4.5.2.12 Factors affecting the nucleation

- **The degree of supersaturation**

Supersaturation is the critical parameter to control the rate of nucleation. At high supersaturation, nucleation dominates crystal growth and nucleation rate is considered a high-order function of supersaturation, as indicated in Fig. 4.15. At higher supersaturation, the adsorbed layer is thicker and results in a large number of nuclei. The size of the critical nucleus decreases with increase in supersaturation. Thus, the probability of the nuclei surviving to form crystals is higher. As the supersaturation increases, the microroughness of the crystal surfaces also increase, resulting in a larger nuclei population [44].

Heterogeneous nucleation usually occurs at a lower level of supersaturation compared to the homogeneous nucleation.

- **Temperature**

Temperature is one of the main factors that affects nucleation rate. Increasing temperature reduces the interfacial tension, decreases the induction time, and thus increases the nucleation rate. However, for other systems the nucleation rate declines with increasing temperature, which can be attributed to

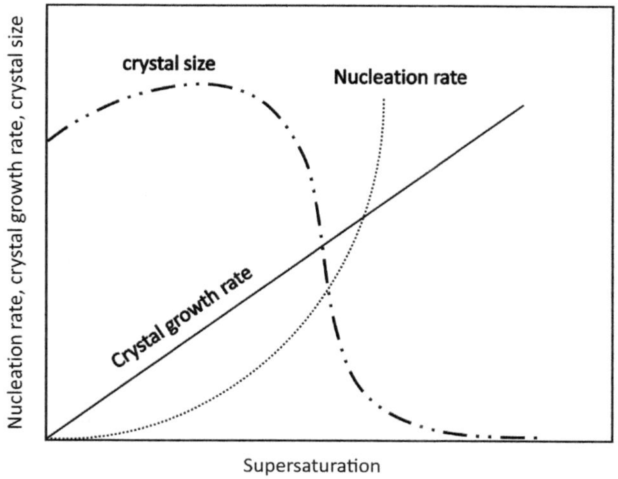

FIG. 4.15

Nucleation and crystal growth rates as a function of supersaturation.

relative decrease in supersaturation, increase in critical size of nucleus, or is due to the faster rate of incorporation of the absorbed layer onto the crystal surface at higher temperatures [56].

Whether the nucleation rate decreases or increases with temperature change basically depends on the kinetic parameters of the system, such as the order of nucleation rate; the order of surface integration rate; and the activation energies of diffusion, surface reaction, and nucleation [1].

The induction time also depends on the temperature, as indicated in Eq. (4.62) [1]:

$$\log t_{ind}^{-1} = A - \frac{E_a}{2.303\,RT} \tag{4.62}$$

where E_a is the molar activation energy for nucleation (J/mol), R is the universal gas constant (8.31,451 J/K)/mol, and A is constant. It indicates that the induction period of the scaling depends strongly on temperature.

- **Impurities**

The presence of impurities (additives) can either enhance or inhibit the solubility of a substance. Some additives with enhanced solubilities would lead to lower supersaturation and lower nucleation rate. Some impurities adsorb on the surface crystals, leading to either one of two opposing effects. In the first, the additive would lower the surface tension and lead to higher nucleation rate. In the second, however, the impurity adsorption blocks potential growth sites and lowers nucleation rate, and this is one of the mechanisms by which chemical inhibition using additives works [44].

In addition, the selectivity of the impurities is different from one solution to another; some ions can accelerate the nucleation and growth process, while others can slow it down or even inhibit it totally. Thus the effect of impurities is complex and unpredictable [44].

Other factors affecting the nucleation include pressure fluid viscosity, turbulence, and degree of agitation.

4.5.3 Step (3) crystal growth

The next stage following the nucleation process is the crystal growth. Crystal growth represents the birth of new crystals from the nuclei formed during the nucleation, by transporting of solute (growth units, atoms, ions, or molecules) molecules from the supersaturated solution and incorporating them into the surface of a crystal, causing an increase in size.

The crystal growth goes through a series of steps that can be categorized into *transport processes*, in which solutes are transported through the solution, and *surface processes*, including solute attachment to the surfaces, solute movement on the surface and solute bonding to the kinks and edges [36,70,71]. Therefore growth can be transported (if solute transport is the slowest) or surface controlled (if any of the surface processes are the slowest) [36].

To understand the mechanism of crystal growth, let's take a look at the crystal surface, as presumed by Kossel [72], which depicts the crystal surface as made of cubic units that form layers of monoatomic height, limited by steps (or edges). These steps contain a number of kinks along their length. The area between steps is referred to as a *terrace*, and it may contain single adsorbed growth units, clusters, or vacancies [36,70]; see Fig. 4.16.

Solutes integrate into the crystal most readily at vacant sites or "kinks" along the edge of the layer, since the solute entering such a kink will be able to form nearest-neighbor bonds with three sides in the crystal. The crystal growth process steps are illustrated in Fig. 4.17A and B.

The process of growth can be considered to occur in the following stages [36,70]:

(1) Transport of solute to the neighborhood of the crystal surface.
(2) Diffusion through a boundary layer, adjacent to the surface, in which a gradient in the solute concentration exists because of depletion of material at the crystal-solution interface.

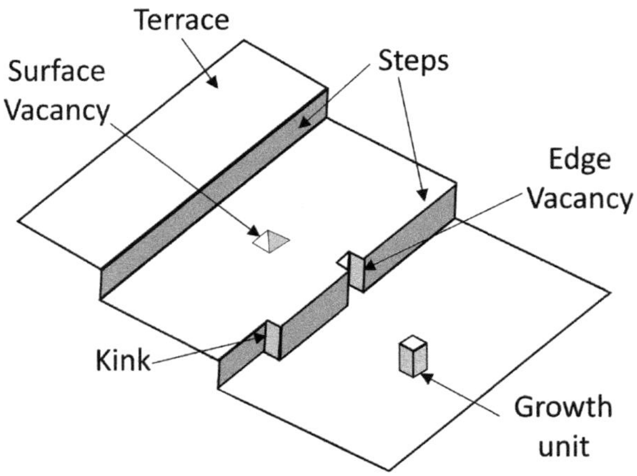

FIG. 4.16

Kossel model of a crystal surface.

Reprinted with permission from P. Cubillas, M.W. Anderson, Synthesis mechanism: crystal growth and nucleation, in: Zeolites and Catalysis: Synthesis, Reactions and Applications, vol. 1, Wiley-VCH, 2010, pp. 1–55, https://doi.org/10.1002/9783527630295.ch1. Copyright (2010), John Wiley and Sons.

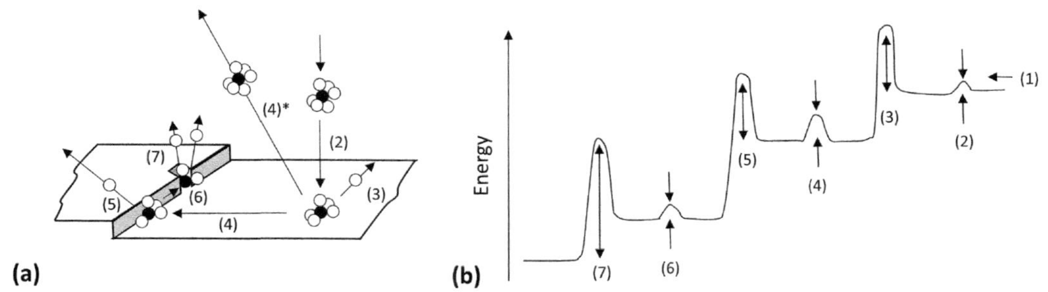

FIG. 4.17

Schematic representation of processes involved in the crystal growth (a), and associated energy changes for the processes depicted in (b).

Reprinted with permission from P. Cubillas, M.W. Anderson, Synthesis mechanism: crystal growth and nucleation, in: Zeolites and Catalysis: Synthesis, Reactions and Applications, vol. 1, Wiley-VCH, 2010, pp. 1–55, https://doi.org/10.1002/9783527630295.ch1. Copyright (2010), John Wiley and Sons.

(3) Adsorption on the crystal surface.
(4) Diffusion over the surface.
(4*) Desorption from the surface; to a step or edge.
(5) Attachment to a step.
(6) Diffusion along the step.
(7) Integration into the crystal at a kink or step vacancy.

As growth rate increases, the growth units fit into less energetically favorable sites (steps) and the crystal typically grows imperfectly. Thus slow growth generally leads to perfect crystals, while fast growth leads to crystals with many more imperfections [34,44,73].

4.5.3.1 Crystal growth theories

The crystal growth theories are well detailed in the literature by Mullin [34], Elwell and Scheel [70], Myerson [74], and Bard [73]. Here a brief outline of each is given for better understanding of that critical process.

- **Surface energy theory**

The principles of surface energy theory are: the total free energy of a crystal in equilibrium with its surroundings at constant temperature and pressure would be a minimum for a given volume [75].

Thus if a crystal is allowed to grow in a supersaturated medium, it should develop into an equilibrium shape, i.e., the development of the various faces should be in such a manner as to ensure that the whole crystal has a minimum total surface free energy for a given volume [34]. Due to the inverse proportion between surface energy of a face and the reticular or lattice density of the respective lattice plane, faces of low density would grow rapidly and eventually disappear [1].

The surface energy theories have no general acceptance and could not explain well-known effects of supersaturation and solution movements on the crystal growth rate [34].

- **Adsorption layer theories**

First suggested by Volmer [76], this concept of a crystal growth mechanism is based on the existence of an adsorbed layer of solute on a crystal face. Upon arrival at the solid-liquid interface, the growth units will not fully integrate into the crystal lattice, but instead they would migrate over the crystal face to form a layer of loosely adsorbed units known as a "third phase" [77].

The growth units adsorb at the active sites (where attraction forces are the greatest) of the crystal until the whole plane face is completed, and before the next crystal face commences, a two-dimensional nucleus (monolayer island nucleus or island), also known as "center of crystallization," must be produced on the adsorption layer to continue crystal growth, as illustrated in Fig. 4.18 [34,77].

Kossel [72] depicted the flat crystal surface as a moving layer of monatomic height that contains one or more kinks, where the loosely adsorbed growth units can be easily incorporated into the crystal face at these kinks. The kink then moves along the surface until the whole layer eventually is complete. A new step can be created by surface nucleation, usually commencing at the corners. The more kinds, the faster the crystal face is complete. Kossel theory is illustrated in Fig. 4.16. One problem associated with this model is due to the dependence on the high level of supersaturation

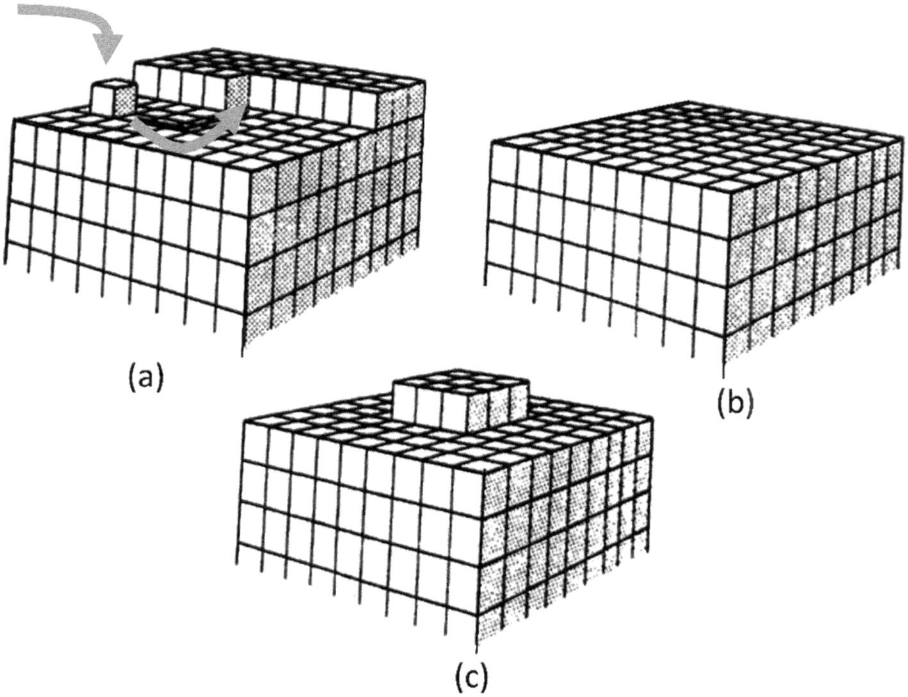

FIG. 4.18

Adsorption layer theory (a) migration towards location, (b) completed layer, (c) surface nucleation.

Reprinted with permission from J.W. Mullin, Crystallization, Butterworth-Heinemann, Oxford, 2001. Copyright (2001), Elsevier.

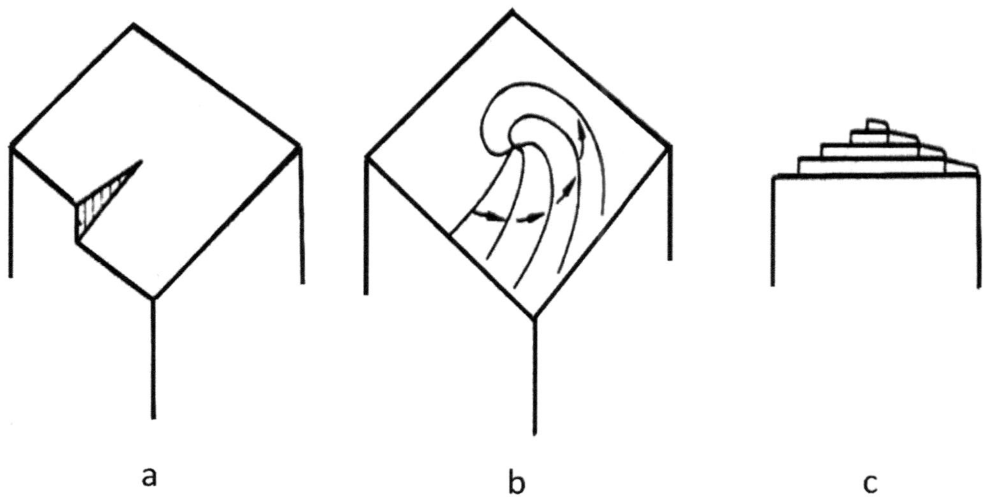

FIG. 4.19

Spiral growth from a dislocation.

Reprinted with permission from J.W. Mullin, Crystallization, Butterworth-Heinemann, Oxford, 2001. Copyright (2001), Elsevier.

for surface nucleation, since it is known that many crystal faces can grow at quite high rates at low supersaturation levels [34].

A solution to this problem was presented by Frank [78], which argued that few crystals grow in the ideal layer-by-layer fashion without developing imperfections, leading to the screw or spiral crystal growth. Spiral growth according to this approach originates from screw dislocations, as shown in Fig. 4.19. The spiral growth is pronounced in several types of minerals, e.g., calcite as indicated in Fig. 4.20 [79]. Spiral growth was also observed in wax deposition.

Burton-Cabera-Frank (BCF) model [80] is another approach in the adsorption theory of crystal growth, in which surface diffusion was considered to be the rate-controlling step, so the model is called the BCF surface diffusion model. According to the BCF model, the relationship between crystal growth rate R_{CG} and supersaturation ratio S is described as:

$$R_{CG} = A(S-1)^2 \tanh\left(\frac{B}{S-1}\right) \tag{4.63}$$

where A and B are constants depending on temperature.

At low supersaturation R_{CG} α $(S-1)^2$, while at high supersaturation R_{CG} α $(S-1)$, i.e., it changes from parabolic relationship to linear when supersaturation increases [1].

- **Kinematic theory**

The kinematic theory, developed by Frank [81], involves two processes of crystal growth on the crystal surface: generation of steps on the crystal surface followed by the movement of layers across the crystal surface. The step velocity depends on the closeness of other steps, since all steps are competing units [34].

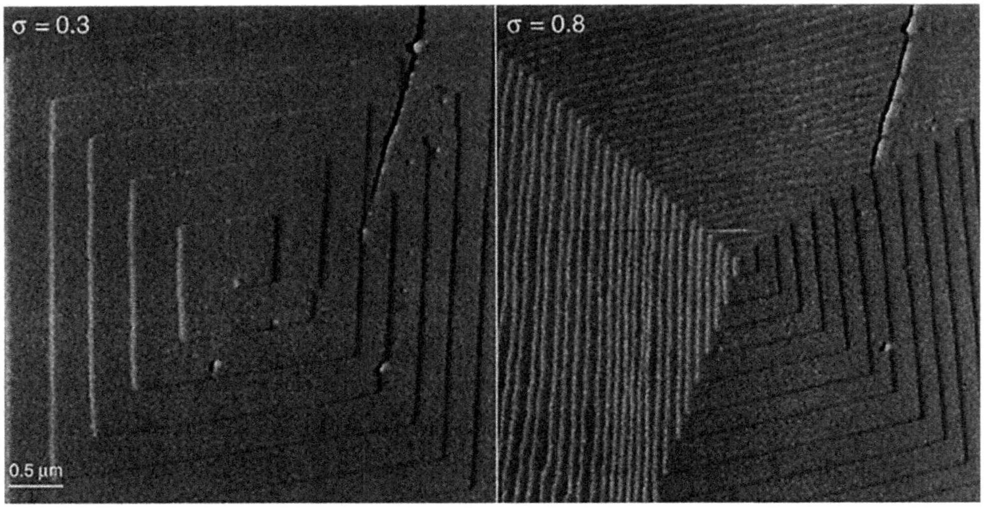

FIG. 4.20

AFM images: spiral growth of calcite at different supersaturations.

Reprinted with permission from H.H. Teng, P.M. Dove, J.J. De Yoreo, Kinetics of calcite growth: surface processes and relationships to macroscopic rate laws, Geochim. Cosmochim. Acta, 64(13) (2000), pp. 2255–2266, https://doi.org/10.1016/S0016-7037(00)00341-0. Copyright (2000), Elsevier.

- **Diffusion theory**

Diffusion theory was first proposed by Noyes and Whitney and Nernst. It assumes that the deposition of a solid on the face of a growing crystal is principally a diffusion process. In this theory, crystallization is the reverse of dissolution and the rates of both processes are controlled by concentration gradients between the bulk solution and the surface. Further modifications to the theory suggested two key steps in the mass deposition: a diffusion process where solute molecules are transported to the surface, and a first-order reaction where the solute molecules get incorporated into the crystal lattice [34]. These can be represented by Eqs. (4.64), (4.65):

$$\text{Diffusion}\ \frac{dm}{dt} = k_d A_s (C - C_i) \tag{4.64}$$

$$\text{Reaction}\ \frac{dm}{dt} = k_r A_s (C_i - C_e) \tag{4.65}$$

where dm/dt is the rate of mass deposited in time t, A_s is surface area, and C and C_i are solute concentrations in the bulk solution and at the interface, respectively; C_e is equilibrium saturation concentration, k_d is coefficient of mass transfer by diffusion, and k_r is rate constant for the surface reaction (integration) process.

Some critics of the diffusion-reaction theory state that it was not consistent with experimental observations; besides, the crystallization is not necessarily the reverse of dissolution [34]. Additionally, it was shown that the rate of crystallization of sparingly soluble electrolytes is mainly controlled by reaction kinetics rather than diffusion [82].

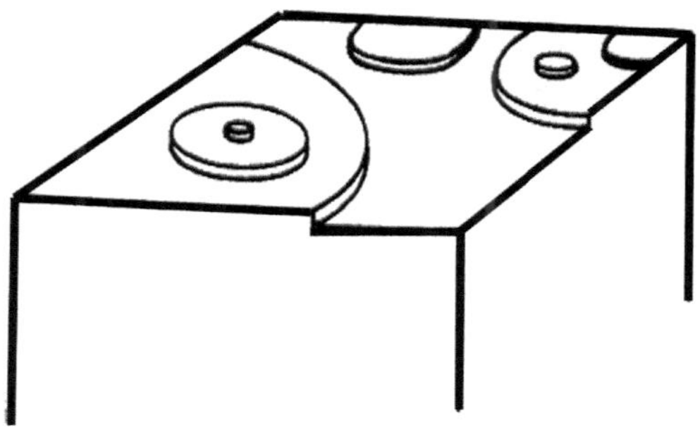

FIG. 4.21

Polynucleation growth in birth and spread theory.

Reprinted with permission from J.W. Mullin, Crystallization, Butterworth-Heinemann, Oxford, 2001. Copyright (2001), Elsevier.

- **Birth and spread theory**

Birth and spread theory (B+S theory), also known as 2D nucleation growth, is based on surface nucleation, followed by spread of the monolayers. Polynucleation and nuclei on nuclei are also observed. This happens when the nucleation rate is faster than the time required for the completion of a whole surface layer. Growth develops from surface nucleation at the corners and other spots on the crystal face, and the formed nuclei will then spread and coalesce to form layers [34,36], as shown in Fig. 4.21. This behavior is observed at high supersaturation values. Fig. 4.22 shows an AFM photo of the birth and spread growth mechanism in calcium carbonate.

4.5.3.2 Crystal growth rate

The growth rate of a crystal is often described by the change in crystals dimensions with time. The change of crystal dimension can be indicated by increasing mass of a crystal and the change in size. The growth rate can be expressed by the following equations:

$$R_{CG} = \frac{dL}{dt} \tag{4.66}$$

where dL is changing size of crystal; and dt is time

$$R_{CG} = K_G \Delta C^g \tag{4.67}$$

where K_G is overall growth rate constant, $\Delta C = C - C_e$, g is the order of the overall growth process.

4.5.3.3 Factors affecting crystal growth

Crystal growth rate is affected by various parameters such as supersaturation, temperature of the system, impurities, solution pH, and degree of agitation.

- The supersaturation is one of the most important parameters affecting rate of crystal growth. Eq. (4.69) indicates the dependence of rate of growth on supersaturation. Different. Spiral growth is dominant at low supersaturation and smoother surface, while 2D nucleation is observed at higher supersaturation, and at high supersaturation surface get rough and adhesive growth is dominant. This is represented in Fig. 4.23.

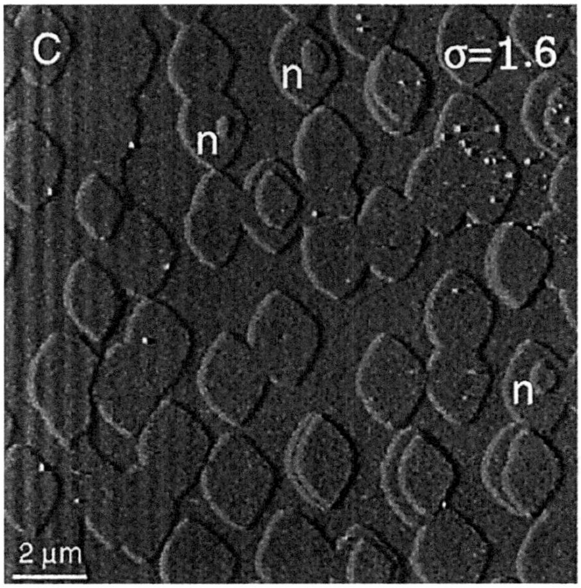

FIG 4.22

AFM of two dimensional nucleation growth.

Reprinted with permission from H.H. Teng, P.M. Dove, J.J. De Yoreo, Kinetics of calcite growth: surface processes and relationships to macroscopic rate laws, Geochim. Cosmochim. Acta, 64(13) (2000), pp. 2255–2266, https://doi.org/10.1016/S0016-7037(00)00341-0. Copyright (2000), Elsevier.

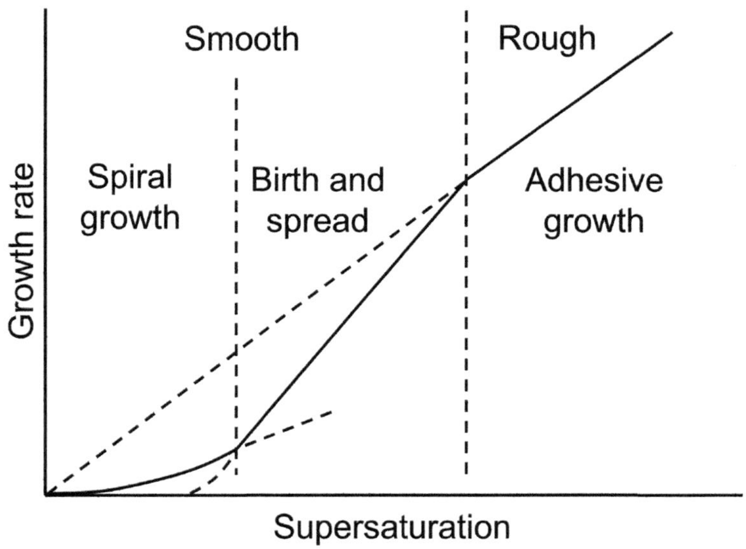

FIG. 4.23

Mechanisms of growth as a function of supersaturation.

Reprinted with permission from P. Cubillas, M.W. Anderson, Synthesis mechanism: crystal growth and nucleation, in: Zeolites and Catalysis: Synthesis, Reactions and Applications, vol. 1, Wiley-VCH, 2010, pp. 1–55, https://doi.org/10.1002/9783527630295.ch1. Copyright (2010), John Wiley and Sons.

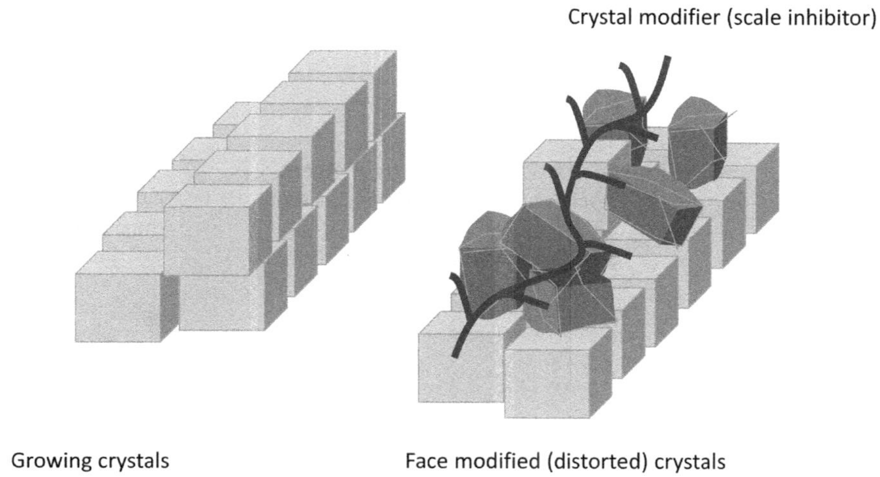

FIG. 4.24

Effect of impurities (e.g., chemical inhibitors) on growing crystals.

- Temperature can change the relative rates of diffusion and surface integration steps for crystal growth. At higher temperatures, the growth may be controlled by diffusion steps while at lower temperature growth it may be controlled by integration steps [53]. The growth rate also increases when temperature increases [83,84], while the crystal size, shape, and crystalline modification may all change with temperature [85].
- The presence of impurities in the solution. Some impurities can suppress growth entirely, while some impurities are selective, where they only affect certain crystallographic faces and modify the crystal habit, as shown in Fig. 4.24. Some impurities may, on the other hand, enhance crystal growth. The presence of impurities can influence crystal growth rate in a variety of ways [34].
  - Changing the properties of solution, the equilibrium saturation concentration, and the supersaturation.
  - Changing the characteristics of the adsorption layer at the crystal-solution interface.
  - It may be incorporated into the crystal's surface.

Generally speaking, the effect of impurities on crystal growth is crucial to understand and develop crystal modifiers that inhibit the growth to control deposits formation. The impurities inhibit the growth by crystal modification when they adsorb across the faces, at a kink, and at a step or ledge between steps blocking the growth sites.

4.5.4 Step (4) adhesion

The next step after nucleation and crystal growth is precipitate adhesion, or sticking to the piping and equipment surface. This is a critical step in solids deposition, since if solids were to flow suspended in the flow fluids without sticking to surfaces, the solids deposition problem would then be temporarily averted.

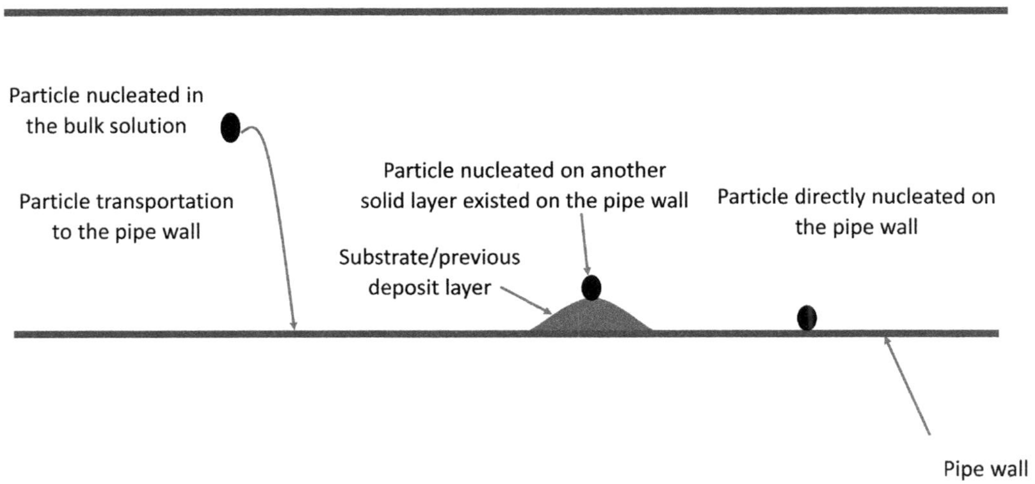

FIG. 4.25

Scenarios of particle deposition on pipeline wall.

Adhesion is generally the molecular attractions that hold the materials (here, the scale and the metallic surfaces) in close contact. Adhesion basically occurs when surfaces cling to one another by different mechanisms, e.g., mechanical interlocking, adsorption, chemical, electrostatic, diffusion, etc. [86]. It is considered a complex phenomenon. It can also be represented in terms of heterocoagulation and homocoagulation of particles [87].

Solids adhesion and deposit layers formation depend on where the solids nucleated and grew in the first place, i.e., on the pipe wall surface or in the bulk solution. These processes are also known as surface deposition or bulk precipitation, as depicted in Fig. 4.25.

In the general sense, the particle adhesion to the surface depends on many factors:

- Fluids properties

Basically fluids composition, level of supersaturation, viscosity, and density affect the adhesion and attachment of the particles.

- Surface properties

Surface properties such as surface energy, surface roughness, surface hydrophobicity, and wettability affect solids formation. The literature mentioned that high surface energy and high surface roughness are associated with high solids deposition, compared to low surface energy surfaces and smooth surfaces. High energy surfaces have high active sites for the solids molecules, and rough surfaces have high surface areas and high contact points with the solid substrate [88]. Yet, recent studies showed that surfaces with low surface energy, e.g., PTFE, can be more susceptible to solids than high surface energy surfaces, e.g., DLC [89]. Furthermore, no strong correlations were found between the surface roughness/hydrophobicity and the scaling deposition [90]. That leads to the fact that the surface properties cannot be seen alone without including the other factors, such as hydrodynamics, fluids composition, etc. Generally speaking, the understanding of such surface properties has been used to design and improve solids-repellent coatings or surfaces.

– Solids properties

Such as size, size distribution, density, shape, charge, etc. Such properties in addition to the fluid hydrodynamics will determine the deposition regime of the solids.

– Hydrodynamics

Type and rate of flow affect the solids adhesion. In turbulent flows like those in industrial operations, two competing events determine whether the solids will adhere to the walls or not: the transport of particles to pipe wall, which favors deposits adherence, and the shear effects that tend to strip off the formed solids and detach the solid layer. Increasing the turbulence was found to increase solids adhesion and deposition rates [91].

Theoretically speaking, the process of solids adhesion to surfaces involves multiple physical particle-fluid, particle-particle, and particle-surface interactions. Substantially, for the bulk precipitation, the solids adhesion to surfaces and deposition take place in two major steps: solid particles transport to a surface, followed by attachment of these solids to that surface, as depicted in Fig. 4.26. The particle transport describes particle-fluid interactions, such as particle-turbulence interactions, whereas the attachment process describes the physicochemical interactions between two bodies [92,93].

4.5.4.1 Particles motion and transport

When a solid particle is nucleated in the bulk of the solution, the common forces that affect particle motion in the fluid and its transport to the walls in a fluid include drag forces, gravitational forces, shear dispersion, shear stripping, lifting forces (Saffman lifting forces), thermophoretic forces, electrical/electrostatic forces, eddy diffusion (turbulence diffusion), Brownian diffusion, particle diffusion, advection, inertial forces, and others [94]. Fig. 4.27 illustrates some of these mechanisms.

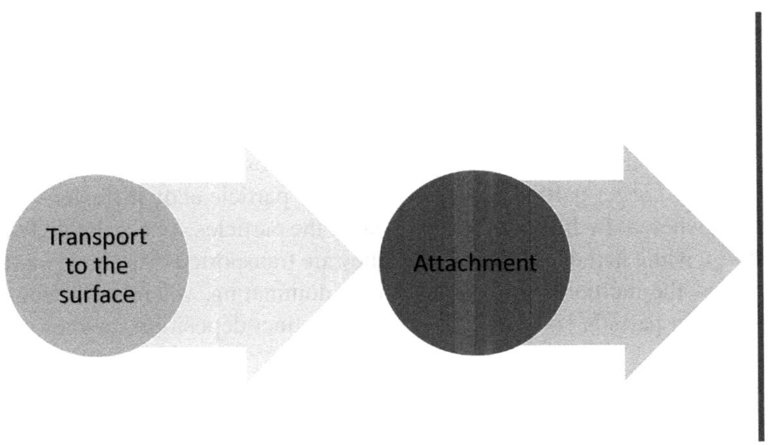

FIG. 4.26

Processes involved in particle adhesion.

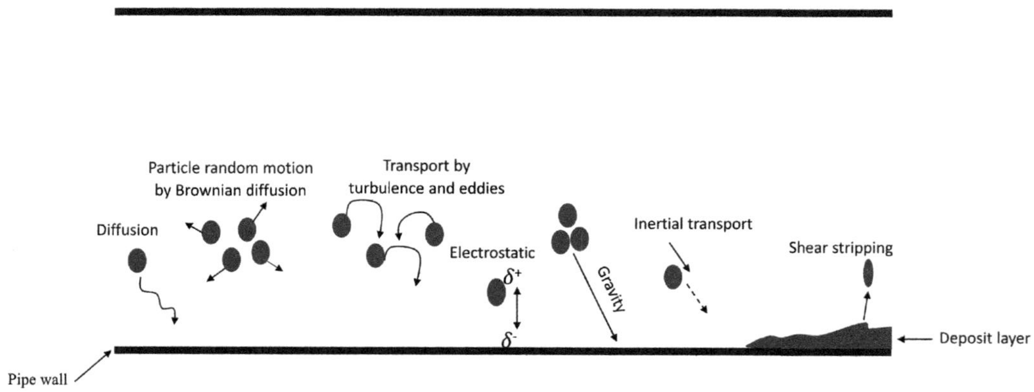

FIG. 4.27

Schematic of the different mechanisms involved in particle transport to the pipe wall.

Collins [93] listed six main forces by which solids transport to the pipe wall and deposit:

1. Particle motion due to electrostatic forces.
2. Deposition owing to gravity forces (sedimentation).
3. Deposition caused by shear dispersion and shear forces.
4. Brownian diffusion.
5. Transport owing to inertial forces.
6. Eddy diffusion.

Some of these forces can apply under a specific set of conditions and a combination of different forces can apply at the same time. For example, in turbulent flow, turbulent diffusion and advection have the heist effect, while diffusion will have the highest effect in laminar flow [91]. Gravitational forces will have a significant impact in horizontal pipes and in surface equipment, especially with high concentrations of suspended solids [93].

The deposition regime depends on the geometry, surface charge, and size of the solids, and the flow regime of the fluid. For very small particles, particle diffusion is the dominant mechanism. In the boundary layer, particles are transported predominantly by diffusion, which is enhanced by the turbulence, whereas in the thin laminar region adjacent to the wall, Brownian diffusivity alone transports the particles toward the wall. Larger particles are transported by particle eddy diffusion-impaction as the dominant mechanism, whereas in the outer boundary layer the particles are accelerated toward the wall by the turbulent eddies, while in the viscous sublayer they are transported by particle inertia to the wall. For large size particles, the inertial deposition regime is dominating, where wall deposition velocity decreases with increasing particle size [95]. These three distinct deposition regimes are presented in Fig.4.28 [96].

Considering a particle in motion with a flowing fluid, the deposition velocity is defined by Eq. (4.68):

$$V_{\text{dep}+} = \frac{J_{\text{wall}}}{\rho_p u} \tag{4.68}$$

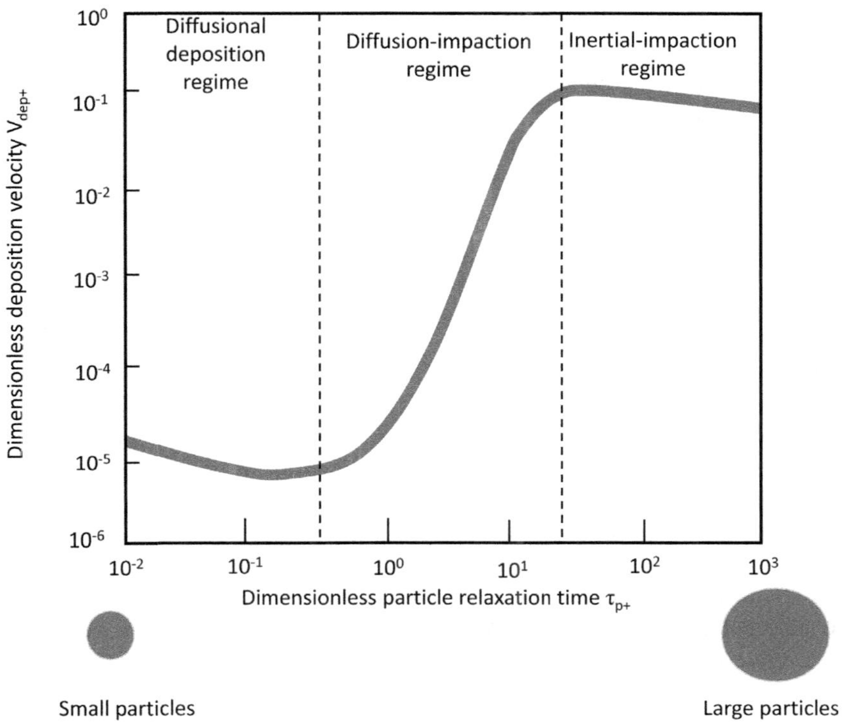

FIG. 4.28

Particle deposition regimes.

Adapted with permission from J. Young, A. Leeming, A theory of particle deposition in turbulent pipe flow, J. Fluid Mech. 340(June) (1997), 129–59, https://doi.org/10.1017/S0022112097005284. Copyrights (1997), Elsevier.

where V_{dep+} is the deposition velocity, J_{wall} is the mass flux of solids, ρ_p is bulk density of the particles, u is the friction velocity ($u^2 = Tw/\rho_l$, where Tw is the wall shear stress and ρ_l is the liquid density). The dimensionless particle relaxation time τ_{p+} is defined by:

$$\tau_{p+} = \frac{\tau_p u^2}{\eta_l} \tag{4.69}$$

where η_l is the liquid viscosity and τ_p is the particle relaxation time, which can be calculated from the following equation:

$$\tau_p = \frac{\rho_p d_p^2}{18\eta_l} \tag{4.70}$$

where ρ_p is the particle material density, d_p is the particle diameter, and η_l is the liquid viscosity. The particle relaxation time is a measure of particle inertia and denotes the time scale with which any slip velocity between the particles and the fluid is equilibrated.

For small particles, net particle flux toward the wall is determined by Fick's law of diffusion. In the diffusion-impaction regime, the deposition rate dramatically increases with increasing the size of the particles, due to the interaction between particles having significant inertia and the fluid turbulent eddies [96]. In this regime, the free-flight or stop-distance model was suggested by Friedlander and Johnstone [97], where particles are transported by gradient diffusion to within one "stop-distance" of the wall, where they acquire sufficient inertia to "coast" across the viscous sublayer [98].

In the inertia-moderated regime, the massive particles acquire sufficient momentum from the eddies in turbulent core and directly reach the pipe wall. In this regime, increasing the particle inertia decreases its response to turbulence [96].

4.5.4.2 Attachment

After the particles are transported to the surface, attachment takes place. At the particle-surface interface, different interactions that are governed by different forces are possible. The attachment usually takes place, leading to deposition, after passing the specific energy barrier for these interactions, such that if the energy barrier is sufficiently high, it may reduce or prevent deposition entirely. The total of the interaction forces between the particle-surface in the fluid medium are the surface forces, while those that work to hold them are known as the adhesive forces [99]. Generally, there are four essential deformation effects in particle-surface contacts: elastic, plastic, viscoelastic, and viscoplastic contact deformations. If the deformation is perfectly elastic, where no energy dissipates during the interaction, the adhesive and surface forces are equal in magnitude [99]. In general, a particle deforms elastically to increase the contact area—and thus the attractive interaction—with the surface [100].

The particle-surface interactions have been studied largely using AFM methods [92]. These interactions include van der Waals, electrostatic, acid-base, solvation, steric forces, and other forces. They are described or categorized under DLVO, and extended DLVO (x-DLVO) theories in addition to heterogeneities; these interactions are illustrated in Fig 4.29.

The DLVO theory is the most common theory to describe the attachment of particles on surfaces. It was named after Derjaguin-Landau and Verwey-Overbeek [101,102]. The DLVO theory considers that interbody interactions are the sum of the van der Waals (VDW) interaction and of the electrostatic double-layer (EDL) interaction [92,103]. Both van der Waals and electrostatic forces are depicted in Fig. 4.30. By convention, attractive forces are negative and repulsive forces are positive.

The van der Waals (VDW) forces are mostly attraction forces, corresponding to low strength interactions between atoms, molecules, or bodies, comprising dipole-dipole and dispersion forces [92]. Two approaches are used to express the van der Waals interactions between two bodies: the Hamaker approach (or microscopic approach) and the Lifshitz approach (also called the macroscopic approach) [92]. Electric double-layer (EDL) is formed when a charged surface exists in an electrolyte solution due to the electrostatic affinity of the ions. EDL can be repulsive (similarly charged particles) or attractive forces (oppositely charged particles). The VDW predominate in small and large separation distances, while EDL repulsions can dominate in medium separation distances, as illustrated in Fig. 4.31. In this figure, the peak in the DLVO curve is the energy barrier, E_{barr}, which can prevent incoming particles from reaching the surface [92].

The DLVO theory successfully describes the long-range interactions but is not suitable for shorter separation distances (a few nanometers) [92]. The extended DLVO theories account for other forces between the bodies, such as solvation forces (or hydration forces in water), acid-base forces, the ion bridging, Brownian motion, magnetic forces, steric forces, capillary forces, and others

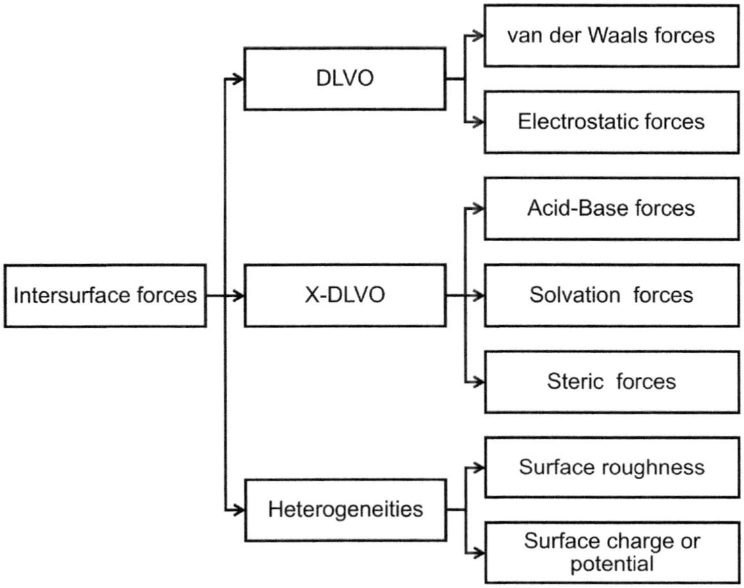

FIG. 4.29

The various components at play in surface forces.

Reprinted with permission from C. Henry, J.P. Minier, G. Lefèvre, Towards a description of particulate fouling: from single particle deposition to clogging, Adv. Colloid Interface Sci. (2012), https://doi.org/10.1016/j.cis.2012.10.001. Copyright (2012), Elsevier.

[92,93]. In bifouling, the adhesion of microorganisms can be strongly dependent on their external appendages [103].

Furthermore, the nature of the surface, e.g., smooth or rough, affects the adhesion process. Surface roughness is the irregularities in the surface. Particles can be entrapped in crevices and cracks, and nonspherical particles can migrate along the surface to find a position of larger optimum contact. On the other hand, if the surface presents a microroughness, the minimal number of contact points may reduce the possibility of adhesion [103]. DLVO theory describes adhesion on perfectly smooth surfaces.

The net effect is a balance between all possible interactions [103]. The total interaction energy ΔG_T between a deposit and a metal surface is given as the sum of the individual interaction parameters, as shown in Eq. (4.71) [93], where adhesion and deposition will take place when ΔG_T is negative.

$$\Delta G_T = \Delta G_W + \Delta G_E + \Delta G_A + \Delta G_B \tag{4.71}$$

where ΔG_w denotes the Lifshitz-van der Waals interaction component, ΔG_E the electrostatic double-layer component, ΔG_A the Lewis acid-base component, and ΔG_B the Brownian motion component.

In addition, the adsorption forces and the cohesive forces must also be taken into consideration. For example, certain systems can have ΔG_T positive and deposition still occurs. That may be attributed to the cohesive forces, which make the particles attract each other, coagulate, and increase in size, thus leading to surface adhesion after converting the ΔG_T value into a negative value [93].

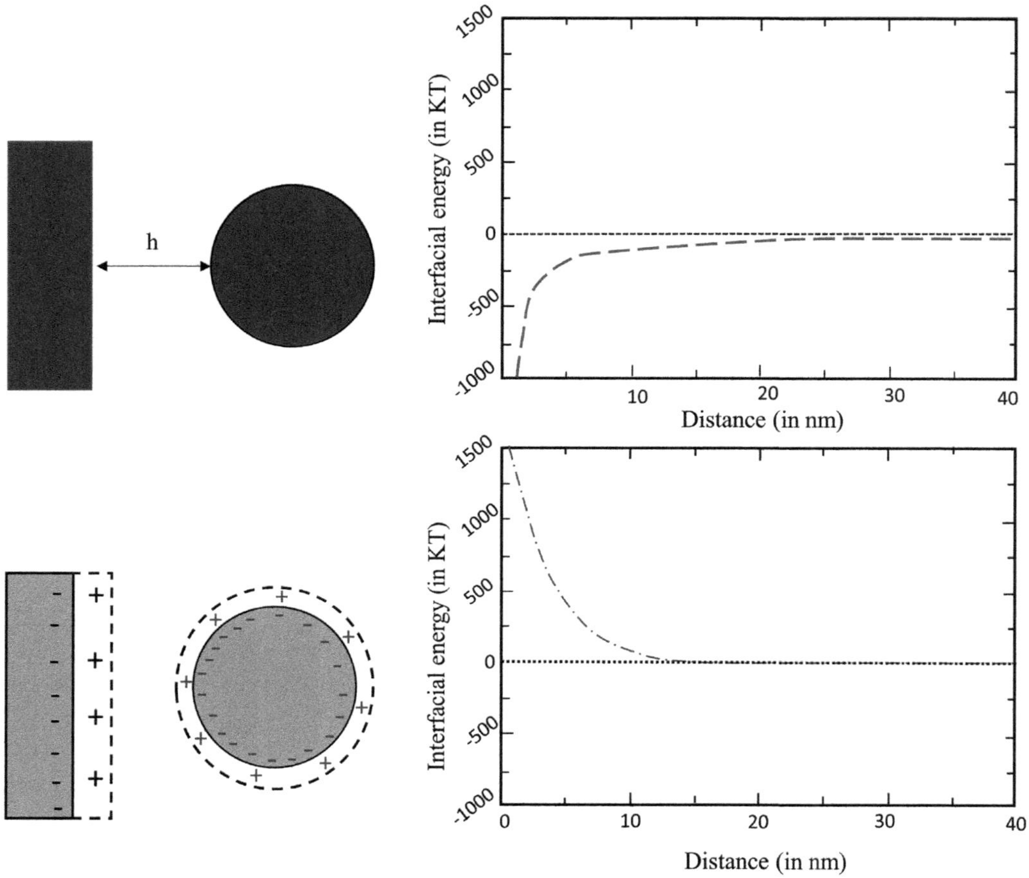

FIG. 4.30

DLVO forces: van der Waals forces (top) and electrostatic double-layer forces (bottom).

Reprinted with permission from C. Henry, J.P. Minier, G. Lefèvre, Towards a description of particulate fouling: from single particle deposition to clogging, Adv. Colloid Interface Sci. (2012), https://doi.org/10.1016/j.cis.2012.10.001. Copyright (2012), Elsevier.

4.5.5 Step (5) aging

After solids have precipitated and deposited on the surface, multiple transformations to the deposited solid crystals take place. The transformations can be changes in the mechanical properties or in the chemical composition.

4.5.5.1 Aging of mineral scales

Aging of mineral scales is controlled by the hydrodynamics and chemical changes that may lead to modifications in texture, morphology, or the composition of the formed scales. Figs. 4.32 and 4.33 illustrate the effect of hydrodynamics on the morphology of scale of the same composition.

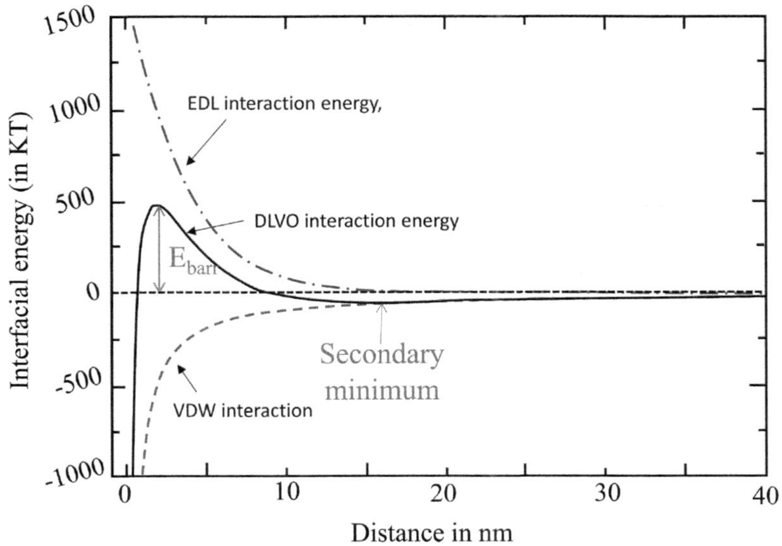

FIG. 4.31

DLVO interaction energy between a particle and a surface.

Reprinted with permission from C. Henry, J.P. Minier, G. Lefèvre, Towards a description of particulate fouling: from single particle deposition to clogging, Adv. Colloid Interface Sci. (2012), https://doi.org/10.1016/j.cis.2012.10.001. Copyright (2012), Elsevier.

FIG. 4.32

Scale deposit collected from well tubing. The scale has two distinct layers (marked with the arrows) that have the same composition: $CaCO_3$, Ba SO_4, Sr SO_4, and $CaSO_4$.$2H_2O$, however their texture and morphology are different, as the outer layer facing the tubing surface (indicated by *blue arrow* (dark gray in print versions)) is amorphous and loose, while the inner layer facing the liquids (indicated by the *yellow arrow* (light gray in print versions)) is dense and extremely hard.

FIG. 4.33

Scale deposits collected from 18″ pipeline both (a) and (b) have the same composition ($CaCO_3$, $BaSO_4$, $SrSO_4$, Fe compounds), with different texture and strength where sample (a) was collected from upstream the pipe (where flow is turbulent and high shear exists) and has extremely high strength, sample (b) was collected from downstream the pipeline (where flow is relatively laminar with low shear), and exhibit layering and less hard than sample (a).

Flowing fluid can erode and slough freshly formed and loose scale layers; it can also dissolve some of the formed scale, ion exchange with flowing fluids, and then reprecipitate them. Cleaver and Yates [104] showed that solids removal depends on wall shear stress. The removal takes place in turbulent bursts depending on the flow rates, and when the shear stress is above critical value, it suggests that scale is preferably formed in low shear locations.

On the other hand, the solids that are formed under high shear flowing fluids are characterized by being compact and hard with improved mechanical properties. Quddus and Allam [105] showed that deposition rates increased with increasing Reynolds numbers (Re), suggesting that once a scale layer is formed, the growth is more rapid when the shear rate is high. Garcia et al. [106] show the same tendency with $CaCO_3$, at least under laminar flow [107].

4.5.5.2 Aging of organic deposits

The formed organic deposits are usually porous and contain amounts of liquids. The aging takes place by decreasing the liquid content in the formed deposit and increasing the solid forming components content. Depending on the type of the deposit, the mechanical properties and the composition of the deposits change during aging.

For gas hydrates, the formed hydrates are usually porous with high volume. The porosity of the formed hydrates depends on the subcooling conditions and temperature gradient between the bulk solution and the surface [108,109]. Under the shearing effect of the flowing fluids, gas hydrates collapse and hydrate sloughing occurs. Sloughing occurs due to a combination of the effects of shear, gravitational forces, and deposit integrity. The debris of collapsed hydrate deposit can then agglomerate, leading to increasing the viscosity, and increasing the pressure drop, ultimately resulting in forming hydrate plugs that clog the flowlines and cause shut in [110]. As the plug sits for a longer period of time, the masses anneal and the plug becomes more solid-like, with less flexibility.

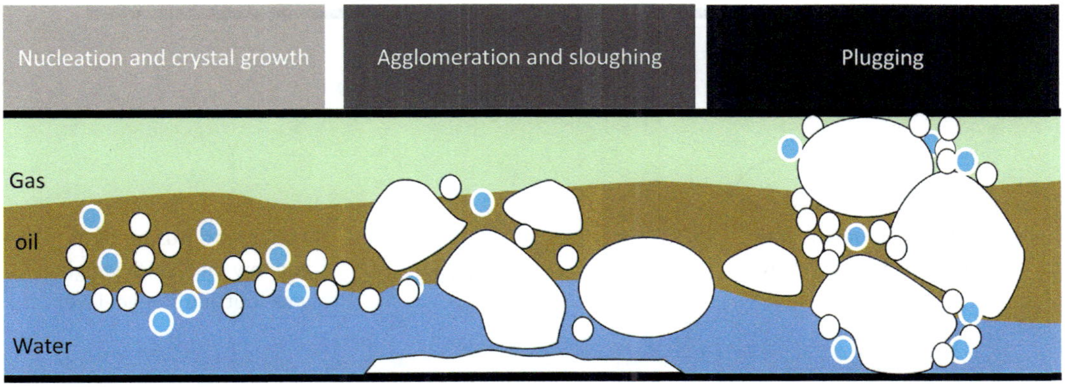

FIG. 4.34

Generalized mechanism of hydrates formation.

That is, both intra- and interparticle growth occurs [60]. Fig. 4.34 illustrates the generalized mechanism of hydrates formation [110].

In paraffin waxes, the aging step means hardening the formed incipient gel layer. That principally occurs by increasing the content of the high molecular weight wax molecules in the formed 3D network or gel layer. Such event takes place by molecular diffusion, which is considered the most acceptable mechanism that explains wax deposition. During wax deposits hardening, the dissolved wax molecules diffuse from the bulk of the solution toward the gel layer, depositing and adding on the gel layer, and diffusing into the gel layer. On the other hand, the liquid oil entrapped in the gel layer and the low molecular weight wax counter diffuse out of the gel layer. This increases the deposit layer wax content and increases its strength. This diffusion process usually depends on the temperature gradient between the bulk oil and the pipe wall surface, oil composition, and the concentration of the high molecular weight wax molecules. The aging of wax deposits is illustrated in Fig. 4.35.

In the case of asphaltenes, it is evident that they undergo aggregation to end up forming primary particles that are on the order of a few hundred nanometers. These primary particles aggregate further, forming microaggregates $> 1\,\mu m$ in size; these microaggregates age and harden to form the solid-like solids. During the aging of asphaltenes, the amorphous particles lose the entrapped oil and liquids inside them and gain some strength [111]. Asphaltenes aging is illustrated in Fig. 4.36. Rogel et al. [112] showed that the precipitation of asphaltenes is a dynamic process in which the composition of the deposit shows it can vary with aging, due to the exchange of molecules between the deposit and fluid. For example, the fractions that are more aromatic and richer in heteroatom molecules are less soluble, and hence they are enriched in the deposit as it ages, whereas those that are less aromatic with fewer heteroatoms redissolve in the oil. This might affect asphaltene solubility during the field cleaning operations, as Wong et al. [113] reported that aged asphaltened deposits are hard to remove using chemical dissolvers like toluene.

In the case of naphthenate soaps, they accumulate at the interface between water and oil, stabilizing the emulsions. These soaps, especially the calcium naphthenates, harden when exposed to air and also when they are cooled down. Besides, the emulsion forms further accumulate more solids in their thick matrix and harden with time, making these soaps difficult to remove.

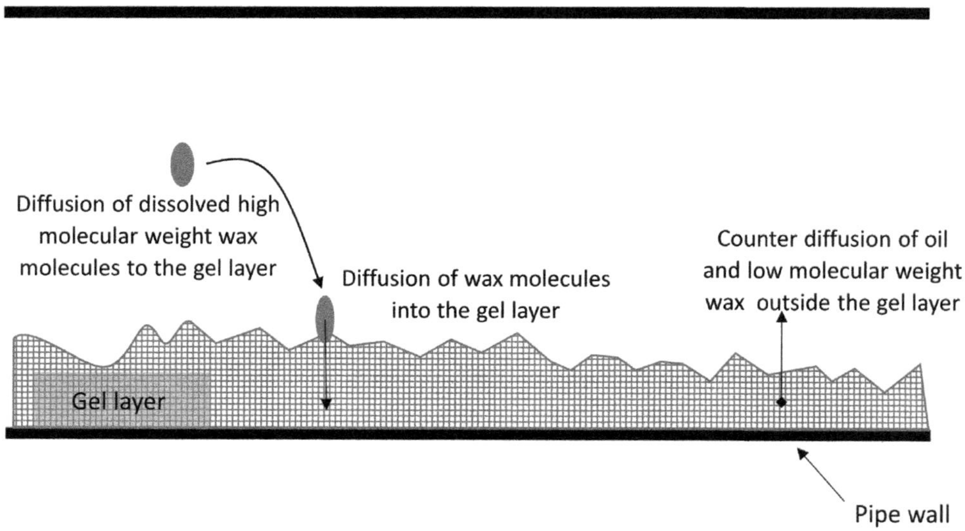

FIG. 4.35

Aging of wax deposits.

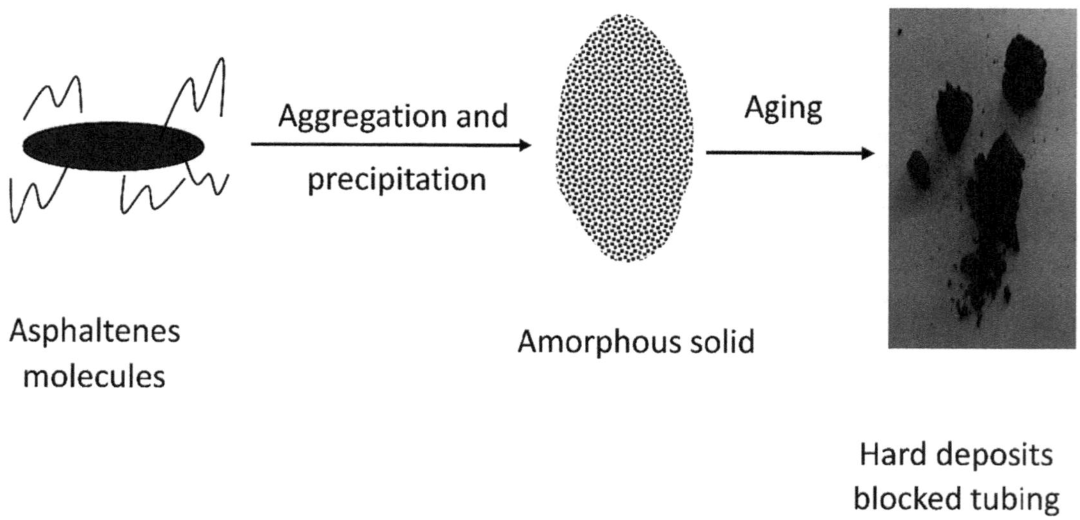

FIG. 4.36

Aging of precipitated asphaltenes particles.

4.5.5.3 Factors affecting aging

The factors affecting the aging process are mainly hydrodynamic parameters. Some of the factors involved include flow rates, flow velocity, flow regimes, shear stress, and hydrodynamic factors that affect heat and mass transfer. In addition, system design and layout can drastically affect these hydrodynamic factors, which consequently affects the aging process.

4.6 Fouling

As mentioned earlier in this chapter, fouling is a general term that describes the build-up of unwanted deposits on the internal surfaces of tubulars, equipment, and other production or manufacturing installations. Sometimes, fouling gets mixed up with scaling, making scaling a general term while fouling is limited to particulate deposition. That actually stems from the conventional use of the word scale to refer to kinds of deposits in oil and gas fields, including organic and inorganic. However, the opposite is true: that fouling is a wider and more generic term of solids precipitation and deposition on surfaces, and that scaling is just a type of fouling. The term is widely used in different industries, including food processing, dairy, pulp and paper, water treatment, pharmaceuticals, oil and gas, etc. Saying that, almost any solid or semisolid material can become a foulant. Some of the industrially encountered foulants include the following: inorganic materials; airborne dusts and grit; waterborne mud and silts; calcium and magnesium salts; iron oxide; organic materials; biological substances, e.g., bacteria, fungi, and algae; oils; waxes and greases; heavy organic deposits, e.g., polymers, tars, and carbon [114].

Generally speaking, fouling can be (1) particulate, (2) precipitation, (3) corrosion, (4) biofouling, and (5) chemical reaction.

- Particulate fouling:

This involves the accumulation of solid particles suspended in the process fluids onto the surface. Examples include ashes, dust, fines, and other types of particulates in the different equipment. Sands and fines that are transported with reservoir fluids to deposit in the tubing and on different production equipment and installations are a common example in the oil and gas fields.

- Precipitation fouling (Sedimentation fouling)

This involves the crystallization process, as mentioned in this chapter, such as scale formation.

- Chemical reaction fouling

In this type, the deposits are formed due to chemical reaction, such as polymerization, and decomposition of organic materials.

- Corrosion fouling

In this type, fouling takes place due to corrosion process.

- Biological fouling

This type takes place when microorganisms, macrorganisms, or their metabolic products accumulate on the surfaces.

4.7 Recent advances in solid deposit formation mechanism research

Solid formation and deposition on surfaces have experienced several updates in the last decades. These advances are associated with improving the mitigation methods of solid deposits. Nonclassical theories of nucleation have emerged to change the classical view of how the clusters of deposits forming species aggregate to form the crystal. It is clear that mineral scales nucleate through intermediate nanosized particles. This can be used for better understanding of the kinetics of the process, and hence developing more effective chemical inhibitors. Similarly, several advanced views of gas hydrates nucleation have emerged as well, changing the typical view of the labile cluster theory. The advances in crystal growth and how the impurities interact with the crystal faces have led to improvements in the crystal modifier types of the different inhibitors. The advances in understanding the intermolecular forces that govern the solids adhesion to surfaces lead to developing different solutions, such as deposit repellent surfaces and the chemical dispersants.

Solids deposition is usually studied in flow loops, static tests, rotating discs, and other deposition equipment. Recently, the use of microfluidic technology has been a game changer, due to its accuracy, low volume samples, availability at high temperature and pressures, and other important advantages.

4.8 Summary

In this chapter, the flow assurance solids formation and deposition were discussed. Solids formation involves supersaturation, nucleation, crystal growth, adhesion, and aging. Supersaturation occurs when the dissolved species exceed their limit of solubility solution. The nucleation is the first step in new phase formation that takes place when a cluster of the chemical species assembles and grows into a critical size known as the critical nuclei. After reaching the critical size, the formed crystal keeps growing by successive addition of the growth units. If the nucleation and crystal growth started on the pipeline wall surface, then it will continue growing until a full layer of solids is formed on the surface. If the solid was formed in the bulk solution, the crystals have to be transported to the pipe wall before they are attached to the pipe wall surface. A good understanding of these steps is critical to develop vital methods for mitigating their formation.

References

[1] T.A. Hoang, Mechanisms of scale formation and inhibition, in: Mineral Scales and Deposits: Scientific and Technological Approaches, Elsevier, 2015, pp. 47–83, https://doi.org/10.1016/B978-0-444-63228-9.00003-6.

[2] J.C. Cowan, D.J. Weintritt, Water Formed Scale Deposits, Gulf Publishing Company, TX, USA, 1976.

[3] A. Kopteva, V. Starshaya, Radioisotope Measuring System for Oil Stream Asphaltene-Resin-Paraffin Deposits ARPD Parameters (Russian), Society of Petroleum Engineers, 2016, October 24, https://doi.org/10.2118/182104-RU.

[4] G.A. Mansoori, Remediation of asphaltene and other heavy in oil wells and in pipelines, J. Reserv. Pet. Eng. (2010).

[5] G. Amy, Fundamental understanding of organic matter fouling of membranes, Desalination 231 (2008) 44–51, https://doi.org/10.1016/j.desal.2007.11.037.

[6] J.S. Gudmundsson, Inorganic scale. The salt-like solids, in: Flow Assurance Solids in Oil and Gas Production, CRC Press, 2017, pp. 159–199, https://doi.org/10.1201/9781315185118-3.

[7] J. Santos, da S.T. dos, Antonio C. Fernandes, and Marco Giulietti., Study of the paraffin deposit formation using the cold finger methodology for brazilian crude oils, J. Pet. Sci. Eng. 45 (1–2) (2004) 47–60, https://doi.org/10.1016/j.petrol.2004.05.003.

[8] F.M. Vargas, J.L. Creek, W.G. Chapman, On the development of an asphaltene deposition simulator, Energy Fuel 24 (2010) 2294–2299, https://doi.org/10.1021/ef900951n.

[9] C. Ruiz-Agudo, E. Ruiz-Agudo, C.V. Putnis, A. Putnis, Mechanistic principles of barite formation: from nanoparticles to micron-sized crystals, Cryst. Growth Des. 15 (8) (2015) 3724–3733, https://doi.org/10.1021/acs.cgd.5b00315.

[10] M.K. Banerjee, Fundamentals of heat treating metals and alloys, in: Comprehensive Materials Finishing, vol. 2–3, Elsevier, 2016, pp. 1–49, https://doi.org/10.1016/B978-0-12-803581-8.09185-2.

[11] P.G. Koutsoukos, Water chemistry and its role in industrial water systems, in: Mineral Scales and Deposits, Elsevier, 2015, pp. 25–45, https://doi.org/10.1016/B978-0-444-63228-9.00002-4.

[12] N.S. Tavare, Industrial Crystallization Process Simulation Analysis and Design, Springer Science+Business Media, Llc, 1995.

[13] N.S. Tavare, Mixing, reaction and precipitation: the barium sulfate precipitation system, in: B.S. Gupta, S. Ibrahim (Eds.), Mixing and Crystallization, Kluwer Academic Publishers, 2000, pp. 191–203.

[14] M. Havenith-Newen, G. Schwaab, F. Sebastiani, Ion hydration and ion pairing as probed by THz spectroscopy, Angew. Chem. Int. Ed. (2018), https://doi.org/10.1002/anie.201805261.

[15] C.C. Patton, Applied Water Technology, first ed., Campbell Petroleum Series, 1987. October 1987.

[16] B. Averill, P. Eldredge, General Chemistry: Principles, Patterns, and Applications 1.0, Flat World Education, 2016, pp. 1–12. http://catalog.flatworldknowledge.com/bookhub/4309?e=averill_1.0-ch24_s05.

[17] F. Civan, Reservoir Formation Damage: Fundamentals, Modeling, Assessment, and Mitigation, third ed., Elsevier, 2015, https://doi.org/10.1016/C2014-0-01087-8.

[18] T.-H. Chung, Thermodynamic Modeling for Organic Solid Precipitation, Society of Petroleum Engineers, 1992, January 1, https://doi.org/10.2118/24851-MS.

[19] C.I. Oseghale, E.J. Akpabio, Managing Paraffin Wax Deposition in Oil Wells—Related Problems in Nigerian Oil Fields, Society of Petroleum Engineers, 2012, January 1, https://doi.org/10.2118/163037-MS.

[20] A. Hammami, A.K. Mehrotra, Non-isothermal crystallization kinetics of binary mixtures of n-alkanes: ideal eutectic and isomorphous systems, Fuel 75 (4) (1996) 500–508.

[21] R.G. Moghanloo, D. Davudov, E. Akita, Formation Damage by Organic Deposition. Formation Damage During Improved Oil Recovery, 2018, pp. 243–273, https://doi.org/10.1016/b978-0-12-813782-6.00006-3.

[22] S.A. Beg, E. Hamad, N.M. Tukur, Solubility studies of paraffin waxes in selected petroleum fractions and toluene, Fuel Sci. Technol. Int. 11 (2) (1993) 363–381, https://doi.org/10.1080/08843759308916073.

[23] M. Freund, R. Csikós, S. Keszthelyi, G.Y. Mózes, Chemical, crystallographical and physical properties of liquid paraffins and paraffin waxes, Dev. Pet. Sci. 14 (C) (1982) 13–140, https://doi.org/10.1016/S0376-7361(08)70146-8.

[24] G.M. Pietraru, D.T. Cramb, Changes in asphaltene microenvironments evidenced by fluorescence solvatochromism, Langmuir 19 (4) (2003) 1026–1035, https://doi.org/10.1021/la026247m.

[25] A. Yen, Y.R. Yin, S. Asomaning, Evaluating asphaltene inhibitors: laboratory tests and field studies, in: Paper SPE #65376-MS Presented at the SPE Int. Symp. Oilfield Chem. 2001,Houston, Texas, 2001.

[26] D. Kashchiev, A. Firoozabadi, Driving force for crystallization of gas hydrates, J. Cryst. Growth 241 (1–2) (2002) 220–230, https://doi.org/10.1016/S0022-0248(02)01134-X.

[27] R.J. Ferguson, B.R. Ferguson, The Chemistry of Strontium and Barium Scales, Association of Water Technologies, 2010, pp. 1–17.

[28] Y. Zhang, E. Isaj, Halite Envelope for Downhole Salt Deposition Prediction and Management, Society of Petroleum Engineers, 2015, June 3, https://doi.org/10.2118/174206-MS.

[29] J. Mand, H.S. Park, T.R. Jack, G. Voordouw, The role of acetogens in microbially influenced corrosion of steel, Front. Microbiol. 5 (June) (2014) 1, https://doi.org/10.3389/fmicb.2014.00268.

[30] R.W. Schurko, Systematic Treatment of Equilibrium, Retrieved from http://chem320.cs.uwindsor.ca/Notes_files/320_l07.pdf. Last visited Jun 2019.

[31] H. Műller-Steinhagen, C.A. Branch, Calcium carbonate fouling of heat transfer surfaces, in: Proceedings of CHEMECA 1988, Sydney, 1988.

[32] A.B. Demir, H.I. Bilgesu, B. Hascakir, The Effect of Clay and Salinity on Asphaltene Stability, Society of Petroleum Engineers, 2016, May 23, https://doi.org/10.2118/180425-MS.

[33] F. Shojaati, S.H. Mousavi, M. Riazi, F. Torabi, M. Osat, Investigating the effect of salinity on the behavior of asphaltene precipitation in the presence of emulsified water, Ind. Eng. Chem. Res. 56 (48) (2017) 14362–14368, https://doi.org/10.1021/acs.iecr.7b03331.

[34] J.W. Mullin, Crystallization, Butterworth-Heinemann, Oxford, 2001.

[35] D. Kashchiev, Nucleation, Basic Theory With Applications, Butterworth-Heinemann, 2000.

[36] P. Cubillas, M.W. Anderson, Synthesis mechanism: crystal growth and nucleation, in: Zeolites and Catalysis: Synthesis, Reactions and Applications, vol. 1, Wiley-VCH, 2010, pp. 1–55, https://doi.org/10.1002/9783527630295.ch1.

[37] R.V. Davis, P.W. Carter, M.A. Kamrath, D.A. Johnson, P.E. Reed, The use of modern methods in the development of calcium carbonate inhibitors for cooling water systems, in: Mineral Scale Formation and Inhibition, Springer, US, 1995, pp. 33–46, https://doi.org/10.1007/978-1-4899-1400-2_4.

[38] M. Crabtree, D. Eslinger, P. Fletcher, M. Miller, A. Johnson, G. King, Fighting scale—removal and prevention, Oilfield Rev. 11 (1999) 30–45.

[39] E. Sanz, C. Vega, J.R. Espinosa, R. Caballero-Bernal, J.L.F. Abascal, C. Valeriani, Homogeneous ice nucleation at moderate supercooling from molecular simulation, J. Am. Chem. Soc. 135 (40) (2013) 15008–15017, https://doi.org/10.1021/ja4028814.

[40] B.C. Knott, V. Molinero, M.F. Doherty, B. Peters, Homogeneous nucleation of methane hydrates: unrealistic under realistic conditions, J. Am. Chem. Soc. 134 (2012) 19544–19547.

[41] D. Gebauer, How can additives control the early stages of mineralisation? Minerals (2018), https://doi.org/10.3390/min8050179.

[42] D. Gebauer, P. Raiteri, J.D. Gale, H. Cölfen, On classical and non-classical views on nucleation, Am. J. Sci. 318 (9) (2018) 969–988, https://doi.org/10.2475/09.2018.05.

[43] S. Karthika, T.K. Radhakrishnan, P. Kalaichelvi, A review of classical and nonclassical nucleation theories, Cryst. Growth Des. 16 (11) (2016) 6663–6681, https://doi.org/10.1021/acs.cgd.6b00794.

[44] E. Ariyanto, Crystallisation and Dissolution Studies of Struvite in Aqueous Solutions, 2013. This Thesis is Presented for the Degree of Doctor of Philosophy of Curtin University.

[45] W. Ke, T.M. Svartaas, D. Chen, A review of gas hydrate nucleation theories and growth models, J. Nat. Gas Sci. Eng. (2019), https://doi.org/10.1016/j.jngse.2018.10.021.

[46] V.A. Garten, R.B. Head, Homogeneous nucleation and the phenomenon of crystalloluminescence, Philos. Mag. A 14 (132) (1966) 1243–1253.

[47] N.F. Otpushchennikov, Determination of the size of the elementary crystal nucleus, from acoustic measurement, Crystallogr. Rep. 7 (1962) 237–240.

[48] T. Adamski, Commination of crystal nucleation by a precipitation method, Nature 197 (4870) (1963) 894.

[49] A.G. Walton, The nucleation of sparingly soluble salts from solution, Anal. Chem. Acta 29 (1963) 434–441.

[50] J.H. van der Merwe, Misfitting monolayers and oriented overgrowth, CRC Crit. Rev. Solid Stat. Mater. Sci. 7 (1978) 209–225.

[51] M.A. Herman, Crystallization of epitaxial layers: fundamentals and basic problems, in: K. Sangwall (Ed.), Elementary Crystal Growth, Saan Publ, Lublin, 1994, pp. 377–407.

[52] D. Kim, J. Kim, H.D. Ryu, S.I. Lee, Effect of mixing on spontaneous struvite precipitation from semiconductor wastewater, Bioresour. Technol. 100 (2009) 74–78.

[53] A.G. Jones, Crystallization Process System, Butterworth-Heinemann, Oxford, 2002.

[54] T.P. Melia, Crystal nucleation from aqueous solution, J. Appl. Chem. 15 (1965) 345–357.

[55] R.F. Strickland-Constable, Kinetics and Mechanism of Crystallization, Academic Press, New York, 1968.

[56] O. Söhnel, J. Garside, Precipitation: Basic Principles and Industrial Applications, Butterworth-Heinemann, Oxford, 1992.

[57] A.D. Randolph, M.A. Larson, Theory of Particulate Processes: Analysis and Techniques of Continuous Crystallization, Academic Press, San Diego, 1971.

[58] L.C. Jacobson, W. Hujo, V. Molinero, Amorphous precursors in the nucleation of clathrate hydrates, J. Am. Chem. Soc. 132 (2010) 11806–11811.

[59] B. Arjun, T. A., & Bolhuis, P. G., Unbiased atomistic insight in the competing nucleation mechanisms of methane hydrates, Proc. Natl. Acad. Sci. USA 116 (39) (2019) 19305–19310, https://doi.org/10.1073/pnas.1906502116.

[60] E.D. Sloan, C.A. Koh, Clathrate Hydrates of Natural Gases, third ed., CRC Press, 2008.

[61] T. Zhu, J.A. Walker, J. Liang, Evaluation of wax deposition and its control during production of Alaska North slope oils, 2008, https://doi.org/10.2172/963363. United States https://www.osti.gov/servlets/purl/963363.

[62] K.S. Wang, C.H. Wu, J.L. Creek, P.J. Shuler, Y. Tang, Evaluation of effects of selected wax inhibitors on wax appearance and disappearance temperatures, in: Petroleum Science and Technology, vol. 21, Marcel Dekker, 2003, pp. 359–368, https://doi.org/10.1081/LFT-120018525.

[63] K. Akbarzadeh, H. Alboudwarej, W.Y. Svrcek, H.W. Yarranton, A generalized regular solution model for the prediction of asphaltene precipitation from n-alkane diluted heavy oils and bitumens, Fluid Phase Equilib. 232 (2005) 159–170.

[64] E.R. Leite, C. Ribeiro, Crystallization and Growth of Colloidal Nanocrystals, Springer, Amsterdam, The Netherlands, 2012.

[65] H.W. Yarranton, J.H. Masliyah, Molar mass distribution and solubility modeling of asphaltenes, AIChE J 42 (12) (1996) 3533–3543.

[66] D.M. Barrera, D.P. Ortiz, H.W. Yarranton, Molecular weight and density distributions of asphaltenes from crude oils, Energy Fuel 27 (5) (2013) 2474–2487.

[67] J.A. Duran, Precipitation, Aggregation, and Settling of Asphaltenes from n-Alkane Diluted Bitumens (Unpublished doctoral thesis), University of Calgary, Calgary, AB, 2019. http://hdl.handle.net/1880/110937.

[68] T.E. Havre, Near-IR spectroscopy as a method for studying the formation of calcium naphthenate, Colloid Polym. Sci. 282 (2004) 270–279.

[69] W.I. Eke, C. Victor-Oji, O. Akaranta, Oilfeld metal naphthenate formation and mitigation measures: a review, J. Pet. Explor. Prod. Technol. 10 (2019) 805–819, https://doi.org/10.1007/s13202-019-00797-0.

[70] D. Elwell, H.J. Scheel, Crystal Growth From High-temperature Solutions, Academic Press, New York, 1975.

[71] A.C. Lasaga, Kinetic Theory in the Earth Sciences, Princeton University Press, Princeton, 1998, p. 811.

[72] W. Kossel, Zur Theorie des Kristallwachstums, Nachrich. Ges. Wiss. Göttingen, Math.-Phys. Klasse (1927) 135–143.

[73] J.P. Bard, Theories of Crystal Growth, Microtextures of Igneous and Metamorphic Rocks, Springer, Dordrecht, 1986, pp. 12–48.

[74] A.S. Myerson, Handbook of Industrial Crystallization, Butterworth-Heinemann, Woburn, USA, 2002.

[75] J. Gibbs, The collected works, in: Thermodynamics, vol. 1, Yale University Press, England, 1948.

[76] M. Volmer, Kinetik der phasenbildung, Steinkopff, Leipzig, 1939.

[77] X. Guan, Kinetics Studies of Reactions at Solid-Liquid Interface: Simulation of Calcification, VDM Verlag, 2009. 3 July 2009.

[78] F.C. Frank, The influence of dislocations on crystal growth, Discuss. Faraday Soc. 5 (1949) 48–54.

[79] H.H. Teng, P.M. Dove, J.J. De Yoreo, Kinetics of calcite growth: surface processes and relationships to macroscopic rate laws, Geochim. Cosmochim. Acta 64 (13) (2000) 2255–2266, https://doi.org/10.1016/S0016-7037(00)00341-0.

[80] W.K. Burton, N. Cabrera, F.C. Frank, The growth of crystals and the equilibrium structure of their surfaces, Philos. Trans. Royal Soc. A, Math. Phys. Sci. 243 (866) (1951) 299–358. http://www.jstor.org/stable/91559. (Accessed 1 October 2020).

[81] Frank, F. C., in: R. Doremus, D. Turnbull (Eds.), Kinematic Theory of Crystal Growth and Dissolution Processes, 1958, pp. 411–420.

[82] G.H. Nancollas, The nucleation and growth of scale crystals, in: R.W. Bryers (Ed.), Fouling of Heat Exchanger Surfaces, Engineering Foundation, New York, 1983.

[83] J. Garside, L. Brečević, J.W. Mullin, The effect of temperature on the precipitation of calcium oxalate, J. Cryst. Growth 57 (1982) 233–240.

[84] K. Maiwa, H. Nakamura, H. Kimura, A. Miyazaki, Effect of temperature and supersaturation on the growth of Sr(No3)2 (1 1 1) face in aqueous solution, J. Cryst. Growth 289 (2006) 303–307.

[85] Y.F. Ma, Y.H. Gao, Q.L. Feng, Effects of pH and temperature on CaCO3 crystallization in aqueous solution with water soluble matrix of pearls, J. Cryst. Growth 312 (2010) 3165–3170.

[86] W.C. Cheong, Biomimetic Approach to Anti-Fouling Surfaces (PhD thesis), University of Leeds, 2010.

[87] TadaoSugimoto, Application to fundamental studies, in: Monodispersed Particles, second ed., Elsevier, Amsterdam, 2001 (Chapter 11).

[88] Z. Wang, A. Neville, A. Meredith, How and why does scale stick—can the surface be engineered to decrease scale formation and adhesion? in: Paper Presented at the SPE International Symposium on Oilfield Scale, Aberdeen, United Kingdom, May, 2005, https://doi.org/10.2118/94993-MS.

[89] M. Förster, M. Bohnet, Influence of the interfacial free energy crystal/heat transfer surface on the induction period during fouling, Int. J. Therm. Sci. 38 (11) (1999) 944–954.

[90] V. Eroini, N. Kapur, A. Neville, M. Euvrard, Preventing scale formation using modified surfaces, 2011. Corrosion 2011.

[91] M.M. Vazirian, T.V.J. Charpentier, M. de Oliveira Penna, A. Neville, Surface inorganic scale formation in oil and gas industry: as adhesion and deposition processes, J. Pet. Sci. Eng. 137 (2016) 22–32, https://doi.org/10.1016/j.petrol.2015.11.005.

[92] C. Henry, J.P. Minier, G. Lefèvre, Towards a description of particulate fouling: from single particle deposition to clogging, Adv. Colloid Interface Sci. (2012), https://doi.org/10.1016/j.cis.2012.10.001.

[93] I.R. Collins, A New Model for Mineral Scale Adhesion, Society of Petroleum Engineers, 2002, January 1, https://doi.org/10.2118/74655-MS.

[94] A. Guha, Transport and deposition of particles in turbulent and laminar flow, Annu. Rev. Fluid Mech. (2008), https://doi.org/10.1146/annurev.fluid.40.111406.102220.

[95] P. Tandon, M.A. Adewumi, Particle deposition from turbulent flow in a pipe, J. Aerosol Sci. 29 (1–2) (1998) 141–156, https://doi.org/10.1016/S0021-8502(97)00039-6.

[96] J. Young, A. Leeming, A theory of particle deposition in turbulent pipe flow, J. Fluid Mech. 340 (June) (1997) 129–159, https://doi.org/10.1017/S0022112097005284.

[97] S.K. Friedlander, H.F. Johnstone, Deposition of suspended particles from turbulent gas streams, Ind. Eng. Chem. 49 (1957) 1151–1156.

[98] J. Escobedo, G.A. Mansoori, Solid Particle Deposition During Turbulent Flow Production Operations, Society of Petroleum Engineers, 1995, January 1, https://doi.org/10.2118/29488-MS.

[99] F.L. Leite, C.C. Bueno, A.L. Da Róz, E.C. Ziemath, O.N. Oliveira, Theoretical models for surface forces and adhesion and their measurement using atomic force microscopy, Int. J. Mol. Sci. (2012), https://doi.org/10.3390/ijms131012773.

[100] S. Ajupet, M. Sow, D.J. Lacks, Particle adhesion to rough surfaces, Phys. Rev. E 102 (1) (2020), https://doi.org/10.1103/PhysRevE.102.012904.

[101] B.V. Derjaguin, Theory of the heterocoagulation, interaction and adhesion of dissimilar particles in solutions of electrolytes, Discuss. Faraday Soc. 18 (1955) 85.

[102] E.J.W. Verwey, J.T. Overbeek, G., Theory of Stability of Lyophobic Colloids, Elsevier, Amsterdam, 1948.

[103] R. Oliveira, Understanding adhesion: a means for preventing fouling, Exp. Therm. Fluid Sci. 14 (4) (1997) 316–322, https://doi.org/10.1016/S0894-1777(96)00134-3.

[104] J.W. Cleaver, B. Yates, The effect of re-entrainment on particle deposition, Chem. Eng. Sci. 31 (2) (1976) 147–151.

[105] A. Quddus, I.M. Allam, BaSO4 scale deposition on stainless steel, Desalination 127 (3) (2000) 219–224.

[106] C. Garcia, G. Courbin, et al., Study of the scale inhibition by HEDP in a channel flow cell using a quartz crystal microbalance, Electrochim. Acta 46 (7) (2001) 973–985.

[107] L.O. Jøsang, M.N. Psarrou, T. Østvold, T. Tjomsland, K. Ramstad, B.M. Hustad, Scale under turbulent flow, in: Tekna 23rd International Oil Field Chemistry Symposium, 18-21 March 2012, Geilo, Norway, 2012.

[108] Z. Liu, Study of Hydrate Deposition and Sloughing of Gas-Dominated Pipelines Using Numerical and Analytical Models (PhD thesis), Colorado School of Mines, 2017.

[109] E.O. Straume, Study of Gas Hydrate Formation and Wall Deposition Under Multiphase Flow Conditions, 2017. (PhD thesis). Departamento De Pesquisa E Pós-Graduação Programa De Pós-Graduação Em Engenharia Mecânica E De Materiais – Ppgem.

[110] E.O. Straume, C. Kakitani, R. Morales, A. Sum, Study of Gas Hydrate Formation and Deposition Mechanisms in Hydrocarbon Systems, Associacao Brasileira de Engenharia e Ciencias Mecanicas - ABCM, 2018, https://doi.org/10.26678/abcm.encit2016.cit2016-0255.

[111] A. Khaleel, M. Abutaqiya, M. Tavakkoli, A.A. Melendez-Alvarez, F.M. Vargas, On the prediction, prevention and remediation of asphaltene deposition, in: Paper Presented at the Abu Dhabi International Petroleum Exhibition and Conference, Abu Dhabi, UAE, November 2015, https://doi.org/10.2118/177941-MS.

[112] E. Rogel, M. Witt, M.E. Moir, Effects of aging on asphaltene deposit composition using ultrahigh-resolution magnetic resonance mass spectrometry, Energy Fuels 33 (10) (2019) 9596–9603, https://doi.org/10.1021/acs.energyfuels.9b01864.

[113] S.M. Campen, S.J. Moorhouse, J.S.S. Wong, Effect of aging on the removal of asphaltene deposits with aromatic solvent, Langmuir 35 (37) (2019) 11995–11998, https://doi.org/10.1021/acs.langmuir.9b01792.

[114] S.N. Kazi, Particulate matter: interfacial properties, fouling, and its mitigation, in: Mineral Scales and Deposits: Scientific and Technological Approaches, Elsevier, 2015, pp. 141–168, https://doi.org/10.1016/B978-0-444-63228-9.00007-3.

CHAPTER

Mineral Scales in Oil and Gas Fields

5

5.1 Introduction

Water is known to be a universal solvent. Water flowing in rivers, streams, oceans, and underground carries various amounts of dissolved salts. Our food and drinks are cooked and brewed, in one way or another, in water. In addition to being a solvent, water is also used for other industrial purposes like heating, cooling, cutting, drilling, washing, diluting, etc. Water is also a natural product that is extracted during operations like oil and gas and geothermal operations. In all these operations, varying amounts of salts (impurities) are dissolved in water. When the changes in operating parameters drive the dissolved salts to their saturation limit, the dissolved salts precipitate and deposit on the surfaces of the operating facilities. These deposits are conventionally called mineral scale deposits or water-formed scale deposits, referring to water as the origin of these deposits.

Mineral scale deposits are a major operational problem in different industries around the world. Pulp and paper, water treatment and desalination, food processing, power generation, pharmaceutical, and energy industries are examples of industries affected by scale formation.

In the petroleum industry, enormous amounts of water carrying significant amounts of dissolved salts are produced and handled during production operations in oil and gas fields. Due to mixing between these waters and the continuous changes in the operating parameters like pressure, temperature, and pH, various types of mineral scales can deposit on the internal surfaces of the producing facilities and equipment, leading to overwhelming operational consequences. Table 5.1 summarizes the major types of mineral scales that form in oil and gas fields. These compounds and their polymorphs and phases are discussed in further detail in this chapter.

The effect of mineral scales is quick and severe, even at very small concentrations. They fiercely affect fluid flow, impair production equipment, increase fuel consumption, and require expensive maintenance and treatments that cost millions of dollars annually, not to mention cause production losses due to scale formation. Several examples of the deleterious effects of mineral scale on production are presented in Chapter 3, involving problems associated with solids formation and deposition.

In this chapter the different types and polymorphs of mineral scales encountered in oil and gas fields, their mechanisms of formation, and the various factors affecting their solubility are discussed.

Essentials of Flow Assurance Solids in Oil and Gas Operations. https://doi.org/10.1016/B978-0-323-99118-6.00004-6

Table 5.1 Common types of oil and gas field mineral scales.

Scale type	Common examples
Carbonate	Calcium carbonate Iron carbonate Barium carbonate Strontium carbonate Zinc carbonate
Sulfate	Barium sulfate Strontium sulfate Calcium sulfate
Sulfide	Iron sulfide Zinc sulfide Lead sulfide
Oxides and hydroxides	Iron oxides Iron hydroxides Calcium hydroxide Magnesium hydroxide
Chloride	Sodium chloride
Silicate	Magnesium silicate Calcium silicate Aluminum silicate
Phosphate	Calcium phosphate

5.2 Calcium carbonate scale

Calcium carbonate is one of the calcareous scales, and it is considered one of the most widespread and difficult problems to deal with, as it affects a range of applications and industries, starting with household appliances like kettles, coffee machines, laundry machines, and faucets, to pipelines, separators, heat exchangers, boilers, cooling towers, and other facilities used in different industries. Fig. 5.1 shows

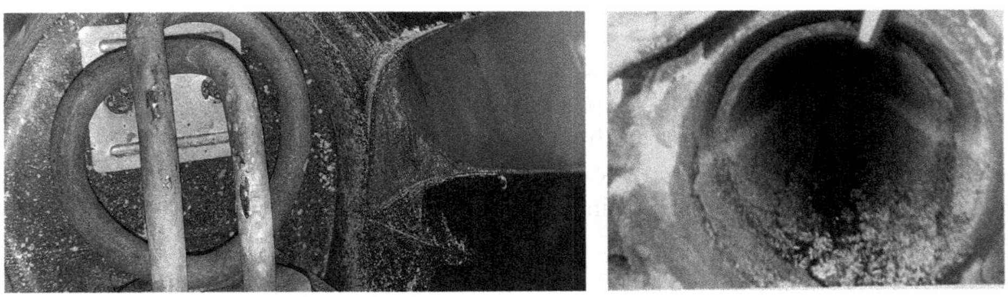

FIG. 5.1

$CaCO_3$ deposits in a kettle (left), and in pipeline (right).

Photo courtesy MERUS oil and gas processing, with permission.

examples of calcium carbonate scale deposition in domestic appliances, e.g., a kettle and an industrial pipeline.

Gross amounts of calcium carbonate scale have been reported in petroleum operations. Vetter and Kandarpa [1] reported a case of complete blockage with calcium carbonate within days.

5.2.1 Mechanism of calcium carbonate formation and its polymorphs

The main mechanism of calcium carbonate precipitation is from the reaction of calcium (Ca^{2+}) ions with either bicarbonate (HCO_3^-) or carbonate (CO_3^{2-}), according to Eqs. (5.1), (5.2) [2]:

$$Ca^{++} + 2\left(HCO_3^-\right) \rightleftharpoons CaCO_3 + H_2O + CO_2 \tag{5.1}$$

$$Ca^{++} + CO3^{--} \rightleftharpoons CaCO_3 \tag{5.2}$$

Carbonate ions are very rare in most produced waters and bicarbonate species are dominant over a wide pH range; therefore Eq. (5.1) represents the principal reaction creating calcium carbonate.

The most common source of dissolved calcium in water is the dissolution of calcium-containing geological features by natural waters [3]. Also, losing calcium-based brines in the formation during drilling and workover provides another source of calcium in the reservoir water. Additionally, the carbonate rocks (calcite and dolomite) are abundant in the petroleum reservoirs. About 45% of the world's petroleum reserves occur in carbonate reservoirs [4], and carbonate can dissolve upon contact with water; higher concentrations of these ions are expected during acid stimulation operations. Another source of carbonate and bicarbonate ions is the dissolution of CO_2 gas in water, which forms the unstable carbonic acid, which dissociates into carbonate (CO_3^{-2}) and bicarbonate (HCO_3^-), according to Eqs. (5.3)–(5.5). This case is associated with miscible gas injection operations using CO_2 gas during enhanced oil recovery [5].

$$CO_2 + H_2O \rightleftharpoons H_2CO_3 \tag{5.3}$$

$$H_2CO_3 \rightleftharpoons HCO3^- + H^+ \tag{5.4}$$

$$HCO3^- \rightleftharpoons CO3^{--} + H^+ \tag{5.5}$$

Microbial activities can also be a rich source of carbonate species, which finally can lead to the precipitation of calcium carbonates [6].

Calcium carbonate is one of the inverse solubility salts, and its solubility in water is very low, at 13 mg/L [7].

Calcium carbonate can deposit in different polymorphs, each polymorph having a different thermodynamic stability and reactivity. These polymorphs may transform or dissolve, or both, as the solution composition approaches equilibrium with the thermodynamically stable phase. Table 5.2 summarizes these different polymorphs of calcium carbonate.

The stability of the $CaCO_3$ polymorphs decreases in the following order: ACC < vaterite < aragonite < calcite [15].

Calcium carbonate crystal growth is believed to take place through the formation of metastable intermediates, which rapidly transform to the thermodynamically stable crystalline forms [16]. Ogino et al. [17] concluded that the amorphous calcium carbonate is initially formed, and then transforms to a mixture of several crystalline $CaCO_3$ polymorphs within several minutes. The polymorphs vaterite and calcite were formed at low temperature (14–30°C), while aragonite and calcite were formed at high

Table 5.2 The different polymorphs of calcium carbonate [8–14].

Polymorph	Formula	Crystal system	K_{sp}	Hardness (Mohs)	sp. gr	Stability
Calcite	Anhydrous $CaCO_3$	Rhombohedral	3.3×10^{-9}	3	2.71	Generally the most thermodynamically stable phase and least soluble between 0°C and 90°C
Aragonite	Anhydrous $CaCO_3$	Orthorhombic	4.47×10^{-9}	3½–4	2.95	Less stable than calcite, reported to be the first phase to precipitate out of solution, particularly at elevated temperatures
Vaterite	Anhydrous $CaCO_3$	Hexagonal	1.23×10^{-8}	3	2.65	A metastable phase of calcium carbonate and forms rather rarely on the surface of the earth at ambient conditions; the most soluble form between 0°C and 90°C
Amorphous calcium carbonate (ACC)	Anhydrous $CaCO_3$ or monohydrate $CaCO_3 \cdot H_2O$	Amorphous	3.98×10^{-7}	–	–	Thermodynamically unstable, acts as a precursor to stable crystalline phases, and hence it spontaneously transforms (crystallizes) to calcite, aragonite, and/or vaterite within minutes to hours at ambient temperatures
Monohydrocalcite	Monohydrate $CaCO_3 \cdot H_2O$	Trigonal	4.68×10^{-8}	2–3	2.95	Thermodynamically unstable and transforms to calcite in Mg-free matrix, or to aragonite when Mg exists
Ikaite	Hexahydrate $CaCO_3 \cdot 6H_2O$	Monoclinic	2.40×10^{-7}	3	1.77–1.83	At room temperature, it evolves quickly towards anhydride forms with a very pronounced increase in volume

temperature (60 – 80°C), and all three polymorphs formed at temperatures from 40°C to 50°C. Aragonite was reported to be the predominant precipitate at high temperature (70°C) and vaterite at low temperature (30°C) [18].

A proposed mechanism of the precipitation starts with the association of hydrated ions Ca^{2+} and CO_3^{2-} in $CaCO_3^0$—shape (ion pairs) that remain partially surrounded by water molecules. The $CaCO_3^0$ precursors group together and lose a part of their hydration cloud and arrange according to an organization sketch close to that of less-hydrated $CaCO_3$ solid form. So, step-by-step, precursors with lower hydration degrees could appear, depending on the different dehydration reaction speeds. If the possibility of precipitating one or several solid forms is then added to this evolution plan, the proposed model can then be written according to Fig. 5.2 [9].

Wang et al. [19] reported a three-stage mechanism of calcite deposition. At the initial stage, the nanoparticles of amorphous $CaCO_3$ initially form on the steel surface. The second stage is the metastable stage where the amorphous $CaCO_3$ grows in the direction perpendicular to the steel surface and forms irregular rhombohedral calcite crystals. Finally, in the stable stage, calcite crystals grow larger and complete, and their surface becomes smooth. In the presence of fluid, the polymorph transitions from aragonite to calcite and from vaterite to calcite or aragonite are rapid [20].

The transformation between the different polymorphs is illustrated in Fig. 5.3 [21]. Initially formed ACC particles transform to the least stable vaterite and the most soluble vaterite undergoes dissolution and crystallization, finally forming the most stable calcite. The solubility of the $CaCO_3$ polymorphs basically decreases in the order of ACC, vaterite, aragonite, and calcite. The ACC transformation into vaterite and calcite was observed below 40°C, while aragonite can be stabilized at elevated temperature above 60°C [17,21–23].

In heterogeneous deposition of calcium carbonate polymorphs, vaterite crystals are observed in direct contact with the substrate, whereas calcite or aragonite crystals are observed on the top layer [16,24]. However, calcite is known to be the most vulnerable form to precipitate, and it is more

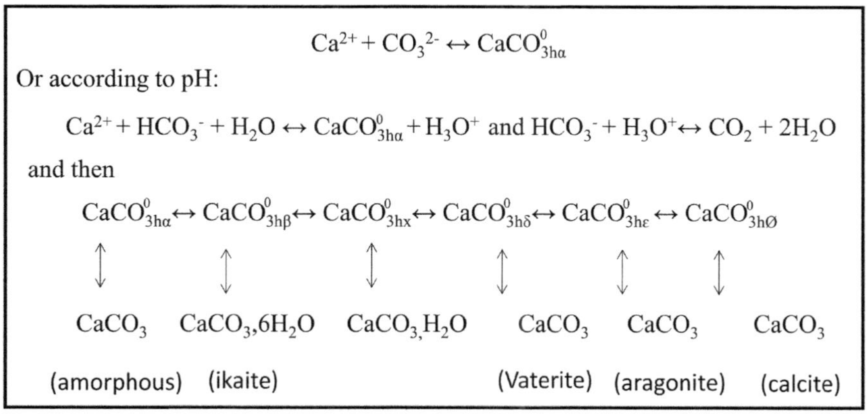

FIG. 5.2

Gal et al. model of the phases involved in calcium carbonate precipitation.

Reprinted from reference J.-Y. Gal, Y. Fovet, N. Gache, Mechanisms of scale formation and carbon dioxide partial pressure influence. Part I. Elaboration of an experimental method and a scaling model. Water Res. 36 (3) (2002) 755–763. https://doi.org/10.1016/s0043-1354(01)00270-6, copyright (2002), with permission from Elsevier.

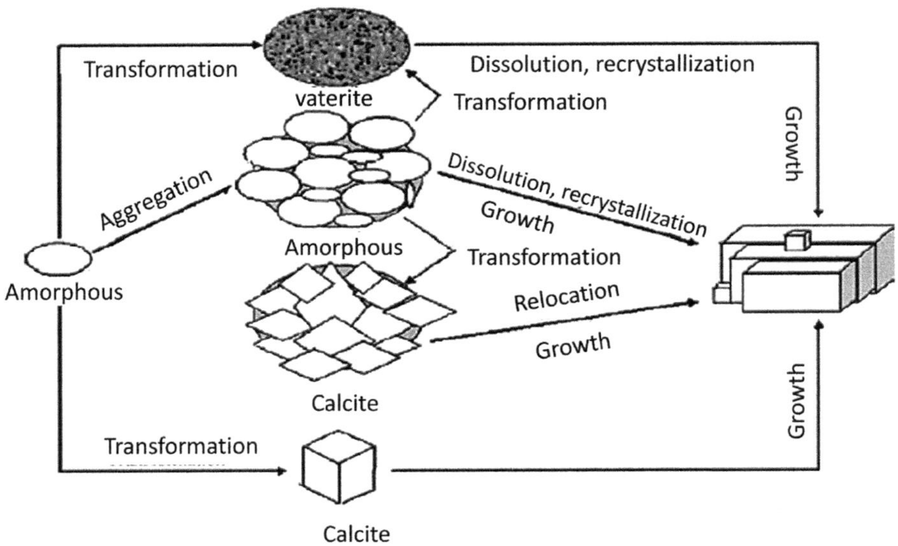

FIG. 5.3

Transformation between different calcium carbonate polymorphs.

Reprinted from reference H. Wei, Q. Shen, Y. Zhao, D.J. Wang, D.F. Xu, Influence of polyvinylpyrrolidone on the precipitation of calcium carbonate and on the transformation of vaterite to calcite, J. Cryst. Growth 250 (2003) 516–524. https://doi.org/10.1016/S0022-0248(02)02484-3, copyright (2003), with permission from Elsevier.

adhesive to the surfaces [19]. Calcite crystals are composed mainly of calcium carbonate but often contain up to 20% iron carbonate and/or magnesium carbonate, so that the average formula for calcite might be represented as $Ca_{0.8-1.0}(Fe,Mg)_{0.2-0.0}CO_3$ [25].

5.2.2 Factors affecting the formation of calcium carbonate scale

A number of factors affect the formation of calcium carbonate, including pressure, temperature, pH, and the presence of particles or impurities.

5.2.2.1 Effect of pH

The higher the pH, the more likely calcium carbonate precipitation will occur [26]. Calcium carbonate is known to precipitate in alkaline media and to dissolve in acidic media. Increasing the solution pH increases the conversion of bicarbonate to carbonate, and hence the tendency for calcite precipitation [27].

The equilibrium between the carbonate species is explained by Eqs. (5.3)–(5.5). These reactions can go in either direction depending on parameters such as pressure, temperature, and pH. Because both the carbonate and bicarbonate ions are conjugate bases of weak acids, they are involved in equilibrium with the hydronium ion (H_3O^+) and their concentrations will be fixed by pH [28]. Fig. 5.4 is a Bjerrum plot that depicts the relative concentrations of carbonate species as a function of pH [29].

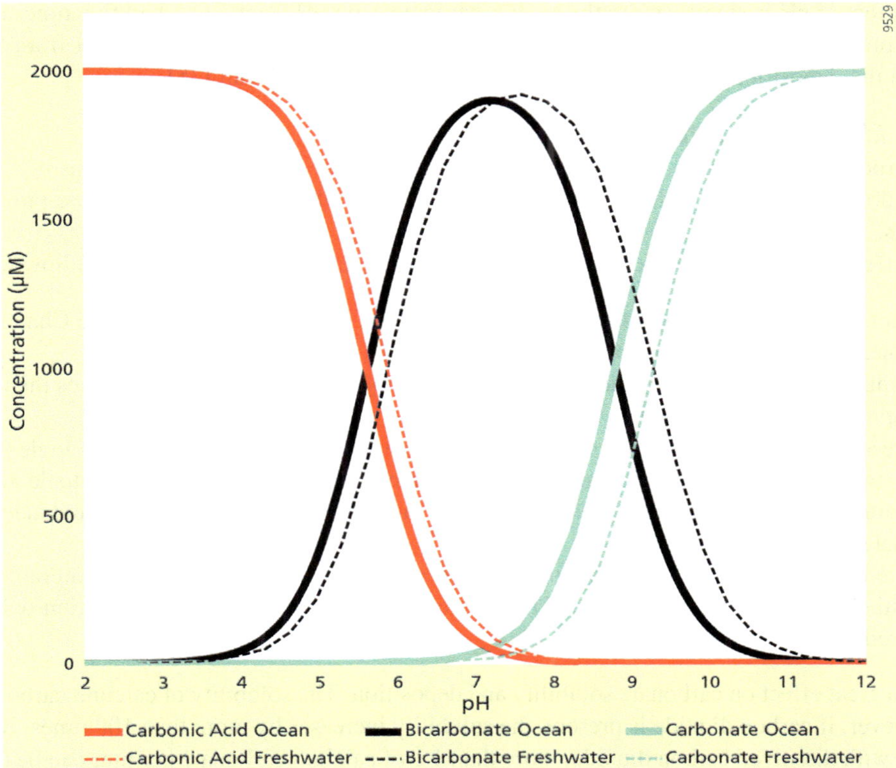

FIG. 5.4

Carbonate species as function of pH.

Source J.J. Middelburg, Biogeochemical processes and inorganic carbon dynamics, 2019, pp. 77–105. https://doi.org/10.1007/978-3-030-10822-9_5.

Treating chemicals that increase the pH of the solution also increase the potential of carbonate scaling; one famous example is the H_2S scavenger triazines. A case was reported on increasing the carbonate scaling after injecting triazine chemicals [30].

Solution pH was also found to affect the type of the formed polymorph. For example, Kitamura et al. [31] suggested that highly alkaline solutions are advantageous for the nucleation of aragonite. Hu et al. [32] reported that at pH 13.4, only ikaite was formed, and at pH 9.0, the precipitate was predominantly vaterite in the absence of phosphate and ikaite in the presence of phosphate. They also concluded that the effect of pH on different polymorphs of calcium carbonate is related to the ratio of HCO_3^-/CO_3^{2-}, which might affect the structure of prenucleation clusters. Oral and Ercan [33] reported that $CaCO_3$ particle size decreased from micron to submicron range and calcite polymorph transformed to vaterite upon decreasing pH of the precursors for all of the precursor solution $[Ca^{2+}]:[CO_3^{2-}]$ ratios investigated in their study.

The effect of pH is significant to the extent that increasing pH from 7 to 8 had five times the effect on scale precipitation against a temperature change of 70°C [34]. Increasing the pH from 8.8 to 10 increased the deposition of $CaCO_3$ from 2 to 12 mg/cm^2, after a 2 h run [35].

5.2.2.2 Effect of pressure

During production operations, deposition of calcium carbonate will occur in locations of significant pressure drop, like chokes, valves, pumps, top parts of the tubing, downstream of the pipeline, and separators.

The effect of pressure on calcium carbonate deposition can be attributed to the following [2]:

- A reduction in pressure favors the precipitation reaction (Eq. 5.1) according to Le Chatelier's principle, as CO_2 is lost from solution.
- Reducing the pressure reduces the concentration of CO_2 in solution, which increases the pH of the solution (becomes less acidic).
- The pressure drop downhole can cause negative Joule-Thomson effect (or reverse Joule-Thomson effect), which increases the temperature of the gas at the perforation, before starting to decrease with expansion as it goes up the tubing, and that leads to localized temperature spike and calcium carbonate deposition [36].
- Pressure drop can induce turbulence in the water, which helps to overcome supersaturation effects and initiate precipitation. The effect of turbulence on calcium carbonate precipitation will be discussed later.

CO_2 has a great effect on carbonate solubility and deposition. The solubility of calcium carbonate is so low, however, if carbon dioxide is present, the solubility increases by more than 100 times. If water is saturated with dissolved carbon dioxide, more than 1 g of calcium carbonate per liter can be dissolved [7]. The amount of CO_2 that can dissolve in the water is determined by Henry's law, which depends on the partial pressure correlated with the total pressure by Eq. (5.6):

$$PCO_2 = \text{Mole fraction of } CO_2 \text{ in gas} \times \text{Total pressure} \tag{5.6}$$

The partial pressure of CO_2 represents the concentration of the gas; thus the concentration can be increased by increasing either the mole fraction or the total pressure. Or, in other words, the higher the system pressure, the higher the CO_2 partial pressure, and the less likely $CaCO_3$ will deposit.

Another way of affecting CO_2 concentration is through the gas lift, which was found to reduce the CO_2 concentration in the water phase by allowing CO_2 to move to the gas phase by agitation (similar to shaking a can of soda and releasing CO_2), which consequently increases carbonate deposition [2].

In the absence of other salts, a shift in the partial pressure of CO_2 from 30 psia to 5 psia can increase the pH from 3.7 to 4.0 at a constant temperature of 77°F [26]. Aragonite is more soluble than calcite at any given temperature and CO_2 pressure but only 10%–15% more soluble [37]. Fig 5.5 shows the effect of CO_2 partial pressure on aragonite and calcite solubility.

5.2.2.3 Effect of temperature

Calcium carbonate is an inverse solubility material, so increasing the temperature promotes calcium carbonate formation [38]. This is the reason why a kettle used in hard water builds up white deposits, and for the same reason calcium carbonate deposits usually form in heat exchangers, heaters, in high-pressure and high-temperature reservoirs and wells, downhole of water injection wells, and on

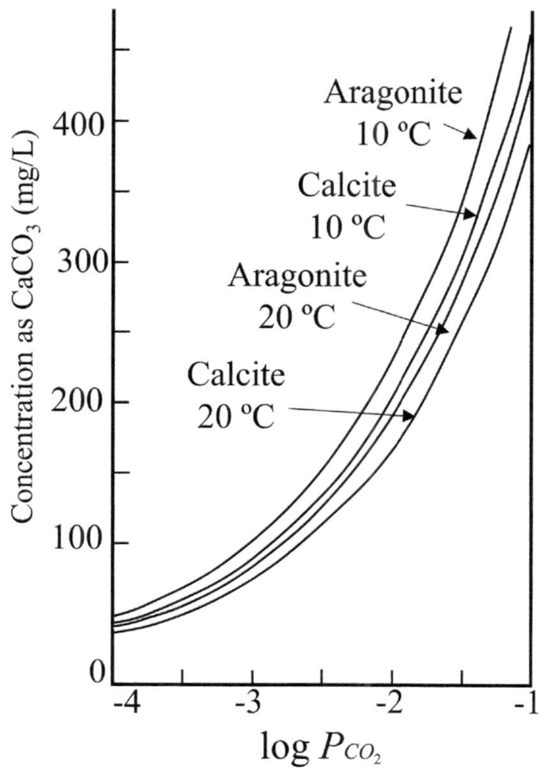

FIG. 5.5

Solubility curves for calcite and aragonite as a function of CO_2 partial pressure. Calculated from thermodynamic data.

Reprinted from reference W.B. White, Speleothems, in: Encyclopedia of Caves (Third Edition), Elsevier, 2019, pp. 1006–1017, copyright (2019), with permission from Elsevier.

geothermal installations [39,40]. Fig. 5.6 shows the precipitation of calcium carbonate in a heat exchanger, and Fig. 5.7 illustrates calcium carbonate deposition in geothermal piping due to the high temperature effect.

The effect of temperature on solubility of $CaCO_3$ is illustrated in Fig. 5.8, indicating its inverse solubility behavior [41].

The K_{sp} of calcium carbonate significantly depends on the temperature; Duggirala [28] reported a relationship for this K_{sp} temperature dependency, empirically determined between 32°F and 180°F, as given by Eq. (5.7):

$$\frac{K_{sp,f}}{K_{sp,i}} = 6.57 \times 10^{-3} \left(T_i - T_f\right) \tag{5.7}$$

where $K_{sp,i}$ and $K_{sp,f}$ are the solubility products at the two temperatures, T_i and T_f (°F). So, at 180°F, the estimated solubility of $CaCO_3$ drops to about 3 mg/L of pure water (a K_{sp} of 1.0×10^{-9} M^2).

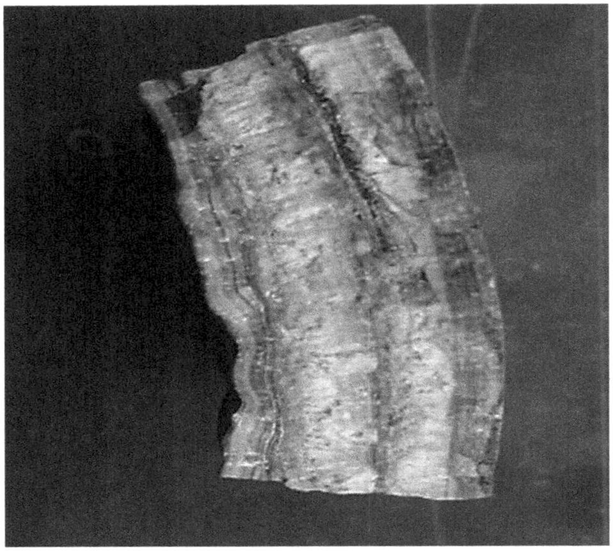

FIG. 5.6

Thick layer of calcium carbonate scale, removed from heat exchanger in oilfield.

FIG. 5.7

Calcium carbonate scale in geothermal pipeline.

Source reference L. Yiman, Calcite Scaling Potential of Kangding Geothermal Field, W-Sichuan Plateau, China, The United Nation University, Reykjavik, Iceland, vol. 16, 2017 with permission from Li Yiman.

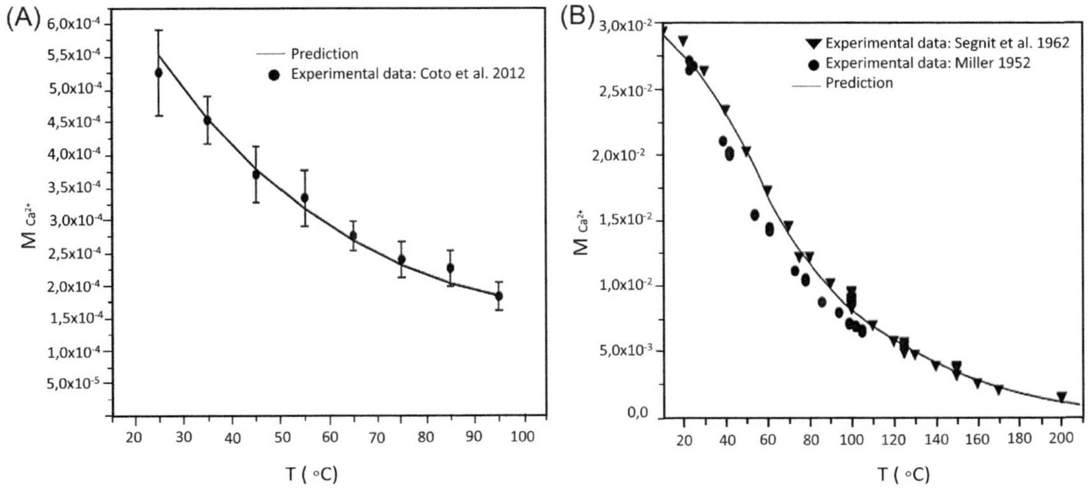

FIG. 5.8

The effect of temperature on CaCO3 solubility (A) at $P = 1$ bar, (B) at $P = 40$ bar.

Reprinted from reference B. Coto, C. Martos, J.L. Peña, R. Rodríguez, G. Pastor, Effects in the solubility of $CaCO_3$: experimental study and model description, Fluid Phase Equilib. 324 (2012) 1–7, https://doi.org/10.1016/j.fluid.2012.03.020, copyright (2012) with permission from Elsevier.

Plummer and Busenberg [42] formulated a correlation for temperatures between 0°C and 90°C (Eq. 5.8):

$$\log K_{calcite} = -171.9065 - 0.077993 \times T + \frac{2839.319}{T} + 71.595 \times \log T \tag{5.8}$$

where $K_{calcite}$ is the equilibrium constant for calcite, T is temperature in °K; similar correlations were formulated for the other polymorphs, vaterite and aragonite.

A formula proposed (Eq. 5.9) by Arnórsson et al. is also widely used at temperatures up to almost 240°C [43]:

$$\log K_{calcite} = 10.22 - 0.0349 \times T - \frac{2476}{T} \tag{5.9}$$

The temperature also affects the polymorphs of calcium carbonates. For example, the initial ACC transforms to a mixture of vaterite and calcite at low temperatures (14–30°C), while vaterite polymorph occurs predominantly at 30–40°C. Overall, the formation of all three polymorphs is observed at intermediate temperatures (40–50°C). Elevating the temperature above 60°C stabilizes the aragonite polymorph of calcium carbonate [44]. The transformation of ACC to thermodynamically stable forms (which ends up with calcite) will be on the order of: ACC → vaterite → aragonite → calcite. At low temperatures (<30°C) ACC transforms to calcite via vaterite intermediate, whereas at high temperatures (>60°C), ACC transforms to calcite via aragonite at higher temperatures) [45]. Fig. 5.9 represents the distribution of $CaCO_3$ polymorphs at different temperatures [17].

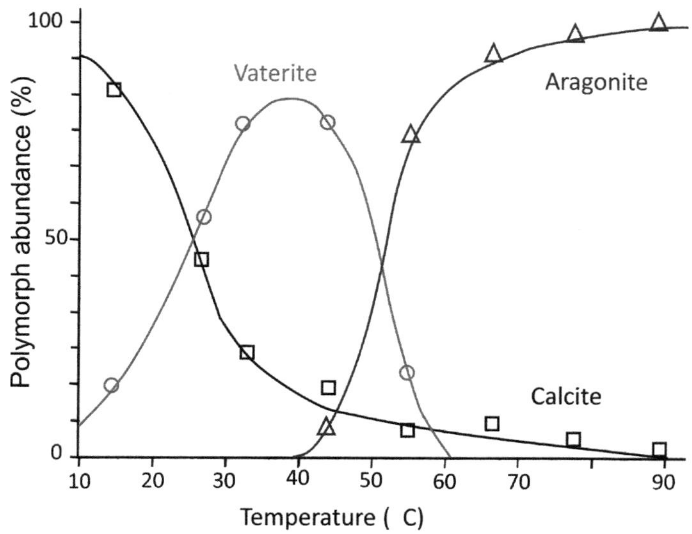

FIG. 5.9

The distribution of calcium carbonate polymorphs versus temperature.

Reprinted from reference T. Ogino, T. Suzuki, K. Sawada, The formation and transformation mechanism of calcium carbonate in water, Geochim. Cosmochim. Acta 51 (1987) 2757–2767. https://doi.org/10.1016/0016-7037(87)90155-4, copyright (1987) with permission from Elsevier.

5.2.2.4 Effect of dissolved salts

Calcium carbonate solubility increases as the salt content of the water increases. The higher the total dissolved solids (not counting calcium or carbonate ions), the greater the solubility of $CaCO_3$ in the water and the lower the scaling tendency up to a maximum of about 200,000 mg/L [26]. The solubility of calcium carbonate at different salinities and temperatures is illustrated in Fig. 5.10 [41].

Evans [46] reported that more vaterite is found among the precipitation as the ionic concentration increases in the solutions.

5.2.2.5 Effect of metal ions

- **Effect of Ca^{++} ion**

Calcium cation concentration has a major influence on the amount of scale formed. Water hardness is a measure of Ca^{2+}, Mg^{2+} cation concentration and it is the primary parameter governing the scaling power of water; it is reported as parts per million (as calcium carbonate). Traditionally, it is a measure of the capacity of the water for precipitating soap [47]. Calcium and magnesium associated with bicarbonate anions are termed *carbonate* or *temporary* hardness, as they can be removed by heating. Calcium and magnesium associated with chloride or sulfate are termed permanent or *noncarbonate* hardness [27]. Table 5.3 summarizes the water hardness ranges [48].

- **Other metal ions**

Metal ions like Mg^{2+}, Ni^{2+}, Co^{2+}, Fe^{3+}, Zn^{2+}, and Cu^{2+} can result in a prolonged induction period leading to slower precipitation; these metal ions were also found to favor the formation of aragonite over calcite,

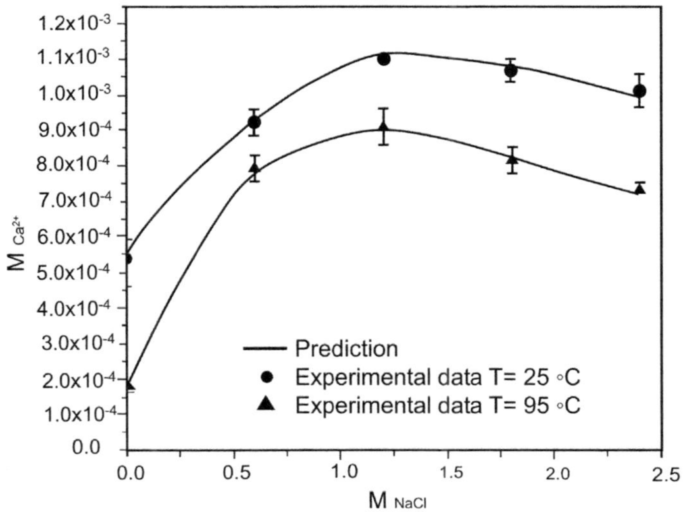

FIG. 5.10

The effect salinity (NaCl) on the $CaCO_3$ solubility at 1 bar air pressure.

Reprinted from reference B. Coto, C. Martos, J.L. Peña, R. Rodríguez, G. Pastor, Effects in the solubility of $CaCO_3$: experimental study and model description, Fluid Phase Equilib. 324 (2012) 1–7, https://doi.org/10.1016/j.fluid.2012.03.020, copyright (2012) with permission from Elsevier.

Table 5.3 Water hardness ranges.

Concentration (mg/L) as $CaCO_3$	Degree of hardness
<75	Soft
75–150	Moderately hard
150–300	Hard
>300	Very hard

Reprinted by permission from [Springer Nature]: [Springer] [International Journal of Environmental Science and Technology [Sources of contaminants and groundwater quality in the coastal part of a river delta, E. Vetrimurugan et al., (copyright) (2013).

whereas ions of Mn^{2+}, Cd^{2+}, Sr^{2+}, Pb^{2+}, and Ba^{2+} favor calcite. Other impurities with a small ionic radius and higher hydration energy than that of calcium Ca^{2+} result in aragonite formation [27,49,50].

5.2.2.6 Effect of alkalinity

The alkalinity of a solution may be defined as the capacity it has for solutes to react with and neutralize acid [51]. The total alkalinity includes all the bases that are titratable with strong acid. It is a measure of the buffering capacity of water, meaning that waters of low alkalinity will be very sensitive to pH changes, while those with high alkalinity content will be more able to resist pH changes. The short description of the total alkalinity is given in Eq. (5.10):

$$A_T = [HCO_3^-] + 2[CO_3^-] + [OH^-] - [H^+] \tag{5.10}$$

where A_T is the total alkalinity in eq/L, [ ] is the molar concentration of the chemical species.

The total alkalinity is a conservative quantity (independent of PH, temperature, and pressure), even though concentrations of individual species are not.

Carbonate species are by far the most important contributor to alkalinity. Phenolphthalein alkalinity (P-alkalinity) is the titratable alkalinity at pH 8.3, while methyl-orange alkalinity (M-alkalinity) is the titratable alkalinity at pH value 4.3 (or the endpoint in the range 4.3–5.1 depending on the carbon dioxide present) [51]. Most produced waters have pH of 4–8. They can be expressed by the M-alkalinity as the total alkalinity [HCO_3^-], and in that case it is termed total alkalinity, which represents the concentration of bicarbonate ion concentration in the water, and the volatile fatty acids are added to that term [HCO_3^-] + [AC] [52].

High total alkalinity waters usually means high bicarbonate/carbonate species concentration, which shifts the reaction towards the precipitation of calcium carbonates (Eqs. 5.1 and 5.2).

5.2.2.7 Effect of bacteria

Microbial carbonate precipitation (MCP) can occur as a by-product of common microbial metabolic activities, including photosynthesis, ureolysis, denitrification, ammonification, sulfate reduction, and methane oxidation [6,53]. In addition, cell walls and extracellular polymeric substances (EPS) are reported to serve as templates for carbonate precipitation [54].

Table 5.4 summarizes reactions and by-products involved in different metabolic pathways leading to MCP [6].

Sulfate reduction by SRB is a common bacterial activity in oil and gas fields. The role of SRB in the precipitation of calcium carbonate is related to three factors [55]: (1) reduction of sulfate increases pH, thus affects the saturation index (SI) and calcium carbonate precipitation, (2) the use of organic acid by SRB increases the availability of free calcium ions, and (3) SRB alters the kinetics of dolomite by removing sulfate ions from the solution.

5.2.2.8 Effect of fluid dynamics

Fluid hydrodynamics have significant impact on the scaling process. High shear rates are expected to strip and detach the formed scale from surfaces. However, flowing fluid turbulence was found to be a major hydrodynamic factor that may promotes scale formation, as more active scale-forming species are adsorbed on the surface, which ultimately adhere and deposit on the substrate. In addition, it also

Table 5.4 Different pathways of calcium carbonate biomineralization by different bacteria, source [6].

Microbial groups	Metabolism	Reactions
Cyanobacteria, algae	Photosynthesis	$2HCO_3^- + Ca^{2+} \rightarrow CH_2O + CaCO_3 + O_2$
Ureolytic bacteria	Ureolysis	$CO(NH_2)_2 + 2H_2O + Ca^{2+} + Cell \rightarrow 2NH_4^+ + cell\text{-}CaCO_3$
Nitrate-reducing bacteria	Denitrification	$CH_2COO^- + 2.6H^+ + 1.6NO_3^- \rightarrow 2CO_2 + 0.8N_2 + 2.8H_2O$ $Ca^{2+} + CO_2(aq) + 2OH^- \rightarrow CaCO_3(s) + H_2O$
Sulfate-reducing bacteria (SRB)	Sulfate reduction	$SO_4^{2-} + 2(CH_2O) + OH^- + Ca^{2+} \rightarrow CaCO_3 + CO_2 + 2H_2O + HS^-$
Methanogens	Methane oxidation	Anaerobic oxidation: $CH_4 + SO_4^{2-} + Ca^{2+} \rightarrow CaCO_3 + H_2S + H_2O$ Anaerobic reduction: $CH_4 + 2O_2 \rightarrow CO_2 + 2H_2O$

provides the necessary activation/surface energy for the nucleation and/or epitaxial growth of scale crystals [56]. Vazirian and Neville [57] concluded that the increase in the level of the turbulence in the bulk would increase the scale formation rate on the surface.

Calcium carbonate is known to be deposited in turbulent locations like elbows, after pumps, chocks, and other production equipment.

Hasson et al. [58] studied the separate effects of flow velocity, scale surface temperature, and water composition in order to determine the mechanism of calcium carbonate scale deposition on heated surfaces. Within the range of 67–85°C and Reynolds numbers 13,000–42,000, they found that the rate of calcium carbonate deposition was controlled mainly by the forward diffusion rate of Ca^{2+} and HCO^{3-} ions toward the scale surface. The backward diffusion rate of dissolved CO_2 gas formed during the reaction is also a rate controlling factor in some instances.

High flow rates increased the deposited scale masses [59]. Carbonate scaling rates increase under turbulent multiphase flow due to increased CO_2 transfer from the water phase to the gas and oil phases [60,61].

5.2.2.9 Geochemical interactions

During thermochemical sulfate reduction (TSR), calcium carbonate is produced as a by-product of the process [62]. These interactions are shown in Eqs. (5.11), (5.12).

$$\text{hydrocarbon} + CaSO_4 \rightarrow \text{altered hydrocarbons} + \text{solid bitumen} + CaCO_3 + H_2S + H_2O \tag{5.11}$$

$$CH_4 + CaSO_4 \rightarrow CaCO_3 + H_2S + H_2O \tag{5.12}$$

Other geochemical reactions include dissolution of calcium-bearing rocks and carbonate-bearing rocks (by the reservoir geochemical reactions, or by injected water) and then precipitating them as calcium carbonate in the near wellbore area.

5.3 Calcium sulfate scale

Calcium sulfate is one of the sulfate deposits of group II metals in the Periodic Table (calcium, strontium, and barium), which commonly occur and impact production operations in oil and gas fields. The precipitation of sulfate scales is basically a consequence of mixing incompatible waters during water flooding operations [26]. Like calcium carbonate deposits, calcium sulfate deposits cause a number of operational problems and need expensive treatments and long downtime. Figs. 5.11 and 5.12 show equipment and tubular blockages due to calcium sulfate scale deposition.

5.3.1 Mechanism of calcium sulfate scale formation and its polymorphs

The chemical reactions in Eqs. (5.13), (5.14) represent calcium sulfate mechanism of formation:

$$Ca^{++} + SO_4^{--} \rightarrow CaSO_4 \tag{5.13}$$

In a brine system, the reaction can be described as follows:

$$CaCl_2 + Na_2SO_2 + xH_2O \rightarrow CaSO_4.xH_2O + 2NaCl \tag{5.14}$$

where $0 \leq x \leq 2$.

FIG. 5.11

Calcium sulfate deposits fully plugging a valve.

FIG. 5.12

Calcium sulfate deposits in a production header.

In the previous equations, the calcium ions are abundantly found in the formation water, while the sulfate ions are coming from the injection of seawater (sulfate concentration in seawater is 2800–3200 ppm). Another source of Ca^{2+} and SO_4^{2-} ions is reservoir rocks, as anhydrite (anhydrous calcium sulfate) is one of the evaporites that forms the cap rocks in the petroleum reservoir [63].

This mechanism involves mixing between produced water and injection water or two incompatible produced waters. Calcium sulfates have been observed in high-temperature fields under seawater flooding where Ca-rich formation water and seawater mix in the production well at elevated seawater fractions, e.g., 70%–80% [62–64]. McCartney et al. [65] reported that $CaSO_4$ can precipitate from injection water by different mechanisms:

- Heating of seawater. Because $CaSO_4$ has retrograde solubility, $CaSO_4$ solubility can be exceeded when seawater is injected into higher temperature reservoirs (>~130°C) leading to $CaSO_4$ deposition.
- Precipitation of brucite from seawater. Deposition of brucite ($Mg(OH)_2$) from seawater can induce calcite dissolution. The released Ca^{2+} ions can cause additional $CaSO_4$ precipitation. Where this reaction occurs, it can proceed until equilibrium between seawater, calcite, brucite, and anhydrite is reached:

$$2CaCO_{3(s)} + Mg^{2+} + SO_4^{2-} + 2H_2O \rightarrow 2CaSO_4(s) + Mg(OH)_{2(s)} + 2HCO_3^- \tag{5.15}$$

- Dolomitization of carbonates by seawater. Dissolution of calcite in the reservoir by seawater can result in precipitation of dolomite (or Mg-calcite) and $CaSO_4$. The reaction proceeds until equilibrium between seawater, calcite, dolomite and anhydrite is reached:

$$2CaCO_{3(s)} + Mg^{2+} + SO_4^{2-} \rightarrow CaSO_4(s) + Ca,Mg(CO_3)_2 \tag{5.16}$$

- Release of Ca^{2+} via ion exchange. During injection of seawater, Ca^{2+} is displaced from ion exchange sites (e.g., on clays, micas, etc.) and this can lead to $CaSO_4$ precipitation if $CaSO_4$ solubility is exceeded.

Calcium sulfate can exist in six different solid phases elucidated from Eq. (5.14): three anhydrites [$CaSO_4$], two hemihydrates ($CaSO_4 \cdot ½H_2O$), and one dihydrate, gypsum ($CaSO_4 \cdot 2H_2O$). Table 5.5 summarizes these polymorphs and their stability.

Classical and nonclassical approaches were used to describe the precipitation of calcium sulfates. The classical picture is based on the classical nucleation theory (CNT) in which the crystallization process takes place in two stages: nucleation and crystal growth. However, the nonclassical picture states that the precipitation of gypsum, for example, is a multistage process that involves one or more precursor/intermediate phases, and the phase(s) formed initially upon precipitation from solution can be quite distinct from the final crystalline state (gypsum) [67,69–72]. Fig. 5.13 summarizes these different pathways.

A new description of the gypsum precipitation process was provided by Stawski et al. [73]. Their study concludes that crystalline gypsum forms via a multistep process, involving four well-defined stages, as illustrated in Fig. 5.14 [67,73]:

- (Stage I) Formation of elongated primary species, <3 nm in length and <0.5 nm in diameter and composed of "anhydrous" Ca-SO_4 cores
- (Stage II) Assembly of these primary species via density fluctuations into loose domains, where the interparticle distance remains still fairly large, i.e., >2 times the radius of gyration of the individual primary particles
- (Stage III) Further condensation of these domains into larger and denser aggregates, still consisting of the same primary units

Table 5.5 Calcium sulfate polymorphs and their properties [66–68].

Polymorph	Other names	Formula	Crystal system	K_{sp} at 25°C	Hardness (Mohs)	sp. gr	Stability
Gypsum (calcium sulfate dihydrate)	Selenite, alabaster	$CaSO_4 \cdot 2H_2O$	Monoclinic-prismatic	3.14×10^{-5} (solubility 0.21 g/100 g at 20°C)	2	2.3	<60–90°C in air <40–60°C in water Precipitate from solution upon mixing Ca^{2+}, SO_4^{2-} ions under ambient conditions
Hemihydrate (calcium sulfate hemihydrate)	Bassanite, Plaster of Paris, calcined gypsum	α-$CaSO_4 \cdot 0.5H_2O$	Deformed monoclinic	3.1×10^{-7} (solubility 0.66–0.88 g/100 g at 20°C)	2	2.69–2.76	Metastable in H_2O Formed by heating gypsum in air at >60–90°C, and can rehydrate into gypsum
		β-$CaSO_4 \cdot 0.5H_2O$	Trigonal				Metastable in H_2O Formed by heating gypsum in water at >90°C; can rehydrate to gypsum
Anhydrite (Anhydrous calcium sulfate)	AIII, soluble anhydrite	γ-$CaSO_4$	Hexagonal	4.39×10^{-5} (solubility 0.27 g/100 g at 20°C)	3–3½	2.98	Metastable in H_2O and air Formed by heating gypsum in air at >100–130°C; rehydrates to hemihydrate or gypsum
	AII, insoluble anhydrite	β-$CaSO_4$	Orthorhombic				Stable <~1200°C Natural anhydrite formed at the earth's surface
	AI	α-$CaSO_4$	Trigonal				Stable >~1200°C Converts immediately into β-$CaSO_4$ on cooling below ~1200°C

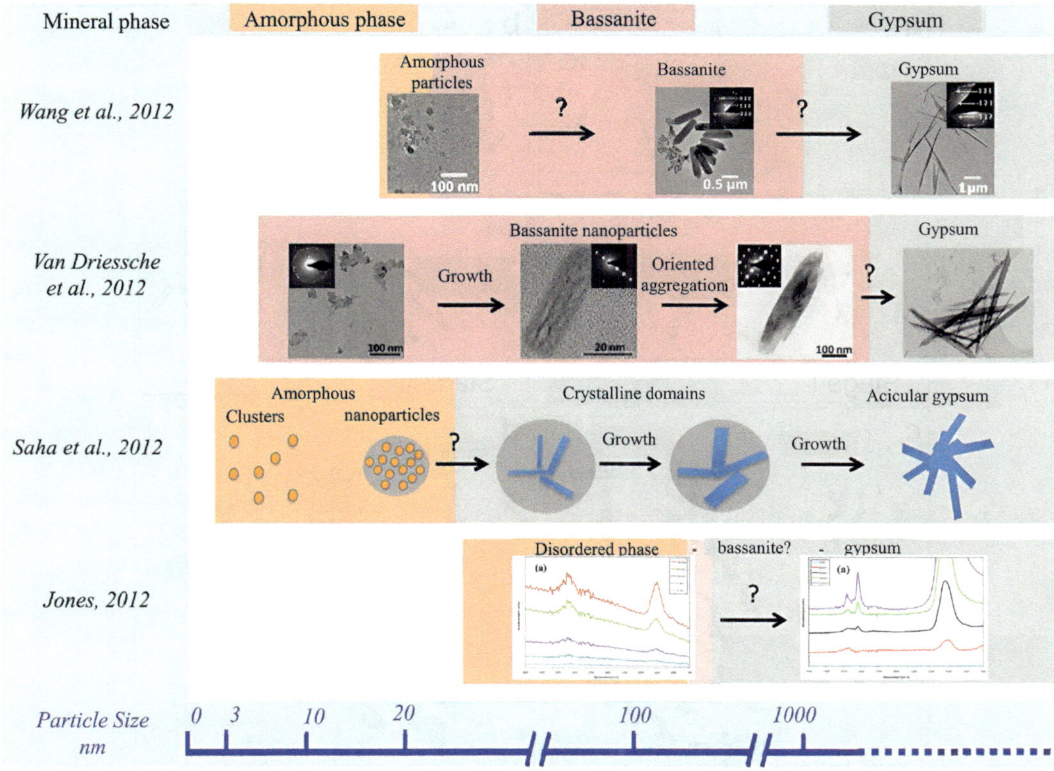

FIG. 5.13

Schematic overview of the various precursor phases and pathways leading to gypsum formation.

*Reprinted/adapted by permission from, Springer Nature, Springer, Calcium Sulfate Precipitation Throughout Its Phase Diagram by Alexander E.S. Van Driessche, Tomasz M. Stawski, Liane G. Benning et al) [**COPYRIGHT**] (2017).*

– (Stage IV) Growth of the primary units within the aggregates, mutual alignment, and eventual collective reorganization/transformation into gypsum, as confirmed through the simultaneously collected WAXS data

The interrelation of all phases, which are partly transformable by changing the water content and temperature, is illustrated in Fig. 5.15 [74].

According to Oddo and Tomson [75], the most likely scale to form from brines will be gypsum at temperatures less than 176°F [80°C]. Between 176°F [80°C] and 250°F [121°C], any of the three types of calcium sulfate may form, with gypsum being more likely at the low end of the temperature range and anhydrite more likely at the high end. Hemihydrate is commonly found in this temperature range in nonturbulent systems with high ionic strengths. Above 250°F [121°C], any calcium sulfate scale formed will almost certainly be anhydrite. It is difficult to precipitate anhydrite directly from solution; it usually transforms from gypsum with time and after dehydration.

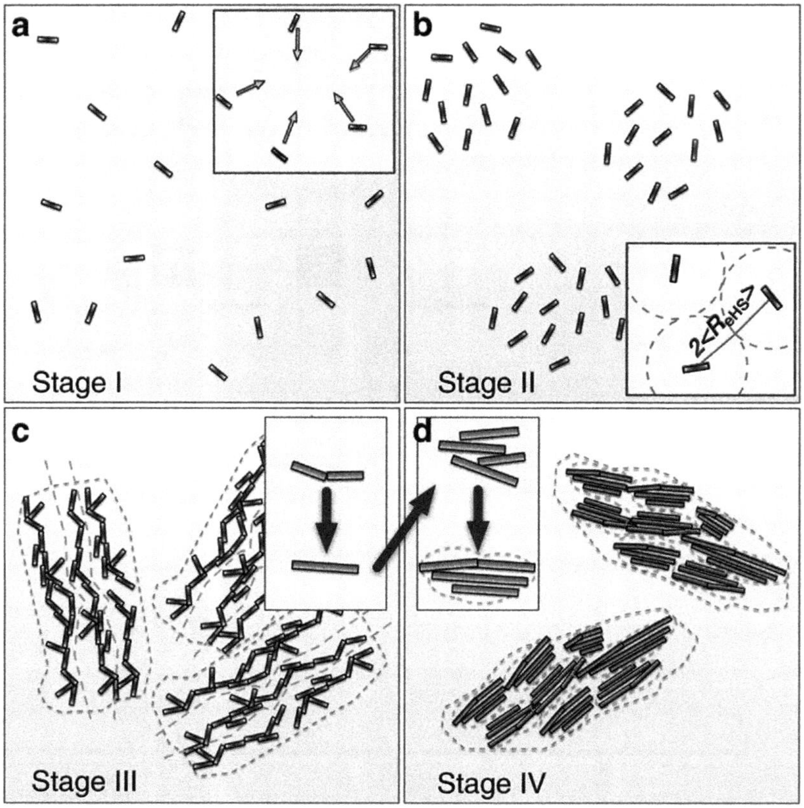

FIG. 5.14

Schematic of the stages of CaSO4 formation.

Source: T. Stawski, A.E.S. Van Driessche, M. Ossorio, J.D. Rodriguez-Blanco, R. Besselink, L.G. Benning, Formation of calcium sulfate through the aggregation of sub-3 nanometre primary species, Nat. Commun. 7 (2016) 10177.

The solubility of gypsum in water is 2530 mg/L at 25°C, which is higher than that of calcium carbonate but still roughly 140 times lower than the solubility of common salt (360 g/L) [76].

5.3.2 Factors affecting precipitation of calcium sulfates

5.3.2.1 Effect of Temperature

Generally, calcium sulfate is a retrograde/inverse solubility salt like calcium carbonate. Higher temperatures provide enough energy to the molecules to overcome the activation energy of the precipitation reaction, increase mass transfer of the scaling species, and reduce the induction time, which speeds up the scaling process [77].

The temperature has a significant effect on the polymorph solubility. Gypsum solubility in pure water increases with temperature up to about 38°C, then decreases with temperature as shown in

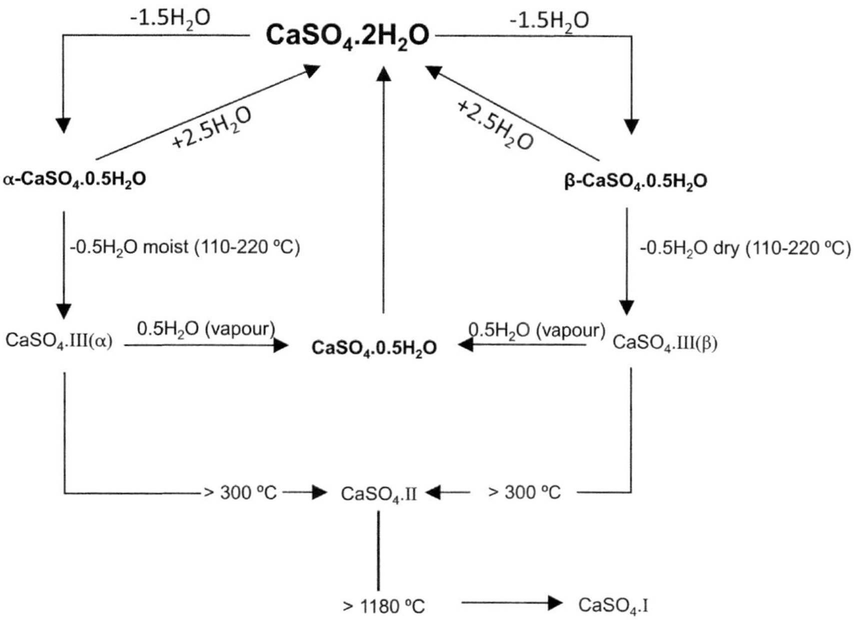

FIG. 5.15

Transformation and modification of calcium sulfates.

Reprinted from reference N.B. Singh, B. Middendorf. Calcium sulphate hemihydrate hydration leading to gypsum crystallization, Prog. Cryst. Growth Charact. Mater. (2007) https://doi.org/10.1016/j.pcrysgrow.2007.01.002, copyright (2007), with permission from Elsevier.

Fig. 5.16 [78]. This is quite different from the temperature-solubility behavior of $CaCO_3$, where an increase in temperature always decreases the solubility [26].

Above 38°C, the anhydrite polymorph becomes less soluble than gypsum, so it could reasonably be expected that anhydrite might be the preferred form of $CaSO_4$ in deeper and hotter wells. Above 100°C, anhydrite will precipitate directly in a stirred or flowing system [26].

The preceding curve for the solubility of anhydrite intersects the solubility curve for gypsum at 42°C. This is the transition point of gypsum-anhydrite [3]. The transition temperatures of gypsum-anhydrite and gypsum-hemihydrate in different studies vary mostly within the ranges from 42°C to 60°C and 80°C to 110°C, respectively [79–82]. Predicting which form of calcium sulfate will precipitate under a given set of conditions is very difficult. For example, gypsum may be found at temperatures up to 100°C [26]. The temperature at which the scale changes form from gypsum to anhydrite or hemihydrate is a function of many factors, including pressure, dissolved salt content, flow conditions, and the speed at which different forms of $CaSO_4$ can precipitate from solution.

Hoang et al. [77] determined the correlation between calcium sulfate scale mass and temperature to be:

$$\log(\text{scale mass}) = -4047.1\left(\frac{1}{T}\right) + 12.141 \tag{5.17}$$

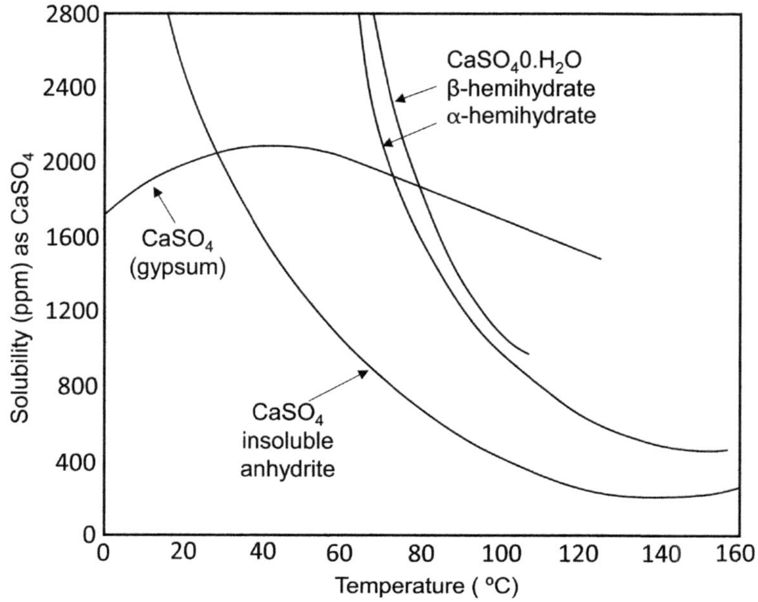

FIG. 5.16

Solubility of calcium sulfate polymorphs vs temperature.

Reprinted (adapted) with permission from D. Hasson, J. Zahavi, Mechanism of calcium sulfate scale deposition on heat-transfer surfaces, Ind. Eng. Chem. Fundam. 9 (1) (1970) 1–10. https://doi.org/10.1021/i160033a001. Copyright {1970} American Chemical Society.

Bansal and Müller-Steinhagen [83] showed that the fouling resistance measured for calcium sulfate fouling of a plate heat exchanger was 50% lower at 85°C than at 90°C.

5.3.2.2 Effect of pressure

Pressure drops are the primary cause of calcium sulfate scaling within a formation, when the formation brine is saturated with calcium sulfate [84]. Under downhole conditions, anhydrite or gypsum deposition is caused mainly by pressure drop, which has a stronger effect than temperature [85]. The greatest pressure change is at the sand face of the producing well. Pressure drop can also occur at the valves, pumps, upwards in the tubing, downstream of the pipelines. Figs. 5.17 and 5.18 illustrate the effect of pressure on the solubility of gypsum and anhydrite [86,87].

The effect of pressure drop on the precipitation and deposition of calcium sulfates can be attributed to:

- The solubility of gypsum or anhydrite in water increases with pressure, due to a slight decrease in total volume when it dissolves [85–88].
- Pressure drops across chokes and valves induce turbulence in the water, which helps to overcome supersaturation effects and initiate precipitation [26].

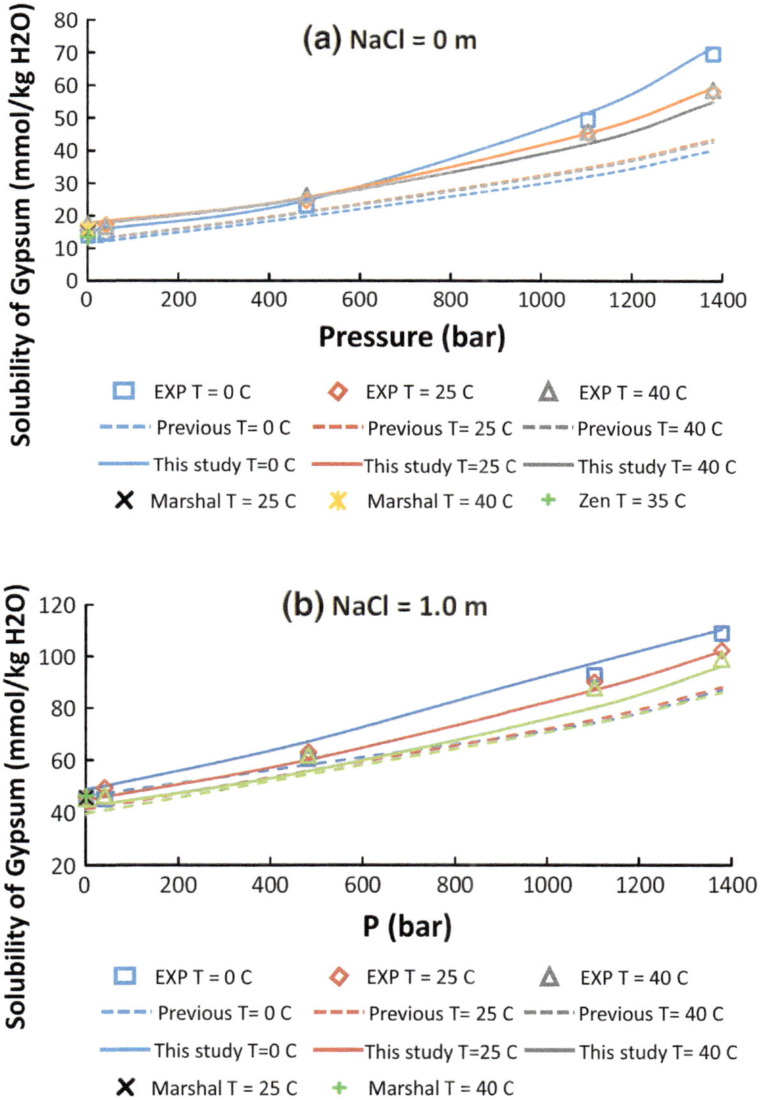

FIG. 5.17

Solubility of Gypsum vs pressure at various ionic strength conditions.

Reprinted by permission from springer Nature: (Springer) Rock Mechanics and Rock Engineering, Z. Dai, A. T. Kan, W. Shi, et al., Solubility measurements and predictions of gypsum, anhydrite, and calcite over wide ranges of temperature, pressure, and ionic strength with mixed electrolytes, copyright (2016). https://doi.org/10.1007/s00603-016-1123-9.

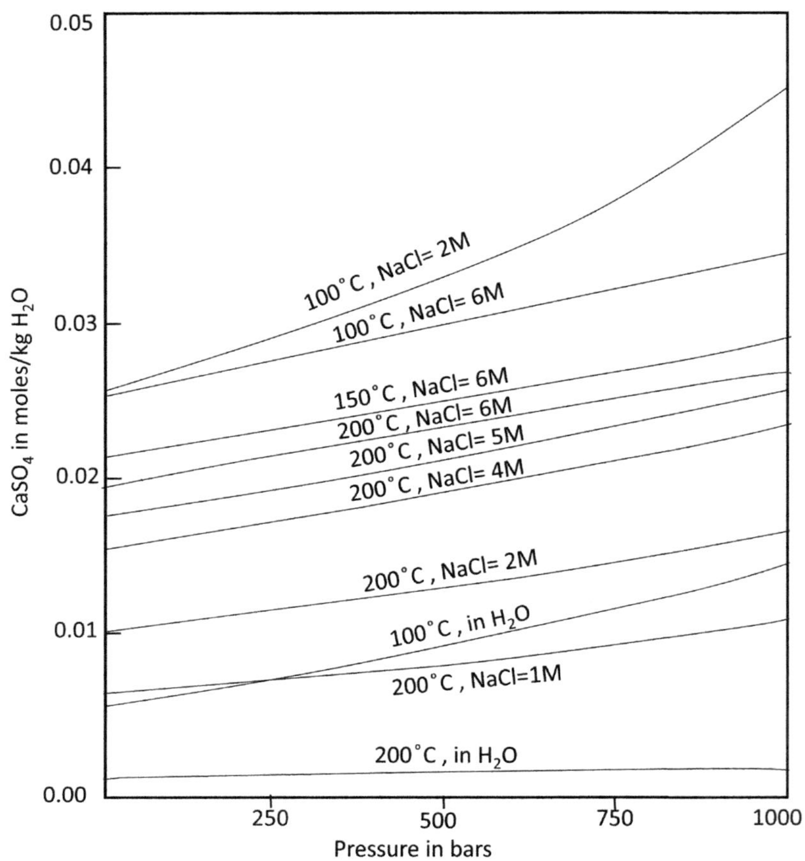

FIG. 5.18

Solubility of anhydrite vs temperature.

Reprinted from reference C.W. Blounot, F.W. Dickson, The solubility of anhydrite ($CaSO_4$) in NaCl-H_2O from 100 to 450°C and 1 to 1000 bars, Geochim. Cosmochim. Acta 33 (2) (1969) 227–245. https://doi.org/10.1016/0016-7037(69)90140-9, copyright (1969), with permission from Elsevier.

For example, at 50°C the solubility of gypsum is about 15 mM at 1 bar and ca. 35 mM at 1000 bar, while at 80°C anhydrite has a solubility of ca. 9 and 25 mM at 1 and 1000 bar, respectively [89].

5.3.2.3 Effect of dissolved salts

The solubility of calcium sulfates in brines is more complicated than pure water. Generally, the presence of NaCl or dissolved salts other than calcium or sulfate ions increases the solubility of $CaSO_4$ just as it does for $CaCO_3$ up to a salt concentration of about 150,000 mg/L. Further increases in salt content decrease $CaSO_4$ solubility [26,90].

Gypsum solubility is highly sensitive to the ionic strength. At 25°C and 1 bar, the gypsum solubility increases from 16.62 mm in pure water to 45.62, 54.97, and 59.27 mm in the presence of 1.0, 2.0, and 4.0 m NaCl, respectively [86]. Fulford [88] concluded that concentrations of sodium chloride above

240 g/L or combinations of sodium chloride and other salts with a total concentration over 240 g/L prevent gypsum scale resulting from pressure decreases.

The dependency of solubility on ionic strength is slightly different for anhydrite, especially at low temperatures, leading to a crossover at high salt concentrations, and suggesting that anhydrite should be in the stable phase at room temperature and high salinity [67].

5.3.2.4 Effect of pH

The solubility of calcium sulfates can increase or decrease with changes in pH of the solution. Kumar et al. [91] found that in high salinity (NaCl) solutions, with pH range of 2.5–8.5 at 35°C, gypsum solubility increased slightly with a decrease in the pH of the solution. At pH 2.5 gypsum solubility was enhanced by nearly 12%. Unexpectedly, gypsum solubility has also increased at pH 8.5. The increase in the solubility of gypsum at low pH values may be because H^+ ions provide localized acidification to dissolve the mineral and also because of the higher hydration of ions.

Bhagawati et al. [92] studied calcium sulfate in the NaCl-H_2O system, in 0.01–0.1 M NaCl concentrations at pH 3–6 and showed that the highest soluble sulfate is observed at pH 5.0 when the concentration of NaCl is 0.05 M at 5 kg/cm^2 pressure.

5.4 Barium sulfate scale

Barium sulfate scale deposits pose a unique and sometimes "unsolvable" problem in production operations. Under most conditions, barium sulfate is considerably less soluble in all solvents than any of the other commonly encountered scale-forming compounds, which makes its chemical removal almost impossible, even with today's advanced technologies. Consequently, barium sulfate deposits most of the time are removed mechanically or the scaled piping/equipment should be discarded [3]. Besides its lower solubility, barite is very rapidly precipitating and rapidly impacting, as one well in Miller field in the North Sea went from 30,000 bpd to zero bpd in just 24 h [93]. Fig. 5.19 shows BaSrSO4 scale in a pipeline [94].

Fig. 5.20 illustrates the distribution of barium sulfate scale (% weight composition of the collected scale) in different oil wells at different depths. From the figure, it is obvious that barium sulfate has formed at all depths in the production tubing.

5.4.1 Mechanism of barium sulfate scale formation

Like calcium sulfates, the primary cause of barium sulfate deposition is the mixing between two sources of water, one of which is rich in calcium ions, i.e., produced water, and the other of which is rich in sulfate ions, i.e., the seawater that is used in water injection or workover and maintenance operations. Commingled production between two different formation waters can cause barite scaling. This can cause formation, near the wellbore and producing string damage. Mixing between two different nonscaling well fluids in the topside headers, risers, and flowlines causes a topside sulfate scale problem [95].

Barium sulfate precipitates according to the following equation:

$$Ba^{++} + SO_4^{--} \rightarrow BaSO_4 \quad (5.18)$$

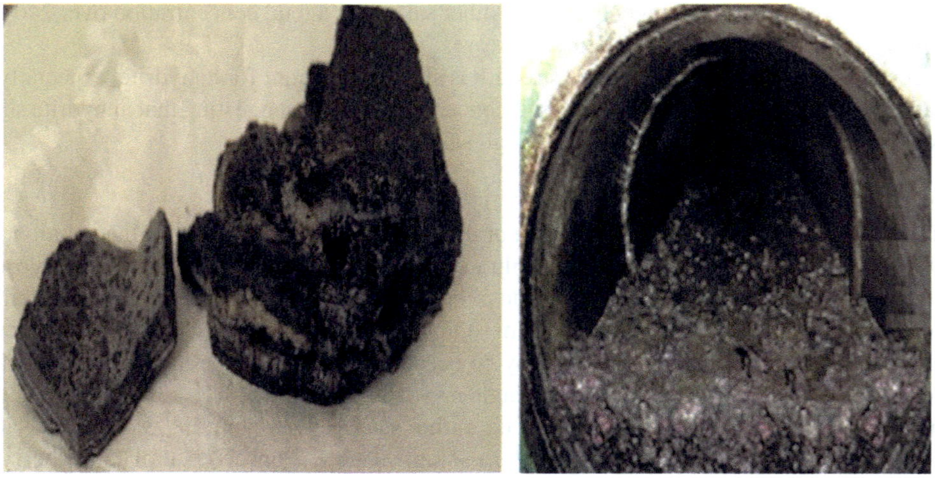

FIG. 5.19

Barium sulfate samples from the field. (left) Collected from a platform. (right) Collected from a transportation pipeline.

Source: Reference H. Gamal, S. Al-Afnan, S. Elkatatny, M. Bahgat, Barium sulfate scale removal at low-temperature, Geofluids (2021), https://doi.org/10.1155/2021/5527818.

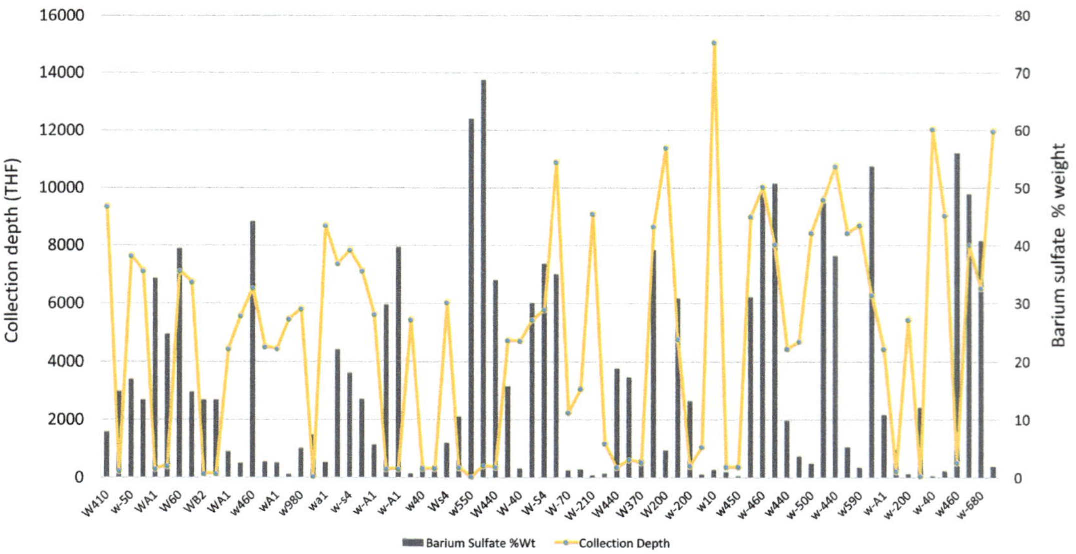

FIG. 5.20

Barium sulfate distribution in the deposits collected from different wells at different depths.

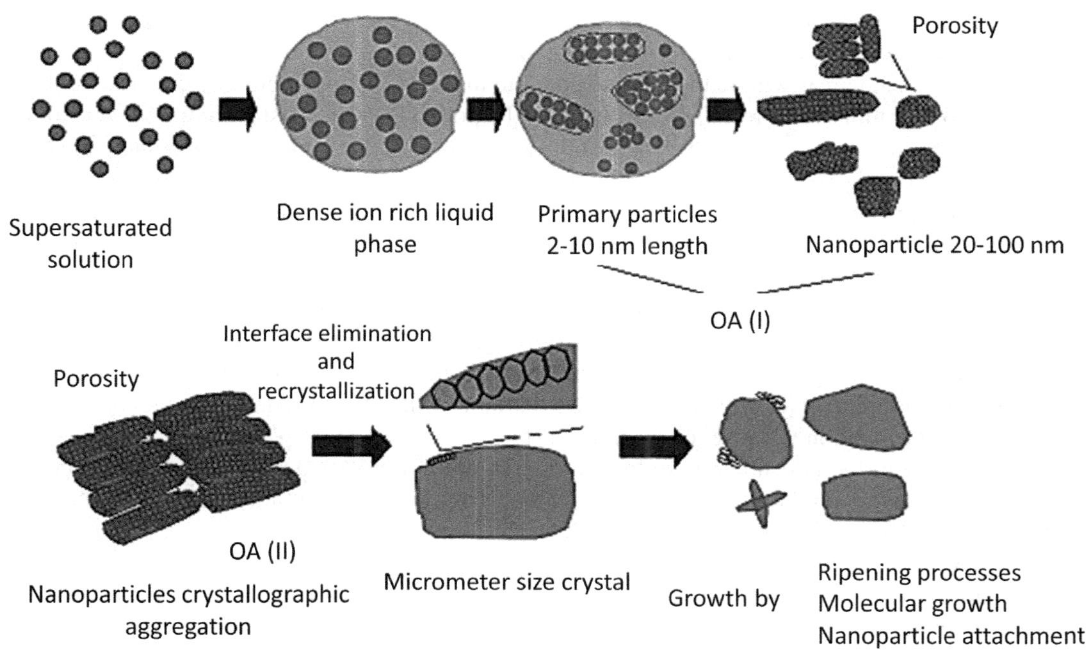

FIG. 5.21

Barite formation proposed model including non-classical and classical mechanisms, OA = oriented aggregates.

Source: Reprinted with permission from C. Ruiz-Agudo, E. Ruiz-Agudo, C.V. Putnis, A. Putnis, Mechanistic principles of barite formation: from nanoparticles to micron-sized crystals, Cryst. Growth Des. 15 (8) (2015) 3724–3733. https://doi.org/10.1021/acs.cgd.5b00315. Copyright {2015} American Chemical Society.

The classical picture of barite precipitation involves the nucleation and crystal growth steps. However, Ruiz-Agudo et al. [96] has reported a nonclassical pathway using electron microscopy. In this nonclassical pathway, the initial stage of the crystallization is a liquid-liquid separation between a denser and less dense phase. The first solid (primary) particles or nuclei seem to form within the initially formed liquid-like precursor phase, which shows crystallographic properties like those of barite. TEM and SEM observations of the nanostructure evolution of samples quenched at successive stages of crystallization indicate two levels of oriented aggregation of nanosized solid particles. The first is the aggregation of crystalline primary nanoparticles of ca. 2–10 nm length to give larger but still nanometer-sized secondary particles (ca. 20–100 nm length). During the second aggregation step of these secondary nanoparticles, most of the porosity in the largest, micron-sized aggregates is annealed, resulting in perfect single crystals [96]. Fig. 5.21 summarizes the mechanisms by which the barium sulfate deposits form:

The authors also concluded that both classical and nonclassical (i.e., oriented aggregation) crystal growth mechanisms are simultaneously operative during barite formation at high supersaturation [96].

In fields under a waterflood with seawater, the barium sulfate scaling tendency at the producer will start low (formation water only), before seawater breakthrough rapidly increases the scaling tendency and it stays high while individual zones or perforations produce a contrast of seawater and formation water. Later in the field life, the scale tendency reduces when seawater dominates and consists of most of the water, while formation water stops being produced [2].

The most affected areas will be the areas exposed to the most mixing, including the near wellbore area and the perforations (or screens). Wylde et al. [97] reported that the worst point for barium sulfate scale is at 200 ft above the perforations, and that the area immediately downstream, including the liner and tailpipe, will be exposed to scaling. The type of reservoir completion most sensitive to blockage will be sand control screens—especially those with low inflow areas, such as cased hole gravel packs. The scaling tendency higher up the well (above the tailpipe) will be lower, as any incompatible fluids will have already mixed and reacted. From field observations, barium sulfate deposits were detected all along the production tubing and on the top locations of the tubing as well, as the mixing causes the precipitation and the temperature, pressure, and flow dynamics determine the location of deposition [3]. Areas of fluid mixing are of high risk of barite deposition, such as the production manifold, where some wells produce mainly formation water (excess barium) while others produce seawater or mixtures of seawater and formation water (excess sulfate) [2].

The solubility of barium sulfate is so low, 2.4 mg/L, which is 150,000 times less than the regular salt, and its solubility product is 1.0810^{-10} at 25°C. That very low solubility allows it to be used medically as a substance taken orally before stomach X-rays are made, because the barium atoms absorb X-rays well. Even though barium ions are quite toxic, barium sulfate can still be taken orally without danger. The compound is so insoluble that it passes through the digestive system essentially unchanged, giving a bright detailed X-ray image [98]. A powder comprising 96% w/w barium sulfate can be reconstituted into a range of concentrations, to be used in both single and air contrast barium enema examinations. The barium enema is shown in Fig. 5.22.

Barium sulfate, also called barite or baryte, exists in a rhombic crystal system, and it is moderately soft mineral (Mohs hardness 2.5–3.5). Its color is crystalline grey, but varies depending on the impurities. Normally opaque, it does occur (rarely) as transparent crystals. Pure barium sulfate occurs as colorless, rhombic crystals with a specific gravity of approximately 4.50. Both commercial barite and barite scale vary considerably in appearance, depending on the deposition conditions and subsequent treatment. Barium sulfate scales were found to contain some strontium, with occasional replacement of Ba by Pb reported to be as high as 30%. A solid solution series between $PbSO_4$ and $BaSO_4$ is possible. Minor replacements of barium by other ions such as mercury, cobalt, and radium may also occur [3].

5.4.2 Factors affecting the formation of barium sulfate scale

5.4.2.1 Effect of temperature

$BaSO_4$ precipitation is affected most strongly by temperature. The solubility of $BaSO_4$ increases with respect to increase in temperature because the dissociation of $BaSO_4$ is an endothermic reaction [99]. Barium sulfate solubility increases with temperature up to 100°C [100–102]. The solubility in water increases from 2.3 mg/L at 25°C to 3.9 mg/L at 95°C, depending on the purity of the sample tested; while the increase is fairly substantial percentage wise, barium sulfate is still insoluble even at this higher temperature. Above 100°C the solubility decreases with temperature in waters with TDS values less than about 50,000 ppm [3,26]. The effect of temperature on solubility of barium sulfate is shown in Fig. 5.23.

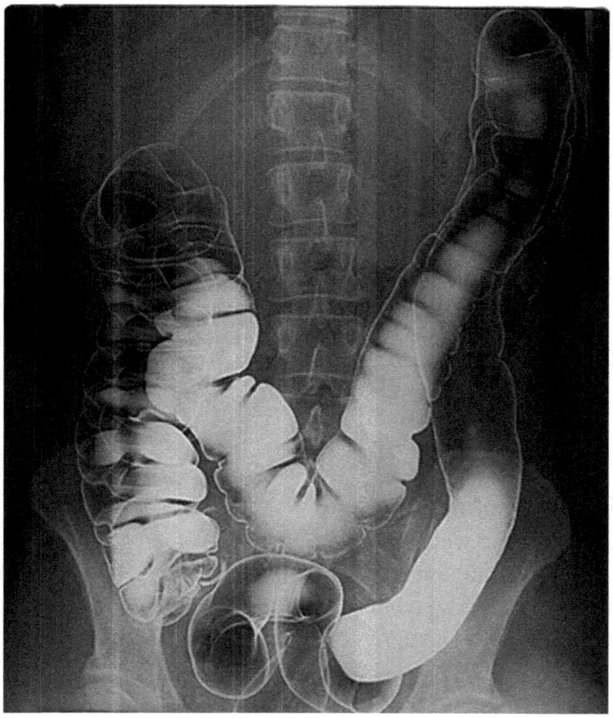

FIG. 5.22

Barium sulfate use in medical X-ray scanning (barium enema).

Source: https://commons.wikimedia.org/wiki/File:Human_intestinal_tract_as_imaged_via_double-contrast_barium_enema.jpg.

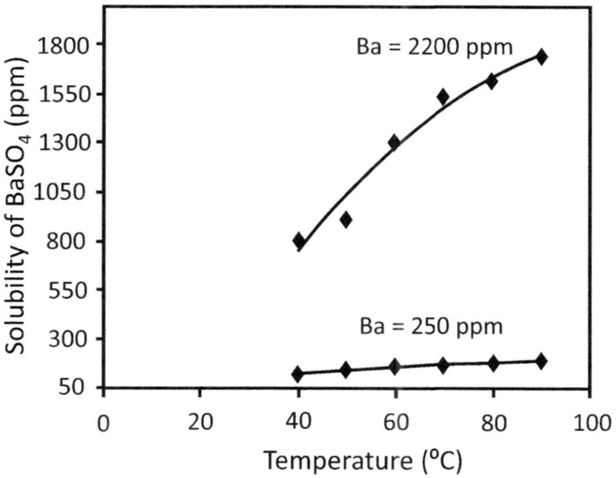

FIG. 5.23

Effect of temperature and salinity on barium sulfate at atmospheric pressure.

Reprinted from reference A.B. BinMerdhah, A.A.M. Yassin, M.A. Muherei, Laboratory and prediction of barium sulfate scaling at high-barium formation water, J. Pet. Sci. Eng. 70 (1–2) (2010) 79–88. https://doi.org/10.1016/j.petrol.2009.10.001, copyright (2010), with permission from Elsevier.

5.4.2.2 Effect of pressure

Pressure drop is one of the primary causes of barium sulfate scale deposition in production systems. Dyer and Graham [103] showed that high temperature and high pressure reduce scaling tendency of barium sulfate in dynamic tube blocking experiments. The tests have also shown that the effect of temperature on scaling tendency was generally greater than the effect of pressure.

Pressure drops across chokes and valves induce turbulence in the water, which helps to overcome supersaturation effects and initiate precipitation [26].

In an oil well in south Louisiana, barium sulfate scale was found at 11,000 ft (possibly due to high mixing downhole), and in over 100 wells operated in this particular field, barium sulfate deposited in the upper 5000 ft (due to pressure and temperature decrease). Also samples of barium sulfate scale were discovered from a 0.5-inch choke that was set at 3652 feet in a 15,500 ft well (due to pressure and temperature decrease), this choke opening was restricted to 0.25 inch in less than a year due to sulfate scale [3].

5.4.2.3 Effect of dissolved solids

The solubility of barium sulfate in water is increased by dissolved salts (ionic strength), just as in the case of calcium carbonate and calcium sulfate. The addition of 100,000 mg/L of NaCl to distilled water increases the solubility of $BaSO_4$ from 2.3 mg/L to 30 mg/L at 25°C. Increasing the temperature to 95°C while maintaining 100,000 mg/L NaCl increases barium sulfate solubility to about 65 mg/L [26].

Some ionic species affect the precipitation of barium sulfate. Zhao et al. [104] found that the deposition rate of the $Ba_xSr_{1-x}SO_4$ solid solution decreased significantly with the increase of strontium ion concentration at a constant barium sulfate supersaturation. The deposition rate decreased up to six times when strontium concentration in aqueous phase increased from $[Sr^{2+}] = 0.5[Ba^{2+}]$ to $[Sr^{2+}] = 8[Ba^{2+}]$.

5.4.2.4 Effect of sulfide scale deposits

Okocha and Sorbie [105] concluded that Fe^{2+}, Pb^{2+}, and Zn^{2+} ions do not cause changes in $BaSO_4$ crystal structure, implying that these ions at low concentrations do not integrate into the $BaSO_4$ crystal lattice and do not affect its morphology. The same results apply for ZnS and PbS, as they show little distortion of the barite lattice when they coprecipitate with it. On the other hand, FeS drastically distorts the crystal lattice, going from a rhombohedral shape to a bipyramidal shape. There also appears to be growth along the crystal lattices enhancing the bipyramidal shape. This indicates that FeS causes extensive intrusion into the $BaSO_4$ crystal lattice, which causes that distortion. This may, in some circumstances, lead to a more easily inhibited barite situation.

5.4.2.5 Effect of fluid hydrodynamics

Field observations have shown that the turbulence found at some locations like chokes, tubing joints, gas lift valves, and safety valves increases the likelihood of scale formation. Increased turbulence also impacts the minimum scale inhibitor concentration required to prevent scale [106].

5.5 Strontium sulfate

That is another hard scale problem usually occurring in oil and gas fields. Strontium sulfate deposits are usually found with barium sulfate deposits, either as separate layers of deposits or cocrystallizing within the barite crystals. Weitreitt and Cowan [107] studied barium sulfate deposits and reported that all of the sulfate deposits analyzed contained strontium sulfate in concentrations ranging from 1.2% to 15.9% and barium sulfate in concentrations ranging from 63.7% to 97.5%. Fig. 5.24 shows a scale sample comprising a mixture of barium and strontium sulfate.

5.5.1 Mechanism of strontium sulfate scale formation

Strontium sulfate precipitates due to the mixing between the formation water that is rich in strontium ion and seawater that is rich in sulfate ion. The following correlation represents the precipitation reaction of strontium sulfate:

$$Sr^{++} + SO_4^{--} \rightarrow SrSO_4 \tag{5.19}$$

Strontium sulfate, also known as celestite or celestine, precipitates as rhombic crystals; the crystals are sky blue in color, which is where the name came from, as celeste means related to the sky. Strontium sulfate solubility is 135 mg/L, its solubility product is 2.23×10^{-7} at 25°C. The specific gravity of strontium sulfate deposits is 3.97, and its hardness is 3–3.5 mohs.

Nucleation and crystal growth of strontium sulfate were found to depend on the degree of supersaturation and presence of impurities, and thus are controlled by a heterogeneous mechanism [108,109].

5.5.2 Factors affecting the formation of strontium sulfate scale

5.5.2.1 Effect of temperature

Like calcium carbonate, strontium sulfate is an inverse solubility salt, as its solubility decreases with increasing the temperature. The solubility of strontium sulfate drops from 135 mg/L at 25°C to 69 mg/L at 125°C [26,110,111]. Fig. 5.25 illustrates the modeled scaling tendency of strontium sulfate for

FIG. 5.24

Sale sample from 18″ pipeline composed mainly of $(BaSr)SO_4$.

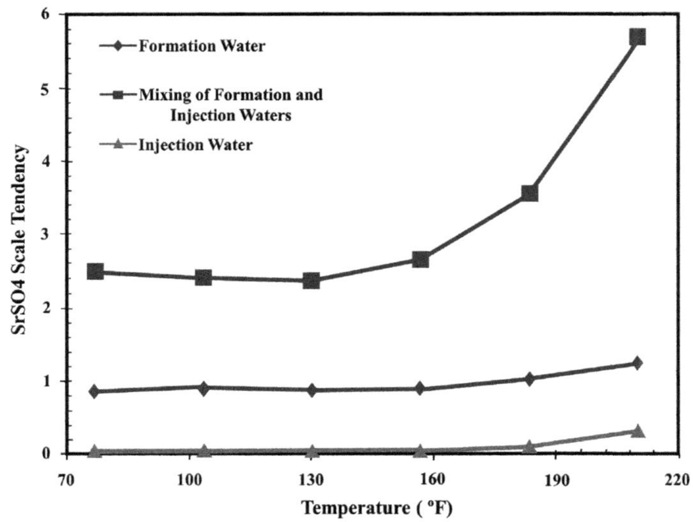

FIG. 5.25

Scaling tendency of strontium sulfate vs temperature.

Reprinted from reference M. Amiri, J. Moghadasi & M. Pordel, The Effect of Temperature, Pressure, and Mixing Ratio of the Injection Water With Formation Water on Strontium Sulfate Scale Formation in the Siri Oilfield, Petroleum Science and Technology, 30:7, (2012) 635-645, https://doi.org/10.1080/10916466.2010.489088 copyright (2012), with permission, from Taylor and Francis.

injection water-formation water mixture in Iranian oilfield, which showed increasing the scaling tendency with temperature [112].

5.5.2.2 Effect of dissolved salts

Like barium sulfate, the solubility of strontium sulfate increases with increasing the ionic strength [113].

Studies showed that strontium sulfate solubility increases as water salinity (as NaCl) increases up to a maximum of approximately 175,000 mg/L, yet further increases in salinity decrease its solubility. Strontium sulfate solubility was found to be greater in brines containing calcium or magnesium than a NaCl brine of equivalent ionic strength [26,114,115].

5.5.2.3 Effect of pressure

Pressure drops are one of the primary causes of strontium sulfate scale deposition in production systems. In addition to decreasing the solubility due to thermodynamic considerations, pressure drops across chokes and valves induce turbulence in the water, which helps to overcome supersaturation effects and initiate precipitation [26].

The effects of temperature and pressure on strontium sulfate and other mineral scales were studied by Amiri et al. [112,116]. The modeled data of injection water-formation water mixture showed relatively decreased strontium sulfate tendency with increasing temperature (Fig. 5.26).

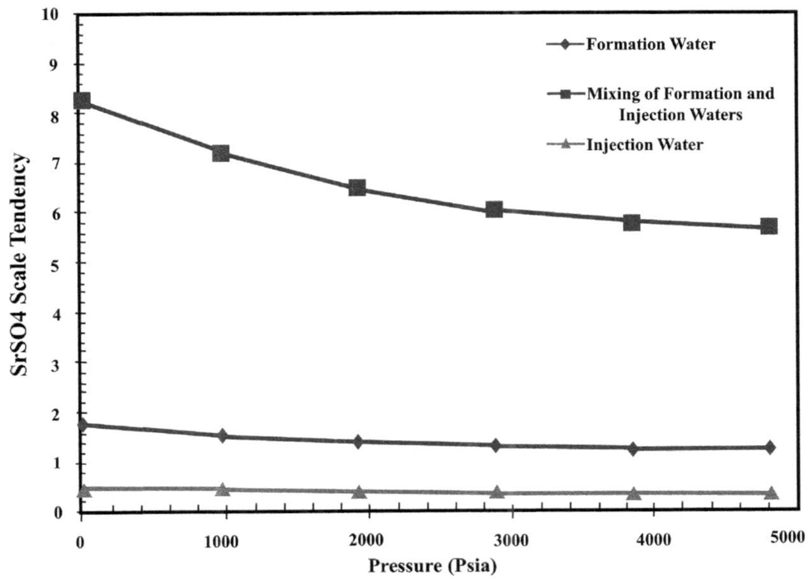

FIG. 5.26

Scaling tendency of SrSO4 vs pressure.

Reprinted from reference M. Amiri, J. Moghadasi, M. Pordel, The effect of temperature, pressure, and mixing ratio of the injection water with formation water on strontium sulfate scale formation in the Siri oilfield, Pet. Sci. Technol. 30 (7) (2012) 635–645, https://doi.org/10.1080/10916466.2010.489088, *copyright (2012), with permission, from Taylor and Francis.*

5.5.2.4 Effect of turbulence

Like barium sulfate, turbulence and highly turbulent locations are more likely to be infested with scale deposits. Additionally, increased turbulence impacts the minimum scale inhibitor concentration required to prevent scale [117].

5.6 Naturally occurring radioactive materials (NORM)

Naturally occurring radionuclides (NORs) or naturally occurring radioactive materials (NORMs) are present at varying concentrations in the earth's crust and consequently will be present in natural concentrations in gas and oil reservoirs. The NORs concentrations in well fluids may become enhanced due to extraction processes and subsequently form NORs enriched deposits within production facilities, thereby forming NORMs [118]. NORs, along with other minerals that are dissolved in the brine, separate and settle out, forming various wastes at the surface [119]:

- Mineral scales inside pipes, which is our main concern in this section.
- Sludges/sediments.
- Contaminated equipment or components.
- Produced waters.

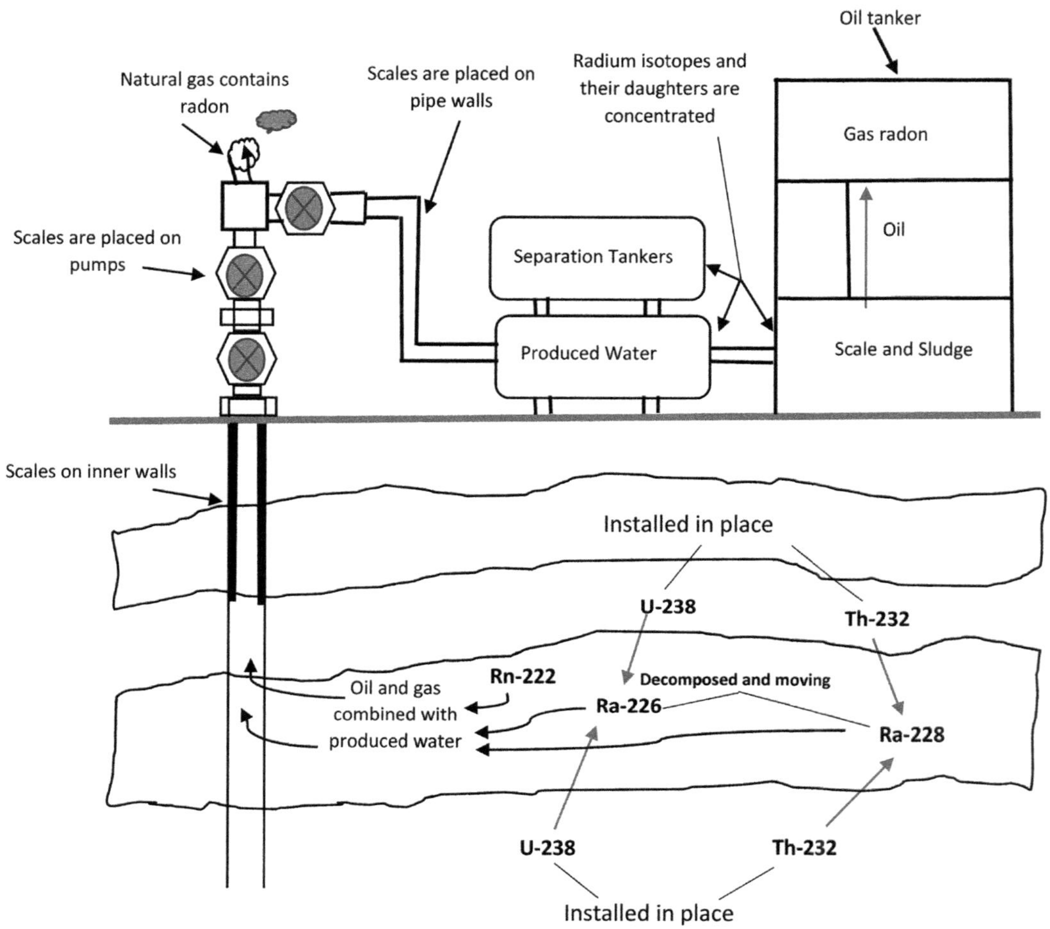

FIG. 5.27

Presence of radioactive materials in oil industry.

Used with permission of [Rightsholder's name], from [Concentrations of TENORMs in the petroleum industry and their environmental and health effects, Ali, M. M. M., Zhao, H., Li, Z., & Maglas, N. N. M., 9, 2019]; permission conveyed through Copyright Clearance Center, Inc.

NORMs can be found in various downhole and surface processing and transport equipment. Flowlines, production tubing, sludge pits, filters, brine disposal/injection wells, and associated equipment may be contaminated with radium NORMs. Radioactive films, coatings, or plating usually form on natural-gas production or processing equipment [120]. Fig. 5.27 illustrates the occurrence of radioactive materials in oil and gas systems [121].

It is estimated that the industry generates about 150,000 m^3 or 260,000 MT of such waste yearly. Field surveys have shown that petroleum pipe scale may have Ra-226 concentrations of 480 pCi/g on average, but can be very high in some cases (more than 400,000 pCi/g) [122]. Some concentrations of radioactive materials in field scale deposits are given in Table 5.6 [121]:

Table 5.6 Concentrations of radioactive materials in field scale.

Radioactive isotope	Concentration in hard scales (Bq kg^{-1})
^{238}U	1–500
^{232}Th	1–2
^{226}Ra	100–15,000,000
^{228}Ra	50–2,800,000
^{224}Ra	–
^{222}Rn	–
^{210}Pb	20–75,000
^{210}Po	20–1500

Used with permission of Royal Society of Chemistry (RSC) Publishing, from [Concentrations of TENORMs in the petroleum industry and their environmental and health effects, Ali, M. M. M., Zhao, H., Li, Z., & Maglas, N. N. M., 9, 2019]; permission conveyed through Copyright Clearance Center, Inc.

5.6.1 Radioactive decay and naturally occurring radionuclide (NOR) formation

Some NORs such as ^{232}Th, ^{235}U, and ^{238}U usually form the starting point of three distinct radioactive decay series. However, due to its low natural abundance ^{235}U, including its decay chain products, is rarely encountered in practice. The ^{232}Th and ^{238}U chains are then the most significant with regards to the origin of oilfield NORs. These "parents" undergo radioactive decay through a series of transformations (Fig. 5.28): loss of an α-particle or a β-particle, or the emission of γ-photons. Each transformation gives rise to a new radioactive "daughter" which is also a NOR that is both lighter and less energetic than its immediate "parent," until the last step when a stable Pb isotope is reached [118,123]. The decay of these unstable radionuclides takes place at different speeds, as some of these elements take millions of years to decay, while others only take seconds to decay. The time span needed for radionuclei to decay is called the half-life period [123].

In a gas/oil reservoir that has not been producing, both the ^{232}Th and ^{238}U decay series are in secular equilibrium (situation in which the quantity of a radioactive isotope remains constant because of its production rate). For reservoirs in production, chemical partitioning equilibrium between NORs present in reservoir rock and reservoir fluids will be dependent on the elemental properties [118].

5.6.1.1 Thorium and uranium NORs

As a gas/oil reservoir is a reducing environment, both Th and U prefer the solid rock phase and do not dissolve in the aqueous or oily phase. Consequently, the top of both series remains with reservoir rock and can only appear in natural concentrations at the surface during drilling operations [118,121].

5.6.1.2 Radium NORs

The partitioning of Ra (^{228}Ra, ^{226}Ra, ^{224}Ra) is special, as they are formed by α-decay from their immediate parent. Ra prefers the aqueous phase, leading to somewhat naturally enhanced concentrations. Therefore upon production the Ra will follow the aqueous produced water stream. Ra is chemically similar to barium (Ba), strontium (Sr), calcium (Ca), and magnesium (Mg) and becomes incorporated in group II sulfate or carbonate deposits and scale [118]. It is the formation of the highly insoluble radium

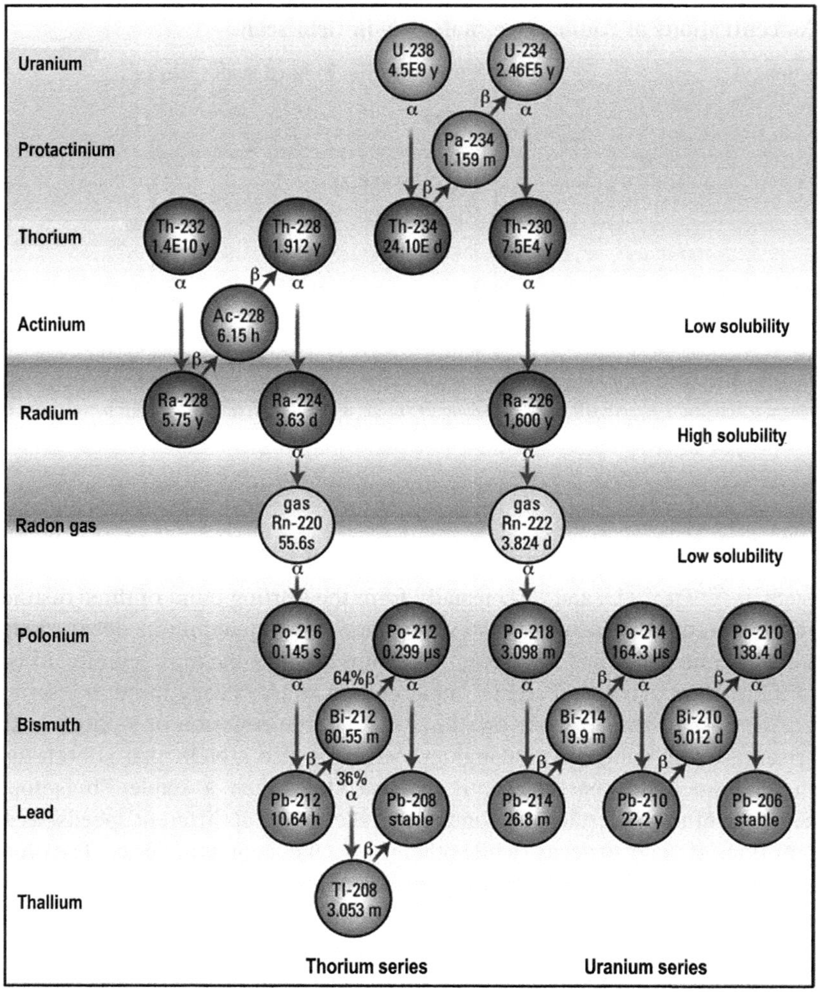

FIG. 5.28

Natural thorium and uranium decay chains. Half-lives and decay, d, days; h, hours; m, minutes; s, seconds; y, years.

Source: A.W. Nelson, E.S. Eitrheim, A.W. Knight, D. May, M.A. Mehrhoff, R. Shannon, R. Litman, W.C. Burnett, T.Z. Forbes, M.K. Schultz, Understanding the radioactive ingrowth and decay of naturally occurring radioactive materials in the environment: an analysis of produced fluids from the Marcellus Shale, Environ. Health Perspect. 123 (7) (2015) 689–696. https://doi.org/10.1289/ehp.1408855.

sulfate within the crystal lattice structures of strontium, but more likely barium sulfate scales, which exhibit radioactive properties [3], as it is considered the primary host of the oil field NORM [124]. Radium may also precipitate to form complex sulfates and carbonates [122]. In practice, any scale or corrosion product may contain coprecipitated radium ions and be radioactive [26]. 226Ra and 228Ra were found in black powder residue in gas pipelines [125]. An example of NORM scale is shown in Fig. 5.29.

FIG. 5.29

Norm scale sample from pipeline.

Once deposited inside installations or produced at the surface, due to their radiation characteristics all three Ra NORMs show different behavior [118,126].

When radium is brought to the surface in produced water, the concentration of radium that is incorporated in barite scale is largely a function of [124]:

(1) The concentration of dissolved radium.
(2) The amount of produced water that moves past the site of barite precipitation.

5.6.1.3 Radon NORs

Ra isotopes prefer the (natural) gas phase for transport from the subsurface to the surface. Hence, scale in gas wells and equipment can also contain the radon progeny ^{210}Pb and ^{210}Po [122]. As ^{220}Rn (also known as thoron) only lives for minutes by the time this NOR reaches the surface, it will have decayed. ^{222}Rn with a half-life of about 4 days may appear in natural gas liquids (NGL) processing installations (particularly the C3 process stream), in the gas cap of crude oil storage tanks (associated gas), or in natural gas (NG) transmission lines. In the absence of NG, ^{222}Rn partitions in the (light) hydrocarbon (NGL) and/or aqueous phase. When produced with the oil and gas, ^{222}Rn will usually follow the gas stream [118,123]. It is worth mentioning that radon itself is not a particularly hazardous material.

Because it is chemically unreactive, it does not accumulate in the body. The health hazards associated with radon exposure are from its decay products [120].

5.6.1.4 Lead NORs

In gas or oil fields where stable Pb is present in produced waters, ^{210}Pb may be incorporated in any formed deposit as well. Such ^{210}Pb deposits were observed in piping, equipment, tank bottoms, and sludge from booster stations, originating from initial ^{222}Rn with high concentrations and supposedly appearing due to prolonged processing. Similarly, $^{210}Pb_{eq}$ can accumulate as thin radioactive films on the inner surfaces of gas processing equipment and pipelines due to prolonged processing times [IOGP 2016] [118]. ^{210}Pb was also observed in black powder residue in gas pipelines and seems to be inversely correlated with the distance to the platforms [125].

Alternative terminology of NORMs can also be commonly encountered. The acronym TENORM—Technologically Enhanced NORM—is mainly used in the United States and is used to discriminate between "natural" NORM and any NORM subject to human driven technological processing. Radioactive scale deposits were initially indicated as low specific activity (LSA) scale. This acronym emerged from the International Atomic Energy Agency (IAEA) transport regulations. Similarly, installation parts that are contaminated at their inner surface with deposited NORMs are sometimes termed surface contaminated object (SCO) [118].

5.6.2 Factors affecting the formation of NORM scale

Changes in operating conditions that can lead to NORM scale formation include [118]:

- Temperature and pressure variations.
- Varying fluid flow (transition between laminar and turbulent) and gas expansion.
- pH variation.
- Increasing sulfate ion concentration during seawater flood injection.
- Mixing incompatible waters or commingled production.
- Water volumes; the amount of NORM material from a producing field generally increases as the amount of water pumped from the formation increases.
- Presence of seed crystals on the inner surface of the equipment and rough surfaces (for example, in sour gas wells, the presence of NORM scale is enhanced by the presence of gypsum).

5.7 Iron compound scales

Iron is the most common metal on earth, and the fourth most common element in Earth's crust by weight. Moreover, much of Earth's core is thought to be composed of iron. Most of the production facilities and equipment in today's industries are made of steel. When fluids flow through these facilities and a corrosion process takes place where the iron ions go into the solution, after a while there is a big chance that these iron ions deposit as iron compound (sulfides, oxides, carbonates, etc.) and cause operational problems. Therefore iron deposits are one of the complicated deposits in all industries, due to their high abundance, the various causes of their formation, and also due to their complex chemistry. Fig. 5.30 represents the diversity of iron compounds.

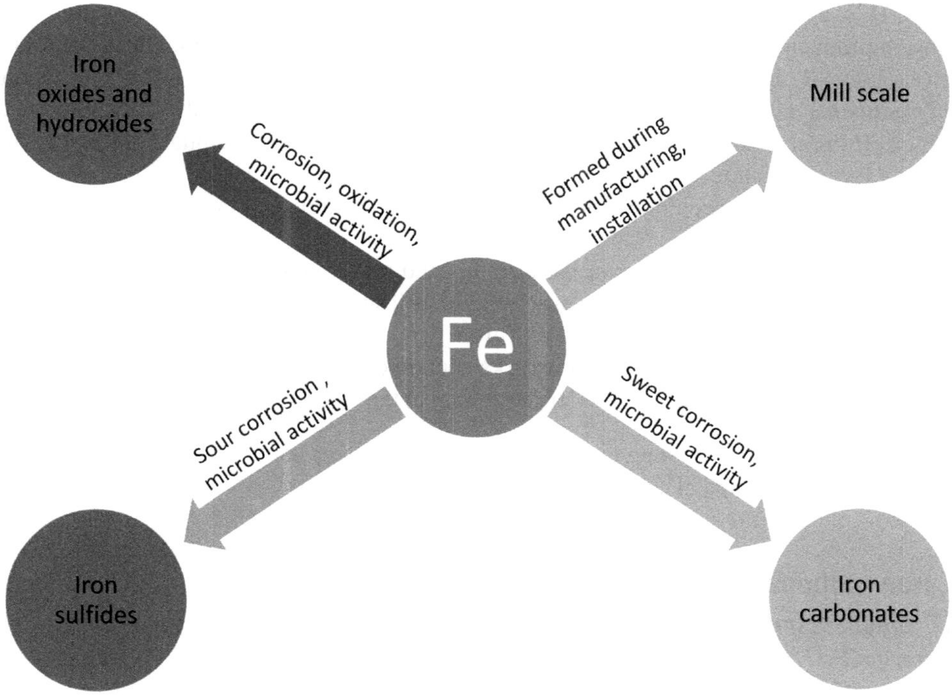

FIG. 5.30

The diversity of iron compounds.

5.7.1 Sources of iron ion in production fluids

Iron deposits can contain compounds of different Fe valences (basically FeII and FeIII) with the possibility of forming various sulfides, carbonates, oxides, and hydroxides. The conversion/transformation between iron compounds of the same group is common (i.e., conversions between sulfides and each other), and also the conversion between the different groups (conversion of iron sulfides to oxides, carbonates to oxides, and vice versa) is another property of iron compounds. These properties of iron compounds make understanding, detecting, and controlling them a daunting task.

Iron has several oxidation states: −2, −1, 0, +1, +2, +3, +4, +5, and +6; while in aqueous solutions, 0, +2, and +3 are most common. At equilibrium conditions, all the iron is present in +3 oxidation state and it happens usually in the surface facilities. At normal reservoir conditions, iron is present in +2 oxidation state. From the analysis of different field samples it was observed that ratio of iron (II) to iron (III) is 5 to 1 but it can be different for different types of well and formation [127–129]. Formation waters normally contain only a few mg/L of natural iron, and values as high as 100 mg/L are rare [26]. Higher iron contents may be invariably attributed to corrosion processes.

Generally speaking, the different sources of iron in formation water are as follows:

1- ***Corrosion***: corrosion is a recurrent problem in oil and gas operations and that makes it a prosperous source of iron species. Among the different corrosion processes comes the corrosion due to gases like O_2, H_2S, and CO_2, as a very common and troublesome issue in oil and gas systems, as their corrosion products are one of the main forms of mineral scales. In addition, bacterial actions (or microbiologically induced corrosion MIC) are also another source of these corrosion products.

2- ***Acid treatments***: after acid cleaning of the pipes or after stimulation jobs, the iron is converted into soluble FeII form, which when the pH reaches a specific value can reprecipitate. Mixing of rust with acids can form a mixture of iron (II) and iron (III) [130].

3- ***Mill scale***: mill scale is steel leftovers that are found after installing a new pipeline or maintaining a used pipeline.

4- ***Drilling fluids***: contaminated or unusual drilling fluid may contain iron. In some unusual drilling fluid, iron oxide or iron-based hydrogen scavenger or iron as a weighing material for drilling fluid are added, which work as another source of iron.

5- ***Reservoir formation***: Different reservoir rocks contain different iron content. Sandstone reservoir usually has high iron content [131]; siderite existing in the reservoir rocks is another source of iron.

As discussed, iron has different sources, and most of these sources and their affiliated deposited compounds are operations related. The different types of iron deposits are discussed in the next sections.

5.7.2 Iron carbonate scale

5.7.2.1 Mechanisms of iron carbonate scale formation

Iron carbonate ($FeCO_3$) or siderite is one of the main corrosion products in the CO_2 corrosion process. CO_2 dissolves in the aqueous phase that is often produced along with crude petroleum, and creates an acidic solution (according to Eqs. 5.3–5.5) that is corrosive to the inner walls of steel pipeline. Sweet corrosion or CO_2 corrosion process is illustrated in Fig. 5.31.

Solid $FeCO_3$ forms on the steel surface if the product of ferrous ion concentration (Fe^{2+}) and carbonate ion concentration (CO_3^{2-}) exceeds the solubility product limit according to the reaction [132,133]:

$$Fe^{++} + CO_3^{--} \rightarrow FeCO_3 \tag{5.20}$$

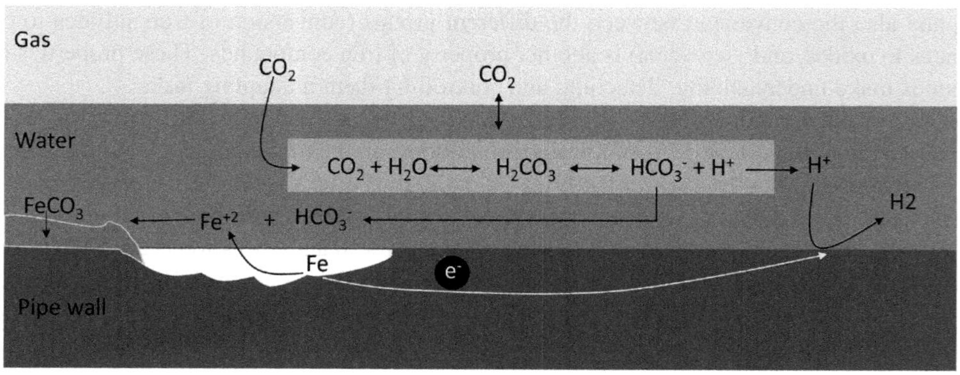

FIG. 5.31

Sweet corrosion process.

Another way of representing the precipitation reaction is as follows [134]:

$$Fe^{++} + 2HCO_3^- \rightarrow FeCO_3 + CO_2 + H_2O \tag{5.21}$$

Other suggested mechanisms include the following [134, 135, 136]:

$$Fe^{++} + 2H_2O \rightarrow Fe(OH)_2 + 2H^+ \tag{5.22}$$

$$Fe(OH)_2 + HCO_3^- \rightarrow FeCO_3\left(\text{or}\left(Fe_2(OH)_2CO_3\right)\right) + H_2O + OH^- \tag{5.23}$$

The source of carbon dioxide (CO_2) gas in the petroleum industry emerges from geological formations and accompanies the crude oil and natural gas that are being brought to the surface via pipelines.

Typically, such aqueous CO_2-induced corrosion scales are assumed to only consist of iron carbonate or siderite ($FeCO_3$). However, in reality, another forms of iron carbonate deposits have been reported, including iron-oxy carbonate ($Fe_2O_2CO_3$) and iron-hydroxycarbonate ($Fe_2(OH)_2CO_3$) Chukanovite, $Fe_6(OH)_{12}(CO_3)$, and $Fe_6(OH)_{12}(CO_3)\cdot 2H_2O$ phases [136], and can also include some forms of iron oxide (Fe_3O_4) [137], at solution temperatures lower than 50°C. This iron oxide, magnetite, is also observed as a sweet corrosion product at higher solution temperatures (>150°C) [138]. Iron can also form a complex carbonate with calcium and magnesium, called $Ca(Fe,Mg)CO_3$ (ankerite).

Once a layer of iron carbonate forms on the surface, it presents a diffusion barrier for the species involved in the corrosion process by covering up a portion of the steel surface and preventing the underlying steel from further dissolution [133].

Another possible source of deposition of the iron carbonate is the bacterial activity (biomineralization). In systems that do not have enough concentration of carbonate or carbon dioxide, siderite deposition is considered strong evidence of bacteria activity or microbial corrosion. The main biogeochemical processes that are believed to have the potential to induce carbonate precipitation are (1) bacterial sulfate reduction, (2) archaeal methanogenesis, and (3) dissimilatory ferric iron reduction. These processes can increase the pH or alkalinity which finally induces the precipitation of the carbonates [139].

In the presence of SRB as products of the microbial corrosion process are the minerals siderite and iron (II) sulfide in a ratio of 3:1 or more [140,141]; the process takes place according to the following equations [142,143]:

$$\text{fatty acid} + SO_4^{2-} + H_2O \xrightarrow{\text{SRB bacteria}} HS^- + HCO_3^- \tag{5.24}$$

$$4Fe + SO_4^{2-} + 3HCO_3^- + H_2O \rightarrow FeS + 5OH^- + 3FeCO_3 \tag{5.25}$$

Methanogens convert the organic compounds into methane gas and CO_2, which dissolves in the water, producing the bicarbonate [144]:

$$CH_3COOH \rightarrow CH_4 + \left[CO_2 + H_2O \rightarrow HCO_3^-\right] \tag{5.26}$$

Also methanogens in the presence of sulfate-reducing prokaryotes can cause corrosion and produce methane and siderite:

$$4Fe + 5HCO_3^- + 2H_2O \rightarrow CH_4 + 5OH^- + 4FeCO_3 \tag{5.27}$$

Siderite formation is also known to be facilitated by both mesophilic and thermophilic iron-reducing bacteria [145]. These bacteria basically reduce Fe III to Fe II and precipitate it, and one of these precipitation forms is siderite, according to the following equation:

$$12FeOOH + CH_3CH(OH)COO^- \text{ (lactate or other organic materials)} + 4H_2O \rightarrow 3HCO_3^- + 22OH^- + 4Fe^{2+} \tag{5.28}$$

As indicated earlier, siderite is generally a product of the sweet and microbial corrosion processes. It precipitates in a trigonal crystal system, with colors ranging from yellowish brown to brown, gray, yellowish gray, to greenish gray. It has a density of ~3.96 g cm^{-3}, hardness 3½–4½, and it has a limited solubility of 0.0067 g/100 cc water at 25°C; its solubility product is 9.3×10^{-12} [3,146]. Siderite has also been identified in extraterrestrial materials, such as meteorites and interplanetary dust particles [147].

5.7.2.2 Factors affecting the formation of iron carbonate scale

The factors discussed in this section are related to different perspectives on precipitation and deposition of iron carbonate: from one perspective, the factors that are known to affect the solubility of iron carbonate, and from a different perspective, how these factors affect the sweet corrosion process that finally leads to siderite precipitation.

5.7.2.2.1 Effect of temperature

Solubility of iron carbonate decreases with increasing temperature, as siderite dissolution is an exothermic reaction [148]. The value of siderite K_{sp} decreases with increasing temperature [149]. Many correlations have been proposed to relate the solubility product of the siderite with the temperature, among them the correlation proposed by Braun [150]:

$$\log K_{sp} = -10.2 - 0.0314T_c \tag{5.29}$$

Temperature generally increases the corrosion rate. The rate of corrosion reaction doubles with every 10°C increase in temperature [151]. At atmospheric pressure, increasing temperature accelerates carbon steel corrosion and sweet corrosion scale formation kinetics [146]. Vuppu and Jepson [152] showed that, for saltwater, the corrosion rate increases with temperature up to approximately 60°C and then decreases due to the formation of protective iron carbonate layers. The corrosion rate increases as the oil composition is increased up to 60%. Above this, the corrosion decreases rapidly, due to the presence of a water layer at the bottom of the pipe for oil compositions up to 80%.

Renpu [153] identified three temperature regions that are included in the carbon dioxide corrosion process:

1. Region of temperature <60°C. In this region, the metal surface may form a small quantity of nonprotective loose $FeCO_3$; hence, uniform corrosion is mainly generated, where the corrosion rate increases with increasing temperature. The maximum corrosion rate of steel is presented at about 40°C for Mn-containing steel and at about 60°C for Cr-containing steel.
2. Region of medium temperature of about 100°C. In this region, the $FeCO_3$ film is irregularly formed on the metal surface, where loose crystal is formed; thus severe localized corrosion is presented.
3. Region of high temperature >150°C. In this region, highly adhesive tight $FeCO_3$ and Fe_3O_4 films are formed, which slow or inhibit corrosion and reduce corrosion rate.

5.7.2.2.2 Effect of pressure

Like other carbonates, increasing the pressure increases the solubility of iron carbonates, because it shifts the equilibrium in Eq. (5.21) to the left. Deposits of iron carbonate type observed in wells in east Texas and south Louisiana were usually situated in the upper portions of the tubing strings and in surface equipment, mainly at areas where pressures drop drastically [3].

On the other hand, experimental and field evidence has often suggested that steel corrosion rate increases with partial pressure of carbon dioxide (P_{CO2}). P_{CO2} raises the concentration of dissolved CO_2 ($CO_{2(aq)}$), which enhances the acid H_2CO_3 formation/dissociation and causes pH to decrease, which encourages iron corrosion and siderite dissolution. Ref. [153] categorized sweet corrosion based on P_{CO2} into three categories:

- Serious corrosion of $P_{CO2} > 0.2$ MPa
- Corrosion if $P_{CO2} = 0.02$–0.2 MPa
- No corrosion if $P_{CO2} < 0.02$ MPa

In a formation water matrix, and when $P_{CO2} = 0.05$–0.1 MPa, the severity of corrosion can be estimated by using the ratio of the molal concentration multiplied by the number of electrovalence for Ca^{2+} to that for HCO_3^- as follows [153]:

- Low corrosion rate if $Ca^{2+}/HCO_3^- < 0.5$
- Medium corrosion rate if $Ca^{2+}/HCO_3^- > 1000$
- High corrosion rate if $0.5 < Ca^{2+}/HCO_3^- < 1000$

The carbon dioxide corrosion of pipe includes uniform corrosion and pitting corrosion.

5.7.2.2.3 Effect of pH

Increasing the pH of the solution increases the potential of the iron carbonate precipitation. The reservoir brines that accompany crude oil can contain significant concentrations of carbonate salts, which adjust solution pH to more alkaline values (~pH 5–8) [146]. The equilibrium of such carbonate species and iron carbonate is also a factor in the pH. The formation of iron-containing scale because of highly dissolved iron content and increased pH (≥8.5) has been observed in a monoethylene glycol regeneration system [95].

On the other hand, decreasing the pH increases the iron metal dissolution, and increases the sweet corrosion rate. Typically, solution pH of CO_2-saturated deionized water (or condensed water on pipeline surfaces) at atmospheric pressure is ~4, although it becomes more acidic at higher CO_2 partial pressure.

pH also affects the texture and hardness of iron carbonate scale. Iron carbonate corrosion scale that was formed at pH 6.0 was less protective and showed higher scale porosity compared with that formed at pH of 6.6 [154].

5.7.2.2.4 Effect of water cut

Increasing the water cut increases the corrosion process and also increases the mass transfer and diffusion of scaling species towards the pipe wall. Generally, when oil and gas lines transport low water cut goods (e.g., water:oil < 30%), corrosion is usually not a concern due to the lack of aqueous phase.

5.7.2.2.5 Effect of dissolved ions

Chokshi et al. [155] demonstrated that by increasing bulk $[Fe^{2+}_{(aq)}]$, the sweet corrosion rate of carbon steel decreased with higher rate of $FeCO_3$ layer formation, or in other words, more siderite deposits and less corrosion rate.

The presence of calcium in solution containing ferrous iron had a large influence on increasing the solubility of siderite. The formation of calcium carbonate salt strongly impaired the affinity of ferrous iron toward carbonate and prevented the further precipitation of siderite at high SI values. In spite of the formation of mixed scale of both calcium and iron carbonate, $Fe_xCa_{1-x}CO_3$, has been observed in several oil wells and water treatment plants [156].

HCO_3^-/CO_3^{2-} ion species have been shown to accelerate both the cathodic and anodic half reactions [157].

The solubility of iron carbonate increases with increasing the ionic strength (I). Silva et al. [158] proposed the following equation for the relationship between ionic strength and solubility product of iron carbonate room temperature (25°C):

$$\log K_{sp} = -10.9 + 2.518 I^{0.5} - 0.657 I \tag{5.30}$$

A more unified equation that relates the ionic strength and temperature of iron carbonate to the solubility product was proposed by Nešić et al. [149]:

$$\log K_{sp} = -59.3498 - 0.041377 T_K - \frac{2.1963}{T_K} + 24.5724 \log(T_K) + 2.518 I^{0.5} - 0.657 I \tag{5.31}$$

On the other hand, some researchers conclude that sweet corrosion is generally accelerated with increasing NaCl concentration with a greater impact on the anodic dissolution kinetics [146].

5.7.2.2.6 Effect of organic acids

When the acids are present in the acid form, they decrease the pH of the solution, increase the corrosion rate, and increase the iron carbonate solubility. However, when these acids are in salt form, they increase the alkalinity of the solution, which affects the carbonate scale formation.

Crolet and Bonis [159] have proposed a classification in which less than 0.1 mM HAc is assumed to give only slightly increased risk of corrosion, while concentrations of more than 1 mM significantly increase the corrosivity. Undissociated HAc concentrations of more than 10 mM have been reported in some fields.

Organic acids were found to increase the CO_2 corrosion rate, and make it harder for protective iron carbonate scales to form due to a "scale undermining" effect. The scale precipitation rate is not directly affected [160].

5.7.3 Iron sulfide scale

Iron sulfide scale is present in sour oil and gas wells and injectors that are contaminated with sulfate-reducing bacteria (SRB). It enhances the corrosion rate of the downhole tubulars, and adversely affects the performance of various wells. It also reduces the efficiency of oil-water separation in various operations [131].

5.7.3.1 Mechanisms of iron sulfide scale formation

There are various mechanisms that can lead to the formation of iron sulfide. However, all of these mechanisms require sources of hydrogen sulfide and iron [131], according to the following equation:

$$Fe^{++} + S^{--} \rightarrow FeS_{(S)} \tag{5.32}$$

where the source of the iron is indicated in Section 5.7.1, and the sulfide can originate from different sources, including the following:

- Microbial activity: SRB and sulfate-reducing archaea (SRA) use sulfate ion in its anaerobic dissimilatory respiration to produce the hydrogen sulfide. The process requires an electron donor and carbon source at optimum temperature. The biochemistry of the sulfate reduction is illustrated in Fig. 5.32 [161].
- TSR: At temperatures above 250°C, aqueous sulfate, derived from seawater, pore water, or from dissolution of solid calcium sulfate (mainly gypsum and anhydrite), can be reduced by a variety of organic compounds, resulting in high concentrations of H_2S in the reservoir fluid [62,162]:

$$\text{Hydrocarbons} + SO_4 \rightarrow H_2S \tag{5.33}$$

- Thermolysis and/or aquathermolysis of organosulfur compounds such as thiophenes in heavy oil also produce H_2S. Aquathermolysis reactions between oil and steam are dominant at temperatures below 240°C, while thermolysis (i.e., in the absence of water) dominates at higher temperatures [162].
- Gas lift operations, as injecting the lift gas that has significant amounts of H_2S into the tubing string leads to sour corrosion and sulfide deposition.
- Acidizing in sour wells.
- Hydrolysis of minerals that are rich in sulfides like pyrite, sphalerite, and galena.
- The decomposition of corrosion inhibitor and drilling compounds can also produce sulfide ions at high temperature when tested in autoclave equipment, but are very unlikely to be the source of sufficient sulfide ions to give scale deposition over many years of production [163].

Bacterial sulfate reduction, thermal cracking, and TSR are the common sources of H_2S and reservoir souring [62]. The degradation of carbon steel in the presence of H_2S proceeds according to the following anodic reactions [164]:

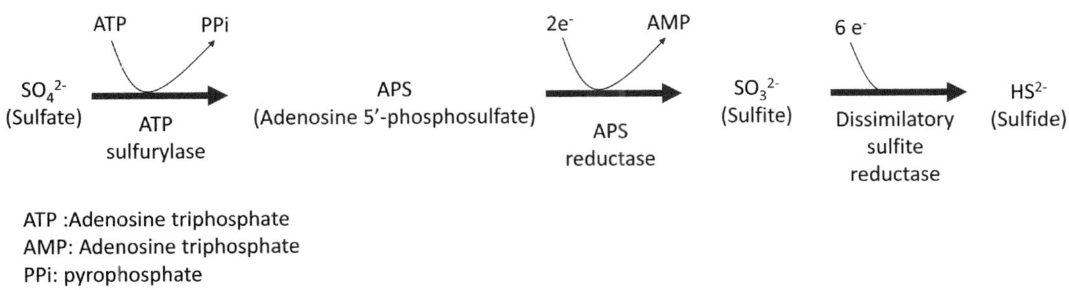

FIG. 5.32

Sulfide production from sulfate reduction by SRB.

$$Fe + H_2S + H_2O \leftrightarrow FeSH^-_{ads} + H_3O^+ \tag{5.34}$$

$$FeSH^-_{ads} \leftrightarrow Fe(SH)_{ads} + e^- \tag{5.35}$$

$$Fe(SH)_{ads} \leftrightarrow FeSH^+ + e^- \tag{5.36}$$

$$FeSH^+ + H_3O^+ \leftrightarrow Fe^{2+} + H_2S + H_2O \tag{5.37}$$

$$2FeSH^+ \leftrightarrow FeS_{ads} + HS^- + H^+ + Fe^{2+} \tag{5.38}$$

$$2nFeS_{ads} \rightarrow nFe_2S_2 \rightarrow 2nFeS_{mackiawite} \tag{5.39}$$

where $FeSH^-{}_{ads}$ and $FeSH_{ads}$ are adsorbed mesophases and FeS_{ads} is the adsorbed amorphous FeS.

The cathodic reactions involve the following:

$$2H_2S + 2e^- \rightarrow H_2 + 2HS^- \tag{5.40}$$

$$HS^- + e^- \rightarrow H + S^{2-} \tag{5.41}$$

$$H^+ + e^- \rightarrow H \tag{5.42}$$

From the preceding equations and depending on its concentration, H_2S plays a dual role. First, it accelerates the corrosion rate by providing an additional cathodic reaction, and second, it favors the formation of an iron sulfide layer by providing more HS^- ions [165].

A common example of the brutal effect of H_2S, even at very low iron sulfide deposition rates, is by considering a pipeline that has hydrogen sulfide at 1 part per million and natural gas flow of 210 TJ/day, with the gas calorific value being 40 MJ/m^3; if all the H_2S is converted to iron sulfide (FeS), then approximately 6600 kg per annum of FeS will be produced in this pipeline [166]. Besides the high volumes of deposits that can potentially form, the solubility of iron sulfide compounds is very low. For example, the solubility of the most soluble stable iron sulfide phase (called mackinawite) is five orders of magnitude lower than the solubility of calcite ($CaCO_3$) [167], which indicates the drastic effect on production operations. The solubility of iron sulfide phases was extensively studied by Rickard [168], Berner [169], and Tewari et al. [170]. Fig. 5.33 shows mixed calcium carbonate and iron sulfide scale in a pipeline, and Fig. 5.34 shows iron sulfide scale collected from a production well.

5.7.3.2 Iron sulfide compounds and polymorphs

The combination of hydrogen sulfide and iron will cause the formation of various types of iron sulfides, with different iron-to-sulfide (Fe/S) ratios, in the same well. Table 5.7 summarizes the iron sulfide polymorphs and their properties.

The Fe/S ratio in these species depends on temperature, pressure, pH, age, H_2S concentration, and the location in the well. This Fe/S ratio determines the iron sulfides solubility and plays a key role in their dissolution and removal. Generally, the Fe/S ratio in the iron sulfide scales decreases as the location of the formed scale is closer to the well head. Hard (low solubility) or insoluble scales (like pyrite FeS_2) exist at shallower depths compared to the soft or soluble scales, which exist at deeper depths (like troilite FeS). Hydrochloric acid can be used to dissolve iron sulfide species that contain iron and sulfur at a molar ratio close to unity. Nonacid formulae can be used to remove iron sulfide scale; however, their ability to dissolve iron sulfide also depends on the molar ratio of Fe/S. Therefore a single treatment cannot be designed for the same well because HCl can remove FeS, but it cannot remove the FeS_2. Fig. 5.35 represents the different forms of iron sulfides and their possible locations and solubility in a production well.

FIG. 5.33

Mixed calcium carbonate and iron sulfide scale layer in pipeline.

FIG. 5.34

Iron sulfide scales collected from production well.

Table 5.7 The different forms of iron sulfide deposits and their properties [131,171–173].

Form/ parameter	Chemical formula	Crystal system	Color	Solubility product	Hardness (Mohs)	sp. gr	Solubility/stability
Mackinawite	Fe_9S_8 ($FeS_{0.93–0.96}$)	Tetragonal	Bronzy	2.88×10^{-18}	2.5	4.30	Metastable phase, found in freshly formed precipitates, can be precursor to other forms, rapidly soluble in acid
Marcasite	FeS_2	Orthorhombic	Tin-white	8.65×10^{-26}	6–6.5	4.88	Metastable phase, FeII disulfide compound, slowly soluble in acid
Pyrite	FeS_2	Cubic	Pale brassy yellow	8.51×10^{-26}	6–6.5	5.01	Very stable phase, FeII disulfide compound, known as "fool's gold"; usually deposits at shallower places downhole. Slow and very difficult to dissolve in acid
Pyrrhotite	Fe_7S_8	Monoclinic	Bronze-yellow	2.70×10^{-19}	3.5–4.5	4.69	Nonstoichiometric stable group where $x > 0.2$; monoclinic form is approximately Fe7S8; hexagonal form is approximately $Fe_{10}S_{11}$. Moderately soluble in acid
Troilite	FeS	Hexagonal	Light grayish brown	6.17×10^{-17}	3.5–4.5	4.85	Stoichiometric end member of the Fe1-xS group. Rapidly soluble in acid
Greigite	Fe_3S_4 ($FeIIFe_2IIIS_4$)	Isometric	Metallic pinkish-tarnishing to a metallic blue	2.99×10^{-55}	4–4.5	4.05	Metastable FeII-FeIII sulfide; the iron member of the thiospinel
Smythite	Fe_9S_{11} $(Fe,Ni)_{3+x}S_4$ ($x = 0–0.3$)	Hexagonal	Brownish black, bronze-yellow	–	4.5	4.32	Metastable phase related to the Fe1-xS group

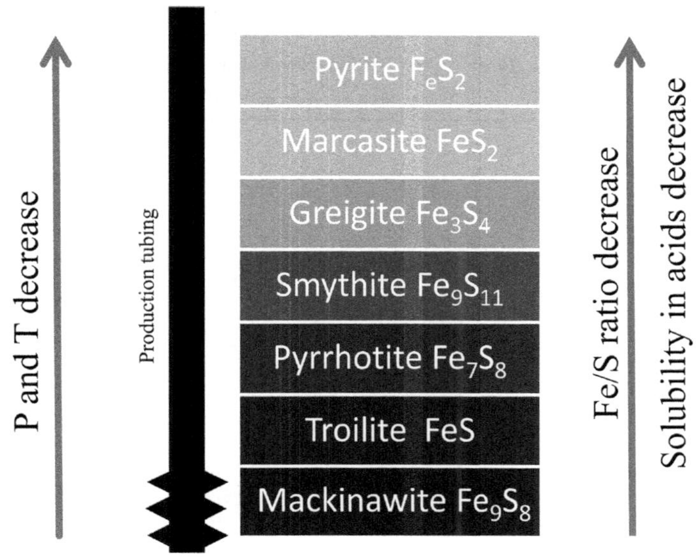

FIG. 5.35

Iron sulfide forms, order of formation, and solubility in a production well.

Among those iron sulfides that form in CO_2/H_2S solutions, it is thought that mackinawite usually forms as a precursor to other types of sulfides, converted to greigite in a time- and pH-dependent manner [174]. Amorphous iron sulfide has also been reported. Pyrite is also transformed from other form phases, which can be mackinawite, pyrrhotite, or troilite. More research is needed to investigate the real transformation processes and the transformation impact factors.

Although mackinawite is the first-appearing iron sulfide phase, in the oilfield pyrrhotite is usually the most common phase [175,176]. It is believed that pyrrhotite is transformed from mackinawite, and the transformation process is related to pH, temperature, and time length [175]. The relative abundance of iron sulfides in one of the Saudi oilfields was found to be 58% pyrrhotite, 13% troilite, 11% mackinawite, 9% pyrite, 6% marcasite, and 3% greigite [176].

5.7.3.3 Factors affecting the formation of iron sulfide scale

As with iron carbonate, since the main cause of sulfide deposition is the sour corrosion process. Therefore the factors discussed here will take into account the factors affecting the iron sulfide solubility as well as the factors affecting the sour corrosion process.

5.7.3.3.1 Effect of temperature

High temperature decreases the solubility of iron sulfides [177] as well as accelerating iron sulfide formation and phase transformation kinetics. Under reservoir high pressure/high temperature conditions (100–250°C, 336.6–1654.7 bar). Yan et al. [178] showed that the solubility of iron sulfides decreases with increasing temperature and increases with increasing pressure. Phases like pyrrhotite and troilite are most common to form at HP/HT conditions, whereas makinawite is common at colder temperature [178].

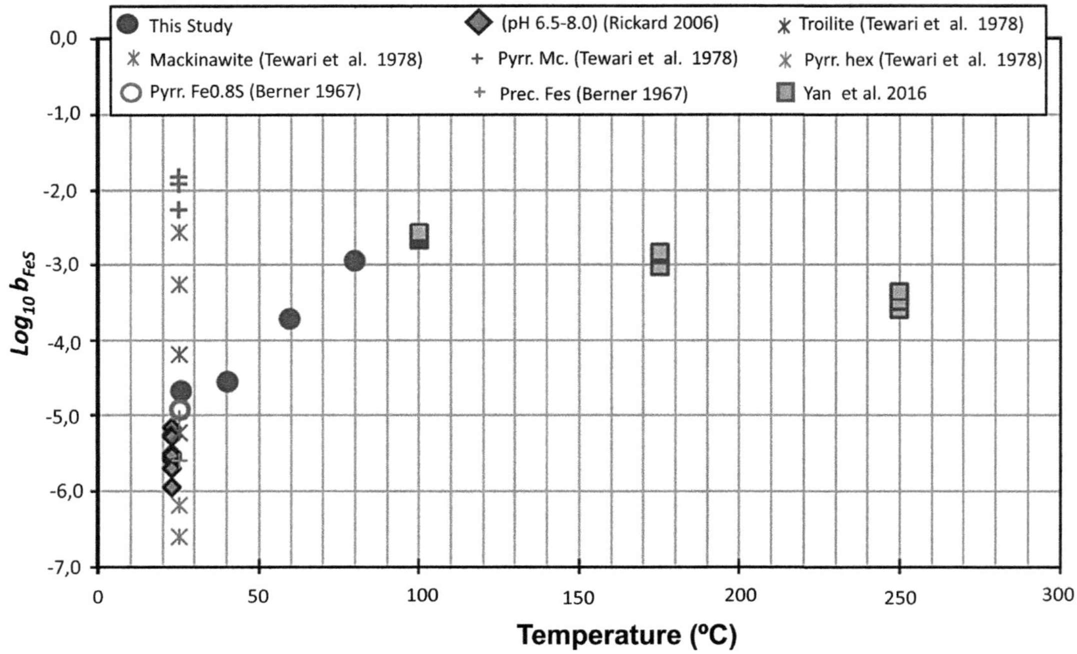

FIG. 5.36

Effect of temperature on FeS solubility in aqueous solution.

Reprint from reference F. Murcia, D. Carolina, P.L. Fosbøl, E.H. Stenby, K. Thomsen, Measurement of iron and lead sulfide solubility below 100°C, Fluid Phase Equilib. 475 (November) (2018) 118–126. https://doi.org/10.1016/j.fluid.2018.07.033, copyright (2018), with permission from Elsevier.

Below 100°C, Thomsen et al. [179] showed that solubility of iron sulfides increased with increasing temperatures. The results are illustrated in Fig. 5.36 and compared with the other studies.

The effect of temperature on sour corrosion is complicated by many factors, such as the metal surface deposits, concentration of acid gases, and even the exposure duration. Generally speaking, corrosion in an H_2S environment increases with rising temperature. Nevertheless, this is counteracted by the formation of a protective iron sulfide layer at elevated temperatures, so the corrosion rate decreases more rapidly at very high temperatures, and at longer exposure times [164]. At temperatures <80°C, temperature accelerates both the corrosion rate and the kinetics of corrosion product layer formation. The corrosion product layer formed is a porous and nonprotective mackinawite layer. At 80°C, a dense and adherent corrosion product layer, composed of mackinawite and pyrrhotite, gives good protectiveness. At high temperature, e.g., 220°C, the corrosion products were found to be pyrrhotite, with traces of pyrite [180].

5.7.3.3.2 Effect of pressure

The comprehensive studies on the effect of pressure on sulfides concluded that pressure drop causes an increase in the solubility of metal sulfides [181,182].

H_2S partial pressure has a strong impact on the process of iron sulfide formation. Generally, sour corrosion occurs in aqueous environments containing acid gases (H_2S/CO_2), but with H_2S concentration sufficiently higher than that of CO_2 to dominate the corrosion behavior and corrosion products. According to NACE, for sour corrosion to take place, the partial pressure of H_2S should be >0.30 kPa (0.05 psi) [183].

5.7.3.3.3 Effect of pH

The pH affects the solubility of iron sulfides. Generally, acidic solutions increase the solubility of iron sulfides. Corrosion also favors the acidic conditions.

Laboratory experiments showed that, when the pH was below 7, typically less than 40% of the added ferrous iron reacted by sulfide precipitation, although sulfide was in excess. However, when the pH was above 8, almost complete precipitation of all the added ferrous iron was observed [184].

5.7.3.3.4 Effect of flow rate

The flow rate has been found to noticeably influence the corrosion behavior of carbon steel in an H_2S environment. High flow rates accelerate corrosion rate due to mass transport processes and affect the formation and precipitation of the protective iron sulfide layers [164].

Other factors that affect the iron sulfides include the dissolved solids or ionic strength. Yan et al. [178] showed that the presence of NaCl in the solution increases the sulfide solubility.

5.7.4 Mixed iron sulfides-iron carbonate scales

In oil and gas production systems, acid gases CO_2 and H_2S are usually coproduced. So, in such systems, sour and sweet corrosion may take place at the same time with relative proportions, where one of them may dominate the other; thus relative proportions of iron sulfides and iron carbonates are expected to form.

Generally, in H_2S/CO_2 systems, H_2S corrosion dominates in an environment with $pCO_2/pH_2S < 20$, while CO_2 corrosion is prevalent when $pCO_2/pH_2S > 500$. An environment with $20 < pCO_2/pH_2S < 500$ exhibits mixed CO_2/H_2S corrosion with mixed sulfides–carbonate scale. Additionally, acid gases can affect each other, as the presence of H_2S can enhance CO_2 corrosion by accelerating anodic dissolution through sulfide adsorption; on the other hand, it can suppress CO_2 corrosion due to the protective sulfide scale layer [185]. Below the NACE sour limit of 3 mbar H_2S, the composition of the corrosion films depends on the CO_2/H_2S ratio, the corrosion rates, and the temperature [185,186]. Fig. 5.37 illustrates corrosion regimes in the presence of H_2S and CO_2 [187].

Dunlop et al. [188] suggested ratios that can be used to estimate the carbonate and sulfide deposition regime at room temperature. According to this suggestion the carbonate regime exists when the ratio of the partial pressure of CO_2/H_2S is above 500 (siderite prevails), while the sulfide regime is reached when this ratio of partial pressures is below 500 (iron sulfides prevail). Although the value was stated to only work at 25°C, temperature was found to have only a minor effect on the ratio within the range of 25–100°C, assuming that the FeS corrosion product is mackinawite. That value of 500 was found to be sensitive to the thermodynamic data, and it can be within a range of 9–2500 (the 2500 value was determined just to account for the change from mackinawite to pyrrhotite) [189]. Calculations based on solubility data lead to theoretical CO_2/H_2S ratios for the transition between FeS and $FeCO_3$ formation in the range of 105–107. Observed CO_2/H_2S ratios of 1000–5000 have been reported in the literature

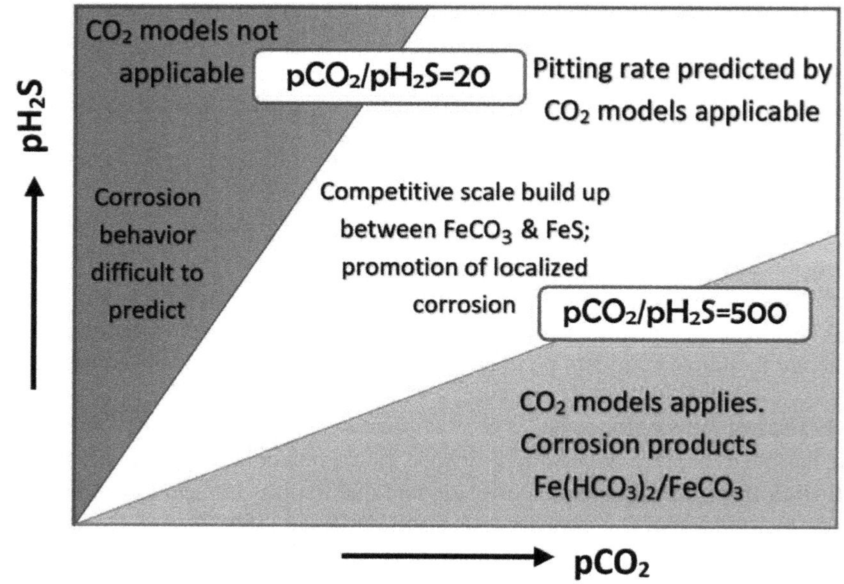

FIG. 5.37

Corrosion regimes in the presence of H_2S and CO_2.

Reprinted from reference M. Askari, M. Aliofkhazraei, R. Jafari, P. Hamghalam, A. Hajizadeh, Downhole corrosion inhibitors for oil and gas production—a review. Appl. Surf. Sci. Adv. 6. (2021) https://doi.org/10.1016/j.apsadv.2021.100128, copyright (2021) with permission from Elsevier.

[190]. New investigations from Smith and Pacheo [191] resulted in more precise formulas for the calculation of critical constants, assuming coexistence of $FeCO_3$ and all forms of Fe_xS_y.

5.7.5 Iron oxides, iron hydroxides, and iron oxy-hydroxides

Iron oxides and hydroxides are very complex, not just because of the immense number of compounds, and the transformations between these compounds, but also because of variable conditions under which these compounds can occur. Generally speaking, iron-oxygen compounds can form under the following conditions:

- Aerobic conditions in the presence or absence of H_2S or CO_2
- Anaerobic conditions in the presence or absence of H_2S or CO_2
- Aerobic microbial activity
- Anaerobic microbial activity
- Conversion of the other corrosion products into iron oxides by heat or oxygen ingress.

Iron-oxygen compounds are formed as a result of corrosion, or conversion of other corrosion products. The formation and transformation pathways for the corrosion products under aerobic and anaerobic conditions are illustrated in Fig. 5.38 [192], and the different forms of iron-oxygen compounds are summarized in Table 5.8.

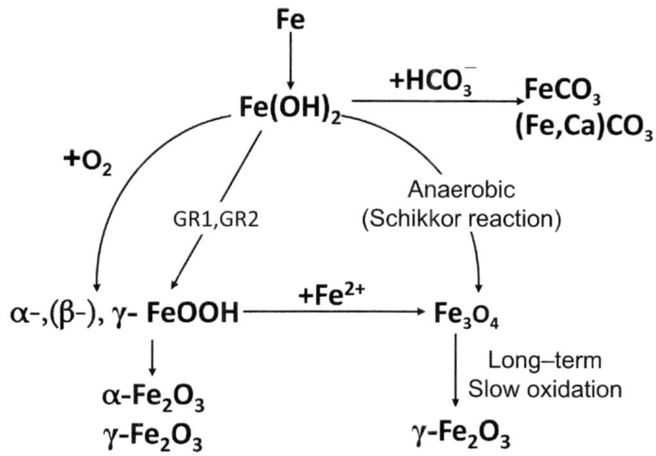

FIG. 5.38

Formation and transformation of corrosion products on carbon steel in aerobic, anaerobic.

Reprinted from reference F. King, C. Padovani, Review of the corrosion performance of selected canister materials for disposal of UK HLW and/or spent fuel, Corros. Eng. Sci. Technol. (2011) https://doi.org/10.1179/1743278211Y.0000000005, copyright (2011) with permission from Tylor and Francis.

Table 5.8 The different forms of iron oxides-hydroxides and their properties.

Form	Chemical formula	Appearance	Crystal system	sp. gr	Hardness (Mohs)
Hematite	α-Fe_2O_3	Reddish gray, black, red	Trigonal	5.26	5.5–6.5
Maghemite	γ-Fe_2O_3	Reddish-brown	Cubic, or tetragonal	4.87	6
Magnetite	Fe_3O_4	Dark grey, Black	Isometric	5.18	5.5–6.5
Wüstite	FeO	Black	Isometric	5.9–5.99	5.5
Iron II hydroxide	($Fe(OH)_2$)				
Bernalite	($Fe(OH)_3$	Dark bottle green	Orthorhombic	3.32	4
Goethite	α-FeO(OH)	Yellow-brown	Orthorhombic	4.26	5–5.5
Akaganeite	β-FeO(OH)	Yellow-brown	Monoclinic	3.56	–
Lepidocrocite	γ-FeO(OH)	Orange	Orthorhombic	4.09	5

5.7.5.1 Formation of iron oxygen compounds under aerobic conditions in the absence of H_2S/CO_2

The mechanism of iron oxide and oxyhydroxide formation on low alloy steel in aqueous solution was studied by Misawa et al. [193,194].

Aerobic corrosion is associated with ferrous/ferric oxides, formed as hardened reddish/orange tubercles [195]. For example, in the case of aerobic conditions (presence of air/oxygen, absence of H_2S, CO_2), the overall reaction of oxygen with steel and water is as follows [196]:

$$O_{2_{aq}} + 2H_2O_{(l)} + 2Fe_{(s)} \rightarrow 2Fe(OH)_{2_{(s)}} \tag{5.43}$$

O_2 also dissolves readily in water and, when in excess, reacts with ferrous hydroxide to produce hydrated ferric oxide (brown rust) [196].

$$4Fe(OH)_{2_{(s)}} + O_{2(aq)} \rightarrow 2H_2O_{(l)} + 2Fe_2O_3 \cdot H_2O_{(s)} \tag{5.44}$$

Also, the formation of $Fe(OH)_3$ takes place according to the following equation [197]:

$$2Fe^{2+} + \frac{1}{2}O_{2(aq)} + 5H_2O \rightarrow 4H^+ + 2Fe(OH)_3 \tag{5.45}$$

Corrosion products normally form a structure consisting of three layers, known as "tubercles" under aerobic conditions, where $Fe[OH]_2$ composes the inner green layer, $Fe[OH]_3$ composes the outer layer, and Fe_3O_4 forms the layer in between [197]. $Fe(OH)_3$ is practically insoluble above pH 4; however, if oxygen can be excluded, 100 ppm of $Fe(OH)_2$ will still be in solution at pH 8.5 [26].

Subsurface waters containing soluble iron oxidize upon contact with air in surface handling equipment. The result is a colloidal form of ferric oxide (Fe_2O_3), which exhibits a red color and is commonly referred to as "red water." Such waters are common in the production of oil and gas, and surface equipment that commonly plug systems with iron oxides. Iron oxide scales are also commonly formed in the heat treatment of iron at elevated temperatures and on the combustion side of boilers [3].

5.7.5.2 Formation of iron-oxygen compounds under aerobic conditions in the presence of H_2S/CO_2

The presence of O_2 can accelerate both sweet and sour corrosion, causing the formation of different corrosion products. In sweet corrosion, several reports have indicated that a trace amount of O_2 can have drastic effects on CO_2 corrosion of mild steel, as low O_2 concentrations in CO_2 saturated aqueous solution usually causes the iron carbonate layer to lose its protectiveness, and hence increases the corrosion rate [196]. Fe^{2+} from siderite is oxidized in the presence of O_2 forming various Fe (III)-rich oxides and oxyhydroxides [198], as shown in Eq. (5.46):

$$4\,FeCO_3 + O_2 \rightarrow 2Fe_2O_3 + 4CO_2 \tag{5.46}$$

When air-free water containing dissolved ferrous iron ions and bicarbonate ions is contacted by air, ferric hydroxide can be formed, as in Eq. (5.47):

$$2Fe^{2+} + 4HCO_3 + H_2O + 1/2O_2 \rightarrow 2\,Fe(OH)_3 + 4CO_2 \tag{5.47}$$

In mixed anion solutions containing carbonate/sulfate/chloride, in which carbonate does not dominate, a mixture of magnetite and siderite is usually formed [195]. XRD and Raman confirm the formation of magnetite (Fe_3O_4), hematite (α-$Fe2O_3$), and goethite (α-FeOOH), along with siderite ($FeCO_3$) in corrosion products [199].

O_2 ingress also accelerates corrosion in sour systems and increases in the severity of localized attacks [196,200]. With the presence of O_2, dissolved H_2S can be further oxidized to elemental sulfur and different sulfur-containing species [200].

Oxygen can oxidize sulfide species according to the following equations [201]:

$$2Fe^{II}S + 2xH_2O + xO_2 \rightarrow 2\,Fe^{II}_{1-3x}Fe^{III}_{2x}S + 2xFe^{2+} + 4xOH^- \tag{5.48}$$

$$4Fe^{II}_{1-3x}Fe^{III}_{2x}S + (2-2x)H_2O + (3-3x)O_2 \rightarrow (4-4x)Fe^{III}OOH + 4S^0 \tag{5.49}$$

$$3Fe^{II}S + S^0 \rightarrow Fe^{II}Fe^{III}_2S_4 \tag{5.50}$$

$$3Fe^{II}Fe^{III}_2S_4 + 2O_2 \rightarrow Fe^{II}Fe^{III}_2O_4 + 6Fe^{II}S_2 \tag{5.51}$$

In the same way, H_2S is a reducing agent for iron oxides, which converts them into metal sulfides:

$$Fe_2O_3 + 2HS^- + 4H^+ \rightarrow 2\,FeS + 3H_2O + S^0 \tag{5.52}$$

With O_2 contamination at high P_{H2S}, the iron sulfides were mainly pyrrhotite and troillite, whereas at lower P_{H2S}, greigite, magnetite, and elemental sulfur were identified [202].

5.7.5.3 Formation of iron-oxygen compounds under anaerobic conditions, in the absence of H_2S/CO_2

Under anaerobic conditions, iron dissolution leads to hydroxide dissolution, according to Eq. (5.53):

$$Fe + 2H_2O \rightarrow Fe(OH)_2 + 2e^- + 2H^+ \tag{5.53}$$

$Fe(OH)_2$ is considered to be the primary corrosion product in the absence of sulfide.

$Fe(OH)_2$ converts into Fe_3O_4 at high temperatures via the "Schikorr reaction" [203], shown in Eq. (5.54):

$$3Fe(OH)_2 \rightarrow Fe_3O_4 + H_2 + 2H_2O \tag{5.54}$$

A passive magnetite layer that grows on the inner metal surface of a boiler provides corrosion protection and, under normal operating conditions, the protective layer is created through reaction between steel and deoxygenated water or steam:

$$3Fe + 4H_2O \rightarrow Fe_3O_4 + 4H_2 \tag{5.55}$$

Fe(II) oxidation can take place by oxic fracture fluids and also can be aided by aqueous bitumen leached from shale by organic compounds in the fracturing fluid [204], leading to the formation of Fe(III)-(oxy)hydroxides. Fe(II) is basically derived from Fe(II)-bearing phases, e.g., primarily pyrite, siderite, and Fe(II) bound directly to organic matter. The effect of bitumen on Fe(II) oxidation is significant, especially under acidic conditions, where bitumen was found to promote Fe(II) oxidation and $Fe(OH)_3$ precipitation that otherwise would not occur [205].

5.7.5.4 Formation of iron-oxygen compounds under anaerobic conditions in the presence of H_2S/CO_2

According to thermodynamic calculations, magnetite (Fe_3O_4) can be formed in sweet as well as sour environments at high temperature. Magnetite was detected in an iron carbonate corrosion product layer formed on steel exposed to a CO_2 containing aqueous solution at $T = 80°C$, pH ranges from 7.1 to 8.1, pCO_2 0.53 bar, NaCl 1 wt%, in stagnant conditions [206]. A recurrent layer of magnetite always forms at all high temperatures with other iron sulfide deposits in sour environments [207]. Fig. 5.39 illustrates a mixture of iron oxides (major) and iron sulfides (minor) from a 12″ spool.

FIG. 5.39

Mixture of iron oxides and iron sulfides from 12″ spool.

5.7.5.5 Formation of iron oxides/hydroxides by bacteria

Iron-oxidizing bacteria (IOB or FeOB) is a common bacteria that has been implicated in microbiologically influenced corrosion (MIC). IOB are aerobic bacteria that convert ferrous iron [Fe^{+2}] to the ferric [Fe^{+3}] state and produce ferric hydroxide [$Fe(OH)_3$], goethite [FeOOH], and ferric oxides [Fe_2O_3, Fe_3O_4], which are a highly insoluble by-product that will damage the formation [208].

These bacteria must oxidize large quantities of iron as an energy source in order to grow. Consequently, even a small number of bacteria can be responsible for precipitating vast amounts of iron oxides and hydroxides [209].

Iron oxides and hydroxides were found to be by-products of anaerobic sulfate reduction by SRB [210].

5.7.5.6 Factors affecting the formation of iron oxides/hydroxides

- **Oxygen concentration**

Iron oxides are a direct product of corrosion of iron in the presence of different oxidants, and most likely in the presence of oxygen. The corrosion potential is controlled by the solution pH and the concentration of dissolved oxygen (DO), so the redox potential (E_H) can be written as follows [211]:

$$E_H(\mathrm{mV}) = 1.189 - 0.059\,\mathrm{pH} + \frac{0.059}{4}\log(\mathrm{DO}) \quad (5.56)$$

where DO is the dissolved oxygen in milligrams/liter.

For a given temperature and water velocity, the rate of oxygen consumption is equivalent to the corrosion rate and is to the first approximation directly proportional to the oxygen concentration in the water, regardless of the concentration of the other constituents dissolved in the water. So oxygen reacts with water, forming different forms of iron oxides depending on the valance of iron [211].

- **pH**

Assuming that the pH and dissolved oxygen concentration in the bulk of the solution are the same at the metal surface, then the possibility of iron corrosion and the product of that corrosion process can be predicted from the Pourbaix diagram. A typical Pourbaix diagram of an iron system is illustrated in Fig. 5.40 [212].

Furthermore, iron oxides, hydroxides, and oxyhydroxides are soluble in acidic solutions. Their rate of dissolution in HCl was found to be on the order of: lepidocrocite > magnetite > akaganeite > maghemite > hematite > goethite [213].

The effect of pH on the iron oxides/hydroxides becomes more critical during acid-involving maintenance operations like acidizing and acid cleanings. During these processes, iron ions are produced from dissolved rust, corrosion products, mill scale, or contaminated acids, which can reprecipitate when the spent acid reaches a specific pH, causing various problems, including permeability plugging. The ratio of Fe^{2+} to Fe^{3+} in spent acidizing fluids is about 5 to 1, which varies from one well to another [128,129]; while Fe^{2+} and Fe^{3+} can lead to precipitate formation under certain conditions, most damage will occur from precipitation of $Fe(OH)_3$ because of its lower solubility in spent acid [214]. Taylor et al.

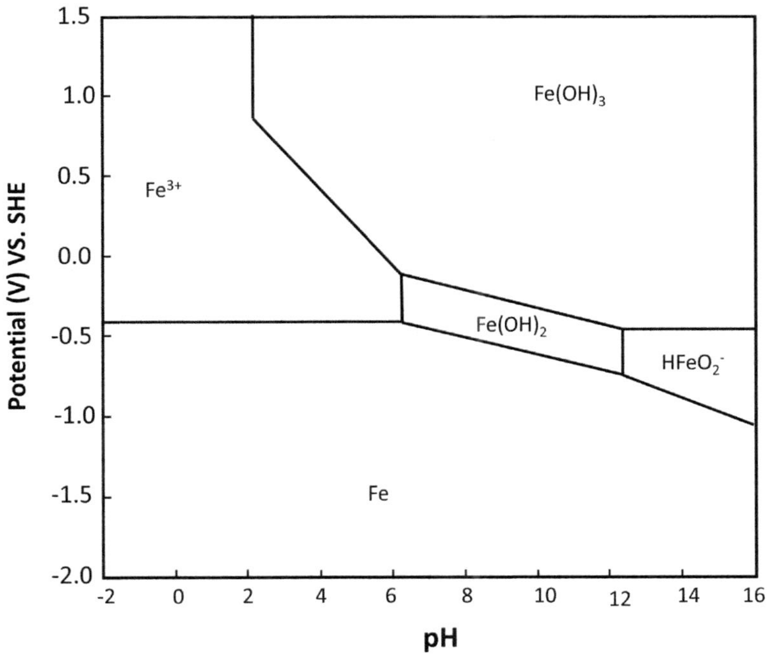

FIG. 5.40

Pourbauix diagram.

Reprinted from reference B.N. Popov, Thermodynamics in the electrochemical reactions of corrosion, in: Corrosion Engineering, Elsevier, 2015, 29–92. https://doi.org/10.1016/b978-0-444-62722-3.00002-1, copyright [2015] with permission from Elsevier.

[214] showed that $Fe(OH)_3$ precipitation can begin at pH 1 and complete at 2 at 25°C, which is lower than the literature reported values, e.g., beginning at pH 2.2 and complete at a pH of 3.3. $Fe(OH)_2$ usually starts precipitating at higher pH values.

- **Temperature**

As with sweet and sour corrosion, increasing temperature increases the corrosion rate. Taylor et al. (2014) showed that as the temperature was increased, Fe^{3+} became significantly less soluble in the spent acid, which indicates higher precipitation and damage potential.

5.7.6 Mill scale

Mill scale, often shortened to just scale, is the flaky surface that is formed on the outer surfaces of hot rolled steel sheets or profiles when they are being produced by rolling red hot iron or steel billets in rolling mills.

Mill scale consists of the iron oxides iron(II) oxide (FeO), iron(III) oxide (Fe_2O_3), and iron(II,III) oxide (Fe_3O_4, magnetite).

Mill scale is bluish black in color, usually less than 0.1 mm (0.0039 in.) thick, and initially adheres to the steel surface and protects it from atmospheric corrosion provided no break occurs in this coating.

Because it is electrochemically cathodic to steel, any break in the mill scale coating will cause accelerated corrosion of steel exposed at the break. Mill scale is thus a boon for awhile, until its coating breaks due to handling of the steel product, or due to any other mechanical cause.

Right after installation and maintenance operations, these mill scales need to be cleaned or they can accumulate and cause serious deposition problems.

5.8 Zinc and lead sulfide scales

Lead and zinc sulfide scale has recently become a concern in a number of oil and gas fields around the globe. These deposits have occurred in the production tubing and also in the topside process facilities [163]. As with barium and strontium sulfate, Zn and Pb sulfides usually occur together. Fig. 5.41 illustrates zinc sulfide from a flowline, and Fig. 5.42 shows mixed zinc and lead sulfide scales from a production well.

5.8.1 Mechanism of zinc and lead sulfide scales formation

The main mechanism of Zn and Pb sulfides is expressed by Eq. (5.57) [95]:

$$\left(Zn^{2+}, Pb^{2+}\right) + S^{2-} \rightarrow (ZnS, PbS)\downarrow \tag{5.57}$$

The potential sources of zinc and lead ions include the following [163]:

- Partial dissolution of reservoir minerals such as sphalerite (zinc sulfide) and galena (lead sulfide) when in contact with connate and aquifer water. Zn and Pb ion concentration within high pressure/temperature fields within the Gulf coast of Mexico were reported as high as 70 ppm Pb and 245 ppm Zn.

FIG. 5.41

Thick layer of zinc sulfide scale in a flow line from Egyptian oilfield.

FIG. 5.42

Mixed zinc and lead sulfide scales from a production well.

- Partial dissolution of the reservoir minerals by injection water, which makes it rich with such heavy metals.
- Loss of heavy brine completion fluids, which contain Zn ions (e.g., zinc bromide) into the formation during drilling and well workover operations. Biggs et al. [215] reported that a loss of 500 bbls of 17.2 ppg zinc bromide completion fluid in a reservoir resulted in significant zinc sulfide scale formation in the presence of only 2 ppm of hydrogen sulfide from the reservoir. Another field in the North Sea UK reported the presence of zinc sulfide on downhole gauges and logging tools after zinc bromide brines had been lost during completion operations. Zinc levels within the produced fluids were in the region of 10–50 ppm for several months during initial water breakthrough.

Sulfide ion sources are the same ones that are discussed in iron sulfide deposits (Section 5.7.3.1).

Keogh et al. [216] demonstrated that the way sulfide scale sticks to surfaces is quite different from other types of scale. While conventional scales deposit by both crystallization (in the bulk and/or on the surfaces) and adhesion onto surfaces, sulfide scale (PbS) undergoes rapid bulk nucleation, and adhesion acts as the dominant mechanism for deposition. They also found that PbS fouling depends on the formation of a stable o/w Pickering emulsion due to PbS wettability and nanometric size, which in a multiphase turbulent flow results in deposition by the impaction on surfaces, suggesting that PbS deposits in the upper regions of production tubing are driven by adhesion after formation of a PbS solid-stabilized Pickering emulsion. In single-phase systems, loose agglomerates of PbS act through turbulent and gravitational deposition. ZnS deposition can also be explained in a similar way.

Zinc sulfide can form in two types, alpha (wurtzite) and beta (sphalerite) forms, while lead sulfide occurs in nature and is known as galena. The chemical and physical properties of zinc and lead sulfide forms are summarized in Table 5.9.

5.8.2 Factors affecting the formation zinc and lead sulfide scales

5.8.2.1 Effect of temperature

Increasing the temperature increases the solubility of both zinc and lead sulfides. Thomsen et al. [220] showed an exponential dependency of the solubility with respect to temperature. An increase of 40°C results in an increase of roughly 12 times for the solubility of ZnS. Thomsen et al. [179] showed that PbS solubility increases with increasing temperature in the range between 25°C and 80°C; however, the increasing trend is not as much as FeS. The temperature effects on zinc and lead sulfide scales are shown in Figs. 5.43 and 5.44.

Solubility data presented by Barrett and Anderson [221] suggest that the solubility of lead and zinc sulfide increases with increasing temperature and increases with increasing brine salinity. Their data showed that an increase in temperature of 15°C produces a solubility increase comparable to or greater than that produced by the 3–5 m NaCl increase. Thus brines cooling during production increases the zinc and lead sulfide scales deposition (Collins and Jordan [222]). This explains the formation of zinc sulfides in the top parts of the production tubing.

Table 5.9 Physical and chemical properties of zinc and lead sulfide minerals [66,217–219].

Name	Formula	Crystal system	Color	K_{sp} at 25°C	Solubility in water (mg/L)	Hardness (Mohs)	Density (g/cm^3)
Sphalerite, Zinc blende	β-ZnS	Cubic	White, gray white, yellow, dark brown	3×10^{-23}	6.5 at 18°C	3.5 – 4.0	4.10
Wurtzite	α-ZnS, (Zn,Fe)S	Hexagonal	Brown, red brown, black	2×10^{-25}	6.9 at 18°C	3.5–4.0	3.98–4.1
Galena	PbS	Cubic	Lead-gray, silvery	3×10^{-28}	124 at 20°C	2.5	7.58

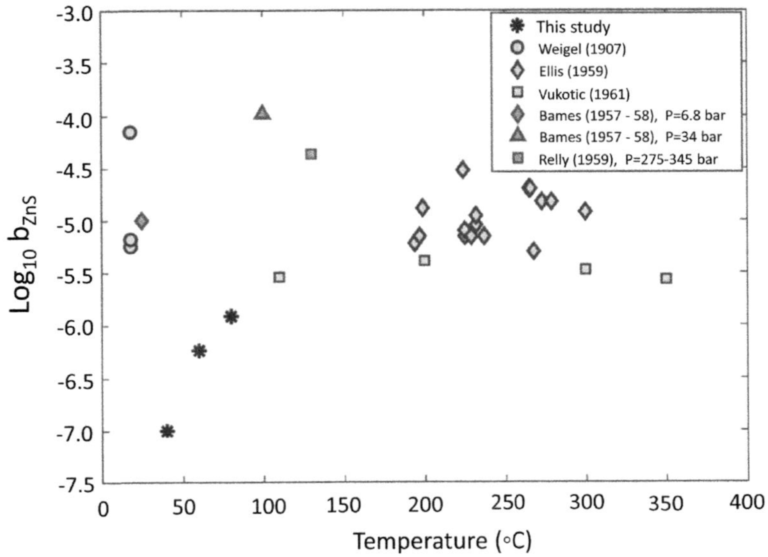

FIG. 5.43

ZnS solubility in molality b_{ZnS} (mol kgwater^{-1}) versus temperature.

Reprinted from reference D. Carolina Figueroa Murcia, P.L. Fosbøl, K. Thomsen, E.H. Stenby, Determination of zinc sulfide solubility to high temperatures, J. Solut. Chem. 46 (9–10) (2017) 1805–1817. https://doi.org/10.1007/s10953-017-0648-1, copyright (2017), with permission from Elsevier.

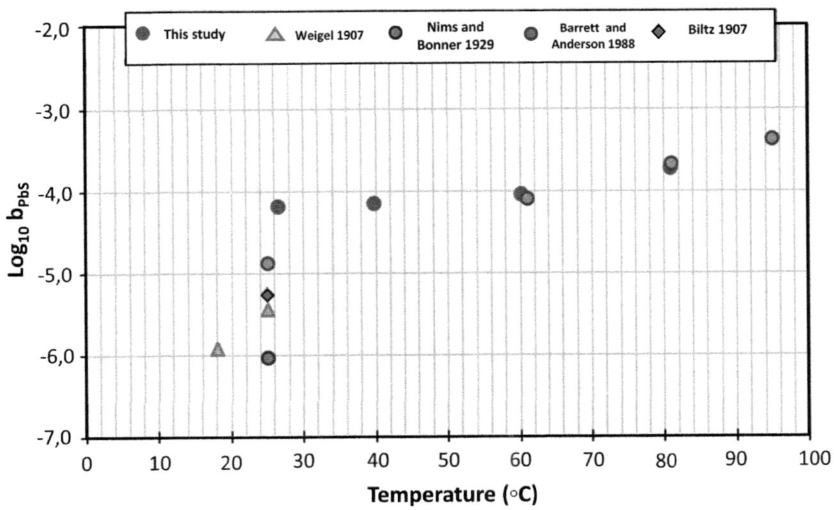

FIG. 5.44

Effect of temperature on PbS solubility in aqueous solution.

Reprint from reference D. Carolina Figueroa Murcia, P.L. Fosbøl, E.H. Stenby, K. Thomsen, Measurement of iron and lead sulfide solubility below 100°C, Fluid Phase Equilib. 475 (November) (2018) 118–126. https://doi.org/10.1016/j.fluid.2018.07.033, copyright (2018), with permission from Elsevier.

Zn and PbS have been reported to deposit in HPHT wells [222–224], as any cooling causes cooling shock, which stimulates sulfide precipitation. Under downhole conditions, the cooling can take place by three different mechanisms:

- Mixing of the formation water with the injection water
- Adiabatic decompression (throttling) when pressure on the fluid approaches the surface changes over a short distance from lithostatic to hydrostatic.
- Conductive heat loss to the walls of the pipe.

Therefore, under any of these conditions, Zn and Pb sulfides can deposit even at the extreme conditions of HPHT wells. Fig. 5.45 shows scaled well tubing where ZnS deposits formed under high temperature and pressure.

5.8.2.2 Effect of pH

As with iron sulfide, the solubility of both lead and zinc sulfide increases with decreasing solution pH. Thus acid washing is conventionally used to remove sulfide scales [222]. The solubility of Fe, Zn, and Pb sulfides on a pH range is illustrated in Fig 5.46 [225].

At pH 5 the solubility of iron sulfide is 65 ppm, whereas lead and zinc sulfides are 0.002 and 0.063 ppm, respectively. Depending on the exact brine conditions, the solubility of zinc sulfide is between 30 and 100 times more soluble than lead sulfide [221,222].

5.8.2.3 Effect of Dissolved solids

The solubility of ZnS and PbS increases by increasing the solution dissolved solids. Barrett and Anderson [221] reported that both ZnS and PbS are about five times more soluble in 5 m than in 3 m NaCl brines. The dilution of a high salinity brine and the increased solution pH caused by injection water mixing with the formation brine could also result in sulfide scale formation [222].

FIG. 5.45

Deposition of zinc sulfide deposits in production tubing under high temperature conditions.

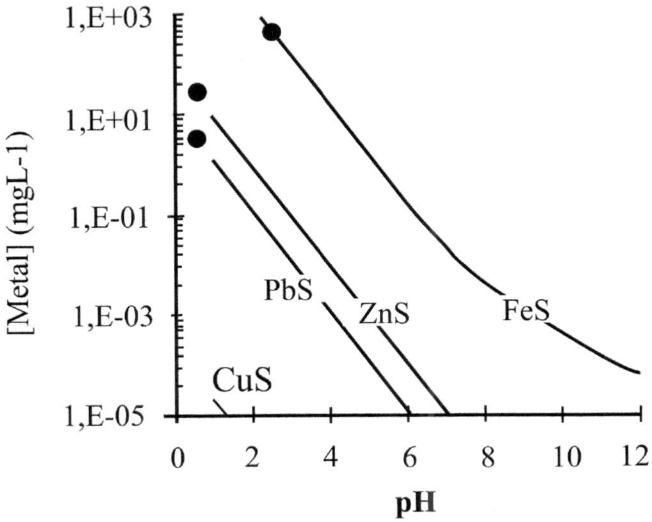

FIG. 5.46

Solubility of iron, Zinc and lead sulfides at different pH.

Reprinted from reference M.M. Marchioretto, H. Bruning, W. Rulkens, Heavy metals precipitation in sewage sludge, Sep. Sci. Technol. 40 (16) (2005) 3393–3405. https://doi.org/10.1080/01496390500423748, copyright (2005), with permission from Taylor and Francis.

5.8.2.4 Effect of pressure

Pressure affects the precipitation of the sulfide deposits in many ways. Pressure drop causes a slight increase in solution pH and decrease in temperature, which causes precipitation of the sulfides [226]. In addition, pressure drop causes turbulence, which allows the chemical species to interact and deposit.

However, only large pressure changes (on the order of 1000 bars) can cause significant changes in the sulfide solubility, so it is seldom a factor in sulfide precipitation except when the pressure leads to boiling. In fact, Hemley et al. [181,182] showed that the solubility of Fe, Zn, Pb sulfides in NaCl increases with pressure drop, so that these metals may be carried over long distances along a decreasing pressure gradient so long as the temperature drop is not enough to offset the pressure effect [227].

5.8.2.5 Mixing incompatible fluids

The effect of mixing incompatible fluids is a major factor in sulfide deposition, as mixing metal ion-containing waters with sulfide-containing waters within the near-wellbore or production tubing during fluid extraction leads to sulfide precipitations; then the well parameters pressure, temperature, pH, flow rate, and hydrodynamics control where they deposit.

Sometimes in a single reservoir interval two different brine types can coexist. For instance, connate water and aquifer water with different composition can both be in thermodynamic equilibrium until mixed within the near-wellbore and production system. Additionally, aquifer water/connate water can be separated from the dissolved hydrogen sulfide gas within crude oil or gas cap until they get mixed during production [222].

5.9 Halite scale

Crude oil is usually produced with varying amounts of water, and in many cases these waters are saturated or supersaturated with sodium and calcium chloride, and carry significant amounts of solid salt – up to several hundred pounds per thousand barrels [3]. Due to the changes in operating parameters, these water-soluble salts can drop out of the solution and deposit. The scale formed in such cases is usually sodium chloride, or halite.

Halite deposition is often a problem for producers operating in high brine matrices, such as those producing from wells located over salt domes [228]. Fig. 5.47 shows halite scale.

5.9.1 Mechanism of halite scale formation

Like other mineral scales, halite precipitation takes place when its concentration exceeds the limit in the solution and becomes supersaturated.

$$NaCl_{(\text{Aqoueous})} \xleftrightarrow{\Delta P,\ \Delta T,\ \text{supersaturation}} NaCl_{(\text{solid precipitate})} \tag{5.58}$$

The first thought that comes to mind is that halite has very high solubility in water, so why is it still precipitating? The answer is that although it has very high solubility, it is precipitating due to two main reasons:

- *First the high potential of precipitating halite*: this is due to high concentration of halite in oilfield brines. Since halite is highly soluble in water, its saturation index, which is a measure of the driving force for precipitation, is small. However, the potential amounts available for precipitation are very large. In simple words, the kinetics of formation of minerals depend on both the driving force and the concentrations of precipitating ions. While the driving force is small for halite, the concentrations of ions are very large indeed. This is the opposite of the situation ordinarily encountered with scales

FIG. 5.47

Halite deposits in Egyptian field.

such as barite [228]. Thus, unlike some scales, halite precipitation occurs soon after saturation; Nieuwland and Collins [229] reported halite deposition with only 5% oversaturation.

- *Second, the operating parameters* that have strong impact on the solubility and deposition of the halite. Change in temperature or pressure can significantly affect the solubility of halite leading to its precipitation. So, unlike the evaporation sequence for seawater under atmospheric pressure, which follows the formation of deposits in the order of: carbonates, gypsum, halite, and potassium salts [230], under the higher pressures of reservoir systems, it is possible that halite precipitates prior to calcium carbonate and gypsum deposits [2].

Knowing this, it is plausible that water becomes supersaturated with halite for a number of reasons, including the following:

- Fluids cooling can occur by many mechanisms: losing heat by conduction to the surrounding environment, mixing cold fluid with the production fluids (water flood, polymer flood, gas lift, fracking, etc.), or by Joule-Thomson throttling (in the reservoir porous media, valves, or perforations).
- Solvent evaporation occurs when dry hot gas escapes the fluids when the pressure drops, leading to water flash-off to the gas phase to equilibrate (saturate the gas with water), leading to supersaturating the solution with halite.

The kinetics of halite precipitation are very fast, such that rapid plugging is expected. Besides, even wells with very low water cut ($<0.5\%$) may rapidly salt up overnight [95]. The factors impacting halite scale formation are illustrated in Fig. 5.48.

To understand these cases, let's consider the halite deposition mechanism in the reservoir, downhole (production tubing), flowlines, and top sides.

In a petroleum reservoir, reservoir brines are in contact with hydrocarbon gases; thus these gases are usually saturated with water under reservoir conditions. However, during production, pressure and temperature changes cause the amount of water that the gases can hold to change, where hotter gases and lower pressure gases hold more water [231]. Therefore, in the reservoir, the dominant mechanism for halite precipitation is pressure reduction combined with gas production and a saturated or near-saturated brine. The gas dehydrates (evaporates) the brine, creating a greater salt concentration and eventual precipitation. With large amounts of mobile water, high amounts of salt are deposited and finally plug the pore throat. In the case of small amounts of residual water, small films of halite usually form and may not affect production. Reducing drawdowns through negative skins will distribute the problem further away, reducing the effect, but also making the salt harder to reach [2]. Low pressure and high temperature locations in the production system (perforation) are expected to be the worst in halite deposition, as the low pressure causes more gas to evolve and high temperature makes the gas drier, which means more water to absorb, more water evaporation, and more halite to precipitate.

In the tubing, flowing from the downhole to the Christmas tree, both the temperature and the pressure decrease. Cooling the fluids down to the crystallization temperature and the pressure dropping to below the bubble point will certainly lead to more halite deposition. Also it is worth mentioning that cooling down causes condensation of water from gas, which decreases water salinity. This beneficial effect will be important if the water-to-gas ratio is very low [2].

In the flowlines, they combine both pressure drop and temperature drop, which makes the halite problematic in subsea flowlines, risers, and manifolds. The problem will be severe in brines with high

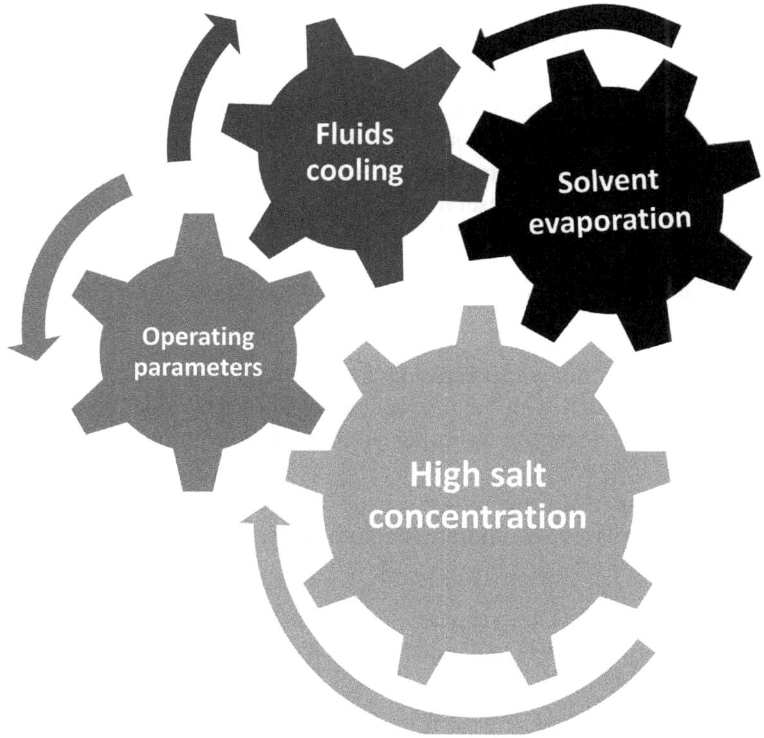

FIG. 5.48

Factors impacting halite deposition.

salt content. Halite plugging in topsides flowlines and risers from brines with TDS reached 380,000 ppm in the Egyptian field.

Sodium chloride precipitates in a cubic crystal system with colors ranging from colorless or white when pure, to gray, yellow, orange, pink, red, purple, and faintly tinted due to impurities. Halite mineral usually has a density of 2.168 g/cm^3, solubility of 360 g/L, and a hardness of 2.5–3.00 mohs. Sodium chloride is often hydroscopic material, which means it has a deliquescence humidity of about 75% RH; therefore it tends to pick up moisture easily in most temperate climates.

5.9.2 Factors affecting halite deposition

5.9.2.1 Effect of temperature

Sodium chloride is much more soluble than the other scales described earlier and its solubility increases with increasing temperature [232]. Fig. 5.49 illustrates the solubility of halite in water at different temperatures.

Strangely, in oil and gas fields a decrease or increase in temperature can lead to halite deposition. When very high salt brines from HPHT reservoirs cool down, they drop out halite crystals, driven by the high concentration of the salt in the produced water and also thermodynamically driven by the

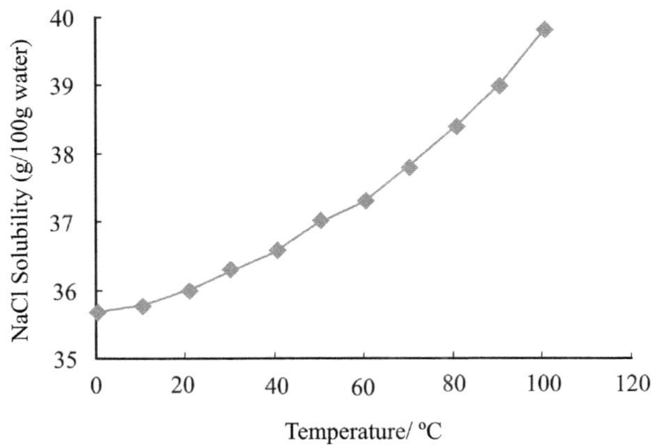

FIG. 5.49

Halite solubility versus temperature.

Source: Y. Tang, R. Yang, X. Kang, Modeling the effect of water vaporization and salt precipitation on reservoir properties due to carbon dioxide sequestration in a depleted gas reservoir, Petroleum 4 (4) (2018) 385–397. https://doi.org/10.1016/j.petlm.2017.12.003, with permission from Tang Yong.

solubility limit when halite reaches the crystallization temperature. Halite precipitation due to reaching the crystallization temperature is independent of the water-to-gas ratio [2]. It is also worth mentioning that cooling will promote condensation of water from the gas, thus decreasing the salinity of the produced water. This beneficial effect will be important if the water-to-gas ratio is very low [2].

On the other hand, in high temperature gas reservoirs, the high temperature makes the gas drier, which drive them towards absorbing more water to become saturated. That leads to more water evaporation, intensifying the halite precipitation problem. Halite deposition is more likely in locations like bottom-hole and at the reservoir interface.

5.9.2.2 Effect of pressure

Pressure also has a major effect on the deposition of halite. The solubility of water in gas increases at lower pressures where gas becomes undersaturated with water; this is indicated in Fig. 5.50, which illustrates the water content of methane gas as a function of pressure [233]. Case studies of halite precipitation of halite in gas-producing wells and dry gas injection wells used for storage were reported [233]. Lower pressures also promote salt crystallization. Two main ways the pressure affects halite deposition are:

- First pressure drop causes water (solvent) evaporation leading to supersaturation and precipitation.
- Second pressure drop causes Joule-Thomson cooling of the fluids, which promotes crystallization.

As pressure decreases during production, water flash-off into the gas phase occurs, which concentrates solutions of sodium chloride, which may eventually also lead to halite precipitation [95]. The problem often becomes significantly worse when reservoir pressure are low, typically less than 150 [234].

Pressure drop is considered the dominant mechanism for halite precipitation in gas reservoir and wells, combined with gas production and a saturated or near-saturated brine [2].

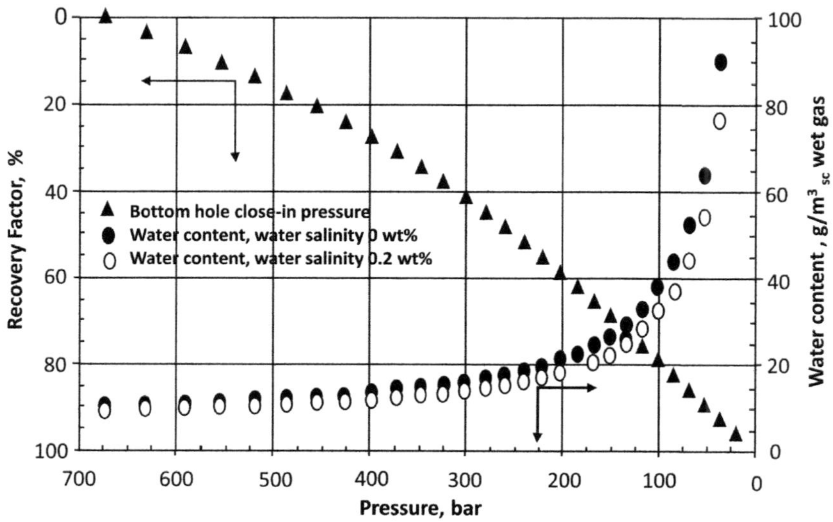

FIG. 5.50

Methane water saturation versus pressure.

Reprinted from reference W. Kleinitz, G. Dietzsch, M. Köhler, Halite scale formation in gas-producing wells, Chem. Eng. Res. Des. 81 (3) (2003) 352–358, https://doi.org/10.1205/02638760360596900, copyright (2003), with permission from Elsevier.

Halite deposition is exacerbated at the bubble point where water-free gas is generated, which is at high temperature, and water will evaporate from the formation water phase into the gas phase to saturate the generated dry gas with water. This leads to water supersaturation with halite followed by its precipitation.

Also, evaporation occurs during CO_2 injection in depleted gas reservoirs. A small part of CO_2 dissolves in the brine, whereas a fraction of the saline water is vaporized by the flowing CO_2. Water evaporates and the salt content contained in the brine is enriched [232,233].

Joule-Thomson cooling also occurs when the pressure drops through porous media, perforation, and valves, leading to instant cooling of the liquids and causing salt precipitation and crystallization.

5.9.2.3 Effect of water/gas ratio

As indicated earlier, gases in the petroleum reservoir are usually saturated with water under reservoir conditions, and their saturation of water depends on the pressure and temperature that leads to vaporization and salt precipitation. It was reported that the location of halite and amount of scale formation was influenced by the water/gas ratio and the drawdown on the well.

Generally speaking, halite deposition is experienced in gas wells with a low water-to-gas ratio (typically less than 5 bbl/MMscf) [235]. With vaporization, low water-to-gas ratios will lead to a greater likelihood of an undersaturated brine becoming saturated. For a saturated brine, precipitation will occur regardless of the water-to-gas ratio [2].

Jordan et al. [236] reported deposition of halite due to the application of gas lift at a low water cut and resulting dehydration of the trace amounts of produced brine.

5.9.2.4 Effect of hydrate inhibitors

Hydrate inhibitors like methanol and glycols have a significant impact on the solubility of the mineral scales [237,238]. The saturation index (SI) of mineral scales increases with the addition of the hydrate inhibitors MeOH, TEG, and MEG; this can be expressed by Eq. (5.59) [238]:

$$SI = SI_{(\text{brine only})} + \Delta SI_{(\text{due to added MeOH, MEG, or TEG})} \tag{5.59}$$

Methanol has a higher effect on the solubility (saturation index) of halite (mineral scales) than MEG or TEG. Halite scale problems can occur at > 210,000 mg/L TDS when 50% (vol.) methanol is present in the production system. At brine TDS of 170,000 mg/L, halite is not expected to form at MeOH percentage approaching 100%, while above 170,000 mg/L great caution should be taken to avoid halite precipitation [238].

5.9.2.5 Effect of flow rate

The flow pattern has a great importance for salts scaling. Flow rate and flow regime affect the mass transfer of the chemical species. Flow rates also affect the deposition regime of the formed crystals.

Additionally, it was found that crystals of salt are scaling when oil and water jointly flow in a laminar flow without mixing or while mixing resulting in formation of "oil-in water" emulsion. If a "water-in-oil" emulsion was formed, halite scaling takes place in individual small micrometer drops of water, where the salt crystals are "trapped" inside the drops and cannot grow into large aggregates. Stable water-in-oil emulsions can then be used to reduce scaling in wells, especially when the water cut is below 40%–50% [239].

5.10 Silica scale

Silica/silicate-based deposits present enormous operational challenges for industrial water systems, including brackish water reverse osmosis, evaporative cooling, oil and gas, and geothermal systems where silica supersaturation levels exceed saturation limits [240]. Silica scale-related problems have been termed the "Gordian knot" of chemical water treatment, highlighting the difficulty to resolve them [241].

To avoid silica-based deposition detrimental problems, water technologists normally maintain silica levels below 180 mg/L in absence of silica inhibitors [240].

5.10.1 Silica-silicate occurrence and chemistry

Silicon is the second most abundant element in the earth's crust. Silica and silicate species usually originate from water dissolving minerals and rocks like quartz, mica, and feldspar, which are found in abundance in the earth's crust [27].

All natural waters contain some dissolved silica [242]. Silica generally occurs at concentrations between 1 and 30 mg/L as SiO_2 in ground waters, but can also occur at concentrations up to 100 mg/L [243]. Silica levels are typically lower in surface waters, but some surface waters may contain elevated silica concentrations, depending on hydrogeologic conditions [243].

Silica can be found in different forms including soluble form (silicic acid, soluble silicates), particulate silica, and amorphous state (colloidal silica). Amorphous silica is further classified into dissolved (reactive), colloidal (nonreactive), and particulate (suspended silica) [27,244]. Reactive silica includes silicic acid, silicates, and polymerized forms with molecules smaller than the colloidal range [245]. Colloidal silica is polymerized silica in which the molecular size is roughly in the range from 0.01 to < 0.5 μm. Colloidal silica may be associated with organic compounds or with other complex inorganic compounds. Particulate silica is larger in size and mostly comprises sand or suspended solids in water with sizes >0.5 μm.

Crystalline silica (quartz) has solubility of 5–6 mg/L at 25°C and pH < 9, while amorphous silica has solubility ranging from 120 to 150 mg/L at 25°C and pH < 8–8.5 (150–180 mg/L solubility was also reported depending on water chemistry and temperature) [241].

Dissolved silica is often described by the formula "SiO_2." When silica is dissolved in water, it hydrolyzes to form silicic acid, and can be expressed as follows [242]:

$$SiO_2 + 2H_2O \rightarrow Si(OH)_4 \tag{5.60}$$

Silicic acid has been reported in different formulas, i.e., H_2SiO_3 (metasilicate), H_4SiO_4 (orthosilicate), which is equivalent to $Si(OH)_4$ [242]. Monomeric silicic acid (the "reactive silica") is the predominant dissolved silica species in natural waters with concentrations varying from 1 to 3 mg/L in mountain lakes to 50–300 mg/L in well waters in oil production fields [246].

Silicic acid can be ionized to form silicate ions. The pK_a of this reaction is reported in the literature in the range of 9.7–9.9 [242]; hence at neutral pH only a very small fraction of silicic acid dissociates (few ppb). As pH is raised above a value of 10, silicate ions become increasingly more abundant. Silicate formation at different pH values is illustrated in Fig. 5.51.

Silicate ions can be combined to form polymeric silica, the polymer is expressed as (H2SiO_3)*n* as presented in Fig. 5.52.

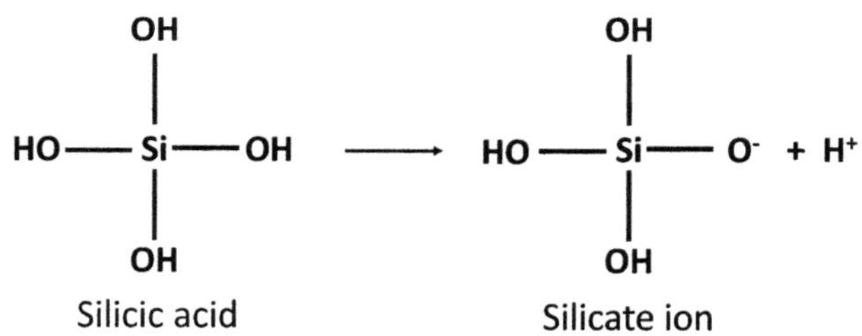

FIG. 5.51

Silicic acid ionization into silicate.

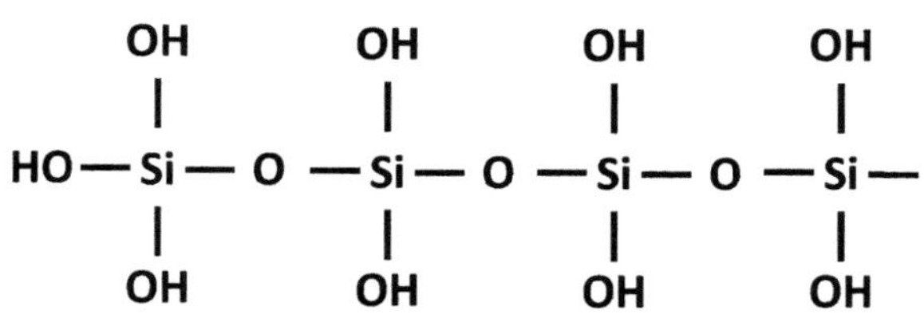

FIG. 5.52

Polymerized silicate.

5.10.2 Mechanism of silica scale formation

Although the chemistry of silica-silicate precipitation and inhibition has been intensively studied, no unified understanding of various processes involved in silica-silicate deposition currently exists [247].

Generally speaking, silica scaling can occur by one of three possible routes [248]:

- Condensation of monomeric silicic acid on solid substrate containing "—OH" group
- Polymerized silicic acid or colloidal deposition schematic of silicate polymerization process is illustrated in Fig. 5.53 and
- Biogenic amorphous silica by living organisms [249]

The product of silica/silicate ion polymerization is different shapes and structures of components, including rings of various sizes, cross-linked polymeric chains of different molecular weights, oligomeric structures, etc. The resulting product is a complex mixture of these components, which continues to grow to form amorphous (colloidal) silica scale [250]. Iler [251] reported that polymers will continue to grow until they reach a critical size for them to be considered colloidal particles, which is normally <50 Å, though sometimes down to 10–20 Å.

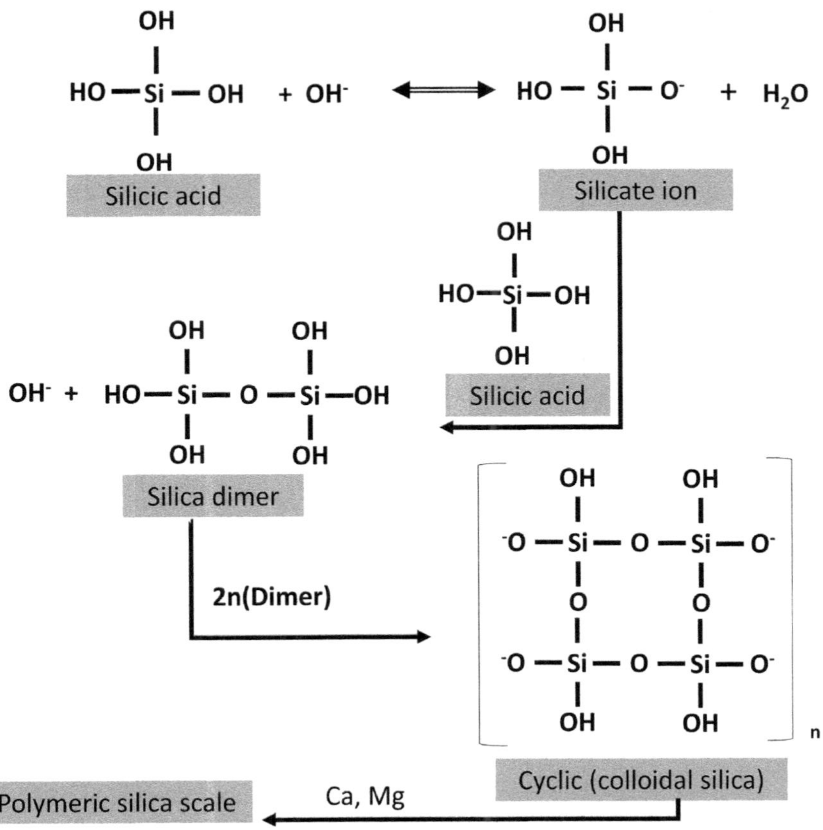

FIG. 5.53

Silicate sale formation by polymerization.

Chan [252] concluded that the polymerization rate is a maximum in the pH range 6 < pH < 9 and will continue until the monomeric silica concentration falls to the solubility of amorphous silica. Magnesium silicate precipitation strongly depends on solution pH and temperature. In systems operating above pH 8.5, the magnesium ion bridges the colloidal silicate particles, leading to the formation of amorphous magnesium silicate scale. As hydroxides, other metals, i.e., Ca, Sr, Na, Fe, Al, form silicates at pH < 8.5, which are more soluble than the magnesium silicates [253–255]. Under high pH conditions in alkaline surfactant polymer (ASP) flooding, the dissolved silica polymerize to form metal colloidal silica; the colloidal silica then becomes amorphous silica scale which, in the presence of ions, forms a metallic silicate scale [253].

Lithium silicate scale was found in boiler tubes in steam for in situ recovery of bitumen [256].

5.10.3 Factors affecting silica scale formation

5.10.3.1 Effect of pH

Silicate precipitation is highly dependent on solution pH. The presence of the different silica species and their relevant concentrations depend on the solution pH. Generally speaking, silicate precipitation is usually favored at a pH of less than 8.5. Magnesium silicate scale forms at a pH of more than 8.5, while sodium silicate polymerize below pH 11. Data suggest silica solubility is largely independent of pH in the range of 6–8 (Demadis [250]). The solubility of amorphous silica with pH at different temperatures is given in Fig. 5.54.

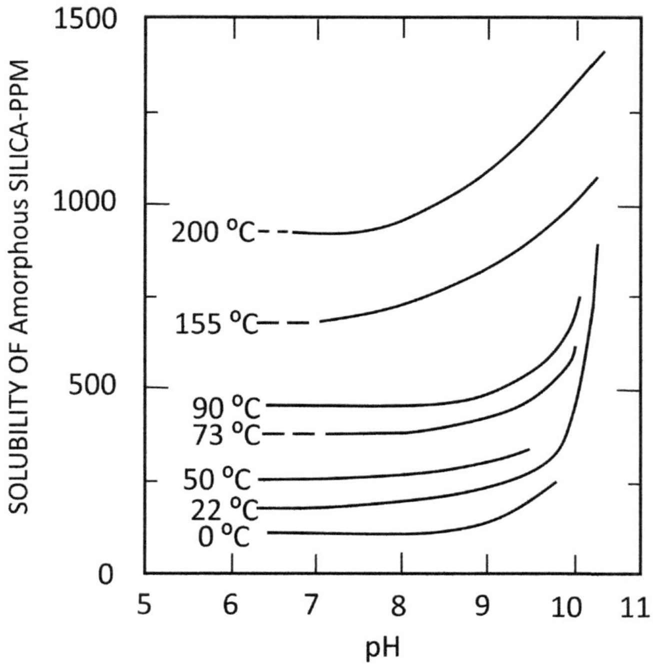

FIG. 5.54

Solubility of amorphous silica as a function of pH at different temperatures.

Reprinted from reference R.K. Iler, The Chemistry of Silica: Solubility, Polymerization, Colloid and Surface Properties and Biochemistry of Silica, Wiley New York, 1979, ISBN: 978-0-471-02404-0, copyright (1979), with permission from Jhon Wiley and Sons.

5.10.3.2 Effect of temperature

Silica and silicate solubility increase proportionally with temperature. Amorphous silica precipitate at low temperatures below its saturation. In contrast, magnesium silicate exhibits inverse solubility (Demadis 2003) [250]. Generally, silicate precipitation begins at a lower pH if the temperature is sufficiently high [257].

5.10.3.3 Effect of dissolved metal ions

Demadis [257] demonstrated that Mg^{2+} ions actually act as a catalyst in the silicic acid condensation reaction for magnesium levels up to 100 mg/L, but not unless the pH is above 9.

The cations were found to decrease silica solubility on the order of $Mg^{2+} > Ca^{2+} > Li^{+} > Na^{+} > K^{+}$ [258], with calcium having a lesser impact compared to magnesium on silica solubility [257].

5.11 Other mineral scales types

5.11.1 Barium and strontium carbonate

Barium and strontium carbonate have also been reported in oil and gas fields. They form by metal ion reaction with carbonate/bicarbonate species. They have very low solubility in water (24 mg/L for barium carbonate at 25°C [259], and 42.7 mg/L for strontium carbonate at 25°C [260]), and are affected by factors such as pH, temperature, and pressure.

5.11.2 Zinc and lead carbonate deposits

The occurrence of Smithsonite ($ZnCO_3$) is mentioned in the context of "exotic" oilfield scales. There have been a few reports of $ZnCO_3$ [261,262], which can also cause flow disruptions. Also Cerussite and $PbCO_3$ have been reported to impair the petroleum process [263].

The mechanism of these deposit formations is similar to that of calcium carbonate and iron carbonate. Another source of these deposits is when they are supplied to the production system for other purposes. For example, zinc carbonate can be used as an H_2S scavenger in water-base muds. It is also used as a bridging agent in cementing operations.

5.11.3 Magnesium-based scale

This includes magnesium hydroxide (brucite), magnesium silicate, and other magnesium-based scale. Magnesium hydroxide and calcium carbonate are commonly known as alkaline scale. Magnesium hydroxide is strongly dependent on other factors such as pH, temperature, release rate of CO_2, and the concentrations of divalent ions. Magnesium silicate was discussed earlier in the silicate scale section. Similar to magnesium silicate, magnesium hydroxide is also a common scale in ASP flooding.

Another type of magnesium-based scale is struvite or magnesium ammonium phosphate hexahydrate ($MgNH_4PO_4 \cdot 6H_2O$, struvite); this type of scale deposit can be a serious operational problem in wastewater treatment.

5.11.4 Calcium phosphate scale

Phosphorus exists in natural waters as particulate phosphate, orthophosphate, polyphosphates, and organic phosphorus. Calcium phosphate is another scale deposit that is commonly encountered in cooling systems and wastewater treatment. It forms by the interaction between calcium ions and phosphate ions. Hydroxylapatite, or calcium hydroxide phosphate ($Ca_5(PO_4)_3(OH)$), is also common in wastewater treatment. Calcium phosphates usually are affected by ion concentration, temperature, and pH. They usually form at high pH values.

5.12 Elemental sulfur deposition (ESD)

Elemental sulfur deposition is an old problem. Sulfur deposition in Canadian oilfields and American refineries dates back to the 1920s. Since then reports from industry have indicated the detrimental presence of ESD in downhole and topsides tubulars and equipment.

Elemental sulfur in gas, liquid, or solid phases has a ring structure with eight atoms linked together (S_8) that will break above 157°C. The solid sulfur exists in rhombic crystalline form at temperatures <96°C, in monoclinic crystalline from up to 118°C, and in gas form above 120°C [264].

5.12.1 Mechanisms of elemental sulfur deposition

There are various mechanisms proposed to explain the formation of solid elemental sulfur in the gas systems. Some of them generally explain the origins of elemental sulfur in the production systems, whether the sulfur will be formed by these deposit mechanisms or not, which depends on the supersaturation level and the system parameters. Nevertheless, the main mechanisms by which elemental sulfur is deposited can be shortlisted into three main mechanisms: chemical or microbial reaction, condensate formation, and desublimation [265,266].

5.12.1.1 Chemical reaction mechanisms

A number of chemical and microbial reactions can lead to the formation of elemental sulfur, including the following:

- Hydrolysis/pyrolysis and/or bacterial leaching of sulfur-containing reservoir rocks: sulfur mineral, pyrite, galena, sphalerite, and others [267,268]
- Decomposition of H_2S, hydrogen polysulfides, as shown in the following equations:

$$2H_2S_{(g)} + O_{2(g)} \rightarrow 2S_{(s)} + 2H_2O_{(g)} \tag{5.61}$$

$$4H_2S_{(g)} + C_6H_{6(g)} \rightarrow 4S_{(s)} + C_6H_{14} \tag{5.62}$$

$$H_2S_{9(g)} \rightarrow S_{8(s)} + H_2S_{(g)} \tag{5.63}$$

- Reaction of mill scale/iron oxides with H_2S [166,269]:

$$3H_2S + Fe_2O_3 \rightarrow S + 3H_2O + 2FeS \tag{5.64}$$

$$4H_2S + Fe_3O_4 \rightarrow S + 4H_2O + 3FeS \tag{5.65}$$

– TSR: TSR can happen at elevated temperatures, where solid $CaSO_4$ reacts with hydrocarbon, generating H_2S, and then solid $CaSO_4$ reacts with generated H_2S gas to form elemental sulfur [265]:

$$CH_4 + CaSO_4 \rightarrow H_2S + CO_2 + Ca(OH)_2 \quad (5.66)$$

$$6H_2S + CaSO_4 \rightarrow S_8 + 4H_2O + Ca(OH)_2 \quad (5.67)$$

– Bacterial sulfate reduction: elemental sulfur can also be formed in the presence of SRB in the crude oil. SRB will reduce SO_4^{2-} to form elemental sulfur and H_2S gas, as shown in the following equation [265,269]:

$$SO_4^{-2} \xrightarrow{\text{bacteria}} S \rightarrow H_2S \quad (5.68)$$

– Reaction with additives: such as methanol, which is a main additive to control gas hydrate formation [166]:

$$CH_3OH + H_2S \rightarrow CH_4 + S_{(s)} + H_2O \quad (5.69)$$

– Thermal decomposition of sulfur compounds during compression or other processing steps [270].

However, field experience does not support the idea that these reactions are a mechanism of deposition of elemental sulfur. Therefore these chemical reactions work as a source for the elemental sulfur in the production stream, and precipitation and deposition of elemental sulfur depends on the next two mechanisms [266].

5.12.1.2 Condensation mechanism

Field observations have shown that the condensation deposition hypothesis matches the physical location of the deposit, i.e., downstream of a pressure reduction facility [266]. Wilkes and Pareek [271] reported condensate formation as a potential cause of the deposit formation, as liquid hydrocarbons form when pressure is reduced. The liquid hydrocarbons can dissolve a part of the vapor sulfur and then remove it from the gas. Evaporation of the condensed liquids might lead to the deposition of sulfur particles [266]. Nevertheless, this condensate mechanism is possible only when the gas is rich in heavy hydrocarbons [272].

5.12.1.3 Desublimation mechanism

This mechanism is the most common acceptable mechanism of sulfur deposition, and it also encompasses the other two submechanisms within it: the chemical/ biological reactions as a source of sulfur, and the condensation as a crucial step in the desublimation.

By definition, desublimation is the passage of substance from the gaseous state directly to the solid state. Solid elemental sulfur has a vapor pressure, at ambient and subambient temperatures [270]. In the sour gas systems, H_2S can act as a solvent and carry elemental sulfur in the stream, or H_2S can react with elemental sulfur to form polysulfides [270]. When the natural gas stream contains sulfur vapor in solution, then with pressure reduction and hence temperature quench, sulfur vapor becomes supersaturated, which furnishes suitable conditions for a sulfur desublimation process to occur. Basically, the transition of the sulfur vapor to solid state as S_8 occurs because, at normal pipeline operating conditions, the partial pressure of the sulfur vapor is well below the triple points [166].

The overall "elemental sulfur" formation and deposition process consists of nucleation, condensation, and coagulation phases, with a fourth phase of agglomeration. The details of the process involve the following steps [166]:

1. Sulfur vapor already in gas stream in sub-ppm levels.
2. The sulfur vapor becomes supersaturated due to cooling of the gas mixture by the rapid gas expansion through common restrictions, e.g., pressure control valve/devices, or nozzles.
3. The supersaturated sulfur vapor molecules start nucleating by forming minute solid particles or nuclei.
4. Due to gas expansion, retrograde condensation can occur for some of the heavier hydrocarbon components in the gas stream.
5. Other molecules (retrograde condensation components) are attracted to the sulfur particle surface through the mechanism of condensation.
6. The resulting larger particles, which have a very high velocity, collide with other particles in the gas stream, forming larger particles, and the process is known as the coagulation process.
7. Other deposits can exist in the system, either sticking to the internal pipe walls and fittings, or suspended and traveling within the gas stream. Collisions between these particles are possible due to the high gas velocities and turbulence, leading to their growth in size and leading to agglomeration of solids.

5.12.2 Factors affecting the formation of elemental sulfur deposits

5.12.2.1 Effect of pressure/temperature reduction

Pressure reduction and temperature quenching are considered the main factors in precipitation and deposition of elemental sulfur. In fact, the main mechanism of deposition (desublimation) is explained in the light of the Joule-Thomson effect, which is basically a pressure drop term [273]. Under reservoir conditions with high pressure and temperature, a large amount of elemental sulfur is present as a vapor species, i.e., typically hydrogen polysulfide (H_2S_x). During production, as the sour gas flows from the reservoir to topsides and processing facilities, pressure reduction and cooling of the flowing gas stream decrease the S_8 solubility, and once S_8 vapor concentration reaches supersaturation level, further decreases in either pressure or temperature promote a thermodynamic instability leading to the precipitation of the solid elemental sulfur [272]. Fig. 5.55 shows the sulfur phase diagram at different pressures and temperatures.

Unless the system has a sudden temperature quench locally, the pressure drop rate appears to be a dominant parameter in S_8 deposition [272].

Zhu et al. [272] concluded that a high initial temperature delays the onset of S_8 desublimation. As a result, a high reservoir pressure and temperature will delay the occurrence of S_8 desublimation.

5.12.2.2 Effect of fluids composition

The produced gas (fluids) has a great impact on the deposition of sulfur. Ref. [166] studied the effect of different components on the deposition process and concluded the following:

- Hydrogen sulfide: Generation of sulfur.
- Oxygen: Generation of free sulfur from sulfur compounds.
- Moisture (water): Generation of free sulfur from sulfur compounds.

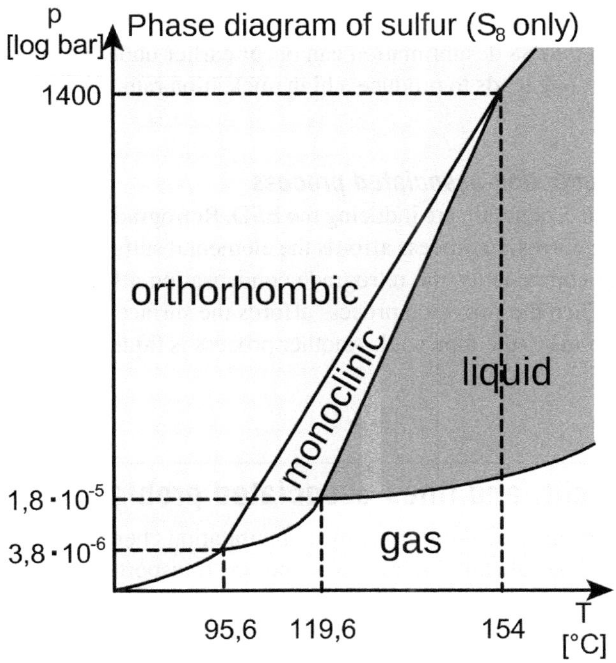

FIG. 5.55

Elemental sulfur phase diagram.

Source: https://commons.wikimedia.org/wiki/File:Sulfur_phase_diagram.svg, photo credit Wnt.

- Aromatics: Generation of free sulfur from sulfur compounds.
- Mercaptans: Reaction with Fe to form H_2S—then potentially S.
- Glycols: Reaction with Fe to form Fe_3O_4—then potentially reaction with H_2S to form S.
- Methanol: Reaction with H_2S to form S.
- Carbonyl sulfide: Potential for conversion to H_2S.
- Sulfur dioxide: Reaction with mercaptans (CH_4) to form S.
- Carbon dioxide: Reaction with H_2S to form COS.
- Solid contaminants in gas stream: Coagulation material.
- Lubricants: Formation of metal sulfides through reaction with H_2S. Possibility of MeS to S.
- Metal particles: Formation of metal sulfides through reaction with H_2S. Possibility of MeS to S.
- Goethite (rust): Reaction with H_2S to form S.
- Hydrocarbons C^{7+}: Liquid generation—coagulation potential.
- Oils—compressor and other coagulation with S particles and other.
- Volatile organic, chlorine, and fluorine compounds: Rapid corrosion.
- Mill scale in pipeline: Generation of S(s).

5.12.2.3 Effect of initial elemental sulfur vapor concentration

A low initial S_8 vapor concentration will postpone S_8 desublimation and have a smaller particle size at the end of the pipeline [272].

5.12.2.4 Effect of production rate

Zhu et al. [272] reported that S8 desublimation can occur earlier under an elevated production rate. A large sour gas production rate tends to produce a high nucleation rate, resulting in a large S8 particle at the end of pipelines.

5.12.2.5 Effect of production-associated process

Processes associated with production are inducing the ESD. Retrograde condensation and corrosion are two examples. While the corrosion process affords the elemental sulfur by reacting the corrosion products with sulfur-bearing components, the retrograde condensation affords the liquids required to start the deposition process. Then the corrosion process affords the surface defects required for the agglomeration/sticking of the solids to the pipe wall. Another process is liquid hold-up, which also influences the EDS.

5.13 Sand, mud, silt, and fines-associated problems

Sand, mud, silt, and other fines are not formed by a precipitation chemical reaction, but are part of the reservoir minerals or the petroleum system rocks and are transported with the fluids as suspended solids. These solids can serve as good nucleating sites for all kinds of deposits, including organic and mineral scales. In addition, when their concentration in the production fluid reaches a specific threshold, depending on the fluids hydrodynamics they can precipitate and plug different locations in the production system.

There are multiple studies and reviews in the literature that discuss the topic of sand, fines, and the different suspended solids [274–276]. Therefore these topics are limited here to concise discussions on their types and causes of transportation with the production fluids.

5.13.1 Sand production

Although approximately 60% of the world's oil and gas production comes from carbonates, 90% of hydrocarbon wells are in sandstone reservoirs. Around 30% of these sandstones may be weak enough to produce sand [277]. Some carbonate reservoirs may also produce solids [2].

Unconsolidated sandstone reservoirs with permeability of 0.5–8 darcies are the most susceptible to sand production, which may start during first flow or later when reservoir pressure has fallen or water breaks through [278].

The forces that restrain and hold the formation sand in place consist of [279]:

- Natural cementation (compressive strength), the primary restraint, is controlled by intergranular cementation. Generally speaking, old sediments are more consolidated than are younger sediments, which have little cementing material and are considered poorly consolidated.
- Friction between sand grains is related to the overburden stresses, including the mechanical stress that results from the overburden and the drag forces associated with viscous flow of fluids through the rock matrix.
- Fluid pressure in the pores of the rock.
- Capillary forces also contribute to sand production.

The main causes of sand production can be either natural or induced.

- *Natural causes of sand production*

Under this category falls the unconsolidated reservoir sands where there is limited weak grain-to-grain cementation. The sand grains are usually held together by osmotic pressure. In such cases, sand production is usually early, sudden, and can sometimes be unpredictable. Capillary pressure control is used to minimize impact on osmotic pressure and to prolong onset of production. Examples of these reservoirs include highly permeable Niger Delta Agbada and Akata formations, and the deepwaters of the Gulf of Mexico and Gulf of Guinea [276].

- *Induced causes of sand production*

The induced causes of sand production are those due to poor drilling, completion, production, and reservoir management strategies. They include:

a) High production rate beyond the critical limit that induces failure and sand migration [276].
b) Water production that dissolves the grain-grain cementation and also breaks down the cohesive osmotic pressure forces. Water production also reduces the capillary bond between water wet grains, and increases drag forces [280].
c) Poor drilling, completion, and production strategies, including high skin, high drawdown, thermal fractures, and shock loading due to repeated startup and blowdown [276].

Sand production strikes with varying degrees of severity, not all of which require action. The rate of sand production may decline over time with constant production conditions and is frequently associated with cleanup after stimulation.

Sand production causes drastic effects on the production system including:

- Causing erosion.
- Causing clogging problems when produced in high amounts.
- It works as nucleating sites for other solid deposits.

Fig. 5.56 shows the distribution of sand and fines deposits (in percent weight of the collected deposits) from oil wells at different depths, and Fig. 5.57 shows a deposit sample comprising mainly sand from a production well.

5.13.2 Fines migration

Fines migration also impacts production when the suspended particles bridge the pore throats near the wellbore, reducing well productivity. Fines can include different materials such as clays (phyllosilicates) and silts (silicates or aluminosilicates). Kaolinite and illite are the most common migrating clays. The clay fines sizes can range from 4 to 64 μm. Fines usually cause damage and permeability decline in the wellbore (within a radius of 3–5 ft of the wellbore), and also in gravel-pack completions [281]. Fines usually present in most sandstones and some carbonates. They remain attached to pore surfaces by electrostatic and van der Waals forces, and they are free to move with the fluid phase that wets them (usually water) [282].

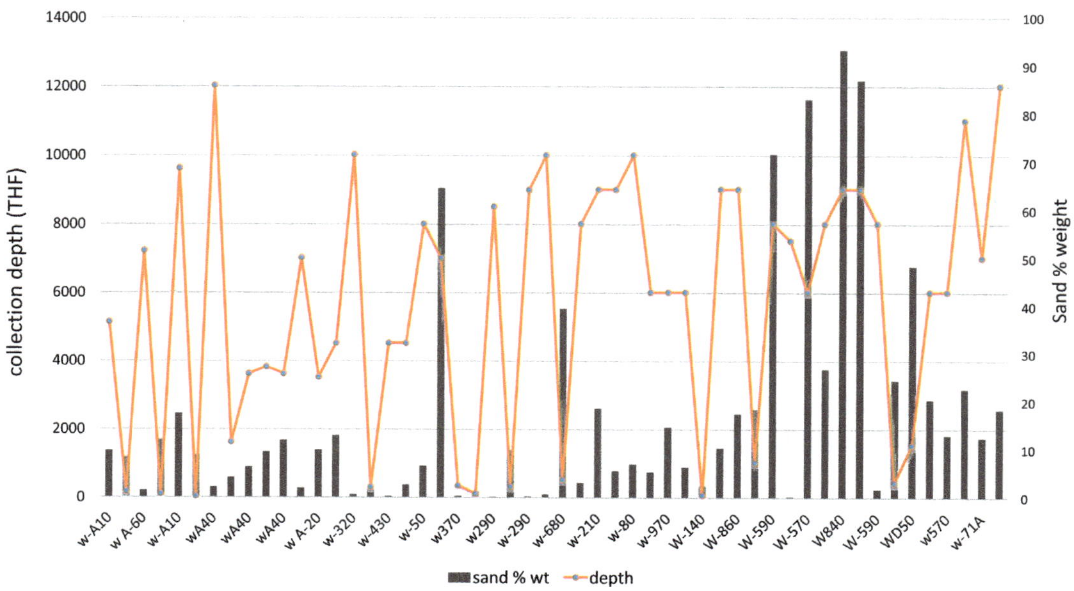

FIG. 5.56

Distribution of sand and fines in deposit samples collected from different wells at different depths.

FIG. 5.57

Deposit sample from oil producer comprises sand and fines.

Fines migration usually results from unconsolidated formation, or from introducing incompatible treatment fluids into the formation, which liberates the fine particles. The different factors affecting fines migration include [283]:

- High flow rate, particularly in wells producing water
- Wettability
- Salinity change. Low salinity is a common cause of fines destabilization and migration. A critical salt solution below which fines separation occurs is in the range of 5000–15,000 ppm (1.5%) sodium chloride [284]; for divalent ions this critical concentration is lower.

– Ion exchange. When clays exchange ions with surrounding water, leading to breaking them apart or swelling.
– pH change and introducing pH altering fluids into a water-sensitive formation. Generally increasing pH induces fines migration (by dissolving silica and other cementing materials). Besides, acidizing jobs can dissolve acid-soluble materials and induce fines migration.

Unlike sand migration that is best stabilized, the material mobilized in fines migration should be produced to avoid near-wellbore damage [281].

5.13.3 Clay swelling

Clay swelling is a common problem that leads to reducing permeability of hydrocarbon reservoirs. It also interferes with other operations such as drilling, completion, workover, fracturing, acidizing, and water flood.

Clays are naturally occurring minerals that exist in the petroleum reservoirs. Clays are phyllosilicates ("phyllo" meaning sheet) or layered minerals, consisting of stacks of negatively charged two-dimensional aluminosilicate layers. Each layer is composed of fused sheets of octahedra of Al^{3+}, Mg^{2+} or Fe^{3+} oxides, and sheets of tetrahedra of Si^{4+} oxides. The ratio and arrangement of the tetrahedral and octahedral sheets indicates the type of the clay. Clay mineral with one tetrahedral and one octahedral sheets is known as a 1:1 clay (and sometimes denoted a T-O clay) such as kaolinite. Clays combining two tetrahedral sheets that sandwich one central octahedral sheet are known as a 2:1 clay (sometimes denoted T-O-T), such as smectites [285]. Clays can also be classified into swelling and nonswelling types or expanding and nonexpanding. The structure of 2:1 clay is illustrated in Fig. 5.58, and the classification of clay types is shown in Fig. 5.59.

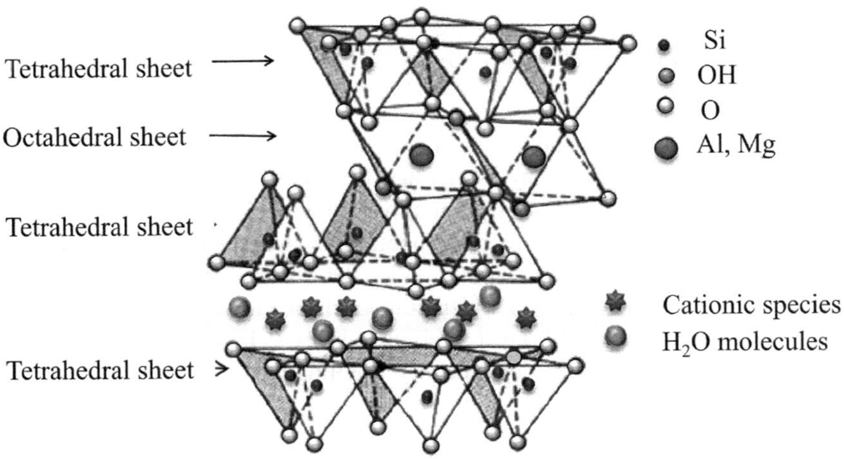

FIG. 5.58

Scheme of of 2:1 smectite clay mineral structure.

Reprinted with permission from Royal society of chemistry, from [Biomedical Applications of Cationic Clay Minerals, Ghadiri, M., W. Chrzanowski, and R. Rohanizadeh, 5, and 2015]; permission conveyed through Copyright Clearance Center, Inc.

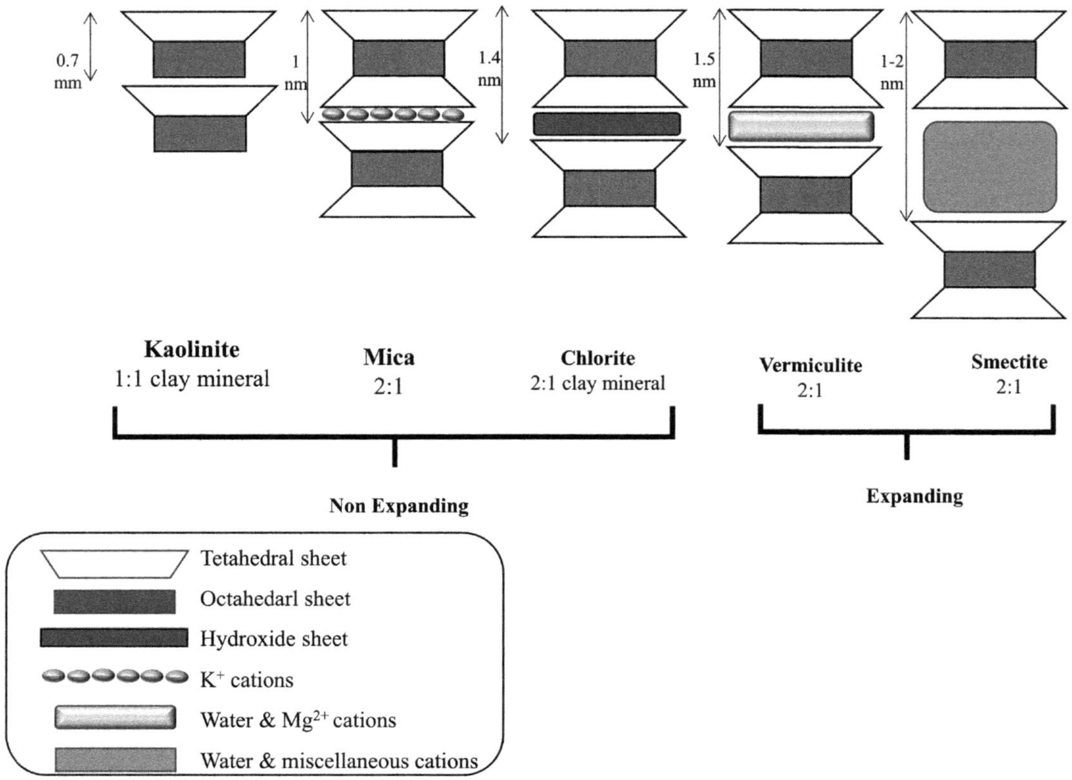

FIG. 5.59

Expanding and non-expanding cationic clay minerals.

Reprinted with permission from Royal society of chemistry, from [Biomedical Applications of Cationic Clay Minerals, Ghadiri, M., W. Chrzanowski, and R. Rohanizadeh, 5, and 2015]; permission conveyed through Copyright Clearance Center, Inc.

Clay minerals, such as smectites and mixed-layer illites, can expand in volume up to 20 times their original volume through adsorption of layers of water between their unit cells.

Crystalline swelling takes place by adsorption of water molecules at clay particle surfaces that occurs at interlayer separations of 10–22 Å. Diffuse double layer swelling occurs at interlayer separations >22 Å. This is driven by interlayer cation water molecules adsorption when exposed to lower ionic strength water, to form thick envelopes of water films over the clay platelets. The tendency of compacted clay to develop osmotic or double layer swelling reduces with increase in solute concentration in bulk solution [286,287]. Fig. 5.60 illustrates the clay swelling.

5.14 Recent advances in mineral scales research

Better understanding of the mechanisms by which mineral scales are formed and the factors affecting them helps in combating this problem and developing feasible mitigation plans.

Mineral scales are usually studied using static tests or using the dynamic flow loop experiments. The flow loop experiments are more accurate than the static tests and are more representative of

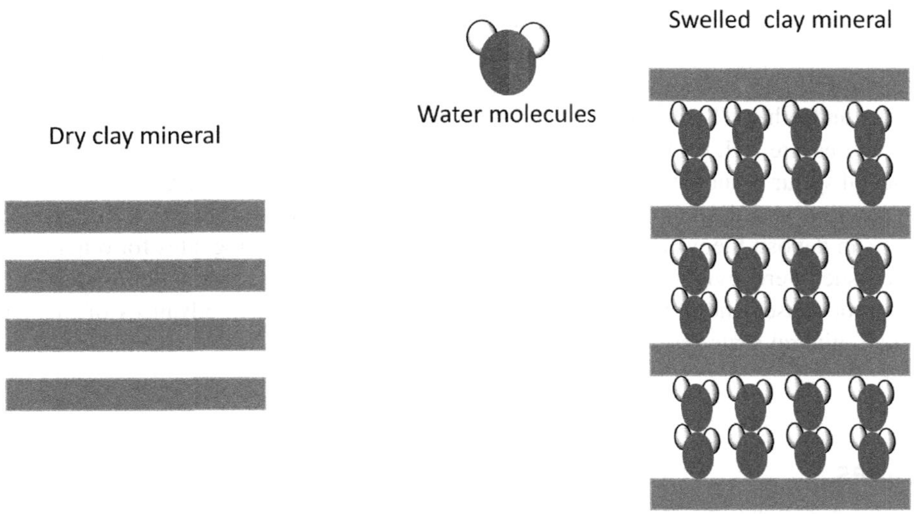

FIG. 5.60

Clay swelling process.

the field conditions. Scale studies have also been implemented using rotating disc equipment, and other electrochemical methods. Recently, microfluidic technology has offered precise and fast methods of studying the different variables that affect mineral scales.

Several advances were revealed recently in the mechanism of formation of different scales. Recent advances in the kinetics of nucleation and crystal growth processes and the findings of the nonclassical theories, which proved the presence of the intermediate phases, have helped in establishing in-depth mechanisms by which scale is formed. This, consequently, improves the way researchers understand mineral scales formation and helps in improving the inhibition and management techniques. Furthermore, the recent studies in particulate deposition on surfaces and surface properties improved the understanding of the scale adherence and deposition on surfaces and consequently that resulted in advancing steps in mitigating scale using manipulated or modified surfaces that are repellent or not adherent to scale or particles in general.

In addition, another factor is understanding the geochemistry of the reservoirs and the effect of the injected fluids to their formation. Moreover, more studies have been reported to investigate the hydrodynamics on scale formation.

These recent results are employed in the scale prediction models, which have witnessed significant advances by employing kinetic and hydrodynamic studies.

5.15 Summary

Mineral scales are a major flow assurance problem in oil and gas operations. Once they are formed in the system, they cause a tremendous number of problems from formation damage, tubing and flowlines blockages, to surface equipment failures. Mineral scales form when the thermodynamic equilibrium of the produced fluids is disturbed during production operations, leading to precipitating the dissolved minerals. Different types of mineral scales can encrust the production system including calcium carbonate, calcium sulfate, barium sulfate, strontium sulfate, iron compounds, zinc and lead sulfides, halite, silicates and others. Such types form due to mixing between incompatible fluids, changes in temperature, pressure, pH, bacterial activity, and other hydrodynamic and geological factors. In addition, elemental sulfur is another mineral deposition problem that can form in relatively different ways, e.g., condensation and desublimation. Furthermore, sand and fines production is another problem where they can deposit themselves or can work as functioning nucleating sites for other scale formation. These issues were reviewed in detail to furnish the necessary background for the reader to understand their root causes and how to combat them, in light of the recent advances of theoretical and experimental mineral scale research.

References

[1] O.J. Vetter, V. Kandarpa, Prediction of $CaCO_3$ scale under downhole conditions, in: Society of Petroleum Engineers—SPE Oilfield and Geothermal Chemistry Symposium, OGC 1980, Society of Petroleum Engineers, 1980, pp. 155–165, https://doi.org/10.2118/8991-ms.

[2] J. Bellarby, Production chemistry, in: J. Bellarby (Ed.), Well Completion Design, vol. 56, Developments in Petroleum Science, 2009, pp. 371–432, https://doi.org/10.1016/S0376-7361(08)00207-0.

[3] J.C. Cowan, D.J. Weintritt, Water-Formed Scale Deposits, Gulf Publishing Company, (TX USA), 1976.

[4] L.T. Ikelle, L. Amundsen, Introduction to Petroleum Seismology, Society of Exploration Geophysicists, 2005, https://doi.org/10.1190/1.9781560803447.

[5] M. Yuan, J. Mosley, N. Hyer, Mineral scale control in a CO_2 flooded oilfield, in: SPE International Symposium on Oilfield Chemistry, 2001, https://doi.org/10.2118/65029-ms.

[6] T. Zhu, M. Dittrich, Carbonate precipitation through microbial activities in natural environment, and their potential in biotechnology: a review, in: Frontiers in Bioengineering and Biotechnology, Frontiers Media SA, 2016, https://doi.org/10.3389/fbioe.2016.00004.

[7] F.W. Tegethoff, Calcium Carbonate: From the Cretaceous Period into the 21st Century, Springer, 2012, https://doi.org/10.1007/978-3-0348-8245-3.

[8] E. Mackenzie, Calcium carbonate hexahydrate, J. Chem. Soc. 188 (1923) 2409–2417.

[9] J.-Y. Gal, Y. Fovet, N. Gache, Mechanisms of scale formation and carbon dioxide partial pressure influence. Part I. Elaboration of an experimental method and a scaling model, Water Res. 36 (3) (2002) 755–763, https://doi.org/10.1016/s0043-1354(01)00270-6.

[10] J. Ihli, W.C. Wong, E.H. Noel, Y.Y. Kim, A.N. Kulak, H.K. Christenson, M.J. Duer, F.C. Meldrum, Dehydration and crystallization of amorphous calcium carbonate in solution and in air, Nat. Commun. 5 (January) (2014), https://doi.org/10.1038/ncomms4169.

[11] J.D. Rodriguez-Blanco, S. Shaw, L.G. Benning, The kinetics and mechanisms of amorphous calcium carbonate (ACC) crystallization to calcite, via vaterite, Nanoscale 3 (1) (2011) 265–271, https://doi.org/10.1039/c0nr00589d.

[12] J.D. Rodriguez-Blanco, S. Shaw, P. Bots, T. Roncal-Herrero, L.G. Benning, The role of Mg in the crystallization of monohydrocalcite, Geochim. Cosmochim. Acta 127 (February) (2014) 204–220, https://doi.org/10.1016/j.gca.2013.11.034.

[13] J.H.E. Cartwright, A.G. Checa, J.D. Gale, D. Gebauer, C.I. Sainz-Díaz, Calcium carbonate polyamorphism and its role in biomineralization: how many amorphous calcium carbonates are there? Angew. Chem., Int. Ed. 51 (2012) 11960–11970.

[14] F. Konrad, F. Gallien, D.E. Gerard, M. Dietzel, Transformation of amorphous calcium carbonate in air, Cryst. Growth Des. 16 (11) (2016) 6310–6317, https://doi.org/10.1021/acs.cgd.6b00906.

[15] Q. Shen, H. Wei, Y. Zhou, Y. Huang, H. Yang, D. Wang, X. Duanfu, Properties of amorphous calcium carbonate and the template action of vaterite spheres, J. Phys. Chem. B 110 (7) (2006) 2994–3000, https://doi.org/10.1021/jp055063o.

[16] P.G. Koutsoukos, T. Chen, Calcium carbonate: polymorph stabilization in the presence of inhibitors, in: The Science and Technology of Industrial Water Treatment, CRC Press, 2010, pp. 61–79, https://doi.org/10.1201/9781420071450.

[17] T. Ogino, T. Suzuki, K. Sawada, The formation and transformation mechanism of calcium carbonate in water, Geochim. Cosmochim. Acta 51 (1987) 2757–2767, https://doi.org/10.1016/0016-7037(87)90155-4.

[18] J.L. Wray, F. Daniels, Precipitation of calcite and aragonite, J. Am. Chem. Soc. 79 (1957) 2031–2034.

[19] H. Wang, V. Alfredsson, J. Tropsch, R. Ettl, T. Nylander, Formation of $CaCO_3$ deposits on hard surfaces—effect of bulk solution conditions and surface properties, ACS Appl. Mater. Interfaces 5 (10) (2013) 4035–4045, https://doi.org/10.1021/am401348v.

[20] G. Zhou, Y. Zheng, Chemical synthesis of CaCO3 minerals at low temperatures and implication for mechanism of polymorphic transition, N. Jb. Miner. Abh. 176 (2001) 323–343.

[21] H. Wei, Q. Shen, Y. Zhao, D.J. Wang, D.F. Xu, Influence of polyvinylpyrrolidone on the precipitation of calcium carbonate and on the transformation of vaterite to calcite, J. Cryst. Growth 250 (2003) 516–524, https://doi.org/10.1016/S0022-0248(02)02484-3.

[22] R. Chang, S. Kim, S. Lee, S. Choi, M. Kim, Y. Park, Calcium carbonate precipitation for CO_2 storage and utilization: a review of the carbonate crystallization and polymorphism, Front. Energy Res. 5 (2017) 17, https://doi.org/10.3389/fenrg.2017.00017.

[23] J. Chen, L. Xiang, Controllable synthesis of calcium carbonate polymorphs at different temperatures, Powder Technol. 189 (2009) 64–69, https://doi.org/10.1016/j.powtec.2008.06.004.

[24] C. Gabrielli, R. Jaouhari, S. Joiret, G. Maurin, In situ Raman spectroscopy applied to electrochemical scaling. Determination of the structure of vaterite, J. Raman Spec. 31 (2000) 497–501.

[25] P. Zhang, A.T. Kan, M.B. Tomson, Oil field mineral scale control, in: Mineral Scales and Deposits: Scientific and Technological Approaches, Elsevier Inc., 2015, pp. 603–617, https://doi.org/10.1016/B978-0-444-63228-9.00024-3.

[26] C.C. Patton, Applied water technology, in: Campbell Petroleum Series, first ed., October 1987.

[27] J. MacAdam, P. Jarvis, Water-formed scales and deposits: types, characteristics, and relevant industries, in: Mineral Scales and Deposits: Scientific and Technological Approaches, Elsevier Inc., 2015, pp. 3–23, https://doi.org/10.1016/B978-0-444-63228-9.00001-2.

[28] P. Duggirala, Formation of calcium carbonate scale and control strategies in continuous digesters, in: CD Del II Coloquio Internacional Sobre Celulosa …, 2005. http://www.eucalyptus.com.br/icep02/prassad_duggirala.pdf.

[29] J.J. Middelburg, Biogeochemical processes and inorganic carbon dynamics, 2019, pp. 77–105, https://doi.org/10.1007/978-3-030-10822-9_5.

[30] M. Sumestry, H. Tedjawidjaja, Case study: calcium carbonate scale inhibitor performance degradation due to H2S scavenger injection in Semoga field, in: Society of Petroleum Engineers—North Africa Technical Conference and Exhibition 2012, NATC 2012: Managing Hydrocarbon Resources in a Changing Environment, vol. 1, 2012, pp. 524–533, https://doi.org/10.2118/150705-pa.

[31] M. Kitamura, H. Konno, A. Yasui, H. Masuoka, Controlling factors and mechanism of reactive crystallization of calcium carbonate polymorphs from calcium hydroxide suspensions, J. Cryst. Growth 236 (2002) 323–332.

[32] Y.B. Hu, M. Wolthers, D.A. Wolf-Gladrow, G. Nehrke, Effect of PH and phosphate on calcium carbonate polymorphs precipitated at near-freezing temperature, Cryst. Growth Des. 15 (4) (2015) 1596–1601, https://doi.org/10.1021/cg500829p.

[33] Ç.M. Oral, B. Ercan, Influence of PH on morphology, size and polymorph of room temperature synthesized calcium carbonate particles, Powder Technology 339 (November) (2018) 781–788, https://doi.org/10.1016/j.powtec.2018.08.066.

[34] D.M. Dawson, A non-chemical water treatment system, Corros. Prev. Control (1990) 61–64.

[35] N. Andritsos, A.J. Karabelas, The influence of particulates on CaCO3 scale formation, J. Heat Transfer 121 (1999) 225–227.

[36] K. Orski, B. Grimbert, C.A. Menezes, E. Quin, Fighting Lead and Zinc Sulphide Scales on a North Sea HP/HT Field, Society of Petroleum Engineers, 2007, https://doi.org/10.2118/107745-MS.

[37] W.B. White, Speleothems, in: Encyclopedia of Caves (Third Edition), Elsevier, 2019, pp. 1006–1017.

[38] G.V. Chilingar, R. Mourhatch, G.D. Al-Qahtani, Scaling, in: The Fundamentals of Corrosion and Scaling for Petroleum and Environmental Engineers, Gulf, 2008, pp. 1–276. https://doi.org/10.1016/B978-1-933762-30-2.50012-4. (Chapter 6).

[39] O.J. Vetter, W.A. Farone, Calcium Carbonate Scale in Oilfield Operations, Society of Petroleum Engineers., 1987, https://doi.org/10.2118/16908-MS.

[40] L. Yiman, Calcite Scaling Potential of Kangding Geothermal Field, W-Sichuan Plateau, China, vol. 16, The United Nation University, Reykjavik, Iceland, 2017.

[41] B. Coto, C. Martos, J.L. Peña, R. Rodríguez, G. Pastor, Effects in the solubility of $CaCO_3$: experimental study and model description, Fluid Phase Equilib. 324 (2012) 1–7, https://doi.org/10.1016/j.fluid.2012.03.020.

[42] L.N. Plummer, E. Busenberg, The solubilities of calcite, aragonite and vaterite in CO_2-H_2O solutions between 0 and 90°C, and evaluation of the aqueous model for the system $CaCO_3$-CO_2-H_2O, Geochim. Cosmochim. Acta 46 (1982) 1011–1040.

[43] S. Arnórsson, S. Sigurdsson, H. Svavarsson, The chemistry of geothermal waters in Iceland I. Calculation of aqueous speciation from 0°C to 370°C, Geochim. Cosmochim. Acta 46 (1982) 1513–1532.

[44] D.B. Trushina, T.V. Bukreeva, M.V. Kovalchuk, M.N. Antipina, $CaCO_3$ vaterite microparticles for biomedical and personal care applications, Mater. Sci. Eng. C (2014), https://doi.org/10.1016/j.msec.2014.04.05.

[45] J.D. Rodriguez-Blanco, K.K. Sand, L.G. Benning, ACC and vaterite as intermediates in the solution-based crystallization of CaCO3, in: New Perspectives on Mineral Nucleation and Growth, Springer International Publishing, 2017, pp. 93–111, https://doi.org/10.1007/978-3-319-45669-0_5.

[46] T. Evans, Influence of Ionic Strength on Calcium Carbonate ($CaCO_3$) Polymorphism, arXiv: Materials Science, 2012.

[47] I.V. Muralikrishna, V. Manickam, Analytical methods for monitoring environmental pollution, in: Environmental Management, Elsevier, 2017, pp. 495–570, https://doi.org/10.1016/b978-0-12-811989-1.00018-x.

[48] E. Vetrimurugan, L. Elango, N. Rajmohan, Sources of contaminants and groundwater quality in the coastal part of a river delta, Int. J. Environ. Sci. Technol. 10 (3) (2013) 473–486, https://doi.org/10.1007/s13762-012-0138-3.

[49] H. Műller-Steinhagen, C.A. Branch, Calcium carbonate fouling of heat transfer surfaces, in: Proceedings of CHEMECA 1988, Sydney, 1988.

[50] J. MacAdam, S.A. Parsons, The effect of metal ions on calcium carbonate precipitation and scale formation, Sustain. Energy Build. (2009) 137–146, https://doi.org/10.1007/978-3-642-03454-1_15.

[51] J.D. Hem, Study and Interpretation of the Chemical Characteristics of Natural Water, US Geological Survey Water-Supply Paper 2254, 1985.

[52] M.B. Tomson, A.T. Kan, G. Fu, L. Cong, Measurement of Total Alkalinity and Carboxylic Acid and Their Relation to Scaling and Corrosion, Society of Petroleum Engineers, 2006, https://doi.org/10.2118/87449-PA.

[53] P. Anbu, C.H. Kang, Y.J. Shin, J.S. So, Formations of Calcium Carbonate Minerals by Bacteria and Its Multiple Applications, SpringerPlus. SpringerOpen, 2016, https://doi.org/10.1186/s40064-016-1869-2.

[54] M. Obst, J. Dynes, J. Lawrence, G. Swerhone, K. Benzerara, C. Karunakaran, et al., Precipitation of amorphous CaCO3 (aragonite-like) by cyanobacteria: a STXM study of the influence of EPS on the nucleation process, Geochim. Cosmochim. Acta 73 (2009) 4180–4198, https://doi.org/10.1016/j.gca.2009.04.013.

[55] H.H. Almahamedh, Sulfate reducing bacteria influenced calcium carbonate precipitation, in: NACE International, 2013.

[56] A. Quddus, L.M. Al-Hadhrami, Hydrodynamically deposited $CaCO_3$ and $CaSO_4$ scales, Desalination 246 (1–3) (2009) 526–533, https://doi.org/10.1016/j.desal.2008.11.005.

[57] M.M. Vazirian, A. Neville, An investigation into the effect of hydrodynamic conditions and surface characteristics on adhesion/deposition processes of carbonate/sulphate scales in the oil and gas industry, 2015.

[58] D. Hasson, M. Avriel, W. Resnick, T. Rozenman, S. Windreich, Mechanism of calcium carbonate scale deposition on heat-transfer surfaces, Ind. Eng. Chem. Fundam. 7 (1) (1968) 59–65, https://doi.org/10.1021/i160025a011.

[59] S. Muryanto, A.P. Bayuseno, H. Ma'mun, M. Usamah, and Jotho., Calcium carbonate scale formation in pipes: effect of flow rates, temperature, and malic acid as additives on the mass and morphology of the scale, Procedia Chem. 9 (2014) 69–76, https://doi.org/10.1016/j.proche.2014.05.009.

[60] C. Garcia, G. Courbin, F. Ropital, C. Fiaud, Study of the scale inhibition by HEDP in a channel flow cell using a quartz crystal microbalance, Electrochim. Acta 46 (7) (2001) 973–985, https://doi.org/10.1016/S0013-4686(00)00671-X.

[61] L.O. Jøsang, M.N. Psarrou, T. Østvold, T. Tjomsland, K. Ramstad, B.M. Hustad, Scale under turbulent flow, in: Tekna 23rd International Oil Field Chemistry Symposium, 18–21 March 2012, Geilo, Norway, 2012.

[62] P. Mougin, V. Lamoureux-Var, A. Bariteau, A.Y. Huc, Thermodynamic of thermochemical sulphate reduction, J. Pet. Sci. Eng. 58 (3–4) (2007) 413–427, https://doi.org/10.1016/j.petrol.2007.01.005.

[63] C.W. Walker, Origin of Gulf Coast salt-dome cap rock, AAPG Bull. 60 (12) (1976) 2162–2166, https://doi.org/10.1306/c1ea3aa0-16c9-11d7-8645000102c1865d.

[64] A.C. Todd, M.D. Yuan, I.R. McCracken, Part 3: Sulphate and carbonate scale prediction in North Sea oil and gas fields, in: E.A. Warren, P.C. Smalley (Eds.), North Sea Formation Water Atlas, vol. 15, Geological Society Memoir, 1994, pp. 105–118.

[65] R.A. McCartney, K. Melvin, R. Wright, E. Sørhaug, Seawater injection into reservoirs with ion exchange properties and high sulphate scaling tendencies: modelling of reactions and implications for scale management, with specific application to the Gyda Field, in: 18th International Oilfield Chemistry Symposium, Geilo, Norway, 26–28 March, 2007.

[66] B.M. Ham, M. Aihui, Analytical chemistry: a chemist and laboratory technician's toolkit, J. Chem. Inf. Model. 53 (2013) 1689–1699.

[67] V. Driessche, E.S. Alexander, T.M. Stawski, L.G. Benning, M. Kellermeier, Calcium sulfate precipitation throughout its phase diagram, in: New Perspectives on Mineral Nucleation and Growth, Springer International Publishing, 2017, pp. 227–256, https://doi.org/10.1007/978-3-319-45669-0_12.

[68] W. Gerhartz (Ed.), Ullmann's Encyclopedia of Industrial Chemistry, fifth ed., Vol. A1, VCH Publishers, Deerfield Beach, FL, 1985, p. VA4558. 1985 to Present.

[69] Y.W. Wang, Y.Y. Kim, H.K. Christenson, F.C. Meldrum, A new precipitation pathway for calcium sulfate dihydrate (gypsum) via amorphous and hemihydrate intermediates, Chem. Commun. 48 (2012) 504–506.

[70] A.E.S. Van Driessche, L.G. Benning, J.D. Rodriguez-Blanco, M. Ossorio, P. Bots, J.M. Gárcia-Ruiz, The role and implications of bassanite as a stable precursor phase to gypsum precipitation, Science 336 (2012) 69–72.

[71] A. Saha, J. Lee, S.M. Pancera, M.F. Bräeu, A. Kempter, A. Tripathi, A. Bose, New insights into the transformation of calcium sulphate hemihydrate to gypsum using time-resolved cryogenic transmission electron microscopy, Langmuir 28 (2012) 11182–11187.

[72] F. Jones, Infrared investigation of barite and gypsum crystallization: evidence for an amorphous to crystalline transition, CrystEngComm 14 (2012) 8374–8381.
[73] T. Stawski, A.E.S. Van Driessche, M. Ossorio, J.D. Rodriguez-Blanco, R. Besselink, L.G. Benning, Formation of calcium sulfate through the aggregation of sub-3 nanometre primary species, Nat. Commun. 7 (2016) 10177.
[74] N.B. Singh, B. Middendorf, Calcium sulphate hemihydrate hydration leading to gypsum crystallization, Prog. Cryst. Growth Charact. Mater. (2007), https://doi.org/10.1016/j.pcrysgrow.2007.01.002.
[75] J.E. Oddo, M.B. Tomson, Why scale forms and how to predict it. SPE production & facilities, J. Sci. 47 (1994).
[76] A. Klimchouk, The dissolution and conversion of gypsum and anhydrite, Int. J. Speleol. 25 (3/4) (1996) 21–36, https://doi.org/10.5038/1827-806x.25.3.2.
[77] T.A. Hoang, H. Ming Ang, A.L. Rohl, Effects of temperature on the scaling of calcium sulphate in pipes, Powder Technol. 179 (1–2) (2007) 31–37, https://doi.org/10.1016/j.powtec.2006.11.013.
[78] D. Hasson, J. Zahavi, Mechanism of calcium sulfate scale deposition on heat-transfer surfaces, Ind. Eng. Chem. Fundam. 9 (1) (1970) 1–10, https://doi.org/10.1021/i160033a001.
[79] S. Casby-Horton, J. Herrero, N.A. Rolong, Gypsum soils-their morphology, classification, function, and landscapes, in: Advances in Agronomy 130, Academic Press Inc., 2015, pp. 231–290, https://doi.org/10.1016/bs.agron.2014.10.002.
[80] K.U.G. Raju, G. Atkinson, The thermodynamics of 'scale' mineral solubilities. 3. Calcium sulfate in aqueous NaCl, J. Chem. Eng. Data 35 (3) (1990) 361–367, https://doi.org/10.1021/je00061a038.
[81] L.A. Hardie, The gypsum-anhydrite equilibrium at one atmosphere pressure, Am. Mineral. 52 (1–2) (1967) 171–200. http://www.minsocam.org/ammin/AM52/AM52_171.pdf.
[82] A.G. Ostroff, Conversion of gypsum to anhydrite in aqueous salt solutions, Geochim. Cosmochim. Acta 28 (9) (1964) 1363–1372, https://doi.org/10.1016/0016-7037(64)90154-1.
[83] B. Bansal, H. Müller-Steinhagen, Crystallization fouling in plate heat exchangers, J. Heat Transfer 115 (3) (1993) 584–591, https://doi.org/10.1115/1.2910728.
[84] A.G. Collins, Geochemistry of oilfield waters, in: Developments in Petroleum Science, vol. 1, ELSEVIER SCIENT.PUBL, Amsterdam, 1975, https://doi.org/10.1016/0012-8252(77)90073-3.
[85] J. Moghadasi, H. Müller-Steinhagen, M. Jamialahmadi, A. Sharif, Scale deposits in porous media and their removal by Edta injection, in: Heat Exchanger Fouling and Cleaning VII, 2007, pp. 59–60. http://dc.engconfintl.org/heatexchanger2007/10.
[86] Z. Dai, A.T. Kan, W. Shi, et al., Solubility measurements and predictions of gypsum, anhydrite, and calcite over wide ranges of temperature, pressure, and ionic strength with mixed electrolytes, Rock Mech. Rock Eng. 50 (2017) 327–339, https://doi.org/10.1007/s00603-016-1123-9.
[87] C.W. Blounot, F.W. Dickson, The solubility of anhydrite ($CaSO_4$) in NaCl-H_2O from 100 to 450°C and 1 to 1000 bars, Geochim. Cosmochim. Acta 33 (2) (1969) 227–245, https://doi.org/10.1016/0016-7037(69)90140-9.
[88] R.S. Fulford, Effects of brine concentration and pressure drop on gypsum scaling in oil wells, J. Petrol. Technol. 20 (06) (1968) 559–564, https://doi.org/10.2118/1830-pa.
[89] C.W. Blount, F.W. Dickson, Gypsum-anhydrite equilibria in systems CaSO4-H2O and CaSO4-NaCl-H2O, Am. Mineral. 58 (1973) 323–331.
[90] B.L. Carlberg, R.R. Matches, Solubility of calcium sulfate in brine, in: SPE Paper 4353 Presented at the Oilfield Chemistry Symposium of the Society of Petroleum Engineers of AIME, Denver, CO, May 24–25, 1973.
[91] J. Shukla, Y.P. Mohandas, A. Kumar, Effect of PH on the solubility of $CaSO_0 \cdot 2H_2O$ in aqueous NaCl solutions and physicochemical solution properties at 35°C, J. Chem. Eng. Data 53 (12) (2008) 2797–2800, https://doi.org/10.1021/je800465f.

[92] D. Bhagawati, U. Kakoty, P.P. Saikia, M.K. Baruah, Study on the effect of pH, ionic strength and pressure on the solubility of $CaSO_4$-NaCl-H_2O system, Int. Res. J. Pure Appl. Chem. 11 (1) (2016) 1–10. Article no. IRJPAC.22661.

[93] M. Brown, Full scale attack, in: Review, The BP Technology Magazine, 1998, pp. 30–32.

[94] H. Gamal, S. Al-Afnan, S. Elkatatny, M. Bahgat, Barium sulfate scale removal at low-temperature, Geofluids (2021), https://doi.org/10.1155/2021/5527818.

[95] M.A. Kelland, Production Chemicals for the Oil and Gas Industry, CRC Press, 2009, https://doi.org/10.1201/9781420092974.

[96] C. Ruiz-Agudo, E. Ruiz-Agudo, C.V. Putnis, A. Putnis, Mechanistic principles of barite formation: from nanoparticles to micron-sized crystals, Cryst. Growth Des. 15 (8) (2015) 3724–3733, https://doi.org/10.1021/acs.cgd.5b00315.

[97] J.J. Wylde, G.D.M. Williams, F. Careil, et al., Deep Downhole Chemical Injection on BP Operated Miller: Experience and Learning, 2006. SPE 92832.

[98] General Chemistry, 5th ed., Whitten, K.W.; Davis, R.E.; Peck, M.L. 824.

[99] O. Bukuaghangin, The kinetics of barium sulphate scale formation and inhibition in the bulk solution and on surfaces, 2017. PhD thesis.

[100] A.B. Merdhah, Bin, and Abu Azam Mohd Yassin., Laboratory study on precipitation of barium sulphate in Malaysia sandstone cores, Open Pet. Eng. J. 2 (2009) 1–11, https://doi.org/10.2174/18748341009020100.

[101] J. Moghadasi, M. Jamialahmadi, H. Muller-Steinhagen, A. Sharif, A. Ghalambor, R.M. Izadpanah, E. Motaie, Scale formation in Iranian oil reservoir and production equipment during water injection, in: The 5th International Oilfield Scale Symposium and Exhibition, January 29–30, Aberdeen, UK, SPE 80406, 2003, pp. 1–14.

[102] A.B. BinMerdhah, A.A.M. Yassin, M.A. Muherei, Laboratory and prediction of barium sulfate scaling at high-barium formation water, J. Pet. Sci. Eng. 70 (1–2) (2010) 79–88, https://doi.org/10.1016/j.petrol.2009.10.001.

[103] S.J. Dyer, G.M. Graham, The effect of temperature and pressure on oilfield scale formation, J. Pet. Sci. Eng. 35 (1–2) (2002) 95–107, https://doi.org/10.1016/S0920-4105(02)00217-6.

[104] Y. Zhao, A.Y.-T. Lu, X. Wang, C. Dai, S. Paudyal, S. Ko, G. Deng, S. Mateen, A.T. Kan, M. Tomson, Strontium effect on Ba-Sr-SO_4 scale deposition rate, in: Paper Presented at the CORROSION 2020, Physical Event Cancelled, June 2020.

[105] C. Okocha, K.S. Sorbie, Effects of Sulphide Scales (PbS, ZnS & FeS) on $BaSO_4$ Crystal Growth and Dissolution, Society of Petroleum Engineers, 2010, https://doi.org/10.2118/130391-MS.

[106] L. Sutherland, C.J. Johnston, W. Taylor, The Influence of Turbulence (or Hydrodynamic Effects) on Barium Sulphate Scale Formation and Inhibitor Performance, Society of Petroleum Engineers, 2013, https://doi.org/10.2118/164070-MS.

[107] D.J. Weintritt, J.C. Cowan, Unique characteristics of barium sulfate scale deposition, J. Pet. Technol. 19 (10) (1967) 1381–1394.

[108] J.R. Campbell, G.H. Nancollas, The crystallization and dissolution of strontium sulfate in aqueous solution, J. Phys. Chem. 73 (6) (1969) 1735–1740, https://doi.org/10.1021/j100726a017.

[109] O. Sühnel, M. Handlířová, Precipitation of strontium sulphate, Cryst. Res. Technol. 19 (4) (1984) 477–490, https://doi.org/10.1002/crat.2170190407.

[110] O.J.G. Vetter, I. Vandenbroek, J. Nayberg, $SrSO_4$: the basic solubility data, in: SPE Paper 11803 presented at the International Symposium on Oil-field and Geothermal Chemistry, Denver, CO (June 1–3), 1983.

[111] A.B. Merdhah, A.A.M. Yassin, Solubility of common oil field scales of injection water and high-barium concentration and high salinity formation water, J. Teknol. 50 (2007) 67–77.

[112] M. Amiri, J. Moghadasi, M. Jamialahmadi, A prediction of the amount of strontium sulfate scale formation in Siri oilfield at different temperatures and pressures, Energy SourcesPart A 36 (1) (2014) 5–14, https://doi.org/10.1080/15567036.2010.535104.

[113] S. Schulien, High-Temperature/High-Pressure Solubility Measurements in the Systems $BaSO_4$-NaCl-H_2O and $SrSO_4$-NaCl-H_2O in Connection With Scale Studies, Society of Petroleum Engineers, 1987, https://doi.org/10.2118/16264-MS.

[114] D.F. Jacques, B.I. Bourland, A Study of Solubility of Strontium Sulfate, Society of Petroleum Engineers, 1983, https://doi.org/10.2118/9625-PA.

[115] J.W. Davis, A. Gene Collins, Solubility of barium and strontium sulfates in strong electrolyte solutions, Environ. Sci. Technol. 5 (10) (1971) 1039–1043, https://doi.org/10.1021/es60057a007.

[116] M. Amiri, J. Moghadasi, M. Pordel, The effect of temperature, pressure, and mixing ratio of the injection water with formation water on strontium sulfate scale formation in the Siri oilfield, Pet. Sci. Technol. 30 (7) (2012) 635–645, https://doi.org/10.1080/10916466.2010.489088.

[117] C. Johnston, L. Sutherland, The Influence of Turbulence (or Hydrodynamic Effects) on Strontium Sulphate Scale Formation and Inhibitor Performance, Society of Petroleum Engineers, 2014, https://doi.org/10.2118/169760-MS.

[118] International Association of Oil and Gas Producers (IOGP), Managing Naturally Occurring Radioactive Material (NORM) in the Oil and Gas Industry Report, 2016.

[119] EPA, TENORM: Oil and Gas Production Wastes, 2019. Retrieved from: https://www.epa.gov/radiation/tenorm-oil-and-gas-production-wastes last visit Dec. 2019.

[120] P.R. Gray, NORM Contamination in the Petroleum Industry, Society of Petroleum Engineers, 1993, https://doi.org/10.2118/22880-PA.

[121] M.M.M. Ali, H. Zhao, Z. Li, N.N.M. Maglas, Concentrations of TENORMs in the petroleum industry and their environmental and health effects, RSC Adv. 9 (2019) 39201, https://doi.org/10.1039/c9ra06086c.

[122] U.S. Environmental Protection Agency, Diffuse NORM Wastes: Waste Characterization and Preliminary Risk Assessment, Draft Report, Environmental Protection Agency, Office of Radiation and Indoor Air, Washington, DC, US, 1993.

[123] A.W. Nelson, E.S. Eitrheim, A.W. Knight, D. May, M.A. Mehrhoff, R. Shannon, R. Litman, W.C. Burnett, T.Z. Forbes, M.K. Schultz, Understanding the radioactive ingrowth and decay of naturally occurring radioactive materials in the environment: an analysis of produced fluids from the Marcellus Shale, Environ. Health Perspect. 123 (7) (2015) 689–696, https://doi.org/10.1289/ehp.1408855.

[124] U.S. Geological Survey, Naturally Occurring Radioactive Materials (NORM) in Produced Water and Oil-Field Equipment—An Issue for the Energy Industry, Fact Sheet, U.S. Geological Survey, 1999, pp. 142–199, https://doi.org/10.3133/fs14299.

[125] Godoy, J.M., Carvalho, F., Cordilha, A., Matta, L.E., & Godoy, M.L. (2005). ^{210}Pb content in natural gas pipeline residues ("black-powder") and its correlation with the chemical composition. J. Environ. Radioact., 83(1), 101–111. https://doi.org/https://doi.org/10.1016/j.jenvrad.2005.02.018.

[126] W. Knaepen, W. Bergwerf, P. Lancée, W. van Dijk, J. Jansen, R. Janssen, KiezenbergW., van Sluijs, R., Tijsmans M., Volkers K., Voors, P., State-of-the-art of NORM nuclide determination in samples from oil and gas production—validation of potential standardisation methods through an Interlaboratory Test Programme, J. Radioanal. Nucl. Chem. 198 (1995) 323–341.

[127] K.C. Taylor, H.A. Nasr-El-Din, M.J. Al-Alawi, Systematic study of iron control chemicals used during well stimulation, SPE J. 4 (1) (1999) 19–24.

[128] C.W. Crowe, Evaluation of agents for preventing precipitation of ferric hydroxide from spent treating acid, J. Petrol. Technol. 37 (4) (1985) 691–695.

[129] C.W. Crowe, Prevention of undesirable precipitates from acid treating fluids, in: Paper SPE 14090 Presented at the International Meeting on Petroleum Engineering, 17–20 March, Beijing, China, 1986.

[130] B.E. Hall, W.R. Dill, Iron control additives for limestone and sandstone acidizing of sweet and sour wells, in: Paper SPE 17157 Presented at the SPE Formation Damage Control Symposium, 8–9 February, Bakersfield, California, 1988.

[131] H.A. Nasr-El-Din, A.Y. Al-Humaidan, Iron sulfide scale: formation, removal and prevention, in: Society of Petroleum Engineers—International Symposium on Oilfield Scale 2001, Society of Petroleum Engineers, 2001, https://doi.org/10.2523/68315-ms.

[132] O.A. Nafday, S. Nesic, Iron carbonate scale formation and CO_2 corrosion in the presence of acetic acid, in: NACE—International Corrosion Conference Series, vol. 2005-April. National Assoc. of Corrosion Engineers International, 2005.

[133] W. Sun, S. Nesic, Basics revisited: kinetics of iron carbonate scale precipitation in CO_2 corrosion, in: NACE—International Corrosion Conference Series, 063651-521, 2006.

[134] A. Krylov, O. Khlystov, T. Zemskaya, H. Minami, A. Hachikubo, Y. Nunokawa, M. Kida, et al., First discovery and formation process of authigenic siderite from gas hydrate-bearing mud volcanoes in fresh water: Lake Baikal, Eastern Siberia, Geophys. Res. Lett. 35 (5) (2008), https://doi.org/10.1029/2007GL032917.

[135] J.G.N. Thomas, J.D. Davies, Influence of hydrogen carbonate and chloride-ions on stability of oxide-films on mild-steel in near-neutral solutions, Br. Corros. J. 12 (1977) 108–114.

[136] D. Marco, Z.T. Roland, B.P. Jiang, E. Poinen, An in situ synchrotron radiation grazing incidence X-ray diffraction study of carbon dioxide corrosion, J. Electrochem. Soc. 152 (10) (2005), https://doi.org/10.1149/1.2007227.

[137] E.W. Chan, Magnetite and its Galvanic Effect on the Corrosion of Carbon Steel Under Carbon Dioxide Environments, PhD Thesis, Curtin University, 2011.

[138] T. Tanupabrungsun, D. Young, B. Brown, S. Nesic, Construction and verification of Pourbaix diagrams for CO_2 corrosion of mild steel valid up to 250°C, in: NACE Corrosion 2012, Paper No. 1418, 2012.

[139] L.M.F. Baumann, D. Birgel, M. Wagreich, J. Peckmann, Microbially-driven formation of Cenozoic siderite and calcite concretions from Eastern Austria, Austrian J. Earth Sci. 109 (2) (2016), https://doi.org/10.17738/ajes.2016.0016.

[140] T.R. Jack, M.J. Wilmott, R.L. Sutherby, Indicator minerals formed during external corrosion of line pipe, Mater. Perform. 34 (11) (1995) 19–22.

[141] H. Venzlaff, D. Enning, J. Srinivasan, K.J.J. Mayrhofer, A.W. Hassel, F. Widdel, M. Stratmann, Accelerated cathodic reaction in microbial corrosion of iron due to direct electron uptake by sulfate-reducing bacteria, Corros. Sci. 66 (January) (2013) 88–96, https://doi.org/10.1016/j.corsci.2012.09.006.

[142] R.J.W. Meulepas, C.G. Jagersma, A.F. Khadem, A.J.M. Stams, P.N.L. Lens, Effect of methanogenic substrates on anaerobic oxidation of methane and sulfate reduction by an anaerobic methanotrophic enrichment, Appl. Microbiol. Biotechnol. 87 (4) (2010) 1499–1506.

[143] O. Oyekola, R.P. van Hille, S.T.L. Harrison, Study of anaerobic lactate metabolism under biosulfidogenic conditions, Water Res. 43 (14) (2009) 3345–3354.

[144] M.J. Whiticar, Carbon and hydrogen isotope systematics of bacterial formation and oxidation of methane, Chem. Geol. 161 (1999) 291–314.

[145] C.L. Zhang, J. Horita, D.R. Cole, J. Zhou, D.R. Lovley, T.J. Phelps, Temperature dependent oxygen and carbon isotope fractionation of biogenic siderite, Geochim. Cosmochim. Acta 65 (2001) 2257–2271.

[146] G. Joshi, Elucidating Sweet Corrosion Scale, PhD Thesis, The University of Manchester, 2015, p. 280.

[147] P. Bénézeth, J.L. Dandurand, J.C. Harrichoury, Solubility product of siderite ($FeCO_3$) as a function of temperature (25–250°C), Chem. Geol. 265 (1–2) (2009) 3–12, https://doi.org/10.1016/j.chemgeo.2009.03.015.

[148] D. Langmuir, Aqueous Environmental Geochemistry, first ed., Prentice Hall, 1996.

[149] W. Sun, S. Nešić, R.C. Woollam, The effect of temperature and ionic strength on iron carbonate ($FeCO_3$) solubility limit, Corros. Sci. 51 (2009) 1273–1276.

[150] R.D. Braun, Solubility of iron (II) carbonate at temperatures between 30 and 80, Talanta 38 (1991) 205–211.

[151] L. Garverick, Corrosion in the Petrochemical Industry, ASM International, Materials Park, OH, 1994. http://books.google.com/books?id=XblTAAAAMAAJ.

[152] A.K. Vuppu, W.P. Jepson, Effect of temperature in sweet corrosion of horizontal multiphase carbon steel pipelines, in: SPE—Asia Pacific Oil & Gas Conference, Society of Petroleum Engineers (SPE), 1994, pp. 635–640, https://doi.org/10.2523/28809-ms.

[153] W. Renpu, Oil and gas well corrosion and corrosion prevention, in: Advanced Well Completion Engineering, Elsevier, 2011, pp. 617–700, https://doi.org/10.1016/b978-0-12-385868-9.00018-x.
[154] F. Mutahhar, G. Aithan, E.V. Iski, M.W. Keller, S. Shirazi, K.P. Roberts, Mechanistic modeling of erosion-corrosion for carbon steel, in: Trends in Oil and Gas Corrosion Research and Technologies: Production and Transmission, Elsevier Inc., 2017, pp. 749–763, https://doi.org/10.1016/B978-0-08-101105-8.00031-0.
[155] K. Chokshi, W. Sun, S. Nesic, M., Technology, iron carbonate scale growth and the effect of inhibition in CO_2 corrosion of mild steel, in: NACE Corros, Paper No. 285, 2005.
[156] H.A. Alsaiari, A. Kan, M. Tomson, Effect of Calcium and Iron (II) Ions on the Precipitation of Calcium Carbonate and Ferrous Carbonate, Society of Petroleum Engineers, 2010, https://doi.org/10.2118/121553-PA.
[157] E.B. Castro, J.R. Vilche, Study of iron electrodissolution in carbonate/bicarbonate solutions by electrochemical impedance spectroscopy, J. Electroanal. Chem. 323 (1992) 231–246.
[158] C.A.R. Silva, X. Liu, F.J. Millero, Solubility of sidertie ($FeCO_3$) in NaCl solutions, J. Solut. Chem. 31 (2002) 97–108.
[159] R. Bonis, J.-L. Crolet, Basics of the Prediction of the Risks of CO2 Corrosion in Oil and Gas Wells, CORROSION/89, Paper No. 466, NACE International, Houston Texas, 1989.
[160] V. Fajardo, C. Canto, B. Brown, S. Nesic, Effect of Organic Acids in CO_2 Corrosion, NACE International, 2007. Paper No. 07319.
[161] H. Cypionka, Sulfate reducing bacteria, in: J. Reitner, V. Thiel (Eds.), Encyclopedia of Geobiology, 2011, https://doi.org/10.1007/978-1-4020-9212-1_197.
[162] M. Basafa, K. Hawboldt, Reservoir souring: sulfur chemistry in offshore oil and gas reservoir fluids, J. Pet. Explor. Prod. Technol. (2019), https://doi.org/10.1007/s13202-018-0528-2.
[163] I.R. Collins, M.M. Jordan, Occurrence, Prediction, and Prevention of Zinc Sulfide Scale Within Gulf Coast and North Sea High-Temperature and High-Salinity Fields, Society of Petroleum Engineers, 2003, https://doi.org/10.2118/84963-PA.
[164] I.B. Obot, M.M. Solomon, S.A. Umoren, R. Suleiman, M. Elanany, N.M. Alanazi, A.A. Sorour, Progress in the development of sour corrosion inhibitors: past, present, and future perspectives, J. Ind. Eng. Chem. (2019), https://doi.org/10.1016/j.jiec.2019.06.046.
[165] S. Gao, B. Brown, D. Young, S. Nesic, M. Singer, A modified thermodynamic model for the prediction of mild steel corrosion product formation at high temperature in an aqueous H_2S environment, in: NACE—International Corrosion Conference Series. Vol. 2019-March. National Assoc. of Corrosion Engineers International, 2019.
[166] J.P. David, Elemental Sulphur Deposition in Natural Gas Transmission Pipelines, PhD thesis, University of Western Australia, School of Oil and Gas Engineering, 2005.
[167] D. Rickard, G.W. Luther, Chemistry of iron sulfides, Chem. Rev. 107 (2) (2007) 514–562.
[168] D. Rickard, The solubility of FeS, Geochim. Cosmochim. Acta 70 (23 SPEC ISSUE) (2006) 5779–5789, https://doi.org/10.1016/j.gca.2006.02.029.
[169] R.A. Berner, Thermodynamic stability of sedimentary iron sulfides, Am. J. Sci. 265 (9) (1967) 773–785.
[170] P.H. Tewari, G. Wallace, A.B. Campbell, The Solubility of Iron Sulfides and Their Role in Mass Transport in Girdler-Sulfide Heavy Water Plants, Atomic Energy of Canada Ltd., 1978. AECL-5960.
[171] S. Elkatatny, New formulation for iron sulfide scale removal, in: SPE Middle East Oil and Gas Show and Conference, MEOS, Proceedings, 2017-March:1809–21. Society of Petroleum Engineers (SPE), 2017, https://doi.org/10.2118/183914-ms.
[172] M.A. Mahmoud, M. Kamal, B.S. Bageri, I. Hussein, Removal of pyrite and different types of iron sulfide scales in oil and gas wells without H_2S generation, in: International Petroleum Technology Conference, IPTC 2015. Vol. 2015-January. Society of Petroleum Engineers (SPE), 2015, https://doi.org/10.2523/iptc-18279-ms.
[173] N.G. Harmandas, E. Navarro Fernandez, P.G. Koutsoukos, Crystal growth of pyrite in aqueous solutions. inhibition by organophosphorus compounds, Langmuir 14 (1998) 1250–1255.

[174] W. Sun, S. Nesic, S. Papavinasam, Kinetics of iron sulfide and mixed iron sulfide/carbonate scale precipitation in CO_2/H_2S corrosion, in: NACE—International Corrosion Conference Series, 066441-426, 2006.

[175] S.N. Smith, B. Brown, W. Sun, Corrosion at higher H_2S concentrations and moderate temperatures, in: Corrosion Conference and Exposition, Houston, TX, 13-17 March. Paper No. 11081, 2011.

[176] Q. Wang, H. Ajwad, T. Shafai, J.D. Lynn, Iron sulfide scale dissolvers: how effective are they? in: SPE Saudi Arabia Section Technical Symposium and Exhibition, Al-Khobar, Saudi Arabia, 19–22 May. SPE-168063-MS, 2013, https://doi.org/10.2118/168063-MS.

[177] G.B. Naumov, B.N. Ryzhenko, I. Khodakovsky, Handbook of Thermodynamic Data. Transl. into ENGLISH of the Publ. "Handbook of Thermodyanic Data" Moscow, Atomizdat, 1971, 1974.

[178] C. Yan, P. Guraieb, R.C. Tomson, Solubility Study of Iron Sulfide FeS under Extremely High Temperature Pressure in Strictly Anoxic, Various Ionic Strength Solutions, in: Society of Petroleum Engineers—SPE International Oilfield Scale Conference and Exhibition. Society of Petroleum Engineers, 2016, https://doi.org/10.2118/179868-ms.

[179] F. Murcia, D. Carolina, P.L. Fosbøl, E.H. Stenby, K. Thomsen, Measurement of iron and lead sulfide solubility below 100°C, Fluid Phase Equilib. 475 (November) (2018) 118–126, https://doi.org/10.1016/j.fluid.2018.07.033.

[180] S. Gao, Thermodynamics and Kinetics of Hydrogen Sulfide Corrosion of Mild Steel at Elevated Temperatures, PhD thesis, The Russ College of Engineering and Technology of Ohio University, 2018.

[181] J.J. Hemley, G.L. Cygan, W.M. D'Angelo, Effect of pressure on ore mineral solubilities under hydrothermal conditions, Geology 14 (5) (1986) 377–379, https://doi.org/10.1130/0091-7613(1986)14<377:EOPOOM>2.0.CO;2.

[182] J.J. Hemley, G.L. Cygan, J.B. Fein, G.R. Robinson, W.M. D'Angelo, Hydrothermal ore-forming processes in the light of studies in rock- buffered systems. I. iron-copper-zinc-lead sulfide solubility relations, Econ. Geol. 87 (1) (1992) 1–22, https://doi.org/10.2113/gsecongeo.87.1.1.

[183] E. Bardal, Corrosion and Protection, Engineering Materials and Processes, Springer-Verlag London Limited, 2004.

[184] A.H. Nielsen, T. Hvitved-Jacobsen, J. Vollertsen, Effects of pH and iron concentrations on sulfide precipitation in wastewater collection systems, Water Environ. Res. 80 (4) (2008) 380–384, https://doi.org/10.2175/106143007x221328.

[185] NACE, MR0175/ISO 15156-1, Petroleum and Natural Gas Industries—Materials for Use in H_2S-Containing Environments in Oil and Gas Production, NACE International/ISO, 2001.

[186] J. Kvarekval, R. Nyborg, M. Seiersten, Corrosion product films on carbon steel in semi-sour CO_2/H_2S environments, in: CORROSION'02, NACE International, Houston, TX, 2002. Paper No. 296.

[187] M. Askari, M. Aliofkhazraei, R. Jafari, P. Hamghalam, A. Hajizadeh, Downhole corrosion inhibitors for oil and gas production—a review, Appl. Surf. Sci. Adv. 6 (2021), https://doi.org/10.1016/j.apsadv.2021.100128.

[188] A.K. Dunlop, H.L. Hassell, P.R. Rhodes, Fundamental considerations in sweet gas well corrosion, in: CORROSION'83, NACE International, Houston, TX, 1983. Paper No. 46.

[189] S.N. Smith, Discussion of the History and Relevance of the CO_2/H_2S Ratio, NACE International, 2011.

[190] G. Schmitt, M. Hörstemeier, Fundamental aspects of CO_2 metal loss corrosion. Part II. Influence of different parameters on CO_2 corrosion mechanisms, in: NACE—International Corrosion Conference Series, 061121-226, 2006.

[191] S.N. Smith, J.L. Pacheco, Prediction of corrosion in slightly sour environments, in: CORROSION'02, NACE International, Houston, TX, 2002. Paper No. 241.

[192] F. King, C. Padovani, Review of the corrosion performance of selected canister materials for disposal of UK HLW and/or spent fuel, Corros. Eng. Sci. Technol. (2011), https://doi.org/10.1179/1743278211Y.0000000005.

[193] T. Misawa, K. Asami, K. Hashimoto, S. Shimodaira, The mechanism of atmospheric rusting and the protective amorphous rust on low alloy steel, Corros. Sci. 14 (4) (1974) 279–289.

[194] T. Misawa, K. Hashimoto, S. Shimodaira, The mechanism of formation of iron oxide and oxyhydroxides in aqueous solutions at room temperature, Corros. Sci. 14 (2) (1974) 131–149.

[195] B. Shera, The Effect of the Environment on the Corrosion Products and Corrosion Rates on Gas Transmission Pipelines, PhD thesis, The University of Western Ontario, 2011.
[196] N.R. Rosli, The Effect of Oxygen in Sweet Corrosion of Carbon Steel for Enhanced Oil Recovery Applications, PhD thesis, The Russ College of Engineering and Technology of Ohio University, 2015.
[197] J.D. Gu, Biofouling and prevention. Corrosion, biodeterioration and biodegradation of materials, in: Handbook of Environmental Degradation of Materials, Elsevier Inc., 2005, pp. 179–206, https://doi.org/10.1016/B978-081551500-5.50011-2.
[198] F. Renard, C.V. Putnis, G. Montes-Hernandez, H.E. King, Siderite dissolution coupled to iron oxyhydroxide precipitation in the presence of arsenic revealed by nanoscale imaging, Chem. Geol. 449 (January) (2017) 123–134, https://doi.org/10.1016/j.chemgeo.2016.12.001.
[199] N.R. Rosli, Y.S. Choi, D. Young, Impact of oxygen ingress in CO2 corrosion of mild steel, in: NACE—International Corrosion Conference Series. National Assoc. of Corrosion Engineers International, 2014.
[200] Y. Song, A. Palencsár, G. Svenningsen, J. Kvarekvål, T. Hemmingsen, Effect of O_2 and Temperature on Sour Corrosion, NACE International, 2011.
[201] J.A. Bourdoiseau, M. Jeannin, C. Rémazeilles, R. Sabot, P. Refait, The transformation of mackinawite into greigite studied by Raman spectroscopy, J. Raman Spectrosc. 42 (2011) 496–504.
[202] M.D.D. Ayagou, G.R. Joshi, T.T.M. Tran, E. Sutter, B. Tribollet, C. Mendibide, C. Duret-Thual, N. Ferrando, J. Kittel, Corrosion and hydrogen permeation in H_2S environments with O_2 contamination. Part 2. Impact of H_2S partial pressure, Corros. J. 75 (4) (2019) 389.
[203] G. Schikorr, On the iron-water system, Zeitsch. Elektrochem. Angew. Phys. Chem. 35 (1929) 62–65.
[204] Q. Li, A. Jew, D. Cercone, J. Bargar, G.E. Brown, K. Maher, Geochemical modeling of iron (hydr)oxide scale formation during hydraulic fracturing operations, in: Paper Presented at the SPE/AAPG/SEG Unconventional Resources Technology Conference, Denver, Colorado, USA, July, 2019, https://doi.org/10.15530/urtec-2019-612.
[205] A.D. Jew, M.K. Dustin, A.L. Harrison, et al., Impact of organics and carbonates on the oxidation and precipitation of iron during hydraulic fracturing of shale, Energy Fuels 31 (4) (2017) 3643–3658.
[206] J. Han, S. Nesic, Y. Yang, B. Brown, Spontaneous passivation observations during scale formation on mild steel in CO2 brines, Electrochim. Acta 56 (2011) 5396–5404.
[207] S. Gao, B. Brown, D. Young, M. Singer, Formation of iron oxide and iron sulfide at high temperature and their effects on corrosion, Corros. Sci. 135 (May) (2018) 167–176, https://doi.org/10.1016/j.corsci.2018.02.045.
[208] R.I. Ray, J.S. Lee, B.J. Little, Iron-Oxidizing Bacteria: A Review o Corrosion Mechanisms in Fresh Water And Marine Environments, NACE International, 2010.
[209] T.D. Brock, D.W. Smith, M.T. Madigan, Biology of Microorganism, Prentice-Hall, Englewood Cliffs, NJ, 1984.
[210] M.K. Khouzani, A. Bahrami, A. Hosseini-Abari, M. Khandouzi, P. Taheri, Microbiologically influenced corrosion of a pipeline in a petrochemical plant, Metals 9 (4) (2019), https://doi.org/10.3390/met9040459.
[211] AWWA Research Foundation, Internal Corrosion of Water Distribution Systems, American Water Works Association. The Foundation, 1996.
[212] B.N. Popov, Thermodynamics in the electrochemical reactions of corrosion, in: Corrosion Engineering, Elsevier, 2015, pp. 29–92, https://doi.org/10.1016/b978-0-444-62722-3.00002-1.
[213] P.S. Sidhu, R.J. Gilkes, R.M. Cornell, A.M. Posner, J.P. Quirk, Dissolution of iron oxides and oxyhydroxides in hydrochloric and perchloric acids, Clays Clay Miner. 29 (4) (1981) 269–276, https://doi.org/10.1346/CCMN.1981.0290404.
[214] K.C. Taylor, H.A. Nasr-El-Din, M.J. Al-Alawi, Systematic study of iron control chemicals used during well stimulation, SPE J. 4 (1) (1999) 19–24, https://doi.org/10.2118/54602-PA.
[215] K. Biggs, D. Allison, W.G.F. Ford, Acid treatment removes zinc sulfide scale restriction, Oil Gas J. 90 (35) (1992) 47–49.

[216] W. Keogh, A. Neville, M. Huggan, V. Eroini, J.H. Olsen, F.M. Nielsen, S. Baraka-Lokmane, et al., Deposition of inorganic carbonate, sulfate, and sulfide scales on antifouling surfaces in multiphase flow, Energy Fuels 31 (11) (2017) 11838–11851, https://doi.org/10.1021/acs.energyfuels.7b02165.

[217] P. Patnaik, Handbook of Inorganic Chemicals, McGraw-Hill, New York, 2003.

[218] OKOCHA EMEKA CYRIL, Mechanistic Evaluation of Sulphide Scale Formation in the Oilfield, PhD thesis, Institute of Petroleum Engineering Heriot Watt University Edinburgh UK, 2011.

[219] J.J. Wylde, C. Okocha, R. Smith, A. Mahmoudkhani, C.J. Kelly, Dissolution of Sulfide Scale: A Step Change With a Novel, High Performance, Non-Mineral Acid Chemical, Society of Petroleum Engineers, 2016, https://doi.org/10.2118/179880-MS.

[220] D. Carolina Figueroa Murcia, P.L. Fosbøl, K. Thomsen, E.H. Stenby, Determination of zinc sulfide solubility to high temperatures, J. Solut. Chem. 46 (9–10) (2017) 1805–1817, https://doi.org/10.1007/s10953-017-0648-1.

[221] T.J. Barrett, G.M. Anderson, The solubility of sphalerite and galena in 1–5 M NaCl solutions to 300C, Geochim. Cosmochim. Acta 52 (1988) 813–820.

[222] I.R. Collins, M.M. Jordan, Occurrence, Prediction and Prevention of Zinc Sulfide Scale Within Gulf Coast and North Sea High Temperature/High Salinity Production Wells, Society of Petroleum Engineers, 2001, https://doi.org/10.2118/68317-MS.

[223] S. Baraka-Lokmane, C. Hurtevent, M. Rossiter, F. Bryce, F. Lepoivre, A. Marais, et al., Design and Performance of Novel Sulphide Nanoparticle Scale Inhibitors for North Sea HP/HT Fields, Society of Petroleum Engineers, 2016, https://doi.org/10.2118/179866-MS.

[224] C. Okocha, K.S. Sorbie, C. Hurtevent, S. Baraka-Lokmane, M. Rossiter, Sulphide Scale (PbS/ZnS) Formation and Inhibition Tests for a Gas Condensate Field with Severe Scaling Conditions, Society of Petroleum Engineers, 2014, https://doi.org/10.2118/169798-MS.

[225] M.M. Marchioretto, H. Bruning, W. Rulkens, Heavy metals precipitation in sewage sludge, Sep. Sci. Technol. 40 (16) (2005) 3393–3405, https://doi.org/10.1080/01496390500423748.

[226] M.H. Reed, J. Palandri, Sulfide mineral precipitation from hydrothermal fluids, Rev. Mineral. Geochem. 61 (2006) 609–631, https://doi.org/10.2138/rmg.2006.61.11.

[227] K.C. Misra, Understanding Mineral Deposits, Springer Netherlands, 2000, https://doi.org/10.1007/978-94-011-3925-0.

[228] J.K. Smith, J.L. Przybylinski, The effect of common brine constituents on the efficacy of halite precipitation inhibitors, in: NACE—International Corrosion Conference Series, 063871-714, 2006.

[229] H.F. Nieuwland, I.R. Collins, Groet 1 Scale Study: Analysis, Diagnosis & Solution Implementation, Society of Petroleum Engineers, 2004, https://doi.org/10.2118/87472-MS.

[230] M.E. Tucker, Sedimentary Petrology, third ed., Blackwell Science, Oxford, UK, 2001, pp. 166–181 (Chapter 5).

[231] Gas Processors Suppliers Association (GPSA), Engineering Data Book, twelfth ed., GPSA, Tulsa, OK, 2004.

[232] Y. Tang, R. Yang, X. Kang, Modeling the effect of water vaporization and salt precipitation on reservoir properties due to carbon dioxide sequestration in a depleted gas reservoir, Petroleum 4 (4) (2018) 385–397,- https://doi.org/10.1016/j.petlm.2017.12.003.

[233] W. Kleinitz, G. Dietzsch, M. Köhler, Halite scale formation in gas-producing wells, Chem. Eng. Res. Des. 81 (3) (2003) 352–358, https://doi.org/10.1205/02638760360596900.

[234] N. Goodwin, G.M. Graham, D.M. Frigo, E. Kremer, Halite deposition—prediction and laboratory evaluation, in: Society of Petroleum Engineers—SPE International Oilfield Scale Conference and Exhibition, Society of Petroleum Engineers, 2016, https://doi.org/10.2118/179861-ms.

[235] The UK Offshore Energies Association Limited OEUK, Management of Salt Deposition in Gas Wells Guideline, OGUK, 2020.

[236] M.M. Jordan, J. Buckman, C.J. Johnston, Direct Monitoring of the Performance of Scale Control Programs Across the Produced Water Life Cycle via Suspended Solids Analysis, Society of Petroleum Engineers, 2005, https://doi.org/10.2118/93892-MS.
[237] A.T. Kan, F. Gongmin, M.A. Watson, M.B. Tomson, Effect of hydrate inhibitors on oilfield scale formation and inhibition, in: SPE Oilfield Scale Symposium, 2002, pp. 83–94, https://doi.org/10.2523/74657-ms.
[238] H.J. Shipley, A.T. Kan, G. Fu, D. Shen, M.B. Tomson, Effect of hydrate inhibitors on calcite, sulfates, and halite scale formation, in: SPE Eighth International Symposium on Oilfield Scale 2006, 2006, pp. 277–289, https://doi.org/10.2523/100522-ms.
[239] E.O. Chertovskikh, A.S. Lapoukhov, V.A. Kachin, A.V. Karpikov, Problems of oil and gas production in the Verkhnechonskoye oil and gas condensate field associated with halite depositing, in: Society of Petroleum Engineers—SPE Arctic and Extreme Environments Conference and Exhibition, AEE 2013, vol. 2, Society of Petroleum Engineers, 2013, pp. 1491–1527.
[240] Z. Amjad, R.W. Zuhl, P.E. Lubrizol, Factors Impacting Silica-Silicate Control Agent Performance in Industrial Water Systems, 2016.
[241] K.D. Demadis, G. Skordalou, I. Aristodemou, Distinct Mechanisms of Silica Scale Formation: Silicic Acid Polycondensation and Silica Particle Growth, NACE International, 2020.
[242] P. Meyers, Behavior of silica in ion exchange and other systems, in: Presented at the International Water Conference, held in Pittsburgh, PA, October 18–20, 1999.
[243] D. Chapman (Ed.), Groundwater, in: Water Quality Assessments—A Guide to Use of Biota, Sediments, and Water in Environmental Monitoring, second ed., UNESCO/WHO/UNEP, 1996 (Chapter 9).
[244] A. Antony, J.H. Low, S. Gray, A.E. Childress, P. Le-Clech, G. Leslie, Scale formation and control in high pressure membrane water treatment systems: a review, J. Membr. Sci. (2011), https://doi.org/10.1016/j.memsci.2011.08.054.
[245] P. Sahachaiyunta, T. Koo, R. Sheikholeslami, Effect of several inorganic species on silica fouling in RO membranes, Desalination 144 (2002) 373–378.
[246] R.Y. Ning, Reactive silica in natural waters—a review, Desal. Water Treat. 21 (1–3) (2010) 79–86, https://doi.org/10.5004/dwt.2010.1229.
[247] Z. Amjad, R. Zuhl, Effects of Thermal Stress on Silica-Silicate Deposit Control Agent Performance, Association of Water Technologies, Inc. Annual Convention and Exposition, 2010.
[248] A.A. Umar, I.M. Saaid, Effects of temperature on silicate scale inhibition during ASP flooding, J. Appl. Sci. 14 (15) (2014) 1769–1774, https://doi.org/10.3923/jas.2014.1769.1774.
[249] D.D. Kostas, S. Aggeliki, K. Antonia, Inhibition and control of colloidal silica: can chemical additives untie the Gordon Knot of scale formation? in: Proceedings of the NACE International Corrosion Conference, March 11–15, Nashville, Tennessee, 2007.
[250] K.D. Demadis, Water treatment's 'Gordian Knot', Chem. Process. 66 (5) (2003) 29–34.
[251] R.K. Iler, The Chemistry of Silica: Solubility, Polymerization, Colloid and Surface Properties and Biochemistry of Silica, Wiley New York, 1979, ISBN: 978-0-471-02404-0.
[252] S.H. Chan, A review on solubility and polymerization of silica, Geothermics 18 (1/2) (1989) 49–56.
[253] Z. Amjad, R.W. Zuhl, An evaluation of silica scale control additives for industrial water systems, in: Paper 08368 Presented at the NACE International Corrosion Conference and Expo, New Orleans, Louisiana, 16–20 March, 2008.
[254] J. Arensdorf, D. Hoster, D. McDaugall, Y. Mingdon, Static and dynamic testing of silicate scale inhibitors, in: Paper SPE 132212 Presented at CPE/SPE International Oil and Gas Conference and Exhibition, Beijing, China, 8–10 June, 2010.
[255] J.S. Gill, Silica scale control, in: Paper 226, Presented at CORROSION 98, NACE 1998 San Diego, California, 22–27 March, 1998.
[256] W. Zalewski, P. Bulkowski, Lithium silicate scale in oilfield steam generation, J. Can. Pet. Technol. 37 (4) (1998) 51–54, https://doi.org/10.2118/98-04-05.

[257] K.D. Demadis, Recent developments in controlling silica and magnesium silicates in industrial water systems, in: The Science and Technology of Industrial Water Treatment, IWA Publishing/CRC Press, 2010 (Chapter 10).

[258] W.L. Marshall, J.M. Warakomski, Amorphous silica solubilities. II. Effect of aqueous salt solutions at 25°C, Geochim. Cosmochim. Acta 44 (1980) 915–917.

[259] Environmental Protection Agency, Barium compounds; toxic chemical release reporting, Federal Reg. 62 (2) (1997).

[260] D.L. Miles, J.D. Burton, The solubility of strontium carbonate in water, J. Chem. Soc., Dalton Trans. 15 (1972) 1691–1692, https://doi.org/10.1039/DT9720001691.

[261] H.J. Ramsay, F.R. Small, Use of carbon dioxide for water injectivity improvement, J. Petrol. Technol. 16 (01) (1964) 25–31, https://doi.org/10.2118/595-pa.

[262] G. Fu, A.T. Kan, M.B. Tomson, Enhanced effect of transition metal ions on inhibitor squeeze treatments, in: SPE International Symposium on Oilfield Chemistry, 2009, https://doi.org/10.2118/121556-ms.

[263] B. Hitchon, Lead and zinc in formation waters, Alberta Basin, Canada: their relation to the Pine Point ore fluid, Appl. Geochem. 21 (1) (2006) 109–133.

[264] Y. Tang, J. Voelker, C. Keskin, Z. Xu, B. Hu, C. Jia, A Flow Assurance Study on Elemental Sulfur Deposition in Sour Gas Wells, Society of Petroleum Engineers, 2011, https://doi.org/10.2118/147244-MS.

[265] N. Yaakob, M. Singer, D. Young, Elemental sulfur corrosion of carbon steel in the presence of sulfur solvent and monoethylene glycol, in: NACE—International Corrosion Conference Series. Vol. 2015-January. National Assoc. of Corrosion Engineers International, 2015.

[266] P. Cézac, J.P. Serin, J.M. Reneaume, J. Mercadier, G. Mouton, Elemental sulphur deposition in natural gas transmission and distribution networks, J. Supercrit. Fluids 44 (2) (2008) 115–122, https://doi.org/10.1016/j.supflu.2007.11.005.

[267] C.G. Ramachandran Nair, A.R. Vasudeva Murthy, Hydrolysis of elemental sulphur in a homogeneous medium, Proc. Indian Acad. Sci. Sect. A 55 (3) (1962) 168–173, https://doi.org/10.1007/BF03045834.

[268] K. Ding, Y. Liu, Hydrolysis of pyrite: a possible pathway for generation of high concentrations of H2S gas in deep-buried reservoirs? Development in Earth Science 5 (2017) 1, https://doi.org/10.14355/des.2017.05.001.

[269] D.J. Pack, A.B. Chesnoy, J. Bosio, Gas flow measurement & the impact of contaminants in the gas stream, in: 16th International Flow Measurement Conference 2013, FLOMEKO 2013, IMEKO-International Measurement Federation Secretariat, 2013, pp. 440–446.

[270] C.L. Kimtantas, M.A. Taylor, An effective solution for elemental sulfur deposition in natural gas systems, in: 2014 Conference Proceedings: Oil and Gas and Chemical Filtration and Separations, AFS 2014, American Filtration and Separations Society, 2014, pp. 216–226.

[271] C. Wilkes, V. Pareek, Sulfur deposition in a gas turbine natural gas fuel control system, Energy-Tech (2001).

[272] Z. Zhu, N. Tajallipour, P.J. Teevens, K. Lepková, R. Gubner, Modeling of elemental sulfur deposition in sour-gas petroleum pipelines, in: NACE—International Corrosion Conference Series, 2011.

[273] J.P.L. Santos, A.K.C.L. Lobato, C. Moraes, L.C.L. Santos, Influence of natural gas composition on the formation and deposition of elemental sulfur in pipelines, Braz. J. Pet. Gas 9 (2) (2015) 45–53, https://doi.org/10.5419/bjpg2015-0006.

[274] S.P. Mathis, Sand management: a review of approaches and concerns, in: SPE European Formation Damage Conference, 13–14 May, The Hague, Netherlands, 2013.

[275] G.E. King, P.J. Wildt, E. O'Connell, Sand Control Completion Reliability and Failure Rate Comparison With a Multi-Thousand Well Database, Society of Petroleum Engineers, 2003, https://doi.org/10.2118/84262-MS.

[276] B. Oyeneyin, Integrated sand management for effective hydrocarbon flow assurance, Dev. Pet. Sci. (2015) 63, https://doi.org/10.1016/B978-0-444-62637-0.00002-6.

[277] I.C. Walton, D.C. Atwood, P.M. Halleck, et al., Perforating Unconsolidated Sands: An Experimental and Theoretical Investigation, SPE 71458, 2001.

[278] J. Carlson, D. Gurley, G. King, C. Price-Smith, F. Waters, Sand control: why and how? Oilfield Rev. (1992).
[279] W.L. Penberthy Jr., Sand control, in: Volume IV—Production Operations Engineering, Society of Petroleum Engineers, 2006, pp. 175–239.
[280] B. Wu, C.P. Tan, N. Lu, Effect of water cut on sand production—an experimental study, in: 2005 SPE Asia Pacific Oil and Gas Conference and Exhibition Proceedings, 2005, pp. 59–67, https://doi.org/10.2118/92715-pa.
[281] Fines Migration, Schlumberger Limited, 2018. Retrieved from http://www.glossary.oilfield.slb.com/Terms/f/fines_migration.aspx, last visited November 2018.
[282] J.G. Speight, Handbook of Hydraulic Fracturing, Wiley, 2016, pp. 1–296, https://doi.org/10.1002/9781119225102.
[283] J. Xiao, J. Wang, X. Sun, Fines migration: problems and treatments, Oil Gas Res. 03 (01) (2017), https://doi.org/10.4172/2472-0518.1000123.
[284] M.M. Sharma, Y.C. Yortsos, L.L. Handy, Release and Deposition of Clays in Sandstones, Society of Petroleum Engineers, 1985, https://doi.org/10.2118/13562-MS.
[285] R.L. Anderson, I. Ratcliffe, H.C. Greenwell, P.A. Williams, S. Cliffe, P.V. Coveney, Clay swelling—a challenge in the oilfield, Earth-Sci. Rev. (2010), https://doi.org/10.1016/j.earscirev.2009.11.003.
[286] S.M. Rao, T. Thyagaraj, P. Raghuveer Rao, Crystalline and osmotic swelling of an expansive clay inundated with sodium chloride solutions, Geotech. Geol. Eng. 31 (4) (2013) 1399–1404, https://doi.org/10.1007/s10706-013-9629-3.
[287] M. Ghadiri, W. Chrzanowski, R. Rohanizadeh, Biomedical applications of cationic clay minerals, in: RSC Advances, Royal Society of Chemistry, 2015, https://doi.org/10.1039/c4ra16945j.

CHAPTER

Gas Hydrates and Diamondoids

6

6.1 Introduction

Two main cage-like classes of molecules are encountered in oil and gas fields: gas hydrates and diamondoids. These two types of molecules are sources of flow assurance troubles in the different production systems, causing production loss and costly maintenance. Gas hydrates have accounted for massive amounts of research in the field of flow assurance, energy resources, and energy storage. The research on diamondoids has gained some momentum in the field of flow assurance, but lately diamondoids have been attracting more research interest due to their superior properties and potentially astounding applications.

In the following sections, gas hydrate and diamondoid flow assurance issues are discussed.

6.2 Gas hydrates

6.2.1 Introduction

Gas hydrates, clathrate hydrates, hydrates, and clathrates are crystalline inclusion compounds, in which "host" molecular cages of water entrap "guest" gas molecules under specific thermodynamic conditions [1]. Fig. 6.1 illustrates a methane hydrate molecule. In order for gas hydrates to form, they basically require relatively high pressures, low temperatures, water, and low molecular weight gases such as methane, ethane, propane, butane, CO_2, H_2S, nitrogen, or chlorine. Even oxygen can create hydrates. It is not necessary that the gas be present as vapor—hydrates can form in liquid systems such as gas condensate reservoirs [2].

Hydrates were discovered back in 1810 by Sir Humphrey Davy, the mentor of Michael Faraday, although it is believed that other scientists may have observed their existence before this date. They are of great importance industrially because of their occurrence in hydrocarbon production systems as one of the major flow assurance problems, presenting high economic loss and ecological risks, as well as potential safety hazards to exploration and transmission personnel. On the other hand, they also have technological importance in separation processes, fuel transportation, and storage, in addition to their potential use as a fuel resource following the discovered natural deposits of predominantly methane hydrate in permafrost and continental margins [3].

Gas hydrates are one of the most reported damaging flow assurance problems in oil and gas production systems. Hydrates are the leading (compared to wax, asphaltenes, scale, and other solids)

Essentials of Flow Assurance Solids in Oil and Gas Operations. https://doi.org/10.1016/B978-0-323-99118-6.00010-1

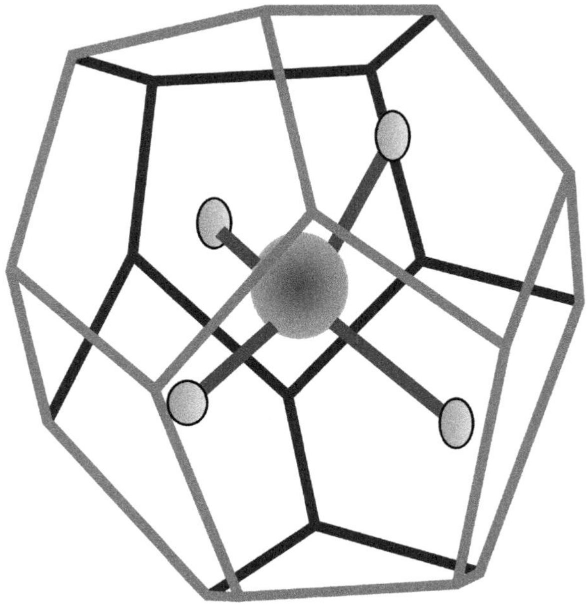

FIG. 6.1

Crystal structure of methane hydrate.

deepwater flow assurance problem by an order of magnitude of 4 and in a survey among 110 energy companies, where flow assurance was listed as the major technical problem in offshore energy development [4,5]. Gas hydrates are most commonly encountered in subsea or cold climate wet gas or multiphase (oil-water-gas) pipelines, as they can interrupt or block the flow in different production systems facilities, tubulars, and equipment. They are also encountered in aqueous chemical injection in gas lift lines; furthermore, gas hydrates interfere with major field operations such as drilling, completion, and workover operations [6]. Gas hydrates are considered the major flow assurance problem in deepwater drilling and completion.

Hydrate crystals look like ice, but unlike ice they can form at much higher temperatures. The gas molecules give them extra attraction and stability. They can be soft, hence not causing damage to tubing, flow line, or vessels. However, in some cases the crystals are hard enough to cause erosion or destroy sensitive parts in the production facilities. In larger quantities, the crystals tend to lump together and cause blockages in the system [7]. Fig. 6.2 shows gas hydrate blockages in pipelines.

Gas hydrates are also found in nature in large quantities. The sediments of deep marine continental margins host nearly 99% of the world's gas hydrates, typically at water depths of 500 m (approximately 1640 ft) and greater, and is also present in areas with permanently frozen ground (permafrost). The sediments on continental margins often contain substantial concentrations of organic carbon, which is used by microbes to produce methane. This methane, as well as microbial produced methane or even methane that migrates upward from deep conventional gas reservoirs, can combine with sediment pore waters to form gas hydrate beneath the sea floor [8]. Fig. 6.3 shows gas hydrate sediments in nature. Estimates of the global resources of natural gas hydrates range from 100,000 to almost 300,000,000

FIG. 6.2

Gas hydrate blockages from Namorado Field, in the Campos Basin.

Source: Reference H. Barbosa, E.C. de Oliveira, S.L. Farias, IBP1233_15 Mapping the State of the Art of Measurement of Hydrocarbon Dew Point in Natural Gas, Rio Pipeline Conference & Exposition 2015, 2015, with permission from Dr. Hall Barbosa.

FIG. 6.3

Gas hydrates in nature: (Right) Scanning electron microscope image of gas hydrate crystals in a sediment sample. (Left) Gas hydrate (white, ice-like material) under authigenic carbonate rock that is encrusted with deep-sea chemosynthetic mussels and other organisms on the seafloor of the northern Gulf of Mexico at 966 m (~3170 ft) water depth.

Right: Courtesy of the U.S. Geological Survey. Left: Courtesy of NOAA.

trillion cubic feet (TCF), while the total estimates of the remaining global reserves and undiscovered resources of conventional natural gas are about 13,000 TCF [9].

6.2.2 Structures of gas hydrates

Gas hydrates are ice-like, crystalline solids in which water molecules form a hydrogen-bonded three-dimensional network enclosing gas molecules in defined cavities of different sizes. The interaction force between host and guest molecules is van der Waals force. The (nonpolar) gas molecules in turn prevent the water cages from collapsing [10–12]. Fig. 6.4 illustrates the difference in structure between gas hydrates and normal ice [13].

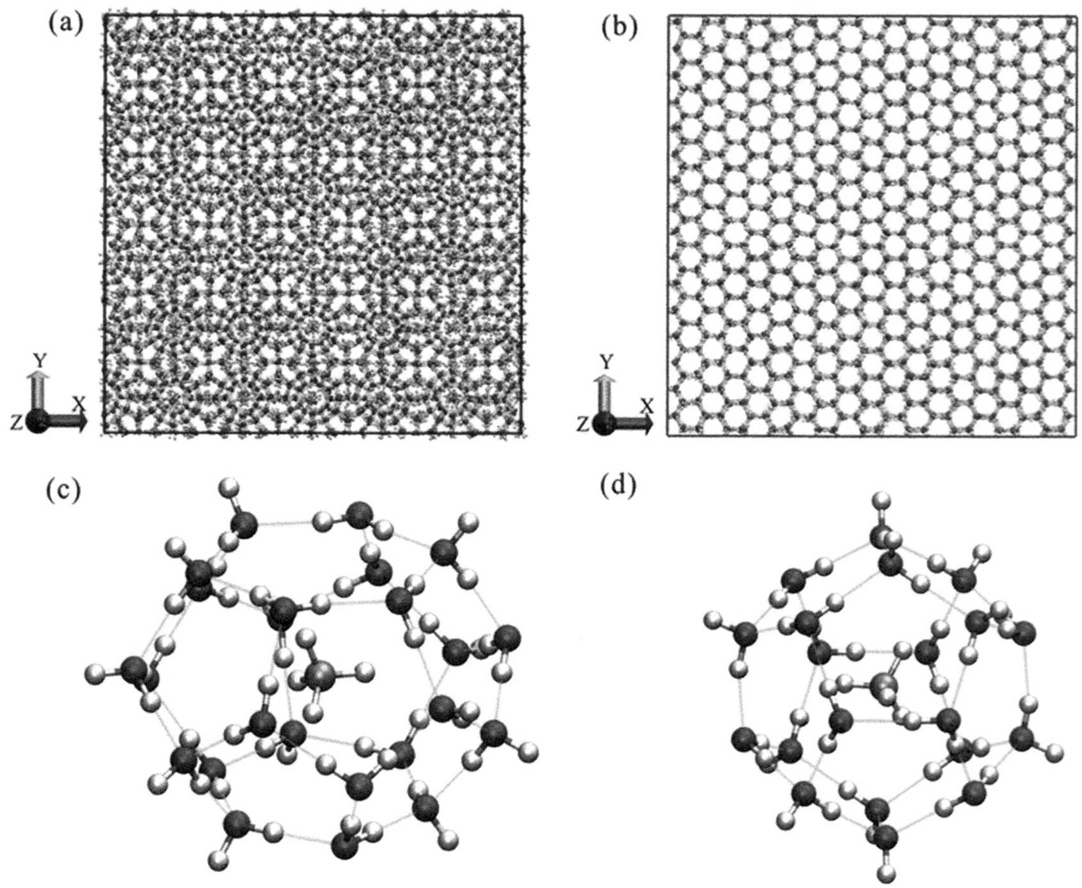

FIG. 6.4

Molecular structures of (a) methane hydrate (sI) and (b) normal ice (Ih). (c) Large ($5^{12}6^2$) and (d) small (5^{12}) methane hydrate cages.

Source: J. Jia, Y. Liang, T. Tsuji, S. Murata, T. Matsuoka, Microscopic origin of strain hardening in methane hydrate, Sci. Rep. 6 (2016). https://doi.org/10.1038/srep23548.

Unlike most inorganic crystalline structures, gas hydrates are nonstoichiometric crystals because not all of the cages will necessarily contain a gas molecule [14,15]. Another major difference is that hydrates are not like the usual chemical compounds in the sense of chemical bonding, since no strong chemical bonds exist between the water and gas molecules [16]. The word "clathrate" has its origin in the Latin word *clathratus*, which means "to encage." It is indeed more precise to use the term "clathrate hydrates" [17].

Three hydrate structures have been confirmed in these natural gas hydrate samples: the cubic structures I (sI) and II (sII) and the hexagonal structure H (sH) [17,18]. Jeffery [19] added four more unusual structures and thus made a list of seven structures of gas hydrates. However, our concern in this chapter will be focused on the three major structures.

To understand these structures, let's begin with understanding the cages or the cavities. First, the simplest of all crystalline cavity structures has 12 regular pentagonal (five-ring) faces (5^{12}), and it can be formed by a pentagonal dodecahedral of 20 water molecules linked together by hydrogen bonds. All the structures will combine this simple 5^{12} cavity with other cavities to form the basic unit cell for these structures. The other cavities are as follows: $5^{12}6^2$ combines 12 pentagons with 2 hexagons, $5^{12}6^4$ combines 12 pentagonal faces with 4 hexagonal faces, $4^35^66^3$ has 3 square faces, 6 pentagons and 3 hexagons, and $5^{12}6^8$ contains 12 pentagons and 8 hexagons and so forth [10]. With the exceptions of cavities containing square faces, all hydrate cavities follow Euler's theorem [20] for convex polyhedra, stated as ($F + V = E + 2$). The number of faces (F) plus the vertices (V) equals the edges (E) plus 2. Euler's theorem is easily fulfilled in cavities, having exactly 12 pentagonal faces and any number of hexagonal faces except one [17].

Understanding the cavities makes perceiving the hydrate structures easier. For example: in structure I, two small pentagonal dodecahedrons (5^{12}) are combined with six tetrakaidecahedrons ($5^{12}6^2$) to form a unit cell. The unit cell of structure II consists of 16 pentagonal dodecahedrons (5^{12}) and 8 hexakaidecahedrons ($5^{12}6^4$). Structure H unit cell combines three pentagonal dodecahedrons (5^{12}), two irregular dodecahedrons ($4^35^66^3$), and one icosahedron $5^{12}6^8$ [10,21].

The ideal unit cell formula for Structure I is: $6X{\cdot}2Y{\cdot}46H_2O$ where: X represents the guest molecule to fill in large cavities and Y represents the guest molecule to fill in small cavities. The ideal unit cell formula for Structure II is: $8X{\cdot}16Y{\cdot}136H_2O$ where: X represents the guest molecule to fill in large cavities and Y represents the guest molecule to fill in small cavities. The ideal unit cell formula for Structure H would be: $1X{\cdot}3Y{\cdot}2Z{\cdot}34H_2O$ where: X represents the guest molecule to fill in large cavities ($4^35^66^3$), Y represents the guest molecule to fill in small cavities (5^{12}), Z represents the guest molecule to fill in the largest cavity ($5^{12}6^8$) [22]. Fig. 6.5 illustrates the different cavity types and gas hydrates structures and Table 6.1 shows the structures and cavities properties.

The structure types of hydrate are generally determined by the guest molecule size, which is affected by key factors such as the shape of guest molecules, pressure and temperature conditions, cage size, and the presence of a hydrate promoter. Smaller guest molecules could enter the various holes of hydrate crystals, and exert different stability effect. In practice, larger guest molecules only stabilize the holes of type II, and exert much more stability than that of smaller guest molecules. Therefore the propane, butane, and *i*-butane tend to form type II of the hydrate [12]. A thermodynamic preference between hydrate structures arises from the ratio of guest size to large cage size. For example, the larger cage of structure I ($5^{12}6^2$ with average cage diameter 5.86 Å) is sufficiently large for ethane (5.5 Å diameter), but too small for propane [24]. Ordinarily the diameter ratio of guest molecules and water cage is near 0.9 [12]. However, in a mixed gas system experimental results showed that it is difficult to judge

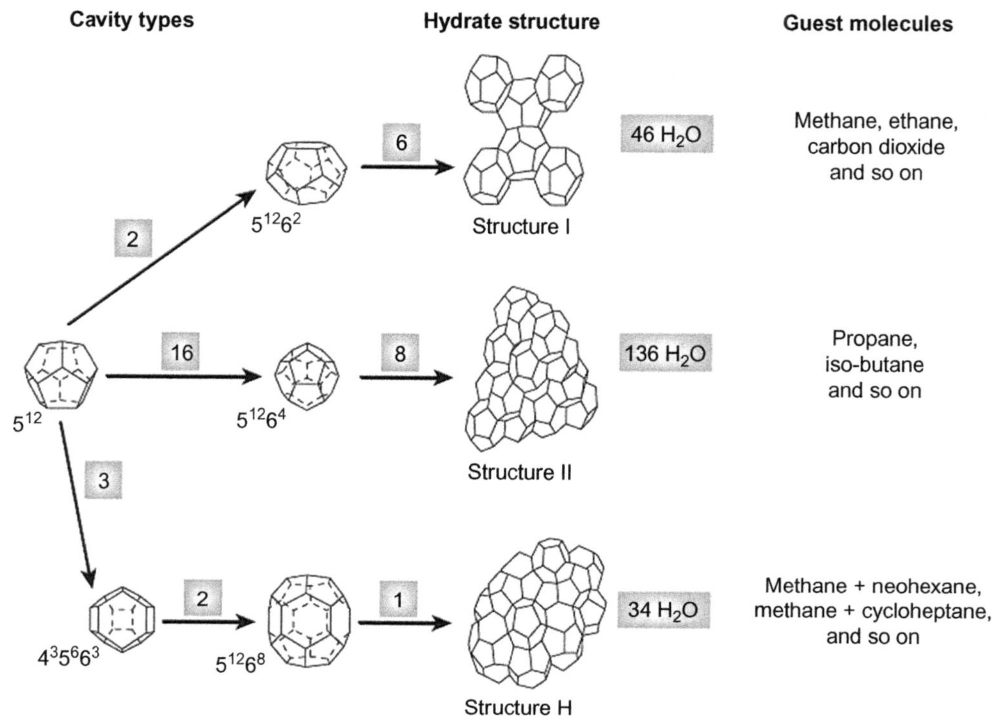

FIG. 6.5

Gas hydrates cavity types and structures.

Reprinted by permission from Springer Nature, Nature, [Fundamental principles and applications of natural gas hydrates, E. Dendy Sloan Jr), [COPYRIGHT] (2003).

Table 6.1 Gas hydrate cavities and structures properties [17,21,23].

Properties of cages	Structure I		Structure II		Structure H		
Cavity	**Small**	**Large**	**Small**	**Large**	**Small**	**Medium**	**Large**
Description	5^{12}	$5^{12}6^2$	5^{12}	$5^{12}6^4$	5^{12}	$4^35^66^3$	$5^{12}6^8$
Cavities per unit cell	2	6	16	8	3	2	1
Coordination number	20	24	20	28	20	20	36
Average cavity radius nm (Å/10)	0.395	0.433	0.391	0.473	0.391	0.406	0.571
Crystal system	Cubic		Cubic		Hexagonal		
Number of faces (H) hexagonal, (P) pentagonal, (S) square	H = 6 P = 48		H = 16 P = 144		H = 7 P = 30 S = 3		
Unit cell formula	$6(5^{12}6^2)\cdot 2(5^{12})\cdot$ $46H_2O$		$8(5^{12}6^4)\cdot 16(5^{12})\cdot$ $136H_2O$		$1(5^{12}6^8)3(5^{12})\cdot 2(4^35^66^3)\cdot$ $34H_2O$		
Water molecules per unit cell	46		136		34		
Density kg/m^3	912		940		1952		

the hydrate structural type, as the gas composition, temperature, and pressure can transfer the hydrate structural type [12,25]. The so-called "hydration number" is the number of water molecules per guest molecule (or in other words the water-gas ratio in hydrates). The host and guest molecules maintain a specific hydration number when water cages are fully filled by guest molecules, which is 5.75 in type I and 5.67 in type II structure. In fact, the guest molecules only fill partial water cages with a filling rate between 0.9 and 1.0 [12,26].

Hydrate structure sI is commonly formed in nature, while sII is often formed in artificial systems, like those formed in oil and gas pipelines. Structure H is rarely encountered in the oil industry, and is only favored when a heavy hydrocarbon such as methylcyclohexane is present with methane and water. It is stabilized by methane in small cages and fairly large hydrocarbons such as methylcyclopentane or benzene in large cages [4,6].

All three main hydrate structures (sI, sII, and sH) are approximately 85% (mol) water and 15% gas when all the cages are filled. This suggests that the mechanical properties of the three hydrate structures are similar to those of ice [23]. However, by standard volume, 1 ft^3 of hydrate equates to 160–180 ft^3 of gas and 0.8 ft^3 of water. This feature can be an advantage as a method of storing and transporting natural gas. As with water ice, hydrates are lighter than water and often sit at the interface between water and oil [2].

Now, before moving on to the mechanism of gas hydrates formation, a critical question must be asked: which gases are able or favorable to form gas hydrates and which are not? The answer is, generally speaking, the gases with high solubility in water such as ammonia or hydrogen chloride cannot form a hydrate regardless of its molecule size. Common guest molecules include hydrocarbon components (CH_4, C_2H_6, C_3H_8, i-C_4H_{10}, n-C_4H_{10}, etc.), acidic component CO_2, H_2S and inert components (N2, Ar, Kr, Xe), etc. [12].

6.2.3 Mechanism of gas hydrate formation

The general chemical equation for the hydrate formation "reaction" can be formulated as follows [27]:

$$\text{Gas} + nwH_2O \leftrightarrow \text{gas} \cdot nwH_2O_{\text{solid}} \tag{6.1}$$

where nw is the number of water molecules, also known as the hydration number, and the expression gas·nwH_2O is regarded as a building unit for a hydrate crystal. It is worth mentioning that the reaction in Eq. 6.1 is not a chemical reaction (no chemical bonding), but rather a physical process.

At temperatures well above 0°C, water molecules exhibit Brownian motion, where the molecules move freely from each other. When temperatures drop approaching the normal freezing point of water, these free movements become slower and the water molecules tend to group themselves in a more stable structure. This is achieved via the hydrogen bonding in which the hydrogen atoms from the water molecules form a loose bond with the oxygen atom of another water molecule, resulting in the formation of clusters of water molecules. During these changes in hydrogen bonding, the formation of unstable crystal-like lattices called "cages" may also occur. In the presence of certain "guest molecules," which have a size to nicely fit in the cavity of a lattice of water molecules, the total structure can thereby be stabilized so as to form a hydrate crystal. As a result hydrate ice crystals can be formed well above the pure water freezing point [7]. Fig. 6.6 shows the hydrate formation process.

Although that basic mechanism is plainly as simple as it seems in Fig. 6.6, the detailed steps of the gas hydrate formation mechanism are a matter of complexity, with many contributing factors under

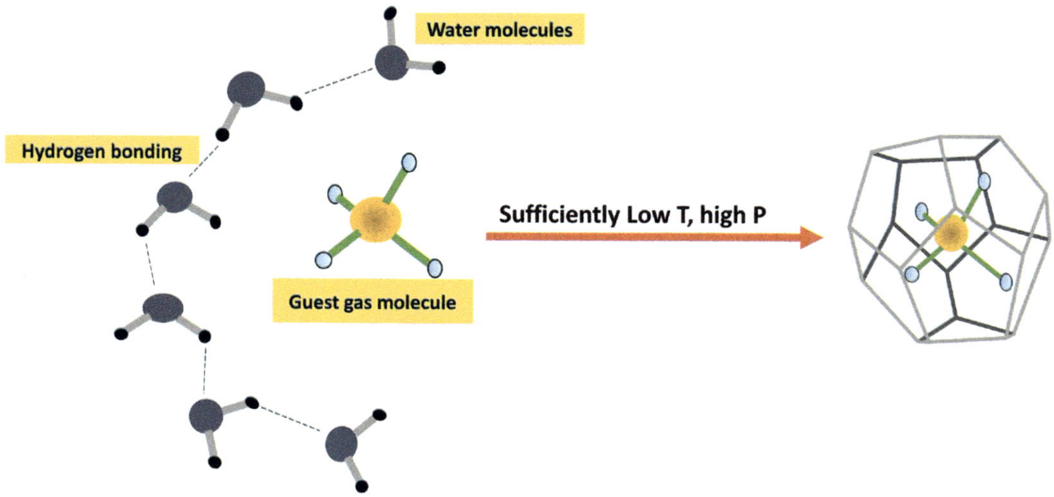

FIG. 6.6

Gas hydrate formation process.

different flow conditions. Basically, the formation of gas hydrates is quite analogous to crystallization processes such as salt precipitation from solution, which includes two main steps: nucleation and then crystal growth [4].

Several mechanisms can be used to explain the nucleation and deposition of gas hydrates, namely the labile cluster hypothesis (LCH), the hypothesis of the nucleation at the interface, the local structuring hypothesis (LSH), the cage adsorption hypothesis (CAH), and the blob mechanism. Ref. [10] reviews these main mechanisms by which gas hydrates can form.

6.2.3.1 Gas hydrates nucleation

In general, nucleation can be either a homogeneous or heterogeneous process. Homogeneous nucleation occurs only in the absence of impurities, while heterogeneous nucleation takes place in the presence of impurities. Homogeneous nucleation is rarely observed in the real world, while heterogeneous nucleation is more commonly applied in industrial applications. The details of the nucleation types and theories are discussed in Chapter 4.

Hydrate nucleation often occurs at a gas-water interface and the phases most probably contain impurities. In addition, foreign surfaces such as a pipe wall or container wall may be involved in the process. Thus hydrate nucleation in real systems is expected to be heterogeneous.

Hydrate nucleation is crucial, as it is literally "the start" of all hydrate studies. Nucleation study is difficult because it occurs at the molecular level and it is stochastic. The next sections discuss the different gas hydrate nucleation theories.

6.2.3.1.1 The labile cluster hypothesis (LCH)

The LCH is considered the first gas hydrate nucleation hypothesis. LCH was proposed by Sloan and Fleyfel [28] and extended by Christiansen and Sloan [29]. It is considered the most common and accepted mechanism for gas hydrate deposits formation. LCH assumes that the hydrate nucleation

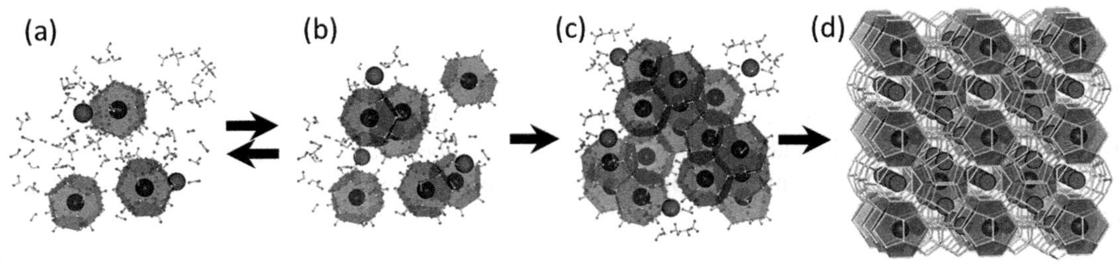

FIG. 6.7

Labile cluster hypothesis of gas hydrate formation: (a) labile clusters, (b) agglomeration of clusters, (c) primary nucleolus, and (d) hydrate crystal.

Republished with permission from Royal Society of Chemistry, from [Gas hydrates in sustainable chemistry. Chemical Society Reviews. *Royal Society of Chemistry. Joonaki et al., 49, 15, 2020]; permission conveyed through Copyright Clearance Center, Inc.).*

process starts from pure water molecules and suggests that cagelike water clusters corresponding to the polyhedral cages of clathrates form around guests in solution, and these combine in the liquid to form the unit cell of the crystal [30–32]. LCH starts with the presumption that the molecules of pure water without any dissolved gas molecules form labile ring structures of pentamers and hexamers. These labile water rings construct labile clusters around gas molecules after gas is dissolved in the water phase. The labile clusters gradually grow, combine, and agglomerate into nuclei of critical size to initiate subsequent growth [10,17,32,33]. Fig. 6.7 illustrates the LCH.

The coordination number of the water molecules depends on the size of the guest molecules, e.g., 20 water molecules surround 1 CH_4 molecule, whereas 24 water molecules surround 1 C_2H_6 molecule and 28 water molecules cover 1 C_3H_8 molecule. The clusters of the dissolved species combine to form unit cells. For the formation of a unit cell, the coordination number of the water molecules surrounding the dissolved gas molecules has to be changed. This transformation of the cluster coordination number of the water molecules needs activation energy, and therefore this step becomes a barrier in the formation process [10,29].

The labile cluster nucleation hypothesis proposes an energy barrier for cluster agglomeration [32]. Using direct molecular dynamic simulations, Walsh et al. [34] could show nucleation steps similar to the LCH. However, using molecular dynamic simulations, Radhakrishnan and Trout [35] showed that the energy barrier for cluster agglomeration is much larger than that for cluster disintegration. As a result, it is actually much more thermodynamically favorable for a labile cluster to decompose than to agglomerate.

6.2.3.1.2 The hypothesis of the nucleation at the interface

The hypothesis of the nucleation at the interface was presented by Rodger [36] and Kvamme [37]. Unlike the labile cluster it assumes that nucleation starts from the vapor side of the gas-water interface [32]. In this hypothesis, heterogeneous nucleation takes place when gas molecules are transported to the water-gas interface, where they are adsorbed at the aqueous surface and diffuse to a suitable location, at which they will be enclathrated into the cavity that was formed from water molecules [10]. These clusters agglomerate and grow on the vapor side of the interface [27].

Observations from microscopic and Raman spectroscopic investigations suggest that the guest molecule is dissolved in the water phase before it is encased into a hydrate cavity. Thereby, the water phase does not necessarily need to be liquid [27]. As described by Wang et al. [38], hydrate formation from ice does not require a conversion of ice into liquid water. Since the guest molecules need to be dissolved first before they are enclathrated, the water phase is initially supersaturated with the gas molecules [27].

The methane hydrate formation is supported by this process; however, it is probably not a satisfying concept for those hydrates with a higher density compared to the density of water, which show a preferential nucleation and growth on the subsurface [10,39].

6.2.3.1.3 Local structuring nucleation hypothesis (LSH)

The local structuring nucleation hypothesis is based on the work performed by Radhakrishnan and Trout [35]. LSH assumes that nucleation is triggered by dissolved guest molecules. LSH states that the key step in the nucleation pathway is the concentration fluctuation that arranges a group of guest molecules in a configuration similar to that of the clathrate crystal. The water molecules around the locally reordered guest molecules will be perturbed, and the rearranged guest molecules may exceed a critical value comparable to that in a critical nucleus. The perturbed water molecules, according to this hypothesis, follow the ordering of the guest molecules, building the cages that constitute the clathrate nucleus [30–32].

6.2.3.1.4 Cage adsorption hypothesis

Guo et al. [40] proposed the cage adsorption mechanism, which features specific multiple steps. At first the gas dissolves into water, and then a cage forms spontaneously; then the cage adsorbs the dissolved gas, and the cages stabilize themselves by adsorbing more gas; the miscellaneous growth of cages then reaches an amorphous hydrate phase, and finally the amorphous phase transforms into a hydrate crystal.

6.2.3.1.5 The blob mechanism

In the blob mechanism, the crystallization proceeds in two steps. First, the guest molecules concentrate in "blobs," amorphous clusters involving multiple guest molecules in water-mediated configurations. These blobs are in equilibrium with the dilute solution and result in the formation of the clathrate cages that eventually transform into an amorphous clathrate nucleus. In a second step, these amorphous clathrate nuclei transform into crystalline clathrate [30]. Eq. (6.2) summarizes the blob mechanism.

$$\text{Solution} \leftrightarrow \text{blob} \rightarrow \text{amorphous clathrates} \rightarrow \text{crystalline hydrates} \tag{6.2}$$

Arjun et al. [31] reported that methane hydrate nucleation at moderate temperature involves two competing mechanisms, amorphous and crystalline polymorph formation. Polymorph selection is thus governed by kinetic accessibility of the correct cage type and, moreover, occurs at precritical nucleus sizes, apparently against Ostwald's step rule.

6.2.3.1.6 Funnel model of crystallization

Hall et al. [41] demonstrated that hydrate nucleation involves funnel-shaped potential energy landscapes, and proposed a funnel model of crystallization. They also showed that the phenomenological similarities between crystallization and protein folding result from underlying similarities between essential features of these microscopic ordering processes.

6.2.3.2 Hydrate growth

After the small hydrate nuclei obtain a critical size, a continuous hydrate growth starts. Regarding this hydrate growth, three important aspects have to be considered: (1) the mass transfer including the transportation of gas and water molecules to the growing crystal surface (and because the hydrate contains up to 15 mol% gas, the mass transport of the gas to the hydrate surface is of major importance); (2) the kinetics of the hydrate growth process; and (3) the heat transfer due to the fact that hydrate formation is an exothermic process (the heat transfers away from the reaction/growth site) [10,17].

Similar to nucleation mechanics, crystal growth of hydrates occurs at the gas–liquid–solid interfaces. A generalized picture for hydrate crystal growth is illustrated in Fig. 6.8, conceptualized from the general crystal growth mechanics taken from Dennis Elwell and Hans J. Scheel [42].

A hydrate/water cluster interacts with the solid hydrate crystal surface (i), and then either attaches to the surface (ii) or diffuses across the surface (iii), and attaches to the hydrate step (iv) or kink sites (v). Water molecules are expelled as the water clusters attach and are incorporated into the hydrate crystal structure [43].

The different types of hydrate crystal growth processes were reported by Sloan and Koh [17] and may be divided into: (1) single crystal growth, (2) hydrate film/shell growth at the water-hydrocarbon interface, (3) multiple crystal growth in an agitated system, and (4) growth of metastable phases.

Hydrocarbon nucleation and growth is believed to be initiated at the hydrocarbon-water interface where the hydrate film (also known as hydrate shell, or crust) growth takes place. The steps of hydrate film formation are summarized in Fig. 6.9. These include formation and propagation of a thin porous

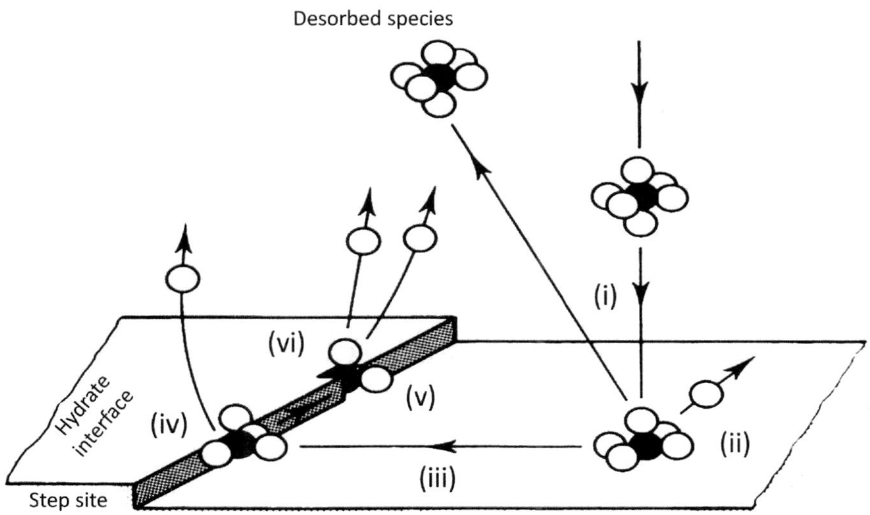

FIG. 6.8

Conceptual model for gas hydrate crystal growth.

Concept from: Z.M. Aman, C.A. Koh, Interfacial Phenomena in Gas Hydrate Systems, Chemical Society Reviews. Royal Society of Chemistry, 2016. https://doi.org/10.1039/c5cs00791g. Original photo reprinted from reference *D. Elwell, H.J. Scheel, Crystal Growth from High-Temperature Solutions, Academic Press, 1975, copyright (1975), with permission from Elsevier.*

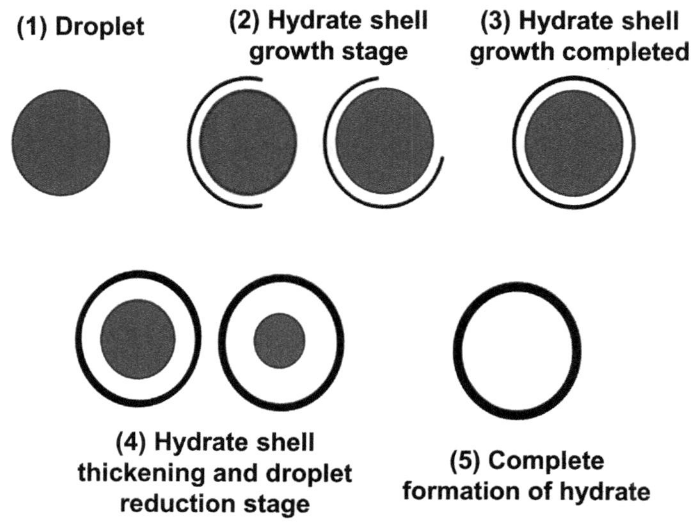

FIG. 6.9

Gas hydrate shell formation steps.

Reprinted by permission from [Springer Nature]: [Springer Nature) [Computational and Experimental Simulations in Engineering] by [Hiroshi Okada, Satya N. Atluri] copyright [2019].

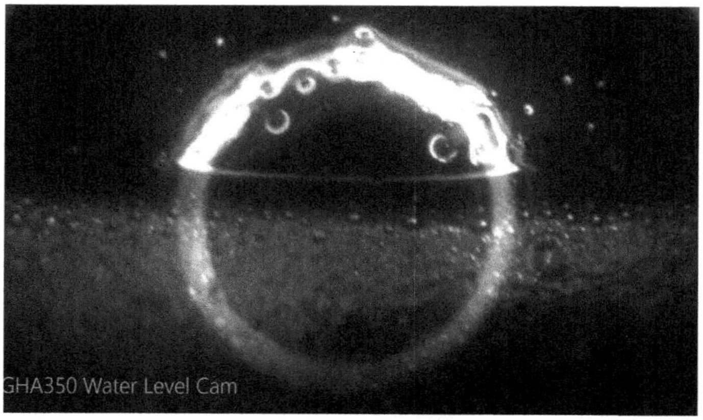

FIG. 6.10

Gas hydrate formation at the water surface.

Photo courtesy PSL Systemtechnik GmbH (https://psl-systemtechnik.com/) reprinted with permission.

hydrate film around the droplet and across the hydrocarbon-water interface (steps 1–3), then film development and thickening (step 4), and eventually hydrate film solidification and conversion to hydrate particles (step 5) [17,44,45].

Fig. 6.10 shows an image of a hydrate shell formed at the water surface in a gas hydrates autoclave, and Fig. 6.11 shows the formation of solid gas hydrate particles in a sapphire rocking cell.

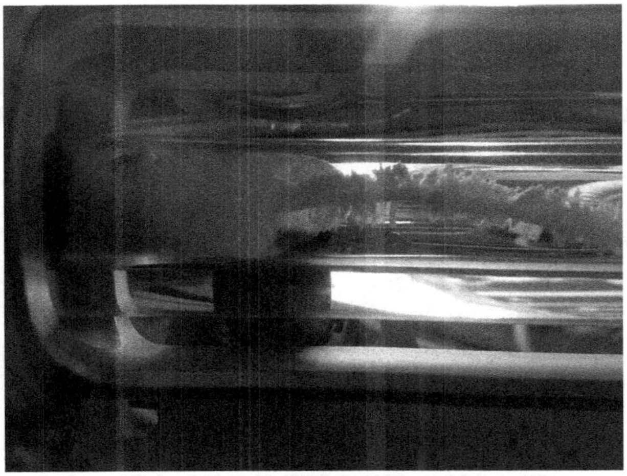

FIG. 6.11

Gas hydrate particles formation in rocking cell.

Photo courtesy PSL Systemtechnik GmbH (https://psl-systemtechnik.com/) reprinted with permission.

The studies using microimaging techniques showed that gas hydrates can grow in the bulk interior droplet (when the gas hydrate former diffuse into the interior droplet), away from the hydrate shell–water interface, by growing out from the hydrate shell, resulting in shell thickening [17].

Besides hydrate shell growth at the water-hydrocarbon interfaces, hydrates can also grow on solid substrates. De Banos et al. [46] studied hydrate growth on (hydrophilic) glass. The hydrates form tenuous halos that emerge from the contact line of a water drop sitting on the substrate. Before the development of the halo, hydrate nucleation occurs at the interface between the primary water drop and the guest phase or at the contact line of the water drop, producing a crust of polycrystalline hydrate covering the drop. As a rule, the halo emerges from the contact line only on completion of the crust in the vicinity, or even complete closure over the drop [47].

Furthermore, it was found that hydrates form in two layers, each with a clearly distinct morphology. As the clathrate aged, this difference in morphology tended to disappear [47].

Molecular experimental studies of the hydrate nucleation and crystal growth processes are challenging due to the small dimensions at which these pathways take place (i.e., nanometer scale), as well as the short timescales (microseconds for nucleation), and the stochastic nature of hydrate formation [32,48].

6.2.3.3 Formation of gas hydrate deposits in multiphase production systems

Knowing the previously described gas hydrate formation mechanisms, the formation of gas hydrates in pipelines is a complex problem and its formation mechanism yet depends on the production fluids composition (gas, oil and water) and the hydrodynamics of these fluids (flow regime).

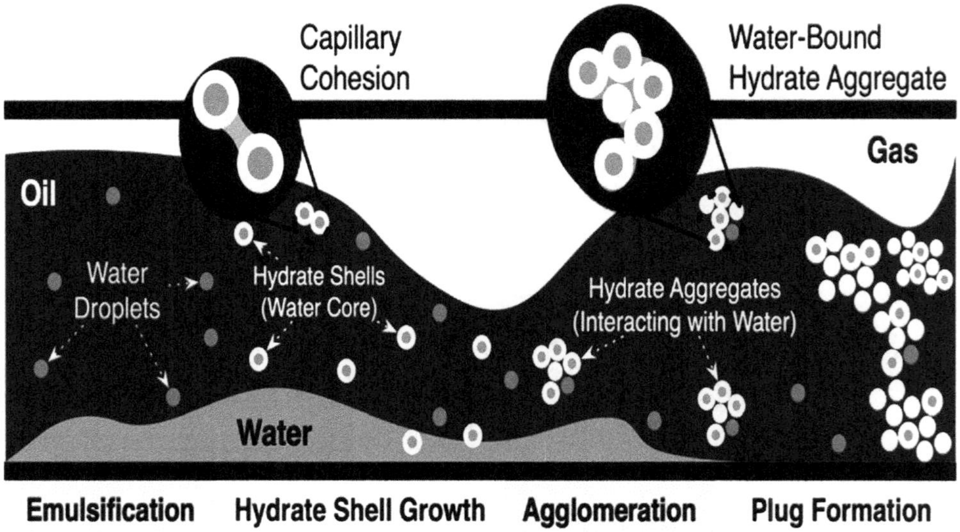

FIG. 6.12

Gas hydrates formation in multiphase flow.

Reprinted from Z.M. Aman, K. Pfeiffer, S.J. Vogt, M.L. Johns, E.F. May, Corrosion inhibitor interaction at hydrate-oil interfaces from differential scanning calorimetry measurements, Colloids Surf. A Physicochem. Eng. Asp. 448 (1) (2014) 81–87, https://doi.org/10.1016/j.colsurfa.2014.02.006, copyright (2014) with permission from Elsevier.

Generally, in a regular gas-oil-water system, the mechanism of hydrate formation is a four-step process, as stipulated by [49,50] and illustrated in Fig. 6.12:

(i) water emulsification in the liquid hydrocarbon phase;
(ii) hydrate nucleation and growth at the water-hydrocarbon interface;
(iii) particle aggregation; and
(iv) blockage formation from aggregate jamming.

In such systems, the water-in-oil emulsion is generally a precursor to hydrate formation.

The hydrate plug formation mechanism is also distinct in the various multiphase flow systems, and it has been studied by many researchers [50]. Due to the changing volumes and ratios of the produced fluids, it is easier to study hydrate formation based on these volumes or according to the dominating phase.

For oil dominant systems, several steps are involved. The water phase is emulsified within the oil phase. A thin malleable ($<6\,\mu m$ thick) hydrate shell grows around the water droplets, from small gas molecules dissolved in the oil phase. Capillary forces of attraction cause the hydrate-encrusted droplets to agglomerate. Further agglomerations of hydrate particles cause viscosity increase and further drop in the system pressure, and when they sufficiently grow in size they form a plug. The plug then undergoes aging and become less flexible, less porous, and more solid-like [17]. Boxall et al. [51] pointed out that hydrates first form at the oil-water interface as a hydrate shell. Then, the hydrate plug is formed mainly due to the shell growth and the decrease of hydrate transportability, which is dependent on the flow

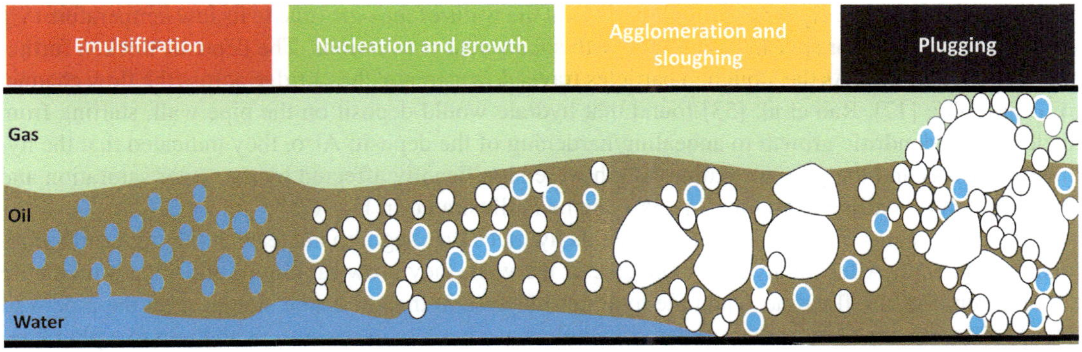

FIG. 6.13

Hydrates formation in oil dominated systems.

velocity and water cut. In addition, they found that plugging tendency is highly dependent on the effective diffusivity and the hydrate-oil slip ratio. Hydrate formation in oil dominated systems is illustrated in Fig. 6.13.

Hydrate blockages in high water cut systems are explained as in oil dominated systems, with the main differences being that the water phase will be partly entrained in the oil phase, while oil droplets will be entrained in the water phase at high water cut, and also there will be a separate free water phase that is not entrained due to the high water cut. Hydrates will form on interfaces where water and hydrate forming hydrocarbon components are present; agglomeration of hydrates leads to increased viscosity and hydrate plug formation. Joshi et al. [52] pointed out that the hydrate plug formation in water dominant systems was a consequence of the increase of hydrate concentration, which would further lead to the formation of hydrate bed and wall deposit. Hydrate formation in water dominated systems is illustrated in Fig. 6.14.

In gas dominant systems, water sources can be both produced water and condensed water from the gas. In such systems, there is much less liquid (both hydrocarbon and water), so that the W/O emulsion

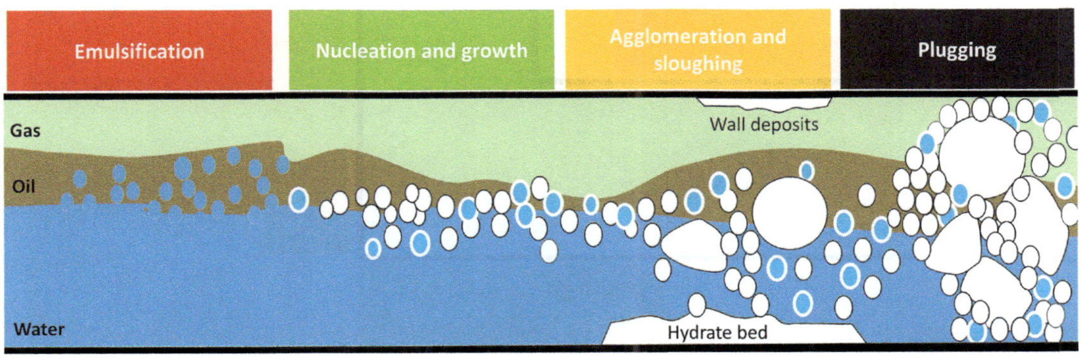

FIG. 6.14

Hydrate formation in water dominant systems.

concept does not apply. Hydrates originally form at the walls of the pipe (due to its low temperature) via vapor deposition and/or splashing of water with subsequent conversions. The deposits thicken further and then they slough. As the sloughed particles travel downstream, they bridge across the flow channel to form a plug [17]. Rao et al. [53] found that hydrate would deposit on the pipe wall, starting from nucleation to dendritic growth to annealing/hardening of the deposit. Also, they indicated that the hydrate thickness and the plug formation length were significantly affected by the water saturation and fluids velocity. Hydrate formation in gas systems is illustrated in Fig. 6.15.

Like gas systems, in gas condensate systems hydrates that begin to nucleate at the pipe surface remain on the wall. After the hydrates nucleate at the wall, they grow rapidly to encompass the entire circumference of the flowline. Then the hydrates increase in thickness and slough from the pipe wall into the flow stream as hydrate particles, which then accumulate in the system and jam to a plug, preventing normal flow. Hydrate formation in gas condensate is illustrated in Fig. 6.16.

Shi et al. [54] studied the hydrate behaviors and the slurry plugging mechanism in different flow patterns using a high pressure flow loop and concluded that the agglomeration degree in order from high to low is: slug flow > stratified flow > bubble flow > annular flow; the deposition degree in order from high to low is: annular flow > slug flow > bubble flow > stratified flow. They also found that the slurry flow in the stratified flow and annular flow conditions was easily blocked. The plugging in the stratified flow was mainly due to the hydrate accumulation and bedding in the liquid phase, while the plugging in the annular flow was mainly caused by the hydrate layer sloughing. The hydrate formation and plugging for each flow pattern is presented in the following text.

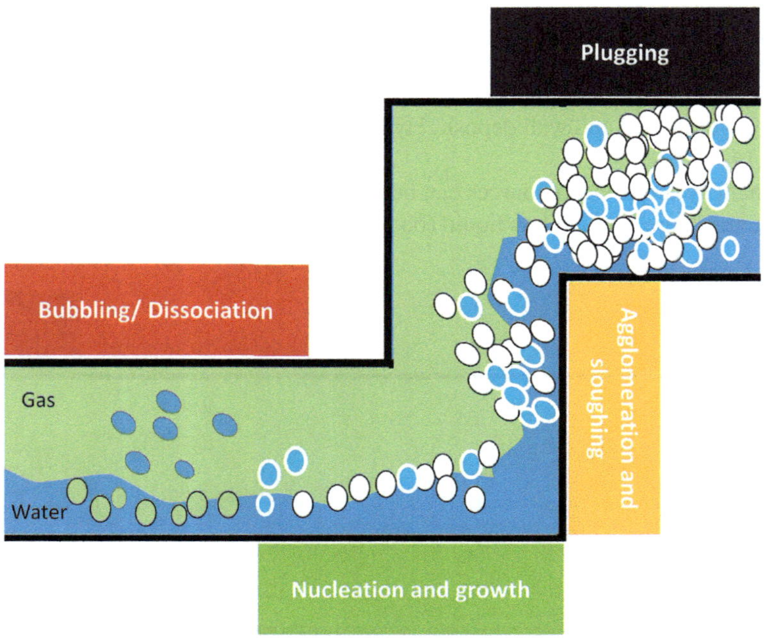

FIG. 6.15

Gas hydrate formation in gas systems.

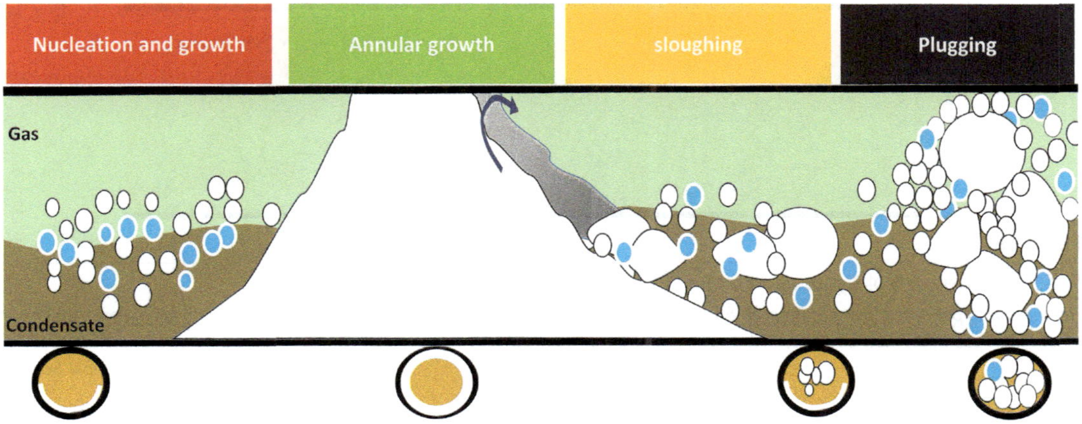

FIG. 6.16

Hydrate formation in condensate systems.

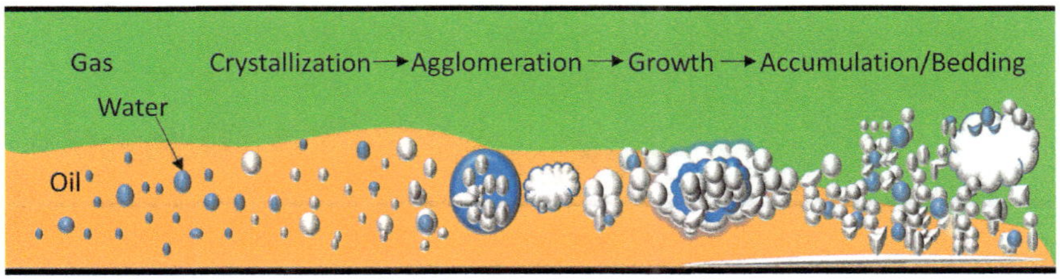

FIG. 6.17

Schematic diagram of hydrates plugging in stratified flow.

Reprinted with permission from L. Ding, B. Shi, X. Lv, Y. Liu, H. Wu, W. Wang, J. Gong, Hydrate formation and plugging mechanisms in different gas-liquid flow patterns, Ind. Eng. Chem. Res. 56 (14) (2017) 4173–4184. https://doi.org/10.1021/acs.iecr.6b02717. Copyright {2017} American Chemical Society.

In the stratified flow, as shown in Fig. 6.17, the system maintains gas-liquid stratified flow with the water phase dispersed in the oil phase. Once the system reaches hydrate formation conditions or a hydrate stable region, hydrate nucleation starts. Then hydrate particles and water droplets begin to agglomerate with each other; the agglomerates grow continuously and eventually the hydrates accumulate in the liquid phase and then bedding occurs on the pipe wall. A plug is then formed and the pipeline is blocked [54].

The hydrate formation and slurry flow process in the bubble flow is shown in Fig. 6.18. The water phase is dispersed in the oil phase, where the system maintains at a stable bubble flow. Upon reaching the onset of hydrate formation conditions, hydrates start forming on the water/oil interface. The hydrate particles and water droplets then agglomerate with each other, forming large agglomerates that grow in the liquid phase. The hydrates and agglomerates deposit on the pipe wall surface, and then the system can maintain a stable flow condition [54].

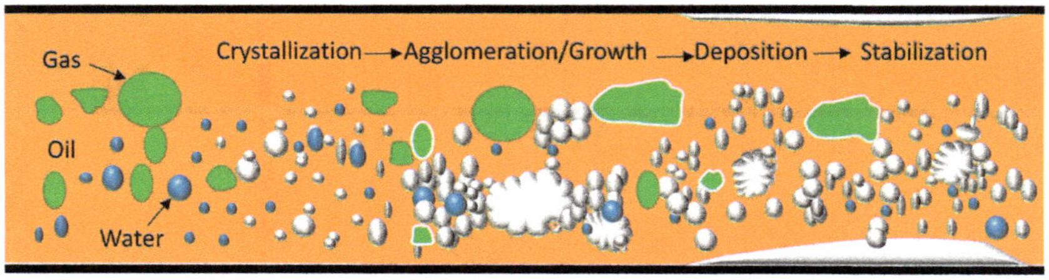

FIG. 6.18

Schematic diagram of hydrates formation in bubble flow.

Reprnted with permission from L. Ding, B. Shi, X. Lv, Y. Liu, H. Wu, W. Wang, J. Gong, Hydrate formation and plugging mechanisms in different gas-liquid flow patterns, Ind. Eng. Chem. Res. 56 (14) (2017) 4173–4184. https://doi.org/10.1021/acs.iecr.6b02717. Copyright {2017} American Chemical Society.

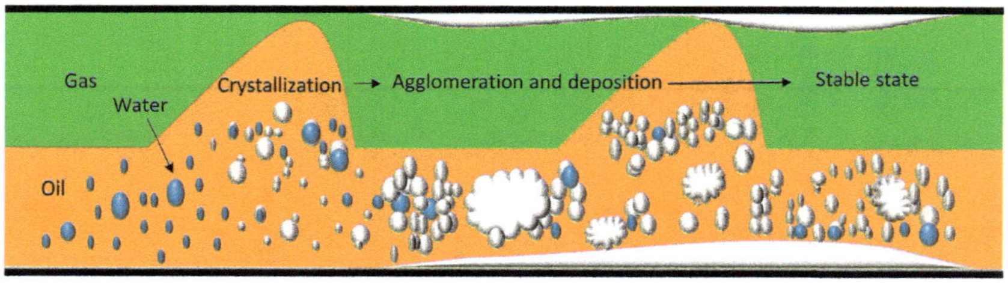

FIG. 6.19

Schematic diagram of hydrate formation and behavior in slug flow.

Reprinted with permission from L. Ding, B. Shi, X. Lv, Y. Liu, H. Wu, W. Wang, J. Gong, Hydrate formation and plugging mechanisms in different gas-liquid flow patterns, Ind. Eng. Chem. Res. 56 (14) (2017) 4173–4184. https://doi.org/10.1021/acs.iecr.6b02717. Copyright {2017} American Chemical Society.

The hydrate formation and slurry flow process in the slug flow are shown in Fig. 6.19. In this regime, the water droplets disperse in the liquid phase, and when the system reaches a hydrate stable region, hydrates start to nucleate on the water/oil interface. Agglomeration of hydrates and water droplets takes place in the liquid phase and deposits on the pipe wall surface, while the shear forces can break some of the agglomerates. The system can maintain a stable flow state, and no plug forms as was observed in the flow loop experiments [54].

The plugging mechanism in annular flow is shown in Fig. 6.20. The system maintains the annular flow before hydrate formation, with some water being dispersed in the oil phase, whereas some water distributes as water film covering the pipe wall. Hydrates start forming on the pipe wall or on the water/oil interface, forming a thick hydrate layer covering the pipe wall. The high shear force then sloughs the formed thick layer, and then the sloughed hydrate fragments accumulate at some uneven section, forming a plug that blocks the flow [54].

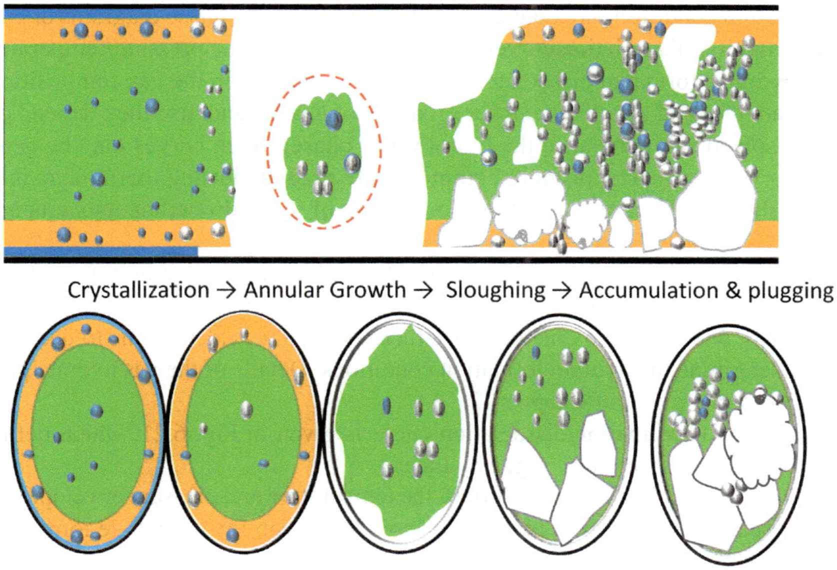

FIG. 6.20

Schematic diagram of hydrate plugging mechanism in annular flow.

Reprinted with permission from L. Ding, B. Shi, X. Lv, Y. Liu, H. Wu, W. Wang, J. Gong, Hydrate formation and plugging mechanisms in different gas-liquid flow patterns, Ind. Eng. Chem. Res. 56 (14) (2017) 4173–4184. https://doi.org/10.1021/acs.iecr.6b02717. Copyright {2017} American Chemical Society.

6.2.4 Factors affecting gas hydrate formation

6.2.4.1 Effect of pressure and temperature

Gas hydrates are formed by combining water and small hydrocarbons at elevated pressures and at lower temperatures. Normally, the stability of hydrates increases with increasing pressure and decreasing temperature [55]. The temperatures at which gas hydrates form are significantly higher than the temperatures at which water ice will form [56]. The temperature below which hydrates can form increases with increasing pressure and can be as high as 25–30°C [6]. For a given pressure, the difference between the thermodynamic hydrate formation/dissociation temperature and the ambient temperature (e.g., seabed or processing facilities surrounding temperature) is known as the subcooling.

The hydrate formation temperature and pressure conditions are different from one gas to another and from pure gas to a mixture of gases. The formation pressure is a function of the guest:cage size ratio. As a molecule better fits the cage, the formation pressure decreases. Methane is a relatively poor fit (0.86) in the 5^{12} cage, so formation pressure of methane hydrate is relatively high (2.56 MPa), whereas since i-C_4H_{10} is considered a good fit (0.98) in the $5^{12}6^4$ cage, its formation pressure is relatively low (0.133 MPa) [57]. In addition, as gas hydrate induction time decreases, nucleation and growth rates increase with decreasing temperature and increasing subcooling [58,59]. Furthermore, the amount of gas consumed during the gas hydrate formation is also dependent on the temperature and the degree of subcooling [60].

Hydrate formation conditions are conventionally represented by phase diagrams or curves known as the hydrate formation/dissociation curves, which are used to define pressure/temperature relationships in which hydrates form and dissociate. These curves depend on the gas composition (different from one gas to another and also different for mixtures of gases), and also they depend on the water composition (dissolved salt content in the liquid water phase). Such curves may be generated by a series of laboratory experiments, or more commonly are predicted using thermodynamics software such as MultiFlash or PVTsim based on the composition of the hydrocarbon and aqueous phases in the system [55].

Predicting gas hydrate formation conditions (T,P) can be relatively straightforward with simple methods such as gas gravity or the *K*-value methods with few inputs, and these methods are known as hand calculations. More complex methods use thermodynamics calculations to describe the phase equilibrium and determine the hydrate formation conditions. Details of the gas hydrate prediction and the industry software are given in Chapter 12.

A typical gas hydrate formation/dissociation curve is given in Fig. 6.21, which defines P and T boundaries of formation/dissociation of gas hydrates.

The hydrate formation curve basically defines the temperature and pressure envelope or boundaries in which the hydrocarbon system must operate, to avoid the possibility of hydrate formation, or it simply defines the hydrate safe zone for operations [55].

Hydrate phase equilibria are normally determined in terms of four variables: (1) pressure, (2) temperature, (3) water-free hydrocarbon phase composition, and (4) the free-water phase composition (excluding hydrocarbons but including salts, alcohols, and glycols) [17]. So, putting more details

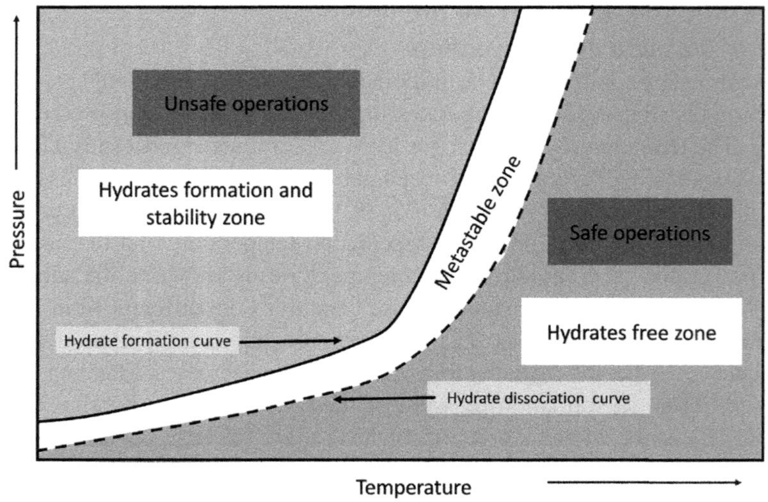

FIG. 6.21

Typical gas hydrate formation/dissociation curve.

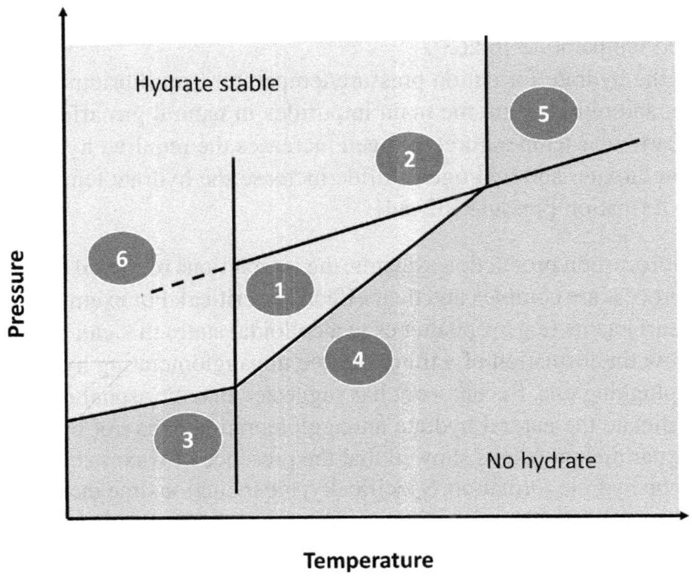

FIG. 6.22

Phase diagram for water/hydrocarbon/hydrate system.

on the preceding diagram, we can find that the general phase diagram for water, hydrocarbon, and solid hydrate has the following essential regions [56], as shown in Fig. 6.22:

(1) Hydrate + gaseous hydrocarbon + liquid water;
(2) Hydrate + liquid hydrocarbon + liquid water;
(3) Ice + gaseous hydrocarbon;
(4) Liquid water + gaseous hydrocarbon;
(5) Liquid water + liquid hydrocarbon;
(6) Hydrate + ice;

Zhang et al. [60] proposed a new method, namely the slope of the gas flow rates against time ($d\nu_g/dt$), to represent the intrinsic driving force for hydrate formation. These slopes represent the acceleration in gas flow rates, and therefore they indicate the magnitude of driving force for hydrate formation.

6.2.4.2 Effect of gas composition

The composition of the natural gas can affect the temperature and pressure required for formation of the hydrate. Even more significant is the effect of impurities in the natural gas on the pressure temperature (PT) curves of the hydrate. Generally the following points have been noticed:

- The relation between hydrate formation temperature and the natural gas molecular weight is directly proportional. The higher the natural gas molecular weight, the higher the hydrate formation temperature will be [61].
- Generally the higher the methane percentage in the gas sample, the higher the hydrate formation pressure at a fixed temperature [62,63].

- Generally the higher the propane concentration in the natural gas, the lower the hydrate formation pressure at a fixed temperature [62,63].
- Impurities affect the hydrate formation pressure/temperature equilibrium line. Carbon dioxide, hydrogen sulfide, and nitrogen are the main impurities in natural gas affecting the hydrate formation. At a particular temperature, nitrogen increases the required hydrate formation pressure while both carbon dioxide and hydrogen sulfide increase the hydrate temperature and lower the required hydrate formation pressure [62–64].

In the multiphase hydrocarbon production systems, the interactions of the oil components with the gas hydrates formation process are complex and their effects are critical. For example, some crude oils have naturally occurring surfactants (e.g., asphaltenes in a colloidal state) that can behave as a hydrate antiagglomerant and allow the formation of a transportable nonagglomerating hydrate slurry. Such crude oils are denoted nonplugging oils. Recent work has suggested that the asphaltene-aggregation state is a parameter that may dictate the natural hydrate antiagglomeration behavior of nonplugging crude oils [65,66]. Recently, experimental results showed that the presence of waxes strengthened the inhibiting effect of asphaltenes on hydrate formation. Specifically, the induction time increased and the amount of formed hydrates decreased with increasing wax content in water-in-oil emulsion containing asphaltenes [67].

The experimental investigation showed that wax tends to increase the induction time of hydrates [68,69], and to inhibit hydrate nucleation [70]. Nonetheless, wax could promote hydrate growth once nucleation is achieved [71]. Simultaneous precipitation of wax and hydrates experiments revealed that the precipitation of wax could significantly enhance hydrate formation [72].

6.2.4.3 Effect of water properties

Water is a basic requirement for gas hydrate formation. Generally, water in free form or emulsified form can cause gas hydrate formation. The quantity of water required for hydrate formation is a key variable in the process. The volume of the formed hydrates usually increases when the pressure and water content increase. The volume of hydrates reaches a maximum as the moles of water increase at fixed pressure and temperature. The water-to-gas ratio depends on the gas composition and system pressure. Generally speaking, the mole ratio of water to natural gas is about 6:1; however, to achieve maximum hydrate formation a gradual increase in water or pressure may be required [63]. Beyond that ratio, further additions of water will result in free-water liquid phase. For a fixed mole ratio, the volume of gas decreases due to the compressibility effect [62].

It is common to express water content in terms of the dew point, and the dew point temperature is usually expressed in °F or °C [73]. The dew point of a gas mixture is the temperature at which the water vapor in the gas mixture starts to condense into liquid water, or, in other words, it is the point at which the gas is in full capacity of water, that is, it cannot hold any more water vapor, and hence the gas mixture is known to be saturated with water vapor. In practice, the water content is plotted on a semi-logarithmic scale against the dew point temperature in the form of isobaric lines. Such diagrams have been widely used in gas processing [74].

If the water dew point of gas is high, the water is likely to condensate during the pressure regulation, leading to the formation of hydrates [75,76]. It is recommended not to exceed water content of 7.0 lbm/MMscf to avoid hydrate formation [77].

It is important not to confuse the gas hydrate formation with condensation. Condensation of water from natural gas under pressure occurs when the temperature is at or below the dew point, whereas the gas hydrates are formed using the free water obtained from that condensation; hence, gas hydrates are formed at temperatures equal to or below the dew point temperature [78].

Besides hydrate formation from free water, a hydrate or ice plug can form in two ways when there is no liquid water phase [6]:

1. Water coalescence when the liquid water saturation curve is crossed (condensation water on top of pipe in shut-in as the pipe cools).
2. Dissolved water in the liquid hydrocarbon phase.

Water composition also affects hydrate formation. The stability of hydrates depends on the salinity of water. At very low salt concentrations, it is likely to be a hydrate promoter to some extent. As the salt solution increases, the solubility of natural gas in the solution decreases; thus the high salt content formation (and completion) fluids inhibit the formation of hydrates [2,79,80]. The salt will not enter the gas hydrate crystal structure, but it will control the chemical activity of the water from which the hydrates form.

6.2.4.4 Effect of hydrodynamics and operating conditions

These are considered the secondary conditions of gas hydrate formation [56]:

- High velocities
- Pressure pulsations
- Any type of agitation
- Introduction of a small crystal of the hydrate
- Flow regimes, as different flow regimes cause hydrate formation at different rates and mechanisms, as mentioned earlier.

Many multiphase production lines are designed to operate safely without hydrate formation, but problems usually emerge during shut-in when the fluids cool, untreated to within the hydrate-forming envelope. Furthermore, if a subsea well is shut-in, there is a high risk of gas hydrate plug formation below the wellhead unless the necessary precautions are taken beforehand [6]. Operating conditions, especially those of shutdowns and restart, must be addressed and specified in the field flow assurance strategy, with the possible mitigation solutions if the gas hydrate problem is estimated to be severe, to avoid extended downtimes and expensive maintenance.

Slow start of production usually creates high pressure and low temperature conditions that are sufficient for gas hydrate formation. Another scenario of gas hydrate risk is during intermittent operations such as aborted start. Production start for a short period at moderate rates and then shutting-in usually creates both high pressure, as the well is full of produced fluids, and low temperatures, which intuitively creates a sufficient environment for gas hydrate formation [2].

Therefore steady-state production cases are less likely to be at risk of hydrate formation unless the rates are very low and the temperature of the flowing fluids is far below the hydrate formation temperature [2].

6.2.4.5 Effect of surfactants

Surfactants and micellar solutions were found to be a chemical hydrate promoter and to increase the gas hydrate formation rate. As hydrate promoters, surfactants generally reduce the induction time required for gas hydrate formation and enhance the rate of hydrate nucleation and growth. Surfactants reduce the surface tension of the gas-liquid interface, which reduces the diffusion resistance between two phases and hence increases the gas molecule solubility [81,82].

Among the three major types of surfactants (anionic, cationic, and nonionic), anionic surfactants such as sodium dodecyl sulfate (SDS) are the most effective in enhancing the rate of hydrate formation and/or reducing the induction time [83].

Zhong and Rogers [81] reported that SDS increases the hydrate formation of ethane and natural gas hydrate. They reported that critical micelle concentration (CMC) of SDS solution was found at 242 ppm at hydrate formation conditions. At concentration of SDS above the CMC, the hydrate formation rate increased by a factor of 700 compared to the system without any added surfactant. The formation of micelles in the presence of surfactants enhanced the gas solubility and also acted as nucleating sites for faster hydrate growth. A similar observation was concluded by Lin et al. [84], as they reported that the maximum storage capacity of methane (170 V/V) was obtained at critical SDS concentration (650). However, this hypothesis was disputed by Di Profio et al. [85], who did not observe any micelle formation under hydrate formation conditions using SDS, including other anionic and cationic surfactants. Zhang et al. [86] concluded that the hydrate nucleation was influenced by the presence of SDS so that the SDS micelle had little influence on the hydrate nucleation. However, SDS increased the hydrate nucleation rate and hydrate growth rate by reducing the interfacial tension between hydrate and liquid, and by increasing the total surface area of hydrate particles and the gas-liquid interfacial area.

Recently, surfactants have been used as gas hydrates crystallization promotors for the industrial uses of gas hydrates, like separation of gas mixtures, CO_2 capture, transportation and sequestration, methane storage and transport, and seawater desalination [83].

6.2.4.6 Effect of surface nature

Gas hydrates are affected by the nature of the surface; it is widely agreed that solid surfaces promote gas hydrate formation by providing nucleation sites. The effect of solid surfaces on hydrate formation can be explained by the changes in the local structure of water molecules near the liquid-solid interface, which alters the pathways of hydrate formation [87].

Hydrophobic surfaces were reported to promote gas hydrate formation, whereas hydrophilic surfaces inhibit hydrate formation. Gas hydrates nucleate easily on less hydrophilic surfaces. This promoting effect of hydrophobic surfaces is attributed to the tetrahedral ordering of water structure and the increase in gas density at hydrophobic solid-water interfaces, whereas hydrophilic surfaces show inhibiting effects owing to the distorted water structure and the decreased gas density at hydrophilic solid-water interfaces [88,89]. Aside from surface hydrophilicity, surface crystallinity shows a weak effect on the local structure of adjacent water molecules and on gas hydrate nucleation [87].

In conclusion, we can say that the general requirements for gas hydrate formation include the following [64,77,78]:

- Natural gas at or below its dew point with free water present;
- Low temperature, specifically temperatures below the hydrate formation temperature for the pressure and gas composition considered;

- High operating pressure;
- High velocity or agitation through piping or equipment;
- Presence of seed crystals of hydrates;
- Presence of highly soluble acid gases like H_2S or CO_2 as they enhance the gas hydrate formation by reducing the required pressure and increasing the required temperature;
- If surfactants are present, they will enhance gas hydrate formation;
- Hydrophobic surfaces also enhance gas hydrate formation.

6.2.5 Recent advances in gas hydrate research

Gas hydrates have received considerable attention due to their important role in flow assurance for the oil and gas industry, their extensive natural occurrence on the earth and extraterrestrial planets, and their significant applications in sustainable technologies [33].

In the past few decades, several experimental techniques and methods have been used to elucidate gas hydrate structures and chemistry and have helped in understanding their properties, which are imperative for controlling their behavior. In situ techniques such as X-ray diffraction (XRD), Raman spectroscopy, NMR, and neutron diffraction at very low temperatures and high pressures allowed real-time monitoring of different structures of clathrates, identification of cage occupancies, and their evolution over time, providing a deeper insight into the kinetic behavior of their crystalline structures. In situ Raman spectroscopy and in situ XRD supplemented with calorimetry and/or NMR are helpful tools to determine hydrate formation and dissociation behavior as well as roles of multiple molecules on structure and cage occupancy, especially for gas hydrates formed from gas mixtures [33].

Furthermore, recently molecular simulation has contributed a great deal towards elucidating microscopic mechanisms of clathrate-hydrate behavior, from equilibrium structural and dynamical properties of the crystalline state, to time-dependent rate phenomena, such as nucleation, crystallization, and dissociation in heterogeneous environments (such as in marine silica) [33].

Besides recent advances in experimental studies of gas hydrates, e.g., flow loops and rocking cells, monitoring systems and other modeling and simulation tools have led to significant advances in predicting hydrate phase equilibria modeling, and these have enabled operators to predict hydrate stability zones, hydrate fractions and compositions, plus other derived thermodynamic properties.

Gas hydrates are also gaining momentum due to their diverse industrial applications. These include but are not limited to:

- Sustainable energy source as gas hydrate reserves are higher than those of fossil fuels.
- Natural gas storage and transportation in solid form with several advantages, including high volumetric storage capacity, nonexplosive nature, moderate storage conditions, and being environmentally benign, safe to handle, and cost effective [90,91].
- Like natural gas, hydrogen gas storage is another use of gas hydrates.
- Their use in refrigeration systems, because of their high temperature zone, suitable latent heat of fusion (more than ice in CO_2 hydrate), and better heat transfer efficiency [33].
- Fighting climate change by reducing methane gas emissions that occur when hydrate deposits trapped in marine and permafrost-associated sediments destabilized due to climate warming.
- Fighting climate change, by using CO_2 capture and storage technology, through CO_2-rich gas injection into methane gas hydrate reservoir. Hydrate-based CO_2 capture (HBCC) has received

increasing attention, due to such advantages as the mild operating pressure and temperature, the ease of regeneration, and its unique separation mechanism [92].
- Clathrate hydrate-based desalination (HyDesal) is a potential technology for seawater desalination [93].
- Gas hydrate formation as a separation technology shows tremendous potential, both from physical feasibility (in terms of effecting difficult separations) as well as an envisaged lower energy utilization criterion [94].

These new applications emerged after extensive research, which means in-depth and better understanding of gas hydrate properties and the factors affecting them, that consequently can be used to optimize their management in oil and gas fields.

6.3 Diamondoids

6.3.1 Introduction

Diamondoids are another cage-like compound that is associated with oil and gas operations. Diamondoids are constituents of petroleum, gas condensate, and natural gas reservoirs [95]. Adamantane (a type of diamondoid) was originally discovered and isolated from petroleum fractions of the Hodonin oilfields in Czechoslovakia in 1933 [96].

Diamondoids in petroleum are believed to be formed from enzymatically created lipids with subsequent structural rearrangement during the process of source rock maturation and oil generation [95]. Studying diamondoids has led to new advancements in different organic geochemistry of petroleum, including maturity, oil-cracking, oil mixing, oil biodegradation, organic facies, oil spill identification, and others [97].

Diamondoids are considered one of the natural gas and gas condensate contaminants, characterized by very high boiling point, melting point, and also high vapor pressure. They usually deposit when the pressure and temperature of the fluid (gas) containing diamondoids decrease, and saturation in the gas is reached, leading to condensation and solidifying a portion of the diamondoid components (Fig. 6.23). Diamondoids usually deposit as waxy white solids, causing fouling and plugging of the gas-handling equipment [98,99].

6.3.2 Diamondoids molecular structure

Diamondoid molecules are cage-like structures of ultrastable, saturated polycyclic hydrocarbons with a molecular formula of $C_{4n+6}H_{4n+12}$ (where n is the number of the diamond cage subunit) [95,97,100–102].

The name "diamondoid" arose from the fact that they are similar to diamonds in their chemical structure, as the carbon-carbon framework in diamondoids constitutes the fundamental repeating unit in the diamond lattice structure [95]. Basically, diamondoids are ringed compounds formed by carbon and hydrogen atoms with sp3-type chemical bonds, which have a diamond-like structure consisting of a number of six-member carbon rings (like cyclohexane) in a "chair" conformation fused together. More explicitly, they consist of repeating units of 10 carbon atoms forming a tetracyclic cage system [95,102–104]. Fig. 6.24 illustrates the diamondoids and their relationship to the diamond structure.

FIG. 6.23

Diamondoids collected from a gas field in the Gulf of Mexico in which they have accumulated.

Reprinted by permission from Springer Nature: Springer Nature, Nature, Diamondoids and oil are not forever, Schoell, M., & Carlson, R. M.K., copyright (1999).

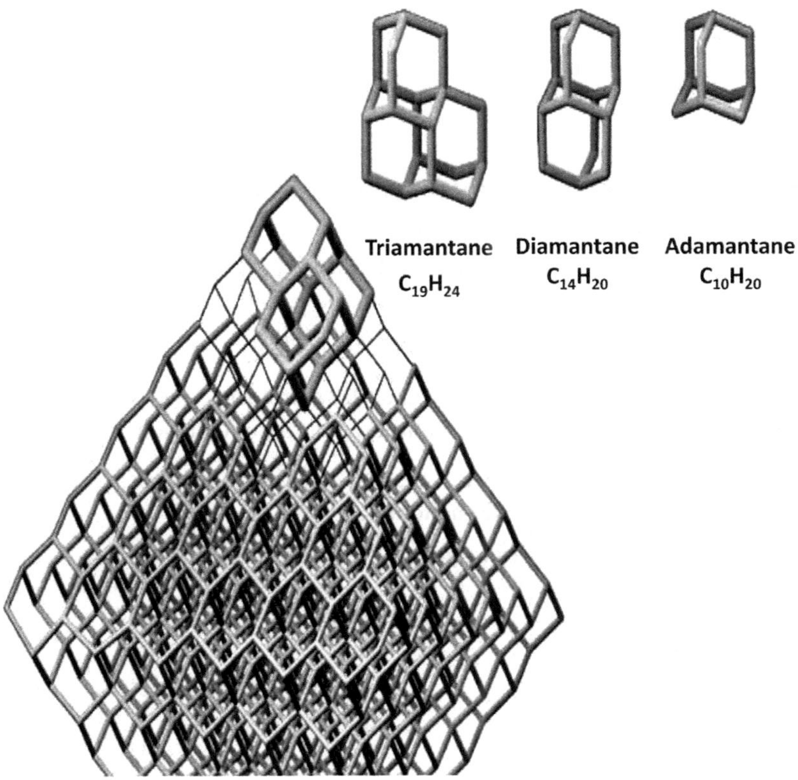

FIG. 6.24

Structures of diamondoids and the carbon framework of triamantane shown superimposed on a 455-carbon diamond lattice.

Reprinted by permission from Springer Nature, Springer, Diamond Molecules Found in Petroleum, in Synthesis, Properties and Applications of Ultrananocrystalline Diamond by R.M.K. Carlson, J.E.P. Dahl, S.G. Liu et al., copyright (2005).

The first and simplest member of the diamondoids group, adamantine ($C_{10}H_{16}$), is a tricyclic saturated hydrocarbon (tricyclo[3.3.1.1]decane), and is formed by 10 carbon and 16 hydrogen atoms. Other diamondoids are formed by face fusing a multiplet of such cages together. Adamantane is followed by its polymantane homologues (adamantologues): diamantine ($C_{14}H_{20}$) and triamantane ($C_{18}H_{24}$), which are known as the lower diammondoids. The others with $n >$ tetramantane, pentamantane, and hexamantane are known as the higher diamondoids or higher polymantanes. The lower diamondoids only have one geometrical isomer, while several geometrical isomers exist in higher diamondoids [95,103,105].

In fossil fuels, the abundance of adamantanes is the highest, while other diamondoids occur in relatively small amounts [104,106].

6.3.3 Diamondoids chemical and physical properties

Diamondoids show unique properties due to their exceptional atomic arrangements. These compounds are characterized by being chemically and thermally stable and strain-free. That make them exhibit high melting points (m.p.) in comparison to other hydrocarbons. They are essentially hydrophobic, are generally soluble in cyclohexane, and their solubility in organic solvents is a function of their hydrophobicity [95,105,107]. Table 6.2 summarizes the different properties of diamondoids [95].

The unique chemical structure and physical properties of diamondoids are advantageous for different applications, including polymer synthesis, nanotechnology, drug delivery, drug targeting, DNA directed assembly, and others [95].

6.3.4 Mechanisms of diamondoids deposits formation

Diamondoids are hydrophobic compounds that are soluble in light hydrocarbons at high pressures and temperatures. They have relatively high melting points, so they basically precipitate during oil and gas production, when the temperature falls below a specific limit. They also deposit when pressure drops, as petroleum fluids expand, resulting in reducing their solubility in the flowing stream and causing their precipitation [95,100].

Diamondoids deposition potential is higher in the "dry petroleum" (petroleum hydrocarbons low in light components) due to pressure and temperature drop during production. They also can deposit in

Table 6.2 The different properties of diamondoids.

Diamondoid	Molecular weight	Melting point (°C)	Boiling point (°C)	Density (g/cm^3)	Crystal structure
Adamantane ($C_{10}H_{16}$)	136.240	269	135.5	1.07	Cubic
Diamantane ($C_{14}H_{20}$)	188.314	236.5	272	1.21	Cubic
Triamantane ($C_{18}H_{24}$)	240.390	221.5	330	1.24	Orthorhombic

Reprinted from reference G. Ali Mansoori, Diamondoid molecules, Adv. Chem. Phys. 136 (2007) 207–258. https://doi.org/10.1002/9780470175422.ch4, copyright (2007), with permission from John Wiley and Sons.

"wet petroleum" streams, as they can partition themselves among the existing phases (vapor, liquid, solid), and consequently deposit upon drastic changes of pressure and temperature [108,109].

Diamondoids are much more soluble in CO_2 than methane, and solid diamondoids are also soluble in xylene and toluene [100]

Diamondoids usually deposit in natural gas, gas condensates, and light crude oil production and processing systems. One main problem of diamondoids is that they are usually ignored due to their low concentration in these fluids. They can easily nucleate on surfaces or other existing deposits upon pressure and temperature change. Thus it is crucial to determine, detect, and quantify diamondoids in reservoir fluid [107].

6.3.5 Factors affecting diamondoids formation and deposition

Like other solid deposits, diamondoids are sensitive to changes in process operating parameters.

Basically, reductions in pressure and temperature are the main factors that affect diamondoid deposition.

Diamondoid concentration and hydrocarbon composition are other significant factors. For this purpose, the separation, detection, and measurement of diamondoids and their bearing produced fluids are crucial.

6.3.6 Recent advances in diamondoids research

Extensive studies have been conducted on diamondoids as flow assurance threats and as a material with extraordinary physical properties. In flow assurance, characterization of the diamondoids in gas transmission lines and understanding their behavior have gained attention from several research institutes.

Diamondoids can also be used in several applications due to their superior properties, such as excellent thermal stability with a high melting point, high hardness and stiffness, as well as excellent chemical resistance.

Applications such as pharmaceutical therapy, catalysis, polymer composites, and optics have been developed. Selective chemical functionalization of diamondoid molecules has also been studied to broaden their applications in the fields of organic and inorganic chemistry, materials science, and biomedical science [110].

6.4 Summary

Two types of cage-like molecules are encountered in oil and gas operations that cause flow assurance issues: gas hydrates and diamondoids.

Gas hydrates form when the gas molecules known as guest molecules are entrapped inside cage-like water molecules formed under high pressure and low temperatures. The cage-like water molecules are stabilized by hydrogen bonds, while the guest gas-cage structure is stabilized by van der Waals forces. Gas hydrates are significantly affected by temperature and pressure, as they form at low temperature and high pressures. Other factors affecting the gas hydrate formation include water cut, surface nature, and fluid hydrodynamics. Gas hydrate deposition in systems is different based on composition (oil dominant, water dominant, gas dominant, gas condensate) and hydrodynamics (laminar flow, turbulent flow), in addition to the other production system factors. The operator should be aware of these formation mechanisms and their variances for safe operations in hydrate-free zones and better management of the problem.

Gas hydrates have received strong research interest for their potential industrial applications including gas storage and transportation, climate change control, gas separation, and water desalination.

Diamondoids are another cage-like deposit encountered in gas transmission lines. They deposit as waxy white crystals due to temperature and pressure changes during production, causing flow assurance issues. Diamondoids also exhibit some superior properties, which have made them a target of extensive recent research for their uses in different industrial applications.

References

[1] Z.M. Aman, C.A. Koh, Interfacial Phenomena in Gas Hydrate Systems, Chemical Society Reviews. Royal Society of Chemistry, 2016, https://doi.org/10.1039/c5cs00791g.

[2] J. Bellarby, Production chemistry, Dev. Pet. Sci. (2009), https://doi.org/10.1016/S0376-7361(08)00207-0 (Chapter 7).

[3] C.A. Koh, Towards a fundamental understanding of natural gas hydrates, Chem. Soc. Rev. 31 (3) (2002) 157–167, https://doi.org/10.1039/b008672j.

[4] L. Jensen, Experimental investigation and molecular simulation of gas hydrates, Ph.D. Thesis, Center for Energy Resources Engineering Department of Chemical and Biochemical Engineering Technical University of Denmark, 2010.

[5] C. Hall Barbosa, E.C. de Oliveira, S.L. Farias, IBP1233_15 Mapping the State of the Art of Measurement of Hydrocarbon Dew Point in Natural Gas, Rio Pipeline Conference & Exposition 2015, 2015.

[6] M.A. Kelland, Production Chemicals for the Oil and Gas Industry, CRC Press, 2009, https://doi.org/10.1201/9781420092974.

[7] R. Hollamby, G. Nuis, Introduction to Production Chemistry, Shell International Exploration and Production BV, 2000.

[8] C.D. Ruppel, Gas Hydrate in Nature, USGS Fact Sheet 2017-3080, 2017, https://doi.org/10.3133/fs20173080.

[9] T.S. Collett, Natural Gas Hydrates-Vast Resource, Uncertain Future, 2001. USGS Fact Sheet FS-021-01.

[10] J.M. Schicks, Gas hydrates: formation, structures, and properties, in: Hydrocarbons, Oils and Lipids: Diversity, Origin, Chemistry and Fate, Springer International Publishing, 2018, pp. 1–15, https://doi.org/10.1007/978-3-319-54529-5_2-1.

[11] E.D. Sloan Jr., Clathrate Hydrates of Natural Gases, Marcel Dekker, New York, 1998.

[12] Z. Wang, B. Sun, Y. Gao, Formation and decomposition of natural gas hydrate, in: Natural Gas Hydrate Management in Deepwater Gas Well, Springer, Singapore, 2020, https://doi.org/10.1007/978-981-15-6418-5_2.

[13] J. Jia, Y. Liang, T. Tsuji, S. Murata, T. Matsuoka, Microscopic origin of strain hardening in methane hydrate, Sci. Rep. 6 (2016), https://doi.org/10.1038/srep23548.

[14] D.W. Davidson, The motion of guest molecules in clathrate hydrates, Can. J. Chem. 49 (1971) 1224–1242.

[15] J.S. Gudmundsson, Flow Assurance Solids in Oil and Gas Production, CRC Press, London, 2018, https://doi.org/10.1201/9781315185118.

[16] L. Krishna, C.A. Koh, Inorganic and methane clathrates: versatility of guest-host compounds for energy harvesting, MRS Energy Sustain. 2 (2015) E8.

[17] E.D. Sloan, C.A. Koh, Clathrate Hydrates of Natural Gases, third ed., CRC Press, Taylor & Francis Group, Boca Raton, FL, 2007.

[18] D.M. Amos, M.E. Donnelly, P. Teeratchanan, C.L. Bull, A. Falenty, W.F. Kuhs, A. Hermann, J.S. Loveday, A chiral gas-hydrate structure common to the carbon dioxide-water and hydrogen-water systems, J. Phys. Chem. Lett. 8 (17) (2017) 4295–4299, https://doi.org/10.1021/acs.jpclett.7b01787.
[19] G.A. Jeffrey, in: J.L. Atwood, J.E.D. Davies, D.D. MacNichol (Eds.), Inclusion Compounds, Academic Press, 1984, p. 135.
[20] L.A. Lyusternik, Convex Figures and Polyhedral, Dover, New York, 1963.
[21] E.D. Sloan, Fundamental principles and applications of natural gas hydrates, Nature (2003), https://doi.org/10.1038/nature02135.
[22] T. Iwamoto, S.-i. Nishikiori, T. Kitazawa, H. Yuge, Mineralomimetic chemistry as a modern aspect of coordination chemistry, J. Chem. Soc. Dalton Trans. 22 (1997) 4127–4136, https://doi.org/10.1039/a702539d.
[23] J.F. Gabitto, C. Tsouris, Physical properties of gas hydrates: a review, J. Thermodyn. 2010 (2010) 1–12, https://doi.org/10.1155/2010/271291.
[24] Z.M. Aman, E.P. Brown, E. Dendy Sloan, A.K. Sum, C.A. Koh, Interfacial mechanisms governing cyclopentane clathrate hydrate adhesion/cohesion, Phys. Chem. Chem. Phys. 13 (2011) 19796–19806.
[25] G.D. Holder, P.F. Angert, Simulation of gas production from a reservoir containing both gas hydrates and free natural gas, in: SPE Annual Technical Conference and Exhibition, Society of Petroleum Engineers, 1982.
[26] T. Maekawa, Experimental study on isotopic fractionation in water during gas hydrate formation, Geochem. J. 38 (2) (2004) 129–138, https://doi.org/10.2343/geochemj.38.129.
[27] J.M. Schicks, M. Luzi-Helbing, Kinetic and thermodynamic aspects of clathrate hydrate nucleation and growth, J. Chem. Eng. Data 60 (2) (2015) 269–277, https://doi.org/10.1021/je5005593.
[28] E.D. Sloan, F. Fleyfel, A molecular mechanism for gas hydrate nucleation from ice, AICHE J. 37 (9) (1991) 1281–1292, https://doi.org/10.1002/aic.690370902.
[29] R.L. Christiansen, E.D. Sloan, Mechanisms and kinetics of hydrate formation, in: E.D. Sloan, J. Happel, M. A. Hnatow (Eds.), International Conference on Natural Gas Hydrates, vol. 715, New York Academy of Science, New Paltz, 1994, pp. 283–305.
[30] L.C. Jacobson, W. Hujo, V. Molinero, Amorphous precursors in the nucleation of clathrate hydrates, J. Am. Chem. Soc. 132 (33) (2010) 11806–11811, https://doi.org/10.1021/ja1051445.
[31] T.A. Arjun, Berendsen, and Peter G. Bolhuis., Unbiased atomistic insight in the competing nucleation mechanisms of methane hydrates, Proc. Natl. Acad. Sci. U. S. A. 116 (39) (2019) 19305–19310, https://doi.org/10.1073/pnas.1906502116.
[32] W. Ke, T.M. Svartaas, D. Chen, A review of gas hydrate nucleation theories and growth models, J. Nat. Gas Sci. Eng. (2019), https://doi.org/10.1016/j.jngse.2018.10.021.
[33] A. Hassanpouryouzband, E. Joonaki, M. Vasheghani Farahani, S. Takeya, C. Ruppel, J. Yang, et al., Gas Hydrates in Sustainable Chemistry, Chemical Society Reviews. Royal Society of Chemistry, 2020, https://doi.org/10.1039/c8cs00989a.
[34] M.R. Walsh, C.A. Koh, E.D. Sloan, A.K. Sum, D.T. Wu, Microsecond simulations of spontaneous methane hydrate nucleation and growth, Science 326 (5956) (2009) 1095–1098, https://doi.org/10.1126/science.1174010.
[35] R. Radhakrishnan, B.L. Trout, A new approach for studying nucleation phenomena using molecular simulations: application to CO_2 hydrate clathrates, J. Chem. Phys. 117 (4) (2002) 1786–1796, https://doi.org/10.1063/1.1485962.
[36] P.M. Rodger, Stability of gas hydrates, J. Phys. Chem. 94 (1990) 6080–6089.
[37] B. Kvamme, A new theory for kinetics of hydrate formation, in: Proceedings of the 2nd International Conference on Natural Gas Hydrates, June 2–6, Toulouse, 1996, pp. 139–146.
[38] X. Wang, A.J. Schultz, Y. Halpern, Kinetics of methane hydrate formation from polycrystalline deuterated ice, J. Phys. Chem. A 106 (32) (2002) 7304–7309.

[39] B. Beeskow-Strauch, J.M. Schicks, R. Naumann, J. Erzinger, The influence of SO_2 and NO_2 impurities on CO_2 gas hydrate formation and stability, Chem. Eur. J. 17 (2011) 4376–4384.

[40] G.J. Guo, M. Li, Y.G. Zhang, W. Chang Hua, Why can water cages adsorb aqueous methane? A potential of mean force calculation on hydrate nucleation mechanisms, Phys. Chem. Chem. Phys. 11 (44) (2009) 10427–10437, https://doi.org/10.1039/b913898f.

[41] K.W. Hall, S. Carpendale, P.G. Kusalik, Evidence from mixed hydrate nucleation for a funnel model of crystallization, Proc. Natl. Acad. Sci. U. S. A. 113 (43) (2016) 12041–12046, https://doi.org/10.1073/pnas.1610437113.

[42] D. Elwell, H.J. Scheel, Crystal Growth from High-Temperature Solutions, Academic Press, 1975.

[43] Z.M. Aman, C.A. Koh, Interfacial Phenomena in Gas Hydrate Systems, Chemical Society Reviews. Royal Society of Chemistry, 2016, https://doi.org/10.1039/c5cs00791g.

[44] P. Englezos, N. Kalogerakis, P.D. Dholabhai, P.R. Bishnoi, Kinetics of formation of methane and ethane gas hydrates, Chem. Eng. Sci. 42 (11) (1987) 2647–2658.

[45] N. Ma, J. Li, K. Luo, S. Liu, M. Wen, Methane hydrate generation model and software development based on P. Englezos method, in: Mechanisms and Machine Science, vol. 75, Springer, 2020, pp. 693–706, https://doi.org/10.1007/978-3-030-27053-7_59.

[46] M.L.M. De Banos, N. Hobeika, P. Bouriat, D. Broseta, E. Enciso, F. Clément, R. Brown, How do gas hydrates spread on a substrate? Cryst. Growth Des. 16 (8) (2016) 4360–4373, https://doi.org/10.1021/acs.cgd.6b00471.

[47] J.G. Beltrán, P. Servio, Morphological investigations of methane-hydrate films formed on a glass surface, Cryst. Growth Des. 10 (2010) 4339–4347, https://doi.org/10.1021/cg1003098.

[48] C.A. Koh, A.K. Sum, E.D. Sloan, Gas hydrates: unlocking the energy from icy cages, J. Appl. Phys. 106 (2009), https://doi.org/10.1063/1.3216463.

[49] D. Turner, K. Miller, E. Sloan, Methane hydrate formation and an inward growing shell model in water-in-oil dispersions, Chem. Eng. Sci. 64 (2009) 3996–4004.

[50] Z.M. Aman, K. Pfeiffer, S.J. Vogt, M.L. Johns, E.F. May, Corrosion inhibitor interaction at hydrate-oil interfaces from differential scanning calorimetry measurements, Colloids Surf. A Physicochem. Eng. Asp. 448 (1) (2014) 81–87, https://doi.org/10.1016/j.colsurfa.2014.02.006.

[51] J.A. Boxall, S.R. Davies, J.W. Nicholas, et al., Hydrate blockage potential in an oil-dominated system studied using a four inch flow loop, in: Proceedings of the Fourth International Conference of Gas Hydrates. Vancouver, British Columbia, Canada, July 6 – 10, 2008.

[52] S.V. Joshi, G.A. Grasso, P.G. Lafond, et al., Experimental flowloop investigations of gas hydrate formation in high water cut systems, Chem. Eng. Sci. 97 (2013) 198.

[53] I. Rao, E.D. Sloan, C.A. Koh, A.K. Sum, Laboratory experiments and modeling for hydrate formation and deposition from water saturated gas systems, in: Proceedings of the 7th International Conference on Gas Hydrates; Edinburgh, U.K., July 17 – 21, 2011.

[54] L. Ding, B. Shi, X. Lv, Y. Liu, W. Haihao, W. Wang, J. Gong, Hydrate formation and plugging mechanisms in different gas-liquid flow patterns, Ind. Eng. Chem. Res. 56 (14) (2017) 4173–4184, https://doi.org/10.1021/acs.iecr.6b02717.

[55] Y. Bai, Q. Bai, Hydrates, in: Subsea Engineering Handbook, Elsevier, 2019, pp. 409–434, https://doi.org/10.1016/b978-0-12-812622-6.00015-4.

[56] J.D. Clegg, Production operations engineering IV, in: Petroleum Engineering Handbook, 2007.

[57] D. Sloan, Introduction, in: Natural Gas Hydrates in Flow Assurance, Elsevier Inc., 2011, pp. 1–11, https://doi.org/10.1016/B978-1-85617-945-4.00001-7.

[58] X. Lv, D. Lu, Y. Liu, S. Zhou, J. Zuo, H. Jin, et al., Study on methane hydrate formation in gas-water systems with a new compound promoter, RSC Adv. 9 (57) (2019) 33506–33518, https://doi.org/10.1039/c9ra06467b.

[59] R.E. Meindinyo, Tariyemienyo, and Thor Martin Svartaas., Gas hydrate growth kinetics: a parametric study, Energies 9 (12) (2016), https://doi.org/10.3390/en9121021.
[60] P. Zhang, W. Qingbai, M. Cuicui, Influence of temperature on methane hydrate formation, Sci. Rep. 7 (1) (2017), https://doi.org/10.1038/s41598-017-08430-y.
[61] J.A. Prajaka, J. Himawan, S.A. Affandy, J.P. Sutikno, R. Handogo, Hydrate formation calculation in the natural gas purification unit, in: IOP Conference Series: Materials Science and Engineering, vol. 543, Institute of Physics Publishing, 2019, https://doi.org/10.1088/1757-899X/543/1/012084.
[62] J.J. Rajnauth, M.A. Barrufet, G. Falcone, Hydrate Formation: Considering the effects of Pressure, Temperature, Composition and Water, Society of Petroleum Engineers, 2010, https://doi.org/10.2118/131663-MS.
[63] J. Rajnauth, M. Barrufet, G. Falcone, Hydrate formation: considering the effects of pressure, temperature, composition and water, Energy Sci. Technol. 4 (1) (2012) 60–67, https://doi.org/10.3968/j.est.1923847920120401.397.
[64] B. Guo, X. Liu, X. Tan, Gas hydrate control, in: Petroleum Production Engineering, Elsevier, 2017, pp. 649–662, https://doi.org/10.1016/b978-0-12-809374-0.00020-9.
[65] D.C. Salmin, H. Stoner, D.T. Wu, L.E. Zerpa, C.A. Koh, K. Mateen, S. Bodnar, P. Prince, A. Teixeira, Effect of alcohols on asphaltene-particle size and hydrate non-plugging behavior of crude oils, in: J.G. Delgado-Linares (Ed.), Paper Presented at the Offshore Technology Conference, Houston, Texas, USA, May, 2020, https://doi.org/10.4043/30916-MS.
[66] D. Costa Salmin, J. Delgado-Linares, D. Wu, et al., Hydrate agglomeration in crude oil systems in which the asphaltene aggregation state is artificially modified, in: Paper Presented at the Offshore Technology Conference, Houston, Texas, 6–9 May. OTC-29484-MS, 2019, https://doi.org/10.4043/29484-MS.
[67] D. Zhang, Q. Huang, R. Li, W. Wang, X. Zhu, H. Li, Y. Wang, Effects of waxes on hydrate behaviors in water-in-oil emulsions containing asphaltenes, Chem. Eng. Sci. 244 (2021), https://doi.org/10.1016/j.ces.2021.116831.
[68] H.M. Zheng, Q.Y. Huang, W. Wang, Z. Long, P.G. Kusalik, induction time of hydrate formation in water-in-oil emulsions, Ind. Eng. Chem. Res. 56 (2017) 8330–8339, https://doi.org/10.1021/acs.iecr.7b01332.
[69] W. Wang, Q. Huang, H. Zheng, Q. Wang, D. Zhang, X. Cheng, R. Li, Effect of wax on hydrate formation in water-in-oil emulsions, J. Dispers. Sci. Technol. (2019), https://doi.org/10.1080/01932691.2019.1637751.
[70] X.F. Song, F. Xin, H.C. Yan, X.A. Li, H.R. Jia, intensification and kinetics of methane hydrate formation under heat removal by phase change of N-tetradecane, AICHE J. 61 (2015) 3441–3450, https://doi.org/10.1002/aic.14867.
[71] Y.S.B.H. Liu, Y. Zhang, L. Ding, Y. Yong, Q.L. Ma, W. Wang, J. Gong, Investigation on the effect of wax on hydrate formation characteristics and flow properties in a high pressure flow loop, in: AIChE Annual Meeting, Minneapolis, MN, 2017.
[72] M.A. Mahabadian, A. Chapoy, R. Burgass, B. Tohidi, Mutual effects of paraffin waxes and clathrate hydrates: a multiphase integrated thermodynamic model and experimental measurements, Fluid Phase Equilib. 427 (2016) 438–459, https://doi.org/10.1016/j.fluid.2016.08.006.
[73] J. Carroll, Water content of natural gas, in: Natural Gas Hydrates, Elsevier, 2020, pp. 307–345, https://doi.org/10.1016/b978-0-12-821771-9.00010-0.
[74] A. Bahadori, Natural gas dehydration, in: Natural Gas Processing, Elsevier, 2014, pp. 441–481, https://doi.org/10.1016/b978-0-08-099971-5.00009-x.
[75] D.Q. Li, M.Y. Ai, Y.B. Wang, et al., Hydrate accident and prevention in sebei-xining-lanzhou gas pipeline, Oil Gas Storage Transport. 31 (4) (2012) 267–269.
[76] C.Y. Sun, Q. Huang, G.J. Chen, Progress of thermodynamics and kinetics of gas hydrate formation, J. Chem. Ind. Eng. (China) 5 (2006) 1031–1039.
[77] A. Hernández, Gas properties, in: Fundamentals of Gas Lift Engineering, Elsevier, 2016, pp. 1–23, https://doi.org/10.1016/b978-0-12-804133-8.00001-4.

[78] J.G. Speight, Gas Engineering: Vol. 1: Origin and Reservoir Engineering, De Gruyter, 2021, ISBN: 9783110690897.
[79] K.-S. Kim, J.W. Kang, S.-P. Kang, Tuning ionic liquids for hydrate inhibition, Chem. Commun. 47 (2011) 6341–6343.
[80] Y. Qi, W. Wu, Y. Liu, Y. Xie, X. Chen, The influence of NaCl ions on hydrate structure and thermodynamic equilibrium conditions of gas hydrates, Fluid Phase Equilib. 325 (2012) 6–10.
[81] Y. Zhong, R.E. Rogers, Surfactant effects on gas hydrate formation, Chem. Eng. Sci. 55 (19) (2000) 4175–4187, https://doi.org/10.1016/S0009-2509(00)00072-5.
[82] K. Inkong, P. Rangsunvigit, S. Kulprathipanja, Effects of mixed surfactants on methane hydrate formation and dissociation, Chem. Eng. Trans. 52 (2016) 151–156, https://doi.org/10.3303/CET1652026.
[83] A. Kumar, G. Bhattacharjee, B.D. Kulkarni, R. Kumar, Role of surfactants in promoting gas hydrate formation, Ind. Eng. Chem. Res. 54 (49) (2015) 12217–12232, https://doi.org/10.1021/acs.iecr.5b03476.
[84] W. Lin, G.J. Chen, C.Y. Sun, X.Q. Guo, Z.K. Wu, M.Y. Liang, L.T. Chen, L.Y. Yang, Effect of surfactant on the formation and dissociation kinetic behavior of methane hydrate, Chem. Eng. Sci. 59 (21) (2004) 4449–4455, https://doi.org/10.1016/j.ces.2004.07.010.
[85] P. Di Profio, S. Arca, R. Germani, G. Savelli, Surfactant promoting effects on clathrate hydrate formation: are micelles really involved? Chem. Eng. Sci. 60 (2005) 4141–4145.
[86] J.S. Zhang, S. Lee, J.W. Lee, Kinetics of methane hydrate formation from SDS solution, Ind. Eng. Chem. Res. 46 (19) (2007) 6353–6359, https://doi.org/10.1021/ie070627r.
[87] D. Bai, G. Chen, X. Zhang, A.K. Sum, W. Wang, How properties of solid surfaces modulate the nucleation of gas hydrate, Sci. Rep. 5 (2015), https://doi.org/10.1038/srep12747.
[88] N.N. Nguyen, M. Galib, A.V. Nguyen, Critical review on gas hydrate formation at solid surfaces and in confined spaces—why and how does interfacial regime matter? Energy Fuel 34 (6) (2020) 6751–6760, https://doi.org/10.1021/acs.energyfuels.0c01291.
[89] Y. Li, M. Chen, H. Song, P. Yuan, D. Liu, B. Zhang, H. Bu, Methane hydrate formation in the stacking of kaolinite particles with different surface contacts as nanoreactors: a molecular dynamics simulation study, Appl. Clay Sci. 186 (2020), 105439.
[90] J. Javanmardi, K. Nasrifar, S.H. Najibi, M. Moshfeghian, Economic evaluation of natural gas hydrate as an alternative for natural gas transportation, Appl. Therm. Eng. 25 (2005) 1708–1723.
[91] W. Hao, J. Wang, S. Fan, W. Hao, Evaluation and analysis method for natural gas hydrate storage and transportation processes, Energy Convers. Manag. 49 (2008) 2546–2553.
[92] H. Dashti, L.Z. Yew, X. Lou, Recent advances in gas hydrate-based CO_2 capture, J. Nat. Gas Sci. Eng. 23 (2015) 195–207.
[93] P. Babu, A. Nambiar, T. He, I.A. Karimi, J.D. Lee, P. Englezos, P. Linga, A review of clathrate hydrate based desalination to strengthen energy-water nexus, ACS Sustain. Chem. Eng. 6 (7) (2018) 8093–8107, https://doi.org/10.1021/acssuschemeng.8b01616.
[94] A. Eslamimanesh, A.H. Mohammadi, D. Richon, P. Naidoo, D. Ramjugernath, Application of gas hydrate formation in separation processes: a review of experimental studies, J. Chem. Thermodyn. 46 (2012) 62–71, https://doi.org/10.1016/j.jct.2011.10.006.
[95] G. Ali Mansoori, Diamondoid molecules, Adv. Chem. Phys. 136 (2007) 207–258, https://doi.org/10.1002/9780470175422.ch4.
[96] S. Landa, V. Machacek, M. Mzourek, M. Landa, Chim. Ind. (Publ. No. 506), 1933.
[97] A. Ma, Advancement in application of diamondoids on organic geochemistry, J. Nat. Gas Geosci. 1 (4) (2016) 257–265, https://doi.org/10.1016/j.jnggs.2016.09.001.
[98] T. Gaskin, Removing diamondoid components from natural gas at reduced temperatures, 2005. US Patent Number US20060111602A1.

[99] M. Schoell, R.M.K. Carlson, Diamondoids and oil are not forever, Nature (1999), https://doi.org/10.1038/19847.

[100] G.D. Holder, R.M. Enick, J.W. Jones, J.A. Reiser, C.R. LeBlond, Dynamic Gas Hydrate Formation and Properties of Diamondoids, Society of Petroleum Engineers, 1993, https://doi.org/10.2118/26164-MS.

[101] J.E. Dahl, S.G. Liu, R.M.K. Carlson, Isolation and structure of higher diamondoids, nanometer-sized diamond molecules, Science 299 (5603) (2003) 96–99.

[102] Y. Zhou, A.D. Brittain, D. Kong, M. Xiao, Y. Meng, L. Sun, Derivatization of diamondoids for functional applications, J. Mater. Chem. C (2015), https://doi.org/10.1039/c5tc01377a.

[103] M. Baumgartner, Electronic and Spatial Structures of Self-Assembled Diamondoid Monolayers, Master-Thesis, Faculty of Science, Physik-Institut (University Of Zurich), 2016.

[104] R.M.K. Carlson, J.E.P. Dahl, S.G. Liu, M.M. Olmstead, P.R. Buerki, R. Gat, Diamond molecules found in petroleum, in: Synthesis, Properties and Applications of Ultrananocrystalline Diamond, 2005, https://doi.org/10.1007/1-4020-3322-2_6.

[105] G. Ali Mansoori, P.L.E. de Araujo, E.S. de Araujo, Molecular Structure and Chemistry of Diamondoids, in Diamondoid Molecules: With Applications in Biomedicine, Materials Science, Nanotechnology & Petroleum Science, World Sci Pub Co, 2012.

[106] R. Lin, Z.A. Wilk, Natural occurrence of tetramantane (C22H28), pentamantane (C26H32) and hexamantane (C30H36) in a deep petroleum reservoir, Fuel 74 (10) (1995) 1512–1521.

[107] D. Araujo, P.L. Barros, G. Ali Mansoori, E.S. De Araujo, Diamondoids: occurrence in fossil fuels, applications in petroleum exploration and fouling in petroleum production. A review paper, Int. J. Oil Gas Coal Technol. (2012), https://doi.org/10.1504/IJOGCT.2012.048981.

[108] D. Vazquez, G.A. Mansoori, Identification and measurement of petroleum precipitates, J. Pet. Sci. Eng. 26 (2000) 49–55.

[109] G.A. Mansoori, Modeling of asphaltene and other heavy organic depositions, J. Pet. Sci. Eng. 17 (1997) 101–111.

[110] K.W. Yeung, Y. Dong, L. Chen, C.Y. Tang, W.C. Law, G.C.P. Tsui, Nanotechnology of diamondoids for the fabrication of nanostructured systems, Nanotechnol. Rev. (2020), https://doi.org/10.1515/ntrev-2020-0051.

CHAPTER

Wax Deposition

7

7.1 Introduction

Waxes are long-chain alkane hydrocarbons that are solid at low to moderate temperatures. They are conventionally called paraffins, waxes, or paraffin waxes by oil and gas operators [1]. For years, the term paraffin has been loosely defined for petroleum productions as anything that inhibits the production of crude, including rust, scale, sand, dirt, ice emulsion, wax, and bent or kinked line [2]. The term has been used in the United States to describe oilfield waxes, although "paraffin wax" is a better terminology [3]. Therefore you might see these terms used interchangeably throughout this chapter; they all refer to the same thing.

Paraffin waxes deposit in oil and gas facilities when the temperature of the flowing stream drops below a specific limit known as the wax appearance temperature (WAT), causing major flow assurance and production troubles. Wax gelation changes the crude rheology, causing flowing problems during operations, shutdowns, and system restart. In addition, wax deposition causes major operational problems, including formation damage, tubing and flowlines blockages, and impairs other production equipment. In the extreme cases, wax can lead to field abandonment, as in the case of UK Lasmo, which abandoned and withdrew its platform in November 1994 due to repeated wax blockage. The US Minerals Management Service published 51 severe wax-related reports between 1992 and 2002 in Gulf of Mexico flowlines [4]. Fig. 7.1 shows wax deposit formation from flowlines during pigging operations in the Egyptian oilfield, and Fig. 7.2 illustrates the composition of deposit samples from a pipeline over 7 months showing the high paraffin content.

Therefore it is of great importance to understand the mechanisms of formation of wax deposits and the factors affecting them in order to be able to proactively counteract this problem. In the next sections, the composition and structure of paraffin waxes are addressed; following that, the mechanisms and factors affecting paraffin wax deposition are examined.

7.2 Paraffin wax composition and structure

Waxes, paraffin waxes, or paraffins are high-molecular-weight long-chain, normal, or branched alkanes with carbon numbers > C15, which are naturally present in crude oils and some condensates and can build up as deposits in the wellbore, feed lines, etc. It worth mentioning that the designation of the carbon numbers in waxes varies in the literature; for example, you can find the wax carbon

Essentials of Flow Assurance Solids in Oil and Gas Operations. https://doi.org/10.1016/B978-0-323-99118-6.00023-X

FIG. 7.1

Paraffin wax deposits from pipelines after pigging.

number is specified as >C15, >C18, >C20, C15–C75, C16–C40, C18–C36, C18–C50, C18–C60, or C20–C80, and others [3,5–10]. This reveals the wide range of paraffin molecules in crude oils and the diversity of crude oils composition. Paraffin wax consists mostly of normal paraffin content (80%–90%), while the rest consists of branched paraffins (isoparaffins) and cycloparaffins [10]. Fig. 7.3 shows the structure of some linear and branched paraffins. The general formula of alkane paraffin

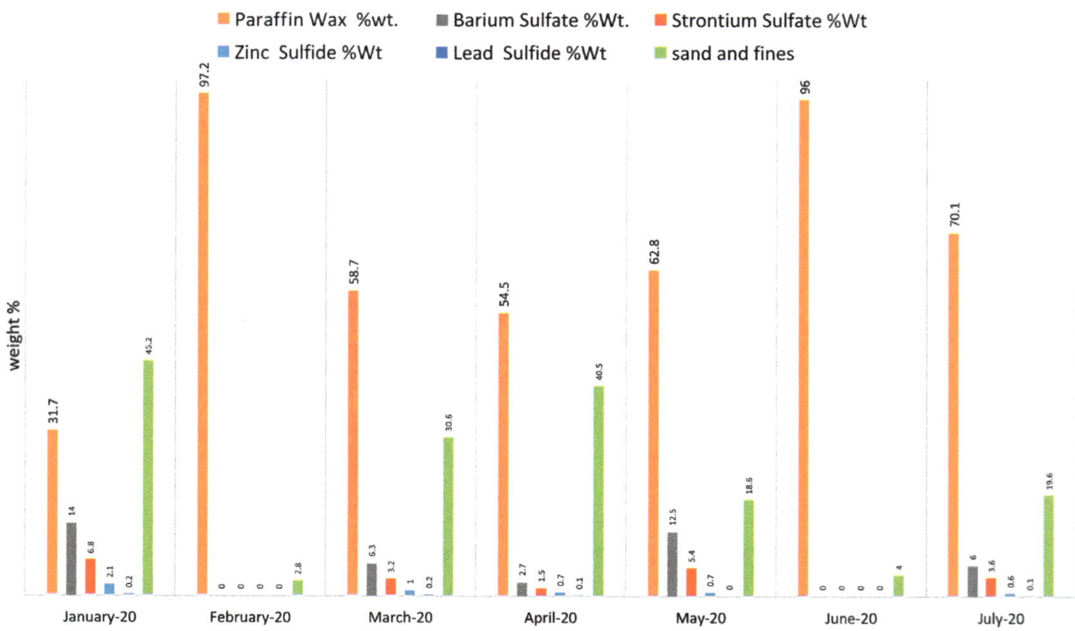

FIG. 7.2

Composition of deposits from 36″ pipeline over 7 months period showing high paraffin content.

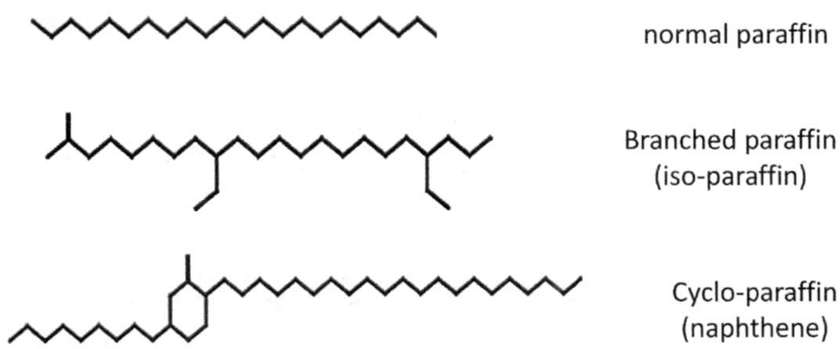

FIG. 7.3

Structure of paraffins.

wax's hydrocarbons is C_nH_{2n+2}. The degree of branching has an important influence on the properties [11]. Pipeline wax deposition and plugging are predominantly caused by normal alkanes (*n*-paraffin waxes) deposition [3]. It is worth mentioning that paraffin wax is not wax in the true sense of the word: true wax is defined as an ester of a long-chain even-numbered fatty acid usually found in nature, while paraffin waxes are hydrocarbon derivatives, mixtures of alkane derivatives usually in a homologous series of chain lengths [2,11].

Waxes can be divided into two groups: macrocrystalline and microcrystalline. The macrocrystalline waxes are composed mainly of paraffins with normal chains (n-alkanes), with carbon numbers between 18 and 30, and melting points ranging between 40°C and 60°C. Microcrystalline paraffins, on the other hand, contain a high content of branched and cyclical paraffins, with chain carbon numbers between 30 and 60, and melting point >60–90°C [12,13].

The microcrystalline waxes can be distinguished operationally from the macrocrystalline waxes. The microcrystalline waxes usually contribute to tank-bottom sludges, while the macrocrystalline waxes cause flow assurance problems in production and transportation operations [14,15]. Microcrystalline wax deposits are usually softer than deposits formed from macrocrystalline waxes. Consequently, wax deposits comprising mainly microcrystalline wax are easier to remediate than the deposits comprising macrocrystalline wax with the same carbon number. It is thus crucial to distinguish microcrystalline wax from macrocrystalline wax during deposit analysis (using modified SARA analysis), for better designing and application of wax treatments and pigging operations [8].

There are three types of wax crystals, namely plate crystals, needle crystals, and amorphous crystals. Needle and plate crystals are formed from paraffinic oils, whereas amorphous crystals usually precipitate from asphaltenic oils. The degree of contact or interaction between crystals and pipeline internal walls increases from amorphous to needle to plates. Thus amorphous crystals are lower in risk during operations than the other two types [16].

The nature and occurrence of paraffin wax deposits have been studied extensively since 1865, and it was 1932 when one of the famous reports written by Reistle and Blade [17] became a classic reference for wax precipitation and deposition [2].

Wax problems can occur with both oil systems (heavy and light) and gas condensates. Wax in crude oils can be up to 32.5%, and indeed there are some condensate reservoirs with wax contents as high as 30% [1,10]. Waxes in crudes are usually harder to control than those in condensates, as they are of longer-chain alkanes [3].

A typical paraffin deposit mainly contains paraffin (normal or branched) besides other different components such as oil, water, other solids, e.g., sand, scale, and corrosion products. Paraffin wax deposits vary in consistency from soft accumulations to hard, brittle deposits. Usually the deposits are firmer and harder as the molecular weight of the deposit increases [6].

7.3 Why do waxes tend to come out of the solution and precipitate?

Waxes deposit as a consequence of temperature and pressure drop during production. In the reservoir, at high temperature and pressures, waxes are dissolved in the solution. As the crude oil temperature drops during its journey from the reservoir traveling towards surface processing facilities, wax begins to precipitate from the crude oil when the temperature drops below WAT. In addition, as the pressure drops during production, loss of low molecular weight hydrocarbons (light ends) to the gas phase reduces the solubility of the waxes in the oil [3]. Paraffin waxes that have the highest melting point and molecular weight separate from the solution and precipitate first, while the lower molecular weight waxes separate as the temperature decreases further. For instance, a C60 alkane with a melting point of about 211°F will deposit first and at a much higher temperature than a C20 alkane with a melting point of 98°F [6].

Generally, the higher the melting point, the higher the tendency of the paraffin to deposit when the temperature drops. The melting point of the paraffin waxes increases as the size of the molecule increases. Table 7.1 summarizes the melting temperatures of some alkanes.

Another way of explaining the driving force of precipitation of paraffin wax is through understanding the surface tension concept. Fig. 7.4 illustrates the structure of a paraffin crystal. Think of the edges of the crystal as tightly packed methyl groups CH_3. The surface tension of these methyl groups is about 22–24 dynes/cm. When compared to the surface tension of oil at 33–35 dynes/cm and water at 72 dynes/cm, once finds that at the low temperatures the crystal is both hydrophobic and oleophobic. This means that the paraffin crystal does not want to go with either the water or the oil. Thus it preferentially precipitates from the solution, adhering to the other surfaces [2].

Principally, the wax deposition is a phase transition process formation and conceptually is similar to the crystallization process. Wax crystallization is an exothermic process, which emits a small amount of heat. Likewise, dissolving these crystals is an endothermic process that requires a certain amount of

Table 7.1 Melting points for different paraffins.

Alkanes	Melting point (°F)
C10	−21.5
C16	64.7
C18	82.9
C20	98.1
C23	117.7
C32	158
C40	179.6
C60	211

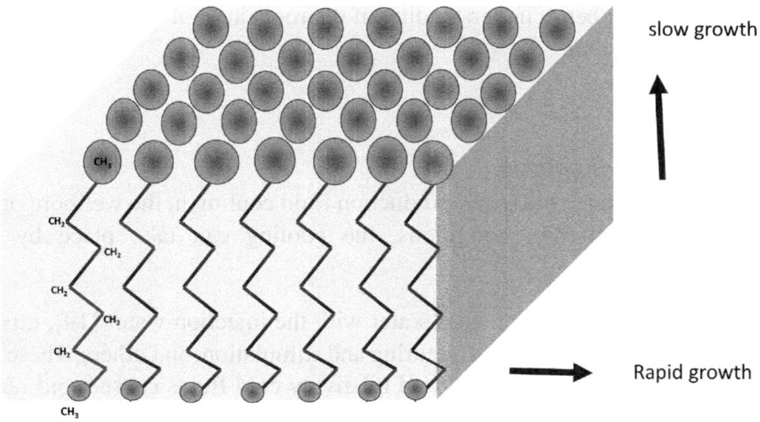

FIG. 7.4

Structure and growth of wax crystal.

heat to dissolve [18]. The crystallization process involves the production of solid ordinate structure from a disordered phase, which typically occurs in thermodynamically unstable solutions.

7.4 Mechanism of wax deposit formation

The precipitation and deposition of paraffin waxes is a complex multistep process. As stated earlier, the hydrocarbons have to cool down below specific temperature (WAT) before the precipitation process begins. Therefore the comprehensive deposition process of paraffin waxes comprises two main steps: the cooling of the fluids, and precipitation and deposition of the paraffins from the solution (Fig. 7.5). Most of the literature focuses on the precipitation and deposition steps without elaborating on the fluid cooling step and the mechanisms included. Therefore it is relevant to include fluid cooling in the mechanism, as it is a critical initial step in wax precipitation, and various mechanisms of fluid cooling must

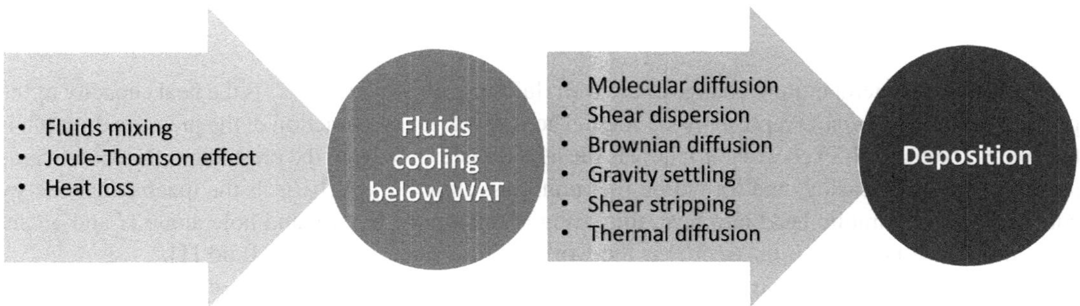

FIG. 7.5

The general mechanism of wax deposition.

be addressed to the reader for better understanding of the root cause of wax deposition in the different locations of a production system.

7.4.1 Fluid cooling

7.4.1.1 Fluid cooling mechanisms

Many mechanisms can lead to heat loss and production fluid cooling in the wellbore or on the topsides during operations. Under different conditions, the cooling can take place by three different mechanisms:

- Mixing of fluids: The mixing of formation water with the injection water [19], mixing of CO_2 gas with the reservoir fluids [20], acidizing, fracturing and stimulation, and others. These types of mixing between high temperature reservoir fluids and relatively cold fluids cause rapid reduction of temperature that can trigger paraffin wax deposition.
- Adiabatic decompression (throttling) when pressure on the fluid approaching the surface changes over a short distance from lithostatic to hydrostatic [21]. The effect, which is also known as the "Joule-Thomson effect" (J-T effect), often causes a temperature decrease as gas flows through pores of a reservoir to the wellbore; it is also noticed after valves, pumps, and other equipment or facilities that encounter throttling and pressure drop. Jamaloei and Asghari [22,23] reviewed Joule-Thomson adiabatic compression and its effects on the production operations. The same effect is expected to happen when dissolved gases are released from liquids. It is worth mentioning that the Joule-Thomson effect can lead to heating or cooling. The J-T coefficient is zero for an ideal gas, and may be either positive or negative for a nonideal gas, depending on both the temperature and pressure. Positive values will lead to cooling on gas expansion while negative values will lead to heating. Typically the J-T coefficient is calculated from equation of state (EoS) models such as Peng-Robinson or Soave-Redlich-Kwong [1].
- Heat loss to the walls of the pipe: several means of heat loss can happen to the produced fluids including conductive heat loss, radiative heat loss, and convective and forced convective heat loss.

Fig. 7.6 summarizes the different cooling mechanisms of the produced fluids in a producing well [1].

In production wells, by ignoring the vertical heat transfer away from the production fluids and fluid condensation/evaporation, the energy balance for a producer is given by Eq. (7.1) [1]:

$$\frac{dT_f}{dz} = C_j\frac{dp}{dz} + \frac{1}{c_p}\left(\frac{Q}{w} - \frac{g\sin\theta}{Jg_c} - \frac{v}{Jg_c}\frac{dv}{dz}\right) \tag{7.1}$$

where dT_f/dz is the temperature change of the well fluids per unit of length, C_p is the heat capacity of the production fluid, $Cj(dp/dz)$ represents the Joule-Thomson effect as a function of the pressure drop (Cj is the Joule-Thomson (J-T) coefficient); Q/w is the heat transfer rate from the production fluids outwards through the annuli, casing, and formation (per unit mass flow); $(g \sin \theta)/Jg_c$ is the mechanical energy conversion for hydraulic head into heat, this being a function of gravity and hole angle (J and gc are conversion factors); and $(v/Jg_c)(dv/dz)$ is the work done by accelerating the fluid [1].

In a pipeline, the temperature as illustrated in Fig. 7.7, is generally expressed by Eq. (7.2) [24]:

$$T_2 = T_u + (T_1 - T_u)\exp\left[\frac{-U\pi d}{mC_p}L\right] \tag{7.2}$$

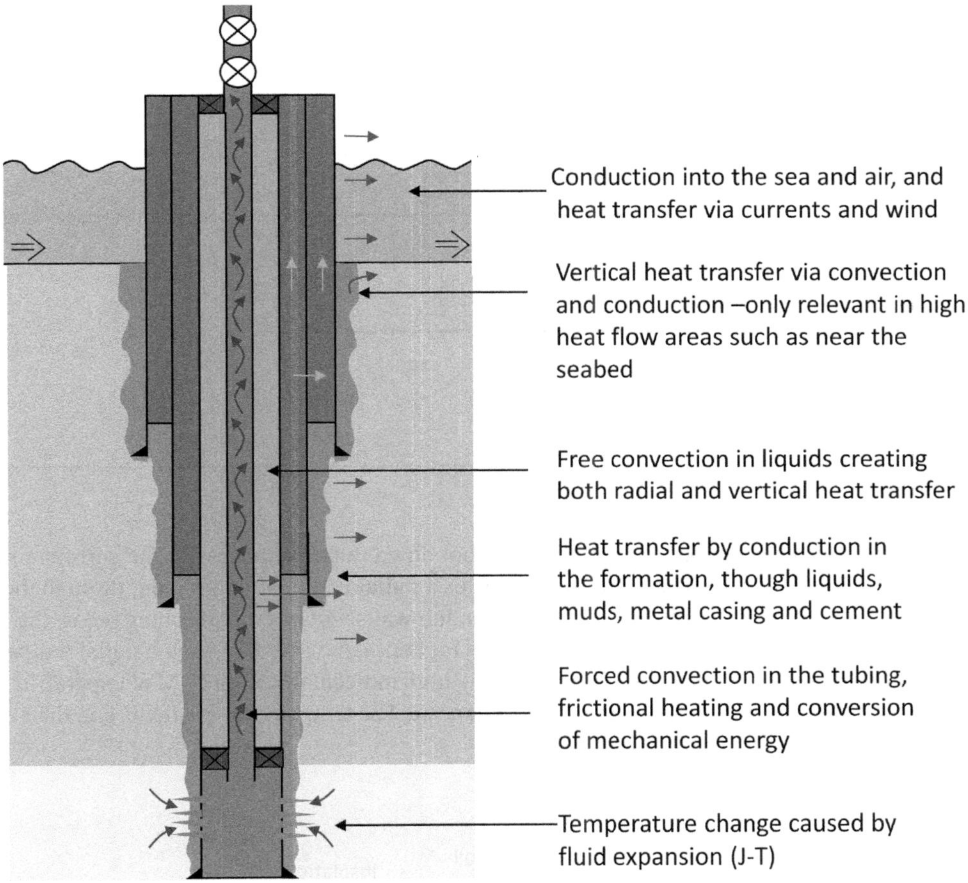

FIG. 7.6

Heat transfer mechanisms in a production well.

Reprinted from J. Bellarby, Well Completion Design, Developments in Petroleum Science, 2009, copyright (2009) with permission from Elsevier.

where T_2 is the fluid temperature at the pipe outlet (°F), T_u is the constant seawater temperature at the outside of the pipeline (°F), T_1 is the fluid inlet temperature (°F), L is the length of the pipe (ft), d is the pipe diameter (ft), m is the mass rate flowing fluid inside the pipe, and C_p is the heat capacity fluid ($\mathrm{BTU \cdot °F^{-1}\ lb^{-1}}$). U is the heat transmission coefficient ($\mathrm{Btu/(ft^2\ h\ °F)}$) and a large U value leads to a rapid cooling [24]. The radial temperature profile of a pipeline is also illustrated in Fig. 7.8.

By any of the previously mentioned cooling mechanisms or a combination of them, the temperature of the flowing fluids decreases down to the WAT, and this is when the first wax crystal starts to precipitate. Besides precipitation, waxes gel in crude oil and thicken to the point that crude oil stops flowing when it reaches another temperature limit known as the pour point (PP).

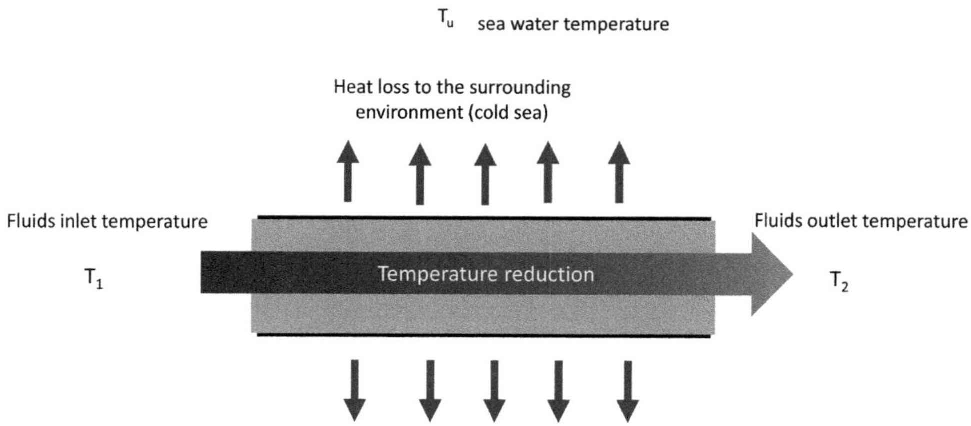

FIG. 7.7

The temperature along the pipeline.

The hot fluids produced from the well slowly cool down due to heat loss to the surroundings. In addition, the cooling effect of the conduit walls creates a radial temperature gradient through the pipe, where the temperature of the fluids adjacent to the conduit walls will be cold, reaching below the WAT, while those in the bulk might still be at temperatures higher than the WAT. Such a radial temperature gradient, as we will see, is responsible for transporting high molecular weight (HMWT) paraffin waxes from the bulk to the pipe wall and causing them to deposit. The temperature gradients and their effects on wax deposition are illustrated in Fig. 7.9.

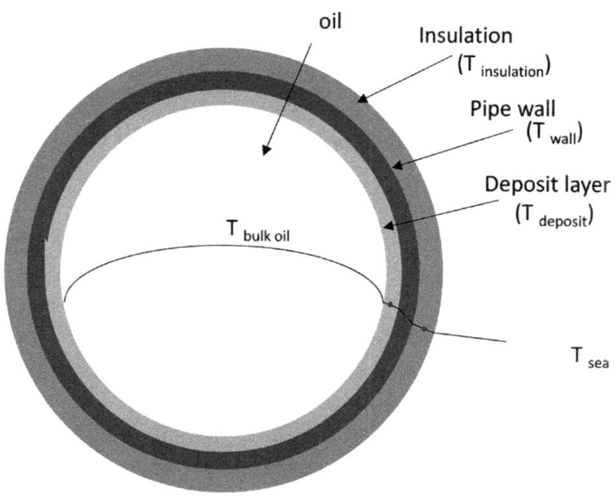

FIG. 7.8

Radial temperature profile in a pipeline.

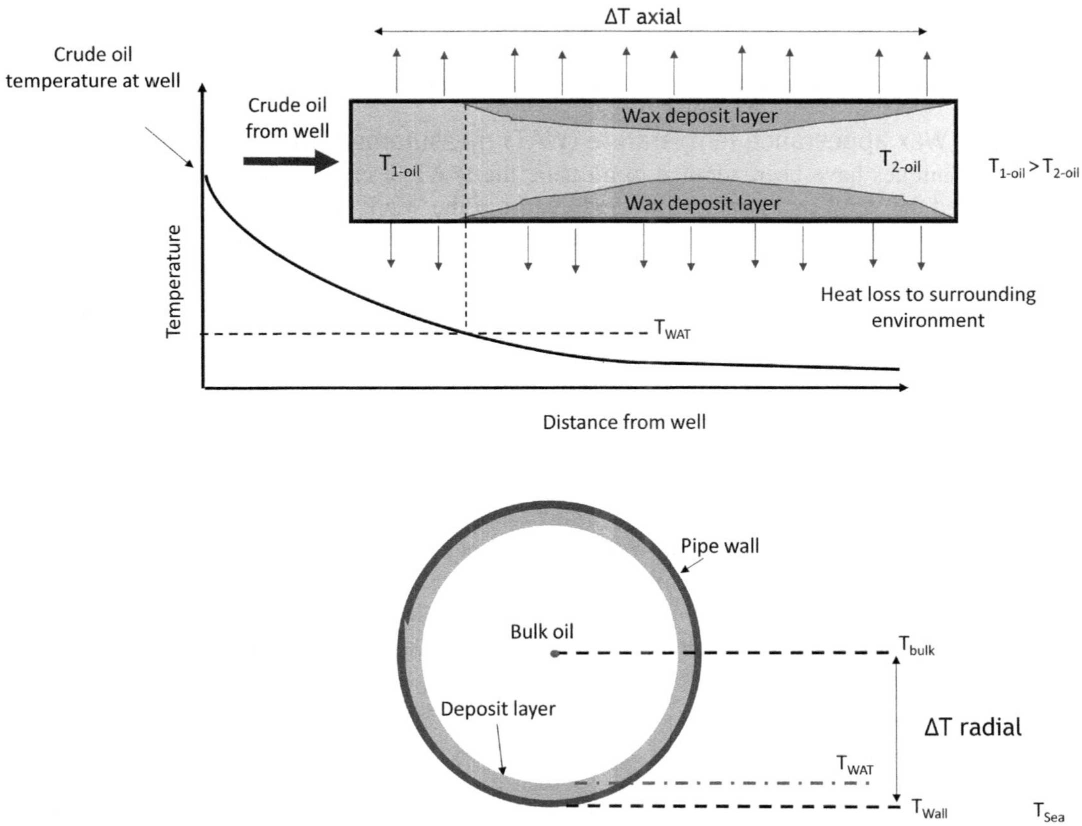

FIG. 7.9

Temperature gradients and their effects on wax deposition.

Since WAT and PP are temperature terms that are related to crude oil cooling, they are discussed in more detail in the following sections.

7.4.1.2 Wax appearance temperature (WAT)

The WAT usually refers to the onset temperature at which the precipitation of wax is observed or, in other words, the temperature at which the first wax crystals begin to precipitate from the crude oil [3,8]. WAT is also known as the cloud point (CP).

Crude oil WAT is a crucial parameter for optimizing the different production operations. It is also important for designing offshore drilling facilities and operations [8]. The WAT can be as high as 50°C for some oils (higher WAT values, e.g. ~66°C), have also been reported [25]. WAT depends on the pressure, oil composition (in particular, the concentration of light ends, asphaltenes surface), and pressure (bubble point) [3].

Surprisingly, two light crude oils with similar chemical composition, e.g., saturates, aromatics, resins, and asphaltenes, can have WAT values that differ by as much as 15°C or more. For live oils, the effect of pressure (amount of dissolved gas) on the WAT can be significant [3].

7.4.1.2.1 Wax appearance temperature (WAT) measurement methods

Various techniques have been adopted to measure the WAT in crude oils. Coutinho and Daridon [26] reviewed the WAT measurement methods. Most of the WAT measurements utilize the change in the physical properties of the oil due to phase transformation during the formation of solid wax crystals. In all techniques, wax precipitation is induced by a controlled cooling of the oil sample, in order for sufficient amount of solid to be present to produce a detectable signal [26]. An optimum WAT measurement technique should detect the very first few crystals that form at a temperature just below and close to the WAT. However, two main problems are associated with the methods currently used:

- No available technique is able to practically detect the very first crystal to appear in a representative fluid sample in thermal equilibrium, so these techniques cannot detect the exact onset of wax precipitation or the thermodynamic WAT, which lies on the solid–liquid phase boundary [8,26].
- Most of these techniques measure the WAT in dead crude oil, where the effects of the pressure and amount of dissolved gas in the live oil are not taken into account [27]. The light ends in live crude often have a bubble point below atmospheric pressure. The loss of light ends upon depressurization was found to increase the WAT. The value of WAT increases by 10°C in North Sea gas condensate due to depressurization and the loss of light ends [27].

- **Visual inspection method**

This technique is based on the fact that the precipitated solid wax crystals can scatter light, so the oil–wax suspension appears opaque. WAT/CP in this technique is determined to be the temperature at which cloudiness of the oil due to wax precipitation can be first observed. Standard testing methods based on this technique, such as ASTM-D2500 and ASTM -D3117, are commonly used [8].

- **Cross polarized microscope (CPM)**

The CPM technique is very similar to that of visual inspection, the difference being that in the CPM method a microscope is used to visualize the onset of wax precipitation instead of the naked eye. In the CPM method the wax crystals are visualized after having been magnified 100 times under the microscope. A drop of liquid oil sample is placed on a glass slide and covered with a top glass to form a ~50 μm liquid film. The liquid film is then typically cooled at a cooling rate of 0.5–1°C/min, and the first sign of crystal formation is monitored by the microscope [8]. CPM is more efficient in detecting WAT and has the following advantages over ASTM-D2500 and ASTM-D3117 [8]: first, CPM can detect wax crystals down to 1 μm in size, and detects the WAT at the very beginning of precipitation, as it is assumed that CPM detects WAT at 0.1 wt% precipitation of wax; second, the application of cross-polarized light allows the detection of solids in dark oil samples, which can be confusing in the visual inspection methods.

- **FT-IR technique**

FT-IR techniques measure the WAT based on the distinct IR absorbance of solid wax compared with liquid oil. It is also regarded as one of the precise methods of detecting, as its sensitivity is comparable to that of CPM [8].

- **Viscosity measurements**

At temperatures slightly below the WAT, the oil usually remains as a Newtonian fluid, but the viscosity increases at a higher rate as temperature decreases compared to crude oil above the WAT. Thus the determination of the WAT based on viscometry utilizes the change in the slope in the viscosity vs. temperature curve. Viscosity measurements detect wax precipitation at 0.3–0.4 wt% [8].

- **Differential scanning calorimetry (DSC)**

As mentioned earlier, wax precipitation is an exothermic process. Thermal techniques such as DSC capture the heat of crystallization released by the sample upon wax precipitation and thus detect the onset of wax precipitation. DSC detects wax crystals at 0.3–0.4 wt% of wax precipitation [8]. Venkatesan et al. [27] introduced high-pressure microdifferential scanning calorimetry (HP-µDSC) and showed that it gave comparable results to the CPM technique.

- **Solid detection system using light transmittance (SDS/LT)**

This device was generally designed to detect when the organic deposition takes place; in other words, it measures the onset conditions of live oil precipitation. The instrument is based on the transmittance of a low intensity laser light through fiber optic transmission probes mounted across a windowed cell [28].

- **Simulation methods**

Methods like cold finger and flow loops are also considered effective in addressing WAT impacting factors, wax deposition mechanisms, developing models to calculate WAT, and testing and screening the wax control chemicals. They basically cool the crude oil down and detect the amounts of wax deposited. A cold finger consists of a refrigerated surface that is inserted into a jar containing a heated crude oil. The temperature gradient formed on the surface promotes wax deposition, and after a determined period of time, the surface is removed from the fluid and the deposit is weighed [29,30]. A flow loop is a closed system of fluid flow in pipes or tubing with a refrigerated measuring section that usually can be removed to analyze the deposit that is formed. The flow loop method is claimed to be the most representative of field conditions due to flow regime similarities and thus shear stress, which impacts the formation of wax deposit [30].

- **Ultrasonic method**

A novel high-pressure ultrasonic detection system was developed to evaluate the WAT and the wax precipitated curve (WPC) for dead crude oil and live oil at pressures up to 22.87 MPa [31]. The results obtained are comparable to results from DSC and rheometry. The technique can give a rough estimate of the amount of precipitated wax as well.

Table 7.2 summarizes the techniques used in WAT measurement. Generally, the WATs determined by FT-IR and CPM are comparable and are higher than the WAT determined either by DSC or by viscometry; for example, the WAT determined from DSC can be ~8°C lower than that determined by the

Table 7.2 WAT measurement techniques.

Technique	Principle	Pros and cons
Visual method	Visual observation of wax precipitation	Simple, cheap, does not require skills Not very accurate, inapplicable for very opaque crudes
CPM	Using CPM to visualize the wax precipitation	Very accurate, small sample volume
FT-IR	Using FT-IR absorption bands to detect the phase change from liquid to solid	Very accurate, small sample volume Expensive technique, requires skilled operator
Viscosity	Significant change in viscosity vs. temperature upon cooling, near the WAT region	Simple, not expensive, different techniques, and the accuracy depends on the technique used
DSC	Capturing the heat of crystallization of wax at WAT	Low sample volume, accurate, and the HP-DSC is very accurate Expensive, requires skilled operator

CPM. Therefore CPM provides the most conservative value (the highest estimation) for the thermodynamic WAT in most cases [8,32]. The WAT measured by CPM is also called onset wax crystallization temperature (OWCT), as it was shown that CPM detects wax crystals at the nucleation stage, while the other techniques ascertain their existence at the subsequent early stage of crystal growth. The WAT measured by CPM for the stock tank oil (STO) samples was found to match field wax deposition temperatures (FWDTs) surveyed from the producing wells, i.e., under flowing live oil conditions, likely within and/or close to the laminar regime. WAT and wax crystals can be undetectable for some techniques such as viscometry, DSC, and/or LT, especially in low wax content crude oils, causing misleading results while the problem is already developing [9].

The melting or dissolution temperatures can also be used as a measure of the equilibrium temperature of solid-liquid coexistence. Ideally, the onset of crystallization occurs at the same temperature as the termination of the melting or dissolution process, when the cooling and heating rates are low enough. They are identified as the equilibrium "cloud point" [33].

Similar to the WAT, the cold filter plugging point (CFPP) is now often used instead of CP as the criterion to predict the low-temperature performance of diesel. The CFPP is the lowest temperature at which fuel will still flow through a specific filter.

7.4.1.3 Pour point

PP is defined as the lowest temperature at which movement of a crude oil test specimen is observed. PP is assumed to occur at a precipitation between 2% and 3% of wax solids [34]. The PP is indicative of the amounts of wax in oil. At low temperatures the wax tends to precipitate and inhibit oil flow, so in crude oil a high PP is generally associated with high paraffin content [35,36]. Generally, the PP is 10–40°C below the WAT. Mixing crudes with different PPs will not necessarily result in a PP between the two [37]. PP is an important parameter, especially during shutdown and system restart. PP measurements are also used in screening and testing wax control chemicals.

According to ASTM, the test is performed after preliminary heating: the test specimen is cooled at a specified rate and examined at intervals of 3°C for flow characteristics. The ASTM standard method for crude oil is ASTM-D5853, and for petroleum products the standard is ASTM-D97. The crude oil standard ASTM-D5853 entails the measurement of the maximum and minimum PP temperatures, which provides a temperature window in which a crude oil, depending on its thermal history, might appear in the liquid as well as the solid state. The maximum (upper) PP, is the PP obtained after the test specimen has been subjected to a prescribed treatment designed to enhance gelation of wax crystals and solidification of the test specimen. The minimum (lower) PP, is the PP obtained after the test specimen has been subjected to a prescribed treatment designed to delay gelation of wax crystals and solidification of the test specimen.

Nikolaev et al. [38] presented a new method of testing PPs of pure substances and multicomponent hydrocarbon mixtures based on the analyses of the path lengths of passing of equal volumes of substances along the tracking grooves on a refrigerated sloping metallic panel. The repeatability of the method surpasses the repeatability of testing PPs by the known international standards (ASTM D5853-95, ASTM D97-09, ISO 3016, GOST 20287-91) by five to seven times.

Although WATs and PP temperatures are specific thermodynamic properties for waxy crude oils, their relative positions/boundaries within pressure-temperature (P-T) diagrams as well as the rate (and amount) of wax deposition and accumulation between them are dependent upon a number of factors, including [9]:

- Oil composition,
- Temperature/cooling rate,
- Pressure,
- Paraffin concentration,
- Molecular mass of paraffin molecules,
- Occurrence of nucleating materials such as asphaltenes, formation fines, and corrosion products,
- Water-oil ratio, and
- Shear environment.

7.4.2 Precipitation and deposition of waxes

After the fluids have been cooled down to a limit (WAT) that triggers their solubility limit, the paraffin waxes start to precipitate and deposit on the pipe internal walls, causing severe flow assurance troubles.

Different theories and mechanisms are mentioned in the literature to describe the deposition of paraffin waxes, including molecular diffusion, shear dispersion, gravity settling, Brownian diffusion, and others. The literature plainly points to molecular diffusion as the main mechanism of paraffin wax deposition.

These theories are discussed in the next section, before delving more deeply into the deposition process.

7.4.2.1 Theories and mechanisms of wax deposition

7.4.2.1.1 Molecular diffusion

Generally speaking, molecular diffusion is a process of transport, where molecules flow from an area of high concentration to areas of low concentration. So, once the pipe wall temperature decreases below the WAT, the dissolved wax precipitates out of the crude oil, forming an incipient gel layer on the internal pipe wall. Hence, it is expected that the inner wall is colder than the bulk fluid, resulting in a high concentration of precipitated wax molecules (crystals) at the pipe wall rather than in the bulk, while the bulk oil will have a higher concentration of dissolved HMWT wax molecules than the area at the pipe wall, creating a radial concentration gradient of the waxy components between the bulk oil and the pipe wall. Therefore, the HMWT dissolved wax molecules diffuse from the bulk oil towards the pipe wall (or the incipient gel layer) where they deposit. The precipitation of the dissolved waxes on the deposit surface leads to the development of the deposit [39].

The wax transport rate is given by Eq. (7.3), the Fick diffusion equation [39,40]:

$$W_m = \rho_m A D_m \frac{dC}{dT}\frac{dT}{dr} \tag{7.3}$$

where W_m is the wax deposition rate (kg s^{-1}), ρ_w is the wax density (kg m^{-3}), A is the surface area available for deposition (m^2), D_m is the diffusion coefficient or diffusivity of wax in oil (m^2 s^{-1}), dC/dT is the variation of dissolved wax concentration with the temperature (°C^{-1}), and dT/dr is the radial temperature gradient (°C m^{-1}) [39].

The molecular diffusion coefficient, D_m, is calculated according to Eq. (7.4), based on Wilke and Chang [41] experimental data:

$$D_m = 7.4 \times 10^{-8} \frac{(x \times MW_{solvent})^{1/2} T}{\mu_{solvent} V_{solvent}^{0.6}} \tag{7.4}$$

where T is the absolute temperature (K), $MW_{solvent}$ is the molecular weight of solvent (g mol^{-1}), x is the Wilke and Chang association parameter ($x = 1$ for nonassociated solvents, and $x = 2.6$ for water), $\mu_{solution}$ is the solution dynamic viscosity (cP), and V_{solute} is the molar volume of solute at normal boiling point (cm^3 mol^{-1}).

The diffusion coefficient of the waxy components in the oil typically ranges from 10^{-10} to 10^{-9} m^2/s [41–43], which is much smaller compared to that for gas systems, which are on the order of 10^{-5} m^2/s [44], not to mention the other convective fluxes of the wax molecules towards the pipe wall due to the fluid flow.

7.4.2.1.2 Brownian diffusion

In Brownian diffusion, the wax solid crystals suspended in crude oil collide with thermally agitated oil molecules, resulting in random Brownian movements of the wax suspended particles, creating irregular movement of wax particles. If a concentration gradient exists for the wax solid particles, it is possible that the net effect of Brownian motion results in the transportation of wax crystals towards the decreasing concentration [39,45]. However, the wax solid particles are highly concentrated at the pipe wall rather than in the bulk oil, and Brownian diffusion stipulates that the particle movements always flow from areas of high concentration to areas of low concentration, which in this case is from the pipe wall to the bulk oil, instead of moving toward the pipe wall and being deposited, which makes Brownian diffusion not a valid postulation [8,46].

For spherical, noninteracting particles, the Brownian diffusion coefficient (m^2 s^{-1}) is given by Eq. (7.5) [39]:

$$D_b = \frac{RT}{N}\frac{1}{6\pi\mu_{solvent} r} \tag{7.5}$$

where R is the gas constant (8.314 J K^{-1} mol^{-1}), T is the temperature (K), r is the radius of the molecule (m), $\mu_{solvent}$ is the solvent dynamic viscosity (Pa s^{-1}), and N is Avogadro's number (6.022×10^{23} mol^{-1}).

7.4.2.1.3 Shear dispersion

This is one of the common mechanisms used to explain wax deposition, and is also used in some commercial wax prediction tools like the OLGA simulator. Shear dispersion postulates that wax particles deposit when they are dispersed toward the wall and reach a region of lower velocity at the pipe wall, where they stick and deposit on the surface of the pipe wall.

The shear dispersion coefficient D_s (cm^2 s^{-1}) was experimentally determined by Eq. (7.6) [47]:

$$D_s = \frac{r^2 \omega \text{Ø}}{10} \tag{7.6}$$

where r is the particle radius (cm) and ω is the mean shear rate of suspending fluid (s^{-1}), and Ø is the volumetric concentration of wax suspended particles (m^3/m^3), for $0 < \text{Ø} < 0.20$; however, for $\text{Ø} > 0.20$, a relation could not be unequivocally established [39].

Some authors have claimed the importance of shear dispersion in modeling wax deposition [48], and state that it predominates over molecular diffusion at a higher velocity of the flow [49], while other reports showed that, for temperatures much lower than WAT and for moderate heat fluxes, the governing process is shear dispersion, and for slightly higher temperatures the dominant process is molecular diffusion [50].

However, experimental studies showed that increasing the shear did not increase the rate of wax deposition [51], and the particle fluid dynamics studies showed that particles in the oil flow are not necessarily dispersed toward the wall, but, on the contrary, the particles located in the viscous layer near the wall tend to be reentrained into the bulk flow by a lifting force generated by the turbulent flow (known as the Saffman lift force) [52,53].

7.4.2.1.4 Gravity settling

Wax deposition is due to the settling of the precipitated particles of the waxy components toward the bottom of the pipe. The settling velocity, Vg, can be computed as the balance of the buoyancy and the viscous forces, as presented in Eq. (7.7):

$$V = \frac{d^2 (\rho_p - \rho) g}{18 \mu} \tag{7.7}$$

where Vg is the settling velocity (m/s), d is the particle diameter (m), ρ_p is the particle density (kg/m^3), ρ is the fluid density (kg/m^3), and μ is the fluid viscosity (Pa s).

However, this postulates that the wax deposits at the 6 o'clock (bottom) of the pipeline should be always higher/thicker than those at the 12 o'clock (top) of the line, which was not reported from the field experience of the wax problems [8]. Gravity settling was ruled out as insignificant in wax deposition, as mentioned in Ref. [54].

7.4.2.1.5 Other theories

Other theories have been reported to explain the deposition of paraffin waxes. Among these theories are shear stripping and thermal diffusion.

Shear stripping is a wax removal mechanism in which wax is removed from the tubing walls, as the flowing stream sloughs pieces of the wax from the deposit layer, due to shear stress induced by the flowing stream on the wax layer [39]. Edmonds et al. [55] expressed the wax shearing rate, J, according to Eq. (7.8):

$$J = c\frac{\delta\tau}{\text{Ø}_m^{2.3}} \tag{7.8}$$

where c is the shear constant, δ is the wax layer thickness (m), Ø_m is the mass fraction of wax in the gel layer, T is the shear stress (N/m^2) and it is obtained by Eq. (7.9):

$$\tau = \frac{1}{2} f \rho_0 u_0^2 \tag{7.9}$$

where f is the Fanning friction factor, ρ_o is the density of the flowing oil (kg/m^3) and u_o is the oil velocity (m/s).

Another approach is the thermophoresis or thermal diffusion, where the temperature gradient is the driving force for wax deposition [39]. The effect of thermal diffusion in wax deposition has been found to be negligible [56]. However, accurate modeling of wax deposition that accounts for thermal effects during diffusion can be obtained if the diffusion is expressed in the form of molecular and thermal diffusion [57]. Under the effect of a temperature gradient, the thermophoretic velocity (m/s) is given by Eq. (7.10) [58]:

$$V_T = -\beta\frac{\mu}{\rho}\frac{\nabla T}{T} \tag{7.10}$$

The negative sign means that the particles move from the hotter zone to the colder one, where the proportionality factor can be determined by Eq. (7.11) [59], which is a function of the thermal conductivity of the fluid, k (W/(m K)), and the thermal conductivity of wax particles, k_p (W/(m K)) [39]:

$$\beta = 0.26\frac{k}{2k + k_p} \tag{7.11}$$

Additionally, emulsification is a mechanism that is usually observed during oil and gas operations. It principally relies on the idea that the emulsified particles in the low-temperature oil come into contact and are combined with each other, and the wax crystal particles are adsorbed on the surface of the dispersed phase of the emulsification system, accumulated, and grown continuously to form a gel network structure and this accelerates the wax deposition process [60].

In a broader sense, molecular diffusion is considered the main mechanism of transporting the HMWT wax molecules into the gel layer at the pipe wall, where they add onto it and replace the entrapped oil and age to form the conventionally encountered hard deposits. It is worth mentioning that while molecular diffusion is generally considered as the main deposition mechanism, the contribution from other forces cannot be neglected.

Now, after overviewing the main theories used to address wax deposition, it is time to go over the deposition process to see how it actually happens, in a stepwise manner.

For simplicity, the precipitation and deposition of paraffin waxes can be summarized in two major processes [46]:

- First: the formation of the incipient gel layer.
- Second: hardening or aging of that gel layer.

The first process is the basic and initial step in paraffin deposition, which involves wax nucleation and crystal growth. The second process is a multistep process that includes molecular diffusion as the main mechanism, in which wax crystals continue growing and hardening.

We can break these two main steps into multiple steps as follows [8,46,61]:

1. Gelation of the waxy oil, or the formation of incipient gel layer on the cold surface.
2. Diffusion of wax molecules with carbon numbers greater than the critical number from the bulk solution towards the gel layer.
3. Internal diffusion of these diffused wax molecules through the trapped oil into the incipient gel layer.
4. Precipitation of the wax molecules in the gel layer, increasing the deposit layer wax content and thickness.
5. Counterdiffusion of dewaxed oil hydrocarbons with carbon numbers lower than the critical number out of the gel layer.

7.4.2.2 First step: The formation of incipient gel layer

The first step in wax crystallization is the formation of an incipient gel layer. The incipient gel layer is a 3D network of wax crystals that entraps crude oil and other production fluids. Wax crystallization and incipient gel layer formation basically involve two distinct stages, namely nucleation and growth [9].

7.4.2.2.1 Wax nucleation

Nucleation can be either homogeneous or heterogeneous, with the heterogeneous being the predominant due to the various impurities in the production fluids, like scale and corrosion products, imperfect surfaces, and others. Nucleation starts when this mixture is cooled below WAT by any of the previously discussed cooling mechanisms. A fraction of hydrocarbons with carbon numbers above a certain value, called the *critical carbon number* (CCN) [46,62] (which are randomly tangled molecules in crude oil) tend to move closer together and form clusters of adjacently aligned chains. These paraffin molecules continue to attach and detach to and from these ordered sites, until the clusters reach a critical size and become stable. These clusters are called nuclei, and those with critical size are called critical nuclei. These nuclei are only stable below the melting/dissolution temperature of the wax, as they are disrupted by thermal motion above this temperature [9].

7.4.2.2.2 Wax crystal growth

After nuclei are formed, while the temperature is kept at or below the WAT, additional molecules are added to the nucleation sites and included in the growing lamellar structure [9]. The crystal growth of the paraffin wax crystals is believed to be spiral growth [63,64], and it is believed to form orthorhombic crystals [8,46]. Fig. 7.10 shows the spiral growth of wax crystals.

The formed orthorhombic subcrystals combine to form a lamellar structure, which is a sheet-like morphology. A closeup of a single packing layer of a multicomponent paraffin wax crystal shows that it has a thickness equal to the average length of the wax molecules in the mixture, as shown in Fig. 7.11 [65]. If a wax molecule has a carbon number greater than the average carbon number, it bends to insert itself inside the layer and associates itself with a molecule with a carbon number lower than the average. The mismatch in length between the molecules and the thickness of the packing layer causes conformational disorders in the interlamellar regions of these crystals. Such disorders impart to these crystals their wax-like properties and may hinder the growth of the sheet-like crystals in a 3D manner [46].

FIG. 7.10

Atomic force microscope photo of spiral crystal growth of paraffin wax.

With permission from Professor M.J. Miles, Bristol University, GB.

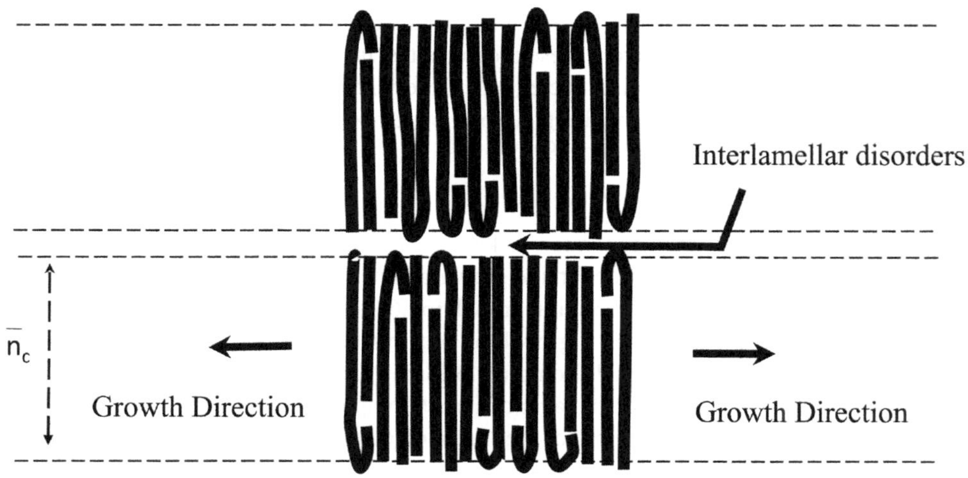

FIG. 7.11

Lamellar structure of paraffin wax source.

Reprinted from P. Singh, R. Venkatesan, H. Scott Fogler, N. Nagarajan, Formation and aging of incipient thin film wax-oil gels, AICHE J. 46 (5) (2000) 1059–1074, https://doi.org/10.1002/aic.690460517 copyright (2000) with permission from John Wiley and Sons.

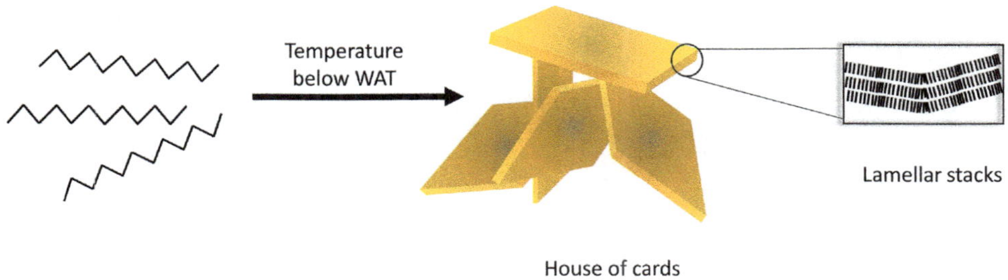

FIG. 7.12

Wax platelets forming house of cards.

With further cooling of the mixture, thin flakes result from the 2D lateral growth of the sheet-like crystals. [46]. When the size and density of these flakes reach a critical value, they form platelets of stacked lamella crystals, where they overlap and interlock with each other to form a house of cards, which is a 3D network that forms the entire wax-oil solution gel [66–68]. Figs. 7.12–7.14 illustrate the house of cards of wax platelets and the 3D network of wax crystals.

The critical size value depends on the mixture composition, shear, and thermal histories of the mixture. For crude oils with $>10\,\text{wt}\%$ of *n*-paraffins, under static conditions, the gelation of the crude oil can take place when approximately $2\,\text{wt}\%$ of waxes have precipitated out in the solution [69]. However, under flow conditions, the weight fraction of the precipitated wax in the incipient gel is a strong function of the shear and thermal histories of the mixture [70]. The incipient gel layer is believed to have 90%–95% of oil entrapped in it [8,46].

Fig 7.15 summarizes the main steps involved in the formation of the incipient gel layer.

7.4.2.3 Second step: Hardening (aging) of the incipient gel layer

As indicated earlier, the incipient gel layer is a 3D network of wax crystals having nearly 90%–96% of oil entrapped in it. The hardening/aging process involves increasing the wax content of the gel deposit with time, leading to hardening of the deposit [46]. This will happen by transporting the wax molecules

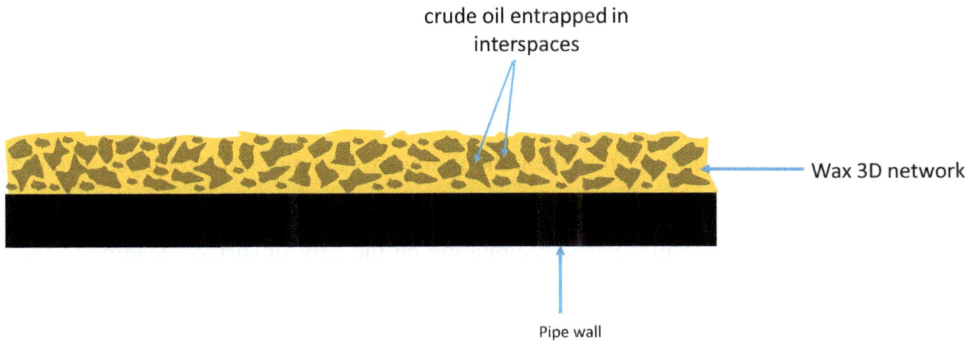

FIG. 7.13

3D network of gel layer.

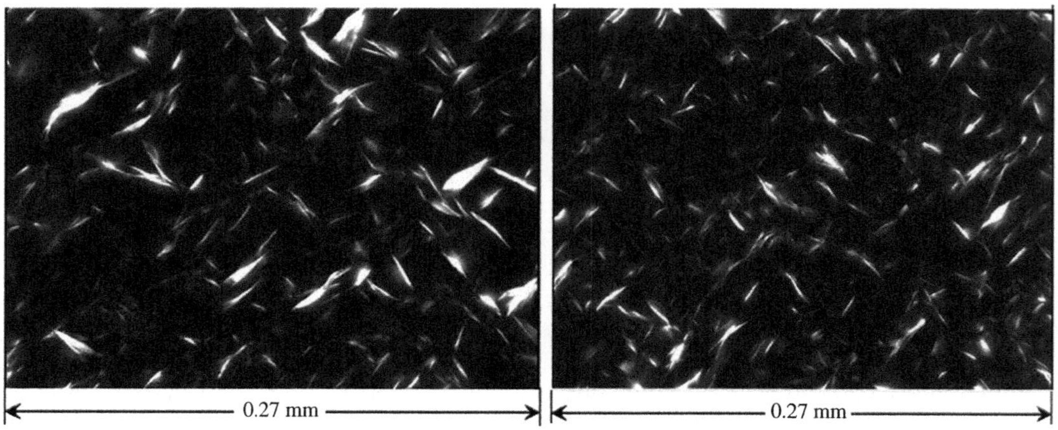

FIG. 7.14

The 3D network of a waxy gel under the microscope.

Reprinted from R. Venkatesan, N.R. Nagarajan, K. Paso, Y.B. Yi, A.M. Sastry, H.S. Fogler, The strength of paraffin gels formed under static and flow conditions, Chem. Eng. Sci. 60 (13) (2005) 3587–3598, https://doi.org/10.1016/j.ces.2005.02.045 copyright (2005) with permission from Elsevier.

(by diffusion) from the bulk of the solution towards the gel layer, and then transporting them from outside the gel layer to inside it through its 3D network.

Then molecular diffusion plays a critical role in hardening and aging of the wax gel deposits [71]. The dissolved HMWT wax molecules diffuse from the bulk of the solution towards the pipe wall (or more specifically towards the incipient gel layer on the pipe wall) to increase the gel layer thickness and wax content.

The molecular diffusion is driven by two main factors:

- The concentration gradient of the dissolved HMWT wax molecules between the bulk oil and pipe wall.
- The temperature or thermal radial gradient between the bulk oil and the pipe wall, which enhances the radial concentration gradient of HMWT wax molecules.

Similarly, the oil diffuses away from the wall due to the concentration gradient of oil. This process is a counterdiffusion of the wax and the oil [46].

Once the waxy molecules diffuse from the bulk oil towards the pipe wall/gel layer, two processes take place:

- First: the dissolved wax molecules keep precipitating on the deposit surface, leading to further growth of the wax deposit, and with the continuous flow of the waxy crude through the pipe, the diffusion of dissolved wax compounds towards the deposit increases, resulting in the buildup of wax deposits [8]. As the deposit thickness grows, the insulating effect of it increases (the deposited gel acts as an insulator to the radial heat transfer), and the driving force for further deposition significantly diminishes. Hence, there is virtually no further increase in the film thickness [46].

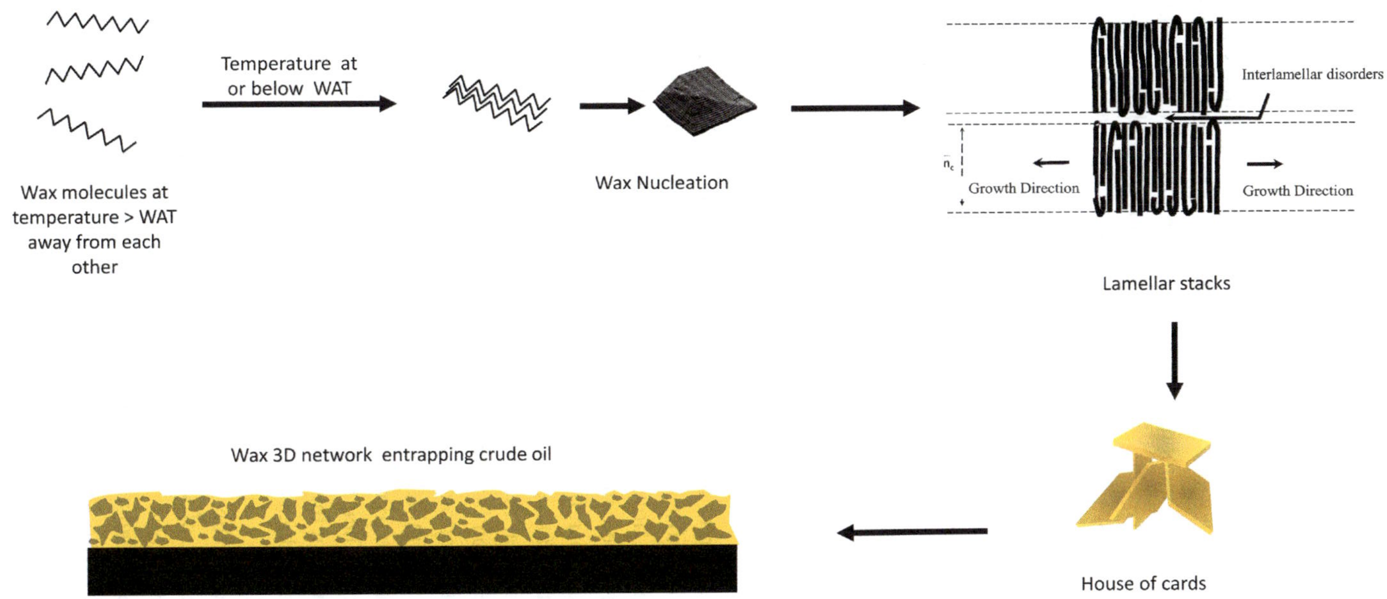

FIG. 7.15

Summary of the incipient gel layer formation steps.

- Second: once they reach the oil-gel layer interface, the wax molecules diffuse into the gel layer and the oil molecules continue to diffuse out of the gel layer (counterdiffusion), until the gel layer hardens with a high wax content [46,71], and this is what is called "aging of the deposit."

Fig. 7.16 illustrates the diffusion and counterdiffusion processes of wax molecules.

The ultimate result of these processes is increasing the wax content in the gel layer, by increasing the fraction of HMWT waxes or the molecules with carbon number greater than the CCN, and decreasing the fraction of molecules with carbon number lower than the CCN in the gel [46].

The wax content in the gel layer also increases by the thermal gradient across the gel layer, even after it stops growing in thickness. This thermal gradient across the gel layer has a greater effect on wax aging than other effects, such as compression or shear forces [46].

From the previous discussion, we can summarize the aging and diffusion process during wax deposition as:

- Diffusion of waxy molecules from the bulk oil towards the pipe wall/gel layer, and counterdiffusion of oil molecules from the pipe wall area towards the bulk oil.
- Diffusion of the waxy molecules from outside the gel layer to inside it, and counterdiffusion of the oil from the wax deposit layer to outside it.
- Diffusion of waxy molecules through the wax deposit layer due to the thermal gradient within when it increases in thickness.
- This is also affected by the shear and compressive forces of the flowing fluid.

Fig. 7.17 summarizes the different fluxes of wax molecules and the forces affecting them.

It is worth mentioning that most of the work done in understanding wax precipitation and deposition mechanisms is based on single flow regimes. In real field operations, multiphase flow is usually encountered and therefore must be considered. In contrast to many studies on single-phase paraffin deposition, gas-oil and oil-water paraffin deposition studies are very limited. There is no published study on three-phase gas-oil-water wax deposition.

Matzain et al. [72] found that the thickness, hardness, and profile of wax deposition in two-phase gas-oil flow show dependence on flow patterns. They concluded that for horizontal flow, the stratified flow gave a soft deposit at the bottom of the pipe, with harder and thicker deposits along the edge of the wavy gas-liquid interface. Intermittent flow resulted in a hard deposit, with increasing hardness from the top to the bottom of the pipe. Finally, annular flow resulted in a very hard deposit, uniform across the circumference of the pipe. Their results for vertical two-phase flow, on the other hand, showed very uniform thickness distribution in the different flow regimes, with very hard deposits for annular flow, deposits of medium to high hardness for intermittent flow, and hard deposits for bubbly flow with high superficial velocity [15,72].

Another way of regarding the wax gelation, especially in hydrocarbon-water multiphase systems, is to consider the effect of water and emulsification. Waxes are known to form gel layers and viscous wax-oil gel emulsions. Visintin et al. [73] hypothesized that the solid paraffin stabilizes the emulsion by being strongly adsorbed at the liquid-liquid interface, forming Pickering emulsions. The growth of the wax gel network is induced by the interactions between wax crystals and the drop surface, and the formed network involves wax crystals with entrapped dispersed water, as shown in Fig. 7.18.

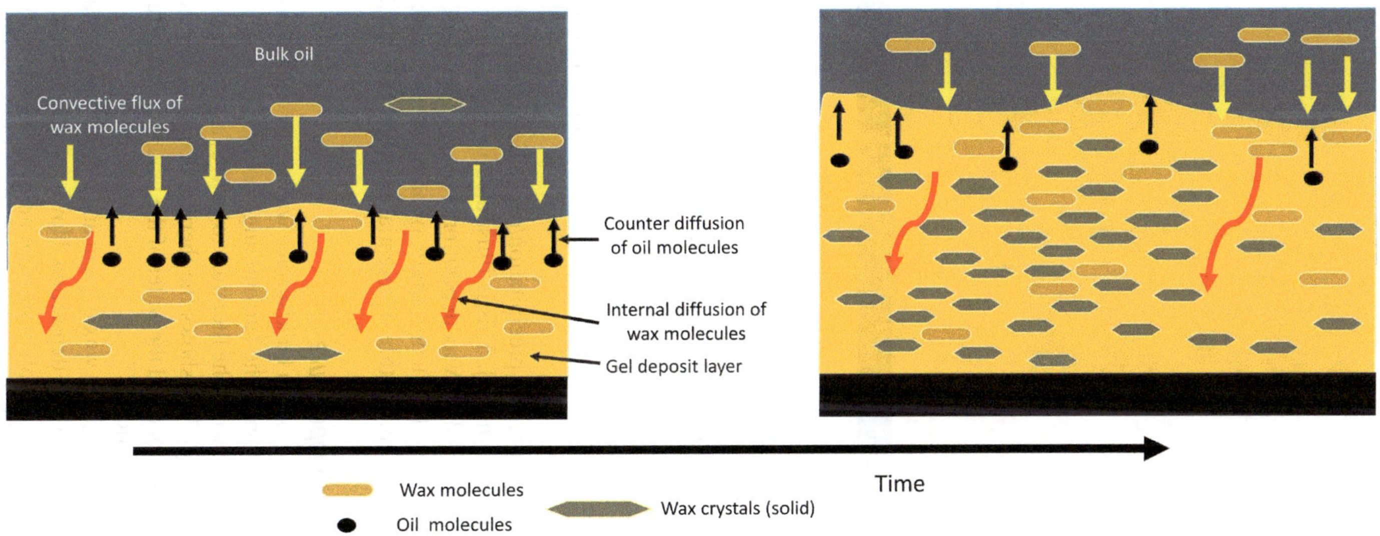

FIG. 7.16

Diffusion processes during wax deposit formation and aging.

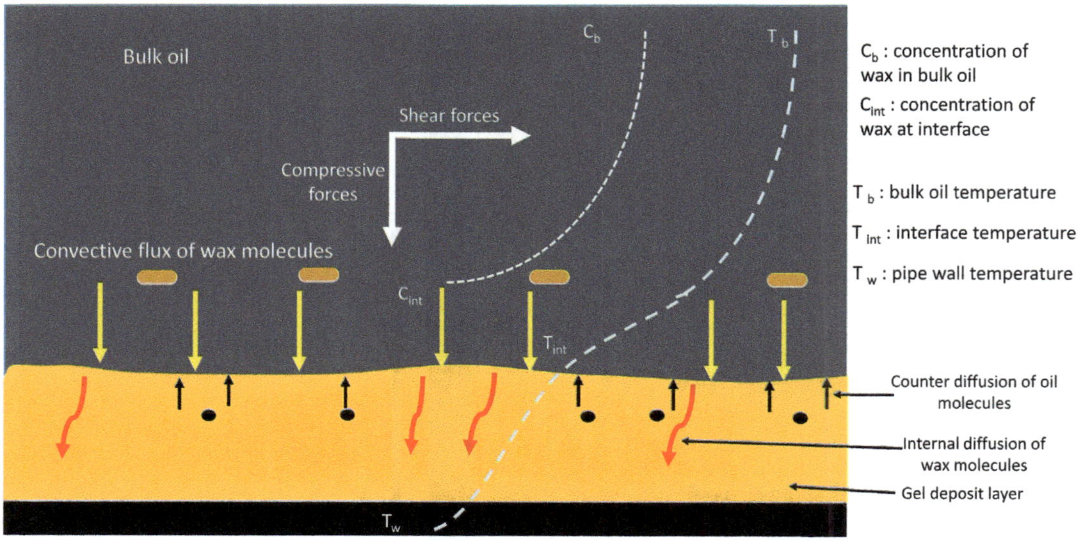

FIG. 7.17

Summary of diffusion processes and forces affecting wax deposit formation and aging.

7.5 Factors affecting the formation of wax deposits

7.5.1 Temperature

Temperature is the major factor that affects paraffin wax gelling and deposition. When the temperature of the fluids drops below the WAT (solubility limit.), wax molecules start to precipitate out of the solution [46].

Compared to gas hydrate deposition, wax precipitation/deposition is normally a problem at a higher temperature than gas hydrate formation [74]. This is illustrated in Fig. 7.19.

Different temperature terms are used for wax deposition. For example, there is the coolant temperature (underwater, ocean floor, land surface temperature, or ambient temperature), the pipe wall temperature, the bulk temperature, and temperature differential between bulk fluids and pipe wall.

7.5.1.1 The effect of coolant temperature

The coolant can be the ocean floor, the ocean bulk/surface water, or the land surface. Production fluids lose their temperature through radiation to the surrounding coolants.

The ocean floor temperature is known in the literature to be 4°C, and it increases with decreasing the depth reaching the water surface. Fig. 7.20 shows the thermocline of ocean temperature vs. depth.

The temperatures of the water surface and the water of the land face are different, depending on the season (summer, winter) and the location.

Generally speaking, crude oils with >2% wax content and a WAT higher than the ocean floor temperature (39°F or 4°C) are vulnerable to wax deposition. However, this is not always the case and such thresholds are only a rough estimation and often provide an overestimation of the problem, as some crude oils with a WAT higher than 39°F did not have wax deposition problems [8].

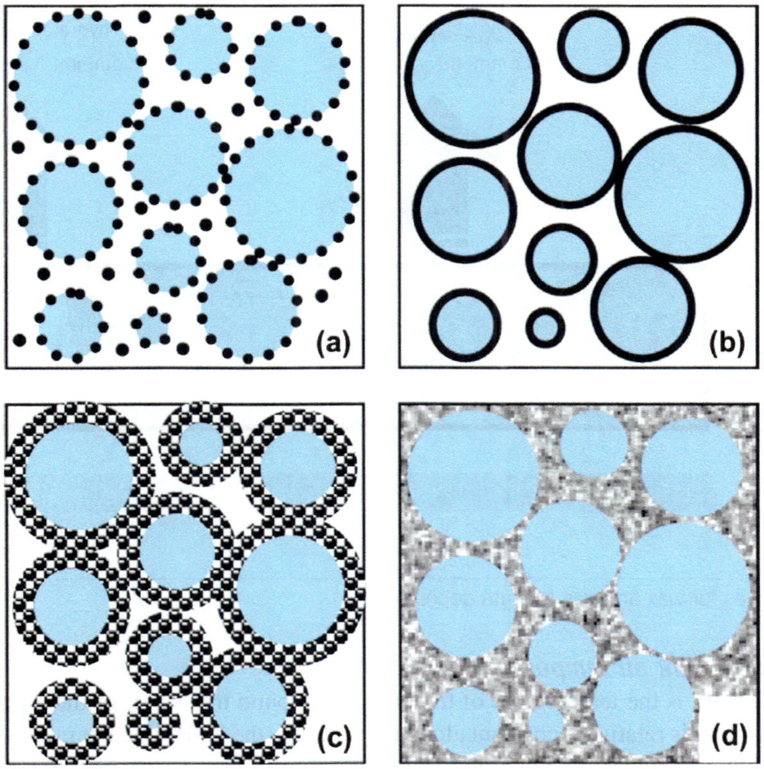

FIG. 7.18

Schematic representation of the gelation of waxy crude oil emulsions. Paraffin crystals that precipitate after a decrease in temperature below the WAT can adsorb on droplet surface (a) or cover it (b), and stabilize the emulsion. Flocs of solid paraffin continuously grow on a drop of water or between them (c). Dispersed water is entrapped by a wax crystal network (d): the system spans the entire volume and the gelation is complete.

Reprinted from reference R.F.G. Visintin, T.P. Lockhart, R. Lapasin, P. D'Antona, Structure of waxy crude oil emulsion gels, J. Non-Newtonian Fluid Mech. 149 (1–3) (2008) 34–39, https://doi.org/10.1016/j.jnnfm.2007.07.008 copyright (2008), with permission from Elsevier.

7.5.1.2 The effect of pipe wall temperature

Wax molecules precipitate on cold surfaces. When the pipe wall temperature falls below the WAT, wax molecules start to precipitate.

Under the same bulk oil temperature, wax precipitation and deposition increase with decreasing of the surface (pipe wall) temperature. This is attributed to enhancing the driving force of the dissolved wax molecule diffusion from the bulk towards the wall, driven by the enhanced radial temperature and concentration gradients [75–77].

Also, the final thickness of the deposit layer is smaller for a higher pipe wall temperature [76,77]. The WAT is reached in a shorter period of time for a higher pipe wall temperature, resulting in slower rate of growth and a thinner deposit layer [46].

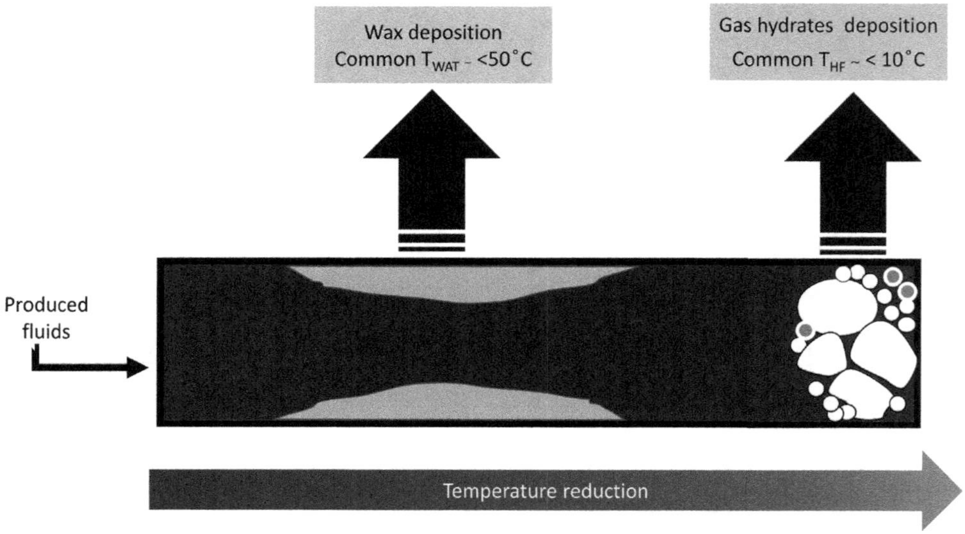

FIG. 7.19

Temperature ranges for wax and gas hydrate deposition.

7.5.1.3 Effect of bulk oil temperature

The bulk temperature is the temperature of the crude oil around the center of the pipeline. While the pipe wall temperature is relatively constant along the pipeline, the bulk oil temperature decreases along the pipe flow, and radially through the pipe diameter.

Higher bulk oil leads to higher radial temperature gradient or higher thermal driving force for deposition, which increases the wax deposition [78]. At constant pipe wall temperature, the higher bulk

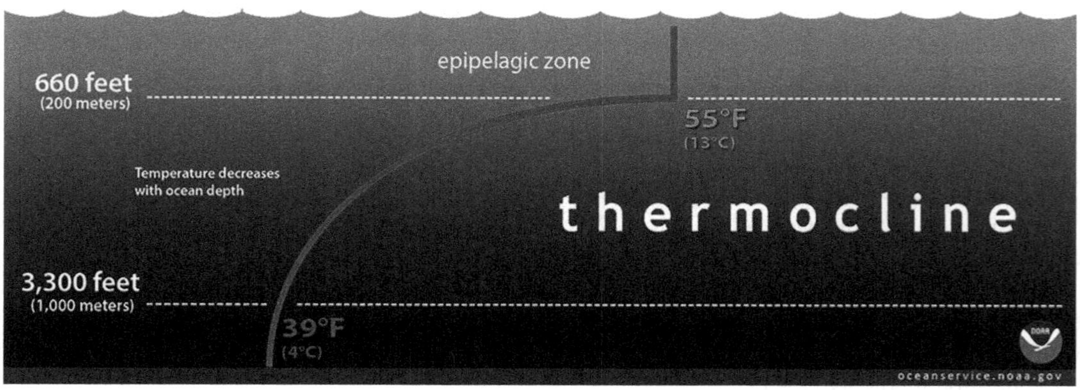

FIG. 7.20

Ocean temperature vs. depth.

Graphic courtesy NOAA.

oil temperature results in HMWT wax molecules dissolving in bulk oil and diffusing to the pipe wall, leading to formation of harder deposits. Lower bulk oil temperature (but high enough to keep the bulk oil flowing) favors relatively lower molecular weight wax molecules dissolving in bulk oil and diffusing to the cold surface, leading to a softer deposit, which is also applied in the "cold flow" technology [75].

Generally, both the wax content and the amount of higher carbon number components in the deposit increase as the bulk oil temperature increases under constant coolant temperature.

7.5.1.4 Effect of differential temperature

The differential temperature, the radial temperature gradient, or the thermal gradient is the difference in temperature between the bulk oil temperature and the pipe wall, where the pipe wall temperature is lower than the WAT. This temperature differential gives rise to a radial concentration gradient of wax molecules. Since wax molecule solubility depends on temperature, the wax concentration in the liquid at the centerline is higher than the wax concentration in the liquid at the wall, owing to the higher centerline temperature. This temperature or concentration gradient works as a driving force for two events [46]:

- Diffusion of wax hydrocarbons with carbon numbers greater than the critical number towards the gel layer from the bulk.
- Counterdiffusion of dewaxed oil hydrocarbons with carbon numbers lower than the critical number out of the gel layer.

These two events are responsible for the aging and hardening of the incipient gel layer and formation of wax deposits.

A generalization was made that the deposit thickness decreases when thermal driving force decreases [79]. Lashkarbolooki et al. [80] found that the deposit thickness depends on the temperature difference, where the thickness obviously decreases but the wax percentage fraction increases at small temperature differences. An increase in the mass of the deposit does not accordingly mean a larger wax percentage fraction. Valinejad and Nazar (2013) [81] revealed that the temperature difference between the oil and pipe wall has the maximum percentage of contribution to the amount of deposited wax.

Cold finger experiments showed an increase in the deposition mass with increase in the differential temperature [76], as illustrated in Fig. 7.21.

Using flow loop, the deposit thickness was shown to increase with increasing the temperature differential in both turbulent and laminar flow [77]. However, the authors also reported some inaccuracy in measuring the deposit thickness due to its loose nature. These results are illustrated in Figs. 7.22 and 7.23.

The amount of lower carbon number components in the deposit decreases and the amount of higher carbon number components (with carbon number higher than the CNN) in the deposit increases as the deposit time increases at a higher temperature range [75]. The CNN of the wax molecules in deposits is an important criteria, since it determines the hardness and solubility of the deposits, which are used to optimize the pigging and chemical dissolution of the deposits. Higher CCN yields a mechanically harder wax with a higher melting point, as well as a lower solubility in typical wax solvents [46,82].

The decrease of the wax solubility in crude oil with temperature reduction can be determined by using the wax precipitation or solubility curve (WPC), which shows the amount of precipitated/

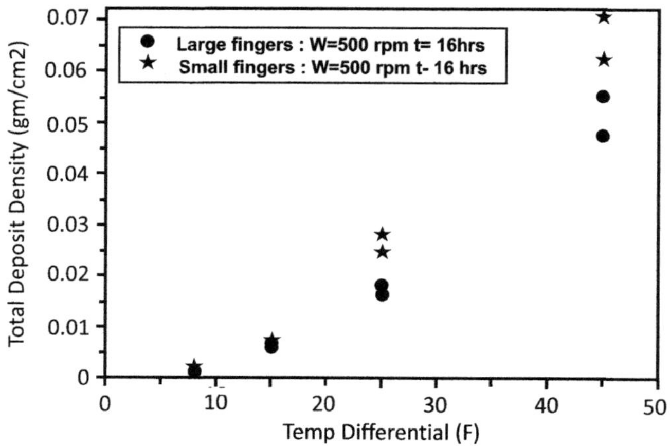

FIG. 7.21

The effect of temperature differential on the amount of wax deposit.

Reprinted with permission from D.W. Jennings, K. Weispfennig, effects of shear and temperature on wax deposition: coldfinger investigation with a Gulf of Mexico crude oil, Energy Fuels 19 (2005) 1376–1386. Copyright {2005} American Chemical Society.

dissolved wax in the crude oil vs. temperature [83]. The different methods of developing and constructing a wax precipitation/dissolution curve were reported by Refs. [83–85].

The WPC characterizes the solubility limit of wax in crude oil and thus affects the concentration driving force for wax deposition [8]. The WAT and WPC are considered significant parameters in studying wax deposition in oil and gas fields. WPC gains significance, as it predicts the amount of the precipitated wax as a function of the operating conditions. Fig. 7.24 illustrates a typical WPC.

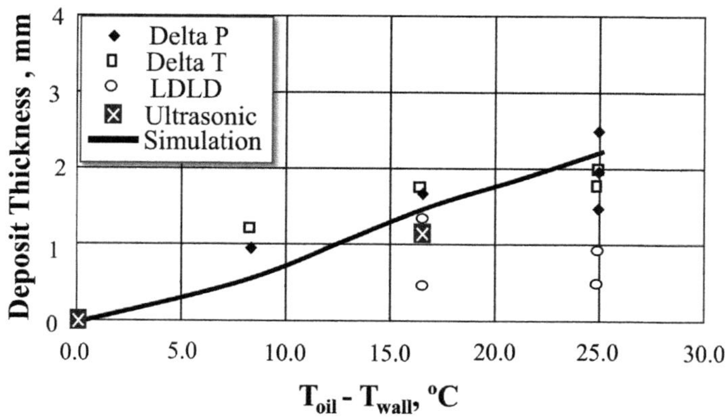

FIG. 7.22

Effect of differential temperature on wax deposit thickness in laminar flow.

Reprinted from J.L. Creek, H.J. Lund, J.P. Brill, M. Volk, Wax deposition in single phase flow, Fluid Phase Equilib. 158 (1999) 801–811 copyright (1999) with permission from Elsevier.

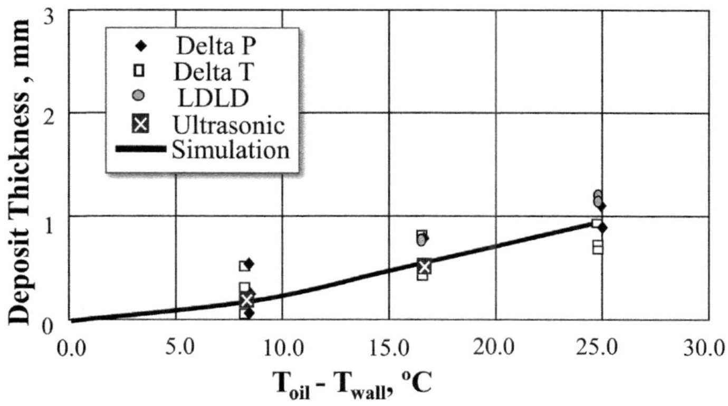

FIG. 7.23

Effect of differential temperature on wax deposit thickness in turbulent flow.

Reprinted from J.L. Creek, H.J. Lund, J.P. Brill, M. Volk, Wax deposition in single phase flow, Fluid Phase Equilib. 158 (1999) 801–811 copyright (1999) with permission from Elsevier.

7.5.2 Effect of pressure

While temperature has the biggest effect on the wax deposition and gelation, pressure was also found to affect the wax deposition. The effect of pressure-temperature is always illustrated in the wax deposition envelope, as indicated in Fig. 7.25 [86], as the solid/liquid-phase boundary is nearly vertical for waxes, illustrating wax precipitation's strong dependence on temperature and weak dependence on pressure [87].

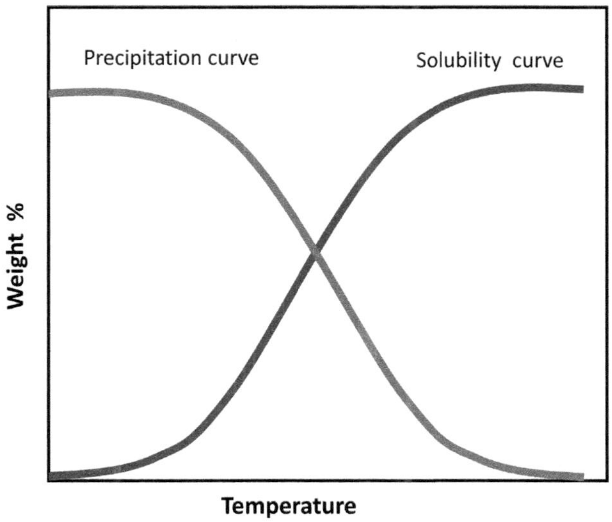

FIG. 7.24

Wax precipitation/dissolution curve.

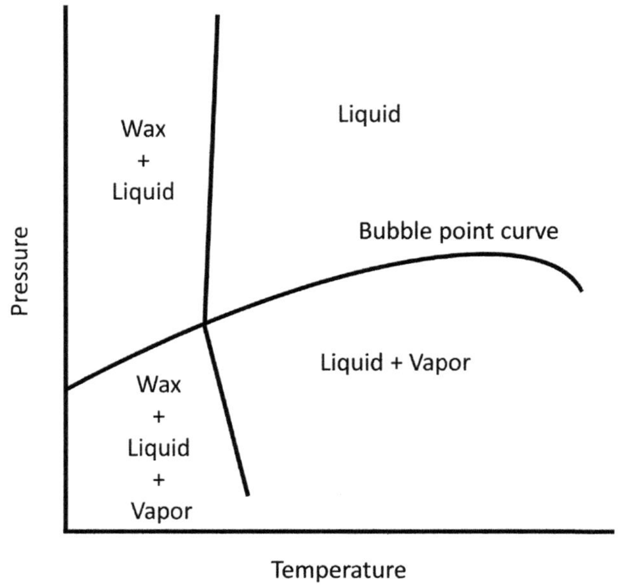

FIG. 7.25

Wax deposition envelope.

Reprinted from K.J. Leontaritis, Asphaltene and wax deposition envelopes, Fuel Sci. Technol. Int. 14 (1–2) (1996) 13–39. https://doi.org/10.1080/08843759608947560, copyright (1996), with permission from Taylor and Francis.

The effect of pressure on wax deposition is represented in many ways:

- As the pressure drops during production, loss of low light ends to the gas phase, which reduces the solubility of the waxes in the oil [3]. The liberation of solution gas from a crude oil as pressure decreases below the bubble point of the fluid has been found to increase the crude oil WAT [88]. This can be understood through Fig. 7.26, which is another representation of the phase diagram for wax precipitation. Point A is the reservoir pressure where fluids are at high temperature and pressure, and the oil is undersaturated. As the fluids are produced, the pressure falls and the light ends expand in greater proportion to the dissolved waxes, increasing their solubility and yielding a lower WAT. And wax precipitation at this stage is driven by fluids cooling below WAT. At the bubble point (B), the volume ratio of light ends to heavy ends is at its greatest, so the WAT is at its lowest, hence the smallest precipitation zone. Reducing the pressure still further (C) results in liberation of dissolved gases and light ends to the gas phase. This reduces the solubility of the wax, and so the WAT increases [3,89].
- Vieira et al. [13] found that increasing the pressure of the crude or condensate using methane gas or natural gas leads to significant decreasing of the WAT. Quan et al. [90] reported that the amount of deposit increases when the pressure is increased from 1 to 2 MPa and decreases thereafter (for pressures ranging from 2 to 6 MPa). The pressure corresponding to the maximum deposition increases with increasing wax content. For example, wax contents of 5%, 7%, and 10% correspond to pressures of 1, 2, and 3 MPa, respectively.
- Pressure drop leads to thermal expansion, which results in localized cooling of the produced fluids. When a fluid passes through a throttling device, the enthalpy remains constant, and the fluid

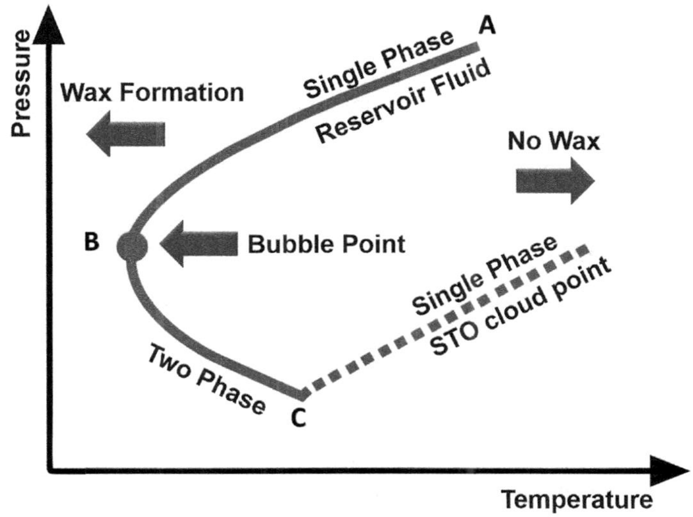

FIG. 7.26

Pressure-temperature relationship of wax appearance in both reservoir fluid and stock tank oil (STO) conditions.

Adapted from M.K. Afdhol, M. Abdurrahman, F. Hidayat, F.K. Chong, H.F. Mohd Zaid, Review of solvents based on biomass for mitigation of wax paraffin in Indonesian oilfield, Appl. Sci. (Switz.) (2019) https://doi.org/10.3390/app9245499.

temperature normally drops. The effect depends on the value of the Joule–Thomson coefficient, C_{JT} [91]:

$$C_{JT} = \left.\frac{dT}{dP}\right|_h \tag{7.12}$$

If C_{JT} is positive, a temperature decrease follows a pressure decrease. If it is negative, a temperature increase results [91]. Saturated fluids experience J-T cooling when pressure decreases below the bubble point, whereas undersaturated or supercritical fluid experiences heating upon pressure decrease, as in well flow [92].

Joule-Thomson cooling can take place in the reservoir, perforation, wellbore, production choke, and any other locations that encounter a throttling mechanism. It is also most likely to occur during restart (after shut-ins), and system depressurization. For example, it is not unusual to see 2000 psig upstream of the choke and 100 psig downstream of the choke. This 1900 psig pressure drop causes a 95°F temperature drop in the choke, or 1°F per 20 psig of pressure drop [93]. The extent of the J-T effect depends on many factors:

- The production fluids; the effect is addressed more in gas and condensate systems than in oil/water systems.
- The system dimensions, design, and operating parameters, like the length of pipe where the fluid can be within the wax equilibrium range due to Joule-Thomson cooling, velocity of the fluids, residence time, all of which determine if the Joule-Thomson is sufficient to cause wax deposition or if the wax

will just be concentrated in limited spots for short times and the fluids can warm up after that to melt the already formed wax deposits.

J-T is one of the main causes of paraffin deposition in oil, gas, and gas condensate facilities. Gas expansion is responsible for the loss of light ends in oil reservoirs and J-T cooling of reservoir fluids in oil and gas reservoirs, which accounts for a number of cases of reservoir damage from paraffin deposition.

7.5.3 Effect of flow rate

The general consensus is that shear environment can either delay or induce wax deposition, depending on the regime, i.e., turbulent or laminar [9].

Wax deposition was found to decrease with increasing flow rates, which is attributed to the shear removal. Fogler et al. [94] showed that the changes in wax deposition with changing flow rates are effects that focus on the heat- and mass-transfer phenomena at the oil–deposit interface. The high flow rates are thought to increase heat transfer rate, which means the fluids reach WAT faster, which does not give the formed gel layer enough time to grow in size and harden enough. In addition, in high shear flow environments, the ability of wax constituents to gel and deposit is diminished as the nucleation process is disrupted [9].

However, some wax deposits are formed at high flow rates; these are harder and more compact as they are capable of firm attachment to the surface, with good cohesion among themselves, and will not be removed from the deposit [95,96]. The wax content in the deposits has shown a gradual increase with an increase in flow rate [97].

Dwivedi et al. [97] showed that both shear stress and Reynolds number were found to be correlating parameters for decrease in deposit thickness. Deposit thickness decreases with increase in wall shear stress/Reynolds number and effective ΔT.

In a low shear environment such as laminar flow regime, the paraffinic molecules tend to be aligned adjacent to one another in the direction of flow, thereby inducing and accelerating the nucleation process [9]. Low flow rates also affect wax deposit, mainly because of the longer residence time of the oil in the tubing [95].

Recently, Daraboina et al. [98], using a state-of-the-art microscopic in situ visualization technique, showed that there exists a direct relation between the flow regime and the particulate deposition mechanism, where the relation is strongly noticed at low Reynolds number. The deposit thickness and growth rate of the deposit in laminar and transition flow regimes were higher than those calculated under the turbulent flow regime. This is principally attributed to the fact that a larger number of crystals is available for deposition under laminar and transition conditions, compared to turbulent flow with high Reynolds numbers, where large shear forces decrease the particles available for deposition.

7.5.4 Effect of time

Studies showed that composition of deposits changes with time, which is related to deposit aging. It is found that the wax content and higher carbon number waxes in the deposit increase as time increases, and thus the hardness of the deposit increases with time [75].

Experimental studies showed that the amount of wax deposition increased with increasing time, reaching a maximum and becoming constant, before gradually tailing off as the time increased to higher values. The drop in paraffin wax deposition weight at longer times could be attributed to the thermal insulation by the deposited wax layer and the variation in the amount of paraffin wax particles available for deposition [95].

The deposit thickness was found to grow initially but to shrink over time as the deposit composition enriched in wax content and the wax concentration in the surrounding oil decreased [99].

7.5.5 Effect of oil composition

The paraffin wax concentration in crude oil is not a direct cause of wax deposition. Some crudes with high wax content were observed flowing smoothly with no problems, while others with low wax content were observed with troubled flowing and plugging issues.

In addition, two oils with similar wax content can widely differ in their wax deposition tendency and in their WAT values [8]. The crude oil has different components, which seem to have an effect on wax deposition and WAT. These are:

a. Light hydrocarbons were found to be good solvents for paraffin waxes. Examples are methane gas and fractions lighter than C3, which were found to be good solvents for wax molecules [13]. As methane, ethane, propane, and butane are lost from the crude, the paraffins increase in volume compared to the rest of the crude. The WAT will also increase as the gas and gas liquids are lost.
b. Gao [100] revealed that gas hydrate particles can facilitate wax precipitation and wax deposition.
c. The isoparaffins tend to delay the formation of wax nuclei and usually form unstable wax solids [9].
d. Aromatics, on the other hand, are known to be good solvents for paraffinic waxes [9].
e. Naphthenes, or cycloparaffins, are stiff and bulky in nature; they tend to disturb and/or disrupt the wax nucleation and growth processes [9,14].
f. Resins do not have a direct effect on wax deposition, unlike their role in asphaltene stability.
g. Asphaltenes: the influence of asphaltenes is inconclusive and sometimes contradictory. This requires some discussion, as follows.

Some studies confirm that asphaltene presence decreases PP, while others affirm that it increases it, and yet a third team showed that asphaltenes have no effect on PP [101]. For those who reported that asphaltenes enhance wax deposition, they reported that asphaltenes induce a wax nucleation process by lowering the energy barrier for forming the critical wax nucleus. Those who reported that asphaltene solids reduced the wax deposition stated that the amorphous nature of asphaltene solids acted as inhibitors for wax deposition [9]. It appears that whether asphaltenes contribute to or prevent paraffin wax deposition depends on their relative concentrations, sizes, and characters. Further investigation needs to be conducted to clarify this issue [9]. The idea that the WAT is a function of asphaltene surface area has been proposed and supported by experiment [102].

Tinsley et al. [103] studied the wax precipitation in the presence of asphaltenes and showed that at low asphaltene concentrations and relatively high wax concentrations, asphaltenes do not have a significant effect on the yield stress of the waxy gel. However, at higher asphaltene concentrations, the asphaltenes behave as wax crystal modifiers, by decreasing the yield stress and distorting the crystal morphology. This can happen at a threshold ratio of wax/asphaltenes. Asphaltenes that are well dispersed at low concentration can reduce the WAT and act as nucleation inhibitors. However, when large asphaltene aggregates were present at higher concentrations, they increased the wax precipitation temperature and served as nucleation sites for wax growth.

Flocculated asphaltenes might coprecipitate with the waxes as well, but they cannot be fully and tightly incorporated in the wax network because of spatial interference, which reduces the yield stress of the formed wax [102].

Another study indicated that asphaltenes may have synergistic effects with a wax deposition inhibitor [104]. Recent technical papers have also shown how asphaltene inhibitors and solvents used to remediate or prevent asphaltene deposition can inadvertently increase wax deposition and/or change the characteristics of the wax in the deposits [105,106].

Yang et al. [107] found that the asphaltenes with lower polarity have higher H/C ratio and can disperse into smaller associating particles in the synthetic waxy oils, which is beneficial for the cocrystallization interactions between asphaltenes and wax molecules, and also can provide more nucleating sites for wax molecules. Decreasing polarity of asphaltenes leads to changes in wax crystal morphologies, as they changed from long needle-like to small needle-like or nearly spherical, which greatly improves the rheological properties of synthetic waxy oils. The addition of asphaltenes with lower polarity can significantly reduce the wax deposition rate but increase the aging rate: that is, a thinner and harder wax deposit tends to form with the decrease of asphaltene polarity.

h. Other components: like N_2 gas, which was found not to act as a solvent of the paraffins present, whether of the macro- or microcrystalline type, but in fact increased the pressure as using the nitrogen gas increased the WAT [13]. Contrary to this, increasing the CO_2 concentration decreased the WAT [108]. Hosseinipour et al. [84] reported a decrease in WAT of the crude oils to about 8.53°C below its original value when CO_2 was injected.

i. Mixing two different oils together: while this occasion is not likely to cause any precipitation issues as in the case of asphaltenes, it may enhance fluids flowing if one of the oils has a high temperature above the WAT, which helps in melting the deposited paraffins and enhances the crude rheology. In this case, however, using asphaltene control chemicals might be necessary if they showed asphaltene precipitation upon mixing.

After the deposit layer is formed, increasing its wax content increases the deposit yield stress, which is an important value for pigging and system restart operations.

Water in waxy crude can affect its gelation and wax deposition. Generally, the emulsion viscosity is higher than either oil or water at single phase. The presence of water above a certain threshold can significantly promote gel formation, modifying both the PP and the yield strength and drastically increasing viscosity [73,109]. They observed a sharp increase in shear viscosity, yield stress, and PP for waxy crude oil emulsions with above 25%–30% volume of dispersed water, as demonstrated in Fig. 7.27.

Nonetheless, very high water cuts (as high as 85%) could lead to the nonformation of wax deposit or the occurrence of a very thin and hard deposit [110]. Zhang et al. [111] were able to confirm that wax deposition rate decreases with the increasing water cut, maintaining the temperature gradient constant. Furthermore, they were also able to conclude that the wax deposition rate decreases with increasing emulsion temperature, if the temperature difference between the emulsion and the cooler surroundings is kept constant (for the same water cut). On the other hand, the same authors observed that, still maintaining the water cut, the wax deposition rate is higher for the higher temperature differences between emulsion and surroundings.

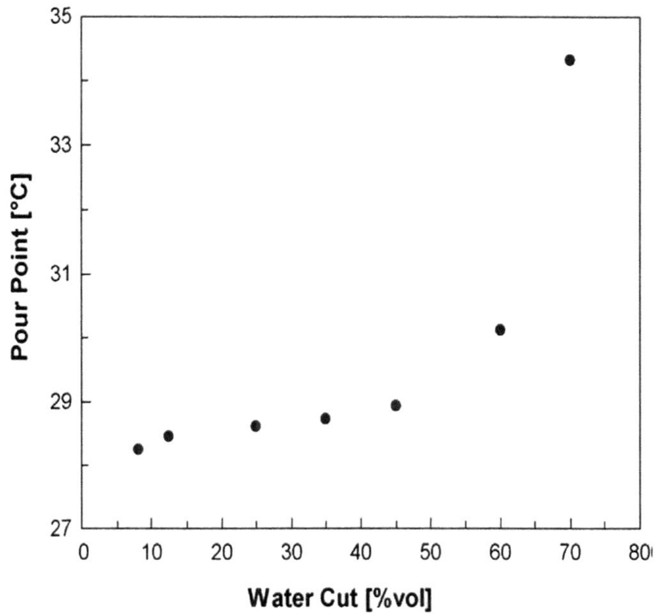

FIG. 7.27

Pour point of the waxy crude oil emulsion upon the water content.

Reprinted from R.F.G. Visintin, T.P. Lockhart, R. Lapasin, P. D'Antona, Structure of waxy crude oil emulsion gels, J. Non-Newtonian Fluid Mech. 149 (1–3) (2008) 34–39, https://doi.org/10.1016/j.jnnfm.2007.07.008 copyright (2008), with permission from Elsevier.

7.5.6 The effect of surface properties

It is proposed that paraffin deposition is initiated by the direct nucleation of wax on or adjacent to the pipe wall and that the deposit grows by the diffusion of wax from solution to the previously deposited wax. It was also reported that paraffin deposits are held in place by surface roughness and not by forces related to the composition of the surface that causes the paraffin to adhere to the surface [112]. Surface defects, high roughness, and corrosion spots are more vulnerable to paraffin deposition than smooth spots.

Experimental tests showed that the deposit does not adhere to a polished stainless-steel surface or a polished cold-rolled steel surface, but does adhere to a sand-blasted stainless-steel surface. The deposit appeared to adhere to the rusty spots of the cold-rolled steel plate but not to the smooth spots. Thus it is evident that the deposits formed on the surface are held in place by surface roughness. Also, some crudes may contain surface active materials that cause the deposit to adhere to the steel surface. Plastic pipe coatings inhibit paraffin deposition, mainly due to their smooth surface rather than their chemical nature [112].

7.5.7 Effect of corrosion inhibitor

It has been observed that the use of corrosion inhibitors can induce wax deposition. A protective corrosion inhibitor film, such as that formed from oleic imidazoline, can generate an ordered layer of long chain alkane molecules with a structure akin to that found in the bulk alkane crystal, which is more likely to act as a nucleation site for wax deposition [113].

7.5.8 Effect of hydrate inhibitor

Methanol is used as an effective way to control gas hydrates in gas wells. Methanol can stop hydrate formation but it can also cause the precipitation of wax. Wax deposition was observed downstream of methanol injection points when methanol was mixed into produced hydrocarbon liquids, as it can precipitate the longer chain, high melting paraffin. The ratio of methanol to condensate that causes wax deposition varies from well to well. In Michigan, methanol was used to inhibit gas hydrates by a gas producer, and a wax deposition problem was experienced in the flowlines of gas wells. Analysis of the problem indicated that the methanol was no longer needed, since the pressures had dropped below hydrate formation pressure. Coincidentally, after shutting off the methanol, the wax problem disappeared. Bottle tests confirmed that methanol was the main cause of wax deposition from the condensate being produced [93].

Recent field studies by four independent US operators who encountered wax deposition showed that the solubility of higher molecular weight waxes was reduced in the presence of methanol and MEG in the hydrocarbon phase. This effect was most pronounced in those paraffins with more than 45 carbons in the chains, where it was found that methanol and MEG forced deposits to precipitate above the original WAT measured for that particular crude. DCS measurements reveal that the WAT is a variable parameter for these oils, which can be raised by increasing the amount of THIs that partition into the hydrocarbon phase [114]. The addition of methanol and MEG to crudes containing C46–C100+ paraffin fraction could significantly change the characteristics of wax precipitation and deposition.

7.6 Recent advances in wax deposition research

Since the early operations of oil and gas, wax congealing and deposition have been in the circle of attention. The first recognition of the wax deposition issue was in 1923, and awareness has been expanding exponentially from the 1980s to currently.

Throughout the wax research studies, different techniques have been used to characterize and understand the behavior of wax and waxy crude oils. Examples of these techniques are DSC, CPM, FT-IR, NMR, GC, gas chromatography, viscometry, XRD, mass spectrometry, elemental analysis, and others [115,116].

The deposition experiments include the conventional cold finger and flow loop experiments, with flow loop trials being dynamic and more representative of field conditions. However, both methods have shown some limitations, like sample volume and test duration, making them impractical for rapid chemical control product performance screening. Some updates were made on this equipment for better representation of the field conditions and avoidance of the limitations of the conventional equipment. One of the recently innovated pieces of equipment is the dynamic paraffin deposition cell (DPDC) for enhanced mixing and multiphase deposition trials. Another recent advance in the testing equipment is the coaxial shear cold finger (CSCF), which gives better representation of the field shear conditions [30].

Wax precipitation and deposition experiments are conventionally conducted on single phase crude oil. However, recently, wax deposition in multiphase systems has been studied by many researchers [72,73,117–120].

Wax modeling has seen immense research and advances in the past years. However, the current commercial and academic models still suffer from uncertainties. Consequently, more work in the future is expected to optimize the current models and innovate more reliable models, by improving the precipitation and deposition experimental equipment, improving the field characterization techniques, and enhancing the collaboration between different institutes and industrial parties.

7.7 Summary

Wax gelation and deposition in oil and gas production systems have been discussed in this chapter. Wax molecules precipitate when the temperature falls below a certain limit, known as the wax appearance temperature, or WAT. Several theories have been frequently used to explain wax precipitation, including molecular diffusion, shear dispersion, gravity settling, Brownian motion, and others. Among those, molecular diffusion is the most acceptable mechanism among academics and industry experts, without denying the possibility of involvement of other mechanisms to some extent in wax precipitation. The factors affecting wax deposition were discussed in some detail. Temperature is the primary factor that controls wax precipitation and deposition; pipe wall temperature, bulk oil temperature, and pipe wall-bulk oil temperature gradient are major temperature parameters that control wax precipitation rate, texture, and hardness of the formed deposits. Pressure affects the formation of wax deposits in one way by stripping the light components, and in another way by causing fluid cooling through the Joule-Thomson effect when a sudden pressure drop takes place. In addition, fluid dynamics affects wax deposition and the formed deposit thickness and texture, where high shear environments experience low deposition. Furthermore, the composition of crude oil affects the wax deposition, with some components like light ends and aromatics working as good solvents for wax, while asphaltenes may or may not induce wax nucleation and growth. Moreover, some additives can influence the wax deposition, i.e., corrosion inhibitors and hydrate inhibitors, which can induce wax precipitation.

References

[1] J. Bellarby, Chapter 7—Production Chemistry, Developments in Petroleum Science, 2009, https://doi.org/10.1016/S0376-7361(08)00207-0.

[2] D.C. Thomas, Paraffin Manual, Petrolite Corporation, Tretolite Division, 1976.

[3] M.A. Kelland, Production Chemicals for the Oil and Gas Industry, CRC Press, 2009, https://doi.org/10.1201/9781420092974.

[4] M.M. El-Dalatony, B.H. Jeon, E.S. Salama, M. Eraky, W.B. Kim, J. Wang, T. Ahn, Occurrence and Characterization of Paraffin Wax Formed in Developing Wells and Pipelines, Energies. MDPI AG, 2019, https://doi.org/10.3390/en12060967.

[5] M. Del Carmen Garcia, Paraffin Deposition in Oil Production, Society of Petroleum Engineers, 2001, https://doi.org/10.2118/64992-MS.

[6] M.M. Sharma, in: J.D. Clegg (Ed.), Formation damage in Petroleum Engineering Handbook, vol. IV, PRODUCTION OPERATIONS ENGINEERING, 2007.

[7] W.D. McCain Jr., The Properties of Petroleum Fluids, PennWell Publishing Company, Tulsa, OK, 1990.

[8] Z. Huang, S. Zheng, H. Scott Fogler, Wax Deposition: Experimental Characterizations, Theoretical Modeling, and Field Practices, CRC Press, 2015, pp. 1–8.

[9] A. Hammami, M.A. Raines, Paraffin deposition from crude oils: comparison of laboratory results with field data, SPE J. 4 (1) (1999) 9–18, https://doi.org/10.2118/54021-PA.

[10] M. Rehan, A.S. Nizami, O. Taylan, B.O. Al-Sasi, A. Demirbas, Determination of wax content in crude oil, Pet. Sci. Technol. 34 (9) (2016) 799–804, https://doi.org/10.1080/10916466.2016.1169287.

[11] J.G. Speight, Hydrocarbons from crude oil, in: Handbook of Industrial Hydrocarbon Processes, Elsevier, 2020, pp. 95–142, https://doi.org/10.1016/b978-0-12-809923-0.00003-5.

[12] M. Dirand, M. Bouroukba, V. Chevallier, D. Petitjean, E. Behar, V. Ruffier-Meray, Normal alkanes, multialkane synthetic model mixtures, and real petroleum waxes: crystallographic structures, thermodynamic properties, and crystallization, J. Chem. Eng. Data 47 (2002) 115–143, https://doi.org/10.1021/je0100084.

[13] L.C. Vieira, M.B. Buchuid, E.F. Lucas, Effect of pressure on the crystallization of crude oil waxes. II. Evaluation of crude oils and condensate, Energy Fuel 24 (4) (2010) 2213–2220, https://doi.org/10.1021/ef900761t.

[14] K.A. Ferworn, D.B.R. Fluid, A. Hammami, D.B. Robinson, H. Ellis, Control of wax deposition: an experimental investigation of crystal morphology and an evaluation of various chemical solvents, in: Proceedings of the SPE International Symposium on Oilfield Chemistry, Houston, TX, USA, 18–21 February, 1997, pp. 1–20.

[15] A. Aiyejina, D.P. Chakrabarti, A. Pilgrim, M. Sastry, Wax formation in oil pipelines: a critical review, Int. J. Multiphase Flow 37 (2011) 671–694.

[16] S. Mokhatab, W.A. Poe, J.Y. Mak, Raw gas transmission, in: Handbook of Natural Gas Transmission and Processing, Elsevier, 2019, pp. 103–176, https://doi.org/10.1016/b978-0-12-815817-3.00003-4.

[17] C.E. Reistle Jr., O.C. Blade, Paraffin and Congealing-Oil Problems. Factors Responsible for Deposition of Paraffin and Congealing Crude Oils; Analysis of Crude Waxes or Paraffin From Crude Oils. United States: N.P., Web, 1931.

[18] N.V. Bhat, A.K. Mehrotra, Measurement and prediction of the phase behavior of wax-solvent mixtures: significance of the wax disappearance temperature, Ind. Eng. Chem. Res. 43 (13) (2004) 3451–3461.

[19] B. Gong, H. Liang, S. Xin, K. Li, Effect of water injection on reservoir temperature during power generation in oil fields, in: Thirty-Sixth Workshop on Geothermal Reservoir Engineering, 2011.

[20] D. Loeve, C. Hofstee, J.G. Maas, Thermal effects in a depleted gas field by cold CO_2 injection in the presence of methane, Energy Procedia 63 (2014) 5378–5393, https://doi.org/10.1016/j.egypro.2014.11.569.

[21] K.C. Misra, Understanding Mineral Deposits, Springer Netherlands, 2000, https://doi.org/10.1007/978-94-011-3925-0.

[22] B. Yadali Jamaloei, K. Asghari, The Joule-Thomson effect in petroleum fields. I. Well testing, multilateral/slanted wells, hydrate formation, and drilling/completion/production operations, Energy Sour. A: Recov. Util. Environ. Effects 37 (2) (2015) 217–224, https://doi.org/10.1080/15567036.2010.551258.

[23] B. Yadali Jamaloei, K. Asghari, The Joule-Thomson effect in petroleum fields. II. CO_2 sequestration, wellbore temperature profiles, and thermal stresses and wellbore stability, Energy Sour. A: Recov. Util. Environ. Effects 37 (3) (2015) 236–244, https://doi.org/10.1080/15567036.2011.557701.

[24] K. Rosvold, Wax Deposition Models, Master Thesis, Norwegian University of Science and Technology (NTNU), 2008.

[25] Z. Jeirani, A. Lashanizadegan, S. Ayatollahi, J. Javanmardi, The possibility of wax formation in gas fields: a case study, J. Nat. Gas Chem. 16 (3) (2007) 293–300, https://doi.org/10.1016/S1003-9953(07)60062-0.

[26] J.A.P. Coutinho, J.L. Daridon, The limitations of the cloud point measurement techniques and the influence of the oil composition on its detection, Pet. Sci. Technol. 23 (9–10) (2005) 1113–1128, https://doi.org/10.1081/LFT-200035541.

[27] P. Juyal, T. Cao, A. Yen, R. Venkatesan, Study of live oil wax precipitation with high-pressure micro-differential scanning calorimetry, Energy Fuel 25 (2011) 568–572, https://doi.org/10.1021/ef101213x.

[28] Solids Detection System SLB Laser Based Solids Detection System, Retrieved from: http://www.vinci-technologies.com/products-explo.aspx?IDM=753562&IDR=113221&IDR2=113124. (last visited November 2019).

[29] D.W. Jennings, K. Weispfennig, Effects of shear and temperature on wax deposition coldfinger investigation with a Gulf of Mexico crude oil, Energy Fuel 19 (2005) 1376–1386, https://doi.org/10.1021/ef049784i.

[30] L. Padula, S. Gon, C. Russell, E. Pohl, I. Littlehales, A. Neilson, Systematic study of temperature and different types of mixing in paraffin deposition test methods, ACS Omega 5 (21) (2020) 12320–12328, https://doi.org/10.1021/acsomega.0c01069.

[31] H. Chen, S. Yang, X. Nie, X. Zhang, W. Huang, Z. Wang, H. Wei, Ultrasonic detection and analysis of wax appearance temperature of Kingfisher Live Oil, Energy Fuel 28 (4) (2014) 2422–2428, https://doi.org/10.1021/ef500036u.

[32] T.G. Monger-McClure, J.E. Tackett, L.S. Merrill, Comparisons of Cloud Point Measurement and Paraffin Prediction Methods, Society of Petroleum Engineers, 1999, https://doi.org/10.2118/54519-PA.

[33] A. Hammami, J. Ratulowski, J.A.P. Coutinho, Cloud points: can we measure or model them? SPE Reprint Ser. 58 (2004) 144–154, https://doi.org/10.1081/LFT-120018524.

[34] B. Coto, C. Martos, J.J. Espada, M.D. Robustillo, D. Merino-García, J.L. Peña, Study of new methods to obtain the n-paraffin distribution of crude oils and its application to flow assurance, Energy Fuel 25 (2011) 487–492, https://doi.org/10.1021/ef100987v.

[35] H.K. Abdel-Aal, M.A. Aggour, M.A. Fahim, Petroleum and Gas Field Processing, CRC Press, 2015, https://doi.org/10.1201/9780429021350.

[36] A. Majhi, Y.K. Sharma, V.S. Kukreti, K.P. Bhatt, R. Khanna, Wax content of crude oil: a function of kinematic viscosity and pour point, Pet. Sci. Technol. 33 (4) (2015) 381–387, https://doi.org/10.1080/10916466.2014.986277.

[37] R. Hollamby, G. Nuis, Introduction to Production Chemistry, Shell International Exploration and Production BV, 2000.

[38] V.F. Nikolaev, A.V. Egorov, I.V. Nikolaev, R.B. Sultanova, A method of testing the pour point of petroleum products on a refrigerated sloping surface, Pet. Sci. Technol. 31 (3) (2013) 276–283, https://doi.org/10.1080/10916466.2010.547900.

[39] A.M. Sousa, H.A. Matos, L. Guerreiro, Wax deposition mechanisms and the effect of emulsions and carbon dioxide injection on wax deposition: critical review, Petroleum (2020), https://doi.org/10.1016/j.petlm.2019.09.004.

[40] D. Burger, T.K. Perkins, J.H. Striegler, Studies of wax deposition in the trans Alaska pipeline, J. Pet. Technol. 33 (1981) 1075–1086, https://doi.org/10.2118/8788-PA.

[41] R. Wilke, P. Chang, Correlation of diffusion coefficients in dilute solutions, AICHE J. 1 (1955) 264–270, https://doi.org/10.1002/aic.690010222.

[42] W. Hayduk, B. Minhas, Correlations for prediction of molecular diffusivities in liquids, Can. J. Chem. Eng. 60 (1982) 295–299.

[43] S. Correra, A. Fasano, L. Fusi, M. Primicerio, F. Rosso, Wax diffusivity under given thermal gradient: a mathematical model, ZAMM Zeitschr. Angew. Math. Und Mech. 87 (2007) 24–36, https://doi.org/10.1002/zamm.200510293.

[44] D.W. Green, Perry's Chemical Engineers' Handbook, eighth ed., McGraw-Hill, New York, 2008.

[45] M.A. Theyab, Wax deposition process: mechanisms, affecting factors and mitigation methods, Open Access J. Sci. 2 (2) (2018). https://doi.org/10.15406/oajs.2018.02.00054.

[46] P. Singh, H. Ramachandran Venkatesan, S. Fogler, N. Nagarajan, Formation and aging of incipient thin film wax-oil gels, AICHE J. 46 (5) (2000) 1059–1074, https://doi.org/10.1002/aic.690460517.

[47] C. Eckstein, D.G. Bailey, A.H. Shapiro, Self-diffusion of particles in shear flow of a suspension, J. Fluid Mech. 79 (1977) 191–208, https://doi.org/10.1017/S0022112077000111.

[48] L. Fusi, On the stationary flow of a waxy crude oil with deposition mechanisms, Nonlinear Anal. Theory Methods Appl. 53 (2003) 507–526, https://doi.org/10.1016/S0362-546X(02)00315-2.

[49] W. Wang, Q. Huang, C. Wang, S. Li, W. Qu, J. Zhao, M. He, Effect of operating conditions on wax deposition in a laboratory flow loop characterized with DSC technique, J. Therm. Anal. Calorim. 119 (1) (2015) 471–485, https://doi.org/10.1007/s10973-014-3976-z.

[50] L.F. Fasano, S. Correra, Mathematical models for waxy crude oils, Meccanica 39 (2004) 441–482, https://doi.org/10.1023/b:mecc.0000046444.98941.3c.

[51] P.A. Bern, V.R. Withers, R.J.R. Cairns, Wax deposition in crude oil pipelines, in: European Offshore Petroleum Conference & Exhibition, Earls Court, London, 1980, p. 571.

[52] P.G. Saffman, The lift on a small sphere in a slow shear flow, J. Fluid Mech. 22 (1965) 385–400.

[53] M. Garcia, F. Lopez, Y. Nino, Characterization of near-bed coherent structures in turbulent open channel flow using synchronized high-speed video and hot-film measurements, Exp. Fluids 19 (1995) 16–28.

[54] F.A. Azevedo, A.M. Teixeira, A critical review of the modeling of wax deposition mechanisms, Pet. Sci. Technol. 21 (2003) 393–408, https://doi.org/10.1081/LFT-120018528.

[55] B. Edmonds, T. Moorwood, R. Szczepanski, X. Zhang, Simulating wax deposition in pipelines for flow assurance, Energy Fuel 22 (2008) 729–741, https://doi.org/10.1021/ef700434h.

[56] D. Merino-Garcia, M. Margarone, S. Correra, Kinetics of waxy gel formation from batch experiments, Energy Fuel 21 (2007) 1287–1295, https://doi.org/10.1021/ef060385s.

[57] R. Banki, H. Hoteit, A. Firoozabadi, Mathematical formulation and numerical modeling of wax deposition in pipelines from enthalpy-porosity approach and irreversible thermodynamics, Int. J. Heat Mass Transf. 51 (13–14) (2008) 3387–3398, https://doi.org/10.1016/j.ijheatmasstransfer.2007.11.012.

[58] J. Buongiorno, Convective transport in nanofluids, J. Heat Transf. 128 (2006) 240, https://doi.org/10.1115/1.2150834.

[59] G.S. McNab, A. Meisen, Thermophoresis in liquids, J. Colloid Interface Sci. 44 (1973) 339–346, https://doi.org/10.1016/0021-9797(73)90225-7.

[60] Z. Liu, Y. Li, W. Wang, G. Song, Z. Lu, Y. Ning, Wax and wax-hydrate deposition characteristics in single-, two-, and three-phase pipelines: a review, Energy Fuel 34 (11) (2020) 13350–13368, https://doi.org/10.1021/acs.energyfuels.0c02749.

[61] J.G. Speight, Wax deposition and fouling, in: Fouling in Refineries, Elsevier, 2015, pp. 155–173, https://doi.org/10.1016/b978-0-12-800777-8.00007-3.

[62] P. Singh, A. Youyen, H. Scott Fogler, Existence of a critical carbon number in the aging of a wax-oil gel, AICHE J. 47 (9) (2001) 2111–2124, https://doi.org/10.1002/aic.690470921.

[63] D.H. Templeton, Crystals: their role in nature and in science (Bunn, Charles), J. Chem. Educ. 42 (7) (1965) A550, https://doi.org/10.1021/ed042pa550.

[64] G. Totten, S. Westbrook, R. Shah, Fuels and Lubricants Handbook: Technology, Properties, Performance, and Testing, ASTM International., 2003, https://doi.org/10.1520/mnl37-eb.

[65] M. Dirand, V. Chevallier, E. Provost, M. Bouroukba, D. Petitjean, Multicomponent paraffin waxes and petroleum solid deposits: structural and thermodynamic state, Fuel 77 (1998) 1253.

[66] G.A. Holder, J. Winkler, Wax crystallization from distillate fuels. I. cloud and pour phenomena exhibited by solutions of binary n-paraffin mixtures, J. Inst. Petrol. 51 (1965) 228.

[67] H.S. Ashbaugh, A. Radulescu, R.K. Prud'homme, D. Schwahn, D. Richter, L.J. Fetters, Interaction of paraffin wax gels with random crystalline/amorphous hydrocarbon copolymers, Macromolecules 35 (18) (2002) 7044–7053, https://doi.org/10.1021/ma0204047.

[68] R. Venkatesan, N.R. Nagarajan, K. Paso, Y.B. Yi, A.M. Sastry, H.S. Fogler, The strength of paraffin gels formed under static and flow conditions, Chem. Eng. Sci. 60 (13) (2005) 3587–3598, https://doi.org/10.1016/j.ces.2005.02.045.

[69] J.M. Létoffé, P. Claudy, M.V. Kok, M. Garcin, J.L. Volle, Crude oils: characterization of waxes precipitated on cooling by D.S.C. and thermomicroscopy, Fuel 74 (1995) 810.

[70] P. Singh, H.S. Fogler, N. Nagarajan, Prediction of the wax content of the incipient wax-oil gel in a flowloop: an application of the controlled-stress rheometer, J. Rheol. 43 (1999) 1437.

[71] M.A. Theyab, P. Diaz, Experimental study on the effect of spiral flow on wax deposition thickness, Society of Petroleum Engineers, 2016, https://doi.org/10.2118/181954-MS.

[72] A. Matzain, M.S. Apte, H. Zhang, M. Volk, J.P. Brill, J.L. Creek, Investigation of paraffin deposition during multiphase flow in pipelines and wellbores. Part 1. Experiments, J. Energy Res. Technol. 124 (2002) 180–186.

[73] R.F.G. Visintin, T.P. Lockhart, R. Lapasin, P. D'Antona, Structure of waxy crude oil emulsion gels, J. Non-Newtonian Fluid Mech. 149 (1–3) (2008) 34–39, https://doi.org/10.1016/j.jnnfm.2007.07.008.

[74] N.F. Carnahan, Paraffin deposition in petroleum production, J. Pet. Technol. 41 (1989) 1024.

[75] Q. Quan, J. Gong, W. Wang, G. Gao, Study on the aging and critical carbon number of wax deposition with temperature for crude oils, J. Pet. Sci. Eng. 130 (2015) 1–5, https://doi.org/10.1016/j.petrol.2015.03.026.

[76] D.W. Jennings, K. Weispfennig, Effects of shear and temperature on wax deposition: coldfinger investigation with a Gulf of Mexico crude oil, Energy Fuel 19 (2005) 1376–1386.

[77] J.L. Creek, H.J. Lund, J.P. Brill, M. Volk, Wax deposition in single phase flow, Fluid Phase Equilib. 158 (1999) 801–811.

[78] H.O. Bidmus, A.K. Mehrotra, Solids deposition during "cold flow" of wax_solvent mixtures in a flow-loop apparatus with heat transfer, Energy Fuel 23 (2009) 3184–3194.

[79] Z. Huang, L. Yingda, R. Hoffmann, L. Amundsen, H. Scott Fogler, The effect of operating temperatures on wax deposition, Energy Fuel 25 (11) (2011) 5180–5188, https://doi.org/10.1021/ef201048w.

[80] M. Lashkarbolooki, A. Seyfaee, F. Esmaeilzadeh, D. Mowla, Experimental investigation of wax deposition in Kermanshah crude oil through a monitored flow loop apparatus, Energy Fuel 24 (2011) 1234–1241.

[81] R. Valinejad, A.R. Nazar, An experimental design approach for investigating the effects of operation factors on the wax deposition in pipelines, Fuel 106 (2013) 843–850.

[82] K.G. Paso, H.S. Fogler, Influence of n-paraffin composition on the aging of wax-oil gel deposits, AICHE J. 49 (12) (2003) 3241–3251.

[83] S. Han, Z. Huang, M. Senra, R. Hoffmann, H. Scott Fogler, Method to determine the wax solubility curve in crude oil from centrifugation and high temperature gas chromatography measurements, Energy Fuel 24 (3) (2010) 1753–1761, https://doi.org/10.1021/ef901195w.

[84] A. Hosseinipour, A.B. Japper-Jaafar, S. Yusup, The effect of CO_2 on wax appearance temperature of crude oils, Procedia Eng. 148 (2016) 1022–1029, https://doi.org/10.1016/j.proeng.2016.06.580.

[85] B. Coto, C. Martos, J.L. Peña, J.J. Espada, M.D. Robustillo, A new method for the determination of wax precipitation from non-diluted crude oils by fractional precipitation, Fuel 87 (10–11) (2008) 2090–2094, https://doi.org/10.1016/j.fuel.2007.12.012.

[86] K.J. Leontaritis, Asphaltene and wax deposition envelopes, Fuel Sci. Technol. Int. 14 (1–2) (1996) 13–39, https://doi.org/10.1080/08843759608947560.

[87] L.X. Nghiem, B.F. Kohse, in: L.W. Lake (Ed.), Asphaltenes and Waxes in Petroleum Engineering Handbook, General Engineering, vol. I, Society of Petroleum Engineers, 2007. John R. Fanchi, Ed.

[88] T.S. Brown, V.G. Niesen, D.D. Erickson, The effects of light ends and high pressure on paraffin formation, in: Paper SPE 28505 Presented at the 1994 SPE Annual Technical Conference and Exhibition, New Orleans, 25–28 September, 1994.

[89] M.K. Afdhol, M. Abdurrahman, F. Hidayat, F.K. Chong, H.F. Mohd Zaid, Review of solvents based on biomass for mitigation of wax paraffin in Indonesian oilfield, Appl. Sci. (Switz.) (2019), https://doi.org/10.3390/app9245499.

[90] Q. Quan, W. Ran, L. Yang, G. Gao, S. Wang, J. Gong, The effect of pressure on wax deposition from wax-solvent mixtures with natural gas, J. Pet. Sci. Eng. (2018), https://doi.org/10.1016/j.petrol.2018.08.040.

[91] B. Zohuri, Thermodynamic relations, in: Physics of Cryogenics: An Ultralow Temperature Phenomenon. Physics of Cryogenics: An Ultralow Temperature Phenomenon, Elsevier Inc., 2017, pp. 245–257, https://doi.org/10.1016/C2017-0-01796-2 (Chapter 10).

[92] T.Y. Makogon, Flow restrictions and blockages in operations, in: Handbook of Multiphase Flow Assurance, Elsevier, 2019, pp. 95–189, https://doi.org/10.1016/b978-0-12-813062-9.00005-1.

[93] K.M. Barker, J.M. Bigler, K. Hake, D.C. Sallee, Paraffin Problems in Gas Systems, Society of Petroleum Engineers, 2003, https://doi.org/10.2118/84827-MS.
[94] Y. Lu, Z. Huang, R. Hoffmann, L. Amundsen, H. Scott Fogler, Counterintuitive effects of the oil flow rate on wax deposition, Energy Fuel 26 (2012) 4091–4097, https://doi.org/10.1021/ef3002789.
[95] E.M. Kelechukwu, H.S.S. Al Salim, A.A.M. Yassin, Influencing factors governing paraffin wax deposition during crude production, Int. J. Phys. Sci. 5 (15) (2010) 2351–2362.
[96] V. Mahto, A. Kumar, Effect of several parameters on wax deposition in the flow line due to Indian waxy crude oil, Int. J. Appl. Eng. Res. Dev. (IJAERD) 3 (4) (2013) 1–10.
[97] P. Dwivedi, C. Sarica, W. Chang, Experimental study of wax deposition characteristics of a waxy crude oil under single phase turbulent flow conditions, in: Paper No., OTC 22953, Presented at the 2011 Offshore Technology Conference, Houston, Texas, USA, 2012.
[98] G. Santos, N. Daraboina, C. Sarica, Dynamic microscopic study of wax deposition: particulate deposition, Energy Fuel 35 (15) (2021) 12065–12074, https://doi.org/10.1021/acs.energyfuels.1c01684.
[99] L.H.A. Mahir, C. Vilas Bôas Fávero, T. Ketjuntiwa, H.S. Fogler, R.G. Larson, Mechanism of wax deposition on cold surfaces: gelation and deposit aging, Energy Fuel 33 (5) (2019) 3776–3786, https://doi.org/10.1021/acs.energyfuels.8b03139.
[100] S.Q. Gao, Investigation of interactions between gas hydrates and several other flow assurance elements, Energy Fuel 22 (2008) 3150–3153, https://doi.org/10.1021/ef800189k.
[101] E. Ariza-León, D.R. Molina-Velasco, A. Chaves-Guerrero, Review of Studies on Asphaltene—Wax Interaction and the Effect Thereof on Crystallization, CTyF—Ciencia, Tecnologia y Futuro. Ecopetrol S.A, 2014, https://doi.org/10.29047/01225383.32.
[102] P. Kriz, S.I. Andersen, Effect of asphaltenes on crude oil wax crystallization, Energy Fuel 19 (2005) 948.
[103] J.F. Tinsley, J.P. Jahnke, H.D. Dettman, K. Robert, Prud'home., Waxy gels with asphaltenes 1: characterization of precipitation, gelation, yield stress, and morphology, Energy Fuel 23 (4) (2009) 2056–2064, https://doi.org/10.1021/ef800636f.
[104] F.G. D'Avila, C.M.F. Silva, L. Steckel, A.C. Ramos, E.F. Lucas, Influence of asphaltene aggregation state on the wax crystallization process and the efficiency of EVA as a wax crystal modifier: a study using model systems, Energy Fuel 34 (4) (2020) 4095–4105, https://doi.org/10.1021/acs.energyfuels.9b04166.
[105] J.I. Aguiar, A.A. Pontifes, A. Nerris, J. Rogers, A. Mahmoudkhani, Impact of solvent treatments for asphaltenes on wax deposition and an efficient alternative with green surfactants, in: Offshore Technology Conference, 2020, https://doi.org/10.4043/30695-MS.
[106] J.I. Aguiar, A. Punase, C. Mazzeo, Influence of asphaltene inhibitors on wax and asphaltene deposition—Are problems associated? in: Offshore Technology Conference, Brazil, 2019, https://doi.org/10.4043/29817-MS.
[107] C. Li, H. Zhu, F. Yang, H. Liu, F. Wang, G. Sun, B. Yao, Effect of asphaltene polarity on wax precipitation and deposition characteristics of waxy oils, Energy Fuel 33 (8) (2019) 7225–7233, https://doi.org/10.1021/acs.energyfuels.9b01464.
[108] H. Pan, A. Firoozabadi, P. Fotland, Pressure and Composition Effect on Wax Precipitation: Experimental Data and Model Results, Society of Petroleum Engineers, 1997, https://doi.org/10.2118/36740-PA.
[109] M.C.K. De Oliveira, R.M. Carvalho, A.B. Carvalho, B.C. Couto, F.R.D. Faria, R.L.P. Cardoso, Waxy crude oil emulsion gel: impact on flow assurance, Energy Fuel 24 (2010) 2287–2293, https://doi.org/10.1021/ef900927g.
[110] A. Bruno, C. Sarica, H. Chen, M. Volk, Paraffin deposition during the flow of water in oil and oil-in-water dispersions in pipes, in: SPE Annu. Tech. Conf. Exhib, 2008, https://doi.org/10.2118/114747-MS SPE 114747.
[111] Y. Zhang, J. Gong, Y. Ren, P. Wang, Effect of emulsion characteristics on wax deposition from water-in-waxy crude oil emulsions under static cooling conditions, Energy Fuel 24 (2010) 1146–1155, https://doi.org/10.1021/ef901065c.

[112] E.B. Hunt, Laboratory Study of Paraffin Deposition, Society of Petroleum Engineers, 1962, https://doi.org/10.2118/279-PA.

[113] M.A. San-Miguel, P.M. Rodger, The effect of corrosion inhibitor films on deposition of wax to metal oxide surfaces, J. Mol. Struct. THEOCHEM 506 (2000) 263–272, https://doi.org/10.1016/S0166-1280(00)00418-8.

[114] J.I. Aguiar, A.A. Nerris, A. Mahmoudkhani, Can paraffin wax deposit above wax appearance temperature? A detailed experimental study, in: Paper Presented at the SPE Annual Technical Conference and Exhibition, Virtual, October, 2020, https://doi.org/10.2118/201297-MS.

[115] L.M. Alghanduri, M.M. Elgarni, J.-L. Daridon, J.A.P. Coutinho, Characterization of Libyan waxy crude oils, Energy Fuel 24 (2010) 3101–3107.

[116] B.J. Musser, P.K. Kilpatrick, N. Carolina, Molecular characterization of wax isolated from a variety of crude oils, Energy Fuel 59 (1998) 715–725.

[117] T.S. Raj, D.P. Chakrabarti, G. Das, Liquid-liquid stratified flow through horizontal conduits, Chem. Eng. Technol. 28 (2005) 899–907.

[118] D.P. Chakrabarti, G. Das, P.K. Das, The transition from water continuous to oil continuous flow pattern, AICHE J. 52 (2006) 3668–3678.

[119] D.P. Chakrabarti, G. Das, P.K. Das, Identification of stratified liquid-liquid flow through horizontal pipes by a non-intrusive optical probe, Chem. Eng. Sci. 62 (2007) 1861–1876.

[120] C. Sarica, E. Panacharoensawad, Review of paraffin deposition research under multiphase flow conditions, Energy Fuel 26 (2012) 3968–3978, https://doi.org/10.1021/ef300164q.

CHAPTER

Asphaltene Deposition

8

8.1 Introduction

Asphaltenes are the heaviest and most polar constituents in crude oil. They increase the crude oil's viscosity and density, hindering its transport, and are also responsible for its color (brown to black) [1].

The term asphaltene was possibly derived from the Accadian word *asphaltu* or *sphallo*, meaning "to split" and later was adapted to the Greek word *asphaltos*, meaning stable or secure, as the Romans used it to seal and secure their baths, reservoirs, and aqueducts. The word asphaltene was coined by the French agricultural chemist Jean Baptiste Boussingault in his 1837 [2] paper, "Memoir upon the Composition of Bitumens." Boussingault described how he distilled bitumen from Pechelbronn oilfield at about 250°C to obtain a distillate and a residue that he named petrolene and asphaltene, respectively [3].

Asphaltenes are defined operationally according to their solubility behavior as the fraction of oil that is insoluble in light aliphatic hydrocarbons such as pentane and heptane, but soluble in aromatic solvents such as toluene. A more generic rule for defining asphaltenes given by Speight [4] is that asphaltenes are insoluble in nonpolar solvents with a surface tension lower than that of 25 dyne cm^{-1} at 25°C (such as low-boiling petroleum naphtha, petroleum ether, *n*-pentane, isopentane, *n*-heptane, and the like), and they are soluble in liquids with a surface tension above 25 dyne cm^{-1} at 25°C, such as pyridine, carbon disulfide, carbon tetrachloride, and benzene.

Chemically, they are the heaviest components in crude oils as they consist of various polyaromatic structures with aliphatic side chains. They also contain heteroatoms such as sulfur, nitrogen, and oxygen, and metals such as nickel, vanadium, and iron, which makes their chemistry very different and more complex than that of waxes [5–7].

Asphaltenes are heavy molecules which have high molecular weight and high densities, are viscous, and tend to stick together and coagulate or aggregate to block the flow of the oil, which is why they often are called the cholesterol of crude oil, like the cholesterol in blood veins when it blocks blood flow.

Asphaltene aggregation and deposition represent a complex and major flow assurance problem in oil and gas fields. Asphaltenes stabilize petroleum emulsions, which interferes with the phase separation operations. They also aggregate and deposit in different locations in the production system, causing formation damage, flow restrictions, and equipment failures leading to production loss and extra expenses for mitigation and maintenance. Fig. 8.1 shows asphaltene deposits from an oil producer from the Egyptian field, and Fig. 8.2 shows asphaltene deposits removed during pigging operations.

Essentials of Flow Assurance Solids in Oil and Gas Operations. http://doi.org/10.1016/B978-0-323-99118-6.00020-4

FIG. 8.1

Asphaltene deposits collected from production well.

FIG. 8.2

Asphaltene deposits after pigging operations.

Furthermore, asphaltene deposits are commonly confused with paraffin wax deposits and vice versa due to their organic nature, although their texture and composition is markedly different [6]. The cost of treatment of asphaltene deposits in the well might be as high as US$3 M, while the financial loss due to lost production can be as high as $1.2 M/day, even in the scenario of a US$30/bbl oil price [8]. Fig. 8.3 shows the severity of the asphaltene problem and treatment history for one of the wells from Hassi Messaoud field [9].

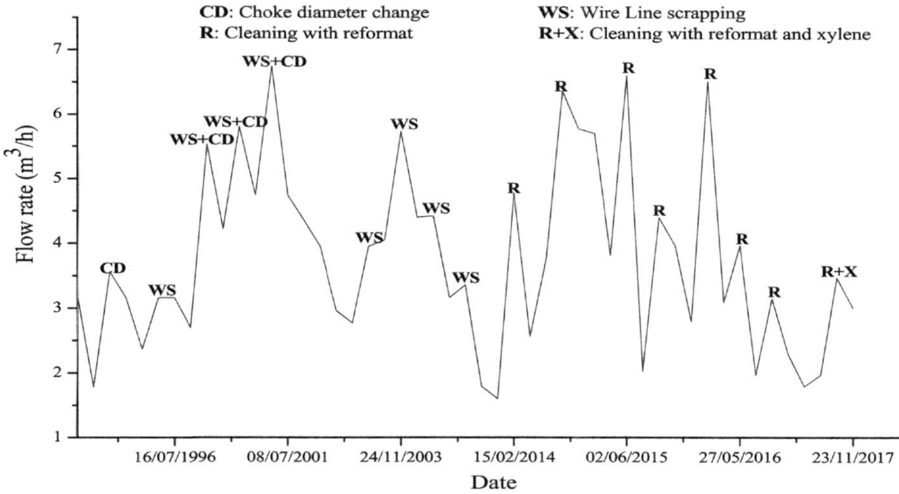

FIG. 8.3

The severity of the asphaltene problems by the drop in productivity and the frequency of intervention operations, in production well from Hassi Messaoud field.

From K. Gharbi, C. Benamara, K. Benyounes, M.A. Kelland, Toward separation and characterization of asphaltene acid and base fractions, Energy Fuel 35 (18) (2021) 14610–14617, https://doi.org/10.1021/acs.energyfuels.1c01999.

8.2 Chemical composition, structure, and nature of asphaltenes

8.2.1 Elemental composition and functional groups

The basic elements in asphaltenes include carbon, hydrogen, oxygen, nitrogen, and sulfur in addition to some metals. The percentage of each element can vary from oil to oil; however, average values can be summarized as follows: 76–86 wt% C, 7.3–8.5 wt% H, 5.0–9.0 wt% S, 0.7–1.2 wt% O, 1.3–1.4 wt% N, and 0.1–0.2 wt% metals (mostly Ni, V, and Fe) [5,10,11]. Various functional groups can be found in asphaltenes. Sulfur is present mostly as sulfidic and thiophenic groups and, to a lesser extent, as sulfoxide [12]. Asphaltene nitrogen is virtually totally present in aromatic groups such as pyrrolic, and to a lesser extent, pyridinic, groups and only very occasionally as tertiary amines [4,13]. Oxygen is present mainly in carbonyl and hydroxyl/phenolic groups. This includes ketones and carboxylic acids. The metals form complexes and impart electrical charge, which in turn may influence asphaltene deposition [5]. Kelland et al. [9] separated asphaltenes from Algerian oil into four fractions: acid, base, neutral, and amphoteric, pointing out that the neutral fraction was the dominant part in the asphaltene sample. They reported that both the acid and basic fractions were aromatic and polar compounds. The acid fraction included groups such as carboxylic acids, sulfoxide groups, and phenols, in addition to aliphatic chains, while the basic fraction has amines in significant amounts, sulfoxide groups, and aliphatic chains in its structure. Fig. 8.4 illustrates the structure of the asphaltene molecule [14].

Not all the asphaltenes have the same composition and properties. In fact, asphaltene composition and properties depend on the method used in extracting and separating them, which is considered one of

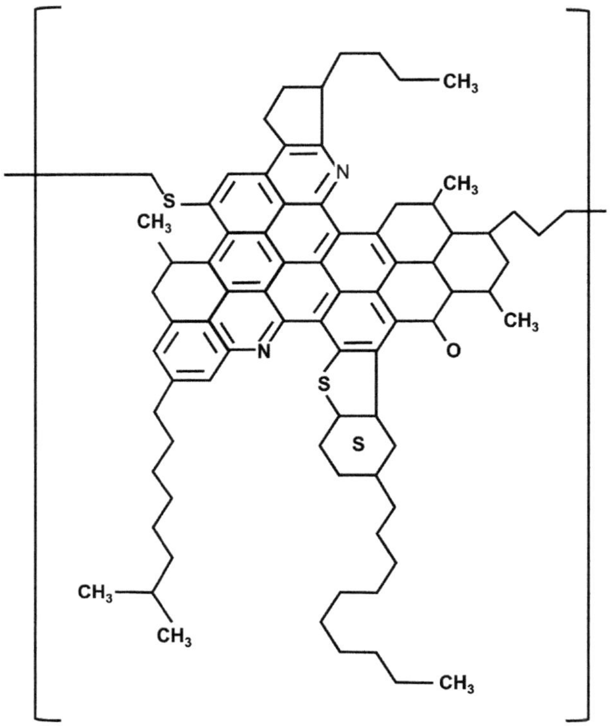

FIG. 8.4

Hypothetical structure of asphaltene molecule.

Reprinted *with permission from J.G. Speight, S.E. Moschopedis, On the molecular nature of petroleum asphaltenes, in: J.W. Bunger, N.C. Li (Eds.), Chemistry of Asphaltenes, in: Advances in Chemistry Series, vol. 195, ACS, 1982, pp. 1–15 (Chapter 1). Copyright 1982, American Chemical Society.*

the main confusions about asphaltenes. Some of the differences between asphaltenes according to the method of their extraction are discussed in the following paragraphs.

- Asphaltenes that are separated/extracted by a solvent: their properties depend on the solvent type (alky chain length), strength, volume, temperature, and the time allowed for precipitation to occur. The use of *n*-heptane as the precipitating medium yields an asphaltene precipitate that is considerably different from those precipitated by *n*-pentane. For example, the hydrogen-to-carbon atomic ratio (H/C) of the *n*-heptane asphaltene precipitate is usually lower than that of the *n*-pentane asphaltenes precipitate, indicating a higher degree of aromaticity in the *n*-heptane precipitate. Other features such as nitrogen-to-carbon (N/C), oxygen-to-carbon (O/C), and sulfur-to-carbon (S/C) ratios are usually higher in the *n*-heptane precipitate, indicating higher proportions of the hetero elements in this material [14,15]. The differences in the asphaltene yield can be attributed to the variation in the solubility parameter of the hydrocarbon solvent [16]. So in the literature asphaltenes are designated by the solvents used to separate them, i.e., C5 asphaltenes means asphaltenes that

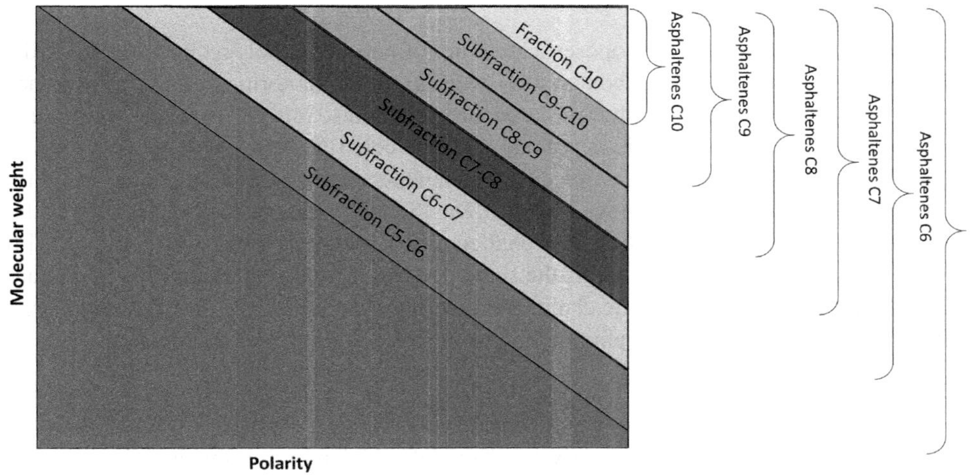

FIG. 8.5

Molecular weight and polarity of different *n*-alkane asphaltenes.

were separated by pentane, C7 asphaltenes are asphaltenes that were separated by heptane, and so on. Fig. 8.5 illustrates the different fractions of asphaltenes separated by different alkanes.

- Asphaltenes precipitated by pressure reduction are different from those precipitated by solvents: Pressure-precipitated asphaltenes are produced basically from live oil, while the alkane-produced asphaltenes are produced using dead oil. Pressure-precipitated asphaltenes were found to have fewer double bonds and more sulfur species than the alkane-separated asphaltenes.
- The asphaltenes precipitated in the lab were found to be different from those deposited in the field. In addition, the lab-precipitated asphaltenes are different in other ways, as it was found that asphaltenes deposited in the organic solid deposition cell (OSDC) are different from those separated by heptane addition [17].
- Asphaltenes are known to be different according to their source: Asphaltenes extracted from crude oil are different from those separated from coal or bitumen. Coal asphaltenes are smaller, more aromatic, and have shorter alkyl chains than petroleum asphaltenes [18]. Coal-derived asphaltenes are half the size of many crude oil asphaltenes [19].

Knowing the structure and the functional groups is of great practical importance, since it allows better understanding of the asphaltene behavior under the various operating parameters, and helps in optimizing the mitigation of asphaltene problems. For example, selecting the proper chemical inhibitor can be achieved through identifying the dominant active group, and thus acidic inhibitors are most suitable for asphaltenes with main active basic groups, and vice versa [9].

8.2.2 Asphaltene molecular structure

Two models have been proposed in the literature for the structure/architecture of asphaltenes: continental and archipelago.

8.2.2.1 The continental structure

The "continental" or "island" model proposes a structure of one single polycyclic aromatic hydrocarbon (PAH) (on average, the core has about six to seven fused aromatic rings) with cycloalkanes and external branched alkyl substituents with some heteroatoms. It is also referred to as the "like your hand" model, which depicts asphaltenes consisting of one aromatic core (the palm) with peripheral alkyl side chains (like fingers) [20]. The monomer molecular asphaltene structure has a molecular weight in the range of ~500–1000 Da, with an average of ~750 Da [5,21,22]. Suggesting this structure has no charged entities, a second island structure of asphaltenes was proposed by Kuznicki et al. [23] based on molecular dynamic simulations, which is the same like the regular one in addition to one anionic terminal carboxylic group on an aliphatic chain, and that model is termed the anionic island structure. Fig. 8.6 illustrates the continental and anionic continental structure of asphaltenes.

8.2.2.2 Archipelago structure

The second model is the archipelago or "rosary-type" model, which proposes that individual asphaltene monomers are composed of clusters of polycondensed groups consisting of five to seven aromatic rings each connected by short aliphatic side chains, possibly containing polar heteroatom bridges [5,24–26]. Fig. 8.7 illustrates the archipelago structure of asphaltenes.

Recent experiments using laser desorption and mass spectral measurements have showed the dominance of island structures [27,28]. Using atomic force and scanning tunneling microscopy [29] proved

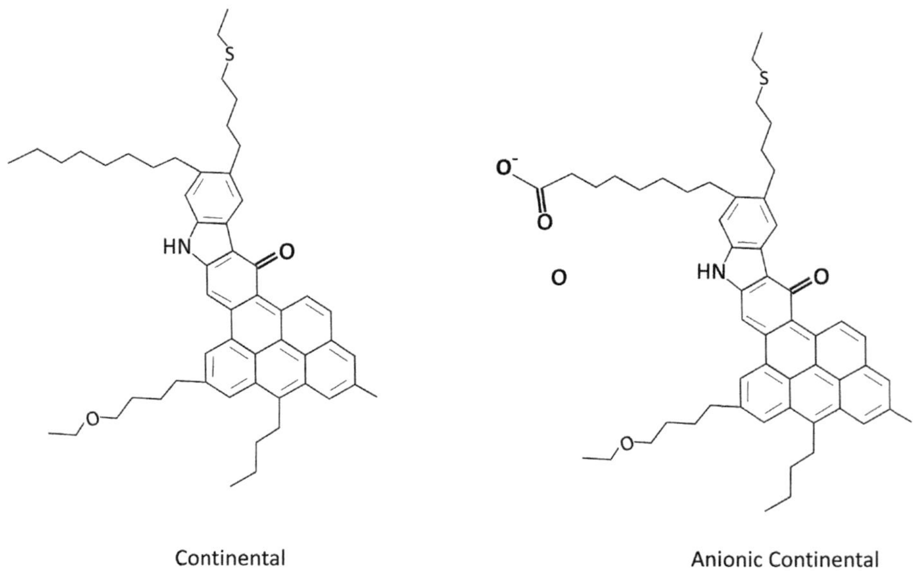

FIG. 8.6

Asphaltene continental structure.

Reprinted with permission from T. Kuznicki, J.H. Masliyah, S. Bhattacharjee, Molecular dynamics study of model molecules resembling asphaltene-like structures in aqueous organic solvent systems, Energy Fuels 22(4) (2008) 2379–2389, https://doi.org/10.1021/ef800057n. Copyright 2008, American Chemical Society.

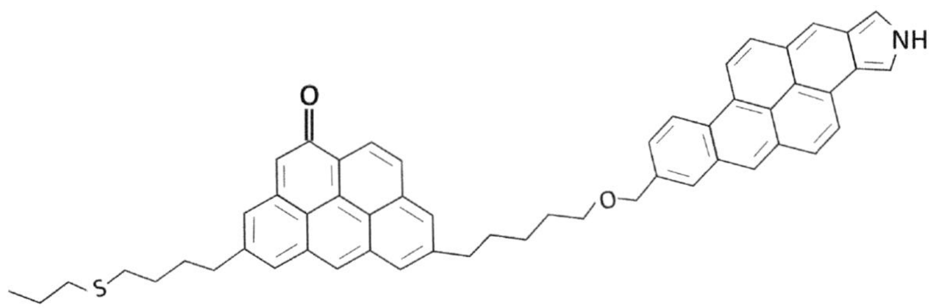

FIG. 8.7

Archipelago structure of asphaltenes.

Reprinted with permission from T. Kuznicki, J.H. Masliyah, S. Bhattacharjee, Molecular dynamics study of model molecules resembling asphaltene-like structures in aqueous organic solvent systems, Energy Fuels 22(4) (2008) 2379–2389, https://doi.org/10.1021/ef800057n. Copyright 2008, American Chemical Society.

the existence of both the island and archipelago structures of asphaltenes, with the dominance of the island structure. Rodgers et al. [30] demonstrated that the asphaltene structure is a continuum of island and archipelago motifs and hypothesized that the dominant structure (island or archipelago) depends upon the asphaltene sample and the analysis technique.

8.2.3 Nature of asphaltene molecules (the asphaltenes continuum)

The nature and chemical structure of asphaltenes have been debated in the scientific community for decades, due to their illusive behavior. Starting with Pliny calling the bituminous materials "asphaltos," and moving along to recent advances, probing the depths of the nature of asphaltenes is still ongoing. Fig. 8.8 summarizes the history of the changes of the asphaltene concept over the years [31–36].

Asphaltenes are described in the literature as a composition continuum of molecular weights and polarities. They also have been identified historically as a solubility class, which is still the operational definition to date. They are also expressed in polymeric terms, so it is common to find asphaltenes described as a polydisperse in the literature, referring to their different molecular sizes [37,38]. Other polymeric terms like monomers and dimers are also common, where the term monomer refers to a single island or archipelago molecule of asphaltenes [38,39]. The theories used to describe the

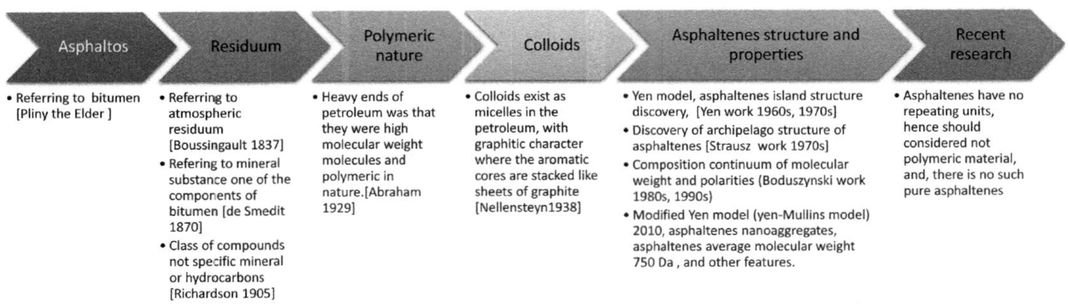

FIG. 8.8

The history of development of the asphaltenes concept over the years.

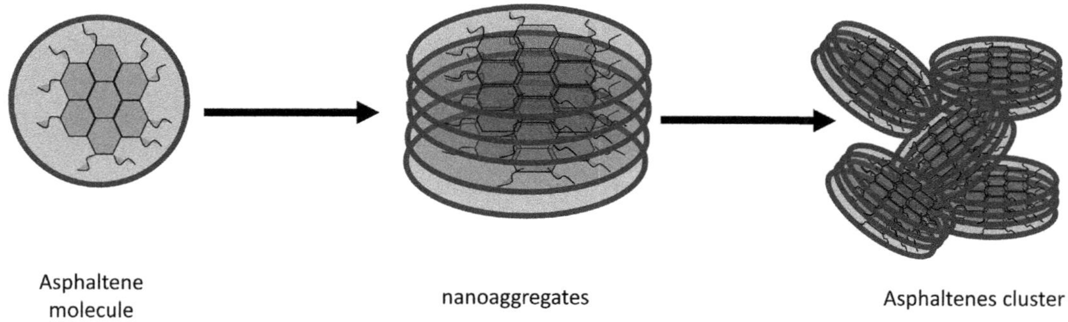

FIG. 8.9

Asphaltene aggregation and clustering.

polymeric solutions, like the Flory-Huggins solution theory, are widely applied in studying and modeling asphaltene behavior and precipitation. These monomers are thought to aggregate by face-to-face stacking of the aromatic cores like graphite sheets [35,40,41], and these aggregates then form clusters that agglomerate later and form the nasty deposits. Other stacking types, including aromatic cores in various configurations having cores perpendicular to stacks, have been reported [42]. Fig. 8.9 illustrates asphaltene aggregation.

However, recent studies by Schuler and coworkers provided evidence of the tremendous variety of molecules that make up the building blocks of asphaltenes. Fig. 8.10 shows a few of the 200 or so molecules that were imaged [43]. Similar results also were reported by Müller et al. [44] when they used molecular dynamics and computer simulations to develop a catalogue of over 100 molecules of asphaltenes and resins. They also concluded that the idea of "the single asphaltene molecule" is likely impossible. Using the molecules elucidated from Schuler's work, Rogel et al. [45] found that the molecules associate to form stable, irregularly stacked assemblies.

Moir [3] concluded that, in light of these recent studies, it is obvious that the language of polymer science and the analogy to the structure of polymers is largely inappropriate in two respects:

- First, the molecular weights of the molecules of the asphaltene fraction are low and are not as high as calculated or determined before.
- Second, and most important, asphaltenes contain no repeating units, or if there is any repetition, it must be purely by chance.

Furthermore, Moir [3] concluded that asphaltenes are amorphous, and the notions of average molecular weight and average molecular structure are largely, if not completely, useless. Also, based on the definition of a pure substance, there are no pure asphaltenes.

8.2.4 Asphaltenes in crude oil

The conflict and confusion about asphaltenes is not just in their composition and structure, but also in studying their existence in crude oil. The question here: is it that important to understand their existence in petroleum? And if so, how do they exist in crude oil?

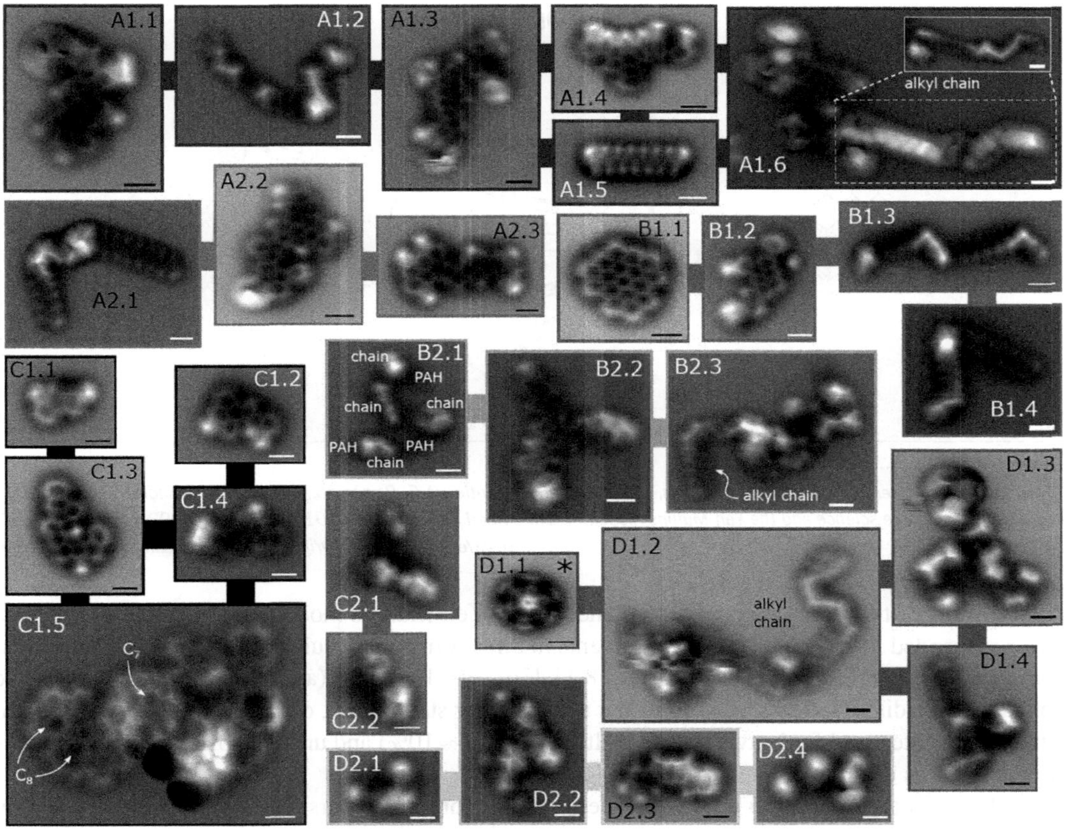

FIG. 8.10

Molecules observed in asphaltene fractions as imaged by AFM.

Reprinted with permission from B. Schuler, S. Fatayer, G. Meyer, E. Rogel, M. Moir, Y. Zhang, ... L. Gross, Heavy oil based mixtures of different origins and treatments studied by atomic force microscopy, Energy and Fuels 31(7) (2017) 6856–6861. https://doi.org/10.1021/acs.energyfuels.7b00805. Copyright 2017, American Chemical Society.

We study how asphaltenes exist in crude oils to understand how they behave under flowing conditions, what stabilizes them and keeps them in solution, and what destabilizes them and makes them precipitate.

Generally speaking, the existence of asphaltenes in crude oil can be classified based on the size distribution of asphaltene, into two types of asphaltenes: dispersed asphaltenes (<2 μm) and aggregated asphaltenes (>2 μm) [46]. Fogler et al. [47] observed three kinds of asphaltenes based on size: stable asphaltene (<0.1 μm), colloidal asphaltene (0.1–1 μm), and flocculated asphaltene (>1 μm).

In the meantime, there are two main approaches that explain the presence of asphaltenes in crude oil. The first is the real solution approach, which makes out the asphaltenes to be solids or a solute that is soluble in the other crude oil hydrocarbons (maltenes); and the second approach is the colloidal approach, which assumes that asphaltenes are suspended in crude oil and peptized/stabilized by resins. These approaches are discussed later under the aggregation and precipitation of asphaltenes.

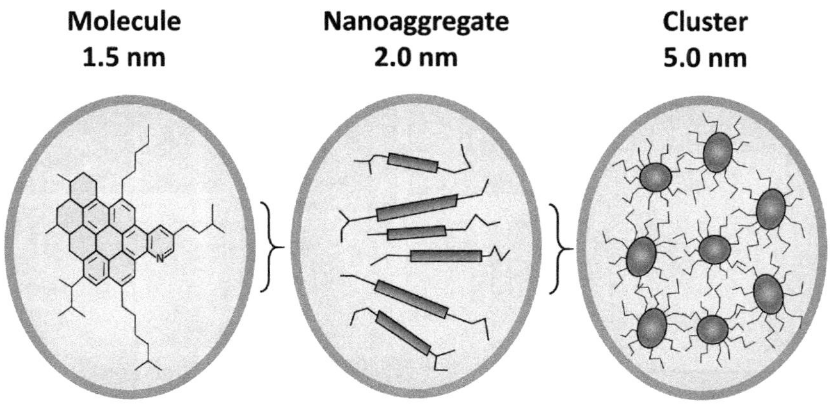

FIG. 8.11

Yen-Mullins model.

Reprinted with permission from O.C. Mullins, H. Sabbah, J. Eyssautier, A.E. Pomerantz, L. Barré, A.B. Andrews, ... R.N. Zare, Advances in asphaltene science and the Yen-Mullins model, in: Energy and Fuels, vol. 26, 2012, pp. 3986–4003. https://doi.org/10.1021/ef300185p. Copyright 2012, American Chemical Society.

According to the recent view of asphaltenes by the Yen-Mullins model, asphaltenes are dispersed and/or suspended in crude oils and/or in solvents in three forms: molecules, nanoaggregates, and clusters of nanoaggregates, as depicted in Fig. 8.11 [36]. For light oils (asphaltene content $\leq 1\%$), the asphaltenes are dispersed as a true molecular solution. For stable black oils, asphaltenes are dispersed as nanoaggregates, and for heavy oils (asphaltene content $\sim 10\%$) and unstable black oils, asphaltenes are dispersed as clusters of nanoaggregates [48,49].

Asphaltenes are thought to exist as single molecules or "monomers" at concentrations below ca. 50 mg/L. As concentrations increase to what is known as the critical nanoaggregate concentration (CNAC) of ca. 100 mg/L (± 50 mg/L), asphaltenes start forming nanoaggregates, which are believed to stabilize water in oil emulsions by creating a "skin" around water droplets. At concentrations of 2–5 g/L, nanoaggregates form clusters, where each cluster is composed of <10 nanoaggregates [50].

8.3 Properties of asphaltenes

The physicochemical properties of asphaltenes have been subjected to extended research. However, like their chemical structure, the asphaltene properties have also been a source of confusion to many researchers. This ambiguity can be attributed to:

- The way the asphaltenes are precipitated and separated. Asphaltenes do exist as a continuity, and the method of precipitation and separation will give different composition.
- The wide variety of techniques that were used in measuring the asphaltene property.

Table 8.1 summarizes the properties of asphaltenes and the techniques used in their characterization [51].

Table 8.1 Asphaltene properties and techniques used to measure them.

Topics	Techniques of the analysis	Reported values
Molecular weight	VPO	2000–10,000 amu
	GPC	10^6 amu
	FIMS, LDMS	500–1000 amu
Size of aggregates	SANS, SAXS	20–500 Å
	GPC	Smaller than 30 nm and larger than 200 nm
Size of molecules	FD, FCS	1–2 nm
Number of aromatic rings in molecules	Optical absorption, FD, FCS, NMR	2–20
Molecular architecture of molecule	RICO, fluorescence spectroscopy, pyrolysis	Archipelago
	FD, FCS, LDMS, LIAD	Island
	DOSY NMR	Both archipelago and island structures

Reprinted with permission from A. Al-Hosani, S. Ravichandran, N. Daraboina, Review of asphaltene deposition modeling in oil and gas production, Energy Fuels (2021, January 21), https://doi.org/10.1021/acs.energyfuels.0c02981. Copyright 2021, American Chemical Society.

8.3.1 Molecular weight

Molecular weight is one property of asphaltenes that has attracted researchers, apparently from an academic point of view rather than an industrial point of view, as it gives evidence on the chemical composition and molecular structure of asphaltenes. However, the molecular weight of asphaltenes was a source of controversy for decades, as the molecular weights of asphaltene constituents spanned a wide range from a few hundred to several million [4,52,53]. It is a matter of fact that in an asphaltenes mixture, there exists a polarity and molecular weight continuum [54]. The source of controversy in estimating asphaltene molecular weight can be explained by the following [4]:

- The difference in the way the asphaltenes are precipitated and separated from the solution, as indicated in Section 8.2.1.
- For the asphaltenes separated by alkane/precipitating agents, their molecular weight varies based on the properties of that solvent: dielectric constant, polarity, etc.
- The tendency of asphaltenes to self-associate.
- The different techniques used to measure the molecular weight. Examples of the different techniques that were used are: vapor pressure osmometry (VPO), gel permeation chromatography (GPC), field-ionization mass spectroscopy (FIMS), laser desorption mass spectroscopy (LDMS), and matrix-assisted laser desorption ionization (MALDI), ESI-FT-ICR and APCI, and fluorescence depolarization.

As a summary, the molecular weight of asphaltenes spans between 500 and 2000 Da, or more specifically between 500 and 1000 Da, with an average of ~750 Da according to the last measurements, and the confusion in measurements is attributed to the previously mentioned reasons [4,55].

8.3.2 Density

Studying the density was used in determining the molecular structure and characterizing the asphaltenes. Asphaltene densities from a variety of sources generally fall between 1130 and 1200 kg/m^3. It is worth mentioning that the density measurements were determined from the density of the asphaltene solubility cuts [26,56].

Asphaltenes from crude oil of unstable nature are characterized by higher densities, lower H/C ratios, and high aromaticities compared with stable crude oil asphaltenes [57].

8.3.3 Solubility parameter

The solubility parameter is a physicochemical property of a substance that provides a numerical estimation of the relative solvency behavior of a specific solvent [58]. The solubility parameter demonstrates the cohesion energy density of a substance, is useful for evaluation of compatibility, wettability, and cohesiveness/dispersibility between substances, and it plays an important role in screening and selecting solvents [59]. The solubility parameter is generally given by:

$$\delta = \left(\frac{\Delta E}{v}\right)^{0.5} \tag{8.1}$$

where ΔE is the difference between the internal energy of the condensed material and that of an ideal gas of the same material at the same temperature, and v is the molar volume of the solute. Solubility parameter is also represented by Eq. (4.21) in Chapter 4.

The solubility parameter is applied in the following realms [60]:

- Studying the aggregation/deposition of asphaltenes in different chemicals/solvents.
- Developing deposition-predicting models that can be used as part of the flow assurance program.
- Studying the solubility of asphaltenes in different solvents, which helps in designing and selecting the proper solvent for removal and control treatments.

Asphaltenes have a solubility parameter between 19 and 24 MPa$^{0.5}$ at ambient conditions [61,62]. The solubility of asphaltenes can be expressed by the following equation:

$$\ln x_a = -M_a/RT\rho_a\left[(\delta_S - \delta_a)^2\right] \tag{8.2}$$

where x_a is the mole fraction solubility of asphaltenes, M_a is the molecular weight of asphaltenes, ρ is the density of asphaltenes, and $(\delta_s - \delta_a)$ is the difference between the solubility parameters of the solute asphaltenes (a) and the solvent (s).

Asphaltene solubility and solubility parameter depend basically on the asphaltene molecular weight, composition (H/C ratio, heteroatoms), polarity, and the solvent properties of molecular weight and polarity [59,60,63].

For a polymer to dissolve in a solvent, the solubility parameter of the solvent should be within about 1.1 (cal/cc)^{½} [1(cal/cc)^{½} = 2.045 MPa$^{1/2}$] of the solubility parameter of the polymer [64]. For a non-polymeric solid material, such as naphthalene, to dissolve in a solvent, the difference in solubility parameters should be less than about 2 (cal/cc)^{½} [65]. And finally, for two liquids to be miscible, the difference in their solubility parameters should be less than about 7 (cal/cc)^{½} [66].

In general, hydrocarbon liquids can dissolve polynuclear aromatic hydrocarbons when the difference between the solvent's lower solubility parameter and the hydrocarbon's solubility parameter is less than three points [4].

8.3.4 Charge

Asphaltenes are charge-bearing molecules, but like the other ambiguous properties of asphaltenes, the nature of that charge is not fully established. While some researchers have claimed that asphaltenes possess predominantly positive charge in most nonaqueous dispersants [67,68], some reported net negative surface charges based on electrodeposition experiments with neat crude oil samples [69,70].

This discrepancy can be tracked down to some factors that include solvent, temperature, and influence of resins [71], as detailed here:

- Solvent: Asphaltenes in aqueous solutions possess a surface charge due to the protonation and dissociation reactions of functional groups enabling electrophoretic mobility [72]. Such solvent induced polarity was found to reverse the electrophoretic mobility of destabilized asphaltene particles; hence, the particles present a negative electrophoretic mobility when dispersed in polar (water) media, whereas in toluene their mobility was positive.
- Dilution: Leon et al. [67] reported different charge of asphaltenes at different dilutions.
- Resins: coprecipitation of resins occurred when they were present in the precipitating medium, indicating a binding process of the resins on the resultant asphaltene particles [73].

The nature of the charge is not the only ambiguity of asphaltenes, but charge reversal is another issue encountered during studying the asphaltenes. Leon et al. [67] reported that the surface charges on asphaltene micelles were predominantly negative in crude oils and at low heptane dilution, but the surface charges were reversed at higher dilution. Taylor [74] observed that stable asphaltene "micelles" in crude oils were not affected by the DC, indicating that they possess little electric field net charge. However, they possessed a small net negative charge upon destabilization of micelles with lower dilutions of *n*-alkanes, and, strangely, higher dilutions resulted in a positively charged deposit. The charge reversal can be attributed to removal of positively charged peptizing resins.

The ambiguities existing on the surface charge of asphaltenes and resins were clarified by the work of Hashmi and Firoozabadi [75], in which they concluded that asphaltene colloids may possess heterogeneous surface charge: some particles possess small net positive charge while others possess negative charge. They concluded that the asphaltene surface charge is derived from the metallic components, pi-electron clouds, and acid-base functional groups on the periphery, and these components may vary from one crude to another.

Salmón-Vega et al. [76] reported that asphaltenes showed an isoelectric point (IEP) at pH 4.5, and that its zeta potential is a function of pH, as it was positive below pH 4.5 and was negative above pH 4.5 and up to pH 10.3. This implies that asphaltene surface charge is positive at lower pH values and negative for high pH values. The effect of pH on the zeta potential and surface charge of asphaltene is attributed to the acidic and basic functional groups, which are pH sensitive and can undergo dissociation and protonation. For instance, a positive zeta potential of asphaltenes is related to the ionization basic groups such as amine, amide, and pyridine, which have pK_a values >2, whereas the negative zeta potential can be attributed to the dissociation of acidic functional groups such as carboxylic, which have pK_a values ≤ 4 [77,78].

8.3.5 Polarity

Polarity is generally defined as the distribution of electrical charge over the atoms joined by the bond. Asphaltenes are polar molecules and it is believed that heteroatomic (N,S,O) functional groups are the source of their polarity and amphoteric characteristics [4,79].

Asphaltene solubility in hydrocarbon media is determined based on the molecular weight, aromaticity, and polarity [80]. However, evidence was found that the asphaltene properties are in fact more related to the polarizability rather than the polarity of the molecules [3].

8.3.6 Thermal stability

Kopsch [81] reported a glass transition temperature for Kirkuk asphaltenes of 313–350°C from differential scanning calorimetry (DSC) for heating rates of 10–50 K/min. Zhang et al. [82] reported glass transition temperatures of 120–130°C for four asphaltenes using DSC at a heating rate of 10 K/min.

The melting points of asphaltenes extracted from five sources were reported to be in the range of 214–311°C [83]. A lower melting point temperature range (125–180°C) was also reported [84]. On heating above 300–400°C, asphaltenes do not melt, but decompose, forming carbon and volatile products [1,85].

The thermal stability of asphaltenes depends on the molecular weight and the content of sulfur and oxygen in various functional groups, whereas the carbon atom distribution over aromatic, naphthenic, and aliphatic fragments of asphaltene molecules exerts no dominant impact on their stability in thermal processes up to 360°C [86].

8.3.7 Refractive index

The refractive indices of asphaltenes (at 25°C), between 1.70 and 1.8, have been extrapolated for asphaltenes from diluted asphaltene solutions in toluene [87].

8.3.8 Asphaltene self-association

Self-association and micellization are used interchangeably in the literature to describe a property of surfactants in aqueous solutions and it is also a property of colloidal solutions. Micellization, self-association, or self-assembly occurs when the surfactants tend to self-assemble into colloidal-like particles for better suspension in the solutions. This phenomenon occurs only when the surfactant concentration exceeds a threshold, known as critical micelle concentration (CMC) or critical aggregation concentration (CAC). Micellization represents a self-association phenomenon in general, yet micellization is conventionally limited to the self-association initiated by the hydrophobic-hydrophilic imbalance, and as a result, it represents a limited class of surface active agents [88].

Asphaltenes have been reported to show some similarities with surfactants, as they apparently can stabilize water-in-oil emulsions. Concentrated asphaltene solutions exhibit some degree of colloidal behavior. For a long time it was believed that asphaltenes in organic solutions had a well-defined CMC, as is the case for surfactants, implying that they remain in a monomeric state in the solution until they reach a certain threshold concentration or CMC, whereas above that concentration, asphaltenes start to aggregate [89]. However, it is evident that the micellization or self-association of regular surfactants cannot be used to describe asphaltene self-association, for the following reasons:

- Asphaltenes do not have well-defined hydrophilic heads and hydrophobic tails and, thus, do not have amphiphilic character and classic surfactant-like behavior [5].
- Asphaltenes can self-associate prior to the region of the commonly reported CMC. Asphaltenes can self-associate at low concentrations (in the ppm region), whereas the commonly reported CMC is in the 1–5 g/L region [89].
- Unlike surfactants, asphaltene self-association is not due to hydrophobic-hydrophilic imbalance [88]. Micellization in aqueous solvents exhibits stronger exterior forces (in the solvent) than the interior forces inside the micelle, while in micellization in nonpolar solvents, the intermolecular forces are higher in the interior of the micelles than at the exterior, as indicated in Fig. 8.12. It is shown that the association of asphaltenes in nonpolar environments is different from these two categories and that new mechanisms must be employed to understand their structure and mechanism for association [90].
- The properties of asphaltene aggregates resulting from the micellization process depend on the properties of the solvent used [89].

The result of asphaltene self-association is the formation of aggregates. The aggregates formed at first are nanoaggregates, and then, under poor solvent conditions or poor production parameters, the aggregates arrange into flocs or clusters, and finally precipitate [91,92].

The self-association and aggregation of asphaltenes take place basically as a result of the intermolecular forces between the asphaltene molecules. Fig. 8.13 summarizes these intermolecular interactions. The intermolecular forces include the following:

- Acid-base interactions: asphaltenes can have amine bases in their composition, so acid-base interactions with other acidic components in the crude oil (such as naphthenic acids) are expected [93]. Barth et al. [94] reported a strong correlation between the asphaltene content and the total

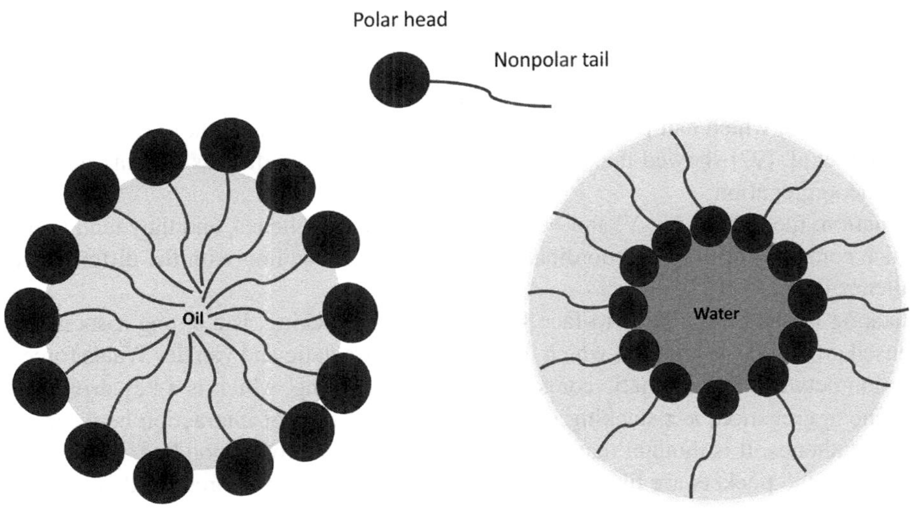

FIG. 8.12

Micellization in aqueous and nonpolar solvents.

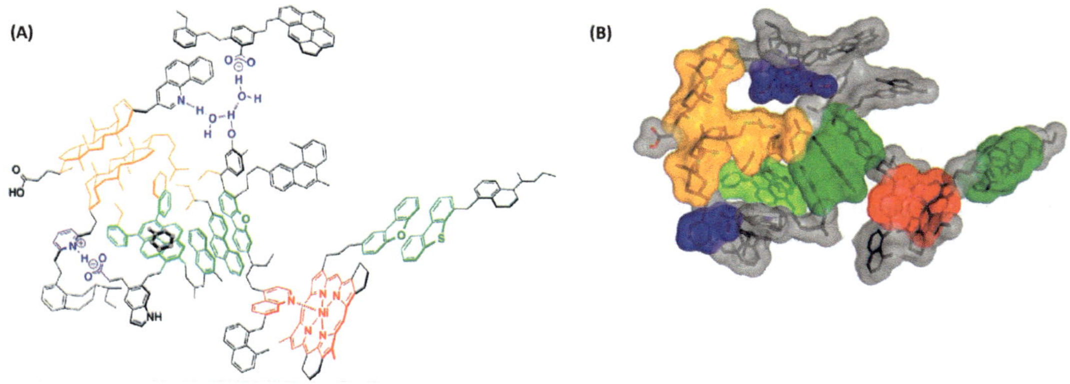

FIG. 8.13

Schematic representation of asphaltene aggregate, showing their intermolecular forces. Associations between molecules are color-coded in (A) the molecular depiction and (B) the space-filling version: acid-base interactions and hydrogen bonding *(blue)*, metal coordination complex *(red)*, a hydrophobic pocket *(orange)*, π-π stacking (face-to-face *dark green*; within a clathrate containing toluene, *light green*).

Reprinted with permission from M.R. Gray, R.R. Tykwinski, J.M. Stryker, X. Tan, Supramolecular assembly model for aggregation of petroleum asphaltenes, Energy Fuels 25 (2011) 3125–3134. https://doi.org/10.1021/ef200654p. Copyright 2011, American Chemical Society.

base number (TBN). They also concluded that asphaltene fractions are not in themselves bases, the petroleum bases stabilize asphaltenes in the bulk crude oil, and that acid-base interaction could be an important mechanism for asphaltene stability. Kelland et al. [9] reported four fractions in crude oil asphaltenes: acid, base, neutral, and amphoteric, where the acid fraction comprises carboxylic acids, phenols, and sulfoxide groups while the basic fraction has amines and sulfoxide groups. Gray et al. [95] stated that ionic association represents the strongest interaction driving asphaltene aggregation.

- Hydrogen bonding: this type of bonding can take place between groups such as hydroxyls and carboxylic acids, which can participate in hydrogen bonding with weak nitrogen bases [96]. Kovalenko et al. [97] showed that hydrogen bonding is as important as π-π interactions for asphaltene aggregation.
- Coordination: metals like Ni, V, and Fe are abundant in asphaltenes, and they tend to interact and influence complex forming by coordination and hydrogen bonding with the different fractions of asphaltenes [96].
- π-π stacking (pi-bonding): Face-to-face stacking, or π-π stacking, is a strong association of condensed aromatics in solution, which is based on electrostatic and van der Waals forces. The π-π interaction between polyaromatic cores leads to their stacking, which is a key driving force for asphaltene aggregation; π-π stacking is also an important type of interaction between resins-asphaltenes. It is thought that acid-base interaction, hydrogen bonding, metal coordination, and hydrophobic pockets are likely to act simultaneously with π-π forces to drive asphaltene aggregation [96]. The aliphatic side chains hinder the stacking due to the steric interferences.
- Hydrophobic pockets: These are driven by van der Waals interactions, such as the association of nonpolar, cycloalkyl, and alkyl groups in hydrophobic pockets, driven similarly to the behavior of steroidal or lipid bilayers [95].

Some of the models that were proposed to explain self-association are peptization models [98], fractal models [99], the Yen model [52], and the modified Yen model [36].

Generally speaking, the single PAH core in asphaltene molecules is assumed to be the primary site of intermolecular attraction due to its polarizability and some degree of charge separation associated with heteroatoms in the aromatic ring system. The asphaltene molecules are thought to dimer or aggregate by face-to-face stacking (π-π stacking) of the aromatic cores like graphite sheets [35,40,41]. However, extensive studies indicate only a fraction of the interactions among asphaltenic molecules may take place through stacking of aromatic sheets [100,101].

The strength of the intermolecular forces is supposedly in the following order: hydrogen bonding > pi bonding > inductive forces [7].

Molecular dynamic simulations of the structure of asphaltene aggregates show that basically asphaltene model compounds can undergo parallel and nonparallel stacking to form three types of aggregates, which are face to face, offset stacking, and T shaped [102]. These shapes are illustrated in Fig. 8.14.

Various morphologies of asphaltene aggregates have been suggested, including disc-like particles, ellipsoids, cylindrical, vesicles, fractals, and monodispersed and polydispersed spheres [103]. The average asphaltene micelle size distribution is around 25 nm, whereas micellar coacervates may have sizes >25 nm [104].

In the recent view of asphaltenes, as depicted by the Yen-Mullins model, asphaltenes aggregate to form nanoaggregates. Therefore the CMC will be replaced by CNAC or critical nanaoaggregation concentration [105], and the CNAC of asphaltenes is in the range of 50–150 mg/L. The dimensions of these nanoaggregates are 2 nm, with an aggregation number of nanoaggregates of ~6 [36]. It is worth noting that the CNAC is not the concentration where aggregation starts, but it is in fact the concentration at which one obtains fully formed nanoaggregates and these nanoaggregates stop growing in size [105].

A generalized hypothetical organization of asphaltene aggregates can be suggested as depicted in Fig. 8.15, which in turn is in close agreement with the modified Yen-Mullins model [106]:

- Core aggregates (20 Å): or 8–10 molecules of asphaltene. It is generally formed due to the π-π stacking of asphaltene molecules.
- Medium aggregates (50–500 Å): These can be considered as secondary aggregates, which generally form via interaction of core aggregates with solvent, media, oil, or maltenes.
- Fractal aggregation (1000 Å): They are tertiary aggregates formed by reaction of medium aggregates. These aggregates are inert with respect to any media.

The key parameters of aggregation include [107]:

- Asphaltene/resin ratio
- The degree of aromaticity
- The effect of solvent systems
- Charge transfer properties
- heteroatoms present in the asphaltenic core

Understanding these effects and counteracting them can help in dispersing and controlling asphaltene behavior in production systems.

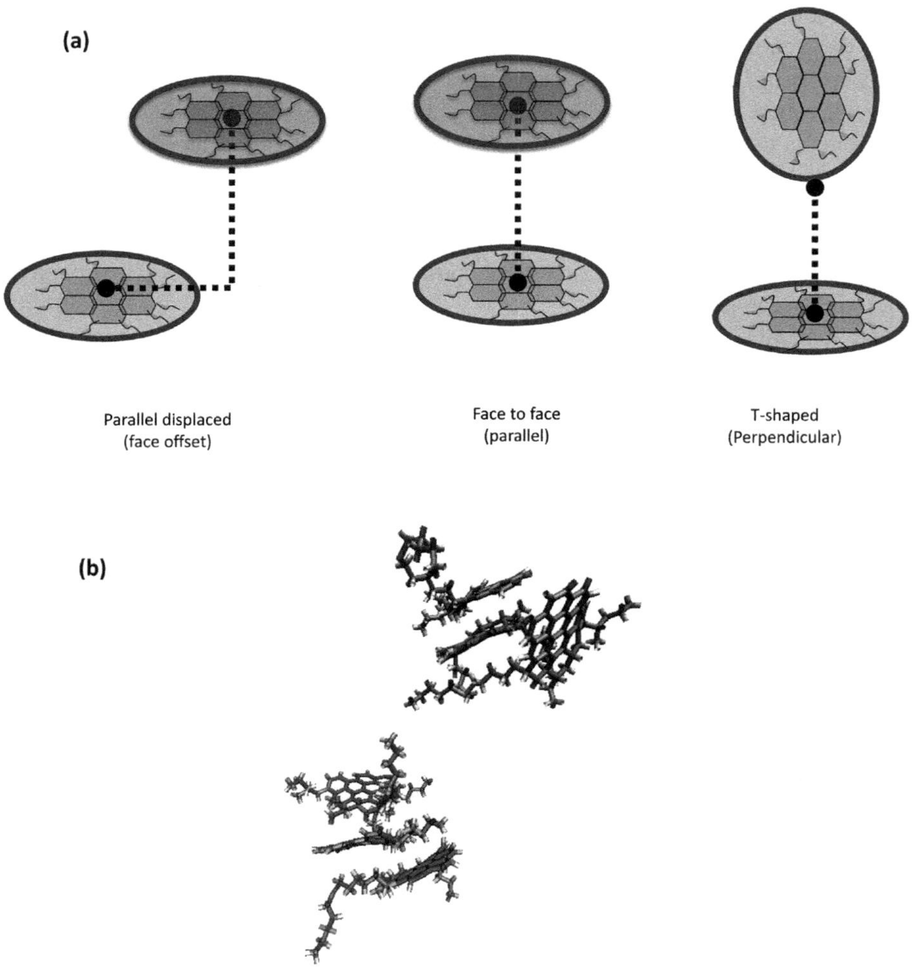

FIG. 8.14

(a) Schematic of the stacking modes of asphalenes. (b) Snapshot of molecular dynamic simulation of asphaltene molecules showing the parallel and nonparallel stacking.

Panel (B) was *reprinted with permission from T.F. Headen, E.S. Boek, N.T. Skipper, Evidence for asphaltene nanoaggregation in toluene and heptane from molecular dynamics simulations. Energy Fuels 23 (2009) 1220–1229. https://doi.org/10.1021/ef800872g.*

8.4 Asphaltene aggregation and deposition

Asphaltene precipitation and deposition is a complex process that has been the subject of research for decades, with many of its puzzles and missing pieces being solved and uncovered. Generally speaking, two basic approaches are currently used to express asphaltene stability and precipitation in crude oil: the solubility model and the colloidal model. Both of these approaches are discussed in some detail in the following sections.

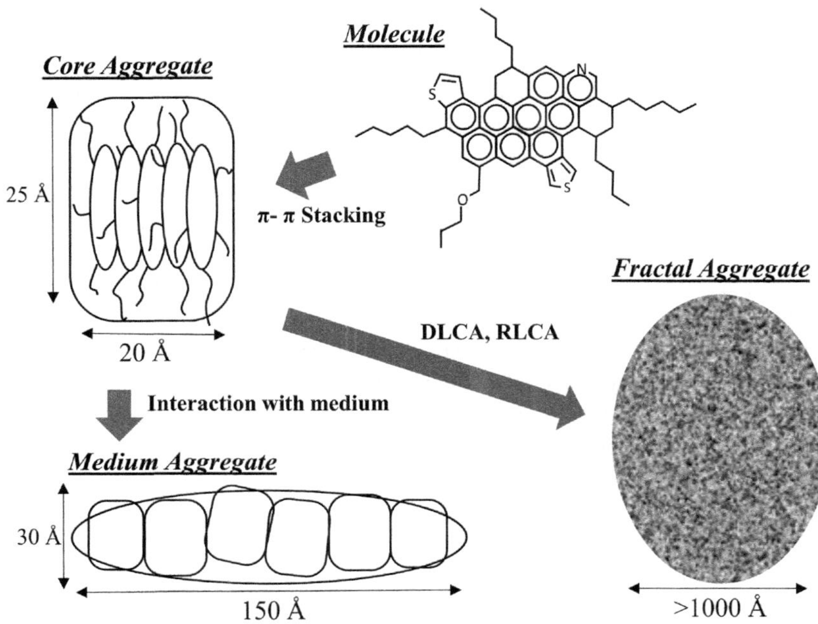

FIG. 8.15

Schematic representation of the phases of asphaltene aggregation based on XRD, SAXS, and SANS data.

Reprinted with permission from R. Tanaka, E. Sato, J.E. Hunt, R.E. Winans, S. Sato, T. Takanohashi, Characterization of asphaltene aggregates using X-ray diffraction and small-angle X-ray scattering, Energy Fuels 18 (2004) 1118–1125. https://doi.org/10.1021/ef034082z. Copyright 2004, American Chemical Society.

8.4.1 Solubility model

Like regular solutions, the solubility approach assumes that the asphaltenes are solutes dissolved in crude oil (as solvent) and the precipitation occurs if the solubility falls below a certain threshold level, or when the oil solvency falls below a specific limit. In that approach, asphaltene precipitation can be treated as liquid-liquid or solid-liquid equilibria. Crude oil components like resins and aromatics solubilize asphaltenes, whereas saturates destabilize them [108]. Examples of this approach are the Hirschberg model [109], Redelius [110], Yarranton [111], Chapman and Vargas [108], and others.

Since this theory and those models treat crude oil as a two-components mixture—asphaltenes as solid phase (solute) and maltenes or deasphalted oil (solvent)—therefore the solubility of asphaltenes is expressed by the use of the solubility parameter, while the properties of the maltenes (solvent) are expressed using thermodynamic calculations and the equation of state [112]. External influences such as alkane addition, depressurization, or gas uptake can change in the solubility parameter to either of the two phases, which results in disturbing the phase equilibria between these two phases (asphaltenes and deasphaltened oils) and eventual asphaltene precipitation. According to this approach, asphaltene precipitation is a reversible process, means asphaltenes can redissolve after precipitation.

8.4.2 Colloidal model

This approach assumes that asphaltenes exist in the colloidal state (in suspension) stabilized by the resin molecules adsorbed at the asphaltene surface; the disturbance of the asphaltene-resin complex/balance causes the asphaltene precipitation. The colloidal theory assumes that the precipitation of asphaltenes is an irreversible phenomenon and that a certain amount of resins is necessary for asphaltenes to be completely peptized in crude oil [112]. The idea of the existence of asphaltenes in colloidal form was first proposed by Nellensteyn [34] and then by Pfeiffer and Saal [98], as illustrated in Fig. 8.16.

The resin molecules interact with the asphaltenes through pi-bonding or π-π stacking to form a complex that is stabilized and kept dissolved in the bulk hydrocarbons by the numerous alkyl chains on the resin surface. Stripping of the resins from the asphaltenes destabilizes the bulky core, rendering it less soluble in the bulk oil, as fewer alkyle chains are present. In addition, the aromatic cores in the asphaltenes can undergo π-π stacking, then aggregating and increasing in size until they precipitate [7].

Examples of this approach include Leontaritis and Mansoori [113], Victorov and Firoozabadi [114], and Pan and Firoozabadi [115].

According to this approach, asphaltene precipitation is an irreversible process, which means the precipitated asphaltenes cannot be redissolved.

This approach was helpful in understanding asphaltene behavior. However, aspects like addition of toluene (not a resin), which can redissolve asphaltenes, showed serious problems in this approach [37]. Also, recent studies by Goual [116] and Mullins et al. [117] have demonstrated that resins are unlikely

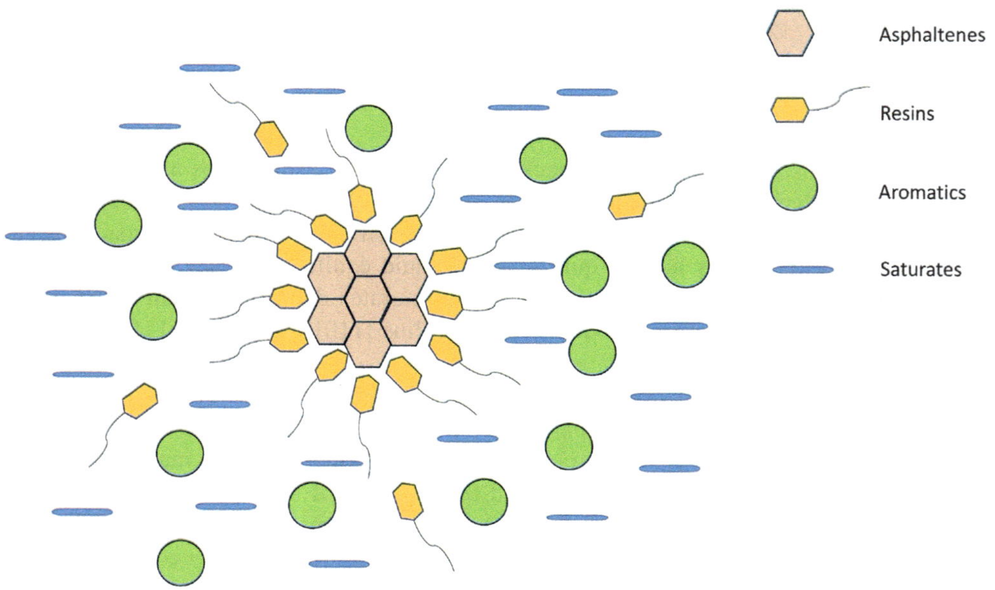

FIG. 8.16

Asphaltene colloidal model by Pfeiffer and Saal.

to coat asphaltene nanoaggregates, and therefore resins are not able to provide the steric stabilizing layer that the colloidal model proposes. All of these have made the asphaltene research community tend towards the solubility approach (first approach).

So now, how exactly so asphaltenes precipitate and deposit? Unfortunately, despite the vast amount of work devoted to understanding asphaltene molecular structure and the mechanisms of asphaltene precipitation and deposition, the latter still represent a challenging problem. Nevertheless, numerous observations in the field and lab studies are pointing at the notion that the precipitation of asphaltenes seems to be a hierarchical process, starting with the asphaltene molecules and scaling up to the particulates that finally accumulate to block the production tubing and flowlines.

8.4.3 General mechanism of deposition

The principal mechanism of asphaltene deposition can be explained in a stepwise hierarchical approach, where asphaltene molecules grow from molecules and aggregates to solid particles that cause the blockages in production systems. The main processes and phase changes involved in this hierarchical mechanism are sensitive to the thermodynamic and hydrodynamic factors (P, T, flow regimes, flow rates, etc.). The hierarchical concept of asphaltene deposition is illustrated in Fig. 8.17.

At reservoir high temperature and high pressure conditions, asphaltenes are dissolved predominantly as nanoaggregates and their clusters. During production, pressure and temperature continuously decrease, which drives these asphaltene nanoaggregates and clusters to agglomerate and form primary (intermediate/precipitated) particles. The primary asphaltene particles can stick together to constitute larger aggregated asphaltene particles [118]. The eventual particles can be carried by the flow or

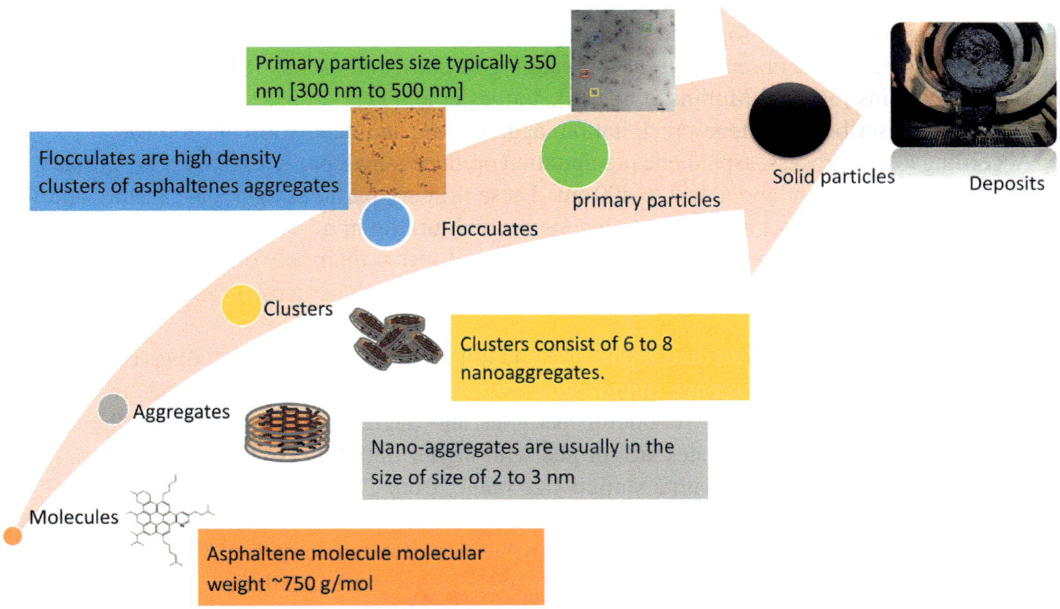

FIG. 8.17

Asphaltene precipitation and deposition mechanism.

transported by some mechanism (these are discussed in Chapter 4) to the pipe wall, where they ultimately stick to the pipe surface and are deposited.

8.4.3.1 Asphaltene self-association and aggregation

Asphaltenes go through self-association and aggregation (once asphaltenes reach the CNAC), forming nanoaggregates with a particle size of 2 to 3 nm [36,119], and then desolubilization due to changes in the environment to which they are subjected. The aggregation of asphaltenes is explained in Section 8.3.8.

8.4.3.2 Clustering of asphaltene nanoaggregates

At higher concentrations than the CNAC, a secondary aggregation process known as the clustering of nanoaggregates takes place. These clusters are several nanometers and larger in size, and have small aggregation numbers (generally consisting of six to eight nanoaggregates) [36,119]. The critical clustering concentration (CCC) is usually observed at higher concentrations, and, as stated by Mullins, the conventional CMC is likely responding to the clustering concentration (CCC) [36,96].

8.4.3.3 Phase change and deposition

Further flocculation of these clusters leads to the formation of high-density clusters or flocculates [120,121], and the agglomeration of these flocculates finally forms microscopic particles known as the primary particles, intermediate particles, or precipitate particles. These primary particles typically have a size of 350 nm, and their size can be different from one type of oil to another, ranging from 300 to 500 nm as reported by Vargas et al. [122], to 200–400 nm in radius as reported by Meng et al. [123]. These primary particles tend to aggregate further to form bigger solid particles with size $>1\ \mu m$ [122], which finally grow in size to form the thick deposits that block production facilities.

Based on experimental observations using scanning electron microscopy (SEM) and previous work reported by Mullins [36] and Mullins et al. [49]. Vargas et al. [122] proposed a conceptual multistep mechanism to describe asphaltene precipitation, aggregation, and deposition phenomena.

According to this process, asphaltene precipitation produces small particles known as "primary particles" in the size range of about 100 to 400 nm. These primary particles aggregate further and form microaggregates, which later age, becoming compact and turning into a solid-like phase. The formed initial particles are amorphous and entrap some crude oil inside their network; thus the aging stage involves diffusion of oil out of these initial particle networks, which hardens the formed particles over time [124].

Recently, Meng et al. [123] studied asphaltene precipitation under diffusive mixing, using an in situ total internal reflection fluorescence microscopic system (TIRF). They showed that two types of asphaltene particles (i.e., primary submicron particles and aggregates) are formed in the dilution-induced precipitation. The aggregates are formed with higher initial precipitant concentration, while the primary submicron particles with a similar size range, from 200 nm to 400 nm in radius.

The formation of asphaltene precipitates/particles is a major step but it does not mean asphaltene deposition will occur. Because of drag, the asphaltene particles can be carried by flow downstream and do not deposit. Nonetheless, other asphaltene particles that are smaller can simultaneously be transported and diffuse radially to the near-wall region, and deposit on the pipe wall (Guan et al. [118]). The particle transportation and deposition mechanisms can vary. Molecular diffusion is the main particle transport mechanism within a thin fluid layer adjacent to the wall, whereas other mechanisms, such as convection, diffusion, thermophoresis, etc., affect the particle transport outside this layer [118,125]. Other mechanisms considered include eddy diffusion and inertia [126]. Three transport mechanisms, Brownian diffusion (for small particles <1 μm), inertia (for medium particles), and medium and impaction (for large particles), have been proposed by Darabi et al. [127]. Other particle transport models were employed in asphaltene particle transport and deposition in the literature.

8.4.3.4 Deposits aging

At the onset of precipitation, asphaltenes form linear and planar aggregate structures, while after sufficient precipitation they become more three dimensional. Therefore the formed asphaltene deposits are amorphous and contain high amounts of crude oil in their formed network. During the aging step, more asphaltene particles accumulate and deposit while the liquid oil diffuses out of the formed deposit layer (network). That leads to increasing the asphaltene content of the deposits and increasing the deposit layer thickness and decreasing the crude oil volume fraction. With time, the formed deposit layer will harden and its mechanical properties will improve.

8.4.4 Asphaltene precipitation under downhole and reservoir conditions

During production, the pressure depletion causes oil expansion and reduces the oil solubility parameter, and hence reduces oil solvency for asphaltenes. Thus the asphaltene precipitation occurs at a certain low pressure known as asphaltene onset pressure, or AOP. According to the Chapman et al. (2010) model, asphaltenes precipitation leads to the appearance of the first particles called microaggregates, which then adhere to one another and undergo aggregation to form larger particles, or can follow a diffusion mechanism to the surface of the tubing, where they stick and build up a deposit. Additionally, both small and big particles are transported upward by the moving fluid, in a process known as advection [108]. Fig. 8.18 summarizes asphaltene precipitation and particle transport in a producing well/pipeline [118].

The rate of asphaltene precipitation is assumed to be proportional to the supersaturation degree of asphaltenes, which is defined as the difference between the actual concentration of asphaltenes dissolved in the oil and the concentration of asphaltenes at equilibrium [108,118]:

$$\frac{dF_A^*}{dV} = r_P = -k\left(C_A^* - C_A^{eq}\right) \tag{8.3}$$

where F_A^* is the rate at which dissolved asphaltenes precipitate, forming microaggregates in units of mass per unit time; V is the system volume; r_P is the precipitation rate in units of mass per unit time and unit volume; k_p is the specific precipitation rate constant; C_A^* is the actual concentration of asphaltenes in the oil; and C_A^{eq} is the concentration of asphaltenes in oil at equilibrium, at the given operating conditions.

The deposit layer thickness D_r can be calculated from Eq. (8.4):

$$D_r = \left(1 - \sqrt{\propto}\right)R \tag{8.4}$$

where R is the original radius of the wellbore/pipeline, α is the crude oil volume fraction given by Eq. (8.5):

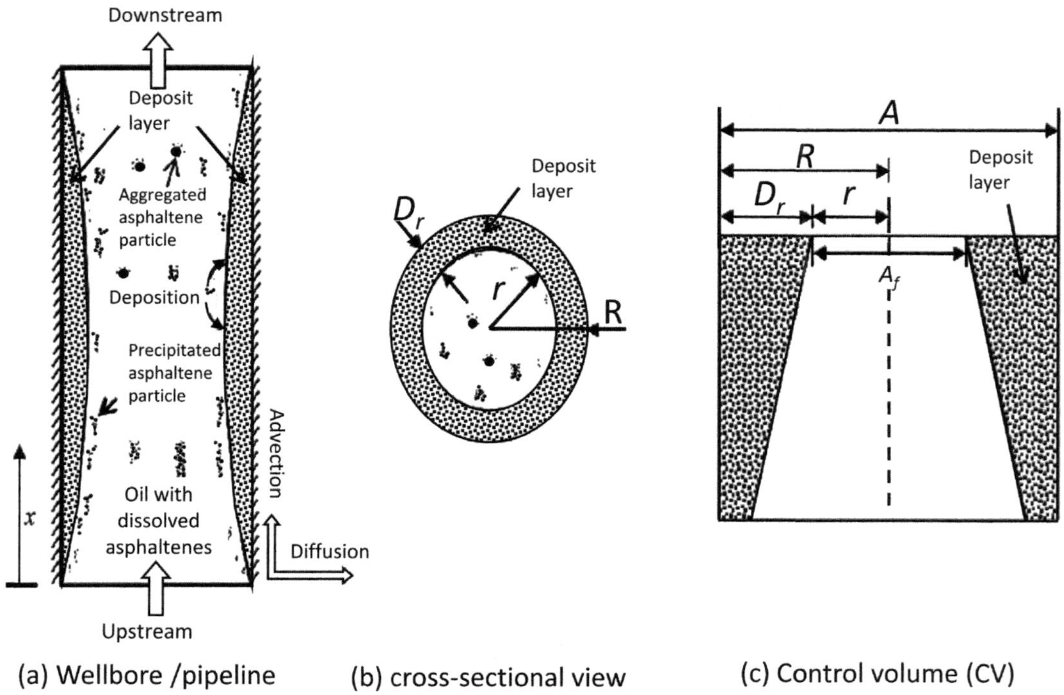

FIG. 8.18

Asphaltene precipitation in a producing well.

Reprinted with permission from Q. Guan, A. Goharzadeh, J.C. Chai, F.M. Vargas, S.L. Biswal, W.G. Chapman ... Y.F. Yap, An integrated model for asphaltene deposition in wellbores/pipelines above bubble pressures, J. Pet. Sci. Eng. 169 (2018) 353–373. https://doi.org/10.1016/j.petrol.2018.05.042. Copyright (2018), Elsevier.

$$\propto = \frac{V_f}{V_{CV}} = \frac{r^2}{R^2} \tag{8.5}$$

where V_f is the volume of the effective flow passage, V_{CV} is the volume of the control volume (CV), and r is the radius of the effective flow passage.

Mohammadzadeh et al. [128] reported four mechanisms of asphaltene deposition in the formation pore throats: surface deposition, entrainment, pore-throat plugging, and pore throat opening, which counterbalances the pore throat plugging mechanism. Such mechanisms result in asphaltene deposition and subsequent permeability reduction as a result of pressure drop across the cores. These mechanisms are indicated in Fig. 8.19 [128].

Asphaltene deposition is often independent of its content in the parent crude oil. Low asphaltene content crudes such as Clyde field in the North Sea, with only 0.5% asphaltenes, encountered serious downhole problems, whereas many Venezuelan fields contain up to 5% or even 10% asphaltenes and produce without any real problems. Here, it is worth mentioning that the aromatic content of the crude oil appears to generally prevent coagulation of the asphaltene molecules [6].

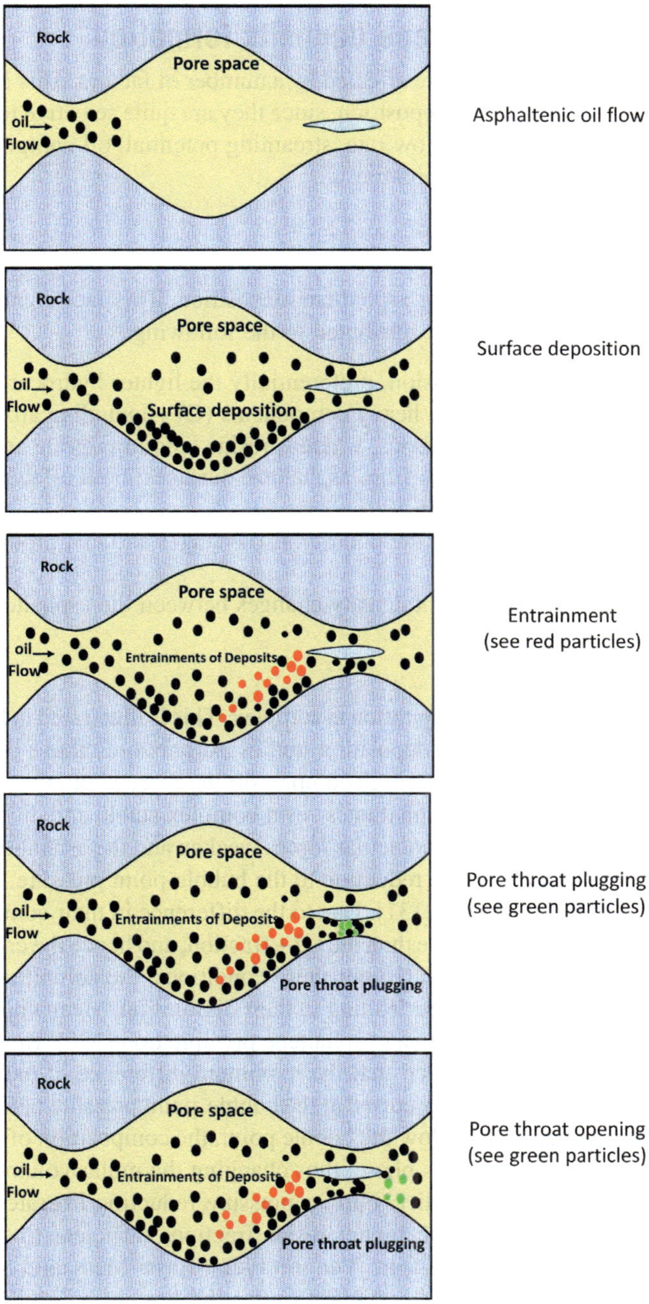

FIG. 8.19

Schematic of different asphaltene deposition and permeability impairment mechanisms in porous media originated from continuous flow of asphaltenic crude oil.

Reprinted with permission from S. Kord, O. Mohammadzadeh, R. Miri, B.S. Soulgani, Further investigation into the mechanisms of asphaltene deposition and permeability impairment in porous media using a modified analytical model, Fuel 117 (2014) 259–268. https://doi.org/10.1016/j.fuel.2013.09.038. Copyright (2014), Elsevier # is 210915-020648.

8.5 Factors affecting asphaltene deposits formation

Asphaltene aggregation and deposition are affected by a number of factors. This is also another way of thinking of the ambiguity in asphaltene deposition, since they are quite sensitive to changes in different parameters. Pressure, temperature, pH, flow rate, streaming potential, oil composition, and corrosion were reported to affect asphaltene precipitation.

8.5.1 Effect of pressure

Pressure reduction is the main reason for asphaltene deposition. This fact comes from various field observations and field cases. This can be attributed to the following:

- Pressure reduction causes fluid expansion, preferentially the lighter hydrocarbons (C6 –), which increase their volume compared to the heavy components (C7 + including aromatics). Asphaltenes are less soluble in such light hydrocarbons, whereas resins tend to migrate and dissolve in these lower-density fluids, which strip resins from asphaltene surfaces; both effects eventually cause asphaltenes to aggregate and deposit [6,129].
- With the changes in fractions volume, the polarity also changes and that promotes asphaltene self-association [5].
- These changes also promote significant density changes between the asphaltenes and the bulk oil, which induce asphaltene aggregation.
- Pressure changes cause pH to shift.

The effect of pressure on asphaltene precipitation is complex. This is illustrated in Fig. 8.20. The bubble point is a crucial parameter in asphaltene deposition studies. At pressures above bubble-point pressure, the undersaturated single-phase oil is under high compressibility forces from the external pressure, and strong intermolecular forces keeps the asphaltenes-resin complex stable and dissolved in the bulk oil [130]. With pressure reduction, the intermolecular forces weaken and the asphaltenes-resins balance is partially disturbed. With further pressure reduction to the bubble-point pressure, the molar volume of the bulk oil increases, and fluids expand [131]. Due to the difference in their compressibility, the light components C6 – increase in volume more than the C7 + fraction, and hence the crude oil acts more like lighter oil, with overall density reduction. Under these conditions, the solubility parameter between resins and evolving lighter ends decreases, leading to their dissolution. Meanwhile the asphaltene micelles solubility decreases, which later causes asphaltene precipitation [130]. Kokal and Sayegh [132] and Hammami et al. [133] have concluded that the maximum difference in molar volume (density) between the asphaltenes and the bulk oil occurred at the bubble-point pressure before the first dissolved gas is released. As pressure decreases below the bubble point, the composition of fluids changes as the lighter hydrocarbons evaporate to the gas phase, thus changing the molar volume of the liquid phase and partially restoring asphaltene solubility. Further pressure reduction liberates more light components, leaving the heavier crude oil with higher resins fraction, sufficient to peptize and stabilize the asphaltenes [131]. Field observations from Ventura Field, Hassi-Messaoud Field, and Lake Maracaibo showed that asphaltene problems diminished after the bottom-hole pressure fell below the bubble-point pressure. There was no deposition problem observed in Ula Field, Norway, below the bubble-point pressure [130].

So, at what exact pressure do asphaltenes start to precipitate?

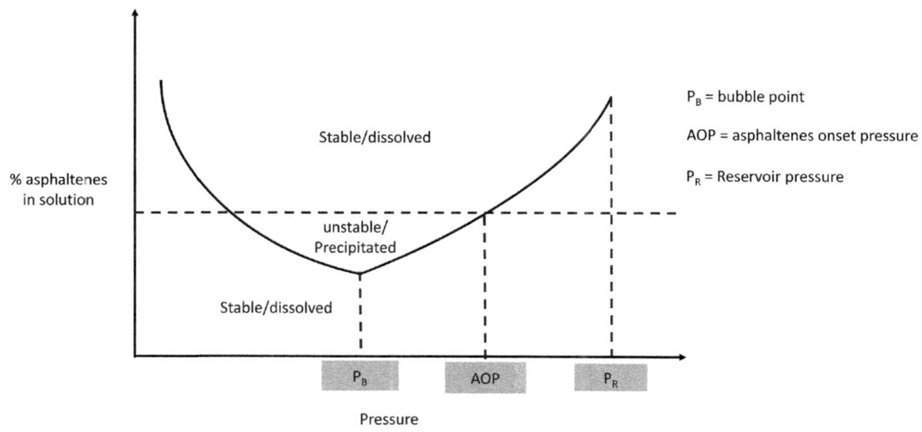

FIG. 8.20

Asphaltene stability and precipitation as a function of the pressure.

The answer is the asphaltene onset pressure, or AOP. It is defined as the pressure at which asphaltenes start to precipitate at a given test temperature during pressure reduction [130].

Operators should always keep the pressure drop at its lowest to assure that the flowing bottomhole pressure (FBHP) or the flowing streams in general are always at pressure higher than the AOP, especially for oil reservoirs with rich asphaltenes. Pressure depletion in reservoirs is usually maintained by gas injection or water flood.

There two main AOPs, the upper AOP and the lower AOP. The point where asphaltenes start to precipitate from zero weight percentage are defined as the upper AOP, or UAOP; then asphaltene precipitations reach their maximum amount at the bubble point, and then below the bubble point the asphaltene deposits starts to redissolve in the crude oil, up to a point called the lower AOP, or LAOP, at which the dissolved asphaltenes reach a constant value [134]. Fig. 8.21A shows the UAOP and LAOP in terms of soluble asphaltenes and Fig. 8.21B shows the UAOP-LAOP curve in terms of the precipitated asphaltenes with the common bell shape.

Many methods have been used to measure the onset of precipitation and precipitation pressure. The different methods that are used in measuring AOP are listed in Table 8.2 [135].

The thermodynamic boundaries (in pressure-temperature space) within which asphaltenes destabilize or flocculate and precipitate is called the asphaltene deposition envelope (ADE). The ADE principally shows the thermodynamic path that must be followed during production operations to avoid or minimize asphaltene problems [136].

Typical ADE is illustrated in Fig. 8.22. It shows that the severity of asphaltene flocculation increases as the fluid conditions move from the boundary to the center of the envelope [136]. For the given example, in moving from reservoir conditions (red dot), pressure decreases with production until it intersects with ADE at the UAOP (upper line in ADE), where asphaltenes start to precipitate. Further decrease in pressure precipitates more asphaltenes, until it reaches its maximum at the bubble-point pressure where the gas normally releases from solution. With additional pressure reduction, gases are removed from the system, and crude oil may begin to redissolve asphaltenes at LAOP (lower line in ADE) [137].

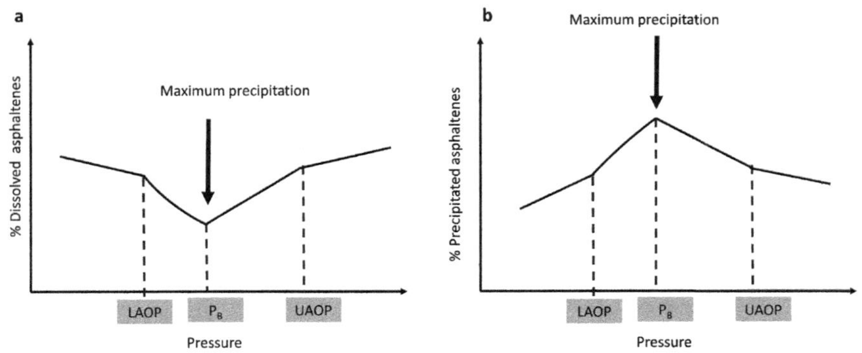

FIG. 8.21

Asphaltene solubility vs pressure curves showing the UAOP and LAOP.

Table 8.2 Methods of measurement of onset of asphaltene precipitation [135].

Method	Principles of AOP detection	Pros and cons
Gravimetric	Detection of AOP during depressurization of live oil, or titrating dead with precipitant, e.g., heptane	Can be applied on dead or live oil Requires enough sample volume, variable accuracy
Viscosity measurement	Oil viscosity measurement during titration with precipitant	Used with live or dead oils, light or heavy oils Depends on the viscosity measurement technique
Optical microscopy	Image processing of captured snapshots of oil during depressurization or titration with precipitant	Can be used with dead or live oil Depends on the opacity of the oil sample, and the accuracy of the microscope
Refractive index (RI) technique	From the plot of RI vs % volume of precipitant, a deviation from the straight line is regarded as the AOP concentration	Used with dead oil, can be applied in HPHT, fast and direct technique Heavy opaque oils can only be estimated by extrapolation
Interfacial tension (IFT) technique	IFT measurements between water and a mixture of oil and precipitant	Used with dead oil Not accurate in samples with low asphaltene content
Heat transfer technique	Detecting the increase in thermal resistance of tested mixture due to asphaltene precipitation	Used with dead oil, relatively fast, can be used in HPHT conditions Its accuracy depends on the temperature sensor and asphaltene content in the sample
Density	Change in density of oil after asphaltene precipitation	Used with dead oil
Electrical technique	Detecting the decrease in conductivity of oil sample after asphaltene precipitation	Used with dead oil, can be used in HPHT Accuracy depends on the sensors and the asphaltene content in the sample

Table 8.2 Methods of measurement of onset of asphaltene precipitation—cont'd

Method	Principles of AOP detection	Pros and cons
Acoustic resonance technique (ART)	Detecting the decrease in acoustic response of oil sample after asphaltene precipitation	Used with live and dead oils, light and dark oils, fast technique, requires small sample volume No mixing is applied, causes some inaccuracy, and less accurate in the lower boundary of asphaltene precipitation envelope
Light scattering technique (LST)	Measuring transmitted light vs pressure, absorbance vs time, or turbidity vs % weight of precipitate to indicate the onset of precipitation	Used with live or dead oils, affordable, can be used in HPHT

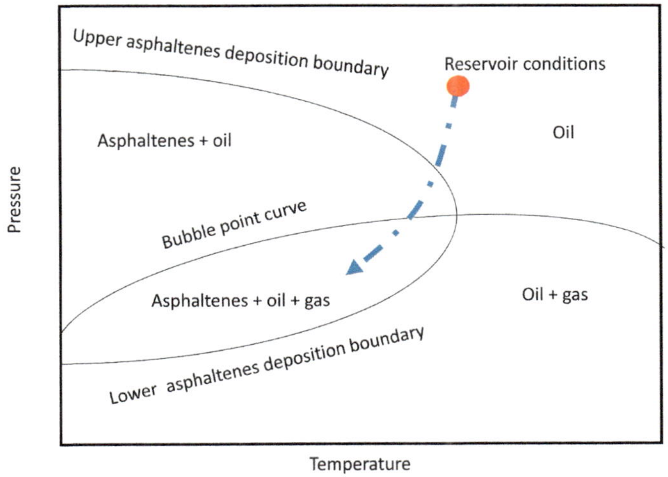

FIG. 8.22

Asphaltene deposition envelope.

Crude oils that exhibit asphaltene precipitation and deposition during primary depletion are typically undersaturated, meaning they exist in the reservoir at pressures higher than the bubble-point pressure. Asphaltene precipitation is quite common in crude oils with low asphaltene content and high gas content.

8.5.2 The effect of temperature

The effect of temperature on asphaltene precipitation is controversial. Some studies have revealed that at low temperatures, asphaltene becomes more concentrated and susceptible to precipitation, while others showed that the stability of asphaltene is higher at lower temperatures [131,138,139].

Generally speaking, it is thought that decreasing temperature reduces interacting energies between asphaltene molecules and solvent (crude oil) molecules, which destabilizes asphaltenes [139], while increasing temperature can break the resin-asphaltene bounds, which destabilizes asphaltenes [138].

Chandio et al. [140] showed that using the Heithaus parameters and the Hildebrand solubility parameter, the solubility of asphaltenes increased with increase in temperature. This suggests that the association between resins and asphaltene particles is stronger, and their tendency to aggregate is lower with increasing temperature. A fairly linear trend up to 60°C is observed, but a significant rise in the solubility parameters has been seen at temperatures above 60°C, presumably due to possible phase transition of asphaltenes above 60°C. This observation is in agreement with Mehranfar et al. [141], who demonstrated that asphaltenes transform from a purely amorphous phase at temperatures below 70°C to a liquid crystalline phase above 70°C. It is also observed that the average asphaltene solubility parameter has largely remained between 18.5 and 22.5 $MPa^{0.5}$.

Zendehboudi et al. [142] reported different behavior of asphaltenes in different crude oil types, as illustrated in Fig. 8.23. They concluded that the increase in temperature leads to an increase in asphaltene precipitation in the light oil, while the opposite behavior is observed for the heavy oil. This is attributed to solubility of asphaltenes in the oil as a function of temperature and pressure. They also reported a lower portion of asphaltenes available in the heavy oil is precipitated compared to the light oil for the same thermodynamic conditions.

The temperature has an effect on the AOP as well. As the temperature decreases, the AOP increases, and the bubble point increases [139].

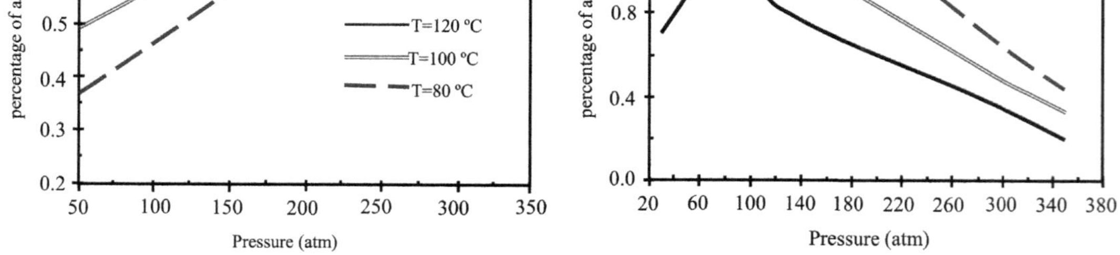

FIG. 8.23

Effect of temperature and pressure on asphaltene deposition in light oil (left) and heavy oil (right).

Reprinted with permission from S. Zendehboudi, A. Shafiei, A. Bahadori, L.A. James, A. Elkamel, A. Lohi, Asphaltene precipitation and deposition in oil reservoirs—technical aspects, experimental and hybrid neural network predictive tools, Chem. Eng. Res. Des. 92(5) (2014) 857–875. https://doi.org/10.1016/j.cherd.2013.08.001. Copyright 2014, Elsevier.

It worth to mention that temperature decrease leads to deposition of paraffin wax as explained in Chapter 7. These wax deposits can entrap some of asphaltenes molecules and deposits within it. That also can nucleate further asphaltenes depositions [143].

8.5.3 Crude oil composition

Extensive field and laboratory data showed that asphaltenes precipitate more easily from light oil as compared to from heavier oil, although the heavier oil may contain higher asphaltene content. Heavier oil mainly comprises higher intermediate components with more resins and aromatics which work as a good solvent to stabilize asphaltenes, whereas lighter oil comprises lighter components with limited asphaltenes solubility in them [130]. This is supported with observations from fields where the Venezuelan Boscan heavy crude oil with 17.2-wt% asphaltenes was producing nearly asphaltenes trouble-free, whereas Hassi-Messaoud light crude oil in Algeria with only 0.15-wt% asphaltenes has numerous asphaltene deposition problems.

8.5.3.1 Effect of light ends

Asphaltenes have low solubility in light hydrocarbons. Besides, in accordance with the colloidal approach, resin molecules tend to desorb from asphaltenes surface and breaking the micelles bond, upon the addition of light hydrocarbons to reservoir fluids for flow improvement or maintenance purposes. Similar issues will arise when condensate diluent is added to heavy oils to reduce viscosity, asphaltenes deposition can occur in surface facilities and pipelines [144].

Many models were used to predict the effect of oil components and paraffinic solvents on asphaltenes deposition. These include Continuous thermodynamic (CT) model, fractal aggregation model (FA), steric colloidal model (SC) and others. Among these de Boer model comes as a conservative screen that is best used to identify nonproblematic oils.

de Boer et al. [145] developed asphaltenes stability plot (de Boer plot) by correlating crude oil density with crude properties, such as solubility parameter, molar volume, and asphaltene solubility in oil. The plot has the difference of reservoir pressure and bubble-point pressure as the *y*-axis and the in situ crude density as the *x*-axis, and classifying crudes oils into 3 main categories: crudes with no asphaltenes problems, crudes with slight asphaltenes problems, and those with severe asphaltenes problems. The difference between reservoir pressure and bubble-point pressure and the crude densities at reservoir conditions can be calculated, which then used to instantly assess the risk level of asphaltene precipitation during production [146]. A simplified de Boer plot is shown in Fig. 8.24.

Based on field experience, de Boer [145] also concluded that asphaltene deposition problems are encountered in light crude oils that are high in C1-C3 ($>$37 mol%) and have a relatively low C7+ content ($<$46 mol%), low asphaltene content ($<$1.0 wt%), and a large difference between reservoir and bubble-point pressures ($>$ 30 MPa). The heavy crude oils that have high asphaltene content tend to have fewer problems with asphaltene deposition as in the examples of Venezuelan crudes mentioned earlier. This also depends on the reservoir pressure and if it was close to the bubble-point pressure [146].

de Boer also concluded that the crude tendency for asphalt precipitation is mainly determined by three parameters: the extent to which the crude is undersaturated with gas, the density of the crude at reservoir conditions, and its saturation with asphalt at downhole conditions [145].

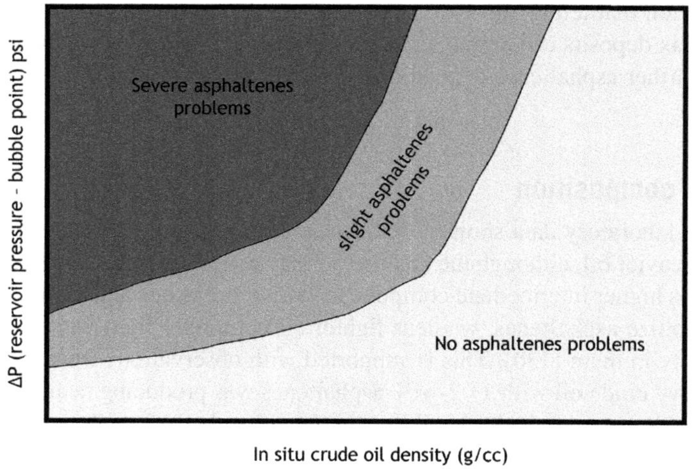

FIG. 8.24

Typical de Boer plot for asphaltenes stability.

8.5.3.2 Effect of naphthenic acids

It was found that naphthenic acids adsorb to asphaltenes and disperse them, but the performance was lower than a commercial AD based on fatty acids and amines [5].

8.5.3.3 Effect of aromatics

The aromatic fraction is also found to disperse asphatenes in solution similar to aromatic solvents such as toluene or benzene [147]. The aromatics (such as toluene) are relatively good solvents for both wax and asphaltenes.

8.5.3.4 Effect of resins

Resins are well known for so long to peptize and stabilize petroleum asphaltenes. The resins are typically composed of a highly polar parts, with heteroatoms such as oxygen, sulfur, and nitrogen, as well as long, nonpolar paraffinic side chains. The resins are attracted to the asphaltene micelles through intermolecular forces, while the paraffinic side chains as a tail that keeps the asphaltene-resin complex dissolved in the bulk oil [129].

However, Mullins et al. [117] showed that this asphaltene-resin micelle model, which has virtually no supporting data, appears to be substantially misleading in treating asphaltene colloidal structure.

In the meantime experimental data showed that asphaltenes precipitations were significantly decreased by the addition of resins into the crude oils, also crude oils with high resins to asphaltenes ratio showed better asphaltenes stability [148,149].

Ashoori et al. [150] showed that the stability of asphaltenes in crude oils is a phenomenon that is related to all of crude oil components (saturates, aromatics, resins and asphaltenes) behavior, and it cannot be associated only with one of them, individually.

Yen et al. [151] used the SARA data and developed the colloidal instability index (CII) of the oil. The CII considers a crude oil as a colloidal system made up of the pseudo components saturates, aromatics, resins and asphaltenes, and asphaltenes stability depends on these components ratios. It is calculated as the ratio of the sum of the saturates and asphaltenes fractions to the sum of the aromatic and resin fractions to assess the stability of asphaltenes. Oils with a CII of below 0.7 are considered stable while those with a CII of above 0.9 are considered very unstable. A CII of 0.7–0.9 shows is indicative of an oil with moderate instability. The CII correlation is as follows:

$$CII = \frac{(\text{saturates} + \text{asphaltenes})}{(\text{resins} + \text{aromatics})} \tag{8.6}$$

Therefore change these ratios by adding or removing one or more of the components can shift this equilibrium and lead to asphaltenes deposition. Fig 8.25 shows the areas of asphaltenes stability and instability according to the CII.

8.5.4 Mixing of crude oil with other fluids

Asphaltenes are held in petroleum in a delicate equilibrium or balance, which can be easily disturbed by changing crude composition through adding saturates or by removing resins or aromatics [152]. Thus, mixing of crude oil with other crudes, solvents and other petroleum fractions is one of the biggest sources of crude oil composition alteration. The different mixing possibilities are:

- Mixing crude oil with paraffinic solvents.
- Mixing crude oil with crude oil.
- Mixing crude oil with miscible gas CO_2.
- Mixing crude oil with rich gas.

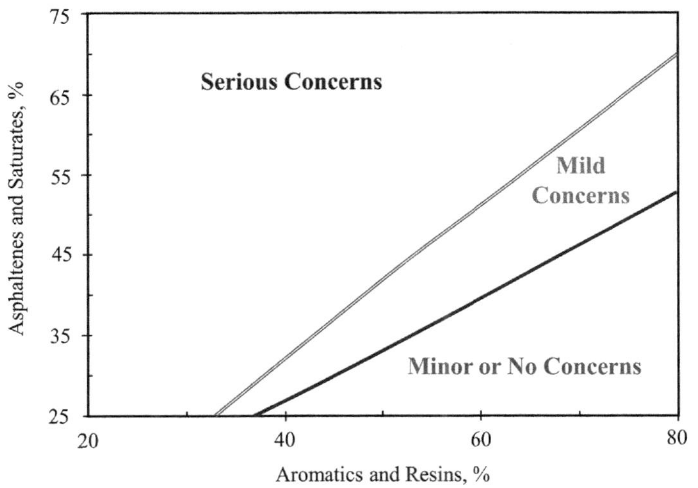

FIG. 8.25

Asphaltenes stability zones determined from the ratio of weight percentage of asphaltenes and saturates versus aromatics and resins percentage.

Reprinted from with permission from S. Zendehboudi, A. Shafiei, A. Bahadori, L.A. James, A. Elkamel, A. Lohi, Asphaltene precipitation and deposition in oil reservoirs—technical aspects, experimental and hybrid neural network predictive tools, Chem. Eng. Res. Des. 92(5) (2014) 857–875. https://doi.org/10.1016/j.cherd.2013.08.001. Copyright 2014, Elsevier.

8.5.4.1 Mixing crude oil with paraffinic solvents

Mixing asphaltenes with paraffinic solvents leads to asphaltenes aggregation and precipitation. Paraffinic solvents are added to the crude as flow improvers or to dissolve organic paraffinic obstructions can lead to the precipitation of asphaltenes.

Experiments showed that paraffinic and other nonparaffinic compounds caused asphaltenes deposition. Cyclohexane, methanol, formic acid, heptane, tetrabutylammonium fluoride, acetonitrile, decane, hexanol-1, furfural, and octene-1, added to the solution of asphaltenes in toluene, were shown to produce a considerable aggregation of asphaltene particles of 10–13nm size. However, decalin, dimethylformamide, dimethylsulfoxide, nitrobenzene, aniline, cyclohexanol, monoethanolamine, triethanolamine, 1-methylnaphtalene, and dichloroethane cause asphaltene aggregate sizes to decrease essentially [153].

8.5.4.2 Mixing different crudes

The blending of crude feed stocks is a common cause for asphaltene precipitation. Light, nonasphaltic crude can be a potential precipitant for heavier crudes, as indicted previously by changing the molecular balance of saturates, resins, aromatic and asphaltenes leading to asphaltenes precipitation [5]. Since resins and asphaltenes always to be associated with each other, the phase behavior and solubility can be represented and determined based on aromatics-saturates balance [152].

Another compatibility criterion proposed by Wiehe [152] which determine the solubility parameter at which asphaltenes precipitate for crudes and crude mixtures using titration of the individual oils with asphaltenes solvent, such as toluene, and a nonsolvent such as n-heptane. This solubility parameter on a reduced n-heptane-toluene scale is called the insolubility number (IN). In addition, the tests measure the solubility parameter of the oil that on a reduced n-heptane-toluene scale is called the solubility blending number (SBN). The criterion for compatibility of any blend is that the volume average solubility blending number is greater than the maximum insolubility number of any component in the blend: (SBN) blend > (IN) max.

These values can be expressed by the following correlations:

$$I_N = 100\frac{(\delta_f - \delta_H)}{(\delta_T - \delta_H)} \tag{8.7}$$

$$S_{BN} = 100\frac{(\delta_{oil} - \delta_H)}{(\delta_T - \delta_H)} \tag{8.8}$$

where δ_f is the flocculation solubility parameter, δ_H is the solubility parameter of n-heptane, δ_T is the solubility parameter of toluene, and δ_{oil} is the solubility parameter of the oil [152].

When oil is completely soluble in n-heptane, hence it speculated to have no asphaltenes and thus IN = 0. However, if the asphaltene-resin dispersion is barely soluble in toluene, here IN = 100. Likewise, an oil with poor solvency like n-heptane has a SBN = 0, and an oil that is as good a solvent like toluene has a SBN = 100 [152].

Fig. 8.26 shows the evaluation of compatibility of Souedie and Forties crudes from the heptane dilution and toluene equivalence tests. The insolubility numbers are the y axis intercepts.

From Fig. 8.26, since IN of each crude is less than its SBN, each crude is compatible with itself. However, since the IN of Souedie, 39, is greater than the SBN of Forties, 27, there will be some mixtures of the two crudes that are incompatible [152].

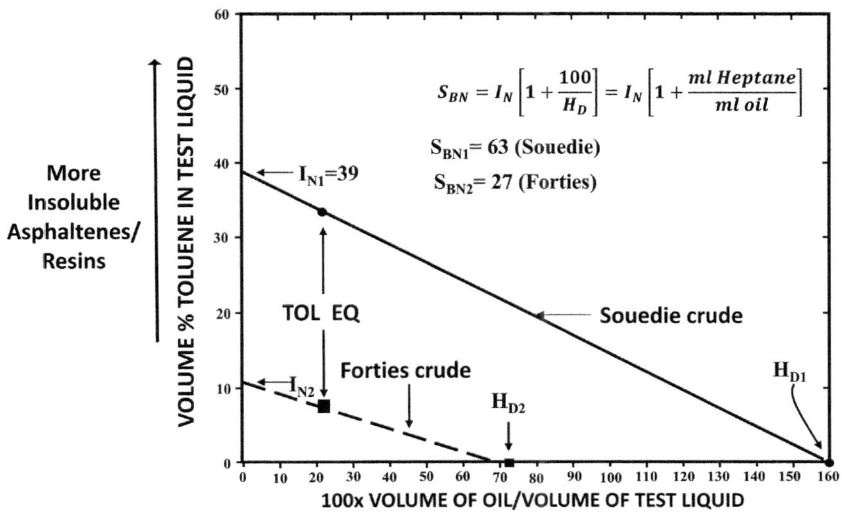

FIG. 8.26

Evaluation of compatibility of crude oils by Wiehe method.

Reprinted with permission from I.A. Wiehe, R.J. Kennedy. Oil compatibility model and crude oil incompatibility, Energy Fuels 14(1) (2000) 56–59. https://doi.org/10.1021/ef990133+. Copyright 2000, American Chemical Society.

That method was found to be practical especially in refineries where crude mixing can cause asphaltenes problems. In field application of this method, blending order is important, as potentially incompatible oils should be blended in the order of decreasing solubility blending number [152].

8.5.4.3 Mixing crude oil with CO_2 gas (CO_2 injection)

Asphaltene precipitation and deposition have been reported to exacerbate by gas injection. Coincidently, light oil reservoirs are the more preferable candidates for gas injection processes, which makes the problem worse. In fact CO_2 have been shown to be the most effective asphaltene precipitant followed by alkanes (C1–C7) [130].

CO_2 injection leads to asphaltenes precipitation due to [130,154]:

- Lowering pH.
- Creating turbulence.
- Most importantly is CO_2 disturbs the solution equilibrium by stripping the resins of the asphaltenes surface and change oil composition which favors the precipitation of asphaltenes.

Experimental work and field observations showed that the amount of precipitated asphaltene increases with increasing CO_2 concentration. The most noticeable primary location of asphalt materials in the CO_2 floods during the early stages of operations is in wellbore and the pumps [130,154,155].

For planning for reservoir management strategy, during CO_2 injection, it is crucial to know [130]:

- The initial temperature and pressure in reservoir condition.
- The bubble-point pressure.
- Asphaltene onset pressure for a reservoir oilfield over different mol% of CO_2 gas injection.

Bahrami et al. [156] showed that while CO_2 is known to cause asphaltenes precipitation at high temperatures, however below a specific crossover temperature, the stability of asphaltene improves. Therefore, CO_2 can behave as an inhibitor or precursor of asphaltene precipitation.

Formation water acts as a CO_2 buffer during CO_2 gas injection, thus it reduces CO_2 concentration in the crude oil reservoir and minimizes asphaltenes flocculation [157].

H_2S gas reduced the risk of asphaltenes deposition, and offsets the the impact of CO_2 gas. Nitrogen was found to show severe effect of on asphaltene precipitation is compared to methane [155,158]. Nitrogen expands the APE boundaries more than the methane gas [156].

8.5.4.4 Mixing crude oil with rich gas

Rich gas flooding destabilizes the asphaltenes, as light hydrocarbons and straight chain hydrocarbons are less solving for asphaltenes. Gas and light components stripping from the oil have been shown to improve the solubility of asphaltenes. The negative effect of rich gas injection peaks near the bubble point, and then decline after the bubble point is reached. This behavior near the bubble point in more pronounced with the rich gas than with CO_2 [143,154].

Asphaltenes flocculation increases when light gases are introduced the reservoir fluids during miscible gas flooding processes. Besides, the bubble-point curves and the asphaltene stability boundary increase when different compositions of methane, associated gas, and N_2 are added to the original fluid [156]. Fig. 8.27 illustrates the asphaltene precipiation bundaries with the addition of different gases (methane, associated gas, N_2, and CO_2) compositions.

8.5.4.5 Effect of gas lift

Gas lift is one of the artificial lift techniques used in production operations. Gas lift is usually used above the bubble-point pressure. Unfortunately, gas lift has been found to cause asphaltene problems since the light components lead to compositional changes that disturb asphaltene stability [159]. Once formed, asphaltene deposits can damage the gas lift system and can leave gas lift check valves stuck open or closed.

This condition becomes severe with low flow rates and the composition of the reservoir is unstable. In general two main concepts need to be considered in the effect of gas lift on asphaltene deposition:

- The composition and percentage of the gas lift. Lighter-component gas lift leads to worse asphaltene deposition problems.
- The flow rate. Optimizing the flow rate slows down asphaltene deposition by gas lift injection.

8.5.5 Effect of pH shift

Asphaltenes have been shown to contain acidic and basic fractions, which makes them amphoteric chemicals that can apparently ionize at either low or high pH, respectively, when they come in contact with water, or other acid or basic fluid consitiuents [160,161]. This basically affects asphaltene hydrophilic behavior and surface charge/activity, which as a result leads to asphaltene aggregation. While this can happen at high or low pH, this effect was found to be stronger in the high-pH range if asphaltenes have more acidic fractions than basic ones. At high pH, it was observed that emulsion is stabilized more by asphaltenes [161]. Strassner [162] found that rigid film emulsions form whenever asphaltene containing crude comes in contact with an aqueous fluid of pH$=$6. This study shows that crude oils with API

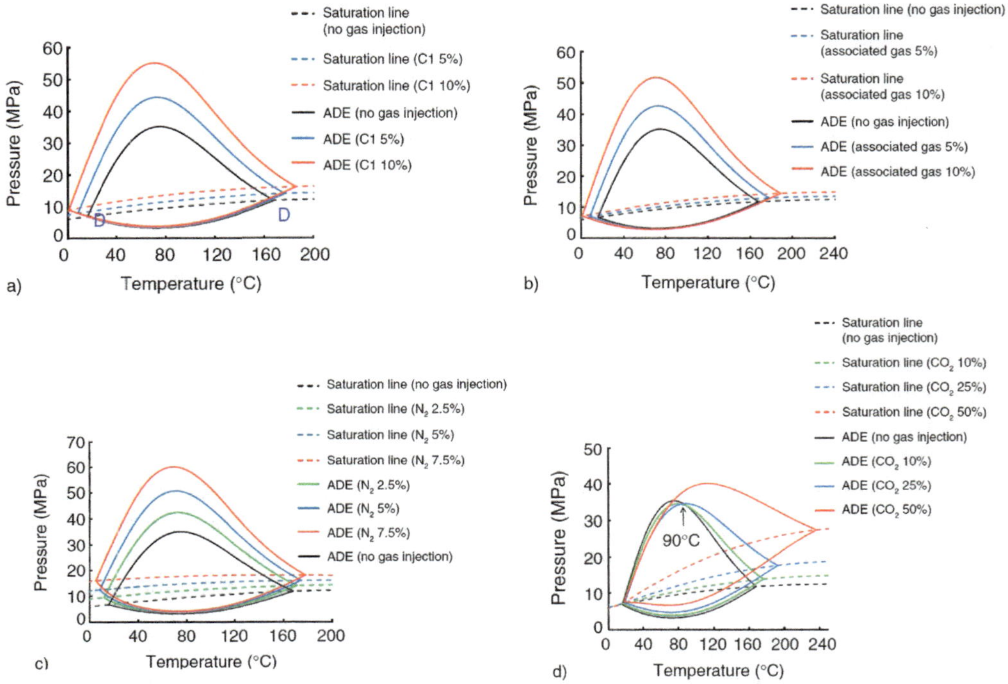

FIG. 8.27

Effect of (a) methane, (b) associated gas, (c) N_2, (d) CO_2 on the asphaltenes precipiation envelope, indicating increasing stability boundaries incase of methane, N_2 and associated gas, while a different behavior observed in case of CO_2.

From P. Bahrami, R. Kharrat, S. Mahdavi, H. Firoozinia, Prediction of the gas injection effect on the asphaltene phase envelope, Oil Gas Sci. Technol. 70(6) (2015) 1075–1086. https://doi.org/10.2516/ogst/2014037.

gravities ~22 and asphaltene contents of >4 typically form rigid film emulsions and subsequently do not aggregate or form sludge.

The pH shift of the crude oil-asphaltene solution may be imposed by CO_2, mineral acid, or naturally occurring or bacterially produced organic acids [143].

8.5.6 Effect of acidizing/acid stimulations

Severe asphaltene precipitation may be caused by well stimulations, such as acidizing, or acid cleaning jobs, etc. Acid treatments lead to the following [143]:

- A drastic shift in local chemical equilibria including pH change.
- CO_2 gas liberation.
- Increase in the concentration of some ions, such as iron, which may influence the formation of asphaltic sludges, flocs, etc.

Houchin et al. [163] concluded that the oxidation potential of the acid system appears to be the primary factor contributing to sludging. They also concluded that acid mixtures HCl:HF seem to cause more severe sludging than HCl acid of the same concentration. The data indicates that sludge can be significantly reduced by adjusting the initial HCl stage with acetic. The use of low surface tension liquids such as diesel can cause sludge. Some acid corrosion inhibitors promote sludge. Generally during acid jobs, reduced acid strength, organic acids and solvent preflushes, comprehensive iron control, and optimized additive packages contribute to reducing asphaltene sludge tendencies.

8.5.7 Effect of metal ions (Fe, Ni, V)

Metal ions cause aggregation and precipitation of asphaltenes. The list includes the metals iron, nickel, vanadium, zinc, copper, lead, arsenic, molybdenum, cobalt, manganese, chromium, etc., all of which have been reported [164].

Iron is thought to cause aggregation by a charge neutralization effect due to coordination of iron with porphyrin, pyrroles, pyridines, or possibly phenolic hydroxyl groups. Another explanation is catalyzing polymerization processes taking place at elevated temperatures in the oil [165]. Iron was found to have higher solubility in the oil phase than other ions, so it can be released from injection or produced water into the oil phase and destabilize the asphaltenes. The mechanism of adsorption of heavy fractions like asphaltenes on iron materials is explained in terms of aromaticity, H/C ratio, and the formation of bonds on heteroatoms, where the heteroatoms and aromaticity were found to decrease the activation energy of the iron-asphaltenes complex [166]. Iron concentrations in natural and produced waters is normally low (normally <100 ppm), but the corrosion process and acid treatments can increase the iron concentrations. Tubing washes and careful cleaning procedures are followed; as little as 1000 ppm iron can be expected [165].

There is a theory concerning the two metals nickel and vanadium, which are found to be present in significant concentrations in many crude oils. Their concentration was found to be higher in asphaltene deposits than it is in the crude oil itself. These metals may be somehow implicated in the mechanism for asphaltene deposition by the formation of organometallic complexes, or imparting surface charge on asphaltene molecule surfaces, which eventually can have impact on asphaltene stability and solubity in the bulk oil [5].

8.5.8 Effect of shear

Aphaltene aggregates are shear sensitive, and can be affected by electric forces. These electrostatic interactions and shear sensitivity can result in stripping the solvent sheathing (resins and maltenes) from the asphaltenes, resulting in destabilizing them and causing their aggregation and precipitation [7].

Stripping of resins that surround asphaltene particles can occur by shearing in pumps, the formation, perforations, valves, and the chokes. Asphaltene deposition in the pumps is very common, such as multistage centrifugal pumps and electrical submersible pumps [167]. Because most ESP applications operate at very close to pump-off conditions, the pump intake pressure is close to bubble point, which makes asphaltene problems worse.

8.5.9 Effect of streaming potential

Streaming potential is the electric potential generated when a liquid (electrolyte) is forced to move by pressure gradient along a stationary channel (pipe, or porous media) with charged surface [71]. It is an electrokinetic effect that influences asphaltenes in flowing crude oil.

This generated electrical potential difference can change asphaltene colloidal particle surface charge and ultimately result in their deposition [167]. This effect can occur in the reservoir porous media, tubing, or pipeline.

The factors influencing this effect are the electrical and thermal characteristics of the conduit, flow regime, flowing oil properties, and characteristics of the polar heavy organics and colloidal particles, e.g., sizes, surface charge, etc.

Leontaritis and Mansoori [113] showed that the net negative charge on the asphaltene colloidal surface is the main cause of their deposition in pipes due to the streaming potential. This feature of asphaltenes can be used to control asphaltene deposition. Since asphaltenes were found to deposit at the anode surface upon application of DC potential, while little or no deposition occurred on cathode surface, the method will be similar to cathodic protection used in corrosion control, where the pipe is converted to a cathode with an anode pair nearby [71]. However, that requires fully addressing fluid physicochemical properties and system parameters, e.g., pressure, temperature, bubble point, AOP, etc.

8.5.10 Effect of water cut

It is generally thought that produced water can help in reducing asphaltene deposition along the production flow path, especially when water cut is high. Explanations for this, arbitrarily, are [168–170]:

- The precipitated asphaltenes tend to accumulate at water-oil interfaces, leaving fewer available asphaltenes to deposit on pipe walls.
- Water can delay asphaltene deposition theoretically due to hydrogen bonds between the water molecules and heteroatoms (N, O, and S) in asphaltenes, thus fewer asphaltene aggregates are expected to accumulate on the pipe walls.
- The reduction of pipe wall surface area available for deposition, due to the bombardment of this surface by water droplets suspended in a turbulent flow.

Distilled water seems to be more efficient in deposition reduction than brines with high ionic strength, since metal ions such as ferric and aluminum ion brines induce asphaltene deposition.

Joonaki et al. [171] showed that, due to the substitution of divalent cations with monovalent ones, asphaltene deposition is repelled and the solid surface becomes more hydrophilic.

On the other hand, Yaseen and Ali Mansoori [41] using molecular dynamics reported that asphaltenes become less soluble in oil when water is partially miscibilized in the oil phase, whereas all model asphaltenes containing nitrogen and/or oxygen atoms in their structures were prone to association. The driving force behind asphaltene aggregation is shown to be the hydrogen bonding between asphaltenes and water molecules. The utilization of brine (3.5 wt% NaCl) instead of pure water induces a slight decrease in asphaltene propensity for aggregation due to the salt-in effect.

8.5.11 Effect of corrosion

Corrosion is a deteriorating process on the production facilities. It also has significant impact on asphaltene deposition, which can be summarized in the following points:

- The main product of metal corrosion is iron ions, which as discussed earlier can enhance asphaltene deposition.
- The corrosion process furnishes the rough surfaces that can work as nucleating and crystal growth sites for asphaltene deposits.

Iron ions form a complex with asphaltenes that stick to steel surfaces. Sung et al. [166] reported that adding iron III sequestering agents like EDTA can prevent the iron III impact on the asphaltene deposition.

8.5.12 Effect of asphaltene composition

As discussed earlier in this chapter, asphaltene molecules can exist over a wide range of compositions and properties. This diverse composition of asphaltenes can give them some stability against deposition, or can induce their aggregation.

Gabrienko et al. [172] concluded that asphaltene molecules comprising polycyclic aromatic hydrocarbons with a minor presence of heteroatoms are more stable compared to those enriched with heteroatoms O, N, and S. This can be explained in terms of intermolecular forces. They also ranked asphaltenes according to their composition and stability as follows:

- Oxygen-containing asphaltenes can be considered as less stable species.
- Asphaltene molecules incorporated by pyrrole and pyridine groups have medium stability.
- Asphaltene molecules consisting of polycyclic aromatic hydrocarbons with the minor presence of heteroatoms have high stability compared to those enriched with O, N, and S.

Besides the factors mentioned, there are other factors that influence asphaltene aggregation and precipitation. Thermodynamic hydrate inhibitors influence hydrate aggregation and precipitation. Alcohols with intermediate-hydrocarbon chains prevented asphaltene aggregation more effectively and reduced their emulsification tendency, compared to short chain alcohols. The effect of alcohols on the asphaltene particle size was also shown to be highly sensitive to the presence of a free-water phase, likely due to observed alcohol partitioning into the water phase [173].

In addition, some wax inhibitors were found to hinder asphaltene aggregation and deposition, while others can aggravate the problem [174].

In the presence of asphaltenic oil, scale inhibitors did not affect the oil/water permeability, whereas the corrosion inhibitors caused a negative effect on the oil/water relative permeability, and a mixture of scale and corrosion inhibitors showed no effect [175].

8.6 Recent advances in asphaltene deposition research

Asphaltenes represent a complex problem that poses multiple flow assurance issues in production operations. They have a complex chemical configuration and a strong precipitating tendency [176].

The interest in asphaltenes started to gain some momentum in the late 1990s, with the developments in analytical techniques and the significant attention given to the unconventional and heavy crudes, along with other production-related issues that made studying asphaltene chemical and physical properties mandatory.

Determining crude oil properties and characteristics is an important step in understanding the phase behavior of crude oil and asphaltene fractions at different stages of production and developing feasible strategies for mitigating and management of asphaltene-associated problems. Some asphaltene physicochemical properties are still subject to controversy in the literature, which is attributed to their complex structure and properties. Several analytical methods have been employed in the characterization of asphaltene aggregates and elucidating their structure and other physicochemical properties, with significant advances in understanding asphaltene composition and behavior. Nonetheless, the current methods have some limitations; they can be lengthy and consume many solvents, and they can be expensive, and inaccurate [176,177].

A number of experimental techniques have been developed to detect the onset of asphaltene precipitation at reservoir conditions, which include: gravimetric technique, acoustic resonance technology (ART), light scattering technique (LST), high-pressure microscopy (HPM), HPHT filtration, and quartz crystal resonator (QCR) technique. Among these techniques, LST and HPM have become the industry's standards to measure asphaltene precipitation onset.

Asphaltene precipitation methods have witnessed significant advances with several proposed models. However, since not every precipitation leads to deposition, the deposition step is considered critical. The deposition step of asphaltenes is not fully understood. Some of the experiments and techniques that are used in studying deposition include the capillary flow loop, the quartz crystal microbalance, the RealView deposition cell, and the packed bed column.

The capillary flow loop is a dynamic method that is capable of estimating the amount and distribution of deposition by the pressure drop across the capillary. However, this technique is always limited to studying dead oil samples under laminar flow regime. The quartz crystal microbalance is highly sensitive to detect the adsorption of asphaltenes on different surfaces by consuming a small amount of sample, yet the technique is not representative to actual field asphaltene deposition. Rear view cells or Organic Solid Deposition and Control Devices (OSDCs) are a recent technique that has been used with ability to simulate pressure, temperature, and hydrodynamic changes. However, the technique is costly and requires high sample volumes. The packed bed column is a newly developed apparatus to investigate asphaltene deposition by direct mass balance using a small amount of oil sample [176]. The Asphaltene Rocking Cell (ARC) test is another recent method of testing asphaltene deposition. It is analogous to the well-known Hydrate Rocking Cell test, but is able to operate in high temperature and high pressure (HT/HP) conditions, up to 15,000 psi and 250°F. ARC is able to simulate live oil conditions by combining synthetic field gas with dead crude oil in HT/HP conditions [178]. Recently, microfluidic technology has been employed in studying asphaltene deposition.

Recent studies showed that asphaltenes in crude oil are different from their corresponding deposits, as the latter are composed of less soluble asphaltenes than those present in the original crude oils, suggesting a complex mechanism of formation that is usually not considered in the tools currently used to study asphaltene deposition. This also indicates that asphaltenes separated using heptane or pentane might not be representative of the asphaltenes found in field deposits [179].

Furthermore, more advances of asphaltene precipitation and deposition under multiphase flow and using live crude oil will step up the game in understanding asphaltene behavior.

On the other hand, asphaltenes can have some benefits in oilfields. For example, Chevron and Schlumberger used the optical properties of asphaltenes to understand reservoir connectivity in the Tahiti field, a deepwater structure in the Gulf of Mexico [137,179]. Besides, while asphaltenes affect the formation wettability, in some wettability models the presence of asphaltene in the oil phase is required for the generation of oil-wetting conditions [180].

Recently, asphaltenes have been studied for potential use in different applications such as biological imaging and drug delivery, display technology, photovoltaics, and others [181]. Asphaltenes have gained a lot of momentum in synthesis of new materials with superior properties that can be used in different applications. For example, asphaltenes were used in synthesis of three-layer graphene sheets [182]. Asphaltene oxide was synthesized and applied as a carbocatalyst avoiding the drawbacks of graphene oxide [183]. They were also investigated for the potential production of carbon-based nanoparticles, electrocatalysis, solar cells, and other energy storage applications [177,180,182].

8.7 Summary

Asphaltenes represent one of the most complicated flow assurance issues. Asphaltenes are the polar and heaviest component in crude oil. They have molecular weight around 750 amu. They interfere with phase separation operations, and precipitate in the production system, causing interrupted operations and production loss. Asphaltene deposits are usually misdiagnosed as paraffins and require special attention from operators. Additionally, asphaltene structure and properties are complex and sensitive to various conditions and their stability depends on many factors. Pressure drop is one of the main factors that cause asphaltene deposition, especially when the pressure falls to near the bubble point, where the highest asphaltene precipitation occurs. Temperature change is a bit controversial with regard to its effect on asphaltene stability; pH changes were found to destabilize asphaltenes, something frequently observed after acid stimulation and cleaning jobs. Moreover, crude oil composition and the interactions between its components are a major affecting factor in asphaltene stability, which can be predicted according to the CII index. Furthermore, mixing incompatible crudes is a main cause of asphaltene deposition. Other factors include streaming potential, surface nature, corrosion process, and water cut.

Asphaltene research is wide and has gained momentum in the last few decades. Recent updates have taken place in asphaltene precipitation and deposition studies. In addition, recent interest has emerged in asphaltenes due to their potential industrial uses.

References

[1] S.O. Honse, S.R. Ferreira, C.R.E. Mansur, E.F. Lucas, G. González, Separation and characterization of asphaltenic subfractions, Quim. Nova 35 (10) (2012) 1991–1994, https://doi.org/10.1590/S0100-40422012001000019.

[2] J.B. Boussingault, Memoire sur la composition des bitumens, Ann. Chim. Phys. 64 (1837) 141–151.

[3] M.E. Moir, Asphaltenes, what art thou?: Asphaltenes and the Boduszynski continuum, in: ACS Symposium Series, vol. 1282, American Chemical Society, 2018, pp. 3–24, https://doi.org/10.1021/bk-2018-1282.ch001.

[4] J.G. Speight, Petroleum asphaltenes—part 1: asphaltenes, resins and the structure of petroleum, Oil Gas Sci. Technol. 59 (5) (2004) 467–477, https://doi.org/10.2516/ogst:2004032 (Editions Technip).

[5] M.A. Kelland, Production Chemicals for the Oil and Gas Industry, CRC Press, 2009, https://doi.org/10.1201/9781420092974.

[6] J. Bellarby, Production chemistry, in: Developments in Petroleum Science, 2009, https://doi.org/10.1016/S0376-7361(08)00207-0. Chapter 7.

[7] J.R. Becker, Crude Oil Waxes, Emulsions, and Asphaltenes, Pennwell Books, Tulsa, Okla, 1997.

[8] J.L. Creek, Freedom of action in the state of asphaltenes: escape from conventional wisdom, Energy Fuel 19 (2005) 1212–1224, https://doi.org/10.1021/ef049778m.

[9] K. Gharbi, C. Benamara, K. Benyounes, M.A. Kelland, Toward separation and characterization of asphaltene acid and base fractions, Energy Fuel 35 (18) (2021) 14610–14617, https://doi.org/10.1021/acs.energyfuels.1c01999.

[10] J.P. Maclean, P. Kilpatrick, Comparison of precipitation and extrography in the fractionation of crude oil residua, Energy Fuel 11 (1997) 570.

[11] S. Asomaning, Test methods for determining asphaltene stability in crude oils, Pet. Sci. Technol. 21 (2003) 581.

[12] N.G. Graham, M.L. Gorbaty, Sulfur K-edge X-ray absorption spectroscopy of petroleum asphaltenes and model compounds, J. Am. Chem. Soc. 111 (1989) 3182.

[13] S. Mitra-Kirtley, O.C. Mullins, J. van Elp, S.J. George, J. Chen, S.P. Cramer, Determination of the nitrogen chemical structures in petroleum asphaltenes using XANES spectroscopy, J. Am. Chem. Soc. 115 (1993) 252.

[14] J.G. Speight, S.E. Moschopedis, On the molecular nature of petroleum asphaltenes, in: J.W. Bunger, N.C. Li (Eds.), Chemistry of Asphaltenes, Advances in Chemistry Series, vol. 195, ACS, 1982, pp. 1–15 (Chapter 1).

[15] J.G. Speight, The Chemistry and Technology of Petroleum, third ed., Marcel Dekker, New York, 1999 (Chapter 10).

[16] D.L. Mitchell, J.G. Speight, The solubility of asphaltenes in hydrocarbon solvents, Fuel 52 (2) (1973) 149–152, https://doi.org/10.1016/0016-2361(73)90040-9.

[17] W.T. Kok, A.J. Tüdös, M. Grutters, A.G. Shepherd, Characterization of asphaltenes by nonaqueous capillary electrophoresis, Energy Fuel 25 (1) (2011) 208–214, https://doi.org/10.1021/ef100921d.

[18] E. Buenrostro-Gonzalez, H. Groenzin, C. Lira-Galeana, O.C. Mullins, The overriding chemical principles that define asphaltenes, Energy Fuel 15 (4) (2001) 972–978, https://doi.org/10.1021/ef0100449.

[19] A.B. Andrews, J.C. Edwards, A.E. Pomerantz, O.C. Mullins, D. Nordlund, K. Norinaga, Comparison of coal-derived and petroleum asphaltenes by 13C nuclear magnetic resonance, DEPT, and XRS, Energy Fuel 25 (2011) 3068–3076, https://doi.org/10.1021/ef2003443.

[20] O.C. Mullins, Petroleomics and structure–function relations of crude oils and asphaltenes, in: O.C. Mullins, E.Y. Sheu, A. Hammami, A.G. Marshall (Eds.), Asphaltenes, Heavy Oils, and Petroleomics, Springer, New York, 2007, pp. 1–16, https://doi.org/10.1007/0–387-68903-6_1 (Chapter 1).

[21] H. Groenzin, O.C. Mullins, Asphaltene molecular size and structure, J. Phys. Chem. A 103 (50) (1999) 11237–11245, https://doi.org/10.1021/jp992609w.

[22] A.E. Pomerantz, M.R. Hammond, A.L. Morrow, O.C. Mullins, R.N. Zare, Energy Fuel 23 (2009) 1162.

[23] T. Kuznicki, J.H. Masliyah, S. Bhattacharjee, Molecular dynamics study of model molecules resembling asphaltene-like structures in aqueous organic solvent systems, Energy Fuel 22 (4) (2008) 2379–2389,-https://doi.org/10.1021/ef800057n.

[24] O.P. Strausz, T.W. Mojelsky, F. Faraji, E.M. Lown, P. Peng, Additional structural details on athabasca asphaltene and their ramifications, Energy Fuel 13 (2) (1999) 207.

[25] O.P. Strausz, T.W. Mojelsky, E.M. Lown, The molecular structure of asphaltene: an unfolding story, Fuel 71 (1992) 1355.

[26] D.M. Barrera, D.P. Ortiz, H.W. Yarranton, Molecular weight and density distributions of asphaltenes from crude oils, Energy Fuel 27 (5) (2013) 2474–2487, https://doi.org/10.1021/ef400142v.

[27] D. Borton, D.S. Pinkston, M.R. Hurt, X. Tan, K. Azyat, A. Scherer, R. Tykwinski, M. Gray, K. Qian, H.I. Kenttämaa, Molecular structures of asphaltenes based on the dissociation reactions of their ions in mass pectrometry, Energy Fuel 24 (2010) 5548–5559.

[28] W. Tang, M.R. Hurt, H. Sheng, J.S. Riedeman, D.J. Borton, P. Slater, H.I. Kenttämaa, Structural comparison of asphaltenes of different origins using multi-stage tandem mass spectrometry, Energy Fuel 29 (2015) 1309–1314.

[29] B. Schuler, G. Meyer, D. Peña, O.C. Mullins, L. Gross, Unraveling the molecular structures of asphaltenes by atomic force microscopy, J. Am. Chem. Soc. 137 (31) (2015) 9870–9876, https://doi.org/10.1021/jacs.5b04056.

[30] M.L. Chacón-Patiño, S.M. Rowland, R.P. Rodgers, Advances in asphaltene petroleomics. Part 1: asphaltenes are composed of abundant island and archipelago structural motifs, Energy Fuel 31 (12) (2017) 13509–13518, https://doi.org/10.1021/acs.energyfuels.7b02873.

[31] H. Abraham, Asphalts and Allied Substances, third ed., D. Van Nostrand, New York, 1929, p. 6.

[32] E.J. Smedt, Improvement in Laying Asphalt or Concrete Pavements or Roads, U.S. Patent 103581, May 31, 1870.

[33] C. Richardson, The Modern Asphalt Pavement, John Wiley & Sons, New York, 1905.

[34] F.J. Nellensteyn, in: A.E. Dunstan, A.W. Nash, H. Tizard (Eds.), The Science of Petroleum, vol. 4, Oxford University Press, London, UK, 1938, pp. 2760–2763.

[35] T.F. Yen, J.G. Erdman, S.S. Pollack, Investigation of the structure of petroleum asphaltenes by X-ray diffraction, Anal. Chem. 33 (1961) 1587–1594.

[36] O.C. Mullins, The modified Yen model, Energy Fuel 24 (2010) 2179–2207.

[37] M. Tavakkoli, S.R. Panuganti, V. Taghikhani, M.R. Pishvaie, W.G. Chapman, Understanding the polydisperse behavior of asphaltenes during precipitation, Fuel 117 (Part A) (2014) 206–217, https://doi.org/10.1016/j.fuel.2013.09.069.

[38] A.I. Victorov, N.A. Smirnova, Description of asphaltene polydispersity and precipitation by means of thermodynamic model of self-assembly, Fluid Phase Equilib. 158–160 (1999) 471–480, https://doi.org/10.1016/s0378-3812(99)00053-9.

[39] F. Alvarez-Ramírez, Y. Ruiz-Morales, Island versus archipelago architecture for asphaltenes: polycyclic aromatic hydrocarbon dimer theoretical studies, Energy Fuel 27 (2013) 1791–1808, https://doi.org/10.1021/ef301522m.

[40] H.C.A. Brandt, E.M. Hendriks, M.A.J. Michels, F. Visser, Thermodynamic modeling of asphaltene stacking, J. Phys. Chem. 99 (26) (1995) 10430–10432, https://doi.org/10.1021/j100026a002.

[41] S. Yaseen, G. Ali Mansoori, Asphaltene aggregation due to waterflooding (a molecular dynamics study), J. Pet. Sci. Eng. 170 (November) (2018) 177–183, https://doi.org/10.1016/j.petrol.2018.06.043.

[42] M. Rapacioli, F. Calvo, F. Spiegelman, C. Joblin, D.J. Wales, Stacked clusters of polycyclic aromatic hydrocarbon molecules, J. Phys. Chem. A 109 (2005) 2487–2497.

[43] B. Schuler, S. Fatayer, G. Meyer, E. Rogel, M. Moir, Y. Zhang, M.R. Harper, A.E. Pomerantz, K.D. Bake, M. Witt, D. Peña, J.D. Kushnerick, O.C. Mullins, C. Ovalles, F.G.A. van den Berg, L. Gross, Heavy oil based mixtures of different origins and treatments studied by atomic force microscopy, Energy Fuel 31 (2017) 6856–6861.

[44] J.C. Law, T.F. Headen, G. Jiménez-Serratos, E.S. Boek, J. Murgich, E.A. Müller, Catalogue of plausible molecular models for the molecular dynamics of asphaltenes and resins obtained from quantitative molecular representation, Energy Fuel 33 (10) (2019) 9779–9795, https://doi.org/10.1021/acs.energyfuels.9b02605.

[45] E. Rogel, K. Hench, R. Dutta, Size Exclusion Chromatography of Crude Oils: From Asphaltene Determination to Modelling 18th Petro Phase Conference Le Havre (Normandie, France), June 1–15, 2017.

[46] Y. Lei, S. Han, J. Zhang, Effect of the dispersion degree of asphaltene on wax deposition in crude oil under static conditions, Fuel Process. Technol. 146 (June) (2016) 20–28, https://doi.org/10.1016/j.fuproc.2016.02.005.

[47] K. Kraiwattanawong, H. Scott Fogler, S.G. Gharfeh, P. Singh, W.H. Thomason, S. Chavadej, Effect of asphaltene dispersants on aggregate size distribution and growth, Energy Fuel 23 (2009) 1575–1582,- https://doi.org/10.1021/ef800706c.

[48] O.C. Mullins, The asphaltenes, Annu. Rev. Anal. Chem. 4 (1) (2011) 393–418, https://doi.org/10.1146/annurev-anchem-061010-113849.

[49] O.C. Mullins, H. Sabbah, J. Eyssautier, A.E. Pomerantz, L. Barré, A.B. Andrews, R.N. Zare, Advances in asphaltene science and the Yen-Mullins model, Energy Fuel 26 (2012) 3986–4003, https://doi.org/10.1021/ef300185p.

[50] A. Svalova, N.G. Parker, M.J.W. Povey, G.D. Abbott, Determination of asphaltene critical nanoaggregate concentration region using ultrasound velocity measurements, Sci. Rep. 7 (1) (2017), https://doi.org/10.1038/s41598-017-16294-5.

[51] A. Al-Hosani, S. Ravichandran, N. Daraboina, Review of asphaltene deposition modeling in oil and gas production, Energy Fuel (2021), https://doi.org/10.1021/acs.energyfuels.0c02981.

[52] J.P. Dickie, T.F. Yen, Macrostructures of the asphaltic fractions by various instrumental methods, Anal. Chem. 39 (1967) 1847–1852.

[53] M.M. Boduszynski, Composition of heavy petroleums. 1. Molecular weight, hydrogen deficiency, and heteroatom concentration as a function of atmospheric equivalent boiling point up to 1400 °F (760 °C), Energy Fuel 1 (1) (1987) 2–11, https://doi.org/10.1021/ef00001a001.

[54] J.F. Schabron, J.G. Speight, The solubility and three-dimensional structure of asphaltenes, Pet. Sci. Technol. 16 (3–4) (1998) 361–376.

[55] O.C. Mullins, B. Martínez-Haya, A.G. Marshall, Contrasting perspective on asphaltene molecular weight: this comment vs. the overview of A. A. Herod, K. D. Bartle, and R. Kandiyoti, Energy Fuel 22 (2008) 1765.

[56] K. Akbarzadeh, A. Dhillon, W.Y. Svrcek, H.W. Yarranton, Methodology for the characterization and modeling of asphaltene precipitation from heavy oils diluted with *n*-alkanes, Energy Fuel 18 (2004) 1434–1441.

[57] S. Elkahky, C. Lagat, M. Sarmadivaleh, A. Barifcani, A comparative study of density estimation of asphaltene structures using group contribution methods and molecular dynamic simulations for an australian oil field, J. Pet. Explor. Prod. Technol. 9 (4) (2019) 2699–2708, https://doi.org/10.1007/s13202-019-0641-x.

[58] J. Burke, Solubility Parameters: Theory and Application, vol. 3, The Book and Paper Group Annual, 1984.

[59] T. Sato, S. Araki, M. Morimoto, R. Tanaka, H. Yamamoto, Comparison of Hansen solubility parameter of asphaltenes extracted from bitumen produced in different geographical regions, Energy Fuel 28 (2) (2014) 891–897, https://doi.org/10.1021/ef402065j.

[60] S. Amjad-Iranagh, M. Rahmati, M. Haghi, M. Hoseinzadeh, H. Modarress, Asphaltene solubility in common solvents: a molecular dynamics simulation study, Can. J. Chem. Eng. 93 (12) (2015) 2222–2232,- https://doi.org/10.1002/cjce.22321.

[61] A. Hirschberg, L.N.J. de Jong, B.A. Schipper, J.G. Meijer, Influence of temperature and pressure on asphaltene flocculation, Soc. Pet. Eng. J. 24 (3) (1984) 283–293.

[62] I.A. Wiehe, Two-dimensional solubility parameter mapping of heavy oils, Fuel Sci. Technol. Int. 14 (1996) 289–312.

[63] F. Mutelet, G. Ekulu, R. Solimando, M. Rogalski, Solubility parameters of crude oils and asphaltenes, Energy Fuel 18 (3) (2004) 667–673, https://doi.org/10.1021/ef0340561.

[64] P.A. Small, Some factors affecting the solubility of polymers, J. Appl. Chem. 3 (1953) 71–80.

[65] G. Scatchard, Equilibria in non-electrolyte solutions in relation to the vapor pressure and densities of the components, Chem. Rev. (1931) 321–333.

[66] H. Burrell, Solubility parameters, in: Interchemical Review, 1955, pp. 3–16.

[67] O. Leon, E. Rogel, G. Torres, A. Lucas, Electrophoretic mobility and stabilization of asphaltenes in low conductivity media, Pet. Sci. Technol. 18 (2000) 913–927.

[68] G. Gonzalez, G.B.M. Neves, S.M. Saraiva, E.F. Lucas, M.A. De Sousa, Electro-kinetic characterization of asphaltenes and the asphaltenes-resins interaction, Energy Fuel 17 (2003) 879–886.

[69] A. Nasser, H. Belhaj, Modeling of crude oil asphaltenes deposition: in-depth colloidal prospective, in: Presented at SPE Production and Operations Conference and Exhibition, Tunis, 8–10 June 2010 (SPE 136087), 2010.
[70] H. Belhaj, H.A. Khalifeh, Asphaltene stability in crude oil during production process, J. Pet. Environ. Biotechnol. 4 (3) (2013) 1–5.
[71] M. Ihtsham Hashmi, B. Ghosh, Dynamic asphaltene deposition control in pipe flow through the application of DC potential, J. Pet. Explor. Prod. Technol. 5 (1) (2015) 99–108, https://doi.org/10.1007/s13202-014-0113-2.
[72] M. Janusz, W. Jablonski, Electrochemical properties of asphaltene particles in aqueous solutions, J. Dispers. Sci. Technol. 21 (6) (2000) 785–802.
[73] H. Parra-Barraza, D. Hernandez-Montiel, J. Lizardi, Z.J. Hernandez, R.H. Urbina, M. Valdez, The zeta potential and surface properties of asphaltenes obtained with different crude oil/nheptane proportions, Fuel 82 (2003) 869–874.
[74] S.E. Taylor, The electro-deposition of asphaltenes and implications for asphaltene structure and stability in crude and residual oils, Fuel 77 (1998) 821–828.
[75] S.M. Hashmi, A. Firoozabadi, Controlling nonpolar colloidal asphaltene aggregation by electrostatic repulsion, Energy Fuel 26 (2012) 4438–4444.
[76] S. Salmón-Vega, R. Herrera-Urbina, C. Lira-Galeana, M.A. Valdez, The effect of ionic surfactants on the electrokinetic behavior of asphaltene from a Maya Mexican oil, Pet. Sci. Technol. 30 (10) (2012) 986–992,- https://doi.org/10.1080/10916466.2010.495966.
[77] A. Jada, M. Salou, Effects of the asphaltene and resin contents of the bitumens on the water–bitumen interface properties, J. Pet. Sci. Eng. 33 (2002) 185–193.
[78] A.W. Marczewski, M. Szymula, Adsorption of asphaltenes from toluene on mineral surface, Colloids Surf. A Physicochem. Eng. Asp. 208 (2002) 259–266.
[79] L. Goual, A. Firoozabadi, Measuring asphaltenes and resins, and dipole moment in petroleum fluids, AICHE J. 48 (11) (2002) 2646–2662.
[80] S. Acevedo, M.A. Ranaudo, G. Escobar, L.B. Gutierrez, X. Gutierrez, in: E.Y. Sheu, O.C. Mullins (Eds.), Asphaltene Constituents: Fundamentals and Applications, Plenum Press, New York, 1995 (Chapter IV).
[81] H. Kopsch, On the thermal behavior of petroleum asphaltenes, Thermochim. Acta 235 (2) (1994) 271–275,- https://doi.org/10.1016/0040-6031(94)85171-9.
[82] Y. Zhang, T. Takanohashi, S. Sato, I. Saito, R. Tanaka, Observation of glass transition in asphaltenes, Energy Fuel 18 (1) (2004) 283–284, https://doi.org/10.1021/ef0301147.
[83] M.R. Gray, G. Assenheimer, L. Boddez, W.C. McCaffrey, Melting and fluid behavior of asphaltene films at 200-500°C, Energy Fuel 18 (5) (2004) 1419–1423, https://doi.org/10.1021/ef049923w.
[84] H. Mahmood Al-Attar, Determination of the functional groups and the melting point of Iraqi asphaltenes, Pet. Sci. Technol. 37 (4) (2019) 486–491, https://doi.org/10.1080/10916466.2018.1552970.
[85] A. Demirbas, Deposition and flocculation of asphaltenes from crude oils, Pet. Sci. Technol. 34 (1) (2016) 6–11, https://doi.org/10.1080/10916466.2015.1115875.
[86] D.S. Korneev, G.S. Pevneva, A.K. Golovko, Study of the thermal stability of petroleum asphaltenes by differential scanning calorimetry, AIP Conf. Proc. 2051 (2018), https://doi.org/10.1063/1.5083377.
[87] J.X. Wang, Predicting Asphaltene Flocculation in Crude Oils (Ph.D. thesis), New Mexico Institute of Mining & Technology, 2000.
[88] E.Y. Sheu, O.C. Mullins, Asphaltenes: Fundamentals and Applications, Plenum Press, New York, 1995.
[89] D. Merino-Garcia, Calorimetric Investigations of Asphaltene Self-Association and Interactions With Resins (Ph.D. thesis), Technical University of Denmark, 2004, ISBN: 87-91435-02-1.
[90] S.E. Friberg, Micellization, in: Asphaltenes, Heavy Oils, and Petroleomics, Springer New York, 2007, https://doi.org/10.1007/0-387-68903-6.
[91] G.P. Dechaine, M.R. Gray, Membrane diffusion measurements do not detect exchange between asphaltene aggregates and solution phase, Energy Fuel 25 (2) (2011) 509–523.

[92] A.H. Alshareef, Asphaltenes: definition, properties, and reactions of model compounds, Energy Fuel 34 (1) (2019) 16–30, https://doi.org/10.1021/acs.energyfuels.9b03291.

[93] J.-A. Stlund, M. Nyden, I.H. Auflem, J. Sjoblom, Interactions between asphaltenes and naphthenic acids, Energy Fuel 17 (2003) 113–119.

[94] T. Barth, S. Høiland, P. Fotland, K. Magne Askvik, R. Myklebust, K. Erstad, Relationship between the content of asphaltenes and bases in some crude oils, Energy Fuel 19 (4) (2005) 1624–1630, https://doi.org/10.1021/ef049750a.

[95] M.R. Gray, R.R. Tykwinski, J.M. Stryker, X. Tan, Supramolecular assembly model for aggregation of petroleum asphaltenes, Energy Fuel 25 (2011) 3125–3134, https://doi.org/10.1021/ef200654p.

[96] Z. Chen, L. Zhang, S. Zhao, Q. Shi, C. Xu, Molecular structure and association behavior of petroleum asphaltene, in: Structure and Modeling of Complex Petroleum Mixtures, Springer, 2015.

[97] L.M. Da Costa, S.R. Stoyanov, S. Gusarov, X. Tan, M.R. Gray, J.M. Stryker, R.D.M. Tykwinski, J.W. Carneiro, P.R. Seidl, A. Kovalenko, Density functional theory investigation of the contributions of π–π stacking and hydrogen-bonding interactions to the aggregation of model asphaltene compounds, Energy Fuel 26 (5) (2011) 2727–2735.

[98] J.P. Pfeiffer, R.N.J. Saal, Asphaltic bitumen as colloid system, J. Phys. Chem. 44 (2) (1940) 139–149.

[99] L. Barré, S. Simon, T. Palermo, Solution properties of asphaltenes, Langmuir 24 (8) (2008) 3709–3717.

[100] G.H.C. Prado, A. de Klerk, Halogenation of oilsands bitumen, maltenes, and asphaltenes, Energy Fuel 28 (2014) 4458–4468.

[101] S.I. Andersen, J.O. Jensen, J.G. Speight, X-ray diffraction of subfractions of petroleum asphaltenes, Energy Fuel 19 (2005) 2371–2377.

[102] T.F. Headen, E.S. Boek, N.T. Skipper, Evidence for asphaltene nanoaggregation in toluene and heptane from molecular dynamics simulations. In, Energy Fuel 23 (2009) 1220–1229, https://doi.org/10.1021/ef800872g.

[103] Z. Hosseini-Dastgerdi, S.A.R. Tabatabaei-Nejad, E. Khodapanah, E. Sahraei, A comprehensive study on mechanism of formation and techniques to diagnose asphaltene structure; molecular and aggregates: a review, Asia Pac. J. Chem. Eng. 10 (2015) 1–14, https://doi.org/10.1002/apj.1836.

[104] S. Priyanto, G. Ali Mansoori, A. Suwono, Measurement of property relationships of nano-structure micelles and coacervates of asphaltene in a pure solvent, Chem. Eng. Sci. 56 (24) (2001) 6933–6939, https://doi.org/10.1016/S0009-2509(01)00337-2.

[105] F. Mostowfi, K. Indo, O.C. Mullins, R. McFarlane, Asphaltene nanoaggregates studied by centrifugation, Energy Fuel 23 (2009) 1194–1200, https://doi.org/10.1021/ef8006273.

[106] R. Tanaka, E. Sato, J.E. Hunt, R.E. Winans, S. Sato, T. Takanohashi, Characterization of asphaltene aggregates using X-ray diffraction and small-angle X-ray scattering, Energy Fuel 18 (2004) 1118–1125, https://doi.org/10.1021/ef034082z.

[107] Z. Rashid, C. Devi Wilfred, N. Gnanasundaram, A. Arunagiri, T. Murugesan, A comprehensive review on the recent advances on the petroleum asphaltene aggregation, J. Pet. Sci. Eng. (2019), https://doi.org/10.1016/j.petrol.2019.01.004.

[108] F.M. Vargas, J.L. Creek, W.G. Chapman, On the development of an asphaltene deposition simulator, Energy Fuel 24 (2010) 2294–2299, https://doi.org/10.1021/ef900951n.

[109] A. Hirschberg, L.N.J. deJong, B.A. Schipper, J.G. Meijer, Influence of temperature and pressure on asphaltene flocculation, SPE J. 24 (03) (1984) 283–293.

[110] P.O. Redelius, Solubility parameters and bitumen, Fuel 79 (2000) 27–35.

[111] H. Alboudwarej, K. Akbarzedeh, J. Beck, W.Y. Svrcek, H.W. Yarranton, Regular solution model of asphaltene precipitation from bitumen, AICHE J. 49 (2003) 2948–2956.

[112] S. Subramanian, S. Simon, J. Sjöblom, Asphaltene precipitation models: a review, J. Dispers. Sci. Technol. 37 (7) (2016) 1027–1049, https://doi.org/10.1080/01932691.2015.1065418.

[113] K.J. Leontaritis, G.A. Mansoori, Asphaltene flocculation during oil production and processing: a thermodynamic colloidal model. SPE 16258, in: SPE International Symposium on Oilfield Chemistry, San Antonio, February 4–6, 1987.

[114] A.I. Victorov, A. Firoozabadi, Thermodynamic micellization model of asphaltenes precipitation from petroleum fluids, AICHE J. 42 (1996) 1753–1764.

[115] H. Pan, A. Firoozabadi, Thermodynamic micellization model for asphaltene aggregation and precipitation in petroleum fluids, SPE Prod. Facil. 13 (1998) 118–127.

[116] L. Goual, Proceedings of the 10th International Conference on Petroleum Phase Behavior and Fouling, Rio de Janeiro, Brazil, 2009.

[117] O.C. Mullins, S.S. Betancourt, M.E. Cribbs, F.X. Dubost, J.L. Creek, A. Ballard Andrews, L. Venkataramanan, The colloidal structure of crude oil and the structure of oil reservoirs, Energy Fuel 21 (5) (2007) 2785–2794, https://doi.org/10.1021/ef0700883.

[118] Q. Guan, A. Goharzadeh, J.C. Chai, F.M. Vargas, S.L. Biswal, W.G. Chapman, Y.F. Yap, An integrated model for asphaltene deposition in wellbores/pipelines above bubble pressures, J. Pet. Sci. Eng. 169 (2018) 353–373, https://doi.org/10.1016/j.petrol.2018.05.042.

[119] K. Indo, et al., Asphaltene nanoaggregates measured in a live crude oil by centrifugation, Energy Fuel 23 (9) (2009) 4460–4469.

[120] S. Leandra, et al., Stress relaxation in quasi-two-dimensional self assembled nanoparticle monolayers, Phys. Rev. E 96 (5) (2018) 052803, https://doi.org/10.1103/PhysRevE.97.052803.

[121] F. Liu, et al., Mixture effect on the dilatation rheology of asphaltenes-laden interfaces, Langmuir 33 (2017) 8, https://doi.org/10.1021/acs.langm uir.6b039 58.

[122] M. Vargas, M. Garcia-Bermudes, M. Boggara, S. Punnapala, M. Abutaqiya, N. Mathew, S. Prasad, A. Khaleel, M. Al Rashed, H. Al Asafen, On the development of an enhanced method to predict asphaltene precipitation, in: Paper Presented at the Offshore Technology Conference, Houston, Texas, May 2014. https://doi.org/10.4043/25294-MS.

[123] J. Meng, J.B. You, H. Hao, X. Tan, X. Zhang, Primary submicron particles from early stage asphaltene precipitation revealed in situ by total internal reflection fluorescence microscopy in a model oil system, Fuel 296 (2021), https://doi.org/10.1016/j.fuel.2021.120584.

[124] A.A. Meléndez Álvarez, Experimental Study of the Effect of Commercial Dispersants on the Precipitation, Aggregation, and Deposition of Asphaltenes (Master thesis), vol. 66, University Rice, 2012, pp. 37–39.

[125] B. Mirzayi, S.A.M. Dehghani, M.B. Chakan, Modeling of asphaltene deposition in pipelines, J. Pet. Sci. Eng. 3 (2) (2013) 15–23.

[126] J. Escobedo, G.A. Mansoori, Heavy-organic particle deposition from petroleum fluid flow in oil wells and pipelines, Pet. Sci. 7 (4) (2010) 502–508.

[127] H. Darabi, M. Shirdel, M.H. Kalaei, K. Sepehrnoori, Aspects of modeling asphaltene deposition in a compositional coupled wellbore/reservoir simulator, in: SPE Improved Oil Recovery Symposium, Society of Petroleum Engineers, 2014.

[128] S. Kord, O. Mohammadzadeh, R. Miri, B.S. Soulgani, Further investigation into the mechanisms of asphaltene deposition and permeability impairment in porous media using a modified analytical model, Fuel 117 (2014) 259–268, https://doi.org/10.1016/j.fuel.2013.09.038.

[129] A. Hammami, J. Ratulowski, Precipitation and deposition of asphaltenes in production systems: a flow assurance overview, in: Asphaltenes, Heavy Oils, and Petroleomics, Springer, New York, 2007, pp. 617–660, https://doi.org/10.1007/0-387-68903-6_23.

[130] T.W. Ru, A.F. Alta'ee, Investigation of asphaltene onset pressure (AOP) in low asphaltenic light oil samples, in: ICIPEG 2014, Springer Singapore, 2015, pp. 23–45, https://doi.org/10.1007/978-981-287-368-2_3.

[131] F. Ameli, A. Hemmati-Sarapardeh, B. Dabir, A.H. Mohammadi, Determination of asphaltene precipitation conditions during natural depletion of oil reservoirs: a robust compositional approach, Fluid Phase Equilib. 412 (March) (2016) 235–248, https://doi.org/10.1016/j.fluid.2015.11.013.

[132] S.L. Kokal, S.G. Sayegh, Asphaltenes: The Cholesterol of Petroleum, Society of Petroleum Engineers, 1995, January 1, https://doi.org/10.2118/29787-MS.

[133] A. Hammami, C.H. Phelps, T. Monger-McClure, T.M. Little, Asphaltene precipitation from live oils: an experimental investigation of onset conditions and reversibility, Energy Fuel 14 (1) (2000) 14–18,- https://doi.org/10.1021/ef990104z.

[134] A. Tabzar, M. Fathinasab, A. Salehi, B. Bahrami, A.H. Mohammadi, Multiphase flow modeling of asphaltene precipitation and deposition, Oil Gas Sci. Technol. 73 (2018) 51, https://doi.org/10.2516/ogst/2018039.

[135] A. Soleymanzadeh, M. Yousefi, S. Kord, O. Mohammadzadeh, A review on methods of determining onset of asphaltene precipitation, J. Pet. Explor. Prod. Technol. (2019, June 1), https://doi.org/10.1007/s13202-018-0533-5.

[136] K.J. Leontaritis, J.O. Amaefule, R.E. Charles, A Systematic Approach for the Prevention and Treatment of Formation Damage Caused by Asphaltene Deposition, Society of Petroleum Engineers, 1994, August 1, https://doi.org/10.2118/23810-PA.

[137] K. Akbarzadeh, A. Hammami, A. Kharrat, D. Zhang, S. Allenson, J. Creek, S. Kabir, et al., Asphaltenes—problematic but rich in potential, Oilfield Rev. 19 (2) (2007) 22–43.

[138] S.I. Andersen, K.S. Birdl, Influence of temperature and solvent on the precipitation of asphaltenes, Fuel Sci. Technol. Int. 8 (6) (1990) 593–615, https://doi.org/10.1080/08843759008915946.

[139] S. Verdier, H. Carrier, S.I. Andersen, J.L. Daridon, Study of pressure and temperature effects on asphaltene stability in presence of CO_2, Energy Fuel 20 (4) (2006) 1584–1590, https://doi.org/10.1021/ef050430g.

[140] Z.A. Chandio, M. Ramasamy, H.B. Mukhtar, Temperature effects on solubility of asphaltenes in crude oils, Chem. Eng. Res. Des. 94 (February) (2015) 573–583, https://doi.org/10.1016/j.cherd.2014.09.018.

[141] M. Mehranfar, R. Gaikwad, S. Das, S.K. Mitra, T. Thundat, Effect of temperature on morphologies of evaporation-triggered asphaltene nanoaggregates, Langmuir 30 (3) (2014) 800–804.

[142] S. Zendehboudi, A. Shafiei, A. Bahadori, L.A. James, A. Elkamel, A. Lohi, Asphaltene precipitation and deposition in oil reservoirs—technical aspects, experimental and hybrid neural network predictive tools, Chem. Eng. Res. Des. 92 (5) (2014) 857–875, https://doi.org/10.1016/j.cherd.2013.08.001.

[143] T.F. Yen, G.V. Chilingarian, Asphaltenes and Asphalts, 1, Developments in Petroleum Science, 40A, Elsevier, 1994, https://doi.org/10.1016/j.profnurs.2003.09.001.

[144] N.A.E. da Silva, V.R. da Rocha Oliveira, G.M. Nunes Costa, Modeling and simulation of asphaltene precipitation by normal pressure depletion, J. Pet. Sci. Eng. 109 (September) (2013) 123–132, https://doi.org/10.1016/j.petrol.2013.08.022.

[145] R.B. de Boer, K. Leerlooyer, M.R.P. Eigner, A.R.D. van Bergen, Screening of Crude Oils for Asphalt Precipitation: Theory, Practice, and the Selection of Inhibitors, Society of Petroleum Engineers, 1995, February 1, https://doi.org/10.2118/24987-PA.

[146] B. Guo, A. Ghalambor, S. Song, T. Ran Lin, Offshore Pipelines: Design, Installation, and Maintenance, second ed., Elsevier, 2013.

[147] A.A. Prakoso, A.D. Punase, B. Hascakir, A Mechanistic Understanding of Asphaltene Precipitation From Varying Saturate Concentration Perspective, Society of Petroleum Engineers, 2015, November 18, https://doi.org/10.2118/177280-MS.

[148] M.A. Anisimov, Y.M. Ganeeva, E.E. Gorodetskii, V.A. Deshabo, V.I. Kosov, V.N. Kuryakov, I.K. Yudin, Effects of resins on aggregation and stability of asphaltenes, Energy Fuel 28 (10) (2014) 6200–6209, https://doi.org/10.1021/ef501145a.

[149] F. Soorghali, A. Zolghadr, S. Ayatollahi, Effect of resins on asphaltene deposition and the changes of surface properties at different pressures: a microstructure study, Energy Fuel 28 (4) (2014) 2415–2421, https://doi.org/10.1021/ef500020n.

[150] S. Ashoori, M. Sharifi, M. Masoumi, M. Mohammad Salehi, The relationship between SARA fractions and crude oil stability, Egypt. J. Pet. 26 (1) (2017) 209–213, https://doi.org/10.1016/j.ejpe.2016.04.002.

[151] A. Yen, Y.R. Yin, S. Asomaning, Evaluating asphalteneinhibitors: laboratory tests and field studies, in: Paper SPE #65376-MS Presented at the SPE Int. Symp. Oilfield Chem. 2001, Houston, Texas, 2001.

[152] I.A. Wiehe, R.J. Kennedy, Oil compatibility model and crude oil incompatibility, Energy Fuel 14 (1) (2000) 56–59, https://doi.org/10.1021/ef990133+.
[153] Y.V. Larichev, A.V. Nartova, O.N. Martyanov, The influence of different organic solvents on the size and shape of asphaltene aggregates studied via small-angle X-ray scattering and scanning tunneling microscopy, Adsorpt. Sci. Technol. 34 (2–3) (2016) 244–257, https://doi.org/10.1177/0263617415623440.
[154] M.G. Trbovich, G.E. King, Asphaltene deposit removal: long-lasting treatment with a co-solvent, in: Proceedings of SPE International Symposium on Oilfield Chemistry, i, 1991, https://doi.org/10.2523/21038-MS.
[155] S. Moradi, M. Dabiri, B. Dabir, D. Rashtchian, M.A. Emadi, Investigation of asphaltene precipitation in miscible gas injection processes: experimental study and modeling, Braz. J. Chem. Eng. 29 (3) (2012) 665–676, https://doi.org/10.1590/S0104-66322012000300022.
[156] P. Bahrami, R. Kharrat, S. Mahdavi, H. Firoozinia, Prediction of the gas injection effect on the asphaltene phase envelope, Oil Gas Sci. Technol. 70 (6) (2015) 1075–1086, https://doi.org/10.2516/ogst/2014037.
[157] R.T. Okwen, Formation Damage by CO_2 Asphaltene Precipitation. SPE International Symposium and Exhibition on Formation Damage Control, Society of Petroleum Engineers, 2006, https://doi.org/10.2118/98180-MS.
[158] H. Yonebayashi, K. Takabayashi, R. Iizuka, S. Tosic, How Acid Gas Affects Asphaltene Precipitation Behavior, Society of Petroleum Engineers, 2014, October 27, https://doi.org/10.2118/170585-MS.
[159] A. Abouie, M. Shirdel, H. Darabi, K. Sepehrnoori, Modeling asphaltene deposition in the wellbore during gas lift process, Soc. Pet. Eng. J. (2015, April 27), https://doi.org/10.2118/174067-MS.
[160] A. Hutin, J.F. Argillier, D. Langevin, Influence of pH on oil-water interfacial tension and mass transfer for asphaltenes model oils. Comparison with crude oil behavior, Oil Gas Sci. Technol. 71 (4) (2016), https://doi.org/10.2516/ogst/2016013.
[161] S. Poteau, J.-F. Argillier, D. Langevin, F. Pincet, E. Perez, Influence of pH on stability and dynamic properties of asphaltenes and other amphiphilic molecules at the oil–water interface, Energy Fuel 19 (4) (2005) 1337–1341, https://doi.org/10.1021/ef0497560.
[162] E. Strassner, Effect of pH on interfacial films and stability of crude oil-water emulsions, J. Pet. Technol. 20 (1968) 303–312. SPE 1939.
[163] L.R. Houchin, D.D. Dunlap, B.D. Arnold, K.M. Domke, The Occurrence and Control of Acid-Induced Asphaltene Sludge, Society of Petroleum Engineers, 1990, January 1, https://doi.org/10.2118/19410-MS.
[164] B.P. Tissot, D.H. Welte, Petroleum Formation and Occurrence, A New Approach to Oil and Gas Exploration, second ed., Springer-Verlag, Berlin, Heidelberg, New York, 1984, p. 363. 1978, 720.
[165] I.C. Jacobs, M.A. Thorne, Asphaltene Precipitation During Acid Stimulation Treatments, Society of Petroleum Engineers, 1986, January 1, https://doi.org/10.2118/14823-MS.
[166] C.A. Sung, M. Tavakkoli, A. Chen, F.M. Vargas, Prevention and control of corrosion-induced asphaltene deposition, in: Proceedings of the Annual Offshore Technology Conference, 2, 2016, pp. 1589–1603, https://doi.org/10.4043/27008-ms (Offshore Technology Conference).
[167] W.A.A. Limanowka, M.J.M. Voytechek, Asphaltene Deposition Problems in Oil Industry With Focus on Electric Submersible Pump Applications, Society of Petroleum Engineers, 1999, January 1, https://doi.org/10.2118/56662-MS.
[168] J. Wang, T. Fan, J.S. Buckley, J.L. Creek, Impact of water cut on asphaltene deposition tendency, in: Offshore Technology Conference. OTC-25411-MS 5-8 May, Houston, Texas, 2014, 2014, May 5, https://doi.org/10.4043/25411-MS.
[169] S. Aslan, A. Firoozabadi, Effect of water on deposition, aggregate size, and viscosity of asphaltenes, Langmuir 30 (13) (2014) 3658–3664, https://doi.org/10.1021/la404064t.

[170] R. Mollaabbasi, D. Eskin, S. Mohammad Taghavi, Analyzing the effects of produced water on asphaltene deposition in a vertical production tubing, Ind. Eng. Chem. Res. 57 (45) (2018) 15450–15459, https://doi.org/10.1021/acs.iecr.8b03336.

[171] E. Joonaki, J. Buckman, R. Burgass, B. Tohidi, Water versus asphaltenes; liquid–liquid and solid–liquid molecular interactions unravel the mechanisms behind an improved oil recovery methodology, Sci. Rep. 9 (1) (2019), https://doi.org/10.1038/s41598-019-47782-5.

[172] A.A. Gabrienko, E.V. Morozov, V. Subramani, O.N. Martyanov, S.G. Kazarian, Chemical visualization of asphaltenes aggregation processes studied in situ with ATR-FTIR spectroscopic imaging and NMR imaging, J. Phys. Chem. C 119 (5) (2015) 2646–2660, https://doi.org/10.1021/jp511891f.

[173] J.G. Delgado-Linares, D.C. Salmin, H. Stoner, D.T. Wu, L.E. Zerpa, C.A. Koh, A. Teixeira, Effect of alcohols on asphaltene-particle size and hydrate non-plugging behavior of crude oils, in: Proceedings of the Annual Offshore Technology Conference, vol. 2020-May, 2020, https://doi.org/10.4043/30916-ms (Offshore Technology Conference).

[174] E. Joonaki, A. Hassanpouryouzband, R. Burgass, A. Hase, B. Tohidi, Effects of waxes and the related chemicals on asphaltene aggregation and deposition phenomena: experimental and modeling studies, ACS Omega 5 (13) (2020) 7124–7134, https://doi.org/10.1021/acsomega.9b03460.

[175] M.M. Amro, Effect of scale and corrosion inhibitors on well productivity in reservoirs containing asphaltenes, J. Pet. Sci. Eng. 46 (4) (2005) 243–252, https://doi.org/10.1016/j.petrol.2005.01.007.

[176] Z. Rashid, C.D. Wilfred, N. Gnanasundaram, A. Arunagiri, T. Murugesan, A comprehensive review on the recent advances on the petroleum asphaltene aggregation, J. Pet. Sci. Eng. (2019), https://doi.org/10.1016/j.petrol.2019.01.004.

[177] S. Enayat, F. Lejarza, R. Doherty, A.T. Khaleel, C. Sisco, J. Kuang, F.M. Vargas, Conclusions, current research, and future directions, in: Asphaltene Deposition, CRC Press, 2018, pp. 327–350, https://doi.org/10.1201/9781315268866-9.

[178] J. Bae, D. Fouchard, S. Garner, J. Macias, Advantages of applying a multifaceted approach to asphaltene inhibitor selection, in: Paper Presented at the Offshore Technology Conference, Houston, Texas, USA, May, 2016, https://doi.org/10.4043/27171-MS.

[179] E. Rogel, C. Ovalles, M. Moir, J. Vien, H. Morazan, Solubility Characterization of Asphaltene Deposits, Society of Petroleum Engineers, 2015, April 13, https://doi.org/10.2118/173757-MS.

[180] A.R. Kovscek, H. Wong, C.J. Radke, A pore-level scenario for the development of mixed wettability in oil reservoirs, AIChE J. 39 (6) (1993) 1072–1085.

[181] S. Enayat, Utilization of Asphaltenes as Inexpensive and Abundant Precursors of Novel Carbon-Based Nanomaterials (Ph.D. thesis), Rice University, 2019.

[182] X. Wu, B. Ma, Y. Xu, J. Xu, H. Zhai, Z. Xiong, X. Liao, X. Shi, X. Chen, Preparation of three-layer graphene sheets from asphaltenes using a montmorillonite template, J. Nanomater. 2019 (2019), https://doi.org/10.1155/2019/2094723.

[183] H. Jung, C.W. Bielawski, Asphaltene oxide promotes a broad range of synthetic transformations, Commun. Chem. 2 (1) (2019), https://doi.org/10.1038/s42004-019-0214-4.

CHAPTER

Naphthenate and Carboxylate Soap Deposition

9

9.1 Introduction

Naphthenate and carboxylate deposits are newcomers to the flow assurance solids group after the strong interest they have gained in recent years. Naphthenate and carboxylate deposits, soaps, or, more specifically, naphthenate soap scale and carboxylate soap emulsion, are different ways of representing these salts, formed when oil-soluble aliphatic carboxylic and naphthenic acids come in contact with metal cations in relatively high-pH produced water [1,2].

Unlike other organic (wax and asphaltenes) and inorganic deposits, naphthenate and carboxylate soaps are not such a widespread problem in the oil production industry; however, they have received considerable research attention in response to increased production in areas such as the North Sea, West Africa, Southeast Asia, and the Gulf of Mexico, where naphthenate deposits are particularly troublesome [1,3,4]. However, the problem is more common than previously anticipated, as naphthenate deposits are present in 10% or more of North Sea crudes, while 30% of Southeast Asian crudes and 20% of West African crudes have encountered soap emulsion or sludge problems. Naphthenates were originally identified in oilfields in the 1960s, but it was not until the mid-1990s that research and development efforts began [5].

Naphthenates and carboxylate soaps accumulate at the oil-water interfaces and interfere with the separation processes, and they also can deposit in the process facilities, causing flow restrictions and blockages.

9.2 Naphthenic acids structure and occurrence in crude oil

Naphthenic acids are mixtures of carboxylic acids with nonsaturated and saturated alkyl-substituted acyclic and cyclic structures, with molecular weights in the range of 200–500, and having the general chemical formula $C_mH_{2m+z}O_2$, where m indicates the carbon number, and z is an even integer that specifies the hydrogen deficiency resulting from ring formation; z can be zero in the case of fatty acids, or negative depending on the number of condensed and/or aromatic rings. Small amounts of nitrogen and sulfur components also occur in naphthenic acids. Although the term "naphthenic" traditionally was used to classify only acids with cycloaliphatic derivatives, in the petroleum industry it refers normally to all the organic acids present in crude oil [6–8]. Fig. 9.1 displays some examples of possible monoprotic naphthenic acid structures for different Z values [9].

Essentials of Flow Assurance Solids in Oil and Gas Operations. https://doi.org/10.1016/B978-0-323-99118-6.00003-4

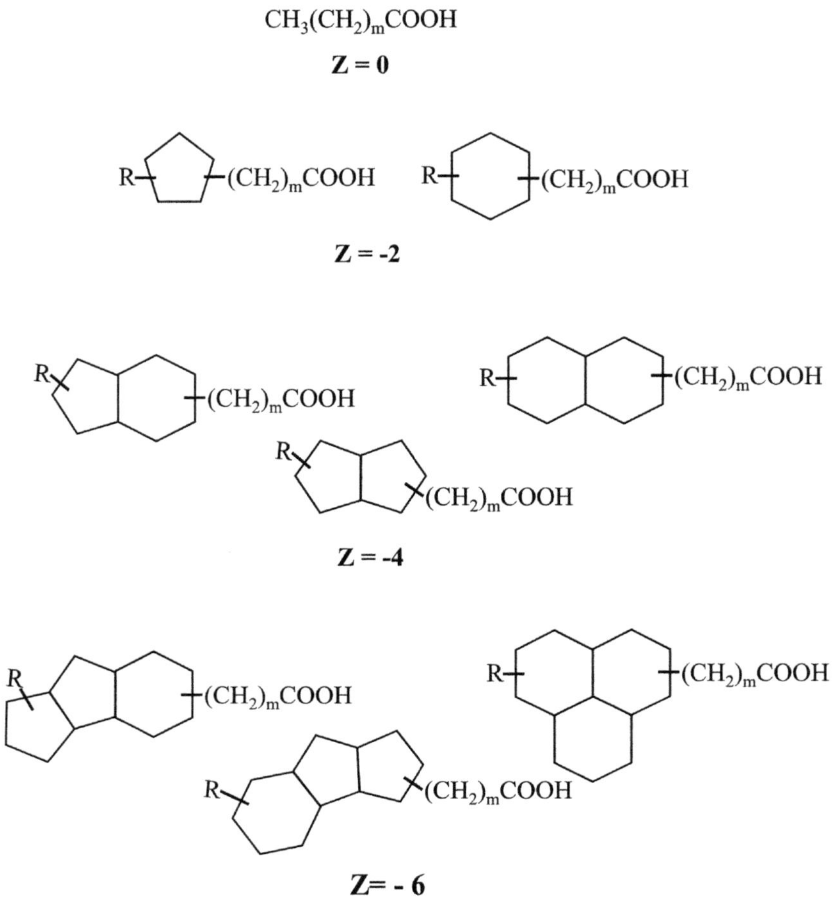

FIG. 9.1

Naphthenic acid structures (R is an alkyl chain, Z is the hydrogen deficiency, and m is the number of CH_2 units). Reproduced *with permission from J.S. Clemente, P.M. Fedorak, A review of the occurrence, analyses, toxicity, and biodegradation of naphthenic acids, Chemosphere (2005), https://doi.org/10.1016/j.chemosphere.2005.02.065. Copyright (2005), Elsevier.*

Naphthenic acids are normal constituents in nearly all crude oils, with their concentration ranging from 0 to 4%wt. Naphthenic acids occur mainly in biodegraded oils and are responsible for high oil total acid number (TAN) [8]. They are predominantly found in immature heavy crudes, whereas paraffinic crudes normally have low acid content [6]. Naphthenic acids are a class of biological markers that were produced during the in-reservoir biodegradation of petroleum hydrocarbons, and hence they are closely linked to the maturity and the biodegradation level of the oil fields [10,11].

Several analytical techniques, including potentiometric titration, LC/MS, NMR, and VPO, have been applied to determine the structure of the main naphthenic acids that are the dominating constituents of these soap deposits, which are typically called ARN. ARN is a family of tetraprotic (4-protic) carboxylic acids containing 4–8 unsaturated sites (rings) in the hydrocarbon skeleton, with molecular

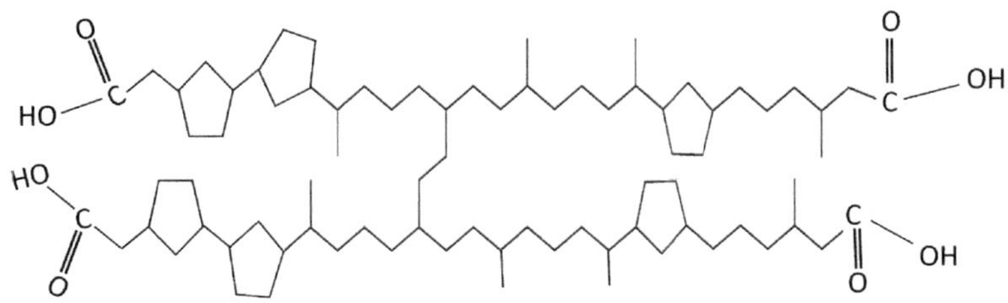

FIG. 9.2

Structure of ARN acid.

weights in the range 1227–1235 g/mol [12]. Fig. 9.2 illustrates the structure of ARN. The molecular weights of the homologous ARN acid series are 1227, 1229, 1231, 1233, 1235 (basic structures) $+ n \times 14$ (n = number of additional CH_2- groups in the hydrocarbon skeleton). The ARN acid with molecular weight 1231 has empirical formula $C_{80}H_{142}O_8$ and has 6 rings, and is considered the most abundant isomer [12–14]. Upon contact with produced waters, the tetraacids form insoluble sticky salts with Na and Ca cations, which interfere with phase separation operations and cause system blockages [1].

Potentiometric titrations are frequently used to measure total acid number (TAN) in crude oils, expressed in (mgKOH/g); the FTIR spectroscopy method has also been used [15]. However, TAN does not necessarily represent the naphthenic acids in crude oils. Naphthenic acids with similar total acid number (TAN) and average molecular weight can have significantly different profiles [16]. Many crude oils with high TAN value showed no naphthenate deposition, while others with low TAN value encountered naphthenates deposits.

The quantification of tetraacids was achieved using a UV-based method [17,18]. Mass spectrometry-based methods have been used by various researchers for the detection and quantification of naphthenic acids in oil and deposits [3,4,19].

9.3 Naphthenic acid equilibrium in oil-water systems

Naphthenic acids are amphiphilic molecules where the carboxylic group represents the hydrophilic or polar part and the hydrocarbon skeleton represents the oil-soluble, hydrophobic or nonpolar part. In oil-water systems, the acids may thus be distributed or partitioned between the different phases. This distribution will depend strongly on conditions like the pH and the salinity of the aqueous phase [6].

The partitioning of naphthenic acid monomers between oil (o) and water (w) may be described by the equilibrium in Eq. (9.1) [20]:

$$(\mathrm{RCOOH})_{\mathrm{O}} \leftrightarrow (\mathrm{RCOOH})_{\mathrm{W}} \tag{9.1}$$

where $(\mathrm{RCOOH})_{\mathrm{O}}$ and $(\mathrm{RCOOH})_{\mathrm{W}}$ represent undissociated naphthenic acid in the oil and water phase, respectively. Here, the preceding equilibrium can be described by the corresponding partitioning coefficient $K_{w/o}$ given by Eq. (9.2) [6,20]:

$$K_{w/o} = \frac{[\mathrm{RCOOH}]_{\mathrm{W}}}{[\mathrm{RCOOH}]_{\mathrm{O}}} \tag{9.2}$$

The partition coefficient is a function of the pH and concentration as well as the structure of the acid [21].

The acid will dissociate in the water phase according to Eq. (9.3):

$$(\mathrm{RCOOH})_{\mathrm{w}} \leftrightarrow (\mathrm{RCOO}^{-})_{\mathrm{W}} + \mathrm{H}^{+} \tag{9.3}$$

The equilibrium constant K_a is given by Eq. (9.4) [6,20]:

$$K_a = \frac{[\mathrm{RCOO}^{-}]_{\mathrm{W}}\,[\mathrm{H}^{+}]}{[\mathrm{RCOOH}]_{\mathrm{w}}} \tag{9.4}$$

Hence, the overall concentration of the naphthenic acids in the water phase, $[\mathrm{RCOOH}]_{w,Tot}$ is given by Eq. (9.5) [6,20]:

$$[\mathrm{RCOOH}]_{\mathrm{W},Tot} = [\mathrm{RCOOH}]_{init,\mathrm{oil}}\frac{V_{\mathrm{oil}}}{V_{\mathrm{water}}} \times \left(1 - \frac{1}{1 + K_{\mathrm{wo}}\left(\frac{V_{\mathrm{water}}}{V_{\mathrm{oil}}}\right) \times \left[1 + \left(\frac{k_a}{[\mathrm{H}^{+}]}\right)\right]}\right) \tag{9.5}$$

where $[\mathrm{RCOOH}]_{init,\mathrm{oil}}$ is the initial concentration of the naphthenic acid in the oil phase, V_{oil} and V_{water} are the volume of the oil and water phases, respectively, and $[\mathrm{H}^{+}]$ is the aqueous concentration of H^{+} [20].

The degree of naphthenic acids partitioning will accordingly depend on their solubility in the two phases. Hence, naphthenic acids with a large hydrocarbon moiety will exhibit lower partition coefficients compared to those with smaller hydrophobic parts [6].

The distribution of acid between the phases is also affected by the pH; with increasing pH, the monomers dissociate and become more water-soluble. Naphthenic acids in crude oil-water systems were found to have $\mathrm{p}K_{\mathrm{a}}$ in the range of 5–6 [22] (or 4.9 as reported by Havre et al. [20]). The extent to which naphthenic acid dissociates depends on the pKa and partitioning coefficient of the individual acids [23].

Generally speaking, two behaviors were found at different pHs. At low pH the two terms believed to be most important are [20]:

- the partitioning of undissociated acid between the phases.
- the dissociation of carboxylic acid in the water phase.

However, at high pH the formation of micelles in the water phase and reversed micelles in the oil phase is encountered and affects the final equilibrium. In addition, dimerization of naphthenic/carboxylic acids is a fairly common property, since they form dimers through hydrogen bonding. Therefore Eq. (9.5) gives a good basic description of the naphthenate concentration at lower pH; however, at higher pH, where formation of micelles and reversed micelles can be significant, Eq. (9.5) may not be able to predict the total acid concentration in the aqueous phase [20].

The micellization must be accounted for when the total concentration exceeds the critical micelle concentration (CMC). CMC is dependent on factors like molecular structure, e.g., the length of the hydrocarbon chain and electrolyte concentration, as well as on physical conditions such as temperature and pH [24]. Reversed micelles may form in the oil phase, which needs a certain dissociation of the acids in the water phase [20]. Fig. 9.3 shows a carboxylic acid system going from a system involving only monomer equilibria, at low pH, to a system at higher pH, with micellization involved. At the

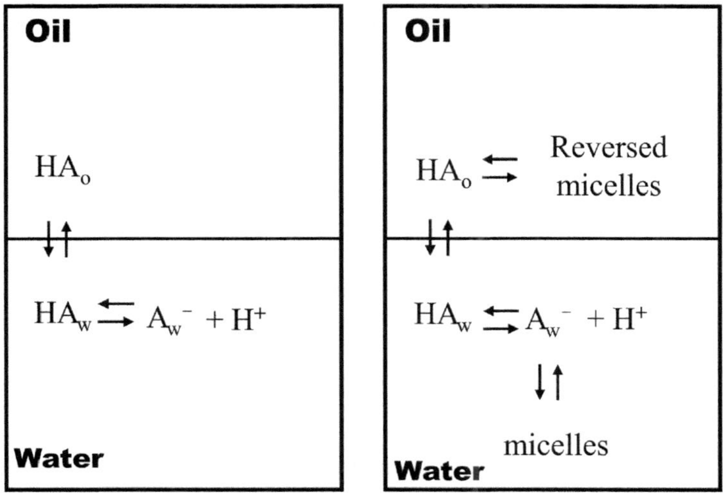

FIG. 9.3

Equilibria in water/oil/naphthenic acid systems; (left) low pH; (right) high pH with formation of micelles and reversed micelles.

Reproduced *with permission from T.E. Havre, J. Sjöblom, J.E. Vindstad, Oil/water-partitioning and interfacial behavior of naphthenic acids, J. Dispers. Sci. Technol. 24(6) (2003) 789–801, https://doi.org/10.1081/DIS-120025547. Copyright (2003), Taylor & Francis.*

oil-water interface, the interfacial tension decreases in the presence of amphiphilic carboxylic acids and with increasing pH. Generally, with increasing pH, the interfacial activity increases gradually due to a higher degree of dissociation of the carboxylic groups [25].

It has been reported that naphthenic acids extracted from deposits have significantly higher interfacial activity compared to those extracted from crude oil, implying that the higher abundance of tetra-acids in naphthenates deposits is related to their greater interfacial activity [26].

9.4 Mechanism of formation of naphthenate and carboxylate soaps

Generally speaking, the naphthenates/carboxylates soaps precipitate when the naphthenic/carboxylic acids (RCOOH) interact with the metal ions (Na^+, Ca^{++}) at an optimum pH. This reaction can take place in two different locations proposed by two different approaches:

- First, in the bulk water where the water soluble acids react with the metal ions [27]. Although this approach is well-known, it is not fully supported due to the limited solubility of most of the naphthenic/carboxylic acids in the water phase under the normal production operations conditions, in addition to the tremendously large interfaces created when coproduced water is emulsified with the crude oil during production [6].
- Second, at the oil-water interface, where different possible interactions between the acids and the metal ions are available including [28]: (1) undissociated acid-metal ion, (2) undissociated

acid-metal hydroxide complexes, (3) dissociated acid-metal ion hydroxide complexes, (4) dissociated acid-free metal ion interactions. The interface mechanism is the most common.

A generalized mechanism for the formation of naphthenate soaps in a stepwise manner can be given as follows:

1. The "naphthenic acids" from the crude oil, represented by the general formula R-COOH, diffuse from the bulk oil and collect at the oil-water interface, due to the chemistry of their functional group [29].
2. The change of the brine pH (from lower to higher) causes the formation of surface active naphthenates $RCOO^-$ anion [8,29] at the oil-brine interface until pH equilibrium is attained [30], as shown in Eq. (9.6):

$$RCOOH + OH^- \leftrightarrow RCOO^- + H_2O \tag{9.6}$$

3. Depending on the cationic composition of the brine, the naphthenate anions will bond with the metal ions to form insoluble substances, i.e., metal naphthenates [29,30]. The most commonly known types are the sodium carboxylates and calcium naphthenates. The possible reaction mechanisms were reported by De Oliveira et al. [30], Ahmed [31], and Turner and Smith [32].

In the case of the Na ion, the reaction is a one-step saponification-style reaction between long-chain fatty acids and monovalent cations (Na+, K+) in brine, as shown in Eq. (9.7):

$$RCOO^- + Na^+ \leftrightarrow RCOONa \tag{9.7}$$

In the case of calcium naphthenate, the reaction supposedly can be as shown in Eq. (9.8):

$$2RCOO^- + Ca^{2+} \leftrightarrow (RCOO)_2Ca \tag{9.8}$$

which is two reaction steps in which the acid monomers sequentially bind to one cation, according to the reactions in Eq. (9.9):

$$2RCOO^- + Ca^{2+} \leftrightarrow (RCOO - Ca)^+ + RCOO^- \rightarrow (RCOO)_2Ca \tag{9.9}$$

The generalized mechanism of naphthenates soaps is illustrated in Fig. 9.4. [33]

While Reactions (9.7)–(9.9) seem to be ionic bonding reaction types, Taylor and Hiu [34] showed these reactions in fact involve coordination-type bonding, including monodentate coordination with one carboxylate O atom and hydrogen-bonded water; bidentate coordination, with both carboxylate O atoms; or bridging, as illustrated in Fig. 9.5 for divalent cations. At higher calcium concentrations, the complexation-style 2:1 bidentate binding of large molecular weight tetraprotic acids (ARN) to Ca^{2+} gives practically insoluble coordination compounds and can be interpreted within the framework of the crystallization theory. The figure also depicts the different modes of bonding depending on the type of metal ion involved, so specifically for Ca^{2+}, monodentate coordination with one carboxylate O atom and hydrogen-bonded water, bidentate coordination, with both carboxylate O atoms, or bridging (Fig. 9.5A–C) are notable. On the other hand, a major difference in bonding mode with carboxylic acids in the case of magnesium is the formation of uncoordinated ionic species of the form shown in Fig. 9.5D.

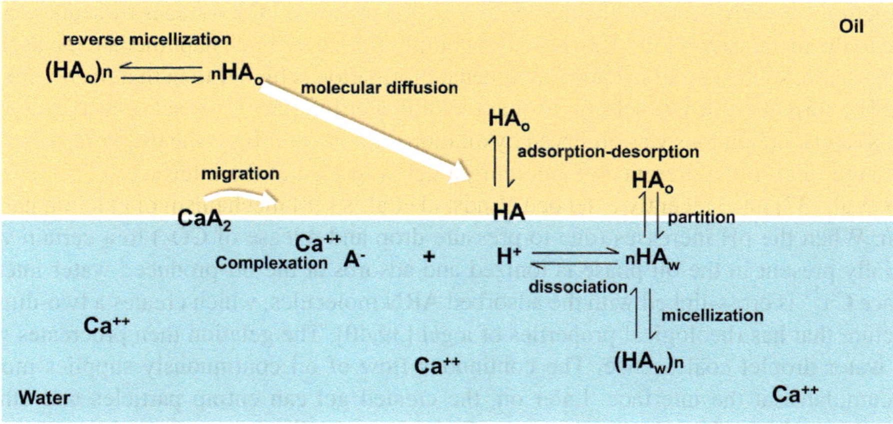

FIG. 9.4

Schematic of the process of metal naphthenates formation and associated acid equilibria at alkaline pH.

From *W.I. Eke, C. Victor-Oji, O. Akaranta, Oilfield metal naphthenate formation and mitigation measures: a review, J. Pet. Explor. Prod. Technol. 10(2) (2020), 805–19, https://doi.org/10.1007/s13202-019-00797-0.*

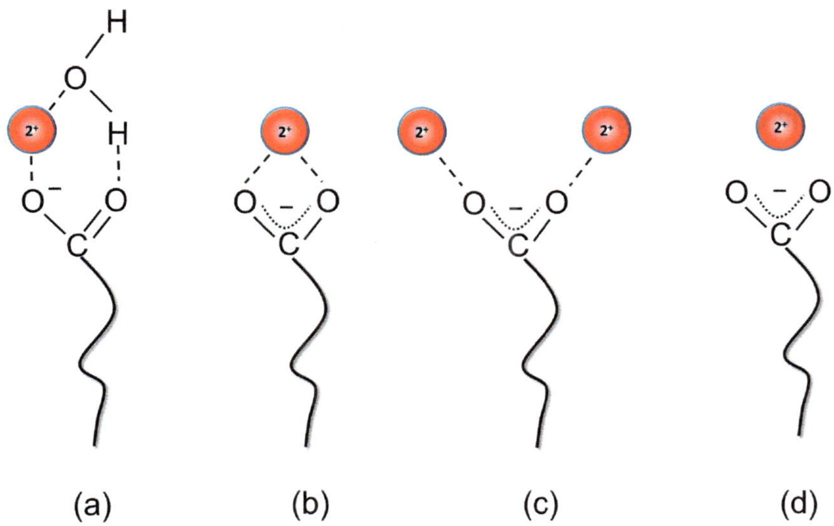

FIG. 9.5

Bonding modes identified for divalent ion interactions with carboxylates: 1:1 coordination is shown in binding modes (a) and (b), with bridging exhibited in (c). (d) shows ionic bonding.

From *S.E. Taylor, T.C. Hiu, Metal ion interactions with crude oil components: specificity of Ca2+ binding to naphthenic acid at an oil/water, Colloids Interfaces 2(40) (2018) 1–21.*

The reaction between the naphthenic acids and Ca ion proceeds in two stages by sequential binding of the acid to the metal cation [35]. Calcium ions complex twice as much naphthenate ion as the monovalent sodium ion, so the rate of calcium naphthenate formation is higher than that of sodium naphthenate [35]. The formation of viscoelastic film at the oil/water interface by reaction between tetrameric acids (ARN) and calcium ions was observed by dilational rheology [36]. If the pH of the aqueous phase is high enough (and in the presence of calcium), a gel is formed at the interface.

Simon et al. [37] have recently extended Vindstad et al.'s [38] mechanism of calcium naphthenate deposition. When the pH increases (due to pressure drop and release of CO_2) to a certain value, the ARN initially present in the oil phase is ionized and adsorbs at the oil-produced water interface. At the interface Ca^{2+} is cross-linked with the adsorbed ARN molecules, which creates a two-dimensional (2D) structure that has rheological properties of a gel [39,40]. The gelation then procreates with time and with water droplet coalescence. The continuous flow of oil continuously supplies more ARN, which accumulates at the interface. Later on, the created gel can entrap particles and other crude oil components, which ultimately leads to the formation of solid deposits, blocking oil production facilities [41].

The two types of naphthenate soaps/deposits described earlier, e.g., calcium and sodium naphthenates, have different features and properties. The calcium-rich deposit is usually referred to as calcium naphthenate and the sodium-rich sample is referred to as sodium carboxylate or sodium naphthenate emulsion, as suggested by Turner and Smith [32]. These two types of soaps are discussed in some detail in the following sections.

9.5 Types of naphthenate and carboxylate soaps

9.5.1 Sodium carboxylate

Sodium naphthenate soap emulsions are monovalent salts formed by the reaction of sodium ion with naphthenic acids that have low to medium molecular weight (C15–C35), with long and mostly unbranched hydrocarbon chains with a polar head group that promotes binding to monovalent sodium ion at the interface [33,42,43]. Other monovalent cations forming carboxylate emulsions are potassium carboxylate [44]. Therefore it is quite common in the literature to refer to these as carboxylate emulsions or soap emulsions.

Sodium emulsion soaps are usually formed from the lighter oil types with a typical °API ranging from 32 to 41, and these oils can have a variable content of organic acids. Such oils usually have a low TAN value, a low sulfur, asphaltene and resin content, and typically high to moderate wax content [45].

Carboxylate soaps are often ultrastable viscous sticky emulsions, which interfere with phase separation operations, e.g., crude oil dehydration and produced water deoiling. These sticky emulsions can also incorporate other materials like clay, paraffins, asphaltenes, or mineral scales. They can also accumulate in the storage tanks and reduce their capacity. In addition, in a reservoir environment, they can affect relative permeability and create emulsion blocks or sludges in the formation, and even when they are treated in the formation, they give up all the particulates they held, causing formation damage or blockages in the flowlines. They are also considered a hazard due to their heavy metals load [2,3,32].

Unlike calcium naphthenate soaps, the carboxylate soaps do not harden upon exposure to air; however, the volume of emulsion soaps can be as high as 10% of the total fluids [32].

9.5.2 Calcium naphthenates

Calcium naphthenate soaps are formed when the high molecular weight naphthenic acids, particularly tetraprotic ARN acids, interact with the calcium cation [43]. However, lower molecular weight naphthenic acids, including acyclic, mono- and bicyclic and alkylbenzoic structures, have been identified in field deposits of calcium naphthenate [3,4].

Calcium is the most likely cation to interact and form the naphthenate soaps, as calcium content of these deposits can be up to 6.2% [13]. However, other metal ions like iron and magnesium may occur in naphthenate deposits [44]. The relatively high amounts of calcium in the naphthenate deposits compared to the other divalent cations is not due to their abundance in the produced water, but rather is due to the following:

- First, selectivity and affinity of ARN toward the Ca^{2+} due to an optimal balance between the ion size and hydration state of the calcium cation [33,35,46,47]. For example, Mg^{2+} with a higher hydration number of 6 compared to 2 for Ca^{2+} experiences greater electrostatic shielding. Therefore Mg^{2+} has a lower tendency to complex with ARN and will barely participate in the formation of an interfacial film [35]. The affinity of tetraacids for metal cations was found to decrease in the order: $Ca^{2+} > Ba^{2+} \approx Sr^{2+} > Mg^{2+} \gg Na^{+}$ [46].
- Second, the solubility product of calcium naphthenate is much lower than that of magnesium naphthenate; thus calcium naphthenate has a greater tendency to exist as a solid precipitate [48].

Vindstad et al. [49] were the first to report on the calcium naphthenate issue in 2003. Calcium naphthenate soaps are commonly encountered from heavy high TAN crude oils, although other crudes with low TAN encountered calcium naphthenate soap problems as well [32].

These soaps cause the same problems as sodium salts by interfering with phase separation operations, and they are also very sticky, so can act as a binding agent for other solids, such as mineral scale, and for formation of fines and sand, which then lump up and drop in the water legs and cause blockages [49–51]. In addition, calcium naphthenates tend to foul the production system by forming hard deposits similar to mineral scales. Calcium naphthenates are known to harden as they cool and, as a result of oxidation upon deposit with exposure to air, to finally form hard, black to greenish-brown deposits. The freshly formed naphthenate soaps are soft and sticky. This phenomena is more likely related to the glass transition temperature of the material [35,51]. A rare case regarding downhole problems with naphthenates was reported from the Kikeh field [52], where calcium-based completion fluids were lost to the formation, promoting a possible naphthenate scale.

Unlike the carboxylate soaps, the formation of calcium naphthenates in the field may typically be quite small, e.g., around 10 mg solid/L of crude oil, as reported. This quantity may appear insignificant, but it should be viewed as a cumulative effect arising from the very large throughput volume of oil and brines during production [31,49,53]. Naphthenates have been reported as the most serious threat to a particular field's (Heidrun) continued production [49].

Although both carboxylate and naphthenate production problems can occur in the same field, it is more common that only one of the two problems dominates. For example, carboxylate problems were encountered in Malaysian fields, whereas naphthenate scale was the major problem in West Africa fields [1].

9.6 Factors affecting naphthenate and carboxylate soap formation and deposition

9.6.1 Effect of oil composition

9.6.1.1 Crude oil acidity and naphthenic acids content

The crude oil composition plays a significant role in the naphthenate formation mechanism. Table 9.1 summarizes the relationship between the crude oil composition and the soap types. Almost all crude oils contain some amount of naphthenic acids in varied amounts [31]. Heavy crudes from geologically young formations or biodegraded crudes (from either formation or reservoir) have high acid content, while paraffinic crudes are associated with low acid content [54].

The key factor in naphthenate formation is the types of naphthenic acid present in the crude oil, and their concentration. The naphthenic acid composition of the crudes varies from one to another. The acid content of crude oil varies from trivial amounts to as high as 4%, depending on the type of oil [55,56]. Naphthenic acid concentration and distribution are thought to be dependent on the level of biodegradation of the crude oil [56].

The acidity of crude oil is usually estimated by the TAN measurement of crude oils, which is usually carried out using either colorimetric or potentiometric titrations (ASTM D974 and ASTM D664) that give an indication of the total acid present in the crude oil. Crude oil TAN has been found to vary in different crudes, depending on many parameters responsible for the crude oil acidity. Sartori et al. [57] reported that crude oils will typically have TAN values in the range of 0.2 to 10 mg KOH/g. Furthermore, crudes with TAN value greater than 0.5 mgKOH/g of oil are considered to have a high total acid number [58].

Field observations showed that lower-TAN crude may present a higher risk of calcium naphthenate formation than one having higher TAN [23]. Some naphthenate-forming crudes were observed to have very high TAN and low naphthenic acid content (8.2 mgKOH/g and 20 ppm), while others show the opposite: i.e., low TAN and high acid content (0.5 mgKOH/g and 5000 ppm) [32,59].

Therefore TAN is actually not indicative of naphthenate potential deposition and there has been no direct correlation between total acid number and naphthenic acid concentration in most crude oils. This is because TAN measures acidic components of crude oil in general, which can be affected by

Table 9.1 Summary of the relationship between crude oil properties and the expected soap types [33].

Property	Sodium carboxylate soap emulsion	Calcium naphthenate soap scale
API gravity	32°–41°	17°–30°
TAN	<0.6 mgKOH/g	>1 mgKOH/g
Wax	4–15 wt%	<4 wt%
Asphaltene	0.05–0.5 wt%	0.5–>3 wt%
Resin	1–6 wt%	>6–17 wt%
Acid type	Fatty acids (C_{15}–C_{35})	Significant ARN; high ARN/non-ARN ratio
Oil classification	Paraffinic	Naphthenic

substances like dissolved H_2S, dissolved CO_2, phenols, and some inorganic acids, and does not measure naphthenic acids particularly and accurately [11,60].

Naphthenate deposits are also associated with heavy crudes with high TAN, with low paraffin concentrations that have formed normally as a result of biodegradation in the reservoir [61].

A more informative index for assessing the calcium naphthenate formation risk is the ratio of ARN acid content to total acid content. Higher ARN acid content relative to non-ARN acids in the oil will favor naphthenates formation [33].

9.6.1.2 Effect of low molecular weight monoacids

Low molecular weight monoacids inhibit the deposition of calcium naphthenates [17,62]. Two main reasons are behind this behavior:

- First, the monoacids dissociate and diffuse to the o/w interface and compete with the naphthenic acids and ARN, and complex with the calcium cations to hinder the cross-linking and gelling of ARN to solid polymer-like material, leading to partially inhibiting or softening of the deposit [63,64].
- Second, the deprotonation of carboxylic acids releases protons that have an inhibitive effect on the dissociation of more naphthenic acids. However, this is counteracted by bicarbonate buffering of brine [33].

9.6.1.3 The effect of asphaltenes

Asphaltenes were found to have an inhibitory effect on naphthenate deposition. This is attributed to the fact that asphaltenes compete with naphthenates at the interface due to their interfacial activity, and asphaltenes dominate more interfacial surface area for naphthenic acid adsorption, which reduces naphthenic acid partitioning. The asphaltene effect is a function of the ionic strength of the aqueous solution [65]. As the ionic strength of brine decreases, competitive adsorption between asphaltenes and naphthenic acids at the interface increases due to higher asphaltene adsorption. Asphaltenes might also complex ARN in bulk oil (through acid-base reactions), reducing the number of tetraacids available at the interface for complexation with calcium [36,66].

Nevertheless, Simon et al. [41] reported that the inhibitory effect of asphaltenes is not due to competition with naphthenate but rather to interactions. They concluded that asphaltenes either inhibit the formation of ARN/Ca^{2+} interfacial gel, desorb it, or disperse it if already formed. Surprisingly, the inhibition seems to result from bulk interactions between ARN and asphaltenes in the oil phase and not from interfacial competition between these two components that was previously anticipated, since asphaltenes are in fact less surface active than ARN (even in the absence of Ca^{2+}). The tendency of asphaltenes to desorb or disperse interfacial gel could also be the result of the same interactions.

9.6.1.4 Effect of resins

Like asphaltenes, resins affect the naphthenate by affecting the interfacial tension. However, asphaltenes were found to predominantly affect the interfacial tension at low divalent salt concentration, whereas resin had higher effect at higher concentrations [67].

9.6.1.5 Effect of light hydrocarbons

Naphthenic acids are dissolved in crude oil, made soluble by the surrounding lighter molecules (solvation effect).

9.6.2 Effect of produced water composition

Water composition plays an important role in naphthenate/carboxylate soaps formation. This includes the following compositions.

9.6.2.1 Effect of carbonate/bicarbonate

From field observations, soap-forming fluids always have a water phase with a high bicarbonate content (typically much higher than 500 mg/L), because they form tight emulsions stabilized by interfacial sodium carboxylate-bicarbonate complexes [68,69]. In the case of naphthenate deposits, bicarbonate is believed to buffer the pH; however, for the carboxylate soaps the bicarbonate is thought to be part of the structure of the soap [32].

However, the effect is not just related to the increase in pH that bicarbonate provides; it can cause naphthenate soap formation at unexpectedly low pH values. Nichols et al. [23] showed that naphthenate may precipitate at untypically low pH values if bicarbonate content (and calcium content) of the brine is high. The low-level dissociation of tetraacid at low pH is sufficient to produce calcium naphthenate deposition due to the emulsion-stabilizing effect of bicarbonate [33].

It has been proposed that HCO_3^- stabilizes the acids at the interface (by hydrogen bonding). Moreover, the species $[Ca(HCO_3)]^+$ is known to be present in significant abundance in solutions containing Ca^{2+} and HCO_3^- ions, and can serve as bridging to bring calcium and naphthenic acids to form calcium naphthenate deposits [23]. Carboxylate salts soaps in sodium form were found to be stabilized by fatty acid-Na-HCO_3 complexation in the Indonesian field [70].

9.6.2.2 Effect of volatile fatty acids (VFAs)

Effect of volatile fatty acid (VFA) concentrations have been reported to be relatively high (greater than 1000 mg/L) in cases where carboxylate soaps are observed, albeit it is unknown whether and how the fatty acids contribute to soap formation, while in the case of naphthenate soaps, VFAs are generally low [32,70].

9.6.2.3 Effect of metal ions

Cations are necessary for the formation of soap deposits/emulsions, and they are abundant in all produced waters. Not only that, but also cations were found to affect the interfacial properties of naphthenic/carboxylic acids.

Generally, the metal cation stabilizes the interfacial naphthenate film, making it less soluble in the bulk phase [71]. Taylor and Hiu [34] showed how the interfacial behavior of naphthenic acids was significantly modified with the presence of metal ions at the *n*-heptane/water interface.

Interfacial activity of naphthenate salts is related to the valency and the hydration number of the cations [20,72]. Generally divalent cations are more stabilizing to the interfacial film than monovalent cations because they decrease the interfacial tension to a greater extent and they can form diacid salts [67,72]. Metal cations with higher hydration number have high interfacial affinity and produce less stable interfacial films because of the greater solubility of these films in the aqueous phase [20,33]. Ca^{2+} with hydration number of 2 complexes with and stabilizes the naphthenate better than Mg^{2+} with a hydration number of 6.

Ca^{2+} and Mg^{2+} were found to form naphthenate soaps, while Na^+ and K^+ ions were found to form carboxylate soaps as well [70]. Ba^{2+} did not form naphthenate or carboxylate soaps, but in fact it reduced the precipitate mass of the linear acids [50].

Also it is worth mentioning that the increase in the ionic strength of brine causes greater partitioning of naphthenic acid into the aqueous phase and consequently increases the rate of metal naphthenate formation [21].

Kherb et al. studied the strength of association of monovalent and divalent cations with carboxylate sites on a polypeptide biopolymer, and identified the divalent ion sequence to be in the order of: $Zn^{2+} > Ca^{2+} > Ba^{2+} > Sr^{2+} > Mg^{2+}$, while monovalent cations showed weaker binding [73].

Besides, the bonding between the carboxylic acids and metal ions is different from one metal to another [34], as depicted in Fig. 9.5.

9.6.3 Effect of pH

The variation of pH of the produced water from reservoir to surface facility pressure is of major importance as a driving force, for both naphthenate and carboxylate soaps formation; pH basically affects the partitioning, dissociation, and interfacial properties of naphthenic and carboxylic acids. Generally speaking, soaps are restricted to produced fluids where the pH of the water is greater than 6–6.2 [32]. Dyer et al. [47] reported that the critical minimum initial brine pH required for the onset of the naphthenate-soap precipitation was determined to be 5.9, while Runham and Smith [44] showed that critical pH to be between 5.9 and 6.3. Even at low calcium concentration in water, CaN formation can still occur substantially if pH is high (> 8) [74]. The relatively high pH is necessary for two things: partitioning and dissociation of the naphthenic/carboxylic acids, as these processes increase with increasing the pH above the mentioned limits, and hence provide a key driver for soap formation [32,75].

A more generalized way of expressing the effect of pH is that concluded by Brandal and Sjöblom [76] that for offshore production, when the pH of the coproduced water exceeds the pK_a values of the naphthenic acids, the acid monomers dissociate, making them even more interfacially active so that they form soaps when they come in contact with brines.

The pH is proven to strongly affect the interfacial properties of the naphthenic acids. As a rule of thumb, interfacial activity of naphthenic acids decreases with pH if pH < neutral, and increases with pH if pH > 7 [77].

9.6.4 Effect of pressure

Pressure affects the formation of naphthenates in many ways:

- Affecting the pH
- Affecting the stability of naphthenates
- Causing shear

When the crude oil is being processed offshore, the pressure decreases. This leads to degassing of carbon dioxide from the formation water and hence an increase in pH. When the pH of the water increases, the concentration of dissociated acid also increases, which in turn favors the formation of calcium naphthenate [10,32]. Also, naphthenate formation and stability was found to be related to asphaltenes, which in turn depend on pressure, especially around the bubble point [10]. Pressure drops cause shear, which leads to smaller water droplets, which creates larger surface area for soap particles to seed and grow, which produces more soap.

Naphthenate sludges, deposits, and emulsions were found to be formed in pressure drop areas like chokes, valves, and pumps, especially ESP, separators, and after shutdowns.

9.6.5 Effect of temperature

Heat is a very important factor in the saponification process of commercial soaps. The higher the temperature, the more soap that is expected to form [78] since increasing the temperature increases the rate of naphthenic acid portioning and dissociation, which increases the rate of the reaction. The partition ratio of a model naphthenic acid reached its highest with increasing temperature [79]. Also, the nucleation and the other kinetics aspects of calcium naphthenate soap scale formation may be affected by the temperature [80]. However, Sodium carboxylate soap emulsion formation responds differently to change in temperature, as temperature elevation can reduce their formation.

9.6.6 Effect of water cut

Water cut affects the naphthenate and carboxylate soap formation. Field observation from oilfields in the North Sea showed that water cut has an impact on soap formation; however, above a certain water cut, soap formation no longer occurs or becomes minor [32].

Generally, metal naphthenate formation is at its peak between 25% and 50% water cut, and then it attains a plateau at about 50%–60% water cut, after which it begins to decline [32,81]. It is thought that is corresponding to the emulsion conversion point of fluids. Other oilfields that generated carboxylate soaps have reached water cut percentages of 70%–85% over nearly 30 years and these fluids still generate significant soaps [32,44]. The effect of water cut on soap formation is illustrated in Fig. 9.6.

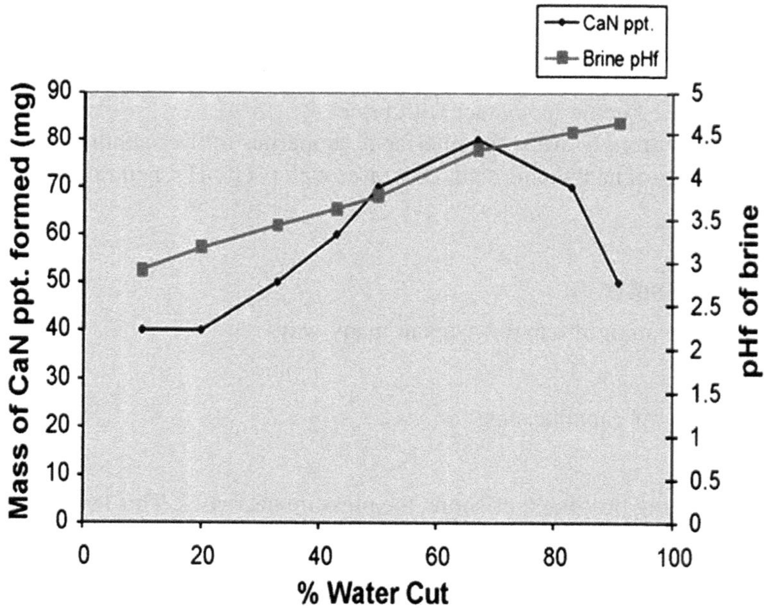

FIG. 9.6

Influence of water cut on amount of calcium naphthenate precipitate and brine final pH.

Reproduced *with permission from M.A. Mohammed, K.S. Sorbie, Thermodynamic modelling of calcium naphthenate formation: model predictions and experimental results, Colloids Surf. A Physicochem. Eng. Asp. 369(1–3) (2010) 1–10. Copyright (2010), Elsevier.*

9.6.7 Mixing of incompatible streams

Commingling incompatible fluids exacerbates soap formation. For example, mixing one stream with high calcium content with another stream with high naphthenic acids at optimum pH will cause soap formation. Also, mixing one stream with high calcium content but low carbonate and an acidic oil with another stream that has a water of high bicarbonate content leads to the formation of soaps. Indeed, such single streams will not form naphthenate/carboxylate deposits by themselves [32].

9.6.8 Effect of electrostatic forces

The dissociated carboxylic/naphthenic acids at the O/W interface can hold a charge, which can interact with the applied electric field, resulting in soap formation. Some field cases reported soap formation in electrostatic grids, which indicated a relationship between the applied electric field and the soap particulates [32].

9.7 Recent advances in naphthenates and carboxylate soap research

Naphthenic acids have been long known for their effect on system integrity due to causing corrosion problems. Yet, naphthenic acid salts (naphthenates/carboxylates) have gained interest from researchers and operators due to their major flow assurance soaps that have been reported in a number of different fields globally.

Several techniques have been used to measure the acidity and detect naphthenic/carboxylic acids in crude oil, including colorimetric techniques, potentiometric techniques, FTIR spectroscopy [15], and UV-based methods [20,21]. In addition, several mass spectrometry-based methods have been used for the detection of naphthenic acids in oil and deposits [3,4,19,63]. There are several challenges to the detailed characterization of naphthenic acid, such as poor peak resolution due to similar homologs, obscured fragmentation patterns due to several co-eluates, cumbersome derivatization, excessive fragmentation, and poor volatility of high molecular weight acids [33,55,71]. Furthermore, naphthenate interfacial properties and gel formation were studied using the NIR technique, pendant drop technique, Langmuir technique, and using a coaxial capillary device fitted to a profile analysis tensiometer [6,48].

Naphthenate deposition studies involve bottle tests, jar tests, flow loop, and recently a naphthenate deposition rig was used. Furthermore, interfacial techniques have been employed to study the mechanism of naphthenate formation and to screen the potential treating chemicals.

9.8 Summary

Naphthenic acids and carboxylic acids are gaining momentum among operators and researchers as major flow assurance impediments. Known as soaps, naphthenate and carboxylate deposits form when crude oil naphthenates and carboxylates interact with metal ions at $pH \geq 6$ at the oil-water interface. Soaps formation is highly affected by pressure change, as pressure decrease is one of the main causes of naphthenate deposition. Besides, pressure drop also causes a pH shift, which induces the formation of such soaps. Soaps were found to form at pH values around 6 and their formation increases with increasing pH, as relatively high pH induces partitioning and dissociation, and induces interaction between the naphthenates and the metal ions. Increasing temperature was also found to increase naphthenate formation as a result of increasing portioning of naphthenates and carboxylates. Mixing incompatible fluids is a main reason for naphthenates formation, i.e., high water cut streams with high Ca content with highly acidic crude oil streams. Furthermore, crude oil and produced water composition also impact soaps formation.

References

[1] M.A. Kelland, Production chemicals for the oil and gas industry, in: Production Chemicals for the Oil and Gas Industry, CRC Press, 2009, https://doi.org/10.1201/9781420092974.

[2] J. Bellarby, Production chemistry, in: Developments in Petroleum Science, 2009, https://doi.org/10.1016/S0376-7361(08)00207-0 (Chapter 7).

[3] M.A. Mohammed, K.S. Sorbie, Naphthenic acid extraction and characterization from naphthenate field deposits and crude oils using ESMS and APCI-MS, Colloids Surf. A Physicochem. Eng. Asp. 349 (2009) 1–18.

[4] M.A. Mohammed, K.S. Sorbie, Spectroscopic determination of naphthenic acid composition from various calcium naphthenates field deposits, in: SPE International Symposium on Oilfield Chemistry, Texas, SPE 121633, 2009.

[5] PT Staff, Techbits: naphthenate deposits, emulsions highlighted in technology workshop, J. Pet. Technol. 60 (2008) 30–31, https://doi.org/10.2118/0708-0030-JPT.

[6] Ø. Brandal, Interfacial (o/w) Properties of Naphthenic Acids and Metal Naphthenates, Naphthenic Acid Characterization and Metal Naphthenate Inhibition in Norwegian University of Science and Technology, Norwegian University of Science and Technology, 2005.

[7] A.C. Scott, M.D. MacKinnon, P.M. Fedorak, Naphthenic acids in athabasca oil sands tailings waters are less biodegradable than commercial naphthenic acids, Environ. Sci. Technol. 39 (21) (2005) 8388–8394, https://doi.org/10.1021/es051003k.

[8] G. Rosseau, H. Zhou, C. Hurtevent, Calcium carbonate and naphthenate mixed scale in deep offshore fields, in: SPE 3rd International Symposium on Oilfield scale, Aberdeen, SPE 68307, 2001.

[9] J.S. Clemente, P.M. Fedorak, A review of the occurrence, analyses, toxicity, and biodegradation of naphthenic acids, Chemosphere (2005), https://doi.org/10.1016/j.chemosphere.2005.02.065.

[10] T.E. Havre, Formation of Calcium Naphthenate in Water/Oil Systems, Naphthenic Acid Chemistry and Emulsion Stability, Department of Chemical Engineering Norwegian University of Science and Technology, October 2002. 71 Pages.

[11] W. Meredith, S.J. Kelland, D.M. Jones, Influence of biodegradation on crude oil acidity and carboxylic acid composition, Org. Geochem. 31 (2000) 1059–1073.

[12] H. Mediaas, N.O. Wolf, T.D. Baugh, K.V. Grande, J.E. Vinstad, The Discovery of High Molecular Weight Naphthenic Acids (ARN Acid) Responsible for Calcium Naphthenate Deposits, Society of Petroleum Engineers, 2005, January 1, https://doi.org/10.2118/93011-MS.

[13] P. Juyal, M.M. Mapolelo, A. Yen, R.P. Rodgers, S.J. Allenson, Identification of calcium naphthenate deposition in South American oil fields, Energy Fuel 29 (2015) 2342–2350.

[14] E.L. Nordgård, A.D. Hanneseth, J. Sjöblom, Inhibition of calcium naphthenate. Experimental methods to study the effect of commercially available naphthenate inhibitors, J. Dispers. Sci. Technol. 31 (5) (2010) 668–675.

[15] R. Chakravarthy, G.N. Naik, A. Savalia, U. Sridharan, C. Saravanan, A.K. Das, K.B. Gudasi, Determination of naphthenic acid number in petroleum crude oils and their fractions by mid-FTIR spectroscopy, Energy Fuel 30 (10) (2016) 8579–8586.

[16] J.A. Brient, Commercial utility of naphthenic acids recovered from petroleum distillates, Prepr. Am. Chem. Soc. Div. Petrol. Chem. 43 (1) (1998) 131–133.

[17] N. Passade-Boupat, M.R. Gonzalez, B. Brocart, C. Hurtevent, T. Palermo, Risk assessment of calcium naphthenates and separation mechanism of acidic crude oil, in: SPE International Conference and Exhibition on Oilfield Scale, Aberdeen, SPE 155229, 2012.

[18] S. Simon, E. Nordgård, P. Bruheim, J. Sjoblom, Determination of C80 tetra-acid content in calcium naphthenate deposits, J. Chromatogr. A 1200 (2008) 136–143.

[19] B.E. Smith, P.A. Sutton, C.A. Lewis, B. Dunsmore, G. Fowler, J. Krane, B.F. Lutnaes, Ø. Brandal, J. Sjoblom, S.J. Rowland, Analysis of ARN naphthenic acids by high temperature gas chromatography and high performance liquid chromatography, Sep. Sci. 30 (2007) 375–380.

[20] T.E. Havre, J. Sjöblom, J.E. Vindstad, Oil/water-partitioning and interfacial behavior of naphthenic acids, J. Dispers. Sci. Technol. 24 (6) (2003) 789–801, https://doi.org/10.1081/DIS-120025547.

[21] T.E. Havre, M. Ese, J. Sjöblom, A.M. Blokhus, Langmuir films of naphthenic acids at different pH and electrolyte concentrations, Colloid Polym. Sci. 280 (2002) 647–652.

[22] J.A. Brient, P.J. Wessner, M.N. Doyle, Naphthenic acids, in: Kirk-Othmer (Ed.), Encyclopedia of Chemical Technology, John Wiley & Sons, New York, 1995, pp. 1017–1029.

[23] D.A. Nichols, F.F. Rosario, M.C.M. Bezerra, S.E. Gorringe, H.L. Williams, G.M. Graham, Calcium naphthenates in complex production systems—evaluation and chemical inhibition challenges, in: SPE International Oilfield Scale Conference and Exhibition, Aberdeen, SPE-169756-MS, 2014.

[24] K. Theander, R.-J. Pugh, The influence of pH and temperature on the equilibrium and dynamic surface tension of aqueous solutions of sodium oleate, J. Colloid Interface Sci. 239 (2001) 209–216.

[25] Ø. Brandal, A.M.D. Hanneseth, P.V. Hemmingsen, J. Sjöblom, S. Kim, R.P. Rodgers, A.G. Marshall, Isolation and characterization of naphthenic acids from a metal naphthenate deposit: molecular properties at oil-water and air-water interfaces, J. Dispers. Sci. Technol. 27 (3) (2006) 295–305, https://doi.org/10.1080/01932690500357909.

[26] Ø. Brandal, J. Sjöblom, G. Øye, Interfacial behavior of naphthenic acids and multivalent cations in systems with oil and water. I. A pendant drop study of interactions between n-dodecyl benzoic acid and divalent cations, J. Dispers. Sci. Technol. 25 (3) (2004) 67–374.

[27] T.E. Havre, Near infrared spectroscopy as a method for studying the formation of calcium naphthenate, Colloid Polym. Sci. 282 (2004) 270–279.

[28] G.A. Wolstenholme, J.H. Schulman, Metal-monolayer interactions in aqueous systems: I. Interaction of monolayers of long-chain polar compounds with metal ions in the underlying solution, Trans. Faraday Soc. 46 (1950) 475–487.

[29] A. Goldszal, C. Hurtevent, G. Rousseau, Scale and naphthenate inhibition in deep offshore fields, in: Presented at the 4th International Oilfield Scale Symposium, Aberdeen, UK, January 30–31, SPE 74661, 2002.

[30] M.C.K. De Oliveira, F.F. Rosario, J.N. Bertelli, R.C.L. Pereira, F.C. Albuquerque, L.C.C. Marques, Flow Assurance Solutions to Mitigate Naphthenates Problems in Crude Oil Production, Society of Petroleum Engineers, 2013, September 30, https://doi.org/10.2118/166235-MS.

[31] M.M. Ahmed, Characterization, Modelling, Prediction and Inhibition of Naphthenate Deposits in Oilfield Production (PhD thesis), Heriot-Watt University, Edinburgh, UK, 2010.

[32] M.S. Turner, P.C. Smith, Controls On Soap Scale Formation, Including Naphthenate Soaps—Drivers and Mitigation, Society of Petroleum Engineers, 2005, January 1, https://doi.org/10.2118/94339-MS.

[33] W.I. Eke, C. Victor-Oji, O. Akaranta, Oilfield metal naphthenate formation and mitigation measures: a review, J. Pet. Explor. Prod. Technol. 10 (2) (2020) 805–819, https://doi.org/10.1007/s13202-019-00797-0.

[34] S.E. Taylor, T.C. Hiu, Metal ion interactions with crude oil components: specificity of Ca2+ binding to naphthenic acid at an oil/water, Colloids Interfaces 2 (40) (2018) 1–21.

[35] J.N. Bertelli, M.M. Rocío, V.P. Renata, F.C. Albuquerque, E.F. Lucas, Shear rheology using De Nouy ring to evaluate formation and inhibition of calcium naphthenate at the water/oil interface, Energy Fuel 28 (2014) 1726–1735.

[36] S. Subramanian, S. Simon, J. Sjöblom, Interfacial dilational rheology properties of films formed at the oil/water interface by reaction between tetrameric acid and calcium ion, J. Dispers. Sci. Technol. 38 (8) (2017) 1110–1116, https://doi.org/10.1080/01932691.2016.1224718.

[37] S. Simon, E. Blanco, B. Gao, J. Sjöblom, N. Passade-Boupat, T. Palermo, M. Rondon-Gonzalez, Rheological properties of gels formed at the oil/water interface by reaction between tetrameric acid and calcium ion under flow condition and at the batch scale, Ind. Eng. Chem. Res. 58 (2019) 15516, https://doi.org/10.1021/acs.iecr.9b02751.

[38] J.E. Vindstad, K.V. Grande, K.M. Høvik, H. Kummernes, H. Mediaas, In calcium naphthenate management, in: 8th International Conference on Petroleum Phase Behavior and Fouling, Pau, France, June 10–14, 2007.

[39] S. Subramanian, S. Simon, J. Sjöblom, Interfacial dilational rheology properties of films formed at the oil/water interface by reaction between tetrameric acid and calcium ion, J. Dispers. Sci. Technol. 38 (2017) 1110, https://doi.org/10.1080/01932691.2016.1224718.

[40] E.L. Nordgård, S. Simon, J. Sjöblom, Interfacial shear rheology of calcium naphthenate at the oil/water interface and the influence of pH, calcium and in presence of a model monoacid, J. Dispers. Sci. Technol. 33 (2012) 1083, https://doi.org/10.1080/01932691.2011.579834.

[41] S. Simon, B. Gao, S. Tofte, J. Sjöblom, N. Passade-Boupat, T. Palermo, M. Rondon-Gonzalez, Influence of asphaltenes on gelation of tetrameric acid with calcium ion at the oil/water interface under flow-model condition, Energy Fuel 34 (5) (2020) 5846–5855, https://doi.org/10.1021/acs.energyfuels.0c00829.

[42] J. Debord, P. Srivastava, Development and field application of a novel non-acid calcium naphthenate inhibitor, in: SPE Annual Technical Conference and Exhibition, Louisiana, SPE 123660, 2009.

[43] M.M. Mapolelo, L.A. Stanford, R.P. Rodgers, A.T. Yen, J.D. Debord, S. Asomaning, A.G. Marshall, Chemical speciation of calcium and sodium naphthenate deposits by electrospray ionization FTICR mass spectrometry, Energy Fuel 23 (2009) 349–355.

[44] G. Runham, C. Smith, Successful naphthenate scale and soap emulsion management, in: SPE International Symposium on Oilfeld Chemistry, Texas, SPE 121522, 2009.

[45] A.-M.D. Hanneseth, An Experimental Study of Tetrameric Naphthenic Acids at w/o Interfaces: Reactivity, Inhibition and Emulsion Formation (Doctoral theses at NTNU), Norwegian University of Science and Technology, 2009.

[46] O. Sundman, S. Simon, E.L. Nordgård, J. Sjöblom, Study of the aqueous chemical interactions between a synthetictetra-acid and divalent cations as a model for the formation of metal naphthenate deposits, Energy Fuel 24 (2010) 6054–6060.

[47] S.J. Dyer, G.M. Graham, C. Arnott, Naphthenate scale formation—examination of molecular controls in idealised systems, in: SPE 80395, Presented at the 5th International Symposium on Oilfield Scale, Aberdeen, UK, January 29–30, 2003.

[48] S. Simon, C. Reisen, A. Bersås, J. Sjöblom, Reaction between tetrameric acids and Ca2+ in oil/water system, Ind. Eng. Chem. Res. 51 (2012) 5669–5676.

[49] J.E. Vindstad, A.S. Bye, K.V. Grande, B.M. Hustad, E. Hustvedt, B. Nerdard, Fighting naphthenate deposition at the Heidrun field. SPE 80375, in: 5th International Symposium on Oilfield Scale, Aberdeen, 2003.

[50] A.G. Shepherd, G.B. Thomson, R. Westacott, K.S. Sorbie, A. Neville, A Mechanistic Study of Naphthenate Scale Formation, Society of Petroleum Engineers, 2005, January 1, https://doi.org/10.2118/93407-MS.

[51] J. Junior, L. Borges, C. Carmelino, P. Hango, J.D. Milliken, S. Asomaning, Calcium naphthenate mitigation at Sonangol's Gimboa field, in: SPE International Symposium on Oilfied Chemistry, Texas, SPE 164069, 2013.

[52] K.C. Hampshire, D. Stokes, N.F. Omar, et al., Kikeh ESS well test—a case history of a deepwater well test, Offshore Malaysia, in: SPE 88564, 2004.

[53] K. Pearson, Separator profiling: Kuito FPSO case study, in: SPE 8847, Presented at the 11th International Petroleum Exhibition and Conference, Abu Dhabi, UAE, October 10–13, 2004.

[54] W.K. Seifert, R.M. Teeter, W.G. Howells, M.J.R. Cantow, Analysis of crude oil carboxylic acids after conversion to their corresponding hydrocarbons, J. Anal. Chem. 41 (1969) 1638–1647.

[55] I. Dzidic, A.C. Somerville, J.C. Raia, H.V. Hart, Determination of naphthenic acids in California crudes and refinery wastewaters by fluoride ion chemical ionization mass spectrometry, Anal. Chem. 60 (1988) 1318–1323.

[56] D.M. Jones, J.S. Watson, W. Meredith, M. Chen, B. Bennett, Determination of naphthenic acids in crude oils using nonaqueous ion exchange solid phase extraction, Anal. Chem. 73 (2001) 703–707.

[57] G. Sartori, C.W. Savage, B.H. Ballinger, D.C. Dalrymple, Process for Extraction of Naphthenic Acids From Crudes, 2001. US Patent 6281328 B1.

[58] W.K. Robbins, Challenges in the characterization of naphthenic acids in petroleum, in: Presented Before the Division of Petroleum Chemistry Inc. 215th National Meeting, American Chemical Society, Dallas, Texas, USA, March 29-April 3, 1998.

[59] P.C. Smith, Soap (naphthenates) scales management from deepwater flow assurance aspects to oil terminal sludge processing, in: Presented at IQPC Flow Assurance. A Holistic Approach Meeting, Kuala Lumpur, Malaysia, December 1-2, 2004.

[60] K.S. Sorbie, A.G. Shepherd, M. Turner, P.C. Smith, R.E. Westacott, Naphthenate formation in oil production: general theories and field observations, in: Proc., Chemistry in the Oil Industry IX, Manchester, UK, 31 October–2 November, 2005.

[61] G.M. Graham, K.B. Melvin, A.E. Gabb, F. Haider, H. Williams, S. Dyer, cummine, C., Simulating Calcium Naphthenate Formation and Mitigation Under Laboratory Conditions, Society of Petroleum Engineers, 2006, January 1, https://doi.org/10.2118/100632-MS.

[62] B. Brocart, C. Hurtevent, Flow assurance issues and control with naphthenic oils, J. Dispers. Sci. Technol. 29 (10) (2008) 1496–1504.

[63] B. Brocart, M. Bourrel, C. Hurtevent, J. Volle, B. Escoffier, ARNtype naphthenic acids in crudes: analytical detection and physical properties, J. Dispers. Sci. Technol. 28 (3) (2007) 331–337.

[64] E.L. Nordgård, J. Ahmad, S. Simon, J. Sjöblom, Oil–water partitioning of a synthetic tetracarboxylic acid as a function of pH, J. Dispers. Sci. Technol. 33 (6) (2012) 871–880.

[65] M. Moradi, E. Topchiy, T.E. Lehmann, V. Alvarado, Impact of ionic strength on partitioning of naphthenic acids in water–crude oil systems—determination through high-field NMR spectroscopy, Fuel 112 (2013) 236–248.

[66] J.A. Östlund, M. Nydén, I.H. Auflem, J. Sjöblom, Interactions between asphaltenes and naphthenic acids, Energy Fuel 17 (1) (2003) 113–119, https://doi.org/10.1021/ef020103y.

[67] M. Lashkarbolooki, S. Ayatollahi, M. Riazi, Effect of salinity, resin, and asphaltene on the surface properties of acidic crude oil/smart water/rock system, Energy Fuel 28 (11) (2014) 6820–6829, https://doi.org/10.1021/ef5015692.

[68] D.L. Gallup, J.A. Curiale, Characterization of sodium emulsion soaps formed from production fluids of Kutel Basin, Indonesia, Energy Fuel 21 (2007) 1741–1759.

[69] A.G. Shepherd, G.B. Thomson, R. Westacott, K.S. Sorbie, M. Turner, P.C. Smith, Analysis of organic field deposit: new types of calcium naphthenate scale or the effect of chemical treatment? in: SPE International Oilfield Scale Symposium, Aberdeen, SPE 100517, 2006.

[70] D.L. Gallup, J. Star, Soap Sludges: Aggravating Factors and Mitigation Measures, Society of Petroleum Engineers, 2004, January 1, https://doi.org/10.2118/87471-MS.

[71] B.F. Lutnaes, Ø. Brandal, J. Sjoblom, J. Krane, Archaeal C80 isoprenoid tetraacids responsible for naphthenate deposition in crude oil processing, Org. Biomol. Chem. 4 (2006) 616–620.

[72] M.A. Mohammed, K.S. Sorbie, A.G. Shepherd, Thermodynamic modeling of naphthenate formation and related pH change experiments, SPE Prod. Oper. 24 (2009) 466–472.

[73] J. Kherb, S.C. Flores, P.S. Cremer, Role of carboxylate side chains in the cation Hofmeister series, J. Phys. Chem. B 116 (2012) 7389–7397.

[74] C. Igwebueze, L. Oduola, O. Smith, P. Vijn, A.G. Shepherd, Calcium naphthenate solid deposit identification and control in offshore Nigerian fields, in: SPE International Symposium on Oilfield Chemistry, Texas, SPE 164055, 2013.

[75] S. Sarac, F. Civan, Mechanisms, Parameters, and Modeling of Naphthenate Soap-Induced Formation Damage, Society of Petroleum Engineers, 2008, January 1, https://doi.org/10.2118/112434-MS.

[76] Ø. Brandal, J. Sjöblom, Interfacial behavior of naphthenic acids and multivalent cations in systems with oil and water. II: formation and stability of metal naphthenate films at oil-water interfaces, J. Dispers. Sci. Technol. 26 (1) (2005) 53–58, https://doi.org/10.1081/DIS-200040145.

[77] J. Bai, W. Fan, G. Nan, S. Li, B. Yu, Influence of interaction between heavy oil components and petroleum sulfonate on the oil-water interfacial tension, J. Dispers. Sci. Technol. 31 (4) (2010) 551–556.

[78] S.J. Dyer, H.L. Williams, G.M. Graham, C. Cummine, K.B. Melvin, F. Haider, A.E. Gabb, Simulating calcium naphthenate formation and mitigation under laboratory conditions, in: SPE Oilfield Scale Symposium, Aberdeen, SPE 100632, 2006.

[79] A. Celesie, J.M. Parnis, D. Mackay, Impact of temperature, pH and salinity changes on the physico-chemical properties of model naphthenic acids, Chemosphere 146 (2016) 40–50.

[80] A.G. Shepherd, M. Dijk, W. Koot, S. Dubey, S. Poteau, G. Zabaras, M. Grutters, Flow assurance in oil systems: on the role of naphthenic acids, in: SPE International Production and Operations Conference and Exhibition, Qatar, SPE 157295, 2012.

[81] M.A. Mohammed, K.S. Sorbie, Thermodynamic modelling of calcium naphthenate formation: model predictions and experimental results, Colloids Surf. A Physicochem. Eng. Asp. 369 (1–3) (2010) 1–10.

CHAPTER

Biofouling

10

10.1 Introduction

Biofouling is a type of fouling that represents the undesirable adhesion and accumulation of micro- and macroorganisms or their excretion products on surfaces, causing flow restriction in tubulars and impairment of the process equipment [1].

Biofouling is basically attributed to bacteria and other microorganism activity. Different types of bacteria are typically encountered in oil and gas fields. Biofouling is often mentioned in the literature as the formation of biomasses composed mainly of microorganisms, biofilm, colonies, or bacterial slimes. Nevertheless, it is common to find other mineral constituents within the biofouling deposits, due to the different interactions of the microorganisms with their environment during their life cycle, such as biocorrosion. Therefore in this chapter the biofouling is classified and categorized into two basic types: biomineralization and biomass fouling. The categories of biofouling are illustrated in Fig. 10.1.

As indicated earlier, biofouling is basically attributed to microorganism activity. Therefore it is intuitive to discuss different types of microorganisms before discussing any of the biofouling processes illustrated in Fig. 10.1.

10.2 Microbial community in oil and gas fields

The tree of life is divided into three primary domains: Bacteria, Archaea, and Eucarya. This classification encompasses the entire realm of living organisms.

A wide variety of microorganisms can be found in oil and gas reservoirs and production and processing facilities. Fig. 10.2 gives an example of the diversity of these microorganisms by illustrating the various types of bacteria isolated from producing and injection wells in southern Algeria [2].

Bacteria require general essentials during their life cycle, such as an energy source (light or chemical compounds), carbon source (CO_2 or organic substances), electron donor (organic or inorganic substances), and electron acceptor (O_2, CO_2, NO_2^-, SO_4^{2-}, NO_3^-) [3].

Bacteria found in oil and gas fields can be classified according to their shape and their oxygen and temperature requirements. Fig. 10.3 illustrates a simple classification of bacteria.

Essentials of Flow Assurance Solids in Oil and Gas Operations. https://doi.org/10.1016/B978-0-323-99118-6.00008-3

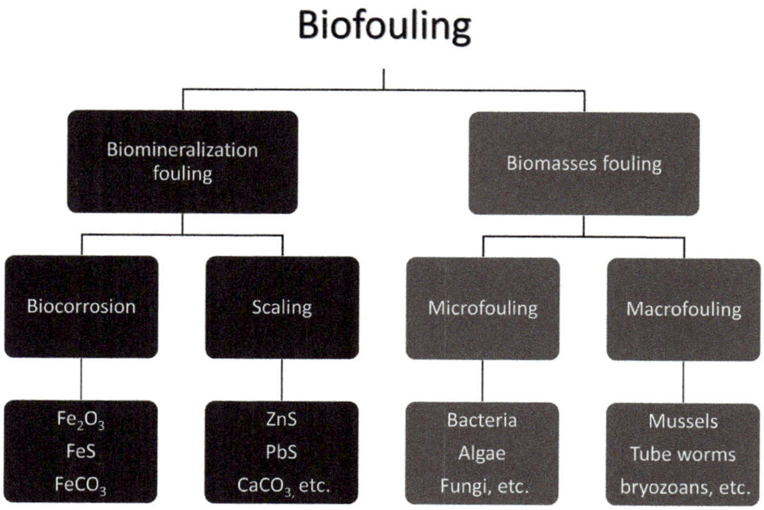

FIG. 10.1

Biofouling types.

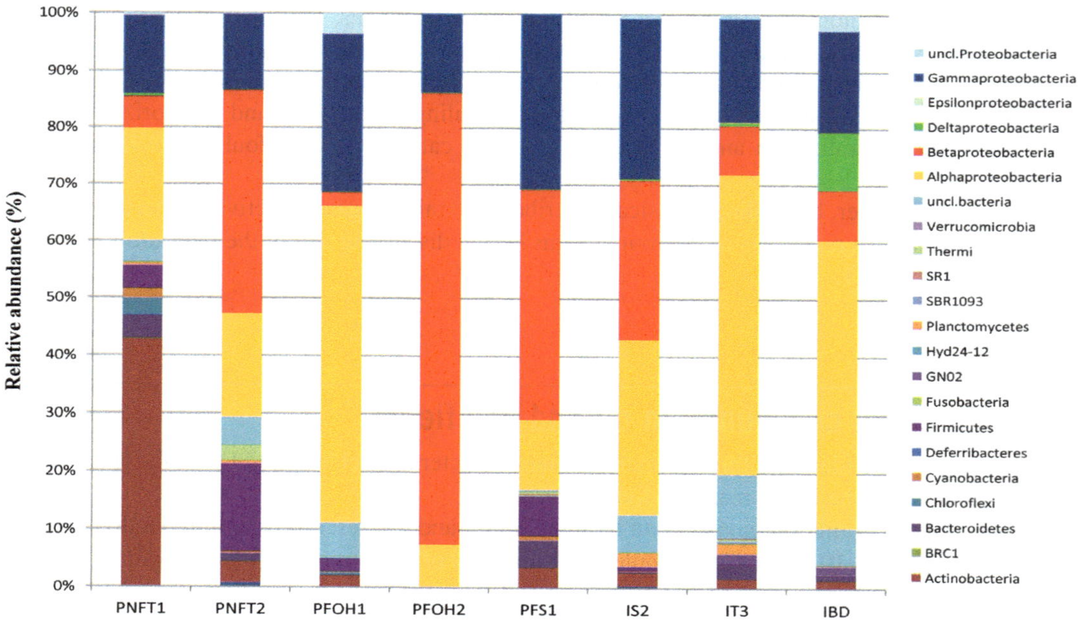

FIG. 10.2

Bacteria phyla from different wells in Algeria (using 16S rRNA gene pyrosequencing).

From *N. Lenchi, Ö. Inceoğlu, S. Kebbouche-Gana, M. Lamine Gana, M. Lliros, P. Servais, T. García-Armisen, Diversity of microbial communities in production and injection waters of Algerian oilfields revealed by 16S RRNA gene amplicon 454 pyrosequencing, PLoS One 8(6) (2013), doi:10.1371/journal.pone.0066588.*

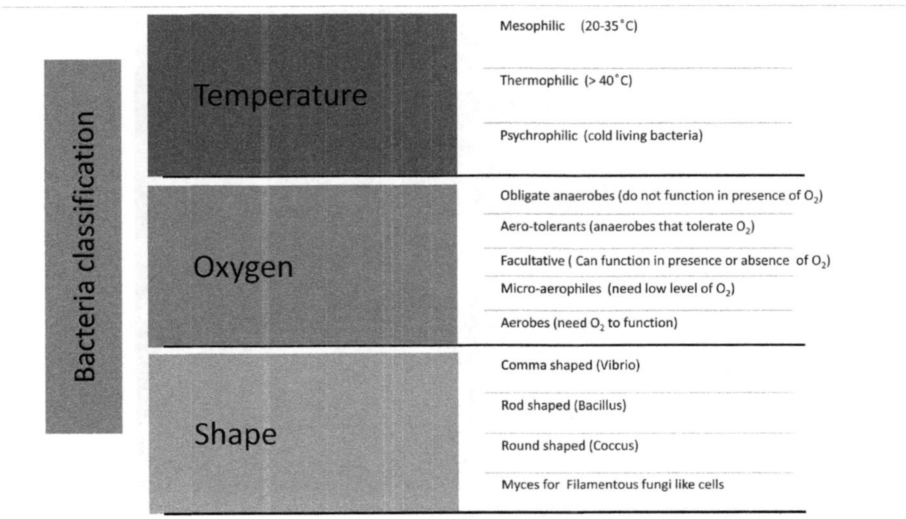

FIG. 10.3

Classification of bacteria.

10.2.1 Sulfate-reducing bacteria

Sulfate-reducing bacteria (SRB) are notorious corrosion-enhancing bacteria in oil and gas fields. They are considered the most studied bacteria in upstream oil and gas operations, due to their serious impacts on production, including severe corrosion of piping, drilling, and pumping machinery and storage tanks, biofouling and tubular plugging, reservoir souring, and contamination of crude oil resulting in increasing the sulfur level of fuels [3–6]. SRB can also be found in the human body, such as the mouth and bowel [3].

Desulfovibrionales, Desulfosporosinus, Desulfotomaculum, Desulfosporomusa, and Desulfobacterales are examples of common SRB, while Thermodesulfobacteria and Thermodesulfobium are examples of thermophilic SRB that are commonly encountered at elevated temperatures. Sulfate-reducing archaea such as Archaeoglobus, Thermocladium, and Caldivirga have been found in oil deposits [7,8]. Fig. 10.4 shows SRB bacteria on a stainless steel surface.

These microorganisms are anaerobic, require sulfate as a main nutrient, grow in a pH range between 4 and 9.5, and can tolerate pressures of up to 500 atm [3]. However, they also can tolerate low levels of oxygen and have been shown to grow in the absence of sulfate by fermenting organic acids and alcohols. Most known SRB are mesophilic; nevertheless, psychrophilic, thermophilic, and hyperthermophilic species have been described [9].

In anaerobic respiration SRB use sulfate instead of oxygen, and oxidize organic substances to organic acids and CO_2 (or oxidize molecular hydrogen) by the reduction of sulfate to sulfide [6–8]. An illustration of the sulfate-reducing bacteria mechanism of action is given in Fig. 10.5 [6].

As seen in Fig. 10.5, SRB attack steel by two main scenarios [8]:

- Chemical MIC (CMIC) of iron by H_2S produced from microbial sulfate reduction process.
- Electrical MIC (EMIC) in which SRB corrode iron by direct utilization of the metal itself; this takes place in only a limited number of recently discovered SRB strains.

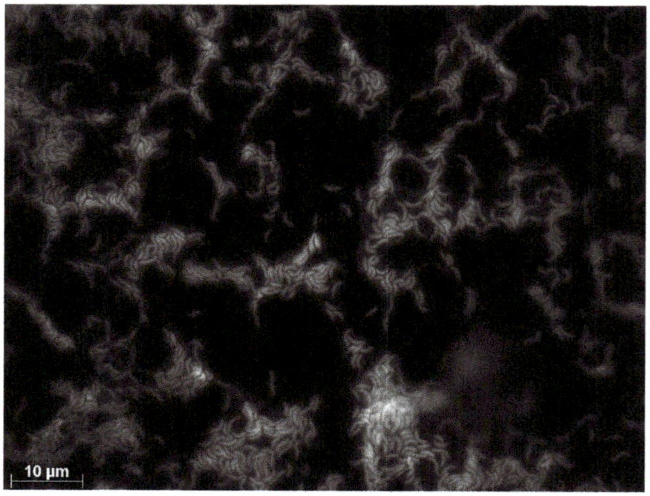

FIG. 10.4

Fluorescent microscopic image of *Desulfovibrio alaskensis* cells on stainless steel 304, 1000× magnification, DAPI staining.

From *https://commons.wikimedia.org/wiki/File:Desulfovibrio_alaskensis_cells_on_stainless_steel_304.jpg, CC BY 4.0, photo credit Vvjbp.*

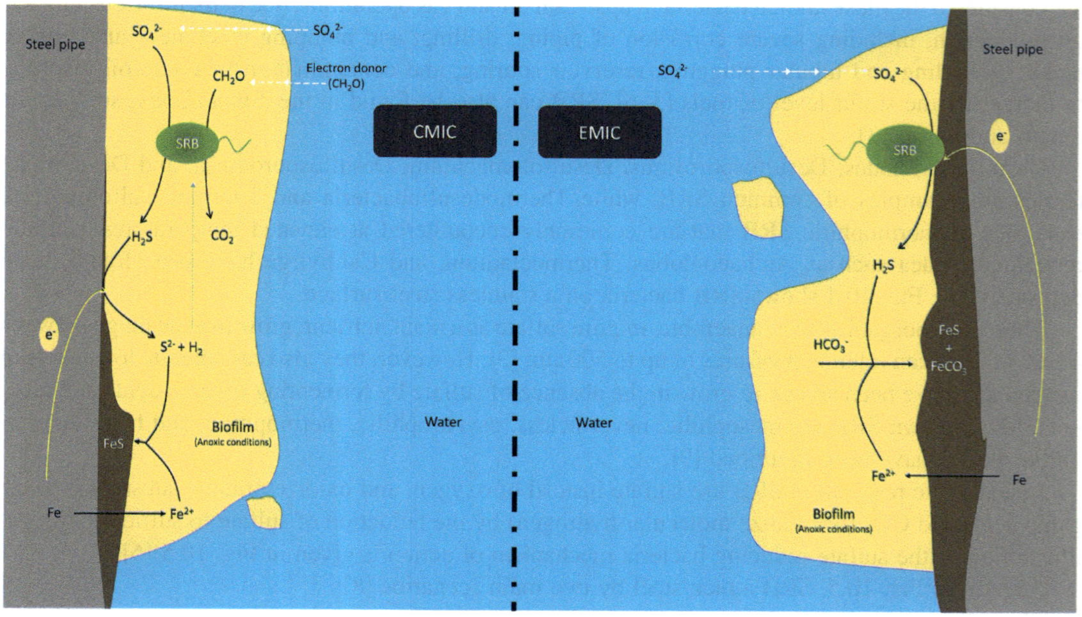

FIG. 10.5

The different mechanisms of iron corrosion by sulfate-reducing bacteria (SRB).

10.2.2 Sulfur-oxidizing bacteria

Sulfur-oxidizing bacteria (SOB) are aerobic bacteria that generally oxidize reduced forms of sulfur to sulfate, producing acids and causing severe corrosion problems [10].

The most investigated genera of SOB are chemolitotrophic sulfur-oxidizing bacteria that include Thiobacillus, Sulfolobus, Thermothrix, Beggiatoa, and Thiothrix. They use the energy of reduced inorganic sulfur compounds (H_2S, $S_2O_3^{2-}$, SO_3^{2-}, S^0), and they can also use sulfur from organic sulfur compounds [11,12]. Most of the studied chemolithotrophic bacteria thrive best in mesophilic or thermophilic conditions. An example, *Thiobacillus ferrooxidans*, is active in the pH range of 1.0–2.5, and is also a thermophile that prefers temperatures of 45–50°C and oxidizes iron- and sulfur-containing minerals [12].

10.2.3 Acid-producing bacteria

Acid-producing bacteria (APB) have been reported as one of the major causes of corrosion, due to their effect of localized production of corrosive acidic metabolites.

APB use organic substrates (carbohydrates) during their fermentation process and produce organic acids. The kinds and amounts of acids produced depend on the type of microorganisms and the available substrate molecules [5]. Common organic acids include acetic and butyric acids. Inorganic acids that are reported to be produced by some bacteria include HNO_3, H_2SO_3, H_2SO_4, HNO_2, and H_2CO_3 [7].

APB can be aerobic like the sulfur-oxidizing bacteria *Acetobacter* and *Gluconobacter* or anaerobic like *Clostridia* [3]. Acetic acid-producing bacteria (*Acetobacter* and *Gluconacetobacter*) are prevalent in environments exposed to fuel grade ethanol [13]. They were also reported to cause biocorrosion in oil and gas fields. *Pseudomonas* has been known to form biofilm, which was reported to be acidic, accelerating the corrosion process [14].

10.2.4 Iron- and manganese-oxidizing bacteria

These are a group of aerobic bacteria that metabolize by utilizing the oxidation of Fe^{+2} and/or Mn^{+2} ions. In the case of iron, IOB convert iron from the ferrous [Fe^{+2}] to the ferric [Fe^{+3}], producing ferric hydroxide [$Fe(OH)_3$] with its well-known reddish brown color. Ferric hydroxide is a highly insoluble compound that can encrust and accumulate in pipes and equipment, causing damage [7]. Fig. 10.6 shows iron oxidation bacteria in surface waters.

Common iron-oxidizing bacteria include *Gallionella*, *Sphaerotilus*, *Crenothrix*, and *Leptothrix* species [15].

IOBs have detrimental effects on metal structures due to the following [7]:

- They cause corrosion.
- They also cover SRB colonies, protecting them from biocides during chemical treatment and, in addition.
- They induce pitting corrosion by oxidizing ferrous to ferric ion form, thus producing extremely aggressive ferric chlorides (very low local pH value: pH $\sim$ 1).
- The formed iron/manganese oxide layer stimulates underdeposit corrosion and increases metal corrosion potential, which finally leads to severe localized pitting.

FIG. 10.6

Iron-oxidizing bacteria in surface waters.

Photo courtesy New Hampshire Estuaries Project, www.nhep.unh.edu. Source https://en.wikipedia.org/wiki/Iron-oxidizing_bacteria#/media/File:Iron_bacteria_in_runoff.JPG.

10.2.5 Iron-reducing bacteria

Iron-reducing bacteria (IRB) are facultative anaerobes that reduce ferric ions to ferrous ions under anaerobic conditions. It has been reported that IRB are able to metabolize other different electron acceptors, including Mn (IV), NO^{3-}, NO^{2-}, $S_2O_3^{-2}$, and SO_3 [5,16,17]. One of the most studied IRB is *Shewanella*. Fig. 10.7 shows iron-reducing bacteria isolated from a uranium mine.

The role and importance of IRB in MIC has been a matter of controversy. Some works indicate that they can be detrimental by directly causing or enhancing corrosion, while other works show that iron reducers may actually give protection against biotic or abiotic aggressive agents [18], depending on the enrichment environment.

It has been postulated that IRB accelerate corrosion by the following actions:

1. The reduction of insoluble ferric ion compounds to soluble ferrous ion that, in turn, removes the protective corrosion scales formed on exposed surfaces.
2. Promotion of the formation of concentration cells among the biofilm [16,19].

On the other hand, IRB could inhibit the corrosion process via different mechanisms, including:

1. Aerobic respiration by removing the oxygen from the system [14,16,20].
2. The presence of IRB along with SRB in the biofilm could be beneficial, as they could destroy the ecological nests of sulfate-reducers within the biofilms formed on metal surfaces [21]. Little et al. [22] reported short-term protection of biofilms composed of a mixed culture of IRB and SRB.

IRB and SRB activity results in the formation of a variety of iron oxides, hydroxides, carbonate, and sulfides as end products, which can accumulate and encrust, causing damage in production systems.

FIG. 10.7

IRB isolated using a roll bottle technique from a former uranium mine near Ronneburg, Germany.

Photo credit: Denise Akob, USGS.

10.2.6 Methanogens

Oil reservoir methanogenic microorganisms metabolize H_2, CO_2, or other small organic compounds with production of methane, in a process known as methanogenesis [23]. The methanogenic pathway of anaerobes generates methane through the reduction of CO_2 with hydrogen or organic molecules as the hydrogen donor if soil pH is very low [24]. Several strains of methanogens (e.g., *Methanosarcina barkeri*, *Methanobacterium bryantii*, and *Methanospirillum hungatei*) can grow in media containing iron or other metals as the only source of electrons [5,25]. Methanogenic Archaea (methanogens) are obligate anaerobes.

Lab and field observations showed that SRB can grow syntrophically with hydrogen- or formate-consuming methanogens on lactate, ethanol, propionate, fumarate, and pyruvate in defined cocultures, eliminating their need to reduce sulfate [26]. Such a microbial community increases the risk of corrosion and produces a variety of corrosion products. The syntrophic association of methanogens with fermentative microorganisms was also found to significantly enhance corrosion [27]. Methanogenic thermophilic Archaea can go through assimilatory reduction of S^0 to H_2S leading to the formation of mineral sulfides [28]. Cooperative interactions between microorganism methanogens and acetogens were found to be responsible for corrosion of infrastructure [29].

10.2.7 Nitrate-reducing bacteria

Nitrate-reducing bacteria use nitrate as the main electron acceptor and reduce it to different products, including nitrite (NO_2), nitrogen dioxide, nitrous oxide (N_2O), and nitrogen gas (N_2). The process is also known as denitrification. Fig. 10.8 illustrates the denitrification process and its impact on MIC.

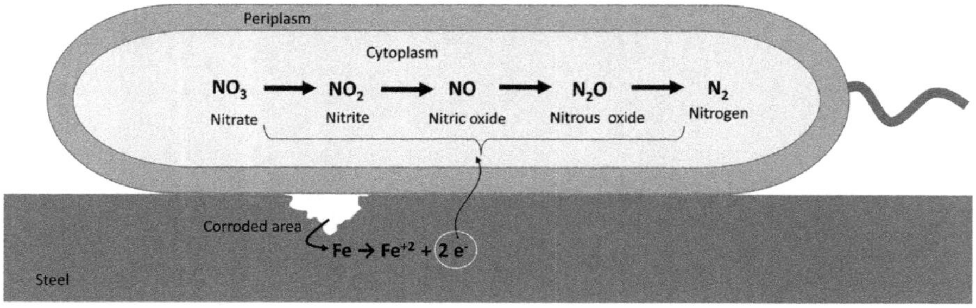

FIG. 10.8

Effect of NRB on steel corrosion.

Nitrate-reducing bacteria are mostly heterotrophic and often facultative anaerobic, with the ability to switch between oxygen and nitrate respiration depending on the environmental conditions [30].

Like SRB and methanogens, electrogenic nitrate-reducing bacteria (NRB) also cause MIC, as they are capable of utilizing electrons from steel for nitrate reduction [31], as illustrated in Fig. 10.8. The theory that explains MIC caused by NRB is known as biocatalytic cathodic nitrate reduction (BCNR) [31].

Nitrate-reducing bacteria are very common in use in oil and gas fields as they are basically used to combat reservoir souring generated by sulfate-reducing bacteria, by injecting nitrates in the reservoir, which promotes respiration by NRB to suppress sulfate respiration by SRB. However, as mentioned earlier, nitrate-reducing bacteria can initiate MIC itself or they can be utilized by SRB.

10.2.8 Other types of bacteria

Different types of bacteria other than those already mentioned can exist and affect production operations. Acetogenic microorganisms utilizing H_2 and CO_2 to anaerobically produce acetate have been reported. Some acetogens can also use different C_1 compounds, amino acids, and sugars [32]. Acetogens were found to exaggerate corrosion problems when they cooperate with methanogens.

Slime-forming bacteria are known to exist in the presence or absence of oxygen and produce the extracellular mats of high-density slime that cover, partially or fully, solid surfaces. This slime layer can cause flow restrictions and plugging in the pipelines. Slime formers protect the other aerobic and anaerobic microorganisms under the slime layer, maintaining their syntrophic living, and also remove the metal ions that are derived from the metal surface [7]. Examples of slime formers are *Pseudomonas*, *Escherichia*, *Flavobacterium*, *Aerobacter*, *Clostridium*, *Bacillus*, *Desulfovibrio*, and *Desulfotomaculum*.

Table 10.1 summarizes the most common organisms that can be found in oil and gas fields [33].

10.3 Biomineralization

Biomineralization is known generally as the process by which living organisms produce minerals [34]. Shells, bones, and teeth are examples of biologically produced materials [35].

The synthesis of minerals by prokaryotes is generally classified into two classes: biologically controlled mineralization (BCM) and biologically induced mineralization (BIM) [36,37]. BCM involves

Table 10.1 Common types of bacteria involved in MIC and biofouling and their essential requirements.

Group	Example of species	pH	Temperature (K)	Nutrient	End product
SRB (anaerobic)	*Desulfovibrio* *Desulfoto-maculum* *Desulfomonas*	4–9.5 6–8	298.15–333.15 283.15–313.15 (some at 319.15–347.15) 283.15–313.15	Organic compounds, hydrocarbons, alcohols, aromatic compounds lactate, and acetate or H_2. SO_4^{2-}, elemental sulfur S^0, sulfide and thiosulfate ($S_2O_3^{2-}$)	H_2S, HS^-, FeS
APB (facultative)	*Clostridium aceticum*	<7		Organic compounds, hydrocarbons, O_2	Organic acids, e.g., formic acid and acetic acid, CO_2
SOB (aerobic)	*Acidithiobacillus thiooxidans*	0.5–8	293.15–323.15	Sulfide, sulfite, thiosulfate and elemental sulfur Organic compounds, CO_2, O_2	Sulfuric acid (H_2SO_4), S^0
IRB (facultative)	*Pseudomonas* sp.	4–9	294.15–313.15	Insoluble ferric iron (Fe^{3+}) O_2, NO_3^-	Soluble ferrous iron (Fe^{2+})
Iron/MnOB (aerobic)	*Acidithiobacillus ferrooxidans* *Gallionella* *Leptothrix*	1–7 7–10 6.5–9	283.15–313.15 294.15–313.15 283.15–308.15	Ferrous iron (Fe^{2+}) Mn^{2+}	Ferric iron (Fe^{3+}), Mn^{4+}
NRB (facultative)	*Azonexus*	7–8	288.15–298.15	Organic compounds, NO_3^-, O_2	NO_2^-, N_2O, NO, N_2
Methanogens (anaerobic)	*Methermicoccus*	5–6	310.15–358.15	Organic compounds, CO_2 (or soluble CO_3^{2-}. HCO_3^-, H_2CO_3) or H_2	Methane (CH_4), CO

Reproduced *with permission from A. Ibrahim, K. Hawboldt, C. Bottaro, F. Khan, Review and analysis of microbiologically influenced corrosion: the chemical environment in oil and gas facilities, Corros. Eng. Sci. Technol. (2018), https://doi.org/10.1080/1478422X.2018.1511326. Copyright (2017), Taylor and Francis.*

minerals that are directly synthesized at a specific location within or on the cell and only under certain conditions, while in BIM the minerals are formed extracellularly as a result of metabolic activity of the organism [34]. Broadly speaking, the BIM type is the dominant process among bacteria and it is the one that leads to the flow assurance issues in the petroleum industry.

Biomineralization is driven by minor changes in the media, such as the introduction of biologically produced metabolic end products (like S^{2-}, OH^-, CO_2, CO_3^{2-}, SO_4^{2-}, ..., etc.), cations released from cells, or the development of a charged surface that can induce the nucleation and growth of minerals with crystal habits similar to those produced by precipitation from regular inorganic solutions [38].

Minerals that form by BIM processes generally nucleate and grow extracellularly when the metabolic byproducts react with ions or compounds in the surrounding environment, leading to the deposition of mineral particles. The minerals formed by BIM in this case can share similar or even indistinguishable crystallochemical features with minerals that are formed by inorganic reactions. Thus BIM is a presumably unintended and uncontrolled consequence of metabolic activities. However, bacterial surfaces or polymeric materials (exopolymers) secreted by bacteria, such as slimes, sheaths, or biofilms, can act as important sites for the adsorption of ions and mineral nucleation [39].

As discussed in Chapter 5, Mineral Scales in Oil and Gas Fields," a number of mineral scales are formed due to various bacterial activities, e.g., reservoir souring, microbiologically influenced corrosion, and other microbial activities and metabolites. Fig. 10.9 illustrates some of the mineral scales that are formed due to microbial activities.

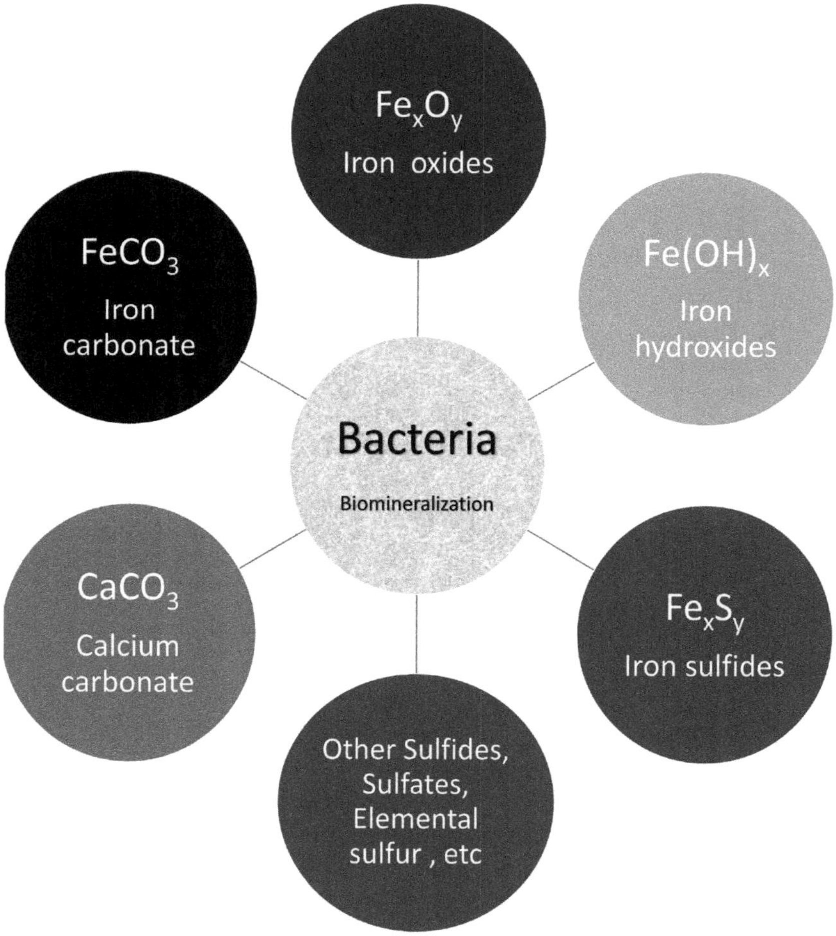

FIG. 10.9

Common mineral scales that can be formed by bacteria in oil and gas fields.

The biomineralization of these compounds in oil and gas operations can be attributed to two basic interactions of the bacteria metabolic byproducts: interactions with the metallic constituents of the system (biocorrosion), and interactions with ions in the flowing fluids whereby they precipitate when they exceed their solubility limit (bioprecipitation or bioscaling).

10.3.1 Biocorrosion

Biocorrosion or microbiologically influenced corrosion (MIC) refers to corrosion caused by the presence and activities of microorganisms. Microorganisms usually influence the kinetics of corrosion by accelerating the corrosion reactions or shifting the corrosion mechanisms.

MIC causes catastrophic damage in downhole and topsides facilities in oil and gas fields; it also damages pipeline systems, storage terminals, and refinery fields. MIC is believed to be responsible for greater than 20% of the total corrosion cost in the oil and gas industry [17]. The products of such biocorrosion are iron compounds (oxides, hydroxides, sulfides, and carbonate) that deposit in the system, causing flow assurance issues. Fig. 10.10 shows the MIC attack on pipeline walls and clogging of the pipeline by the biofoulants [7].

10.3.1.1 Mechanisms of biocorrosion

Despite the great number of studies on tracing and understanding MIC, there is no universal mechanism underpinning MIC. Instead of a universal mechanism, several mechanisms whereby microorganisms affect the rates of corrosion have been reported. Such effects of microorganisms on the corrosion rate stem from the microbial modifications to the chemical environment near metal surfaces [40–42]. Therefore MIC cases require a thorough and rigorous approach to each case individually and an assessment of the role of microorganisms in a specific medium is possible only by analyzing and simulating the corrosion parameters found in the field [43].

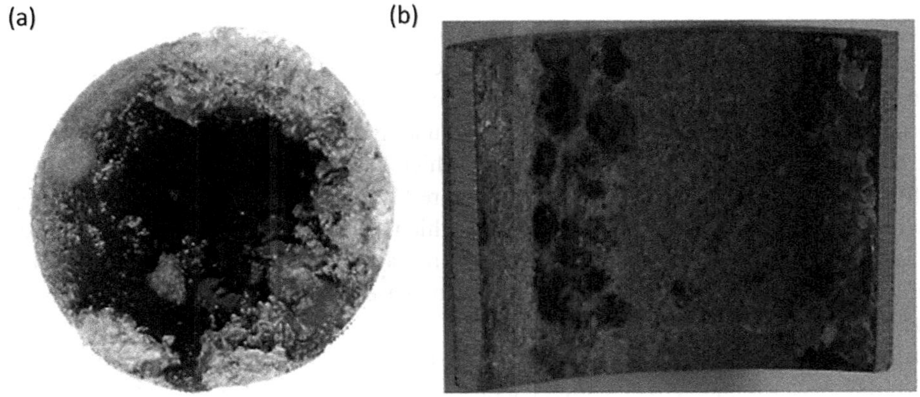

FIG. 10.10

Effect of MIC on pipeline: (A) clogging of pipeline by biofoulants, (B) localized attacks on pipeline walls.

Reproduced *with permission from J. Telegdi, A. Shaban, L. Trif, Microbiologically influenced corrosion (MIC), in: Trends in Oil and Gas Corrosion Research and Technologies: Production and Transmission, Elsevier, 2017, pp. 191–214, https://doi.org/10.1016/B978-0-08-101105-8.00008-5. Copyright (2017), Elsevier.*

Despite the lack of consensus on the MIC mechanism, when studying MIC one can find five basic mechanisms that are always mentioned [40–42]:

- Biofilms create oxygen heterogeneities on the metal surfaces.
- Biofilm matrix increases the mass transport resistance near the metal surface, thus changing the kinetics of the corrosion process.
- Metabolic reactions in biofilms generate corrosive substances.
- Metabolic reactions in biofilm generate substances that serve as cathodic reactants.
- A complex consortia of bacteria.

The first mechanism suggests that when biofilm is developed it accelerates corrosion as it creates a differential aeration cell where areas covered with biofilm exhibit lowered oxygen concentration and become anodic, while those with less biofilm exhibit higher oxygen concentration and become cathodic.

The second mechanism presumes that the EPS (extracellular polymeric substances) layer of the biofilm initiates and accelerates localized corrosion in many ways: first, by crosslinking with the corroding metal ions, which affects the equilibrium between the corroding metal and its ions and, second, by hydrolyzing the metal ions by water to form metal hydroxides and producing H+, which decreases the local pH. Third, the EPS layers create a good anaerobic environment for anaerobic microorganisms, and also create a differential aeration cell as indicated in the first mechanism. Therefore the treatment in this case must involve removing the biofilms, besides killing the microorganisms, to stop MIC [40–42].

The third mechanism is the classical picture of MIC in which the microorganisms secrete different products (H_2S, organic acids) that initiate or accelerate corrosion. The way these metabolic products cause corrosion is a matter of controversy and encompasses different mechanisms. For example, MIC by SRB bacteria involves different approaches, including cathodic depolarization by hydrogenase enzyme, anodic depolarization, a three-stage mechanism proposed by de Romero [44], and sulfide-induced stress-corrosion cracking. Enning and Garrelfs [6] and Enning et al. [45] defined two fundamentally different SRB corrosion mechanisms: chemical MIC or CMIC, which was used to describe the corrosive effects of biogenic H_2S, and electrical MIC or EMIC, which is only observed in the presence of specialized lithotrophic SRB, capable of accepting electrons directly from iron. The EMIC mechanism is also known as the biocatalytic cathodic sulfate reduction (BCSR) theory [46]. The corrosion products according to this third approach include iron hydroxides and iron sulfides that in the presence of oxygen convert into elemental sulfur, which accelerates the corrosion even more.

The fourth mechanism refers to ennoblement, where MIC changes the electrochemical properties of metal by increasing the metals corrosion potential, which accelerates corrosion. An example of ennoblement is when IOB in biofilms oxidize dissolved iron and manganese to form mineral-biopolymer encrustations, which work as efficient cathodes and increase cathodic current density of stainless steel by two to three orders of magnitude [40–42].

The fifth mechanism is the complex nature of the microbial community that can grow in a biofilm where different bacteria interact/cooperate, resulting in aggravating corrosion: for example, the interactions between IRB and SRB. Another example occurs when some bacteria syntrophically live on each other (crossfeeding between the bacteria) like the relationship between methanogens and SRB, which accelerates the corrosion and leads to the formation of iron sulfides and iron carbonate [47]. Such a syntrophic relationship allows some strains of SRB to survive and cause MIC, even in the absence of sulfate, based on fermenting organic compounds like lactate, ethanol, propionate, fumarate, and pyruvate, or by proton reduction to form H_2 gas in syntrophic cocultures with hydrogen and/or formate scavenging methanogens, thereby eliminating the inhibitory effect of these end products on the SRB [48].

10.3.1.2 Factors affecting biocorrosion

- **Fluids chemistry**

Crude oil and water chemistry include the constituents that the microorganisms can use as nutrients during their life cycle, and other constituents that can retard/inhibit the microorganism growth.

- Sulfate, carbon source (volatile fatty acids, other organic compounds) are necessary for SRB growth [49]. H_2S was found to have direct and reversible toxicity on SRB [50]. IRB require Fe^{3+}, while IOB require Fe^{2+}. Oxygen is required for obligate aerobes, while low or no oxygen conditions are required for anaerobes.
- Total dissolved solids (TDS): High TDS (or ionic strength) accelerates corrosion in general. Generally speaking, high salt concentration is used to limit bacterial growth [3]. However, different bacteria can live in solutions with different salt concentrations. SRB were reported to live in waters of all salinities, from near zero to saturation, depending on the genus. *Desulfotomaculum* are seldom found in waters containing $>$20,000 ppm NaCl; for waters of $>$20,000 ppm, the population is almost always *Desulfovibrio*. *Desulfovibrio* are salt tolerant and can grow in NaCl concentrations as high as 100,000 ppm; however, *Desulfovibrio vulgaris* lives in freshwater with NaCl concentrations $<$10,000 ppm [49]. *Desulfovermiculus halophilus* has recently been recovered from a Russian oil-field and was found to grow at NaCl concentrations up to 23% (230,000 ppm) [51].
- pH: Acidic solutions are basically corrosive, while neutral and basic solutions are noncorrosive. Besides solutions, pH supports the growth of bacteria. Most bacteria favors neutral conditions (neutrophils), however, this depends on the bacteria type. For example, SRB were found to grow in solutions with pH between 5 and 10 [52]. Iron-oxidizing bacteria typically live in acidic, aerobic environments, and they grow poorly at pH values greater than 4 (low pH is critical in keeping Fe(II) from being spontaneously oxidized to Fe(III)) [53,54]. Methanogens generally operate at a pH range of 6 to 8 [55]. If the pH drops below 5.5, activity is significantly hindered and can stop completely and irreversibly.

- **Temperature**

Increasing the temperature usually leads to increasing corrosion rates. The ambient temperature encountered in oil-handling facilities is also suitable for microbial growth of mesophilic bacteria. Different bacteria types grow at different conditions of temperature. Different strains of thermophilic bacteria were isolated from oil and gas reservoirs and surface facilities with temperatures reported up to 130°C [56]. In addition, hyperthermophilic bacteria and archaea were also detected in oil and gas fields [57].

- **Pressure**

The effect of pressure on corrosion depends on the partial pressures of acid gases (e.g., H_2S and CO_2), which may influence the dissolution of corrosive species into the solution. SRB can tolerate pressures of up to 500 atm [3]. Many microorganisms can withstand a 100-fold or greater variation in pressure

[5]. Li et al. [58] revealed that the application of negative pressure may slow down the microbial growth rate and inhibit biofilm development.

- **Flow**

Flow usually affects MIC in different ways: by removing the protective layer of corrosion products, and by affecting the biofilm formation and the mass transfer of nutrients required for microorganism growth.

- **System design components**

System design features such as dead ends, dead legs, and other stagnant locations are more vulnerable to MIC and biofouling than other locations. Engineering designs must incorporate drains, eliminate traps for stagnant water, reduce the number of dead legs, bends, and elbows, to reduce the system vulnerability for MIC [5]. Welding spots are also vulnerable to MIC and require attention from operators [3].

10.3.2 Bioprecipitation

Biocorrosion leads to precipitation of various types of iron compounds. But consider what happens when the microbiological metabolic end products (CO_3^{2-} or S^{2-}) are transported into the bulk solution, where they interact with counter ions (Ca^{2+}, Zn^{2+}, or Pb^{2+}) and finally precipitate when they exceed their solubility limit. This is known as bioprecipitation.

Bioprecipitation is usually pictured by the operators and researchers through the secretion of scaling anion, e.g., S^{2-}, CO_3^{2-}, OH^-. However, delving deeper into the bacterial interactions you can find different mechanisms by which bacteria lead to mineral precipitation, including the following:

- Secretion of scaling anions (S^{2-}, CO_3^{2-}, OH^-, SO_4^{2-}).
- Secretion, accumulation, and concentration of metal ions on the surface of bacteria.
- Working as nucleating sites for precipitation.
- Increasing fluid viscosity in the vicinity of the biofilm.

For example, CO_3^{2-} or SO_4^{2-} ions can be produced as byproducts of multiple bacterial activity, which, in the high metal ions produced waters, can easily form carbonate and sulfate deposits. Aside from iron sulfides scale that forms by biocorrosion, zinc and lead sulfides scale usually forms when they exceed their supersaturating limit due to increasing sulfide ion concentration from SRB activity.

Bacteria also affect the metal ion concentration leading to scale formation. One way is by changing metal ion valence to a state that favors precipitation: a common example is the production of Fe^{3+} from the oxidation of Fe^{2+} by IOB, and then Fe^{3+} precipitates as hydroxide or oxides. Another way is by binding and accumulating metal ions into their surfaces, since bacteria have large surface area to volume ratio with cell components anionic at pH between 5 and 8, which gives them the ability to sequester and chelate metal ions into their surfaces. Biofilms are also considered ideal cation scavengers [59].

Furthermore, the bacteria also work as nucleating agents/sites for precipitation due to their peculiar characteristics [60]:

- Bacterial cells are very small.
- They have the highest surface area-to-volume ratio.

- The presence of charged chemical groups on their cell surface.
- From the polysaccharides in bacteria and algae cell walls, biofilms can work as templates for the formation of the mineral phase of organic matter [61].

The nucleation on the bacterial surface is a stepwise process. In the first step, metal ions present in the aqueous surroundings of the cell interact and bind with charged groups in the surface cellular polymers, which lowers the total free energy of the system and initiates further metal deposition. The next step involves precipitate formation at the nucleation site between metal ions and excess counterions from the fluid phase. The formed mineral phases depend on microenvironmental geochemistry, and negatively charged counterions (e.g., sulfate, phosphate, carbonate, sulfide, or silicate ions) [60,62]. Fig. 10.11 depicts precipitate nucleation on a microbial cell surface [62].

Another mechanism of bioprecipitation involves changing the viscosity of the surrounding fluids by the chemicals secreted for biofilm and the other chemicals used during the bacterial life cycle.

10.3.3 Carbonates biomineralization

Microbiologically induced carbonate precipitation (MICP) is a well-known process used by various types of bacteria in different natural environments. Table 5.4 in Chapter 5 summarized these interactions to produce calcium carbonates. Microorganisms induce carbonate precipitation through different metabolic pathways, such as photosynthesis, ureolysis, ammonification, denitrification, sulfate reduction, anaerobic sulfide oxidation, and methane oxidation, by increasing either the pH or the dissolved inorganic carbon. Furthermore, most microbial cells provide nucleation sites for carbonate formation [62]. Fig. 10.12 illustrates these different processes.

It is worth mentioning that SRB, a well-known and notorious oilfield bug, can also lead to the formation of carbonates in several ways. First, by changing the pH, it affects the saturation index and thus directly the precipitation of carbonate minerals. Second, when SRB use low molecular weight organic acids (e.g., lactate, acetate) as electron donors for growth, the availability of free calcium ions may increase due to the removal of carboxylic acids binding calcium. Third, by removing sulfate ions from solution, SRB can alter the kinetic inhibition of dolomite formation. On top of all that, the SRB cells can be nucleating sites for calcium carbonate precipitation. In addition, the SRB EPS was found to have high buffer capacity and high Ca binding capacity [63].

The biomineralization of $CaCO_3$ has been reported to occur in different environments such as seawater, freshwater, wastewater, and soil by a large variety of bacteria, including cyanobacteria, sulfate-reducing bacteria, nitrate-reducing bacteria, methanogens, heterotrophs, and urea-degrading bacteria [64,65].

A recent research trend is growing that deals with biologically induced calcium carbonate deposition as a way of enhancing oil recovery. The technique is based on the idea of simple plugging of the high-permeability “thief zones” of oil reservoirs, which allows better sweeping of the hydrocarbons [66,67].

Iron carbonate is a direct product of biocorrosion from many types of bacteria. The actions of SRB lead to the formation of a mixture of iron sulfides and iron carbonate, where in the case of EMIC the ratio is 1 (sulfides): 3 (carbonates), while in the case of CMIC they can form at different ratios, as per the following equations [6]:

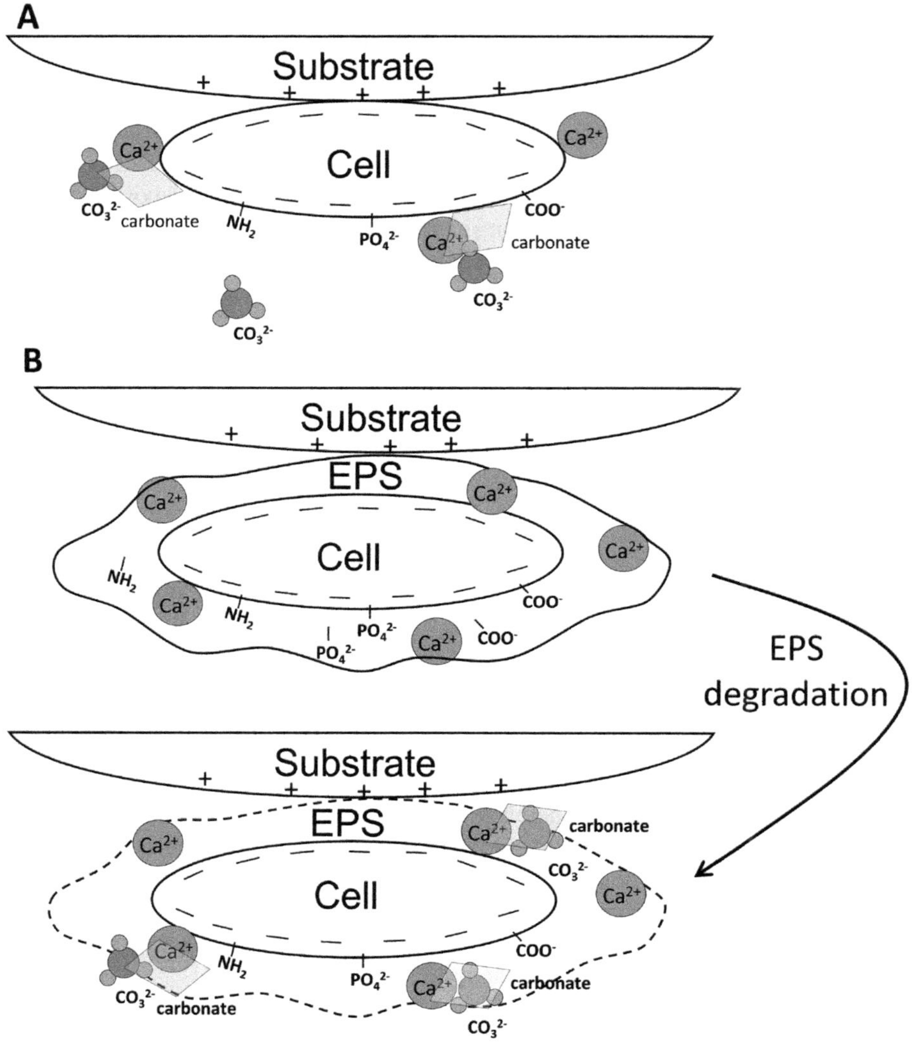

FIG. 10.11

Nucleation of carbonates on microbial surfaces: (A) the cell wall with negatively charged functional groups adsorbs Ca^{2+}, and thus carbonates precipitate on the cell surface when carbonate ions are available; (B) EPS with functional groups entraps Ca^{2+} ions, and then when EPS is degraded, high concentration of Ca^{2+} is released locally, which in presence of carbonate ions results in the precipitation of calcium carbonate.

From *T. Zhu, M. Dittrich, Carbonate precipitation through microbial activities in natural environment, and their potential in biotechnology: a review, Front. Bioeng. Biotechnol. 2016, https://doi.org/10.3389/fbioe.2016.00004.*

$$\text{EMIC}\quad 4Fe^{o} + SO_4^{2-} + 3HCO_3^- + 5H^+ \rightleftharpoons FeS + 3FeCO_3 + 4H_2O \tag{10.1}$$

$$\text{CMIC}\quad 3CH_2O + 5Fe^{o} + 2SO_4^{2-} + H^+ \rightarrow 3HCO_3^- + 2FeS + 2H_2O \tag{10.2}$$

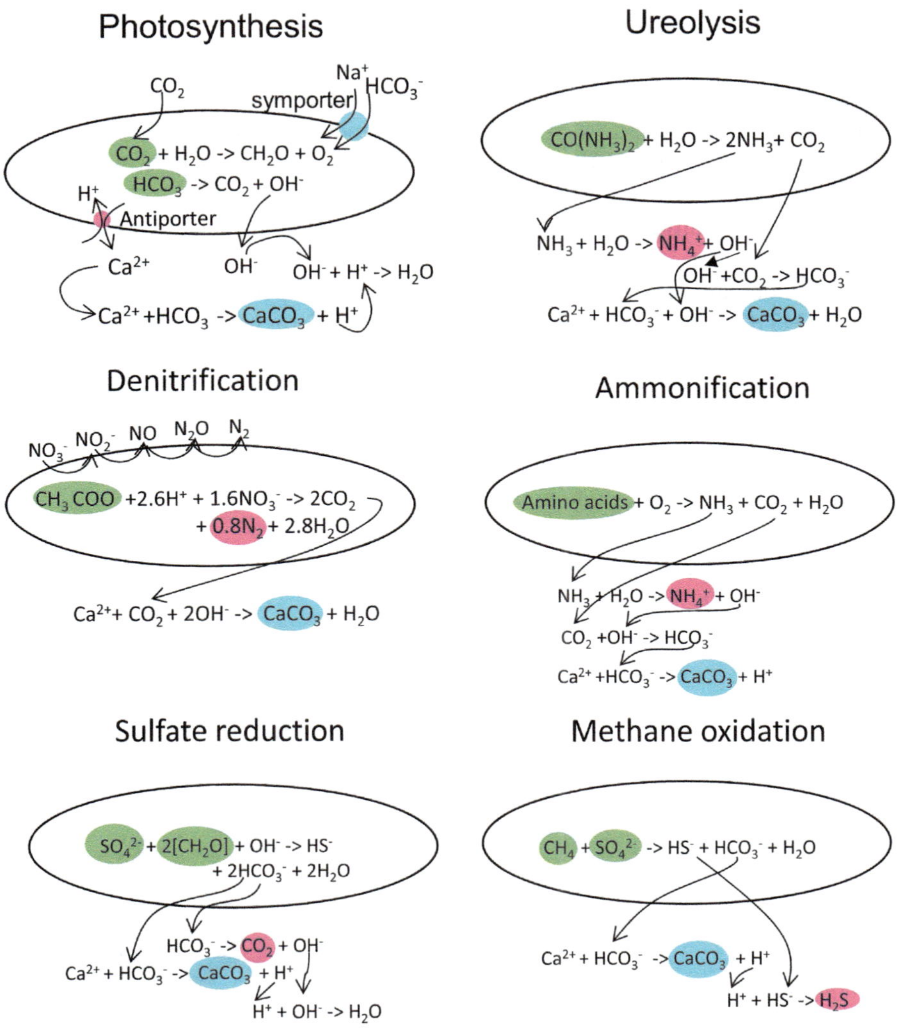

FIG. 10.12

Microbial carbonate precipitation induced by different metabolisms.

From *T. Zhu, M. Dittrich, Carbonate precipitation through microbial activities in natural environment, and their potential in biotechnology: a review, Front. Bioeng. Biotechnol. (2016), https://doi.org/10.3389/fbioe.2016.00004.*

Iron carbonate is also formed due to methanogenic archaea activities, including biocorrosion and syntrophic growth with SRB [47].

$$4Fe^o + 5HCO_3^- + 2H_2O \rightarrow CH_4 + 4FeCO_3 + 5OH^- \tag{10.3}$$

Solid calcium, magnesium, and iron carbonates were reported in a water injection system in the Umm Shaif field in the UAE due to carbonate-generating bacteria (CGB) in association with SRB [68].

10.3.4 Sulfides biomineralization

Sulfide solids are the best-known biofouling minerals to oil and gas operators. They are also known to be produced as a result of sulfate reduction by SRB bacteria. Sulfide minerals in nature are largely formed by microorganisms, as up to 97% of the sulfide produced on earth is ascribed to the activity of sulfate-reducing prokaryotes (SRP) in low-temperature environments [69], while the remaining 3% is produced in volcanoes and deep-sea hydrothermal vents [70].

Different types of iron sulfides were reported to form due to microbial activities in oil and gas fields [6,39,71].

While most sulfides are formed extracellularly (BIM), some reports have shown that some magnetotactic bacteria (MTB) of the *Desulfovibrio magneticus* strain form intracellular iron sulfides (mostly Fe_3S_4) ranging in size from 35 to 120 nm, contained in bilayer membranes within an organelle called the magnetosome [72].

Two mechanisms of extracellular BIM have been established, as depicted in Fig. 10.13 [73,74]:

1. SRP, and possibly other microorganisms, produce free sulfide, which reacts abiotically with dissolved metal ions to form sulfide minerals.

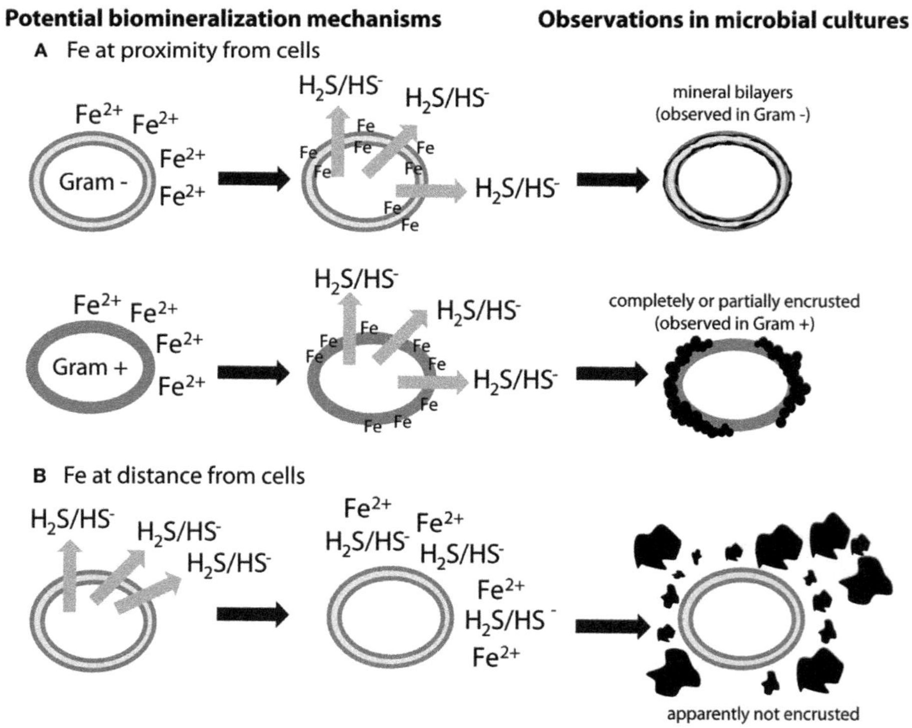

FIG. 10.13

Different mechanisms of biomineralization patterns observed in sulfate-reducing microorganisms.

From *A. Picard, A. Gartman, P.R. Girguis, What do we really know about the role of microorganisms in iron sulfide mineral formation?, Front. Earth Sci. 4 (June) (2016), https://doi.org/10.3389/feart.2016.00068.*

2. The association of iron with their cell walls and EPS (by interacting with anionic cell surface charges), and then reacting with sulfide on these surfaces (biogenic H_2S) to form metal sulfide minerals, which are believed to be amorphous FeS, which then through mineral transformations cause the production of other FeS types and eventually pyrite [74,75].

The minerals precipitated in the presence of SRP showed different characteristics from the abiotic ones. The mackinawite particles formed in the presence of SRP were larger and they tended to aggregate more than the nanocrystalline abiotic mackinawite. Live SRM were found to accelerate the transformation of mackinawite to greigite. The cellular surfaces of live and dead SRM and EPS were found to act as templates for the nucleation of iron sulfide minerals [76]. However, Rickard [77,78] previously concluded that extracellular, biogenic iron sulfide minerals could not be distinguished from abiogenic (inorganic) minerals.

Besides iron sulfides, zinc and lead sulfides are also a product of interaction of biogenic sulfide with Zn and Pb ions present in the formation water. Zinc and lead sulfides are known for their low solubility and quick plugging actions inside the process tubulars [79].

10.3.5 Oxides biomineralization

The microbial precipitation of ferrihydrite is very common in nature [38]. Different strains of bacteria produce iron oxides, hydroxides, or oxyhydroxides at metabolic product. Ferrihydrite also serves as a precursor to more stable iron oxides, such as goethite and hematite [80].

Iron hydroxide formation by bacteria can occur either passively or actively. The passive pathway occurs by oxidation, and hydrolysis of cell-bound iron or the binding of cationic colloidal species ($Fe(H_2O)_5(OH)_5(OH)^{2+}$) can induce the transformation to insoluble hydroxide forms. Another way of passively precipitating these hydroxides is by working as a template for nucleating sites for the mineral formation, whether the cells are alive or dead; in fact, many cases reported biological remains and biomasses within the iron minerals [38,81].

The active means of iron hydroxide formation is by a direct effect of bacteria that leads to precipitation. Iron-oxidizing bacteria (FeOB) oxidize FeII ions as an energy source, leading to iron hydroxide precipitation [38]. Because very little energy is generated in the oxidation of FeII to FeIII ions, these bacteria must oxidize large quantities of iron in order to grow. Consequently, even a very small number of bacteria can be responsible for precipitating significant amounts of iron [38,82].

Magnetite is known to form biologically controlled (intracellular) and biologically induced (extracellular). Magnetite is formed intracellularly by magnetotactic bacteria, which use it to orient and swim along geomagnetic field lines to get to optimum oxygen levels [83]. Extracellularly produced magnetite can be produced by different bacteria. Iron-reducing bacteria are known to produce magnetite as a byproduct during the oxidation of organic compounds coupled to the reduction of poorly crystalline, synthetic FeIII oxide [84]. In addition, magnetite and other iron oxide biomineralization was reported by SRB bacteria [85] and also by denitrifying bacteria [86].

10.3.6 Sulfates biomineralizatioon

Sulfate biomineralization is based on the production of sulfate ion by bacterial action, or by the conversion of bacterial metabolic products.

Sulfate is conventionally produced by the action of sulfur-oxidizing bacteria, which oxidizes reduced forms of sulfur (S^{-2}, H_2S, S_0, and $S_2O_3^{-2}$) into sulfate. In the presence of produced water that is rich in metal ions, different types of sulfates can precipitate. Nitrate stimulates nitrate-reducing, sulfide-oxidizing bacteria (NR-SOB) and heterotrophic nitrate-reducing bacteria (hNRB) that compete with SRB, which uses sulfate as the main nutrient, thus keeping sulfate ion concentrations high enough to cause precipitation [87]. Ansari et al. [64] reported precipitation of simultaneous biomineralization of $CaCO_3$ and $CaSO_4$ in the presence of three different microbial communities from freshwater, wastewater, and seawater.

Another way of producing sulfate is by introducing chemicals that are used to mitigate the bacteria and then they convert into sulfate chemically (by decomposition or oxidation) or biologically by the action of bacteria. Examples are THPS biocides that introduce sulfate ions to the produced fluids, and sulfite oxygen scavengers that are easily converted to sulfate by bacteria with oxidation.

Barite precipitation by bacteria is very common in marine environments. Ba ions can be accumulated on marine bacterial biofilms and particularly EPS substances, providing adequate conditions for barite precipitation. Bacteria also work as templates or nucleating sites for the precipitation [88].

10.3.7 Factors affecting biomineralization

The factors discussed here are the general factors that affect biomineralization, which are added to the list of factors discussed earlier that affect biocorrosion.

- **Temperature**

The solubility of some minerals increases with temperature, such as sulfides and sulfates, while others decrease with temperature, like the inverse solubility carbonates. Another example of the temperature effect on biomineralization is affecting the bacterial activity.

- **pH**

Like temperature, pH affects the solubility of the minerals. Carbonates, sulfides, and oxides dissolve in acidic conditions and precipitate in alkaline conditions, while sulfates are less sensitive to the pH effect. The formation and distribution of magnetite and siderite phases during microbial Fe(III) reduction is, in part, a function of pH and oxidation-reduction potential (Eh) [89].

- **Organic matrix**

The organic matrix from microorganisms, composed of proteins, phospholipids, collagen, carbohydrates, and so on, acts as a mediator for the mineralization of biological systems. These organic matrices, especially proteins, play a significant role in biomineralization, controlling the nucleation, growth, crystalline form, and trend of inorganic crystals. This process is also called molecular recognition [90,91].

10.4 Biomass fouling

The other type of biofouling is fouling by the biomasses or organic materials produced by the bacteria during their life cycle. They are known for their sticky nature and quick impact, and diverse bacteria are involved.

10.4.1 Biofilm and slim layers

Microorganisms can live either in free-floating form, called planktonic, or attached to some sort of surface, called sessile.

During the course of their growth, the bacteria produce various types of organic materials that they use to adapt and survive under the different environment conditions they exist in. One of the common materials is the glycocalyx, or what is known as the extracellular polymeric substances or EPS, which are used by microorganisms to adapt to the surrounding environment in different forms, like capsules, slime layer, and biofilm [92,93].

EPS produced by the microorganisms can exist as tightly bound (capsular), loosely adhered (slime type) to cells, or as free dissolved matter (component of dissolved organic matter, DOM). The capsule has a more organized polymeric structure, more densely packed, less diffusible, and more tightly bound to the cell than slimes [92].

The biofilm, on the other hand, is a community of microorganisms that are irreversibly attached to a surface and encased in a self-produced EPS with compact structures, and having increased resistance to chemical and environmental effects [94]. It is considered an adaptation and survival strategy employed by the bacteria. Fig. 10.14 depicts the biofilm, slime layer, and capsule.

Biofilms and slime layers are the root cause of many problems in industrial systems, due to their role in biocorrosion and their biomass growing in size and accumulating other corrosion products, causing plugging in pipelines and heat transfer equipment.

10.4.2 Stages of biofilm formation

Studying the biofilm formation and its properties is crucial for MIC and biofouling research. The biofilm usually forms in five main steps: conditioning film formation, adhesion, attachment, growth, and then detachment and dispersal to other locations (Fig. 10.15).

- **Stage 1: Conditioning film formation**

Biofilm formation begins with adsorption of macromolecules (proteins, polysaccharides, and humic acids) and smaller molecules (fatty acids and lipids) at surfaces, to form what is known as a

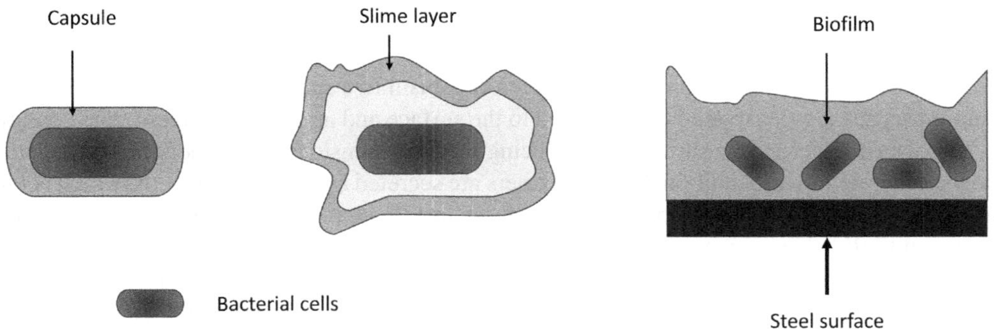

FIG. 10.14

The capsule, slime layer, and biofilm.

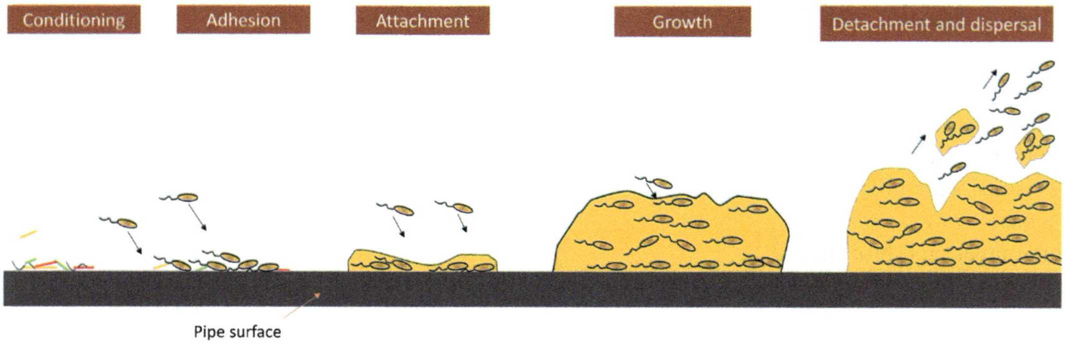

FIG. 10.15

Biofilm formation stages.

conditioning film. This film tends to change physiochemical properties of the interface, such as surface hydrophobicity and electrical charge [95].

- **Stage 2: Adhesion**

In this step, the microorganisms approach the conditioned surface. Planktonic bacteria in the bulk of the solution approach the conditioning film and become sessile bacteria [3]. The bacteria can reach the surface if they are motile or with the aid of flowing fluids. The physicochemical properties of the bacterial cell surface (e.g., cell-surface hydrophobicity and charge, cell size, presence of flagella and pili) and the pipe surface (chemical composition, surface roughness, surface defects, and coverage by oxide films or organic coatings) in addition to the composition and strength of the aqueous medium and the hydraulic flow regime are crucial in determining whether the cell can attach or not during the initial attachment step [5,96].

The adhesion of bacterial cells to the pipe surface can be explained by the DLVO theory, as discussed in Chapter 4. Bacteria possess negative charge and have colloidal dimensions at regular natural waters pH levels, and yet they still approach a surface by van der Waals forces [96,97]. Desorption may occur during this stage as a result of the release of reversibly adsorbed cells, due to fluid shear forces or a hostile environment [96,97].

- **Stage 3: Attachment**

In this step the irreversible attachment of the microorganism to a substratum occurs [96]. During this step more bacteria are transported and supplied to the surface and attach themselves to the conditional film [7]. Bacteria adhere to the surface by producing surface extensions or appendages like flagella and exopolysaccharide fibrils. Additional EPS polymers are secreted by the attached cells, which act to cement the cells to the surface and stabilize the colony against fluctuations or effects from the surrounding environment [7,96]. The EPS produced by bacteria provides the structural framework for the biofilm, besides exhibiting other physical, chemical, and electrical responses to environmental stimuli [92,93].

Further adhesion as well as EPS production and microbial growth result in a thick, gelatinous biofilm that mainly consists of water, organisms of different species, and other components from the environment [7].

- **Stage 4: Cell growth**

Once they adsorb to the surface, the attached bacteria use the macromolecules for growth and reproduction. The later steps in biofilm development involve microbial cell-cell interactions, between the main colonizing bacteria and other microorganisms with different nutritional needs, knowing that different bacteria types can exist in the biofilm, living syntrophically together. The different interactions, growth properties, and metabolic activity of the microorganisms in the biofilm determine its properties, which ultimately lead to mature biofilms. In a mature biofilm, the EPS contain sufficient amounts of polysaccharides, proteins, nucleic acids, and lipids to maintain the structural integrity of the biofilm and furnish suitable conditions for bacterial growth. Thus the biofilm continues to develop until it reaches a plateau stage with a certain thickness and metabolic capacity [96]. The diverse growth rates of the biofilm bacteria and the accumulation of solids from the environment also contribute to increase the biofilm thickness.

- **Stage 5: Biofilm detachment and dispersal**

The last step occurs when the biofilm bacteria detach, disperse, and colonize new locations to achieve propagation. Some microorganisms differentiate and produce spores that are spread to new locations during their life cycle [96].

10.4.3 Properties of biofilms

- **Complexity of biofilms**

Biofilms are remarkably heterogeneous and widely diverse. They exhibit different internal structures and chemical, electrical, and other properties. Besides, they often contain a wide spectrum of microorganisms, as in naturally occurring biofilms (as opposed to those grown in a laboratory for experimental reasons) there are nearly always a large number of different kinds of microorganisms living together [98]. Fig. 10.16 shows an SEM image of mixed culture biofilm. Images of biofilms taken by confocal laser microscopy (CLM) showed that microorganisms in biofilms were aggregated in microcolonies, not uniformly distributed as previously thought [99].

This diversity in microorganisms and properties contributes to the characteristics of the biofilm as a whole that make it peculiar and different (e.g., hard to kill), compared to dealing with planktonic microorganisms [98].

- **Biofilm bacteria behavior**

The same bacteria type can behave dramatically differently in the biofilm, compared to when they are planktonic. For example, genetic expression is different in biofilm bacteria when compared to planktonic bacteria. In addition, the biochemical signaling molecules are thought to be more concentrated in the biofilm, sufficient to cause changes in cellular behavior compared to the planktonic state.

- **EPS**

Flemming et al. [100] defined biofilm as a "city of microbes," while the EPS represents the "house" of the biofilm cells.

The extracellular polymeric substances, or EPS, are biologically generated biopolymers in which biofilm microorganisms are embedded. They are generally composed of polysaccharides, a wide

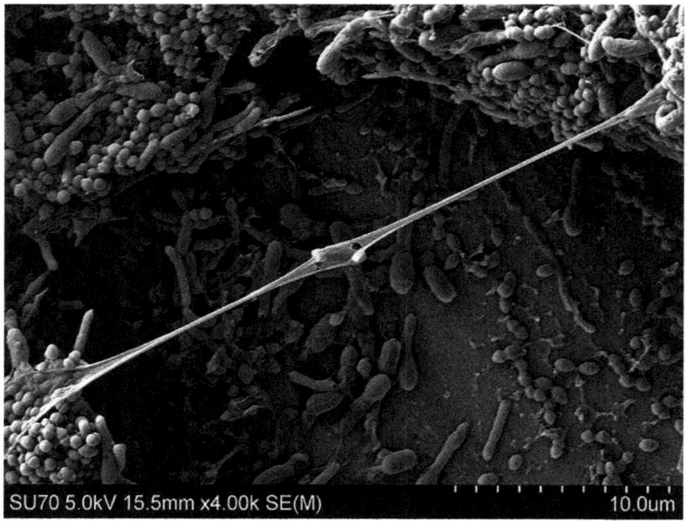

FIG. 10.16

Scanning electron micrograph of mixed-culture biofilm, demonstrating in detail a spatially heterogeneous arrangement of bacterial cells and extracellular polymeric substances.

From *https://commons.wikimedia.org/wiki/File:Mixed-culture_biofilm.jpg. Photo credit: Krzysztof A. Zacharski.*

variety of proteins, glycoproteins, glycolipids, and extracellular DNA (e-DNA) [100,101]. Cations like Ca^{2+} and Mg^{2+} have also been widely reported [92]. All the EPS biopolymers are very hydrated and form a matrix, which keeps the biofilm cells together and retains water [100,101]. The polysaccharides are high molecular weight polysaccharides (10 to 30kDa) and generally have heteropolymeric composition [92]. In the biofilm, the bacteria cells make <10% of the dry mass, while the matrix accounts for >90%, where EPS makes up to 50% to 90% (wt) of the organic matter in a biofilm [102]. The EPS composition is different from one bacteria to another. For example, EPS secreted by *Vibrio* sp. bacteria contains 51.1% carbohydrates, 1.8% proteins, and 14.9% uronic acid, while that of *Marinobacter* sp. is composed of 16.9% carbohydrates, 43.9% proteins, 17.6% uronic acid, and 2.7% sulfates [92].

The EPS exhibits different functionalities in the biofilm [103]:

- Responsible for adhesion to surfaces.
- Structural functions, e.g., scaffolding cells together, and maintaining the three-dimensional architecture of the biofilm.
- Protection functions: protects the bacterial cells against various stresses, such as antimicrobials, host immune systems, oxidation, and metallic cations.
- Cell communication and signaling functions, as the EPS inside the biofilm retains quorum sensing (QS) signaling molecules, extracellular enzymes, and metabolic products.

Thus the EPS affects different characteristics and parameters of the biofilm such as porosity, density, water content, charge, sorption properties, hydrophobicity, and mechanical stability, which ultimately affect and determine the immediate conditions of life of the living cells in the biofilm [104]. Some of these key functions and their associated components are listed in Table 10.2.

Table 10.2 EPS component functions.

Function of EPS in biofilm	EPS component evolved
Structural and protection	Polysaccharides, proteins
Adhesion	Polysaccharides, proteins, eDNA, amphiphilic compounds
Aggregation of bacteria cells	Polysaccharides, proteins, DNA
Sorption	Charged or hydrophobic polysaccharides, proteins
Retention of water	Hydrophilic polysaccharides and some proteins
Enzymatic activity	Proteins
Exchange of genetic information	DNA
Binding of enzymes	Polysaccharides and enzymes
Export of cell components	Membrane vesicles containing nucleic acids, enzymes, lipopolysaccharides, and phospholipids

Reproduced *with permission from H.C. Flemming, J. Wingender, The biofilm matrix, Nat. Rev. Microbiol. 8 (2010) 623–633. https://doi.org/10.1038/nrmicro2415. Copyright (2010), Springer Nature: Macmillan Publishers.*

- **Mechanical properties of biofilm**

The EPS matrix is a network that provides sufficient mechanical stability to maintain spatial arrangement for microorganisms over a longer period of time, and also determine their tendency to spread and colonize more spots. The forces that stabilize and keep the biofilm EPS matrix together include physicochemical interactions such as hydrogen bonds, van der Waals forces, and electrostatic interactions. Crosslinking between cations like Ca^{+2} and EPS material also increases its mechanical stability. These forces are illustrated in Fig. 10.17 [101,105].

While these individual forces are low, the gross overall binding forces are strong enough and can exceed that of covalent bonds. Therefore, in response to shear forces, biofilms first show characteristics of viscoelastic bodies, or they show viscoelasticity, in which to dissipate energy coming from external forces and to withstand external mechanical stress; when a breaking point is exceeded, they have properties of viscous liquids [106,107].

Consequently, when designing chemical cleaning or killing of bacteria in a biofilm, the designed chemical should overcome or at least weaken the EPS forces with the aid of the shear force from flowing fluids before reaching the bacteria to achieve the killing.

- **Permeability of biofilm**

The physiochemical properties of the biofilm matrix affect the exchange of nutrients, dissolved gases, molecules, and cells between the environment and the biofilm. The transport of solute or dissolved nutrients/waste and water depends on the structure and composition of the biofilm constituents. Due to the restrictions the biofilm matrix imposes on materials transport, concentration gradients between the biofilm and its surroundings arise, and thus diffusion is the dominant mechanism of materials transport of materials through the biofilm. Molecules can move by random walk through the porous matrix or physically interact with matrix components. In other structural arrangements where channels form throughout the biomass, flow is introduced and a combinatory effect of advection within the channels and diffusion within the microcolonies influences mass transport [108].

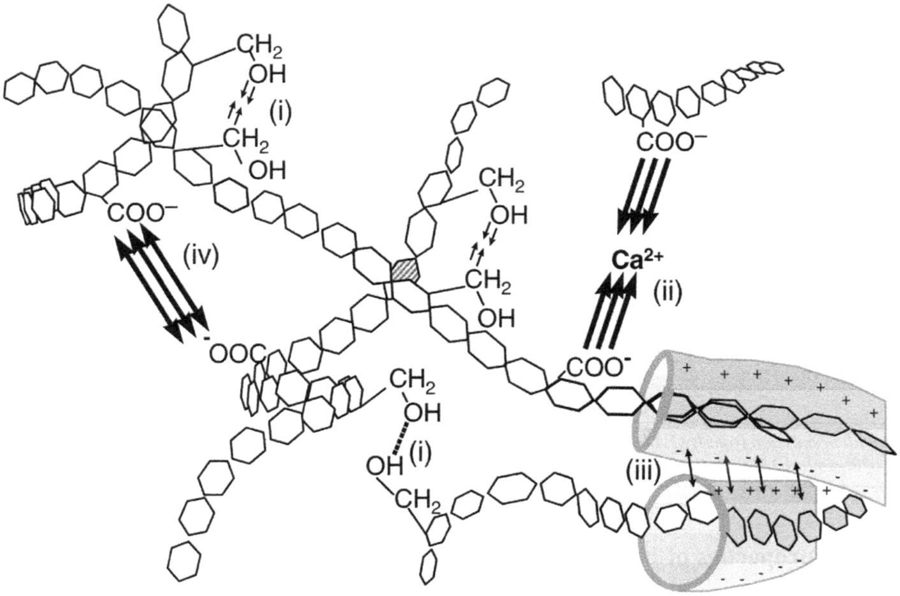

FIG. 10.17

Forces that keep the EPS matrix together: (i) hydrogen bonding, (ii) cations bridging, (iii) van der Waals forces, (iv) repulsive forces.

Reproduced *with permission from H.-C. Flemming, Why microorganisms live in biofilms and the problem of biofouling, in: Marine and Industrial Biofouling, Springer, Berlin, Heidelberg, 2008, pp. 3–12, https://doi.org/10.1007/978-3-540-69796-1_1. Copyright (2009), Springer Nature: Springer.*

10.4.4 Factors affecting biofilm formation

There are several factors that influence biofilm formation. The type of bacteria and the availability of nutrients within the biofilm are the most important factors affecting the growth and properties of the biofilm. The natural environments that most bacteria inhabit are typically complex and dynamic. The different interactions between the species (competing or syntrophic), their distribution, the biomass and metabolic products, contribute to the organization of multispecies biofilms, and also affect the different properties of the biofilm [109].

10.4.4.1 Effect of pH

The pH value affects the growth of different bacteria, which affects the development of the biofilm, as indicated in Section 10.3.1.2. Results showed that increasing pH correlated with an increase in biofilm production by all microorganisms analyzed. Significantly higher biofilm production was observed at pH 7.5 and 8.5 compared with pH 5.5 [110]. The thickness of SRB biofilm at pH 4 was found to be much lower than the thickness of specimens at higher pH [111].

10.4.4.2 Effect of temperature

Temperature affects cell physiology and growth, where an adequate temperature is required for the growth rate of each bacterial strain, and if not maintained, bacterial growth will decrease. Properties of EPS components, such as the viscosity of the polysaccharides, are affected by temperature [112].

Polysaccharides can transform from an ordered state at lower temperatures and in the presence of ions, to a disordered state at elevated temperature under low ionic environments. Thus the more uniform properties of polysaccharides at lower temperatures could stimulate biofilm formation [113].

10.4.4.3 Effect of system design

System design is a significant factor in bacterial activity and biofilm formation. System designs that involve stagnant locations and more welds are more vulnerable to biofilm growth. Stagnant locations are kinetically favored by the bacterial cells to attach and spread the cells. Welds give the bacterial cells the preferred conditions for biofilm growth.

10.4.4.4 Effect of surface nature

The adhesion to the substratum is dependent on the physicochemical properties of the surface, such as texture (rough or smooth), wettability, stiffness, topography, hydrophobicity, and surface charge [114,115].

Surfaces with imperfections (cracks, scratches, corrosion, and roughness) are more vulnerable to bacterial cell attachment and biofilm growth. While rough surfaces are readily colonized because shear forces are diminished and surface area is increased in rougher surfaces [109,114], some studies have also indicated that bacteria can colonize smooth surfaces equally [109,116].

The more hydrophobic cells adhere more strongly to hydrophobic surfaces, while hydrophilic cells strongly adhere to hydrophilic surfaces [117,118]. However, the heterogeneity of the microbial population should be considered, as it is expected to observe the presence of both hydrophilic and hydrophobic cells. Another factor is the behavior microorganisms show when they switch between hydrophobic and hydrophilic phenotypes in response to changes in environmental conditions (temperature, composition of nutrients, etc.) [119].

Studies have also demonstrated that microorganisms typically attach more rapidly to hydrophobic surfaces such as plastics than to hydrophilic glass or metals [109,114].

10.4.4.5 Effect of hydrodynamics

Hydrodynamics and nutrient availability are two key factors influencing biofilm growth in MIC investigations. Generally speaking, biofilm can form under both stagnant and flowing conditions. Field observations showed that less shear flow and stagnant environments are preferred by bacterial activity and biofilm formation. Low turbulence locations, dead legs, improper hydrotesting conditions, maintenance downtimes, and shut downs are more favored for biofilm formation. Furthermore, particulates are places for bacteria to attach. They settle in low velocity locations providing deposits under which biofilms can form and grow. Particulates also increase the demand for biocides and cleaning solutions, due to their high surface area, which consumes some of the used chemicals and gives the biofilm some protection [120].

Fluid flow can enhance mass transfer but it may also produce a high shear that inhibits cell attachment and even causes detachment of an established biofilm [121]. Experimental results in glass cells inoculated with *D. desulfuricans* indicated that at 3000 rpm (roughly equivalent to 3.5 m/s in pipe flow), sessile SRB cells could not adhere on the coupon surface to form an SRB biofilm [122], suggesting using high shear as a mitigation method for bacterial biofilm.

Fluid flow or hydrodynamics influences biofilm structure and can have dramatic effects on the type of biofilm that is formed [109]. The biofilm cell density, strength of attachment, and other physical properties can be affected by fluid shear [109,123–125]. *Pseudomonas aeruginosa* biofilms grown under high shear tended to form filamentous streamers (filamentous biofilms usually clog piping) and more strongly attached to surfaces, whereas those grown under low shear formed an isotropic pattern of mound-shaped microcolonies [124].

Others speculate that turbulent flow may enhance bacterial adhesion and biofilm formation by impinging cells on the surface [116].

Some other factors include oxygen levels, presence of specific ions, osmolarity, the presence of specific ions, nutrients, and factors derived from the biotic environment [109].

10.5 Recent advances in biofouling research

Biofouling is an ancient problem in different industries globally. The impact of biofouling in oil and gas operations has been shown to be severe and costly. Better understanding of the biofouling mechanisms is the key to fighting it and developing proper mitigation plans.

Recent advances have been shown in the methods of identifying the diverse types of bacteria that can be found in oil and gas fields. Highly sensitive techniques like molecular microbiological methods (MMM) methods have been shown to reveal the various types of bacteria that exist in petroleum reservoir and production systems. Consequently, the various traits, behavior, nutrients, and most importantly the interactions between these different bacteria types can be disclosed and this helps in studying biocorrosion and biofilm-forming mechanisms. Some techniques were developed and others were adopted from other industries to monitor biocorrosion.

However, some gaps still exist in this discipline of science. Detection and diagnosis are not always accurate and attempts to correlate numbers and types of microorganisms to corrosion have produced misleading results. The monitoring techniques also are not always accurate, with two main drawbacks: first, most of them can't distinguish biocorrosion from other types of corrosion, and second, current methods only detect the biocorrosion in the location of probe insertion, whereas the problem could be somewhere else.

In addition, further work is needed to optimize the MIC experiments to represent the field conditions and the different consortia of microorganisms, such as those overserved in the environment.

10.6 Summary

Bacterial activity is a major, complex problem in oil and gas fields, causing various types of damage to the production system. Bacteria cause biocorrosion, produce hazardous H_2S gas, degrade chemicals, stabilize emulsions, and their final product of interaction with system facilities is biofouling, which results in flow restrictions and blockages. A wide range of microorganisms can be found in oil and gas fields, e.g., SRB, NRB, IOB, IRB, APB, methanogens, and others. The activity of these microorganisms depends on microorganism type, fluids composition, availability of nutrients, pressure, temperature, hydrodynamics, and system design.

Biofouling was discussed in this chapter. Biofouling can be attributed to biomineralization or biomass deposition. The former is due to deposition of the mineral byproducts that result from biocorrosion or bioprecipitation, whereas the latter is a result of biofilm formation and biomasses secreted by the bacteria during their life cycle. Usually the biofouling deposits combine both of these types of deposits at the same time, which complicates diagnosing and mitigating the problem.

Biomineralization leads to deposition of mineral scales in the system. The process and the type of the deposits formed depend principally on the type of the bacteria, fluids composition, pressure, temperature, fluid hydrodynamics, and the nature of the surface.

Biomass deposition depends on the formation of biofilms, slime layers which under the optimum conditions of growth can lead to severe blockages in the system. Biofilm growth depends on the type of the bacteria present in the medium, availability of nutrients, temperature, hydrodynamics, surface nature, and properties.

References

[1] K.A. Natarajan, Biofouling and microbially influenced corrosion, in: Biotechnology of Metals, Principles, Recovery Methods, and Environmental Concerns, Elsevier, 2018, pp. 355–393.

[2] N. Lenchi, Ö. Inceoğlu, S. Kebbouche-Gana, M.L. Gana, M. Lliros, P. Servais, T. García-Armisen, Diversity of microbial communities in production and injection waters of Algerian oilfields revealed by 16S RRNA gene amplicon 454 pyrosequencing, PLoS One 8 (6) (2013), https://doi.org/10.1371/journal.pone.0066588.

[3] R. Javaherdashti, Microbiologically influenced corrosion (MIC), in: Microbiologically Influenced Corrosion (MIC)—An Engineering Insight, Springer, 2017, pp. 29–79.

[4] M. Sharma, G. Voordouw, MIC detection and assessment: a holistic approach, in: Microbiologically Influenced Corrosion in the Upstream Oil and Gas Industry, CRC Press, 2017, pp. 177–212, https://doi.org/10.1201/9781315157818.

[5] B.J. Little, J.S. Lee, Microbiologically Influenced Corrosion, John Wiley & Sons, 2007.

[6] D. Enning, J. Garrelfs, Corrosion of iron by sulfate-reducing bacteria: new views of an old problem, Appl. Environ. Microbiol. (2014), https://doi.org/10.1128/AEM.02848-13.

[7] J. Telegdi, A. Shaban, L. Trif, Microbiologically influenced corrosion (MIC), in: Trends in Oil and Gas Corrosion Research and Technologies: Production and Transmission, Elsevier, 2017, pp. 191–214, https://doi.org/10.1016/B978-0-08-101105-8.00008-5.

[8] R. Javaherdashti, F. Akvan, Microbiologically influenced corrosion (MIC), in: Failure Modes, Effects and Causes of Microbiologically Influenced Corrosion, Elsevier, 2020, pp. 43–104, https://doi.org/10.1016/b978-0-12-818448-6.00003-x.

[9] M. Bomberg, M. Arnold, P. Kinnunen, Characterization of the bacterial and sulphate reducing community in the alkaline and constantly cold water of the closed Kotalahti mine, Minerals 5 (3) (2015) 452–472, https://doi.org/10.3390/min5030452.

[10] N. Zaveri, R. Sun, N. Zufelt, A. Zhou, Y.Q. Chen, Evaluation of microbially influenced corrosion with electrochemical noise analysis and signal processing, Electrochim. Acta 52 (19) (2007) 5795–5807, https://doi.org/10.1016/j.electacta.2007.02.084.

[11] D. Pokorna, J. Zabranska, Sulfur-oxidizing bacteria in environmental technology, in: Biotechnology Advances, Elsevier, 2015, https://doi.org/10.1016/j.biotechadv.2015.02.007.

[12] D. Pokorna, J. Zabranska, Sulfur-oxidizing bacteria in environmental technology, Biotechnol. Adv. 33 (6 Pt 2) (2015) 1246–1259.

[13] C.H.D. Williamson, L.A. Jain, B. Mishra, D.L. Olson, J.R. Spear, Microbially influenced corrosion communities associated with fuel-grade ethanol environments, Appl. Microbiol. Biotechnol. 99 (16) (2015) 6945–6957, https://doi.org/10.1007/s00253-015-6729-4.

[14] S. Chongdar, G. Gunasekaran, P. Kumar, Corrosion inhibition of mild steel by aerobic biofilm, Electrochim. Acta 50 (24) (2005) 4655–4665.
[15] S.K. Shukla, T. Atif Khan, S. Rao, Microbial fouling in water treatment plants, in: Microbial and Natural Macromolecules Synthesis and Applications, Elsevier, 2021, pp. 589–622 (Chapter 22).
[16] B. Mishra, D. Apelian, Corrosion of advanced steels: challenges in the oil and gas industry, in: Energy Materials 2014, Conference Proceedings, John Wiley and Sons, 2014, pp. 71–79, https://doi.org/10.1002/9781119027973.ch8.
[17] R. Javaherdashti, Microbiologically Influenced Corrosion: An Engineering Insight, Springer-Verlag, London, 2008.
[18] H.A. Videla, S. Le Borgne, C. Panter, R.K. Singh Raman, Mic of Steels by Iron Reducing Bacteria, NACE International, 2008, January 1.
[19] H.A. Videla, Corrosion inhibition by bacteria, in: Manual of Biocorrosion, CRC Lewis Publishers, Boca Raton, Florida, 1996, pp. 121–135.
[20] M. Dubiel, C.H. Hsu, C.C. Chien, F. Mansfield, D.K. Newman, Microbial iron respiration can protect steel from corrosion, Appl. Environ. Microbiol. 68 (2002) 1440–1445.
[21] H.A. Videla, Corrosion inhibition in the presence of microbial corrosion, in: CORROSION/96 NACE International, Houston Paper No. 223, 1996.
[22] B. Little, P. Wagner, K. Hart, R. Ray, D. Lavoie, K. Nealson, et al., The role of metal-reducing bacteria in microbiologically influenced corrosion, in: CORROSION/ 97 Paper No. 215. NACE International, Houston, TX, 1997.
[23] S. Lahme, J. Mand, J. Longwell, R. Smith, D. Enning, Severe corrosion of carbon steel in oil field produced water can be linked to methanogenic archaea containing a special type of [NiFe] hydrogenase, Appl. Environ. Microbiol. 87 (3) (2020) e01819–e01820, https://doi.org/10.1128/AEM.01819-20.
[24] R.M. Flores, Origin of coal as gas source and reservoir rocks, in: Coal and Coalbed Gas, Elsevier, 2014, pp. 97–165, https://doi.org/10.1016/b978-0-12-396972-9.00003-3.
[25] N. Belay, L. Daniels, Elemental metals as electron source for biological methane formation from CO2, Antonie Van Leeuwenhoek 57 (1990) 1–7.
[26] L. Raskin, B.E. Rittmann, D.A. Stahl, Competition and coexistence of sulfate-reducing and methanogenic populations in anaerobic biofilms, Appl. Environ. Microbiol. 62 (10) (1996) 3847–3857, https://doi.org/10.1128/aem.62.10.3847-3857.1996.
[27] R. Liang, R.S. Grizzle, K.E. Duncan, M.J. McInerney, J.M. Suflita, Roles of thermophilic thiosulfate-reducing bacteria and methanogenic archaea in the biocorrosion of oil pipelines, Front. Microbiol. 5 (MAR) (2014), https://doi.org/10.3389/fmicb.2014.00089.
[28] Y. Liu, L.L. Beer, W.B. Whitman, Sulfur metabolism in archaea reveals novel processes, Environ. Microbiol. (2012), https://doi.org/10.1111/j.1462-2920.2012.02783.x.
[29] P.A. Palacios, O. Snoeyenbos-West, C.R. Löscher, B. Thamdrup, A.E. Rotaru, Baltic sea methanogens compete with acetogens for electrons from metallic iron, ISME J. 13 (12) (2019) 3011–3023, https://doi.org/10.1038/s41396-019-0490-0.
[30] V.M. Luque-Almagro, A.J. Gates, C. Moreno-Vivián, S.J. Ferguson, D.J. Richardson, M. Dolores Roldán, Bacterial nitrate assimilation: gene distribution and regulation, Biochem. Soc. Trans. 39 (6) (2011) 1838–1843, https://doi.org/10.1042/BST20110688.
[31] T. Gu, D. Xu, Why are some microbes corrosive and some not? in: Paper Presented at NACE Corrosion/2013, Paper No. C2013-0002336, Orlando, FL, March 17–21, 2013.
[32] B. Schiel-Bengelsdorf, P. Dürre, Pathway engineering and synthetic biology using acetogens, FEBS Lett. 586 (2012) 2191–2198.
[33] A. Ibrahim, K. Hawboldt, C. Bottaro, F. Khan, Review and analysis of microbiologically influenced corrosion: the chemical environment in oil and gas facilities, Corros. Eng. Sci. Technol. (2018), https://doi.org/10.1080/1478422X.2018.1511326.

[34] N.K. Dhami, M. Sudhakara Reddy, M. Sudhakara Mukherjee, Biomineralization of calcium carbonates and their engineered applications: a review, Front. Microbiol. (2013), https://doi.org/10.3389/fmicb.2013.00314.
[35] L.A. Estroff, Introduction: biomineralization, Chem. Rev. (2008), https://doi.org/10.1021/cr8004789.
[36] H.A. Lowenstam, Minerals formed by organisms, Science 211 (1981) 1126–1131, https://doi.org/10.1126/science.7008198.
[37] H.A. Lowenstam, S. Weiner, On Biomineralization, Oxford University Press, New York, NY, 1989.
[38] K.O. Konhauser, Bacterial iron biomineralisation in nature, FEMS Microbiol. Rev. 20 (1997) 315–326, https://doi.org/10.1016/S0168-6445(97)00014-4.
[39] R.B. Frankel, D.A. Bazylinski, Biologically induced mineralization by bacteria, Rev. Mineral. Geochem. 54 (1) (2003) 95–114, https://doi.org/10.2113/0540095.
[40] Z. Lewandowski, H. Beyenal, Mechanisms of microbially influenced corrosion, in: Marine and Industrial Biofouling, 2008, pp. 35–64, https://doi.org/10.1007/7142_2008_8.
[41] A.K. Lee, D.K. Newman, Microbial iron respiration: impacts on corrosion processes, Appl. Microbiol. Biotechnol. (2003), https://doi.org/10.1007/s00253-003-1314-7.
[42] S. Kakooei, M. Che Ismail, B. Ariwahjoedi, Mechanisms of microbiologiacally influenced corrosion. A review, World Appl. Sci. J. 17 (4) (2012) 524e531.
[43] J. Starosvetsky, D. Starosvetsky, R. Armon, Identification of microbiologically influenced corrosion (MIC) in industrial equipment failures, Eng. Fail. Anal. 14 (8 SPEC. ISS) (2007) 1500–1511, https://doi.org/10.1016/j.engfailanal.2007.01.020.
[44] M. de Romero, The mechanism of SRB action in MIC based on sulfide corrosion and iron sulfide corrosion products, in: Corrosion/2005, NACE International, Houston, TX, (2005), No. 05481, 2005.
[45] D. Enning, H. Venzlaff, J. Garrelfs, H.T. Dinh, V. Meyer, K. Mayrhofer, et al., Marine sulfate-reducing bacteria cause serious corrosion of iron under electroconductive biogenic mineral crust, Environ. Microbiol. 14 (2012) 1772–1787.
[46] T. Gu, X. Dake, Why are some microbes corrosive and some not? in: CORROSION'2013, Orlando, FL, March (2013), 2013, Paper No. C2013-0002336, 2013, p. 17e21.
[47] T.L. Skovhus, J.S. Lee, B.J. Little, Predominant MIC mechanisms in the oil and gas industry, in: Microbiologically Influenced Corrosion in the Upstream Oil and Gas Industry, CRC Press, 2017, pp. 75–85, https://doi.org/10.1201/9781315157818-4.
[48] K.A. Brileya, L.B. Camilleri, G.M. Zane, J.D. Wall, M.W. Fields, Biofilm growth mode promotes maximum carrying capacity and community stability during product inhibition syntrophy, Front. Microbiol. 5 (DEC) (2014), https://doi.org/10.3389/fmicb.2014.00693.
[49] C.C. Patton, Applied Water Technology, first ed., Campbell Petroleum Series, 1986.
[50] M.A.M. Reis, J.S. Almeida, P.C. Lemos, M.J.T. Carrondo, Effect of hydrogen sulfide on growth of sulfate reducing bacteria, Biotechnol. Bioeng. 40 (5) (1992) 593–600, https://doi.org/10.1002/bit.260400506.
[51] N. Youssef, M.S. Elshahed, M.J. McInerney, Microbial processes in oil fields. Culprits, problems, and opportunities, in: Advances in Applied Microbiology, Elsevier, 2009 (Chapter 6).
[52] R. Javaherdashti, A review of some characteristics of MIC caused by sulphate reducing bacteria: past present and future, Anti-Corros. Methods Mater. 46 (1999) 173–180.
[53] F. Walsh, R. Mitchell, A PH-dependent succession of iron bacteria, Environ. Sci. Technol. 6 (9) (1972) 809–812, https://doi.org/10.1021/es60068a001.
[54] C.F. Earhart, Iron Metabolism in Encyclopedia of Microbiology, Elsevier, 2018.
[55] M. Henze, Basic biological processes, in: M. Henze, P. Harremoes, J. la Cour Jansen, E. Arvin (Eds.), Wastewater Treatment: Biological and Chemical Processes, third ed., Springer, 2002, pp. 65–129.
[56] G. Grassia, A.J. Sheehy, Thermophilic bacteria from petroleum reservoirs, Dev. Pet. Sci. 39 (C) (1993) 429, https://doi.org/10.1016/S0376-7361(09)70081-0.

[57] C. Jeanthon, O. Nercessian, E. Corre, A. Grabowski-Lux, Hyperthermophilic and methanogenic archaea in oil fields, in: Petroleum Microbiology, ASM Press, 2014, pp. 55–69, https://doi.org/10.1128/9781555817589.ch4.

[58] T. Li, G. Wang, P. Yin, et al., Effect of negative pressure on growth, secretion and biofilm formation of Staphylococcus aureus, Anton. Leeuw. Int. J. Gen. Mol. Microbiol. 108 (2015) 907–917.

[59] G.G. Geesey, L. Jang, J.G. Jolley, M.R. Hankins, T. Iwaoka, P.R. Griffiths, Binding of metal ions by extracellular polymers of biofilm bacteria, Water Sci. Technol. 20 (11–12) (1988) 161–165, https://doi.org/10.2166/wst.1988.0279.

[60] S. Douglas, T.J. Beveridge, Mineral formation by bacteria in natural microbial communities, FEMS Microbiol. Ecol. (1998), https://doi.org/10.1016/S0168-6496(98)00027-0.

[61] P.D. Natsi, P.G. Koutsoukos, Calcium carbonate biofouling, in: Paper Presented at the Corrosion 2019, Nashville, Tennessee, USA, March 2019.

[62] T. Zhu, M. Dittrich, Carbonate precipitation through microbial activities in natural environment, and their potential in biotechnology: a review, Front. Bioeng. Biotechnol. (2016), https://doi.org/10.3389/fbioe.2016.00004.

[63] O. Braissant, A.W. Decho, C. Dupraz, C. Glunk, K.M. Przekop, P.T. Visscher, Exopolymeric substances of sulfate-reducing bacteria: interactions with calcium at alkaline PH and implication for formation of carbonate minerals, Geobiology 5 (4) (2007) 401–411, https://doi.org/10.1111/j.1472-4669.2007.00117.x.

[64] A. Ansari, J. Peña-Bahamonde, S.K. Fanourakis, H. Yandi, D.F. Rodrigues, Microbially-induced mineral scaling in desalination conditions: mechanisms and effects of commercial antiscalants, Water Res. 179 (July) (2020), https://doi.org/10.1016/j.watres.2020.115863.

[65] T. Onal Okyay, D. Frigi Rodrigues, Optimized carbonate micro-particle production by Sporosarcina pasteurii using response surface methodology, Ecol. Eng. 62 (2014) 168–174.

[66] M. Nemati, G. Voordouw, Biologically formed calcium carbonate, a durable plugging agent for enhanced oil recovery, J. Can. Pet. Technol. 43 (2004), https://doi.org/10.2118/04-02-TN1.

[67] C. Song, C. Song, Y. Chen, J. Wang, Plugging high-permeability zones of oil reservoirs by microbially mediated calcium carbonate precipitation, ACS Omega 5 (24) (2020) 14376–14383, https://doi.org/10.1021/acsomega.0c00902.

[68] P.A. Lapointe, M.A. Muhsin, A.F. Maurin, Microbial corrosion and biologically induced new products example from a seawater injection system, Umm Shaif Field, U.A.E, in: Paper Presented at the Middle East Oil Show, Bahrain, November 1991, https://doi.org/10.2118/21367-MS.

[69] D.T. Rickard, D.T. Rickard, Microbial sulfate reduction in sediments, in: Developments in Sedimentology, Elsevier, Oxford, 2012, pp. 319–351.

[70] R. Andres, A. Kasgnoc, Atime-averaged inventory of subaerial volcanic sulfur emissions, J. Geophys. Res. 103 (1998) 25251–25261.

[71] H.A. Nasr-El-Din, A.Y. Al-Humaidan, Iron sulfide scale: formation, removal and prevention, in: Paper Presented at the International Symposium on Oilfield Scale, Aberdeen, United Kingdom, January 2001, https://doi.org/10.2118/68315-MS.

[72] C.T. Lefèvre, D.A. Bazylinski, Ecology, diversity, and evolution of magnetotactic bacteria, Microbiol. Mol. Biol. Rev. 77 (2013) 497–526.

[73] A. Picard, A. Gartman, P.R. Girguis, What do we really know about the role of microorganisms in iron sulfide mineral formation? Front. Earth Sci. 4 (June) (2016), https://doi.org/10.3389/feart.2016.00068.

[74] D. Fortin, G. Southam, T.J. Beveridge, An examination of iron sulfide, iron-nickel sulfide and nickel sulfide precipitation by a desulfotomaculum species: and its nickel resistance mechanisms, FEMS Microbiol. Ecol. 14 (1994) 121–132.

[75] D. Fortin, T.J. Beveridge, Mechanistic routes to biomineral surface development, in: E. Bäuerlein (Ed.), Biomineralization: From Biology to Biotechnology and Medical Application, Wiley-VCH, Weinheim, Germany, 2000, pp. 7–24.

[76] A. Picard, A. Gartman, D.R. Clarke, P.R. Girguis, Sulfate-reducing bacteria influence the nucleation and growth of mackinawite and greigite, Geochim. Cosmochim. Acta 220 (January) (2018) 367–384, https://doi.org/10.1016/j.gca.2017.10.006.

[77] D.T. Rickard, The microbiological formation of iron sulfides, Stockh. Contrib. Geol. 20 (1969) 50–66.

[78] D.T. Rickard, The chemistry of iron sulfide formation at low temperatures, Stockh. Contrib. Geol. 20 (1969) 67–95.

[79] I.R. Collins, M.M. Jordan, Occurrence, prediction, and prevention of zinc sulfide scale within gulf coast and north sea high-temperature and high-salinity fields, SPE Prod. Facil. 18 (2003) 200–209, https://doi.org/10.2118/84963-PA.

[80] R.M. Cornell, R. Giovanoli, P.W. Schindler, Effect of silicate species on the transformation of ferrihydrite into goethite and hematite in alkaline media, Clay Clay Miner. 35 (1) (1987) 21–28, https://doi.org/10.1346/CCMN.1987.0350103.

[81] F.G. Ferris, K. Tazaki, W.S. Fyfe, Iron oxides in acid mine drainage environments and their association with bacteria, Chem. Geol. 74 (3–4) (1989) 321–330, https://doi.org/10.1016/0009-2541(89)90041-7.

[82] T.D. Brock, M.T. Madigan, J.M. Martinko, J. Parker, Biology of Microorganisms, seventh ed., Prentice-Hall, 1994.

[83] D.A. Bazylinski, Structure and function of the bacterial magnetosome, ASM News 61 (1995) 337–343.

[84] D.R. Lovley, Magnetite formation during microbial dissimilatory iron reduction, in: Iron Biominerals, Springer, US, 1991, pp. 151–166, https://doi.org/10.1007/978-1-4615-3810-3_11.

[85] T. Sakaguchi, J. Grant Burgess, T. Matsunaga, Magnetite formation by a sulphate-reducing bacterium, Nature 365 (6441) (1993) 47–49, https://doi.org/10.1038/365047a0.

[86] S. Hauck, M. Benz, A. Brune, B. Schink, Ferrous iron oxidation by denitrifying bacteria in profundal sediments of a deep lake (lake constance), FEMS Microbiol. Ecol. 37 (2) (2001) 127–134, https://doi.org/10.1016/S0168-6496(01)00153-2.

[87] S. Lahme, D. Enning, C.M. Callbeck, D.M. Vega, T.P. Curtis, I.M. Head, C.R.J. Hubert, Metabolites of an oil field sulfideoxidizing, nitrate-reducing Sulfurimonas sp. cause severe corrosion, Appl. Environ. Microbiol. 85 (3) (2019), https://doi.org/10.1128/AEM.01891-18.

[88] M.T. Gonzalez-Muñoz, F. Martinez-Ruiz, F. Morcillo, J.D. Martin-Ramos, A. Paytan, Precipitation of barite by marine bacteria: a possible mechanism for marine barite formation, Geology 40 (8) (2012) 675–678, https://doi.org/10.1130/G33006.1.

[89] P.E. Bell, A.L. Mills, J.S. Herman, Biogeochemical conditions favoring magnetite formation during anaerobic iron reduction, Appl. Environ. Microbiol. 53 (1987) 26102616.

[90] Y. Chen, Y. Feng, J.G. Deveaux, M.A. Masoud, F.S. Chandra, H. Chen, D. Zhang, L. Feng, Biomineralization forming process and bio-inspired nanomaterials for biomedical application: a review, Minerals (2019), https://doi.org/10.3390/min9020068.

[91] H. Ding, H. Pan, X. Xu, R. Tang, Toward a detailed understanding of magnesium ions on hydroxyapatite crystallization inhibition, Cryst. Growth Des. 14 (2014) 763–769, https://doi.org/10.1021/cg401619s.

[92] P.V. Bhaskar, N.B. Bhosle, Microbial extracellular polymeric substances in marine biogeochemical processes, Curr. Sci. 88 (1) (2005) 45–53.

[93] H.C. Flemming, EPS-then and now, Microorganisms 4 (4) (2016) 41, https://doi.org/10.3390/microorganisms4040041.

[94] S. McCarty, E.M. Jones, S. Finnegan, E. Woods, C.A. Cochrane, S.L. Percival, Wound infection and biofilms, in: S.L. Percival, D.W. Williams, J. Randle, T. Cooper (Eds.), Biofilms in Infection Prevention and Control, Academic Press, Boston, 2014, pp. 339–358 (Chapter 18) http://www.sciencedirect.com/science/article/pii/B9780123970435000189.

[95] J.F. Banfield, K.H. Nealson, Geomicrobiology: Interactions Between Microbes and Minerals, De Gruyter, Berlin, Boston, 2018, https://doi.org/10.1515/9781501509247.

[96] T.S. Rao, Biofouling in industrial water systems, in: Mineral Scales and Deposits: Scientific and Technological Approaches, Elsevier, 2015, pp. 123–140, https://doi.org/10.1016/B978-0-444-63228-9.00006-1.

[97] T.S. Rao, R. Kesavamoorthy, C. Babu Rao, K.V.K. Nair, Influence of flow on ordering characteristics of a bacterial biofilm, Curr. Sci. 73 (1) (1997) 69–74.

[98] A.B. Cunningham, J.E. Lennox, R.J. Ross, Biofilms: The Hypertextbook, A Web-Based Active-Learning Approach for Rapid Infusion of Emerging Knowledge Into Undergraduate STEM Curricula, Montana State University, Bozeman, 2010. Http://Www.Hypertextbookshop.Com/Biofilmbook/V005/R001/.

[99] Z. Lewandowski, Mic and biofilm heterogeneity, in: NACE—International Corrosion Conference Series. Vol. 2000-March. National Assoc. of Corrosion Engineers International, 2000.

[100] H.C. Flemming, T.R. Neu, D.J. Wozniak, The EPS matrix: the 'house of biofilm cells, J. Bacteriol. (2007), https://doi.org/10.1128/JB.00858-07.

[101] H.-C. Flemming, Why microorganisms live in biofilms and the problem of biofouling, in: Marine and Industrial Biofouling, Springer, Berlin, Heidelberg, 2008, pp. 3–12, https://doi.org/10.1007/978-3-540-69796-1_1.

[102] H.C. Flemming, J. Wingender, The biofilm matrix, Nat. Rev. Microbiol. 8 (2010) 623–633, https://doi.org/10.1038/nrmicro2415.

[103] M. Toyofuku, T. Inaba, T. Kiyokawa, N. Obana, Y. Yawata, N. Nomura, Environmental factors that shape biofilm formation, in: Bioscience, Biotechnology and Biochemistry, Japan Society for Bioscience Biotechnology and Agrochemistry, 2016, https://doi.org/10.1080/09168451.2015.1058701.

[104] S.S. Branda, A. Vik, L. Friedman, R. Kolter, Biofilms: the matrix revisited, Trends Microbiol. 13 (2005) 20–26.

[105] C. Mayer, R. Moritz, C. Kirschner, W. Borchard, R. Maibaum, J. Wingender, H.C. Flemming, The role of intermolecular interactions: studies on model systems for bacterial biofilms, Int. J. Biol. Macromol. 26 (1999) 3–16.

[106] V. Körstgens, H.C. Flemming, J. Wingender, W. Borchard, Uniaxial compression measurement device for investigation of the mechanical stability of biofilms, J. Microbiol. Methods 46 (1) (2001) 9–17, https://doi.org/10.1016/S0167-7012(01)00248-2.

[107] B.W. Peterson, et al., Viscoelasticity of biofilms and their recalcitrance to mechanical and chemical challenges, FEMS Microbiol. Rev. 39 (2015) 234–245.

[108] N. Billings, A. Birjiniuk, T.S. Samad, P.S. Doyle, K. Ribbeck, Material properties of biofilms-a review of methods for understanding permeability and mechanics. Reports on progress in physics, Phys. Soc. 78 (3) (2015) 036601, https://doi.org/10.1088/0034-4885/78/3/036601.

[109] C.C. Goller, T. Romeo, Environmental influences on biofilm development, Curr. Top. Microbiol. Immunol. (2008), https://doi.org/10.1007/978-3-540-75418-3_3.

[110] E.M. Jones, C.A. Cochrane, S.L. Percival, The effect of pH on the extracellular matrix and biofilms, Adv. Wound Care 4 (7) (2015) 431–439, https://doi.org/10.1089/wound.2014.0538.

[111] T.T.T. Tran, K. Kannoorpatti, A. Padovan, S. Thennadil, Effect of pH regulation by sulfate-reducing bacteria on corrosion behaviour of duplex stainless steel 2205 in acidic artificial seawater, R. Soc. Open Sci. 8 (1) (2021), https://doi.org/10.1098/rsos.200639.

[112] A. Villain-Simonnet, M. Milas, M. Rinaudo, A new bacterial exopolysaccharide (YAS34). II. Influence of thermal treatments on the conformation and structure. Relation with gelation ability, Int. J. Biol. Macromol. 27 (2000) 77–87, https://doi.org/10.1016/S0141- 8130(99)00119-1.

[113] T.R. Garrett, M. Bhakoo, Z. Zhang, Bacterial adhesion and biofilms on surfaces, Prog. Nat. Sci. 18 (2008) 1049–1056, https://doi.org/10.1016/j.pnsc.2008.04.001.

[114] R.M. Donlan, Biofilms: microbial life on surfaces, Emerg. Infect. Dis. 8 (2002) 881–890.

[115] S. Zheng, M. Bawazir, A. Dhall, H.E. Kim, L. He, J. Heo, G. Hwang, Implication of surface properties, bacterial motility, and hydrodynamic conditions on bacterial surface sensing and their initial adhesion, Front. Bioeng. Biotechnol. (2021), https://doi.org/10.3389/fbioe.2021.643722.

[116] R.M. Donlan, J.W. Costerton, Biofilms: survival mechanisms of clinically relevant microorganisms, Clin. Microbiol. Rev. 15 (2002) 167–193.

[117] V. Kochkodan, S. Tsarenko, N. Potapchenko, V. Kosinova, V. Goncharuk, Adhesion of microorganisms to polymer membranes: a photobactericidal effect of surface treatment with TiO_2, Desalination 220 (2008) 380–385, https://doi.org/10.1016/j.desal.2007.01.042.

[118] E. Giaouris, M. Chapot-Chartier, R. Briandet, Surface physicochemical analysis of natural *Lactococcus lactis* strains reveals the existence of hydrophobic and low charged strains with altered adhesive properties, Int. J. Food Microbiol. 131 (2009) 2–9, https://doi.org/10.1016/j.ijfoodmicro.2008.09.006.

[119] A. Krasowska, K. Sigler, How microorganisms use hydrophobicity and what does this mean for human needs? Front. Cell. Infect. Microbiol. (2014), https://doi.org/10.3389/fcimb.2014.00112.

[120] R. Heidersbach, Metallurgy and corrosion control in oil and gas production, second edition, in: Metallurgy and Corrosion Control in Oil and Gas Production, second ed., Wiley, 2018, pp. 1–354, https://doi.org/10.1002/9781119252351.

[121] P. Stoodley, I. Dodds, J.D. Boyle, H.M. Lappin-Scott, Influence of hydrodynamics and nutrients on biofilm structure, J. Appl. Microbiol. Symp. Suppl. 85 (1999) 19–28.

[122] J. Wen, T. Gu, S. Nesic, Investigation of the effects of fluid flow on SRB biofilm, in: NACE—International Corrosion Conference Series, 2007, pp. 075161–0751611.

[123] P. Stoodley, K. Sauer, D.G. Davies, J.W. Costerton, Biofilms as complex differentiated communities, Annu. Rev. Microbiol. 56 (2002) 187–209.

[124] P. Stoodley, R. Cargo, C.J. Rupp, S. Wilson, I. Klapper, Biofilm material properties as related to shear-induced deformation and detachment phenomena, J. Ind. Microbiol. Biotechnol. 29 (2002) 361–367.

[125] M.C. Van Loosdrecht, J.J. Heijnen, H. Eberl, J. Kreft, C. Picioreanu, Mathematical modelling of biofilm structures, Antonie Van Leeuwenhoek 81 (2002) 245–256.

CHAPTER

Chemical Incompatibility Deposits (Pseudoscale) 11

11.1 Introduction

Flow assurance solids in oil and gas operations are known to originate predominantly from the produced fluids themselves. Natural gas hydrates, paraffins, asphaltenes, naphthenates, and mineral scales are principally derived from the produced gas, oil, and water streams due to disturbance of the produced fluids' thermodynamic equilibrium during production operations. However, a different type of deposit can emerge when specialty chemicals and additives (such as scale inhibitors, corrosion inhibitors, hydrate inhibitors, demulsifiers, etc.) are introduced to the production system. These chemicals are added to the system at various doses and in locations where they interact with the produced streams, in order to treat specific production-related problems or to improve the process performance. Nonetheless, some undesirable interactions between the added chemicals themselves, the added chemicals and produced fluids, and the added chemicals and system components are not uncommon. Such interactions can lead to incompatibility issues and end up forming solid deposits usually known as pseudoscale, pseudodeposits, secondary deposits, or chemical incompatibility deposits [1]. The interactions between corrosion inhibitors (CIs)-scale inhibitors (SIs), biocides-oxygen scavengers (OSs), scale inhibitors-metal ions, and acidic chemicals-pipeline materials are common examples of incompatibilities that lead to the formation of pseudoscales.

In a general sense, the problems associated with treating chemical incompatibility include the following:

- Decline in the injected chemical performance and loss of its efficiency.
- Solid deposits formation.
- Emulsion stabilization and phase separation issues. Emulsions may also be thickened, accumulate more solids, and turn into sludge or solid deposits.
- Sludge formation.
- System material integrity issues, i.e., corrosion, swelling, and other types of deformation.
- Product quality issues, which also affect the downstream operations, i.e., the refinery.

11.2 Treatment chemicals and treatment chemical formulations

Chemical agents are extensively used throughout the life cycle of oil and gas fields. They are commonly known as chemical additives, specialty chemicals, process treatment chemicals, or treating chemicals for short. These chemicals are added with different doses at different locations in the production

Essentials of Flow Assurance Solids in Oil and Gas Operations. https://doi.org/10.1016/B978-0-323-99118-6.00013-7

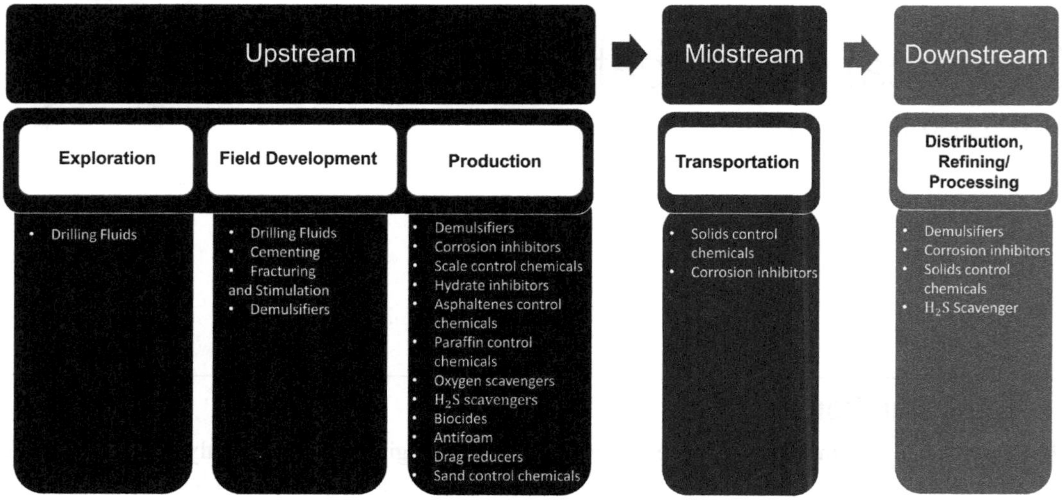

FIG. 11.1

The use of production treatment chemicals in different oil and gas sectors.

system (reservoir formations, wellbore, and topsides) and during different field operations (drilling, completion, production, intervention, workover, and abandonment), to impart unique properties and functionality to the produced fluids to mitigate the production problems and/or to improve the process performance. Fig. 11.1 illustrates the different treatment chemicals that are used in the different oil and gas industry sectors.

These chemicals cover a wide range of chemistries from simple molecules to complex polymeric compounds. They are usually supplied as a homogeneous formulation (mixture) that contains an active material as the main functioning compound. Besides the active materials, and depending on their technical application and environmental safety aspects, the treating chemical formulations can contain one or more of the following ingredients:

- The solvent (organic, aqueous)
- Viscosity adjustor
- Density adjustors
- pH adjustor
- Detergent
- Catalysts/intensifiers
- Other malfunctioning compounds if needed (demulsifiers, antifoams)

Considering the different chemistries and functionalities of these compounds, although they are compatible within the final formulated chemical blend, one can expect this list of ingredients to show incompatibilities when introduced with different chemical formulations of different types of ingredients under the process operating conditions.

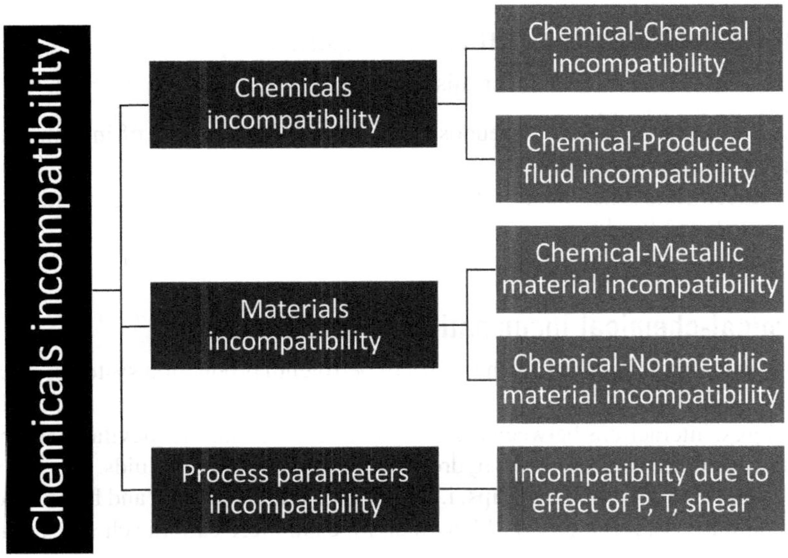

FIG. 11.2

Summary of the types of chemical incompatibility deposits (pseudoscales).

Generally speaking, the chemical incompatibility and pseudoscales can fall into three categories, as illustrated in Fig. 11.2:

- Chemical incompatibility: this type includes the incompatibility of treatment chemicals with each other and with the produced fluids.
- Material incompatibility: this type includes the incompatibility of the treatment chemicals with the system facilities materials (metallic, nonmetallic).
- Process operating parameters incompatibility: this type arises due to the response of the different chemicals to the process parameters, i.e., pressure, temperature, pH, etc., leading to chemical decomposition, or polymerization byproducts that affect the chemical performance and compatibility.

The treating chemicals are principally screened for efficiency prior to applying them in the field. After efficiency screening, the chemicals are applied upstream of the problem to prevent its damage. Methods of deploying the chemicals involve continuous injection, batch injection, squeeze treatments, or slow release treatments. The treating chemical injection systems are discussed in Chapter 22.

Besides the conventional chemical efficiency testing and prior to their application, comprehensive testing must be conducted under the expected production system operating parameters (P, T, shear, etc.) to evaluate their compatibility with the currently applied chemicals, and with the process fluids and materials, to proactively avoid any potential incompatibility issues that may arise. Some details of these incompatibilities are covered in the following sections.

11.3 Treatment chemicals-fluids incompatibility

Two types of pseudoscale can form under this category:

- Chemical-chemical incompatibility pseudoscale, where different types of incompatible treating chemicals interact with each other.
- Chemical-produced fluids incompatibility pseudoscale, where the treating chemicals interact with incompatible produced fluids.

11.3.1 Chemical-chemical incompatibility pseudoscale

Due to the complexity of the production chemistry of petroleum production systems, a number of treating chemicals are added at the same time at different locations in the system. Although they may be added in small doses, interactions between these chemical ingredients are inevitable if they are added in close vicinity to each other and due to the hydrodynamics of the flowing fluids. The active ingredients in these chemicals contain functional groups, i.e., cationic, anionic, acidic, and basic functional groups used to fulfill their purpose. It is then very likely that, for example, a cationic chemical interacts with an anionic chemical, forming a complex or salt that is insoluble under the process operating conditions. Another form of incompatibility can arise from the other ingredients in the chemical formulations (solvent, density or viscosity adjustor, etc.) potentially due to miscibility issues, i.e., aqueous-based solvent in one chemical versus oil-based solvent in another chemical. Some examples of these incompatibilities are given in the following paragraphs.

A common example of functional group interaction is the corrosion inhibitor-scale inhibitor interaction. Film-forming corrosion inhibitors are generally cationic species (i.e., quaternary ammoniums and imidazolines) whereas scale inhibitors are anionic in nature (i.e., phosphates, polyphosphates, phosphate ester, phosphonate, polyphosphonates, phosphino, polyphosphinates, polycarboxylates, polysulfonates). Complexation between these different species in solution would effectively reduce their activity and lead to unwanted precipitations [2–5]. Fig. 11.3 shows the precipitation between scale inhibitor and corrosion inhibitor.

Also, polymeric and nonpolymeric scale inhibitors when present in low concentrations in water form insoluble complexes/salts with cationic charged polymeric flocculants/coagulants, biocides, biodispersants, and antifoams. The driving forces in the formation of such insoluble complexes are electrostatic in nature (anionic-cationic interactions of the polymeric additives). These interactions are strongly dependent on the polymer composition, polymer charge (i.e., cationic, anionic, and neutral), polymer molecular weight, polymer concentration, and solution pH [6–9].

Conventional oilfield biocides tetrakis(hydroxymethyl)phosphonium sulfate (THPS) and glutaraldehyde were reported to show incompatibility issues with bisulfite (sodium or ammonium) based oxygen scavengers [10–12]. Fig. 11.4 shows an example of pseudoscale formation as a result of interaction between glutaraldehyde and bisulfite chemicals during a scale squeeze job. In this case study, biocides and oxygen scavengers were additives employed during the squeeze job to prevent bacterial contamination and dissolved oxygen ingress. The operator was supposed to flush the chemical delivery hoses with fresh water in between the oxygen scavenger and the biocide batches, and when he forgot to, the precipitate built up and was only noticed when pressure built up in the injection system and the delivery hose blew up, releasing some of the formed deposits. In addition, THPS was also found to have incompatibility issues with corrosion inhibitors [13].

FIG. 11.3

A precipitate formation indicates the incompatibility between phosphonate-based scale inhibitor and amine-based corrosion inhibitor.

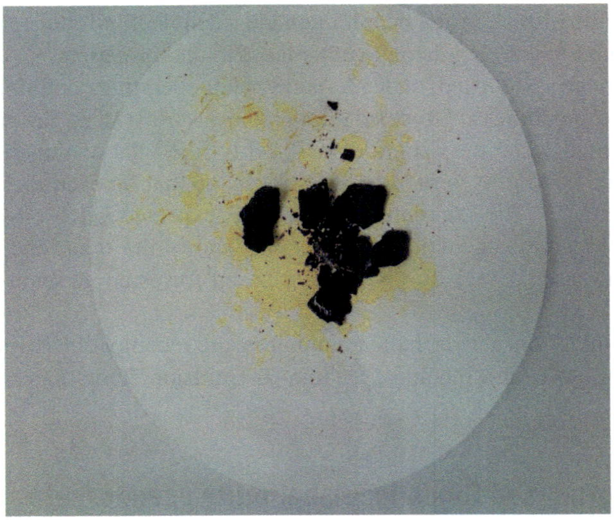

FIG. 11.4

Pseudoscale formed when biocide (glutaraldehyde) was pumped immediately after oxygen scavenger (bisulfite) without inbetween solvent (water) flushing, during a scale inhibitor squeeze job, which lead to blockage of the pumping hoses.

Incompatibility issues between corrosion inhibitors and gas hydrate inhibitors were reported, as hydrate inhibitors can negatively influence the performance of corrosion inhibitors. This issue has been raised in the literature due to absorption of corrosion inhibitor onto the hydrate inhibitor [14,15]. For example, the effect of kinetic hydrate inhibitors (KHIs) on corrosion inhibitors (CI) may be due to preferential adsorption of KHI on the walls of the pipe or due to CI-KHI surfactant–polymer interactions [14–16].

Interferences of H_2S scavengers with corrosion inhibitors are rare. However, during their working process, H_2S scavengers undergo degradation, forming various byproducts, presenting a risk of chemical incompatibility with other additives resulting in undesirable products [17].

Corrosion inhibitors are known to interfere with demulsifiers and stabilize emulsions. Their effect was shown to be stronger than scale inhibitors and antifoams [18]. The emulsions caused by corrosion inhibitors are referred to as "chemically stabilized emulsions" [19]. These emulsions can be a flow assurance issue themselves, or they can get tighter and thickened over time and collect more solids, and collectively result in sludge and deposition problems.

Wax inhibitors were reported to have negative impact on the gas hydrate inhibitors at low dosage, while the effect was negligible at higher dosage [20].

Methanol has incompatibility issues with scale inhibitors, corrosion inhibitors, asphaltene inhibitors, and wax inhibitors. In fact, the thermodynamic hydrate inhibitors methanol and glycol affect scale formation kinetics and can have incompatibility issues with scale inhibitors and affect their performance [16].

Another incompatibility arises due to the difference in miscibility of the chemicals when oil-based chemicals and water-based chemicals are used in close vicinity. This type of incompatibility usually leads to the formation of thick emulsions, which can be redissolved in the bulk of the produced fluids or they can collect more suspended solids and, under the right operating conditions, transform into sludge or precipitate.

A major incompatibility between a scale inhibitor and a demulsifier is due to the former being water based and the latter oil based; mixing them together leads to the formation of a thick emulsion that can accumulate solids from the system and, under certain system operating conditions, turns into sludge or a deposit. Demulsifiers sometimes are injected through the gas lift system. In one of the field cases, the gas lift valve was blocked when the operators mistakenly filled the demulsifier chemical tank with scale inhibitors instead, leading to formation of thick sludge that blocked the gas lift system, which required costly maintenance. The same incompatibility is expected when oil-based demulsifiers are mixed with water-based corrosion inhibitors. Fig. 11.5 shows the thick and turbid emulsions that formed after mixing oil-based demulsifier with water-based corrosion inhibitor and water-based scale inhibitor.

Oil-based paraffin inhibitors and asphaltene inhibitors are incompatible with water-based scale inhibitors, corrosion inhibitors, and biocides, leading to emulsion formation and precipitation in the worst case.

11.3.2 Chemical-produced fluids incompatibility pseudoscale

Although the treating chemicals are designed to solve problems in the flowing streams, if they are incompatible, they can interact with the produced fluids and form precipitates that generate additional problems.

FIG. 11.5

The emulsions formed due to incompatibility between oil based demulsifier and water based corrosion inhibitors (left), oil based demulsifier and water based scale inhibitor (right).

Anionic scale inhibitors interact with metal ions in the produced water, like Ca, Mg, and Fe [21]. In a case from the Egyptian oil field, a malfunctioning downhole chemical injection system overly dosed a phosphonate-based scale inhibitor that reacted with the high calcium content produced water and plugged the production tubing, which required costly workover operations with mills to remove the formed deposit. Further details on the chemical analysis of the deposits in this case are discussed in Chapter 14.

Halite inhibitor hexacyanoferrate (HCF) is known to have incompatibility issues with iron ion. HCF reacts with Fe(III) salts in the produced water to produce a Prussian blue color; therefore an iron sequestering agent must be added to the formulation [16].

THPS was found to serve as a potential source of sulfate [22,23]. High sulfate ion concentration in the produced water in addition to gypsum and barite deposits were detected in the pigging debris of a pipeline with increased dosage of THPS. Upon decreasing the THPS dosage, the amount of sulfate ion in the produced water as well as the sulfate deposits subsequently decreased [13].

Triazines, which are amine-based H_2S scavengers, interact with H_2S and precipitate dithiazine solids in systems [24,25]. In some cases, an MEA-dithiazine product has been found causing fouling in the sales gas in downstream transmission and distribution pipeline systems [26]. In addition to this, triazines are known to increase the system pH to the level that causes calcium carbonate deposition [27]. In the reported BP-operated Bruce produced water reinjection plant, the 6″ discharge pipework was 50% choked and the suction pipework 30% choked. Analysis later confirmed this scale to be mostly mackinawite ($FeS_{0.93–0.96}$) with minor amounts of siderite ($FeCO_3$) and strontium-bearing barite ($Ba(Sr)SO_4$). Approximately 1.8 mT was removed from one 40-m run of pipework. The root cause of scale deposition was the alkaline H_2S scavenger [28].

Sulfite-based oxygen scavengers are known to give off sulfate ions as an end product after their reaction with oxygen. In the presence of metal ions, sulfate scale can form. Furthermore, field observations showed that iron sulfide and sulfur were found only downstream of the bisulfite injection point, suggesting a contribution of the bisulfite-disproportionating and sulfate-reducing bacteria in the scale to iron corrosion [29].

High pH solvents, production chemicals, are definitely incompatible with produced waters with high Ca, Fe, and Mg ions.

Acids used in downhole stimulations and in cleaning jobs cause asphaltene destabilization and precipitation. Case study (1) in Chapter 14 illustrates the effect of acids on asphaltene stability. The acid job was performed due to improper analysis of previously collected deposit samples, which stated that the main deposit composition was carbonate scale and misdiagnosed asphaltenes as paraffins.

Compatibility issues may arise between HF-based stimulation fluids that are used in sandstone with metal ions, like Ca ions. Furthermore, the metal ions will precipitate the dissolved silica, forming silicate deposits. Therefore a preflush and overflush of an ammonium salt is often used to remove incompatible ions [16,30].

Paraffinic dissolvers (pentane, hexane, heptane) are highly incompatible with asphaltenic crude oils, causing immediate aggregation and deposition of asphaltenes.

Richter et al. [31] reported black tar-like fouling material in a highly sour system. The analytical work indicated that the nitrogen-containing corrosion inhibitor polymerized with sulfur compounds (polysulfides, elemental sulfur, and/or H_2S) in a type of a vulcanization process, resulting in a hard-to-clean insoluble fouling product.

Thermodynamic hydrate inhibitors, i.e., methanol and MEG, are known to induce solids deposition. They change the WAT of paraffin waxes, induce asphaltene aggregation, and affect the supersaturation of mineral scales, making them deposit faster and in higher amounts.

Some KHIs have cloud points and deposition points in water. For example, PVCap has a cloud point of about 35–40°C in fresh water and would be expected to precipitate above this temperature [16]. They were reported to deposit in hot-produced high saline waters. The problem relates to the salting-out effect of electrolytes according to the Hofmeister series. Most salts show a salting-out effect, including sodium chloride [16,32–34]. Precipitating KHIs renders them inactive and leads to gunking at the injection point, which is usually at the wellhead where the produced fluids are hot. Therefore most operators have to check KHI compatibility with the hot produced fluids at the injection point [16].

Drag reducers based on partially hydrolyzed polyacrylamide (PHPA) or other polymers containing acrylate monomer have low compatibility with high-calcium brines [16].

Some viscoelastic surfactants used in acid stimulations have compatibility issues with high brine concentration. Water shut-off chemicals that are based on silicates also have compatibility issues with high calcium content brines [16].

Also, many classes of surfactants (nonionic, cationic, and amphoteric) can accelerate hydrate formation, especially if they have good foaming or emulsifying properties giving better contact of the gas (or gas dissolved in oil) with the water phase [16].

11.4 Treatment chemicals-materials incompatibility

The production chemicals not only have to be compatible with each other and the produced fluids, but also with the different materials, e.g., metals, alloys, and polymers, that constitute the production system tubulars, equipment, umbilical lines, and capillaries. In offshore applications, particularly in deepwater where many components of the production system are not readily accessible, the system compatibility of any production chemical is an essential part of the flow assurance process [35].

The incompatibility between the chemicals and the system materials can be in the form of metal corrosion, in the case of nonmetal or polymeric materials swelling, clumping, dissolution, or shrinkage. The corrosion products and deformed polymeric materials may finally pose a threat of flow assurance plugging in the production system. A few examples are given in the following paragraphs.

Highly acidic production chemicals cause pipeline corrosion. Mineral acids, acidic scale dissolvers, are highly corrosive and usually are mixed with corrosion inhibitor to prevent corrosion.

Scale inhibitors like phosphate esters, phosphonates, and some polymer scale inhibitors all are known to be corrosive to low alloy steels (LAS) [36].

Recently, an 18″ low-pressure production header on a producing platform in the Gulf of Mexico failed under normal working pressure. The failure ruptured the header, opening a 10-ft long split and releasing production fluids at a pressure of 250 psi. Three workers were sprayed with production fluid and approximately 25 bbls were released. The event resulted in the evacuation of the platform [37]. The main cause was found to be corrosive loss of pipe integrity at chemical injection point due to improper injection of acidic demulsifier.

Biocides were also reported to be corrosive. For example, THPS biocide is corrosive to mild steel [38] and 316 stainless steel [39], and its corrosivity increases with the addition of catalysts, like ammonium chloride, and with increasing temperature.

Overly dosed corrosion inhibitors can promote further corrosion, and under the effect of temperature and pressure can lead to deposit formation and plugging in the production system. Some cases of corrosion inhibitors overdosing deposits are mentioned in Chapter 14.

Methanol was reported to cause severe corrosion problems under sour conditions, including localized corrosion, sulfide stress cracking (SSC), and stress orientated hydrogen-induced cracking (SOHIC). In addition, methanol interferes with the corrosion inhibitors and removes their surface passive film [40]. Methanol was reported to cause damage to pipelines in under 4 years of service after permeating through HDPE, which, under the operating conditions and in the presence of water and formic acid from treatments, the problem was exacerbated and corrosion products were accumulated between the liner and the steel pipe [41]. Water-based drilling muds were also reported to be corrosive [42].

A few examples of the incompatibility of the treating chemicals with the nonmetallic (polymer) systems are:

- LDHI (low dosage hydrate inhibitor) is often incompatible with HNBR and Viton.
- Corrosion inhibitor is usually incompatible with Viton at temperatures exceeding 140°F. Also, corrosion inhibitors that contain amines are usually incompatible with certain Viton grades.
- Methanol, paraffin inhibitors, and demulsifiers can permeate some polymers and elastomers. Methanol does not appear to be compatible with nylon. In one example, methanol migrated through a nylon umbilical and into a scale inhibitor in an adjacent umbilical, with the result that the scale inhibitor precipitated [16].
- Compatibility of the asphaltene solvents with elastomers and plastic materials that are contained within the system must be checked before use [16].
- Pyridines are good asphaltene dissolvers, but they are toxic and incompatible with many elastomers [16].

The compatibility between the treating chemicals and the materials was and still is an issue of conflict in the literature, due to the differences in purity of chemicals, concentration, application conditions, and also the materials compositing and manufacturing conditions [35].

11.5 Treatment chemicals-process conditions incompatibility

Process operational conditions can cause physical and chemical changes in the treating chemical. It is common to watch chemicals aggregate, degrade, polymerize, or junk under high temperature or high pressure, which finally leads to precipitation. A few examples are given in the following paragraphs.

Scale inhibitors are known to degrade under HPHT conditions [43,44].

Sulfite-based oxygen scavengers break down at pressures as low as 600 psig (41 bar abs), resulting in the formation of sulfur dioxide or hydrogen sulfide; they can also be a source of sulfate ions upon heating, which in total might cause sulfate and sulfide deposits [45,46].

Inappropriate handling, dosing, or production chemicals in the system usually lead to problems of junking, clumping, and deposition in the pipelines, injection points, and umbilical lines. In one case in the Egyptian field, the overly dosed corrosion inhibitor, under the effect of high temperature from a heater treater, promoted corrosion and gunking, which led to severe depositions in the system. In another case, the corrosion inhibitor, under high pressure and high temperature, caused sulfide deposits in the gas system.

Glycols were found to carry over in the system and accumulate in laminar and dead spots, causing major deposition problems. In one case in the Egyptian field, the glycol unit suffered from corrosion that resulted in glycol leaks in the gas system, which under system pressure and temperature led to a severe deposition problem. They also can polymerize at high temperature to polyethylene glycol, causing plugging. In severe cases, high temperature can transform glycol into charred material that can deposit or serve as nucleating sites for other solid deposits.

Solvent evaporation or flashing is common in downhole injection systems, and gas lift lines can lead to chemical deposition if the solvent system evaporates too rapidly. This has been known to occur with scale inhibitors and other production chemicals [16].

11.6 Screening and testing of treatment chemicals compatibility

Treating chemicals are usually screened for their efficiency and performance before application in the field. Besides, extensive testing must also be applied to explore their compatibility with the produced fluids, the other injected chemicals, and process materials under the process operating conditions.

In a wider sense, the treating chemical tests include the following:

- Static precipitation (bottle test, jar test)
- Turbidity tests
- Dynamic blockage tube/flow loop tests
- Thermal aging
- Total suspended solids
- Foaming tendencies

- Viscosity tests
- Other physical properties

Depending on the complexity of the system, the number of chemicals applied in the system, and the operating conditions, some of the previous tests can be used to evaluate the chemical compatibility. Usually these tests are applied on a single chemical, a mixture of chemicals, or chemical/s+produced fluids under room temperature and system operating conditions (p, T) for a specific period of time.

Static precipitation tests are simple and commonly used. To screen for compatibility issues, it is also recommended to apply static tests while mixing different chemicals to test their efficiency under process conditions, using different time intervals to add the chemicals to each other and different total testing times (for example, testing a scale inhibitor+corrosion inhibitor mixture, in synthetic brine or real produced water at T°C, adding the corrosion inhibitor to each scale inhibitor at different times of 30 s, 1 min, and 5 min intervals, to understand their reaction times and incompatibility, and their effects on performance).

Dynamic tests are effective in screening the chemical performance and also indicative in the case of chemical gunking. Turbidity tests are very effective in detecting small precipitations and incompatibilities between chemicals and are highly recommended. Thermal aging involves aging the chemical/chemical mixture in an autoclave at a specific temperature for a specific time and evaluate its performance and gunking or residues. Viscosity and other physical properties are critical for application in different chemical injection systems. In addition, chemicals that are used in squeeze treatments may require core flooding tests for better addressing of their efficiency and compatibility. Furthermore, the state of the art and newly applied chemicals may require stricter testing protocols due to the lack of compatibility history.

Total suspended solids (TSS) are recommended for chemicals once received, after thermal aging, and can also be applied during other tests such as static tests.

Moreover, after compatibility tests are concluded, interaction maps can be constructed for pairs of additives of different chemical types. These maps indicate regions of concentrations at which one or both additives are not effective [47].

Taking scale inhibitors as an example, scale inhibitors used in squeeze treatments require further studies besides the regular efficiency testing, to account for their chemical compatibility. First, compatibility tests are needed to check that the scale inhibitor does not precipitate when mixed with formation brines, causing formation damage (in some instances, precipitation can be advantageous for enhancing inhibitor retention in a squeeze treatment). Second, and especially for high-temperature reservoirs, thermal aging tests are needed to make sure the inhibitor is stable at the reservoir temperature for the expected squeeze lifetime. The inhibitor solution is aged in a static bottle and then performance-tested against a nonaged sample. Then turbidity testing of the scale inhibitor (aged and nonaged) is carried out with the other chemicals injected in the same well or pipeline, by mixing them at different ratios and seeing the change in turbidity. Furthermore, core flood tests are used to assess the performance of the chemical and the potential of formation damage [16].

Special attention must be given to the chemicals applied in high pressure/high temperature environments (for possible thermal degradation and byproduct interactions), those applied in low temperatures like subsea and deepwater systems (for possible thickening and gunking), and those applied through umbilical lines (for chemical thickening and gunking) and gas lift systems (for solvent evaporation and chemical gunking).

To check the compatibility between the chemicals and the system materials, some basic tests can be done, including:

- Corrosivity test. Some standards that can be referred to are NACE TM0169-2021/ASTM G31-21 Standard Guide for Laboratory Immersion Corrosion Testing of Metals and standard NACE SP0775, Standard Practice Preparation, Installation, Analysis, and Interpretation of Corrosion Coupons in Oilfield Operations as reference for the acceptance criteria.
- For nonmetallic materials, some standard methods that might be useful include ASTM F2331-11 Standard Test Method for Determining Chemical Compatibility of Substances in Contact with Thermoplastic Pipe and Fittings Materials, ASTM D543-20 Standard Practices for Evaluating the Resistance of Plastics to Chemical Reagents, NORSOK Standard M-710 Edition 3—Qualification of Nonmetallic Materials and Manufacturers—Polymers, and ISO 23936—Petroleum, Petrochemical and Natural Gas Industries—Nonmetallic Materials In Contact With Media Related To Oil And Gas Production.

A suggested procedure to deal with the chemical treatment issues is as follows:

- Rigorously discuss and check the compatibility issues with the chemical vendor, especially with the state of art and new emerging chemicals.
- Review the chemical's MSDS, and refer to the safety, chemical composition, physical properties, and handling and storage sections. Try to collect as much information as possible about the main ingredients and their percentages (the main ingredient of some chemicals is a property of the manufacturer). In addition, the supplier usually mentions other chemicals that should be avoided, such as oxidizing agents, strong acids, or strong alkaline chemicals.
- Refer to the chemical's compatibility charts. Although these are mostly applicable to commodity chemicals, the specialty chemicals may have a major component that these charts are applicable for. However, these main ingredients are the property of the manufacturer. Nonetheless, the compatibility of some common chemicals like acids, alcohols, glycols, and others may be covered.
- It is highly recommended to do fingerprinting tests for the received chemicals with fingerprinting techniques like FT-IR or gas chromatography and compare the spectra/chromatogram with the successive received batches for any changes in composition, decomposition, or contaminations, and if any changes in composition are detected, the vendor must be notified and compatibility issues should be reevaluated.
- Perform the previously mentioned compatibility tests. It is also recommended to construct an interaction map to determine the windows/borders of incompatibility and efficiency of the different chemicals, and to prepare the possible solution in case the compatibility issues are significant.
- Adhere to the industry standards for chemical injection points to avoid the possible interactions. For example, during system design, subsea injection points are placed at least three pipe diameters away from each other to avoid incompatibilities that arise from mixing the injected chemicals. Furthermore, pumping or injecting chemicals must be applied in a safe way by pumping/injecting a proper solvent spacer between the different incompatible chemical batches. During long-term shutdowns, production chemicals in the umbilical tubes are usually replaced with solvent packages to ensure their integrity [16].
- Plan the chemical injection points to avoid the incompatibility that might arise between two interacting chemicals by choosing which chemical to inject first after adjusting the distance

between the two chemicals, to give the first injected chemical enough retention time to do its function in the produced fluids before injecting the other chemical. For example, when SI and CI are injected in close vicinity in a pipeline, it is recommended to inject SI before CI, since SI interacts with the scale-forming species, preventing the scale formation before it reacts with the CI. If, however, the CI was injected prior to SI, the film-forming CI will interact with SI, which results in diminishing both chemicals' performance. The same thing is expected with oxygen scavengers and corrosion inhibitors, as oxygen scavengers should be injected first to interact with the dissolved oxygen before it reaches the corrosion inhibitor injection point. This phenomena is illustrated in Fig. 11.6.

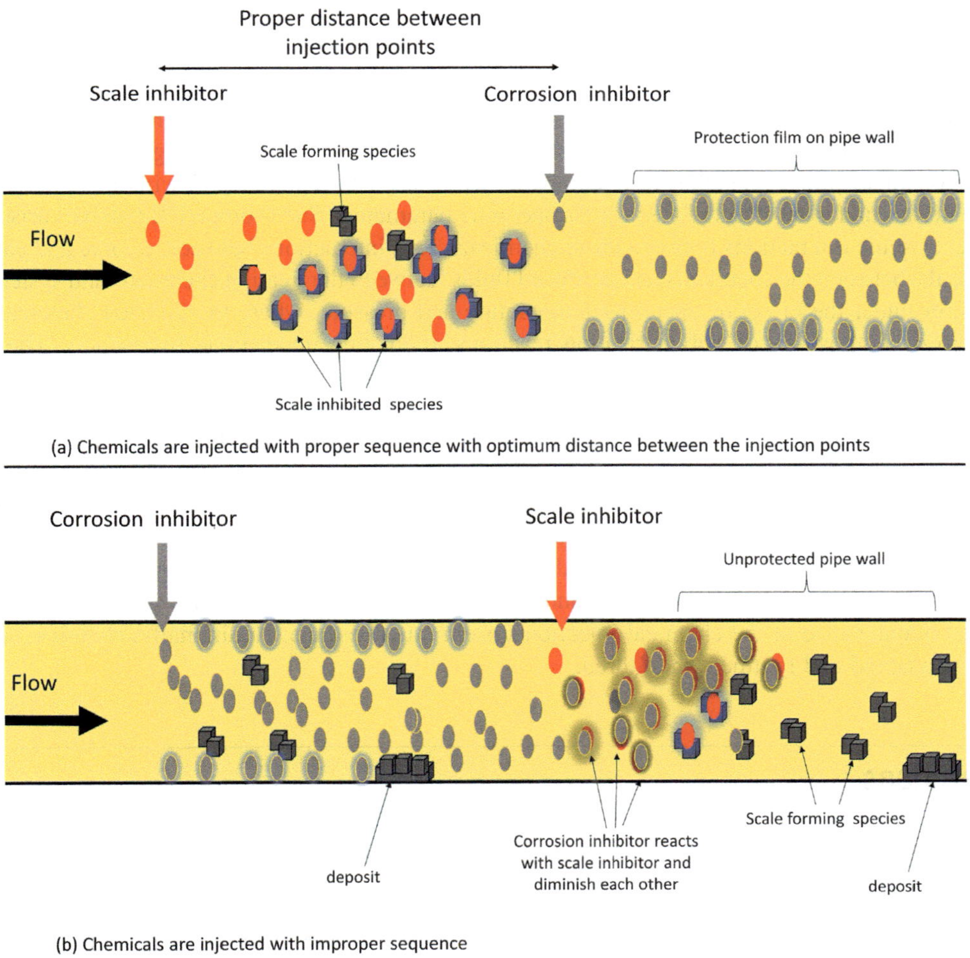

FIG. 11.6

(A) Proper and (B) improper chemicals injection scenarios.

  - In the case of severe incompatibility issues, dilution of the chemicals is considered a valid solution; however, the dose must be optimized to show good performance.
  - Another solution is changing the chemical for an alternative chemical with similar efficiency and better compatibility, i.e., *N,N*-didecyl-*N,N*-dimethylammonium chloride can be used intead of conventional biocides, to avoid incompatibility with oxygen scavengers, replacing triazine H_2S scavengers with more compatible oxidizing ones.
- Rigorously maintain and inspect the chemical injection systems, with continuous monitoring for changes in the chemical performance and chemical injected doses.
- Rigorously check on the stored chemicals to make sure their storage conditions are valid and do not exceed their shelf life.

11.7 Summary

Treating chemicals have been used in oil and gas fields to solve major production operation and flow assurance problems, and to improve process performance. However, these treating chemicals can pose a flow assurance threat when they have compatibility issues with each other, or with the process fluids and materials under the effect of the process operating parameters. Such compatibility issues arise from the electrostatic interactions between these chemicals, e.g., interactions between anionic and cationic chemicals, between the chemicals and produced fluids, between the treating chemicals and system materials, or under the effect of operating temperature and pressure, where the injected chemicals can undergo decomposition or polymerization reactions, producing undesirable byproducts. The main and direct impacts of such interactions are rendering the applied chemicals inefficient and affecting the system integrity. In addition, such interactions can lead to phase changes and precipitations in the system, known as pseudoscale or incompatibility deposits, which can clog the production system. Furthermore, emulsion formation and stabilization and sludge formation are other side effects of chemical incompatibility. Therefore a rigorous plan has to be in place for testing and applying these chemicals, to assure proper deployment and satisfying results, with lower costs and lower flow assurance problems. This includes chemical compatibility testing, besides the regular screening tests, and mapping the incompatible chemicals and their border concentrations. Moreover, routine monitoring and maintenance of the chemical injection system must be done. Additionally, chemical storage must be adapted and routinely monitored.

References

[1] W.C. Lyons, G.J. Plisga, M.D. Lorenz, Standard Handbook of Petroleum and Natural Gas Engineering, Elsevier Inc, 2015, pp. 1–1822, https://doi.org/10.1016/C2010-0-64695-5.

[2] G.M. Graham, D.M. Frigo, I.R. McCracken, G.C. Graham, W.J. Davidson, S. Kapusta, P. Shone, The influence of corrosion inhibitor/scale inhibitor interference on the selection of chemical treatments under harsh (HP/HT/HS) reservoir conditions, in: Society of Petroleum Engineers—International Symposium on Oilfield Scale 2001, Society of Petroleum Engineers, 2001, https://doi.org/10.2523/68330-ms.

[3] C. Meric, A. Sanchez, Compatibility of corrosion and scale inhibitors in simulated petroleum production sweet corrosion system, Rev. Tec. Intevep. 5 (2) (July 1985) 189–195.

[4] G.M. Graham, K.S. Sorbie, L.S. Boak, K. Taylor, L. Blilie, Development and application of accurate detection and assay techniques for oilfield scale inhibitors in produced water samples, in: SPE 28997, Presented at the SPE Int. Symposium on Oilfield Chemistry, Held in San Antonio, TX. 14–17 February, 1995.

[5] M. Rondon Gonzalez, N. Passade-Boupat, P. Bascoul, S. Baraka-Lokmane, C. Hurtevent, Selection of Anti-scale and Anti-corrosion Products: How to Avoid Interactions?, Society of Petroleum Engineers, 2012, January 1, https://doi.org/10.2118/155235-MS.

[6] Z. Amjad, R.W. Zuhl, A new look at industrial water system fouling caused by the interaction of anionic deposit control agents and cationic water treatment chemicals, in: NACE—International Corrosion Conference Series. Vol. 2002-April. National Assoc. of Corrosion Engineers International, 2002.

[7] Z. Amjad, R.W. Zuhl, The influence of polymer architecture on inhibition of amorphous calcium phosphate precipitation, Phosphorus Res. Bull. 13 (2002) 51–57, https://doi.org/10.3363/prb1992.13.0_51. Japanese Association of Inorganic Phosphorus Chemistry.

[8] Z. Amjad, Calcium phosphate inhibition by polymeric inhibitors the influence of aluminum (111), iron (11), and iron (111), Phosphorus Res. Bull. 9 (1999) 119–125.

[9] Z. Amjad, J.F. Zibrida, R.W. Zuhl, The effect of biocide on deposit control polymer performance, in: Association of Water Technologies' Annual Conference, Honolulu, HI, 2000.

[10] D.B. McIlwaine, Oilfield application for biocides, in: Directory of Microbicides for the Protection of Materials, Springer, Netherlands, 2004, pp. 157–175, https://doi.org/10.1007/1-4020-2818-0_10.

[11] J. Moore, V. Keasler, B. Bennett, Compatibility of Tetrakis (Hydroxymethyl) Phosphonium Sulfate (Thps) and Ammonium Bisulfite (Abs), NACE International, 2010.

[12] B. Folwell, Incompatibility of biocides and oxygen scavengers, an overlooked corrosion risk? in: Paper Presented at the CORROSION 2017, New Orleans, Louisiana, USA, March 2017.

[13] D. Andrade, A. Cynthia, E.G.M. Lopes, E.M.O. De Paiva, M.O. De Penna, The side effects of THPS treatment in oil pipelines, in: Rio Pipeline Conference and Exposition, Technical Papers. Vol. 2009-September. Instituto Brasileiro de Petroleo, 2009.

[14] B. Fu, Development of non-interfering corrosion inhibitors for sour gas pipelines with co-injection of kinetic hydrate inhibitors, in: NACE—International Corrosion Conference Series, 076661-626, 2007.

[15] J. Moore, L.V. Vers, P. Conrad, SS: flow assurance: understanding kinetic hydrate inhibitor and corrosion inhibitor interactions, in: Offshore Technology Conference, No. OTC 19869, Houston, TX, Paper Number 19869-MS, 2009, pp. 1–21.

[16] M.A. Kelland, Introduction and environmental issues, in: Production Chemicals for the Oil and Gas Industry, CRC Press, 2009, pp. 19–38, https://doi.org/10.1201/9781420092974-4.

[17] V.S. Saji, Sulfide scavengers and their interference in corrosion inhibition, in: Corrosion Inhibitors in the Oil and Gas Industry, Wiley, 2020, pp. 421–432, https://doi.org/10.1002/9783527822140.ch18.

[18] M. Yang, A.C. Stewart, G.A. Davies, Interactions Between Chemical Additives and Their Effects on Emulsion Separation, Society of Petroleum Engineers, 1996, January 1, https://doi.org/10.2118/36617-MS.

[19] J. Jackson, R. Harrington, D. Manko, Emulsion Tendency Studies—Understanding Method, Inhibitors and Water Cut, NACE International, 2012.

[20] X. Lang, P. Lv, S. Xu, B. Li, S. Fan, Y. Wang, Investigation of the performance of low-dosage hydrate inhibitors combined with wax inhibitors for a mixed CH_4-CO_2 gas hydrate formation, Can. J. Chem. 93 (9) (2015) 992–997, https://doi.org/10.1139/cjc-2014-0595. National Research Council of Canada.

[21] D. Shen, D. Shcolnik, R. Perkins, G. Taylor, M. Brown, Evaluation of Scale Inhibitors in Marcellus High-Iron Waters, Society of Petroleum Engineers, 2012, October 1, https://doi.org/10.2118/141145-PA.

[22] Q. Wang, et al., Laboratory assessment of tetrakis (hydroxymethyl) phosphonium sulfate as dissolver for scales formed in sour gas wells, Int. J. Corros. Scale Inhib. 4 (3) (2015) 235–254.

[23] D.O. Trahan, Method and Composition to Remove Iron and Iron Sulfide Compounds From Pipeline Networks, U.S. Patent No. 8,673,834, 2014.

[24] H.T. Madsen, E.G. Søgaard, Fouling formation during hydrogen sulfide scavenging with 1,3,5-tri-(hydroxyethyl)-hexahydro-s-triazine, Pet. Sci. Technol. 32 (18) (2014) 2230–2238, https://doi.org/10.1080/10916466.2013.783066.

[25] C.W. Titley, P.H. Wieninger, Method and Composition for Removing Sulfides From Hydrocarbon Streams, 2006. U.S. patent 7 (115), 215.

[26] S. Wang, M. Madekufamba, K.L. Lesage, F. Bernard, B.S. Gelfand, P.M. Davis, K. Tittemore, R.A. Marriott, K.K. Botros, Dithiazine vapour saturation in natural gas mixtures, J. Nat. Gas Sci. Eng. 78 (June) (2020), https://doi.org/10.1016/j.jngse.2020.103286. Elsevier B.V.

[27] M. Sumestry, H. Tedjawidjaja, Case Study—Calcium-Carbonate-Scale Inhibitor Performance Degradation Because of H_2S-Scavenger Injection in Semoga Field, Society of Petroleum Engineers, 2013, February 1, https://doi.org/10.2118/150705-PA.

[28] J.J. Wylde, H. McAllister, A.W. Duthie, Root cause failure analysis, removal and mitigation of iron sulfide scale deposition in the Bp Bruce produced water reinjection plant, in: Paper Presented at the CORROSION 2008, New Orleans, Louisiana, March 2008.

[29] H.S. Park, I. Chatterjee, X. Dong, S.H. Wang, C.W. Sensen, S.M. Caffrey, T.R. Jack, J. Boivin, G. Voordouw, Effect of sodium bisulfite injection on the microbial community composition in a brackish-water-transporting pipeline, Appl. Environ. Microbiol. 77 (19) (2011) 6908–6917, https://doi.org/10.1128/AEM.05891-11.

[30] M.U. Shafiq, H.B. Mahmud, Sandstone matrix acidizing knowledge and future development, J. Pet. Explor. Prod. Technol. (2017), https://doi.org/10.1007/s13202-017-0314-6. Springer Verlag.

[31] S. Richter, J. Locklear, M. Achour, T. Baugh, P. Singh, R. Goff, Control of Corrosion Inhibitor Fouling in a Highly Sour System, NACE International, 2018, July 20.

[32] M.A.J. Kelland, E. Fuels, Additives for kinetic hydrate inhibitor formulations to avoid polymer fouling at high injection temperatures: part 1, Energy Fuel 34 (3) (2020) 2643–2653, https://doi.org/10.1021/acs.energyfuels.9b04040.

[33] M.A. Kelland, K. Monig, J.E. Iversen, K. Lekvam, Feasibility study for the use of kinetic hydrate inhibitors in deep-water drilling fluids, Energy Fuel 22 (4) (2008) 2405–2410, https://doi.org/10.1021/ef800109e.

[34] M.A. Kelland, J.E.J.E. Iversen, Fuels, kinetic hydrate inhibition at pressures up to 760 bar in deep water drilling fluids, Energy Fuel 24 (5) (2010) 3003–3013, https://doi.org/10.1021/ef9016152.

[35] L.M. Frostman, C.G. Gallagher, S. Ramachandran, K. Weispfennig, Ensuring Systems Compatibility for Deepwater Chemicals, Society of Petroleum Engineers, 2001, January 1, https://doi.org/10.2118/65006-MS.

[36] M. Bhandari, I. Littlehales, S. Linares-Samaniego, R. Christmas, S. Mancuso, Development of low-corrosivity scale inhibitors for low alloys steels, in: Offshore Technology Conference, 2017, May 1, https://doi.org/10.4043/27816-MS.

[37] J. Mathews, Chemical Injection Point Corrosion Causes Production Header Failure, Bureau of Safety and Environmental Enforcement (BSEE), 2016. Safety Alert No. 319.

[38] Q. Wang, S. Shen, H. Badairy, T. Shafai, Y. Jeshi, T. Chen, F.F. Chang, Laboratory assessment of tetrakis(hydroxymethyl)phosphonium sulfate as dissolver for scales formed in sour gas wells, Int. J. Corros. Scale Inhib. 4 (3) (2015) 235–254, https://doi.org/10.17675/2305-6894-2015-4-3-235-254. IFHAN.

[39] A.E. Hussein, H.S. Mohamed, Studying the use of tetrakis(hydroxymethyl)phosphonium sulfate (THPS) as zinc sulfide and lead sulfide scales dissolver and the factors influencing the dissolution, Int. J. Corros. Scale Inhib. 6 (3) (2017) 349–358, https://doi.org/10.17675/2305-6894-2017-6-3-9. Russian Association of Corrosion Engineers.

[40] N.G. Park, L. Morello, J.E. Wong, S.A. Maksoud, The Effect of Oxygenated Methanol on Corrosion of Carbon Steel in Sour Wet Gas Environments, NACE International, 2007.

[41] L. Simon, R. MacDonald, K. Goerz, M. Grigis, Corrosion failure in a lined sour gas pipeline—part 1: case history of incident, in: Proceedings of the NACE NAWC, Regina, SK, Canada, February 2010, pp. 15–18.

[42] Z. Wang, D. Yin, Y. Luo, H. Liang, L. Yang, L. Hu, Corrosion and its inhibition in water-based drilling fluid used in onshore oilfied, Nat. Resour. 04 (07) (2013) 456–459, https://doi.org/10.4236/nr.2013.47056. Scientific Research Publishing, Inc.

[43] W. Wang, A.T. Kan, W. Shi, C. Fan, M.B. Tomson, Scaling Control for High-Temperature and High-Pressure Oil Production in Mineral Scales in Biological and Industrial Systems, Taylor and Francis, 2010.

[44] K.D. Demadis, A. Ketsetzi, Degradation of phosphonate-based scale inhibitor additives in the presence of oxidizing biocides: 'collateral damages' in industrial water systems, Sep. Sci. Technol. 42 (2007) 1639–1649, https://doi.org/10.1080/01496390701290532.

[45] A. Jaffer, D. Fulmer, J. Ensslen, Recent Developments in Organic Oxygen Scavenger Technology, Corrosion NACE, 2006. Paper Number 06678.

[46] Arkema, Oxygen Scavengers, Selection Guide for Oxygen Scavengers, QC173 7/01, LITHO IN USA, 2001.

[47] M.R. Gonzalez, N. Passade-Boupat, P. Bascoul, S. Baraka-Lokmane, C. Hurtevent, Selection of anti-scale and anti-corrosion products: how to avoid interactions? in: Paper Presented at the SPE International Conference on Oilfield Scale, Aberdeen, UK, May 2012, https://doi.org/10.2118/155235-MS.

CHAPTER

Flow Assurance Solids Prediction and Modeling

12

12.1 Introduction

Flow assurance solids are challenging problems in petroleum production systems. The management strategy to control such solids usually comprises fluids characterization, modeling, mitigation, and monitoring. Thus modeling is considered a crucial and central step in the solids management strategy, as it is a key tool in the deposits risk assessment. Furthermore, the recent advanced versions of solids modeling packages are able to calculate and propose chemical inhibitors and their proper doses.

A good modeling job during the early stages of field development assists in modifying and optimizing the design concept and the operating parameters, so that the system will be less prone to solids formation, and in adding solids treatment solutions to the system, such as electric heating, internal coating, physical treatment devices, smart completions, monitoring devices, chemical injection lines and points, and others.

While solids prediction software packages seem like a black box to most of their users, their basic concepts are not that mysterious. A prediction model basically uses the fluids chemical and physical properties and the system parameters to calculate the tendency towards solids precipitation and deposition. Such calculations are based on thermodynamic, kinetic, and hydrodynamic data, which accordingly are based on a good understanding of the mechanisms by which the deposits form and the factors affecting them, in addition to the field observations from the different case studies. These data are then all put together in interactive algorithms, which are introduced to the end user through sophisticated user-friendly interfaces, to assess the risks of deposit formation. Fig. 12.1 illustrates the basic data used in constructing a prediction model.

12.2 General terms

In this section, the general terms that are frequently used in simulation and modeling studies are briefly reviewed.

- **System**

The word system is generally used to describe a separate part of the world subjected to study. A system can be anything physical or virtual, real or hypothetical, and can range from the size of a box containing a few molecules, to entire planets. Anything not part of the system is referred to as its surroundings or

Essentials of Flow Assurance Solids in Oil and Gas Operations. https://doi.org/10.1016/B978-0-323-99118-6.00018-6

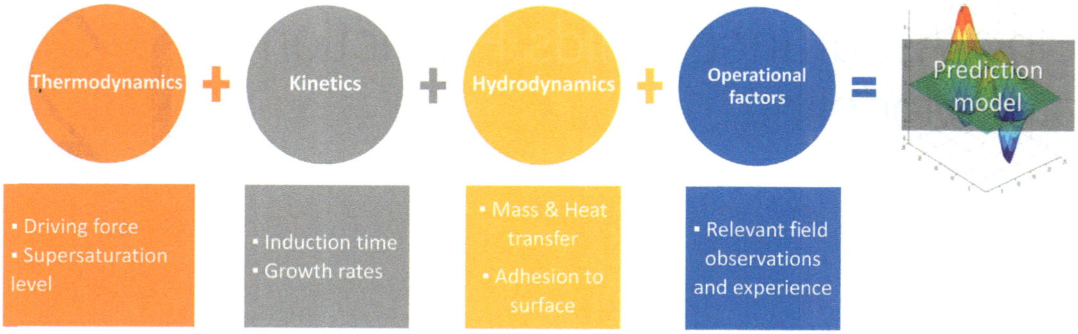

FIG. 12.1

Basic components of solids deposition prediction models.

environment and is separated from the system by a wall or boundary [1]. An open system allows exchanging mass and energy; a closed system allows exchanging energy but not mass, whereas an isolated system does not allow exchanging energy or mass.

A system in this chapter will necessarily mean a package of production fluids: water + oil + gas that being tested; this should be distinguished in the context from a "production system" where the main operations take place.

- **State**

A system thermodynamic state is defined by the values of the system properties, including among others pressure, volume, and temperature. A system state is uniquely defined by its property values, and if any of them change, the state of the system changes. A mathematical expression that defines the value of a state property is called a state function [1,2]. An equation of state (EOS) is a functional relationship between state functions describing the equilibrium state of a system [3].

EOS is a thermodynamic expression that relates pressure (P), temperature (T), volume (V), and composition of a system (or a fluid system in our case). In this chapter, different EOSs are generally used to understand the phase behavior and other properties of complex liquid mixtures at high pressures and temperatures. The first reliable EOS for relating pressure, P, volume, V, and temperature, T, of dilute gases was the ideal gas law. Subsequent equations of state had to be developed to describe their behavior, including cubic and noncubic equations of state [4].

- **Thermodynamics**

Thermodynamics studies energy and work and their effect on a system or the properties of matter. Chemical thermodynamics studies the change in phase or state of material and disposition of a chemical reaction. Thermodynamics usually is used to predict if the reaction is spontaneous, the heat absorbed or released in a chemical reaction, the partitioning of species in different phases, the density of fluids, viscosity of fluids, molar volumes, and to predict other "thermophysical" properties of matter at a known temperature and pressure.

- **Kinetics**

Kinetics is the study of actions of forces in producing or changing the motion of masses. In chemistry, chemical kinetics is concerned with chemical reaction rates and how the molecular processes change with time [5]. In this chapter, kinetics describes the growth rates of solids, which can also include mass transfer and hydrodynamic terms.

- **Phase equilibrium**

Multicomponent mixtures are not always homogeneous and might divide into two or more phases. In a multiphase system, the number of phases coexisting in equilibrium is determined by a stability analysis. The system is said to be in equilibrium when the total Gibbs energy (*G*) is minimum; therefore a fluid mixture will continuously split into additional phases until Gibbs energy cannot be reduced any further. Fig. 12.2 shows the stability analysis that determines the number of phases by comparing the total Gibbs energy in one, two, and three phases. For this example, if G_1 is lower than G_2, then only one phase exists. However, if G_2 is lower than G_1, then the fluid would be more stable by creating a second phase. Similarly, a fluid exists in at least three phases when G_3 is lower than G_2 and G_1 [3].

This means that, among the many states that may satisfy the equilibrium relations, only the state that minimizes the Gibbs energy is the true equilibrium state [4].

A system at a state of equilibrium is a thermodynamic state, wherein the different phases of the substance having common boundary surfaces do not vary quantitatively. The condition of phase

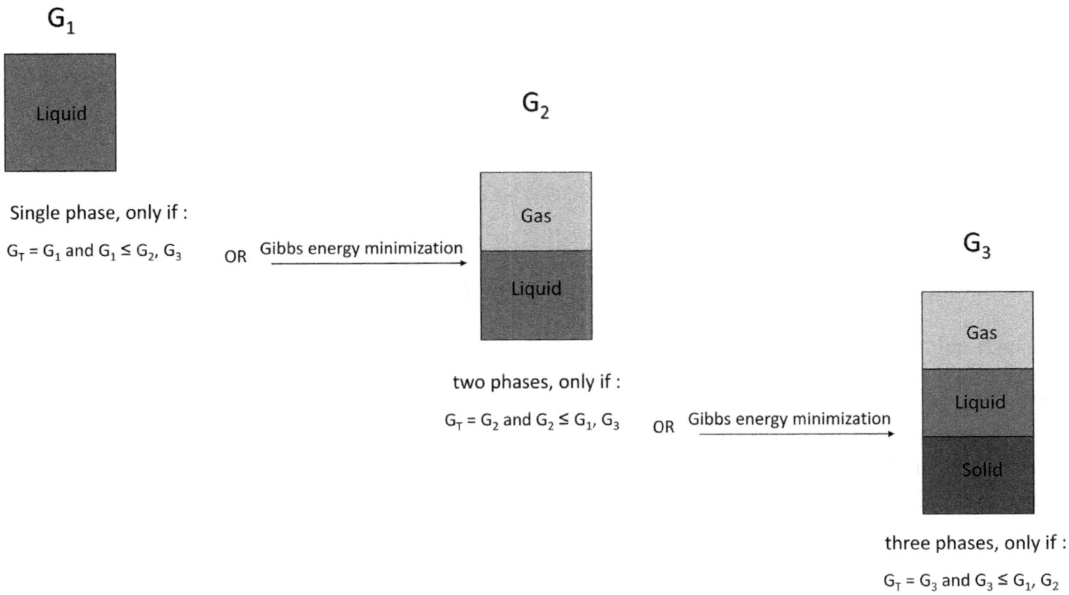

FIG. 12.2

A representation of system stability, which determines the number of phases (gas, liquid, solid) considering that the total Gibbs energy is minimized.

equilibrium is the equality of chemical potentials for each component in all phases; for component i the chemical potential μ (J/mol) at equilibrium is governed by Eq. (12.1):

$$\mu_i^{oil} = \mu_i^{gas} = \mu_i^{Aq} \tag{12.1}$$

or in fugacity *f* (Pa) terms

$$f_i^{oil} = f_i^{gas} = f_i^{Aq} \tag{12.2}$$

For multicomponent systems such as petroleum, it is a complex procedure to determine the equilibrium for every single component. Thus it is feasible to lump similar fractions into pseudocomponents in order to simplify the calculations. The less complicated the system is, the easier is the use of the model. This will be applied during modeling wax and asphaltenes [6].

- **Gibbs energy**

Gibbs free energy ΔG is the amount of energy available for work for any chemical reaction. The Gibbs free energy is also known as the free energy, the Gibbs function, and originally called "*available energy*" by Gibbs in 1873. ΔG is usually used to determine the spontaneity of a reaction where:

$$\begin{array}{c} \Delta G^{\circ} < 0 \rightarrow \textit{spontaneous reaction} \\ \Delta G^{\circ} > 0 \rightarrow \textit{nonspontaneous reaction} \end{array} \tag{12.3}$$

A minimum in the Gibbs free energy will determine the stable state or phase.

Gibbs free energy of component *i* with activity a_i in a mixture is given by the difference in the chemical potential between the current state μ_i and the standard state μ_i^o [7]:

$$\mu_i - \mu_i^o = \Delta G_i = RT \ \ln \frac{a_i}{a_i^o} \tag{12.4}$$

Related to the Gibbs free energy is the excess Gibbs free energy ΔG^E, which is used to express the difference between a real and ideal solution, where [7]:

$$\Delta G^E = \Delta G^{Real} - \Delta G^{ideal} \tag{12.5}$$

$$\Delta G^E = RT \ \ln \left(a_i^{Real} - a_i^{ideal} \right) = RT \ \ln \left(C_i \gamma_i - C_i \right) = RT \ \ln \gamma_i \tag{12.6}$$

where a is the activity, γ is the activity coefficient, and C is the molar concentration.

- **Flash calculations**

For a multicomponent mixture, calculating the number of phases at equilibrium and the distribution of components between those phases at specific pressure and temperature conditions is known as flash calculations, or PT flash [4].

For example, for a specific mixture of fixed overall composition, flash calculations are used for processes with vapor/liquid-equilibrium (VLE) to determine the number of phases and the components distribution between the phases at equilibrium (vapor at fixed pressure and temperature). For example, Fig. 12.3 shows a flash tank where a typical process requiring flash calculations occurs when a feed stream (*F*) is separated into a vapor (*V*) and liquid (*L*) product.

In a production system of a mixture of fluids, the gas starts to expel from liquids when the pressure falls below the bubble point.

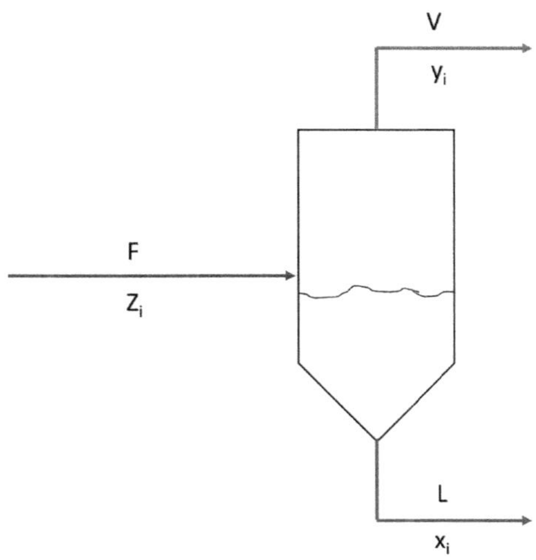

FIG. 12.3

Flash tank and phase separation, F=fluids mixture, V=vapor, L=liquid.

- **Concentration**

Material concentration is defined as the amount of a given component related to the amount of surrounding matrix or amount of component related to the total amount of material [8]. It can be expressed in molarity, M=moles/liter (mol/L), which is normally represented by the two brackets [ ], normality, N =equivalents/liter (Eq. L^{-1}), molality, m=moles/kg (mol kg^{-1}). Lower concentrations are usually expressed by parts per million, ppm (mg L^{-1} or mg kg^{-1}), parts per billion, ppb (μg L^{-1} or μg kg^{-1}). Relatively high concentrations are expressed in percentage, % (%wt or %vol), or by the dimensionless mole fraction, X. For reactions or systems that involve gas or a mixture of gases, the gas concentration is usually expressed by its partial pressure.

Materials concentration/amount is a crucial variable in the prediction models, since when the dissolved material component exceeds its solubility limit it starts to precipitate.

- **Activity**

Activity (symbol a) is a measure of the "effective concentration" of a species in a mixture. The concentration C can be related to activity a by using the activity coefficient γ, where:

$$a = \gamma \cdot C \tag{12.7}$$

The activity of pure substances in condensed phases (solid or liquid) is normally taken as unity. Activity depends on temperature, pressure, and composition of the mixture, among other things. For gases, the activity is the effective partial pressure, and is usually referred to as *fugacity*.

- **Fugacity**

The fugacity of a real gas is an effective partial pressure. Fugacity is also a quantitative measure of the "escaping" tendency of a chemical from one phase to another [9]. When the fugacities of a chemical in two different compartments that are in contact are unequal, the chemical will be redistributed between the compartments, with net transfer of the chemical from the higher fugacity compartment to the lower fugacity compartment. When the fugacities are equal, the net transfer is zero and equilibrium exists [10].

Fugacity can be used to replace chemical potential in equilibrium calculations, as it makes the calculations easier. The fugacity is used to define the thermodynamic equilibrium in a hydrocarbon multicomponent mixture where gaseous hydrocarbons are in equilibrium with liquid hydrocarbons.

The fugacity f_i of component i is given by:

$$f_i = x_i \varnothing_i \gamma_i P \tag{12.8}$$

$\varnothing_i$ is the fugacity coefficient, x_i is the mole fraction (or molar concentration), γ is the activity coefficient, P is the pressure (kPa).

Generally speaking, fugacities and partition coefficients are usually needed for determining the equilibrium state of a mixture.

- **Chemical potential**

Chemical potential principally determines the stability of substances, such as chemical species, compounds, and solutions, and their tendency to chemically react to form new substances, to transform to new physical states, or to migrate from one spatial location to another [11]. Chemical potential (μ) is defined as the chemical energy (or in other words the Gibbs free energy, G) possessed by 1 mol of the substance as expressed by Eq. (12.9):

$$\mu = \frac{G}{n} \tag{12.9}$$

where n is number of moles of the substance.

The reactivity of the chemical species described by the chemical potential is determined by how the Gibbs free energy (ΔG) of the system changes when the number of moles or molecules of the reactant in the system changes at constant pressure and temperature. The system tends to minimize its Gibbs free energy by lowering its chemical potential.

In other words, the chemical potential is the driving force that causes the dissolved substance to do phase transformation and precipitation. Chemical species usually tend to migrate from high chemical potential regions to lower chemical potential regions, until the difference in chemical potential is ~0. At thermodynamic equilibrium, the chemical potentials of each component in liquid, solid, and vapor should be equal.

The chemical potential is related to the activity (a) and fugacity (f) according to Eq. (12.10):

$$\mu = \mu^o + RT \ln\frac{f}{f^o} = \mu^o + RT \ln a \tag{12.10}$$

where μ^o is the chemical potential at the standard state, R is the gas constant, T is the temperature, and f^o is fugacity at the standard state.

- **Theory of solution**

The theory of solution describes, using statistical mechanics, how the free energy varies with the composition of the solution, and with the chemical composition of its components. Such an EOS provides the key to the determination of phase transitions, critical points, solubilities, immiscible regions, metastable and unstable regions, etc. [12].

Electrolyte theory or theory of electrolytes is used to address the properties of aqueous electrolyte solutions, such as NaCl. The mixed-solvent electrolyte (MSE) model is an advanced excess free energy theory that expands the prediction range for solutions containing electrolytes and nonelectrolytes. MSE is commonly applied in the scale prediction models.

- **Mass transfer**

Mass transfer is a net movement or transport of substance from one location to another in a fluid (liquid or gas). Some of the modes of mass transfer are convection, diffusion, or advection.

- **Heat transfer**

Heat transfer is a process of transfer of heat (thermal energy) from hotter spots to colder spots. Main mechanisms of heat transfer include convection, conduction, and radiation.

In the following sections the main simulation and modeling tools are discussed in brief.

12.3 Fluid flow modeling

Computational fluid dynamics modeling or CFD is based on the principles of fluid mechanics, utilizing numerical methods and algorithms to solve problems that involve fluid flows. The equations governing fluid motion are the three fundamental principles of mass, momentum, and energy conservation. The fundamentals of CFD are well furnished elsewhere in the literature, but here it is mentioned because it is considered a crucial tool for more realistic representation of solids deposition.

Models usually simulate multiphase flow and the interaction of liquids and gases; they can also integrate chemical reactions, the flow of solids through a system, and other processes associated with fluid flow, to holistic 3D depiction of the flowing fluids in a conduit/vessel.

CFD has been used for different applications in oil and gas fields, as mentioned in Chapter 2, Flow Assurance. It is important to employ CFD in solids formation and deposition modeling since it gives realistic representation of the precipitated solids transportation in the flowing fluids and the chances of their adherence to the surfaces, so deposition thickness, masses, and growth rates will be more realistic.

12.4 Chemical and geochemical reactions simulation

Various types of interactions can take place in a geological system like a petroleum reservoir, leading to different transformations and changes in phase equilibria of the reservoir fluids, which consequently affects the fluids modeling and solids prediction calculations. Continuous pressure, temperature changes, and fluids injection in the reservoir result in reservoir rock-fluids and fluids-fluids interactions. It is crucial to identify and quantify the effect of such geochemical reactions on the phase behavior of any real reservoir fluid mixture.

Geochemical simulation usually utilizes reactive transport modeling to understand the chemical reaction and fluids flow in porous media.

12.5 Reservoir simulation

In reservoir modeling, physics, mathematics, engineering, and geological data are used to simulate the reservoir fluids behavior over time under operating strategies. It is one of the necessary tools used to evaluate and select the future recovery options to maximize revenue.

Reservoir simulation can be used in solids deposition to provide crucial data such as flow in porous media, injection water breakthrough, possible geochemical reactions, fluids flow in the wells, and others. Such data can be used to improve the solids deposition simulation results.

12.6 Mineral scale prediction

Scale prediction is very common in many industries: e.g., oil and gas, power plants, paper and pulp, water and wastewater treatment, and others. The prediction software applications range from simple straightforward equations with a few inputs to complex multifunctioning integrated commercial packages. Some basics of the principles that are used in these prediction tools are discussed in the next few sections.

12.6.1 Basic principles of scale prediction

12.6.1.1 Thermodynamics of mineral scale prediction

Thermodynamics of scale prediction is principally based on the solution chemistry and solubility equilibria.

Assume a sparingly soluble salt in water *MX*: its equilibrium is represented by Eq. (12.11):

$$MX \leftrightarrow M^{z+} + X^{z-} \tag{12.11}$$

Following the law of mass action, and given that the concentration of the solid *MX* is the unit, then the solubility product constant is expressed by Eqs. (12.12)–(12.14):

$$K_{sp,MX,(T,P)} = a_{M^{z+}} \times a_{X^{z-}} \tag{12.12}$$

$$K_{sp,MX,(T,P)} = IAP \tag{12.13}$$

$$SR = \frac{IAP}{K_{sp,MX,(T,P)}} \tag{12.14}$$

where K_{sp} is the salt solubility product (or the equilibrium constant, dimensionless), a_{Mz+}, a_{xz-} are the activities of the ionic species (mol/L), IAP is the ion activity product, and SR is the saturation ratio. Here, the equilibrium status of the solution can be one of three cases: when $\mathrm{SR} < 1$ the solution is undersaturated and no precipitation occurs; when $\mathrm{SR} = 1$ the solution is saturated or at equilibrium; while when $\mathrm{SR} > 1$ the solution is supersaturated and precipitation occurs.

From Eq. (12.14), the saturation index (SI) can be represented as:

$$\mathrm{SI} = \log_{10}\mathrm{SR} = \log_{10}\frac{IAP}{K_{sp,MX,(T,P)}} \tag{12.15}$$

where SI is the saturation index, and like SR, SI has three cases: SI = 0 at equilibrium, SI < 0 at under-saturation and no precipitation, and SI > 0 represents supersaturation and salt precipitation.

SI is a measure of the degree of supersaturation, and thus the precipitation driving force, the larger the value of SI, the greater the likelihood that scale will occur [13].

While these equations look straightforward (measure the ions concentrations and compare it to the solubility constant), they still have countless of thermodynamic work to represent real solutions. To realize this, let's take the activity in Eq. (12.12) and represent it in terms of activity coefficient and concentration in Eq. (12.16):

$$K_{sp} = \gamma_{M^{z+}} C_{M^{z+}} \times \gamma_{X^{z-}} C_{X^{z-}} \tag{12.16}$$

where $\gamma_{M^{z+}}$, $\gamma_{C^{z-}}$ are the activity coefficient of M^{z+}, and X^{z-} (dimensionless), $C_{M^{z+}}$, and $C_{X^{z-}}$ are the concentration of M^{z+} and X^{z-} respectively (mol/L).

In the same way, Eq. (12.14) will be re-written to be as:

$$\mathrm{SR} = \frac{\gamma_{M^{z+}} C_{M^{z+}} \times \gamma_{X^{z-}} C_{X^{z-}}}{K_{sp,MX,(T,P)}} \tag{12.17}$$

Similarly, the saturation index of the common mineral scales can be represented by Eqs. (12.18)–(12.22)

$$\mathrm{SI}_{\mathrm{calcite}} = \log_{10}\left(\frac{\gamma_{\mathrm{Ca}^{2+}}\left[\mathrm{Ca}^{2+}\right] \times \gamma_{\mathrm{CO}_3^{2-}}\left[\mathrm{CO}_3^{2-}\right]}{K_{sp\mathrm{CaCO}_3}}\right) \tag{12.18}$$

$$\mathrm{SI}_{\mathrm{barite}} = \log_{10}\left(\frac{\gamma_{\mathrm{Ba}^{2+}}\left[\mathrm{Ba}^{2+}\right] \times \gamma_{\mathrm{SO}_4^{2-}}\left[\mathrm{SO}_4^{2-}\right]}{K_{sp\mathrm{BaSO}_4}}\right) \tag{12.19}$$

$$\mathrm{SI}_{\mathrm{celestite}} = \log_{10}\left(\frac{\gamma_{\mathrm{Sr}^{2+}}\left[\mathrm{Sr}^{2+}\right] \times \gamma_{\mathrm{SO}_4^{2-}}\left[\mathrm{SO}_4^{2-}\right]}{K_{sp\mathrm{SrSO}_4}}\right) \tag{12.20}$$

$$\mathrm{SI}_{\mathrm{gypsum}} = \log_{10}\left(\frac{\gamma_{\mathrm{Ca}^{2+}}\left[\mathrm{Ca}^{2+}\right] \times \gamma_{\mathrm{SO}_4^{2-}}\left[\mathrm{SO}_4^{2-}\right]}{K_{sp\mathrm{CaSO}_4}}\right) \tag{12.21}$$

$$\mathrm{SI}_{\mathrm{halite}} = \log_{10}\left(\frac{\gamma_{\mathrm{Na}^{+}}\left[\mathrm{Na}^{+}\right] \times \gamma_{\mathrm{Cl}^{-}}\left[\mathrm{Cl}^{-}\right]}{K_{sp\mathrm{NaCl}}}\right) \tag{12.22}$$

where the brackets [ ] represents the molar concentration of the ions, and γ is the activity coefficient of the ions, K_{sp} is the solubility product. K_{sp} and γ of the different species are basically calculated by the scale prediction software.

For the following produced water (Table 12.1) chemistry calculate the saturation index of calcium carbonate and barium sulfate, knowing that the K_{sp} of calcium carbonate = 3.36×10^{-9}, and that of barium sulfate = 1.08×10^{-10}.

Solution:

Using Eqs. (12.18), (12.19)

$$\mathrm{SI}(\mathrm{CaCO}_3) = \log_{10}\left(\frac{[0.029 \times 0.253]\left[3.3 \times 10^{-5} \times 0.093\right]}{3.36 \times 10^{-9}}\right) = 0.83$$

Table 12.1 Produced water ions concentration, and their activity coefficient.

Cation	Concentration (mol/L)	Activity coefficient (25°C, 1 atm)	Anion	Concentration ppm (mol/L)	Activity coefficient (25°C, 1 atm)
Na^+	0.76	0.665	Cl^-	0.7418	0.681
K^+	0.026	0.612	SO_4^{--}	0.021	0.087
Ba^{++}	2.18×10^{-5}	0.207	HCO_3^-	0.0043	0.555
Ca^{++}	0.029	0.253	CO_3^{--}	3.3×10^{-5}	0.093

$$\mathrm{SI}(\mathrm{BaSO_4}) = \log_{10}\left(\frac{\left[2.18\times10^{-5}\times0.27\right]\left[0.021\times0.087\right]}{1.08\times10^{-10}}\right) = 1.99$$

Speculate doing these calculations without taking the activity coefficient in consideration. The results will be different indeed. Therefore it is then really important to take the activity coefficient in consideration to represent the real solution and to adjust for the deviation from ideality.

Both K_{sp}, γ are temperature, pressure and ionic strength dependent and technically, their data are available either in low salinity matrix at different temperatures and pressures or in high salinity matrix (usually using simple background electrolyte KCl) but at STP which does not represent the high ionic strength reservoir waters which are produced at high temperature and pressure. Hence, the main essence of the scale prediction software is to calculate K_{sp}, γ or to correlate the experimentally determined effect of temperature, pressure and ionic strength on both K_{sp} and γ [14]. So we will discuss these two terms in brief.

The solubility constant $K_{sp(T,P)}$ at a constant temperature and pressure can be described by Eq. (12.23) [15]:

$$K_{sp(T,P)} = e^{\left(-\frac{\Delta G^\circ}{RT}\right)} \tag{12.23}$$

where ΔG° is the change in standard-state Gibbs energy for equilibrium ($\mathrm{kJ\,mol^{-1}}$) at the temperature T (K), and R is the gas constant $8.3\,\mathrm{J\,K^{-1}\,mol^{-1}}$.

The temperature effect on the solubility constant is also addressed using the relationship between the change in Gibbs free energy (ΔG) for a reaction, the enthalpy ΔH and the entropy ΔS at a specific temperature T, and reference pressure P_o, according to Eq. (12.24) [15–17]:

$$\ln K_{sp(T,P_o)} = -\frac{\Delta H^o_{T,P_o}}{RT} + \frac{\Delta S^o_{T,P_o}}{R} \tag{12.24}$$

where $\Delta H^o_{T,Po}$ is the standard state enthalpy change in $\mathrm{kJ\,mol^{-1}}$, $\Delta S^o_{T,Po}$ is the standard state entropy change in $\mathrm{J\,K^{-1}\,mol^{-1}}$, T in K, and P in Pa.

The relationship between the solubility product at pressure P (Pa) and at the reference pressure P_o is described according to Eq. (12.25) [15–17]:

$$\ln K_{sp(T,P)} = \ln K_{sp(T,P_o)} - \frac{\Delta V^o_{P_o}}{RT}(P-P_o) + \frac{\Delta k^o}{2RT}(P-P_o)^2 \tag{12.25}$$

where $\Delta V^{o}_{P_o}$ is the change in standard partial molal volume at infinite dilution for a given reaction [m^3/mol], and Δk^{o} is change in standard partial molal compressibility [$cm^3\ mol^{-1}\ bar^{-1}$].

The other term that is crucial for scale prediction is the activity coefficient. It is a measure of the deviation from ideality in solutions. In diluted solutions, the activity of species is affected by the long-range forces (repulsion between same charge molecules and the attraction between oppositely charged molecules), and at infinite dilution the activity coefficient is unity. With increasing of the ionic strength, the short-range forces (attraction and repletion) increase. This increase in the intermolecular forces decreases the ion energy available for reactions (affects ion activity). The activity coefficient is then a measure of the solvent-solute and solute-solute and solvent-solvent intermolecular interactions.

One of the common theories to calculate the activity coefficient is the Debye-Hückel rule: the activity coefficient is determined according to Eq. (12.26) [18]:

$$\log\gamma = -Az^2\sqrt{I_s} \tag{12.26}$$

where A is a temperature-dependent constant (0.5092 at 25°C), Z is the ions electric charge, I_s is ionic strength given by Eq. (12.27):

$$I_s = \frac{1}{2}\sum C_m z^2 \tag{12.27}$$

where C_m is the ions molal concentration.

Later, the Debye-Hückel rule was extended by many others, like Robinson and Stokes, Davis, and Helgeson (B dot form). A comparison between the different models is depicted in Fig. 12.4.

The activity coefficient in the Debye-Hückel equations and their modified forms are derived based on NaCl matrix brines; however, when dealing with other brines, such as NaBr-based completion fluids, other methods are more appropriate. The Pitzer model is one of the commonly accepted models for high ionic strength solutions, over a wide range of temperatures and pressures. Pitzer theory

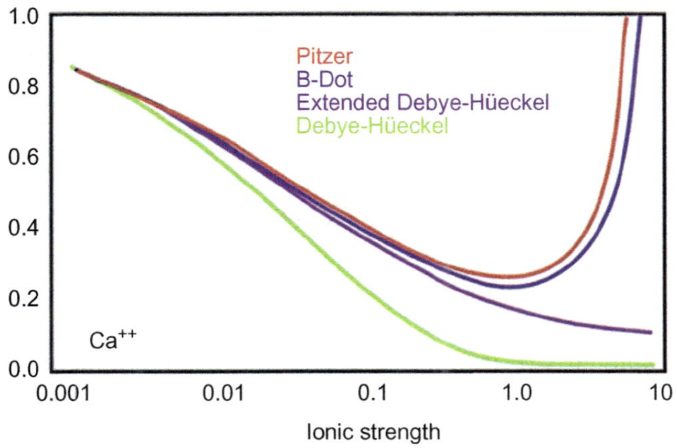

FIG. 12.4

Activity coefficient models.

Reprinted from R.J. Ferguson, Scaling indices: types and applications, in: Mineral Scales and Deposits, 2015, pp. 721–735. Elsevier. https://doi.org/10.1016/b978-0-444-63228-9.00030-9, copyright (2015), with permission from Elsevier.

basically uses viral coefficients to represent the binary and ternary species interactions as functions of temperature, pressure, and compositions; thus it takes into consideration the short-range interactions and the long-range interactions [19].

The mean activity coefficient for an anion and a cation can be calculated from the excess Gibbs energy by using Eq. (12.28) [20–22].

$$\begin{aligned}\ln \gamma_M &= \left(\frac{\partial G^{Excess}_{pitzer}/RT}{\partial n_M}\right)_{T,P,n_j} \\ &= z_M^2 F(I) + \sum_a m_a(2B_{Ma} + ZC_{Ma}) + \sum_c m_c\left(2\Phi_{Mc} + \sum_a m_a \Psi_{Mca}\right) + \sum_{a<}\sum_{a'} m_a m_{a'} \Psi_{Maa'} \\ &+ |z_M| \sum_c \sum_a m_c m_a C_{ca} + 2\sum_n m_n \lambda_{nM}\end{aligned} \tag{12.28}$$

where Z is the ion charge, m denotes molality of the ions, the subscripts M and c denote cations and a and a' denote anions, and n denotes neutral molecules. The Pitzer coefficients B_{Ma}, C_{Ma}, Φ_{Mc}, Ψ_{Mca}, $\Psi_{Maa'}$, C_{ca}, and λ_{nM} are functions of temperature, pressure, ionic strength, and composition. Φ represents the interaction between the independent ions, which are not common with the substances, and Ψ represents the interaction between the common ions. F is a function of the ionic strength and molalities of cations and anions [20,22].

The Pitzer model is applied to determine the solubility and activity coefficient of minerals allowing a wide range of temperatures, pressures, ionic strength, and cosolvents [22].

Now, the ion concentrations in the preceding equations are assumed to be free species, meaning they are free to react, which is not true, as these ions do not tend to exist totally as free ions. Calcium, for example, may be paired with sulfate, bicarbonate, carbonate, phosphate, and other species, forming complexes even in low ionic strength waters [23]. The direct impact of this in scale prediction is overestimation of scaling risks. While this overestimation might seem negligible in low ionic strength waters, it can lead to significant errors in scaling prediction in high ionic strength waters, knowing that oil and gas produced waters are known for their high ionic strength nature.

For example, the concentration of the sulfate ion is determined using the formula in Eq. (12.29):

$$C_{SO4} = \underbrace{[CaSO_4^o] + [MgSO_4^o] + [SrSO_4^o] + [BaSO_4^o]}_{pairedsulfate} + \underbrace{[SO_4^{--}]}_{Freesulfate} \tag{12.29}$$

Oddo and Tomson [24] calculated the free sulfate concentration by defining the complex stability constant K_{st} according to Eq. (12.30):

$$\log K_{st} = +1.86 + 4.5 \times 10^{-3}\mathrm{T} - 1.2 \times 10^{-6}\mathrm{T}^2 + 10.7 \times 10^{-5}\mathrm{P} - 2.38\sqrt{I_s} + 0.58 I_s - 1.3 \times 10^{-3}\sqrt{I_s}\, T \tag{12.30}$$

where T is the temperature in °F, I_s is the ionic strength (mg/L), P = total absolute pressure (psia).

Using the stability constant, the free sulfate ion concentration is expressed by Eq. (12.31):

$$[SO_4^{--}] = \frac{-\left\{1 + K_{st}\left(\sum C_M - C_{SO4}\right)\right\} + \left\{\left[1 + K_{st}\left(\sum C_M - C_{SO4}\right)\right]^2 + 4K_{st}C_{SO4}\right\}^{0.5}}{2K_{st}} \tag{12.31}$$

where

$\sum C_M = C_{Ca} + C_{Mg} + C_{sr} + C_{Ba}$, where C refers to the total molar concentration of the metal ion. In the same way, the free metal ion concentration is calculated according to Eqs. (12.32)–(12.35):

$$[\mathrm{Mg}^{++}] = \frac{C_{Mg}}{1 + K_{st}[\mathrm{SO}_4^{--}]} \tag{12.32}$$

$$[\mathrm{Ca}^{++}] = \frac{C_{Ca}}{1 + K_{st}[\mathrm{SO}_4^{--}]} \tag{12.33}$$

$$[\mathrm{Sr}^{++}] = \frac{C_{Sr}}{1 + K_{st}[\mathrm{SO}_4^{--}]} \tag{12.34}$$

$$[\mathrm{Ba}^{++}] = \frac{C_{Ba}}{1 + K_{st}[\mathrm{SO}_4^{--}]} \tag{12.35}$$

So far we have discussed the interactions and phase equilibria within the aqueous phase: the mineral solubility constants, activity coefficient, ion interactions and associations, which are relatively sufficient to do some valid calculations in water systems. However, in complex multiphase petroleum production systems that include gas, oil, and water phases, the equilibrium between the different phases must be solved to estimate the partitioning of gases like H_2S or CO_2 between the phases. The partitioning of such gases can affect certain water properties like pH, which is a critical parameter in scale prediction calculations, affecting the scaling tendency of carbonates and sulfides scales (Fig. 12.5).

For this purpose, the vapor-liquid equilibrium (VLE) for every component in the mixture is calculated by the quality of the component fugacities in the liquid and in the gas phases [25]. Such calculations are known as the flash calculations, as introduced in Section 12.2.

Various types of models are used to calculate the component fugacity; however, a few are accurate for all phases and compositions. The EOS (such as the Soave-Redlich-Kwong or Peng-Robinson EOS) is used for hydrocarbon gas and oil calculations, and Henry's law is usually applied to the aqueous phase, since EOS is not suitable for fugacity calculations in aqueous high salt phases. Gibbs excess

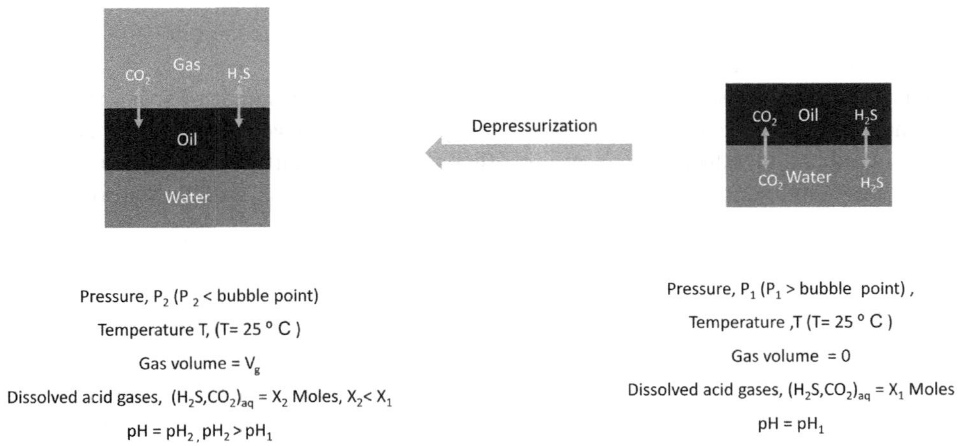

FIG. 12.5

Gases partitioning between phases and their effects on pH upon depressurization.

models are also much better in the aqueous phase [7]. Thus a combination of these models is usually used in scale prediction packages to describe the phase behavior of the liquid and gas [25].

The result of the VLE flash calculations are the fugacities, bubble point, and other properties of the modeled fluids, which ultimately are used with other formulas to determine the phase equilibrium and partitioning of gases between the phases, which affects the final results of scale prediction.

Using these calculations, a proper scaling tendency can be achieved; for example, based on Eq. (5.1) (Chapter 5) the saturation index in case of calcium carbonate in presence of free gas containing CO_2 can be rewritten to be:

$$SI_{calcite} = \log_{10}\left(\frac{K_2\,\gamma_{Ca^{2+}}\,[Ca^{2+}] \times \left(\gamma_{HCO_3^-}\right)^2 \times \left[HCO_3^-\right]^2}{K_{spCaCO_3} \times K_1 \times K_{H,CO_2} \times P_{CO_2} \times \varnothing_{CO_2}}\right) \tag{12.36}$$

where K_1 and K_2 represent the first and *sec*ond ionization constants of carbonic acid; $\varnothing_{CO_2}$ represents the fugacity coefficient of CO_2 in the gas; and $K_{H,CO2}$ represents Henry's law constant for the partition of CO_2 between gas and water (molal atm^{-1}) [22].

Another issue that affects the thermodynamics of scale formation is the effect of hydrate inhibitors on the salts scaling tendencies, as it increases their saturation index, as mentioned in Chapter 4. Many models have been proposed to study the effect of glycols and methanol on the different types of scales [7,22,26].

12.6.1.2 Kinetics of scale prediction

The thermodynamics of scale modeling indicates the scaling tendency of minerals formation, while the scaling rate—and thus the time required to cause plugging—is basically determined by kinetics. Kinetics can also refer to the mass transfer and hydrodynamic terms that affect the scale growth rates.

The kinetics of scale prediction is fundamentally concerned with the induction time and precipitation reaction rates, specifically nucleation and crystal growth rates. If the scale induction time is shorter than the residence time in a production system, then deposit formation is inevitable. Such data are crucial not only for scale prediction but are used in scale control, as the scale inhibitor application is based on extending the induction time and retarding the growth rates.

The induction time, as discussed in Chapter 4, Principles of Flow Assurance Solids Formation Mechanisms, is the inverse of nucleation rate ($t = 1/J_n$). Many researchers have proposed different models and formulas to calculate the induction time under different testing conditions. For example, Eq. (12.37), which represents a correlation for barite scale induction time, was formulated by Tomson et al. [26].

$$\log\left(t_{ind}^0, \sec\right) = -2.235 + \frac{[1087.2 - 0.297T]^2}{T^2 SI} - 0.123\frac{[1087.2 - 0.297T]^3}{T^3 SI^3} \tag{12.37}$$

where t_{ind}^0 is the predicted induction time (s); T is the temperature (T absolute); and SI is the calculated saturation index at a certain temperature, pressure, and composition.

They also predicted the impact of applying scale inhibitors on the nucleation time in a semiempirical manner by assuming a separation of the effect of the inhibitor from that of the uninhibited mineral nucleation, according to Eq. (12.38) [27]:

$$\log_{10}\left(t_{ind}^{inh}\right) = \log_{10}\left(t_{ind}^0, \sec\right) + b_{inh}.C_{inh} \tag{12.38}$$

where t_{ind}^{inh} is the nucleation time in the presence of inhibitors; t_{ind}^{0} is the nucleation time in the absence of inhibitors; C_{inh} is the concentration of inhibitor (mg/L) added, and b_{inh} is inhibitor effectiveness (L/mg).

The rate of scale deposition is basically determined by both transport and surface reaction.

12.6.1.3 Hydrodynamics of scale prediction

The previous kinetics and thermodynamics discussions can be used to indicate the tendency, rate, and mass of formed precipitates in a solution. But what affects the mass transport of the chemical species in the flowing fluids, and are the nucleation and crystal growth the same in both laminar flow and turbulent flow? What are the factors affecting them? All these aspects are usually addressed by the hydrodynamics.

Solids can nucleate, grow, and deposit right on the metal surface, or they can precipitate in the bulk of the solution and then be transported to the metal surface where they deposit.

Generally speaking, for scaling on surfaces, the growth of deposits can be limited thermodynamically (by favorable nucleation sites) or kinetically (when the material supply limits the growth). In laminar flow, the scale deposition is thermodynamically dominant, where scale is formed in favored nucleation sites, whereas in turbulent flow the growth is limited by the kinetics of influx of scaling ions to the surface. The turbulence usually increases the ionic influx towards the wall [28].

Bulk precipitated solids are transported to the surface by many mechanisms (or regimes), such as electrostatic, gravitational, diffusion, shear dispersion, or inertial forces. Furthermore, the transported particles have to attach and interlock with the surface. Thus whether the formed precipitate will stick and form a deposition layer is basically a hydrodynamics-based process, based on the fluids flow regimes in the different media (reservoir porous media, vertical or horizontal tubulars), in addition to the particle deposition regimes.

Hydrodynamics are also taking into consideration the effect of surface roughness, where a smooth newly fabricated pipeline is expected to experience lower deposition rates than a worn surface, which will agitate turbulence in the fluid flow and reduce energy requirements for surface deposition, thereby reinforcing scale deposition [28].

12.6.1.4 Geochemical terms of scale prediction

Reservoir fluids are subjected to various changes during production operations, which lead to changes in overall equilibrium in the reservoir. The reservoir fluids phase equilibrium is critical information that is extensively studied in scale prediction, reservoir simulation, and other tools. The geochemical interactions, such as fluids-reservoir rock interactions, are also a significant factor that can seriously affect the final calculations of scale prediction. For example, the interactions between the reservoir fluid-reservoir rocks under different pressure and temperature and interaction between injected fluids-reservoir rocks can subsequently create a new phase equilibrium. Such interactions can create new scaling conditions or eliminate existing ones which, if not taken into consideration, might lead to underestimation or overestimation of the scaling tendencies.

One of the common geochemical simulators is PHREEQC, which is a geochemical modeling program developed by the United States Geological Survey and written in C and C++ to model a variety of geochemical reactions and transport calculations.

12.6.2 Thermodynamic saturation and stability indices

The saturation and stability indices are thermodynamic tools that are used to estimate the risk of scale formation. These are simple, straightforward methods of estimating the risk of scale with fewer variables and input data. Table 12.2 summarizes the different scaling and stability indices [29–33].

The Langelier saturation index is the most famous calcium carbonate scaling index. It was developed by Langelier in 1936 to describe the ability of a natural oxygen-containing water to form protective carbonate layers on an iron surface or to be corrosive [13,34]. This index indicates the tendency of a water to precipitate calcium carbonate, but it does not indicate the amount of precipitate [13]. The Langelier index is valid in solutions with pH between 6.5 and 9.0, up to 4000 ppm, and temperatures up to 95°C [35,36].

Ryznar [37] developed an empirical method for predicting the stability of a water in terms of its corrosiveness or tendency to deposit $CaCO_3$ scale. The RSI was meant to be semiquantitative in the sense that higher index means high corrosion rates, while the lower the index, the larger the amount of scale that is formed.

The Puckorius (or Practical) scaling index (PSI) was developed by the work of Puckorius and Brooke [38], taking into account some of the shortfalls of LSI and RSI. PSI attempts to further quantify the relationship between saturation state and scale formation by incorporating an estimate of buffering capacity of the water into the index [39].

The Caldwell-Lawrence diagram is a graphical method consisting of a family of curves for different alkalinity, calcium, and pH values.

Stiff and Davis [40] first introduced a graphical method for predicting the formation of calcium sulfate scale in oil field waters, taking common ion effects like Na and Mg into consideration. They then extended the Langelier method for calcium carbonate to apply to oilfield brines with high total dissolved solids, commonly known as the Stiff and Davis stability index (SDSI). It is valid between temperatures from 0 to 100°C and ionic strengths up to 3.6 [41]. Inputs needed are pH, alkalinity, calcium concentration, and ionic strength. Values of K in their formula are obtained from a graphical correlation with ionic strength and the temperature of the water. Figs. 12.6 and 12.7 illustrate the curves that are used to determine the K value [42].

The Larson-Skold Index (L&SkI or LS) is used to describe the corrosivity of water towards mild steel and was developed based on in-situ measurements of corrosion in steel lines transporting Great Lakes water [43]. The index describes the ratio of concentration of chloride and sulfate ions to the concentration of bicarbonate and carbonate ions [44].

Skillman et al. [45] developed a method to predict the solubility of gypsum scale (NOT the Saturation index OR the total possible scale formed), and it is valid between temperatures of 10–80°C and ionic strengths from 0 to 6. Inputs needed are ionic strength and sulfate and calcium ion concentrations. The solubility constant, K_{sp} is calculated using the graph of K_{sp} vs. temperature and ionic strength. This method can also be used to predict the solubility of barite and celestite scale, if the K_{sp} variation with temperature and pressure are known [46].

The Oddo-Tomson method is another way of predicting the formation of calcium carbonate and various sulfate scales by calculating SI values. Their initial work was published in 1982 and the method was subsequently modified to include the effects of additional parameters. The reader is referred to their published paper in 1994 [24], which summarized the main formulas. It is valid between temperatures of 0–200°C (32–392°F), ionic strengths of 0–4.0, and pressures of 1–1380 bar (0–20,000 psig).

Table 12.2 Summary of common mineral scale saturation and stability indices.

Index	Equation	Variables	Index criteria and Water conditions
Langelier saturation index (LSI)	$LSI = pH_{meas} - pH_s$	pH_{meas} = water measured pH pH_s = pH at saturated $CaCO_3$ $pH_s = (9.3 + A + B) - (C + D)$ $A = \frac{(\log_{10}[TDS] - 1)}{10}$ $B = -13.12 \times [\log_{10}(^{\circ}C + 273) + 34.55]$	$LSI < 0$: No $CaCO_3$ precipitation, water is corrosive $LSI > 0$: Water is supersaturated, $CaCO_3$ precipitates $LSI = 0$: Borderline scale potential, water is saturated.
Ryznar stability (RSI)	$RSI = 2pH_s - pH_{meas}$	$C = \log_{10}[Ca^{2+}$ as ppm $CaCO_3] - 0.4$ $D = \log_{10}[Alkalinity\ as$ ppm $CaCO_3]$	$RSI < 6$ Supersaturated, tend to precipitate $CaCO_3$ $6 < RSI < 7$ Saturated, $CaCO_3$ is in equilibrium $RSI > 7$ Undersaturated, tend to dissolve solid $CaCO_3$
Puckorius scaling (PSI)	$PSI = 2pH_s - pH_{eq}$	pH_{eq} is the equilibrium pH based on the value of total alkalinity $pH_{eq} = 1.465 \times \log\ (T.\ Alk) + 4.54$, where $T.Alk$ is the total alkalinity, pHs is measured as LSI and RSI.	$PSI < 6$ Scaling is likely to occur $PSI > 7$ It likely tends to dissolve scale
The Aggressive Index (AI)	$AI = pH + \log(Alk)(H)$	*Alk* is the alkalinity and *H* is the calcium hardness, both expressed in mg/L as $CaCO_3$	$AI > 12$ Scaling, or nonaggressive (noncorrosive) $10 < AI < 12$ It is moderately aggressive $AI < 10$ It is very aggressive
Larson ratio (LS)	$LS = \frac{[Cl^-] + 2[SO_4^{--}]}{[HCO_3^-]}$	The square braces indicate the concentrations of respective ions in mol/L	$LS < 0.8$ Scaling, chloride and sulfate are unlikely to interfere with the formation of protecting film $0.8 < LS < 1.2$ Corrosion rates may be higher than expected $LS < 1.2$ High rates of localized corrosion may be expected
Stiff and Davis	$SDSI = pH + \log[Ca^{+2}] + \log[Alk] - K$	*K* is a constant that is a function of salinity, composition, and water temperature. Values of *K* are obtained from a graphical correlation with ionic strength and the temperature of the water in the range of: *T* (0°C, 30°C, 50°C), *P* (1 atm), *I* (0–3.6). All curves outside of this data range are extrapolated	$SDSI < 0$ water is undersaturated, no $CaCO_3$ precipitates $SDSI > 0$ water is supersaturated, $CaCO_3$ will precipitate

Continued

Table 12.2 Summary of common mineral scale saturation and stability indices—cont'd

Index	Equation	Variables	Index criteria and Water conditions
			SDSI = 0 equilibrium state, water is saturated
Skillman Index	$I = \frac{S_{actual}}{S_{theritical}}$	S_{actual} = actual solubility, $S_{throretical}$ = theoretical solubility, $S_{theoretical} = 1000 \times \left(\sqrt{(X^2 - 4K_{sp})} - X\right)$ X is the excess common-ion concentration of the calcium and sulfate ions $x = \{2.5 \times [Ca^{2+}]$ or[Ba] or $[Sr] - 1.04 \times [SO_4^{2-}]\} \times 10^{-5}$ M/L	$S_{theoretical} = S_{Actual}$, water is saturated with sulfate deposit $S_{theoretical} > S_{Actual}$, water is undersaturated, no sulfate precipitations $S_{theoretical} < S_{Actual}$, water is supersaturated, Sulfate precipitates

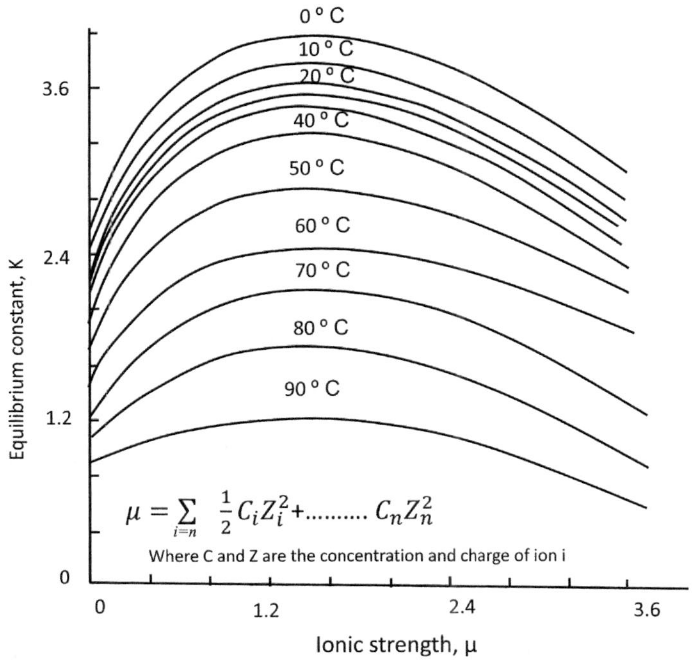

FIG. 12.6

Values of K as a function of ionic strength.

Reprinted from Z. Ahmad, Selection of materials for corrosive environment, in: Principles of Corrosion Engineering and Corrosion Control, Elsevier, 2006, pp. 479–549. https://doi.org/10.1016/b978-075065924-6/50010-6, copyright (2006), with permission from Elsevier.

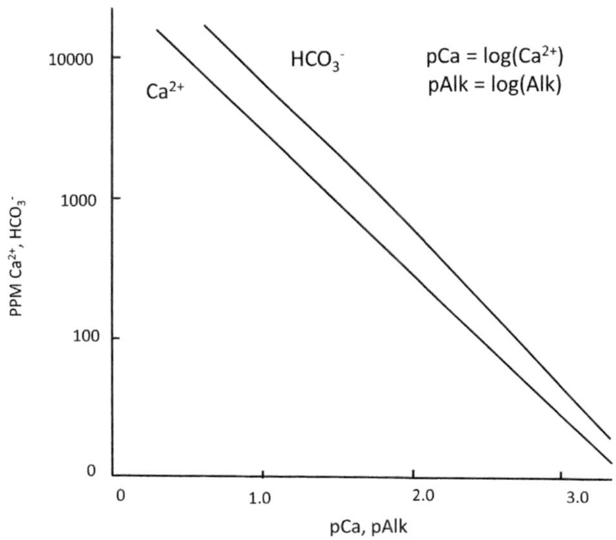

FIG. 12.7

Determination of pCa and pAlk from the ionic concentration.

Reprinted from Z. Ahmad, Selection of materials for corrosive environment, in: Principles of Corrosion Engineering and Corrosion Control, Elsevier, 2006, pp. 479–549. https://doi.org/10.1016/b978-075065924-6/50010-6, copyright (2006), with permission from Elsevier.

The model considers the effect of total pressure, ionic strength, and varying CO_2 partial pressures [13,46]. The model needs simple data for ion analysis (Ca^{2+}, Ba^{2+}, Sr^{2+}, HCO_3^-, CO_3^{2-}, and SO_4^{2-}), temperature in °F, pressure in psia, and CO_2 concentration in the gas and aqueous phases. If the amount of carbon dioxide is unknown but there is an accurate pH measurement, the method uses the pH to calculate the saturation indices [46]. The model also is capable of calculating the pH based on the amount of CO_2 gas and bicarbonate in the water, which allows greater accuracy in calculating the actual saturation index of a water sample, since pH measurements decline in accuracy very quickly after the sample is taken out of production systems at high pressure and temperature.

The Oddo-Tomson model is more accurate than the Stiff-Davis and other models, and also represents the basic scaling index module in some of the recently developed commercial scale prediction software applications, which are discussed in the following sections.

12.6.3 Industrial and academic scale prediction software packages

12.6.3.1 Basic principles and calculations of prediction software

Computerized scale prediction software was first published in the early 1980s. Since then, there have been many computer programs and software packages developed either for in-house use or commercial purposes [47,48].

Examples of these software packages are OKSCALE, SCALE, SCALED, DOWNHOLE SAT, and GEOCHEMIST'S WORKBENCH [48]. Recently a variety of scale prediction software has become available on a commercial and academic basis. Examples are Scalesoftpitzer, MultiScale, SCALE2000, ScaleChem, and others.

The main organization of the current scale prediction software consists of inputs, processing, and outputs. The inputs include fluids physicochemical properties and operating conditions. The processing involves flash calculations to determine the phase equilibria, and also theory of electrolytes calculations to determine the activity coefficient and solubility product of the scale compounds. The output is basically the scaling index of each type of deposit, amount of scale, recommended dose of chemical inhibitors, and other outputs that are relevant to the case. Fig. 12.8 summarizes the main organization of scale prediction software.

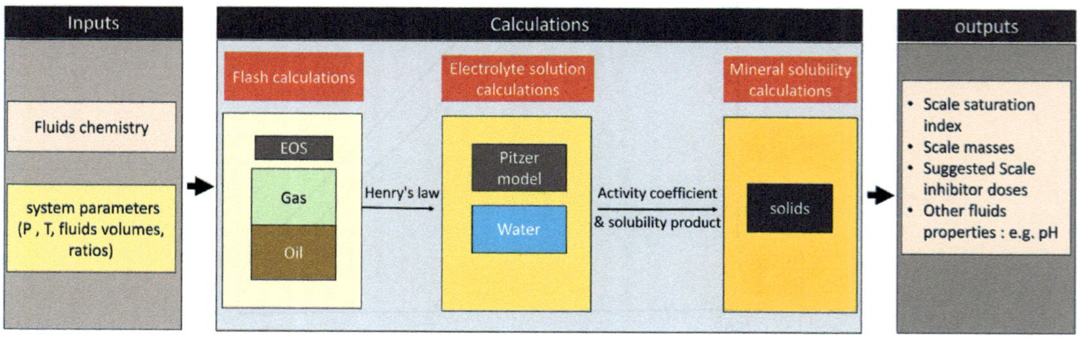

FIG. 12.8

Scale prediction software data and calculations.

- **Inputs**

The detailed inputs can be different from one software package to another; however, they all share basic necessary inputs, including:

- Water composition cations (Na, K, Ca, Mg, Zn, Pb, Ba, Sr, Mn, etc.) and anions (Cl, SO_4, CO_3, HCO_3, Br, silica, F, etc.), fatty acids, TDS, ammonia, dissolved H_2S and CO_2, pH, specific gravity, resistivity.
- Hydrocarbons analysis (gas/oil), C1–C7+ or higher, H_2S, CO_2, gas and oil specific gravity/density (recommended for flash calculation).
- Injected hydrate inhibitor volumes per day.
- Operating temperature test range, e.g., downhole and well head temperatures for wells, upstream and downstream temperatures for a pipeline.
- Operating pressure range, e.g., downhole and well head pressure for wells, or upstream and downstream for a pipeline.
- Production rate, gas, oil, and water volumes per day.

- **Calculations**

The calculations are based on the vapor liquid equilibrium using EOSs. These determine the bubble point, fugacity, and the partitioning of H_2S and CO_2 between the phases. Besides, the Pitzer (or any other model, e.g., Bromley-Zemaitis activity model or UNIQUAC model) is used to determine the solubility product and the activity coefficient of the scaling species.

- **Output**

The main calculated outputs are phase equilibrium of phases, calculated pH of the water, the scaling tendencies, amount of scales, the recommended doses of inhibitor mitigation, and other phase properties.

12.6.3.2 Examples of industrial scale prediction software

- **OKSCALE**

OKSCALE is scale prediction software developed at Oklahoma University. This software includes $BaSO_4$, $SrSO_4$, $CaSO_4$, $CaCO_3$, and $MgSO_4$ scale prediction. The theory is mainly based on Pitzer's model for calculating scaling ion activity coefficients and mineral solubilities. The scaling potential is expressed as the amount of precipitation of a mineral, and no saturation index is available. It has been shown to work at the Miller downhole conditions (120°C and 492 bar) [48].

- **SCALE**

SCALE is scale prediction software marketed by Petrotech of Norway. The scales included in this software are $BaSO_4$, $SrSO_4$, $CaSO_4$, $CaCO_3$, and $FeCO_3$. The scaling potential is expressed as saturation ratio and amount of precipitation of a mineral. The model is applicable up to 150°C and 400 bar [48].

- **SCALEUP**

SCALEUP is scale prediction software developed by the Oilfield Scale Research Group at Heriot-Watt University, Edinburgh, Scotland. Only sulfates, i.e., $BaSO_4$, $SrSO_4$, and $CaSO_4$, are considered in this model. The theory is based on the Pitzer model. The model is applicable up to 150°C and 500 bar [48].

- **Multiscale**

Multiscale was developed by NTNU and distributed by Expro Petrotech, Norway. It is a thermodynamic software application, which uses the Pitzer model for calculating scaling ion activity coefficients and mineral solubilities, and SRK or PR with Peneloux and Huron-Vidal mixing rule for the PVT model. [49].

The software working range is 0–300°C and 1–1000 bar, and it is capable of mixing up to six waters, six oils, and six gases simultaneously. MultiScale can be used to predict the scaling tendency of the following minerals: NaCl, KCl, $BaSO_4$, $SrSO_4$, $CaSO_4$ (a, g, h), FeS, $CaCO_3$ (c, a, v), $FeCO_3$, $BaCO_3$, $SrCO_3$, $NaHCO_3$, $KHCO_3$, Na2CO_3 (a, d, m), K_2CO_2 (h, a), NaAc (t, a), $Mg(OH)_2$ [50].

It has modules for scale calculations in gas condensate systems in the presence of hydrate inhibitors. Also it includes modules for pH simulations in the presence of triazine H_2S scavengers.

- **ScaleSoftPitzer (SSP)**

This is another thermodynamic software package developed by the Brine Chemistry Consortium at Rice University. The last developed versions of the software included kinetic and hydrodynamic calculations [51].

The software uses Pitzer electrolyte theory to calculate activity coefficients and mineral solubility; PengRobinson EOS for gas fugacity coefficients; and ethylene glycol and methanol effect on activity coefficients for calcite, barite, gypsum, celestite, and halite. The model is applicable up to about 300 °C, around 20,000 psi, and about 8 to 10 m ionic strength. The program is used to predict $CaCO_3$, $BaSO_4$, $CaSO_4$ (g, a, h), $SrSO_4$, CaF_2, ZnS, FeS (a, m, t), $FeCO_3$, NaCl, and others [14,22,52,53].

The pH is calculated based on the alkalinity (carbonate, bicarbonate, and volatile fatty acids), dissolved gases partial pressure, and their concentration in water [22].

- **ScaleChem**

ScaleChem was developed and distributed by OLI Systems, United States. The model is based on the combined work of the Bromley-Zeimaitis activity model combined with Pitzer, Meissner work for complex high ionic strength systems, the Helgeson EOS for equilibrium constant values, and the SRK EOS for nonaqueous vapor-liquid fugacity calculations. Recent version uses mixed solvent electrolyte (MSE)-SRK model to predict high temperature/high pressure oil and gas environments. The model predicts and considers all of the true species in solution in the range of −50 to −350°C, up to 3000 bar, and 0–30 m ionic strength [54].

ScaleChem has standard scales that include $CaSO_4$, FeS, $BaSO_4$, NaCl, $CaCO_3$, $SrSO_4$, $CaMg(CO_3)_2$, $FeCO_3$, and other uncommon types of scale deposits [54].

- **SCALE2000**

SCALE2000 is a thermokinetic geochemical calculation-based software. SCALE2000 was specifically developed at BRGM for the study of scaling problems and is now largely used in geothermal energy, oil production, storage of acid gases, and in other industrial systems involving salt-rich electrolytic solutions [55].

The conceptual model is a thermokinetic geochemical calculation program based on Pitzer's ion interaction model for Na-K-Ca-Mg-Sr-Ba-Fe-Cl-SO_4-OH-H-Ac-CO_3-HCO_3-CO_2-AcH-SiO_2-H_2O systems under 25–250°C and 1–1000 bar conditions, and it takes into account the fluid flow effect using a specific approach based on a scheme of serially connected homogeneous reactors. The minerals currently included in the software database are carbonates (calcite, dolomite, strontianite, magnesite, siderite, witherite, aragonite, and vaterite); the most common salts (halite, sylvite); sulfates (anhydrite, gypsum, celestite, barite); silicas (quartz, chalcedony, amorphous silica); and sulfides (mackinawite and pyrrhotite) [55,56].

The kinetic law considered in SCALE2000 is the general kinetic law derived from the transition state theory (TST) applicable to all minerals [55,56].

In order to simulate the feedback between chemical interactions and fluid flow, the "reactors" option of SCALE2000 was implemented. It enables the simulation of fluid flow through a set of four perfectly mixed reactors connected in series [55,56].

The basic equations used in the reactor's model can be split into two categories: first, the geochemical equations, which are based on mass balance and law of mass action formulations, including kinetics for dissolution and precipitation of mineral phases, which are based on the complete Pitzer formalism; second, water flow and mass transport equations, which are based on the classical equations of mass balance for perfectly mixed reactors [55,56].

Another good feature of this software is the ability to determine the scale mass and thickness in the tubing or pipeline.

- **DOWNHOLE SAT (DHSAT)**

DOWNHOLE SAT (DHSAT) is scale prediction software developed by the French Creek Software Inc. This model is a quasigeochemical model. It takes into consideration all the possible sulfates and carbonates, plus some hydroxides in the calculation of scale potentials. The scaling potential is expressed as saturation level and amount of precipitation of a mineral. For calcium carbonate, several other saturation indices, such as the Stiff-Davis index and Oddo-Tomson index, are also calculated. It has shown to be applicable at the Miller downhole conditions (120°C and 492 bar). The prediction results are displayed onscreen either as 2D or 3D charts or as text [48,57].

Other models have been used, including those models that have been developed for geochemical uses.

Recently, Abouie et al. [58] developed a comprehensive thermal compositional wellbore simulator (UTWELL) coupled with the USGS state-of-the-art geochemical package, IPhreeqc, to generate a robust, accurate, and flexible integrated simulator (i.e., UTWELL IPhreeqc), to monitor scale problems throughout the wellbore, which is able to model geochemical reactions, transportation, and consequent deposition in the wellbore.

12.6.3.3 Critiques of the current scale prediction software

Scale prediction software has witnessed major advances in the last two decades; however, they still have some weaknesses that make them in some instances nonrepresentative of the field cases. These include the following:

- The models depend on the quality of water sampling and analysis. Water chemistry is the core inlet of these software packages: water must be sampled, preserved, and analyzed properly. Water/fluids must be sampled upstream of the problem. While this is achievable in the surface facilities, it is unachievable for downhole samples, since the common practice is to sample the fluids from the wellhead, which in this case represents spent water, where the scaling species might be already deposited in the near wellbore region or in the production tubing, which leads to underestimation in the prediction results. To improve the calculations, downhole fluid sampling can be used, or analog field water chemistry can be used that already used downhole sampling.

- The models are mainly thermodynamic, where the effect of kinetics is not fully integrated into these modules.
- The system hydrodynamics are not fully incorporated into these software packages. The current software modules assume that all the precipitated solids will deposit on the pipe wall. However, this is not true since other scenarios can take place, e.g., they can be redissolved in the bulk fluids, or dispersed and carried away with the flowing fluids. These processes depend on the system hydrodynamics. Ignoring the hydrodynamics can basically lead to overestimation of the scale problem.
- Related to the hydrodynamics, these modules do not consider the pipes internal surface roughness conditions.
- These models do not consider the effect of the other system and phase properties, such as oil volumes, oil viscosity, wax deposits, asphaltenes deposits, sand and fines production, and other physical properties of scale particles deposition. Wax and asphaltenes deposits are known to include mineral scales in their deposits. Sand and fines are known to work as nucleating sites for mineral scales.
- The current scale prediction modules cannot accurately determine the location and thickness of the scale layer, and this is related to the system hydrodynamics, besides thermodynamic and kinetic factors.
- The current modules do not take geochemical or hydrogeochemical processes into consideration, which can lead to scale overestimation or underestimation. For example, the injected seawater, aquifer rock minerals, and gas combined with advective transport processes can alter the final chemical composition of the injected seawater, and thus reservoir minerals can dissolve in the final mixture and then redeposit in the production tubing. This leads to underestimation of the problem since this geochemical process was not originally considered; on the other hand, a mineral that showed tendencies in the original mixture may not deposit due to the changes in the final mixture composition, which leads to overestimation of the problem [59].

A case study is given here wherein the scaling tendency of an oil producer was performed using ScaleSoftPitzer, where the prediction results showed a tendency of forming barite scale only. However, the well suffered from multiple obstructions that were detected during multiple gauging jobs (over 5 months). The composition of the collected scale deposit samples showed carbonates, sulfates, sulfides, and chlorides, as indicated in Tables 12.3 and 12.4.

It is clear from the analysis results that the scale composition showed consistency of formation of specific types of scales at different depths. Some factors that are thought to contribute to the inconsistent scaling prediction results include sampling issues, since the sampled water showed a pH of 3.9 and sulfate ion concentration was 30 ppm, which indicates that the surface sampled water was spent water, where the dissolved gases were expelled upon depressurization, and most of the scaling species have already deposited downhole. In this case, downhole sampling, proper preservation, and analysis was

Table 12.3 Scaling tendency of oil producer.

T	*P*	Calcite	Barite	Gypsum	Anhydrite	Hemihydrate	Celestite
(°F)	(psi)	mg/L	mg/L	mg/L	mg/L	mg/L	mg/L
60	300	0	45	0	0	0	0
89	878	0	37	0	0	0	0
118	1456	0	27	0	0	0	0
147	2033	0	19	0	0	0	0
176	2611	0	13	0	0	0	0
204	3189	0	8	0	0	0	0
233	3767	0	1	0	0	0	0
262	4344	0	0	0	0	0	0
291	4922	0	0	0	0	0	0
320	5500	0	0	0	0	0	0

Table 12.4 Composition of different scale samples collected at different depths of the well.

Collection depth	% wt								
	Paraffins + asphaltenes	NaCl	$CaCO_3$	FeS	ZnS	PbS	$BaSO_4$	$SrSO_4$	Sand
8000’THF	18.2	2.5	6.3	2.4	25.8	10.8	25.3	4.5	4.2
1505’THF	21.9	4	2	1	25	17	1.7	0.3	27.1
452’ THF	6.4	2		1	19.4	4.3	35.1	7	24.8
20’ THF	4	4	2.5	0.6	10	2.5	41	7.2	28.2

thought to be a key factor. Also the presence of sand and fines could be another secondary factor that might be contributing to scale nucleation.

12.6.4 Scale prediction using artificial intelligence methods

One of the recent advances in the field of scale prediction is the incorporation and usage of artificial intelligence methods. The artificial neural network (ANN), or neural network (NN), is a component of artificial intelligence that has been used in various industrial applications. Like the brain, ANN learns by experience and example without prior knowledge of mathematical models. Thus the ANN is trained using sets of data consisting of input data and their corresponding outputs.

Falode et al. [60] has applied the ANN and the time series neural network (TSNN) in sulfate scale prediction, and they were found to be more robust than other methods. Machine-learning techniques have been used to predict $CaCO_3$ scale formation and inhibition in oil wells [61]. Multilayer perceptron (MLP) and probabilistic neural network (PNN) methods were used to predict scaling under severely scaling conditions that were amplified to test their accuracy, which showed good and consistent results [62].

12.6.5 Prediction of elemental sulfur deposition

Modeling sulfur deposition basically relies on understanding the phase behavior of sulfur in sour gas systems under different temperatures and pressures; from this the solubility of elemental sulfur can be estimated, and its deposition can be modeled.

Two main approaches are used to the modeling of sulfur solubility in H_2S and sour gas mixtures: the EOS modeling, and the use of analytical expressions, incorporating empirical constants [63].

Equations of state such as Peng-Robinson EOS may be used to determine the fugacity of sulfur in the liquid or vapor phases. Several models were developed based on the PR EOS, such as Tomcej et al. [64], Karan et al. [65], Gu et al.[66], and others.

One of the simplest correlations for sulfur solubility was developed by Chrastil [67], which is empirical correlation based on the associative law and entropy principle, as presented in Eq. (12.39):

$$C_r = \rho^k \exp\left(\frac{a}{T} + B\right) \tag{12.39}$$

where C_r, is the concentration of the solid component at reservoir temperature and pressure (g/m^3), ρ is the fluid density (kg/m^3), and T is the fluid temperature (K). The parameters k, a, and B are empirical constants determined by matching experimental solubility measurements [63].

Another correlation by Brunner and Woll [68] was developed as shown in Eq. (12.40):

$$C_r = \rho^4 \exp\left(\frac{-4666}{T} - 4.5711\right) \tag{12.40}$$

Based on the phase behavior of sulfur in gas systems, different models have been developed to estimate sulfur deposition and the resulted formation damage by Roberts [69] and Kuo [70]; the rate of sulfur-saturation buildup, *dSs/dt*, is given by Eq. (12.41):

$$\frac{dS_s}{dt} = \frac{q^2\left(\frac{dc}{dp}\right)Bu}{4\pi^2 k_a k_r h^2 \varnothing (1 - S_{wi}) r^2} \tag{12.41}$$

where q is the gas flow rate (m^3/s), u is the gas viscosity (pa s), Ø is the porosity, k_a is the absolute permeability (m^2), k_r is the relative permeability, h is pay thickness (m), B is gas formation volume factor, S_{wi} is the initial water saturation, and r is the radial distance from the well (m) [63].

12.6.6 Recent advances in scale prediction

Scale prediction has witnessed many advances lately. One of the main advances was to extend the range of the experiments that are used to build or to be fitted with the theoretical data, to have a wider range of temperature and pressure; for example, ScaleSoftPitzer can be applied up to 300°C and 20,000 psi and up to 10 M ionic strength with the use of mixed electrolytes background. Also, the solubility of different minerals at high and ultrahigh pressures and temperatures in complex brine systems has been validated.

Another important development was the incorporation of kinetics into these models, which were originally based on thermodynamic calculations. Furthermore, hydrodynamic models have also been incorporated. Another groundbreaking factor is the incorporation of hydrogeochemical calculations, such as the models developed by Fu et al. [59] and Abouie et al. [58].

Also, with the recent research advances and the immense amounts of data from theoretical, experimental, and field experience and case studies, it is expected that the use of artificial intelligence in scale prediction will play a major role in the future.

12.7 Gas hydrate prediction

To avoid the expensive and hazardous impact of gas hydrates, a robust management strategy must be developed, which usually is based on gas hydrate predictions to assess their risk and propose mitigation solutions. Various commercial and academic software packages have been developed based on the thermodynamics of the phase equilibrium, and the mass and heat transfer of the process.

12.7.1 Basic principles of gas hydrates prediction

12.7.1.1 Gas hydrates phase equilibrium

A good understanding of gas hydrates formation and prediction of their formation relies on understanding gas hydrates phase equilibria. By determining the possible phases in a hydrocarbon-water mixture at different pressures and temperatures, the hydrate formation and temperature can be estimated. Fig. 12.9 illustrates a gas hydrates phase diagram, defining the hydrate stable zone (gray area (light gray in print versions)) and hydrate unstable zone (green area (dark gray in print versions)).

*Q*1 and *Q*2 represent quadruple points where four phases coexist. At *Q*1, liquid water (Lw), ice (*I*), hydrate (*H*), and hydrocarbon vapor (V_{HC}) coexist, while at *Q*2 liquid water (L_W), hydrate (*H*), hydrocarbon vapor (V_{HC}), and liquid hydrocarbon (L_{HC}) coexist. Such a phase diagram is usually developed by determining the equilibrium of the three phases (L_w-H-V_{HC}) [71]. These diagrams allow one to define under what conditions of pressure and temperature hydrates can form for a given gas and water composition.

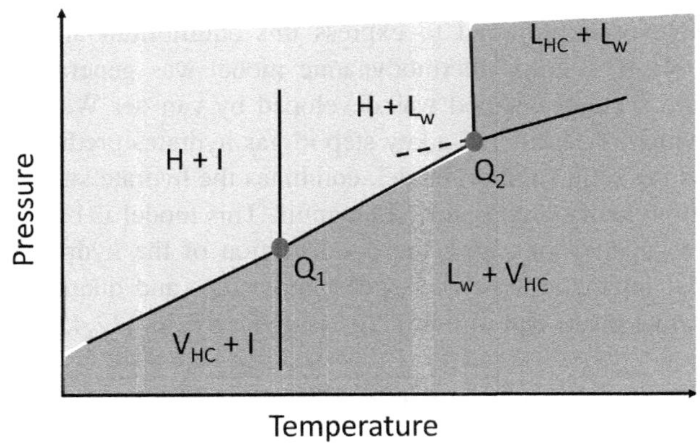

FIG. 12.9

Phase equilibrium of hydrate-hydrocarbon-water system.

Generally speaking, there are five major variables in phase equilibria: pressure, temperature, concentrations of the gas, liquid(s), or hydrate phases, volume or density, and phase amounts. The availability of information about these major variables makes the phase equilibria calculations easier [72].

Two common methods for determining the equilibrium of the three phases are the gas gravity method and the phase equilibrium constant method (conventionally known as the *K*-value method). These methods are simple, straightforward, and commonly known as hand calculations. A third way to calculate the gas hydrates phase equilibrium conditions is based on statistical thermodynamic methods. These methods will be discussed shortly.

12.7.1.2 Thermodynamics of gas hydrates prediction

Equilibrium thermodynamics calculations basically describe the pressure and temperature conditions of formation and dissociation of gas hydrates. These calculations are based on the standard thermodynamic phase equilibrium criteria, which states that the equilibrium between the different phases (hydrate, liquids, vapor) is obtained by minimizing the Gibbs free energy or by equalizing the chemical potentials (or fugacities) of each component in all their different phases, as represented by Eq. (12.42):

$$\mu^H = \mu^w \tag{12.42}$$

where μ^H is the chemical potential of water in the hydrates phase (J/mol), and μ^w is the chemical potential of water (J/mol). Since thermodynamically hydrate formation takes place in two steps—forming a cage and then filling the hydrate lattice [73]—by using the chemical potential of the unoccupied lattice μ^β, the equilibrium can be rewritten as [74]:

$$\mu^\beta - \mu^H = \mu^\beta - \mu^w \tag{12.43}$$

Or in other words:

$$\Delta\mu^H = \Delta\mu^w \tag{12.44}$$

where $\Delta\mu^H$ is the difference in the chemical potential of water in the unoccupied hydrate lattice and in the hydrate phase (J/mol) and $\Delta\mu^w$ is the difference in the chemical potential of water in the unoccupied hydrate lattice and in the water rich phase (J/mol).

Various models have been proposed to express this equilibrium and to define the terms in Eqs. (12.43) and (12.44). The initial thermodynamic model was generated by Barrer and Stuart (1957), and then a more accurate method was developed by van der Waals and Platteeuw (1959), the (VdW-P model), which is considered a key step in gas hydrates prediction. The VdW-P model, which is also known as the solid solution model, combines the hydrate structural characteristics and gas isothermal adsorption theory proposed by Langmuir. This model is based on several postulates, including: the single occupancy of cages, the nondistortion of the hydrate lattice under pressure, and the assumption that interactions between guest molecules and quantum effects are negligible [72]. In the case of hydrate phase equilibrium, $\Delta\mu^H$ is given by Eq. (12.45) [75]:

$$\Delta\mu^H = RT\sum_{i=1}^{2} \nu_i \ \mathrm{ln}\left(1 - \sum_{j=1}^{Nc} \theta_{ij}\right) \tag{12.45}$$

$$\theta_{ij} = \frac{C_{ij} f_j}{1 + \sum_{j=1}^{Nc} C_{ij} f_j} \tag{12.46}$$

where T shows the system temperature, °C; ν_i stands for the number of i-type holes in the hydrate phase; Nc is the number of the molecules in the mixture for hydrates formation; θ_{ij} represents the proportion of guest molecules j to i-type holes, %; f_j refers to the fugacity of guest molecules j in each phase MPa; C_{ij} represents the Langmuir constant of guest molecules j in i-type holes, which could be solved by Eq. (12.47) [75]:

$$C_{ij} = \frac{A_{ij}}{T} \exp\left(\frac{B_{ij}}{T}\right) \tag{12.47}$$

where A_{ij}, B_{ij} show the empirical constant.

When gas hydrate and nongas hydrate phases are in equilibrium, this is expressed by Eq. (12.48) [75]:

$$\mu^w + RT\sum_{i=1}^{2} \nu_i \ln\left(1 - \sum_{j=1}^{Nc} \theta_{ij}\right) = \mu_o^w + RT \ln\left(\frac{f^w}{f_o^w}\right) \tag{12.48}$$

where f^W is the fugacity of water in water-rich phase, MPa; μ_0^W represents the chemical potential of pure water under T and P in reference states, J/mol; f_0^W represents the fugacity of pure water under T and P in reference states, MPa.

In addition, the chemical potential of water can be expressed as per Eq. (12.49) [75]:

$$\frac{\mu^w - \mu_o^w}{RT} = \frac{\Delta\mu_0}{RT_0} - \int_{T_0}^{T} \frac{\Delta H_0 + \Delta C_P(T - T_0)}{RT^2} dT + \int_{P_0}^{P} \frac{\Delta V}{RT} dP \tag{12.49}$$

From Eqs. (12.48) and (12.49), the phase equilibrium conditions can be obtained as in Eq. (12.50) [75]:

$$\frac{\mu^w - \mu_o^w}{RT} = \frac{\Delta\mu_0}{RT_0} - \int_{T_0}^{T} \frac{\Delta H_0 + \Delta C_P(T - T_0)}{RT^2} dT + \int_{P_0}^{P} \frac{\Delta V}{RT} dP = \ln\left(\frac{f^w}{f_o^w}\right) - \sum_{i=1}^{2} \nu_i \ln\left(1 - \sum_{j=1}^{Nc} \theta_{ij}\right) \tag{12.50}$$

where $\Delta\mu_0$ is the chemical potential difference of water in hydrate structure and pure water under standard conditions, J/mol; T_0, P_0 refers to the temperature (K) and pressure (MPa) under standard conditions, respectively; and $T_0 = 273.15\,K$, $P_0 = 0\,MPa$; ΔH_0, ΔV, ΔC_P represent the specific enthalpy difference (J/mol), specific volume difference (m^3/kg), and specific heat difference (J/(kg K)) between hydrate structure and pure water, respectively; $\ln(f^W/f_0^W) = \ln x_W$, and if the hydrate inhibitor is added, then $\ln(f^W/f_0^W) = \ln(y_W x_W)$; x_W, y_W give the mole fraction and activity coefficient of water in the water-rich phase [75].

The preceding equations were a simple way to elucidate the phase equilibrium conditions of gas hydrates formation. For example, if the system pressure is completely known, the equilibrium temperature of hydrate can be calculated as follows [75]:

(1) Use the basic data and phase equilibrium temperature of the hydrate under one pressure;
(2) Calculate the fugacity of gas-phase with a proper EOS;
(3) Determine the hydrate structure type and calculate the θ_{ij} and C_{ij} constants;
(4) Solve the phase equilibrium temperature of hydrate, adopting the iterative approach.

Likewise, if the system temperature is fully known, the solution method for hydrate phase equilibrium pressure is the same as the preceding steps.

Other models mentioned in the literature include Parrish-Prausnitz [76], Ng-Robinson [77], and others.

12.7.1.3 Mass and heat transfer

Gas hydrate growth is controlled by three main factors: kinetics of crystal growth, mass transfer towards the growing hydrate crystal, and heat transfer of the exothermic heat of hydrate formation away from the growing crystal surface [72].

Some of the early kinetic models are Englezos et al. [78,79] and Malegaonkar et al. [80]. However, the intrinsic kinetics were suggested to play a small role in the growth of gas hydrates.

Most of the modern models of gas hydrate growth include mass transfer of the components from the bulk phases to the growing hydrate. These models assume that the diffusion of gas molecules to the hydrate formation interface determines the hydrate formation rate. Common mass transfer models include the Yapa et al. [81], Mori [82], and Skovborg and Rasmussen [83] models. The Skovborg and Rasmussen model is presented in Eq. (12.51):

$$\frac{dn}{dt} = K_L A_{(g-1)} C_{wo} (x_{int} - x_b) \tag{12.51}$$

where $A_{(g-1)}$ is the gas-liquid interface area (m^2); C_{wo} is the initial concentration of water (mol/m^3); K_L is the mass transfer coefficient of the liquid layer (m/s); x_{int} is the mole fraction of gas under interfacial gas-liquid phase equilibrium; and b shows the mole fraction of gas under the solid-liquid phase equilibrium.

Heat transfer models are based on the fact that gas hydrate formation is an exothermic reaction. Thus during lab experimentation on gas hydrate formation, the temperature increases within the reactor, causing an increase in the bulk fluid and gas temperature above the set experimental temperature. This causes a local drop in subcooling and, thus, limits the hydrate growth rate, which necessitates the removal of that released heat (by circulating a coolant) for hydrate formation to continue. This is enabled by circulating a cooling agent around the shell of the reactor. The best result is that rate of hydrate formation invariably becomes dependent upon the rate of removal of produced heat from the formation of new hydrate mass from the system [84].

Various heat transfer models have been developed to express the hydrate growth, such as Uchida et al. [85], Mori [82], Mochizuki and Mori [86], and Freer et al. [87]; these models are illustrated in Fig. 12.10.

The Uchida et al. [85] heat transfer model simplifies the temperature gradient on the surface of the hydrate layer and purports that hydrate growth rate is directly proportional to the driving force, as indicated in Eq. (12.52):

$$\frac{\nu_f}{\lambda} = \frac{1}{(L\rho_h r_c)} \Delta T \tag{12.52}$$

where ν_f is the hydrate growth rate (m/s); λ is the thermal conductivity of the surrounding phases, J (s mm K)$^{-1}$; L is the latent heat of hydrate formation (J/mol); r_c refers to the curvature radius of hydrate layer (mm); ΔT refers to the driving force of system temperature, K; ρ_h represents the mole density of hydrate, (mol/m^3).

Mori [82] developed a convective heat transfer model where the growth rate is related exponentially to the heat driving force; the model is simplified as Eq. (12.53):

$$\nu_f \xi = \text{Ç}\, \Delta T^{3/2} \tag{12.53}$$

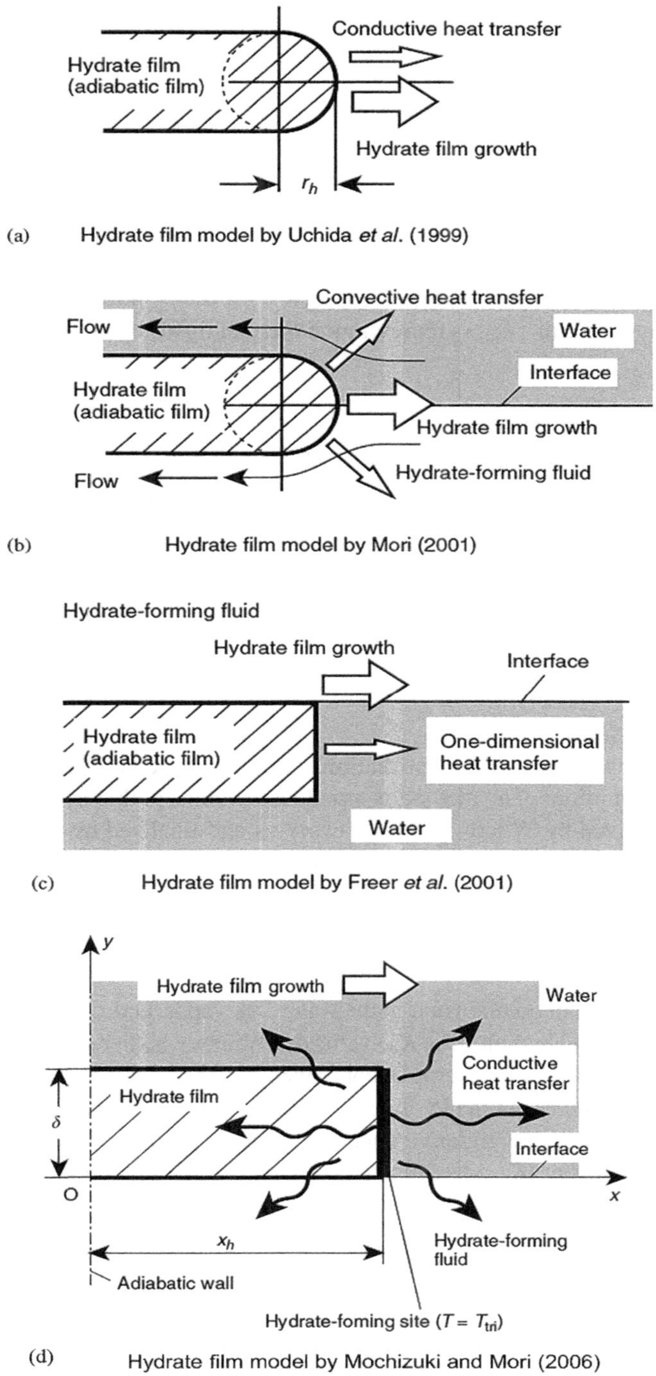

FIG. 12.10

Physical models of hydrate film growing along an interface between liquid water and a hydrate-forming fluid (guest fluid).

Reprinted from T. Mochizuki, Y.H. Mori, Clathrate-hydrate film growth along water/hydrate-former phase boundaries-numerical heat-transfer study, J. Cryst. Growth, 290 (2) (2006) 642–652. https://doi.org/10.1016/j.jcrysgro.2006.01.036, copyright (2006) with permission from Elsevier.

where ν_f shows the hydrate growth rate, m/s; ξ refers to the hydrate thickness, m; ΔT is the subcooling degree, K; and $\c{C}$ is given by Eq. (12.54):

$$\c{C} = \frac{\pi \c{A}}{4} \frac{1}{\rho_h \Delta h_h} \left(\rho_w c_{p,w} k_w^{2/3} + \rho_g c_{p,g} k_g^{2/3} \right) \tag{12.54}$$

$\c{A}$ stands for the dimensionless parameter; Δh_h is the latent heat of hydrate formation, J/kg; $c_{p,w}$, $c_{p,g}$ show the heat capacity of water and guest, J/(kg K); ρ_h, ρ_w, ρ_g refer to the respective density of hydrate, water, and guest, kg/m^3; k_w, k_g represent the thermal diffusion coefficient of water and guest molecules, m^2/s.

12.7.2 Methods of predicting gas hydrates

12.7.2.1 Correlations for prediction of gas hydrate formation conditions

Different correlations have been developed to determine the hydrate formation conditions: pressure and temperature. Table 12.5 summarizes some of these correlations:

Other formulations that allow quick calculation of the hydrate formation temperature and pressure include Berge [92], Bahadori and Vuthaluru [93], Ghiasi [94], Kobayashi et al. [95], Østergaard et al. [96], among others.

12.7.2.2 K-value method

K-value method utilizes the vapor-solid distribution coefficient (or equilibrium constants) for predicting hydrate-forming conditions. The method is based on the concept that hydrates are solid solutions. The method was conceived by Wilcox and his coworkers and finalized by Carson and Katz.

Considering the gas hydrates as solid solution, the vapor-solid distribution coefficient K_{vsi} for each component is calculated by Eq. (12.55):

$$K_{vsi} = \frac{y_i}{x_i} \tag{12.55}$$

where y_i is the mole fraction of component i in the water free vapor, and x_i is the mole fraction of component i in the water free solid hydrate. The K_{vsi} value determines whether a component prefers residing in the hydrate or the vapor phase; for example, components with $K_{vsi} > 1$ (e.g., CH_4, N_2) preferentially concentrate in the vapor rather than the hydrate, whereas components with propane or isobutene have $K_{vsi} < 1$ (C_3H_8, i-C_4H_{10}) are preferentially concentrated in the hydrate phase [72].

The distribution coefficient (K_{vsi}) is a function of temperature and pressure, and either calculated from the charts where the K_{vsi} values are presented as a function of temperature and pressure or from the equation obtained by fitting all the K_{vsi} values from these charts [97].

There is a chart available for each of the components commonly encountered in natural gas that is a hydrate former: methane, ethane, propane, isobutane, n-butane, hydrogen sulfide, and carbon dioxide. The K_{vsi} charts are usually used in three methods: (1) given the temperature and pressure, calculate the composition of the phases, (2) given the temperature, calculate the hydrates formation pressure and the composition of the hydrate, and (3) given the pressure, calculate the hydrate formation temperature and the composition of the hydrate [73].

The K_{vsi} charts or equation are then used to estimate the conditions of three-phase (LW-H-V) hydrate formation. The technique used in the calculation of the equilibrium conditions involves the calculation of

Table 12.5 Correlations for determining gas hydrate formation conditions.

Equation	Variables	Reference
$T=8.9P^{0.285}$	T (°F) and P (psi) are temperature and pressure of hydrate formation	Hammerschmidt [88]
$P=\exp\left(a+\frac{b}{T}\right)$	P is hydrate dissociation pressure, kPa; T, temperature in K depends on two coefficients that are functions of temperature range for each gas. The coefficients can be determined by curve fitting, for methane, Structure I; $a=38.9803$, $b=-8533.90$	Holder et al. [74]
$\log P=\beta+0.0497(T+kT)$-1	The P-explicit correlation, where: T in °C, P is in MPa $\beta=2.681-3.811\gamma_g+1.679\gamma_g^2$ $k=-0.006+0.011\gamma_g+0.011\gamma_g^2$ $\gamma_g=$ gas specific gravity $\gamma_g=MW_{Gas}/MW_{Air}$	Makogon [89]
$T=-\ 283.24469+78.99667\log(P)-5.352544\log(P)^2+349.473877\gamma_g-150.854675\gamma_g^2-27.604065\gamma\log(P)\gamma_g$	T is the hydrate temperature in K P is the pressure in kPa	Motiee [90]
$T=13.47\ \ln(P)+34.27\ \ln(\gamma_g)-1.675\ \ln(P)\ln(\gamma_g)-20.35$	T is the hydrate temperature in °F P is the pressure in psi	Towler and Mokhatab [91]

the K_{vsi} value, such that the sum of mole fraction of each component in the vapor phase divided by the K_{vsi} value equals unity at the three-phase equilibrium conditions [72], according to Eq. (12.56):

$$\sum_{i=1}^{n} \frac{y_i}{K_{vsi}} = 1 \tag{12.56}$$

A full discussion of the K-value method is provided by Sloan and Koh [72] and Carrol [73].

The K_{vsi}-value method is both more accurate and slightly more complex than the gravity method that is discussed next.

12.7.2.3 Gas gravity plot method (Katz's method)

The gas gravity method is a simple graphical method for estimating the hydrate formation conditions. The method is based on Katz gas gravity charts, where gas gravity is defined as the ratio of the molecular mass of the hydrate-forming gas to that of air (e.g., 16.04 g/mol for CH_4 by that of air, 28.966 g/mol) [97].

The method is very simple in use, as illustrated in the example in Fig. 12.11. The gas mixture gravity is calculated (summation of the product of each component mole fraction multiplied by its specific gravity) and either temperature or pressure is specified; then the second unknown variable (either pressure or temperature) at which hydrates will form is read directly from the chart [72].

The gas gravity method for predicting hydrate formation was generated from a limited amount of data and should be used with caution, as it has been proven to show considerable deviations. In the specific gravity method, errors of up to 50% have been reported [98], so it should be considered as an approximate or initial estimate of hydrate formation conditions [72,99]. In addition, the original charts were generated for hydrates containing only hydrocarbons, so they should be used with care for hydrates that contain a considerable amount of other gases like CO_2, H_2S, and N_2 [97].

The Baillie-Wichert method is another graphical method used for gas gravities between 0.6 and 1, with an advantage over the gas gravity and K-factor methods in that it can be used for sour gases [100].

12.7.2.4 The charts of permissible expansion

This method involves the charts of permissible expansion that a natural gas can undergo without risk of hydrate formation. These charts were redrawn with the aid of the gas gravity charts using the Joule-Thomson cooling curves [101]. This method is more suitable for rough design of valves, chokes, and flow provers where gas expansion normally occurs downstream of these flow restrictions. The charts are limited to natural gases composed predominantly of methane. An average error of 10% should be expected with this method. Higher errors may result when significant amounts of nonhydrocarbons or hydrocarbons heavier than ethane are present [102].

12.7.2.5 Artificial intelligence methods

One of the recent advances in gas hydrates prediction is the use of artificial intelligence in gas hydrates prediction.

Elgiblay and Elkamel [102] proposed four ANN models using a total of 2389 data entries obtained for phase equilibrium of pure, binary, and multicomponent systems, where the data were divided into single, binary, ternary, and multicomponent systems. These systems were composed of hydrate-forming and nonhydrate-forming constituents with and without inhibitors. In model-A, the pressure P depends on temperature T and specific gas gravity γ_g, or:

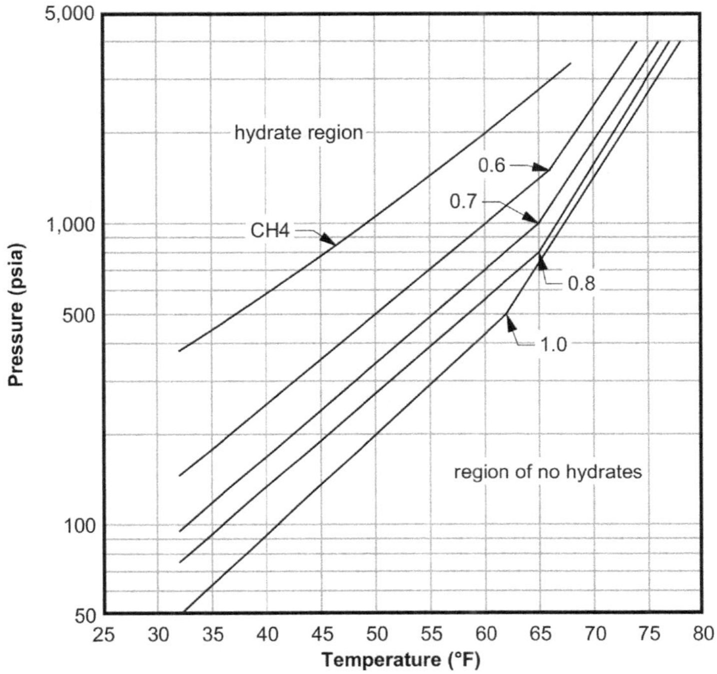

FIG. 12.11

Hydrate prediction for sweet natural gas using the gas gravity method—American Engineering Units.

Reprinted from J.J. Carroll, Natural Gas Hydrates: A Guide for Engineers, Elsevier Inc., 2014, https://doi.org/10.1016/C2013-0-13425-3, copyright (2014), with permission from Elsevier.

$$P = f_1(T, \gamma_g) \tag{12.57}$$

In model-B, the correlation is expressed in terms of the composition of pure hydrocarbon components and their mixtures, i.e.,

$$P = f_2(T, y_{c1}, y_{c2}, y_{c3}, y_{i-c4}, c_{n-c4}) \tag{12.58}$$

where y is the mol% of each component.

In model-C, the nonhydrocarbon hydrate formers carbon dioxide, nitrogen, and hydrogen sulfide are also taken into account so that:

$$P = f_3(T, y_{c1}, y_{c2}, y_{c3}, y_{i-c4}, c_{n-c4}, y_{CO_2}, y_N, y_{H2S}) \tag{12.59}$$

Finally, in model-D, the hydrocarbon-nonhydrate-formers such as i-pentane, n-pentane lumped as pentane, and hexane plus, and the inhibitors, including methanol, sodium chloride, ethylene glycol, calcium chloride, and ethanol, are collectively included in the previous model, i.e.,

$$P = f_4\left(T, y_{c1}, y_{c2}, y_{c3}, y_{i-c4}, c_{n-c4}, y_{CO_2}, y_N, y_{H2S}, y_5, y_{c_6^+}, y_{MeOH}, y_{EG}, y_{EtOH}, y_{NaCl}, y_{CaCl2}\right) \tag{12.60}$$

Zahedi et al. [103] have reported on a prediction of hydrate formation temperature with assessment using two different methods. The assessment included statistical models and an ANN. Statistical models used were engineering equation solver (EES) and statistical package for social sciences (SPSS), whereas a MLP was used to predict the hydrate formation temperature by ANN. The statistical methods used correlation to predict hydrate formation temperature, such as Berge correlation and Kobayashi and Sloan correlation. The MLP used 70% of experimental data for training purposes and the rest of the data to prove the prediction. From 203 experimental data points, it was found that the ANN could well predict the 30% of the data for the hydrate formation temperature.

Ghavipour et al. [104] also studied the use of NNs in predicting hydrate formation conditions. Experiments using six different gas mixtures were carried out, with more than 130 hydrate equilibrium points in the pressure range of about 450–3000 psia. The inputs for ANN training are the pressure and specific gravity of the gas mixture and in this NN prediction 70% of data were used to train the network and the remaining was used to test the network.

Recently, Landgrebe and Nkazi [105] used deep learning regression to develop a universal gas hydrates predictor. They trained a NN MLP by means of backpropagation to model gas hydrate equilibrium.

12.7.2.6 Equation of state models

EOS models refer to models that are based on or use equations of state as part of the calculations for the model. So, there is no such thing as a "PR Hydrate Model" [73]. The properties of fluid phases are described by the EOS: for example, for the gas and nonaqueous liquids, the fugacity, which is a critical parameter in the thermodynamic models, is usually calculated using EOS, and different EOS have been used in different models. The Parrish-Prausnitz model used the SRK EOS, while the Ng-Robinson model used the PR EOS.

Yoon-Chun used the SRK EOS to calculate the fugacity of the components in a gas-liquid mixture. Zhang et al. [106] used the PRSV EOS to calculate the fugacity of the liquid phase. Moradi and Khosravani [107] developed a thermodynamic model based on the PRSV2 EOS and van Laar binary parameter mixing rules.

12.7.2.7 Kinetic models of gas hydrates formation

The kinetics is used to understand the rate of growth of gas hydrates. As mentioned before, the hydrates growth is explained by intrinsic kinetics and mass and heat transfer.

Various models based on intrinsic kinetics have been developed. A review of the gas hydrates kinetic models is provided by Linga et al. [108]. One of the earliest models is the Vysniauskas and Bishnoi [109] model, which assumes the hydrates growth follows three chemical reactionsteps: initial cluster reaction, formation of critical nucleus, and growth of the hydrate [75]. They used the rate of gas consumption to express the hydrate formation rate, according to Eq. (12.61) [109]:

$$r_H = A_c a_s \exp\left(\frac{\Delta E_a}{RT}\right) \exp\left(-\frac{a}{\Delta T^b}\right) P^\gamma \tag{12.61}$$

where r_H is the consumption rate of methane, cm^3/min; A_c shows the comprehensive preexponential constant, cm^3/(cm^2 min bar$^\gamma$), generally taking 4.554×10^{-26} cm^3/(cm^2 min bar$^\gamma$); ΔE_a stands for the activation energy, KJ/mol, taking 106.204 kJ mol^{-1}; R represents the gas constant, 8.314 J/(mol K);

T refers to the system temperature, K; P refers to the system pressure, kPa; ΔT stands for the subcooling degree, K, that is the difference of hydrate equilibrium temperature and system temperature; a, b, γ show the experimental constants, $a = 0.0778$ Kb, $b = 2.411$, $\gamma = 2.986$; a_s is the area of gas-liquid interface, cm^2 [75].

Englezos et al. [79] describes gas hydrate growth based on intrinsic kinetics. The model is shown in Eq. (12.62):

$$\left(\frac{dn_i}{dt}\right)_p = K^* A_p \left(f_i^b - f_i^{eq}\right) \tag{12.62}$$

With

$$\frac{1}{k^*} = \frac{1}{k_r} + \frac{1}{k_d} \tag{12.63}$$

where $(dn_i/dt)_p$ is rate of gas consumption (mol/s); A_p = surface area of particle (m^2); $(f_i^b - f_i^{eq})$ is the driving force, as f_i^b is the fugacity of component i in bulk liquid (bar), and f_i^{eq} is the equilibrium fugacity of component i in the liquid at the hydrate interface (bar); K^* = hydrate formation growth rate constant (s^{-1}); k_r is the reaction rate constant (adsorption processes) (s^{-1}); k_d is the mass-transfer coefficient through the film around the particle (s^{-1}).

12.7.3 Examples of industrial gas hydrate prediction software

- **Colorado School of Mines Hydrate Kinetics Model (CSMHyK)**

This model was developed by the Center for Hydrate Research of the Colorado School of Mines; it is a gas hydrate model especially designed for oil-dominated systems, which has been incorporated as a plug-in module in a transient multiphase flow simulator [110–112].

CSMHyK was developed for oil-dominated systems based on the mechanism of hydrate plug formation and is divided into four main stages [112]:

1. Water entrainment: The water is dispersed in the oil phase as a water-in-oil emulsion.
2. Hydrate growth: A hydrate shell forms at the oil-water interface.
3. Agglomeration: The hydrate-encrusted water droplets then agglomerate, and grow in size.
4. Plugging: The agglomerated hydrates increase the slurry viscosity, and form a plug.

The current version of the CSMHyK software is composed of three submodels, including a kinetics model, transport model, and cold flow model. An overview of the CSMHyK software and these submodels is furnished by Zepra et al. 2012 [112].

The CSMHyK model has been integrated into the OLGA multiphase simulator, and tested against three different industrial scale flowloops [111].

- **Colorado School of Mines Hydrate Flow Assurance Simulation Tool (CSMHyFAST)**

Another model developed by the Colorado School of Mines is the Colorado School of Mines Hydrate Flow Assurance Simulation Tool, or CSMHyFAST. It is a hydrate formation model for water-dominated systems as a part of understanding and evaluating the hydrate plugging risk for water-dominated systems. This tool is benchmarked against flow loop data available from ExxonMobil's flowloop facility for high water cut systems. CSMHyFAST employs a simple and speedy algorithm compared to CSMHyK-OLGA to calculate the hydrate-related phenomena for high water cut systems. It has been shown that the heat transfer model employed in CSMHyFAST captures the cooling of the loop accurately for various cases [113]. This model is available in the form of a user-friendly interface for performing hydrate-related calculations.

- **Colorado School of Mines Gibbs Energy Minimization model (CSMGem)**

Another model that was developed by Colorado School of Mines is the CSMGem. It is a thermodynamic model that uses the Soave-Redlich-Kwong (SRK) EOS for fluids phase calculations. The first piece of software generated was CSMHYD for the prediction of hydrate equilibrium. Following CSMHYD, the next-generation hydrate equilibrium tool CSMGem was developed [114].

The library of components for CSMGem includes 30 hydrocarbon species (hydrate formers and nonformers), four nonhydrocarbons (hydrate formers), and one user-definable petroleum fraction. It includes the following thermodynamic inhibitors: methanol, ethanol, ethylene glycol, NaCl, KCl, and $CaCl_2$ [73]. Fig. 12.12 illustrates the development procedure for CSMGem [115].

- **DbrHydrate**

Formerly known as EQUI-PHASE Hydrate and renamed as DBRHydrate, this program was developed by DB Robinson & Assoc. DBRHydrate* fluid analysis software is designed to predict incipient gas-to-

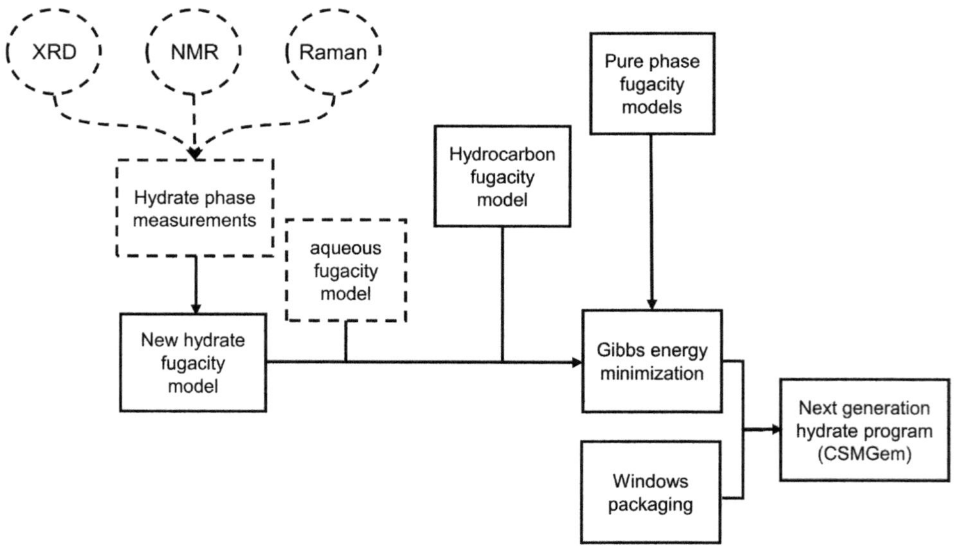

FIG. 12.12

Schematic of the development procedure for the CSMGem program.

Reprinted from A.L. Ballard, E.D. Sloan Jr, The next generation of hydrate prediction: an overview, J. Supramol. Chem. 2 (4–5) (2002) 385–392, https://doi.org/10.1016/S1472-7862(03)00063-7, copyright (2002), with permission from Elsevier.

liquid hydrate formation conditions in multicomponent hydrocarbon systems, in the absence or presence of thermodynamic hydrate inhibitors [73]. The software is currently supplied by Schlumberger.

Combining the Peng-Robinson equation-of-state with the Ng-Robinson hydrate-forming model, DBRHydrate software offers the capability of predicting the effect of common thermodynamic hydrate inhibitors.

- **HydraFLASH**

HydraFLASH is another thermodynamic model that is based on using van der Waals and Platteeuw (vdW-P) theory. It is designed to calculate the phase equilibria and physical properties of petroleum reservoir fluids (including oil, gas, water, salt, alcohols, glycols, hydrates, and ice) over a wide range of pressure and temperature conditions. It was developed through numerous industry and government sponsored R&D projects by the Heriot-Watt Hydrate and Reservoir Fluids research groups.

It incorporates the Valderrama-Patel-Teja (VPT) EOS with nondensity dependent mixing rules (NDD) used in modeling fluid phases. The hydrate phase is modeled by the solid solution theory of vdW-P. The Kihara model for spherical molecules is applied in order to calculate the potential function for compounds forming the hydrate phases [116].

Other commercial products, such as PVTsim, PIPESIM, and HYSYS, include hydrate modeling modules in their packages.

12.7.4 Recent advances in gas hydrate deposits prediction

Gas hydrates phase equilibria modeling has witnessed significant advances that have enabled predicting hydrate stability zones, hydrate fractions, and compositions, plus other derived thermodynamic properties, by employing the basic thermodynamics models and different EOS. Another rising trend in gas hydrates modeling is employing artificial intelligence methods, which have proven to be fast and reliable.

However, some challenges are still encountered in the current models. The current thermodynamic models are accurate in systems with no or low concentrations of inhibitors; however, prediction errors increase when high concentrations of inhibitors are used. In addition, modeling phase equilibria at high pressures or with very small guest molecules is associated with significant errors. Furthermore, for systems containing CO_2 or H_2S in the presence of second guest molecules, uncertainties increase significantly [117].

12.8 Wax prediction

Wax modeling and prediction are crucial for wax deposition management. Various models have been developed to predict the formation of wax deposits. They can be generally classified into thermodynamic models and deposition models. Thermodynamic models study the phase equilibrium of wax-oil systems and calculate the wax precipitation conditions (WAT), the wax precipitation curve, and other oil properties. The deposition models use hydrodynamics, mass transfer, and heat transfer data to calculate the rate of growth of the wax deposits on the pipe walls depending on the expected deposition mechanism. The two types of models are combined and used to calculate the wax deposits growth and thickness, as illustrated in Fig. 12.13.

Various mechanisms have been mentioned in the literature to explain wax deposition, such as molecular diffusion, shear dispersion, gravity settling, Brownian diffusion, and others, with a general

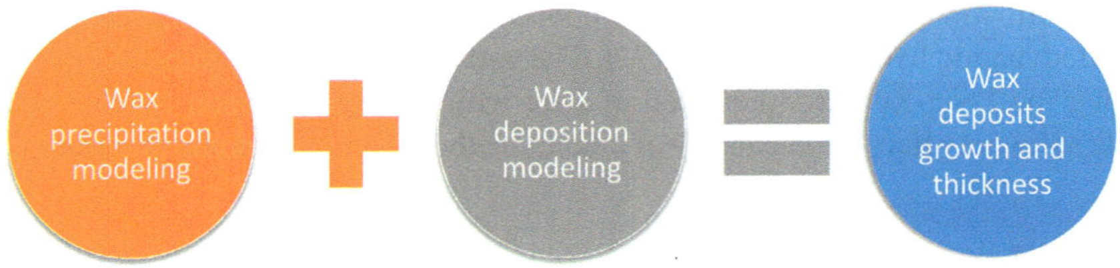

FIG. 12.13

The estimation of wax deposits using precipitation and deposition models.

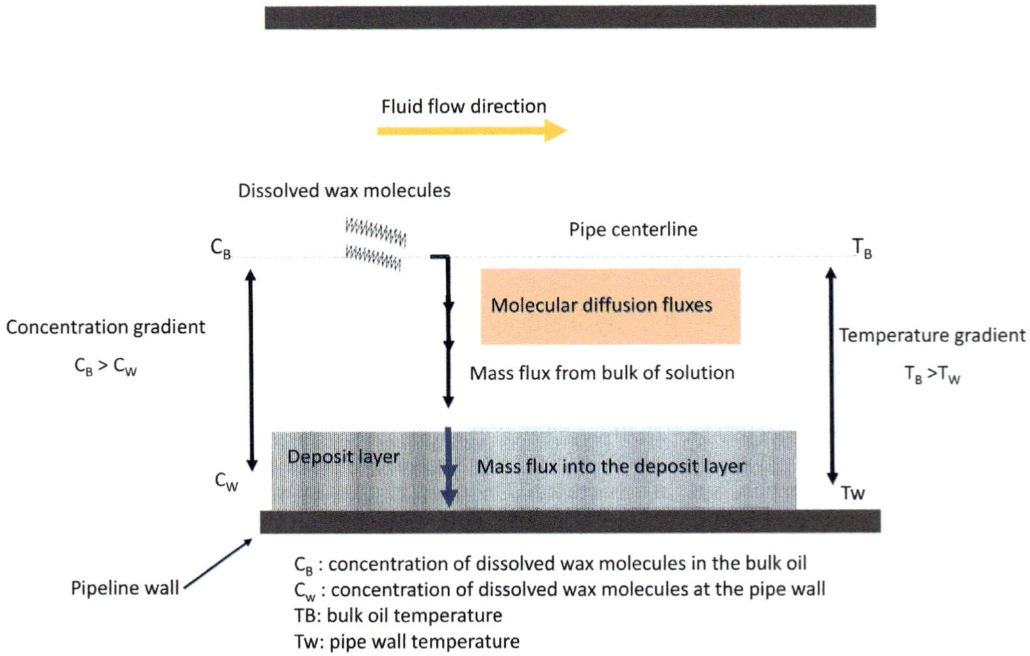

FIG. 12.14

Schematic of wax deposition mechanism by molecular diffusion.

consensus that molecular diffusion is the main mechanism of wax deposition. Fig. 12.14 depicts the molecular diffusion mechanism.

12.8.1 Basic principles

12.8.1.1 Phase equilibria of wax precipitation

Modeling of wax precipitation depends on understanding the phase equilibrium of hydrocarbon systems. It is principally based on the description of equilibrium between the "solid" wax and the hydrocarbon "liquid" phases and thermophysical prediction that relates thermophysical properties of *n*-paraffin in wax solutions with the thermophysical properties of its pure components under standard conditions [118].

Wax-oil is a complex hydrocarbons system composed of wax or paraffinic components (high molecular weight alkanes > C15 that precipitate below wax appearance temperature (WAT)) and nonparaffinic components (the rest of crude oil components). Fig. 7.25 in Chapter 7 illustrates a phase diagram of crude oil, where the boundaries between the phases are identified. Both of these groups of components can have hundreds or thousands of compounds. During wax precipitation modeling, both vapor-liquid and solid-liquid phase equilibrium are addressed. This involves determining the amount of each phase and the mole fraction of the components in each phase, which makes the modeling a complex problem. Therefore a way to simplify this and reduce calculation intensity is by lumping components with neighboring carbon numbers into pseudocomponents [119].

The main outcomes of phase equilibrium calculations are the wax precipitation conditions and the boundaries between the phases at the different conditions.

12.8.1.2 Thermodynamics of wax precipitation

At equilibrium, the fugacity of a component i must be equal in vapor, liquid, and solid phases:

$$f_i^v = f_i^l = f_i^s \tag{12.64}$$

where f is the fugacity of component i (Pa), and the subscripts v,l,s refer to vapor, liquid, and solid, respectively.

In the vapor-liquid equilibrium, the fugacity is usually described using the EOS; however, this is not applicable in the case of solid phase, so the fugacity of the pure solid should be calculated, and then the solid-liquid or wax-oil fugacities for a nonideal mixture are defined, since the assumption of ideal solid-liquid phase behavior usually leads to overestimation in the calculations [119].

The fugacity of a pure solid is calculated based on the fugacity of a pure liquid and the Gibbs free energy change from pure liquid to pure solid, according to Eq. (12.65) [119]:

$$\ln\frac{f_i^{0l}}{f_i^{0s}} = \frac{\Delta H_i}{RT} 1 - \frac{T}{T_i^f} + \frac{\Delta Cp_i}{R} 1 - \frac{T_i^f}{T} + \frac{\Delta Cp_i}{R} \ln\frac{T_i^f}{T} + \int_{P_0}^{P} \frac{\Delta V_i}{RT} dp \tag{12.65}$$

where f_i^{0l}, f_i^{0s} is the fugacity of component i in pure liquid and pure solid phases (Pa); ΔH_i is the heat of fusion of component i (J/mol); R is gas constant (J/mol/K); T is equilibrium temperature (K); ΔCp_i is change of heat capacity of component i due to phase transition (J/mol/K); T_i^f is melting point of component i (K); and ΔV_i is change of molar volume of component i at phase transition (L/mol).

For a nondeal solid/liquid mixture, the fugacity is the product of the mole fraction, activity, and phase fugacity of a pure component, thus [119]:

$$f_i^l = x_i \gamma_i^l f_i^{0l} \tag{12.66}$$

$$f_i^s = y_i \gamma_i^s f_i^{0s} \tag{12.67}$$

where x_i, y_i is the mole fraction of the component i in the liquid and solid phases, and γ_i^l, γ_i^s is the activity coefficient of component i in the liquid and solid phases,

Substituting Eqs. (12.66), (12.67) in Eq. (12.65) yields [119]:

$$\ln\frac{f_i^l}{f_i^s} = \ln\frac{x_i \gamma_i^l}{y_i \gamma_i^s} + \frac{\Delta H_i}{RT} 1 - \frac{T}{T_i^f} + \frac{\Delta Cp_i}{R} 1 - \frac{T_i^f}{T} + \frac{\Delta Cp_i}{R} \ln\frac{T_i^f}{T} + \int_{P_0}^{P} \frac{\Delta V_i}{RT} dp \tag{12.68}$$

The various developed thermodynamic models use different theories to analyze the liquid-solid phase nonidealities and estimate the activity coefficients based on composition. Besides the nonidealities, the models also discuss the phase transition and multiple solid phases in equilibrium with each other. Thus the developed models can be single-component or multicomponent, single-phase or multiphase solid deposits [119]. The majority of published wax precipitation models make simplifying assumptions, typically with regard to the ideality of the liquid and solid phases, or are limited to a single solid phase. The different models are briefly discussed in Section 12.8.2.

The main goal of such thermodynamic models is to predict the wax precipitation conditions, or WAT, the wax deposition curve (WPC), and the composition of precipitated wax, all of which are used as input in the deposition models when the experimental data are not known or are inaccurate.

12.8.1.3 Hydrodynamics of wax deposition

Determining molecular diffusion, as the main mechanism of wax deposition, involves determining the diffusive flux of dissolved wax molecules reaching the pipe wall/deposit-fluid interface. In order to obtain the diffusive flux of wax, three sequential steps need to be performed [120,121]:

1- Hydrodynamic calculations to obtain the velocity profile and eddy diffusivity (in turbulent flow regime),
2- Heat transfer calculations to obtain the temperature profile, and
3- Mass transfer calculations to obtain the concentration profile.

The hydrodynamic calculations are the crucial step, as they are essential to the mass transfer and heat transfer calculations. Multiphase flow parameters, like flow patterns, pressure gradients, and liquid holdups, basically influence the heat and mass transfer characteristics. For example, the heat capacities of gas and liquid phases are drastically different, so the liquid holdup has significant impact on the temperature profile in the pipe. A gas-liquid flow with higher liquid holdup generally exhibits a more gradual decrease in temperature during cooling [119].

In single-phase flow, hydrodynamic calculations are used to determine friction factors in order to calculate the pressure drop across the pipe. In multiphase flow, empirical correlations are used to estimate velocity, pressure, liquid holdup, and flow regimes; this is usually initiated by numerically solving the continuity equation and the Navier-Stokes equation to generate the radial velocity profiles [120].

The pipeline pressure drop or, equivalently, the shear stress at the pipe wall/deposit-fluid interface is a critical parameter for pipeline design, besides the fact that the prediction of the deposit growth rate is achieved by comparing the shear stress imposed by the fluid and the dynamic yield stress of the deposit. The shear stress in laminar flow is calculated using pipe diameter, flow rate, and fluids properties; however, in turbulent flow the calculations are more complex. Under the pseudo-steady state and incompressible fluid assumptions, the continuity equation and the Navier-Stokes equation lead to a balance between the pressure drop in the streamwise direction and the local shear stress, shown in Eq. (12.69) [120,121].

$$-\frac{\partial P}{\partial z} = \frac{2\tau}{r} \tag{12.69}$$

where z is the axial position (m), τ is the local shear stress (Pa), and r is the radial position (m).

Eq. (12.69) suggests that the local shear stress profile varies linearly with radial position and equals the wall shear stress at the effective pipe radius. Therefore the hydrodynamic calculation is started with an initial guess of the wall shear stress, and the linear radial shear stress τ (r) distribution can be expressed by Eq. (12.70) [120,121]:

$$\tau(r) = \tau_w \frac{r}{R} \tag{12.70}$$

where τ_w is the guessed wall shear stress (Pa), r is the radial position (m), R is the effective pipe radius (m), which equals the pipe inner radius at $t = 0^+$ when the deposit has yet to form, and equals the radius minus the deposit thickness after the deposit has formed.

From the relationship between shear stress Eq. (12.70) and viscosity, the shear rate (ϑ), i.e., the radial gradient of the velocity profile can be calculated [120,121]:

$$\vartheta = \frac{\tau}{\eta} = \frac{\tau\left[1 - \left(\frac{A\varnothing\tau^{-x}}{\varnothing_M}\right)\right]^2}{\eta_L(1 - A\varnothing\tau^{-x})} \tag{12.71}$$

where η is the viscosity (Pa s), η_L is viscosity of the liquid phase in the absence of suspended solid crystals and is a function of temperature (Pa s), A and X are rheological parameters that can be obtained from experimental measurements, ϕ is solid volume fraction, and ϕ_M is maximal possible solid volume fraction.

Numerical integration of the shear rate profile with respect to radial position using the no-slip boundary condition at the pipe wall generates the entire radial velocity profile [120,121].

12.8.1.4 Heat transfer

Wax deposition is a temperature-dependent process. Heat loss from the bulk of the solution to ambient temperature at the pipe wall creates a temperature gradient that is responsible for the wax precipitation. Therefore the temperature profile across the system is mandatory for wax deposition prediction. The initial step in this is solving the energy balance equation (Eq. 12.72) in both the radial and axial directions with appropriate boundary conditions, to obtain the radial and axial temperature profiles [119–121]:

$$\rho_{oil} Cp_{oil}\left(\frac{\partial T}{\partial t} + V_z \frac{\partial T}{\partial z}\right) = \frac{1}{r}\frac{\partial}{\partial r}\left(r k_{oil}\frac{\partial T}{\partial r}\right) \tag{12.72}$$

where Cp_{oil} is the heat capacity of the oil (J/kg/K), T is the temperature (K), k_{oil} is the thermal conductivity of oil (W/m/K), V_z is the axial velocity, ρ_{oil} is the density of the oil (kg/m^3), and t is the time (s).

Under production, two boundary conditions are specified: the inlet temperature profile is specified based on the operating conditions and the heat flux due to conduction from the fluid to the pipe wall is equated to the heat loss due to the convective cooling by ambient conditions. These are listed in Eqs. (12.73), (12.74) [119–121]:

$$T_{oil} = T_{inlet}\, at\, z = 0 \tag{12.73}$$

Therefore

$$k_{oil}\frac{\partial T_{oil}}{\partial r} = -h\left(T_{oil} - T_{ambient}\right) at\, r = R \tag{12.74}$$

where h is the eternal convective heat transfer coefficient (W/m^2/K).

Some of the heat transfer correlations used in different wax deposition models include Sieder and Tate [122], Gnielinski [123], Dittus and Boelter [124], and Deen [125].

12.8.1.5 Mass transfer

In the molecular diffusion mechanism, the high molecular weight dissolved wax molecules (not the bulk precipitated ones), under the effect of temperature gradient driven by the heat transfer effects, diffuse to the pipe wall to deposit and diffuse internally into the formed incipient gel layer, leading to its aging and hardening.

Generally, the mass flux of the dissolved wax molecules from the bulk oil J_{bo} is given by Fick's law and can be rewritten as [119–121]:

$$J_{bo} = -D_{wax}\frac{dC}{dr} = D_{wax}\frac{C_{oil} - C_{wall}}{\delta_{MT}} = D_{wax}\frac{C_{oil}(T_{oil}) - C_{wall}(T_{wall})}{\delta_{MT}} \tag{12.75}$$

where C is the concentration of the dissolved waxy component (kg/m^3), D_{wax} is the diffusion coefficient of wax in oil (m^2/S), T is the temperature (K), δ_{MT} is the thickness of the mass transfer layer (m). The $C_{oil} - C_{wall}$ is the mass transfer driving force and can be obtained from the thermodynamics modeling (by inverting the wax precipitation curve) [119].

Notably, from Fig. 12.14, two mass fluxes of wax molecules are involved in the process: one is from the bulk oil J_{bo} (mentioned earlier), and the other is the internal diffusion into the deposit layer (or incipient gel layer) J_{dl}, and this one is responsible for increasing the wax fraction in the deposit layer (aging). The difference between these two ($J_{bo} - J_{dl}$) corresponds to the growth of the deposit layer and this is given by Eq. (12.76) [119]:

$$\rho_d F_w \frac{d\delta_d}{dt} = (J_{bo} - J_{dl}) \tag{12.76}$$

where ρ_d is the density of the deposit (kg/m^3); F_w is the wax mass fraction in the wax deposit (all wax molecules entrapped as oil in the gel layer or precipitated as crystals); and δ_d represents deposit thickness (m). Eq. (12.76) also represents the mass balance on dissolved waxy components at the oil-deposit interface. The internal diffusion flux is given by Eq. (12.77) [119]:

$$\frac{\rho_d\left(R_p^2 - (R_p - \delta_d)^2\right)}{2\pi(R_p - \delta_d)}\frac{dF_w}{dt} = J_{dl} \tag{12.77}$$

where ρ_d is the density of the deposit (kg/m^3), R_p (m) represents the radius of the pipe; t represents time (s), and dF_w/dt represents the aging rate of wax deposits, which indicates its hardening with time. Eq. (12.77) also represents the mass balance on the dissolved waxy components in the deposit.

In most of the deposition models, the internal diffusion term (J_{dl}) is neglected and the deposition growth is described by J_{bo}, and the wax mass fraction is represented by the deposit porosity (φ); thus the molecular diffusion is represented by Eq. (12.78) [119]:

$$\rho_d(1 - \varphi)\frac{d\delta_d}{dt} = J_{bo} \tag{12.78}$$

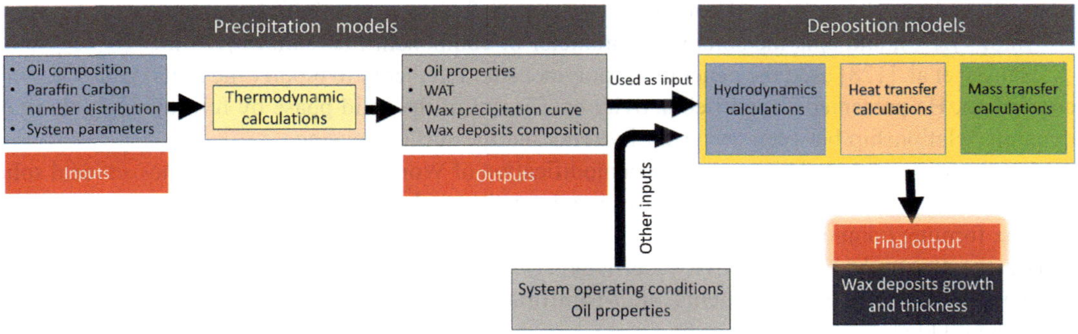

FIG. 12.15

Wax deposits prediction: the precipitation model results are incorporated into the deposition model to ultimately determine the wax deposits growth.

12.8.2 Wax precipitation and deposition models

As mentioned earlier, the holistic wax deposition modeling involves thermodynamics modeling besides the deposition modeling. The thermodynamic modeling predicts WAT, WPC, wax deposits composition, and other oil properties to be used as input in the deposition modeling, which uses hydrodynamics, heat transfer, and mass transfer to calculate the deposit growth rates and deposit thickness. The main organization of wax modeling is illustrated in Fig. 12.15.

12.8.2.1 Wax precipitation models

Currently, there are two types of thermodynamic wax precipitation models, namely the solid solution model, and the multisolid phase model. In the solid solution model, the precipitation is considered as a single solid solution and the activity model and EOS are used for solid-liquid and vapor-liquid equilibrium, respectively. In the multisolid phase model, the precipitated hydrocarbon that consists of pure components is assumed to be unstable and segregates from other precipitated species.

Won's models such as Won [126,127], Hansen et al. [128], and Pedersen et al. [129] are characterized by adopting the regular solution theory for the description of the solid phase nonideality and assuming that all compounds are crystallizable [130]. Won and Hansen defined the K values (equilibrium constant) to determine the solid-liquid equilibrium state, so for component i and by assuming that the liquid- and solid-phase heat capacities are equal, the equilibrium constant can be defined as [129]:

$$K = \frac{y_i}{x_i} = \frac{\gamma_i^l}{\gamma_i^s} \exp\left[\frac{\Delta H_{fi}}{RT}\left(1 - \frac{T}{T_i^f}\right)\right] \tag{12.79}$$

where y_i and x_i are the mole fractions of component i in the solid phase and liquid phase, ΔH_{fi} is the heat of fusion, T_i^f is the temperature of melting or fusion.

Won defined the activity coefficient by the solubility parameter using the Scatchard-Hildebrand model, as in Eq. (12.80):

$$\ln \gamma_i = \frac{V_i(\bar{\delta} - \delta_i)^2}{RT} \tag{12.80}$$

where V_i is the molar volume of component i (L/mol), $\bar{\delta}$ is average solubility parameter of the phase ($J^{0.5}/mol^{0.5}/L^{0.5}$), and δ_i is the solubility parameter of component i ($J^{0.5}/mol^{0.5}/L^{0.5}$).

Hansen et al. [128] used the generalized polymer solution theory to modify the liquid phase activity coefficient. He presented a modified activity coefficient of the solid phase by formulating a thermodynamic relationship between the viscosity term and the surface tension.

Erickson's model [131] is considered a modification of Won's model, but it assumes an ideal solid phase; thus the activity coefficient is considered equal to 1. In this model, they distinguish wax and nonwax forming components by assuming an ideal solution of pure components in the liquid and solid phases, thereby neglecting the activity coefficient. Hence the model requires huge efforts in experimental work to determine the amounts of *n*-paraffinic, isoparaffinic, and naphthenic parts of each component [118].

Pedersen and his coworkers introduced thermodynamic models with several modifications in different research papers [132,133]. They improved Won's model by including the terms containing ΔCpi in the phase equilibrium. They also improved the empirical correlations to estimate physical properties by fitting model predictions with experimental data of 17 North Sea crude oils. Pedersen [134] modeled VLE using SRK EOS and treated the wax phase as an ideal solution to calculate the mole fraction of the potentially wax-forming fraction. The later modified version of the Pedersen model is implemented in the PVTsim software.

The Lira-Galeana model [135] is a multisolid model based on the grouping of hydrocarbon components using average properties assigned to component or pseudocomponent fractions in oil mixtures. They used the Won's model description of the wax fugacity and the pure-liquid fugacity was obtained using EOS [118].

Coutinho's model is a multisolid model that considers both solid and liquid phases to be nonideal solutions. The nonideality is obtained by the predictive UNIQUAC model with use of the predictive free energy model. It is considered a theoretically comprehensive model. It was based on extensive work of Coutinho and coworkers [130,136]. The Coutinho wax model is the default wax model in Multiflash software.

Fig. 12.16 compares the total amount of solid precipitate as a function of temperature for the models with experimental data performed by Pauly et al. [137].

12.8.2.2 Wax deposition models

Various models have been developed to model wax deposition. The two main inputs utilized during the calculations are oil properties and operating conditions. The oil properties include viscosity, density, specific heat, thermal conductivity, WAT, WPC, and others, with WAT and WPC being the most important parameters, obtained from the thermodynamic models discussed earlier. The operating conditions include geometry, dimensions, temperature, pressure, flow rates, etc. [119].

The RRR (Rygg, Rydahl, and Rønningsen) [138] wax model is a multiphase wax model that implements molecular diffusion and shear dispersion effects as the mechanisms responsible for wax deposition. The RRR model functions as an integrator of submodels that account for the pressure drop and regime changes, viscosity of the fluid mixture, heat and energy balances, wax thermodynamic properties, and the wax transport mechanism itself. The model is implemented in OLGA and the commercial wax deposition simulator with the Matzain model.

The Matzain model [139] incorporates shear stripping alongside molecular diffusion and shear dispersion as potential wax deposition models. The model predicted wax thickness with an acceptable

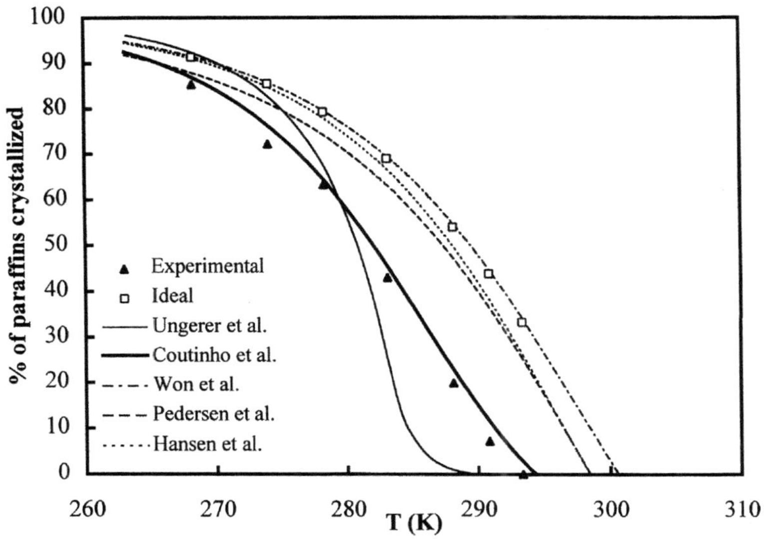

FIG. 12.16

Amount of solids precipitated as a function of temperature for a mixture of n-alkanes predicted with various models.

Reprinted from J. Pauly, C. Dauphin, J.L. Daridon, Liquid-solid equilibria in a decane + multi-paraffins system, Fluid Phase Equilib. 149 (1–2) (1998), 191–207. https://doi.org/10.1016/S0378-3812(98)00366-5, copyright (1998), with permission from Elsevier.

accuracy, especially at high oil flow rates. It also incorporated the effect of flow regimes on wax deposition. The model is implemented in the OLGA wax simulator.

Singh et al. [140] is a wax model based on the thin-film concept. The model considers the aging of the wax deposits and the internal diffusion mechanism for waxy gel layers inducing physical gelation on the wall. The model is implemented in the Michigan Wax Predictor (MWP).

Other models include the Lindeloff and Krejbjerg [141] model, which is implemented in the DepoWax module in the PVTsim software package, and Edmonds et al. [142], which is implemented in FloWax and is a wax deposition model developed by Infochem.

Table 12.6 summarizes some of the features of these models. Thorough discussions of thermodynamic and deposition models are furnished in [118,143,144].

12.8.2.3 Artificial intelligence methods

Artificial intelligence methods have been implemented in wax deposition prediction. Commonly used AI algorithms are the evolutionary algorithm (EA), ANN, swarm intelligence (SI), and fuzzy logic (FL).

An ANN model based on mathematical modeling techniques using regression analysis was developed to predict the wax depositional potential of petroleum reservoirs. The model showed good agreement with the field data records when used to evaluate the waxing tendencies of some crude reservoirs in Nigeria [145].

Table 12.6 Wax deposition models.

Model	Features
RRR	- Deposition mechanism: molecular diffusion and shear dispersion - Does not account for multiphase flow, laminar flow - Implemented in OLGA
Matzain	- Deposition mechanism: shear stripping, molecular diffusion, and shear dispersion - Accounts for flow regimes - Implemented in OLGA
Singh	- Deposition model: molecular diffusion - Considers the aging of wax deposits - Considers wax precipitation kinetics - Implemented in Michigan Wax Predictor (MWP)

An adaptive neural-fuzzy inference system (ANFS) model was developed to predict wax deposition thickness in single-phase turbulent flow and was found to be more accurate than other conventional models [146].

An ANN model trained using different combinations of thermodynamic properties of 12 different hydrocarbon fluids and validated with experimental data was used to predict the WAT for various hydrocarbon mixtures, which showed accurate predictions of the WAT compared to the existing models [147].

Machine learning methods based on MLP optimized with the Levenberg-Marquardt algorithm (MLP-LMA) and the Bayesian regularization algorithm (MLP-BR) were developed to predict wax deposition under production conditions, where the MLP-LMA achieved the best performance [148].

12.8.3 Examples of industrial wax deposition prediction software

- **Wax Deposition Model from the University of Tulsa**

The wax deposition model from the University of Tulsa (TUWAX) was developed through the Tulsa University Paraffin Deposition Projects (TUPDP) [119]. Several versions of TUWAX are available to the members of the TUPDP, and it has been applied in some of the field wax remediation practices [149,119].

The information summarized in the study about TUWAX for single phase is based on the study of Hernandez et al. [150]. The information on wax deposition modeling for the gas-oil two-phase flow is based on the study of Apte et al. [151], while the model's transport equations for the oil-water two-phase flow are likely to be based on the study of Couto ([152] and Bruno et al. [153].

- **Wax Deposition from the University of Michigan (Michigan Wax Predictor)**

The MWP was developed by the University of Michigan Porous Media Research Group [119]. The model predicts the trends of deposition rates with varying operating conditions, and contains no tuning parameter for upper and lower limits of deposit thickness. The model has been verified with 50+ flow-loop experiments from different facilities around the world [154].

The MWP first evaluates the transport characteristics in a pipe flow system and then applies molecular diffusion as the deposition mechanism for wax deposition prediction [155]. The MWP originated from the study of Singh et al. [140], which was the first wax deposition model that accounts for the phenomenon of aging [119]. This software predicts the wax deposition thickness and wax fraction in the deposit under different flow conditions.

- **DepoWax**

The DepoWax module is found in the commercially available software PVTsim, by Calsep International Consultants.

The deposition model is based on the Lindeloff and Krejbjerg [141] model for numerical discretization, heat transfer, energy balances, thermodynamic equilibrium, and wax deposition [156]. The DepoWax is well suited for pipelines with multiple inlets. It contains two flow models, one for single-phase flow and another for two-phase flow [157]. This software calculates pressure drops, temperature profile, and position and thickness of wax deposited along the pipeline [158].

- **FloWax**

This is a wax deposition module that is implemented by KBC. The module is based on the study of Edmonds et al. [142] and Venkatesan [159]. The thermodynamic wax model is based on the Coutinho et al. [136] solid solution model. The model is considered a powerful modeling tool to estimate the deposition of waxes along pipe branches for multi- or single-phase flows in flow assurance and production applications [160].

12.8.4 Recent advances in wax deposits prediction

Wax deposition have witnessed many advances, due to the improvements in the lab testing and simulation tools. Improvements in flow loop testing and incorporating wax characterization methods has removed some of the ambiguities. Using artificial intelligence methods is projected to make a big contribution in the future. However, some of the current models still suffer from critiques, including [156]:

- Some models overestimate the wax deposition, while others underestimate the problem, since most of these models are based on lab experiments, and upscaling these tests is costly.
- Some models do not account for the aging step of wax deposition.
- The models present a number of arbitrary or not well defined values of parameters that are often used to adjust predictions with respect to a specific experimental result.

12.9 Asphaltene prediction

Asphaltene prediction is a key step in managing its flow assurance challenges. Asphaltene prediction spans from quick screening methods, like the colloidal instability index (CII) and deBoer plot, to fully developed methods [161]. Generally speaking, asphaltene modeling can be categorized into two basic types: thermodynamics and kinetic, as depicted in Fig. 12.17.

Thermodynamic methods predict asphaltene precipitation by describing the phase equilibria of the oil-asphaltenes system to determine the phase boundaries, the asphaltene onset pressure, asphaltene

FIG. 12.17
Asphaltene prediction models.

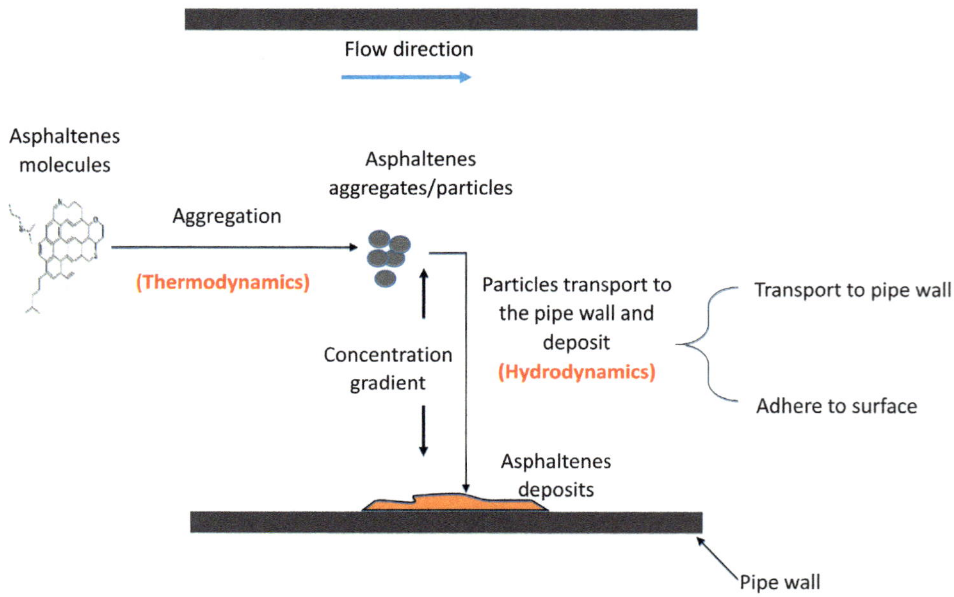

FIG. 12.18
Asphaltene aggregation and deposition mechanism.

solubility, and saturation points, besides other oil properties. The kinetic methods use hydrodynamic and mass transfer calculations to determine the deposition of the precipitated/aggregated particles on the pipe walls to finally determine their growth rates. These processes are depicted in Fig. 12.18.

12.9.1 Basic principles

12.9.1.1 Thermodynamics of asphaltene precipitation

Fig. 12.19 illustrates the phase equilibria of crude oil-asphaltenes. At A the crude oil exists as a single phase where asphaltenes are stable. At point B the crude oil starts to become thermodynamically unstable, where asphaltenes exceed their solubility limit and start to aggregate/precipitate, forming a second phase (or asphaltene rich phase), and the boundary line at this point is known as the upper asphaltene onset pressure (UAOP). Point C marks the bubble point at which asphaltene solubility

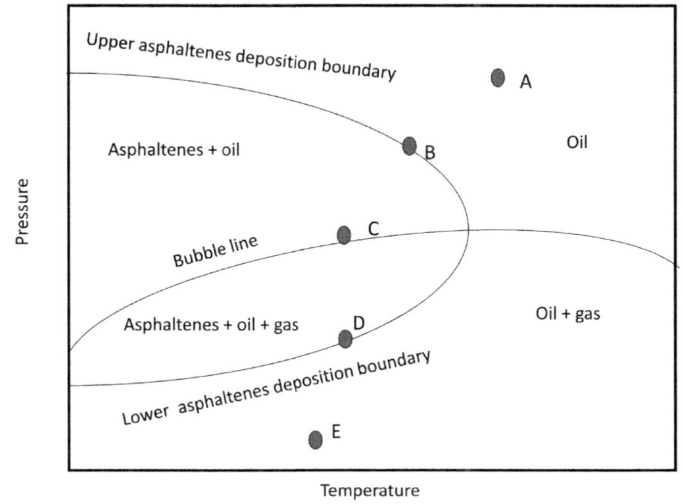

FIG. 12.19

The phase equilibria of asphaltene-crude oil system.

in the bulk oil is lowest and the amount of asphaltenes precipitated is largest. Below the bubble point gases expel from the solution, so below this point three phases exist: asphaltenes (or asphaltene rich phase), gas, and oil (or asphaltene lean phase). At point D, the majority of the light components that destabilize asphaltenes are expelled to the gas phase, so asphaltenes can redissolve in the oil phase. The boundary at point D is known as the lower asphaltene onset pressure (LAOP). At point E gas and oil phases coexist. Hence, thermodynamic modeling of asphaltenes is basically used to predict these boundaries of asphaltene instability (between UAOP and LAOP), calculate the asphaltene onset pressure, estimate the asphaltene solubility. Which then is incorporated in deposition models to predict asphaltene deposit growth rate.

As for the wax modeling, in multicomponent systems such as petroleum, the first step in thermodynamic modeling is the lumping of similar fractions into pseudocomponents in order to simplify the calculations of phase equilibrium [6]. Second, using the rule of thumb that for any two phases at equilibrium, the chemical potentials of each component in these phases are equal, or more beneficial in the asphaltene case, the fugacities of each component in existing phases are equal and they are given by Eq. (12.81) [162]:

$$f_i = \gamma_i x_i f_i^0 \exp\left[\int_0^P \frac{v_i dP}{RT}\right] \tag{12.81}$$

from which the equilibrium ratio is defined as:

$$K_i^{L2,L1} = \frac{x_i^{L2}}{x_i^{L1}} = \left(\frac{\gamma_i^{L1}}{\gamma_i^{L2}}\right)\left(\frac{f_i^{0L1}}{f_i^{0L2}}\right)\exp\left[\int_0^P \frac{v_i dP}{RT}\right] \tag{12.82}$$

where K is the equilibrium ratio (or the equilibrium partition coefficient), i is the component, L_1 is the oil phase (or asphaltene lean phase) and L_2 is the asphaltenes (or asphaltene rich phase), x is the mole fraction, f_0 is the standard fugacity (Pa), v is the molar volume (m^3/mol), P is the pressure (Pa), R is the gas constant, and T is the absolute temperature (K).

For liquid-liquid equilibrium, the terms $\left(\frac{f_i^{0L1}}{f_i^{0L2}}\right)$ and $\exp\left[\int_0^P \frac{v_i^* dP}{RT}\right]$ are equal to 1, so the equilibrium partition coefficient can be reduced to Eq. (12.83):

$$K_i^{L2,L1} = \frac{x_i^{L2}}{x_i^{L1}} = \left(\frac{\gamma_i^{L1}}{\gamma_i^{L2}}\right) \tag{12.83}$$

Several approaches are employed to define the activity coefficient and the other terms in Eq. (12.83) to calculate the onset of asphaltene precipitation UAOP. They assume the oil and asphaltene as two pseudocomponents: one component representing the deasphalted oil and the other the asphaltenes.

Generally speaking, asphaltene precipitation is usually explained through two main thermodynamic approaches, namely the solubility and the colloidal approaches. In the solubility approach, asphaltene molecules are solubilized in oil due to their interactions with other oil molecules dominated by the weak van der Waals forces. The colloidal approach assumed that asphaltenes exist in crude oil as micelles that are stabilized by the polar forces between asphaltenes and resin molecules. Thermodynamic equilibrium calculations for all the components and/or pseudocomponents of the system are performed to predict the phase behavior of oil at specified conditions [163]. More details about these two approaches and their models are discussed in the next few sections.

12.9.1.1.1 Solubility approach and solubility models

The solubility approach assumes that the asphaltenes are dissolved (partially or fully) in the bulk crude oil and the precipitation starts when the solubility falls below a certain threshold level. The models in this approach use the solubility parameter concept, considering that petroleum crude consists of two phases, asphaltenes and the deasphalted oils (maltenes). Therefore the equilibrium can be studied as solid-liquid (S-L) or liquid-liquid (L-L), where asphaltene stability is a consequence of dispersion forces between it and the other components present in the oil [4,164] Regular solution models are based on solving Eq. (12.83). Table 12.7 summarizes the different solubility models. Asphaltene solubility parameters are usually calculated using either the Scatchard [165] or Hildebrand [166] equations or by calculating the interaction parameters between the asphaltenes and the remaining oil constituents. The properties of deasphalted oil are generally calculated using cubic equations of state. The difference between the asphaltene present in the oil and the solubility under given conditions is used for calculating the amount of asphaltene precipitated [6,164].

Models in the solubility approach can be based on the regular theory or EOS, where the difference between them is the way they treat the nonasphaltene components of crude oil. Under regular solution theory, asphaltenes are considered as a solute in a bulk oil solvent, as if crude oil is a binary mixture, whereas EOS models do not consider crude oil as a simple binary mixture. Therefore in the solubility approach asphaltene solubility is quantified by calculating either the solubility parameters of asphaltenes and crude oil, or the binary interaction parameters (BIPs) between asphaltene and the other oil constituents [4].

Hirschberg et al. [167] developed a simple thermodynamic model based on Flory-Huggins polymer solution theory to describe the flocculation tendency and calculate the amount of precipitated asphaltenes in light crudes upon changes in pressure, temperature, or composition. They assumed that asphaltene flocculation is reversible, but only with very slow redissolution kinetics [168]. Chung et al. [169]

Table 12.7 Examples of asphaltenes solubility models [164, 168–174].

Model	Equation	Variables
Hirschberg et al.	$\ln(\phi_a)_{\max} + \left(1 - \frac{v_a}{v_m}\right) + \frac{v_a}{RT}(\delta_a - \delta_m)^2 = 0$	$(\phi_a)_{\max}$ is maximum volume fraction of asphaltenes soluble in crude oil, v is the molar volume, δ is the solubility parameter. The subscripts a and m refer to asphaltenes and oil mixture (excluding asphaltenes)
Thomas et al.	$\ln(K_i^S) = \ln\left(\frac{x_i^S}{x_i^L}\right) = \frac{1}{RT}\left[v_i^L(\delta^L - \delta_i^L)^2 - v_i^S(\delta^S - \delta_i^S)^2\right] + \frac{\Delta H_f}{RT}\left(1 - \frac{T}{T_f}\right) + \frac{\Delta C_p}{R}\left[1 - \frac{T_f}{T} + \ln\frac{T_f}{T}\right] + \frac{1}{RT}\int_0^P (v_i^L - v_i^S)dp$	The superscripts S and L refer to solid and liquid, v_i and δ_i refer to the mole fraction, molar volume and solubility parameter for ith fraction, respectively. δ is the average solubility parameter of the mixture, C_p is the heat capacity, ΔH_f is the heat of fusion, and T_f is the temperature of fusion
Chung et al.	$x_a^L = \exp\left[-\frac{\Delta H_a^f}{RT}\left(1 - \frac{T}{T_f}\right) - \frac{v_a^L}{RT}(\delta_m^L - \delta_a^L)^2 - \ln\left(\frac{v_a^L}{V_m}\right) - 1 + \frac{v_a^L}{V_m}\right]$	x_a^L is the mole fraction of asphaltene in solution, the superscript L stands for liquid, δ_a is the solubility parameter for asphaltene, δ_m is the average solubility parameter, v_a is the molar volume of asphaltenes, and V_m is the average molar volume of the liquid phase
Cimino et al.	$\ln(1 - \phi_a^*) + \left(1 - \frac{v_m}{v_a}\right)\phi_a^* + \frac{v_m}{RT}(\delta_a - \delta_m)^2 \phi_a^{*2} = 0$	ϕ_a^* is the asphaltene volume fraction in the nucleating phase at the cloud point/onset of flocculation, v_a is the molar volume of asphaltenes, and V_m is the average molar volume of the liquid phase, δ_a is the solubility parameter for asphaltene, δ_m is the average solubility
De Boer et al.	$\frac{\Delta S}{S} = \int_{p=pb}^{pr} \left(\frac{\partial S}{S\partial p}\right)_T dp \cong \left(\frac{\partial S}{S\partial p}\right)_T (p_r - p_b)$	$(\Delta S/S)$ is the maximum supersaturation, p_b and p_r are the bubble-point and reservoir pressures, respectively
Yarranton and coworkers	$K_i = \frac{\chi_i^S}{\chi_i^l} = \exp\left[\frac{\Delta H_i^f}{RT}\left(1 - \frac{T}{T_i^f}\right) + 1 - \frac{v_i^L}{V_m} + \ln\left(\frac{v_i^L}{V_m}\right) + \frac{v_i^L}{RT}(\delta_m^L - \delta_i^L)^2\right]$	K_i is the solid-liquid equilibrium ratio of component i, χ_i^S and χ_i^l are the solid phase and liquid phase mole fractions of component i, ΔH_i^f is the enthalpy of fusion, and T_i^f is the melting point temperature of component i, v_i^L and v_m are the molar volumes of component i and solvent, respectively, δ_i^l and δ_m are the solubility parameters for component i and solvent, respectively
Buckley and Wang	$\delta = 52.042\, F_{RI} + 2.904$	δ is the solubility parameter in solutions of non-polar molecules, The function of refractive index (F_{RI}): $F_{RI} = \frac{n^2-1}{n^2+2}$ where n is the refractive index

developed their model for organic solid precipitation based on the thermodynamic principle for solid-liquid equilibrium. Thomas et al. [170] developed a multicomponent model based on Won's [126] liquid-solid solution theory. Cimino et al. [171] proposed a precipitation model utilizing polymer-solution thermodynamics; they hypothesized that phase separation leads to obtaining a heavy liquid phase containing both asphaltenes and a fraction of solvent. Their model also relies on cloud point measurement for investigating the phase behavior. De Boer et al. [172] developed a thermodynamic model based on the work of Hirschberg et al. They established a simple screening criteria for potential asphaltene precipitation from crude oils, where the tendency of asphaltenes to precipitate depends on the extent of crude undersaturation with gas, extent of crude saturation with asphaltenes, and crude density [164]. The Yarranton and Masliyah [173] model was built by considering asphaltenes as a series of polyaromatic hydrocarbons with randomly distributed associated functional groups and molar volume, and the solubility parameter distributions are calculated from experimental measurements of molar mass and density. The solubility was modeled by solid-liquid equilibrium calculations, with the K values derived from Scatchard-Hildebrand solubility theory, incorporating the Flory-Huggins entropy of mixing [164]. Buckley et al. [174] developed a model on the assumption that London dispersion forces dominate aggregation and precipitation of asphaltenes; they also proposed an alternate method for predicting the onset of asphaltene precipitation using refractive index (RI) measurements.

The EOS models, combined with flash calculations, are used to produce any set of desired properties, including solubility of asphaltene in oil in terms of volume fraction or mass per volume [4].

There are three major EOS models used for asphaltene modeling [161]:

1. Cubic Equations of State (CEOS): Peng-Robinson (PR EOS) and Soave-Redlich-Kwong (SRK EOS) with/without Peneloux volume correction factors lie in this category. Exmples of models in this category are: Thomas et al. [170]; Nghiem et al. [175]; Nghiem and Coombe [176]; Gupta [177], Kohse et al. [178].

2. Cubic Equations of State plus Association (CPA): This category includes cubic EOS models that account for self-association between asphaltenes and aromatics/resins. Examples of models in this category are: Du and Zhang [179]; Sabbagh et al. [180]; Li and Firoozabadi [181]; Shirani et al. [182].

3. Statistical EOS: Statistical EOS are the most recent advancements in the EOS modeling field, which have been developed to deal with asymmetric mixtures. SAFT (Statistical Associating Fluid Theory) and PC-SAFT (Perturbed Chain SAFT) belong to this category. Examples of models in this category are: Ting et al. [183]; Gonzalez et al. [184]; Panuganti et al. [185].

SAFT and cubic EOS models are often used to model asphaltene precipitation behavior (Abouie et al.) [186,187], while EOS models (in general) are considered the most widely used [164]. These models assume the asphaltene is treated as one homogeneous component.

12.9.1.1.2 Colloidal approach and the colloidal models

The colloidal theory assumes that the asphaltenes exist as colloidal particles stabilized by resins adsorbed onto their surfaces. The partitioning of resins between the colloidal surface and the surrounding medium controls the asphaltene solubility. If a sufficient amount of resins desorb, the asphaltenes will be destabilized and precipitate [188]. The colloidal form was first proposed by Nellensteyn and then by Pfeiffer and Saal, and it was considered the main theory of asphaltenes until recently (2000s).

Leontaritis and Mansoori [189] proposed a thermodynamic colloidal model capable of predicting the onset of flocculation of colloidal asphaltene in oil mixtures due to either changes in composition (i.e., solvent addition) or electrical phenomena (i.e., streaming potential). They concluded that asphaltenes are kept stabilized in solution by the adsorbed resins on their surfaces. They defined critical resin concentration (CR) as the concentration of resin in the oil phase that is just enough to keep the colloidal asphaltene particles completely covered with resins. The critical chemical potential of resins in oil phase is then equal to the chemical in solid phase (asphaltene phase). When the resin concentration (CR) falls below the critical resin concentration (CR), asphaltenes start to flocculate and aggregate. Victorov and Firoozabadi [190] developed an alternate thermodynamic colloidal model assuming asphaltenes exist as micelles comprising cores of aggregated asphaltenes surrounded with resin molecules adhered to the surface. Another thermodynamic model within the framework of colloidal theory was proposed by Pan and Firoozabadi [191,192]; their approach is also known as a reverse micelle approach. In this approach, the precipitated phase is assumed to consist of asphaltenes and resins that do not associate with each other, while the liquid phase is assumed to consist of a mixture of asphaltenes and resin monomers, asphaltene-resin micelles, and other oil species.

12.9.1.2 Asphaltene deposition modeling

Once asphaltenes have aggregated and precipitated in the solution, they are transported to the pipe wall or rock surface under the turbulent flow conditions, and then stick to the surfaces where they deposit. The process involves hydrodynamic and thermodynamic concepts.

Generally speaking, the asphaltene deposition models are based on the following information [193]:

(1) Particle transport towards the surface,
(2) Particle attachment process to the surface, and
(3) Particle concentration gradient between fluid bulk and the wall surface.

The first two pieces of information are related to the hydrodynamics of the deposition process, while the last one is thermodynamics related and is obtained from the thermodynamic modeling discussed previously.

Mathematically representing these data gives us Eq. (12.84) [193]:

$$\dot{m} = K_t \times SP \times (C_b - C_s) \tag{12.84}$$

where $\dot{m}$ is the mass deposition flux (kg/s m^2), K_t is the transport coefficient that considers the macroscopic and microscopic mechanisms, SP is the sticking probability, and $(C_b - C_s)$ is the concentration gradient between fluid bulk and wall surface.

Particle transport to the pipe wall takes place under the influence of different mechanisms, including diffusion, turbophoresis, thermophoresis, inertial impaction, gravitational settling, electrical forces, and the effects of surface roughness and particle interception [194]. Under these forces, and when optimum conditions are attained, the particle will reach a minimum velocity below which it cannot be lifted in the body of the fluids; thus it will settle on the surface. These mechanisms are discussed in Chapter 4.

Escobedo and Mansoori [195] developed an asphaltene deposition model by accounting for both diffusive and convective mechanisms for transport of asphaltenes based on a previous model used for aerosol (microscopic liquid or solid particles dispersed in air currents). Ramirez-Jaramillo et al. [196] hypothesized that the asphaltenes are transported to the wall due to molecular diffusion. Soulgani et al. [197] concluded that the deposition rate on the tubing surface is controlled by the chemical reaction mechanism. Vargas et al. [198] described asphaltene deposition in a multistep transportation using diffusion and convection. Kurup et al. [199] modified the multistep Vargas et al. model. Eskin et al. [200] concluded that the precipitated particles grow due to agglomeration driven by Brownian motion and the turbulent dispersion and shear removal, where the major mechanism of particle transport to the wall is Brownian motion. The Kor and Kharrat (2016) model considered both diffusional and inertial transport mechanisms. A more rigorous analysis of the asphaltene deposition model can be made using a computational fluid dynamic (CFD) approach [201]. Other models have also been reported in the literature.

Two of the main factors in particle transport are relaxation time and the stopping distance. The stopping distance is the distance a particle will travel in free flight through a stationary fluid (the viscous sublayer close to the wall) before it stops because of drag forces [193,202], while the relaxation time is a measure of particle inertia and denotes the time scale with which any slip velocity between the particles and the fluid is equilibrated [194]. The stopping distance is given by particle relaxation time multiplied by its velocity. A dimensionless relaxation time (T_p^+) is given by Eq. (12.85) [[193,203]:

$$T_p^+ = \frac{\rho_p d_p^2}{18\eta_d} \times \frac{\frac{f_f}{2} V_{avg}^2}{\eta_k} \tag{12.85}$$

where ρ_p is the particle density (kg/m^3), d_p is the particle diameter (m), η_d is the dynamic viscosity (kg/ms), V_{avg} is the average fluid velocity (m/s), η_k is the kinematic viscosity (m^2/s), and f_f is the fanning friction factor.

The dominating transport mechanisms are diffusion regime ($T_p^+ < 0.1$), inertia regime ($0.1 < T_p^+ < 10$), and impaction regime ($T_p^+ > 10$) [Epstein N (1997)] [196].

Another important term in transport is the transport coefficient. Many models have been applied to obtain the transport coefficient, such as Friedlander and Johnstone [193], Cleaver and Yates [204], Jamialahmadi et al. [205], and Escobedo and Mansoori [1995].

The developed asphaltene deposition models can be classified into those that modeled deposition in the porous media and those that modeled deposition in flow lines.

The models describing deposition in porous media are based on asphaltene adsorption on the reservoir rock pores forming a deposit layer, blockage of the pores by the deposited particles, and shear removal of the deposited asphaltene particles, causing partial increase in permeability [206]. One of the earliest models of asphaltene deposition was by Gruesbeck and Collins [207], which combined surface deposition, entrainment, and clogging. Wang and Civan [208] modified the Gruesbeck and Collins model and described a diffusivity model for asphaltene deposition. The mass balance of the asphaltene phase was represented by Eq. (12.86) [206]:

$$\frac{\partial}{\partial t}\left(\varphi^* \rho_a C_{np} + \varphi^* \rho_L w_{aL}\right) + \nabla\left(\varphi^* \rho_a C_{nP} u_f + \varphi^* \rho_L w_{aL} u_f\right) = \rho_a \frac{\partial w_{aL}}{\partial t} \tag{12.86}$$

where φ^* is the rock porosity, ρ_a is the density of asphaltenes (kg/m^3), C_{np} is the concentration of nanoparticles (kg/m^3), ρ_L is the density of the liquid phase (kg/m^3), w_{aL} is the mass fraction of dissolved asphaltene in the liquid phase, and u_f is the mean fluid velocity represented by the Darcy law (m/s).

The deposition by adsorption, deposition removal by drag force, and clogging of pores is represented by Eq. (12.87) [206]:

$$\frac{\partial w_{aL}}{\partial t} = \propto \varphi^* C_{nP} - \beta w_{aL}\left(u_L - u_{L,crit}\right) + \zeta C_{nP} u_f \tag{12.87}$$

where u_L is the interstitial velocity of the liquid phase (m/s), $u_{L,crit}$ is the critical interstitial velocity of the liquid phase (m/S), and α, β, and ζ are empirical constants for deposition, removal of the deposit, and plug clogging, respectively; these constants were experimentally obtained by core flooding experiments and are independent of the particle size. The Wang and Civan model was later modified by others like Kord et al. [209] and Behbahani et al. [210].

The flowlines asphaltene deposition models are based on the particle transportation mechanisms discussed earlier. One of the earliest models was Ramirez-Jaramillo et al. [196], a model based on molecular diffusion. In this model the overall mass flux of asphaltenes deposited at any time is the product of molecular diffusion minus the shear removal fluxes, as in Eq. (12.88) [206]:

$$\frac{1}{A_{pipe}}\frac{dM_s}{dt}=J_{MD}-J_{sr} \tag{12.88}$$

where M_s is total mass deposited (kg), A_{pipe} is the surface area of the pipe (m^2), J_{MD} represents molecular diffusion, and J_{sr} represents shear removal; they are represented in Eqs. (12.89), (12.90) [206]:

$$J_{MD}=-D_m\rho_m\left[-\frac{\partial w_s}{\partial T}+\frac{(1-w_s)}{\rho_m}\frac{\partial\rho_m}{\partial T}\right]\frac{\partial T}{\partial r} \tag{12.89}$$

where D_m refers to the molecular diffusivity of dissolved asphaltene molecules (m^2/S), w_s refers to the weight fraction of asphaltene present in the system, and ρ_m refers to the mixture density of asphaltene and oil (kg/m^3).

$$J_{sr}=\bar{\mathrm{A}}\,\tau_w M_s\exp\left(\frac{-\underline{\mathrm{B}}}{T_i}\right) \tag{12.90}$$

where $\bar{A}$ and $\underline{B}$ are constants and τ_w is the shear stress at the wall (Pa).

The Vargas et al. [198] model is a multistep transportation mechanism (diffusion and convection). This model solved the component mass balance of asphaltenes considering diffusion, convection, precipitation, and aggregation, and it predicted the concentration profile in two dimensions (radial and axial), along with the transient nature of the deposition. Kurup et al. [199] modified this model by incorporating second-order kinetics for aggregation and dissolution of precipitated particle axial diffusion in the flow instead of radial diffusion. The modified precipitation rate (Rp) is given by Eq. (12.91):

$$Rp=\begin{cases}K_p(C_A-C_{eq})\text{ if }C_A>C_{eq}\\ -k_{dis}(C_A)\text{ if }C_A<C_{eq}\end{cases} \tag{12.91}$$

where K_p is the precipitation rate constant (S^{-1}), C_A is concentration of asphaltene microaggregates (kg/m^3), C_{eq} is asphaltene concentration at equilibrium (kg/m^3), and k_{dis} is kinetics of dissolution factor (S^{-1}).

Kor and Kharrat described asphaltene deposition based on diffusional and inertial mechanisms. They used the relaxation time of Eq.(12.85) to define the range of deposition regimes, and they also defined the transport coefficient for the diffusion and inertial regimes using different models. They also defined the particle sticking possibility (SP) according to Eq. (12.92) [193]:

$$SP=K_d\frac{e^{-\frac{E_a}{RT_s}}}{V_{avg}^2}C_d \tag{12.92}$$

where E_a is activation energy (kJ), K_d is deposition rate constant (m^2/s^2), C_d is drag coefficient, and V is average fluid velocity (m/s).

12.9.1.3 Artificial intelligence methods

More recently, artificial intelligence-based models have been proposed to overcome the complexity of existing processes/phenomena. Models based on ANN, Bayesian Belief Network (BBN), and other advanced machine learning algorithms have been used to predict asphaltene precipitation [211–213].

An ANN model was proposed to predict formation damage due to asphaltene deposition. The model was trained on experimental data and showed good agreement with the experimental results [214]. Artificial intelligence methods have been developed to predict asphaltene precipitation and deposition. A feed-forward ANN optimized by the imperialist competitive algorithm (ICA) to predict asphaltene precipitation was proposed [215]. A genetic algorithm-support vector regression (GA-SVR) was also proposed and applied to predict the amount of asphaltene precipitation. The model was trained and tested on the experimental datasets reported in the literature and compared with two scaling equation models, and it showed highly reliable performance [216].

12.9.2 Examples of industrial asphaltene prediction software

- **ADEPT Deposition Model**

A deposition model that has been under development is the Rice University Asphaltene Deposition Tool (ADEPT), which resulted from a collaboration between engineering professors at Rice University in Houston and the DeepStar Global Research and Development Consortium [217].

ADEPT comprises thermodynamic and deposition modules. The thermodynamic module describes the phase behavior of asphaltenes using PC-SAFT EOS. The main inputs for the thermodynamics model include oil chemical and physical properties (live oil composition, live oil/STO density, bubble point, and onset pressure at different temperatures) [217].

The deposition module describes the transport of precipitated asphaltene particles based on advection and dispersion mechanisms, as well as three kinetic processes, precipitation, aggregation of precipitated particles, and deposition, and each process is modeled as first-order kinetics. The deposition model is basically a modification of Vargas et al. [198,217].

The deposition module requires three input parameters [217]:

- Asphaltene solubility as a function of pressure, temperature, and composition (obtained from thermodynamic module)
- Operating parameters
- Constants describing the kinetics of precipitation, aggregation, and deposition

- **ASIST Predictive Model**

ASIST is an empirical method to predict asphaltene stability at reservoir conditions using ambient experimental results and PVT data. The predictions depend heavily on the measured *PRI* (refractive index of the mixture of oil/precipitant at onset of asphaltene precipitation), which is limited by the accuracy of the precipitation onset measurements [4].

ASIST estimates the conditions for instability in a reservoir through a linear extrapolation of the onset solubility parameter vs. the square root of the partial molar volume of a precipitant for a series of *n*-paraffins [218].

ASIST helps predict the asphaltene onset pressures during oil production through titration experiments using stock tank oil with liquid n-paraffins and alkane solubility parameter estimates, based on measurements of the refractive index, which come from an automatic refractometer at a constant temperature [218].

12.9.3 Recent advances in asphaltene deposit prediction

The modeling of asphaltene precipitation and deposition has witnessed multiple advances. Different thermodynamic and kinetic models have been developed to predict asphaltene deposit formation at different conditions, including high pressure and high temperature conditions, to wellbore and pipeline conditions. The experimental methods used to develop the models have also improved. Asphaltene deposition has been studied using capillary deposition tests, realView cell, packed bed column, micromodels, and others. Asphaltenes have also been studied in a variety of geometries and under different flow conditions (temperature, flow regimes, shear rates). Besides, the new revolution in artificial intelligence methods will be game changing in the near future, based on training the theoretical and accumulated field data.

However, the current models still suffer from other drawbacks: for example, they require AOP data as input, which makes them not completely predictive. In addition, the methods depend on crude analysis (SARA analysis), which can suffer from inaccurate measurements leading to inaccuracies in calculations [4].

12.10 Naphthenate deposition prediction

Naphthenate soaps deposition takes place when the naphthenic acids dissociate at the oil-water interface and interact with the metal ions. The deposition model is then based on understanding such interactions at the oil-water interface and the factors affecting them, such as pH, temperature, and pressure.

12.10.1 Basic principles

12.10.1.1 Thermodynamics of naphthenate deposition

Naphthenate and carboxylate soaps (MA) form when the cations (M) interact with the naphthenate/carboxylate anion (A) at the oil-water interface. Similar to mineral scales, the solubility product is expressed by Eqs. (12.93), (12.94):

$$MA = M + A \tag{12.93}$$

$$K_{sp} = [M][A] \tag{12.94}$$

However, the mass transfer of this process depends on the partitioning of the naphthenic acids between the oil-water phases and the dissociation of the acid in the water phase, knowing that naphthenic acids are insoluble in water, with the exception of low-molecular weight acids, which are relatively soluble in water. Soap formation also depends on the naphthenic acids interfacial properties. Therefore the thermodynamic basics of naphthenate deposition is based on understanding the phase equilibria of naphthenic acids and their interfacial properties [219].

12.10.1.2 Phase equilibria of naphthenate deposition

Consider a naphthenic acid HA in oil-water systems: the acid may be distributed between the oil and water phases according to:

$$HA_{f,o} \leftrightharpoons HA_{f,w} \tag{12.95}$$

The corresponding partitioning coefficient K_{ow} is given by Eq. (12.96):

$$K_{ow} = \frac{[HA_{f,w}]}{[HA_{f,o}]} \tag{12.96}$$

where $[HA_{f,w}]$ is the equilibrium (final) concentration of the naphthenic acid in the water phase, and $[HA_{f,o}]$ is the final concentration of the naphthenic acid in the oil phase.

Once partitioned into the water acid dissociates, according to Eq. (12.97), the acid dissociation constant is given by Eq. (12.98):

$$HA_{f,w} \leftrightharpoons H_f^+ + A_f^- \tag{12.97}$$

$$K_a = \frac{[H_f^+][A_f^-]}{[HA_{f,w}]} \tag{12.98}$$

where the $[H_f^+]$, $[A_f^-]$ are the molar concentration of the species, and the subscript f is the equilibrium (final) concentration at equilibrium, and w is the water phase.

Summarizing, the key parameters required for modeling possible naphthenate formation include the naphthenic acid dissociation constants (K_a), the water-oil partitioning coefficient (K_{ow}), and the solubility product (K_{sp}).

At higher pH, formation of micelles in the water phase and reversed micelles in the oil phase is believed to be of importance. Other equilibria involved in the system are dimerization in both water and oil and formation of different metal soaps. The micellization must be accounted for when the total concentration exceeds the critical micelle concentration (CMC). However, most of the developed models neglect the effect of self-association (micellization) of naphtheninc acids [220].

As a consequence of the amphiphilic nature of the carboxylic acids, they prefer the oil-water interface. The pH at the interface, pH_{int}, can be related to the bulk phase pH_{bulk} by assuming a Boltzmann distribution of the counterions in the electrical double-layer, i.e. [220]:

$$pH_{int} = pH_{bulk} + \frac{e\psi}{2.3kT} \tag{12.99}$$

where e is the electronic charge, ψ is the surface potential, k is the Boltzmann constant, and T is the temperature.

12.10.2 Naphthenate deposition models

Several models have been proposed to predict naphthenic acid behavior and naphthenate soaps formation.

Mohammed and Sorbie [219] proposed a model wherein they considered soap-forming species as only a single "pseudonaphthenic acid" species denoted as HA. A schematic of the napththenate-oil-water system is depicted in Fig. 12.20.

The model is based on the key parameters K_{sp}, K_{ow}, K_a, and K_w ($K_w = [H^+][OH^-]$), and the mass balance of each species is then identified. The initial mass of A species $m_{i,A}$ in moles is:

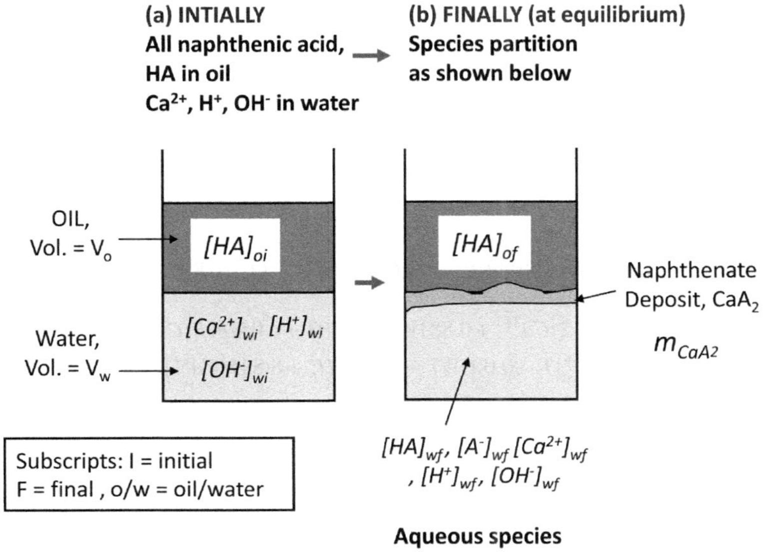

FIG. 12.20

Distribution of chemical species in naphthenate-oil-water system.

Reprinted from M.A. Mohammed, K.S. Sorbie, Thermodynamic modelling of calcium naphthenate formation: model predictions and experimental results, Colloids Surf. A Physicochem. Eng. Asp., Elsevier, 2010. https://doi.org/10.1016/j.colsurfa.2010.08.034, copyright (2010), with permission from Elsevier.

$$m_{i,A} = [HA_{i,o}]V_o \tag{12.100}$$

and at equilibrium

$$m_{f,A} = \left[HA_{f,o}\right]V_o + [HA_w]V_w + [A^-]V_w + 2m_{CaA2} \tag{12.101}$$

where $m_{f,A}$ is the final mass of naphthenate anion A at equilibrium in moles, $[HA_{i,o}]$ and $[HA_{f,o}]$ are the initial and final concentration of the HA in oil phase, V_o is the oil volume, V_w is the water volume, and m_{CaA} is the naphthenates soap mass.

The initial mass of Ca species in moles $m_{i,ca}$ is:

$$m_{i,ca} = \left[Ca^{2+}_{i,w}\right]V_w \tag{12.102}$$

and the final mass at equilibrium is given by:

$$m_{f,ca} = \left[Ca^{2+}_f\right]V_w + m_{CaA2} \tag{12.103}$$

The final charge balance at equilibrium is defined by:

$$C_f = \left(2\left[Ca^{2+}_f\right] + \left[H^+_f\right] - \left[OH^-_f\right] - \left[A^-_f\right]\right)V_w \tag{12.104}$$

By solving the corresponding equations, the main unknowns can be obtained to estimate the mass of calcium naphthenate.

Sulaimon et al. [221] proposed a model to calculate the final pH of the solution and to calculate the precipitate mass. The final solution pH is determined by Eq. (12.105):

$$pH_f = 6.79698 + 0.66140pH_i - 1.21572[Ca^{2+}] + 0.01075W - 0.21425T + 0.01608pH_i[Ca^{2+}] + 0.00108pH_iW - 0.00818pH_iT - 0.00110[Ca^{2+}]W + 0.03088[Ca^{2+}]T - 0.00003WT - 0.01744pH_i^2 - 0.07938[Ca^{2+}]^2 - 0.00006W^2 + 0.00227T^2 \quad (12.105)$$

and the calcium naphthenate precipitate m_p is given by Eq. (12.106):

$$m_p = -59.895 + 26.033pH_i + 33.912[Ca^{2+}] + 0.881W - 2.279T + 0.052pH_i[Ca^{2+}] - 0.005pH_iW - 0.131pH_iT + 0.028[Ca^{2+}]W + 0.252[Ca^{2+}]T - 0.005WT - 1.089pH_i^2 - 8.845[Ca^{2+}]^2 - 0.006W^2 + 0.035T^2 \quad (12.106)$$

where pH_f is the final pH of brine-model oil mixture, pH_i is the initial pH of the brine, $[Ca^{2+}]$ is the concentration of calcium cation, W is the water cut in percentage, T is the temperature, and m_p is the mass of the solid precipitate.

Sarac and Civan [222] proposed a model for the formation damage due to soap formation. The precipitation of naphthenate soap is represented by a power-law equation adopted from the theory of emulsion decomposition/formation reaction, which is given by Eq. (12.107):

$$\frac{dm}{dt} = k_d(m_f - m)^n \quad (12.107)$$

where m_f is the upper limit for the amount precipitated (g), m is the precipitate amount (g) at a given instant, k_d is the precipitation rate constant (1/min/g^{n-1}), and n is the order of reaction.

12.10.3 Recent advances in naphthenate soaps prediction

Naphthenate studies have improved due to the reported cases from different fields. This consequently required a quick and robust risk assessment method to estimate soap threat. Although there is not such a fully developed naphthenate soap deposition model, flow loop, naphthenate rig tests, interfacial tests, and innovative pH sensitivity tests have witnessed improvements that have led to an important step in assessing the risk of naphthenates. Important advances to come in soaps prediction and modeling are projected, with the accumulation of field data and improvements in lab experimental designs.

12.11 Biofouling prediction

Biofouling is another challenging flow assurance issue that depends on the microbiological activity. Good management of the microbial activity requires robust risk assessment and monitoring methods. These can be based on microbiologically influenced corrosion (MIC) prediction methods combined with good analytical methods.

12.11.1 Basic principles

Biofouling modeling relies on understanding the activity and behavior of the different bacteria and their contribution to biocorrosion or MIC. Some of the main factors and aspects that are used include [223]:

- Biofilm development kinetics: microbial cell transport from bulk fluid to the metal surface, and their attachment, accumulation (biofilm growth), and detachment,
- Mass transfer phenomena: diffusion of chemical species (e.g., substrate, metabolites, buffering species) from bulk fluid through the bulk-biofilm interface to the biofilm metal interface, and
- Chemical, biochemical, and electrochemical reactions: within the biofilm, biofilm metal interface, and on the surface of the metal.

12.11.2 Biocorrosion prediction models

Many types of MIC models exist, but they can be classified into three main categories: (1) empirical models, (2) mechanistic models, and (3) risk-based models [224].

- Empirical MIC models are based on a best fit or multivariate correlation to experimental datasets (either laboratory or field based). In these models, the key measured input parameters include measured biological activity, fluid chemistry, process conditions (e.g., temperature, pressure, and flow rate) and physical properties (e.g., pipe material and geometry) [224]. Examples of these models include Fatah et al. [225], Grzelak [226] models, and others.
- Mechanistic models are based on simulating actual physical, chemical, and/or biological processes within an MIC environment, such as a biofilm or under solid deposits. They are semiempirical, and give the user crucial information on the effect of select input variables [224]. Examples of this type of model include Peng et al. [227], Al-Darbi et al. [228], and others.
- Risk-based MIC models are used to predict and identify the potential magnitude and location of MIC threats in oil and gas infrastructure such as production facilities or pipeline systems. Risk-based methods are most often used in industry for planning inspection and maintenance activities of assets to ensure optimal safety and reliability in resource-constrained operations. Risk-based approaches can be classified as qualitative, quantitative, and semiquantitative [224].

12.11.3 Biofouling prediction models

As described in Chapter 10, biofouling involves biomass deposition and biomineralization. These processes can be interconnected and confusing, especially industrially when dealing with such problems caused by bacteria in oil and gas fields. The models describing the biofilm formation are commonly used to describe membrane biofouling rather than on pipe walls. Some of those used to describe biofilm formation on pipe walls belong to the biocorrosion models. The models used to describe membrane biofouling use some key factors, including EPS production rates, membrane permeability, and attempts at modeling the specific mechanisms responsible for the decrease in membrane permeability, such as pore blockage, pore constriction, and cake filtration [229]. On the other hand, biomineralization models are used extensively in clinical applications (to study tissues classification), and in geological applications.

12.11.4 Recent advances in biofouling prediction

Recently, molecular modeling has been used for in-depth investigation of the different and complex interactions of chemical and biological reactions occurring on a material surface. It can also be used to investigate the impact of operation conditions such as pH and temperature on MIC, giving more realistic and effective predictions, thus more effective treatments. Combining quantum chemical models with the molecular models helps in understanding the different reactions at the metal surface that induce MIC. Furthermore, incorporation of the data from molecular microbiological methods can improve the current methods [224]. Additionally, the use of ANN and machine learning is another way to improve the MIC and biofouling risk assessment.

12.12 Summary

In this chapter, the various methods for predicting the different types of flow assurance solids have been discussed. The prediction models use thermodynamic, kinetic, and hydrodynamic data. Mineral scale prediction is based on describing the vapor liquid equilibrium using equations of state to determine the portioning of gases between vapor and liquids, and using electrolyte solution theory, e.g., Pitzer theory, to determine the activity coefficient and solubility product of the minerals to estimate their scaling tendency. Gas hydrates are predicted by resolving the thermodynamic equilibrium between the hydrate-liquid-vapor phases. Gas hydrate prediction models range from simple hand calculation methods to sophisticated thermodynamic methods. Wax prediction models are based on solving the thermodynamic equilibrium between solid wax and liquid oil phases. In wax prediction, the precipitation models predict the WAT and WPC, which then are used in the deposition models to predict the deposition rates and deposit thickness. Similarly, asphaltene precipitation models are used to predict the oil properties and solution saturation with asphaltenes or asphaltene solubility, which then are used in the deposition models to predict the rate of deposition. Naphthenate deposition is based on the phase equilibrium and partitioning of naphthenic acids between the oil and water phases. Finally, the biofouling relies on the understanding of the bacteria behavior, biofilm formation kinetics, and the factors affecting the MIC.

References

[1] J. Poort, Solving Vapor-Liquid Flash Problems Using Artificial Neural Networks, Master Thesis, Delft University of Technology, 2018. http://resolver.tudelft.nl/uuid:220ebd33-4755-466f-9d3f-b62298e55c0e.

[2] J. Smith, H. Van Ness, M. Abbott, Introduction to Chemical Engineering Thermodynamics, sixth ed., McGraw-Hill Higher Education, New York, 2003.

[3] K.G.G. Abad, Development of a Compositional Reservoir Simulator for Asphaltene Precipitation Based on a Thermodynamically Consistent Model, Master Thesis, Texas A&M University, TX, 2013.

[4] C. Sisco, M.I.L. Abutaqiya, F. Wang, J. Zhang, M. Tavakkoli, F.M. Vargas, Asphaltene precipitation modeling, in: Asphaltene Deposition, CRC Press, 2018, pp. 111–159, https://doi.org/10.1201/9781315268866-4.

[5] W.C. Gardiner, Rates and Mechanisms of Chemical Reactions, W.A. Benjamin, New York, 1969.

[6] S.I. Andersen, J.G. Speight, Thermodynamic models for asphaltene solubility and precipitation, J. Pet. Sci. Eng. 22 (1–3) (1999) 53–66.

[7] B. Kaasa, K. Sandengen, T. Ostvold, Thermodynamic predictions of scale potential, pH and gas solubility in glycol containing systems, in: Paper Presented at the SPE International Symposium on Oilfield Scale, Aberdeen, United Kingdom, May 2005, https://doi.org/10.2118/95075-MS.

[8] R. Dybkaer, The meaning of 'concentration', Accred. Qual. Assur. 12 (12) (2007) 661–663, https://doi.org/10.1007/s00769-007-0316-z.

[9] F.A.P.C. Gobas, S. Xu, G. Kozerski, D.E. Powell, K.B. Woodburn, D. Mackay, A. Fairbrother, Fugacity and activity analysis of the bioaccumulation and environmental risks of decamethylcyclopentasiloxane (D5), Environ. Toxicol. Chem. 34 (12) (2015) 2723–2731, https://doi.org/10.1002/etc.2942 (Wiley Blackwell).

[10] B.L. Murphy, Chemical partitioning and transport in the environment, in: Introduction to Environmental Forensics, third ed., Elsevier Inc., 2015, pp. 165–197, https://doi.org/10.1016/B978-0-12-404696-2.00007-2.

[11] L.-Q. Chen, Chemical potential and Gibbs free energy, MRS Bull. 44 (7) (2019) 520–523, https://doi.org/10.1557/mrs.2019.162.

[12] K.E. Gubbins, The theory of non-electrolyte solutions: an historical review, Mol. Phys. (2013), https://doi.org/10.1080/00268976.2013.831140.

[13] C.C. Patton, Applied Water Technology, Campbell Petroleum Series, 1986.

[14] A.T. Kan, X. Wu, G. Fu, M.B. Tomson, Validation of Scale Prediction Algorithms at Oilfield Conditions, Society of Petroleum Engineers, 2005, January 1, https://doi.org/10.2118/93264-MS (SPE-93264-MS 2005).

[15] C. Fan, A.T. Kan, P. Zhang, H. Lu, S. Work, J. Yu, M.B. Tomson, Scale prediction and inhibition for oil and gas production at high temperature/high pressure, SPE J. 17 (2) (2012) 379–392, https://doi.org/10.2118/130690-PA (Society of Petroleum Engineers).

[16] G. Atkinson, K. Raju, R.D. Howell, Thermodynamics of Scale Prediction, Society of Petroleum Engineers of AIME, 1991, pp. 209–215, https://doi.org/10.2118/21021-ms.

[17] G. Atkinson, M. Mecik, The chemistry of scale prediction, J. Pet. Sci. Eng. 17 (1–2) (1997) 113–121, https://doi.org/10.1016/S0920-4105(96)00060-5 (Elsevier B.V.).

[18] R.J. Ferguson, Scaling indices: types and applications, in: Mineral Scales and Deposits, Elsevier, 2015, pp. 721–735, https://doi.org/10.1016/b978-0-444-63228-9.00030-9.

[19] Z. Dai, A.T. Kan, F. Zhang, F. Yan, G. Ruan, N. Bhandari, Z. Zhang, Y. Liu, H.A. Al-Saiari, M.B. Tomson, A thermodynamic model for the solution density and mineral solubility predictions up to 250°C, 1,500 bars for Na-K-Mg-Ca-Ba-Sr-Cl-CO_3-HCO_3-SO_4-CO_2 Aq systems, in: Society of Petroleum Engineers—SPE International Oilfield Scale Conference and Exhibition, Society of Petroleum Engineers, 2016.

[20] A. Lassin, C. Christov, L. André, M. Azaroual, A thermodynamic model of aqueous electrolyte solution behavior and solid-liquid equilibrium in the Li-H-Na-KCl-OH-H_2O system to very high concentrations (40 molal) and from 0 to 250 C, Am. J. Sci. 815 (8) (2015) 204–256.

[21] K. Pitzer, Activity Coefficients in Electrolyte Solutions, second ed., Taylor & Francis Group, USA, 1991, pp. 75–146 (Reissued 2018 by CRC Press).

[22] A. Kan, M. Tomson, Scale prediction for oil and gas production, SPE-132237-MS, in: International Oil and Gas Conference and Exhibition in China, Society of Petroleum Engineers, Beijing, China, 2010.

[23] R.J. Ferguson, A.J. Freedman, G. Fowler, A.J. Kulik, J. Robson, D.J. Weintritt, The practical application of ion association model saturation level indices to commercial water treatment problem solving, in: Z. Amjad (Ed.), Mineral Scale Formation and Inhibition, Springer, Boston, MA, 1995, https://doi.org/10.1007/978-1-4899-1400-2_26.

[24] J.E. Oddo, M.B. Tomson, Why scale forms and how to predict it, SPE Prod. Facil. 9 (1) (1994) 47–54,- https://doi.org/10.2118/21710-PA.

[25] A.S. Ribeiro, D. Silva, E.J. Mackay, K.S. Sorbie, The impact of vapour-liquid equilibria VLE calculations on scale prediction modelling, in: Paper presented at the SPE International Oilfield Scale Conference and Exhibition, Aberdeen, Scotland, UK, May 2016, https://doi.org/10.2118/179885-MS.

[26] M.B. Tomson, G. Fu, M.A. Watson, A.T. Kan, Mechanisms of mineral scale inhibition, Soc. Petrol. Eng. J. 18 (2002) 192–199.
[27] P. Zhang, A.T. Kan, M.B. Tomson, Oil field mineral scale control, in: Mineral Scales and Deposits: Scientific and Technological Approaches, Elsevier Inc., 2015, pp. 603–617, https://doi.org/10.1016/B978-0-444-63228-9.00024-3.
[28] I.A. Løge, J.R. Bentzon, C.G. Klingaa, J.H. Walther, B.U. Anabaraonye, P.L. Fosbøl, Scale attachment and detachment: the role of hydrodynamics and surface morphology, Chem. Eng. J. 430 (2022), 132583, https://doi.org/10.1016/j.cej.2021.132583.
[29] H. Khorsandi, A. Mohammadi, S. Karimzadeh, J. Khorsandi, Evaluation of corrosion and scaling potential in rural water distribution network of Urmia, Iran, Desalin. Water Treat. (2015), https://doi.org/10.1080/19443994.2015.1042058.
[30] M. Taghavi, M.H. Mohammadi, M. Radfard, Y. Fakhri, S. Javan, Assessment of scaling and corrosion potential of drinking water resources of Iranshahr, MethodsX 6 (January) (2019) 278–283, https://doi.org/10.1016/j.mex.2019.02.002 (Elsevier B.V.).
[31] M. Yousefi, H. Najafi Saleh, A. Hossein Mahvi, M. Alimohammadi, R. Nabizadeh, A. Akbar Mohammadi, Data on corrosion and scaling potential of drinking water resources using stability indices in Jolfa, East Azerbaijan, Iran, Data Brief 16 (February) (2018) 724–731, https://doi.org/10.1016/j.dib.2017.11.099. Elsevier Inc.
[32] G. Ozair, An overview of calcium carbonate saturation indices as a criterion to protect desalinated water transmission lines from deterioration, Nat. Environ. Pollut. Technol. 11 (2) (2012) 203–212.
[33] H.L. Skillman, J.P. McDonald Jr., H.A. Stiff Jr., A simple, accurate, fast method for calculating calcium sulfate solubility in oil field brine, in: Paper No. 906-14-I, Spring Meeting of the Southwastern District, Division of Production, American Petroleum Institute, Lubbock, TX, 1969.
[34] AWWA Research Foundation, Internal Corrosion of Water Distribution Systems, American Water Works Association The Foundation, 1996, p. 586.
[35] Dow Chemical Company, filmTec Reverse Osmosis Membraness Technical Manual, Form no. 609-00071-0705, Dow Liquid Separations, 2007.
[36] Cowan and Weintritt, Water Formed Scale Deposits, Gulf Publishing Company, 1976.
[37] J.W. Ryznar, A new index for determining the amount of calcium carbonate scale formed by water, J. AWWA 36 (1944) 472.
[38] P.R. Puckorius, J.M. Brooke, A new practical index for calcium carbonate scale prediction in cooling systems, Corrosion 47 (4) (1991) 280–284.
[39] DownHole SAT Copyright 1998 French Creek Software, Inc. Kimberton, PA 19442 All rights reserved.
[40] H.A. Stiff Jr., L.E. Davis, A method for predicting the tendency of oil field waters to deposit calcium Sulfate, Trans. AIME 195 (1952) 25 (lour. Pet. Tech., April, 1952.).
[41] H.A. Stiff, L.E. Davis, A method for predicting the tendency of oil field water to deposit calcium carbonate, Pet. Trans. AIME 195 (1952) 213–216.
[42] Z. Ahmad, Selection of materials for corrosive environment, in: Principles of Corrosion Engineering and Corrosion Control, Elsevier, 2006, pp. 479–549, https://doi.org/10.1016/b978-075065924-6/50010-6.
[43] T.E. Larson, R.V. Skold, Corrosion and tuberculation of cast iron, J. AWWA 49 (10) (1957) 1294.
[44] L. Frank, G. Katie, Water Chemistry Analysis for Water Conveyance, Storage, and Desalination Projects, U. S. Department of the Interior, 2013, https://doi.org/10.1103/PhysRevB.69.014501. August 12.
[45] H.L. Skillman, J.P. McDonald Jr., H.A. Stiff Jr., A simple, accurate, fast method for calculating calcium sulfate solubility in oil field brine, in: Presented at the Spring Meeting of the Southwestern District, API, Lubbock, Texas, March 12–14, 1969.
[46] M. Cather, P.I. Robert Lee, I. Gundiler, A. Sung, N. Davidson, A. Kumar Reddy, M. Wei, A Produced Water Quality and Infrastructure GIS Database for New Mexico Oil Producers, DE-FC26-02NT15134, New

Mexico Petroleum Recovery Research Center New Mexico Institute of Mining and Technology Socorro, 2003 (NM 87801 (505) 835-5142).

[47] O.J. Vetter, V. Kandarpa, A. Harouaka, Prediction of scale problems due to injection of incompatible waters, J. Pet. Technol. 34 (1982) 273–284.

[48] M. Yuan, Water scaling in the north sea oil and gas fields and scale prediction: an overview, in: Oil Field Chemicals 7th International Symposium 17–20 MARCH 1996 Dr Holms Hotel Geilo, Norway, 1996.

[49] R. Børeng, T. Schmidt, O. Vikane, L.A. Tau, B. Dybdahl, T.I. Dale, O. Thowsen, Downhole Measurement of pH in Oil & Gas Applications by Use of a Wireline Tool, Society of Petroleum Engineers, 2003, January 1, https://doi.org/10.2118/82199-MS.

[50] Expro Petrotech® Analysis Services, Multiscale brochure A3020—MultiScale ver. 0714, Retrieved from: https://81422d14c097702b6a9f-fdabdf1ed13d1990275f510cf3764dd3.ssl.cf3.rackcdn.com/Datasheets/Fluids/MultiScale.pdf.

[51] M. Tomson, C. Fan, H. Lu, P. Zhang, H.A. Alsaiari, A.T. Kan, Integration of Kinetics into Scale Prediction Software, Society of Petroleum Engineers, 2009, https://doi.org/10.2118/121681-MS.

[52] A.T. Kan, Z. Dai, F. Zhang, N. Bhandari, F. Yan, Z. Zhang, et al., Scale Prediction and Control at Ultra HTHP, Society of Petroleum Engineers, 2015, https://doi.org/10.2118/173803-MS.

[53] M.B. Tomson, A.T. Kan, G. Fu, M. Al-Thubaiti, NORM scale formation, control, and relation to gas hydrate control, in: Report From 10th Annual International Petroleum Conference, IPEC, Tulsa, 2003.

[54] Getting Started with Hysys™ OLI™, OLI Systems, Inc, 2021. Retrieved from: https://support.olisystems.com/Documents/Manuals/Getting%20Started%20with%20Hysys%20OLI.pdf. (Last Visited August 2021).

[55] M. Azaroual, C. Kervévan, M.V. Durance, P. Durst, SCALE2000: reaction-transport software dedicated to thermokinetic prediction and quantifiction of scales. Applicability to desalination problems, Desalination 165 (Suppl) (2004) 409–419, https://doi.org/10.1016/j.desal.2004.06.046.

[56] M. Azaroual, C. Hurtevent, C. Kervevan, S. Brochot, M.V. Durance, Quantitative prediction of scale depositions induced by oil production: application of the thermo-kinetic software SCALE2000, SPE-68303-MS, in: International Symposium on Oilfield SCALE, Society of Petroleum Engineers, Aberdeen, United Kingdom, 2001.

[57] S.A.T. DownHole, DownHole SAT Copyright 1998, French Creek Software, Inc., Kimberton, PA, 2011 (All rights reserved).

[58] A. Abouie, A.K.N. Korrani, M. Shirdel, K. Sepehrnoori, Comprehensive Modeling of Scale Deposition by Use of a Coupled Geochemical and Compositional Wellbore Simulator, Society of Petroleum Engineers, 2017, August 1, https://doi.org/10.2118/185942-PA.

[59] Y. Fu, W. Van Berk, H.M. Schulz, Temporal and spatial development of scale formation: one-dimensional hydrogeochemical transport modeling, J. Pet. Sci. Eng. 112 (2013) 273–283, https://doi.org/10.1016/j.petrol.2013.11.014.

[60] O. Falode, C. Udomboso, F. Ebere, Prediction of oilfield scale formation using artificial neural network (ANN), Adv. Res. 7 (6) (2016) 1–13, https://doi.org/10.9734/air/2016/27008.

[61] N.M. Al-Hajri, A. Al-Ghamdi, Z. Tariq, M. Mahmoud, Scale-prediction/inhibition design using machine-learning techniques and probabilistic approach, SPE Prod. Oper. 35 (2020), https://doi.org/10.2118/198646-PA. 0987-1009.

[62] P.A. Paz, J.D. Caprace, J.F. Cajaiba, T.A. Netto, Prediction of calcium carbonate scaling in pipes using artificial neural networks, in: Proceedings of the International Conference on Offshore Mechanics and Arctic Engineering-OMAE, vol. 5A-2017, American Society of Mechanical Engineers (ASME), 2017, https://doi.org/10.1115/OMAE2017-61233.

[63] B.E. Roberts, Flow impairment by deposited sulfur—a review of 50 years of research, J. Nat. Gas Eng. 2 (1) (2017) 84–105, https://doi.org/10.7569/jnge.2017.692504.

[64] R.A. Tomcej, H. Kalra, B.E. Hunter, Prediction of sulfur solubility in sour gas mixtures, in: Paper Presented at the 39th Annual Laurance Reid Gas Conditioning Conference, Norman, Oklahoma, USA, March 6–9, 1989.
[65] K. Karan, R.A. Heidemann, L.A. Behie, Sulfur solubility in sour gas: predictions with an equation of state model, Ind. Eng. Chem. Res. 37 (5) (1998) 1679–1684.
[66] M.-X. Gu, Q. Li, S.-Y. Zhou, W.-D. Chen, T.-M. Guo, Experimental and modeling studies on the phase behavior of the high H_2S content natural gas mixtures, Fluid Phase Equilib. 82 (1993) 173–182.
[67] J. Chrastil, Solubility of solids and liquids in supercritical gases, J. Phys. Chem. 86 (15) (1982) 3016–3021.
[68] E. Brunner, W. Woll, Solubility of sulfur in hydrogen sulfide and sour gases, Soc. Pet. Eng. J. 20 (5) (1980) 377–384.
[69] B.E. Roberts, The effect of sulfur deposition on gas well inflow performance, SPE Reserv. Eng. 12 (2) (1997) 118–123.
[70] C.H. Kuo, On the production of hydrogen sulfide-sulfur mixtures from deepformations, J. Pet. Technol. 24 (9) (1972) 1142–1146.
[71] Susana Leon Caceres, Safety Problems Caused by Hydrate Formation in Deepwater Production Operation, Master Thesis, Texas A&M University, 2017.
[72] E.D. Sloan, C.A. Koh, Estimation techniques for phase equilibria of natural gas hydrates, in: Clathrate Hydrates of Natural Gases, third ed., 2007, pp. 189–256, https://doi.org/10.1201/9781420008494.ch4.
[73] J.J. Carroll, Natural Gas Hydrates: A Guide for Engineers, Elsevier Inc, 2014, https://doi.org/10.1016/C2013-0-13425-3.
[74] G.D. Holder, S.P. Zetts, N. Pradhan, Phase behavior in systems containing clathrate hydrates: a review, Rev. Chem. Eng. 5 (1–4) (1988) 1–70, https://doi.org/10.1515/REVCE.1988.5.1-4.1.
[75] Z. Wang, B. Sun, Y. Gao, Formation and decomposition of natural gas hydrate, in: Natural Gas Hydrate Management in Deepwater Gas Well, Springer, Singapore, 2020, pp. 9–50, https://doi.org/10.1007/978-981-15-6418-5_2.
[76] W.R. Parrish, J.M. Prausnitz, Dissociation pressures of gas hydrates formed by gas mixtures, Ind. Eng. Chem. Process. Des. Dev. 11 (1972) 26–35 (Correction Ind. Eng. Chem. Process Des. Dev., 11, 462, (1972).
[77] H.-J. Ng, D.B. Robinson, The prediction of hydrate formation in condensed systems, AICHE J. 23 (1977) 477–482.
[78] P. Englezos, N. Kalogerakis, P.D. Dholabhai, P.R. Bishnoi, Kinetics of formation of methane and ethane gas hydrates, Chem. Eng. Sci. 42 (11) (1987) 2647–2658, https://doi.org/10.1016/0009-2509(87)87015-X.
[79] P. Englezos, N. Kalogerakis, P.D. Dholabhai, P.R. Bishnoi, Kinetics of gas hydrate formation from mixtures of methane and ethane, Chem. Eng. Sci. 42 (11) (1987) 2659–2666, https://doi.org/10.1016/0009-2509(87)87016-1.
[80] M.B. Malegaonkar, P.D. Dholabhai, P.R. Bishnoi, Kinetics of carbon dioxide and methane hydrate formation, Can. J. Chem. Eng. 75 (1997) 1090–1099, https://doi.org/10.1002/cjce.5450750612.
[81] P.D. Yapa, L. Zheng, F. Chen, A model for deepwater oil/gas blowouts, Mar. Pollut. Bull. 43 (2001) 234–241, https://doi.org/10.1016/S0025-326X(01)00086-8.
[82] Y.H. Mori, Estimating the thickness of hydrate films from their lateral growth rates: application of a simplified heat transfer model, J. Cryst. Growth 223 (1–2) (2001) 206–212, https://doi.org/10.1016/S0022-0248(01)00614-5.
[83] P. Skovborg, P. Rasmussen, A mass-transport limited model for the growth of methane and ethane gas hydrates, Chem. Eng. Sci. 49 (8) (1994) 1131–1143.
[84] R.E.T. Meindinyo, T.M. Svartaas, T.N. Nordbø, R. Bøe, Gas hydrate growth estimation based on heat transfer, Energy Fuel 29 (2) (2015) 587–594, https://doi.org/10.1021/ef502366u.
[85] T. Uchida, T. Ebinuma, J. Kawabata, H. Narita, Microscopic observations of formation processes of clathrate-hydrate films at an interface between water and carbon dioxide, J. Cryst. Growth 204 (3) (1999) 348–356, https://doi.org/10.1016/S0022-0248(99)00178-5.

[86] T. Mochizuki, Y.H. Mori, Clathrate-hydrate film growth along water/hydrate-former phase boundaries-numerical heat-transfer study, J. Cryst. Growth 290 (2) (2006) 642–652, https://doi.org/10.1016/j.jcrysgro.2006.01.036.

[87] E.M. Freer, M. Sami Selim, E. Dendy Sloan, Methane hydrate film growth kinetics, Fluid Phase Equilib. 185 (1–2) (2001) 65–75, https://doi.org/10.1016/S0378-3812(01)00457-5.

[88] E.G. Hammerschmidt, Formation of gas hydrates in natural gas transmission lines, Ind. Eng. Chem. 26 (1934).

[89] Y.F. Makogon, Hydrates of Hydrocarbons, PennWell, Tulsa, OK, 1997.

[90] M. Motiee, Estimate possibility of hydrate, Hydrocarb. Process. 70 (7) (1991) 98–99.

[91] B.F. Towler, S. Mokhatab, Quickly estimate hydrate formation conditions in natural gases, Hydrocarb. Process. 84 (2005) 61.

[92] B.K. Berge, Hydrate predictions on a microcomputer, in: Paper Presented at the Petroleum Industry Application of Microcomputers, Silvercreek, Colorado, June, 1986, https://doi.org/10.2118/15306-MS.

[93] A. Bahadori, A. Vuthaluru, A novel correlation for estimation of hydrate forming condition of natural gases, J. Nat. Gas Chem. 18 (2009) 453–457.

[94] M.M. Ghiasi, Initial estimation of hydrate formation temperature of sweet natural gases based on new empirical correlation, J. Nat. Gas Chem. 21 (2012) 508–512.

[95] R. Kobayashi, K.Y. Song, E.D. Sloan, in: H.B. Bradley (Ed.), Petroleum Engineers Handbook, SPE, Richardson, TX, 1987, pp. 25-1–25-28.

[96] K.K. Østergaard, B. Tohidi, A. Danesh, A.C. Todd, R.W. Burgass, A general correlation for predicting the hydrate-free zone of reservoir fluids, SPE Prod. Facil. 15 (2000) 228–233.

[97] N. Garapati, Determination of Mixed Hydrate Thermodynamics for Reservoir Modeling, 2009 (Graduate Theses, Dissertations, and Problem Reports. 2068) https://researchrepository.wvu.edu/etd/2068.

[98] E.D. Sloan, Phase equilibrium of natural gas hydrates, in: Proceedings of 63rd GPA Convention, 1984, pp. 163–169.

[99] M.R.N. Loh, E.J.H. James, New hydrate formation data reveal differences, Oil Gas J. 81 (1983) 96–98.

[100] C. Baillie, E. Wichert, Chart gives hydrate formation temperature for naturalgas, Oil Gas J. 85 (1987) 37–39.

[101] D.L. Katz, A look ahead in gas storage technology, in: Proceedings AGA Trans. Conference, 1981, pp. T283–T289.

[102] A.A. Elgiblay, A.M. Elkamel, A new correlation for predicting hydrate formation conditions for various gas mixtures and inhibitors, Fluid Phase Equilib. 153 (1998) 10–22.

[103] G. Zahedi, Z. Karami, H. Yaghoobi, Prediction of hydrate formation temperature by both statistical models and artificial neural network approaches, Energy Convers. Manag. 50 (8) (2009) 2052–2059.

[104] M. Ghavipour, M. Ghavipour, M. Chitsazan, S.H. Najibi, S.S. Ghidary, Experimental study of natural gas hydrates and a novel use of neural network to predict hydrate formation conditions, Chem. Eng. Res. Des. 91 (2) (2013) 264–273, https://doi.org/10.1016/j.cherd.2012.08.010.

[105] M.K.B. Landgrebe, D. Nkazi, Toward a robust, universal predictor of gas hydrate equilibria by means of a deep learning regression, ACS Omega 4 (27) (2019) 22399–22417. Published 2019 Dec 19 https://doi.org/10.1021/acsomega.9b02961.

[106] Y. Zhang, P.G. Debenedetti, R.K. Prud'Homme, et al., Accurate prediction of clathrate hydrate phase equilibria below 300 K from a simple model, J. Pet. Sci. Eng. 51 (1–2) (2006) 45–53.

[107] G. Moradi, E. Khosravani, Modeling of hydrate formation conditions for CH_4, C_2H_6, C_3H_8, N_2, CO_2, and their mixtures using the PRSV2 equation of state and obtaining the Kihara potential parameters for these components, Fluid Phase Equilib. 338 (2) (2013) 179–187.

[108] Z. Yin, M. Khurana, H.K. Tan, P. Linga, A review of gas hydrate growth kinetic models, Chem. Eng. J. (2018), https://doi.org/10.1016/j.cej.2018.01.120.

[109] A. Vysniauskas, P.R. Bishnoi, A kinetic study of methane hydrate formation, Chem. Eng. Sci. 38 (7) (1983) 1061–1072.

[110] K.H. Bendiksen, D. Maines, R. Moe, S. Nuland, The dynamic two-fluid model OLGA: theory and application, SPE Prod. Eng. 6 (2) (1991) 171–180.

[111] D. Turner, J. Boxall, S. Yang, D. Kleehammer, C. Koh, K. Miller, E.D. Sloan, Z. Xu, P. Matthews, L.D. Talley, Development of a hydrate kinetic model and its incorporation into OLGA2000s transient multiphase flow simulator, in: Proceedings of the 5th International Conference on Gas Hydrates, Trondheim, Norway, 2005, pp. 1231–1240.

[112] L.E. Zerpa, E.D. Sloan, A.K. Sum, C.A. Koh, Overview of CSMHyK: a transient hydrate formation model, J. Pet. Sci. Eng. (2012), https://doi.org/10.1016/j.petrol.2012.08.017.

[113] P.N. Chaudhari, Development of Hydrate Risk Quantification in Oil and Gas Production, PhDT, 2016.

[114] Colorado School of Mines, Center of Hydrates Research, 2021. http://hydrates.mines.edu/CHR/Software.html. (Last Visited July 2021).

[115] A.L. Ballard, E.D. Sloan Jr., The next generation of hydrate prediction: an overview, J. Supramol. Chem. 2 (4–5) (2002) 385–392, https://doi.org/10.1016/S1472-7862(03)00063-7.

[116] Hydrafact Website, 2021. https://www.hydrafact.com/index.php?page=software. (Last Visited July 2021).

[117] A. Hassanpouryouzband, E. Joonaki, M. Vasheghani Farahani, S. Takeya, C. Ruppel, J. Yang, et al., Gas hydrates in sustainable chemistry, Chem. Soc. Rev. (2020), https://doi.org/10.1039/c8cs00989a. Royal Society of Chemistry.

[118] O.E. Ajayi, Modelling of Controlled Wax Deposition and Loosening in Oil and Gas Production Systems, M. Sc. Thesis, Norwegian University of Science and Technology NTNU, 2013.

[119] Z. Huang, S. Zheng, H.S. Fogler, Wax Deposition, CRC Press, 2016, https://doi.org/10.1201/b18482.

[120] S. Zheng, M. Saidoun, K. Mateen, T. Palermo, Y. Ren, H.S. Fogler, Wax deposition modeling with considerations of non-Newtonian fluid characteristics, in: Proceedings of the Annual Offshore Technology Conference, vol. 1, 2016, pp. 548–565, https://doi.org/10.4043/26914-ms.

[121] S. Zheng, M. Saidoun, T. Palermo, K. Mateen, H.S. Fogler, Wax deposition modeling with considerations of non-Newtonian characteristics: application on field-scale pipeline, Energy Fuel 31 (5) (2017) 5011–5023,-https://doi.org/10.1021/acs.energyfuels.7b00504.

[122] E.N. Sieder, G.E. Tate, Heat transfer and pressure drop of liquids in tubes, Ind. Eng. Chem. 28 (1936) 1429–1435.

[123] V. Gnielinski, New equations for heat and mass transfer in turbulent pipe and channel flow, Int. J. Chem. Eng. 16 (1976) 359–368.

[124] F. Dittus, L. Boelter, Heat transfer in automobile radiators of the tubular type, Int. Commun. Heat Mass Transfer 12 (1985) 3–22.

[125] W.M. Deen, Analysis of Transport Phenomena, second ed., Oxford University Press, 1998.

[126] K.W. Won, Thermodynamics for solid solution-liquid-vapor equilibria: wax phase formation from heavy hydrocarbon mixtures, Fluid Phase Equilib. 30 (1986) 265–279, https://doi.org/10.1016/0378-3812(86)80061-9.

[127] K.W. Won, Thermodynamic calculation of cloud point temperatures and wax phase compositions of refined hydrocarbon mixtures, Fluid Phase Equilib. 53 (1989) 377–396, https://doi.org/10.1016/0378-3812(89)80104-9.

[128] J.H. Hansen, A. Fredenslund, K.S. Pedersen, H.P. Rønningsen, A thermodynamic model for predicting wax formation in crude oils, AICHE J. 34 (12) (1988) 1937–1942, https://doi.org/10.1002/aic.690341202.

[129] K.S. Pedersen, P. Skovborg, H.P. Rønningsen, Wax precipitation from North Sea crude oils. 4. Thermodynamic modeling, Energy Fuel 5 (6) (1991) 924–932, https://doi.org/10.1021/ef00030a022.

[130] J.A.P. Coutinho, J. Pauly, J.L. Daridon, Modelling phase equilibria in systems with organic solid solutions, in: Computer Aided Chemical Engineering, vol. 19, Elsevier, 2004, pp. 229–249, https://doi.org/10.1016/S1570-7946(04)80012-9 (Chapter 10).

[131] D.D. Erickson, V.G. Niesen, T.S. Brown, Thermodynamic measurement and prediction of paraffin precipitation in crude oil, in: Paper presented at the SPE Annual Technical Conference and Exhibition, Houston, Texas, October 1993, https://doi.org/10.2118/26604-MS.

[132] K. Schou Pedersen, P. Skovborg, H.P. Roenningsen, Wax precipitation from North Sea crude oils. 4. Thermodynamic modeling, Energy Fuel 5 (6) (1991) 924–932, https://doi.org/10.1021/ef00030a022.

[133] W.B. Pedersen, A.B. Hansen, E. Larsen, et al., Wax precipitation from North Sea crude oils. 2. Solid-phase content as function of temperature determined by pulsed NMR, Energy Fuel 5 (6) (1991) 908–913, https://doi.org/10.1021/ef00030a020.

[134] K. Schou Pedersen, Prediction of cloud point temperatures and amount of wax precipitation, SPE Prod. Oper. 10 (1) (1995) 46–49. SPE-27629-PA https://doi.org/10.2118/27629-PA.

[135] C. Lira-Galeana, A. Firoozabadi, J.M. Prausnitz, Thermodynamics of wax precipitation in petroleum mixtures, AICHE J. 42 (1996) 239–248.

[136] J.A.P. Coutinho, B. Edmonds, T. Moorwood, R. Szczepanski, X. Zhang, Reliable wax predictions for flow assurance, Energy Fuel 20 (2006) 1081–1088, https://doi.org/10.1021/ef050082i.

[137] J. Pauly, C. Dauphin, J.L. Daridon, Liquid-solid equilibria in a decane + multi-paraffins system, Fluid Phase Equilib. 149 (1–2) (1998) 191–207, https://doi.org/10.1016/S0378-3812(98)00366-5.

[138] O.B. Rygg, A.K. Rydahl, H.P. Rønningsen, Wax deposition in offshore pipeline systems, in: BHRGroup 1st North American Conference Multiphase Technology, Baniff, Canada, 1998.

[139] A. Matzain, M.S. Apte, H.-Q. Zhang, M. Volk, C.L. Redus, J.P. Brill, J.L. Creek, Multiphase flow wax deposition modeling, in: ETCE 2001: Petroleum Production Technology Symposium, Houston, TX, 2001.

[140] P. Singh, R. Venkatesan, H.S. Fogler, N. Nagarajan, Formation and aging of incipient thin film wax-oil gels, AICHE J. 46 (2000) 1059–1074.

[141] N. Lindeloff, K. Krejbjerg, A compositional model simulating wax deposition in pipeline systems, Energy Fuel 16 (2002) 887–891.

[142] B. Edmonds, T. Moorwood, R. Szczepanski, X. Zhang, Simulating wax deposition in pipelines for flow assurance, Energy Fuel 22 (2007) 729–741.

[143] K. Rosvold, Wax deposition models, M.Sc. Thesis, Department of Petroleum Engineering and Applied Geophysics, Norwegian University of Science and Technology, Trondheim, 2008.

[144] T. Zhu, J.A. Walker, J. Liang, Evaluation of Wax Deposition and its Control during Production of Alaska North Slope Oils—Final Report, United States Department of Energy and National Energy Technology Laboratory, 2008.

[145] E.O. Obanijesu, E.O. Omidiora, Artificial neural network's prediction of wax deposition potential of Nigerian crude oil for pipeline safety, Pet. Sci. Technol. 26 (16) (2008) 1977–1991, https://doi.org/10.1080/10916460701399485.

[146] M.J. Jalalnezhad, V. Kamali, Development of an intelligent model for wax deposition in oil pipeline, J. Pet. Explor. Prod. Technol. 6 (1) (2016) 129–133, https://doi.org/10.1007/s13202-015-0160-3.

[147] B.J. Adeyemi, A.A. Sulaimon, Predicting wax formation using artificial neural network, in: Paper Presented at the Nigeria Annual International Conference and Exhibition, Lagos, Nigeria, August, 2012, https://doi.org/10.2118/163026-MS.

[148] M. Nait Amar, A. Jahanbani Ghahfarokhi, C. Shang Wui Ng, Predicting wax deposition using robust machine learning techniques, Petroleum (2021), https://doi.org/10.1016/j.petlm.2021.07.005.

[149] A. Singh, H.S. Lee, P. Singh, C. Sarica, Flow assurance: validation of wax deposition models using field data from a subsea pipeline, in: Paper Presented at the Offshore Technology Conference, Houston, Texas, USA, May 2011, https://doi.org/10.4043/21641-MS.

[150] O.C. Hernandez, H. Hensley, C. Sarica, J.P. Brill, M. Volk, E. Delle-Case, Improvements in single-phase paraffin deposition modeling, in: SPE Annual Technical Conference and Exhibition, Society of Petroleum Engineers, Denver, CO, 2003, pp. 1–9.

[151] M.S. Apte, A. Matzain, H.-Q. Zhang, M. Volk, J. Brill, J.L. Creek, Investigation of paraffin deposition during multiphase flow in pipelines and wellbores—part 2—modeling, J. Energy Resour. Technol. 123 (2001) 150–157.
[152] G. Couto, Investigation of Two-Phase Oil-Water Paraffin Deposition, M.S. Thesis, University of Tulsa, Tulsa, OK, 2004.
[153] A. Bruno, C. Sarica, H. Chen, M. Volk, Paraffin deposition during the flow of water-in-oil and oil-in-water dispersions in pipes, in: ACTE 2008 Proceedings: Proceedings of SPE Annual Technical Conference and Exhibition, Society of Petroleum Engineers, Denver, CO, 2008 (SPE 114747, September 21–24, 2008).
[154] MWP—The Michigan Wax Predictor, 2021. Retrieved from: http://cheresearch.engin.umich.edu/fogler/mwp.html. (Last Visited July 2021).
[155] Z. Huang, H.S. Lee, M. Senra, H. Scott Fogler, A fundamental model of wax deposition in subsea oil pipelines, AICHE J. 57 (11) (2011) 2955–2964, https://doi.org/10.1002/aic.12517.
[156] R. Bagatin, C. Busto, S. Correra, M. Margarone, C. Carniani, Wax modeling: there is need for alternatives, in: Society of Petroleum Engineers—SPE Russian Oil and Gas Technical Conference and Exhibition 2008, vol. 1, 2008, pp. 362–369, https://doi.org/10.2118/115184-ms.
[157] P. Montalvo, Modeling of Wax Deposition in a Crude Oil Carrying Pipeline, Master of Science, Department of Chemical Engineering The University of Utah, 2012.
[158] Calsep, PVTsim Technical Overview, 2021. https://www.calsep.com/uploads/Modules/Footerbrochure/technical.pdf. (Last Visited Accessed July 2021).
[159] R. Venkatesan, H.S. Fogler, Comments on analogies for correlated heat and mass transfer in turbulent flow, AICHE J. 50 (7) (2004) 1623–1626.
[160] KBC, FloWax: Wax Deposition Modelling, KBC Advanced Technologies Limited, 2021. Retrieved from: https://www.kbc.global/uploads/files/software/KBC%20FloWax.pdf. (Last visited July 2021).
[161] N. Anand, Review of Asphaltene Properties and Precipitation Modeling, Master of Science in Engineering, The University of Texas at Austin, 2018.
[162] H. Alboudwarej, K. Akbarzadeh, J. Beck, W.Y. Svrcek, H.W. Yarranton, Regular solution model for asphaltene precipitation from bitumens and solvents, AICHE J. 49 (11) (2003) 2948–2956, https://doi.org/10.1002/aic.690491124.
[163] A.E. Shoukry, A.H. El-Banbi, H. Sayyouh, Enhancing asphaltene precipitation modeling by cubic-PR solid model using thermodynamic correlations and averaging techniques, Pet. Sci. 17 (1) (2020) 232–241, https://doi.org/10.1007/s12182-019-00377-1.
[164] S. Subramanian, S. Simon, J. Sjöblom, Asphaltene precipitation models: a review, J. Dispers. Sci. Technol. 37 (7) (2016) 1027–1049, https://doi.org/10.1080/01932691.2015.1065418 (Taylor and Francis Inc.).
[165] G. Scatchard, Equilibria in non-electrolyte solutions in relation to the vapor pressures and densities of the components, Chem. Rev. 8 (2) (1931) 321–333.
[166] J.H. Hildebrand, R.L. Scott, Regular Solutions, Prentice-Hall, Englewood Cliffs, NJ, 1962.
[167] A. Hirschberg, L.N.J. deJong, B.A. Schipper, J.G. Meijer, Influence of temperature and pressure on asphaltene flocculation, Soc. Pet. Eng. J. 24 (3) (1984) 283–293, https://doi.org/10.2118/11202-PA.
[168] A. Pina, P. Mougin, E. Béhar, Characterisation of asphaltenes and modelling of flocculation—state of the art, Oil Gas Sci. Technol. (2006), https://doi.org/10.2516/ogst:2006037a.
[169] T.-H. Chung, Thermodynamic Modeling for organic solid precipitation, in: SPE Annual Technical Conference and Exhibition, Society of Petroleum Engineers, Washington, DC, 1992.
[170] F.B. Thomas, D.B. Bennion, D.W. Bennion, B.E. Hunter, Experimental and theoretical studies of solids precipitation from reservoir fluid, J. Can. Pet. Technol. 31 (1) (1992) 22–31.
[171] R. Cimino, S. Correra, P.A. Sacomani, C. Carnniani, Thermodynamic modelling for prediction of asphaltene deposition in live oils, in: SPE 28993, Presented at SPE International Symposium on Oilfield Chemistry, 14–17 February, San Antonio, TX, USA, 1995.

[172] R.B. de Boer, K. Leerlooyer, M.R.P. Eigner, A.R.D. Van Bergen, Screening of crude oils for asphalt precipitation: theory, practice, and the selection of inhibitors, SPE J. 10 (1) (1995) 55–61.

[173] H.W. Yarranton, J.H. Masliyah, Molar mass distribution and solubility modeling of asphaltenes, AICHE J. 42 (12) (1996) 3533–3543.

[174] J.S. Buckley, G.J. Hirasaki, Y. Liu, S. Von Drasek, J.X. Wang, B.S. Gill, Asphaltene precipitation and solvent properties of crude oils, Pet. Sci. Technol. 16 (3–4) (1998) 251–285.

[175] L.X. Nghiem, M.S. Hassam, R. Nutakki, et al., Efficient modelling of asphaltene precipitation, in: SPE Annual Technical Conference and Exhibition, Sigma, Houston, Texas, 1993, https://doi.org/10.2118/26642-ms.

[176] L.X. Nghiem, D.A. Coombe, Modeling asphaltene precipitation during primary depletion, SPE J. 2 (2) (1997) 170–176, https://doi.org/10.2118/36106-PA.

[177] A.K. Gupta, A Model for Asphaltene Focculation Using an Equation of State, University of Calgary, Calgary, 1986.

[178] B.F. Kohse, L.X. Nghiem, H. Maeda, et al., Modelling phase behaviour including the efect of pressure and temperature on asphaltene precipitation, in: SPE Asia Pacifc Oil and Gas Conference and Exhibition, Brisbane, Australia, 2000.

[179] J.L. Du, D. Zhang, A thermodynamic model for the prediction of asphaltene precipitation, Pet. Sci. Technol. 22 (7) (2004) 1023–1033, https://doi.org/10.1081/LFT-120038724.

[180] O. Sabbagh, K. Akbarzadeh, W.Y. Svrcek, et al., Applying the PR-EoS to asphaltene precipitation from n-alkane diluted heavy oils and bitumens, Energy Fuel 20 (2) (2006) 625–634.

[181] Z. Li, A. Firoozabadi, Cubic-plus-association equation of state for asphaltene precipitation in live oils, Energy Fuel 24 (5) (2010) 2956–2963, https://doi.org/10.1021/ef9014263.

[182] B. Shirani, M. Nikazar, A. Naseri, et al., Modeling of asphaltene precipitation utilizing association equation of state, Fuel 93 (2012) 59–66, https://doi.org/10.1016/j.fuel.2011.07.007.

[183] P.D. Ting, G.J. Hirasaki, W.G. Chapman, Modeling of asphaltene phase behavior with the SAFT equation of state, Pet. Sci. Technol. 21 (3–4) (2003) 647–661, https://doi.org/10.1081/LFT-120018544.

[184] D.L. Gonzalez, P.D. Ting, G.J. Hirasaki, et al., Prediction of asphaltene instability under gas injection with the PC-SAFT equation of state, Energy Fuel 19 (4) (2005) 1230–1234, https://doi.org/10.1021/ef049 782y.

[185] S.R. Panuganti, F.M. Vargas, D.L. Gonzalez, et al., PC-SAFT characterization of crude oils and modeling of asphaltene phase behavior, Fuel 93 (2012) 658–669, https://doi.org/10.1016/j.fuel.2011.09.028.

[186] A. Abouie, M. Rezaveisi, S. Mohebbinia, et al., Static and dynamic comparison of equation of state solid model and PC-SAFT for modeling asphaltene phase behavior, in: SPE Western Regional Meeting, Anchorage, Alaska, 2016, https://doi.org/10.2118/180480-ms.

[187] A. Abouie, H. Darabi, K. Sepehrnoori, Data-driven comparison between solid model and PC-SAFT for modeling asphaltene precipitation, in: Ofshore Technology Conference, Houston, TX, 2016, https://doi.org/10.1016/j.jngse.2017.05.007.

[188] K.J. Leontaritis, G. Ali Mansoori, Asphaltene deposition: a survey of field experiences and research approaches, J. Pet. Sci. Eng. 1 (3) (1988) 229–239, https://doi.org/10.1016/0920-4105(88)90013-7.

[189] K.J. Leontaritis, G.A. Mansoori, Asphaltene flocculation during oil production and processing: a thermodynamic collodial model, in: SPE International Symposium on Oilfield Chemistry, 4–6 February, Society of Petroleum Engineers, San Antonio, Texas, 1987.

[190] A.I. Victorov, A. Firoozabadi, Thermodynamic micellizatin model of asphaltene precipitation from petroleum fluids, AICHE J. 42 (6) (1996) 1753–1764.

[191] H. Pan, A. Firoozabadi, Thermodynamic micellization model for asphaltene precipitation inhibition, AICHE J. 46 (2) (2000) 416–426.

[192] H. Pan, A. Firoozabadi, Thermodynamic micellization model for asphaltene precipitation from reservoir crudes at high pressures and temperatures, in: SPE Annual Technical Conference and Exhibition, 5–8 October, San Antonio, Texas, 1997.

[193] P. Kor, R. Kharrat, Modeling of asphaltene particle deposition from turbulent oil flow in tubing: model validation and a parametric study, Petroleum 2 (4) (2016) 393–398, https://doi.org/10.1016/j.petlm.2016.08.010.

[194] A. Guha, Transport and deposition of particles in turbulent and laminar flow, Annu. Rev. Fluid Mech. (2008), https://doi.org/10.1146/annurev.fluid.40.111406.102220.

[195] J. Escobedo, G.A. Mansoori (Eds.), Asphaltene and other heavy-organic particle deposition during transfer and production operations, SPE Annual Technical Conference and Exhibition, Society of Petroleum Engineers, 1995.

[196] E. Ramirez-Jaramillo, C. Lira-Galeana, O. Manero, Modelling asphaltene deposition in production pipelines, Energy Fuel 20 (2006) 1184–1196.

[197] B.S. Soulgani, D. Rashtchian, B. Tohidi, M. Jamialahmadi, Integrated modelling method for asphaltene deposition in well string, J. Jpn. Pet. Inst. 52 (2009) 322–331.

[198] F.M. Vargas, J.L. Creek, W.G. Chapman, On the development of an asphaltene deposition simulator, Energy Fuel 24 (2010) 2294–2299.

[199] A.S. Kurup, F.M. Vargas, J. Wang, J. Buckley, J.L. Creek, H.J. Subramani, W.G. Chapman, Development and application of an asphaltene deposition tool (ADEPT) for well bores, Energy Fuel 25 (10) (2011) 4506–4516, https://doi.org/10.1021/ef200785v.

[200] D. Eskin, J. Ratulowski, K. Akbarzadeh, S. Pan, Modelling asphaltene deposition in turbulent pipeline flows, Can. J. Chem. Eng. 89 (3) (2011) 421–441, https://doi.org/10.1002/cjce.20507.

[201] N. Babu, P. Rajan, J. Lin, M.T. Zhang, F.M. Vargas, Modeling methods for prediction of asphaltene deposition, in: Asphaltene Deposition, CRC Press, 2018, pp. 203–241, https://doi.org/10.1201/9781315268866-6.

[202] S. Friedlander, H. Johnstone, Deposition of suspended particles from turbulent gas streams, Ind. Eng. Chem. 49 (1957) 1151–1156.

[203] N. Epstein, Elements of particle deposition onto nonporous solid surfaces parallel to suspension flows, Exp. Thermal Fluid Sci. 14 (1997) 323–334.

[204] J. Cleaver, B. Yates, A sub layer model for the deposition of particles from a turbulent flow, Chem. Eng. Sci. 30 (1975) 983–992.

[205] M. Jamialahmadi, B. Soltani, H. Müller-Steinhagen, D. Rashtchian, Measurement and prediction of the rate of deposition of flocculated asphaltene particles from oil, Int. J. Heat Mass Transf. 52 (2009) 4624–4634.

[206] A. Al-Hosani, S. Ravichandran, N. Daraboina, Review of Asphaltene deposition modeling in oil and gas production, Energy Fuel (2021), https://doi.org/10.1021/acs.energyfuels.0c02981. American Chemical Society.

[207] C. Gruesbeck, R.E. Collins, Entrainment and deposition of fine particles in porous media, Soc. Pet. Eng. J. 22 (6) (1982) 847–856, https://doi.org/10.2118/8430-PA.

[208] S. Wang, F. Civan, Productivity decline of vertical and horizontal wells by asphaltene deposition in petroleum reservoirs, in: Proceedings of the SPE International Symposium on Oilfield Chemistry, Houston, TX, February 13–16, 2001, https://doi.org/10.2118/64991-MS.

[209] S. Kord, O. Mohammadzadeh, R. Miri, B.S. Soulgani, Further investigation into the mechanisms of asphaltene deposition and permeability impairment in porous media using a modified analytical model, Fuel 117 (2014) 259–268.

[210] T. Jafari Behbahani, C. Ghotbi, V. Taghikhani, A. Shahrabadi, Experimental study and mathematical modeling of asphaltenes deposition mechanism in core samples, Oil Gas Sci. Technol. 70 (6) (2015) 1051–1074.

[211] J. Sayyad Amin, A. Alamdari, N. Mehranbod, S. Ayatollahi, E. Nikooee, Prediction of asphaltene precipitation: learning from data at different conditions, Energy Fuel 24 (7) (2010) 4046–4053, https://doi.org/10.1021/ef100106r.

[212] G. Zahedi, A.R. Fazlali, S.M. Hosseini, G.R. Pazuki, L. Sheikhattar, Prediction of asphaltene precipitation in crude oil, J. Pet. Sci. Eng. 68 (3–4) (2009) 218–222, https://doi.org/10.1016/j.petrol.2009.06.023.

[213] S. Zendehboudi, A. Shafiei, A. Bahadori, L.A. James, A. Elkamel, A. Lohi, Asphaltene precipitation and deposition in oil reservoirs—technical aspects, experimental and hybrid neural network predictive tools, Chem. Eng. Res. Des. 92 (5) (2014) 857–875, https://doi.org/10.1016/j.cherd.2013.08.001.

[214] A. Rezaian, A. Kordestany, M. Haghighat Sefat, An artificial neural network approach to formation damage prediction due to asphaltene deposition, in: Paper Presented at the Nigeria Annual International Conference and Exhibition, Tinapa—Calabar, Nigeria, July, 2010, https://doi.org/10.2118/140683-MS.

[215] M.A. Ahmadi, Prediction of asphaltene precipitation using artificial neural network optimized by imperialist competitive algorithm, J. Pet. Explor. Prod. Technol. 1 (2–4) (2011) 99–106, https://doi.org/10.1007/s13202-011-0013-7.

[216] M. Ghorbani, G. Zargar, H. Jazayeri-Rad, Prediction of asphaltene precipitation using support vector regression tuned with genetic algorithms, Petroleum 2 (3) (2016) 301–306, https://doi.org/10.1016/j.petlm.2016.05.006.

[217] A.S. Kurup, J. Wang, H.J. Subramani, J. Buckley, J.L. Creek, W.G. Chapman, Revisiting asphaltene deposition tool (ADEPT): field application, Energy Fuel 26 (9) (2012) 5702–5710, https://doi.org/10.1021/ef300714p.

[218] S. Whitfield, Modeling the behavior of asphaltenes, Oil Gas Facil. 4 (1) (2015) 20–27, https://doi.org/10.2118/0215-0020-ogf.

[219] M.A. Mohammed, K.S. Sorbie, Thermodynamic modelling of calcium naphthenate formation: model predictions and experimental results, Colloids Surf. A Physicochem. Eng. Asp. (2010), https://doi.org/10.1016/j.colsurfa.2010.08.034. Elsevier.

[220] A. Bertheussen, S. Simon, J. Sjöblom, Equilibrium partitioning of naphthenic acids and bases and their consequences on interfacial properties, Colloids Surf. A Physicochem. Eng. Asp. 529 (2017) 45–56, https://doi.org/10.1016/j.colsurfa.2017.05.068.

[221] A.A. Sulaimon, A.N. Masri, U.A.A. Jamil Sabri, B.J. Adeyemi, Predicting naphthenate precipitation and evaluating the effect of ionic liquids on its deposition, J. Pet. Sci. Eng. (2021), https://doi.org/10.1016/j.petrol.2021.109865.

[222] S. Sarac, F. Civan, Mechanisms, parameters, and modeling of naphthenate-soap-induced formation damage, SPE J. 14 (2) (2009) 259–266, https://doi.org/10.2118/112434-PA.

[223] A. Marciales, Y. Peralta, T. Haile, T. Crosby, J. Wolodko, Mechanistic microbiologically influenced corrosion modeling—a review, Corros. Sci. (2019), https://doi.org/10.1016/j.corsci.2018.10.004. Elsevier Ltd.

[224] J. Wolodko, R. Eckert, T. Haile, S.J. Hashemi, F. Khan, A.M. Ramirez, et al., Modeling of microbiologically influenced corrosion (MIC) in the oil and gas industry—past, present and future, in: NACE—International Corrosion Conference Series (vol. 2018-April), National Association of Corrosion Engineers International, 2018.

[225] M. Fatah, M. Ismail, B.A. Wahjoedi, Empirical equation of sulphate reducing bacteria (SRB) corrosion based on abiotic chemistry approach, Anti-Corros. Methods Mater. 60 (4) (2013) 206–212.

[226] L.A. Grzelak, A Statistical Approach to Determine Microbiologically Influenced Corrosion (MIC) Rates of Underground Gas Pipelines, Thesis, Delft University of Technology, Delft, Netherlands, 2006.

[227] C.G. Peng, S.-Y.A.P.J. Suen, Modeling of anaerobic corrosion influenced by sulfate-reducing bacteria, Water Environ. Res. 66 (5) (1994) 707–715.

[228] M. Al-Darbi, K. Agha, M. Islam, Modeling and simulation of the pitting microbiologically influenced corrosion in different industrial systems, in: NACE CORROSION 2005, Houston, Texas, 2005.

[229] T. Opher, M. Rom, L. Kronaveter, E. Friedler, A. Ostfeld, Some observations on biofouling prediction in pipelines using model trees and artificial neural networks versus logistic regression, Urban Water J. 9 (1) (2012) 11–20. To link to this article: https://doi.org/10.1080/1573062X.2011.633611.

CHAPTER

Monitoring of Flow Assurance Solids in Oil and Gas Fields

13

13.1 Introduction

Solids deposition has a deleterious impact on oil and gas production operations. Solids cause loss of productivity, equipment impairment, and expensive maintenance, costing millions of dollars every year. Some examples of the drastic impacts of solids formation are summarized in Chapter 3—problems associated with flow assurance solids in production.

Due to their rapid and drastic effects, early warnings of solids formation and factors affecting them before the problem worsens are valuable to operators. An example of an extreme case was reported by Brown [1], in which the production fell from 30,000 BPD (4770 m^3/d) to zero in just 24 h in one of the wells in the Miller field, due to mineral scales. Therefore one of the major steps in a flow assurance deposits control plan is to monitor their formation. The early detection of solid deposits permits applying the proper mitigation method, i.e., the injection of chemical inhibitors at the appropriate time and location, before the problem worsens. Moreover, continued monitoring following treatments enables practical evaluation and optimization of the mitigation methods, i.e., optimizing scale inhibitor dose or changing chemical type [2].

One thing needs to be mentioned here, since some operators use solids prediction and modeling as the primary method of monitoring. While deposits prediction methods provide a theoretical estimation of their risks, deposits monitoring methods can furnish an actual and real-time evaluation of the problem. Monitoring methods have some advantages over prediction methods, including real-time monitoring and evaluation of the problem including the changes in process parameters, and these methods can provide the location and size (thickness) of the deposits. Prediction methods definitely depend on the proper sampling and analysis of the system fluids, and how accurately these theoretical calculations represent the mechanisms of formation and system operations. Any changes in the system parameters, or in the nature of the sampled fluids, that are not accounted for in the prediction model will lead to nonrepresentative results. Furthermore, with all the advances in the prediction methods, the field operators still find some gaps between the real field results and the modeled results. On the other hand, monitoring methods are not always applicable, some locations are inaccessible for placement of a monitoring device, some monitoring methods are extremely expensive, and others require strong safety precautions, like those that use gamma rays. Finally, the accuracy of some of the monitoring methods is still questionable, especially when it comes to identifying the deposit type. It is strongly recommended to establish a deposits mitigation plan based on both the prediction and monitoring methods

Essentials of Flow Assurance Solids in Oil and Gas Operations. http://doi.org/10.1016/B978-0-323-99118-6.00002-2

together. It is also recommended to include the monitoring device ports, sampling points, and chemical injection points in the system during building of the flow assurance main concepts.

An ideal deposits monitoring system/program should be capable of the following [3]:

- Determining whether or not deposits are present in the system
- Detecting deposits buildup in their early stages, i.e., the method has to be sensitive to thin layers of deposits before the problem worsens
- Determining the thickness of the deposit layer
- Detecting the location of solids buildup
- Identifying the type of deposit.

13.2 Classification of flow assurance solids monitoring methods

Generally speaking, methods used in process monitoring systems can be classified into inline, online, or at-line (or offline). These monitoring systems are also known as process analyzers or industrial analyzers. Fig. 13.1 illustrates the different types of process analyzers.

- For inline systems, a sensor can be placed in a process vessel or stream of flowing material to monitor its specific property/composition in real time. Inline methods provide fast, direct measurements without sample collection or product waste, and with very low manpower needs, and they permit continuous process control [4–6].

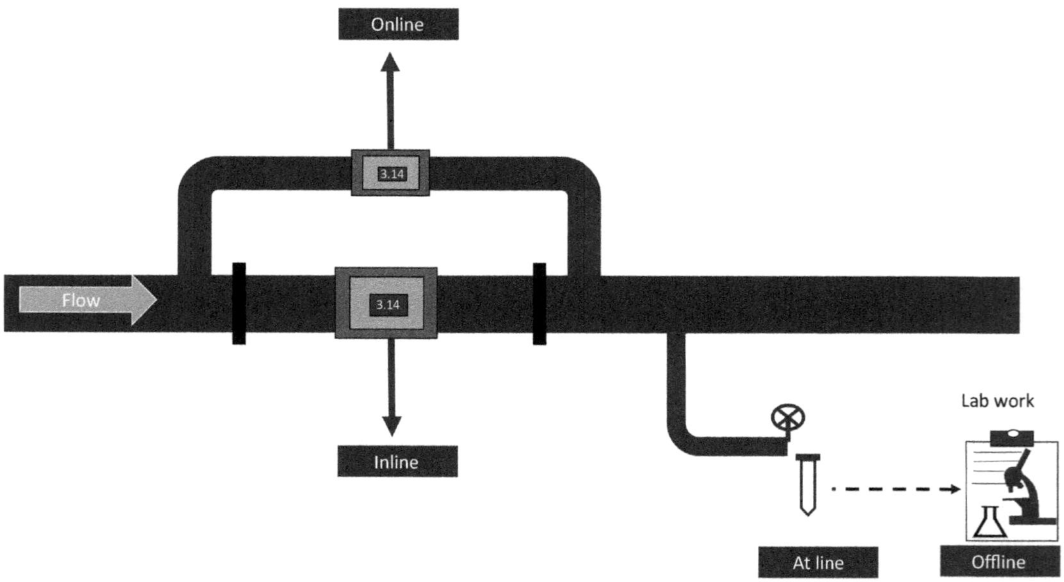

FIG. 13.1

Types of process analyzers.

- Online (on-line) analyzers are also directly connected to a process and conduct automatic sampling occasionally through a side stream or bypass, and measure its specific property/composition. They are also used when the analyzer requires preconditioning before taking measurements. With online analyzers, the fluid samples are circled back to the process. Online analyzers are automated with fast feedback of results, limited manpower needs, and also permit continuous process control [4–6].
- Offline and at-line analyzers, on the other hand, are characterized by manual sampling of the process fluids, followed by discontinuous sample preparation, measurement, and evaluation. The fluid properties can change during the time between sampling, analysis, and availability of the results, so direct process control is not possible. A bypass line is usually suggested in terms of process control, to prevent a shutdown or lost product. In cases when cleaning, calibration, and/or validating the analyzer are frequent, then a bypass is the best solution [4–6].

Aside from the location of the analyzer in the process, process monitoring can also be classified into three main categories [7]:

- Process parameters, e.g., pressure, temperature, flow rate, etc.
- Fluid-specific composition, e.g., suspended solids, deposited solids, cations, gases, etc.
- Fluid physical properties, e.g., pH, density, viscosity, refractive index, etc.

Furthermore, the monitoring methods can be destructive or nondestructive. In the destructive methods, the sample is modified and its properties are changed after performing the test, like chemical analysis that requires adding reagents to the sample, changing temperature or pressure. Nondestructive methods do not modify the sample or change its properties, like electromagnetic and acoustic methods. In non-destructive methods, if the monitoring device is on a side stream or bypass, it can be returned to the process main stream without changing its properties.

A wide variety of methods can be used in the field to monitor the solids deposition in its earliest stages (Fig. 13.2). System parameter changes, e.g., pressure, temperature and flow rate, are considered early warnings of deposit formation in the system. Similarly, produced fluid changes are indicative of solids formation, so conventional chemical analysis methods can be applied to monitor the change in the scaling or deposit-forming species. Additionally, pigging and gauging operations are routinely used to monitor the rate and size of deposit buildup. Furthermore, sensors and devices based on electric, electromagnetic, acoustic, radiographic, or radioactive methods are commonly used to detect signals that are representative of solids buildup in the system. Some of these devices can give a general estimation of solid deposits formed at the location of measurements, while other devices can specify with high accuracy the type of deposit formed and its thickness. However, with all the advances in these sensors, there is no universal sensor that can detect with high accuracy and distinguish between all types of solid deposits in oil and gas fields.

13.2.1 Process parameters monitoring methods

The primary and authentic warning signs that have always been used as system stability monitoring methods are the system parameters themselves. A solids deposition problem (or any production-related problem) in most cases leads to a change in the system parameters. Analogous to when the body gets

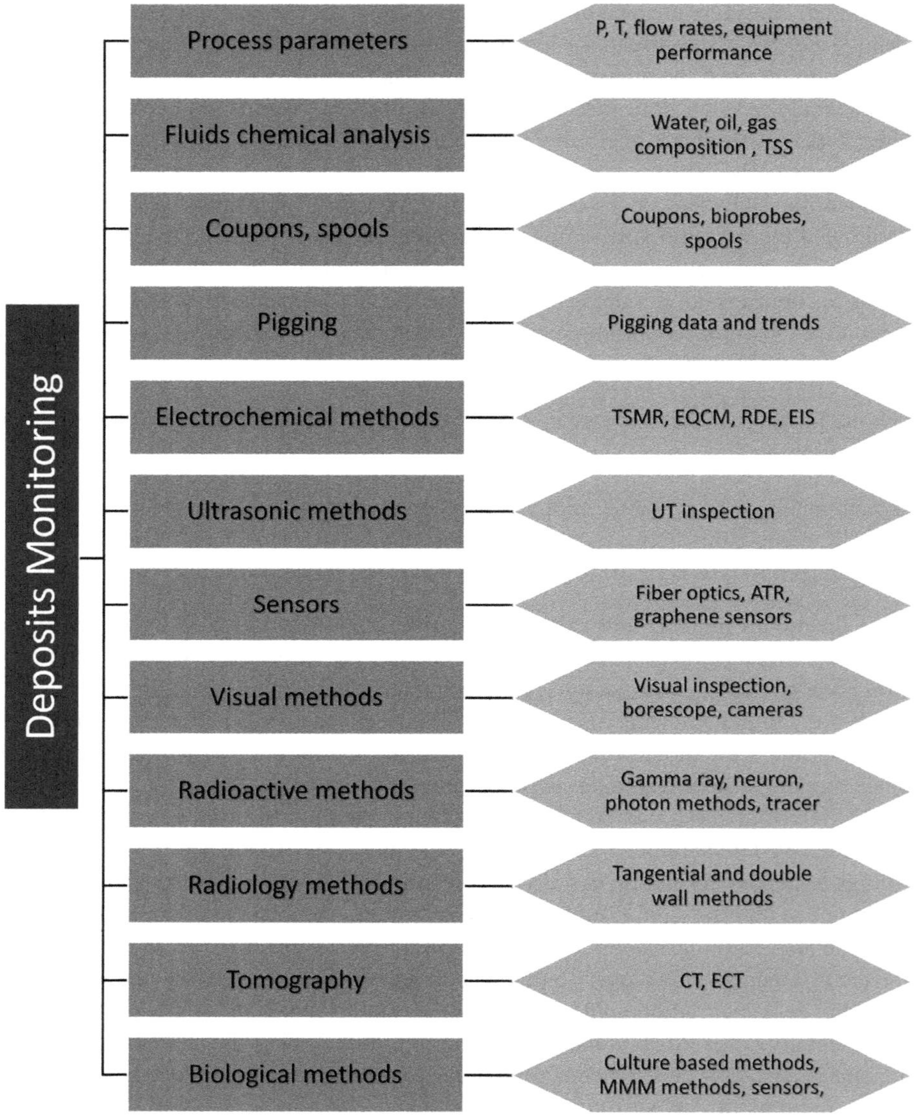

FIG. 13.2

Methods of monitoring solid deposits in oil and gas fields.

sick, signs and symptoms start to appear, such as inflammation, fever, or sweating. Deposits formation in the system is usually accompanied by pressure drop, temperature change, decrease in flow rate, and/ or equipment failure. A vigilant production operator/engineer should report any changes in the system parameters and investigate more fully to identify the root cause, before the problem escalates and causes significant issues.

13.2.1.1 Pressure

Solids deposition usually leads to pressure drop in the system. Rigorous monitoring of the system pressure and early detection of pressure drops are very effective ways of monitoring deposits formation in the system. Pressure monitoring in the different system locations is part of the routine measurements and calculations that are usually carried out by operators. The estimation of pressure drop due to solid deposition in the system is discussed in detail in Chapter 3.

According to the Poiseuille formula, pressure drop and the volumetric rate of flow through a section of a scaled pipe are related, and can be expressed as in Eq. (13.1) [8,9]:

$$\Delta p = \frac{8\eta LQ}{\pi(r-t)^4} \tag{13.1}$$

where Q is the volumetric rate of flow, P is the pressure difference across a sectional length L, r is the radius of the pipe, t is the thickness of the scale deposit (assumed to be uniform over L), and η is the coefficient of viscosity of the fluid. Similar correlations were developed for wax and asphaltenes [10–12].

As can be seen from Eq. (13.1), the volumetric flow is largely dependent on the effective radius $(r - t)$. The higher the scale thickness, the smaller the effective radius, and consequently the higher the pressure drop. Chen et al. [13] showed that the friction pressure drop can increase 31.5% for laminar flow and 37.8% for smooth pipe turbulent flow, respectively, for a wax layer with a uniform thickness of 0.0787 in. (2 mm) deposits in a pipe section with inside diameter of 2 in. (50.8 mm).

Knowing the pressure drop, the productivity index, *PI*, can be estimated as well, according to Eq. (13.2):

$$PI = \frac{Q}{\Delta p} \tag{13.2}$$

Therefore knowing the productivity index, the pressure drop, and the volumetric flow rate over a length of pipeline, the thickness of deposits can be estimated.

The productivity index can be calculated from downhole gauge measurements, estimated reservoir pressure, and hydraulics models. Once calculated, it can be compared with historical PI and checked for declines, which might be an indication of deposits formation. Tjomsland et al. [14] reported another approach to monitor scale deposits in wells producing from two zones. The method involves performing multirate well tests where produced water samples are collected at each rate. Depending on the conditions for each test, either qualitative or quantitative integrated analysis of the well test results and produced water compositions can be performed. Information obtained from the tests for each zone includes pressure, productivity index, water cut, produced water composition, and seawater fraction, which collectively can be used to monitor and control scale formation.

On the other hand, by knowing the deposit thickness, the actual pressure drop and the expected production rate can be calculated. Haindade and Mundhe [15] mechanically estimated the thickness of paraffin deposits using cutters to infer the actual tubing diameter, which then was used in modeling the vertical lift performance (VLP) and inflow performance (IPR) curve matching, for determining the actual pressure drop and the actual behavior of the producing well.

Mohamed et al. [16] used the skin factor to detect the formation of naphthenate deposits downhole of a production well.

Another way of using pressure to monitor deposits formation is pressure pulse technology, in which the pipeline pressure profile is used to detect and monitor solid deposits. The pressure profile originates from the combined effects of frictional pressure drop and changes in the mixed fluids properties. The pressure profile is obtained from pressure measurements at one location, immediately upstream of a quick-acting valve in a multiphase flow pipeline. When the valve is activated, the upstream pressure is measured, resulting in a pressure-time log. The pressure-time log is then converted into a pressure-distance log. The pressure-distance log gives the location and extent of deposits in a pipeline [17]. The method was found to be suitable for routine production testing and allocation measurements, and it was applied to monitor and locate solid deposits in multiphase wells in the North Sea on the Oseberg and Gullfaks [18].

Sarmento et al. [19] reported on a pressure-based method for detection of wax deposits in pipelines. In that method, the wax depositions are detected by pressurizing the pipeline and measuring the pipeline external diameter. Due to wax buildup, significant increase in the external diameter is induced due to the high pressure upstream of the wax blockage, with no changes in the pipe diameter along the downstream of the wax accumulation.

Bonuccelli et al. [20] proposed wax deposition maps, applicable both for single flowline as well as integrated (well-flowline) systems, correlating the pressure drop across the flowline (also taking into account natural depletion), the oil flowrate, and the wax deposition thickness. The wax deposition takes into account the real pipeline section where the conditions for the deposition are satisfied, on the basis of the wall temperature being lower than the WAT value, and the heat transfer higher than zero.

Additionally, routine system pressure monitoring is necessary to avoid pressure affected solids deposition. For instance, pressure drop below bubble point, or below asphaltenes onset pressure (AOP) is foreshadowing of asphaltenes deposition problem.

13.2.1.2 Temperature

When deposits stick to the internal walls of a pipeline, they act as an extra internal insulation layer, which affects the heat transfer to the pipe wall and from the pipe wall to the surrounding environment; therefore a decrease in heat flow rate in the solids buildup locations is usually observed [21].

Assuming a clean pipe, and before deposit is formed, the heat transfer resistance is accounted for by the convective heat transfer from the flowing fluids to the pipe wall, heat conduction through the pipe wall and any insulation or coatings, and the heat transfer from the pipe wall to the surroundings. The addition of deposit layer/layers to the internal pipe wall adds resistance to the total heat transfer resistance, and that resistance depends on the type of the deposit (most of the deposits have very low thermal conductivity) and also depends on the thickness of the deposit layer [13].

By addressing the heat transfer from the internal flowing fluid to the outside wall of the pipe, not only the deposits formation can be monitored but also the deposits thickness can be calculated as described by Eq. (13.3) [13]:

$$\frac{T_f - T_0}{q_o} = \frac{1}{h_d}\frac{r_0}{r_i - t} + \frac{r_0}{k_d}\ln\frac{r_i}{r_i - t} + \frac{r_0}{k_p}\ln\frac{r_0}{r_i} \tag{13.3}$$

where T_f is the bulk fluid temperature in the pipe, T_0 is the outside pipe wall temperature, q_o is the heat flux through the outside pipe wall, r_0 and r_i are the outside and inside diameters of the pipe, respectively, h_d is the film heat transfer coefficient from the flowing fluid to the deposit layer, k_p, k_d are the thermal conductivity of the pipe wall and the deposit layer, and t is the thickness of the deposit layer.

In production systems, the bulk fluids temperature is usually monitored on a routine basis. Nevertheless, using temperature for monitoring solid deposits requires measuring the outer pipe wall temperatures as well. Unusual cold spots on the external pipe wall will be a good indication of a developing solid deposits problem.

Deposit layers were also found to affect the temperature of the bulk produced fluids. Almutairi and Davies [22] reported that the low thermal conductivity of scale deposits increases the temperature of the producing fluid in the scaled region. This temperature increase becomes greater with increasing the thickness of the deposit layer, due to the increased insulation provided by the scale. Thus an increase in the bulk fluids temperature is another indicator for solids buildup.

Thus when monitoring temperature profile of a conduit, areas infested with solid deposits will have an unusual decrease in the external pipe wall temperature followed by an increase in the temperature in the deposit-free areas. This observation should not be confused with Joule-Thomson cooling (cooling of fluids due to gas expansion), which is not usually followed by an increase in temperature.

That effect of temperature will be more addressed in fluids produced with high temperatures, and/or in downhole applications.

Additionally, routine monitoring of system temperature provides early warning signs of potential problems and if intervention is necessary. For instance, a system temperature that is below wax appearance temperature (WAT) is a preindication of developing wax deposition problems. Similarly, a system temperature that is below the hydrates formation temperature or dew point temperature is a sign of potential hydrates problems.

13.2.1.2.1 Temperature-based sensors

Temperature and heat sensors are extensively used in oil and gas operations with wide variety, including resistance temperature detectors (RTDs), thermocouples, thermistors, and distributed temperature sensors (DTSs).

DTSs measure temperatures by means of optical fibers. These optoelectronic devices provide a continuous profile of the temperature distribution along the fiber cable. DTSs have wide applications in oil and gas fields. The main measuring principles are based on detecting the backscattering of light, e.g., using the Rayleigh, Raman, and Brillouin principles [23].

Almutairi and Davies [22] employed this technique in studying the scale deposition in a conventional producing wall. The thermal insulation provided by the scale causes a unique temperature profile. By determining DTS depth-temperature profile data, the presence of scale was detected and scale thickness as well as inside radius of scaled area could also be estimated through an algorithm.

The inside diameter of a scaled tubing can be calculated according to Eq. (13.4):

$$r_{si} = \frac{r_{ti}}{\exp\left(\frac{K_{scale}}{r_{to}}\left[U_{scale}^{-1} - U_{noscale}^{-1}\right]\right)} \tag{13.4}$$

where r_{si} is the inside radius of the scaled-up region of the tubing (ft), r_{ti} is the outside radius of the scaled-up region of the tubing (ft), k_{scale} is the scale layer thermal conductivity (Btu/(hr.ft.°F)), r_{to} is the outside radius of the tubing (ft), U_{scale} and $U_{noscale}$ are heat transfer coefficients for the well when there is no scale deposited and when the scale is deposited at a particular time (Btu/(hr.ft^2.°F)).

Adebayo et al. [24] reported the use of combined electrical resistivity and temperature logs for monitoring scale in carbonate formations during carbon dioxide sequestration in a saline carbonate aquifer or during CO_2-EOR projects. They also concluded that low to moderate production rate environments with low water production yield the greatest increase in flowing fluid temperature when scale is present. In general, scale detection using DTS temperature data can be best achieved when the well is experiencing large heat losses to the formation. DTS is not an independent tool for scale detection. However, it can serve as a complementary tool.

Guzman [25] reported the use of DTS in monitoring wax buildup in a production tubing. The DTS provided temperature trends that showed temperature decline in a section of the tubing string, caused by the additional layer of wax between the crude oil and the gas-filled annulus. The paraffin layer has its own heat transfer coefficient, so thickness can be estimated from temperature data for unchanged flow rate and fluid composition.

13.2.1.2.2 Infrared thermography

Another method that depends on the temperature measurement is infrared thermography. Principles of infrared thermography (or the IR camera shown in Fig. 13.3) are based on the fact that infrared radiation is the energy radiated by the surface of an object whose temperature is above absolute zero. The emitted radiation is a function of the temperature of the material; the higher the temperature, the greater the intensity of the infrared energy emitted [26]. Thermal imagers make pictures from heat, not visible light (infrared or thermal energy). The imaging is done through specially designed thermal imaging cameras and equipment that are able to measure temperature variation by quantifying the sent, detected, and emitted IR energy [27].

FIG. 13.3

IR camera.

As the deposits stick to the internal pipeline wall, they reduce the apparent temperature of that spot, which can be detected by an IR camera that produces two-dimensional images of invisible infrared or "heat" radiation.

Different types of thermographic imaging are available on the market. These include:

- Noncontact temperature sensors.
- Handheld scanners.
- High-resolution infrared imaging systems.
- Thermal wave interferometer systems.
- Radiometers and pyrometers.
- Contact temperature sensors.

Some of these give a direct temperature reading, while others are integrated with computers running preprogrammed software to perform the analysis and produce thermal images of the object.

Dulce et al. [28] used an IR camera to monitor silica deposits and to determine their thickness. An equation for the calculation of the thickness of silica scales inside pipe walls was derived by Villena and de Lara [29] (Eq. 13.5):

$$t = r_i \left[1 - \frac{\left(\frac{r_0}{r_i}\right)^{\left(\frac{k_{scale}}{r_{pipe}}\right)}}{e^{\left(\frac{k_{scale}}{r_0 h}\right)\left(\frac{T_1 - T_2}{T_2 - T_{air}}\right)}} \right] \tag{13.5}$$

where t, r_i, r_o, k_{pipe}, k_{scale}, h, T_2, T_1, T_{air} are thickness of silica, inside radius of pipe (without scale), outside radius of pipe, thermal conductivity of pipe, thermal conductivity of scale, convection coefficient, temperature of inside surface of scale, temperature of outside surface of pipe, and ambient temperature of air, respectively.

Saifi et al. [30] used infrared thermography to study solids deposition and proposed a model for scale thickness measurement.

Al-Ghamdi et al. [27] used an IR gun to monitor scale deposits in the Saudi Aramco giant El Ghawar field. The method was successfully applied on several wells. They utilized a thermal infrared gun pointed toward the production flowlines to observe the thermal transmission signature through different materials. All tests were conducted in conjunction with actual flowline dismantling to confirm the findings of the IR guns with respect to scale accumulation and to collect scale samples for further investigations.

Anomalies in the temperature readings indicated abnormalities inside the flowlines and generally they were found to be related to scale and solids deposition. Comparing the temperature profiles of scaled flowlines to a base model containing a thermal image of the same spot in a healthy flowline revealed that scale accumulation inside flowlines is associated with thermal signatures very different from that of the base case.

13.2.1.2.3 Heat pulse wax monitoring technology

This technology was developed at Statoil's Research Centre in Porsgrunn. The heat pulse wax monitoring technique exploits the fact that a wax layer on the inner pipe wall acts as thermal insulation [31].

The measurement equipment consists of a heat source, e.g., an electric heating element, and a temperature sensor, which are installed at a fixed point on the pipeline, preferably where most wax deposition occurs. By applying a short external heat pulse to the pipeline and measuring the transient thermal response, the wax thickness can be derived [31].

If such a system could also be installed in a subsea pipeline, it would make continuous measuring of the wax buildup possible. This would in turn allow for a much more efficient use of wax control techniques, e.g., by sending a pig only when a certain thickness threshold has been passed [31].

13.2.1.3 Decreased injectivity

Usually seawater is injected into the reservoir to maintain reservoir pressure and improve secondary recovery. Also produced water can be used, known as produced water reinjection (PWRI). Poor injected water quality, solid particulates carryover, solids deposition due to incompatibility between injected water and reservoir fluids, and biofouling cause blocking of pore throats and injectivity decline.

Injectivity decline is evidenced by increasing injection pressures to maintain the injection rate. Injectivity decline is a firm indicator of formation plugging, which may be attributed to scale particles [32].

Siri field in Iran encountered a drop in injectivity of approximately 7000 bbl/day (from 9100 bbl/day to only 2200 bbl/day) over a period of 6 years, due to scale formation [33,34].

Other cases of injectivity decline with time were reported by [35–38].

13.2.1.4 Emulsion problems

Solid deposits are known to stabilize emulsions, and emulsions that are stabilized by solids are called Pickering emulsions. Exaggerated fluids separation, emulsion problems, and crude oil desalting may be signs of solids deposition, especially in the case of steady production and when demulsifier chemicals are already applied.

In such situations, initially an investigative testing and root cause identification is required to determine whether solids precipitation is causing the emulsion problem or not. Sampling from key locations in the system, i.e., from the wellhead through the entire process to water disposal allows the determination of which solids will stay in solution, precipitate out, stay suspended in the flow, or deposit out on the piping and vessels. Then deposit characterization takes place after collecting the deposits using total suspended solids (TSS) filters, or by means of corrosion or scale coupons to determine their composition and causes of their formation. Finally, after identifying the deposits and their causes, mitigating solids deposition takes place alongside emulsion mitigation [39].

Barium sulfate scale particles were reported to stabilize the emulsion, resulting in reduced separation efficiency in an oil production platform in the North Sea [40,41]. Iron sulfides have helped to stabilize the oil/water emulsions as well as increase chemical treatment and other operating costs [42].

Asphaltenes are well-known natural surfactants that cause emulsion stabilizations [43]. Paraffin waxes are also known to stabilize emulsions in crude oil [44]. Stark and Asomaning [45] showed that the asphaltene stabilizer—demulsifier combination does a better job at demulsifying crude oils with asphaltene-stabilized emulsions than when demulsifiers are used alone, keeping in mind the careful choice of such combinations through initial testing.

Naphthenate soaps and emulsions are another type of production problem that most operators do not pay enough attention to, often confusing them with regular emulsions. Sometimes these require a combination of naphthenate inhibitors/demulsifiers in addition to acids for treatment and better process performance [46].

13.2.1.5 Water cut in gas systems

Gas hydrates are formed by combining gas and water at high pressure and low temperature. Intuitively, if the arrival of water to the separator decreases significantly, hydrates can be formed in line. Several field tests have shown that the earliest sign of hydrate formation in a pipe is a decrease or interruption in water production. Increasing the water hold-up in the pipe is attributed to the formation of nontransportable hydrates. Unfortunately, the rate of water production is not monitored closely and continuously enough, and therefore this early warning sign is often not noticed [47].

13.2.1.6 Equipment performance

Equipment fouling is one of the main factors that affect the different production equipment performance. In some cases, changes in pumps, heat exchangers, valves, flowmeters, control systems, separators, and other production equipment performance can be accounted for by possible fouling problems. Some examples of equipment failure due to solid deposits are demonstrated in Chapter 3.

13.2.2 Conventional analytical methods

Conventional analytical methods depend on offline monitoring of the flowing stream chemical composition at specific locations over time. Measuring the solid-forming species in produced fluids, i.e., ion content in water, paraffin content, and asphaltenes content in the crude oil, are usually used as indicators for solid precipitation when their concentrations fall over time.

13.2.2.1 Complete water analysis

Complete water analysis is usually used in scale tendency calculations utilizing one or more of the commercial software, where the types and amounts of deposits can be estimated. Additionally, the scaling ions concentration can be plotted over time and the pattern can be used to monitor the likelihood of scale formation [48]. A drop in Ba^{++}, Ca^{++}, Sr^{++}, Zn^{++}, Pb^{++}, etc. concentration is usually interpreted as a possibility of scale formation. The same trends can be elucidated for HCO_3^-, CO_3^{--}, and SO_4^{--}, keeping in mind that these anion concentrations might also fluctuate due to bacterial activity. A baseline concentration of the ion concentration is also necessary at the start of production from the reservoir formation and also when major changes or maintenance in the production system occurs, e.g., reperforation, changing the sidetracks, or any major changes taking place in the reservoir formation. Fig. 13.4 illustrates the reduction of Ba and SO_4 ion concentrations in produced water with increasing the scaling of $BaSO_4$. Similarly, in pipelines, water analysis is performed on upstream and downstream samples to measure ion concentrations and assess the potential scale formation problem.

Samples are frequently collected from the wellheads, flowlines, separators, and other locations and preserved using EDTA, acid, or scale inhibitors and sent to the lab for analysis.

Knowing the process materials and process fluids composition can help in correlating chemical analysis results with the encountered problems to identify their root cause. Correlating Fe content in the produced water with the system materials is a good example. If the system is mainly duplex steel,

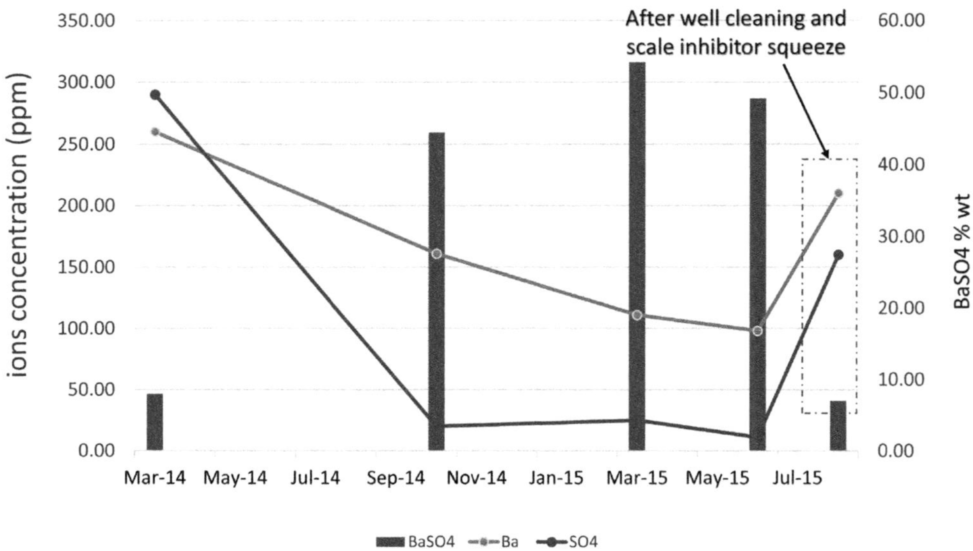

FIG. 13.4

The reduction of scaling ions (Ba, SO_4) concentration with increasing scale formation (% $BaSO_4$) from a production well in the Egyptian field.

then the high Fe content is likely from the reservoir water, while if the system consists primarily of carbon steel then high Fe content is likely from corrosion problems, especially if supported by the Fe/Mg ratio [49].

In addition, detection of the residual scale inhibitor's concentration is another indirect means. When there is a drop in inhibitor concentration, potential scaling can occur since the missing inhibitor is consumed by scale particles [21], or wasted if the produced water rates are increased. An inhibitor concentration above the minimum inhibitory concentration (MIC) means the system is safe from scale.

The main issue in such offline samples is that they may not be representative of the system, and the problem is worse when the scale has been deposited downhole, while the water samples are collected from the well head, which means the operator is collecting the spent water after scale formation is already done, giving a false representation of the scaling ions. In such cases a downhole or reservoir sampling is necessary. Other concerns are that composition changes occur to the samples during deploying them to the lab and until they are analyzed. Samples may encounter loss of dissolved gases and precipitations, which affect the fluid physicochemical properties and consequently can affect the analysis results. Such wrong practice during sampling and analysis can be costly, either from overestimating scale deposits leading to unnecessary chemical inhibitor treatments, or underestimating them, leading to detrimental effects on the production system [21].

13.2.2.2 Crude oil analysis

As with water analysis, analysis of crude oil components like paraffin and asphaltenes can give an indication of a deposition problem. It is worth mentioning that high paraffin and asphaltene contents do not necessarily mean a potential deposition problem. Paraffin and asphaltene deposits were observed in crude oils with low levels of these components.

However, crude oil paraffin and asphaltene contents can be alarming signs if they are done correctly by choosing the right location (locations where different crudes are mixing, high pressure drop, high corrosion rates, high streaming potential for asphaltenes, cold spots, Joule-Thomson throttling locations for wax) and by collecting sets of samples over a period of time where a decreasing trend of asphaltenes/wax content over time (concentration vs. time plot) can be a valid sign of a deposition problem. Samples can also be collected from upstream and downstream of a location of interest, where a decrease in paraffin and asphaltene content at the downstream side is a sign of a deposition problem. The asphaltenes content is conventionally measured according to standard methods IP-143 or ASTM D-3279, while paraffin wax content is measured according to UOP-46. This monitoring method using asphaltene content is a simple, straightforward method of monitoring asphaltenes in produced crude oil.

Besides the regular crude oil components measurement, the oilfield practice usually follows rigorous speciation of the conditions at which paraffin and asphaltenes can form, including [50]:

- PVT analysis.
- SARA analysis (saturates, aromatics, resins, asphaltenes), or modified SARA analysis.
- Asphaltenes onset precipitation tests and asphaltenes onset pressure tests.
- Screening for asphaltene stability using the de Boer criteria, colloidal instability index (CII), and SARA plots.
- Crude mixing compatibility using the Wiehe method [51].
- Atmospheric WAT on dead oil sample using the cross polar microscope (CPM) technique, DSC, and viscosity or infrared methods. Live WAT at high pressure can be measured using the SDS technique.
- Crude oil pour point.
- Carbon distribution for crude oil using gas chromatography improves the investigation of paraffin problem.
- Crude oil density, viscosity, and API are used for investigating multiple problems, including asphaltenes, wax, and naphthenates.
- Total acid number (TAN) analysis is used for naphthenate problem investigations.
- Determination of naphthenic acid contents is crucial for corrosion, naphthenate deposits, and naphthenate emulsion problems.

13.2.2.3 Gas analysis

Gas composition analysis in gas-producing facilities is commonly performed using gas chromatography. Other routine analysis includes water cut, water vapor, and dew point. The formation of gas hydrate will change the hydrocarbon composition in both the gas phase and aqueous phase. Therefore gas hydrate formation will result in a reduction in the concentration of the gas components (i.e., C3 and i-C4) in the gas phase, while hydrate dissociation will result in an increase in the concentration of those sII hydrate formers in the aqueous phase. Online gas analyzers like GasPT can also be used to give real-time results [52].

13.2.2.4 Suspended solids analysis

Suspended solids are defined as the nonwater, undissolved particles/materials suspended in the water. These typically include, but are not limited to, mineral scales and corrosion products (e.g., iron sulfides and oxides, carbonates, and sulfates), sands, silts, reservoir fines, oils, paraffins and asphaltenes, and materials of biological origin [32,53].

TSS is routinely measured as part of the injection water quality parameters to avoid formation plugging and injectivity impairment. Additionally, it was found to be a valuable method of monitoring solids deposition when it was performed on the produced fluids, with a number of reported successful case studies.

NACE Standard method TM0173 is used to determine injection water quality, and it is also applicable to produced water. In this method, injection water is filtered through a membrane filter, which is then dried and weighed to represent the concentration of the suspended solids in milligrams per liter (weight of suspended solids on filter/total volume of filtered water). In the case of produced water, produced fluids are collected, and then produced water is separated and a specific volume is filtered through a membrane filter (after being forced through the membrane using a specific syringe or pump). The solids collected on the membrane surface are then analyzed to identify their full composition and to represent their concentration. Furthermore, this test is performed over specific periods of time, e.g., daily or weekly, to generate a concentration versus time plot that is indicative for the rate of solids formation (excluding reservoir fines and sand); this plot can also include produced water volumes per day for more precise results.

The main advantage of this method is that it offers the direct identification of the actually formed deposits where the samples are collected, avoiding the vague estimations using cation analysis or software predictions.

A major breakthrough in the TSS method was achieved by Jordan et al. [40,41] by conducting suspended solids on produced waters and then measuring the composition, amount, and texture of suspended solids in produced fluids via environmental scanning electron microscopy (ESEM) combined with energy dispersive analysis (EDX). This information can be combined with water production rate data to create a risk matrix to monitor solids deposition and to evaluate the effectiveness of the applied scale control programs. The general procedure for TSS for solids identification is illustrated in Fig. 13.5.

In this test, a sample of produced fluid is collected from wellhead, flowline, or process vessel, and then the produced water is separated and a specific volume is filtered through a membrane filter (0.45 μm). The filters can then be transported to the laboratory where they are opened and the filter papers prepared for analysis. The amount of scale is assessed based on the coverage of the filter surface, i.e., from totally clean filter paper to a paper totally covered. The more scale, the higher the risk. The composition of the solids is then assessed using ESEM-EDX. The amount of scale on the paper is quantified along with the amount of reservoir minerals and corrosion/erosion products. The amount and type of scale is then plotted against the amount of water produced or injected. The larger the water volume, the more potential there is for scale formation and so the higher the risk. From these data, a risk matrix is built to assess the scaling problem and efficiency of the treatments being applied. A typical Boston square type plot is then used to visually display the risk, thus helping to prioritize well mitigation [40,41]. A typical risk matrix is illustrated in Fig. 13.6.

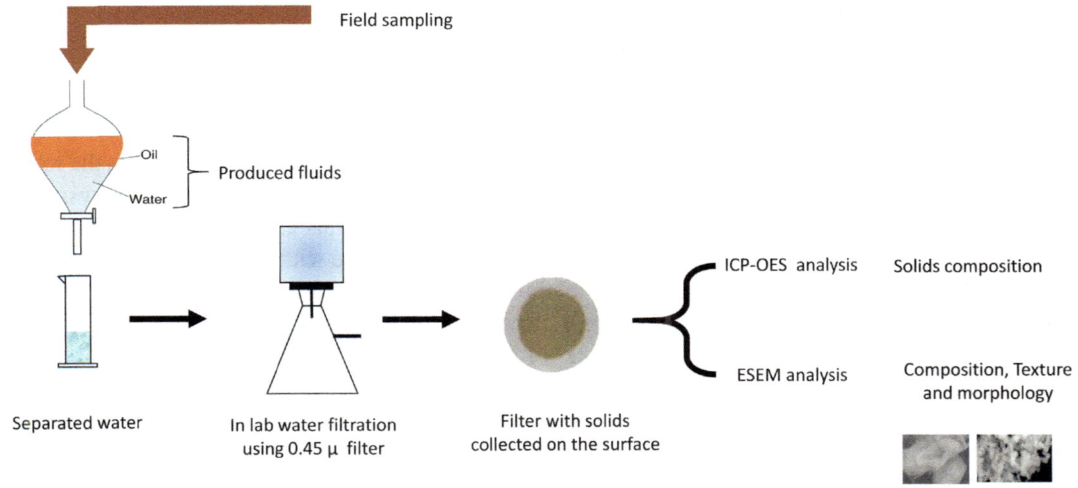

FIG. 13.5

General scheme of TSS test and its use in monitoring solids formation.

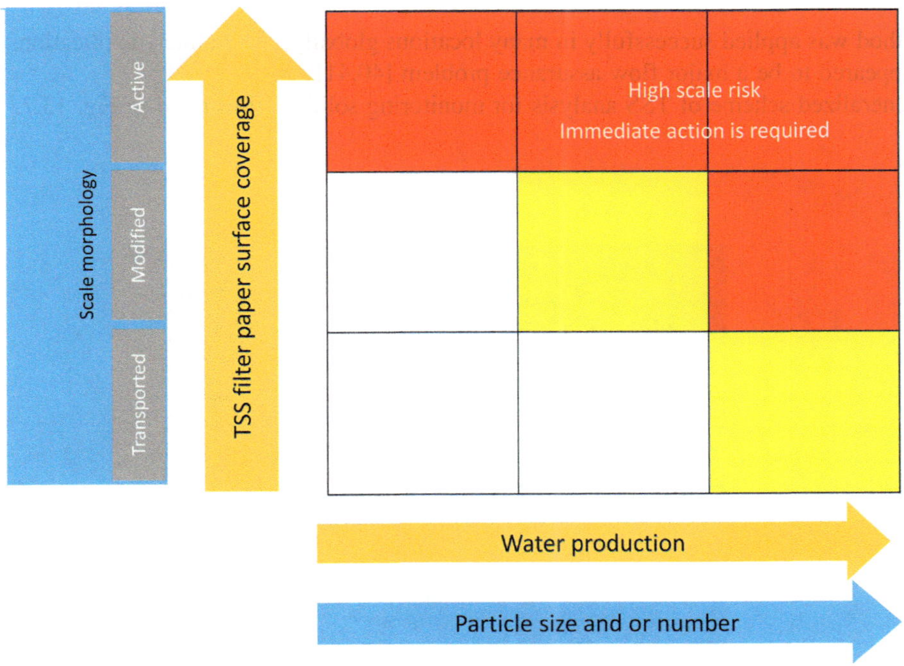

FIG. 13.6

Mineral scales risk matrix constructed by complete TSS analysis, and water production.

Some advantages of this method are:

- It detects actually formed deposits.
- Its capability to differentiate between the different phases of the solids, for example, scale deposits, drilling mud solids, reservoir rocks.
- Its capability to differentiate between the types of scale deposits based on their texture, as the results are reported as "Active," "Modified," or "Transported" scales. Active scales means scale crystals are actively growing in the produced fluids. Modified means the crystals are manipulated due to the effect of a scale inhibitor. Transported means scale that has a texture with chipped and abraded grain edges; these scales formed upstream of the inhibitor injection point or eroded from already formed scale prior to inhibitor application, and were transported with the fluids to the sample point. Usually active scale poses a threat and requires quick mitigation.
- Applicable at very low water cut wells, and in emulsions. Jordan et al. [40] reported application of this method in $<0.5\%$ water cut wells, and in cases of scale-stabilized emulsions.
- The method has been effectively used in evaluating scale inhibition treatments and scale inhibitor levels, and avoiding the over- or underestimations of the other analytical techniques and the scale prediction software.
- Also, the method can be used to identify and measure both inorganic and organic deposits like wax and asphaltenes in crude oil; in this case the crude oil will be filtered through an 8-μm membrane filter [40,41].

The method was applied successfully in many locations globally, and in many applications when deposits appeared to be a major flow assurance problem [40,41].

A generalized scheme of TSS analysis for monitoring solids is illustrated in Fig. 13.7.

FIG. 13.7

Different types of high-pressure corrosion and scale coupons.

Courtesy of Metal Samples Company, reprinted with permission.

13.2.3 Coupons and spools

One of the conventional methods of monitoring solids deposition is by using corrosion and scale coupons and spools. They are installed in and exposed to the flowing streams. These parts are then pulled off the system for further analysis and evaluation.

13.2.3.1 Coupons

Corrosion coupons are usually inserted inline for a specific period of time, exposing the coupon material (which should be the same as the pipeline material) to the flowing fluids and process conditions, and then the corrosion rate can be assessed through the weight loss of the coupon material. Sometimes these corrosion coupons gain some weight due to solids deposition on their surface; once this occurs, the deposits are separated from the coupon, weighed, and identified according to NACE Standard RP0775.

Scale coupons are similar to corrosion coupons, except that they are perforated with a minimum of six to eight holes. The coupon is placed in the system for a specific period of time with the flat side facing the direction of flow, to emphasize any possibility of scale deposition. Average scale growth rate can be calculated by the gain in the weight after a certain period of time, after any hydrocarbons and impurities have been removed from the coupon [32]. Also, the severity of the problem can be determined from the accumulation of the deposits in the holes: the greater the scaling tendency of the fluids, the larger the hole size that is likely to accumulate scale [54,55]. Scale coupons can be strip or cylindrical coupons, but strip scale coupons are usually used. Scale coupons can also be used to evaluate and assess scale treatment methods [54,56,57]. Figs. 13.7 and 13.8 show the different types of coupons and Fig. 13.9 shows the case of a scale coupon with deposits covering its surface.

The success of this scale-monitoring method depends on the location and accessibility of the coupons, how rapidly scale forms on the coupons compared to the process equipment, and how often the coupons can be weighed to determine the amount of scale formed per time and/or capacity units [58]. Fig. 13.10 illustrates the mounting of the coupons in the pipeline.

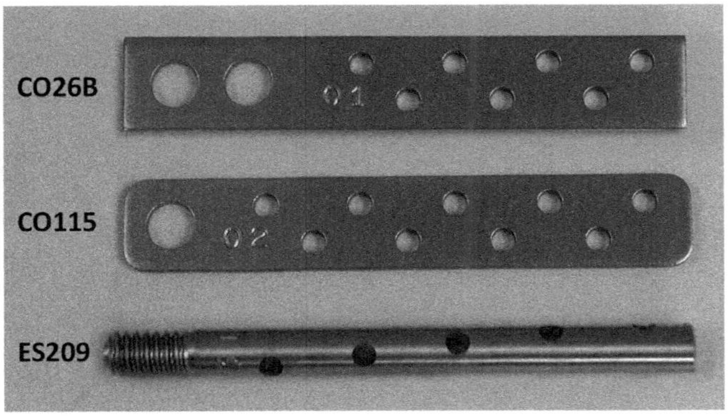

FIG. 13.8

Different types of scale/deposition coupons.

Courtesy of Metal Samples Company, reprinted with permission.

FIG. 13.9

Scale coupon removed from production system with deposits accumulating on its surface.

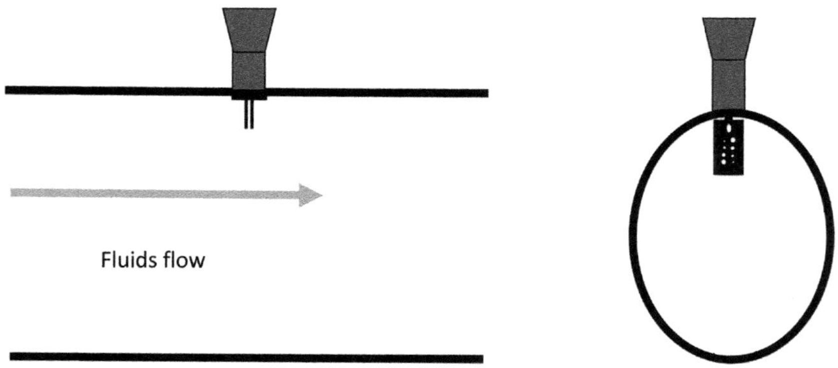

FIG. 13.10

schematic of proper method of mounting scale coupons in the pipelines.

Scale coupons should be located where scaling is occurring or most likely to occur. Examples are main stream mixing locations, pressure and temperature drop locations, and downstream of scale inhibitor injection points (for treatment assessment). The coupon locations can be different from one system to another, depending on the system design, produced fluid volumes, and the system parameters.

Coupons also have been used to monitor asphaltene deposition and the efficiency of asphaltene dispersants and inhibitors, as reported by [59,60].

13.2.3.2 Spools

Spools are like coupons but instead of installing a small strip of metal, a short section of the whole pipeline is installed for a specific period of time, and then the system is depressurized or bypassed, the spool is removed, and the deposits are investigated (Fig. 13.11).

The use of a test spool is also referred to as the "direct method" or the "take-out section method," and it is considered simple and reliable, representing the system with a great deal of potential for studying many parameters associated with deposits formation. Because of these advantages, it is widely used in experimental studies of wax deposition in low-pressure and single-phase flow. However, it is time consuming and needs piping to be bypassed or the entire system to be shut down. In addition, for wax deposits it will not give reliable results with high-pressure single-phase and multiphase systems, since the testing pipes must be depressurized and phase equilibrium is consequently changed [13].

Another way of using spools is to employ what is called liquid displacement-level detection, or the LD-LD method [13,61]. This technique allows estimating the wax deposit volume and thickness by measuring the volume difference of the test section before and after wax deposit inside the pipe. In this process, a clean pipe, with the same geometrical specifications of the spool piece, is filled up with water, and then the water is displaced into the spool piece; the difference between the water levels is a measure of the thickness of the deposit inside the spool piece. The measurement is done multiple times to verify repeatability [13,61,62].

The limitation of this technique comes when it is used to measure the thickness of soft or thin wax deposits, as this method is based on the subtraction of two large volumes (the volumes of the pipe before and after wax deposition). Another error is expected for thin deposits, from the residual oil layer on top of the deposit layer. For soft deposits, part of the deposit can be removed during the measurement and thus is not accounted for [63].

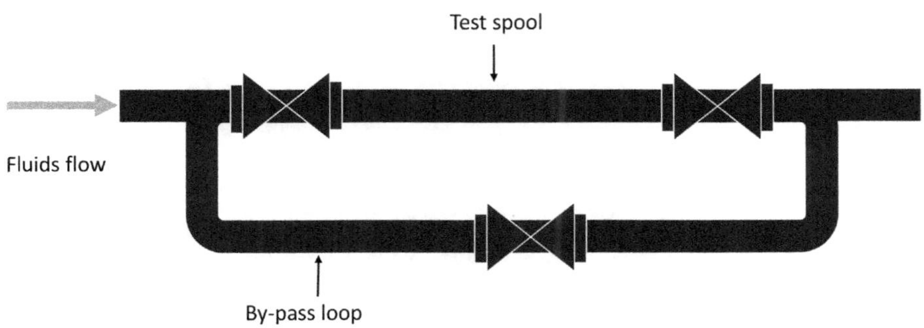

FIG. 13.11

schematic of a pipe spool installed in bypass loop, which used to investigate solids deposition.

13.2.4 Pigging

A pig is a device that travels along the inside of the pipeline to perform different tasks like internal cleaning, physical separation between fluids flowing through the pipeline, visualizing the inside of the pipeline, or inspections [64,65] (Fig. 13.12). Pigging and pig applications are discussed in detail in Chapter 21.

Pigging operations are performed routinely on pipelines to assure a clean path for the produced fluids. It is an effective, cheap, and simple method that has been used for decades in the petroleum industry.

Generally speaking, pigs and pigging operations are valuable in deposit monitoring in many ways, including:

- Deposits that are scrubbed from the internals of the pipelines are sent for chemical analysis and further evaluation. Understanding the root cause of deposits and their mitigation depends on their composition and characteristics. Pigging is the main source of deposit samples in topsides flowlines.

FIG. 13.12

Cleaning pig with organic deposits received from the system after cleaning run.

- Evaluation of the masses of the deposits versus time can develop deposition rate trends that are useful in evaluating the deposition problem and assessing the efficiency of the applied mitigation plans.
- Some types of pigs are able to determine the internal diameter of the pipeline and give a log of the pipe diameter profile where the location of obstruction can be spotted. Examples are [66]:
  - Caliper pigs: used to measure pipeline internal bore.
  - Internal navigation pigs: used to accurately map a pipeline route.
  - Shock logger pigs: used to identify the location of blockages or restrictions in a pipeline.
  - Intelligent pigs and in-line inspection (ILI) tools.
  - Some deposition in-line inspection (DILI) tools were also manufactured, which can determine the scale location, thickness, and type (Fig. 13.13).
- Smart pigs showed promising performance in detecting deposits in pipelines. Smart pig technology was able to identify, locate, and assess the severity of deposits in [67]. In this technology, each form of debris results in characteristic signals in the acquired data, in particular, the vibration and differential pressure data, and changes in the pig inclination within the pipeline as it pitches over debris [68]. Other types of smart pigs are the ultrasonic and magnetic flux pigs, although they are currently used extensively in corrosion-monitoring applications.

FIG. 13.13

Deposition ILI tool developed by Rocsole, based on the electrical tomography method, which is able to locate the deposits and determine their type and thickness.

Photo courtesy of Rocsole LTD, (www.rocsole.com) reprinted with permission.

13.2.5 Electrochemical methods

The use of electrochemical techniques has been reported in lab studies and field applications for deposit detection and monitoring. Among the reported techniques are thickness shear-mode resonator (TSMR), rotating disc electrode (RDE), and conductometric scaling potential test. These methods are used in monitoring deposit formation, in addition to evaluating and optimizing treatment programs; however, except for TSMR, they are not commonly practiced [21].

13.2.5.1 Thickness shear-mode resonator (TSMR)

The TSMR is a mass-sensitive device, which are based on the piezoelectric effect, discovered by Pierre and Jacques Curie in 1880. When crystals are subjected to mechanical stress, an electrical potential on the surface appears, possessing a polar axis without symmetry center (e.g., quartz) and the contrary is also true: the so-called inverse piezoelectric effect [69]. This phenomena is applied in most of the quartz crystal microbalances (QCMs), where a microbalance measures micro quantities of materials. QCM does not act only as a microbalance, but responds also to fluid properties, such as the viscoelastic characteristics of the deposits.

A TSMR is typically constructed from a piezoelectric wafer with thin-film metal electrodes on each side of the wafer. The application of an electrical potential across the wafer causes shear deformation at a resonant frequency, dependent on the properties of the wafer. Deposition on the surface of the wafer will hence alter the resonant frequency, which then can be converted to mass per unit area [70,71]. TSMR is one of the most universally applicable mass-sensing methods, with less interference from the analyte [70]. Usually the results are reported in μg (scale)/cm^2 (probe crystal surface) over a specific time.

Successful application of the technique in monitoring the scaling tendency of produced fluids, monitoring and assessing the applied treating chemicals, and also optimizing the chemical dose when needed were reported by Refs. [70–72].

The early devices were used offline [70,72]. However, the latest version allows online real-time scale monitoring [71].

13.2.5.2 Electrochemical quartz crystal microbalance (EQCM)

EQCM is the combination of QCM and the electrochemical technique. In this technique, electrochemically induced precipitations lead to variations of the total mass that forms at the surface of the material, which can be quantitatively measured by a shift in the resonant frequency of the EQCM [73]. The EQCM technique allows real-time monitoring of the film or deposit layer growth.

The precipitation reaction induced electrochemically by the cathodic reduction of molecular oxygen to hydroxide ion (Eq. 13.6) can be expressed as:

$$O_2 + 2H_2O + 4e^- \rightarrow OH^- \tag{13.6}$$

The OH^- ions increase the pH, which enhances the precipitation of calcium carbonate (Eq. 13.7):

$$Ca^{2+} + HCO_3^- + OH^- \rightarrow CaCO_3 + H_2O \tag{13.7}$$

Successful lab and field applications of this technique in monitoring scale deposit, studying scaling process kinetics, and screening and evaluating scale inhibitors have been reported [74–76].

Problems associated with this technique include the fact that the electrochemical reaction greatly accelerates the scaling process, which sometimes leads to overestimating of the problem. Other disadvantages with this technique are that it does not specify the deposit type, different electrode materials give different responses, and its heat sensitivity and tolerance are not good [77].

13.2.5.3 Chronoamperometry

In the chronoamperometry method, electrochemically accelerated scaling processes are studied wherein the current flow against time is recorded during the scaling process while the electrode is polarized [74]. Normally the current decreases when scale is deposited on the active surface. Often scaling time is the main parameter in this method, as the scaling time represents the time required to reach the lowest and stable value of the current. A large scaling power indicates a smaller scaling time [74,77].

Problems with this method are that it does not indicate scale type or thickness, and also it was found that the scale deposition is underestimated by this technique [74,77].

13.2.5.4 Rotating disc electrode (RDE)

The RDE is an electrochemical technique for scale monitoring as a function of time. This technique is based on the analysis of mass transport to an unscaled and scaled metallic surface. Like the electrochemical process described in Eqs. (13.6), (13.7), the scale deposition on metallic surfaces can be estimated through an assessment of the kinetics of the oxygen reduction to hydroxide ions process, under varying hydrodynamic conditions. The method uses the Levich equation to describe the diffusion, mass transport, and flow conditions at an RDE. The use of the rotating disk ensures that the mass transport to the surface is well defined and may be controlled by changing the disk rotation speed [77–79]. Normally, the scaling problem severity is indicated by the percentage of the working electrode covered by mineral scale deposits.

The technique was reported to be a promising monitoring method for $BaSO_4$ and $CaCO_3$ in the early trials [78,79]. However, some limitations are its pressure and temperature tolerance, the fact that it requires specific skills to interpret the data, and it is time-consuming [77].

13.2.5.5 Conductometric method

In this method, brine scaling is stimulated (by temperature change) and the change in specific resistivity or the brine is measured. Thus a conductivity bridge is used to indicate the stability of produced water and it can also give a quantitative measure of the carbonate scale precipitated. The scaling potential is obtained from measured resistivity-change data using a graphical precalibration method. The method can also be used to screen scale inhibitors [80]. The method was reported to successfully determine quantitatively the capacity of water to precipitate $CaCO_3$, and it was used for evaluating the effectiveness of different carbonate scale inhibitors [80].

13.2.5.6 Electrochemical impedance spectroscopy (EIS)

EIS has been used widely to monitor corrosion processes and has been used extensively for characterizing the performance of protective polymer coatings.

The same technique was adopted to monitor and assess scale formation in aqueous solutions. Calcium carbonate, calcium sulfate, and calcium phosphate deposition tendency were evaluated by EIS

[81]. EIS-based sensors have also been developed for detecting gas hydrates and biofilm-forming bacteria.

As an in situ technique, the EIS monitors the scaling process by measuring the electrical impedance at the solid-liquid interface. It is composed of a working electrode, a counter electrode, and a reference electrode. The working electrode provides a surface for mineral deposition and detects the changes induced by the growth of mineral deposits [82].

13.2.6 Ultrasonic methods

Ultrasonic inspection is a nondestructive technique (NDT) that is commonly used in inspecting steel and other alloys. In this technique, high-frequency sound waves are introduced into materials to detect surface flaws. The sound waves travel through the material with some loss of energy (or attenuation) and then are reflected at interfaces, where the reflected beam is analyzed to detect and estimate the surface flaws [83], as illustrated in Fig. 13.14. The method is widely used in oil and gas system integrity inspections as one of the nondestructive testing methods.

However, the conventional ultrasonic methods may not be applicable for the quantification of deposits, for two main reasons: first, the signals reflected off the scale-oil boundary may be disturbed or masked by the reverberating echoes within the pipe wall, and second, the velocity of sound within the different deposit types is unknown [84].

Gunarathne and Keatch [85,86] suggested a method for indirectly monitoring and characterizing deposits by extracting features from ultrasound and estimating sound velocity from stored data. To estimate deposit thickness, this approach enhances weak echoes from deposits using a number of methods, including preferential filtering in the frequency domain. Scale nucleation in clean pipes is

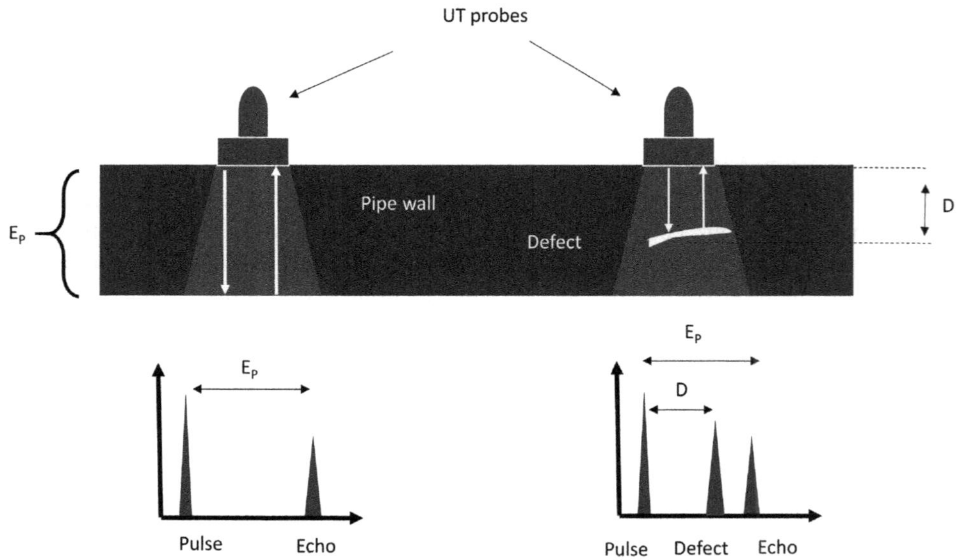

FIG. 13.14

Basic principles of UT inspection.

monitored by evaluating on-line data in the amplitude and frequency domains and comparing the characteristics to acoustic templates stored in the system's database. They also claimed that at a frequency of 5 MHz, their approach could detect 0.5 mm of barium sulfate.

Their work was extended further to its application online and real-time identification of the scale deposit types, in addition to using ultrasound to enhance scale dissolution [84,87–90]. Deposits identified include calcium carbonate, wax, barium sulfate, and strontium sulfate. A significant increase of $BaSO_4$ dissolution rate from 4 to 17 times for an ultrasound intensity of 10%–55% was claimed.

With all the advantages of this technique, however, it is still expensive, with many interferences from the produced fluids, insulation, and polymers, so optimum cleanliness of the pipeline must be attained, in addition to production shutdown to halt fluids flow. Also, the technique requires a skilled person to perform the test.

13.2.7 Fiber optics methods

Optical fibers are flexible transparent fibers made of glass or plastic and having a tiny diameter, capable of guiding light based on total internal reflection. They are employed as extrinsic and intrinsic components of chemical and physical sensors [91]. They are commonly used to monitor temperature, heat flux, strain, pressure, fluids flow, vibrations in deepwater completions, corrosion, gas hydrate plugs, determining amount and location of paraffin deposits, pig detection, and other applications in the oil and gas industry [92,93].

Exposed core fiber optics (has a removed section of the cladding part) were used to monitor $CaCO_3$ precipitation, employing monochromatic laser light and a detector [94]. When solid particles deposit on the exposed fiber surface, part of the guided light will be absorbed by the scale crystal instead of reflecting back into the core part due to the higher refractive index than the fiber core. That significantly reduces the radiation transmitted through the end of the fiber and gives an estimation of the formation of heterogeneous scale [91,95]. This configuration was first proposed by Philip-Chandy et al. [96] and was capable of measuring silver chloride crystallization. This configuration is also termed here an "intrinsic exposed-core optical fiber sensor" (IECOFS).

The technique was used in geothermal waters and was able to detect scale deposits and evaluate the applied scale inhibitors [97]. Furthermore, Means [98] developed a fiber optics-based sensor for monitoring asphaltene deposition in crude oil. The novelty of this probe is that it is combined with asphaltene chemical injection control, which provides a novel method for the control of asphaltene deposition in crude oil production systems. The probe can be used either with discrete samples or as an in situ, online measurement and it is designed to withstand high pressure/high temperature process conditions up to 300°F in temperature and up to 3000 psi.

13.2.8 Attenuated total reflectance

The attenuated total reflectance (ATR) technique is usually used in FT-IR spectroscopy methods for chemical analysis (fingerprinting). An ATR probe is a direct surface analysis technique that can be used as an in situ scale sensor using infrared light. The sensor basically measures the energy loss of an incident light and the change in the refractance of the material that is in contact with the surface of the probe, allowing the detection of deposits formation [99,100].

The technique cannot identify the type of the deposit. However, it was applied in the field and showed promising performance in monitoring calcium carbonate and helped in assessing the applied scale inhibitor [99,100]. A fiber optic ATR probe was developed for sensing asphaltenes [101].

13.2.9 Graphene-based sensors

Graphene has emerged as one of the most promising nanomaterials, having superior properties that make it an excellent material for sensors. A specially designed graphene-based sensor for scale monitoring was proposed by Bouchalkha et al. [102]. The sensor was designed and tested for inline measurement of strontium ion concentrations. The sensor showed good sensitivity in response to a strontium concentration ranging from 0.01 to 0.25 m/L.

13.2.10 Turbidity methods

Shih et al. [103] used a technique based on continuous in situ monitoring of precipitate formation via a back-light scattering turbidity probe to monitor gypsum deposition and evaluate scale inhibitors.

13.2.11 Borescope inspection

Like endoscopy for the human body, a borescope can be used to inspect the inside of pipes. It is a long, tubular optical device (camera) that illuminates and allows the visual inspection of surfaces inside tubulars and chambers. It usually comes in a variety of lengths and diameters, and provides the necessary optical connection between the camera and viewing end [83]. The recently developed borescopes can be viewed through mobile phones with the use of special apps (Fig. 13.15).

Langen and Healy [104] reported a fiber optics borescope that can be used for many purposes in oil and gas fields. Borescope inspection was the main method of monitoring and assessing the mitigation of naphthenate deposits/emulsions in the Heidrun field [105].

The borescope can be accessed through narrow points in the pipes, which makes it a very useful tool to monitor plugged locations where pipelines, tubing, or process equipment are inaccessible for inspection or cleaning by other methods. It allows a visual inspection of the internals of the pipeline and assessing the problem. It also can be used after cleaning jobs to evaluate them.

13.2.12 Radioactive techniques

Several radioactive techniques have been used to monitor deposits in oil and gas fields. Their application covers a wide spectrum of deposit types, at different locations starting from reservoir, production tubing, wellhead, to the topsides flowlines. Gamma ray and neutron-based methods are more involved in scale identification applications due to their penetration ability, where gamma ray technology is easier to detect than neutron and can be used in different ways [21]. These methods are based on the interaction of the incident radiation with the inspected pipe and material, which is illustrated in Fig. 13.16. The gamma radiation log relies on the emitted gamma ray from the NORM scale.

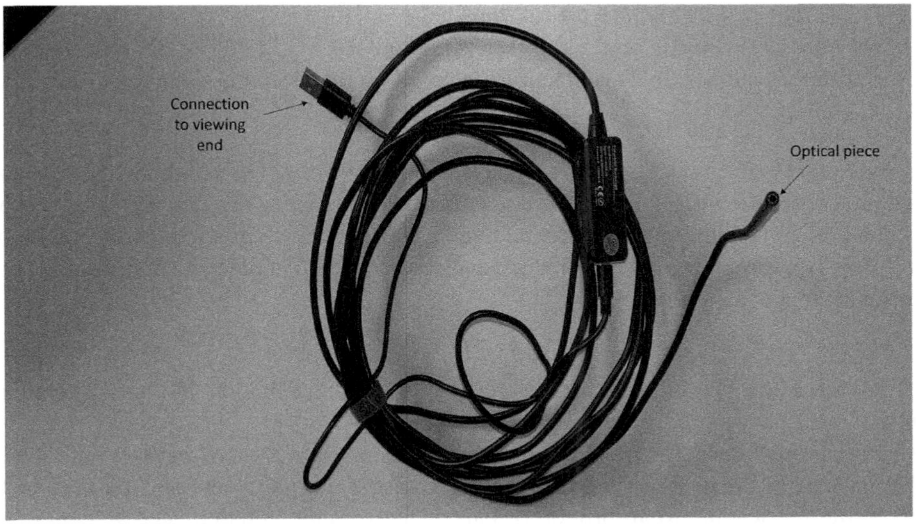

FIG. 13.15

Industrial borescope. Other longer and more feasible versions for oil and gas applications are available commercially.

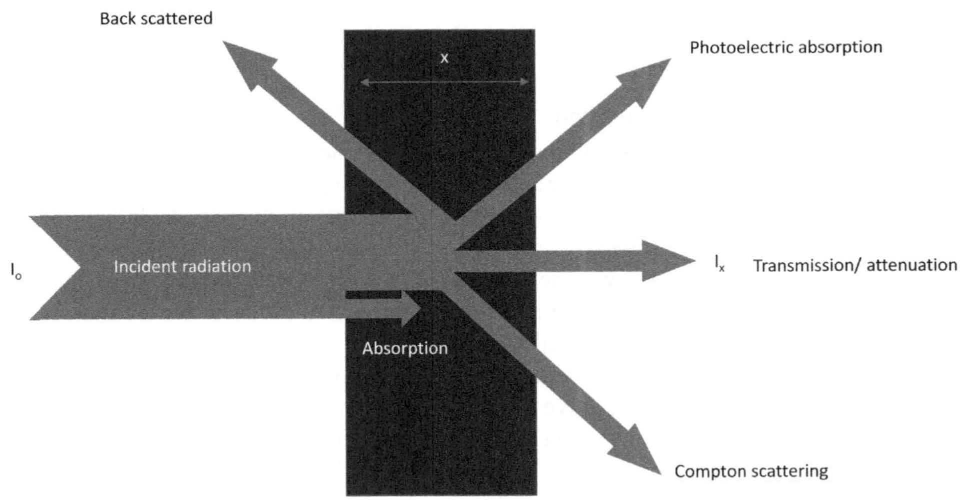

FIG. 13.16

Some of the mechanisms of interaction between radiation and matter.

13.2.12.1 Gamma ray transmission techniques

The different interactions using gamma ray transmission are NDTs, like ultrasonic, used to inspect pipelines in different industries. The technique is based on the Beer-Lambert law [106] (Eq. 13.8):

$$I = I_0 \cdot e^{-\mu \rho x} \tag{13.8}$$

where I_0 and I refer to the intensity of gamma rays initially and after adsorption, while μ, ρ, and x refer to the radiation absorption coefficient (linear attenuation coefficient), density, and thickness of the absorber (radiation path length in cm^2/g), respectively. The preceding equation is true for narrow-beam geometry. However, broad-beam geometry is usually encountered in practical applications. For broad-beam geometry this equation becomes:

$$I = BI_0 \cdot e^{-\mu \rho x} \tag{13.9}$$

where B is called the "buildup factor." B represents the contribution of the scattered radiation [107,108].

When gamma rays strike the pipeline surface, the gamma ray photons interact with the electrons inside the material of the pipeline via different mechanisms, e.g., photoelectric, pair production, and Compton scattering, leading to absorption of part of the gamma radiation. The amount of radiation absorbed depends on the material density and thickness, with the greater the thickness and density, the higher the radiation absorption. Different density materials give different absorption patterns. The thickness of the deposit on the pipeline can be estimated from the extent of the radiation absorption [21,108,109].

Sources of gamma rays can be ^{60}Co, ^{137}Cs, ^{133}Ba, ^{170}Th, or ^{75}Se, and specific sources are used with specific piping diameters and thickness [107].

Oliveira et al. [109] reported the efficiency of using a gamma transmission in finding and quantifying the thickness of the scale conducted in oil exploration pipelines. Cesium-137 was used as the radioactive source and coupled with a NaI(Tl) scintillation detector.

The method has some disadvantages, such as the geometry of the transmission, as the pipe may not be accessible from both sides, and it may not be applicable in large-diameter pipework [3].

13.2.12.2 Gamma ray logging method

Production logging, or production logging tool (PLT), is a technique used in measuring and collecting some wellbore data. It is a process whereby geophysical and petrophysical parameters along a borehole are continuously recorded. Different methods are used in production logging, including temperature, acoustic, gamma ray, and others [110].

The *gamma ray* log (commonly given the symbol *GR*) measures the total natural gamma radiation (γ) emanating from a formation, which originates from radioactive decay of the uranium-radium and rhorium series. A gamma ray detector is usually used to detect this naturally emitted gamma ray [111].

Using this technique in monitoring scale downhole is based on recordings of the radioactivity from the NORM (Naturally Occurring Radioactive Material) scale. The radioactive detector is inserted into a system to measure and record natural gamma radiation emitted from radium-266, radium-288, and bismuth. These radioactive elements usually are coprecipitated with $BaSO_4$ scale, causing the scale to be radioactive. An increase in radioactive levels in the system indicates NORM scale buildup in the reservoir or on the walls of the tubing [112]. Scale precipitation can be detected in the tubing and casing by measuring any decrease in the tubing and casing inner diameters [113].

Gamma ray logs are a valuable tool in detecting and monitoring scale formation. Bamforth et al. [112] reported using gamma ray logs to detect radioactive mineral scale deposited on tubing and pipeline surfaces. Mahmud and Ermila [113] reported using a gamma ray PLT log to detect the formation of barite scales at the wellbore of three production wells; the technique was also used to monitor the effectiveness of the applied mitigation method.

Unfortunately, this technique is only suitable for NORM associated scale, i.e., $BaSO_4$ and $SrSO_4$, and it cannot detect other scale types such as $CaCO_3$, which usually does not contain any radioactive elements. However, some reports showed that NORM can form with black powder and some other deposits rather than sulfate scale. This technique is generally not sensitive enough to detect a low level of radiation emitted by the scales [77].

13.2.12.3 Dual and triple energy spectral gamma ray

In this technique, the scale is detected by an additional increase of attenuation in the path of gamma rays, which is due to the differences in composition between scale and flowing fluids (metals and anions that are heavier than hydrogen and carbon). The method uses radioactive sources on one side of the material being examined, which emits gamma rays at different energy levels, and then the attenuation of the gamma rays at two or three different energy levels is detected with a detector placed on the other side of the material [2].

Bertrand et al. [2] developed a multipurpose dual-energy spectral gamma ray/Venturi multiphase flowmeter used to monitor the gas, oil, and water flow rates in a combination of dual/triple gamma ray attenuation measurements to detect mineral scale on the component surfaces. The application of this method enabled the early detection, type identification, and estimation of the thickness of the scale deposit and its rate of accumulation over time. Scales detected were $CaCO_3$, $BaSO_4$, $SrSO_4$, and ZnS. The data from this flowmeter was used to define the type of acid treatment required and to evaluate its efficiency. Another method was developed based on continuous triple-energy gamma-ray attenuation measurements, which can detect thin scale layers at a dedicated, instrumented mandrel or spool piece. The method can identify the scale types and can also be used in evaluating scale inhibitor treatments [114].

13.2.12.4 Photon-scattering technique

Unlike transmission techniques, the use of photon scattering (incoherent scattering) requires access to the measurement volume from only one side, but this can mean that the attenuation of the scattered photons will be much lower than that of those transmitted through the pipework in the case of large diameter pipes [3].

Meric et al. [3] developed a photon scatter gauge for scale monitoring and characterization that, as opposed to transmission gauges, eliminates the need for access to both sides of the pipework for the required measurements. The scale detection method was based upon using several detectors in a certain geometry that would allow acquisition of data on both photon backscattering and attenuation properties of the process fluids and scale minerals. In this study, simulations were performed with the pipe filled with water, oil, gas, $CaCO_3$, $BaSO_4$, and $SrSO_4$.

13.2.12.5 Gamma ray backscatter technique

Abdul-Majid [115] reported a gamma ray backscatter method based on the fact that, when a point source is put between a gamma detector from one side and material of a certain thickness from the other, the Compton interaction causes some of the incident gamma rays interacting with the material to scatter backwards to the detector. The technique does not require surface preparation or cleanliness, and can work at high temperatures where ultrasonic methods fail.

Margret et al. [116] reported that collimated Compton backscattered gamma rays from a radioactive source have been used to inspect scales by automatic scanning in steps along the axial direction of different pipes.

13.2.12.6 Neutron moderation method

Abdul-Majid et al. [117] described a method for organic deposits detection based on neutron moderation, where neutrons from a neutron source interact with the scale inside the pipe. A fast neutron beam penetrates the pipe wall without significant interaction; the neutron then is scattered elastically with the H and C atoms in the hydrocarbon deposits and slowed down. Some of the slowed-down neutrons diffuse backward and are measured by a detector. The method was used to detect and monitor asphaltenes and wax deposits.

13.2.12.7 Neutron back diffusion technique

In this method, fast neutrons from a source penetrate the pipe with little absorption, since iron has a small absorption cross section for fast neutrons, whereas such fast neutrons interact elastically with the hydrogen and carbon atoms of the organic scale and are slowed down. Some of the slowed-down neutrons will move backwards and are detected by the BF3 neutron detector, the count rate of which increases with the amount of scale. A thicker scale leads to more moderation and backscattering of the slow neutrons. The method is reported by [118,119] and was employed in determination of scale in flare lines [120].

13.2.12.8 Radioactive tracer technique

Tracer technology has been used for various applications in the oil and gas industry, like drilling, well completion, reservoir evaluation, enhanced oil recovery, and determination of residual oil saturation [121]. Tracer technology works by injecting certain volumes of tracers that are selective with oil, water, or gas phases, at known rates, and then, by measuring the dilution and the tracer pulse generated at the far downstream, the individual phase flow rates and velocity of fluid at a particular location along the flow path can be determined. Tracers used can be dye, fluorescent, or radioactive tracers [122].

The radioactive tracers have been reported to be used in many applications [123]:

- Determining flow rates and flow profiles
- Diagnosing completion problems
- Evaluating treatment effectiveness

The radioactive tracer logs comprise a radioactive material and a gamma ray detector. The methods of using radioactive materials together with gamma ray detectors include [123]:

- Treating the well with a radioactive injectant
- Using a tracer ejector tool
- Monitoring the deposit of radioactive salts on the casing from water production

Radioactive tracers are effective, nonintrusive, real-time evaluating tools; however, they require a sensitive detector and good choice of tracer material to be similar to the investigated scale [21].

Martini et al. [124] and Brichart et al. [125] reported the use of time-resolved fluorescent (TRF) tracer in monitoring and evaluating scale squeeze jobs. The characteristic luminescence signature (lifetime value, emission, and excitation spectra) of TRF tracers allows the reliable tracking of the scale inhibitors.

13.2.13 Radiography techniques

Radiography is a common industrial inspection method that has been applied for decades to identify deposits, corrosion, and other different pipe defects. The technique is nondestructive and basically depends on using electromagnetic radiation of high energy (usually X-ray or gamma ray) to penetrate and interact with the material under study. From the variation of radiation intensity due to the emitted radiation, a radiographic image can be created [21]. Radiography techniques can be divided into two types: tangential radiography and double wall radiography (Fig. 13.17); the latter can be further divided into single and double image. Tangential is preferred for thickness measurements, while double wall radiography covers a larger area, but is more subject to errors [21]. Al-Shamari et al. [55] reported that radiography has been used to locate and quantify scale depositions.

If a monochromatic radiation beam is passed through an insulated pipe with deposit, the formula of transmitted radiation intensity (I) is as shown in Eq. (13.10) [126]:

$$I = BI_0 \cdot e^{-(\mu c\, xc + \mu s\, xs + \mu d\, xd + \mu m\, xm)} \tag{13.10}$$

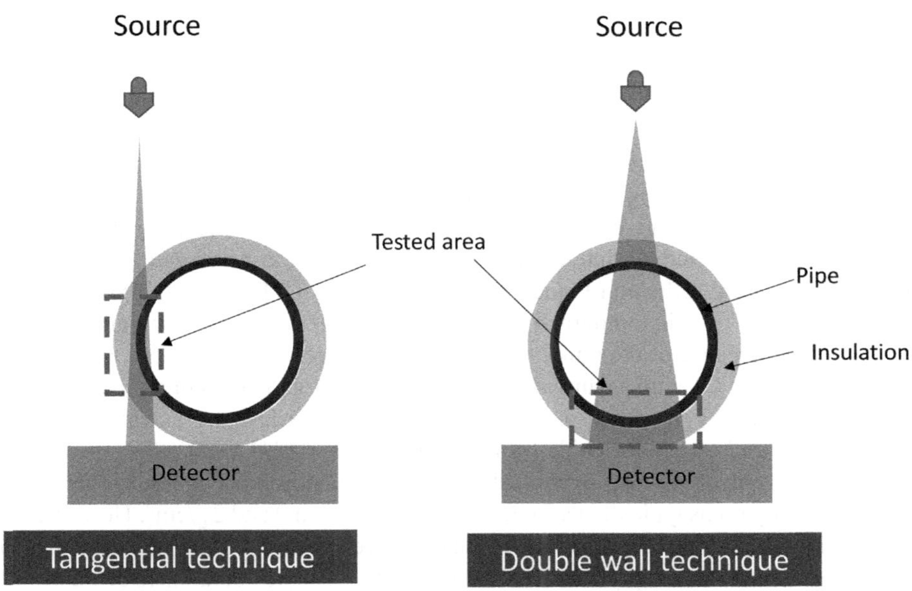

FIG. 13.17

Tangential and double wall radiographic testing.

where $\mu_c, \mu_s, \mu_d, \mu_m$ are the linear absorption coefficients of the insulator, steel, deposit, and transported matter, and x_c, x_s, x_d, x_m are the transmitted thickness of the insulation, steel, deposit, and transported matter.

13.2.13.1 Tangential radiography

In the tangential method, radiation interacts and penetrates the pipe sidewall where the parts of the radiograph, which lie behind the tangential location of the pipe, are interpreted. This method offers many advantages over conventional methods for determination of the thickness of the insulation, pipe, and deposit as well as detecting internal and external corrosion [127].

The method was reported by [126–128]. Harara [119] showed that using the tangential method was sensitive to almost all kinds of deposit at very minimum thickness of the deposit layer; the method is also fast compared to neutron capture gamma ray and neutron back diffusion techniques, which are sensitive for certain type of deposits.

13.2.13.2 Double wall technology

This technique is suitable for larger size pipes or lower X-ray radiation energies, where the tangential projection technique is not applicable. The source is kept perpendicular with respect to the pipe axis. Usually two exposures are taken by rotation around the pipe of 90 degrees [128].

The double wall technique can be single image or double image; the single image is mostly suitable for pipes of diameter above 85 mm, while double image is adopted for smaller size pipes of diameter up to 85 mm. [129].

In general, several improvements have been carried out in the field of radiography in terms of sources, image capturing method, portability, and size. However, it is still considered an expensive and slow technique, and might be inapplicable in some instances; it requires two-side access, it measures a small fraction of the pipe, and is hazardous due to the use of a high-energy source [21].

13.2.14 Tomography

Tomography is another scanning and imaging technique. An image or density mapping is produced by introducing physical stimulus, e.g., radiation source, to the material under test, and then the material response signals are detected and, by the use of computers and mathematical models, the tomographic reconstruction is achieved and the internal material distribution can be visualized [21,130].

X-ray, attenuated total reflection Fourier transform infrared (ATR-FTIR), NMR, and gamma ray are the different sources used in tomography scanning. X-ray electron beam tomography has remained as indoor equipment because of the bulky hardware of X-ray generation devices. By replacing X-ray devices of electron beam CT with a gamma-ray source, a tomographic system can be a portable device. In addition, a tomographic system with a gamma-ray source can be designed as a clamp-on type system because it does not need a vacuum guide for an electron beam [131].

Different density ranges are used to evaluate flowline contents and deposits. For flowline contents evaluation, the density maps typically use a density scale between 0 and 2 g/cm^3. However, for deposit buildup evaluation, density maps typically use a scale between 0 and 3.8 g/cm^3 in order to separate the color of the scale from that of the pipe walls. The density maps enable operators to evaluate the contents and deposits types. The expected deposit types include [132]:

- Wax (0.8 g/cm^3)
- Hydrate (0.9–0.96 g/cm^3)

- Asphaltene (1.1 g/cm^3)
- Scale (1.92–2.65 g/cm^3).

13.2.14.1 Computed tomography

Computed tomography (CT) technology is used to facilitate the inspection of subsea pipelines from the outside, without the need to remove the protective coating.

The method uses a gamma ray (137Cs) source and was successfully used to monitor and evaluate flow assurance and pipeline integrity issues for subsea pipeline flow assurance by providing high-resolution images of pipeline contents and pipe wall thickness [132].

The system is applicable under various conditions, and is available for scanning subsea pipelines since it can be operated by an ROV when scanning subsea pipelines, and is rated for use at 10,000 ft. (3048 m) of water. It also can be used with piggable and unpiggable pipelines, with no pipeline or system modifications, and without production interruptions or downtimes. The system has been proven in the field, and was recently deployed to survey an unpiggable pipeline system in the Gulf of Mexico [132].

13.2.14.2 Electric capacitance tomography (ECT)

Electrical capacitance tomography is an industrial imaging technique used for monitoring and evaluating processes and production system components. It basically measures the external capacitance of the enclosed objects to determine the internal dielectric permittivity distribution, which is then used to reconstruct an image of the process [133,134].

Preliminary investigation for the use of ECT showed that it can be used as a real-time monitoring method for wax deposition in pipelines, and to measure the amount of wax fractions [135].

ECT sensors are available commercially and can be installed in the system with different setups to provide real-time and online measurements of deposits. Sensors manufactured by Rocsole can be used to identify and measure the thickness of different organic and inorganic deposits, besides their use in other flow assurance applications (Fig. 13.18) [136].

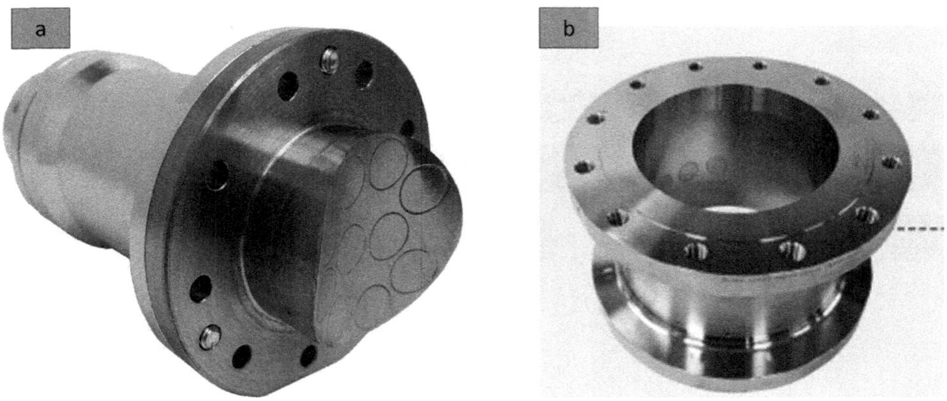

FIG. 13.18

ECT tomography sensors: plug sensor (a), pipe sensor (b).

Photos courtesy Rocsole LTD, https://www.rocsole.com/, reprinted with permission.

13.2.15 Biofouling monitoring

Biofouling and microbiologically influenced corrosion (MIC) are complex problems in oil and gas operations, which are associated with microbial activity. Early detection and monitoring of microbial activity is critical for flow assurance and system integrity.

Microbial activity can be detected online or offline, where offline methods are commonly employed in oil and gas fields. Online methods are based on using sensors. Offline methods depend on detecting and quantifying microorganisms in the produced fluids (mainly water) or in the formed deposits using biological techniques. Usually, diagnosing and quantifying the extent of microbial activity in deposit samples is more accurate and representative of the problem than testing in water samples.

Javaherdashti [137] distinguished between recognition and detection of MIC, as "recognition" does not use biological techniques while "detection" uses biological techniques. Little and Lee [138] listed biological and chemical techniques to "diagnose MIC," and the physical sensors and tools used to "monitor MIC." In this section, the words recognition, diagnosis, detection, and monitoring will be put in the same bucket and used to determine the origins of biofouling and microbial activity. A summary of methods of detecting and investigating MIC in oil and gas fields is furnished in NACE standard TM0212.

Monitoring microbial activity can be online or offline. Fig. 13.19 summarizes these different methods.

13.2.15.1 Online biofouling monitoring

13.2.15.1.1 Sensors

Janknecht and Melo [139] reviewed the online methods of monitoring bacteria and biofilms. All online biofilm monitoring techniques are based on introducing a signal to the surface under test, wherein the

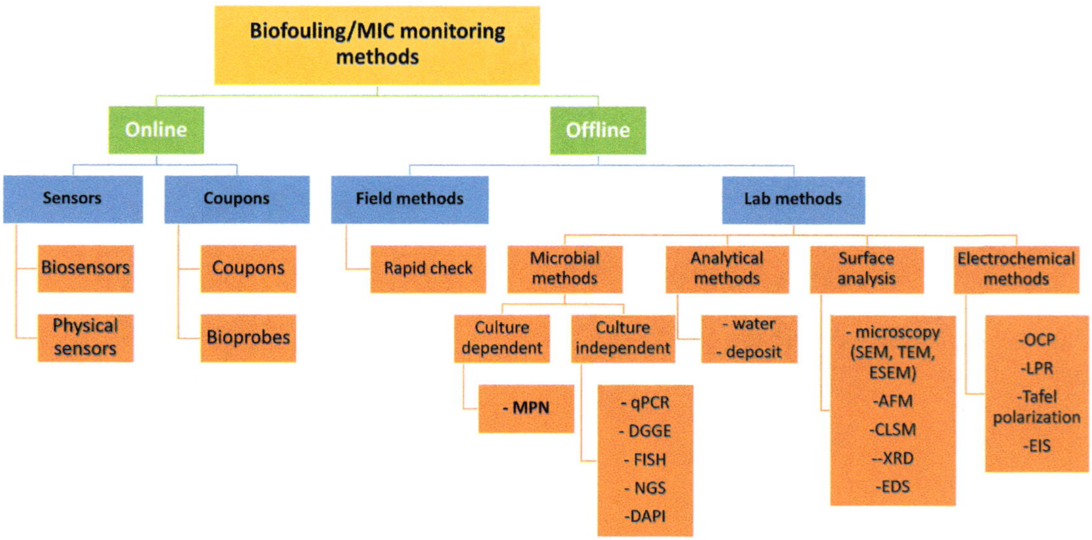

FIG. 13.19

Summary of the methods used in detecting and studying MIC and biofouling.

signal is modified by the biofilm and then the modified signal that contains the characteristics of the biofilm is detected and analyzed. Some examples of these signals include:

- Intensity of light (differential turbidity)
- Intensity of sound (ultrasonic frequency—domain reflectometry)
- Color/wavelengths (bioluminescence, fluorometry, spectroscopy)
- Mechanical resonance frequencies (QCM)
- Electrical capacitance (dielectric sensor)
- Electrical conductivity (electrochemical electrodes)
- Light refraction indices (surface plasmon resonance)
- Friction (pressure drop)
- Thermal resistance (heat transfer coefficient)
- Optical input signals that are being modified into acoustic output signals (photoacoustic spectroscopy).

Some of these signals are actively emitted by the biofilm, e.g., bioluminescence, but the majority of them are passive response to input signals that were generated by the monitoring equipment [139]. Probes should be chosen based on their accuracy and compatibility with fluids and resistance to the harsh environment in oil and gas systems.

Several trials for developing online sensors for monitoring SRB activities have been reported. Sooknah et al. [140] reported a sulfide oxidase enzyme electrode for monitoring of biogenic sulfide resulting from the activities of sulfate-reducing bacteria (SRB). Haile et al. [141] reported a sulfide oxidase (SO) based biosensor for online monitoring of sulfate-reducing bacteria (SRB) activity. Pang et al. [142] constructed an amperometric sensor for on-line detection of sulfide. Haile [143] developed a four-probe sensor for simultaneous monitoring of SRB activity and corrosion, which is based on biosensor technology in which sulfide oxidase was used as the indicating element.

13.2.15.1.2 Coupons and bioprobes

Corrosion and scale coupons can be used for microbiological analysis. Biofilm formation and MIC can be monitored using metal coupons especially designed for biofilms and installed either in-line in the pipeline or in a side stream. After installation, the coupons are then removed periodically (after a certain time) and tested for weight loss, microbiological analysis, or microscopic examination [138]. A wide range of tools has been developed, all of which present multiple small studs to the fluid under test, and they can be classified or fall into three categories: direct probes, side-stream probes, and laboratory rigs.

A bioprobe is a coupon specially designed to study biofilm growth within the pipe system. Coupons, commonly known as "studs," with a surface area of 1 cm^2 fit into the bioprobe holder; these are inserted into the bioprobe monitor so that the coupon surface is flush with the pipe wall. The coupon holders are made of Acetal (polyemer formaldehyde) to avoid any galvanic corrosion through dissimilar metal contact [144]. Figs. 13.20 and 13.21 show a bioprobe and biofilm mesh coupon.

Biofilm, or deposits accumulated on these studs, are then collected and analyzed for bacterial identification and counting.

Other devices are designed to monitor the biofilm development in water systems, including the Pedersen device and the Robbins device, lately known as the modified Robbins device; these devices have been used extensively in studying biofilm and its management strategies in oil and gas fields.

FIG. 13.20

Bioprobe for online collection of bacteria samples, and biofouling monitoring.

Courtesy of Metal Samples Company, reprinted with permission.

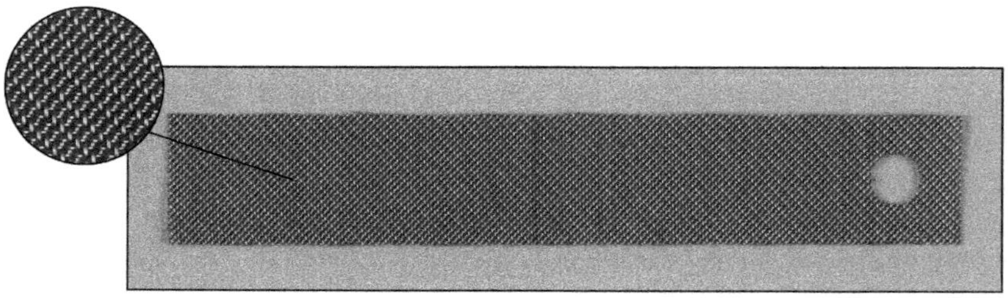

FIG. 13.21

Biofilm mesh coupon, used to monitor biofilm formation in production systems.

Courtesy of Metal Samples Company, reprinted with permission.

13.2.15.2 Offline methods

Offline methods are based on collecting samples from specific locations in the system and sending them to the lab for investigation, to determine the extent of microbial activity. The samples can be mixed production fluids, water samples, and deposit samples from pig returns or from cleaning, pipe section, coupons, or bioprobes.

The collected sample must follow the standard protocols in sampling and preserving the biological samples. ASTM D4057 provides standard practices and procedures for the collection of petroleum-related liquid samples. It is worth noticing that

- Deposit samples are more representative of the problem than water samples, since solid samples contain biofilm and sessile bacteria, which are more damaging than the planktonic bacteria in water samples.
- Sampling from the failure site is more representative of the problem.

- Sterility of sample containers and preservation of the samples is mandatory.
- The use of highly sensitive methods, e.g., molecular microbiological methods (MMMs), is recommended, especially in severe cases where the conventional culture-based methods may underestimate the problem.

13.2.15.2.1 Culture-based methods

The culture methods are based on allowing certain types of bacteria to grow in a media (artificially made) that contains the necessary nutrients and the temperature for their growth for certain periods of time, known as incubation time [137]. The most commonly used method is most probable number (MPN) and serial dilution method (or extinction dilution technique); they are also conventionally known as check bottles or bug bottles in the oil and gas industry. Fig. 13.22 illustrates the MPN or serial dilution method procedure, and Fig. 13.23 shows typical results of bacteria growth in the growth media. Another method similar to culture methods but somewhat faster is the BART method.

API RP 38 provides the recommended practice for biological analysis of surface injection waters.

NACE TM-0194 furnishes the standard test method for monitoring of bacterial growth in oilfield systems; it describes collection and analysis of water and solid samples, also mentioning the ability of using direct emulsion for bacterial isolation in low water cut samples. Xue et al. [145] recently

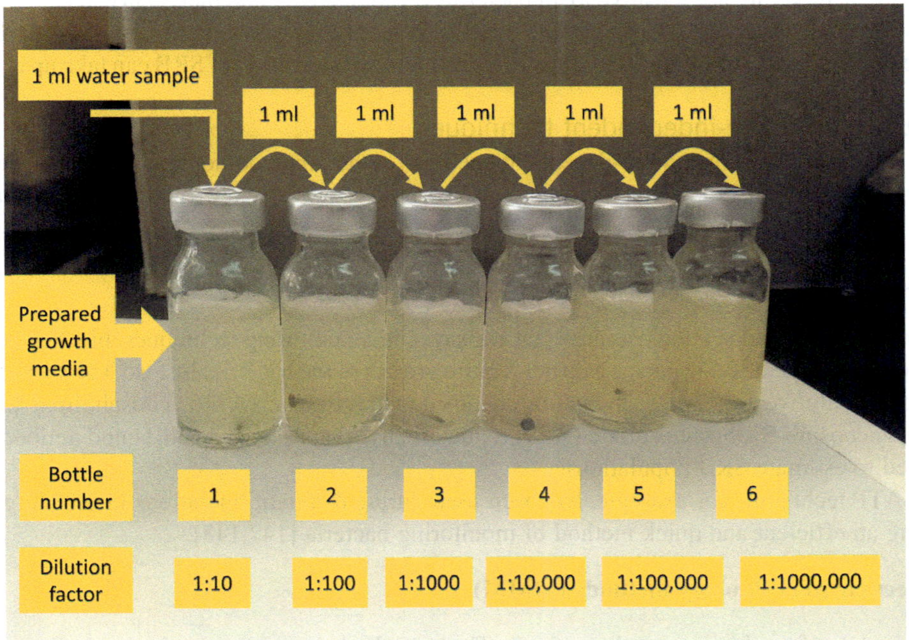

FIG. 13.22

Representation of serial dilution technique.

FIG. 13.23

Results from serial dilution technique shows positive results (a), negative results (b).

documented a step-by-step methods of field sample collection, water chemistry analyses, microbiological analyses, and laboratory incubation strategies to investigate microbial souring.

Culture methods are not sensitive to all bacteria (very selective), and their data can be irrelevant, as underestimation is expected (only accounts for 10% or less of the viable bacteria present and 1% may be more typical) [146], in addition to being time consuming (incubation of SRB can take up to 28 days).

13.2.15.2.2 Culture-independent techniques

Due to the limitations of culture-dependent methods, other methodologies have been developed. Biochemical assay and molecular microbiology methods (MMMs) are two common examples of culture-independent methods.

- **Biochemical assay:**

Biochemical assay is also known as enzymatic assay. Unlike culturing techniques, biochemical assays for detecting and quantifying bacteria do not require growth of the bacteria. Instead, biochemical assays measure constitutive properties, including adenosine triphosphate (ATP), phospholipid fatty acids (PLFA), adenosine-5′-phosphosulfate (APS) reductase, hydrogenase, and cell-bound antibodies, have been used to estimate SRB populations.

The ATP technique, for example, has been used during different operations in oil and gas fields, providing an efficient and quick method of monitoring bacteria [147,148].

- **Molecular microbiology methods (MMM)**

These are also referred to as genetic methods. These methods have completely revolutionized microbial ecology research and the ways microbes are enumerated in technical systems. A short summary of these methods is provided in NACE standard TM0212. These methods are based on the direct extraction of nucleic acids (DNA, RNA) from natural samples, e.g., water, soil, and sediments (both organic and inorganic materials) [149]. These methods have the capacity to [137]:

- Highlight the dominant bacteria in a natural sample (an ecological system) that contains other species as well, with no worries regarding the serious limits of common viable counting methods
- Determine the relative proportion of MIC-related bacteria within the whole bacterial community,
- Identity the bacteria that are resistant to biocides,
- Evaluate the population size changes imposed using biocides or nutrient modifications,
- Be counted as a more reliable method for sampling biological samples, which is not affected by time or transportation factors,
- MMMs accept very small amounts of any type of sample (liquid, biofilm, solid, and dry) with or without live bacteria.

Techniques like PCR, fluorescent in situ hybridization (FISH), DNA microarray, DAPI, denaturing gradient gel electrophoresis (DGGE), and stable isotope probing (SIP) have been reported to add more value to the microbiological analysis in different oil and gas operations. Fig. 13.24 illustrates the emergence of MMM methods [150].

13.2.15.2.3 Other analytical methods

Other analytical methods are used to diagnose biofouling deposits and troubled source water. Methods including water analysis, deposits analytical methods, and surface and microscopic methods are usually used to investigate biofouling and MIC-related problems.

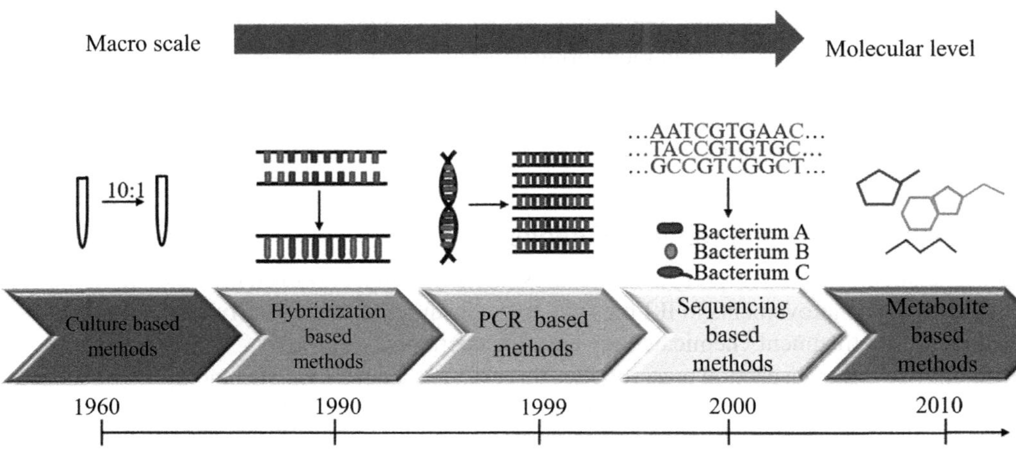

FIG. 13.24

The emergence of the use of molecular techniques in MIC and biofouling investigations.

Reprinted from Kotu S.P., Mannan M.S., Jayaraman A., Emerging molecular techniques for studying microbial community composition and function in microbiologically influenced corrosion. Int. Biodeterior. Biodegrad. 2019, copyright (2019) with permission from Elsevier.

A new sensor based on the concept of electron paramagnetic resonance (EPR) was developed for asphaltenes real-time monitoring. The sensor was applied in field cases and showed promising results [151].

13.3 Recent updates and future advances

Deposits monitoring is a key step in flow assurance strategies. The currently used monitoring methods have been improved over decades of research to cover a wide spectrum of applications in oil and gas fields. Some of the gaps in the current methods include accurately identifying all the types of the deposits, applicability in reservoir and downhole conditions, and working accurately under the harsh conditions encountered in oil and gas systems. Hence, some advances in the monitoring systems have emerged and will continue in the future.

13.3.1 Improving current technology

Some methods, although frequently used in other applications like corrosion or flow characterization, are still limited for use in deposits identification. Examples are the NDT methods of ultrasonics, radiology, and radioactivity (gamma ray transmission and neutron methods), which, despite their extensive and successful use in tubular thickness and corrosion monitoring and also despite the several trials to employ them in deposits monitoring as reported earlier in this chapter, are not practicaly applied as much in the realm of deposits monitoring.

13.3.2 Application of fiber optics

Fiber optics have been used in many applications for reservoir, tubing, and topsides monitoring. However, the application of fiber optics in monitoring deposits is still limited. Fiber optics sensors exhibit some advantages over the other sensors, like small sizes, electrical passivity, electromagnetic immunity, compactness and lightweight, wide dynamic range, greater sensitivity, and remote operation over several kilometer lengths without any lead [152].

13.3.3 Improving ILI and robotic methods

ILI such as smart pigs are extensively used in corrosion monitoring. However, their use is still limited in solids monitoring. Improvements in ILI pigs will make locating, sizing, identifying the deposits types, and delivering the treatment chemicals very easy for operators.

In addition to inline inspection pigs, robotics inspection is a developing technology. The robot usually features a multisensor platform with a variety of condition assessment tools inside the pipeline in a single deployment. Remotely operated vehicles (ROVs) (Fig. 13.25) are robotic tools that have been used in underwater inspections. ILI robots are another emerging technology, usually known as robotic crawler ILI. They exhibit multiple advantages including operating in hard access pipelines like unpiggable pipelines, deepwater pipelines, and buried pipelines, and they also are very accurate [153].

Recently, the Eindhoven University of Technology has developed, in collaboration with universities and the business community, robotic marbles powered by artificial intelligence to inspect

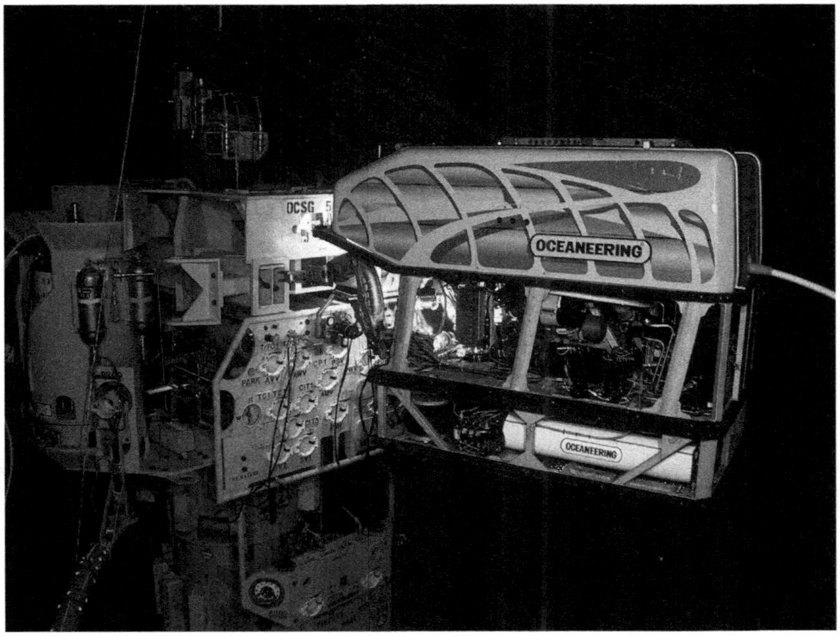

FIG. 13.25

Remotely operated vehicle.

Source: https://commons.wikimedia.org/wiki/File:ROV_working_on_a_subsea_structure.jpg.

production systems. They are very small (the size of a golf ball), allowing them to inspect difficult to inspect or even inaccessible environments. They are inserted in the pipe and float freely to perform measurements and inspections with the sensors and cameras fitted in. They are capable of detecting leaks and obstructions, and mapping pipeline networks.

13.3.4 Using wireless sensors

One common disadvantage of using regular electronic sensors for continuous monitoring in pipelines in remote locations is their need for power and communications. Recently, the use of a low-power wireless sensor network to overcome these challenges was reported. Such a network allows sensors to be powered by locally installed energy-harvesting elements (i.e., thermoelectric, solar). The result is a robust and easily deployable network of sensors that spans large distances [154] (Fig. 13.26).

13.3.5 Smart installations

Smart installations are those manufactured and embedded with sensors and electronics that enable digital tracking of them over their life cycle. Smart pipelines, completions, and production equipment are examples of smart installations that have been reported in oil and gas fields.

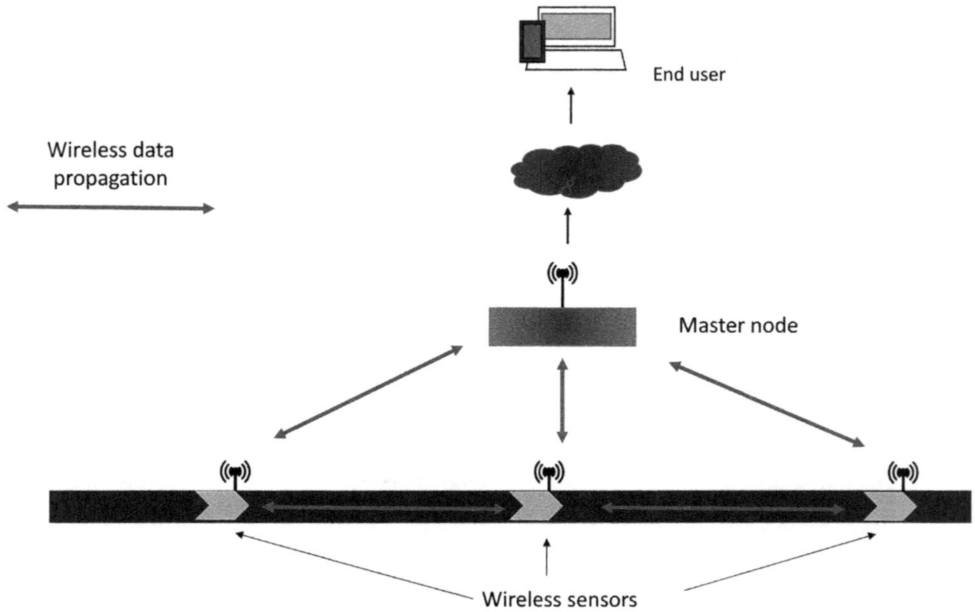

FIG. 13.26

Typical pipeline wireless network.

The technology allows easy and rapid automatic detection, monitoring, controlling, and management during process. The technology makes use of Big Data, cloud computing, and IoT (Internet of Things) to improve the system of data acquisition and storage capacity [155].

Manufacturers of smart pipelines include Tri-D Dynamics Inc., Japan-based global steel distributor Marubeni-Itochu Steel Inc. (MISI), The Smart Pipe, and others.

13.3.6 Application of artificial intelligence

The application of AI in the oil and gas industry is rapidly developing, as the concept of AI gradually penetrates various stages of the industry, with intelligent drilling, intelligent production, intelligent pipeline, intelligent refinery, etc., and it is a future research direction. Artificial intelligence techniques like the artificial neural network (ANN), fuzzy logic, support vector machine (SVM), hybrid intelligent system (HIS), genetic algorithm (GA), particle swarm optimization (PSO), and others have been reported in the literature [156].

Massive amounts of data have been and are still being collected from pipelines in the different operations. These data are integrated in machine learning and artificial intelligence in the monitoring systems or into the smart installations mentioned in the previous section, to allow monitoring, prediction, and optimization of the performance of the pipelines used [157].

Artificial intelligence and machine learning have been applied in oil and gas monitoring systems, including [156,158].

13.4 Mineral scales monitoring

13.4.1 Mineral scales monitoring strategies

Monitoring scale formation in reservoirs is different from that in downhole, topsides, or process facilities. It is recommended that the operator set up a specialized monitoring strategy based on system components, accessibility, and availability of monitoring techniques. A generalized strategy for mineral scales monitoring that can be easily applied in a production system, regardless of its complexity, will involve the following steps:

- Continuous monitoring of water chemistry. Sampling and analysis of produced water at different points in the system, and watching for changes in cations and anions content. A drop in the cations or anions content is indicative of solids precipitation.
- Use scale prediction calculations, especially after changes or maintenance in the system.
- If water sampling is not available, use analogous field data (similar formation with expected water chemistry) as a guide, especially if the fields had a history of deposits formation.
- Continuous monitoring of the water TSS. This involves monitoring TSS content trends, analyzing the solids collected on the filter (preferably using SEM or ESEM).
- Similar to brine water chemistry, analysis of petrographic cuttings can be used to detect mineral scales formation.
- Routine monitoring of the system parameters, e.g., pressure, temperature, flow rates, etc. In addition, monitoring of other process operations, injectivity, and phase separation is indicative of deposits problems.
- Production data such as gamma ray logs can be used to monitor the formation of scale in the reservoir and downhole.
- Monitor the performance of production equipment, e.g., pumps, ESP, separators, metering systems, etc.
- Monitor using scale and corrosion coupons, sidestreams, or spools.
- Routine pigging is an effective mitigation and monitoring method, to estimate the accumulation rates of deposit, sample, and characterize the deposits.
- Monitor the well gauging, wireline, and slickline operations and other intervention methods. If deposits are encountered, obtain a sample, identify the scale type, and set up a mitigation plan.
- If chemical inhibitor is used, monitor its concentration to make sure it is not below MIC.
- Correlate with the outcomes from corrosion monitoring data. Usually corrosion problems are accompanied by deposits problems, and vice versa (corrosion under deposits). Checking the corroded spots and analyzing the process fluids is usually helpful in assessing the scaling problem. MIC usually requires extra biological work for microbe type identification.
- Use one or more of the monitoring sensors, e.g., electrochemical, ultrasonic, radioactive, etc., based on its availability and accessibility to the scaling location. IR guns are commonly used for surface pipelines, borescopes can perform quick inline visual inspection, DTS fiber optics can be installed in downhole systems, microbalance and TSMR are easily used in pipelines, and pressure pulse and CT tomography are used in subsea pipelines.

Fig. 13.27 summarizes mineral scales monitoring methods.

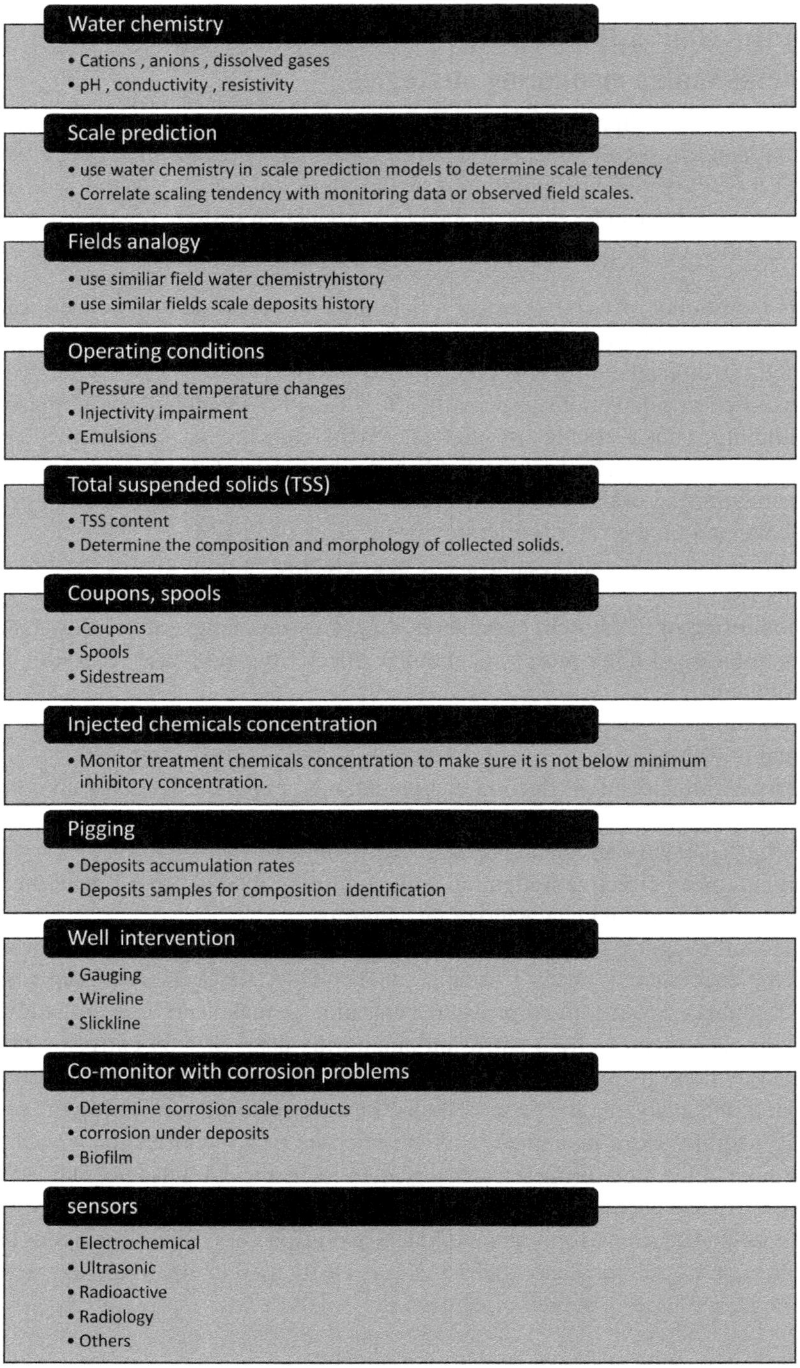

FIG. 13.27

Summary of mineral scale monitoring methods.

13.4.2 Case studies

Case study 1: Scale stabilized emulsions problems in North Sea.

An oil well in the North Sea field with 50% BS&W produced a very tight emulsion, causing severe separation problems in the production platform. Emulsion breakers were evaluated and found to require very high treatment rates for this single well. A root cause analysis was done by investigating the nature of the emulsifying agents using TSS and ESEM methods. The collected solids were identified to be $BaSO_4$ scale and particles were actively growing and stabilizing the emulsion, resulting in reduced separation efficiency. Based on the suspended solids information, the well was scale squeezed with a $BaSO_4$ inhibitor. Following the scale squeeze, this production well has not been the source of separation problems for the process. Regular ESEM samples are taken to ensure that the produced fluids do not contain evidence of scale solids [40]. The same reference, Jordan et al. [40,41], contains multiple case studies where TSS and ESEM were efficiently used to monitor scale problems and to assess the applied chemicals.

Case study 2: Calcite scaling in South Burgan field, Kuwait [55].

This case study involves a combination of corrosion and scale-monitoring methods that are applied methodically to monitor the scale problems and to evaluate the efficiency of the applied treatments. The monitoring methods involve the following:

- operational parameters,
- water chemistry,
- scale prediction,
- corrosion and scale coupons,
- UT inspection,
- Radiography, and
- visual inspection methods.

Calcite scale has formed in several wells and flowlines alongside a reported increase in water cut and decrease in oil production. Several monitoring methods were applied to evaluate the problem, including the methods already mentioned. It was found that field observations such as pressure drop, choke inspection, corrosion, and scale coupons were more realistic than the underestimated results by scale prediction using Langelier Saturation Index (LSI) and Ryznar Stability Index (RSI). A mitigation method was designed that included inhibited acid to dissolve the scale, and scale inhibitor injection into the wellhead upstream of the choke to prevent scale formation. During the treatments, a rigorous monitoring strategy was applied to monitor scale buildup/dissolution and corrosion rates. Online internal corrosion monitoring coupons were installed on the flowlines to monitor the internal corrosion during acid cleaning and routine operation. Similarly, scale coupons have been installed to evaluate the scale inhibitor injection efficiency. Fluid sampling was carried out during acid cleaning to check dissolved iron (Fe) content, as a corrosion monitoring indicator, and scaling tendency. Radiography was used to locate and quantify the scale depositions. Ultrasonic thickness scanning was carried out to ensure the integrity of flowlines during acid cleaning activities.

13.5 Gas hydrates monitoring

13.5.1 Gas hydrates monitoring strategies

The general measures to monitor gas hydrates include:

- Gas composition analysis. GC analysis of the produced fluids will indicate the early signs of gas hydrates formation. A decline in the composition of one of the gas components is a significant sign of hydrates formation.

- Gas hydrates prediction to estimate the risk of gas hydrates. This is a common practice that most operators follow.
- Water content monitoring. A decline in water content is another sign of hydrates formation.
- Pressure monitoring. Pressure drop is an early sign of a buildup of a gas hydrate layer. This is a common observation in gas/condensate lines that operate in the annular flow regime at a relatively low water cut [4].
- Monitor gas rates. Gas hydrates cause pressure drop, which is accompanied by a decreasing gas production rate.
- Monitor temperature. Temperature drop is another sign of hydrates formation. Temperature drop below hydrate formation temperature or below dew point is a preindication of potential hydrates problem.
- Usually pipelines operating in fields with hydrates risk are equipped with DTS temperature monitoring systems to monitor the outer temperature of the pipeline.
- Pigging operations. Pigging returns should be examined carefully for evidence of hydrate particles. Hydrate masses are stable even at atmospheric pressure (metastable equilibrium) in pig receiver discharges.
- The use of sensors. An IR gun can be used for surface pipelines. Pressure pulse and tomography scans can be used for subsea pipelines. DTS sensors can be affixed on well tubing or pipelines for temperature measurements.

Fig. 13.28 summarizes gas hydrates monitoring methods.

13.5.2 Case studies

Case study 1: Hydrate problems in offshore field in Qatar [159].

The field with a hydrate problem (A-south) is an offshore field comprising a single remotely operated minimum-facility wellhead platform (A-S WHP) with two wells (A-1 and A-2), which were completed in two different reservoirs and were producing into a central production (PDP) via a 6″ × 14 km subsea flowline.

Hydrate formation incidents had been encountered seasonally in the winters since 2011. The problem was also observed when producing from both wells A1 + A2, whereas producing from A2 alone didn't show any problems. Hydrate control chemicals were applied at the central production; however, no hydrate inhibition method was applied in the A-South production system.

An initial way to mitigate the problem is to shut in the well A1 during the winter or produce with the lowest rates, and produce from both wells in the summer when the temperature is high enough to avoid hydrates formation.

In addition, a flow assurance plan was developed to monitor and mitigate the A1 hydrates formation problem. The method is based on controlled flow and mixing of the two well streams.

The monitoring plan was established adopting the following steps:

- steady-state simulation software package to predict hydrate conditions at varying well flow ratios
- modeling of the system under inhibitor injection
- monitoring system temperature
- monitoring system pressure
- monitoring fluids arrivals.

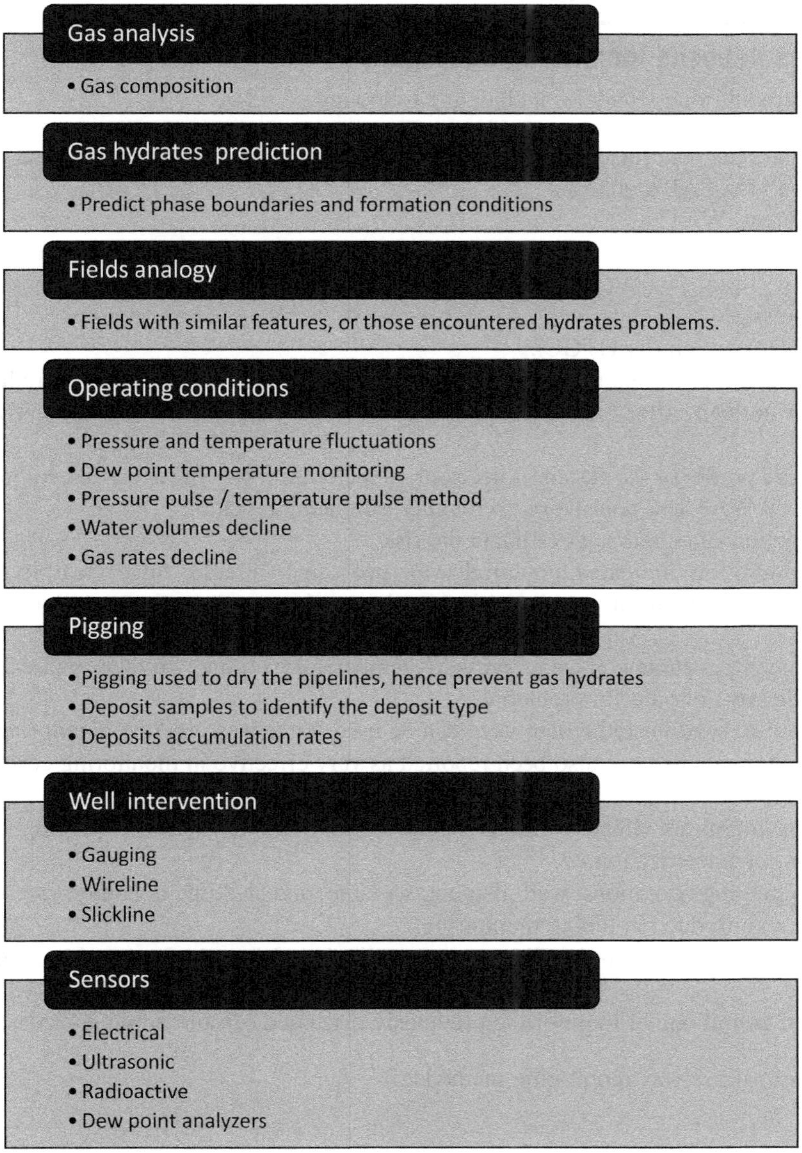

FIG. 13.28

Summary of gas hydrates monitoring methods.

13.6 Wax deposits monitoring

13.6.1 Wax deposits monitoring strategies

Wax deposits monitoring strategies include the following:

- Create a rigorous plan for crude oil sampling and characterization. Common tests include:
  - PVT analysis, SARA analysis
  - wax content, WAT,
  - pour point,
  - viscosity,
  - yield stress
  - carbon distribution (by HTGC).

  Frequent crude oil analysis can indicate wax deposition if a significant reduction in wax content is noticed, after excluding other factors. Viscosity increase is indicative of crude gelling.
- Temperature profile of the system is necessary to determine the cold spots and compare them with the crude oil WAT and pour point, to roughly estimate the risk.
- Wax prediction calculations to estimate the risk.
- If wax prediction indicates potential wax problems, a cold finger test is used to simulate wax deposition problems in the lab and to screen the possible treating chemicals.
- Production system parameters, e.g., pressure, temperature, production rates, phase separation, etc. can be indicative of paraffin problems.
- Pressure pulse, heat pulse/thermal wave can be used for subsea pipeline monitoring. Tomography methods have also been reported as very effective in monitoring corrosion and wax deposits.
- Pigging operations are effective mitigation and monitoring methods. After pigging, collect deposit samples for characterization.
- Similar to pigging operations, well gauging, wireline, and slickline operations are vital for downhole and production tubing monitoring.
- Use side streams and spools. These are usually helpful tools for topsides deposition problems.
- Use sensors. Install one or more of the previously discussed sensors on the topsides.

Fig. 13.29 summarizes wax monitoring methods.

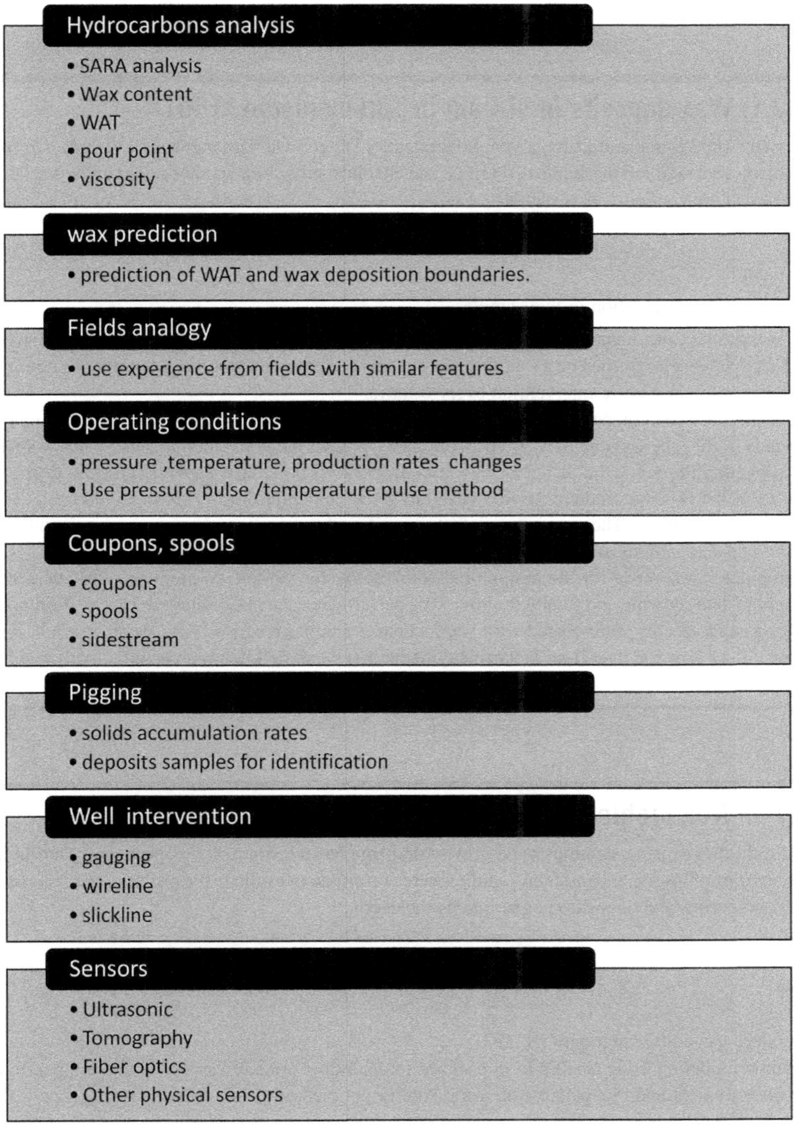

FIG. 13.29

Summary of wax monitoring methods.

13.6.2 Case studies

Case study 1: Wax deposits in oilfield in Turkmenistan [160].

Two offshore fields (Dzheitune and Dzhygalybeg) operated by Dragon Oil Turkmenistan Ltd. suffered from severe wax deposition problems that required mechanical, thermal, and chemical mitigation methods. Part of the surveillance and monitoring plan included conventional tests such as:

- WAT
- Pour point
- Cold finger tests
- Effect of shear on wax deposition in live samples.

Besides these methods, mechanical methods of surveillance were carried out using slickline operations (SLK) and wax cutting jobs (WC) that were performed on a routine basis due to the severe wax problems. Some wells needed to be attended daily, whereas others might require intervention every 60 days.

The data collected from these mechanical (SLK, WC) jobs were organized in spreadsheets, including the frequency of the jobs, efficiency of the job, wax depth, wax type (soft, hard), if other methods (thermal, chemical) were applied. Putting all these data altogether improved the SLK jobs by going deeper than previously reported (1000 m deep). A station point was added for all of the flowing gradient surveys (FGs) to get a temperature value closer to the wellhead, for better understanding of the wax problem. The outcome was that a significant temperature drop occurs between the seabed and the tree (5C), which explains why most of the wells showed wax formation between 0 and 50 m.

Similar compliance was made for the thermal methods (hot water, hot oil) to monitor its efficiency and frequency, which improved the data showing and finally improved the performance, and also improved the performance of the desanders in separating sand, as they were coated with wax, which is another bonus of wax management.

This method shows how the data from mechanical methods such as slicklines can be effectively used to monitor and manage wax deposition.

Case study 2: Wax problems in SAGD field [161].

A redeveloped field that is using steam-assisted gravity drainage has a history of wax problems. During the field development stages, a series of testing and modeling studies were performed to evaluate the problem and its possible mitigation. Wax deposition properties and deposition experiments included:

- Cloud point
- Pour point
- Wax content
- Yield stress
- High temperature gas chromatography (HTGC)
- Thermodynamic modeling using Shell's EOS package to determine critical wax deposition temperature.
- Deposition measurement and rate calculation using cold finger method.

The experiments showed high wax content, relatively high cloud point, and high yield stress, suggesting wax gelling and deposition, troubled startups, and unplanned shutdowns.

The downhole precipitation was observed during the startup phase and resulted in IPR for a number of wells. Low reservoir temperatures (close to the cloud point) and Joule-Thomson effects during drawdown were the main causes for wax deposition. Downhole treatments were deployed and good results were achieved with extended soak times (>3 days) with optimum dissolver. With the failed solvent treatments and unresponsive SAGD effects, it was speculated that EOR activity resulted in wax precipitation in the formation itself.

After treatment, monitoring wax content and wax properties, in addition to production data, were used to manage the problem in the field.

13.7 Asphaltenes monitoring

13.7.1 Asphaltenes monitoring strategies

Asphaltenes monitoring strategies include the following:

- Fluid sampling and analysis. Several tests are performed on crude oil, including:
  - PVT analysis
  - SARA analysis
  - Asphaltenes content
  - Asphaltenes onset precipitation, asphaltenes onset pressure (AOP)
  - Viscosity measurements
  - Density measurements
- The data from the crude analysis is used in asphaltenes prediction modules to estimate the risk of asphaltenes and determine the asphaltenes deposition envelope (ADE). Other simple methods of estimating the risk include colloidal stability index (CII index) and de Boer diagrams
- Static precipitation tests using heptane can also be performed, and the Wiehe method can be used to estimate the risk of mixing different crudes.
- System pressure profile should be known and compared with the measured or calculated bubble point and asphaltenes onset pressure.
- Monitoring system parameter is very crucial in asphaltenes case. Pressure drop is the most significant parameter. On the one hand, pressure drop is an indicator of an asphaltene deposition problem. Even if asphaltenes did not deposit, pressure drop is an alert that asphaltene deposition is possible if the pressure fell to below AOP, or near the bubble point. Similarly, temperature monitoring can be indicative of a deposition problem.
- Frequent monitoring and measurement of asphaltenes content can be meaningful, since a drop in asphaltene content can be indicative of a deposition problem. However, the test is time consuming and may not be applicable for frequent measurements.
- TSS analysis was reported to be applicable for organic deposits by using 8 μm filter paper with crude oil to collect organic and inorganic deposits [40,41].
- For subsea pipelines, tomography or pressure pulse methods can be used.
- Pigging operations: routine and rigorous pigging operations are used to estimate the deposition rates and collected deposit samples are characterized.
- Locations where different crudes are mixing need attention, as incompatible crude mixing is one of the main causes of asphaltene deposition.
- Locations of high corrosion rates, high streaming potential, and high shear are the most vulnerable to asphaltene precipitation, so they will need frequent inspection.
- Use sensors for inline or online monitoring of deposition.

Fig 13.30 summarizes asphaltene monitoring methods.

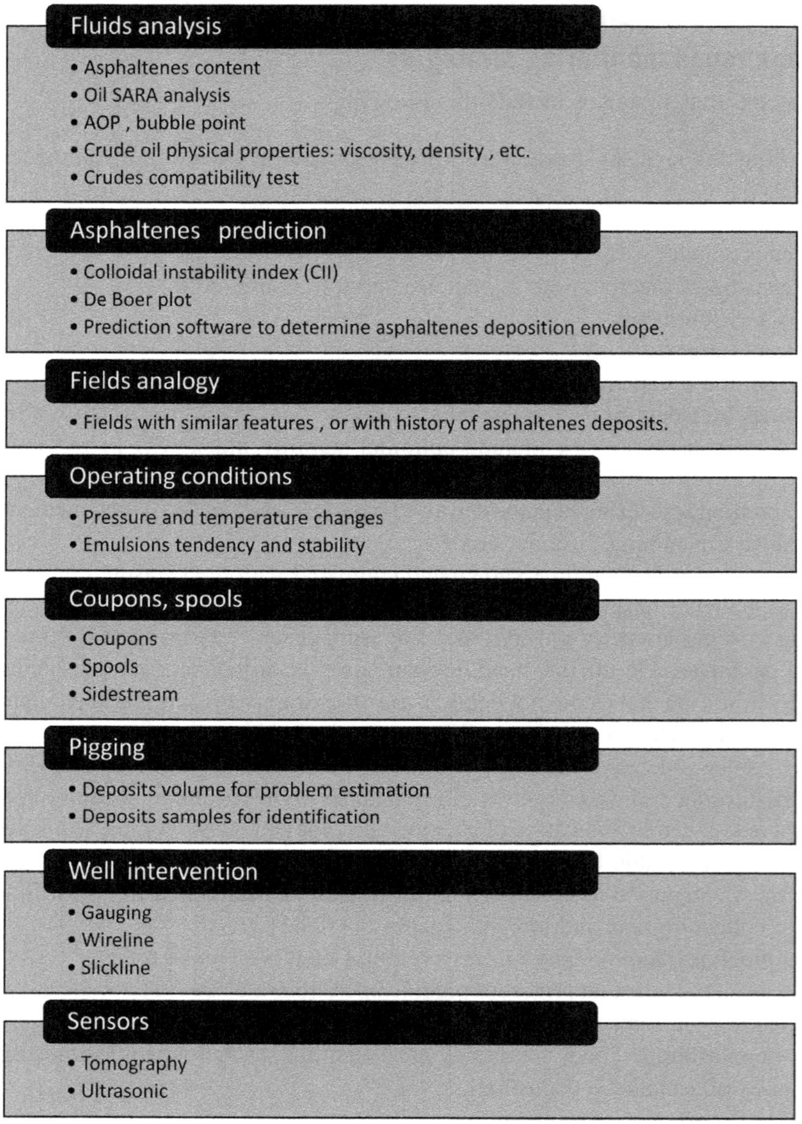

FIG. 13.30

Summary of asphaltene monitoring methods.

13.7.2 Case studies

Case study 1: Asphaltene issues in the production system of a Gulf of Mexico Deepwater subsea development [50].

Asphaltene problems were identified in the Gulf of Mexico Deepwater field, causing production loss and high-cost maintenance.

The investigation and surveillance strategy included the following:

- Reservoir fluids sampling and analysis, which involves:
  - PVT studies
  - SARA analysis
  - Viscosity
  - AOP
  - de Boer asphaltenes stability plot
  - Particle size distribution
  - WAT on dead oil and live WAT at relatively high pressure.
- The data from analysis was then used in modeling and predicting analysis, and then used in modeling and predicting the stability of asphaltenes using de Boer plot and asphaltenes stability/deposition envelope.
- Observation of the effect of chemicals on solids buildup during field trials.
- Surveillance of damage near wellbore using pressure transient analysis (PTA) and multirate test (MRT).
- Real-time monitoring of normalized nodal analysis based on frictional pressure losses.

Case study 2: Monitoring asphaltenes in West Kuwait field [162].

Asphaltene deposition in oilwell tubing has been a serious problem in some of the West Kuwait Marrat Jurassic wells. They developed a method to determine the onset of asphaltene deposition using the flowing wellhead pressure (FWHP) data based on a programmable data logger. The monitoring method is applied on a well-by-well basis and analyzed with the following well information:

- Fluid PVT Analysis.
- Well Gradient Sensitivity Analysis.
- Well Test Analysis.
- GOR and Oil Rate Production.

To achieve accurate FWHP data, a battery-powered digital pressure gauge system was installed on the wellhead before the choke. The gauge is connected via a cable with a sensitive transducer. The system contains an output 0–1 V dc signal from the gauge head to the data logger device. The data logger can record conditions or parameters such as temperature, pressure, fluid level, voltages, etc. and can be programmed to record data ranging from once a day to once per second. The data are downloaded into a computer and analyzed by the software.

The FWHP monitoring method is based on the following well preparation steps:

- Clean the well to remove asphaltenes.
- Run caliper log to measure production tubing after cleaning job.
- Perform production/GOR tests, PLT, and pressure building survey.
- Set the data logger, as desired, to record the FWHP.
- Calculate and analyze the FWHP losses.
- Run caliper log to measure asphaltene buildup inside well tubing.
- Compare the theoretical results with the caliper log.

By plotting these data, the asphaltene onset pressure will appear as a sudden drop in the FWHP. The data were also optimized and used to calculate the asphaltenes thickness.

13.8 Naphthenate deposits monitoring

13.8.1 Naphthenate deposits monitoring strategies

General naphthenate soaps monitoring strategies include:

- Rigorous water sampling and analysis to determine the water chemistry, cations, anions, alkalinity, and pH are of great importance. Waters with high Na and Ca content and pH > 6.2 will have higher potential of developing naphthenates soaps.
- Use TSS technique to monitor the rate of formation of solids and collect the solids to identify them.
- Rigorous crude oil sampling and analysis. Crude oil analysis is the most crucial step in naphthenate monitoring and detection. Runham and Smith [163] summarized the possible tests that can be used to estimate the risk of naphthenate deposits and soaps:
  - SARA analysis
  - TAN, ASTM D664
  - FT-IR analysis of oils and filtered solids to determine naphthenic and carboxylic acid presence
  - Mass spectrometric methods are to identify the presence of ARN acids in oil samples
  - Deposition tests such as autoclave tests, flow loop, or naphthenates rig tests.
- Monitor the operational parameters. Pressure, temperature, level controls, and flow rate drops are signs of a deposition problem. Pressure drop is indicative of deposit formation, and if no deposits were formed yet, then pressure drop is a warning that deposits/soaps are about to form.
- Thermographic methods have been used successfully to determine evidence of naphthenate fouling in vessels. An infrared photograph indicates cool spots within a vessel or pipe where solids have accumulated, which may be naphthenate. Level control malfunctioning is a also good indication of naphthenate/carboxylate soaps formation.
- Locations associated with pressure drop are more vulnerable to naphthenates soap, so special attention should be given to these locations.
- Monitor the performance of separation equipment. Changes in the performance of separators, electrostatic coalescers, desalters, heat exchangers, filters, and hydrocyclones are an indication of soap emulsions.
- Nucleonic density profilers provide active monitoring of internal fluid levels, and they are also capable of detecting foam, oil, emulsion, water, and solids levels through their density differences. They can be used to detect soap solids or emulsions [163].
- Monitor the performance of the conventional demulsifiers. If demulsifier efficiency decreased with no change in crude oil composition, then there is a possible naphthenate emulsion in the system.
- Subsea and deepwater pipelines can be monitored by tomography or pressure pulse methods.

Fig. 13.31 summarizes naphthenates monitoring methods.

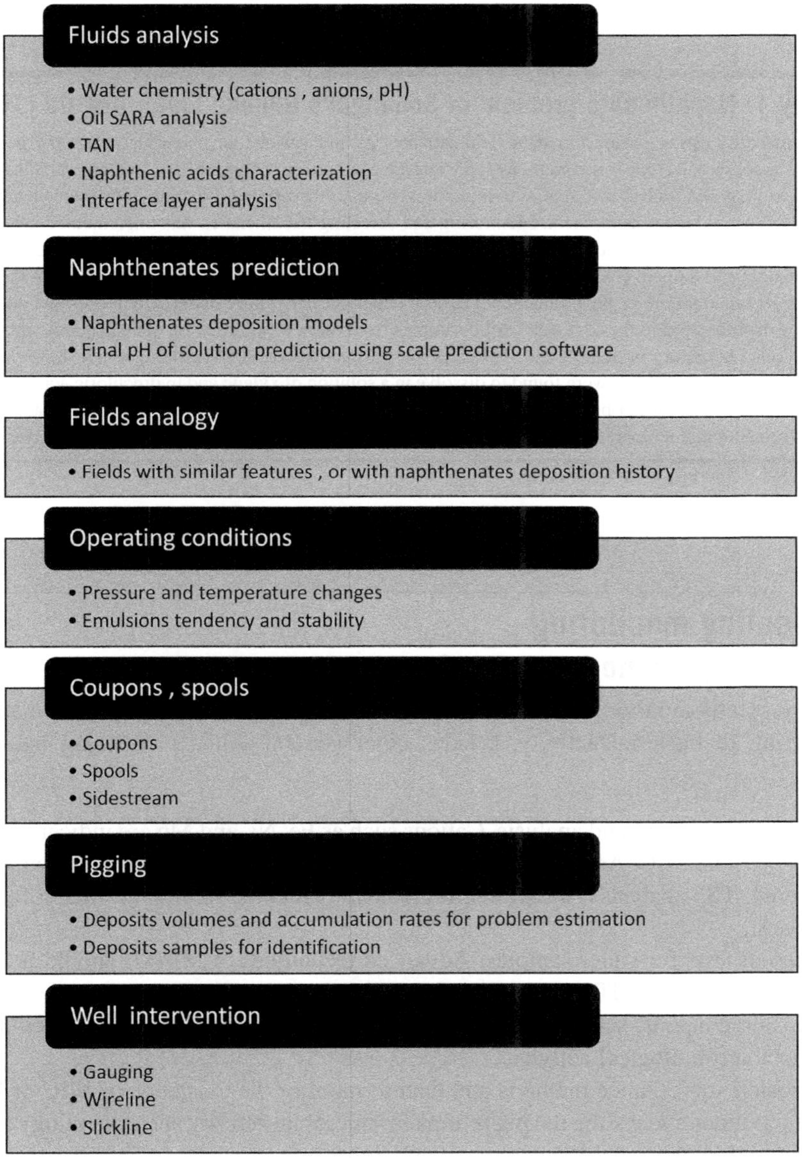

FIG. 13.31

Summary of naphthenates monitoring methods.

13.8.2 Case studies

Case study 1: Naphthenate problem in Sonangol's Gimboa Field, Angola [164].

Calcium naphthenate soaps began to form in the field with increase in the water cut, about 5 months after first oil. The soaps were observed as solids and emulsion formations at the interface, in the separators, bulk oil treaters, and the hydrocyclones, which started to plug and foul in different system equipment, e.g., the bridles, flotation unit, desalter, pumps, heat exchangers, and other production equipment. Monitoring and detecting naphthenates was implemented using an inline filtration technique, using a uniquely designed Millipore filter apparatus instituted by the Nalco engineers in collaboration with their Sonangol and Saipem counterparts. Filtration was performed at the most vulnerable points in the process train, including the inlet separator outlet, the production separator outlet, the dehydrator outlet, and the desalter outlet to establish the presence of naphthenates in these vessels. Solid samples from these locations were collected and sent to Nalco's Diagnostic Solutions Laboratory in Sugar Land, Texas for analysis and solids identification, which showed the main composition's naphthenates. The solids were found to dissolve in a solution of xylene and hydrochloric acid, resulting in a top oily phase and a hazy milky water phase—the telltale signs of calcium naphthenates. Both chemical (acid + demulsifier) and physical (pressure and temperature controls) measures were put in place to mitigate solids formation.

13.9 Biofouling monitoring

13.9.1 Biofouling monitoring strategies

Biofouling is typically monitored by many means, including corrosion/biocorrosion methods and continuous detecting of bacterial activity, besides other general solids monitoring methods. General methods are:

- Water chemistry sampling and analysis. Cations such as Fe, Ni, and Mn are indicators of a corrosion process and possible MIC. An increase in S^- and CO_3^- and a decrease in SO_4^- can also be indicators of SRB activity. TSS analysis is a very helpful indicating method for monitoring biofouling minerals and biomasses.
- Microbiological tests for water samples. Advanced techniques of MMMs should be applied, in addition to conventional ATP and culture-dependent methods.
- Install corrosion coupons, scale coupons, and bioprobes in the system, inspect them regularly, and test them for microbiological activity.
- Use of corrosion surveillance methods and data to monitor the existence of MIC or biofouling.
- Pigging the pipelines and testing the pig returns for microbial activity and mineralogy is often used to monitor bacterial activity and biomass accumulations.
- Well intervention (gauging, slickline, and wireline) returns must be analyzed for mineralogy and bacterial activity.
- Using sensors, e.g., biosensors or physical sensors.

Fig. 13.32 summarizes the biofouling monitoring methods.

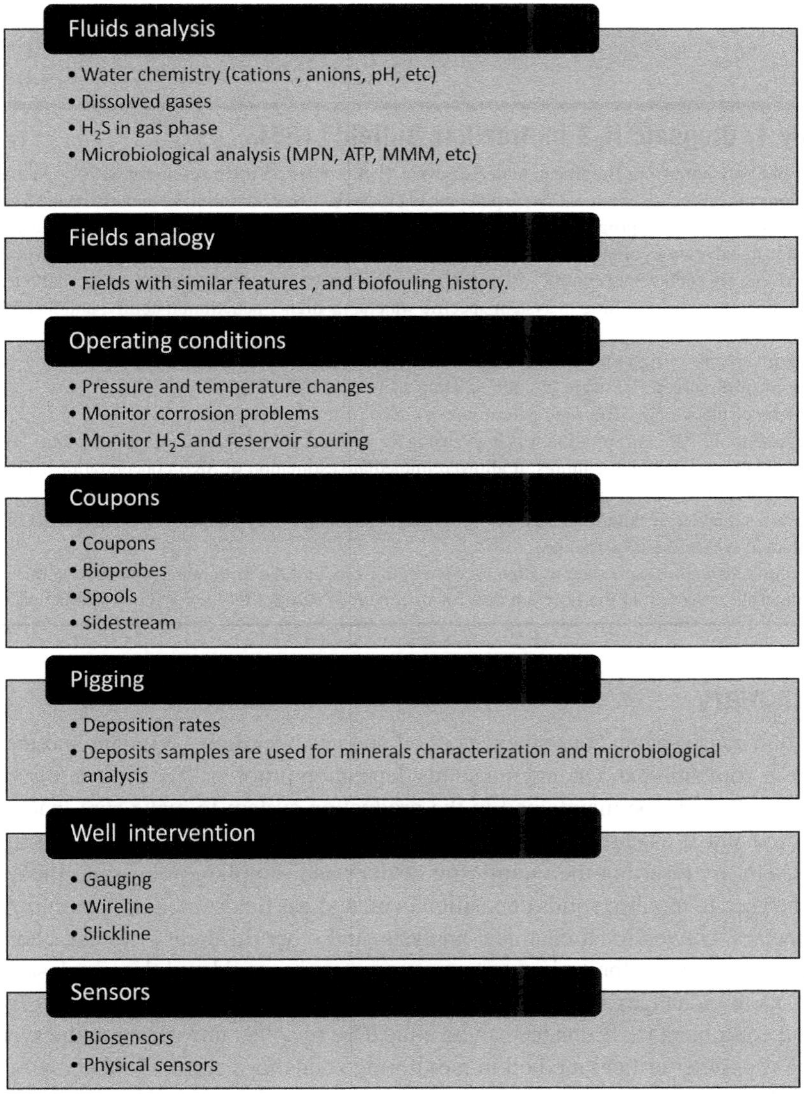

FIG. 13.32

Biofouling monitoring methods.

13.9.2 Case studies

Case study 1: Biogenic H_2S in Brazilian oilfield [165].

A Petrobras FPSO offshore from Brazil encountered severe H_2S problems due to persistent SRB populations in the produced water slop tanks. Biocide treatments were planned, to kill the bacteria and clean the tanks from the formed biofouling minerals and biomasses. A first program of glutaraldehyde was applied; however, it failed to achieve any progress due to the configuration and dynamics of the vessels being treated and partly due to the inability of glutaraldehyde to penetrate the oily layer to effectively contact and control sessile SRB. Another treatment was planned based on THPS as the biocide of choice, based on its efficacy of killing in system-specific waters, its higher penetration ability, and its ability to dissolve sulfides.

During the treatments, a rigorous monitoring plan was put in place, including:

- concentrations of dissolved H_2S were monitored using an HACH H_2S field kit.
- Concentrations of planktonic SRB were determined by serial dilution technique.
- Quick assessment of SRB concentration was made using Rapid Chek II kits. Although this method is less sensitive and less accurate than serial dilution cultures, it allowed immediate estimation of SRB concentrations. The two methods were used in parallel.
- Active THPS residuals were determined by iodometric titration using the THPS field kit to ensure that biocide residuals were maintained within the target range.

With these monitoring measures the treatments were optimized, and the mitigation costs were reduced with time. A greater than ninefold reduction in the program cost for treatment of sulfides on the FPSO was achieved.

13.10 Summary

Solids deposition causes harsh consequences in oil and gas operations. Vigilant and methodical operations are key in combating and managing solids deposition problem. To achieve this, different solids monitoring methods must be maintained in the production system to make sure that the problem location is detected and at its earliest stage. Monitoring starts in the earliest operation stages, during design of the system, by inserting the monitoring devices and sampling points into the design. Various methods can be used to monitor solids deposition in oil and gas fields, including monitoring the various process parameters, conventional chemical analysis, and other different physical, chemical, and mechanical based techniques. Monitoring process parameters is considered the simplest and most indicative method, as any changes in pressure, temperature, flow rates, fluids rheology, hydrodynamics, injectivity, and equipment performance can be indicative of solids infestation in the system. Produced fluids analysis is another critical method in monitoring solids formation. Fluid composition is usually utilized in solids prediction calculations and other system design aspects. In addition, analysis of the produced fluids and checking for composition changes over time are indicative of solids formation. Furthermore, suspended solid counts and chemical composition of these solids can work as a realistic method of monitoring all kinds of solid deposition. Mechanical methods used to clean and gauge the system tubulars, e.g., pigs, wireline, and slickline, are conventionally used to monitor the solids accumulation in production systems. The solids return from these mechanical methods can be analyzed and identified to prepare the proper mitigation methods. Methods such as coupons and spools are realistic means of monitoring solids deposition in the system. Moreover, a wide variety of physical and chemical sensors and devices are used, including electrochemical, ultrasonic, radiographic, radioactive, optical, and other sensors. The main goal of these devices and sensors is to detect the solids in very small thickness/concentration and to give indicative signals representing the solids thickness and type if possible. These devices need to be installed in the proper location in the system to assure genuine detection of the problem, with fewer interferences from the production environment. Comprehensive steps to monitor solid deposits include:

- Monitor process parameters, e.g., pressure, temperature, flow rates, other parameters.
- Analyze produced fluids and monitor changes in composition over time.
- Analyze suspended solids and monitor their counts and chemical composition. An increasing trend is indicative of developing solids problems.
- Coupons are a very handy method of monitoring solids.
- Routinely use mechanical methods, e.g., pigging, slickline, and wireline, and analyze their solids returns.
- Use physical sensors and devices in the system in the proper location.

References

[1] M. Brown, "Full Scale Attack," REview, vol. 30, The BP Technology Magazine, October-December 1998, pp. 30–32.

[2] B. Bertrand, G. Ségéral, P. Ove Moksnes, Detection and identification of scales using dual energy/Venturi subsea or topside multiphase flow meters, in: Society of Petroleum Engineers (SPE), 2001, https://doi.org/10.4043/13152-ms.

[3] I. Meric, G. Johansen, S. Melo, E. Lima, J.M. Jimenez, C. Dantas, Monitoring of scale deposition in petroleum pipelines by means of photon scattering—a preliminary study, in: 2013 International Nuclear Atlantic Conference—INAC 2013 Recife, PE, Brazil, November 24–29, 2013 Associação Brasileira De Energia Nuclear—Aben, 2013, ISBN: 978-85-99141-05-2.

[4] L. Dagge, K. Harr, M. Paul, G. Schnedl, Classification of process analysis: offline, atline, online, inline, Cem. Int. 7 (2009) 72–81.

[5] Aavos, Online, Inline, Atline and Bypass (Analysers), Aavos International, 2017. https://aavos.eu/glossary/online-inline-atline-bypass-analysers/. (Last Visited February 2021).

[6] Metrohm, Industrial Process Analyzers, Metrohm AG, Herisau, Switzerland, 2019.

[7] P. Wilks, Process monitoring: in-line, at-line, or slip-stream? Spectroscopy 21 (2) (2006).

[8] G.P.P. Gunarathne, Measurement and monitoring techniques for scale deposits in petroleum pipelines, in: Conference Record—IEEE Instrumentation and Measurement Technology Conference, vol. 2, Institute of Electrical and Electronics Engineers Inc., 1997, pp. 841–847, https://doi.org/10.1109/IMTC.1997.610200.

[9] K. Alhammadi, A. Bouchalkha, S. Sowwan, Scale detection using light sensing technique, in: 2015 IEEE Jordan Conference on Applied Electrical Engineering and Computing Technologies, AEECT 2015, Institute of Electrical and Electronics Engineers Inc., 2015, https://doi.org/10.1109/AEECT.2015.7360561.

[10] J.J.C. Hsu, M.M. Santamaria, J.P. Brubaker, Wax deposition of waxy live crudes under turbulent flow conditions, in: SPE Annual Technical Conference and Exhibition 1994, Society of Petroleum Engineers, New Orleans, Louisiana, 1994.

[11] K.A. Lawal, J.P. Crawshaw, E.S. Boek, V. Vesovic, Experiemntal investigation of asphaltene deposition in capillary flow, Energy Fuel 26 (4) (2012) 2145–2153.

[12] J. Buckley, J. Wang, Estimate Thickness of Deposit Layer from Displacement Test, 2010.

[13] X.T. Chen, T. Butler, M. Volk, J.P. Brill, Techniques for Measuring Wax Thickness during Single and Multiphase Flow, Society of Petroleum Engineers, 1997, January 1, https://doi.org/10.2118/38773-MS.

[14] T. Tjomsland, B. Sandøy, F.H. Fadnes, Application of multi-rate well tests to scale management, in: Paper Presented at the SPE International Conference on Oilfield Scale, Aberdeen, UK, May 2010, https://doi.org/10.2118/131011-MS.

[15] Z.M.W. Haindade, P.A. Mundhe, A novel approach to estimate severity of paraffin deposition inside production tubing for vertical lift performance matching in well models, in: Offshore Technology Conference Asia 2016, OTCA 2016, 2016, pp. 2939–2944. Offshore Technology Conference https://doi.org/10.4043/26766-ms.

[16] A.S. Mohamed, A. Goyal, M.D. Mohd Ismail, G. Munainni, J.K. Amar Singh, M.Z. Ismail, M. Anwar, Detection and removal of naphthenate deposit in oil well, in: Paper Presented at the Offshore Technology Conference-Asia, Kuala Lumpur, Malaysia, March, 2014, https://doi.org/10.4043/25003-MS.

[17] J.S. Gudmundsson, I. Durgut, H.K. Celius, K. Korsan, Detection and monitoring of deposits in multiphase flow pipelines using pressure pulse technology, in: 12th International Oil Field Chemistry Symposium, April 1–4, 2001, Geilo, Norway, 2001.

[18] J.S. Gudmundsson, H.K. Celius, Gas-liquid metering using pressure pulse technology, in: SPE 56584, Annual Technical Conference and Exhibiton, Houston, 3–6 October, 1999.

[19] R.C. Sarmento, G.A.S. Ribbe, L.F.A. Azevedo, Wax blockage removal by inductive heating of subseapipelines, Heat Transfer Eng. 25 (7) (2004) 2–12.

[20] M. Bonuccelli, V. Faluomi, C. Carniani, Flow maps development for wax deposition monitoring, in: International Conference on Multiphase: Extending the Boundaries of Flow Assurance, 2003, pp. 273–280.

[21] P. Rostron, Critical Review of pipeline scale measurement technologies, Indian J. Sci. Technol. 11 (17) (2018) 1–18, https://doi.org/10.17485/ijst/2018/v11i17/121459 (Indian Society for Education and Environment).

[22] F.H. Almutairi, D.R. Davies, Detection of scale deposition using distributed temperature sensing, in: SPE International Oilfield Scale Conference, Society of Petroleum Engineers, 2008, pp. 1–13.

[23] A. Ukil, in: IEEE, H. Braendle, P. Krippner (Eds.), Distributed Temperature Sensing: Review of Technology and Applications, July 2011.

[24] A.R. Adebayo, H.Y. Al-Yousef, M. Mahmoud, A.S. Sultan, The prospect of using permanent downhole sensors for monitoring oilfield scales, in: Society of Petroleum Engineers—International Petroleum Technology Conference 2014, IPTC 2014—Innovation and Collaboration: Keys to Affordable Energy, vol. 2, Society of Petroleum Engineers, 2014, pp. 1447–1455, https://doi.org/10.2523/iptc-17856-ms.

[25] M. Guzman, Use of distributed temperature sensing for wax detection and treatment optimization, in: SPE Production and Operations Symposium, Proceedings, vol. 1, Society of Petroleum Engineers (SPE), 2012, pp. 366–374, https://doi.org/10.2118/156077-ms.

[26] R. Usamentiaga, P. Venegas, J. Guerediaga, L. Vega, J. Molleda, F.G. Bulnes, Infrared thermography for temperature measurement and non-destructive testing, Sensors (Basel, Switzerland) 14 (7) (2014) 12305–12348, https://doi.org/10.3390/s140712305.

[27] A.A. Al-Ghamdi, I. Mohamad El-Zefzafy, M.H. Al-Malki, A.A. Al-Mulhim, First Flowline Scale Inspection in Ghawar Field Using a Robust Infrared Technology Offers a Reliable and Cost Effective Approach, Society of Petroleum Engineers, 2016, September 26, https://doi.org/10.2118/181356-MS.

[28] R.G. Dulce, A.D. Baltazar, F.E.B. Bayon, An evaluation on the applicability of infrared thermography in identifying silica deposition in geothermal brine lines, in: World Geothermal Congress 2015 April, 2015, p. 7.

[29] J.P. Villena, D.N.G. de Lara, A Model for Finding the Thickness of Silica Deposit, PNOC-EDC Internal Report, 2006.

[30] A. Saifi, H. Halloula, A. Elamiri, A. Elhassnaoui, A. Obbadi, Y. Errami, S. Sahnoun, Infrared thermography for detecting the deposit nature in steel water pipes, in: QIRT Council, 2016, https://doi.org/10.21611/qirt.2016.111.

[31] R. Hoffmann, L. Amundsen, R. Schüller, Online monitoring of wax deposition in sub-sea pipelines, Meas. Sci. Technol. 22 (7) (2011), https://doi.org/10.1088/0957-0233/22/7/075701 (Institute of Physics Publishing).

[32] C.C. Patton, Applied Water Technology, Campbell Petroleum Series, 1986.

[33] J. Moghadasi, M. Jamialahmadi, H. Müller-Steinhagen, A. Sharif, Formation damage due to scale formation in porous media resulting from water injection, in: Proceedings—SPE International Symposium on Formation Damage Control, Society of Petroleum Engineers (SPE), 2004, pp. 581–591, https://doi.org/10.2523/86524-ms.

[34] J. Moghadasi, M. Jamialahmadi, H. Müller-Steinhagen, A. Sharif, M.R. Izadpanah, Formation damage in Iranian oil fields, in: Proceedings—SPE International Symposium on Formation Damage Control, 2002, pp. 757–765, https://doi.org/10.2523/73781-ms.

[35] C.A. Todd, D.M. Yuan, Barium and Sbontium Sulfate Solid-Solution Scale Formation at Elevated Temperatures, SPE Prodrrtion Engineering, 1992, pp. 85–92.

[36] K. Asghari, R. Kharrat, Alteration of permeability by fine particle movement-a water Injectivity problem, in: The SPE International Symposium on Oilfield Chemistry-February 14–17, 1995, pp. 655–666.

[37] I.A. Voloshin, V.V. Ragulin, E.N. Tyabayeva, L.I. Diakonov, J.E. Mackay, Scaling problems in Westem Siberia, in: The SPE fifth International Symoosium on Qilfield Scale, January 29–30, 2003, pp. 1–8.

[38] P. Bedrikovetsky, E. Mackay, R.P. Montelro, F. Patricio, F.F. Rosário, Injectivity impairment due to Sulfate scaling during PWRI: analytical model, in: SPE Eighth International Symposium on Oilfield Scale 2006, 2006, pp. 184–198, https://doi.org/10.2523/100512-ms.

[39] B. Long, Technology update: preventing emulsions by mitigating solids, J. Pet. Technol. 69 (1) (2017) 19–21, https://doi.org/10.2118/0117-0019-jpt (Society of Petroleum Engineers (SPE).

[40] M.M. Jordan, J. Buckman, C.J. Johnston, Direct Monitoring of the Performance of Scale Control Programs Across the Produced Water Life Cycle via Suspended Solids Analysis, Society of Petroleum Engineers, 2005 January, https://doi.org/10.2118/93892-MS.

[41] M.M. Jordan, C.J. Johnston, M. Robb, Evaluation Methods for Suspended Solids and Produced Water as an Aid in Determining Effectiveness of Scale Control both Downhole and Topside, Society of Petroleum Engineers, 2005 January, https://doi.org/10.2118/92663-MS.

[42] J. Brownlee, J.A. Kirk, T.S. Dougherty, R.H. Hausier, Solving iron Sulfide problems in an offshore gas gathering system, in: NACE—International Corrosion Conference Series, vol. 2000-March, National Association of Corrosion Engineers International, 2000.

[43] L. Xia, S. Lu, G. Cao, Stability and demulsification of emulsions stabilized by asphaltenes or resins, J. Colloid Interface Sci. 271 (2) (2004) 504–506, https://doi.org/10.1016/j.jcis.2003.11.027.

[44] R. Aafaqi, Flow Behaviour of Water-in-Oil Emulsions Stabilized by Wax Crystals, 2009 (Theses and Dissertations. Paper 1069).

[45] J.L. Stark, S. Asomaning, Synergies between Asphaltene stabilizers and demulsifying agents giving improved demulsification of asphaltene-stabilized emulsions, Energy Fuel 19 (4) (2005) 1342–1345,- https://doi.org/10.1021/ef0498014.

[46] C. Hurtevent, S. Ubbels, Preventing naphthenate stabilised emulsions and naphthenate deposits at Ekoundou oil field in Cameroon, in: Society of Petroleum Engineers (SPE), 2006, https://doi.org/10.2118/100430-ms.

[47] B. Chandragupthan, Detecting and Dealing with Hydrate Formation Digital Refining, 2010, Retrieved from: https://www.digitalrefining.com/article/1000465/detecting-and-dealing-with-hydrate-formation#.XxMn-J5KiyJ. (Last Access July 2020).

[48] B.E. Moriarty, On-Line Monitoring of Water Treatment Chemicals Moriarty and Associates, Mineral Scales and Deposits, Fairfax, VA, USA, 2015, pp. 737–745 (Chapter 31).

[49] J.J. Wylde, Sulfide scale control in produced water handling and injection systems: best practices and global experience overview, in: Society of Petroleum Engineers—SPE International Conference and Exhibition on Oilfield Scale 2014, Society of Petroleum Engineers, 2014, pp. 366–380, https://doi.org/10.2118/169776-ms.

[50] D. Gonzalez, F. Gonzalez, M. Pietrobon, M. Haghshenas, M. Shurn, A. Mees, C. Stewart, C. Ogugbue, Strategies to monitor and mitigate asphaltene issues in the production system of a Gulf of Mexico deepwater subsea development, in: Proceedings of the Annual Offshore Technology Conference, vol. 3, 2016, pp. 2679–2694, https://doi.org/10.4043/27123-ms (Offshore Technology Conference).

[51] I.A. Wiehe, R.J. Kennedy, The oil compatibility model and crude oil incompatibility, Energy Fuel 14 (2000) 56–63.

[52] S. Mazloum, A. Chapoy, J. Yang, B. Tohidi, Developing a robust hydrate early warning system, in: ICGH 2011, 2011.

[53] NACE TM0173-2005 Standard Test Method Methods for Determining Quality of Subsurface Injection Water Using Membrane Filters n.d.

[54] Caproco Special Purpose Coupons & Coupon Holder Assemblies, Caproco Ltd.

[55] A.R. Al-Shamari, S. Kuthe, S. Al-Sulaiman, S. Londhe, Integrity management of oil wellheads and flowlines having scaling, in: NACE-2017-9323 CORROSION 2017, 26–30 March, New Orleans, Louisiana, USA 2017, NACE International, 2017, April 27.

[56] L. Castillo, E. Torin, J.A. Garcia, M. Carrasquero, M. Navas, A. Viloria, New product for inhibition of calcium carbonate scale in natural gas and oil facilities based on aloe vera: application in Venezuelan oilfields, in: Latin American and Caribbean Petroleum Engineering Conference, 31 May–3 June, Cartagena de Indias, Colombia, Society of Petroleum Engineers, SPE-123007-MS, 2009, January 1, https://doi.org/10.2118/123007-MS.

[57] M. Sumestry, H. Tedjawidjaja, Case study: calcium carbonate scale inhibitor performance degradation due to H_2S scavenger injection in Semoga field, in: North Africa Technical Conference and Exhibition, 20–22 February, Cairo, Egypt Society of Petroleum Engineers, SPE-150705-MS, 2012, January 1, https://doi.org/10.2118/150705-MS.

[58] C. Greer, S. Shevchenko, P. Duggirala, Inorganic scale in ECF bleach plants: Rapid test method for prediction and control, in: Tappi Pulping Conference, Seattle, WA, 2001.

[59] J. Bae, D. Fouchard, S. Garner, J. Macias, Advantages of applying a multifaceted approach to Asphaltene inhibitor selection, in: OTC-27171-MS Offshore Technology Conference, 2–5 May, Houston, Texas, USA, 2016, May 2, https://doi.org/10.4043/27171-MS.

[60] D.M. Kaimaki, B.E. Smith, C. Durkan, On the use of nanomechanical atomic force microscopy to characterise oil-exposed surfaces, RSC Adv. 8 (12) (2018) 6680–6689, https://doi.org/10.1039/c7ra12209h.

[61] H.-J. Lund, Investigation of Paraffin Deposition during Single-Phase Liquid Flow in Pipelines (M.S. Thesis), University of Tulsa, 1998.

[62] M.A. Theyab, Experimental methodology followed to evaluate wax deposition process, J. Pet. Environ. Biotechnol. 9 (1) (2018), https://doi.org/10.4172/2157-7463.1000357. OMICS Publishing Group.

[63] Z. Huang, S. Zheng, H.S. Fogler, Wax Deposition: Experimental Characterizations, Theoretical Modeling, and Field Practices, Taylor & Francis Group. LLC, USA, 2015.

[64] R. Davidson, An Introduction to Pipeline Pigging, PPSA Aberdeen Seminar, UK, 2002, pp. 1–104.

[65] M.S. Chowdhury, M.F. Abdel-Hafez, Pipeline inspection gauge position estimation using inertial measurement unit, odometer, and a set of reference stations, ASME J. Risk Uncertainty Part B 2 (2) (2016), https://doi.org/10.1115/1.4030945. American Society of Mechanical Engineers (ASME).

[66] C. Reed, A.J. Robinson, D. Smart, Techniques for Monitoring Structural Behaviour of Pipeline Systems, American Water Works Association, USA, 2004, pp. 1–246.

[67] R. Case, S. Hare, B. Snodgrass, Smart pigging proves useful inspecting deepwater tiebacks, Pipeline Gas J. 230 (12) (2003) 24–28.
[68] B. Nicholson, Smart Utility Pig Technology in Pipeline Operations, Pigging Products and Services Association, 2004.
[69] N. Iqbal, P.A. Lieberzeit, Artificial receptors for mass-sensitive sensors: targeting analytes from surfaces, nanoparticles, and bioanalytes by molecular imprinting, in: Molecularly Imprinted Sensors, Elsevier, 2012, pp. 195–235, https://doi.org/10.1016/B978-0-444-56331-6.00009-8.
[70] D.H. Emmons, G.C. Graham, S.P. Holt, M.M. Jordan, B. Locardel, On-Site, Near-Real-Time Monitoring of Scale Deposition, Society of Petroleum Engineers, 1999 January, https://doi.org/10.2118/56776-MS.
[71] M.M. Jordan, Deployment of Real-Time Scale Deposition Monitoring Equipment to Optimize Chemical Treatment for Scale Control during Stimulation Flowback, Society of Petroleum Engineers, 2008 January, https://doi.org/10.2118/113804-MS.
[72] N.D. Feasey, E. Freiter, M.M. Jordan, R. Wintle, Field Experiences with a Novel near Real Time Monitor for Scale Deposition in Oilfield Systems, NACE International, 2000 January.
[73] C. Valero Vidal, A. Igual Muñoz, Influence of protein adsorption on corrosion of biomedical alloys, in: Bio-Tribocorrosion in Biomaterials and Medical Implants, Elsevier Publishing, Amsterdam, 2013, pp. 187–219.
[74] C. Garcia, G. Courbin, C. Noïk, F. Ropital, C. Fiaud, Development of the electrochemical quartz crystal microbalance to control carbonate scale deposit, in: NACE—International Corrosion Conference Series. Vol. 1999-April, National Association of Corrosion Engineers International, 1999.
[75] D. Peronno, H. Cheap-Charpentier, O. Horner, H. Perrot, Study of the inhibition effect of two polymers on calcium carbonate formation by fast controlled precipitation method and quartz crystal microbalance, J. Water Process Eng. 7 (September) (2015) 11–20, https://doi.org/10.1016/j.jwpe.2015.05.002 (Elsevier Ltd).
[76] Z. Ma, B. Brown, S. Nesic, M. Singer, Investigation of FES precipitation kinetics by EQCM, in: NACE—International Corrosion Conference Series. Vol. 2019-March, National Association of Corrosion Engineers International, 2019.
[77] T.H. Teh, A Novel Electrochemical Technique for Mineral Scale Coverage and Scaling Tendency Quantification (Doctor of Philosophy), Heriot-Watt University School of Engineering and Physical Sciences, 2011.
[78] A. Neville, T. Hodgkiess, A.P. Morizot, Electrochemical assessment of calcium carbonate deposition using a rotating disc electrode (RDE), J. Appl. Electrochem. 29 (4) (1999) 455–462, https://doi.org/10.1023/A:1026431519148 (Kluwer Academic Publishers).
[79] A. Neville, A.P. Morizot, A combined bulk chemistry/electrochemical approach to study the precipitation, deposition and inhibition of $CaCO_3$, Chem. Eng. Sci. 55 (20) (2000) 4737–4743, https://doi.org/10.1016/S0009-2509(00)00046-4 (Elsevier Science Ltd).
[80] H.E. Englander, Conductometric Measurement of Carbonate Scale Deposition and Scale Inhibitor Effectiveness, Society of Petroleum Engineers, 1975 July, https://doi.org/10.2118/5001-PA.
[81] H. Li, D. Dzombak, R. Vidic, Electrochemical impedance spectroscopy (EIS) based characterization of mineral deposition from precipitation reactions, Ind. Eng. Chem. Res. 51 (7) (2012) 2821–2829, https://doi.org/10.1021/ie201802n.
[82] R.D. Vidic, W. Liu, H. Li, C. He, Water treatment chemicals: types, solution chemistry, and applications, in: Mineral Scales and Deposits: Scientific and Technological Approaches, Elsevier Inc., 2015, pp. 169–191, https://doi.org/10.1016/B978-0-444-63228-9.00008-5.
[83] ASM, ASM Handbook Vol. 17 – Nondestructive Evaluation and Quality Control, ASM International, Metals Park, OH, 1992, p. 1608.
[84] G.P.P. Gunarathne, Online monitoring and characterisation of mineral deposits in petroleum pipelines, in: IEE Seminar on On-line Monitoring Techniques for the Off-Shore Industry (Ref. No. 1999/143), Aberdeen, UK, 1999, 1999, pp. 6/1–6/7, https://doi.org/10.1049/ic:19990725.

[85] G.P.P. Gunarathne, R.W. Keatch, Novel techniques-for monitoring and enhancing dissolution of mineral deposits in petroleum pipelines, in: Presented at the Offshore Europe'95 Conference and Exhibition, July 1995, Aberdeen UK and in the Proceedings of the Conference, 1995, pp. 507–522.

[86] G.P.P. Gunarathne, R.W. Keatch, Novel techniques for monitoring and enhancing dissolution of mineral deposits in petroleum pipelines, in: Ultrasonics Sonochemistry, vol. 34, Elsevier Science, September 1996, pp. 411–419.

[87] G.P.P. Gunarathne, Q. Zhoy, K. Christidis, Ultrasonic feature extraction techniques for characterisation and quantification of scales in petroleum pipeline, in: Presented at the IEEE Ultrasonic International Symposium (US98), Sendai, Japan, Oct 98, and in the Conference Proceedings 1′01.1, 1998, pp. 859–864.

[88] G.P.P. Gunarathne, et al., Analysis and optimisation of transmission characteristics of cables for down-hole tools and data acquisition systems, in: Paper Accepted for Presentation at the IEEE, Instrumentation and Measurement Technology International Conference (IMTC99) in Venice, Italy, May 1999.

[89] K. Christidis, G.P.P. Gunarathne, K. Peteineri, Intelligent instrumentation for identification of scales in petroleum pipelines in situ, Trans. Inst. Meas. Control. 35 (4) (2013) 457–463, https://doi.org/10.1177/0142331212453598.

[90] M.H. Skjelvareid, Y. Birkelund, Y. Larsen, Internal pipeline inspection using virtual source synthetic aperture ultrasound imaging, NDT&E Int. 54 (2013) 151–158, https://doi.org/10.1016/j.ndteint.2012.10.005.

[91] M. Boerkamp, D.W. Lamb, P.G. Lye, C.M. Fellows, A. Al-Hamzah, A.D. Wallace, Detecting and monitoring industrial scale formation using an intrinsic exposed-core optical fiber sensor, Ind. Eng. Chem. Res. 49 (10) (2010), https://doi.org/10.1021/ie901471p.

[92] C. Baldwin, Fiber optic sensors in the oil and gas industry: current and future applications, in: Opto-Mechanical Fiber Optic Sensors: Research, Technology, and Applications in Mechanical Sensing, Elsevier Inc., 2018, pp. 211–236, https://doi.org/10.1016/B978-0-12-803131-5.00008-8.

[93] D.V. Brower, C.N. Prescott, J. Zhang, C. Howerter, D. Rafferty, Real-time flow assurance monitoring with non-intrusive fiber optic technology, in: Proceedings of the Annual Offshore Technology Conference, 2005-May, 2005, pp. 1054–1061, https://doi.org/10.4043/17376-ms.

[94] T. Okazaki, K. Imai, S.Y. Tan, Y.T. Yong, F.A. Rahman, N. Hata, H. Kuramitz, Fundamental study on the development of fiber optic sensor for real-time sensing of $CaCO_3$ scale formation in geothermal water, Anal. Sci. 31 (3) (2015) 177–183, https://doi.org/10.2116/analsci.31.177.

[95] M. Boerkamp, D. Lamb, P. Lye, An intrinsic exposed core optical fiber sensor as a quantitative surface crystallization monitoring sensor, Sens. Actuators B Chem. 177 (2013) 964–969.

[96] R. Philip-Chandy, P. Scully, D. Thomas, A novel technique for on-line measurement of scaling using a multimode optical fibre sensor for industrial applications, Sens. Actuators B Chem. 71 (1) (2000) 19–23.

[97] T. Okazaki, T. Orii, A. Ueda, A. Ozawa, H. Kuramitz, Fiber optic sensor for real-time sensing of silica scale formation in geothermal water, Sci. Rep. 7 (1) (2017), https://doi.org/10.1038/s41598-017-03530-1.

[98] M. Means, Novel method for monitoring and controlling asphaltene deposition in crude oil, in: SPE Western Regional Meeting, Society of Petroleum Engineers, 2001, p. 10, https://doi.org/10.2523/68850-ms.

[99] J.K. Smith, M. Yuan, T.H. Lopez, M. Means, J.L. Przybylinski, Real-time and detection of calcium carbonate scale in a West Texas oilfield, in: Paper Presented at the International Symposium on Oilfield Scale, Aberdeen, United Kingdom, January 2003, https://doi.org/10.2118/80372-MS.

[100] J.K. Smith, M. Yuan, T.H. Lopez, M. Means, J.L. Przybylinski, Real-Time and In-Situ Detection of Calcium Carbonate Scale in a West Texas Oil Field, Society of Petroleum Engineers, 2004, May 1, https://doi.org/10.2118/87376-PA.

[101] C. Gallagher, M.C. Means, Asphaltenes Monitoring and Control System, United States Patent 6467340, 2002.

[102] A. Bouchalkha, R. Karli, K. Alhammadi, A. Anjum, I. Saadat, A. Al Ghaferi, Graphene based sensor for scale monitoring, in: SEG/AAPG/EAGE/SPE Research and Development Petroleum Conference and

Exhibition 2018, RDP 2018, Society of Exploration Geophysicists, 2018, pp. 187–189, https://doi.org/10.1190/RDP2018-41995526.1.

[103] W.Y. Shih, K. Albrecht, J. Glater, Y. Cohen, A dual-probe approach for evaluation of gypsum crystallization in response to antiscalant treatment, Desalination 169 (3) (2004) 213–221.

[104] M.G. Langen, G. Healy, The development and application of a fiber optic inspection device for subsea use, in: Offshore Technology Conference, 1991 January, https://doi.org/10.4043/6741-MS.

[105] J.E. Vindstad, A.S. Bye, K.V. Grande, B.M. Hustad, E. Hustvedt, B. Nergård, Fighting naphthenate deposition at the Heidrun field, in: Paper presented at the International Symposium on Oilfield Scale, Aberdeen, United Kingdom, January, 2003, https://doi.org/10.2118/80375-MS.

[106] J.B. Abdullah, R. Alami, H. Ben Abdelouahed, P. Brisset, W.P. Calvo, S. Charlton, A. Hills, J.H. Jin, Q. Nguyen Huu, J. Thereska, V. Yelgaonkar, Industrial Applications of Sealed Radioactive Sources, IAEA-TECDOC-1925, IAEA TECDOC Series, International Atomic Energy Agency, 2020.

[107] R. Drai, A. Benchaala, M. Zergoug, A. Badidi Bouda, Y. Kabir, J. Hadj Djilani, International Atomic Energy Agency, Development of Protocols for Corrosion and Deposits Evaluation in Pipes by Radiography, Atomic Energy, 2005, pp. 1–114 (April).

[108] R.M. Zain, A.M. Terry, N.S. Dahing, N.H. Yusoff, R. Yahya, L.N.S. Dahing, M.F.A. Rahman, I. Mustapha, N.P. Hassan, E.M.F.E. Chik, A.Q.A. Bakar, J. Abdullah, M.R. Shari, A.A. Mahmood, M.A.K. Adnan, Technical aspect on procedure of gamma-ray pipeline inspection, in: Konvensyen Teknikal Nuklear Malaysia 2015 (NTC2015), 2015.

[109] D.F. Oliveira, J.R. Nascimento, C.A. Marinho, R.T. Lopes, Gamma transmission system for detection of scale in oil exploration pipelines, Nucl. Instrum. Methods Phys. Res., Sect. A 784 (2015) 616–620,-https://doi.org/10.1016/j.nima.2014.11.030.

[110] M.S. Chowdhury, H. Al Tanjil, S. Akter, M. Al Amin, S. Kumar Pal, Production logging and its implementation: a technical review, Int. J. Pet. Petrochem. Eng. 5 (2) (2019), https://doi.org/10.20431/2454-7980.0502004 (ARC Publications Pvt Ltd.).

[111] P. Glover, Petrophysics, GLG—66565, Département de Géologie et de Génie Géologique Université Laval Canada, 2014.

[112] S. Bamforth, et al., Revitalizing production logging, Oilfield Rev. 8 (1996) 44–60.

[113] W.M. Mahmud, M. Ermila, Case study: barium sulfate scale removal by dynamic underbalance, in: Proceedings—SPE International Symposium on Formation Damage Control, vol. 2020-February, Society of Petroleum Engineers (SPE), 2020, https://doi.org/10.2118/199232-ms.

[114] J.-P. Poyet, G. Segeral, E. Toskey, Real-Time Method for the Detection and Characterization of Scale, Society of Petroleum Engineers, 2002 January, https://doi.org/10.2118/74659-MS.

[115] S. Abdul-Majid, Determination of scale thickness and general and localized metal removal in desalination plants by the gamma-ray interaction method, Desalination 91 (1) (1993) 35–49, https://doi.org/10.1016/0011-9164(93)80045-O.

[116] M. Margret, V. Subramanian, R. Baskaran, B. Venkatraman, Detection of scales and its thickness determination in industrial pipes using compton backscattering system, Rev. Sci. Instrum. 89 (11) (2018), https://doi.org/10.1063/1.5039579 (American Institute of Physics Inc.).

[117] S. Abdul-Majid, A. Melaibari, B. Malki, Hydrocarbon scale deposits measurements by neutron moderation and capture gamma methods, Nucl. Instrum. Methods Phys. Res., Sect. B 119 (3) (1996) 433–437, https://doi.org/10.1016/0168-583X(96)00296-0.

[118] S. Abdul-Majid, Determination of wax deposition and corrosion in pipelines by neutron back diffusion collimation and neutron capture gamma rays, Appl. Radiat. Isot. 74 (2013) 102–108, https://doi.org/10.1016/j.apradiso.2013.01.012.

[119] W. Harara, Deposit thickness measurement in pipes by tangential radiography using gamma ray sources, Russ. J. Nondestruct. Test. 44 (2008) 796, https://doi.org/10.1134/S1061830908110090.

[120] S. Al-Smairat, M.M. Damoom, M.S. Al-Johani, S. Abdul-Majid, Determination of scale deposition in a flare line by neutron back-diffusion, in: Applied Radiation and Isotopes, Elsevier Ltd., 2021 February, https://doi.org/10.1016/j.apradiso.2020.109424.

[121] T. Bjornstad, K. Garder, I. Hundere, O.B. Michelsen, Tracer tests in oil appraisal and reservoir evaluation: state of the art, in: North Sea Oil and Gas Reservoirs—II, Graham & Trotman, 1990, pp. 261–270, https://doi.org/10.1007/978-94-009-0791-1_22.

[122] S.A. Ebenezer, J. Gudmundsson, Tracer Behaviour in Pipelines with Deposits and Analysis of Natural Gas Pressure Functions (Diploma), Norwegian University of Science and Technology, Norway, 2006.

[123] R.M. Bateman, Cased-Hole Log Analysis and Reservoir Performance Monitoring, 2015, https://doi.org/10.1007/978-1-4939-2068-6_6.

[124] M. Martini, T. Brichart, A. Marais, A. Moussaron, O. Tillement, S. Baraka-Lokmane, How to Monitor Scale Inhibitor Squeeze Using Simple TRF Tracers, Society of Petroleum Engineers, 2015 April, https://doi.org/10.2118/173768-MS.

[125] T. Brichart, A. Moussaron, A. Marais, M. Martini, O. Tillement, C. Hurtevent, S. Baraka-Lokmane, The use of fluorescent tracers for inhibitor concentration monitoring useful for scale inhibitor squeeze evaluation, in: International Petroleum Technology Conference, 2014 December, https://doi.org/10.2523/IPTC-17933-MS.

[126] J. Zirnhelt, I. Einav, S. Infanzón, Radiographic evaluation of corrosion and deposits: an IAEA co-ordinate research project, in: 3rd PAN American Conference for Non Destructive Testing-PANNDT, Brazil, 2003, pp. 1–10.

[127] K. Edalati, N. Rastkhah, A. Kermani, M. Seiedi, A. Movafeghi, The use of radiography for thickness measurement and corrosion monitoring in pipes, Int. J. Press. Vessel. Pip. 83 (10) (2006) 736–741, https://doi.org/10.1016/j.ijpvp.2006.07.010.

[128] U. Zscherpel, U. Ewert, S. Infanzon, M. Aendur, U.N. Rastkhan, P.R. Vaidya, I. Einav, S. Ekinci, Radiographic evaluation of corrosion and deposits in pipelines: results of an IAEA co-ordinated research programme, in: Proceedings of the European Conference on NDT, Mo.2.4.1, Berlin, 2006.

[129] J. Cowan, Radiographic testing: increased detection sensitivity using optimum source to object distance, in: 18th World Conference on Nondestructive Testing, April 2012, Durban, South Africa (WCNDT 2012), 2012.

[130] D. Benson, Non-intrusive Pipeline Inspection Techniques for Accurate Measurement of Hydrates and Waxes within Operational Pipelines, Society of Petroleum Engineers, 2007 January, https://doi.org/10.2118/108361-MS.

[131] J. Kim, S. Jung, J. Moon, G. Cho, A feasibility study on gamma-ray tomography by Monte Carlo simulation for development of portable tomographic system, Appl. Radiat. Isot. 70 (2) (2012) 404–414, https://doi.org/10.1016/j.apradiso.2011.09.019.

[132] J. Bramlett, Computed tomography offers effective tool for subsea pipeline inspection, Offshore Mag. 76 (3) (2016) 49–52.

[133] E. Al Hosani, M. Zhang, M. Soleimani, A limited region electrical capacitance tomography for detection of wax deposits and scales in pipelines, in: 5th International Workshop on Process Tomography (IWPT-5), Jeju, South Korea, 2014.

[134] K.H.-Y. Wei, C.-H. Qiu, K. Primrose, Super-sensing technology: industrial applications and future challenges of electrical tomography, Phil. Trans. R. Soc. A 374 (2016), 20150328, https://doi.org/10.1098/rsta.2015.0328.

[135] I.L.S. Mei, I. Ismail, A. Shafquet, B. Abdullah, Real-time monitoring and measurement of wax deposition in pipelines via non-invasive electrical capacitance tomography, Meas. Sci. Technol. 27 (2) (2015), https://doi.org/10.1088/0957-0233/27/2/025403.

[136] Pipeline Monitoring Solutions, 2022. Retrieved from: https://www.rocsole.com/solutions-for-pipelines. (Last Visited March 2022).

[137] R. Javaherdashti, Microbiologically Influenced Corrosion, an Engineering Insight, second ed., Springer, 2017, pp. 99–124, https://doi.org/10.1007/978-3-319-44306-5.

[138] B.J. Little, J.S. Lee, Microbiologically influenced corrosion, in: Kirk-Othmer (Ed.), Encyclopedia of Chemical Technology, vol. 18, John Wiley & Sons, Inc., 2007, p. 38, https://doi.org/10.1002/047011245X.

[139] P. Janknecht, L.F. Melo, Online biofilm monitoring, Rev. Environ. Sci. Biotechnol. 2 (2–4) (2003) 269–283, https://doi.org/10.1023/B:RESB.0000040461.69339.04.

[140] R. Sooknah, S. Papavinasam, R.W. Revie, A Sulfide Oxidase Biosensor For Monitoring Sulfide, NACE International, 2008 January.

[141] T. Haile, R. Sooknah, S. Papavinasam, W.D. Gould, O. Dinardo, Simultaneous Online Monitoring of Srb Activity and Corrosion Rate, NACE International, 2010 January.

[142] X. Pang, C. Shi, M. Arafin, R. Zhang, A Highly Sensitive Amperometric Sensor for Monitoring Sulfide and Microbiologically Influenced Corrosion, NACE International, 2015 May.

[143] T. Haile, Evaluation of a Portable Online 4-Probe Sensor for Simultaneous Monitoring of Corrosion Rate and Sulfate Reducing Bacteria Activity, NACE International, 2013 March.

[144] M.W. Lechevallier, Microbial Impact of Biological Filtration, Amer Water Works Association, June 1998, p. 140.

[145] Y. Xue, G. Voordouw, L.M. Gieg, Laboratory protocols for investigating microbial souring and potential treatments in crude oil reservoirs, in: T.J. McGenity, K.N. Timmis, B. Nogales Fernandez (Eds.), Hydrocarbon and Lipid Microbiology Protocols, Springer Protocols Handbooks, Springer-Verlag, Heidelberg, 2015, pp. 1–28.

[146] P.J.B. Scott, et al., Expert consensus on MIC: prevention and monitoring—part 1, Mater. Perform. 43 (3) (2004) 50.

[147] S.A. Hernandez, X. Zhu, G.A. Al-Humaid, F.S. Al-Attallah, A.A. Al-Humam, ATP Bioluminescence Assay for Onsite Microbial Assessment in Drilling Rig Operations, Society of Petroleum Engineers, 2018 January, https://doi.org/10.2118/189391-MS.

[148] J. Chalut, G.A. Small, J.M. Payton, Control and Monitoring the Effectiveness of Different Biocides with the Use of Free ATP, NACE International, 1996 January.

[149] C. Whitby, T.L. Skovhus (Eds.), Applied Microbiology and Molecular Biology 3 in Oilfield Systems, Springer, 2010, https://doi.org/10.1007/978-90-481-9252-6_1.

[150] S.P. Kotu, M.S. Mannan, A. Jayaraman, Emerging molecular techniques for studying microbial community composition and function in microbiologically influenced corrosion, Int. Biodeterior. Biodegrad. (2019 October), https://doi.org/10.1016/j.ibiod.2019.104722 (Elsevier Ltd.).

[151] D. Abdallah, S. Punnapala, K. Al Daghar, O. Kulbrandstad, M. Godoy, J. Lovell, S. Madem, A. Babakhani, From ideation to commercialization—development of the first real-time monitoring device for Asphaltene deposition, in: Paper Presented at the SPE Gas & Oil Technology Showcase and Conference, Dubai, UAE, October, 2019, https://doi.org/10.2118/198615-MS.

[152] N. Sabri, S.A. Aljunid, M.S. Salim, S. Fouad, Fiber optic sensors: short review and applications, in: Springer Series in Materials Science 204, Springer Verlag, 2015, pp. 299–311, https://doi.org/10.1007/978-981-287-128-2_19.

[153] J. Minder, Robotic inspection of deep well booster pumps, in: Pipeline Pigging and Integrity Management Conference 2017, 2017 (Papers. Clarion Technical Conferences).

[154] M. Cullin, T. Petersen, R. Srinivasan, C. Forbes, Development of a low-power wireless sensor network of conductivity probes for the detection of corrosive fluids in pipelines, in: NACE—International Corrosion Conference Series, vol. 2020-June, NACE International, 2020.

[155] G. He, Y. Li, Y. Huang, L. Sun, K. Liao, A framework of Smart pipeline system and its application on multi-product pipeline leakage handling, Energy 188 (December) (2019), https://doi.org/10.1016/j.energy.2019.116031 (Elsevier Ltd.).

[156] H. Li, H. Yu, N. Cao, H. Tian, S. Cheng, Applications of artificial intelligence in oil and gas development, Arch. Comput. Methods Eng. (2020), https://doi.org/10.1007/s11831-020-09402-8. Springer.
[157] A. Hochstein, E. Horn, M. Palomino, Intelligent pipeline solution: leveraging breakthrough industrial internet technologies and big data analytics for safer, more efficient oil and gas pipeline operations, in: Pipeline Technology Conference 2015, 2015.
[158] L. Alegre, Potential applications for artificial intelligence in the petroleum industry, J. Pet. Technol. 43 (11) (1991) 1306–1309, https://doi.org/10.2118/21138-PA.
[159] A. Kanu, N. Al-Hajiri, Y. Messaoud, N. Ono, Mitigating hydrates in subsea oil flowlines: consider production flow monitoring & control, in: Paper presented at the International Petroleum Technology Conference, Doha, Qatar, January 2014, https://doi.org/10.2523/IPTC-17492-MS.
[160] S. Macary, N. Mahtumov, H. Muhamadiyev, A. Akyyev, G. Mashadov, A. Al-Hassan, J. Terry, A. AlWazzan, Wax management: comprehensive approach to assure flow in harsh climate-Brown field conditions, in: Paper Presented at the SPE Russian Petroleum Technology Conference, Moscow, Russia, October 2018, https://doi.org/10.2118/191541-18RPTC-MS.
[161] A.G. Shepherd, T. Vandeweyer, D.R.S. Van Kins, E. Baksteen, A. Cheong, Risks, mitigation and management of wax production chemistry in a SAGD environment: downhole to surface and beyond, in: SPE—European Formation Damage Conference, Proceedings, EFDC, vol. 1, 2013, pp. 262–272, https://doi.org/10.2118/165109-ms.
[162] S.F. Alkafeef, F. Al-Medhadi, A.D. Al-Shammari, A simplified method to predict and prevent asphaltene deposition in oilwell tubings: field case, SPE Prod. Facil. 20 (2) (2005) 126–132, https://doi.org/10.2118/84609-PA.
[163] G. Runham, C. Smith, Successful naphthenate scale and soap emulsion management, in: Proceedings—SPE International Symposium on Oilfield Chemistry, vol. 1, Society of Petroleum Engineers (SPE), 2009, pp. 326–337, https://doi.org/10.2118/121522-ms.
[164] J. Junior, L. Borges, C. Carmelino, P. Hango, J.D. Milliken, S. Asomaning, Calcium naphthenate mitigation at Sonangol's Gimboa field, in: Paper Presented at the SPE International Symposium on Oilfield Chemistry, The Woodlands, Texas, USA, April, 2013, https://doi.org/10.2118/164069-MS.
[165] J.E. Penkala, N. Shioya, E. Costa Bastos, C. De Azevedo Andrade, H.A. Baldotto, T. Salma, E.D. Burger, A cost effective treatment to mitigate biogenic H_2S on a FPSO, in: NACE Meeting Papers, 2004.

CHAPTER

Flow Assurance Solids Chemical Analysis and Characterization 14

14.1 Introduction

Deposits analysis and characterization are crucial steps in a flow assurance management strategy. The details of deposit composition are normally used to track down the root cause of the problem and to develop the mitigation method.

The nature of deposits formed in oil and gas fields is complex. Considering the various produced fluids phases and composition, and the continuously changing process parameters, one can expect that no pure single component deposit will be formed, but rather a composite of different components. Other aspects of the deposits like their texture, hardness, and the location where they deposit can vary, depending on the fluids composition, system design and hydrodynamics. Fig. 14.1 illustrates different shapes and types of oil field-formed deposits.

Deposit sample analysis is typically routine work for production chemists and lab technicians. Due to their complex nature, and in order to acquire as much information as possible from their analysis, deposit samples should be handled with utmost care during all of the analysis steps (sampling, labeling, analysis, and reporting).

Proper analysis and characterization of the deposit samples provide information that can assist in understanding the root cause of the deposits problem and developing the right treatment strategy for these deposits. The use of the right dissolver, converter, and inhibitor, or even switching between treatment techniques (chemical to mechanical or physical), are major process decisions that are made based on the correct deposit characterization, in addition to the economic, safety, and engineering data (Fig. 14.2). Also, knowing the deposit compositions and the factors affecting them allows the operators to optimize the process parameters for operating in the deposit-free zones. Furthermore, regular analysis of deposits is necessary to generate a fouling/deposition profile, which is a map of the types of deposits, their rates of formation, locations, and how they are related to the process. Once the map is in place, less costly analysis is required and the time between shutdown and cleaning is reduced [1,2]. This map is also useful with new developed fields, where deposits formation may be expected, but deposits modeling or sampling is not possible. It is also useful when deposit predictions are overestimating/underestimating the problem.

Essentials of Flow Assurance Solids in Oil and Gas Operations. https://doi.org/10.1016/B978-0-323-99118-6.00016-2

The layering or banding in scale sample from 18" pipeline; the main composition is $BaSO_4$, $SrSO_4$ with minor $CaCO_3$, sulfides (FeS, ZnS, PbS), paraffins and asphaltenes

Deposit sample collected from oil well at 6000 THF, main composition is paraffin, asphaltenes, with minor $BaSO_4$, $SrSO_4$ and sand residues

Scale sample with two distinct layers with different textures but having the same composition, collected from downhole oil well. The main composition is $CaCO_3$, $BaSO_4$, $SrSO_4$, and $CaSO_4 . 2H_2O$.

Two samples from the same well at different depths and different composition, (d) At 12000 THF, main composition is sand (e) At 4000 THF main composition is FeS, ZnS and PbS.

FIG. 14.1

The different shapes, textures, and composition of deposit samples.

Besides being a key step in a deposits management plan, deposit characterization can help in better understanding of other production chemistry aspects of the system, such as:

- Reservoir fluids phase equilibrium and reservoir geochemical interactions.
- Injection water breakthrough, interaction with reservoir fluids and the reservoir rocks.
- Sand and fines production.
- Production fluids—treating chemicals compatibility.

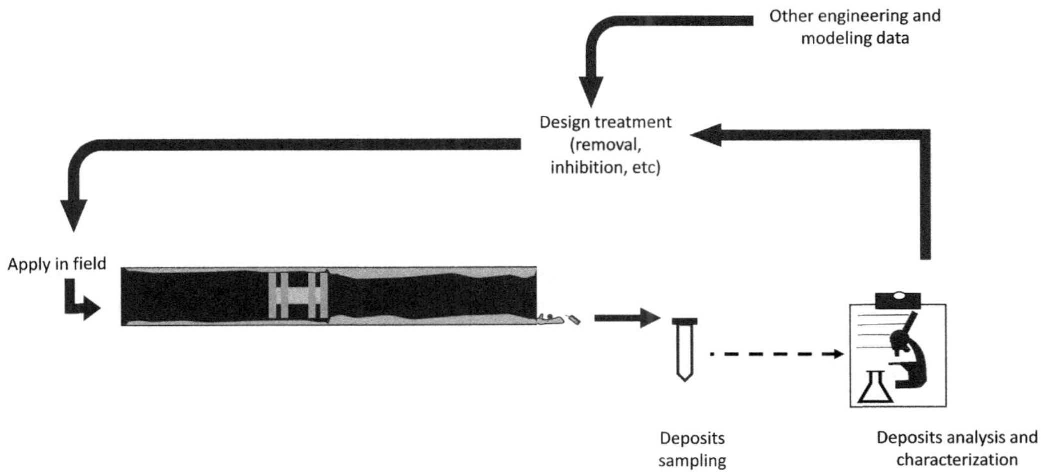

FIG. 14.2

The use of deposits analysis data in designing the treatment.

- Corrosion problem root causes and mechanisms.
- System microbiological activities.
- Treatment chemicals efficiency and optimization.

14.2 General sample analysis flow chart

Some general steps are always applied to samples during the course of their chemical analysis. These include: (1) sampling, (2) sample preparation, (3) chemical analysis, (4) calculations, and (5) results presentation.

Sampling means a portion (subset) of a bulk material is removed in order to be assayed. This portion must be representative, meaning that the sample should reflect the properties of the entire body from which it was collected. Sometimes multiple samples are necessary for accurate analysis; they can be analyzed either individually or as a composite after being mixed together. After sampling, sample fractionation is required to analyze separate phases, e.g., sampling produced water from produced fluids mixture, or separating mineral scales from composite scale.

Following sampling, the next step is sample preparation. This means converting the laboratory sample (by chemical or physical means) into a form that is suitable for chemical analysis by the different analytical techniques and accessible to the analytical instruments. Sample preparation involves different chemical, physical, and mechanical treatments, e.g., dissolution, extraction, masking, surface treatment, pulverizing, filtering, dilution, and other methods.

Chemical analysis is the step in which small test portions of the lab-prepared sample (aliquots) are used for qualitative and quantitative analysis. Proper analysis method and technique must be chosen. A combination of techniques is usually used to fully characterize field deposits. Calibration and validation of the device used is necessary before starting analysis.

The calculations step refers to correlating the obtained results from the analyzed aliquots to the bulk sample size and evaluating them: for example, correlating the elements/groups to their expected mineral/compound stoichiometrically, and finally calculating the final percentage of each component in the bulk sample.

Finally, the results are interpreted and presented in the analysis report, which should be meaningful to the operator and correlated to the actual running processes.

These analytical procedures are usually organized into so-called standard operating procedures (SOPs), which represent an optimized procedure adapted to the available standard methods, i.e., ASTM, and with the adopted analysis techniques that can be applied in the lab. The SOP should elaborate the details and safety concerns of each step and should be updated according to the latest advances in the adopted analytical techniques.

Fig. 14.3 shows a flowchart of the chemical analysis process. For a complex problem like solid deposits, it is important to incorporate as much data as possible about the running processes where the samples were taken. Information such as process design, operating parameters, injected chemicals, corrosion status, microbiology, recent maintenance, and others are helpful in making the analysis results dynamic and related to the system operations, which assists in understand the actual root cause of the problem. Analysis results that are very simple and undetailed, not process related, are insufficient and can sometimes be misleading.

14.3 Solids sampling

14.3.1 Collecting the sample

Unlike the routine sampling of produced fluids (gas, water, and oil), which usually takes place from specific sampling points already designed in the system (e.g., wellhead, separator, flowline sampling points), solid deposit samples are collected from the location where they form in the system after growing into a significant size that is detectable by the system monitoring methods or removable during routine cleaning/maintenance operations (e.g., pigging, wireline, etc.) (Figs. 14.4 and 14.5). Some spots in the system are known to be more vulnerable to deposit formation than others, such as locations of fluids mixing and locations of pressure and temperature drop. These locations must be monitored for deposits buildup and deposit samples must be collected when the opportunity arises during the routine maintenance or cleaning operations.

Generally speaking, the solid deposits can be sampled by the following:

- Pigging operations, i.e., pig returns, and other similar pipeline mechanical cleaning methods.
- Well intervention methods such as slickline, wireline, and methods using bailers, cutters and brushes.
- Scale and corrosion coupons.
- Total suspended solids, where a sizable amount of the solids suspended in the production fluids is collected on a membrane filter which can be characterized afterwards. This method can be applied for any location in the system and also for locations of higher potential of solids formation or locations that are inaccessible for obtaining deposit samples. It is an effective method of monitoring solids buildup.

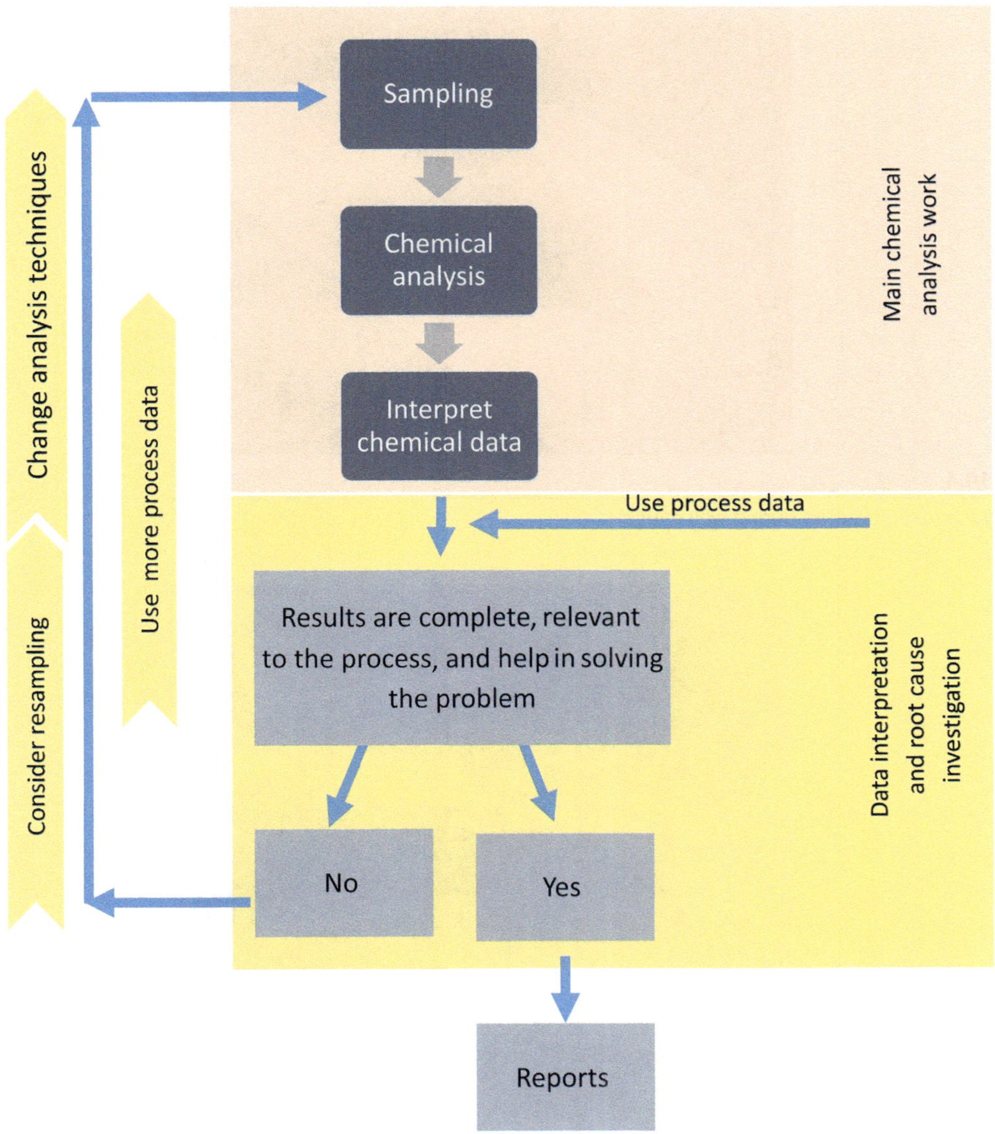

FIG. 14.3

Flowchart of deposit sample analysis procedure in oil and gas operations.

- Pipelines, completions, and other production equipment that are removed/opened for maintenance or for replacement. In this case, solid samples should be collected once the pipe or equipment is removed/opened to avoid solids oxidation and other transformation which may affect the analysis results.
- Solids can also be collected during shutdown.

FIG. 14.4

Solids sample collected during well gauging.

FIG. 14.5

Solids collected during pigging.

Currently, there is no standard method specified for sampling and analysis of all kinds of solid deposits in oil and gas fields. However, a number of standards can be adopted to provide the essential procedures that can be used in oil and gas fields. These include ASTM, NACE, UOP, IP, API, EPA, and others. ASTM D887 [3] provides standard practice for sampling water-formed deposits from boiler tubes and turbine components. NACE standard SP0775 [4] provides standard guidelines for using corrosion coupons in oilfield operations, in which it mentions the visual inspection and quantitative analysis of scale deposit adherent to the corrosion coupon surface.

Some guiding points for collecting a representative sample include:

- The sampling step depends on the experience of the investigator and depends on the intended use of the sample, the system design and accessibility, type of the deposit, and the nature of the problem (scaling, biofouling, corrosion scale, organic deposition).
- There is no specific amount of sample that has to be collected and submitted. The amount of deposit should be consistent with the type of analysis to be performed. However, it is recommended to collect a sample size sufficient to be used in chemical analysis and in chemical dissolution studies, and to send deposit samples to third-party labs for more specified analysis or dissolution studies.
- It is highly recommended to collect the sample in preserved shape, i.e., large pieces or chunks. Powder samples are still applicable, but samples with preserved shape give more information about the deposit thickness and layers, which is crucial for designing the cleaning and removal procedure.
- Samples with distinct layers must be treated with caution, since each layer can have different composition; in this case, analysis of each layer separately is critical in the cleaning and removal procedure.
- The sample that is collected from production tubing is representative of the depth from where it was collected, and that depth must be mentioned in the label/report. It is common to see scale samples with different composition from different depths in the same well. In one well in Egypt, a deposit sample collected at 11,000’ THF was mainly $FeCO_3$ and sand, while the deposit sample from the same well at 230’ THF was mainly $(Ba,Sr)SO_4$. Therefore samples from different depths should be analyzed separately.
- Pig return sample represents a composite of the deposits along the cleaned pipeline. It is important to take a representative composite sample, since deposit types, texture, and hardness can vary along the pipeline. In the case of significant variations of the solid phases, separate samples can be obtained from each phase, in addition to the composite solid sample.
- For biofouling and biological testing samples, the recommended method for the sampling of pig returns involves the collection of a minimum of two composited samples from the pigging solids and from three different locations on the body of the pigging tool [5]. For biological testing, sterilized equipment and containers should be used to collect samples [6]. Ostroff [7] recommends that this type of sample be refrigerated during storage and shipment. It is also recommended that, after a sample has been obtained, it should be kept as close to its original condition as possible.
- Some transformations may occur to the deposit sample, e.g., drying, oxidation, leaching and recrystallization, dissolution, and polymorphism, so the samples must be handled (collected, packed, shipped, prepared) prior to analysis in a manner that safeguards against change in the particular constituents or properties to be examined. Iron compound transformation and polymorphism are a common example of these transformations. High-salinity water associated with scale can evaporate, forming halide solids. Gypsum can transform to anhydrite and vice versa. Paraffinic crude oil associated with solid samples can form wax sludge when it cools below WAT. These transformations lead to misleading analysis and, hence, false root cause and improper treatments.
- Collecting produced fluids samples from where the deposits were collected is valid. Ion concentrations, bacteria types, and paraffin and asphaltenes content are measured and will be used in the root cause analysis and in developing the treatments.

- Pseudoscale is formed due to treating chemical incompatibility issues, so it is crucial to collect samples of the treating chemicals that are injected in the system, as they may be needed during the analysis.
- Various sample containers can be used during sampling. Dry samples are sometimes placed directly in plastic or paper containers. Sample envelopes are convenient for this purpose. Wet samples can be collected in clean glass bottles. Sterilized equipment and containers should be used to collect samples containing biological or organic deposits [6]. Avoid metal containers, as any corrosion within it can cause misleading analysis.
- After sampling, samples should be inspected for any sources of contamination. Metal pieces, rubber pieces, gravel, equipment internal parts, coating, lubricant, gelled crude oil, and broken parts from the cleaning tools are common examples.
- Samples should be preserved adequately after collection: samples can be stored in the refrigerator, icebox, cooler, or desiccator.

14.3.2 Sample labeling, sample and process data required, and why these data are important

After collecting the sample, clear and detailed information on the label should be affixed to the sample container or attached with the sample in a separate document with the following data:

- Name of sample collector
- Plant location
- Sample collection location
- Number of sample/s
- Date and time of sampling
- Collection method
- Appearance of deposit (color, uniformity, texture, odor, oily matter)
- Basic system operating conditions: temperature, pressure, rate of flow

These data are the basic and essential information normally included on sample labels. However, additional information may be considered crucial for accurate chemical analysis and process-related interpretation of analysis results. Such information can be included on the label or sent in a separate document attached to the sample container. This information includes

- History of deposits in that location and the applied treatments.
- Types of chemical/physical treatments applied in the system.
- Any system design changes, maintenance or workover operations, or reservoir formation changes that might cause changes in the produced fluids composition and properties.
- Good description of problems and discrepancies that arose in the system due to the deposits, and it is useful to attach a diagram of the system if possible.
- Any specific test necessary to be done (other than the routine analysis).
- Basic information about the system construction materials.
- Fluids composition; if fluids composition data is not available, then a fluid sample should be submitted along with the deposit sample, as indicated earlier in the sample collection steps.

Such an amount of data might seem to be excessively large to be submitted with a solid sample; however, field experience shows that these data can be required at different stages of sample analysis. Therefore it is useful to have them beforehand, which helps prioritize work, reduce errors, and save time. In addition, such information makes the analysis results more related to the running system.

14.4 Field examination of solid deposits

Generally, a simple on-site analysis is used for early qualitative identification of the sample components. It also helps prioritize the lab analysis work to some extent, and is also useful when urgent field operations might need quick identification of the deposits and cannot wait for the full analysis. The inorganic scale deposits generally can be analyzed qualitatively, and sometimes quantitatively, in the field. On the other hand, deposits containing large amounts of organic materials or biological products usually have to be sent in for laboratory analyses [6].

Simple and quick tests that are indicative for solids composition are listed in Table 14.1.

Some of the qualitative tests for deposits are listed in methods ASTM D1245, NACE standard TM0173, and ASTM 2331.

Table 14.1 Qualitative and field tests for solid deposits.

Testing for	Treating the sample with	Observation
Paraffins and asphaltenes	Heptane	The part that dissolved in the heptane is paraffin while the residues are asphaltenes
Gas hydrates	Methanol	Solubilization of the solids
$CaCO_3$	HCl (15%) wt	Strong effervescence (liberation of CO_2 gas) and will dissolve immediately at ambient temperature
NaCl	Fresh water	Sample will dissolve in fresh water, and will give white precipitate with the addition of $AgNO_3$
(Fe, Zn, Pb) sulfides	HCl (15%) wt	Liberation of H_2S gas with rotten eggs odor, which will turn the lead acetate paper into varying shades from dark brown to black
	Acid arsenite (1.0% sodium arsenite and 0.05% liquid detergent to a 15% HCl)	A bright yellow, readily identifiable precipitate of arsenic sulfide will form
FeO (iron oxides) and also used generally for other iron compounds	Magnet	In a powdered sample the sample will attach to the magnet
$FeCO_3$	Hot HCl (15% wt)	Slow reaction with cold HCl, while gives mild effervescence with hot HCl
$CaSO_4$ $BaSO_4$	Potassium glycolate (25% wt)	Within 24 h the sample, if it is gypsum or anhydrite, will disintegrate into a milky suspension that is readily soluble in dilute hydrochloric acid If it is barium sulfate, no noticeable change will occur

14.5 Sample preparation

Once the sample has been collected, it must be prepared in such a way that it can be safely analyzed using the various analysis techniques. The basic sample preparation includes: removing any source of hazards and contamination, initial screening, and treatment physically and chemically to convert it to a form that is accessible to the analytical devices.

14.5.1 Hazard identification and exclusion

Solid deposit samples can contain hazardous materials such as radioactive scale (NORM) and pyrophoric deposits.

First, and for health and safety reasons, the deposit sample should be examined for radioactivity immediately upon receipt. NORM scale are widespread in oil and gas fields and some of them can have high levels of radioactivity that pose a significant health hazard to operators. Usually radioactivity is measured using a portable or benchtop style radiation counter. If the samples have high levels of radioactivity, stop the analysis and inform the HSE officer to dispose of the sample or send it to a third-party lab specialized in handling such samples. Samples with no radioactive signals are ready for the next step of sample preparation, while samples that have radioactivity within accepted limits should be handled using proper PPE, reduced work hours, switching between analysts, etc.

Pyrophoric deposits are those that can combust in air. Iron sulfides are pyrophoric and they release a tremendous amount of heat that can be enough to make the iron deposits glow red. Careful handling of pyrophoric deposits is required, keeping deposits to a minimum, in a desiccator, and removing all combustible materials from the surrounding area.

14.5.2 Initial preparation and screening

Initial screening involves inspecting and recording the apparent physical properties of the deposits, and performing some simple qualitative tests to primarily identify some major chemical species in the deposit.

Some of the preliminary tests are listed here (Fig. 14.6):

- Visual inspection of the sample to identify the apparent physical features.
- Microscopic examination for structure, color, odor, texture, oily matter, and other characteristics of note.
- Qualitative chemical tests similar to those in Table 14.1, which are applied in the field test to verify the presence of carbonates, sulfides, sulfates, or oxides.
- FT-IR fingerprinting to check for the origin of the organics, produced hydrocarbons (paraffin, asphaltenes, naphthenates), biomass, treating chemicals, processing chemicals.
- Check for magnetic properties.

Record the physical properties and features of the sample. Table 14.2 represents some of the physical features recommended to monitor and report in oilfield deposits.

After identifying the main features of the sample, the next step is eliminating and excluding any contaminants or confusing materials that are not actual solid deposit. Rocks, stones, gravel, metal pieces, plastics, and rubbery materials should be excluded and not mistaken for mineral scales.

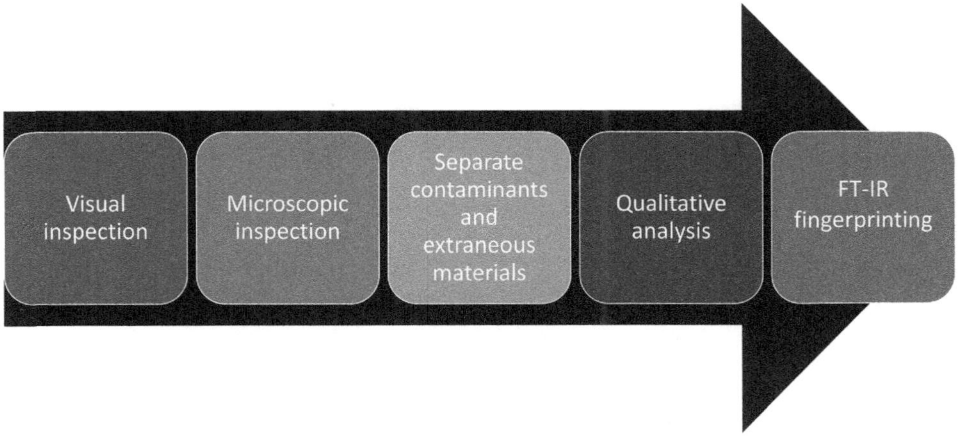

FIG. 14.6

Initial screening steps for deposit samples.

Table 14.2 Different properties of solid samples.

Appearance	Wet
	Moist
	Oily
	Dry
Texture	Fine powder
	Sand
	Chunks
	Flakes
	Oil
Hardness	Hard
	Soft
	Sticky
Color	White gray
	Dark gray
	Reddish brown
	Black
Extraneous materials	Metal pieces
	Coating
	Rubber/plastics
	O-rings
	Paint lumps

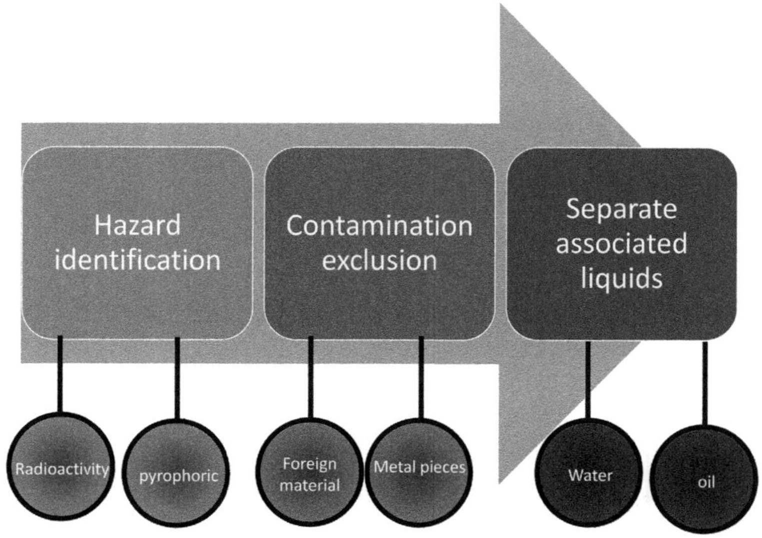

FIG. 14.7

Initial stages of sample preparation.

Residual crude oils can gel during pigging and other maintenance, and should not be mistaken for actual hard wax deposits. Such contaminants should be excluded from the sample and set aside, and not analyzed but to be reported in the final analysis report as is, with mentioning the expected source of contamination.

If the sample includes an appreciable quantity of separated water, remove the solid material by filtration. Save the filtrate, undiluted, pending a decision as to whether or not its chemical examination is required. Samples with oily or chemical matter should be scanned using FT-IR to identify the oily material.

After the contaminations are removed, the sample can be dried. Air drying takes place by spreading the sample in a thin layer on a nonreactive, impervious surface. Air drying should be avoided if the sample contains easily oxidizable materials, such as sulfide; in this case the samples can be dried using nitrogen or any other inert gas or the analysis for these materials should be completed before air drying. The weight loss after air drying is often reported. Deposit samples with high chlorides content will absorb water from the air due to its hygroscopic properties, and will have significant weight loss when air dried; thus these samples should be analyzed right away.

The initial steps of sample preparation are illustrated in Fig. 14.7.

14.5.3 Sample fractionation and preparation

Deposit samples are usually a mixture of organic and inorganic compounds. These compounds cannot be analyzed all together, but need to be separated/fractioned from each other. Fractionation can be done at different levels. At first organic constituents are separated from inorganic constituents; the next level is fractionating organic deposits, e.g., paraffins from asphaltenes, and then fractionating inorganic

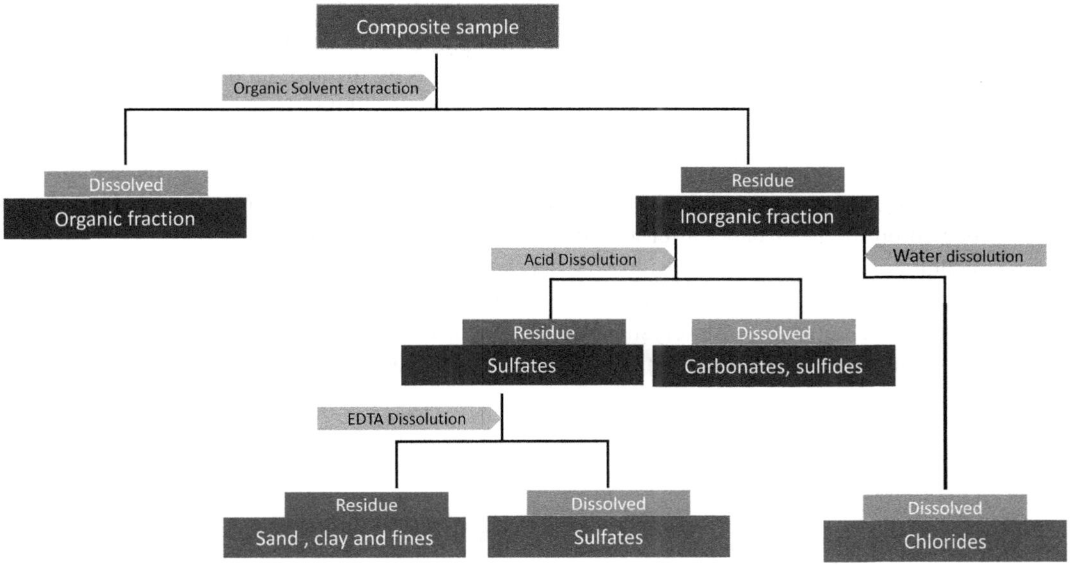

FIG. 14.8

General solids sample fractionation and preparation scheme.

constituents as acid soluble and insoluble, and so on. Standard method ASTM D2331 furnishes the standard practices for preliminary preparation and testing of water-formed scale deposits and a generalized scheme of fractionation is given in Fig. 14.8.

Using proper solvent, organic deposits can be washed off the inorganic scale. Extract solvent in soxhlet at high temperature for the specific period of time (until the solvent in the extraction chamber becomes clear).

Once the organics are washed off from the inorganics, inorganic scale can be dried in the oven at 105°C. After drying, inorganic scale is ground using mortar and pestle and screened using a proper sieve size.

After this, the inorganic samples are prepared for analysis by the different chemical analysis techniques, e.g., wet methods, spectroscopic, chromatographic, etc. These techniques can vary in their sample preparation requirements. Techniques like ICP, atomic absorption, and IC analyze the samples in liquid form after being dissolved in a proper dissolver, whereas others such as XRD and microscopic methods can analyze the sample in solid form after being converted into fine powder or compressed into discs or pellets. Mineral scales can be prepared according to standard method ASTM D2331 or as per method USEPA Method 3050B (Acid Digestion of Sediments, Sludges and Soils).

The inorganic scales usually occur as layers of one or more of water solubles (chlorides), acid solubles (carbonates, sulfides, oxides) and acid insolubles (sulfates). Thus the three portions of the sample are used during scale dissolution. The water solubles are dissolved by treating a small sample volume (~0.5 g) with distilled water with vigorous mixing for a specific period of time. The acid solubles are dissolved by treating a small volume of the sample (~ 0.5 g) with HCl or HNO_3 alternatively or combined at high temperatures until the sample is dissolved. The acid insolubles are usually sulfate scales.

These are usually dissolved by treating a small sample volume (~0.5 g) with aminocarboxylates solutions (EDTA, CyDTA, or DTPA) with heating and mixing for a minimum of 2 h. Another method to dissolve the sulfates is by converting the sulfates to carbonates by fusing a small volume of the sample with Na_2CO_3, and then dissolving the melt in distilled water or diluted HCl.

Silica compounds and sand components in the solids samples are prepared by dissolving the sample in $HF/H_2SO_4/HNO_3$. Another way of preparing the sample for silica species is by the alkali fusion method, by fusing the sample with NaOH; then the melt is dissolved in HCl. Fig. 14.9 shows sand residues collected on the filter paper after acid dissolution of scale sample.

It is recommended that the sand residues be checked with the microscope before proceeding to analyze it further or before reporting it. In some instances, sulfide scale can be oxidized during the acid digestion procedure, forming sulfur crystals that can be mistaken for sand or fines. Fig. 14.10 shows a case where sulfur particles formed after acid digestion (using $HCl + HNO_3$ with heat) of sulfide scales.

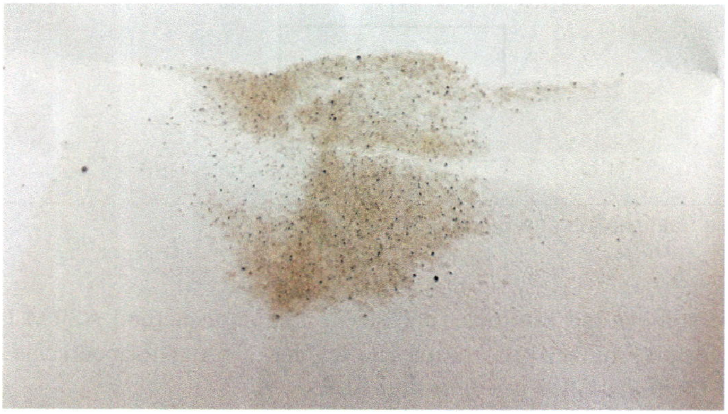

FIG. 14.9

Sand residue collected on filter paper after solid sample dissolution.

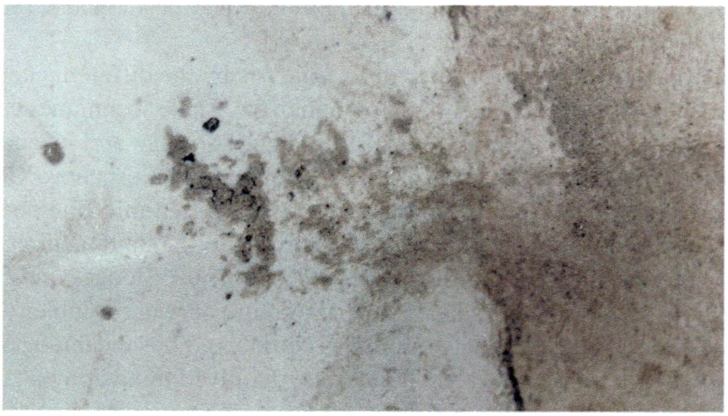

FIG. 14.10

Sulfur residue formed due to oxidation during acid digestion of the solid sample and collected on filter paper.

Another way of preparing the samples was reported by Frenier and Ziauddin [8]. A sample of the deposit is ashed in a high-temperature furnace to drive off any hydrocarbon-containing chemicals. Then the remaining solids are extracted using HCl or an HCl/HF mixture.

14.6 Sample analysis

In this step the quantitative measurements are performed to determine the concentration of elements and components in the prepared deposit sample. Typically, the lab chemist follows a standard operating procedure that encompasses analysis methodology for the different components.

14.6.1 Analysis methodology

The analysis methodology shows the flow of the analysis procedures, depending on the complexity of the sample composition. This analysis methodology depends on the sample data (obtained from initial screening steps) and system data (system parameters and operations). A few points to keep in mind during methodology development are

- The objectives and intentions of the analysis. This determines how far the analysis should go and whether further analysis is necessary. For example, to address the root cause and design inhibition methods, scale phases and polymorphs are necessary, microbiological analysis is crucial, and distinguishing micro- and microcrystalline wax in addition to carbon distribution of waxes is necessary.
- Knowing the sample physical properties, source, and the operations running in the system helps the analysts to prioritize their work and use more sensitive techniques if necessary, to avoid ambiguous analysis results. For instance, organic deposits can occur in water systems due to biofouling, treating chemicals pseudoscale, processing chemicals carryover, lubricant carryover, etc. This requires careful analysis to precisely characterize the sample and disclose the root cause of the problem.
- Analytical methods: some techniques are qualitative more than quantitative, and vice versa, hence it is highly recommended to use a combination of analytical techniques to obtain as much data as possible.

14.6.1.1 Composite deposits

A composite deposit usually contains a mixture of organic and inorganic deposits. Fig. 14.11 shows a typical composite deposit sample. These deposits can be hard-layered, or consist of a sticky, slimy material called schmoo. Schmoo usually comprises sand fines, sulfur, scale, naphthenate soaps, corrosion products, biomass, and other materials, all bound together with oil.

Analyzing these samples starts with FT-IR scans to qualitatively identify the components and distinguish between them, e.g., hydrocarbons (paraffin, asphaltenes, naphthenates), biomass, production chemicals, lubricating oil, etc.

If treating chemicals are detected, then that may be a sign of pseudoscales; in this case, the possible chemicals that are used in the process have to be fingerprinted to find a match. Usually, field-based labs have a library of the FT-IR spectra of these chemicals. Sample manipulations, e.g., evaporating the solvent or reacting with other chemicals at process pressure and temperature, may be necessary to

FIG. 14.11

A composite deposit sample from downhole of a well; main composition: 17% organics, 66% (Fe, Zn, Pb) S, 5% NaCl, 4% $CaCO_3$, 5% (Ba, Sr)SO_4 and 3% sand residue.

obtain a full match between the solid sample and the tested chemical/s spectra after excluding the regular IR technique interferences. The pseudoscale is then quantified after being dissolved in a proper solvent. Similarly to this, deposits formed by glycol, amine, or carryovers can be identified using FT-IR.

Detecting biomasses in the deposits requires microbial testing besides detecting the percent weight of the biomass. Testing for bacteria from the collected deposit samples is a common practice in oil and gas fields.

Naphthenates deposits can be detected in the sample using FT-IR, and in this instance, it can be estimated according to Section 14.6.1.2.4.

If no peculiarity is detected in the FT-IR, then regular hydrocarbon organic deposits and mineral scales measurements are sufficient. Organic deposits and mineral scales analysis are discussed in the following sections. Fig. 14.12 summarizes the composite sample analysis methodology.

14.6.1.2 Organic deposits

Organic deposits can be separated and quantified by many methods, including loss on ignition, solvent extraction, SARA, modified SARA analysis, or by employing the standard methods specified for measurement of each component separately. Fig. 14.13 illustrates these methods.

Loss on ignition is based on oxidizing the organic compounds at high temperatures >750°C to CO_2 and ashes. The difference in weight before and after heating determines the organic and carbonate components.

The general extraction is carried out by extracting the organic in soxhlet extractor using a suitable solvent, first by using heptane or hexane, which dissolves paraffins and leaves asphaltene residues, and then in a second step, asphaltenes are extracted using toluene. The weight loss after each step determines the percentage of each component.

SARA analysis is usually used to measure the crude oil fractions and has been modified to measure organics in deposit samples. Alian et al. [9] used SARA to characterize organic deposits and extended it

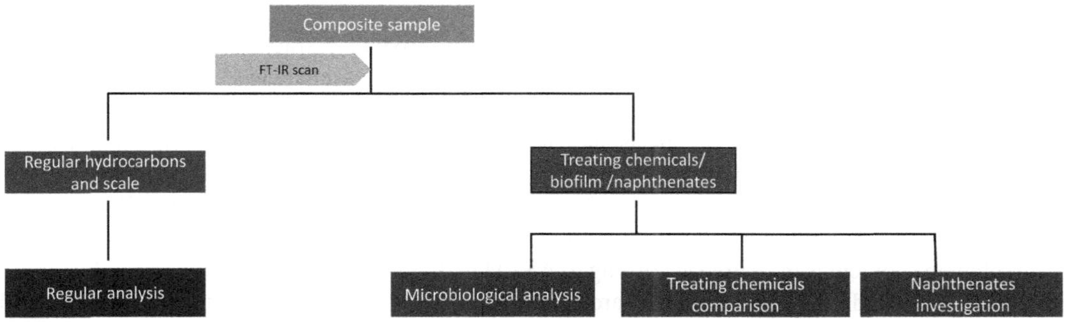

FIG. 14.12

Composite sample screening and analysis methodology.

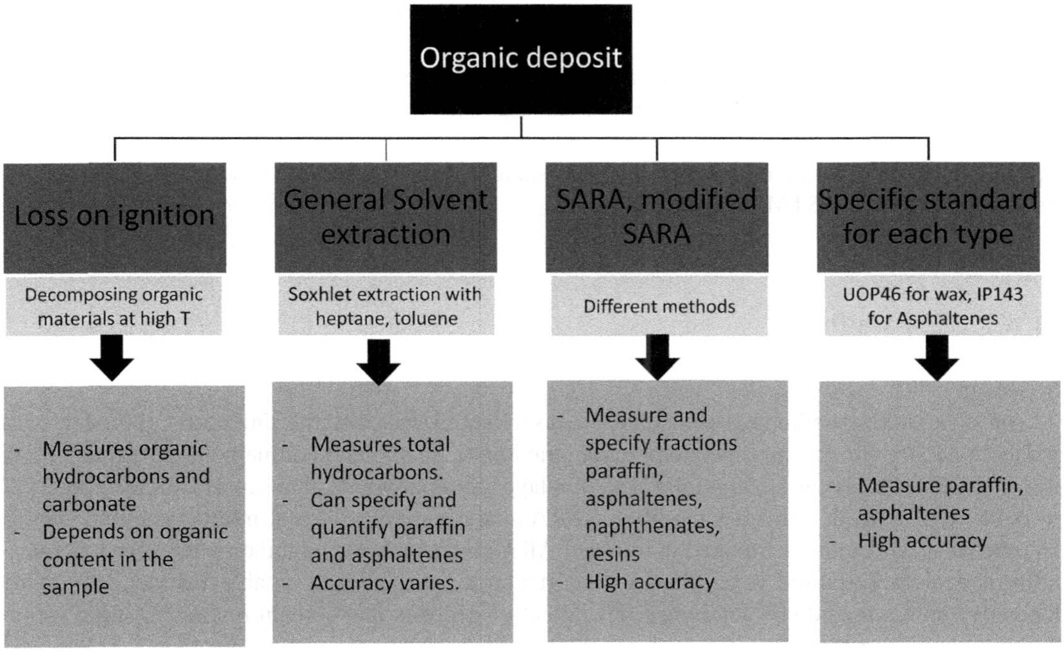

FIG. 14.13

Methods used in determining the organic deposits composition.

to modified SARA analysis to detect microcrystalline wax and naphthenates, which are overlooked in routine SARA procedures.

The specific methods include standard methods designed to measure specific fraction of crude oil and also applicable to organic deposits, e.g., UOP46 used to measure paraffins in crude oil, and IP143 used to measure asphaltenes. These methods and more in-depth characterization of organic deposits are discussed with each type in the following sections.

14.6.1.2.1 Asphaltenes

As mentioned already, asphaltenes can be quantified by Soxhlet extraction. Cosultchi et al. [10] reported a procedure for separating deposit fractions and estimating asphaltenes by washing the deposit with toluene within a Soxhlet unit according to ASTM D473 to separate the toluene-insoluble fraction (mineral scales and debris), and then the asphaltenes were prepared by *n*-heptane precipitation from the toluene soluble fraction according to ASTM 4124.

Multiple standards can be adopted for SARA analysis, including ASTM 2007 and IP-469. SARA analysis has been modified and adopted for solids analysis [11].

Asphaltene content is measured according to ASTM D-3279, ASTM D6560 methods, or Institute of Petroleum Standard IP143. These standards quantify the asphaltenes content generally in crude petroleum and petroleum products, and yet they were adopted for use in deposit samples as well. Other uncommon methods include IFP 9313 and AFNOR T60-115.

Furthermore, other techniques can be used to fully characterize the deposits. Thermogravimetric analysis (TGA) can be used to determine the weight loss as a function of temperature; FT-IR fingerprint is used to indicate the main functional groups; and ESI-Mass is used to determine the degree of non-saturation [12,13].

Measuring the metal content of asphaltene deposits is important, as it may help in identifying the root cause of the deposition. Ni, V, Fe, Al, Na, Ca, and Mg have been reported to accumulate in asphaltenes, with Ni and V being the most abundant. Moreover, Fe was detected in asphaltene deposits, especially in locations with high corrosion rates. Metal ions can be measured using ICP-OES following the standard method ASTM D5708.

14.6.1.2.2 Paraffins

Like asphaltenes, solvent extraction can be used to measure paraffin content using hexane or heptane as the solvent.

The standard test method for wax content measurement is based on the Universal Oil Products LLC method UOP 46, which is a solvent extraction gravimetric method. Other methods of wax determination are DSC [14], gas chromatography [15], thin layer chromatography [16], and NMR technique [17].

Alian et al. [18] developed a modified SARA analysis that can detect microcrystalline wax and naphthenates, which are overlooked in routine SARA procedures. Differentiating between macrocrystalline and microcrystalline waxes plays an important role in designing a suitable treatment, due to their distinctive structure and characteristics. Treatment of deposits that contain microcrystalline wax required higher temperature compared to macrocrystalline waxes.

Knowing the carbon distribution (and also what is known as critical carbon number) of the paraffin deposits using a gas chromatography technique helps in electing the optimum chemical inhibitor. The optimum length of the side chains in paraffin inhibitors depends on the length of alkanes in the waxes deposited and not the oil, since the carbon chain lengths in oils or in lab-generated deposits (cold finger) may not be representative of those in the field deposits, and thus they are not adequate for selecting the chemical inhibitor [19].

Deposit or gel strength is another important feature necessary in order for the operators to design suitable pigging operations and to design system start-up pressure [20].

14.6.1.2.3 Gas hydrates

Multiple modern analyzing instruments and techniques, including Raman, XRD, X-CT, SEM, NMR, UV-visible, and high pressure DSC, have been applied in the study of structure, formation mechanisms, phase equilibrium, and the thermal and physical properties of gas hydrates [21].

Gas hydrate structure has been studied by XRD, SEM, UV-visible, and 129Xe NMR, whereas the kinetics of formation and dissociation have been studied using Raman techniques, ^{13}C NMR, and MIR [22].

Diamondoids are another type of caged molecule. Vazquez and Mansoori [13] reported GC-MS analysis of diamondoids in petroleum fluids can be performed on the saturate fraction of SARA separation.

14.6.1.2.4 Naphthenates

Naphthenate characterization involves identification of the metal ion and the naphthenic acid. Oduola et al. [23] used XRF to measure the metal ion concentrations and used mass spectroscopy to identify the naphthenic acid. Rodgers et al. [24] used high-resolution Fourier transform ion cyclotron resonance mass spectrometry (FT-ICR MS) to characterize naphthenate deposits and determine the naphthenic acid molecular weight. Moreira and Teixeira [25] used FT-IR and thermogravimetry (TG) to characterize naphthenate deposits.

Gallup et al. [26] reported multiple techniques to characterize naphthenate and carboxylate soap from the Indonesian field, including gas chromatography-mass spectroscopy (GCMS), inductively coupled plasma-atomic emission spectroscopy (ICP-AES), ion chromatography (IC), pH, high-performance liquid chromatography (HPLC), electrospray mass spectroscopy (ESMS), Fourier transform infrared (FTIR) spectroscopy, scanning electron microscopy coupled with energy-dispersive X-ray analysis (SEM/EDXA), X-ray diffractometry (XRD), nuclear magnetic resonance (NMR), and high-temperature gas chromatography (HTGC).

Mohamed et al. [27] employed modified SARA analysis and XRD to characterize naphthenate deposits in the offshore Malaysian oilfield.

14.6.1.3 Mineral scale

Standard method ASTM D2331 furnishes the standard practices for preliminary preparation and the different techniques used to determine the composition of water-formed scale deposits. Conventionally, mineral scale flakes are ground together, screened, dissolved, and then measured using the proper technique. Analytical techniques used in mineral scales analysis have been reviewed by East et al. [1] and Woodward et al. [28].

Metal ions are usually measured by ICP-OES, ICP-MS, AA, IC, UV-visible spectrometry, potentiometry, or wet chemistry methods. These methods require the sample to be dissolved in a proper dissolver before analysis and prepared to fit the technique. ICP-OES and ICP-MS are the more common methods and have better detection limits. The anions can be measured either by ion chromatography technique (IC), UV-visible spectrometry, potentiometric methods, or by classical wet chemistry methods. For example, chlorides can be measured using volumetric titration with $AgNO_3$ or potentiometric titration to determine halite scale percentage.

X-ray-based techniques are generally used in mineral scale analysis and they are nondestructive techniques. XRF, XPS, and EDX techniques are basically quantitative elemental analysis techniques.

XRD is another technique that is more qualitative, which is crucial in identifying mineral phases and other crystal structure parameters. Normally a combination of the X-ray qualitative and quantitative techniques provides the necessary data on the scale composition, which then are used in designing the treatments. For example, iron compounds can exist in scale deposits in different phases, which are basically associated with different problems, such as oxides, carbonates, and sulfides are associated with general corrosion, sweet corrosion, sour corrosion, and microbial activities, respectively. Such problems have different treatment regimes, e.g., corrosion inhibitors, H_2S scavengers, biocides, or scale inhibitors, or some combination of these. Differentiating and identifying the exact phases and their ratios is crucial in identifying the root cause of the problem and then designing the optimal treatment. Also, iron sulfide phases are important to identify and distinguish, since the different phases show different solubility, e.g., pyrite is acid insoluble, while mackinawite can easily dissolve in acids.

Another way of identifying the phases and their structure properties is by using combined techniques like EDS with SEM (EDX/SEM) or EDX/ESEM.

FT-IR can also be used to identify the molecules and the compounds and their polymorphs. The deposit spectra might show some differences from the FT-IR from IR libraries or other databases, since these were made using pure compounds under optimal lab conditions. The organic residues in the scale layers can also cause some interference.

Ivakhnenko et al. [29] developed a magnetic method of identifying mineral scales. Scales can be classified into magnetic classes comprising diamagnetic, paramagnetic, and ferrimagnetic scales. A method based on laser-induced breakdown spectroscopy (LIBS) was developed by Siozos et al. [30] to identify the main components of the mineral scales, with the ability of applying it in the field analysis and process monitoring.

During scale analysis, it is critical to analyze each layer separately. Sometimes each layer can have different composition from the others, and from the bulk in general. Knowing the composition of each layer can be crucial for designing cleaning jobs. For multilayer scale with different compositions, a multistep chemical cleaning for the different layers should be considered, especially if they have different solubility behavior, such as an acid soluble layer followed by an acid insoluble layer. For example, if the scale layer that is thought to be facing the fluid in the pipe (the inner layer) is acid soluble and the successive layer is acid insoluble, the cleaning should consider an acid-based dissolver at the first step of cleaning, followed by a proper dissolver for the next layer, like EDTA or similar products for sulfate scales. This first layer must be dissolved to allow dissolver penetration to the next layers. The dissolver can also be designed to target the cementing materials between the layers, hence causing their collapse and eventual dissolution. Fig. 14.14 illustrates a scale deposit sample with different layers, where the inner and outer layers contain mainly acid-insoluble sulfate scale, while the in-between layer contains a mixture of acid solubles and acid insoluble scales.

Using the layer analysis can also provide perspective on the factors controlling the deposition mechanism, like hydrodynamics and induction time, which will be obvious when layers have the same composition but different texture and strength.

14.6.1.4 Biofouling samples

Biofouling samples must be handled with optimum care. The sample container must be sterilized before use, the operator must be trained on handling such samples, and sample preparation and preservation must follow the industry standards.

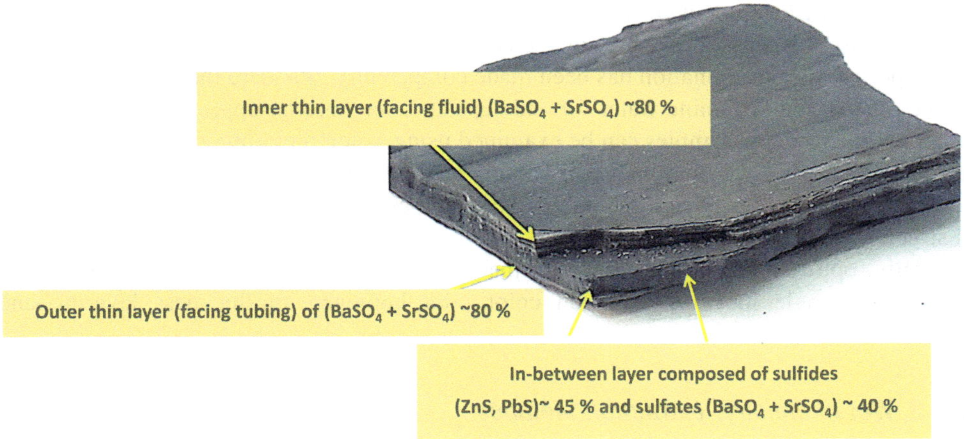

FIG. 14.14

A flake of scale sample showing multiple layers of different composition.

The biofouling can be identified using FT-IR during the initial screening of the sample. Microbial growth tests are performed according to standard method (NACE standard TMO194-2014), which describes the culture-based method (MPN), considered the most common way to test for biofilm-forming microorganisms in the oil and gas industry. Also, standard method API-RP 38 is still a useful standard in use for the biological analysis of surface waters. The new revolutionary molecular biological methods (MMM) were found to give better identification of the types of bacteria, better counting, and a smaller effect of sampling and preservation on the analysis.

14.6.2 Analysis techniques

Many analytical techniques are usually utilized for analysis and characterization due to their complex nature. A combination of these purposely selected techniques and a knowledge of the process operations and fluids where the deposits originally formed are required for full understanding of the deposit formation root cause and planning for efficient mitigations.

The analytical methods/techniques used must be validated to ensure that they do what they are intended to do. Validation provides an assurance of reliability during normal use, ensuring that an analytical methodology is accurate, specific, reproducible, and rugged over the specified range that a target analyte will be analyzed.

Analytical procedures are characterized by a number of parameters known as "figures of merit," such as selectivity, sensitivity, accuracy, precision, limit of detection (LOD), limit of quantification (LOQ), dynamic range, robustness and ruggedness. Other parameters that need specific attention during analysis are interferences, matrix effect, and dilution effects.

14.6.2.1 Visual and microscopic inspection

Visual and microscopic inspections are conducted on the sample as received to ensure that the selection of sample portions for examination has been made competently. Standard method ASTM D1245 furnishes the procedure to examine water-formed deposits using microscopy.

In optical microscopy, samples can be examined using a stereomicroscope or a compound optical microscope. Polarized light illumination is another powerful tool for particle classification. Confocal laser scanning microscopy gives a quantitative three-dimensional image of the scale surface or the internal structures [1,28].

Generally, optical microscopy gives information about sample features like texture, crystalline habit, extinction angle, index of refraction, color, odor, homogeneity, thickness, layering, and other features.

14.6.2.2 Wet chemistry, extraction methods

Qualitative and quantitative wet chemistry methods are very common in oilfields. Qualitative chemistry tests are used in the field and lab tests for early identification of the deposit main components. Quantitative methods like gravimetric and volumetric methods are used to quantify the amounts of the individual components of the deposit. These methods are easy, cheap, and do not need special skills from the analyst; however, they are not as accurate or fast as the instrumental techniques. Wet chemistry methods are commonly used to determine halite scale (as chlorides) by titrating dissolved sample with $AgNO_3$.

14.6.2.3 Electron microscopy

Electron microscopy (EM) is used where a higher resolution is required than that obtainable by light microscopy techniques [31]. The general principle of electron microscopy techniques is the interaction between an electronic beam and the atoms of a target sample, which usually emits different signals that are collected by specific detectors and converted into an image of the sampled area [32].

Unlike the light microscope, electron microscopes like SEM are used to give quantitative analysis of the scale samples. Scanning electron microscopy gives a high-resolution image of the sample surface, showing fine details down to the 100 nm scale [1]. Crystal structure, cross sections, layering, and other features of deposits can be deduced from SEM. Due to the fact that scale deposition is a surface process, SEM is widely used in deposits analysis, and is also used in surface processes related to deposition and fouling, including underdeposits corrosion (UDC) [33] and biofouling or microbial corrosion (MIC) [34].

Jordan et al. [35,36] used the environmental scanning electron microscope (ESEM) combined with energy dispersive analysis (EDX) to analyze scale deposits collected on membrane filters as suspended solids from produced waters to investigate scaling problems. The technique allows drilling related solids, reservoir rocks, to be distinguished from scale formed within produced brine; it also allows the scale deposits to be distinguished as actively growing deposits or distorted due to the effect of an inhibitor, which helps in evaluating the applied treatments.

While SEM focuses on the sample's surface and its composition, TEM provides details about internal composition. Therefore TEM can show many characteristics of the sample, such as morphology, crystallization, stress, or even magnetic domains. Generally, TEM has much higher resolution than SEM. SEM also provides a three-dimensional image, while TEM provides a two-dimensional picture.

14.6.2.4 X-ray diffraction

X-ray diffraction (XRD) gives qualitative and quantitative information about minerals and can be used for the identification of mineral phases. XRDs are working according to Bragg's law, where the samples are irradiated with X-ray beams of short wavelength, and the X-rays interact with the atoms in the crystal and are scattered in a unique diffraction pattern, which produces a fingerprint of the crystal's structure [1]. [ASTM D934] covers standard practices for identification of crystalline compounds in water-formed deposits by X-ray diffraction. XRD can be coupled with elemental analysis like EDX, XRF, or similar methods to improve its accuracy and remove erroneous phase identification [1].

Due to the valuable details that can be obtained from XRD, it has been used extensively in the literature in scale deposits analysis [33,37]; it was also used to understand the occurrence of naphthenate deposits [27,38], asphaltenes characterization [39,40], understanding corrosion under deposits [41], and to understand biofouling, reservoir rocks characterization [42], biofilm structure, and MIC mechanisms [43].

14.6.2.5 X-ray fluorescence

X-ray fluorescence (XRF) is a bulk elemental analysis technique that uses X-rays as the excitation source. XRF usually requires minimum sample preparation. XRF is superior in characterizing amorphous deposits that are unsuitable for XRD or microscopic techniques [1].

XRF has been used to identify geological in addition to production deposits [37,44]. ASTM D2332 covers standard practice for analysis of water-formed deposits by wavelength-dispersive X-ray fluorescence.

14.6.2.6 X-ray photoelectron spectroscopy

X-ray photoelectron spectroscopy (XPS) is much more a surface rather than a bulk analysis technique (normally penetrates a few nanometers into the material). A major advantage of this technique is that minimal sample size is required (grams or even milligrams); however, measurements can only be made under high vacuum conditions [1]. XPS is used to analyze geological samples in addition to scale deposits and also to study corrosion scales, where analysis of a very thin surface layer may be required [45], along with underdeposit corrosion [46], biofouling, and MIC [43].

14.6.2.7 Inductively coupled plasma based techniques

Inductively coupled plasma (ICP) based techniques use high-temperature discharge generated by flowing an ionizable gas through a magnetic field. The samples in the form of solutions are introduced to such flame and go through vaporization, atomization, excitation, and ionization. In atomic emission techniques like ICP-AES or ICP-OES, the light emitted by the element after their relaxation is detected to quantify their concentration. The detection limit of ICP-OES is in the range of ppm, and ppb for some elements. In ICP-MS the ions are extracted from the plasma and feed into a mass spectrometer, where the ions are separated by their mass-to-charge ratio and the detector receives a signal proportional to the ion concentration. The detection limits for ICP-MS are in the ppb and ppt ranges for some elements. ICP-OES or ICP-MS are used for elemental analysis and further analysis for anion content must be carried out to elucidate the full composition of the deposit sample.

ICP techniques are bulk destructive techniques, as the samples must be ground and dissolved in a proper solvent. ASTM 2331 covers the step-by-step procedure to digest and dissolve the water-formed scale deposits using the proper dissolver.

14.6.2.8 Infrared spectroscopy

Infrared spectroscopy (IR) is a widely used qualitative technique that is used to give information about the functional groups, and the bonding between the atoms in the compound (known as the fingerprint). It can also be used quantitatively. Fourier transform infrared (FT-IR) and attenuated total reflectance FT-IR (ATR-FTIR) are advanced techniques known to be precise, sensitive, fast, and have small sample size and less sample preparation (samples are analyzed in solid form).

FT-IR is extensively used for qualitative purposes in deposits analysis, during the initial screening and to distinguish between compounds and phases. The technique is highly recommended to be the qualitative starting point for most deposits analysis, as it gives precise qualitative information about the nature of the sample: inorganic/organic, treating chemicals, corrosion, and biofouling, thus reducing analysis time [47–49].

Tay [50] used FT-IR to identify and analyze clusters of chemically different compounds in crude oil deposits from the refinery, such as asphaltenes, carbonates, sulfates, sulfoxides, oxalates, and even “coke-like” materials. Synchrotron radiation-based Fourier transform infrared (SR-FTIR) technology can be used to distinguish bacteria from archaea within the biofilm by comparing spectral features of their lipids in the C-H region due to differences in cell envelope compositions [51].

FT-IR is a very helpful tool in identifying and characterizing pseudoscale or treating chemical incompatibility scales. Fig. 14.15 shows IR spectra of pseudoscales collected from low pressure gas compressor, which showed matching with a corrosion inhibitor; thus the samples were identified (precipitated corrosion inhibitor), root cause defined (chemical overdose), and the problem later was solved (dose optimized, injection point fixed).

It is worth mentioning that, in some instances, the compared chemical has to be treated/manipulated (evaporating the solvent, or reacting with another chemical, produced water, or oil and then evaporating the solvent) to get a complete matching in FT-IR spectra between the tested chemical and the deposit sample.

14.6.2.9 Raman spectroscopy

Raman spectroscopy is based on the Raman effect, which is inelastic scattering of photons by the molecular species. It is a nondestructive technique that provides detailed information about chemical structure, phases and polymorphy, crystallinity, and molecular interactions. Raman spectroscopy is ideal for examining inorganic surfaces, and while not usually as effective in resolving specific chemical species as FTIR, can readily distinguish between, for example, sulfates and carbonates in a heterogeneous scale sample of corrosion product scale samples [52]. Raman spectroscopy is used as a quantitative tool to determine the composition of mixed gas hydrates [53] and asphaltenes deposits [54].

14.6.2.10 Thermogravimetric analysis

Thermogravimetry (TG) is a thermal analysis technique in which the mass of the sample is measured as a function of temperature over time, while the substance is subjected to a controlled temperature and atmosphere [55,56]. In scale-forming minerals, these changes will be associated with water content, loss of waters of hydration, and decomposition of anions (e.g., loss of CO_2 from CO_3^{--} and SO_2 from SO_4^{2-} to give oxides) [1]. Moreira and Teixeira [25] used FT-IR and thermogravimetry (TG) to characterize mineral scales and naphthenate deposits.

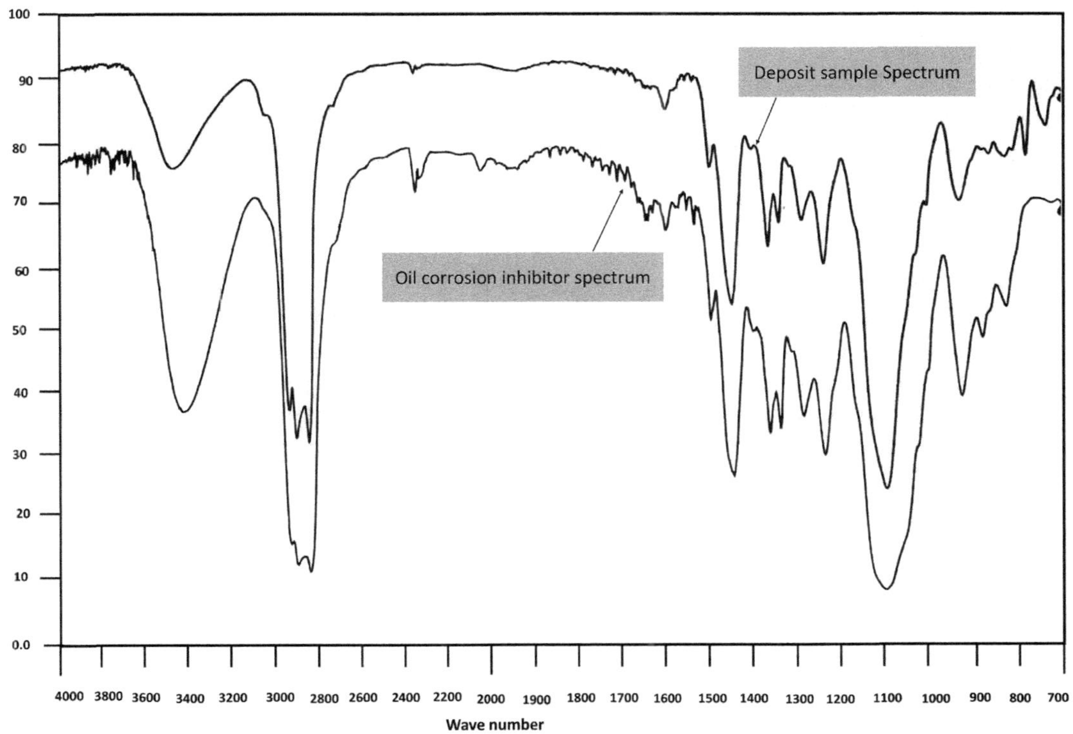

FIG. 14.15

Using IR technique to identify the deposit sample composition. A match was found between the deposit sample and corrosion inhibitor that was injected in the system.

14.6.2.11 Atomic absorption spectroscopy

Atomic absorption spectrometry (AAS) is another leading technique in elemental analysis, like ICP-OES and ICP-MS. Unlike the ICP methods, however, in AAS the amount of light absorbed (not emitted) is measured to quantify the element, after the sample is dissolved and vaporized in a flame or furnace [57,58]. The element to be measured must be known beforehand. AAS has a lower detection limit than ICP-OES and ICP-MS and a longer analysis time, but still it is one of the cheapest and most sensitive instruments.

14.6.2.12 Gas chromatography

Gas chromatography (GC) is one of the leading techniques in characterizing crude oil and its compositions. GC is also used to characterize paraffin wax deposits by measuring their carbon distribution, which is critical in designing and choosing chemicals [19].

GC is also used in characterizing asphaltenes and asphaltene deposits. In addition, it is used to solve complex samples where treating chemicals or an unknown hydrocarbon source is introduced to the system. GC-MS is a more valuable technique, since it combines GC and mass spectrometry features. GC-MS has been used in analyzing asphaltene and naphthenate deposits [26].

14.6.2.13 Ion chromatography, liquid chromatography, high-performance liquid chromatography

These are other chromatographic techniques. high-performance liquid chromatography (HPLC) and liquid chromatography (LC) were used to characterize naphthenate deposits [59,60], and ion chromatography (IC) is used to measure ion concentration, which can be applied in the case of mineral scales after the samples are dissolved in the proper solvent [61].

14.6.2.14 Other techniques

Other techniques that are not so common in practice but have been successfully used to characterize deposits include EDX, which is another elemental analysis method that is usually combined with an electron microscopy method like SEM or TEM, time-of-flight secondary ion mass spectrometry (ToF-SIMS) [1], ion electrospray ionization Fourier transform ion cyclotron resonance mass spectrometry (ESI-FT-ICR-MS) [24], and particle size analysis. Table 14.3 summarizes the different analytical techniques.

Table 14.2 summarizes the different types of analytical techniques and their uses.

Table 14.3 Summary and comparison between the different analytical techniques.

Technique	Information	Pros and cons
Wet chemistry	Elemental, composition	Simple Time consuming Variable accuracy depending on the operator and method limits of detection
Microscopy	Morphology, size	Provide details of surface structure, high cost of SEM, TEM
ICP methods	Elemental	High analysis speed after sample preparation High sensitivity Better limits of detection High cost Suffers from matrix effects Requires skilled operator
FT-IR	Composition	Compounds fingerprinting Can analyze solid or liquid forms Suffers from interferences Requires skilled operator
XRD	Composition and surface analysis	Allows phase identification High cost Requires skilled operator
Thermal methods	Composition	Allows phase identification Requires skilled operator
HPLC	Composition	Identification of organics Requires skilled operator

14.7 Interpretation of data

After the measurements are complete, the data are collected, verified, and stoichiometric calculations are done to obtain the percent weight of each component or its polymorphs in the original raw deposit sample. The results are then outlined in a form that is process related and understandable to the operators; eventually the data are presented in the final report.

14.7.1 Importance of interpretation of the data

The main aim of chemical analysis is to determine the composition of the solid sample. However, that is not achieved blindly, but has to use some process data to interpret the data from chemical analysis and formulate it into a form that is meaningful to the operators, in such way that they can use the final analysis results and combine it with other chemistry and engineering data to understand the root cause of the problem and design the mitigation plans (Fig. 14.16). This creates harmony between the lab work and the other engineering departments. Hence, it is the main goal of data interpretation to be useful to the operators. A few examples are given in the following section.

Iron compounds can be formed due to particular problems, e.g., general corrosion, sweet corrosion, sour corrosion, microbial activity, or reservoir geochemical interactions, where each problem has an indicative or distinctive end product, such as sulfides for sour corrosion or SRB biocorrosion, carbonate for sweet corrosion, etc. In this case, a good interpretation of the data is used to understand the problem and to track down its root cause, and this generally involves using the data from phase or polymorph

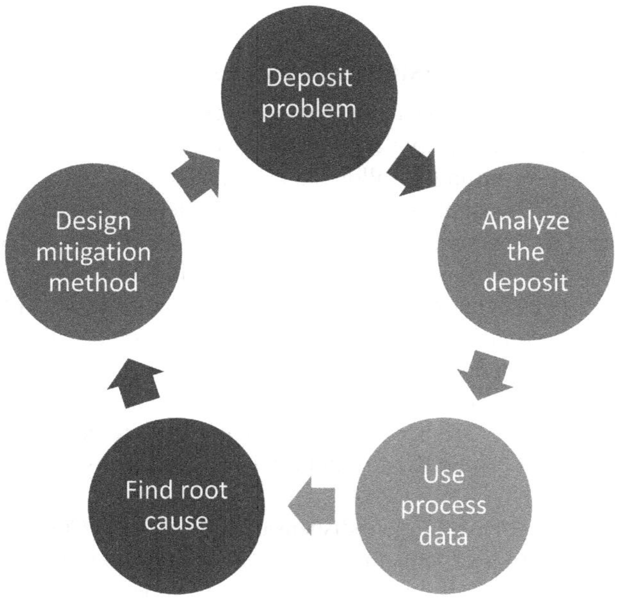

FIG. 14.16

Integrating process data with analysis data to identify the root cause and mitigate the process problems.

identification techniques like SEM, XRD, or FT-IR combined with process data such as acid gases concentrations and microbiology data.

$BaSO_4$ is common in fields; however, it can be a contaminant from drilling operations. Data from SEM or XRD can distinguish $BaSO_4$ from drilling fluids or from scaling. Hence, this can distinguish and identify the root cause of the problem.

Extending paraffin analysis to include the carbon distribution of the deposits and the gel strength is valuable in understanding the root cause of the problem and designing pigging operations, startup conditions, and choosing the chemical inhibitors.

Similarly, asphaltene content is preferably accompanied by analysis data that can specify its composition, including polarity and hetero atoms, which helps in selecting the optimum inhibitor; however, such analysis is not available in most of the field labs. Also, the metal ions Fe, Ni, V, for example, in asphaltenes can be crucial for identifying the root cause.

Additionally, it is the main and direct use of the analysis results to select and design the optimum chemical cleaning treatment. Identifying the different phases and polymorphs is crucial if they have different solubilities. A common example is iron sulfides, which are known to have different solubilities, with mackinawite known to be acid soluble and pyrite known to be acid insoluble. Also, calcium sulfates (gypsum, anhydrite, hemihydrates) have different solubilities. In addition to phase identification, dissolution studies are important, especially if they determine the dissolution rate (amount dissolved/time) under process conditions.

Another use of good data interpretation is that by knowing the root cause, the operators can optimize their system operating parameters, to operate the system within a deposits-free window.

14.7.2 The use of production system and process data during results interpretation

Analysts usually require process data that helps them solve critical problems during analysis, especially when the problem is complex with many variables. Process data are also used in interpretation of the results. A few examples are given in the following paragraphs.

Typical information such as sample source, location, updated system design, and operating conditions is helpful to investigate the main cause of the problem. The root cause can be pressure, temperature or pH change, chemical carryover, or mixing incompatible fluids.

Knowing the system design helps to understand where fluids mixing takes place, where corrosion is possible, where chemicals are injected, and where stagnancy is possible, and this later can be correlated with the location of solids formation to determine the root cause.

Knowing the composition of system materials of construction can be useful to identify the origin of iron scale. For example, carbon steel is more vulnerable to corrosion than duplex steel, so the source of Fe ions can be identified, whether it is the corrosion process or from natural reservoir iron.

Knowing process fluids composition is indispensable. Acid gases CO_2 and H_2S concentration is very important. In the Egyptian field, mineral scales composition showed mixed iron sulfide and iron carbonate scales, thought to be sweet/sour corrosion. However, the acid gases composition in the system was very low and did not support this hypothesis. After performing microbiological analysis using an advanced technique (qPCR) on the collected deposits, the results showed the presence of methanogens/SRB bacteria causing the formation of such mixed deposits.

Crude oil composition is indicative of possible organic deposits when correlated with the operating conditions and water chemistry. Gas composition is another factor in gas hydrates deposits.

Geochemical data of the reservoir are of great help in investigating reservoir deposits.

Knowing the system applied chemicals, their doses, injection points, and their SDS is necessary in deposits involving pseudoscale. Knowing these chemical pathways, the compatibility with the produced fluids and the other chemicals, and their stability under the operating conditions helps in coming to a conclusion when the analysis is performed. This is also related to the system design, especially in the case of carryover of chemicals like glycols, amines, or lubricating oils and depositing them in specific locations in the system.

14.8 Results reporting and follow-up

After results are interpreted, the next step is to present them in a report. It is very important to represent the data correctly and clearly, rather than "stylishly." Standard test method ASTM D 933-84 furnishes the procedure to report the results of analysis of water-formed deposits. It provides useful details about the proper reporting of deposits analysis results.

A summary of the important data to report is as follows:

- Sample supplier name/department, source of sample, field name, time and date of sampling and date of analysis, number of samples, and method of sampling.
- Sample properties: approximate sample size, texture, color, odor, hardness, thickness, number of layers if possible, magnetism, and hygroscopic.
- Whether there is any anomaly, metal pieces, rubber parts, lumpy rocks, or gravel.
- Final % weight of every compound, phase, or polymorph.
- The analysis techniques used.
- Whether biomasses detected and the approximate time that biological test results will be available; if already available, report bacterial type, count and severity, and refer to the biocide kill test if there is one, or conduct one if no biocides have been applied.
- Possible root cause for the deposit type: for example, fluids mixing, water breakthrough, pressure change, temperature change, corrosion, biocorrosion, geochemical, treating chemical compatibility, system hydrodynamics, maintenance operations (acidizing, fracturing, redrilling, reperforating).
- Dissolution study for each type of deposit, and possible inhibition chemicals (based on the literature data and lab studies).
- Common prevention methods for each type.

14.9 Case studies

CASE STUDY 1. Incomplete deposits sample analysis from Egyptian field goes wrong

A deposit sample was collected from a producing well in an Egyptian oilfield during well gauging operations. The sample analysis is shown in Table 14.4.

Table 14.4 Composition of scale sample collected from the well.

Compound	$CaCO_3$	FeS	ZnS	PbS	Organic deposits	Sand residues
% weight	55.0	6.3	8.7	2.0	21.0	7.0

As shown in Table 14.4, the organic deposits were not specified as to whether they were paraffins or asphaltenes. Based on this analysis, the cleaning job was designed based on removing the 21% hydrocarbons using a diesel and xylene mixture, followed by acid cleaning to remove the 72% acid solubles (calcite and sulfides). During the cleaning, some of the acid was lost into the formation. Just days after the cleaning job was performed, the production rate started to gradually decline. Upon performing well gauging, obstructions were detected at multiple points in the well, and deposit samples were collected from these locations. The analysis of the collected samples showed that the deposit samples were composed of asphaltenes and paraffin, as given in Table 14.5.

Fig. 14.17 shows part of the asphaltene deposit samples collected from the well at depth 6000' THF. Apparently the asphaltenes problem was induced in the well right after the acid job, due to the significant effects on asphaltenes stability of the acids and the high iron content in the fluids after the acid job. These deposits were later removed using a combination of diesel, xylene, and other specialty chemicals.

Surprisingly, when the old deposits sample was accurately analyzed [the one that had 21% organic deposits without specifying them (Table 14.4)], it was found that the organic fraction in that deposit was mostly asphaltenes. In this case, the operator could have been alerted that the crude has a tendency for asphaltene deposition and utmost care could have been taken during the cleaning to avoid acid contact with the crude oil, or another safe cleaning solution could have been chosen,

Table 14.5 Analysis of deposit samples from well at different obstruction depths.

	% weight	
Compound	Paraffin	Asphaltenes
Sample #1 (6000' THF)	80%	20%
Sample #1 (2600' THF)	52%	48%

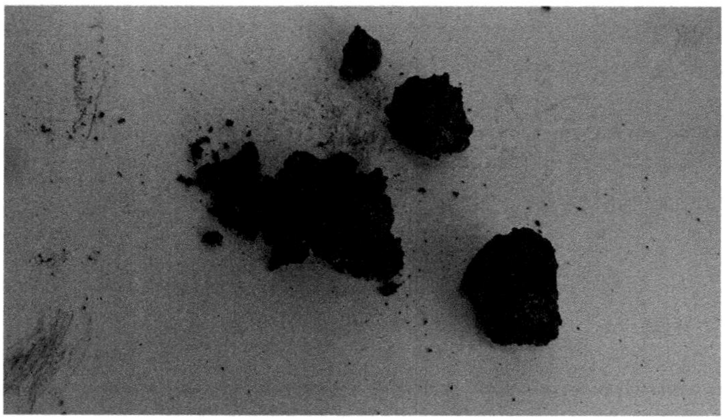

FIG. 14.17

Asphaltene deposits from producing well in an Egyptian oilfield.

and thus a significant deposition problem could have been averted if proper and complete analysis of the deposits sample had been performed. Further, lab investigation confirmed that, while the crude has low asphaltenes content of 2.5%, it has a high tendency for asphaltene deposition.

CASE STUDY 2. Sample from gas lift compression system

Two deposit samples were collected from the gas lift compression system from an offshore oilfield in Egypt. The sample was a composite deposit that included mineral scales and organic scales. The mineral part was crushed powder and the organic part was loose sludge material. The operator mentioned that the gas lift system showed some fluctuating behavior. Preliminary examination of the samples showed sulfides scale, with the existence of some source of chemical contamination. FT-IR fingerprinting of both samples confirmed the chemical contamination; therefore a request was made to bring samples from all the treating chemicals used on that platform. After scanning all the chemicals, a partial match was found between the first sample and the gas corrosion inhibitor, which gave a full match upon evaporating the corrosion inhibitor solvent. The sample was then solvent extracted in toluene to determine the corrosion inhibitor concentration and the mineral residue was dissolved and analyzed using ICP-OES. The analysis results of the first sample are shown in Table 14.6.

Table 14.6 Composition of first deposit sample from gas lift system.

Compound	Precipitated gas corrosion inhibitor	FeS	ZnS	PbS	NaCl	Sand residues
% weight	86	6.0	0.3	0.7	0.5	6.5

The FT-IR spectra of the second sample did not match any of the treating chemicals, so a sample of the compressor lubricating oil was requested and FT-IR fingerprinting was performed, which showed a complete match, except for a few peaks, which were identified later to be lubricant degradation by-products (sulfation). The deposit was solvent extracted using an organic solvent, and the mineral scales were dissolved and analyzed. The full composition of the deposit is given in Table 14.7.

Table 14.7 Composition of the second deposit sample from gas lift system.

Compound	Degraded lube oil	$CaCO_3$	NaCl	Sand residues
% weight	40	3	32	25

Based on these results, it is apparent that a severe corrosion problem was encountered in the gas compression system. The main investigation results pointed at an overdose of the corrosion inhibitor, leading to its precipitation from one side and exaggerating corrosion from another side. That was confirmed by checking the chemical pump, which was found to be adjusted at a higher dose than planned. Also, the supply gas was found to have an impact, causing corrosion and corrosion products due to the relatively high H_2S.

CASE STUDY 3. Deposition of oil corrosion inhibitor in high-pressure gas pipeline in Egypt

Similar to Case study 1, a deposit sample was collected from a high-pressure gas line. The sample was organic in nature and dissolved in methanol. The sample was dried and fingerprinted using an IR technique, which showed a complete match with an oil corrosion inhibitor after evaporating the corrosion inhibitor solvent. Fig. 14.18 shows the IR spectra of the deposit sample, which completely matches with the corrosion inhibitor.

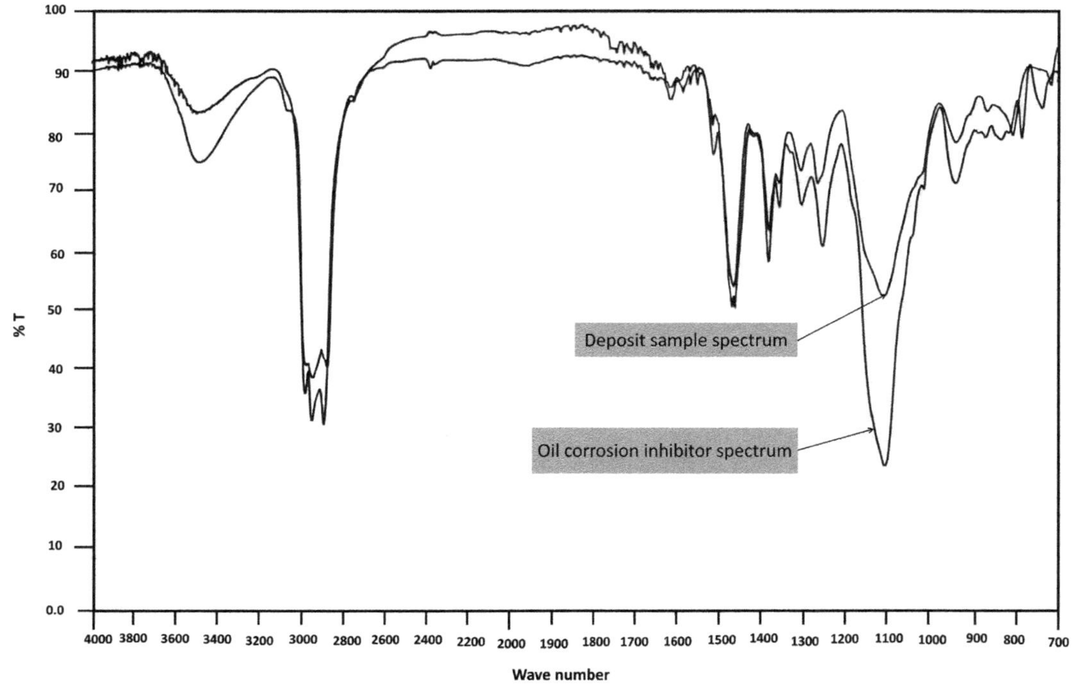

FIG. 14.18

IR spectrum of deposit sample, which matched with the spectrum of the corrosion inhibitor injected in the system.

Further communication and investigation with the engineering team showed that the corrosion inhibitor was diluted using methanol and injected in the high-pressure line. That leads to evaporating or flashing the solvent and thus depositing of the corrosion inhibitor. Some solutions were suggested to the engineering team, including changing the injection point, optimizing the injection dose, and changing the dilution solvent to a less volatile solvent, e.g., solar.

A similar case of corrosion inhibitor deposition was also reported in another high-pressure line and the deposit was identified in a similar way. The causes were quite the same, with an additional cause of overdosing the line with corrosion inhibitor during pigging operations.

Furthermore, a similar case was reported of depositing demulsifier in high-pressure gas, where the sample was identified in similar ways and the causes were solvent evaporation and chemical overdose, in addition to a wrong injection point.

CASE STUDY 4. Scale inhibitor deposition downhole in a production well in Egypt

A deposit sample was collected from downhole in a production well. The deposit was causing severe flow restriction in the production tubing, which required the use of a workover rig to mill the deposits. The deposits were collected from near the location of downhole chemical injection of scale inhibitor.

Regular sample analysis after dissolution in distilled water, acid, and hydrocarbon extraction showed 3% organic deposits, 25% (Fe, Zn, Pb)S, besides a high Ca content compound that composes the rest of the deposit. Therefore FT-IR fingerprinting of the deposit was performed, which showed the presence of a treating chemical. Several trials were conducted to find a match with the injected chemicals, yet no complete match between the deposit and any of the chemicals was attained, although a partial match was found between the deposit and the scale inhibitor. Since the blockage was near the scale inhibitor chemical injection point (injection through capillary line), the focus was on the assumption that some sort of chemical reaction between the inhibitor and the produced fluids could lead to the formed deposit. The chemical data showed that the produced water has high calcium content [26,000–31,000 ppm]. Therefore the lab work involved reacting the scale inhibitor with artificial high Ca content water, followed by evaporating the solvent at high temperature close to the downhole temperature of 198°F; then the product of this reaction (Ca-scale inhibitor precipitate) was fingerprinted, which showed complete matching with the deposit sample. Further investigation showed that the scale inhibitor was overdosed due to a faulty injection system; also no monitoring or maintenance were done to the umbilical lines system since it was installed. Preventive maintenance work was done to the injection system and a maintenance schedule was planned for the umbilical line to solve the problem and to avoid future complications.

14.10 Summary

Deposits analysis and characterization represent a crucial step in the flow assurance management plans. Addressing the detailed composition of the formed deposits helps in identifying the root cause of the problem, maintaining the removal procedure, and planning further mitigation and management strategies. Furthermore, the detailed composition of the deposits is used to understand other aspects of the production system, including the following:

- Fluids chemical compatibility
- Fluid hydrodynamics
- Geochemical interactions
- Corrosion root causes and mechanisms
- Microbiological activity

Samples collection and labeling with the right sample and system information is the initial crucial step. Usually the deposit samples are a composite mixture of organic and inorganic deposits. That requires fractionation and preparation by separating organics from inorganics, followed by dissolving the sample in a proper dissolver before introducing it to the proper analytical technique.

The proper technique must be used to give the necessary information that will be interpreted to give sample composition components in percent weights. Furthermore, production system details and parameters should be considered during analysis and results interpretation, so that the final results are dynamic and linked to the field operations.

References

[1] C.P. East, T.L. Schiller, C.M. Fellows, W.O.S. Doherty, Analytical techniques to characterize scales and deposits, in: Mineral Scales and Deposits: Scientific and Technological Approaches, Elsevier, 2015, pp. 681–699, https://doi.org/10.1016/B978-0-444-63228-9.00028-0.

[2] D.W. Rackemann, W.O.S. Doherty, C.P. East, Development of descriptor tools for the characterization of Australian sugar mill evaporator scale, in: Proceedings of the 27th International Society of Sugar Cane Technologists Congress, 2010.

[3] ASTM D 887–82, Reapproved, Standard Practices for Sampling Water-Formed Deposits, ASTM International, West Conshohocken, PA, 1999.

[4] NACE International, Preparation, Installation, Analysis, and Interpretation of Corrosion Coupons in Oilfield Operations, NACE-SP0775-2013, NACE International, 2013, January 1.

[5] J. Kilbane, Testing for Microbiologically Influenced Corrosion in Pipelines, Corrosionpedia, August 14, 2017.

[6] J.C. Cowan, D.J. Weintritt, Water-Formed Scale Deposits Houston, Gulf Publishing Company, Texas, 1976.

[7] A.G. Ostroff, Introduction to Oilfield Water Technology, Prentice-Hall, Englewood Cliffs, NJ, 1965.

[8] W.W. Frenier, M. Ziauddin, A multifaceted approach for controlling complex deposits in oil and gas production, in: Proceedings—SPE Annual Technical Conference and Exhibition, vol 1, Society of Petroleum Engineers (SPE), 2010, pp. 285–308, https://doi.org/10.2118/132707-ms.

[9] S.S. Alian, K. Singh, A. Saidu Mohamed, M.Z. Ismail, M.L. Anwar, Organic Deposition: From Detection and Laboratory Analysis to Treatment and Removal, Society of Petroleum Engineers, 2013, October 22, https://doi.org/10.2118/165912-MS.

[10] A. Cosultchi, E. Garciafigueroa, B. Mar, A. García-Bórquez, V.H. Lara, P. Bosch, Contribution of organic and mineral compounds to the formation of solid deposits inside petroleum wells, Fuel 81 (4) (2002) 413–421, https://doi.org/10.1016/S0016-2361(01)00187-9.

[11] A.A. Al-Taq, S.M. Abou Zeid, H.H. Al-Haji, J.A. Saleem, Removal of organic deposits from oil producing Wells in a sandstone reservoir: a lab study and a case history, in: SPE Middle East Oil and Gas Show and Conference, MEOS, Proceedings, vol. 3, Society of Petroleum Engineers (SPE), 2013, pp. 1889–1902, https://doi.org/10.2118/164410-ms.

[12] J. AlAamri, M. AlDahlan, S. AlDarwesh, Efficiency of selected terpene-based and conventional solvents in the dissolution of asphaltene deposits, in: Society of Petroleum Engineers—SPE Kingdom of Saudi Arabia Annual Technical Symposium and Exhibition 2018, SATS 2018, Society of Petroleum Engineers., 2018, https://doi.org/10.2118/192249-ms.

[13] D. Vazquez, G.A. Mansoori, Identification and measurement of petroleum precipitates, J. Petrol. Sci. Eng. 26 (2000) 49–55, https://doi.org/10.1016/S0920-4105(00)00020-6.

[14] J. Chen, J. Zhang, H. Li, Determining the wax content of crude oils by using differential scanning calorimetry, Thermochim. Acta 410 (1–2) (2004) 23–26, https://doi.org/10.1016/S0040-6031(03)00367-8.

[15] P. Jokuty, S. Whiticar, Z. Wang, M. Landriault, L. Sigouin, J. Mullin, A new method for the determination of wax content of crude oils, Spill Sci. Technol. Bull. 3 (4) (1996) 195–198, https://doi.org/10.1016/S1353-2561(97)00011-X.

[16] X. Lu, B. Kalman, P. Redelius, A new test method for determination of wax content in crude oils, residues and Bitumens, Fuel 87 (8–9) (2008) 1543–1551, https://doi.org/10.1016/j.fuel.2007.08.019.

[17] H. Saxena, A. Majhi, B. Behera, Prediction of wax content in crude oil and petroleum fraction by proton NMR, J. Pet. Sci. Technol. 37 (2) (2019) 226–233, https://doi.org/10.1080/10916466.2018.1536713.

[18] S.S. Alian, K. Singh, A.S. Mohamed, M. Zaki Ismail, M.L. Anwar, Organic deposition: from detection and laboratory analysis to treatment and removal, in: Society of Petroleum Engineers—SPE Asia Pacific Oil and Gas Conference and Exhibition, APOGCE 2013: Maximising the Mature, Elevating the Young, vol. 2, Society of Petroleum Engineers, 2013, pp. 1263–1275, https://doi.org/10.2118/165912-ms.

[19] M.A. Kelland, Production Chemicals for Oil and Gas Industry, CRC Press, 2014, pp. 259–287.

[20] C.Y. Bai, J.J. Zhang, New method to determine the strength of wax deposits in field pipelines, Adv. Pet. Explor. Dev. 10 (2) (2015) 31–36. Available from: http://www.cscanada.net/index.php/aped/article/view/7816. https://doi.org/10.3968/7816.

[21] N.-y. Wu, C.-l. Liu, X.-l. Hao, Experimental simulations and methods for natural gas hydrate analysis in China, China Geol. 1 (1) (2018) 61–71, https://doi.org/10.31035/cg2018008.

[22] F. Rauh, B. Mizaikoff, Spectroscopic methods in gas hydrate research, Anal. Bioanal. Chem. (2012), https://doi.org/10.1007/s00216-011 5522-1.

[23] L. Oduola, C.U. Igwebueze, O. Smith, P. Vijn, A.G. Shepherd, Calcium naphthenate solid deposit identification and control in offshore Nigerian fields, Soc. Pet. Eng. (2013, April 8), https://doi.org/10.2118/164055-MS.

[24] M.M. Mapolelo, L.A. Stanford, R.P. Rodgers, A.T. Yen, J.D. Debord, S. Asomaning, A.G. Marshall, Chemical speciation of calcium and sodium naphthenate deposits by electrospray ionization FT-ICR mass spectrometry, Energy Fuels 23 (1) (2009) 349–355. 2009.

[25] A.P.D. Moreira, A.M.R.F. Teixeira, An investigation on the formation of calcium naphthenate from commercial naphthenic acid solutions by thermogravimetric analysis, Braz. J. Pet. Gas 3 (2) (2009) 051–056.

[26] D.L. Gallup, J.A. Curiale, P. Colin Smith, Characterization of sodium emulsion soaps formed from production fluids of Kutei Basin, Indonesia, Energy Fuel 21 (3) (2007) 1741–1759, https://doi.org/10.1021/ef060198u.

[27] A.S. Mohamed, A. Goyal, M.D. Mohd Ismail, G. Munainni, J.K. Amar Singh, M.Z. Ismail, M. Anwar, Detection and removal of naphthenate deposit in oil well, in: Offshore Technology Conference Asia, 2014, https://doi.org/10.4043/25003-ms.

[28] V.P. Woodward, R.C. Williams, Z. Amjad, Analytical techniques for identifying mineral scales and deposits, in: The Science and Technology of Industrial Water Treatment, CRC Press, 2010, pp. 425–445, https://doi.org/10.1201/9781420071450.

[29] O.P. Ivakhnenko, R. Abirov, A. Logvinenko, New method for characterisation of petroleum reservoir fluid-mineral deposits using magnetic analysis, Energy Procedia 76 (2015) 454–462, https://doi.org/10.1016/j.egypro.2015.07.877.

[30] P. Siozos, A. Philippidis, M. Hadjistefanou, C. Gounarakis, D. Anglos, Chemical analysis of industrial scale deposits by combined use of correlation coefficients with emission line detection of laser induced breakdown spectroscopy spectra, Spectrochim. Acta, Part B 87 (September) (2013) 86–91, https://doi.org/10.1016/j.sab.2013.05.008.

[31] R.M. Twyman, Microscopy applications | Food, in: Encyclopedia of Analytical Science, second ed., Elsevier, 2005, pp. 50–57.

[32] R. Marassi, F. Nobili, Measurement methods | Structural and chemical properties: scanning electron microscopy, in: Encyclopedia of Electrochemical Power Sources, Elsevier, 2009, pp. 758–768, https://doi.org/10.1016/B978-044452745-5.00071-X.

[33] H.H. Almahamedh, Under-deposit caustic corrosion on sodium carbonate pipeline, Procedia Eng. 114 (2015) 34–37, https://doi.org/10.1016/j.proeng.2015.08.026.

[34] J.S. Lee, B.J. Little, Diagnosing microbiologically influenced corrosion, in: Microbiologically Influenced Corrosion in the Upstream Oil and Gas Industry, CRC Press, 2017, pp. 157–175, https://doi.org/10.1201/9781315157818.

[35] M.M. Jordan, J. Buckman, C.J. Johnston, Direct monitoring of the performance of scale control programs across the produced water life cycle via suspended solids analysis, in: SPE European Formation Damage Conference, 2005, https://doi.org/10.1016/j.jbiosc.2014.10.022.

[36] M.M. Jordan, C.J. Johnston, M. Robb, Evaluation methods for suspended solids and produced water as an aid in determining effectiveness of scale control both downhole and topside, Soc. Pet. Eng. (2005), https://doi.org/10.2118/92663-MS.

[37] H. El Hajj, O. Pal, B. Zoghbi, Compositional analysis and treatment of oilfield scales, Soc. Pet. Eng. (2015), https://doi.org/10.2118/178002-MS.

[38] A.G. Shepherd, G.B. Thomson, R. Westacott, K.S. Sorbie, M. Turner, P.C. Smith, Analysis of organic field deposits: new types of calcium naphthenate scale or the effect of chemical treatment? Soc. Pet. Eng. (2006), https://doi.org/10.2118/100517-MS.

[39] R. Tanaka, E. Sato, J.E. Hunt, R.E. Winans, S. Sato, T. Takanohashi, Characterization of asphaltene aggregates using X-ray diffraction and small-angle X-ray scattering, Energy Fuel 18 (4) (2004) 1118–1125, https://doi.org/10.1021/ef034082z.

[40] F. Trejo, J. Ancheyta, T.J. Morgan, A.A. Herod, R. Kandiyoti, Characterization of asphaltenes from hydrotreated products by SEC, LDMS, LAMDI, NMR, and XRD, Energy Fuel 21 (4) (2007) 2121–2128, https://doi.org/10.1021/ef060621z.

[41] K. Allan, N. Obeyesekere, D. Fell, T. Brown, M. Bluth, Field Sidestream Testing to Evaluate the Effectiveness of Chemicals to Mitigate Under Deposit Corrosion, NACE International, 2012, January 1.

[42] A.J. Mandile, A.C. Hutton, Quantitative X-ray diffraction analysis of mineral and organic phases in organic-rich rocks, Int. J. Coal Geol. 28 (1) (1995) 51–69, https://doi.org/10.1016/0166-5162(95)00004-W.

[43] M. Sharma, G. Voordouw, MIC detection and assessment: a holistic approach, in: Microbiologically Influenced Corrosion in the Upstream Oil and Gas Industry, CRC Press, 2017, pp. 177–212, https://doi.org/10.1201/9781315157818.

[44] R.A.J. Wust, X. Cui, From X-ray Fluorescence (XRF) to Mechanical Profiling for Better Well Completion. Valuable Analytics for the Exploration Toolbox!, 2018. https://csegrecorder.com/articles/view/x-ray-fluorescence-xrf-to-mechanical-profiling-for-better-well-completion.

[45] M.J.L. Felipe, Extended Understanding of Inhibition Mechanism of New Corrosion Inhibitor Via Electrochemical Measurements and X-ray Photoelectron Spectroscopy (XPS), NACE International, 2017, April 27.

[46] C.M. Menendez, V. Jovancicevic, Z. Zhu, M. Morton, D. Stegmann, New Method for Assessing Corrosion under Iron Sulfide Deposits and CO2/H2S Conditions, NACE International, 2011, January 1.

[47] University of North Dakota, FT-IR Spectroscopic Investigation of Fireside Deposits in a Pilot Scale Combustor, Technical Report for US Department of Energy, Grand Forks, ND, US, 1995.

[48] J. AlAamri, M. AlDahlan, S. AlDarwesh, Efficiency of selected terpene-based and conventional solvents in the dissolution of Asphaltene deposits, Soc. Pet. Eng. (2018), https://doi.org/10.2118/192249-MS.

[49] G. Curtis, C. Weaver, The application of Fourier transform infrared spectrophotometry in industrial chemical cleaning, in: Paper 337 Presented at Corrosion/98, NACE International, Houston, TX, 1998, January 1.

[50] F.H. Tay, Characterisation of Deposited Foulants and Asphaltenes Using Advanced Vibrational Spectroscopy, 2009. PhD.

[51] A.J. Probst, H.Y.N. Holman, T.Z. Desantis, G.L. Andersen, G. Birarda, H.A. Bechtel, C. Moissl-Eichinger, Tackling the minority: sulfate-reducing bacteria in an archaea-dominated subsurface biofilm, ISME J. 7 (3) (2013) 635–651, https://doi.org/10.1038/ismej.2012.133.

[52] Z. Zeng, K. Natesan, M. Grimsditch, Effect of Oxide Scale Compositions on Metal Dusting Corrosion of Fe-Based Alloys, NACE International, 2004, July 1.

[53] J. Qin, W.F. Kuhs, Quantitative Raman Scattering of Guest Molecules in Gas Hydrates and Its Relevance to Gas Exchange Processes Involving N2, International Society of Offshore and Polar Engineers, 2013, January 1.

[54] J.S. Riedeman, N.R. Kadasala, A. Wei, H.I. Kenttämaa, Characterization of asphaltene deposits by using mass spectrometry and Raman spectroscopy, Energy Fuels 30 (2) (2016) 805–809, https://doi.org/10.1021/acs.energyfuels.5b02002.

[55] M.L.A. Gonçalves, D.A. Ribeiro, D.A.P. Mota, A.M.R.F. Teixeira, M.A.G. Teixeira, Investigation of petroleum medium fractions and distillation residues from Brazilian crude oils by thermogravimetry, Fuel 85 (2006) 1151–1155.

[56] M.C. Mayoral, M.T. Izquierdo, J.M. Andrés, B. Rubio, Different approaches to proximate analysis by thermogravimetry analysis, Thermochim. Acta 370 (2001) 91–97.

[57] M.M. Houck, J.A. Siegel, M.M. Houck, J.A. Siegel, Light and matter, in: Fundamentals of Forensic Science, 2010, pp. 99–122 (Chapter 5).

[58] E.H. Ivanova, Atomic absorption spectrometry | Principles and instrumentation, in: Encyclopedia of Analytical Science, second ed., Elsevier, 2005, pp. 149–156. 2005.

[59] M. Bataineh, A.C. Scott, P.M. Fedorak, J.W. Martin, Capillary HPLC/QTOF-MS for characterizing complex naphthenic acid mixtures and their microbial transformation, Anal. Chem. 78 (24) (2006) 8354–8361, https://doi.org/10.1021/ac061562p.

[60] B. Brocart, M. Bourrel, C. Hurtevent, J.-L. Volle, B. Escoffier, ARN-type naphthenic acids in crudes: analytical detection and physical properties, J. Dispers. Sci. Technol. 28 (2007) 331–337.

[61] R.A. Durst, W. Davison, K. Toth, J.E. Rothert, M.E. Peden, B. Griepink, Analysis of wet deposition (acid rain): determination of the major anionic constituents by ion chromatography, Pure Appl. Chem. 63 (6) (1991) 907–916, https://doi.org/10.1351/pac199163060907.

CHAPTER

Mineral scale management

15

15.1 Introduction

Mineral scale formation is one of the major challenges in oil and gas fields. The effects of scale are detrimental to production operations, requiring early detection and management.

Scale control, management, or mitigation are terms that are often used interchangeably to refer to all the measures and methods for eliminating or reducing the impact of mineral scales (or generally all the deposits) on production operations. These measures and methods work in two main ways: to *prevent* the formation of scale deposits and/or to *remove* the scale deposits that have already formed. These ways of controlling mineral scales must be:

- Applicable and effective
- Economically feasible
- Time saving
- Environmentally friendly.

Fig. 15.1 summarizes the different ways of controlling mineral scales in oilfields.

15.2 Mineral scales prevention

The phrase "prevention is better than cure" is a fundamental principle of modern healthcare. The same approach is applied in oil and gas fields, where preventing scale formation averts its deteriorating effects. Although scale prevention seems to be simple and possible theoretically, field experience has shown that it is a complex process requiring cooperation of different science and engineering disciplines to achieve satisfactory results. Scale prevention technologies in the different industries have witnessed numerous advances and improvements.

Scale prevention falls into three basic categories: *operational*, *chemical*, and *nonchemical* methods. The operational methods are based on optimizing the production parameters to be within scale-free zones by eliminating the fators that lead to scale formation. The chemical methods are implemented by adding chemical additives to the flowing stream to inhibit scale formation in its early stages (nucleation, crystal growth). The nonchemical methods use physical or mechanical means to manipulate the water properties, leading to scaling prevention.

Essentials of Flow Assurance Solids in Oil and Gas Operations. https://doi.org/10.1016/B978-0-323-99118-6.00009-5

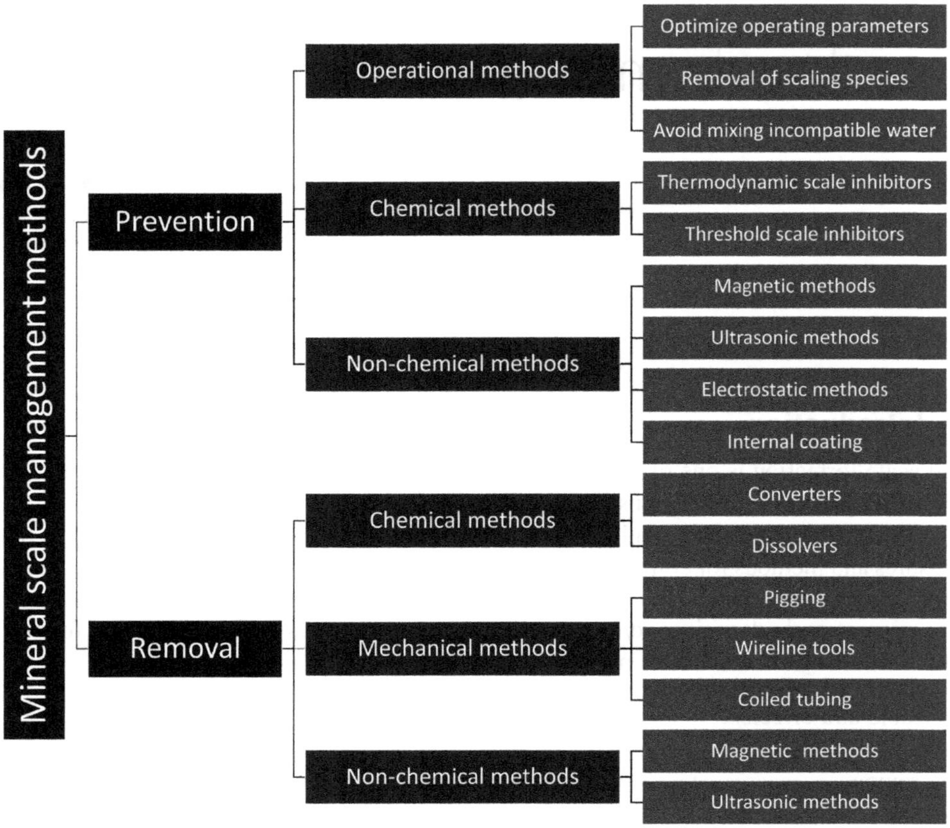

FIG. 15.1

Mineral scale management methods.

15.2.1 Operational methods of scale prevention

15.2.1.1 Controlling operating parameters and fluids mixing

Scale deposits are basically formed due to changes in temperature, pressure, flow rates, pH, or mixing of different streams. Thus avoiding significant changes in these parameters and optimizing them within a scale safe zone can be the first line of defense against scaling issues, as long as manipulating/optimizing these parameter will not lead to production loss. Major steps to avoid scale formation include:

- Avoid mixing of incompatible waters as much as possible. This can be achieved by choosing compatible water for injection, proper system design, production streams re-routing and splitting. If splitting streams is not possible, chemical inhibitors should be injected upstream the mixing point to prevent scale formation. Produced water reinjection is a common option to control scale if sea water injection will cause deposition troubles, such as hard barium sulfate scale [1].

- Water dilution can be generally used to attenuate the scaling risk of injection sea water when mixed with produced water. It is also a feasible mitigation method in the case of halite deposition, where fresh water is basically injected through the annulus between the production tubing and casing.
- Avoid significant pressure drops in the system.
- Avoid adding chemicals that increase pH, such as caustic compounds, triazine-based H_2S scavengers, and others; also, pressure drop can cause pH changes.
- Operate at flow rates sufficiently higher than the average settling velocity of the solid particulates. It is common to observe scaling downstream in long pipelines while the upstream is scale free, due to pressure and friction causing the solids suspended in the solution to settle and adhere to the pipe wall. Also, higher flow rates can strip off the solids attached to the pipe walls.
- Adjust the design of the system to reduce the dead legs, bends, dead ends, and points of stagnant flow rates to avoid suspended solids accumulation.
- Moisture control in gas pipelines helps reduce black powder formation.
- Good commissioning and installation practices of the pipelines and equipment, improving hydrotesting, proper dewatering and drying, and proper cleaning of the lines, removing mill scale and other particulates that may induce corrosion or act as nucleating sites for scale.

15.2.1.2 Injection water pretreatment

The main purpose of water treatment is to remove or reduce the concentration of the scale-forming and scale-inducing species, like dissolved gases and scaling cations and anions.

- **Dissolved gases removal**

Free and dissolved H_2S, CO_2, and O_2 gases in water have tremendous impacts on the integrity of the production facilities. They cause corrosion and scaling, as detailed in Chapter 5, "Mineral scales in oil and gas fields." Removal of these dissolved gases from oilfield water can be achieved by two methods: mechanical or chemical.

Mechanical methods involve thermal degassing, gas stripping, and vacuum degasification. Thermal degassing is achieved by heaters; deaerating heaters are used to remove oxygen for some plant operations, but are not usually used in the oilfield [2]. Gas stripping and vacuum degasification techniques are usually used in removing dissolved gases in injection water. In gas stripping, counter-current stripping towers contain packing or perforated trays, where the water flows into the top of the tower and the stripping gas (usually natural gas) is put in at the bottom, thereby increasing the bubbling gas concentration (pressure), and the dissolved gas partial pressure decreases. In vacuum towers, which also can be packed or trayed towers, by decreasing the total pressure by vacuum, the partial pressure of the dissolved gas consequently decreases [2,3]. Fig. 15.2 illustrates the mechanical degasification methods. Other new technologies of water degasification have also been reported.

Chemical methods of degasification are based on the application of oxygen scavengers and H_2S scavengers. Oxygen scavengers like bisulfites are applied in injection sea water and also during the various chemical injection processes downhole, e.g., scale inhibitor squeeze, fracturing, and workover maintenance. H_2S scavengers like triazines have been applied in the downhole and topsides applications. However, these chemicals must be used with caution since they can be potentially scaling, i.e., bisulfites can be a source of sulfate (by oxidation), causing sulfate scale formation, while triazines increase the pH, causing carbonates precipitation.

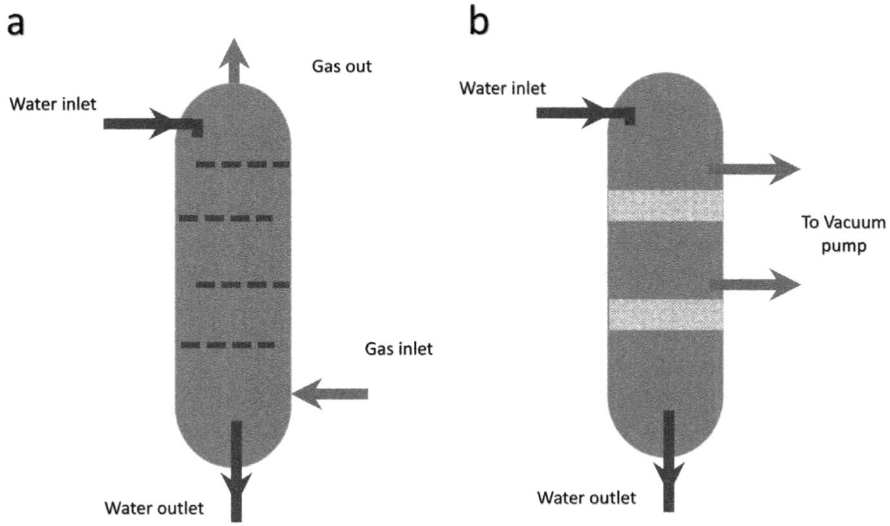

FIG. 15.2

Mechanical methods of removing dissolved gas: (a) gas stripping, (b) vacuum degasification.

Other techniques for removing dissolved gases in membrane technology are emerging. H_2S, CO_2, and O_2 removal from produced waters using membranes were reported [4,5].

- **Water softening**

Water softening is used to remove the scaling ions such as Ca, Mg, Ba, and others. Water softening can be applied by precipitation (cations stripping) or by ion exchange.

Precipitation softening processes are used to reduce raw water hardness, alkalinity, silica, and other constituents. The water is treated with a precipitating agent to form insoluble compounds, which can be separated by sedimentation and filtration [2,6]. Lime or lime soda (cold and hot process) softening, and hot process softening are common types of softening. Cogan [7] reported the removal of Ba^{+2} (by sulfation), Sr^{+2} (coprecipitation with Ca and Ba), Ca^{+2} (by lime soda), and Mg^{+2} (by hydrolysis) from synthetic produced water solutions. Precipitation softening skids and facilities are commercially available and offered by different suppliers; however, they are not very commonly in use in oil and gas fields.

In an ion exchange system, undesirable scaling ions in the water supply are replaced with nonscaling ions. For example, in a sodium zeolite softener, scale-forming calcium and magnesium ions are replaced with sodium ions [6]. Ion-exchange units have been widely used for softening boiler feed water in gas processing plants and for steam generators used in thermal recovery operations. They also are used to soften water in several of the enhanced oil recovery processes [8,9].

- **Filtration**

Filtration is a separation technique that has been used in various applications. Microfiltration, ultrafiltration, nanofiltration, and reverse osmosis have long been used to treat and clarify different oilfield

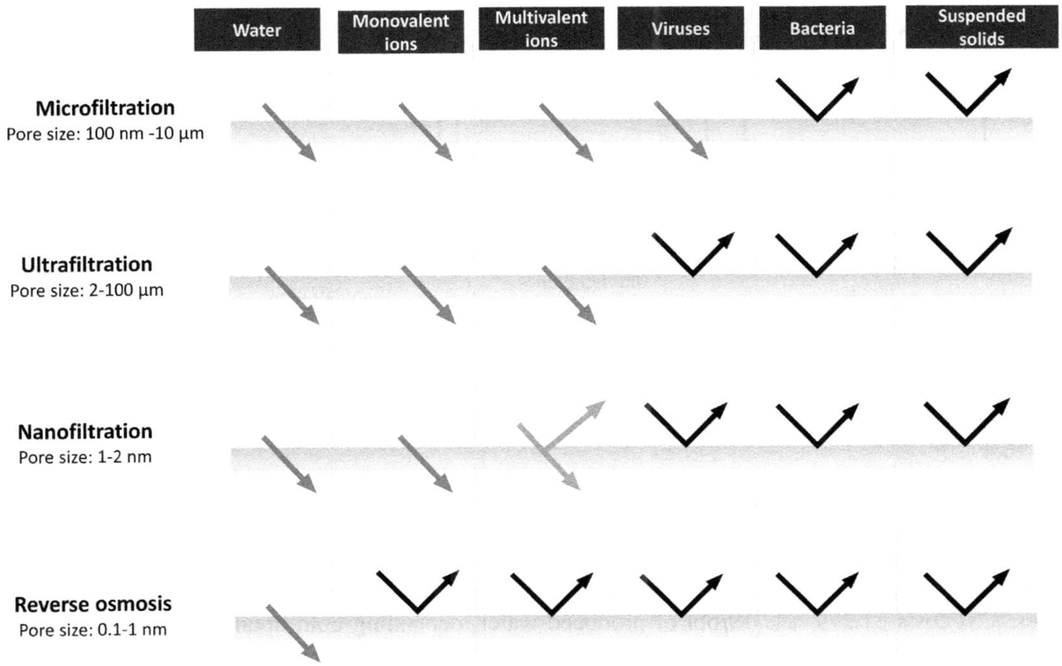

FIG. 15.3

Comparison between membrane techniques in terms of pore size and retained species.

waters (injection water, produced water, and fracturing fluids water); they basically remove oil, particulates, and scaling ions like Ca, Mg, and SO_4 [10,11]. Fig. 15.3 summarizes the different membrane techniques.

Sulfate removal/reduction processes (SRPs) are well-known examples of successful use of filtration techniques to inhibit scale deposition and reduce reservoir souring. SRP is a crossflow filtration system. SRPs were first commercialized in the 1980s, with the first sulfate removal facility (SRF) installed in 1988, on the Marathon Oil UK in the North Sea, where high concentrations of barium (1500–2000 mg/L) were discovered in the reservoir formation water, and seawater injection for pressure maintenance was likely [12]. Sulfate removal efficiency was reported to be as high as 99% [2,13].

The system comprises a pretreatment of the water source for removal of suspended solids and contaminants prior to entering the sulfate removal nanofilter membrane system. Typical pretreatment technologies have included cartridge filters, multimedia filters, or a combination of both, while more recently ultrafiltration has been gaining momentum with oil and gas operations. The nanofiltration membranes have a pore size around 1.0 nm, which is slightly larger than that of RO membranes and smaller than ultrafiltration (UF) membranes [14]. Fig. 15.4 illustrates an SRP module.

SRP is preferred, especially when the chemical inhibition is not sufficiently effective or is very costly, particularly in treating deepwater fields and installations [12,13].

Speaking of the economics, the total discounted cost for the $25/bbl oil cases indicated low sulfate sea water (LSSW) costs $0.112/bbl of water produced ($0.037/bbl of LSSW injected) and was

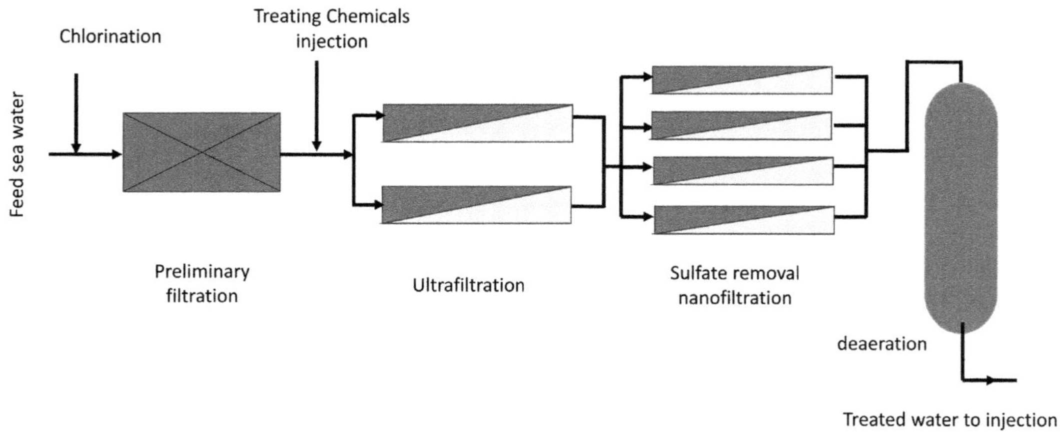

FIG. 15.4

Schematic of sulfate removal process.

insensitive to both oil price and producing scenario. For the bullheaded case, the chemical squeeze inhibition (CSI) and LSSW costs are very nearly the same, with a slight edge to CSI. For the deepwater cases, CSI costs \$1.279–\$2.719/bbl of produced water, depending both upon oil price and CSI frequency; thus LSSW is less costly by an order of magnitude [13].

SRPs have been reused in Heidrun and South Arne fields in the North Sea [15], Ewing Bank in the Gulf of Mexico (GOM), and the Girassol and Plutonio fields offshore West Africa [16,17] and offshore Brazil [18].

Other filtration techniques used are membrane distillation (MD) and liquid-phase precipitation (LPP) technology, nanofiltration (NF) and reverse osmosis (RO) in conjunction with LPP or compressed-phase precipitation (CPP) technologies [19].

Other methods of sulfate removal that can be used in different industries are:

- Sulfate removal by crystallization
- Biological sulfate removal
- Precipitation sulfate removal

However, these methods are usually applied in acid mine drainage treatments and are not in use in oil and gas operations.

15.2.1.3 Optimizing gas lift operations

Mineral scales have been reported to damage gas lift components. Optimizing gas lift operations can protect the gas lift system and other production systems against scale problems. For example, gas assisted plunger lift (GAPL) solutions can reduce the rate of solids deposition and attenuate their impact.

Produced gas reinjection after increasing the CO_2 content by mixing produced gases from different sources has led to reducing carbonate scale deposition in the Siri field in the Danish sector of the North Sea [1].

Specifically designed gas lift valves were reported to reduce the likelihood of mineral scale formation on the latch and in the vicinity of the check valve. They have a track record of successful field cases in the North Sea [20].

15.2.1.4 Scale ions stripping in the reservoir

Both calcium and barium sulfate can precipitate in the reservoir through brine/brine interactions and brine/rock interactions. This reduces the scaling tendency of the produced brines at the production wells and the need for scale squeeze treatments.

In chalk reservoirs, as the injected seawater is warmed in the reservoir and cation exchange takes place with the reservoir rocks, calcium ions are released, which react with the high concentration of sulfate ions in the heated seawater, precipitating $CaSO_4$. This leads to sulfate stripping in the reservoir, which limits the available sulfate for precipitation reactions such as $BaSO_4$ and $SrSO_4$ when mixed with reservoir brine. Sulfate stripping was found to occur in in high temperature chalk reservoirs (>120°C) due to the inverse solubility of calcium sulfate, and also in lower temperature chalk reservoirs at 90°C [21].

15.2.1.5 Sand and fines control

Sand and fines can deposit in the production system and can also serve as nucleating sites for mineral scales and other deposits. Thus removing and controlling such solids reduce their impact on the production system. Sand and fines are generally controlled either in passive or active ways. The passive methods are nonintrusive methods like [22]:

- Oriented perforation
- Selective perforation
- Sand management

While the active methods include the following [22]:

- Standalone screens (slotted liner, wire-wrapped screen, prepacked screen and premium screen)
- Expandable sand screen
- Gravel pack and frack pack
- Chemical consolidation

15.2.2 Chemical scale prevention

Chemical scale prevention is achieved by injecting chemical additives (known as scale inhibitors) in small amounts into the production fluid streams, which interferes with the scaling process, leading to delay and reducing or preventing mineral scales formation. Kelland [23] defined scale inhibitors as water-soluble chemicals that prevent or retard the nucleation and/or crystal growth of inorganic scales.

Early methods of scale inhibition, during the 19th and early 20th century, were limited to and focused on preventing scaling in boiler and cooling waters using natural products like potato leftovers from worker lunches, vegetable products containing tannin and starch, hay infusions, glue, gums, alginates, lignins, humic acids, and others [24,25]. A huge breakthrough in scale inhibition technology occurred in the 1960s when polymer scale inhibitors were introduced to replace the natural lignins, tannin, and starch, followed by the introduction of phosphonate as a more hydrothermal scale inhibitor than inorganic phosphates [25]. In the oil and gas industry, scale inhibitors have witnessed huge advances with the introduction of new technologies like emulsion scale inhibitors (ESIs) [26], oil miscible scale inhibitors [27], encapsulated scale inhibitors [28], and green scale inhibitors [29].

15.2.2.1 Classification of scale inhibitors

Scale inhibitors are diverse in their chemistry, mechanism of action, properties, and applications, and therefore many classifications of scale inhibitors are available. Some of these classifications are discussed in the following paragraphs.

- According to mechanism of action: thermodynamic and kinetic scale inhibitors.

Thermodynamic inhibitors are basically chelating and sequestering agents added to the system in stoichiometric amounts relevant to that of the potential scaling species; they are also called stoichiometric scale inhibitors. Examples are aminocarboxylates like ethylene diamine tetraacetic acid (EDTA) and diethylenetriamine pentaacetate (DTPA). Also, some organophosphorous compounds should be used in stoichiometric amounts to be effective [30]. Kinetic scale inhibitors are those that inhibit scale formation when added to the system in a markedly lower concentration than the stoichiometric concentration of the scaling species, usually in the range of a few parts per million. Such inhibitors are also called substoichiometric or threshold scale inhibitors. Their actions are understood in terms of stereospecific and nonspecific mechanisms [30]. Most of the scale inhibitors used in the oil industry currently are threshold inhibitors like phosphonates and other polymeric scale inhibitors.

- According to their chemistry: inorganic and organic scale inhibitors; also, another common classification is polymeric and nonpolymeric scale inhibitors.

Inorganic scale inhibitor examples are phosphate salts and inorganic polymer-condensed phosphates, such as polymetaphosphates; organic scale inhibitor examples are polyacrylic acid (PAA), phosphinocarboxylic acid, sulfonated polymers, and phosphonates [30,31]. Common polymeric scale inhibitors are polyphosphonates, polyphosphinates, polycarboxylates, and polysulfonates. They are considered good nucleation inhibitors and dispersants. Common nonpolymer scale inhibitors are phosphates, phosphate esters, nonpolymeric phosphonates, and amminophosphonates; they are good at preventing crystal growth [23,32].

- According to their manufacturing method: synthetic inhibitors and extracted inhibitors.

Synthetic scale inhibitors are those designed, synthesized, and modified in specific labs to function as inhibitors. Natural or extracted scale inhibitors are those extracted from natural sources, like potato, corn stalks, olive leaves, and tobacco rob extract, and they may or may not be modified to function as inhibitors [33,34].

- According to their application: Aqueous (water soluble), nonaqueous (oil soluble/miscible), emulsion, capsulated/solid, gelled, foamed, and nanoparticle scale inhibitors.

Most of the scale inhibitors are designed to be water soluble, because scale deposits are primarily a water-based problem, but in some instances miscibility of a scale inhibitor in an oil phase is needed, either due to problems associated with the conventional aqueous scale inhibitors (wettability, clay swelling) or for better performance of the inhibitor [35]. Microencapsulated scale inhibitors are designed to offset problems associated with conventional batch treatments [36]. The same thing applies with inverse emulsion scale inhibitors, which were introduced to overcome the main problems

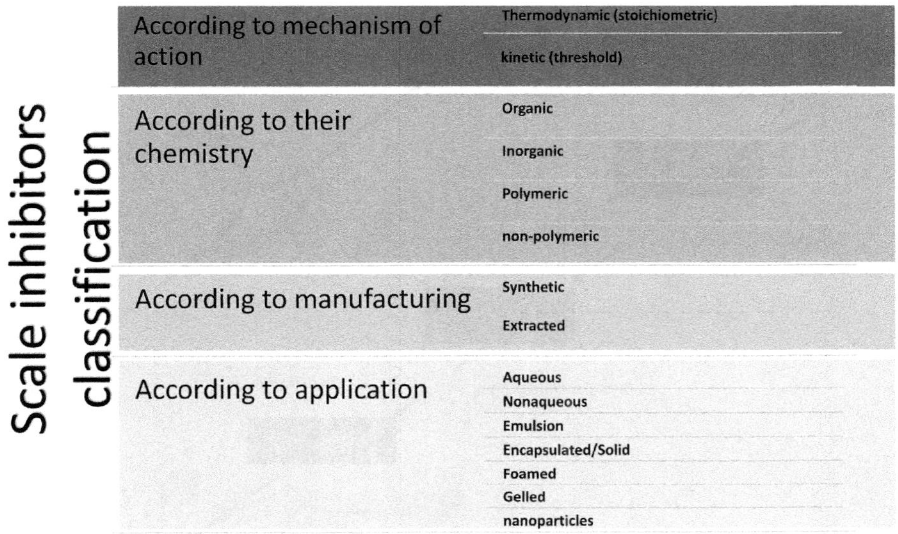

FIG. 15.5

classification of scale inhibitors in oil and gas industry.

associated with normal conventional scale inhibitors [37]. Foamed and gelled scale inhibitors are used to deliver the main treatment to the lower permeability zone. The use of nanoparticles is one of the recent advances in scale inhibition, which is used to extend the scale squeeze lifetime. Fig. 15.5 summarizes the classification of scale inhibitors.

15.2.2.2 Mechanisms of chemical scale inhibition

How exactly do scale inhibitors work? The answer is brimful of ambiguities and complexities!

The fact of the matter is that each inhibitor probably functions in several ways, depending on the inhibitor, concentration, type and concentration of solids in the water, temperature, and other factors [24]. The known scale inhibitors are diverse and quite specific in their performance and a universal inhibitor for all scale types does not exist, as each scale type has its own idiosyncrasies and scale inhibitors are tested mostly on a trial and error basis [25].

Different terms have been used in the literature to describe the scale inhibition mechanisms including: chelation, sequestration, complexation, antiprecipitation, protective colloid, threshold treatment, monomolecular film former, dispersant, deflocculant, antinucleation agent, colloidal stabilization, modification, peptization, crystal/surface modification, flocculation/coagulation, and sludge conditioning [24,38].

To simplify the matter, we can say that, since the scaling process undergoes four basic steps—supersaturation, nucleation, crystal growth, and adhesion—thus scale inhibitors work by interfering with these steps. Sequestering or chelating agents interfere with the supersaturation step, while threshold inhibitors, crystal modifiers, and dispersants interfere with nucleation, crystal growth, and adhesion steps by some type of "surface mechanism." Some of the main mechanisms are discussed in the following paragraphs. A summary of these mechanisms is depicted in Fig. 15.6.

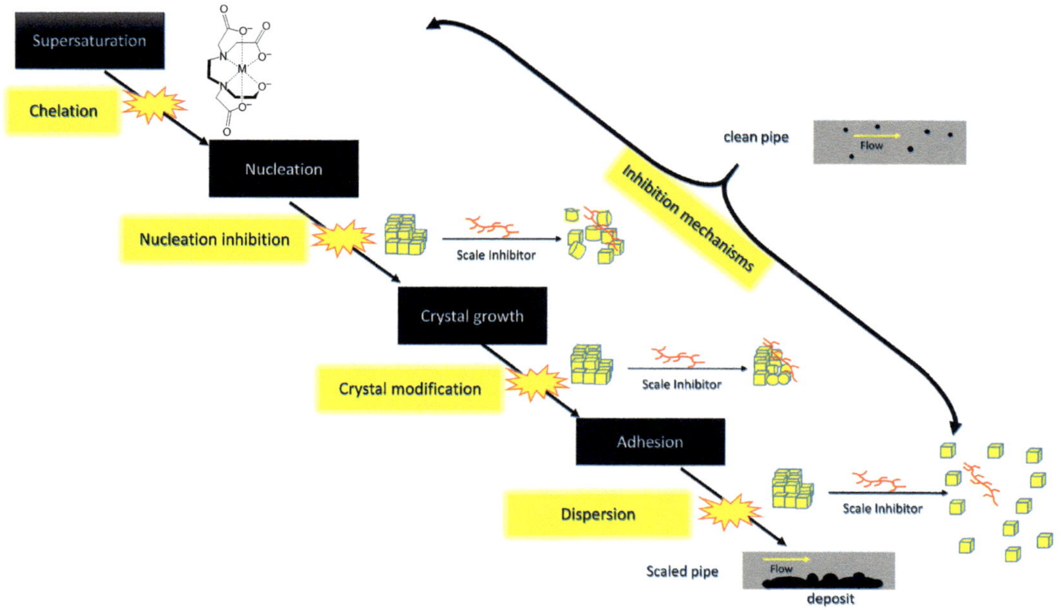

FIG. 15.6

General scheme of the different scale inhibition mechanisms.

- **Chelation mechanism**

Chelants, chelating agents, complexing agents, or sequestering agents are terms to express these organic compounds that contain two or more electron donating groups (such as carboxyl, amine, ether, nitro, and hydroxyl), which function as Lewis bases and form coordinate bonds with a central metal atom through electron donation. This formation of multiple coordinate bonds from a single molecule results in the formation of one or more heterocyclic rings, or *chelate* rings, hence the name *chelating agents*. This method of bonding with the metal ion to form a chelate compound (complex) makes the metal ion unavailable to react and precipitate with other counter anions, hence reduces the actual supersaturation level, thus reducing the possibility of scale formation in that water [39]. Examples of chelating chemicals are aminopolycarboxylates, organic acids, quaternary ammonium salts, and phosphonates. An EDTA chelate with metal ion is illustrated in Fig. 15.7.

- **The nucleation inhibition mechanism**

In surface mechanisms, unlike the sequestering agents, scale inhibitors do not stop the initial formation of scale crystal nuclei, but keep them in the submicroscopic range by inhibiting their growth. Inhibitors in the surface mechanisms perform at a threshold concentration; thus the mechanism is called a threshold or minimum suppression mechanism.

These surface mechanisms are explained in light of the adsorption theory of crystal growth by Burton, Cabrera, and Frank. The scale inhibitors adsorb to the critical sites (preferred sites) on the scale crystal surfaces, blocking (poisoning) them and preventing them from reaching the active sites where they can grow, hence they hinder their further growth, even with stable nuclei [31,40]. Thus the amount

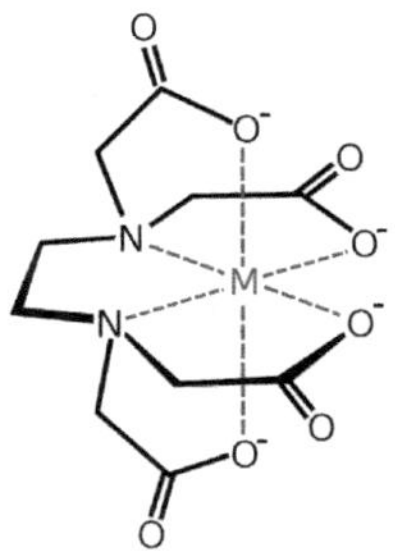

FIG. 15.7

EDTA chelate with metal ion.

From: https://commons.wikimedia.org/wiki/File:Metal-EDTA.svg.

of chemical required for scale inhibition is closely proportional to the surface area of the precipitates. In the case of the stable nuclei (like seed crystals), it has been shown that complete crystal growth inhibition can be achieved with as little as 3%–5% coverage of the total surface area for carbonate or sulfate scale crystal by some polymeric inhibitors [23,25]. However, to completely inhibit barite formation, the inhibitor needed is approximately equal to 16% surface coverage [40]. In the case of unstable nuclei, i.e., below the critical size, if the inhibitor has blocked its active sites and prevented its further growth to a critical size, it will finally dissociate [25,41].

- **Crystal modification mechanism**

In this mechanism, the inhibitors can dramatically affect particle shape and size [42]. The inhibitor will be adsorbed on the crystals, forming complex surfaces or nets, causing morphological changes to the scale deposit crystal lattice [31,43]. The changes are the results of either stress/strain on the crystal lattice, or an uneven growth of different faces of the crystal (when the inhibitor blocks growth of the face/corner it is adsorbed on, while allowing growth of other faces), or both. The face-selective adsorption of the inhibitor can be explained by "lattice matching," where the anionic moieties replace lattice anions at the crystal surface [42] or by "charge matching" [44]. The lattice matching may contribute to the additive's ability to "charge match" more efficiently [44].

The final result of crystal modification is the distortion of the crystal faces (transforming them from sharp active edges to spherical or less active sites) or converting them from a less soluble form to a more soluble form (e.g., from calcite to vaterite). Fig. 15.8 shows the SEM micrograph of mineral scales after crystal modification by scale inhibitors [45,46].

- **Dispersion mechanism**

In this mechanism the chemical additives prevent any crystals from agglomerating into large clusters and then settling onto surfaces [47]. That takes place by electrostatic repulsion when the inhibitors create an electric double layer in the boundary at which the crystal nucleation grows, preventing scaling ions from coagulating on the surface [48,49]. Another way of explaining the dispersion mechanism is that the scale inhibitor can form a thin monolayer that alters the interfacial relationship between the surface and scale deposits, consequently preventing the scale deposits from adhering to the surfaces [24,50]. Examples of this type are the poly-carboxylate and polyacrylic scale inhibitors.

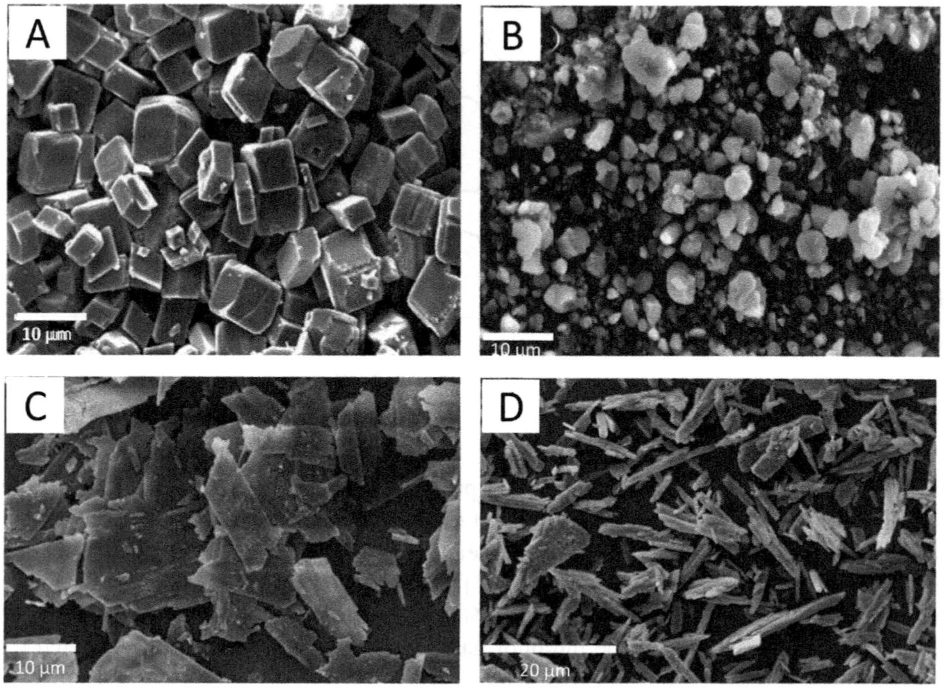

FIG. 15.8

SEM images of morphology of mineral scales in the presence and absence of chemical inhibitors. (A) $CaCO_3$ crystals in the absence of inhibitor, and (B) $CaCO_3$ crystals in the presence of inhibitor, and (C) $CaSO_4$ crystals in the absence of inhibitor, (D) $CaSO_4$ crystals in the presence of inhibitor.

Figures (A) and (B) from: Y. Zuo, W. Yang, K. Zhang, Y. Chen, X. Yin, Y. Liu, Experimental and theoretical studies of carboxylic polymers with low molecular weight as inhibitors for calcium carbonate scale, Crystals 10 (5) (2020), https://doi.org/10.3390/cryst10050406; figures (C) and (D) from: P. Santoso, M.R. Setiawan, Suharso., Piper betle leaf extract as a green inhibitor of calcium sulphate (CaSO4) scale formation, IOP Conference Series: Earth and Environmental Science, vol. 258, Institute of Physics Publishing, 2019, https://doi.org/10.1088/1755-1315/258/1/012038.

- **Film formation mechanism**

Scale inhibitors and corrosion inhibitors are both surface active chemicals with their performance relying upon adsorption on metal surfaces (film-forming corrosion inhibitors) or adsorption onto crystal growth sites and/or on metal surfaces (film forming scale inhibitors). These film-forming species may restrict heterogeneous nucleation at the metal surface and thus prevent scale formation [51,52].

Halite scale inhibitors have been claimed to work through film formation, in which the scale inhibitor absorbs on the surface of substrates and prevents the adhesion of scale crystals [38].

Some scale inhibitors show one or more mechanisms of performance, like those of ATMP (sodium salt of amino trimethylene phosphonic acid), as it shows sequestering, crystal modification, and dispersing behavior in inhibiting barium sulfate deposits, as reported by many researchers [53].

15.2.2.3 Design of scale inhibitor chemical

Now we know how inhibitors work on scale in general. For the chelating scale inhibitors, we know the chelation takes place through a chemical reaction, but for surface-mechanism scale inhibitors, how do they sneak onto the formed scale's crystal surface in the first place?

The answer is that the scale inhibitor will interact with the produced water and get a hold on scale particle surfaces to interact with the lattice ions on the crystal surface. To achieve this binding, the inhibitor molecule must have oppositely charged functional groups that interact with the cations (Ca^{2+}, Br^{2+}) or the anions $\left(SO_4{}^{2-}, CO_3{}^{2-}\right)$ on the crystal surface [23]. Cationic groups like quaternary ammonium, phosphonium, or sulfonium groups are used to bind to the anions in the scale crystal, while anionic groups like phosphate ions, phosphonate ions, phosphinate ions, carboxylate ions, and sulfonate ions are used to bind to the cations on the scale crystal surface [23]. After the inhibitor finds its way to the crystal surface, the adsorption properties of the inhibitor will do the rest of the job. Thus some of the properties considered during design of the scale inhibitor include [54]:

- Proper functional groups to bind to the scale crystal surface.
- Strong-enough binding energies between the inhibitor and the scale surface.
- Anionic groups were found to be more effective, as they have high affinity and binding energy.
- If the binding energy is low for a stereospecific inhibitor, the expression of a stereochemical effect on morphology of crystals will require higher doses.
- Although some inhibitors may lack high affinity, stereospecific binding still may be an effective antiscalant, as the lower-affinity, weaker binding could lead to nonspecific inhibition and produce greater morphological variability, including porous and other high-surface-area crystals.
- Proper spacing of the functional groups that interact with the lattice ions of the crystal surfaces.
- The groups have to have specific spatial orientation when binding with crystal surface.
- Adding some hydrophobic character to the inhibitor will help in crystal disruption and inhibit crystal growth.
- The inhibitor has to be economical.
- It has to be compatible with its application medium.

15.2.2.4 Common oil and gas scale inhibitor chemicals

15.2.2.4.1 Thermodynamic scale inhibitors

Different types of thermodynamic inhibitors are used to treat mineral scales (Table 15.1). These chemicals are usually added to the system in stoichiometric ratios with the scale-forming species.

- **Acids**

Acids are usually used to remove deposits like oxides, carbonates, and sulfides, to stimulate formations (acidizing) for productivity improvement [55].

Acids (organic, such as formic, acetic, citric, and inorganic, such as HCl, H_2SO_4, HF) and acid-forming salts (i.e., iron chloride) and acid-forming gases (CO_2) are also used to control or prevent the deposition of alkaline scales like calcium carbonate by adjusting the water pH in seawater evaporators and oilfield steam generators [24,30]. HCl acid was formulated into a scale inhibitor to reduce the surface tension between the oil and the aqueous scale inhibitor, which can be used to prevent calcite formation in oil reservoirs [56].

Table 15.1 Thermodynamic scale inhibitors.

Chemical type	Examples
Acids	*Mineral acids* HCl, H_2SO_4, HF *Organic acids* Acetic acid, formic acid, citric acid, etc.
Chelating agents	*Aminopolycarboxylates* Ethylenediaminetetraacetic acid (EDTA) Cyclohexylenedinitrilotetraacetic acid (CDTA) Diethylenetriaminepentaacetic acid (DTPA) *Organic acids* Gluconic acid Citric acid Tartaric acid Some phosphonic acids

Acids have to be used with the utmost caution, by adding proper corrosion inhibitors and studying the system metallurgy, temperature, and acid-metal contact time [55].

- **Chelating agents and sequestrants**

One of the common chelating agent classes is the aminopolycarboxylates like NTA, EDTA, DTPA, CDTA, HEDTA, etc. (Fig. 15.9). These compounds are regarded as scale dissolvers more than scale inhibitors [57,58]. Aminocarboxylates were reported to have the lowest efficiency between different calcium carbonate and calcium sulfate inhibitors [59], while they were reported by Billman [60] to have satisfactory to high performance in inhibiting and dissolving iron sulfide scale, with the drawback that inhibition and dissolution effects of EDTA are kinetically "poisoned." NTAA was also reported to inhibit iron sulfide scale in moderate temperature wells. One main drawback of these chemicals is that when they are used as scale inhibitors, they are applied in a stoichiometric ratio, which limits their use.

Another group of chelating agents is the organic acids. Gluconic and tartaric acids and their derivatives were reported to be effective scale dissolvers and inhibitors [24,61–65]. Citric acid and its salts were found to be effective as ferrous and calcium carbonate scale inhibitors [62,63]; citrate also was found to enhance the inhibition efficiency of phosphonates in high dissolved iron water [61]. Polycitric acid was found to be effective on calcium sulfate scale [64]. Environmentally friendly scale inhibitor formulations containing citric acid were also introduced [65]. Citric acid outperformed tartaric acid and maleic acid in inhibiting calcium sulfate phases; the addition of citrate-stabilized bassanite changed the final gypsum habit from typical needle-like crystals in the pure $CaSO_4$ system to plates [66].

Phosphonates are used as chelating agents in many applications, e.g., in pulp, paper, textile, and medical uses. In oil and gas operations, phosphonates are known to inhibit scale by many mechanisms, including sequestering, crystal modification, and dispersion.

Quaternary ammonium salts and quaternary phosphonium salts are another class of sequestering agents. Tetrakis(hydroxymethyl)phosphonium sulfate (THPS) was reported to work as a sulfide scale dissolver and inhibitor [67,68]. It can be used alone or along with aminocarboxylate, phosphonates, or ammonia to improve its inhibitory effect. However, Li et al. [69] concluded that THPS may not be as effective as needed for FeS inhibition benefits.

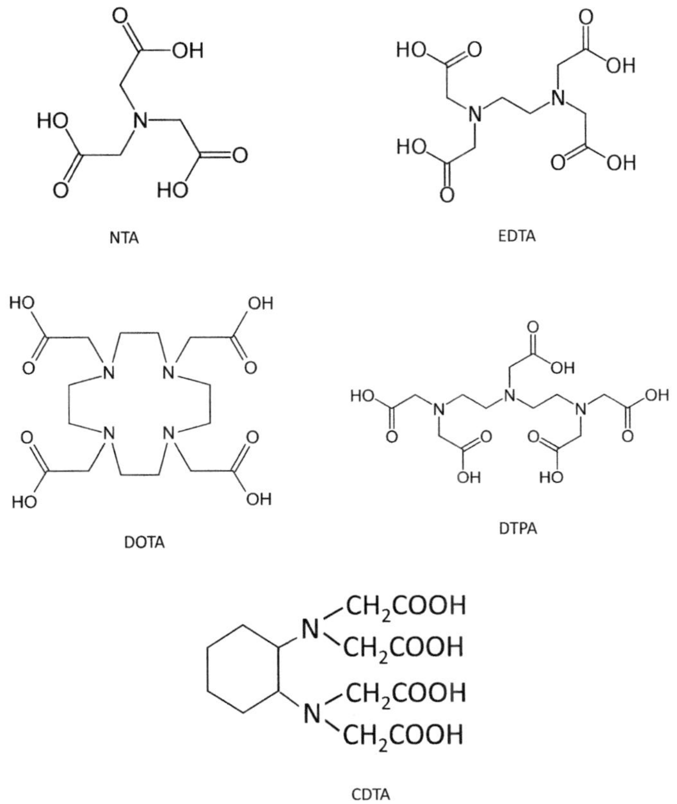

FIG. 15.9

Aminopolycarboxylate chelating agents.

Trace amounts of chelating agents, such as EDTA, citric acid, or gluconic acid, may lower the efficiency of scale inhibitors [70]. However, in high iron waters the addition of sequesterants enhanced the efficiency of phosphonate scale inhibitors [61]; also, sequestering agents such as the trisodium salt of NTAA and alkali metal citrates are used to sequester iron to prevent a halite inhibitor from reacting with the iron [71].

The complex formed by chelating/sequestering agents has a certain stability; its stability is expressed using the stability constant K (K represents the ratio of the chelated metal ion to the free metal ion). The log K value, or *stability constant*, is commonly used to describe the effectiveness of various sequestrants. The complexing is influenced by pH, temperature, nature of electrolytes, etc. More cations can be complexed under one set of conditions than under another [72].

15.2.2.4.2 Threshold scale inhibitors

Oilfield scale inhibitors are threshold inhibitors that are used in relatively low doses to inhibit scale formation. Their exact inhibition mechanism is not fully understood, and it is often believed to be a combination of different mechanisms. Various types of polymeric and nonpolymeric scale inhibitors have been used in oil and gas fields (Table 15.2).

Table 15.2 Common oil and gas fields scale inhibitor chemicals.

Scale inhibitor type	Examples
Inorganic phosphate and polyphosphate	Sodium tripolyphosphate
	Sodium hexametaphosphate
Phosphate ester	Triethanolamine phosphate ester
	Polyhydric alcohol phosphate ester (PAPE)
Phophonates	1-Hydroxyethane-1,1-diphosphonic acid (HEDP)
	2-Hydroxyphosphono acetic acid (HPA or HPAA)
	2-Phosphonobutane-1,2,4-tricarboxylic acid (PBTCA)
	Nitrilotrimethylenephosphonic acid (NTMP)
	Ethanolamine-*N*,*N*-bis(methylenephosphonate) (EBMP)
	Aminotris(methylenephosphonic acid) (ATMP)
	1,2-Diaminoethanetetrakis(methylenephosphonic acid) (EDTMP)
	Diethylenetriaminepentakis(methylenephosphonic acid) (DTPMP or DETPMP)
	Hexamethylenediamine-tetra(methylenephosphonic) acid (HDTMP)
	Bis(hexamethylenetriamine-penta(methylenephosphonic acid)) (BHMTMP)
	Octa-methylene tetra-amine hexa (methylene-phosphonic acid)
	Polyamino polyether methylene phosphonate (PAPEMP)
	N-Phosphonomethylated amino-2 hydroxypropylamine polymer
	Vinyl phosphonic acid (VPA), or vinyl diphosphonic acid (VDPA)
Phosphino polymers and polyphosphinates	Polyphosphinocarboxylic acid (PPCA)
Polycarboxylate	Polyacrylic acid
	Polymethacrylic acid
	Polymaleic acid
Poly sulfonate	Sulfonated polycarboxylate
	Polyvinyl sulfonate (PVS)
	Acrylamido(methyl)propylsulfonic acid (AMPS)

- **Phosphate and phosphate ester scale inhibitors**

The inorganic phosphates including pyrophosphates, metaphosphates, and polyphosphates [24] are widely used in scale prevention and are commonly referred to as "condensed phosphates" or "molecularly dehydrated phosphates," or phosphate glass. They are common as scale and corrosion inhibitors [23,24,73–75]. Fig. 15.10 illustrates some phosphate scale inhibitors [76]. However, phosphate-based scale inhibitors are mainly used in boiler water treatment at low calcium concentrations, as there are more thermally stable and more compatible products for oilfield scale inhibition [23].

Generally, phosphate esters are much more familiar to petroleum engineers as corrosion inhibitors, as well as being used as gelling and lubricant additives [77,78]. Phosphate esters are more tolerant of acid conditions than polyphosphates and are generally compatible with high-calcium brines [2,23]. Phosphate esters have limited thermal stability. They are not recommended for application above

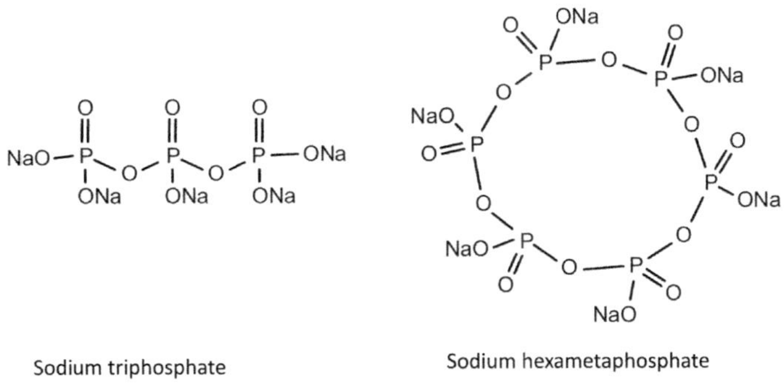

FIG. 15.10

Inorganic phosphates.

From: M. Mpelwa, S. Fa Tang, State of the art of synthetic threshold scale inhibitors for mineral scaling in the petroleum industry: a review, Pet. Sci. (2019), https://doi.org/10.1007/s12182-019-0299-5.

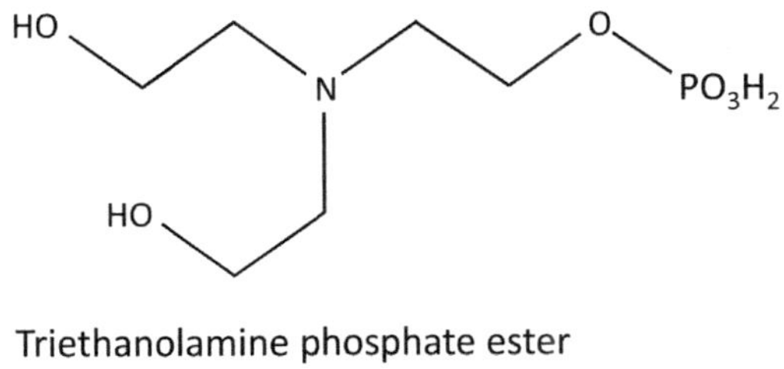

FIG. 15.11

Example of phosphate ester.

approximately 175°F [79°C], as there is a danger of hydrolysis, resulting in a loss of effectiveness [2,77]. Phosphate esters are well known as environmentally friendly scale inhibitors [77]. An example of a phosphate ester is given in Fig. 15.11.

- **Phosphonate scale inhibitors**

Since patented by Ralston in 1969, phosphonate chemistry got off the ground as a premier water treatment chemical for different applications [25]. Phosphonates are complexing agents that contain one or more C—$PO(OH)_2$ groups, which also have a very strong interaction with surfaces [77]. It is known that phosphonates tend to have a lower "cutoff" temperature than many polymeric scale inhibitors, below which they are much less effective [23]. Low biodegradability, low toxicity, bioaccumulation, and low calcium compatibility are the main problems with phosphonates [23,77]. Fig. 15.12 illustrates the structure of some phosphonate scale inhibitors.

Aminophosphonates are considered the most common effective nonpolymeric phosphonates. Common aminophosphonates are ATMP, EDTMP, HEDP, and DTPMP (also known as DETPMP). DETPMP is an excellent carbonate and sulfate scale inhibitor and possibly the most often used scale inhibitor in the oil and gas production sector [77,79]. Also, it was reported to inhibit sulfide scale

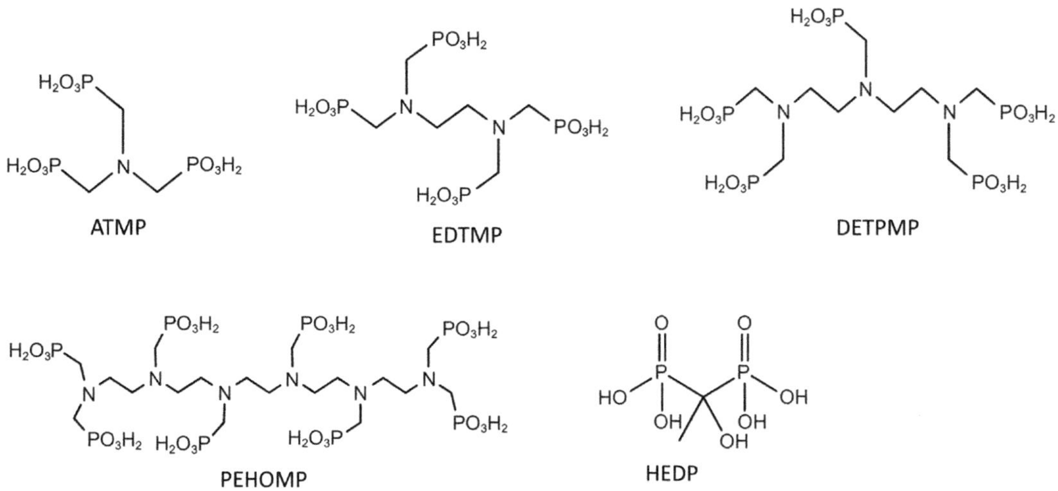

FIG. 15.12

Structure of phosphonate scale inhibitors.

From: M. Mpelwa, S. Fa Tang, State of the art of synthetic threshold scale inhibitors for mineral scaling in the petroleum industry: a review, Pet. Sci. (2019), https://doi.org/10.1007/s12182-019-0299-5.

(mainly by chelation rather than a threshold mechanism) [23]. DTPMP was reported to have good thermal stability when aging at 130°C, while it showed slight instability after aging at 150°C and 170°C, with reduction in performance at 170°C [80,81]. DETPMP is affected by the Ca (it has very low calcium tolerance), Fe cations, and pH [23]. (BHMTMP).is another highly effective aminophosphonate scale inhibitor. This molecule has excellent tolerance to high calcium ion concentrations and has good stability at high temperatures up to 140°C so can be useful for high-temperature applications. In fact, BHPMP showed better thermal stability than DTPMP, with more than 50% remaining active upon preheating at 200°C [80,82]. PAPEMP has high calcium tolerance and good scale inhibition effects, especially on silica scale, calcium carbonate, calcium sulfate, and calcium phosphate. It stabilizes metal ions such as Zn, Mn, and Fe [83,84].

Amjad ranked the phosphonate performance against gypsum and calcite in the order PAPEMP > AMP > PBTC > HEDP > HPA for gypsum, and PAPEMP > PBTC ≈ AMP ≈ HPA ≈ HEDP for calcium carbonate [85]. Hamdona et al. ranked phosphonate effectiveness on strontium sulfate crystal growth inhibition to be in the order HEDP> TENTMP> NTMP> ENTMP [86].

- **Polymeric scale inhibitors**

PPCA is a common phosphino polymer scale inhibitor. The presence of the phosphorus makes PPCA easier to analyze, and enhances the inhibition performance and calcium compatibility. Moreover, it increases the attachment of PPCA to formation rocks, resulting in an extended squeeze lifetime [32]. However, phosphinate groups do not bind as well to rock as phosphonate acid groups. At 60°C, PPCA superseded DETPMP and other phosphonate-based inhibitors in inhibiting $BaSO_4$ [87]. PPCA is also an effective inhibitor of sulfide scale.

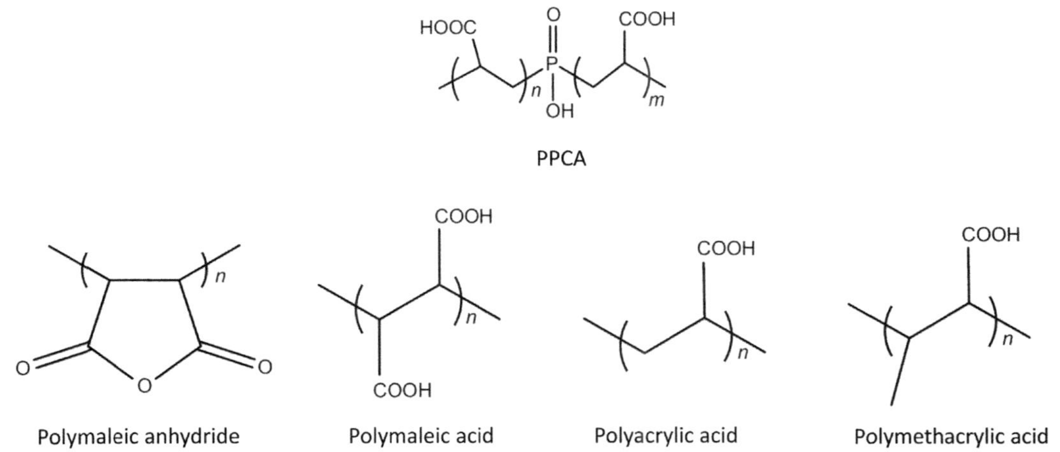

FIG. 15.13

Structures of some polymeric scale inhibitors.

From: M. Mpelwa, S. Fa Tang, State of the art of synthetic threshold scale inhibitors for mineral scaling in the petroleum industry: a review, Pet. Sci. (2019), https://doi.org/10.1007/s12182-019-0299-5.

The most common classes of polycarboxylic acids are based on polyacrylic acid, polymethacrylic acid, and polymaleic acid [23] (Fig. 15.13). Copolymers of acrylic acid and maleic acid are very common as polymeric scale inhibitors, which have superior performance against barium and strontium sulfate scales [77]. Wang et al. introduced carboxylate-sulfonate copolymers, which exhibited excellent hydrolytic and thermal stability up to 350°F (177°C) and were tolerant to waters containing high levels of divalent metal ions [88]. Fig. 15.14 illustrates some of the copolymeric scale inhibitors.

Common types of polysulfonates are SPCA and PVS (Fig. 15.15). PVS bears some advantages over other inhibitors: they have high thermal stability, which makes them applicable for high temperature oil wells [23]; they also were reported to work well at low temperatures (4–5°C); they can be applied in high Ca^{2+} and Mg^{2+} concentration brine [30]; and they can work well in low pH solution [30]. However, polysulfonates do not adsorb as strongly to rock, so their squeeze lifetimes will be shorter.

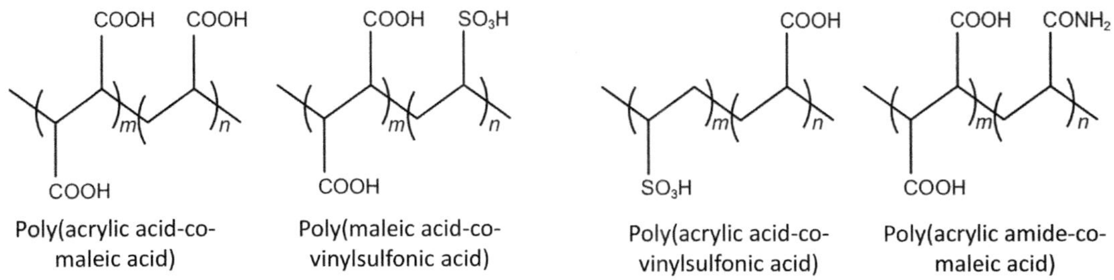

FIG. 15.14

Structure of copolymeric scale inhibitors.

From: M. Mpelwa, S. Fa Tang, State of the art of synthetic threshold scale inhibitors for mineral scaling in the petroleum industry: a review, Pet. Sci. (2019), https://doi.org/10.1007/s12182-019-0299-5.

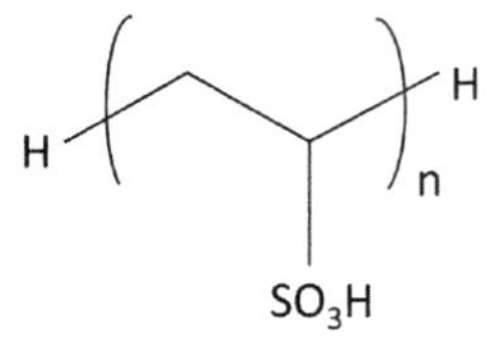

FIG. 15.15

Polyvinyl sulfonate (PVS) scale inhibitor.

The performance of the polymeric inhibitors is dependent on the functional groups present in the molecules (best performance with 15–20 functional groups), the molar mass (best performance with molecular masses between 1000 and 30,000) [89], and their charge density; their effectiveness may also be influenced by external factors such as temperature, pH, ionic strength, flow rates, shear forces, nucleation sites, and residence times [90].

Generally speaking, to inhibit sulfide scale using the common scale inhibitors, 10 times the dosage for sulfate scale inhibition will be required [91]. Another class of polymers are commonly used to disperse iron sulfide scale, known as dispersants. A polymeric compound containing amide or its derivative functionalities, poly(2-ethyl-2-oxazoline and polyacrylamide (PAM), at threshold quantities significantly influenced the growth and deposition behaviors of the FeS particles by dispersing nanoparticulates for an extended period of time [92,93]. High molecular weight sulfonated copolymers were also introduced as iron sulfide dispersants [94].

Due to the fact that halite (sodium chloride) deposit only contains monovalent ions, the previously described scale inhibitors will not work on halite scale. Fresh water is usually used to remove halite deposits, and also to dilute the produced water salt concentration, which helps in inhibiting halite formation [95]. However, that requires extra expense for water use and disposal, besides the water effects on oil production. Polysaccharidic polyelectrolytes (carrageenan and sodium alginate) were studied as halite inhibitors [96]. Another class of chemicals is hexacyanoferrate (HCF) salts, such as potassium HCF, which has been used as a common additive for drilling through salt and an anticaking agent for cooking salt [97], and is also proven to inhibit halite deposition at a concentration of about 250 mg/L [23]. Due to their poisoning by ferrous ions, they are formulated with a chelating agent such as citric acid or nitrotriacetic acid [98]. Nitrilotrialkanamides, like nitrilotriacetamide and nitrilotripropionamide, have been reported in the field to successfully reduce halite scale formation, both topside and downhole in squeeze treatments [23]. Nitrilotrialkanamides have successfully inhibited halite formation, but 250 to 500 mg/L is required to do so. While the molecule is expensive to manufacture, it is stable at high temperatures and has a relatively favorable biodegradation profile [98].

- **Environmentally friendly scale inhibitors (green inhibitors)**

Most of the scale inhibitors used in oil and gas fields have the potential for acute or long-term effects on aquatic organisms when they are discharged into the marine environment. Of particular concern are phosphorus-based inhibitors, which can serve as nutrients leading to eutrophication difficulties [65]. The number of chemicals allowed to be used as inhibitors has thus been limited mainly to a three-criteria tier level of biodegradability, bioaccumulation, and toxicity. An ideal green inhibitor should have: (1) excellent scale inhibition; (2) low aquatic and human toxicity;

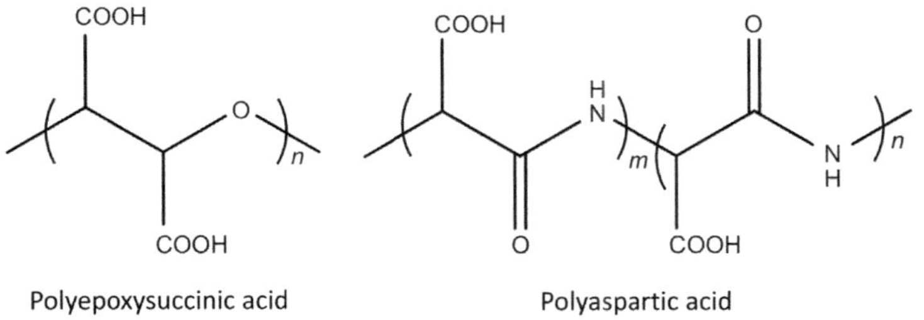

FIG. 15.16

Structure of polyaspartic acid and polysuccinic acid.

From: M. Mpelwa, S. Fa Tang, State of the art of synthetic threshold scale inhibitors for mineral scaling in the petroleum industry: a review, Pet. Sci. (2019), https://doi.org/10.1007/s12182-019-0299-5.

(3) high biodegradability; (4) no bioaccumulation; (5) good price/performance ratio; and (6) no phosphorus, nitrogen, or heavy metals [99,100].

Most of the toxic phosphorus-based scale inhibitors have been replaced by less toxic organic phosphorus compounds, such as dibutylphosphorodithoic acid and carboxyhydromethylphosphonic acids.

Currently, synthetic green scale inhibitors are mostly polyaspartic acid (PASP), polyepoxysuccinic acid (PESA), and so on (Fig. 15.16). PESA and PASP are internationally accepted as highly efficient green inhibitors. Hence, researchers have studied and developed a large number of efficient green inhibitors from derivatives [101].

Polyaspartate is readily biodegradable. It is used to inhibit both carbonate- and sulfate-based scales, and has a dual functionality, acting as both a scale and corrosion inhibitor under oil field conditions [102]. It has been used in squeeze treatments at up to ca. 85°C, although there is an improved polyaspartate for squeezing at up to 120°C. It has also been claimed to reduce fines migration after a squeeze treatment [23]. PASP is used particularly in regions where the environmental regulations normally require >20% biodegradability, such as the North Sea basin.

Another class of green inhibitors is carboxymethylinulin CMIs, which are reported to be effective sulfates and carbonate inhibitors in oilfield applications [103,104]. Inulin is an inherently biodegradable scale inhibitor with an excellent ecotoxicity profile for fresh and seawater species. CMI is particularly effective in sequestration of hard water cations, and thus serves as an antiscalant that could find uses in food processing [102]. Polyepoxysuccinic acid (PESA) is a green corrosion scale inhibitor that has been used in many cooling water systems and shows excellent scale performance [105]. One set of tests showed that PESA performed better than polyaspartate on $CaCO_3$ and $SrSO_4$ scale, but not on gypsum and barite scale [23].

Other inhibitors that are considered to be green chemicals are chelating agents, including sodium iminodisuccinate, disodium hydroxyethyleneiminodiacetic acid, sodium gluconate, and sodium glucoheptonate [30].

Natural products or extracts are emerging types of green inhibitors. Raw potato, hay infusion, glue, albumen, starch, and others are among the naturally used scale inhibitors [24]. Apart from these, fig leaf extract was reported to inhibit $CaCO_3$ [106] and olive leaf extract was tested as a corrosion and scale inhibitor [107]. Wang et al. [33] used a natural extract of tobacco rob as a scale ($CaSO_4$, $CaCO_3$) and corrosion inhibitor in artificial seawater. Zhang et al. [108] synthesized sulfonate (PS-NAEP) using heteropolysaccharide, corn stalks, and cheap raw materials. Polysaccharide derivatives of aloe with carboxylic and alcohol groups have also been claimed as scale inhibitors. An aloe gel dissolved in water has been used as a scale inhibitor [30,31]. Pilot field tests of an *aloe vera*-based scale inhibitor for inhibition of $CaCO_3$ in Venezuelan oilfield wells were successful as a scale inhibitor and more efficient than commercial products used in those production wells [109]. A series of natural proteins and partially hydrolyzed proteins (casein peptones and tryptones) from various animal and plant sources have been tested for the ability to prevent the formation of barium sulfate and calcium carbonate scaling, as they often contain a substantial amount of carboxylic acid groups from aspartic acid or glutamic acid residues [110].

- **Nonaqueous scale inhibitors**

Nonaqueous (oil soluble) scale inhibitors have been used since the late 1990s in different fields in Colombia, the UK, and Norwegian sectors of the North Sea, and in North America [111]. The oil soluble scale inhibitors are used in situations where conventional methods are inapplicable or are not efficient enough, such as [111,112]:

- Permeability effects
- Wettability changes
- Sand production
- Water blocking
- Formations sensitive to water
- Fluid lifting issues
- Deep chemical penetration of the near-well formation
- High-value wells at low water-cut

Thus a major driver for the selection of nonaqueous products is efficiently placing an inhibitor having the minimum effect on the formation [111].

Nonaqueous scale inhibitor delivery systems can be divided into the following main types [111]:

- Oil soluble
- Water-free materials
- Invert emulsions
- Amphiphilic solvent systems
- Microemulsions
- Part-aqueous systems

Oil soluble scale inhibitors (OSSIs), or oil-miscible scale inhibitors, are scale inhibitors (e.g., DTPMP) with low water content (5%–10%) in hydrocarbon solvents (e.g., diesel, kerosene, crude, paraffin, and xylene) with or without mutual solvents that have been developed to avoid the practice of injecting aqueous solutions or calcium-sensitive scale inhibitors [23]. They can be used in continuous injection (with other chemicals), in gas lift injection, and in squeeze applications [23,35]. During squeeze treatments, the overflush used for deeper penetration of the inhibitor can be hydrocarbon based, such as diesel [23]. A larger hydrocarbon spearhead should be considered for wells with a 10% to 25% water cut [35].

Water-free materials are products that do not contain water within the manufactured formulation [111]. Phosphorus-tagged, polymeric, sulfonated scale inhibitors have been found to dissolve in blends of organic solvents without any water present at all [23]. They work by partitioning back into the formation water on squeezing [23].

Invert emulsions are water-in-oil macroemulsions that comprise scale inhibitors (DETPMP, PCCA, PMPA), brine, organic solvent (kerosene, etc.), and surfactant/emulsifier (glycol ester, sorbitan ester blend, and carboxymethylate). After downhole injection of the emulsion, it breaks down during shut-in, releasing the inhibitor chemical and contacting the rock structure, promoting effective retention [113]. This process is promoted by time, temperature, and salinity changes and can be controlled through careful selection of the surfactant components and also the product rheology [111].

Amphiphilic solvent systems (or oil dispersible systems) are composed of a traditional aqueous scale inhibitor and an amphiphilic solvent. The nonionic amphiphiles (NIAs) were found to remove water blocks, prevent water block/relative permeability effects, and extend squeeze life by enhancing inhibitor adsorption [114].

In microemulsions, scale inhibitors exist as the dispersed phase, and then the microemulsion is destabilized by the extreme downhole conditions, e.g., pH and temperature. The use of microemulsions was found to extend the squeeze lifetime and also enable reaching regions of the near wellbore region, which cannot be reached using conventional methods [115].

Part-aqueous systems include both aqueous and nonaqueous stages. Usually the spearhead and overflush stages are nonaqueous, since these constitute the largest proportion of the squeeze by volume, while the main treatment is aqueous. Examples include treatments using microemulsion scale squeeze enhancers and also treatments using mutual solvent or diesel stages to minimize formation damage in water-sensitive reservoirs [116].

- **Solid scale inhibitors**

Solids particles are also mentioned in the nonaqueous scale inhibitors. There are various ways of producing particles containing scale inhibitor, including 100% solid products, powder, briquettes, balls, encapsulated products, and highly porous materials that can capture the inhibitor. These are placed in locations wherein the solid inhibitors are released upon contact with produced water. Thus they can be used in perforated baskets, placed behind screens in gravel packs, in by-pass feeders, placed behind sliding sleeves, added to proppants in fracture operations, or dumped downhole [23,24]. They were applied successfully in the field where the conventional methods were not efficient or needed to be frequently used [117,118].

- **Nanomaterial scale inhibitors**

Nanomaterials have been gaining much attention in the oil and gas industry due to their numerous applications in enhanced oil recovery, stimulation, drilling, and processing. Recently, nanoparticle scale inhibitors have been reported, known as nanomaterial scale inhibitors or scale inhibitor nanomaterials (SINMs). Numerous papers have been published on the development of nanosized scale inhibitors for squeeze treatments [119–121]. An example of developing SINMs is illustrated in Fig. 15.17 [119]. Zhang [122] reviewed SINMs and categorized them into the following:

- Metal-inhibitor SINM, e.g., amorphous Ca-DTPMP SINM, crystalline Ca-DTPMP SINM, reverse micelle Ca-DTPMP nanofluid, Ca-DTPMP inhibitor nanoparticles with remaining synthesis fluid and ca-DTPMP SINM for bulk water process calcite scaling control.
- Scale inhibitor nanoparticle capsule, e.g., SiO_2-PAH-DTPMP nanoparticle capsule.

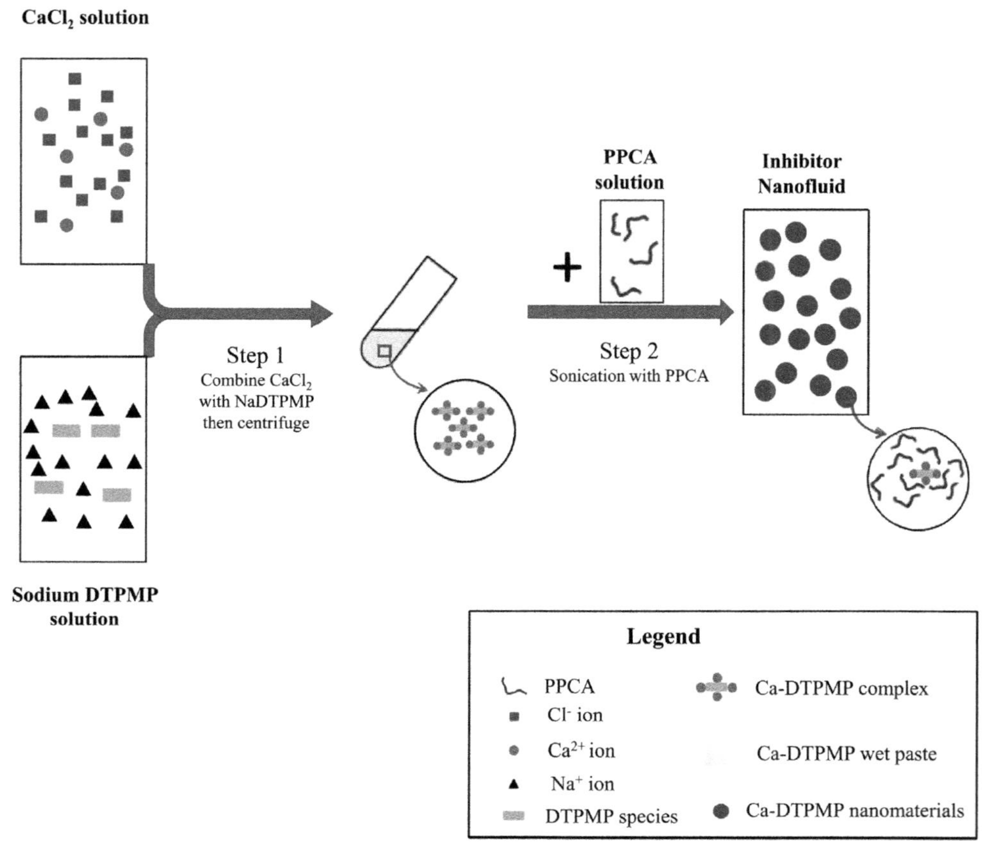

FIG. 15.17

Schematic diagram of the synthesis procedure of PPCA modified Ca-DTPMP nanofluid.

Reproduced with permission from P. Zhang, D. Shen, G. Ruan, A.T. Kan, M.B. Tomson, Phosphino-polycarboxylic acid modified inhibitor nanomaterial for oilfield scale control: synthesis, characterization and migration, J. Ind. Eng. Chem. 45 (2017b) 366–374, https://doi.org/10.1016/j.jiec.2016.10.004, Copyright (2017), Elsevier.

- Cross-link scale inhibitor nanomaterials, e.g., Al cross-linked sulfonated polycarboxylic acid (Al-SPCA) SINM, boehmite [γ-AlO(OH)] cross-linked polymeric SINM.
- Metal oxide-based nanomaterials, e.g., Al_2O_3 nanofluid in scale inhibitor, silica nanoparticles on DTPMP.
- Carbon-based nanomaterials.

The main advantages of SINMs include extended lifetime of squeeze treatments, superior transportability in the formation porous media, and improvements in the scale inhibitor performance.

- **Foam and gel scale inhibitors**

Foam scale inhibitor treatment is used to push the treatment fluid into the low permeability and potentially the most productive zone, especially in horizontal wells. One more advantage of this technology is that since the injection fluids depend mostly on gas to create the foam, the technology is seen as suitable for wells with low reservoir pressure and no gas lift system in place [123].

Gelled scale inhibitors involve a shear thinning gelling agent, xanthan gel, and breaker that have been formulated into a squeeze treatment [124]. They are also useful to deliver the chemical treatment to the low permeability zones and for high water production horizontal wells [125].

15.2.2.5 Factors affecting scale inhibitor performance

- **Brine composition (cations)**

Cations in produced water affect the scale inhibitor performance in many ways. Ca^{2+} and Fe^{2+} are the predominant cations that can precipitate with scale inhibitors. The precipitation of the inhibitor can lower its performance. However, intended precipitation of inhibitors during scale squeeze jobs was shown to improve the inhibitor performance and enhance its lifetime. The solubilities of Ca-SI increase in the order of Ca-NTMP $\sim$ Ca-DTPMP $<$ Ca-PPCA $<$ Ca-BHPMP [126]. The presence of Ca^{2+} and Mg^{2+} affects the inhibition of barite using phosphonate inhibitors. Fe(II) and Fe(III) can significantly impair performance of common scale inhibitors such as DTPMP, PPCA, and PVS. The inhibition time of DTPMP for barite can drop more than 90% in the presence of 1 mg/L Fe(II) or Fe(III) at 70°C [127]. However, some other reports showed that inhibitors can have some tolerance towards FeII [23]. Zn^{2+} ions were found to improve scale inhibitor (BHPMP and DTPMP) performance and extend squeeze lifetime due to the low solubility of Zn-SI salts [128].

- **Ionic strength**

Ionic strength also impacts the inhibitor performance and scale solubilities. The adsorption/release of inhibitors is affected by the ionic strength of produced water.

- **Temperature**

Temperature affects both the inhibitor performance and the scale solubility and supersaturation. Low temperature affects inhibitor adsorption: for example, DTPMP performance decreases at low temperature, whereas PVS is improved. High temperature also affects the inhibitor thermal stability, and solubility of some scales decreases with increasing temperature, i.e., $CaSO_4$, $SrSO_4$, and $CaCO_3$. Thus the temperature effect must be studied thoroughly during the inhibitor selection step.

- **Compatibility issues**

The inhibitor has to be compatible with the produced fluids and with the other chemicals. Phosphonates are the least compatible with high Ca/Mg concentrations, carboxylates are intermediate, and sulfonates are the most compatible [23]. Another compatibility issue is the other injected chemicals. Cationic corrosion inhibitors are known to precipitate anionic scale inhibitors, leading to diminishing their efficiency. Thermodynamic hydrate inhibitors can also interfere with scale inhibitors and reduce their efficiency.

- **pH**

pH has a strong impact on the inhibitor's performance. A common example is the polyacrylica acid performance, which is effective at high pH as a calcium carbonate inhibitor, as in ash sluice and some mining applications; shows mediocre performance near a neutral pH, as in cooling water applications;

and very low activity in an acid pH range, as in gypsum control in a pH range from 2 to 4 [129]. That effect is attributed to the dissociation and distribution of the chemical species at different pH, and is also due to the charge these species hold. The pH also affects the inhibitor precipitation, adsorption, and release mechanisms.

15.2.2.6 Screening of scale inhibitors

Selecting the proper scale inhibitors for field application depends on some factors, including the following:

- Scale type and severity.
- Available inhibitor types, their mechanism of action, factors affecting their performance, efficiency at low doses, and compatibility with the process fluids and other chemicals.
- Inhibitor thermal stability.
- The side effects of the inhibitor on the system, i.e., can it cause formation damage, corrosion, induce emulsions, etc.
- Availability of monitoring and detecting inhibitor concentrations at low concentrations.
- Costs and economics of the chemical inhibitor.
- Deliverability and method of application of the inhibitor in the field.
- Environmental impacts of the inhibitor, i.e., toxicity, biodegradability.

A number of methods and techniques are used to screen chemical inhibitors to evaluate their efficiency and stability, including:

- Compatibility studies
- Static bottle tests
- Kinetic turbidity tests
- Tube-blocking tests
- Thermal aging (in solution or on rock)
- Static adsorption tests
- Dynamic adsorption tests—core flooding, including permeability changes

15.2.2.6.1 Static precipitation tests

Also known as static bottle test or jar test, these tests are used as a rough screening method to rank the performance of scale inhibitors. The procedure of static bottle testing involves mixing synthetic cation and anion solutions based on the field's water analysis along with scale inhibitor at different inhibitor concentrations, incubating samples at temperature for a specified length of time, and filtering and submitting the solution for ion analysis [130].

NACE standard TM0197 furnishes the standard methods for static precipitation tests for screening scale inhibitors for barium sulfate and strontium sulfate, and the standard TM0374 for calcium sulfate and calcium carbonate.

While the static bottle test is a low cost, quick test to evaluate inhibitor performance on scale formation control in bulk solution, static tests were found to underestimate the amount of scale that can potentially form compared with using stirred or flowing test cells. Therefore it is recommended to test scale inhibitors systematically with regards to performance under turbulent conditions during qualification for field application [131].

15.2.2.6.2 Kinetic turbidity test

The kinetic turbidity test (KTT) employs a spectrophotometer with temperature control, to provide information not only on scale inhibitor performance but also on scale formation kinetics, inhibition mechanism, inhibition efficiency, and inhibitor-brine compatibility [130]. The method was applied for screening scale inhibitors for barite and calcite [130], halite, celestite, gypsum [132], and silica/silicate [133].

A combination of static jar, posttest SEM, and KTT were reported to obtain robust selection and improved chemical dosage recommendations for different field applications without the need for DSL tests for mild scaling brines [134].

15.2.2.6.3 Dynamic tubing blockage test

This is a dynamic method of testing scaling and screening scale inhibitors; it is also known as dynamic scale loop testing (DSL) or differential scale loop, where individual cationic and anionic brines are mixed and the scale inhibitor is dosed while the change in differential pressure is monitored. Fig. 15.18 illustrates a schematic of DSL and Fig. 15.19 shows a typical DSL unit with its capillary component [135].

In the dynamic tube-blocking test, inhibitors are tested at decreasing concentrations until a given pressure drop occurs across the tube, as illustrated in Fig. 15.20. Prescaling of the tube is often carried out to get more repeatable results. Using this technique, the minimum inhibitory concentration (MIC) or minimum effective dose (MED) necessary for complete inhibition can be ascertained, and thus the performance of inhibitors can then be compared. In the case of hard scale, like barium sulfate, it is more convenient to run inhibited solutions first and incrementally lower the inhibitor dose until a pressure increase is seen. At this point the tube can be flushed and the difficult problem of removing barium sulfate is minimized [23].

Blockage tube testing is preferred over the static tests for several reasons, including:

- Automated technique
- Temperature controlled
- More accurate and realistic results
- Suitable to screen inhibitors and estimate MIC in low and mild scaling brines

15.2.2.6.4 Adsorption studies

Two main types of adsorption studies are used to assess the extent of adsorption/desorption behavior of the inhibitor on a given substrate: static and dynamic studies.

Static adsorption tests are performed by placing a known weight of mineral separates or crushed formation rocks (as adsorbent) and a measured volume of inhibitor solution of specific concentration in a beaker or a bottle to evaluate the coupled adsorption behavior. After a certain time (sufficient to attain adsorption), the adsorbent mineral is separated and the liquid is analyzed for the equilibrium inhibitor concentration, where the difference between the initial and final inhibitor concentrations represents the adsorbed inhibitor quantity. The adsorption isotherm is obtained by plotting the amount of inhibitor adsorbed as a function of the final equilibrium inhibitor concentration in solution [136]. Dynamic studies are also known as coreflood tests, which can utilize a column packed with sand, dolomite, or any other rock type, or by using an actual core sample. This is discussed in a separate section that follows. For static adsorption/compatibility tests, the main factors influencing the inhibitor

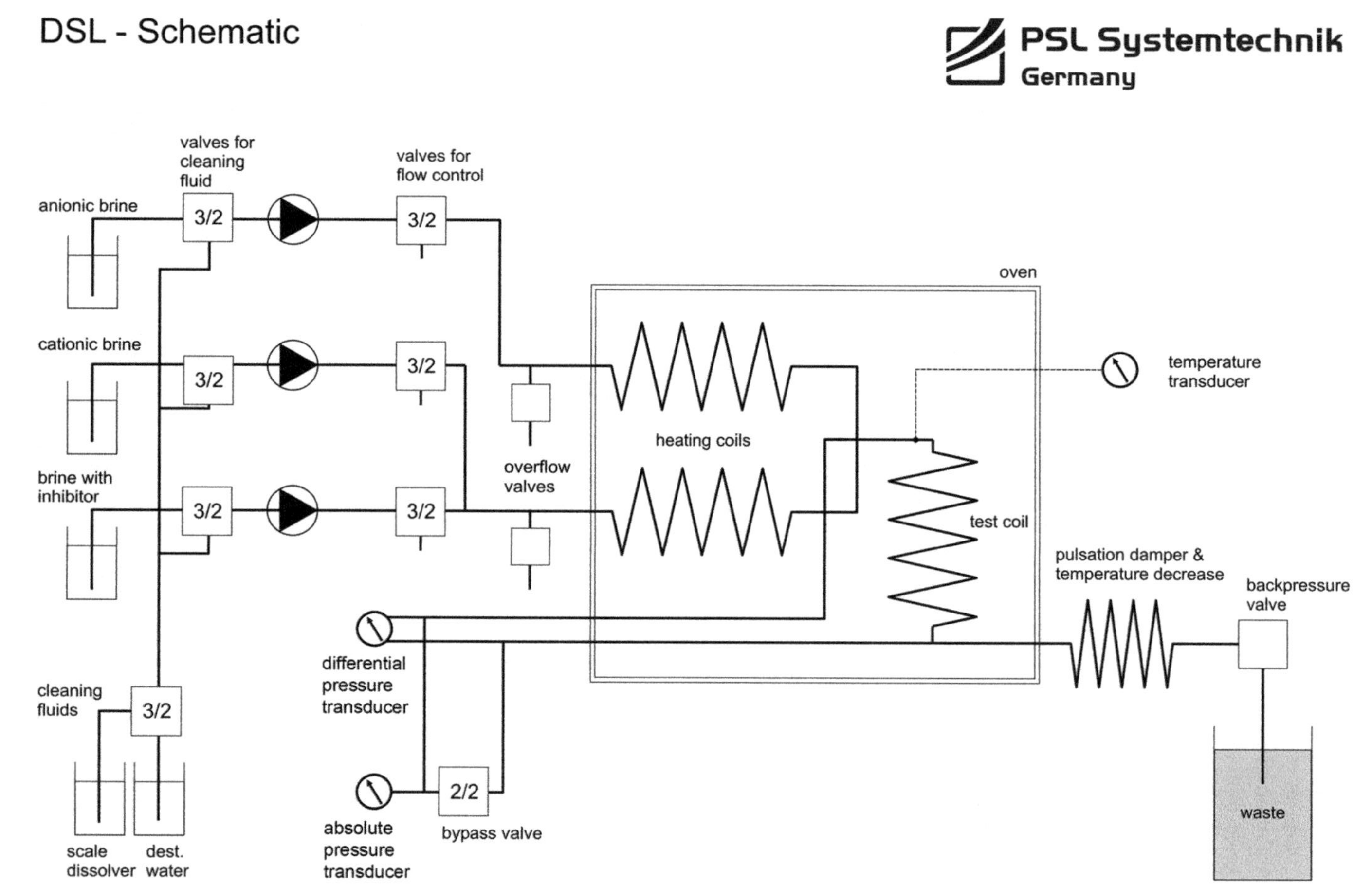

FIG. 15.18

Schematic of Dynamic Scale Loop.

Courtesy of PSL Systemtechnik GmbH, https://psl-systemtechnik.com/en/, reprinted with permission.

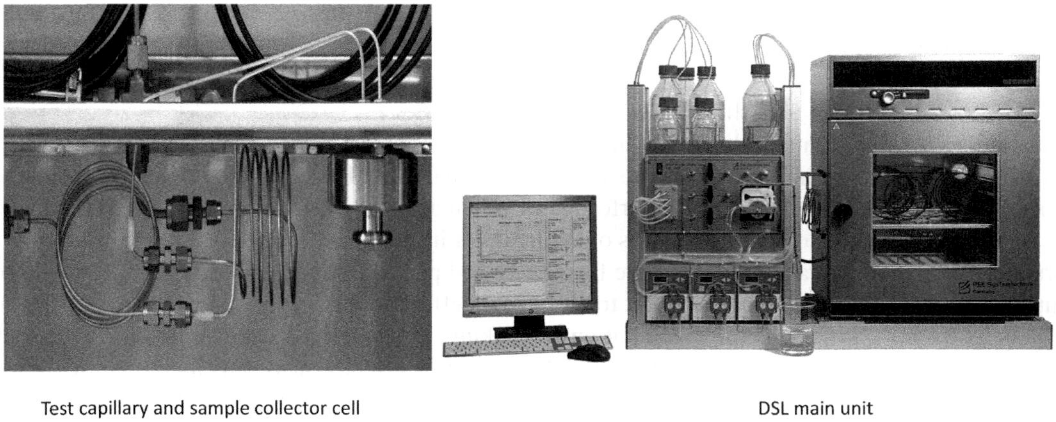

FIG. 15.19

Dynamic Scale Loop unit, and the capillary component.

Courtesy of PSL Systemtechnik GmbH, https://psl-systemtechnik.com/en/, reprinted with permission.

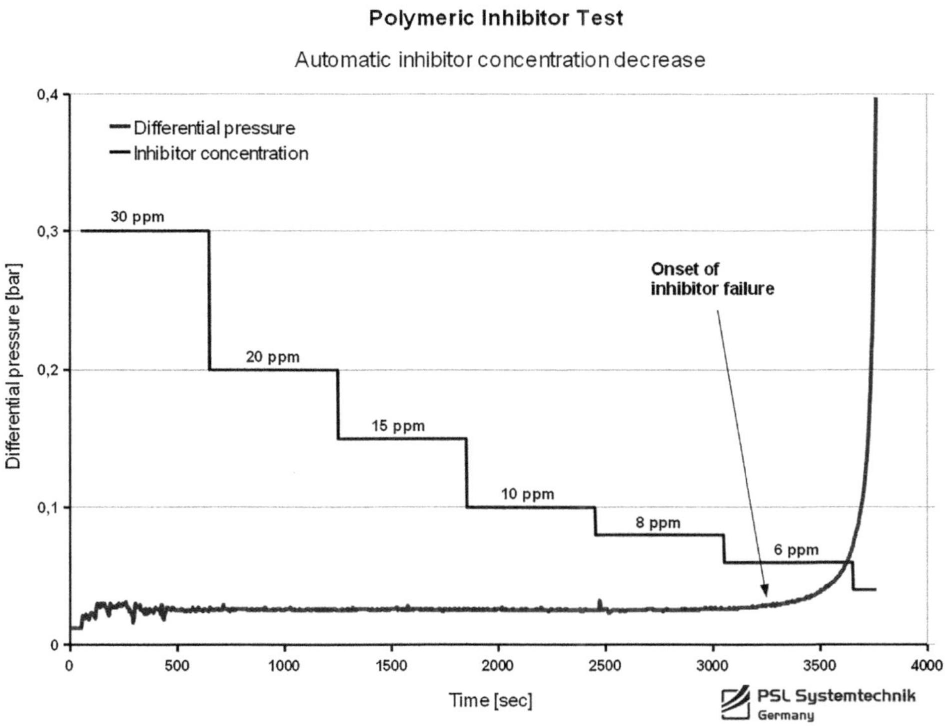

FIG. 15.20

Scale inhibitor testing in DSL. It shows the injection of the different doses of the chemical inhibitor and the changes in pressure.

Courtesy of PSL Systemtechnik GmbH, https://psl-systemtechnik.com/en/cas, reprinted with permission.

adsorption/desorption properties are pH, inhibitor type, mineral substrate type, brine composition, time, and temperature [136].

15.2.2.6.5 Thermal stability tests

Thermal aging tests are needed to make sure the inhibitor is stable at the reservoir temperature for the expected squeeze lifetime. The inhibitor solution is thermally aged in a static bottle and then characterized using chemical techniques and performance-tested against a nonaged sample [23]. Usually autoclave tests are carried out with solutions of the inhibitor in synthetic formation water in contact with crude oil at high temperature and pressure for a described period of time, and then the autoclaved solution of SI is used in a performance test to check if its effectiveness is impaired. The aged scale inhibitor is visually inspected for precipitation or junking, characterized using Raman, NMR, mass spectrometry, FT-IR, and other techniques, and then tested for effectiveness. Any abnormalities or loss in performance will eliminate the chemical from further consideration for squeeze treatments due to its diminished efficiency or possibility of formation damage [137].

15.2.2.6.6 Compatibility tests

Static precipitation tests and dynamic blockage tests can be used to evaluate the compatibility of the scale inhibitor with the produced fluids and with the other applied chemicals. Metal ions, corrosion inhibitors, hydrate inhibitors, and other chemicals can be added to the test solutions to evaluate the percent decline in the scale inhibitor performance or the amount of formed precipitate due to the added chemical.

15.2.2.6.7 Coreflood tests

Coreflood testing is another type of adsorption tests. Fig. 15.21 illustrates coreflooding equipment [56]. The inhibitor is flooded into a core sample at reservoir conditions and back-produced using formation water to simulate inhibitor squeeze treatments.

FIG. 15.21

Coreflooding testing equipment.

From: A. Khormali, D.G. Petrakov, Laboratory investigation of a new scale inhibitor for preventing calcium carbonate precipitation in oil reservoirs and production equipment, Pet. Sci. 13 (2) (2016) 320–327, https://doi.org/10.1007/s12182-016-0085-6.

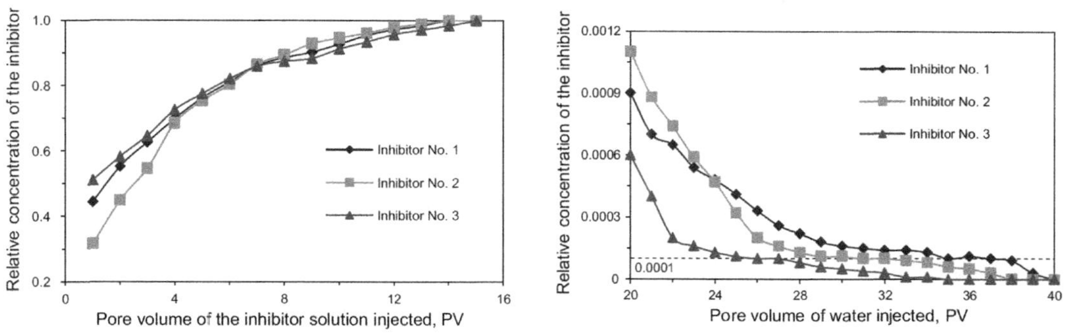

FIG. 15.22

Changes of inhibitor concentration during adsorption (left) and desorption (right).

From: A. Khormali, D.G. Petrakov, Laboratory investigation of a new scale inhibitor for preventing calcium carbonate precipitation in oil reservoirs and production equipment, Pet. Sci. 13 (2) (2016) 320–327, https://doi.org/10.1007/s12182-016-0085-6.

An inhibitor main treatment (also called slug or pill) is injected, followed by a postflush slug and then the effluent composition is monitored as a function of time or produced cumulative pore volumes (PV) of fluid. The concentration of inhibitor in the produced water (the inhibitor effluent profiles, in particular in the inhibitor desorption region or tail region) is monitored until it drops below the MIC. From these data, one can plot an adsorption isotherm and determine the expected squeeze lifetime compared with other products. Permeability changes can also be monitored with a core flood [23]. Typical inhibitor concentration curves obtained from coreflooding are represented in Fig. 15.22 [56]. Some case studies reported that corefloods can be misleading in determining the proper scale inhibitor for squeeze [138,139].

Apart from the preceding methods, other methods that can be used to assess the scale inhibitor include quartz crystal microbalance (QCM) and electrochemical QCM, thickness of shear mode resonator (TSMR), conductometry, rotating cylinder technique, electrochemical methods, electrochemical impedance spectroscopy, and the other methods mentioned in Chapter 13, "Monitoring of flow assurance formation in oil and gas fields."

15.2.2.7 Field application of scale inhibitors

There are a variety of methods for applying scale inhibitors in the field, including the following:

- Continuous treatments
- Batch treatments (squeeze treatments)
- Solid, slow-release scale inhibitor compositions

15.2.2.7.1 Continuous injection treatments

The continuous treatments are based on the presence of scale inhibitors at the time of nucleation, precipitation, and crystal growth to interfere with them and inhibit scale formation [24]. Scale inhibitors can be continuously injected on the topside facilities, or downhole of the production wells.

- **Continuous injection in topside facilities**

This includes wellheads, flowlines, risers, headers, and so on. There are two main points when injecting on topsides:

– The first point regarding the scale inhibitor on the topsides is to choose the proper location of the injection point and apply it upstream of the scale problem location, e.g., upstream of mixing spots and locations of significant pressure and temperature changes.
– The second point is that the scale inhibitor should be compatible with the other production chemicals.

The maximum fluid velocity is usually at the center of the line. Therefore the most effective position for injection is generally at the center of the pipe in the direction of the product flow. Injecting the scale inhibitor in the midpoint stream is important, as adequate mixing can occur; this might also help to eliminate the compatibility issues between concentrated scale inhibitor and brine components like Ca and Fe [24].

- **Continuous injection in the production wells**

Continuous injection is a method to protect the upper part of the wells that do not need to be squeezed because of low scaling potential in the near wellbore, or in cases where scale squeezing may be difficult and costly to perform on a regular basis, e.g., tie-in of subsea fields [140].

Scale inhibitors can be injected downhole through capillary string, the gas lift injection system, and/or continuous injection into the well annular space [23,24,56]. Fig. 15.23 depicts downhole chemical injection methods using downhole injection systems and gas lift.

Downhole chemical injection systems (DCIS or DHCI), conventionally known as macaroni string or umbilical lines, are chemical injection systems that have been implemented in subsea and downhole production systems [140–142]. The design and characteristics of the downhole chemical injection system was discussed by Guerra and Oliveira [142], and the downhole injection systems are discussed in Chapter 22, "Chemical injection systems." They basically comprise chemical tanks, pumps, control and metering systems, and umbilical/capillary lines.

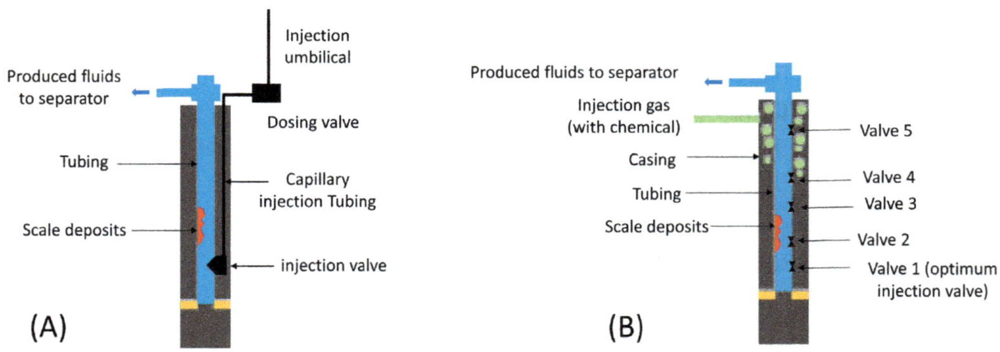

FIG. 15.23

Methods of continous injection of chemical inhibitor in production wells. (A) Downhole chemical injection system, (B) chemical injection through gas lift.

Chemical inhibitors can be deployed downhole the wells by injection with the gas in a gas lift operation. Gas is injected down the annulus through gas lift valves (GLVs), which allow flow into the production tubing. In this case, the scale inhibitors are added to water, alcohol, or other suitable solvent and added to the gas lift system. Usually gas lift systems (mandrels) have multiple valves at different depths, and the injection valve must be upstream of the scale problem. Problems associated with using gas lift are gunking of the scale inhibitor due to solvent evaporation or flash off, particularly where the gas has increased in temperature with depth, and due to the compression required to raise its pressure for reinjection, which plugs the gas lift valves. Another problem is the hydrate formation in the gas lift system. Solutions for such cases involve:

- Substantial scale inhibitor water/solvent dilution
- The addition of low-vapor-pressure solvent (vapor pressure depressant, VPD) such as a glycol to the aqueous scale inhibitor solution to avoid "gunking" [143].
- If VPD did not prevent gunking, the solid scale inhibitor can be dissolved in a very high boiling solvent [23].
- Glycol or some other hydrate inhibitor may be needed to suppress gas hydrate formation.
- A mixture of dissolver blended with scale inhibitor can be used to lower the risk [23].

Other problems associated with gaslift injection include disturbing of wellbore dynamic stability, either by insufficient pressure gradients at the GLV or by localized slugging initiated in the casing. Hence, chemical accumulation in the casing and intermittent pressure build-up upstream of the GLV were responsible for the nonuniform injection into the well tubing. The system's dynamic stability can be restored by either increasing the casing pressure to a level high enough so the GLV tolerates normal variations in the casing pressure, or by manipulating the flow pattern in the casing to avoid slug flow [144].

15.2.2.7.2 Scale inhibitor squeeze treatments

The common batch treatment method is the "squeeze technique" or "squeeze job," which is conventionally used by oil and gas operators. The method was first used for treating a well with a corrosion inhibitor and Stone et al. [145] reported one of the first uses of an organophosphonate scale inhibitor squeeze in 1966 [24].

Squeeze treatment technique involves "squeezing" or pumping a liquid inhibitor down a well into the formation under low, nonfracturing pressures and allowing them to be retained in the reservoir [24,146].

Scale squeeze treatments are basically implemented to protect the well downhole from scale deposition and formation damage, yet field observations have shown that the inhibitor can continue to work above the wellhead, protecting the pipeline and topsides from scaling. However, a further dose of a scale inhibitor may be needed [23].

Generally, scale squeeze treatments include the following main steps [23,147], as depicted in Fig. 15.24:

- Placement (injection) of inhibitor: the scale inhibitor is pumped into the well and introduced into the near-well formation rock pores.
- Retaining: after placement, the well is shut in for a period of time to allow the inhibitor to be retained in the formation rock matrix.

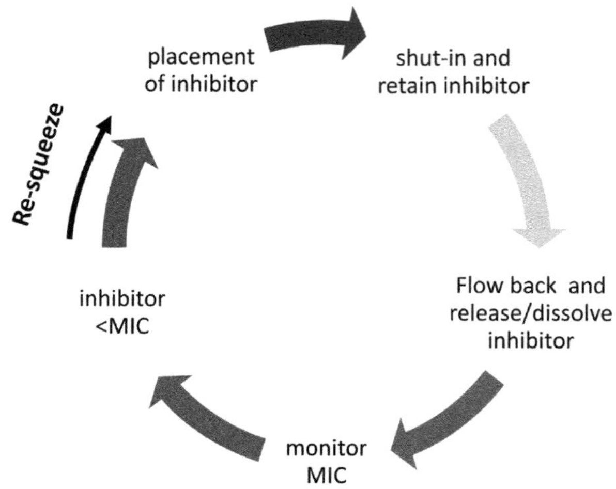

FIG. 15.24

The cycle of scale inhibitor squeeze treatments.

- Dissolution: The well is put on production and the flowing produced water will dissolve the retained chemical, releasing enough chemical (above MIC) to protect against scale formation.
- Monitoring: The concentration of residual scale inhibitor is monitored and when the concentration of the inhibitor falls below the MIC, the well should be resqueezed.

An important factor in squeeze treatment is the squeeze lifetime, which is measured either in terms of the time it takes for the scale inhibitor concentration to drop below MIC or, more commonly, on how many barrels of produced water are "protected" in this period. In terms of time, such treatments may often last from 3 months to 1 year and in terms of protected volume of produced water, they may be from 250 Mbbl to 3 MMbbl [136].

15.2.2.7.2.1 Placement of scale inhibitor. Typically, a squeeze placement consists of the following stages: preflush/spearhead, scale inhibitor main treatment, and overflush.

- **Preflush (spearhead)**

The objectives of the preflush are to [148–150]:

- Displace the wellbore fluids.
- Work as a spacer between these fluids and the main treatment stage.
- Clean the formation (by removing the adsorbed oil, change rock wettability, and lowering oil-water interfacial tension) to be ready to receive the main treatment for enhanced chemical inhibitor retention.
- To cool the formation to allow the inhibitor to improve transportability of inhibitor in the formation.

An acid wash may precede the preflush to clean the scale and debris out of the wellbore to "pickle" the tubing (this fluid should not be pushed into the formation) [151]. Surfactants and cosurfactants are used to wash oil leftovers, and help in changing the wettability and extending the squeeze lifetime. Nonionic surfactants were found to enhance squeeze treatment lifetime by as much as 240% compared with conventional treatment [152]. The mutual solvent chemicals are used in the preflush formula to avoid emulsions and other formation damage, improve inhibitor adsorption, improve fluids mobility, and speed up well cleanup and flowback [148]. The preflush will have a low concentration of inhibitor to avoid incompatibility between the injected chemicals and formation brine. Other additives are demulsifier oxygen scavenger and biocides to avoid emulsions, oxygen ingress, and bacteria contamination.

- **The main chemical treatment**

This includes the main chemical that is used for the inhibition, also called the inhibitor slug or the inhibitor pill. The main scale inhibitor volume is injected, which typically contains inhibitor chemical in the concentration range of 2.5% to 20% in a makeup water of ~1% KCl or filtered produced water. More developed software packages have recently been developed to design the squeeze jobs and calculate the amounts of chemicals.

- **Overflush**

The overflush step is applied to push the inhibitor slug into the desired depth in the warmer regions away from the wellbore and allows the inhibitor achieve a higher level of retention by contacting a larger volume of rock. Various fluids can be used in overflush including brines, diesel and crude oil or combinations of the three fluids [150,153]. Squeeze treatments using aqueous overflush are known as conventional squeeze treatments, where non-conventional refers to treatments where the overflush is splitting to aqueous and non-aqueous stages, typically diesel being used for the non-aqueous stage, in case of split overflush is seems more beneficial to inject the water stages first [154].

If brine is used it should be compatible with reservoir brines i.e. similar salinity, and no scaling ions. Chelating agents and scale inhibitors may be added to assure brines compatibility. Non-aqueous overflush is applied when aqueous treatment are not desirable e.g. where water blocking, fluid lifting, or hydrate formation area concern, hence diesel or non-aqueous overflush are applied [153].

The volume of overflush required to maximize adsorption, inhibitor concentration (volume) and the adsorption capacity of the reservoir.

Ionic polymer additives such as a poly amino acid or poly quaternary amine were found to improve squeeze treatment lifetimes as part of the overflush by working as bridging agents for inhibitor retention [155,156]. They are also used for clay stabilization. The treatments on two wells in an HP/HT field at 165°C demonstrated improved chemical retention and scale inhibitor returns compared to treatments without the ionic polymer additive in the overflush [155].

Various methods are used to place the squeeze treatment chemicals. Among these, bullheading and coiled tubing are the most common. Bullheading is the simplest method, whereby the chemicals are pumped downhole with less control of placement. For better control of the scale inhibitor placement, two or more inhibitor squeeze stages may be separated by diversion stages, which can be used to provide zonal isolation and thus control the subsequent inhibitor placement and assure its propagation into the targeted location. Two possible diverting agents are thermally degrading wax beads and thermally breaking shear-thinning gels [157–159]. Another way to make sure the chemical has reached enough concentration in the targeted locations or low permeability ones is by overdosing the higher permeability regions until sufficient treatment volume enters the low permeability zone. Coiled tubing can also be applied for better control of the placement, and for subsea fields and long completions. Remotely operating vehicles (ROVs) can also be used in subsea fields [160].

Some of the issues that affect the efficiency of the squeeze treatment chemicals placement include [161,162]:

- Fluids mixing in the tubing. These mixing events basically depend on tubing dimensions and pumping rates. For example, if the tubing volume is larger than the main treatment and overflush stage volumes, they can be completely mixed up, and this can affect the overall mass of injected scale inhibitor, causing lower than expected inhibitor concentration contacting the reservoir rocks, which certainly affects the squeeze lifetime and efficiency.
- Reservoir heterogeneity, wherein most of the bullheaded squeeze chemicals are placed in the higher permeability zones, leaving other low permeability zones untreated. The effect can be worse if the untreated zones are the water producing zones.
- Pressure gradients and crossflow, wherein injection being favored in the lower pressure zones, injection at lower flow rates may even fail to penetrate the higher pressure zone, especially in wells with a strong crossflow. Increasing the viscosity of the chemical slug was shown to overcome crossflow and improve chemical slug penetration.
- Wellbore friction, which is strong in the horizontal wells and increases with decreasing tubing diameter, and the presence of sand screens exaggerates the friction resistance.
- Other factors include layer pressures, properties of the fluids in place, and difference in mobility ratios between different zones.

Fig. 15.25 shows a schematic of the squeeze treatment placement stages.

15.2.2.7.2.2 Shut-in. After the chemical treatment is injected into the well, it is shut-in for a period of no flow prior to returning the well to production, to allow the retention of the injected chemical inhibitor. Depending on the reservoir brines, composition, reservoir temperature, and other factors, shut-in times can be between 6 and 24 h. Shut-in times are affected by the kinetics of chemical inhibitor retention. Since it removes the well from production, shut-in times should not extend beyond economic/technical acceptable periods. The chemical inhibitor retention will be discussed in a separate section.

- **Scale inhibitor retention**

Scale inhibitor retention refers to any mechanism whereby the scale inhibitor chemical is held back or "retained" in the porous medium [136]. Retention of SIs in the reservoir-pore spaces is achieved basically by phase trapping, adsorption, or precipitation. Phase-trapped inhibitor is that amount left mixed with the reservoir water after squeeze, and the majority of this fraction is produced back after the well is put back on flow, so only adsorption and precipitation contribute to the long-term retention of the inhibitor.

- *Adsorption*

In adsorption the chemical inhibitor adsorbs on the mineral surface through physical and/or chemical interaction. Adsorption is believed to occur through van der Waals interactions, hydrogen bonding, and other electrostatic interactions between the inhibitor and formation minerals. For example, the retention of DETPMP is believed to be based on at least four mechanisms: (i) acid/base dissolution of the mineral surface; (ii) adsorption to the surface as the result of acid/base dissolution in step (i),

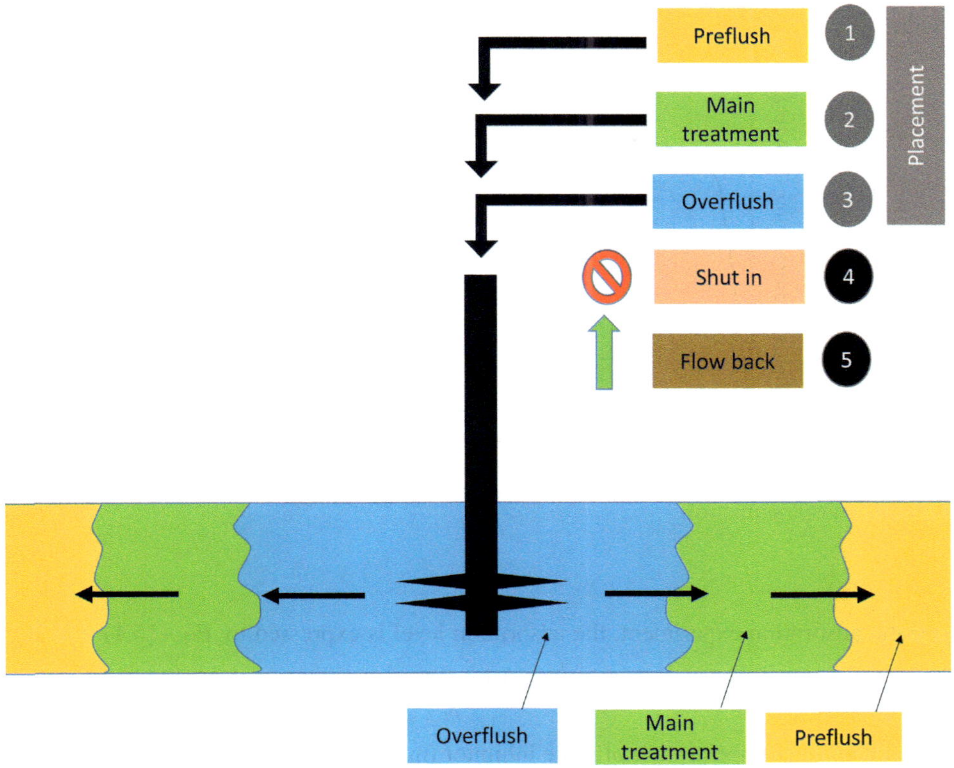

FIG. 15.25

Schematic of scale inhibitor squeeze procedure.

(iii) mass-transport molecular diffusion of inhibitor in solution to the solid surface; and (iv) solid phase maturation towards a thermodynamically stable phase, as the solid surface material interacts with the solution [136,163]. "Adsorption" squeezes often perform in certain noncarbonate (sand) reservoirs, whereby a neutralized phosphonate pill is injected into a sandstone formation [126].

The adsorption may be an equilibrium or nonequilibrium (kinetic) process that, in either case, is governed by an adsorption isotherm, which is the key characteristic of inhibitor adsorption/desorption [164–166]. The isotherm $\Gamma(C)$ in simple terms is a relationship that quantitatively relates the amount of inhibitor adsorbed on a formation rock surface, as a function of inhibitor concentration in the bulk formation fluid at equilibrium at specified temperature. Fig. 15.26 depicts a typical scale inhibitor adsorption isotherm curve.

The adsorption isotherm curve is the most important factor in squeeze treatments. The adsorption isotherm normally is measured using static or dynamic (coreflooding tests) adsorption tests, as discussed in Sections 15.2.2.6.4 and 15.2.2.6.7. The adsorption isotherm can be expressed using empirical expressions, such as the Langmuir and Freundlich forms, which have been employed for representing the inhibitor/rock interaction isotherms.

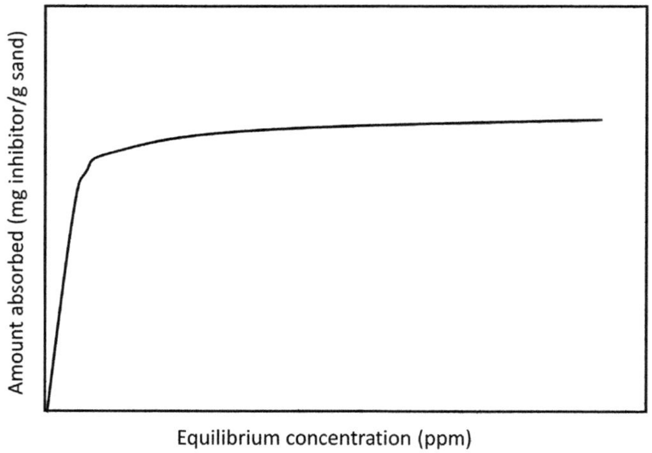

FIG. 15.26

Typical adsorption isotherm curve.

In a static adsorption experiment, the adsorption level is expressed by Eq. (15.1): [167]:

$$\Gamma = \frac{V(c_0 - C_{eq})}{m} \tag{15.1}$$

where c_0 is the initial concentration of scale inhibitor (mg/L, ppm) in volume V (L), of bulk solution; C_{eq} is the equilibrium concentration; and m (g) is the mass of the crushed rock material (sand or lime).

In practice, if coreflood data are not available, the isotherm can be derived using field return data by history matching the measured field squeeze returns assuming that the isotherm is a Langmuir or a Freundlich expression [168].

Scale inhibitor adsorption is a function of many factors including:

- pH

Phosphonate inhibitor adsorption level is higher at lower pH values. Nonequilibrium adsorption is also evident in the low pH floods [169].

- Temperature

The adsorption of inhibitors increases with increase in temperature utilized [170].

- Mineral substrate

Adsorption/desorption depends on the mineral type (sand, kaolinite, siderite, etc.), the nature of the surface (smooth or rough), and the solution [170]. Gdanski and Funkhouser classified minerals into three groups: (a) strongly adsorbing (siderite), (b) moderately adsorbing silica-like minerals (silica and kaolinite), and (c) weakly adsorbing alumina-like minerals (illite, smectite, and alumina) [171,172].

- Molecular weight

The adsorption/desorption characteristics of polymeric scale inhibitors (PAA, PPCA, PVS) have showed that preferential adsorption of the higher molecular weight components occurs [173].

- Scale inhibitor functional groups

Phosphonate groups generally adsorb better than carboxylate groups, which are better than sulfonate groups. Thus squeeze lifetimes with sulfonated inhibitors such as PVS may be lower than with inhibitors with many phosphonate groups, owing to low retention in the rock matrix [23].

- Cation concentration

Ca^{2+} ions can enhance phosphonate scale inhibitor retention (due to a SI-Ca precipitation mechanism, or Ca ions bridging between inhibitor and mineral surface [166,167,169]). The squeeze lifetime using phosphonate SI was reduced by one-half when the calcium in the postflush brine decreased from 5450 to 1000 ppm with the same in situ brine [174]. Polymeric inhibitors are effective at low Ca^{2+} concentrations. Phosphonate performance was drastically reduced in the presence of Mg^{2+} whereas polymeric species were much less affected [81,175]. Iron ions also affect the inhibitor retention.

- Other factors

These include the ionic strength, which affects the mineral surface potential. The preflush surfactants change the wettability of the rocks and can enhance the retention of the chemical inhibitors.

– *Precipitation*

The second mechanism of scale inhibitor retention is the precipitation mechanism in which the chemical inhibitor reacts with metal ions, e.g., Ca^{+2} or Fe^{+2}, and precipitates. The process usually starts with the adsorption stage followed by precipitation (gel-like SI-cation complex), which is controlled by temperature, cation concentration (e.g., $[Ca^{2+}]$) and/or pH [152,164]. The metal ions can be part of the formation fluids, formation minerals, or can be introduced during the squeeze treatment prior to injecting the chemical inhibitor. A coupled adsorption/precipitation mechanism is also common. Common precipitation squeeze occurs when an acidic phosphonate pill is injected into a carbonate formation to cause the precipitation of calcium phosphonate [126]. The precipitated inhibitor then slowly dissolves in the formation water, releasing enough concentration to protect the wellbore area. The concentration of inhibitor in the flow back return depends on the solubility and dissolution rate of the inhibitor/calcium complex [164]. The solubility products of some SI-Ca complexes have been reported in Ref. [176].

Precipitation squeeze treatments improve the retention of the chemical inhibitor and enhance the squeeze lifetime, especially in formation where adsorption squeeze treatments are short-lived or less effective. Precipitation treatments depend on many factors, among them the water volumes, water composition, pH, and reservoir mineral types.

15.2.2.7.2.3 Production (flowback). After enough shut-in time, the well is brought back on production. The well flowback releases the retained chemical inhibitor sufficiently to protect the wellbore and production tubing from scale risks. During the production phase the inhibitor return concentrations are monitored to ensure they are still above the minimum inhibitor concentration. Normally, following the squeeze treatment, a fair percentage (25%–35%) of the scale inhibitor is produced back immediately

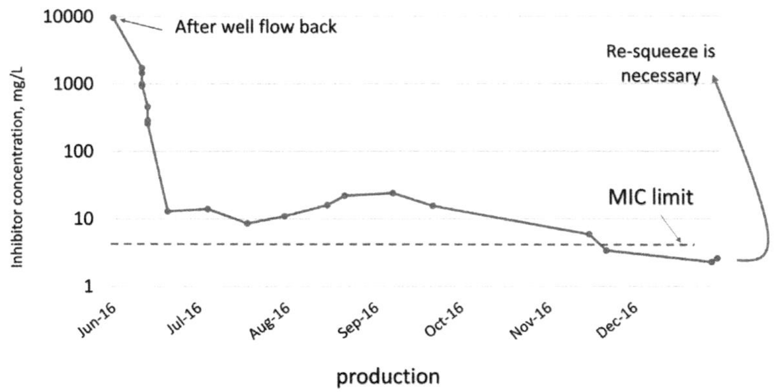

FIG. 15.27

Scale inhibitor returns after scale squeeze from Egyptian field.

after the well is put on production; therefore the return scale inhibitor concentration increases rapidly and peaks to some value and then declines, typically within a few days, to a low plateau concentration, which comprises the bulk of the squeeze duration, as illustrated in Fig. 15.27. It is crucial that during steady-state production the plateau flowback inhibitor concentration is sufficient to inhibit scale ($\geq$ MIC)] [126].

When the scale inhibitor concentration falls below the MIC, the operator must plan for a resqueeze to keep the wells safe from scale accumulations.

Some methods to improve the squeeze treatment lifetime include the following:

- Improve preflush process and preflush chemicals, e.g., mutual solvent, proper surfactant, cosurfactant, cationic polymers, for better conditioning of the formation to enhance the water wettability. Using proper surfactants leads to enhancing squeeze lifetime by 240% [149].
- In the same way, improving the overflush is also a key factor to extending the lifetime of the squeeze job.
- The use of Zn^{2+} ions in squeeze formulations has been shown in laboratory studies to significantly increase the retention of the inhibitor.
- Microcrystalline kaolinite can be combined with the fixation agent and scale inhibitor as a means of mechanically altering near wellbore mineralogy and surface property characteristics within clean, high permeability sandstones, to enhance the squeeze lifetime [177].
- Manipulating the pH to simulate inhibitor precipitation leads to better retention and extended lifetime.
- Using nanomaterial scale inhibitors.

There are other means to batch treat the production wells besides the squeeze treatment. Scale inhibitors can be deployed mixed with other production operations, providing more efficient and extended scale prevention. The scale inhibition can be added to conventional fracture treatments, thereby eliminating a separate inhibitor squeeze treatment [178,179]. The scale inhibitor can be added as [180]:

- A liquid squeeze separate from a hydraulic fracture
- A liquid inhibitor in a hydraulic fracture fluid
- A solid inhibitor in a hydraulic fracture fluid

In this way the placement of the inhibitor is improved: since the inhibitor is mixed into the fracturing fluid during the treatment, the inhibitor is well mixed and dispersed throughout the entire fracture. The propped fracture also reduces pressure drawdown, which reduces scaling potential [178].

The inhibitor should be compatible with the fracturing fluids and not adversely affect the fracturing fluids/proppant pack conductivity, whereas for the acid stimulations, compatibility issues may show up between the acid, the acid additives, the scale inhibitor, and the spent acids with high load of cations [23,180]. Normally, scale inhibitors are deployed under acidic conditions (pH 1–5) to ensure brine compatibility and optimal retention characteristics, whereas common fracking gels (guar or derivatized guar crosslinked with borate) require alkaline conditions (pH ~9) for viscosity development [181]. Such combined simultaneous treatment has been applied to the Gulf of Mexico and West Africa wells with good squeeze life [182–184].

Other operations that scale inhibitor can be mixed with are water shutoff treatments and clay stabilization treatments [185]. Another feasible method of deploying scale inhibitors is during water injection operations, especially for injection wells that are close enough to production wells. If the inhibitors can make it through the formation flowing to reach the production zone without being totally adsorbed/precipitated in the formation, they can protect a large area in the reservoir, wellbore, tubing, and possibly surface facilities as well. Inhibitors with weak adsorption abilities are recommended in this case [126].

15.2.2.7.3 Solid scale inhibitor/controlled release scale inhibitors

To avoid the complications of aqueous scale inhibitor treatments, some scale inhibitors are applied in solid forms, such as powders, briquettes, or balls, wherein they can be applied in perforated baskets, bypass feeders, gravel pack, sliding sleeves, fracturing operations, or are dumped downhole [24]. The various ways of producing particles containing scale inhibitor include 100% solid products, encapsulated products, and highly porous materials that can capture the inhibitor [23].

The solid scale inhibitors can be deployed into the production wells by different means, including:

- Batching downhole
- With fracturing proppant: scale inhibitor impregnated proppants (SIIP)
- With gravel pack system: scale inhibitor impregnated gravel (SIIG)

In the first method, a small volume of ~2 barrels is preflushed with conventional scale inhibitor and then the encapsulated chemical solids (microencapsulated scale inhibitor in permeable polymer membrane) can be placed in the bottom of the well, in the rathole (70%–100% of the rathole should be filled with solid inhibitor) and the well is put back on production; as fluids pass over the charge a small amount of phosphate dissolves, and this in turn provides scale protection to the subsurface equipment [28]. Hydrolytic and thermal stability as well as dissolving rate are important in this type of application [24]. Encapsulated scale inhibitors are applied for weak wet oil, high water cut, and low reservoir pressure wells, with 100 field cases in Saudi Aramco, where they were found to be efficient, cutting costs by 71% [28,186].

Problems with such treatments include the following [24]:

- The solids become coated with oil or surface reaction products that interfere with their dissolution rates.

- Downhole annular or tubing treatments with solids are sometimes impossible due to completion techniques. Packers, slim hole completions, storm chokes, etc. prevent the introduction of solids by normal batch procedures.

The release of the solid inhibitor depends on temperature (the higher the temperature, the greater the release of active material), brine composition, pH, inhibitor concentration in the brine, and the extent of turbulence [118].

Another way of deploying the treatment is by delivering solid inhibitor through hydraulic fracturing, by adding solid inhibitors into the fracturing fluid through the porous scale inhibitor impregnated proppant-SIIP or placing solid scale inhibitor with the proppant-laden fluid [187,188]. The maximum concentration of scale inhibitor that could be incorporated into the solid matrix was approximately 12 wt%/v [189].

Solid scale inhibitors impregnated into porous ceramic particles used with sand control, conventionally called scale inhibitor impregnated gravel (SIIG), can be used. This has a higher concentration of scale inhibitors than SIIP, to be suitable for gravel pack application and to offer more protection. The technology offers considerable cost benefits for scale control when compared to scale inhibitor squeeze treatments and negates the need for preemptive squeeze treatments in anticipation of seawater breakthrough; the technology was applied in the Heidrun field with success [189]. However, the technique is suitable for low scaling wells, as inhibitor loss can occur and the released inhibitor residuals are difficult to detect.

15.2.2.8 Scale inhibitor treatment monitoring and assessment

After treatments, the scale inhibitor must be monitored and its efficiency in preventing scale formation evaluated. Monitoring and assessment of scale inhibition treatments can be achieved either by lab techniques or by online and real-time techniques. These methods are discussed in detail in Chapter 13, "Monitoring of flow assurance solids formation in oil and gas fields."

Lab methods are based on measuring the ions and scale inhibitor residuals in the produced water, and as long as the measured inhibitor residuals are equal to or higher than the MIC, then the system is theoretically protected against scale formation. Methods like inductively coupled plasma optical emission spectroscopy (ICP-OES), inductively coupled plasma mass spectroscopy (ICP-MS), ion chromatography (IC), liquid chromatography (LC) or LC-MS, fluorescence techniques, and high-performance liquid chromatography (HPLC) have been reported. Although these methods are common, they suffer from some issues:

- Inaccuracies due to sample collection and preservation.
- Inaccuracies due to technique sensitivity to the inhibitor chemical, limit of detection, limit of quantification, matrix interferences, and other chemicals in the produced water.
- Issues due to technique selectivity, e.g., ICP methods are elemental and suitable for phosphorus based chemicals; polymer chemicals need more selective techniques like HPLC or the dedicated hyamine method.
- On top of all this, knowing that the inhibitor residual is above the MIC does not guarantee that the system is protected against scale. Some scale chemicals that showed success in lab screening have been failing in field application. Additionally, the continuous changes in produced water volume and composition, besides the changes in pressure and temperature, can affect the efficiency of the injected inhibitor.

Measuring the ions in the produced water is conventionally conducted for treatment monitoring and evaluation. One offline method is effectively used: the total suspended solids followed by evaluating the solids collected on the membrane using ESEM was reported to be highly effective in assessing the scale inhibitor effectiveness and the activity of scale formation. The method determines if scale is actively formed, transported, or modified due to the action of the scale inhibitor, then relates these data with the produced water volumes to build a risk matrix used to determine if the well needs resqueeze or not [190].

Online and real-time monitoring of scale in the system is preferred for rapid and relatively accurate results. Methods like microbalance, thickness of shear resonator, tomography, and thermography are usually used. In addition, routine pigging, well gauging, wireline logging, and other mechanical methods of maintenance and intervention can be used to detect the formation of scale, which indicates the efficiency of the treatment.

15.2.2.9 Recent advances in chemical scale inhibition methods

Recently, chemical scale inhibition has witnessed a number of advances at all levels. New emerging chemical types and synergists have been introduced. Development of new products and introducing them in new forms, e.g., aqueous, nonaqueous, emulsion, encapsulated, solid impregnated on porous materials, foamed, and gelled scale inhibitors, have all been reported. Additionally, scale inhibitor nanomaterials (SINMs) and ionic liquid scale inhibitors are new emerging technologies that are environmentally friendly and efficient (especially SINMs, in improving squeeze lifetime). Green inhibitors have witnessed a number of advances using natural products and extracts. Viscofied fluids, surfactants, and cosurfactants were introduced to improve the squeeze lifetime and efficiency. Nonaqueous inhibitors, to allow their application in low water cut and water sensitive formation, and foamed and gelled inhibitors were introduced, to improve squeeze treatments to reach targeted zones. Solid inhibitors were game changing for high water cut and low reservoir pressure wells. Advances in inhibitor testing have allowed rapid testing using kinetic tests in combination with static tests. More advances have been shown in scale squeeze simulation. Scale inhibitor squeeze simulation software has been developed, Heriot-Watt University's SQUEEZE 10 modeling software is considered the industry standard.

15.2.3 Nonchemical scale prevention

In some instances, chemical scale control is not feasible economically or technically, depending on the system design, capacity, fluids properties, and severity of the scale problem. For example, in deepwater and subsea fields with extremely harsh scaling problems, scale squeeze treatment can be rendered uneconomical.

Hence, operational methods (discussed in Section 15.2.1) physical or nonchemical devices (NCDs) like magnetic fields, electric fields, or scale preventing surfaces and internal liners can be a useful alternative. These physical methods are generally based on the bulk solution precipitation of solids rather than surface precipitation and dispersing the formed solids.

15.2.3.1 Magnetic scale prevention

Antiscale magnetic water treatment devices (AMTs) are one of the most common and controversial solutions for scaling dilemma since first patented in 1873. For more than a century, this technique has been a matter of controversy on its efficiency and mode of action, although evidence of its efficacy has been provided on countless occasions [191].

In this method, either permanent magnets or electromagnets in different configurations are used to produce a magnetic field (MF) within the conduit wherein the fluid is flowing, generating a magnetic

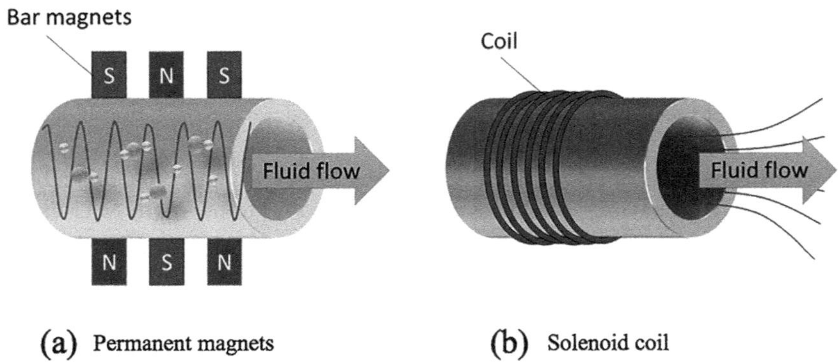

FIG. 15.28

Typical EMF devices with different configurations.

From: L. Lin, W. Jiang, X. Xu, et al., A critical review of the application of electromagnetic fields for scaling control in water systems: mechanisms, characterization, and operation, npj Clean Water 3 (2020) 25, https://doi.org/10.1038/s41545-020-0071-9.

field that can be static or pulsating, orthogonal to fluid flow, or generated in the direction of fluid flow (Fig. 15.28). Manufacturers usually have specifications for their devices, specifying the location on a piping system and the flow regime [191,192]. The configurations involve the kind of magnetic circuit and the space between the magnets, and the solution to be treated [193].

As mentioned earlier, there is no consensus on the mechanism of AMT at the moment, due to the inconsistencies in the results obtained by different researchers, and also due to the reported ineffective application of ATM [191]. Some studies refer the action to the magnetohydrodynamic (MHD) effects, while others refer it to the agglomeration of ferromagnetic compounds [193]. Throughout the literature, there seem to be two possibilities as to how MF devices prevent scale formation:

- By altering the structure or chemical composition of the water itself, or
- By altering the structure of the deposit.

Changes in water characteristics and structure include changing the water surface tension, water conductivity, evaporation, and hydrogen bond stability [191,194,195], which favors the dispersion of water molecule aggregates and decreases the aggregation of their associated ions.

One way MF could bring about scale prevention is by modifying the particle zeta potential, size, and surface charge of burgeoning seed crystals, thus achieving reduced tendency for the deposition of particles on walls [196,197]. This also enhances faster agglomeration/aggregation of the species in larger size, which promotes bulk precipitation rather than adherence to the walls of pipes and vessels [198]. It also involves changes in crystallinity in general (morphology, crystal phase, solubility, and rate of precipitation) [199]. The effects on crystallinity can be explained by the formation of prenucleation clusters called DOLLOPs (dynamically ordered liquid-like oxyanion polymers), which implicate the formed solid morphology [200]. A field study in a power station gave an alternative explanation that the crystallization of carbonates was blocked due to the initiation of another competitive process, i.e., the activation of colloidal silica by deforming the diffuse layer mainly consisting of water molecules [201].

The magnetic field technique depends on many factors, including:

- Supersaturation
- Magnetic field strength
- Time of exposure to the magnetic field
- Water salinity, pH, dissolved gas concentration
- Concentration of impurities and particulates
- Temperature
- Fluid flow rate
- Pipe material
- Flow regimes

Magnetic water treatments have been used in municipal applications with many suppliers worldwide. In industrial applications, the technique has been proven to prevent scale deposits on many occasions. Szkatula et al. [201] implemented MF treatment on industrial water scaling in 25 kW heat exchangers and a 1 GW power plant with an observed reduction in scale formation. Kobe et al. [202] used magnetic treatment to prevent scaling in heat exchangers for 2 years in a pilot plant that treats tab water. Grutsch and McClintock [203,204] reported that a magnetic treatment device was employed successfully on a cooling tower at their Texas City refinery. Donaldson and Grimes [199] used the magnetic field to prevent scale in heat exchangers. Magnetic field technique was applied in a Brazilian field in different onshore and offshore applications, including downhole magnetic subs installed at a tubing string and surface electromagnetic/magnetic devices at production separation facilities and water, and showed promising results in preventing/controlling scale formation [205].

Contrary to this, Pritchard et al. [206] assessed magnetic treatment for use in oilfields and concluded that the magnetic treatment did not show significant efficiency in reducing scale deposits [206]. In another field study of three (two magnetic and one electronic) alternative devices, to determine if they would reduce or prevent scale formation under field conditions typical of those found at US Army installations, the tested devices did not prevent nor appreciably reduce mineral scale formation in comparison to a control [207].

15.2.3.2 Electrostatic scale prevention

This method can be applied in many ways, e.g., solenoid coil electronic device, RF electric field, or capacitor systems, where the created electric field disperses the scale and fouling species. For example, a high-voltage DC power supply and a capacitor-like cell can produce an electric field. The water to be treated flows through this field, so the electric field has an effect on the water, possibly by changing the quantities and size of colloids, which reduces the rate or scaling. Sun Jinghui et al. [208] developed an electrostatic descaling device and applied it to a condenser in a power plant, which showed descaling functioning besides the ability to reduce corrosion. An ion-rod electrostatic water processor was also found to have promising scale inhibition effect. Ye and Liu [209,210] reported that a small current generated by a low-voltage electrostatic processor caused the water to dissolve into more bubbles of hydrogen and oxygen. The resulting bubbles wrapped around the $CaCO_3$ particles and inhibited the growth of $CaCO_3$ microcrystals, to achieve the effect of scale inhibition. A self-powered intelligent water meter based on triboelectric nanogenerators (TENGs) for electrostatic scale prevention was proposed [211].

15.2.3.3 Ultrasonic scale prevention

Ultrasonic scale prevention (USP) is another physical method of inhibiting scale formation. it is believed that USP works by bulk precipitation compounds rather than encrusting on the surface. Furthermore, the ultrasound induces high-frequency vibrations that propagate over the surface, preventing precipitated particles from sticking to the surface. In the event some particles adhere to the surface, they will be split, broken by the hydrodynamic cavitation effect of the ultrasonic waves, which have been used frequently in scale removal.

15.2.3.4 Scale prevention surfaces

This method is based on preventing the formed solid particles from adhering to the surface. In developing an antiscale/antifouling engineered surface, crucial points must be understood, including the following [212–214]:

- The surface parameters: surface roughness, surface energy, hydrophobicity, wettability.
- Fluid hydrodynamics: flow velocity, flow regime, system dimensions, and parameters (T, P, etc.).
- Fluid properties: composition, degree of supersaturation, viscosity, suspended solids.
- The kinetics of surface deposition: induction time for scaling and the link between the time constant for bulk and surface deposition.

A range of chemically and morphologically modified coatings to prevent/reduce mineral scale surface fouling have been reported in the literature, including glass, polytetra fluoroethylene (PTFE), siloxane polymer, acrylate polymer, superhydrophobic nanosilica, sol-gel, modified ceramics, diamond-like carbon (DLC) [1,215–217], and, more recently, titanium dioxide modified with ethylenediamine tetra methylene phosphonic acid (EDTMPA), which has been incorporated into epoxy resins [218], has become available and is used in the oil and gas industry to protect high-risk areas like valves, chokes, and electrical submersible pumps (ESPs) from corrosion and scale deposition [1]. They found that these coatings are specific for each type of deposit.

An alternative method for scale protection of valves and chokes and other key equipment has been proposed in which an insert is installed in wells, flowlines, risers, and topside facilities to change the localized flow rate around valves, etc., to promote bulk scale deposition and reduce scale adhesion. The concept of this technology has been around for a number of years, but has yet to be fully proven in the field [1].

15.2.3.5 Smart well completions

Smart wells are designed to maximize production and recovery factors in both new and existing oil and gas fields and are equipped with permanent downhole measurement equipment and control valves that provide the ability to directly monitor and control each segment of the well automatically [1]. They also can be equipped with physical methods to combat scale formations, e.g., magnetic, acoustic, or electrostatic methods that can be built in the completion.

Smart well technology can be used to control gas or water breakthrough, to select which zone can be brought to production, to solve sand production problems, to manage water injection for pressure maintenance, and to shut off water to manage scale and corrosion problems and minimize topside handling problems [1,219].

15.2.3.6 Other nonchemical methods

The radiofrequency method is another nonchemical method that has been used in wells and reported successful in case studies in the onshore and offshore facilities [220,221]. In this method, a surface-mounted device is connected to the well and an electromagnetic pulsed current is injected along metal pipework to the target well tubulars. These signals induce controlled bulk microprecipitation in which the formed microcrystals remain dispersed in the fluid and are transported downstream with the flowing fluid rather than depositing on the tubular surfaces. Fig. 15.29 illustrates the CLEARWELL technology in a production well and equipment.

Another method of controlling scale deposition is by using catalytic metal ions or alloys. As discussed in Chapter 5, "Mineral scales in oil and gas fields," some metal ions can interrupt the nucleation and crystal growth of mineral scales. Adding these ions/alloys to water can reduce scaling rates or form less soluble forms of the mineral, e.g., vaterite instead of calcite. Metal ions that can be used include Zn, Cu, Fe, and Mg, where Zn was found to be the most effective in inhibiting $CaCO_3$ scaling [222].

15.2.3.7 Pros and cons of nonchemical scale prevention methods

Nonchemical scale prevention methods provide many advantages, including:

- Less well intervention once installed
- Less downtime and less production loss
- Green methods of treatment
- Smaller footprint, no storage area needed and no shelf life
- Low operating costs
- Easily optimized, monitored, and maintained (usually provided by the supplier)

These advantages are compared to the chemical treatments, which need well intervention, downtime during squeeze, and dose optimization requiring several lab trials. However, chemical methods have been proven more efficient in all fields and in some instances can be more economical compared to other methods. An old but useful comparison between these different methods in controlling $CaCO_3$ scale was given by MacAdam and Parsons [223], as shown in Table 15.3.

Some drawbacks of the nonchemical methods include:

FIG. 15.29

Radiofrequency technology (CLEARWELL) application in ESP well (left) and surface production equipment (right).

Photo courtesy ClearWELL Energy, www.clearwellenergy.com, reprinted with permission.

Table 15.3 Comparison of the effectiveness of scale control methods.

Method	Reported % effectiveness
Ion exchange softening	100
Acid dosing	100
Chemical inhibitor	100
Metal ions	80
Magnetic conditioner	80
Electronic conditioner	40
Electrolytic method	30
Ultrasound	65
Surface modification	90

Reprinted by permission from Springer Nature, Springer, Reviews in Environmental Science and Bio/ Technology, calcium carbonate scale formation and control, Jitka MacAdam et al., Copyright (2004).

- Conflicting reports of effectiveness.
- Type of field: these methods cannot be used in subsea and deepwater fields.
- May be limited due to system design and accessibility: for example, dimensions of the pipes where devices will be fitted, accessibility to that location, some of the devices are metal-type dependent, and they may not be suitable for downhole applications (although some devices were reported downhole).
- Availability of power source.
- May be sensitive to fluid type and composition.
- They may not be effective in extremely scaling environments.

15.3 Mineral scales removal

Scale removal methods must be efficient and time saving, nondamaging to the production formation, wellbore, tubing, piping, or equipment, and must prevent reprecipitation or reaccumulation in the installations after removing the deposits [41]. It is very common in the field to use a combination of methods to remove scale.

Many techniques can be used to remove scale, including chemical and nonchemical methods. The selection of the removal technique depends on many aspects, including:

- Type of the deposits and its detailed composition.
- The physical properties of the scale, e.g., thickness, hardness, strength, and age.
- Location of the deposit and the accessibility of the removal method to that location.
- Economic cost of the removal method.
- Time frame required/availability to finish the removal job efficiently.

15.3.1 Chemical scale removal

Chemical scale removal is considered an efficient and low-cost approach, especially when scale is not easily accessible or exists where conventional mechanical removal methods are ineffective or expensive to deploy [102].

Table 15.4 Scale types according to their solubility.

Solubility	Scale type
Water soluble	Halite
Acid soluble	Iron sulfides, iron oxides, iron carbonate, calcium carbonate, strontium carbonate, zinc carbonate, zinc sulfide, lead sulfide, magnesium hydroxide, calcium hydroxide, and partially calcium sulfate
Acid insoluble/chelating agent soluble	Barium sulfate, strontium sulfate, calcium sulfate

The chemical methods obviously depend on the solubility behavior of the formed scale deposits. Deposits can be water soluble (like halite), acid soluble (like sulfides, oxides, and carbonates), or acid insoluble (like barite). Since these types are frequently deposited mixed together, there is no such universal dissolver known to dissociate and remove all these scale types. Table 15.4 summarizes the different types of chemical removal methods.

The success of the chemical scale removal methods depends on some factors:

- The scale type. $CaCO_3$ can be easily removed using an acid dissolver or chelating agent, while $BaSO_4$ is challenging and most of its dissolvers require long dissolution time, high temperature, and mechanical aids.
- The detailed composition of the scale deposit. Layerwise analysis to identify the main component and the polymorphs or phases is essential, especially when different phases or polymorphs have different solubilities. Based on these data, layerwise dissolution or selective dissolution, targeting the main component or the cementing material, can be efficiently applied at lower cost.
- Physical properties of the scale. Scale hardness, strength, texture, thickness, and age affect the efficiency of the cleaning. Thick and aged scales are usually hard to dissolve compared to freshly formed brittle deposits. Porous scale layers dissolve faster due to large reactant surface areas, whereas smaller surface-area-to-volume ratio in thick, nonporous sheets of scale react slowly with dissolvers [41]. One reason scale deposits in tubing are hard to remove is because they exhibit a small surface area for a large total deposited mass, which makes their dissolution too slow during practical removal methods [41].
- System design and parameters. A system with high temperature, good accessibility, and ability to circulate/agitate the dissolver will have better chances of dissolving deposits than cold, stagnant locations. Cleaning downhole may have the benefit of the high temperature and pressure, but it lacks agitation, whereas topsides can have better access and better chances for circulation and agitation, but they might lack the high temperature advantage, which can be applied using heating equipment or circulating the dissolver through a heat exchanger. Agitation downhole can be implemented by applying pressure at the surface and then bleeding it (this will basically move the dissolver up and down in the tubing), or by gas bubbling.
- Time available for the removal job. The lab trials of chemical cleaning usually take a few hours to fully dissolve the scale; however, field cases showed that it might take days to fully remove scale depending on their composition and physical properties. The well cleaning time must be assigned, while pipelines and equipment can be bypassed during cleaning, reducing the downtime.
- Cost of the chemicals. The chemical expenses can switch the mitigation plan from chemical methods to mechanical or nonchemical methods, if cheaper and more time saving.

15.3.1.1 Removing calcium carbonate scales

Carbonate scales are one of the most abundant types in oil and gas fields, and they are also one of the easiest scales to remove using acids. In addition, chelating agents like polyaminocarboxylates offer another corrosion-safe option for dissolving calcium carbonates. The dissolution in acid takes place according to Eq. (15.2):

$$2H_3O^+ + CaCO_3 \longrightarrow CO_2 + 3H_2O + Ca^{++} \tag{15.2}$$

where the exothermicity of the reaction and the release of CO_2 gas can both speed up the rate of reaction [23].

HCl is very common in use in scale cleaning and acid stimulating jobs in carbonate reservoirs. Two major problems associated with acid use are: first, its corrosivity to metals and alloys, and second, its loss of efficiency when the acid becomes spent, as the concentration of metal ions (e.g., Ca, Fe ions) in solution increases and slows or retards the reaction with the potential to reform scale [23]. Therefore some additives that must be used in an acid dissolver package include:

- Corrosion inhibitors like amines and imidazoline are used to reduce acid corrosivity. They are added depending on the acid type and strength, pipe metallurgy, temperature, contact time, and inhibitor solubility in the acid.
- Iron control agents, like EDTA, erythorbic acid, citric acid, or combinations of these are used to keep the iron ions in the solution.
- Surface active agents are used for different purposes during the cleaning, including wetting agents, penetrating agents, and dispersants.

Other additives can be employed in the acid formulations, depending on their use in downhole cleaning, including mutual solvents, clay stabilizers, demulsifiers, and oxygen scavengers.

Due to the corrosivity concerns of HCl and other mineral acids, especially with chrome completions, organic acids have also been used in removing calcium carbonate scale. Formic acid [224], acetic acid [225], citric acid [226], sulfamic acid, and gluconic acid [24] have been used, and long chain organic acids have been reported by Huang et al. [227]. These acids are less reactive with carbonate scale than HCl (slower and not to completion); nonetheless they are less corrosive to metals, which makes them a good alternative to HCl acid. However, organic acids are more expensive than HCl. They have typically been deployed in fields where operating temperatures do not exceed 100°C [228]. Acetic acid has showed unexpected high performance in dissolving carbonate rocks in acid stimulations at high temperatures [225]. Gelled organic acids were found to be efficient in removing $CaCO_3$ in horizontal openhole wells in the Heidrun field, where the regular bullheading plain HCl only gave temporary relief, and the gelled HCl breaking time was too short for bullheading applications [229].

Chen et al. [228] introduced a low corrosive, environmentally friendly calcium carbonate scale dissolver for HTHP wells. New acid formulations to dissolve a calcium carbonate oil-based filter cake were introduced [230,231].

Another problem associated with the acid treatments is the destabilization of asphaltenes. In the treating of oil reservoirs with acids potential, asphaltene sludging is possible. Acetic acid was found to introduce traces or no asphaltenes sludge, even in the presence of FeIII up to 3000 ppm [232,233].

Another safe way to remove calcium carbonate scales is using the chelating agents, such as aminopolycarboxylates, which have been reported to be used in removing different scale deposits in oil and gas fields. The chemistry and application of the aminopolycarboxylates is discussed in detail in the sulfate scales removal section, due to the fact that it is considered one of the main methods for removing them. One of the first EDTA oil field applications was to remove calcium carbonate scale from a sandstone formation in the Prudhoe Bay field [234,235], where the HCl treatments were shown to be short-lived. EDTA was used to remove sulfate and carbonate mineral scale from clay assemblages [236].

Like acids, the EDTA relation to asphaltene aggregation must be investigated before using it in downhole applications. EDTA only led to traces of precipitates, due to the lack of acidity and due to the chelation of Fe ions in the solution [233]. EDTA was able to mitigate the extent of corrosion-induced asphaltene deposition on metallic surfaces by sequestering the iron ions; however, it significantly increased the amount of deposit collected on the PTFE surface [237].

Phosphonates are another type of chelating agent. Phosphonate scale inhibitors with carboxylic acid groups, such as 2-phosphonobutane-1,2,4-tricarboxylicacid, can also be used [238]. However, they are limited to scale inhibition use.

CO_2 affects the solubility of $CaCO_3$. Dissolved CO_2 causes the water to be more acidic, thus dissolving the carbonate scale. Several successful laboratory and field tests have been performed by the US Army Construction Engineering Research Laboratory using carbon dioxide treatment for scale removal and control in potable water systems [239,240].

15.3.1.2 Removing barium, strontium, and calcium sulfate scales

Sulfates of group 2 in the periodic table, or the alkaline earth metals, are the most challenging hard scales in oil and gas fields. Calcium sulfate (gypsum, anhydrite, hemihydrate) is the easiest sulfate scale to remove chemically, and barium sulfate (barite, $BaSO_4$) is the most difficult [23]. Pure barium sulfate is normally of low porosity and is extremely impervious to chemical removal, only slowly removable by most established mechanical techniques [41].

The solubility of $CaSO_4$ in diluted HCl is low (0.1211 mol kg^{-1} [241] as only 1.8%–2% of the solid will dissolve at 25°C [57]. Similar low solubility of $SrSO_4$ in diluted HCl was reported (2338 mg/L at 25°C) [242], while $BaSO_4$ is insoluble in dilute acids and soluble in hot concentrated H_2SO_4.

The most common method of dissolving alkaline earth metal sulfate scale deposits in oil and gas fields is use of the chelating agents, such as aminopolycarboxylates or aminopolycarboxylic acids (APCAs). EDTA, DTPA, NTA, CDTA and others are common examples of this group of compounds. They have one or more nitrogen atoms connected through carbon atoms to two or more carboxyl groups, which work as multiple arms that can pick up the metal ion from a solid state and bring them into the solution as a water-soluble complex [243].

The stability of the complex depends on the nature of the metal ion, nature of the ligand, pH, temperature, pressure, nature of electrolytes, and others [244] and is expressed by the stability constant *K*. The higher the *K* value, the more effective the chelating agent [24]. Table 15.5 summarizes the values of some aminopolycarboxylates with the alkaline earth metal ions.

Table 15.5 Stability constant of some aminocarboxylates with metal ions.

	Stability constant (*K*)			
Metal ion	**NTA**	**EDTA**	**DTPA**	**CDTA**
Ca^{2+}	6.4	10.7	10.9	12.5
Ba^{2+}	4.8	7.7	8.6	8.7
Sr^{2+}	5.0	8.6	9.7	10.5

The applications of chelating agents in the oil and gas industry were reviewed [58,245,246]. Aminopolycarboxylates have long been used in oil fields. In practice, a real breakthrough occurred when [234] used EDTA successfully to improve producibility of oil wells in Alaska. Since then, many case studies have reported the use of aminopolycarboxylates in dissolving oilfield hard scales.

One of the earliest cases was reported by Smith et al. [247] when EDTA was used to remove calcium sulfates from waterflood producing wells, which increased production from an average 15 to 20 BOPD to over 100 BOPD. EDTA was used to stimulate wells scaled with calcium sulfate, with a significant production increase sustained for over 6 months, where HCl acid stimulations only gave 50% success with less than 3 months sustained production. Also the cost of EDTA treatments was only 60% of the cost of the conventional acid treatments [248]. Severe calcium sulfate damage experienced in a high-temperature gas-condensate well of Kalinovac field was removed using an 8 wt% tetrasodium EDTA treatment [249]. EDTA was used to remove calcium sulfate scale from a number of wells and production units, leading to significant increase in production [250]. EDTA was applied to dissolve calcium sulfate deposits for plugged ESP wells in the Rumaila oilfield [251]. Chelating agents were applied to remove carbonate and sulfate scale in electric submersible pumps in offshore oil wells in the Gulf of Mexico [252].

Barium sulfate dissolution using chelating agents is one of the most studied topics due to its limited solubility and the limited chemical solutions options other than chelating agents. De Vries and Arnaud [253] reported one of the industry's first removals of barium sulfate scales from the upper part of the tubing in a well on the Alwyn North field in the North Sea using a chelating agent. The chemical was able to dissolve 200 kg of barium sulfate scale at a total cost of £65,000, preventing full workover with a huge total cost of about £1 million. Rhudy [254] reviewed the use of EDTA formulations to remove calcium, barium, and strontium scales from reservoir cores. DTPA is commonly applied in removing $BaSO_4$ and $SrSO_4$ scale [255,256].

The dissolution of mineral scales by aminopolycarboxylates depends on many factors, including the following:

- The type of polyaminocarboxylate: while EDTA is the common APCA used in the industry, DTPA is reported as the most efficient APCA scale dissolver [255–261]. In fact, combining DTPA with other chelating agents such as DTPA/GLDA or DTPA/MGDA resulted in lower performance than DTPA alone [243]. Lakatos et al. [258,259] ranked the sequences of technical and cost efficiency of APCAs to be in the order NTA < HEDTA < DCTA < EDTA < TTHA < DOCTA < DTPA.
- APCA salt type and concentration: usually potassium salts of the APCA are more efficient than the sodium salts [255,256,261–264]. Also, increasing the concentrations up to a specific limit increases the dissolution efficiency, and then the performance declines with higher concentrations due to steric hindrance, viscosity, and dissolver solubility issues [255,256,260,261]. While the literature mentions low concentrations, down to 0.05 M, and also high concentrations up to 40 wt%, it is believed that in field applications the concentration of APCA will be based on the scale type, physical properties, and the economic aspects of the cleaning job.
- pH: generally, APCA has shown improved dissolving efficiency in alkaline pH solutions for many reasons. First, the alkaline pH is required for their solubility in water since their acid form is less soluble in water. Second, the high pH improves their efficiency, as this deprotonation at alkaline pH allows all the carboxylate groups to ionize and interact with metal ions to form a more stabilized chelant complex. Technically, it was proven for some APCA that a pH ~12 is required for prominent dissolution performance, and the chelating agent-metal ion complex was proven to have the greatest

stability [255,256,265–268]. Finally, alkaline pH maintains a high pH and attenuates the effect of downhole acid gases (CO_2, H_2S) or organic acids, which can render the treatment ineffective [269].

- Temperature: dissolution of mineral scales using aminopolycarboxylates increases with temperature. One important issue to consider with increasing the temperature is the thermal stability of these aminopolycarboxylates [255,256,260,261,263,265].
- Catalysts: Some additives were found to enhance the dissolution efficiency of these APCAs. These include sodium/potassium hydroxides, potassium chloride, formate, oxalate, citrate, succinate, malonate, maleate, dithionate, thiosulfate, potassium carbonate, phosphonates, aminophosphonates, sodium polyacrylate, and others. Their synergistic activity is attributed to the mineral scale being chemically modified by an exchange reaction between the sulfate and the catalyst anions [255,256,265,270,271].
- Soaking time: The dissolution of mineral scales by APCAs is a slow process. While the full dissolution of pure minerals, e.g., barite, calcite, or gypsum, can be achieved within hours in the lab experiments, the dissolution of actual field-formed scale deposits may take days [260,261].
- Solution matrix: The matrix effect and the higher stability of calcium, magnesium, and iron complexes may limit considerably the dissolution capacity of barite [265]. Al-Khaldi et al. [268] reported that the presence of both magnesium and iron (III) ions had a negative effect on gypsum dissolution in EDTA fluid. Compared to magnesium, iron (III) ion resulted in a significant decrease in gypsum solubility in EDTA. Dissolved magnesium ions in EDTA solutions could reprecipitate as magnesium sulfate when gypsum is dissolved. This reprecipitation is more in low pH EDTA solutions. In addition, the compatibility of the alkaline APCA solutions with the reservoir water is very important. Some reservoir waters instantly precipitate solids upon contact with high alkaline solutions.

Another type of chelating agent is the crown ethers. Crown ethers are cyclic compounds that possess an inner cavity, generally consisting of oxygen atoms linked by ethylene [272]. Despite their early studies in the 1970s, their usage is still limited due to their expense. Early patents involving crown ethers were reported by Van Zon et al. [273] and De Jong et al. [274]. $BaSO_4$ dissolution properties of macropa were dramatically found to surpass those of the state-of-the-art ligands DTPA and DOTA [275]. Using macropa, complete dissolution of a molar equivalent of $BaSO_4$ was reached within 30 min at room temperature in a pH 8 buffer, conditions under which DTPA and DOTA only achieved 40% dissolution of $BaSO_4$. Fig. 15.30 illustrates the use of DTPA + phosphonate mixture to dissolve a mixture of sulfate and sulfide scales.

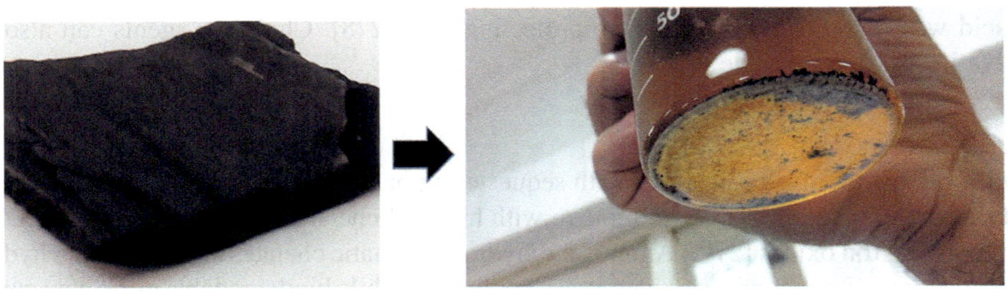

FIG. 15.30

Using aminopolycarboxylate (DTPA) in the presence of phosphonate as catalyst to dissolve sulfate + sulfide composite scale.

Other types of chelating agents include phosphonates; however, their use is limited to scale inhibition or as synergists for the other chelating agents. Gluconic acid/sodium gluconate is another sequestering agent that has been applied in other industries. Glycolic acid is another complexing agent, which was studied to remove gypsum deposits from oilfields [24].

Another way of dissolving the insoluble sulfates is by converting them to soluble form using converters. Two major converters have been used in oil and gas fields: carbonates and hydroxides.

Carbonate converters attack scale by surface reaction, as the carbonate in solution exchanges with the deposited sulfate ions and leaves a carbonate deposit, which is then removed by an acid. Carbonate converters are inexpensive and have been successfully used in many wells, but they work better on light porous/permeable scales, and are not effective on the majority of dense laminated scales commonly found on most downhole equipment, because the scales lack permeability and porosity. Hydroxide converters are more effective for these less-permeable gypsum scales, because they penetrate only the surface, cause physical disintegration (sloughing) of the crystal lattice, thus continuously exposing new surface to the chemical. Although the first converter acid treatment on calcium sulfate scales may be quite effective, each successive retreatment may become less effective, since the remaining scale becomes less permeable as a result of continuing deposition. This scale left behind by the incomplete removal treatment "seeds" a scale deposit that is considerably less subject to further conversion treatment [247]. Field experience has shown that potassium hydroxides and potassium carbonates are more efficient than the sodium hydroxide and sodium carbonate [247].

The use of soda ash (sodium carbonate) to "convert" the gypsum to calcium carbonate and subsequently use an HCl wash to dissolve this salt is a low-cost, two-stage process that has often been used for gypsum remediation, leading to an increase in production by about 2000 BOPD [276].

15.3.1.3 Removing iron compounds scales

Iron compounds are very diverse and different in their solubility. However, they share some common ground by being more soluble in low pH solutions. The next few sections will discuss the dissolution of the different iron compounds.

15.3.1.3.1 Removing iron carbonate

Like calcium carbonate, iron carbonate will dissolve in acids like HCl; however, it will require higher temperature to achieve complete dissolution, as the reaction rate is slower than for calcium carbonate. The rate of dissolution of siderite has been measured at 25°C, 60°C, 80°C and 100°C in 0.1 M (NaCl + HCl) solutions and pH from 1 to 4.6 [277]. Organic acids can also be used to remove siderite scale. Acetic acid was found to remove siderite protecting film [278]. Chelating agents can also be used to dissolve iron carbonate scale [23].

15.3.1.3.2 Removing iron oxides

Iron oxides can be dissolved using HCl with sequestering or reducing agents to avoid precipitation of by-products. Iron oxides are harder to remove with high pH chelants. Organic acids and EDTA were used in removing iron oxides [279]. Amino acid/hydroxyaromatic chelates and sulfonated hydroxyaromatics are better at dissolving iron oxides at pH 7 or higher, while biodegradable citric acid can be used at low pH [280]. Phosphonates also show good performance in removing iron oxides and leave the surface passivated against corrosion [281].

15.3.1.3.3 Removing iron sulfides

Iron sulfides are the most studied iron compounds for chemical removal. Iron sulfides have different forms, which have different solubilities, besides the ability of converting from one form to another. Therefore many chemical methods have been tested and used as dissolvers/removers for iron sulfides. These dissolvers are summarized in Table 15.6.

Dissolution of iron sulfide scale downhole presents unique challenges compared to many other inorganic scales [282], due to the following reasons:

- Formation of H_2S during acidic dissolution of iron sulfides provides additional safety and operational risks, including increased corrosion rates.
- Dissolution of iron sulfide is often reversible, and in sour environments iron sulfide growth initiates from calcium carbonate surfaces starting at a largely unspent pH of only ~1 [283,284].
- Different forms of iron sulfide range from high to very low solubility in acid, existing in a number of forms (such as mackinawite, marcasite, pyrite, and troilite) [285,286].
- Finally, iron sulfide is also oil-wet, and its coating protects the inorganic mineral from simple acid/chelating agent dissolution [287].

- **Acids**

Generally speaking, iron sulfides dissolve in acid solutions. The literature mentions the use of HCl and organic acids (including acetic acid, formic acid, maleic acid, glutamic acid, succinic acid, gluconic acid, thioglycolic, and glyoxalic acid) [23,67,288,289], with HCl usually more effective in dissolving sulfide scales than the organic acids. Elkatatny [289] developed a formulation that contains a mixture of organic and mineral acids, where iron sulfides showed a solubility of 78 g/L at 100°C.

The acid dissolution of iron sulfides is associated with some problems:

Table 15.6 Chemical methods used to dissolve iron sulfides.

Acids	Biocides	Chelating agents	New products
• Mineral and organic acids • Efficient except in certain acid insoluble sulfides such as pyrite • Cheap • Produce H_2S • Corrosive • Reprecipitation • Not environmentally friendly	• THPS, acroline, etc. • Efficient • Work as H_2S scavenger • Kill bacteria • Work as scale dissolver and scale inhibitor • Corrosive (corrosivity less than HCl) • Hazardous and not environmentally friendly	• Aminopolycarboxylates • pH, time, temperature dependent • Possible reprecipitation • Can dissolve sulfate scale as well • Environmentally friendly	• Oxiders • Does not produce H_2S • Possible reprecipitation • Still not common in use

- First, recent studies showed that iron sulfide solubility in acids is diverse and inconclusive. Iron sulfide scale is easily removed by conventional acid systems if the iron-to-sulfur ratio is close to unity [285,286,290]. In iron-to-sulfur ratios close to 0.5, iron sulfide removal can become difficult to accomplish through conventional acid systems. This is because sulfur will shield iron atoms and prevent dissolution [58]. When the iron scale depositions contain low sulfur content, they exhibit a higher degree of solubility in HCl acid [290]. Leal et al. [291] reported that pyrite and marcasite iron sulfides are acid-insoluble, pyrrhotite showed a slow-pace solubility, while mackinawite and troilite are highly soluble. Wang et al. [288] concluded that HCl can be used in general to remove the soft type of iron sulfide scale (FeS).
- Second, acids are corrosive and affect the integrity of the facilities, as discussed in calcium carbonate removal.
- Third, acid dissolution of sulfides produces high levels of H_2S, which is a hazardous and corrosive gas.
- Fourth, the spent acid turned inefficient with increasing cation concentrations.
- Finally, and due to these facts already stated, more additives are required to optimize the acid cleaning and preserve the facility integrity, including corrosion inhibitors, iron control chemicals, sequestering agents, H_2S scavengers, gelling agents, and others, all of which make the acid cleaning more expensive. Even worse, corrosion inhibitors were found to decrease thte rate of sulfide dissolution [292]. The same observation was found with H_2S scavengers and iron control agents like citric acid, which also decreased the sulfide dissolution rate [23].

- **Biocides**

Recently, a number of different oxidizing and nonoxidizing biocides were found to effectively dissolve iron sulfide scales, supported by lab and field studies.

Acrolein, which is a strong oilfield biocide and H_2S scavenger, has also been shown to dissolve iron sulfide scales [293,294]. Penkala et al. [295] reported an increase of 30% capacity of an injection well, 400%–500% improvement in Millipore rate tests, decreased levels of SRB, greater than 90% decrease in soluble H_2S, and a 65% decrease in filter replacements after acrolein squeeze stimulations. Horaska et al. [296] reported that a ninefold improvement in water quality and a 16% increase in injection water rates were achieved after acrolein batch applications. However, acrolein has high acute toxicity and is a suspected carcinogen, so it needs to be handled carefully.

Another type of biocide that is widely used to dissolve iron sulfides is the tetrakis(hydroxymethyl) phosphonium sulfate (THPS). This chemical is an environmentally friendly biocide with high thermal stability and low foaming potential, which is widely used in downhole and topsides applications in oil and gas fields [67]. Recently THPS has gained attention by working as an effective iron sulfide scale dissolver. THPS-based iron sulfide dissolvers (particularly in combination with ammonium chloride or organic phosphonate scale inhibitors) were reported to be comparable to or even better than the uninhibited HCl acid, with the ability to dissolve pyrite, which is generally considered insoluble in hydrochloric acid [297,298]. Many additives were studied as synergists for THPS-based dissolvers including ammonium chloride [297], other nonoxidizing ammonium salt, primary amine [299], amino carboxylic acid (NTA and EDTA) [67], amino phosphonate acids (BHTPMP, EDTMP, DTPMP, ATMP, and EBMP), hydroxy alkyl amines (ethanolamine (MEA), diethanolamine (DEA), triethanolamine (TEA), and aminoethylethanolamine (AEEA)) [300], and organic acids [formic acid, acetic acid, thioglycolic acid, citric acid, and EDTA (acid form)] [300]. However, the most common use of THPS in removing iron sulfide scales is in combination with ammonium chloride. The dissolution reaction depends on many factors, including temperature, soaking time, pH, and scale composition. Higher temperature, longer soaking time, and lower pH gave better results. The mechanism by which it dissolves iron sulfides is not fully understood, but iron ion chelation to a nitrogen-phosphorus ligand, giving the water a red color, and in situ production of HCl by the decomposition of ammonium chloride at high temperatures were suggested as competing mechanisms [67,297,298]. The in situ production of HCl was observed when a composite scale sample was dissolved in THPS-ammonium chloride, with pyrite, marcasite, and anhydrite left almost insoluble, whereas iron oxyhydroxides calcite, siderite, and pyrrhotite were dissolved [67]. One main drawback of using a THPS-NH_4Cl blend is their corrosivity, especially at high temperatures, which requires adding a corrosion inhibitor to the blend, which may impact its dissolution efficiency.

In one field, use of THPS with a surfactant killed the SRB, removed iron sulfide scale, reduced the corrosion rate, and reduced H_2S levels in the gas phase, leading to increasing the production in several wells by 20% to 300% (44% average) [301]. In another field application, a complex damage combination of polymer residue, biomass, and FeS plugging the screen liner was resolved in multiple stages. THPS was used to remove iron sulfide and hydrogen sulfide and control the growth of bacteria. Then, in the second stage, formic acid was added to remove residual iron sulfide and some polymer residue in the near-wellbore area. The third stage was based on a two-part oxidative treatment and was designed to remove polymer damage deep in the formation [302].

Some drawbacks of THPS are that it fails in the presence of triazine H_2S scavenger. Besides, pH changes influence THPS, as high pH renders THPS ineffective as a chelant. Additionally, THPS introduces sulfate ion to the production system, which can have compatibility issues with produced waters, not to mention the corrosivity of THPS-NH_4Cl, especially at high temperature.

Another approach to remove iron sulfide scales is by the conversion of the insoluble Fe(II) to soluble Fe(III) using oxidizing agents such as chlorites/chlorine dioxide or permanganates [23].

Chlorine dioxide reacts with FeS, oxidizing it to sulfate and hydroxide according to Eqs. (15.3), (15.4) [303]:

$$5FeS + 9ClO_2 + 2H_2O \longrightarrow 5Fe^{3+} + 5SO_4^{2-} + 4H^+ \tag{15.3}$$

$$5H_2S + 8ClO_2 + 4H_2O \longrightarrow 8Cl^- + 5SO_4^{2-} + 18H^+ \tag{15.4}$$

Also, depending on the stoichiometry of chlorine dioxide and S^{-2} ion, other possible products are $SO_3{}^{2-}$ and elemental sulfur S^0 [304].

Early field trials of chlorine dioxide as a stimulation fluid were completed in late 1987 specifically for crosslinked polyacrylamide damage removal, where the acid treatments failed to remove it. Successful removal of iron sulfide damage in oil wells and injection wells was reported [304]. A total of 4500 bbl of ClO_2 was pumped in three phases to clean 66 miles of a waterflood-distribution system. Results indicated that ClO_2 was effective in cleaning the well guard screens, the injection lines, and the injection wells [303].

In the same way, potassium permanganate will react with pyrophoric iron sulfide to form either iron oxides or iron sulfate, as in Eqs. (15.5), (15.6) [305]:

$$9FeS + 26KMnO_4 + 4H_2O \longrightarrow 3Fe_3O_4 + 26MNO_2 + 9SO_4^{2-} + 5OH^- + 26K^+ \tag{15.5}$$

$$FeS + 2KMnO_4 \longrightarrow 2MNO_2 + FeSO_4 + 2K^+ \tag{15.6}$$

A number of facilities have found that use of potassium permanganate results in increased productivity, improved safety, and significant cost savings [306].

Hydrogen peroxide, which is another oxidizing agent and biocide, was also used in removing iron sulfides. H_2O_2 was used in removing iron sulfides from a twisted tube exchanger [307]. Such oxidizing agents are so strong that corrosion would be expected and they must be combined with corrosion inhibitors.

- **Chelating agents**

The use of chelating agents in removing iron sulfides has been intensively studied. Aminopolycarboxylates (EDTA, DTPA, NTA, GLDA, MGDA, HEDTA) were studied as iron sulfide scale dissolvers.

EDTA at pH 8–10 dissolved 80% of iron sulfide scale after 20h at 175°F [308]. Low pH DTPA showed high iron sulfide scale dissolution capability [191,287]. DTPA with a converting agent (K_2CO_3) at pH >11 dissolved over 85% of iron sulfide scales with high concentration of sulfur at 155°F and prevented the regeneration of H_2S [309]. De Wolf et al. [310] have shown that GLDA and MGDA can be used to remove FeS scale. Other synergists were studied with APCA potassium chloride (or KCl), potassium iodide (or KI), potassium formate (or HCOOK), sodium fluoride (or NaF), and potassium citrate (or K-citrate) [311]. Elkatatny et al. [312] developed a low-temperature noncorrosive sulfate/sulfide scale dissolver composed of a combination of NTA + ethylenediamine + chloroacetic acid.

- **New dissolvers**

A polymer-based dissolver was developed which, when combined with THPS-based dissolver, showed improved performance [313]. Another polymeric product was proposed that showed promising efficiency of dissolving iron sulfide scales of high pyrite and marcasite content at low to neutral pH, with low H_2S production [314]. A new borax-based formulation for the removal of pyrite scales was developed [315]. Solugan Company developed multiple types of iron sulfide dissolvers based on biochelation and oxidizers (oxidized dextrose).

Gelled fluids were also reported to remove black powder from gas pipelines. In this process, gels that can hold heavy deposits in suspension for days are pumped between mechanical pigs, which scrape off the debris from the pipe wall, distribute it through the volume of the gelled fluid, and contain the gel to give 360-degree coverage. The gels may be based on water or hydrocarbon fluids and can be selected to suit the deposits and binders, and appropriate binder solvents may also be included in the pig train if required [316].

15.3.1.4 Removing zinc and lead sulfides

Like iron sulfides, zinc and lead sulfides can be basically removed using acid solutions. HCl will remove ZnS but not PbS. PbS scale will only dissolve in very hot concentrated oxidizing acids, such as nitric acid, which poses the risk of corrosion problems. Acid treatment was used to remove wurtzite, a form of zinc sulfide scale, from a well in the Gulf of Mexico [317].

Similar to iron sulfides, biocides were also reported to dissolve ZnS and PbS. One of the first reported studies was using THPS-based dissolver to remove ZnS/PbS mixed scale [68]. They showed that the THPS solution alone did not dissolve the mixed sulfide scale, while a THPS-NH_4Cl mixture

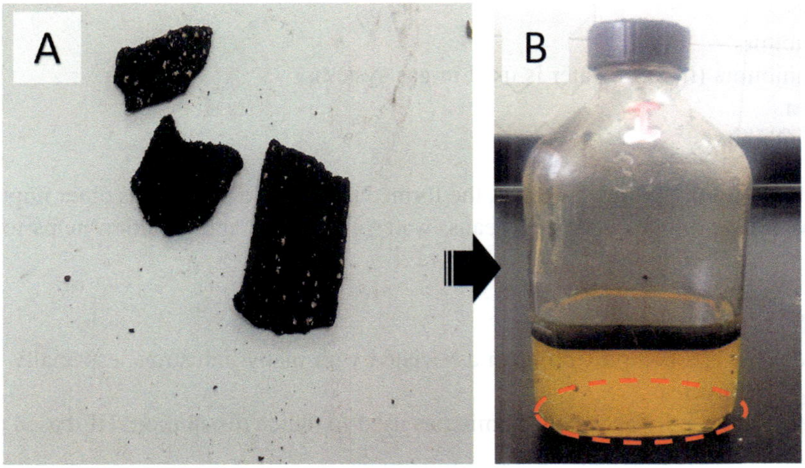

FIG. 15.31

ZnS and PbS scale dissolution by THPS+NH_4Cl mixture at 85°C. (A) Raw scale sample before dissolution, and (B) scale sample after dissolution (*circled in red*). The top organic layer in (B) is the hydrocarbon deposit in the scale that has released after scale dissolution and washed further with organic solvent.

showed the best dissolution efficiency at 85°C after 24 h soaking time, whereas other additives including potassium chloride, ammonium oxalate, ammonium citrate, ammonium bisulfite, EDTA, and sodium azide did not improve the dissolution efficiency. Fig. 15.31 illustrates lab dissolution of zinc and lead sulfides using a THPS-NH_4Cl mixture.

Another biocide that was found effective in removing lead sulfide is peracetic acid, a blend of acetic acid (CH_3COOH) and hydrogen peroxide (H_2O_2), which generates the more oxidizing peracetic acid (CH_3COOOH) in situ, and was effective in dissolving PbS [318]. Similar lead dissolution results were obtained using potassium permanganate ($KMnO_4$) as the oxidant; however, this led to MnO_2 precipitates. The MnO_2 precipitate could be removed using solutions of citric acid or other chelates [23].

15.3.1.5 Removing halite scales

Halite and other chloride scales are water soluble deposits. The most common economic and efficient method of managing the deposition of halite scale is the injection of fresh or less saline water [319].

Wash water can be applied to the affected zones through many ways [320]:

(a) Bullhead water-wash.
(b) Free-fall batch wash.
(c) Coiled tubing wash.
(d) Capillary string with continuous water injection.

Wash water usually requires some additives to enhance the dissolution process and to avoid other side effects. These additives include:

- Oxygen scavengers
- Foaming agents

- Biocides
- Halite inhibitors
- Hydrate inhibitors (if wash water is used in gas systems)
- Demulsifiers
- Corrosion inhibitors

Although halite is highly soluble in water, the formed halite scale always has other impurities and can be aged; thus using some mechanical means, water circulation and agitation helps to speed up the dissolution.

15.3.1.6 Removing silicate scale

Silica/silicate scale is a complex problem associated with many industries, especially in geothermal and oil and gas industries.

There are two generally accepted chemistries used for silica dissolution: HF-based chemistry and strong alkali method.

The industry standard and the one most often used in chemical cleaning is HF (hydrofluoric acid), either in the form of pure HF, or as fluoroboric acid, or created in situ by combining HCl with ammonium bifluoride (ABF) NH_4HF_2 [321]. Although this approach is effective, it also requires meticulous attention to issues such as health hazard potential (generation of HF in situ) and acid-driven metallic corrosion (since cleanings must be done at low pH). Also, calcium (which is present in most of the silicates) can decrease the efficiency of HF cleaning (HF spending). Fluoride ions attack calcium and produce an insoluble toxic by-product. The risk of forming CaF_2 is particularly high, as calcium carbonate is also often found in conjunction with silicate scales [321].

The second method is to react with strong alkali—NaOH or KOH. This is more typically used in industrial settings for the production of sodium or potassium silicates. Alkaline dissolution is thought to be more effective than acidic dissolution, but the reactions are complex and the mechanism is not clear [321].

Gharaibeh et al. [321] improved the alkaline method efficiency by using chelating agents. The addition of the chelating agents prevents the secondary reactions of the divalent cations with OH- at the reaction surface allowing the OH to react with the silicate backbone of the deposit [321].

Other methods that may be used to control silica/silicate scaling in steam flood operations include [322]:

- dilution with fresh water
- reducing the pH of the water
- treating the water with reducing, complexing, and sequestering agents
- removing silica from water by lime softening
- precipitation of silica in water with metals or cationic surfactants
- treating the water with geothermal silica scale inhibitors/dispersants.

15.3.1.7 Removing elemental sulfur deposits

Elemental sulfur dissolves in a number of organic solvents: DADS (diaryl disulfides), DMDS (dimethyl disulfides), CS_2 (carbon disulfide), amines, and/or hydrocarbons (diesel, kerosene, toluene, xylene, benzene), chloroform, and other solvents. DMDS and CS_2 are the most efficient solvents used;

however, they entail some safety hazards and should be used with caution by limiting the operator exposure time.

Laboratory tests proved the suitability of a nonionic sulfur solvent based on tar-oil distillation components (alkylnaphthalenes) and a mineral oil. With this solvent system, a new phase is introduced into the production fluid, because the produced sour gases are completely free of condensable hydrocarbons [323].

A sulfur removal dissolver consisting of alkalines (sodium hydroxide), sodium sulfite, and sodium bisulfite and about 10 ppm nonionic surfactant wetting agent (ethylene oxide adduct of nonyl phenol) was claimed [324].

15.3.1.8 Removing mixed scale composite

Removing composite scale is a major challenge for oil and gas operators. As mentioned before, these mineral scales have diverse solubility behavior, hence no universal dissolver is commercially available that can be used to remove all of them at once, but a specific dissolver is required for each type. However, some attempts have been made recently to find a formula that can be used to dissolve most mineral scales.

Elkatatny et al. [312] reported a novel noncorrosive dissolver for sulfate and sulfide composite scale in alkaline pH that works at low-temperature conditions, composed of a combination of NTA + ethylenediamine + chloroacetic acid. It showed prominent dissolution efficiency, more so than chelating agents and acids.

Gamal et al. [325,326] developed a new dissolver (mixture of acids (acrylic and lactic), nonionic surfactant, and ethylene oxide at pH 9) that is able to dissolve complex scale with a high solubility rate, achieving 100% for the scale sample at two temperature levels of 160 and 210°F for 6 h.

Generally speaking, the most common practice in dissolving mineral scales is based on:

- Multistep dissolution, targeting each component separately, e.g., first wash using organic solvent to remove organic debris, followed by acid to remove acid solubles, and then chelating agent for sulfates.
- Selective dissolution by targeting the main component or the cementing material in the scale layer.

This dissolution process is based on the details of the composition of scale samples, i.e., the composition of each individual layer and composition of layers composite, minerals polymorphs, and phases. Based on the proper chemical analysis of the deposits layerwise, with full details of the polymorphs and phases, a good and effective scale removal program can be planned.

15.3.2 Nonchemical scale removal

15.3.2.1 Magnetic scale removal methods

Magnetic field treatments are typically used in scale inhibition; however, few trials have achieved removal of formed scale.

The magnetic field has many effects on the scale materials, including changes in particle size, crystallinity, crystal morphology, crystal phase, solubility, and the rate of precipitation [327]. Under the effect of a magnetic field, hard scale is turned into "soft gunge" that sticks loosely to the pipes and can easily be removed [327]. Magnetic treatment has shown significant removal of scale from the pipe walls, by 30%. Higher magnetic strength showed higher removal of scale compared to lower strength

[328]. A magnetic field was found to recrystallize already formed calcite scale into aragonite, which is a more soluble, less adherent, and easier to remove scale [329].

15.3.2.2 Electrostatic scale removal methods

Electronic devices are characterized by a solenoid coil or coils wrapped around a pipe and a signal generator, which supplies current to the coil, producing induced magnetic and electric fields within the pipe, which interfere with the scaling process.

Tijing et al. [330] found that applying an electric field reduced $CaCO_3$ fouling, producing blunt rather than sharp crystals that were easily washed. Tijing et al. [331] demonstrated that calcium contents of fouling deposits dropped by 4%–49%, and asymptotic fouling resistances decreased by 88% in high-frequency electric fields [332]. As the applied voltage rose, the structure of $CaCO_3$ fouling changed from aragonites to the more soluble spherical vaterites.

An electronic antifouling device based on a radiofrequency alternating electric field inductance was found to prevent scale formation and to remove existing hard scale in plumbing systems [333].

15.3.2.3 Ultrasonic scale removal methods

Ultrasonic cleaning uses high-frequency sound waves and combines the phenomenon of cavitation, which dislodges contaminants, and agitation (microstreaming), which accelerates the dissolution of contaminants by supplying fresh solution to the surface being cleaned [334].

In fact, ultrasonic cleaners are used more frequently than you think, as they are used to clean many different types of objects, including jewelry, scientific samples, lenses and other optical parts, watches, dental and surgical instruments, tools, coins, fountain pens, golf clubs, fishing reels, window blinds, firearm components, car fuel injectors, musical instruments, gramophone records, industrial machine parts, and electronic equipment [334].

The mechanism of ultrasonic cleaning can be understood as a combination of effects as a result of the collapse of cavitation bubbles near the surface of an object through the formation of a corresponding shock wave and reentrant microjet [335]. The cavitation describes the life cycle of a transient cavity or bubble from formation via inception and nucleation, its expansion, and its eventual implosion. Finally, the rapid collapse of the cavitation bubbles produces an extreme transient ($<2\,\mu s$) pressure/shock wave (102 MPa) and temperatures (3500–8000 K) inside the bubble during collapse, which contributes to the velocity of the microjet and heating of the liquid [335,336]. Its observable effect, when near a solid boundary, is material damage known as cavitation erosion. Fig. 15.32 depicts the growth and collapse of a cavitation bubble.

Kunanz and Wölfel [337] reported the first test and research results using ultrasonic waves to remove scaling in the borehole. Experimental results show the ultrasonic waves play a critical role in the removal of inorganic scales in low permeability core plugs, subsequently increasing the efficiency of water. While water cannot remove these depositions from core plugs solely [338,339], field tests carried out in low permeability reservoirs in northern Shaanxi and in the Daqing oilfield showed that ultrasonic treatment can increase the production and injection remarkably, with the advantages of ultrasonic being simple operation, no damage to reservoirs and the environment, low overall operation costs, among others [340]. They showed that transducer frequency, transducer power, ultrasonic processing time, and core initial permeability all influence the plugging removal effect, which improves with the increase of transducer frequency and power. The optimal processing time is about 80 to 120 min.

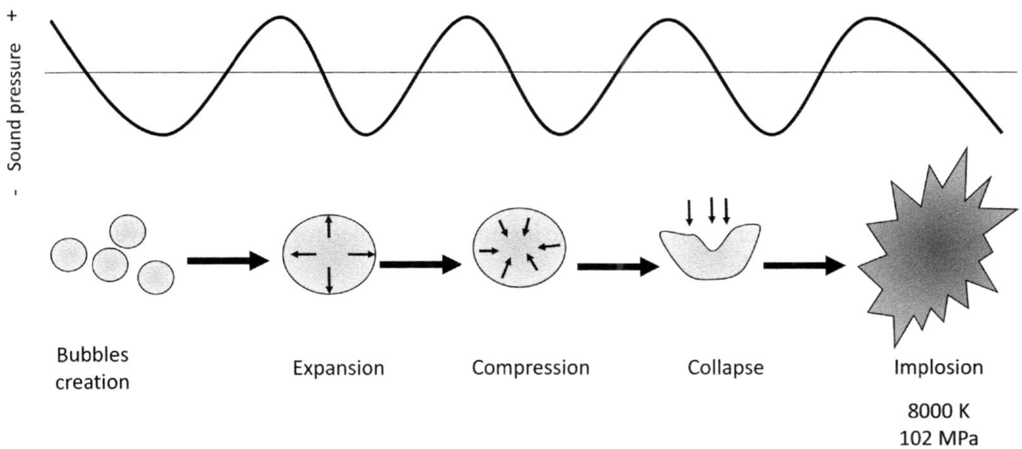

FIG. 15.32

Principle of ultrasonic cavitation.

Kieser et al. [335] reported using an ultrasonic method to effectively clean heat exchangers on both the inside of the tubes (i.d.) and the outside of the tubes (shell side, o.d.) to bare metal and return to service condition in less than 1 day, whereas the chemical and mechanical cleaning usually becomes ineffective. The method successfully removed 150 kg of deposits from one unit and around 1000 kg from the other unit.

15.3.2.4 Mechanical scale removal methods

Different mechanical methods are used to remove mineral scales from oil and gas fields. Pigging is the most common method for topside flow lines, whereas different wireline, slickline, and coil tubing operations can be used for production tubing and downhole cleaning. The details of the different mechanical methods of removal are discussed in Chapter 21, "Mechanical removal of flow assurance solids."

Besides the previously mentioned physical and mechanical methods, filtration and separation methods are used to remove fine solids like black powder, elemental sulfur, and other fines from the system. Typical methods include gas particle filter, filter separator, cyclone separator, and cyclo-filter. Magnetic separators have also been used to control black powder in gas systems. These physical means must be selected properly based on the size of the solid grains and their load in the system. Frequent maintenance and replacement of filtration system parts is very common.

15.3.3 Tips for an efficient scale removal job

Removing mineral scale is not an easy job. Scale deposits in production tubulars and equipment can be hard as rocks, and the chemicals that are commonly applied are not magic. Therefore a methodology for scale removal must be established. That methodology will rely on some factors:

- The detailed chemical composition of the scale
- Scale layer thickness
- Scale layer texture and hardness/strength

- System conditions: temperature/pressure
- Robustness of system facilities integrity, i.e., corrosion status
- Accessibility of the different treating methods to the scaled part
- Compatibility of the chemicals used with the produced fluids
- Economics of the method used

Chemical and mechanical removal methods are the most used methods for removal of mineral scale deposits. They are used either solely or, in some instances, combined together to improve the efficiency in the case of very hard scale.

- The scale dissolver is usually tested on the collected scale sample, whereby scale samples are soaked in the dissolver at specific times, under specific temperature (tubing temperature/flowline temperature).
- Chemical composition of mineral scale should be detailed, and each individual layer, phases, and polymorphs must be identified.
- The thickness of the scale can be determined, e.g., manually, using gauging, caliper pigging, or other methods. The thickness is used to determine the scale weight, and then to determine the required dissolver volume. Fig. 15.33 illustrates a general schematic of these calculations. The ratio of the solid (scale weight) to liquid (dissolver volume) is also used to determine the required dissolver volume, and to theoretically evaluate if many rounds of cleaning are needed. For example, if the dissolution studies showed that the dissolver performance is 5 g scale/100 mL dissolver, and the solid/liquid ratio was determined to be 10 g/100 mL, then two rounds of dissolution are necessary: this applies to the case in which the scales have the same thickness all over the pipe. Higher thickness reduces the effective diameter and effective volume of the pipe, thus requiring more rounds of cleaning using the dissolver.

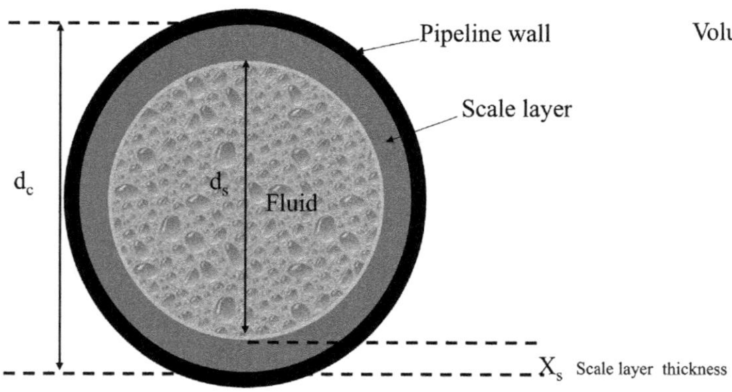

$$\text{Weight of solid scale} = \frac{\pi}{4}(d_c^2 - d_s^2)\rho_s$$

$$\text{Volume of fluid/dissolver in scaled pipe} = \frac{\pi}{4}d_s^2$$

$$\text{Ratio of } \frac{\text{solid}}{\text{liquid}} = \frac{(d_c^2 - d_s^2)\rho_s}{d_s^2}$$

d_c : clean pipe inside diameter
d_s: scaled pipe inside diameter
ρ_s: density of scale

FIG. 15.33

Schematic of scaled pipe and the required calculations to estimate the scale weights and dissolver volumes.

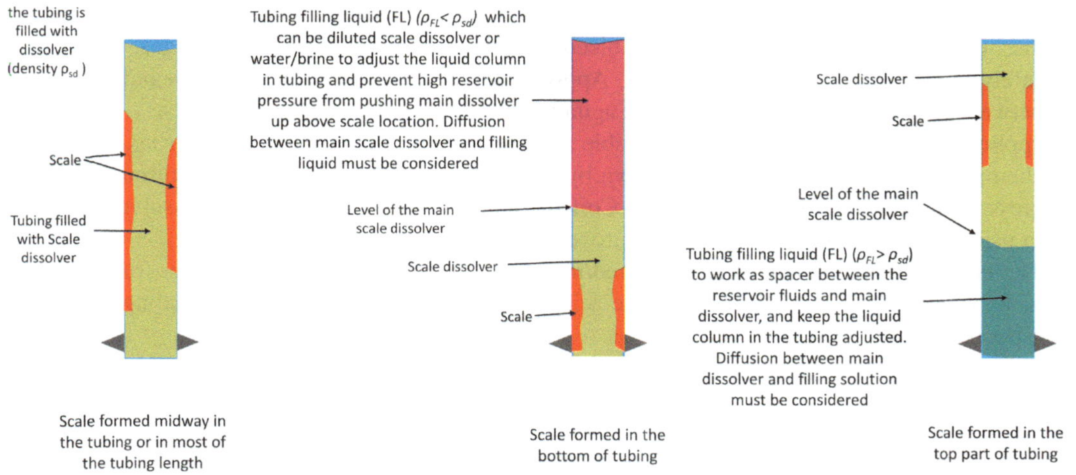

FIG. 15.34

Tubing cleaning and the different scenarios to reduce the cost of the cleaning job.

- Most of the cleaning in production tubing is achieved by filling the tubing either completely, or sufficiently above the location of the scale obstruction. In some instances, scale is formed only in the top part of the well or in the bottom part, and filling the whole tubing with dissolver is very costly, especially with the high cost of scale dissolvers ($50/gal or more). To reduce the volumes and the costs of the dissolver, the targeted location of scale is fully and sufficiently covered, while tubing filling solutions (F_L) can be used to fill the rest of the tubing. These filling solutions must be compatible with the reservoir fluids in case they come in contact with them. Fig. 15.34 shows some of these scenarios.
- The texture and strength of the scale layer added to its thickness are of great importance. Dense, thick, and hard scale may not respond to chemical treatments instantly.
- The dissolution time is determined by many factors: dissolver efficiency (efficient dissolvers act faster), robustness of system integrity (acidic dissolvers are applied for very short time, neutral and alkaline dissolvers can be applied for longer times), production loss (very low production wells, or those that were shut in for a long time, or pipelines that can be bypassed without interrupting production are good candidates for long treatment time), and availability of dissolution aids (like mechanical means to speed up the dissolution).
- Temperature is a big factor in dissolving scale. High temperature/high pressure (HTHP) wells furnish a good environment for chemical dissolvers that require high temperature. The chemical dissolution is usually based on the well static temperature, not the flowing temperature. Sometimes, the depth at the scaled location has lower temperature than the recommended for dissolution; in this case one of the options is to push the dissolver to the hotter depths to increase its temperature and then bleed the pressure to optimize the warm dissolver at the scaled location depth. For topsides, heaters and heat exchangers are commonly used.
- Agitation is another factor in the chemical dissolution operations. Agitation can be achieved in many ways, such as circulation of the dissolver in the flowlines, or through the system heat exchangers.

Another way to provide some agitation is by bubbling gas inside the pipeline filled with the dissolver; this should be intermittent to bleed the pressure buildup after using the gas, with extreme safety precautions for applying the gas. Applying the cleaning solution with a cleaning pig or between two pigs gives the system some agitation and improves the cleaning efficiency. In downhole applications, agitation is nearly impossible and the possible way around it is by shaking the chemicals inside the tubing up and down by applying pressure to the wellhead (through the high-pressure gas lines) and then bleeding it. Coiled tubing can be used to apply hot dissolver in addition to using its mechanical aids, which improves the dissolution.

- The tanks that the chemicals are prepared/mixed in should be clean and rust-free. Iron from rusty tanks will decrease the efficiency of the acid, and the same thing applies to chelating agents, especially aminopolycarboxylates.
- The removal program starts with an organic solvent wash to remove oil residues and organic debris.

15.4 Generic strategies to control mineral scales in oil and gas fields

Scale management is a discipline that entails scientific and engineering investigation to analyze the risk and build a mitigation plan that is efficient and cost effective. The various steps involved in scale management are discussed in the following sections.

15.4.1 Risk assessment

During this step engineers and operators use theoretical and field data to analyze and assess the risk of scale formation.

15.4.1.1 Brine sampling and analysis

The next step in scale management is water sampling and analysis. Water samples must be collected properly in proper sample bottles and preserved using optimum preservation methods, to avoid secondary precipitation and other water properties changes. Fig. 15.35 illustrates the water preservation methods. After sampling, water chemical and physical properties are measured using proper techniques. Cations, anions, pH, alkalinity, density, specific gravity, resistivity, dissolved gases, total dissolved solids, total suspended solids, and other properties are usually measured.

15.4.1.2 Scale prediction and modeling

After water chemistry is addressed, these data in addition to system pressure and temperature are utilized in scale prediction software packages to determine the scaling tendency. Commercial and academic software packages are available, e.g., ScaleSoftPitzer, MultiScale, ScaleChem, and others. Scale prediction is discussed in Chapter 12, "Flow assurance solids prediction and modeling." It is important to understand the limits of each software application: temperature and temperature ranges and brine systems. More than one software package can be used to assure precise results; however, the different software can have only small differences in their results.

Most of the scale prediction software packages are thermodynamic tools that do not take scale formation kinetics, reservoir geochemistry, and fluid hydrodynamics into consideration. They also do not give information about location of scale, or its thickness. Thus reservoir simulations, chemical reaction simulations, and fluid hydrodynamics simulations (CFD) are also recommended to be performed.

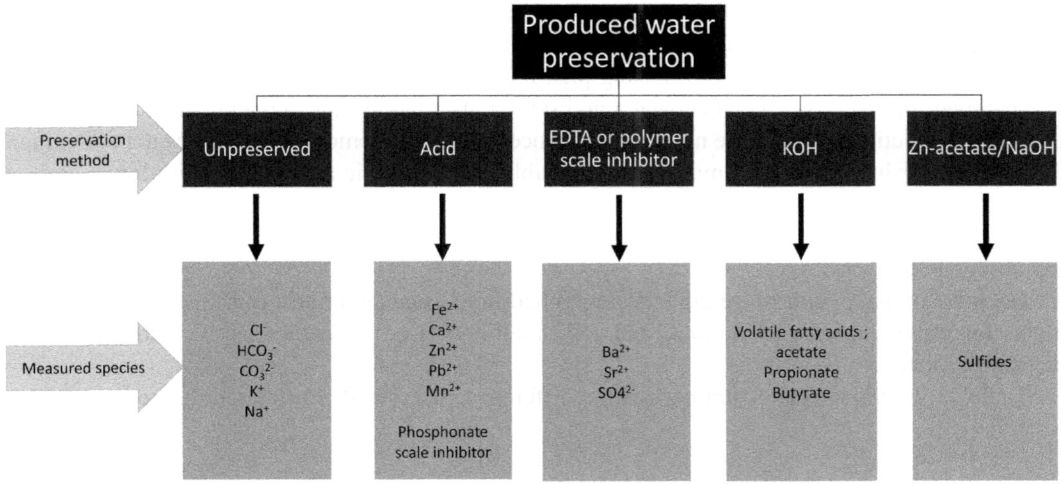

FIG. 15.35

Methods of water sample preservation.

15.4.1.3 Process monitoring

Process performance is monitored as a method of monitoring the scale formation during production. Changes in pressure, temperature, and productivity are indicative of solids formation in the system. Continuous monitoring of these parameters can be used to determine location and severity of scale deposits. That can also change the course of brine sampling and scale prediction from one location to another, depending on where the severity of the problem lies.

15.4.1.4 Analogous fields and cases

Analogous fields are basically used for two main reasons: first, if brine sampling is unavailable, e.g., very low water cut, and secondly, to compare with the obtained results, sharing knowledge and learning from previous practical experience. However, analogous fields should be used with care, as different fields with similar brines can act differently.

Based on the brine analysis, scale prediction, and analogous fields data, the chemist can estimate the risk of scale deposits. The risk assessment should also involve the production rates, water volumes, and system design.

15.4.2 Selecting the proper control method

The proper method of mitigation, whether prevention or removal, depends on:

- System design
- Production rates and water volumes
- Fluids composition
- Scale type and its severity
- Economics of the treatment method

Onshore and shallow offshore fields are easy to treat with conventional methods like scale squeeze treatments. It is also convenient to apply scale inhibitor in pipelines and to frequently scrub the small amounts of formed scale in the pipes using pigging.

However, in severely scaled deepwater and subsea installations, the conventional scale squeeze or continuous injection methods are not feasible; hence, the sulfate removal methods, water softening, or produced water injection are common more feasible and economic alternatives with lower scale potential than using chemical inhibitors during injecting sea water. Magnetic and electronic devices are also used when scale squeeze and chemical injection are inapplicable and in cases such as protecting unpiggable pipelines.

Unconventional chemical treatments are applied in unusual situations: nonaqueous squeeze treatments for water sensitive formations, and solid inhibitors used in water sensitive and low reservoir pressure formations.

Table 15.7 compares and summarizes the different mineral scale mitigation methods.

Table 15.7 The different methods of mitigating mineral scales.

Mitigation method	Description
Operational methods	– Generally used in mineral scales prevention – Require system predesign or system modification – Avoid mixing incompatible fluids, and removing scaling species is very efficient and low cost – Sulfate removal method has high initial costs, but has economic operating costs, with efficient prevention of mineral scales especially in severe scaling cases, and when chemical methods are expensive or require high volumes of inhibitors – Case dependent
Chemical methods	– Used in prevention and removal – Wide variety of chemicals for different applications, e.g., downhole, topsides – Generally, do not require system modification – Generally cost effective, but can be expensive in severe cases that require high volumes of chemicals – Can be applied combined with mechanical methods – Very effective for use during shutdowns to protect the system – Chemicals can have side effects, e.g., incompatibility with other chemicals or produced fluids, or can enhance other production chemistry problems
Mechanical methods	– Used mainly in scale removal – Can be used in topsides (pigging) or downhole (wireline, coiled tubing) – Can be combined with chemical methods to enhance dissolution and removal – System design and accessibility dependent
Other nonchemical methods	– Internal coatings have shown success in field applications – Magnetic methods have been reported having success in the fields, despite the controversy in results – Ultrasonic, electric, radiofrequency methods have been reported with successful cases studies – They may require system modification for installation of physical devices – These methods are not generally common in field application compared to chemical methods – They can be case dependent

15.4.3 Monitoring and assessment

After the mitigation method is applied, the applied method is monitored to evaluate its efficiency. Monitoring process parameters and performance, water chemistry, total suspended solids, scale inhibitor residuals, and other physical devices is used to assess the actively applied scale mitigation method. Chapter 13, "Monitoring of flow assurance solids formation in oil and gas fields," discusses the methods of monitoring scale and scale mitigation methods. Among these, the total suspended solids methods have been proven as quick and effective methods for monitoring scale formation and evaluating scale prevention methods.

A general scale control management plan is depicted in Fig. 15.36.

15.5 Recent advances in scale management methods

An immense number of advances have been achieved in the last few years in the realm of scale management. These have stemmed from deeper understanding of scale formation mechanisms and employing this in early control of the problem.

Deposit monitoring has been shown to improve in accuracy and reliability. While some fields still rely on classic water chemistry, advanced sensors for TSMR, tomography, and radiography have been shown to provide reliable real-time monitoring means. These methods are expected to show improvements by applying smart completions, which are supplied with advanced monitoring devices connected to the surface stations (using cables or wireless), allowing early detection and management of problems.

Scale prediction packages have also improved. While most of them are thermodynamics based, other new emerging packages have included kinetics and hydrodynamics in their calculations, allowing more realistic results. They also improved their temperature, pressure, and ionic strength limits to be able to model high temperature, high pressure, and high salinity fields. Combining big data and artificial intelligence to the realm of scale prediction is expected to give scale prediction a powerful boost towards more realistic and applicable results.

The mitigation methods have witnessed several advances on different sides. On the operational side, the fluid mixing is improved by softening the injection water. Sulfate reducing processes have been applied in many fields with proven success rates, most importantly in deepwater fields, leading to major cost savings. Similarly, low salinity water injection (LOWSAL) has been proven to improve oil recovery and also eliminate the hard scale formation. On the chemical side, many advances have been witnessed, including improving the existing chemicals, developing new chemicals, and using nanomaterials and other natural extracts as green alternatives. Also, further improvements have been gained in deploying these chemicals by improving scale squeeze placement using nonaqueous treatments and employing foamed, gelled, and solid inhibitors. On the nonchemical side, different physical methods have been successfully applied in field operations. Magnetic, radiofrequency, ultrasonic, and internal coating methods have been used to prevent and remove scale from different installations. While their usage is not that frequent, the higher success rates reported in field applications pushes towards more frequent usage of these methods. With the rapid advances in materials science, it is expected that the use of nanomaterials, carbon nanotube composites, graphene, and others, which can be employed in chemical methods, in internal coatings, or in smart monitoring devices, will continue to increase.

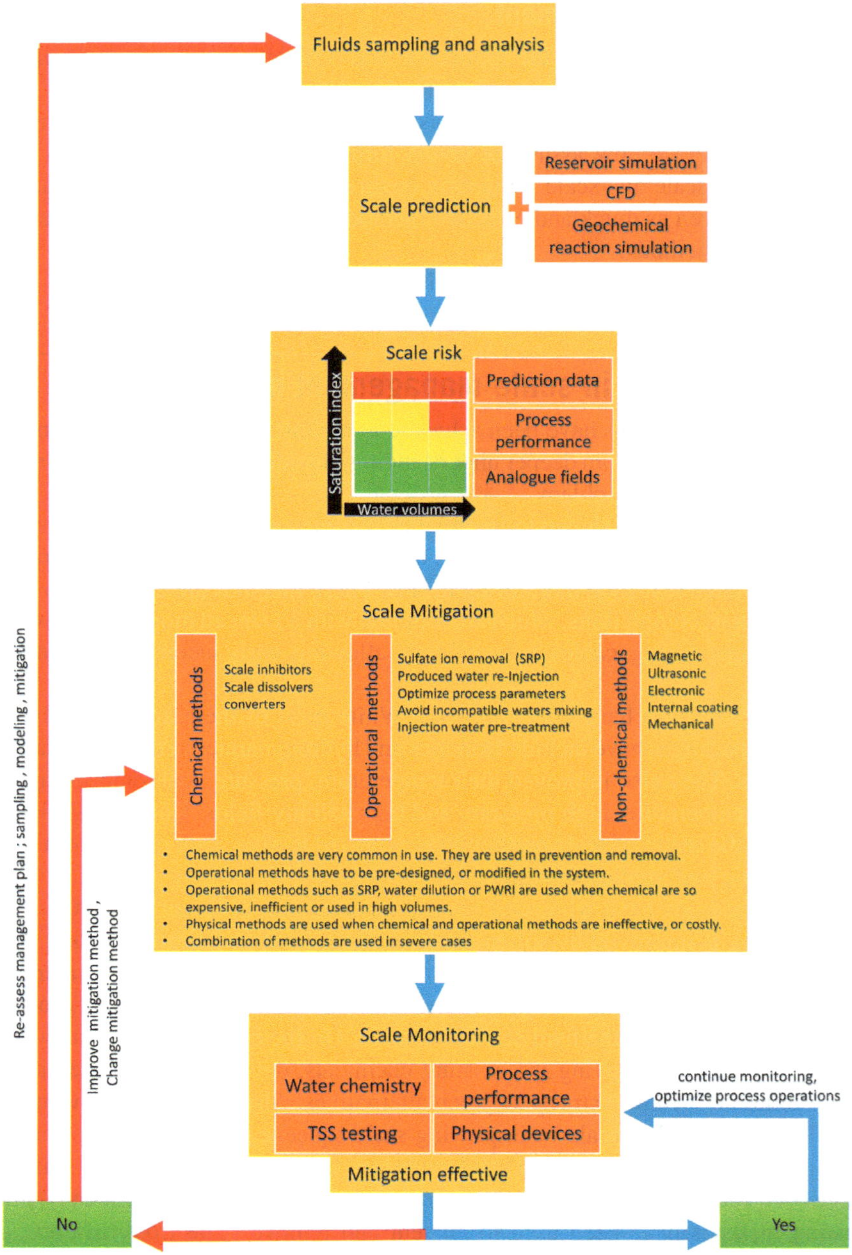

FIG. 15.36

Mineral scale management strategy.

15.6 Case studies

CASE STUDY 1: SULFATE SCALE IN EGYPTIAN FIELD

An oil producer in the Egyptian field encountered composite scale deposition at different depths. The well started production in 2007 and encountered scale formation in 2010, after significant increase in water cut. The scale sample collected from the obstruction at depth 5000' THF composed of organic deposits (8%), halite (4%), $CaCO_3$ (2%), (Fe,Zn,Pb)S (20%), $BaSO_4$ (49%), and $SrSO_4$ (17%), and a relatively similar composition of the scale that was collected at depth 6' THF, comprising organic deposits (4%), halite (4%), $CaCO_3$ (8%), (Fe,Zn,Pb)S (23%), $BaSO_4$ (45%), and $SrSO_4$ (16%). The well was not on scale inhibitor injection treatment, because the water analysis in the water analysis and scale prediction early in 2010 showed no potential scale risks in the well. The water chemistry of the well is summarized in Table 15.8.

Another water sampling and analysis session was performed, which showed no big differences from the concentrations cited in Table 15.8. Hence, the corresponding scaling tendency calculations showed no scaling potential in the well, which was not consistent with the gauging operations, which always came out with scale obstructions.

Unfortunately, no downhole water sampling was available at this time to make sure the water was representative of the actual water composition upstream of the obstruction problem. Hence, the treatment was planned based on the actual scale found in the well, and the scale prediction calculations were ruled out.

While this case study seems simple, it highlights the importance of proper sampling and proper scale tendency calculation, and matching that with the actual field cases. The obvious explanation for the inconsistent results from water analysis and scale prediction is that the well has high downhole pressure >5000 psi and high temperature of 205°F, and upon production the fluids are subjected to pressure and temperature drop, leading to scale formation in the production tubing, and thus the water collected at the well head is in fact the spent water (water after scaling species already precipitated, and dissolved gases are lost) rather than a downhole representative water. The formed scale deposits were not detectable earlier at lower water cut; however, when the water cut increased (from 27% to 87%) more scale masses formed, leading to the observed obstructions.

Furthermore, this case also highlights that the scale prediction in some cases requires a combination of scale prediction, reservoir simulation, and geochemical simulation. The well in this case study was not on water injection; however, it showed sulfate deposits, suggesting geochemical interactions in the reservoir or commingled production from different zones.

After lab tests to screen the available inhibitors, phosphonate-based inhibitor was squeezed into the well. During a lifetime of approximately 6 months, the well performance was restored and well gauging did not detect significant scale on the tubing. Fig. 15.37 illustrates the phosphonate residuals in the well.

Table 15.8 Water chemistry of the well with scale problems.

Na^+ (ppm)	K^+ (ppm)	Ca^{2+} (ppm)	Mg^{2+} (ppm)	Fe^{2+} (ppm)	Mn^{2+} (ppm)	Ba^{2+} (ppm)	Sr^{2+} (ppm)	Zn^{2+} (ppm)	Pb^{2+} (ppm)	Cl^- (ppm)	HCO_3^- (ppm)	SO_4^{2-} (ppm)	CO_3^{2-} (ppm)
50,000	4500	38,100	6100	930	500	80	800	105	2	169,900	0	11	0

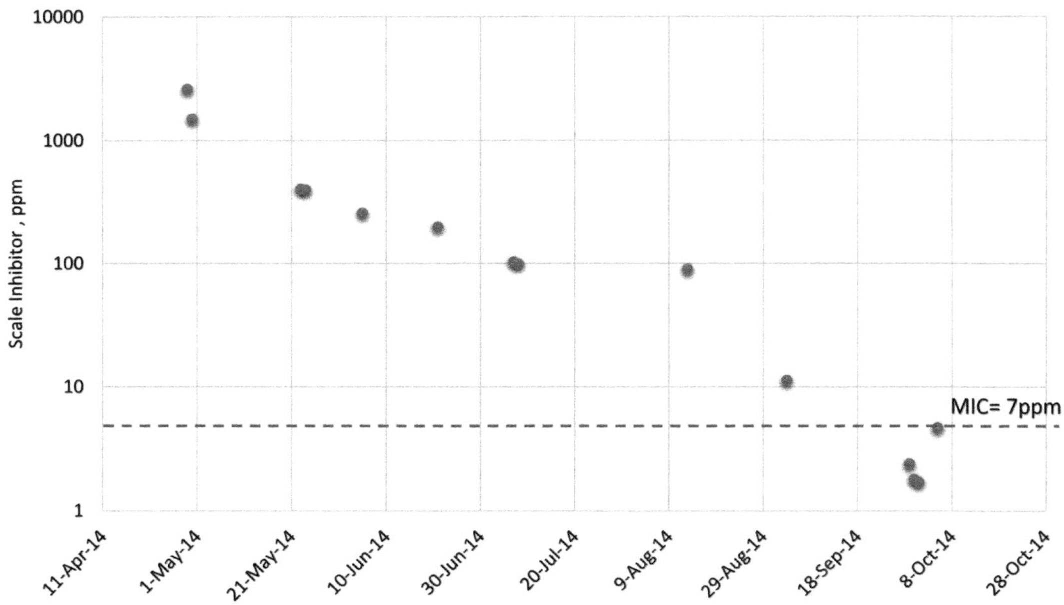

FIG. 15.37

Residual scale inhibitor concentration from the treated well in Case study 1.

CASE STUDY 2: TUBING CLEANING USING CHEAP CHEMICAL DISSOLVER

In this case study, two wells were successfully treated with a formulation comprising aminopolycarboxylate and phosphonate mixture to remove hard carbonate and sulfate scale from two producing wells. The aminopolycarboxylate used are EDTA and DTPA, while the phosphonates used are penta-phosphonate DETPMP and di-phosphonate HEAMBP.

The phosphonates were used as catalyst for the aminopolyphosphonates, enhancing their performance. Such formulations are effective, cheap, and safe for use without side effects on the well integrity.

The first treated well was producing from a sandstone formation, where mineral scales were detected at well depth of 10,100' THF during a wireline job. The scale was hard, dense, and multilayered with a thickness of 0.2–0.3 in., and its chemical composition was found to be mainly $BaSO_4$ and $SrSO_4$, as shown in Table 15.9.

Table 15.9 Composition of scale sample from well#1, in Case study 2.

Compound	$BaSO_4$	$SrSO_4$	$CaCO_3$	FeS	ZnS	PbS	NaCl	Hydrocarbons	Sand residues
% wt	50.4	19.1	1.2	0.8	4.5	15.0	3.2	5.0	0.8

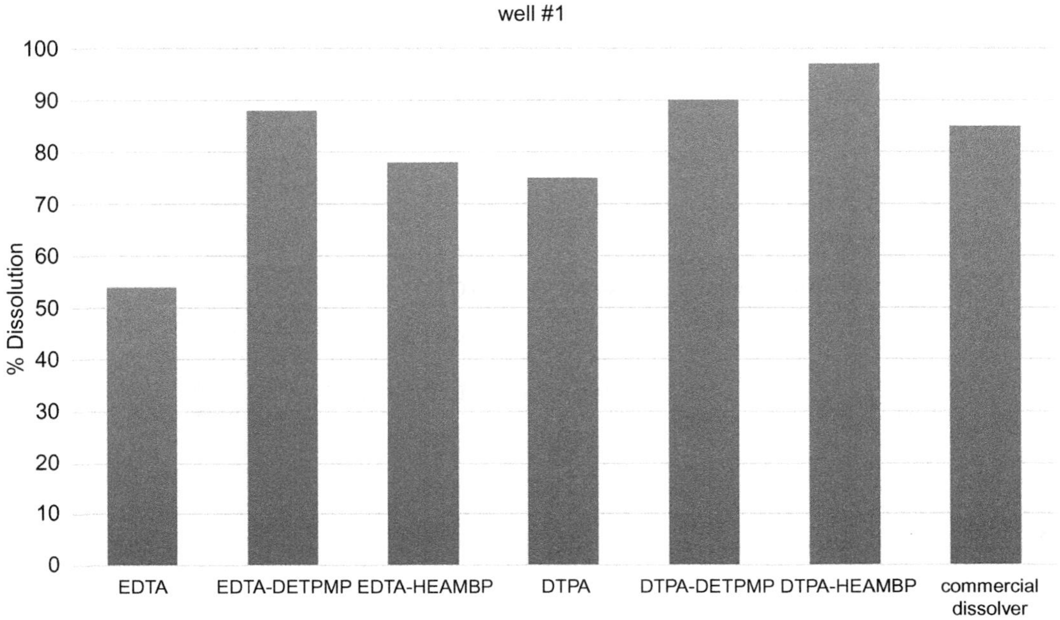

FIG. 15.38

The dissolution efficiency of different formulations for scale deposits of well #1.

Lab trials were performed to assess the proper dissolver. An initial screening test was performed by dissolving 1 g of the scale sample in 50 mL dissolver at 90°C for a specific soaking time (up to 5 days). The various formulations and their efficiency are illustrated in Fig. 15.38.

As shown in Fig. 15.38, the formulations comprising DTPA-DETPMP and DTPA-HEAMBP gave the best performance, and they even superseded the commercial dissolver. Although the DTPA-HEAMBP gave the best results, this formulation showed compatibility issues with the well produced water, which can cause formation damage. Thus the decision was made to use the DTPA-DETPMP formulation, which comprises 20% wt DTPA-10% vol. DETPMP+ KOH (for pH adjustment at ~12), and the soaking time was assigned to be a minimum of 4 days.

The cleaning job started by pumping 50 barrels of diesel to clean the production string from any oil or organic deposits leftovers; then the formulated dissolver (50 barrels) was bullheaded down the production tubing to cover the area from the perforation and up, to cover ~1000 ft above the obstruction. The rest of the tubing was filled with a filling solution comprising a 1:1 diluted dissolver formulation (Fig. 15.39). Then the well was shut-in to let the scale be soaked in the dissolver for 4 days. Once the dedicated soaking time had elapsed, the well was put back into production and a well gauging job was performed to assess the removal of the obstruction. The tubing gauging confirmed that the mineral scales were successfully dissolved, and the obstruction at 10100' THF was removed, which helped retrieve one of the fishing tools that was stuck because of the deposits; however, the other tools that were left stuck in the well were attributed to other mechanical reasons rather than scale deposits.

The total cost of the job was in the range of $17,000, including the chemicals, labor, and pumping expenses, compared to $100,000 estimated cost if the commercial scale dissolver had been used.

The second successful case study was performed in a cretaceous oil well, which was drilled in 1990 and is producing from sandstone formation. The scale deposits were detected during a gas lift valve change (GLVC) job and the obstruction was found to be at 10935' THF; the scale samples analysis showed that the main composition was gypsum and calcium carbonate, as shown in Table 15.10.

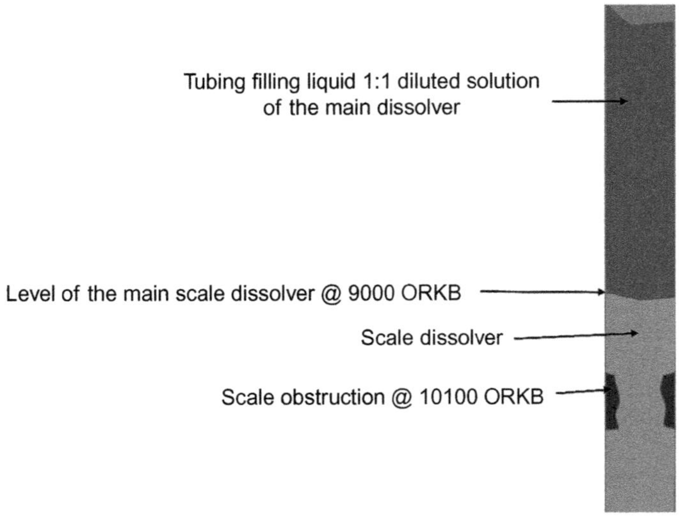

FIG. 15.39

Design of cleaning job for well #1.

Table 15.10 Well#2 scale composition.

Compound	$CaCO_3$	$CaSO_4{\cdot}2H_2O$	FeS	ZnS	PbS	NaCl	Hydrocarbons	Sand residues
% wt	32.8	28.2	3.7	0.1	0.2	1.6	16.9	16.5

After lab trials, all the formulations showed good dissolution efficiency (Fig. 15.40). Hence, based on the availability of chemicals and their economic feasibility, the decision was made to use the EDTA-DETPMP formulation, which comprises 20% wt EDTA-10% vol. DETPMP-KOH, to be soaked in the well for a minimum of 3 days. This solution was chosen based on the cost of the job and the availability of soaking time. The job started by pumping 50 bbl of diesel into the production string to clean the oil and organic deposit leftovers. Then 50 bbl of main treatment chemical was pumped to contact the obstruction and cover 1000 ft. above the obstruction. Then the tubing was filled with filling solution comprising 1:4 diluted dissolver solution. The well was then shut-in and soaked for the dedicated time. After the soaking time, the well was gauged and it was found that the obstruction was successfully removed. After dissolving the mineral scale obstruction, the well production increased by ~580 bpd, in addition to a significant improvement in the API of the produced oil, reaching 48 API. The job was economically feasible, as the total cost was $13,000, including the chemical, pumping, and the labor expenses, which was cheaper than the conventional acid cleaning job. This cleaning job, besides being successful, is considered economically feasible and technically safer to the tubing than the conventional acid jobs that had been regularly applied to remove acid-soluble calcite.

These reported formulations were shown to be effective, cheap, and safe methods for removing mineral scales from the production wells. Their only drawback is the long soaking time, which luckily was available in these cases. However, the operator can manage the operations to afford the required long soaking times. Besides, some other cleaning operations can take days and still fail.

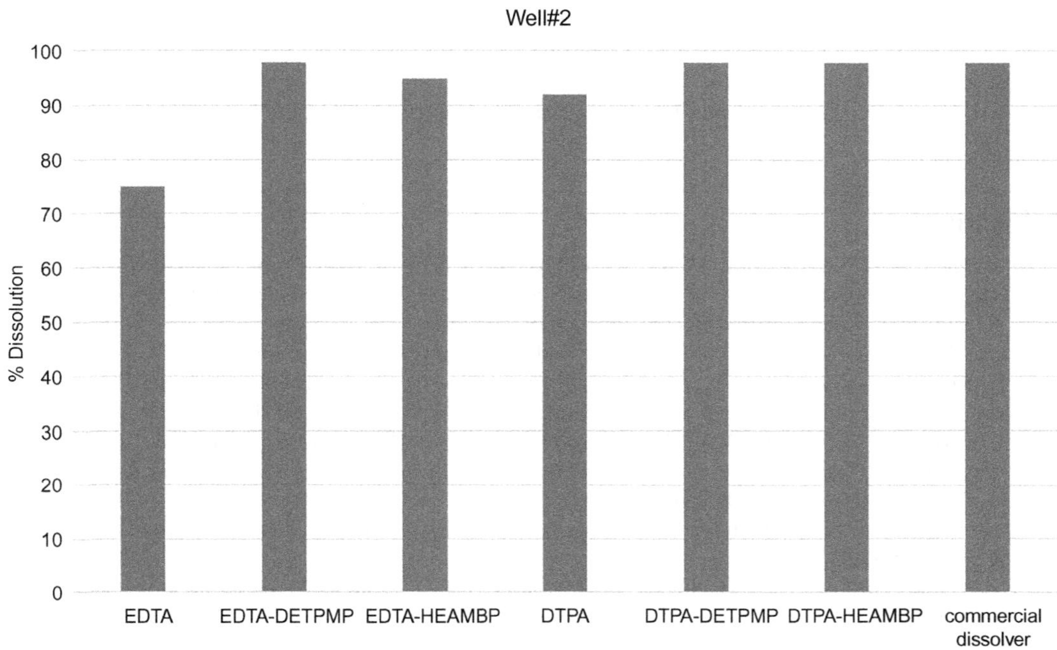

FIG. 15.40

Dissolution efficiency of different formulations for scale deposits in well #2.

15.7 Summary

Mineral scale is a major flow assurance problem in the oil industry. In this chapter, the different methods of scale mitigation were reviewed. Scale mitigation methods involve operational, chemical, and nonchemical methods.

Operational methods involve a number of factors related to scale formation, such as avoiding the mixing of incompatible waters, avoiding significant pressure and temperature drops, and removing the scaling species such as dissolved gases and scaling ions (e.g., sulfate ions). Chemical methods involve the use of chemical agents (polymeric and nonpolymeric chemicals) that can interfere with the scale formation process and slow down or prevent scale formation. These chemicals can be batched into the producing reservoir or continuously injected in the system. The nonchemical methods involve the use of physical, mechanical, or biological methods that can prevent scale formation.

Similarly, scale removal can be achieved by chemical methods known as dissolvers or converters, or by the use of physical methods and mechanical methods to disturb and diminish the strength of the formed hard scale layer.

Scale management strategy involves rigorous produced-fluid sampling, preferably upstream of the expected scale problem; accurate scale prediction calculations by using a combination of scale prediction, reservoir simulation, and geochemical simulators; and application of the mitigation plans, whether chemical or nonchemical, followed by rigorous monitoring of the system.

References

[1] S. Heath, M.Z. Bin Ruslan, E. McKay, O. Ishkov, Non-Chemical Methods for Downhole Control of Carbonate and Sulphate Scales—An Alternative Approach to Scale Management?, Society of Petroleum Engineers, 2018, June 20, https://doi.org/10.2118/190706-MS.

[2] C.C. Patton, Applied Water Technology, second ed., Campbell Petroleum Series, 1995.

[3] K. Juniel, H. Rawlins, Water-treating facilities in oil and gas operations, in: Volume III—Facilities and Construction Engineering, in Petroleum Engineering Handbook, SPE, 2006, pp. 123–183 (Chapter 4).

[4] S.H. Macklin, W.E. Haas, W.S. Miller, Carbon Dioxide and Dissolved Oxygen Removal From Makeup Water By Gas Transfer Membranes, General Electric Company, 2010. TP1060EN.doc Jan-10.

[5] G. George, N. Bhoria, S. Alhallaq, A. Abdala, V. Mittal, Polymer membranes for acid gas removal from natural gas, in: Separation and Purification Technology, Elsevier, 2016, https://doi.org/10.1016/j.seppur.2015.12.033.

[6] suez water technologies, Handbook of Industrial Water Treatment, Retrieved from: https://www.suezwatertechnologies.com/handbook/chapter-07-precipitation-softening#:~:text=Precipitation%20softening%20processes%20are%20used,%2C%20silica%2C%20and%20other%20constituents.&text=The%20water%20is%20treated%20with,water%20to%20form%20insoluble%20compounds. Accessed 11 August 2020.

[7] J.D. Cogan, The Removal of Barium, Strontium, Calcium and Magnesium From Hydraulic Fracturing Produced Water Using Precipitation With Traditional and Alternative Reactant Feedstocks, Environmental Studies, April 2016.

[8] R.B. Reyes, Softening of Oilfield Produced Water by Ion Exchange for Alkaline Flooding and Steamflooding, Society of Petroleum Engineers, 1983, January 1, https://doi.org/10.2118/11706-MS.

[9] C. Brown, M. Sheedy, A New Ion Exchange Process for Softening High TDS Produced Water, Society of Petroleum Engineers, 2002, January 1, https://doi.org/10.2118/78941-MS.

[10] T. Bilstad, E. Espedal, A.H. Haaland, M. Madland, Ultrafiltration of Oily Wastewater, Society of Petroleum Engineers, 1994, January 1, https://doi.org/10.2118/27136-MS.

[11] S. Das, E.J. Alvarez, N. Lange, Offshore ultrafiltration with multi element vessels, in: Offshore Technology Conference, 2016, May 2, https://doi.org/10.4043/27116-MS.

[12] M.W. Heatherly, M.E. Howell, J.E. McElhiney, Sulfate removal technology for seawater waterflood injection, in: Offshore Technology Conference, 1994, January 1, https://doi.org/10.4043/7593-MS.

[13] J.E. McElhiney, Deepwater project economics demand sulfate removal to ensure scale-free operation, Offshore Magazine (2003).

[14] M. Boczkowski, P. Eriksson, M. Simionato, Water injection and sulfate removal in the offshore oil & gas industry, Technical Paper, TP1203EN, General Electric Company, 2015.

[15] Davis, et al., The advancement of sulphate removal from seawater in offshore water flood operations, in: Paper 02314 Presented at NACE International Conference on Corrosion 2002, 2007.

[16] V. Koi Vu, et al., Eliminating the need for scale inhibition treatments for Elf exploration's girassol field, in: SPE Conference on Oilfield Scale 2000, Paper 60,220, Aberdeen, United Kingdom, 2000.

[17] Collins, et al., Sulphate removal for barium sulphate scale mitigation in a deepwater subsea production system, in: SPE Conference on Oilfield Scale 2004, Paper 87465, Aberdeen, United Kingdom, 2004.

[18] Mota, et al., Forecasts and alternative analysis for sulphate removal or chemical treatments for barium and strontium scale deposition—offshore Brazil, in: SPE Conference on Oilfield Scale 2004, Paper 87465 Aberdeen, United Kingdom, 2004.

[19] M.S.H. Bader, Sulfate removal technologies for oil fields seawater injection operations, J. Pet. Sci. Eng. 55 (1–2) (2007) 93–110, https://doi.org/10.1016/j.petrol.2006.04.010.

[20] Petroleum Technology Company, Cost Savings and Improved Production Uptime Due to the Use of Dura-Lift 'Anti Scale' Gas Lift Valves, 2016, Retrieved from: https://ptc.no/_extension/media/493/orig/attachment/Cost%20Savings%20when%20using%20Duralift%20GLV's%20.pdf. (Accessed November 2021).

[21] Ishkov, et al., Impact of temperature on scale formation in chalk reservoirs, in: SPE Europec Featured at 79th EAGE Annual Conference & Exhibition 2017, Paper 185772, Paris, France, 2017.

[22] Basic Sand Control Methods in Oil and Gas Industry, 2016. Retrieved from: https://www.drillingformulas.com/basic-sand-control-methods-in-oil-and-gas-industry/. Accessed 8 November 2021.

[23] M.A. Kelland, Production Chemicals for the Oil and Gas Industry, CRC Press, 2014, https://doi.org/10.1201/9781420092974.

[24] J.C. Cowan, D.J. Weintritt, Water Formed Scale Deposits, Gulf Pub. Co., 1976.

[25] J.S. Gill, Development of Scale Inhibitors, NACE International, 1996, January 1.

[26] R.N. Smith, T.A. Lawless, H.M. Bourne, A. Brunger, D. Kerr Nicoll, C. Hurtevent, A. Ainsworth, Planning and execution of a field trial utilising new invert emulsion squeeze technology, in: International Symposium on Oilfield Scale c: 7, 2000. https://www.onepetro.org/conference-paper/SPE-60210-MS.

[27] J.J. Wylde, E.K. McAra, Optimization of an oil soluble scale inhibitor for minimising formation damage: laboratory and field studies, in: Paper Presented at the SPE International Symposium and Exhibition on Formation Damage Control, Lafayette, Louisiana, February, 2004, https://doi.org/10.2118/86477-MS.

[28] J.S. Al-Thuwaini, B.J. Burr, Encapsulated Scale Inhibitor Treatment, Society of Petroleum Engineers, 1997, January 1, https://doi.org/10.2118/37790-MS.

[29] S. Baraka-Lokmane, K. Sorbie, N. Poisson, N. Kohler, Can green scale inhibitors replace phosphonate scale inhibitors?: carbonate coreflooding experiments, Pet. Sci. Technol. 27 (4) (2009) 427–441, https://doi.org/10.1080/10916460701764605.

[30] J. Fink, Petroleum Engineer's Guide to Oil Field Chemicals and Fluids, Elsevier, 2012, pp. 253–275, https://doi.org/10.1016/C2009-0-61871-7.

[31] A. Viloria, L. Castillo, J.A. Garcia, J. Biomorgi, Aloe Derived Scale Inhibitor, 2010. US Patent 7,645,722, assigned to Intevep, S.A., Caracas, VE, January 12, 2010.

[32] P. Zhang, A.T. Kan, M.B. Tomson, Oil field mineral scale control, in: Mineral Scales and Deposits: Scientific and Technological Approaches, Elsevier, 2015, pp. 603–617, https://doi.org/10.1016/B978-0-444-63228-9.00024-3.

[33] H. Wang, M. Gao, Y. Guo, Y. Yang, R. Hu, A natural extract of tobacco rob as scale and corrosion inhibitor in artificial seawater, Desalination 398 (November) (2016) 198–207, https://doi.org/10.1016/j.desal.2016.07.035.

[34] H. Hang, F. Wang, X. Jin, Y. Zhu, A botanical polysaccharide extracted from abandoned corn stalks: modification and evaluation of its scale inhibition and dispersion performance, Desalination 326 (October) (2013) 55–61, https://doi.org/10.1016/j.desal.2013.07.015.

[35] D. Denney, Oil-Soluble Scale Inhibitors, Society of Petroleum Engineers, 1999, March 1, https://doi.org/10.2118/0399-0072-JPT.

[36] D. Ruiz, M. Jackson, O. Fabian Torres, M. Nohra, Application of Encapsulated Scale Inhibitors—Understanding the Field Success Factors, NACE International, 2017, April 27.

[37] T.A. Lawless, R.N. Smith, New Technology, Invert Emulsion Scale Inhibitor Squeeze Design, Society of Petroleum Engineers, 1999, January 1, https://doi.org/10.2118/50705-MS.

[38] K. Ho, T. Chen, P. Chen, T. Hagen, H. Montgomerie, R. Benvie, J. Lewis, Development of Test Methods and Inhibitors for Halite Deposition in Oilfield Water Treatment, NACE International, 2013, March 17.

[39] A. Suzuki, T. Ide, T. Fukaya, T. Imada, Chelating Agent, and Scale Inhibitor, Detergent and Water Treatment Method, US Patent US 2016/0060145 A1, Minato, JP, March 3, 2016.

[40] M.B. Tomson, G. Fu, M.A. Watson, A.T. Kan, U. Rice, Mechanisms of mineral scale inhibition, in: Oilfield Scale Symposium, Society of Petroleum Engineers, Aberdeen, 2002, pp. 1–12, https://doi.org/10.2118/84958-PA.

[41] M. Crabtree, D. Eslinger, P. Fletcher, M. Miller, A. Johnson, G. King, Fighting scale—removal and prevention, Oilfield Rev. (1999) 30–45, https://doi.org/10.2307/3921.

[42] F. Jones, M.I. Ogden, Controlling crystal growth with modifiers, CrystEngComm 12 (4) (2010) 1016–1023, https://doi.org/10.1039/b918849e.

[43] K.S. Sorbie, N. Laing, How scale inhibitors work: mechanisms of selected barium sulphate scale inhibitors across a wide temperature range, in: Proceedings of SPE International Symposium on Oilfield Scale, Society of Petroleum Engineers, 2004, https://doi.org/10.2523/87470-MS.

[44] F. Jones, M.I. Ogden, G.M. Parkinson, A.L. Rohl, Comparison of the effect growth inhibitors have on an isostructural pair of salts, CrystEngComm 7 (April) (2005) 320–323, https://doi.org/10.1039/b501965f.

[45] Y. Zuo, W. Yang, K. Zhang, Y. Chen, X. Yin, Y. Liu, Experimental and theoretical studies of carboxylic polymers with low molecular weight as inhibitors for calcium carbonate scale, Crystals 10 (5) (2020), https://doi.org/10.3390/cryst10050406.

[46] P. Santoso, M.R. Setiawan, Suharso., Piper betle leaf extract as a green inhibitor of calcium sulphate (CaSO4) scale formation, in: IOP Conference Series: Earth and Environmental Science, vol. 258, Institute of Physics Publishing, 2019, https://doi.org/10.1088/1755-1315/258/1/012038.

[47] M. Schwamborn, Chemical synthesis of polyaspartates: a biodegradable alternative to currently used polycarboxylate homo- and copolymers, Polym. Degrad. Stab. 59 (1998) 39–45.

[48] R. Erany, Study of Scale and Scaling Potential During High Salinity and Low Salinity Water Flooding (PhD), University of Stavanger, 2016.

[49] D. Liu, Research on Performance Evaluation and Anti-Scaling Mechanism of Green Scale Inhibitors by Static and Dynamic Methods (Ph.D.), Environmental Engineering. Ecole nationale supérieure d'arts et métiers, ParisTech, 2011.

[50] L.S. Herbert, U.J. Sterns, Saline water conversion-11, in: Advances in Chemistry Series, vol. 38, 1963, p. 52.

[51] S. Labille, A. Neville, G.M. Graham, L.S. Boak, An Assessment of Adhesion of Scale and Electrochemical Pre-Treatment for the Prevention of Scale Deposition on Metal Surfaces, Society of Petroleum Engineers, 2002, January 1, https://doi.org/10.2118/74676-MS.

[52] D. Hasson, et al., Influence of the flow system on the inhibitory action of $CaCO_3$ scale prevention additives, Desalination 108 (1996) 67–79.

[53] D.J. Weintritt, J.C. Cowan, Unique characteristics of barium sulfate scale deposition, J. Pet. Technol. 19 (1967) 1381–1394.

[54] C.S. Sikes, A. Wierzbicki, Mechanisms of regulation of crystal growth in selected biological systems, in: Mineral Scale Formation and Inhibition, Springer, US, 1995, pp. 183–206, https://doi.org/10.1007/978-1-4899-1400-2_16.

[55] E.D. Burger, G.R. Chesnut, Screening corrosion inhibitors used in acids for downhole scale removal, Mater. Perform. 31 (7) (1992) 40–44.

[56] A. Khormali, D.G. Petrakov, Laboratory investigation of a new scale inhibitor for preventing calcium carbonate precipitation in oil reservoirs and production equipment, Pet. Sci. 13 (2) (2016) 320–327, https://doi.org/10.1007/s12182-016-0085-6.

[57] J. Moghadasi, H. Müller-Steinhagen, M. Jamialahmadi, A. Sharif, Scale deposits in porous media and their removal by Edta injection, in: Heat Exchanger Fouling and Cleaning VII PR5 (10), 2007, pp. 57–70. http://dc.engconfintl.org/heatexchanger2007/10.

[58] T. Almubarak, J. Hong Ng, H. Nasr-El-Din, Oilfield scale removal by chelating agents: an aminopolycarboxylic acids review, in: SPE Western Regional Meeting, Society of Petroleum Engineers, 2017, https://doi.org/10.2118/185636-MS.

[59] L.W. Jones, Development of a mineral scale inhibitor, Corrosion 17 (1961) 232t–236t.

[60] J.A. Billman, Antibiofoulants: a practical methodology for control of corrosion caused by sulfate reducing bacteria, Mater. Perform. 36 (1997) 43.

[61] D. Shen, D. Shcolnik, R. Perkins, G. Taylor, M. Brown, Evaluation of Scale Inhibitors in Marcellus High-Iron Waters, Society of Petroleum Engineers, 2012, October 1, https://doi.org/10.2118/141145-PA.

[62] S. Yean, H. Al Saiari, A.T. Kan, M.B. Tomson, Ferrous carbonate nucleation and inhibition, in: Paper SPE 114124 Presented at the SPE International Oilfield Scale Conference, Aberdeen, 28–29 May, 2008, https://doi.org/10.2118/114124-MS.

[63] K. Blanco, H. Aponte, E. Vera, Effect of the temperature and the CO_2 concentration on the behavior of the citric acid as a scale inhibitor of $CaCO_3$, J. Phys. Conf. Ser. 935 (2017) 012053.

[64] Y. Zhao, L. Jia, K. Liu, P. Gao, H. Ge, L. Fu, Inhibition of calcium sulfate scale by poly (citric acid), Desalination 392 (August) (2016) 1–7, https://doi.org/10.1016/j.desal.2016.04.010.

[65] D. Hasson, H. Shemer, A. Sher, State of the art of friendly 'Green' scale control inhibitors: a review article, Ind. Eng. Chem. Res. 50 (12) (2011) 7601–7607, https://doi.org/10.1021/ie200370v.

[66] T. Rabizadeh, C.L. Peacock, L.G. Benning, Carboxylic acids: effective inhibitors for calcium sulphate precipitation? Mineral. Mag. 0026-461X, 78 (6) (2014) 1465–1472, https://doi.org/10.1180/minmag.2014.078.6.13.

[67] Q. Wang, S. Shen, H. Badairy, T. Shafai, Y. Jeshi, T. Chen, F.F. Chang, Laboratory assessment of tetrakis(hydroxymethyl)phosphonium sulfate as dissolver for scales formed in sour gas wells, Int. J. Corros. Scale Inhib. 3 (2015) 235–254, https://doi.org/10.17675/2305-6894-2015-4-3-235-254.

[68] A.E. Hussein, H.S. Mohamed, Studying the use of tetrakis(hydroxymethyl)phosphonium sulfate (THPS) as zinc sulfide and lead sulfide scales dissolver and the factors influencing the dissolution, Int. J. Corros. Scale Inhib. 2017 (3) (2017) 349–358, https://doi.org/10.17675/2305-6894-2017-6-3-9.

[69] W. Li, G. Ruan, N. Bhandari, X. Wang, Y. Liu, H. Dushane, M. Tomson, Development of novel iron sulfide scale control chemicals, in: SPE International Oilfield Scale Conference and Exhibition, 2018, https://doi.org/10.2118/190755-ms.

[70] R.T. Barthorpe, The Impairment of Scale Inhibitor Function by Commonly Used Organic Anions, Society of Petroleum Engineers, 1993, January 1, https://doi.org/10.2118/25158-MS.

[71] N.H. Zaid, U.S. Patent 5396958, 1995.

[72] D. Licker, McGraw-Hill Dictionary of Chemistry, second ed., McGraw-Hill, US, 2003.

[73] D.E. Abd-El-Khalek, B.A. Abd-El-Nabey, Evaluation of sodium hexametaphosphate as scale and corrosion inhibitor in cooling water using electrochemical techniques, Desalination 311 (2013) 227–233, https://doi.org/10.1016/J.DESAL.2012.11.017.

[74] H.K. Moudgil, S. Yadav, R.S. Chaudhary, D. Kumar, Synergistic effect of some antiscalants as corrosion inhibitor for industrial cooling water system, J. Appl. Electrochem. 39 (8) (2009) 1339–1347, https://doi.org/10.1007/s1080 0-009-9807-4.

[75] C. Curless, J. Baclaski, R. Sachdev, Phosphate glass as a phosphate source in high cell density *Escherichia coli* fermentations, Biotechnol. Prog. 12 (1) (1996) 22–25, https://doi.org/10.1021/bp9500369.

[76] M. Mpelwa, S. Fa Tang, State of the art of synthetic threshold scale inhibitors for mineral scaling in the petroleum industry: a review, Pet. Sci. (2019), https://doi.org/10.1007/s12182-019-0299-5.

[77] H.A. Craddock, Oilfield Chemistry and Its Environmental Impact, John Wiley & Sons, 2018, pp. 21–233. https://onlinelibrary.wiley.com/book/10.1002/9781119244233.

[78] M.M. Jordan, G. Seland, N.D. Feasey, C.J. Graff, Latest advances in bio-degradation of scale inhibitors. Laboratory and field evaluation of a green carbonate and sulphate scale inhibitor with deployment histories in the North Sea and Alaska, in: Presented at Tekna Oilfield Chemistry Symposium, Geilo, Norway, 2000.

[79] Z. Amjad, R.T. Landgraf, J.L. Penn, Calcium sulfate dihydrate (gypsum) scale inhibition by PAA, PAPEMP, and PAA/PAPEMP blend, Int. J. Corros. Scale Inhib. 3 (1) (2014) 035–047, https://doi.org/10.17675/2305-6894-2014-3-1-035-047.

[80] W. Wang, A.T. Kan, F. Zhang, C. Yan, M. Tomson, Measurement and Prediction of Thermal Degradation of Scale Inhibitors, Society of Petroleum Engineers, 2014, December 1, https://doi.org/10.2118/164047-PA.

[81] S.J. Dyer, G.M. Graham, Thermal stability of generic barium sulphate scale inhibitor species under static and dynamic conditions, J. Pet. Sci. Eng. 37 (3–4) (2003) 171–181, https://doi.org/10.1016/S0920-4105(02)00351-0.

[82] S.J. Dyer, C.E. Anderson, G.M. Graham, Thermal stability of amine methyl phosphonate scale inhibitors, J. Pet. Sci. Eng. 43 (3–4) (2004) 259–270, https://doi.org/10.1016/j.petrol.2004.02.018.

[83] Z. Amjad, R.T. Landgraf, J.L. Penn, Calcium sulfate dihydrate (gypsum) scale inhibition by PAA, PAPEMP, and PAA/PAPEMP blend, Int. J. Corros. Scale Inhib. 3 (1) (2014) 35–47, https://doi.org/10.17675/2305-6894-2014-3-1-035-047.

[84] J. Gill, Scale control in geothermal brines—new inhibitors for calcium carbonate and silica control, in: GRC Transactions, Vol. 32, 2008. Geothermal Resources Council 2008 Annual Meeting, October 5–8, 2008, Reno, Nevada, 2008.

[85] Z. Amjad, A Multifunctional Additive for Controlling Inorganic Foulants in Industrial Water Systems, Materials Performance, NACE International, 2016.

[86] S.K. Hamdona, S.M. Hamza, A.H. Mangood, The influence of polyphosphonates on the precipitation of strontium sulfate (celestite) from aqueous solutions, Desalin. Water Treat. 24 (1–3) (2010) 55–60,-https://doi.org/10.5004/dwt.2010.1176.

[87] A.B. BinMerdhah, Inhibition of barium sulfate scale at high-barium formation water, J. Pet. Sci. Eng. 90–91 (2012) 124.

[88] S. Wang, A. Miller, C. Haugen, C. Okocha, A. Mahmoudkhani, A. Kaiser, Multifunctional polymeric scale inhibitor for high brine systems: development, implementation and evaluation, in: SPE Symposium: Production Enhancement and Cost Optimisation, 2017, https://doi.org/10.2118/189246-ms.

[89] A. Jada, R. Ait Akbour, C. Jacquemet, J.M. Suau, O. Guerret, Effect of sodium polyacrylate molecular weight on the crystallogenesis of calcium carbonate, J. Cryst. Growth 306 (2007) 373–382.

[90] S. Carvalho, L. Palermo, L. Boak, K. Sorbie, E.F. Lucas, Influence of terpolymer based on amide, carboxylic, and sulfonic groups on the barium sulfate inhibition, Energy Fuel 31 (10) (2017) 10648–10654, https://doi.org/10.1021/acs.energyfuels.7b01767.

[91] B. Wang, T. Chen, P. Chen, H. Montgomerie, T. Hagen, X. Liu, X. Yang, Development of test method and environmentally acceptable inhibitors for zinc sulfide deposited in oil and gas fields, in: SPE 156005 (Paper Presented at the SPE International Conference on Oilfield Scale, Aberdeen, UK, 30–31 May, 2012.

[92] N. Bhandari, G. Ruan, Y. Liu, et al., Identification of new chemical formulation for control of metal sulfides FeS, ZnS and PbS scale deposition, in: SPE International Conference on Oilfield Chemistry, 3 – 5 April, Montgomery, Texas, USA, 2017.

[93] W. Li, G. Ruan, N. Bhandari, X. Wang, Y. Liu, H. Dushane, A.T. Kan, Development of novel iron sulfide scale control chemicals, in: Society of Petroleum Engineers—SPE International Oilfield Scale Conference and Exhibition 2018, Society of Petroleum Engineers, 2018, https://doi.org/10.2118/190755-ms.

[94] Y.K. Alduailej, K.S. Sorbie, Iron sulfide scale inhibition: limitations at sour conditions, in: Society of Petroleum Engineers—SPE International Oilfield Scale Conference and Exhibition, OSS 2020, Society of Petroleum Engineers, 2020, https://doi.org/10.2118/200706-ms.

[95] N. Zhang, D. Schmidt, W. Choi, D. Sundararajan, Z. Reisenauer, J. Freeman, M. Tomson, Halite challenges and mitigation in the bakken-experience of managing high saline produced water from hydraulically fractured wells, in: SPE International Oilfield Scale Conference and Exhibition, 2018, https://doi.org/10.2118/190739-ms.

[96] J.D. Birchall, R.J. Davey, The crystallization of sodium chloride from aqueous solution in the presence of polysaccharides, J. Cryst. Growth 54 (1981) 323–329.

[97] J. Bellarby, Production chemistry, in: Developments in Petroleum Science, Elsevier, 2009, https://doi.org/10.1016/S0376-7361(08)00207-0.

[98] J.J. Wylde, J.L. Slayer, Halite scale formation mechanisms, removal and control: a global overview of mechanical, process, and chemical strategies, in: SPE International Symposium on Oilfield Chemistry, 2013, https://doi.org/10.2118/164081-ms.

[99] S. Taj, S. Papavinasam, R.W. Revie, Development of green inhibitors for oil and gas applications, in: Proceeding of the CORROSION 2006, San Diego CA, March 12-16, 2006, pp. 1–9.

[100] M. Schweinsberg, W. Hater, J. Verdes, New stable biodegradable scale inhibitor formulations for cooling water: development and field tests, in: Presented at the 64th International Water Conference, Pittsburgh, PA, October 19–23, 2003.

[101] J. Li, M. Tang, Z. Ye, L. Chen, Y. Zhou, Scale formation and control in oil and gas fields: a review, J. Dispers. Sci. Technol. 38 (5) (2017) 661–670, https://doi.org/10.1080/01932691.2016.1185953.

[102] A.A. Olajire, A review of oilfield mineral scale deposits management technology for oil and gas production, J. Pet. Sci. Eng. (2015), https://doi.org/10.1016/j.petrol.2015.09.011.

[103] B. Bazin, N. Kohler, A. Zaitoun, T. Johnson, H. Raajimakers, A new class of mineral scale inhibitors for squeeze treatments, in: Presented at the SPE International Symposium on Oilfield Scale Conference, Aberdeen, UK, May 26-27, 2004.

[104] C.E. Inches, K.S. Sorbie, C. Christophe, L. Papirer, Thermal stability of selected green scale inhibitors, in: Presented at the International Oil Field Chemistry Symposium, Geilo, Norway, March 25-28, 2007.

[105] D. Liu, W. Dong, F. Li, F. Hui, J. Lédion, Comparative performance of polyepoxysuccinic acid and polyaspartic acid on scaling inhibition by static and rapid controlled precipitation methods, Desalination 304 (2012) 1–10, https://doi.org/10.1016/j.desal.2012.07.032.

[106] A.M. Abdel-Gaber, B.A. Abd-El-Nabey, E. Khamis, D.E. Abd-El-Khalek, Investigation of fig leaf extract as a novel environmentally friendly antiscalent for $CaCO_3$ calcareous deposits, Desalination 230 (1–3) (2008) 314–328.

[107] A.M. Abdel-Gaber, B.A. Abd-El-Nabey, E. Khamis, H. Abd-El-Rhmann, H. Aglan, A. Ludwick, Green anti-scalent for cooling water systems, Int. J. Electrochem. Sci. 7 (12) (2012) 11930–11940.

[108] H. Zhang, W. Feng, X. Jin, Y. Zhu, A botanical polysaccharide extracted from abandoned corn stalks: modification and evaluation of its scale inhibition and dispersion performance, Desalination 326 (5) (2013) 55–61.

[109] L. Castillo, E. Torin, J.A. Garcia, M. Carrasquero, M. Navas, A. Viloria, New Product for Inhibition of Calcium Carbonate Scale in Natural Gas and Oil Facilities Based on Aloe Vera: Application in Venezuelan Oilfields, Society of Petroleum Engineers, 2009, January 1, https://doi.org/10.2118/123007-MS.

[110] M.F. Mady, M.A. Kelland, Study on various readily available proteins as new green scale inhibitors for oilfield scale control, Energy Fuel 31 (2017) 5940–5947.

[111] A.F. Miles, O. Vikane, D.S. Healey, I.R. Collins, J. Saeten, H.M. Bourne, R.G. Smith, Field experiences using 'oil soluble' non-aqueous scale inhibitor delivery systems, in: SPE International Symposium on Oilfield Scale, 2004, https://doi.org/10.2118/87431-ms.

[112] H. Guan, K.S. Sorbie, E.J. Mackay, The comparison of non-aqueous and aqueous scale inhibitor treatments: experimental and modeling studies, in: SPE International Symposium on Oilfield Scale, 2004, https://doi.org/10.2118/87445-ms.

[113] T.A. Lawless, R.N. Smith, New technology, invert emulsion scale inhibitor squeeze design, in: SPE International Symposium on Oilfield Chemistry, 1999, https://doi.org/10.2118/50705-ms.

[114] M.M. Jordan, C.J. Graff, K.N. Cooper, Development and Deployment of a Scale Squeeze Enhancer and Oil-Soluble Scale Inhibitor to Avoid Deferred Oil Production Losses During Squeezing Low-Water Cut Wells, North Slope, Alaska, Society of Petroleum Engineers, 2000, January 1, https://doi.org/10.2118/58725-MS.

[115] I.R. Collins, J.A. Hewartson, Extending squeeze lifetimes using miscible displacement, in: International Symposium on Oilfield Scale, 2002, https://doi.org/10.2118/74650-ms.

[116] N. Poynton, R. Tidswell, J. Steele, J. Wood, R. Early, Squeezing Aqueous Based Scale Inhibitors Into a Water Sensitive Reservoir—Development of a Squeeze Strategy, Society of Petroleum Engineers, 2000, January 1, https://doi.org/10.2118/60219-MS.

[117] J.S. Al-Thuwaini, et al., Encapsulated scale inhibitor treatment, in: Paper 37790 Presented at the 1997 Middle East Oil Show, Bahrain, 1997.

[118] D. Ruiz, M. Jackson, O. Fabian Torres, M. Nohra, Application of Encapsulated Scale Inhibitors—Understanding the Field Success Factors, NACE International, 2017, April 27.

[119] P. Zhang, D. Shen, G. Ruan, A.T. Kan, M.B. Tomson, Phosphino-polycarboxylic acid modified inhibitor nanomaterial for oilfield scale control: synthesis, characterization and migration, J. Ind. Eng. Chem. 45 (2017) 366–374, https://doi.org/10.1016/j.jiec.2016.10.004.

[120] W. Wang, A. Kan, C. Yan, M. Tomson, The use of inhibition kinetics and NMR spectroscopy in thermal stability study of scale inhibitors, in: SPE International Symposium on Oilfeld Chemistry, 2013/4/8/, The Woodlands, Texas, USA, 2013, https://doi.org/10.2118/164047-MS.

[121] G. Ruan, A.T. Kan, F. Yan, F. Zhang, N. Bhandari, H.A. Alsaiari, et al., Citrate assisted metal phosphonate colloidal scale inhibitors for long-term squeeze treatments, in: SPE International Oilfeld Scale Conference and Exhibition; 2016/5/11/; Aberdeen, Scotland, UK, 2016, https://doi.org/10.2118/179898-MS.

[122] P. Zhang, Review of synthesis and evaluation of inhibitor nanomaterials for oilfield mineral scale control, Front. Chem. (2020, November 19), https://doi.org/10.3389/fchem.2020.576055.

[123] O.M. Selle, O.-K. Nygård, E. Storås, A. Moen, P. Chen, R. Matheson, B. Juliussen, T. Håland, A. Mebratu, I. Melien, Foamed scale inhibitor and scale dissolver treatments for improved placement in a subsea norne well, in: Paper Presented at the SPE International Conference on Oilfield Scale, Aberdeen, UK, May, 2012, https://doi.org/10.2118/155407-MS.

[124] O.M. Selle, M. Springer, I.H. Aufem, P. Chen, R. Matheson, A.A. Mebratu, et al., Gelled Scale inhibitor treatment for improved placement in long horizontal wells at Norne and Heidrun fields, SPE Prod. Oper. 24 (3) (2009) 425–438, https://doi.org/10.2118/112464-PA.

[125] P. Chen, H. Montgomerie, T. Hagen, O. Vikane, B. Juliussen, T. Haaland, et al., Meeting the challenges in downhole scale control for high-water-production horizontal wells in an environmentally sensitive part of the Norwegian North Sea, SPE Prod. Oper. 27 (04) (2012) 382–389, https://doi.org/10.2118/135249-PA.

[126] A.T. Kan, Z. Dai, M.B. Tomson, The state of the art in scale inhibitor squeeze treatment, Pet. Sci. (2020, December 1), https://doi.org/10.1007/s12182-020-00497-z.

[127] Z. Zhang, Y. Liu, Z. Dai, N. Bhandari, F. Zhang, F. Yan, G. Ruan, H. Alsaiari, Y.-T. Lu, G. Deng, A. Kan, M. Tomson, Impact of FeIII/FeII on scale inhibition, in: Paper Presented at the SPE International Oilfield Scale Conference and Exhibition, Aberdeen, Scotland, UK, May, 2016, https://doi.org/10.2118/179905-MS.

[128] A.T. Kan, G. Fu, H.A. Al-Saiari, M.B. Tomson, D. Shen, Enhanced scale-inhibitor treatments with the addition of zinc, SPE J. 14 (2009) 617–626, https://doi.org/10.2118/114060-PA.

[129] J.R. Ferguson, pH impact on inhibitor performance, Analyst 21 (3) (2014) 11–17.

[130] T.D. Baugh, J. Lee, K. Winters, J. Waters, J. Wilcher, A fast and information-rich test method for scale inhibitor performance, in: Offshore Technology Conference, 2012, April 30, https://doi.org/10.4043/23150-MS.

[131] L. Sutherland, C.J. Johnston, W. Taylor, The influence of turbulence (or hydrodynamic effects) on barium sulphate scale formation and inhibitor performance, in: SPE 164070 (Paper Presented at the SPE International Symposium on Oilfield Chemistry, The Woodlands, TX, 8–10 April, 2013.

[132] H. Lu, J. Penkala, J. Brooks, C. Haugen, C. Harbaugh, C. McAfee, L. Kalfayan, Novel Laboratory and Field Approaches to Control Halite and other Problematic Scales in High Salinity Brine From the Bakken, Society of Petroleum Engineers, 2015, September 28, https://doi.org/10.2118/174788-MS.

[133] H. Lu, J. Brooks, R. Legan, S. Fritz, Novel Laboratory Test Method and Field Applications for Silica/Silicate and Other Problematic Scale Control, Society of Petroleum Engineers, 2018, June 20, https://doi.org/10.2118/190705-MS.

[134] M. Barber, S. Heath, A new approach to testing scale inhibitors in mild scaling brines—are dynamic scale loop tests needed? in: Proceedings—SPE International Symposium on Oilfield Chemistry (vol. 2019), Society of Petroleum Engineers (SPE), 2019, https://doi.org/10.2118/193580-ms.

[135] V. Kartnaller, F. Venâncio, F.F. do Rosário, J. Cajaiba, Application of multiple regression and design of experiments for modelling the effect of monoethylene glycol in the calcium carbonate scaling process, Molecules 23 (4) (2018), https://doi.org/10.3390/molecules23040860.

[136] J. Ibrahim, Establishing Scale Inhibitor Retention Mechanisms in Pure Adsorption And Coupled Adsorption/Precipitation Treatments (Ph.D. thesis), Heriot-Watt University, 2011, pp. 1–308.

[137] R. Wat, L.-E. Hauge, K. Solbakken, K.E. Wennberg, L.M. Sivertsen, B. Gjersvold, Squeeze chemical for HT applications—have we discarded promising products by performing unrepresentative thermal aging tests? in: SPE 105505 (Paper Presented at the SPE International Symposium on Oilfield Chemistry, Houston, TX, 28 February–2 March, 2007.

[138] G.M. Graham, S. Kidd, R. Stalker, R. Wright, Effect of coreflood test methodology on appropriate simulation of field treatments, in: 151779 (Paper Presented at the SPE International Symposium and Exhibition on Formation Damage Control, Lafayette, LA, 15–17 February, 2012.

[139] K. Ramstad, H.C. Rohde, T. Tydal, D. Christensen, Scale squeeze evaluation through improved sample preservation, inhibitor detection, and minimum inhibitor concentration monitoring, SPE Prod. Oper. 24 (4) (2009) 530. (SPE 114085).

[140] B.M. Hustad, O.G. Svela, J.H. Olsen, K. Ramstad, T. Tjomsland, Downhole Chemical Injection Lines—Why Do They Fail? Experiences, Challenges and Application of New Test Methods, Society of Petroleum Engineers, 2012, January 1, https://doi.org/10.2118/154967-MS.

[141] D.J. Peng, A. Low, Quantitative Risk Assessment of Downhole Continuous Chemical Injection for Subsea Multiphase Production System, BHR Group, 2015, September 4.

[142] L.A.D.O. Guerra, G.H. Oliveira, Designing Downhole Chemical Injection Systems, Society of Petroleum Engineers, 2014, October 27, https://doi.org/10.2118/170772-MS.

[143] G. Poggesi, C. Hurtevent, J.L. Brazy, Scale inhibitor injection via the gas lift system in high temperature block 3 fields in Angola, in: Paper Presented at the International Symposium on Oilfield Scale, Aberdeen, United Kingdom, January, 2001, https://doi.org/10.2118/68301-MS.

[144] M.A. Daas, T.S. Golczynski, A.A. Oliveira Magalhaes, E. Bastos, P.A. Ferreira, Downhole chemical injection through gas lift: options and consequences, in: Brasil Offshore, 2011, https://doi.org/10.2118/142951-ms.

[145] J.D. Stone, New solutions to an old problem: calcium sulfate deposition in Permian basin, Baroid News Bull. 18 (I) (1966) 11–14.

[146] A.G. Ostroff, Introduction to Oilfield Water Technology, second ed., Texas, Houston, 1979.

[147] O. Vazquez, E.J. Mackay, K.S. Sorbie, Development of a non-aqueous scale inhibitor squeeze simulator, in: SPE International Oilfield Scale Symposium, 2006, https://doi.org/10.2118/100521-ms.

[148] O. Vazquez, E.J. Mackay, M.M. Jordan, K.S. Sorbie, Impact of Mutual Solvent Preflushes on Scale Squeeze Treatments: Extended Squeeze Lifetime and Improved Well Clean-Up Time, Society of Petroleum Engineers, 2009, January 1, https://doi.org/10.2118/121857-MS.

[149] B. Ghosh, M.Y. Alklih, X. Li, Pre-flush design for extended scale inhibitor squeeze lifetime in limestone reservoir—laboratory investigations and cost benefit analysis, in: Offshore Technology Conference Asia, 2016, https://doi.org/10.4043/26604-ms.

[150] J.A. Patroni Zavala, E.J. Mackay, O. Vazquez, L.S. Boak, M.A. Singleton, G. Ross, The Cost and Value of Field, Laboratory, and Simulation Data for Validating Scale Inhibitor Treatment Models, Society of Petroleum Engineers, 2008, January 1, https://doi.org/10.2118/114106-MS.

[151] D. Blumer, Y. Xu, G. Mclelland, J. Waters, Preflush Chemicals For Scale Inhibitor Squeeze, 2017. Patent International Publication Number WO 2017/156538 Al. US.

[152] B. Ghosh, X. Li, M.Y. Alklih, Preflush design for oil-wet carbonate formations: key to enhance scale-inhibitor-squeeze lifetime, SPE J. 21 (03) (2016) 0776–0785, https://doi.org/10.2118/178910-pa.

[153] O. Vazquez, E.J. Mackay, K.H. Al Shuaili, K.S. Sorbie, M.M. Jordan, Modelling a Surfactant Preflush with Non-Aqueous and Aqueous Scale Inhibitor Squeeze Treatments, Society of Petroleum Engineers, 2008, January 1, https://doi.org/10.2118/113212-MS.

[154] O. Vazquez, P. Herrero, E. Mackay, M. Jordan, Non-aqueous vs aqueous overflush scale inhibitor squeeze treatment in an oilfield offshore norway, J. Pet. Sci. Eng. 138 (February) (2016) 1–10, https://doi.org/10.1016/j.petrol.2015.11.033.

[155] S.M. Heath, B. Juliussen, P. Chen, T. Chen, R. Benvie, Novel Scale Squeeze Technology and Treatment Designs for Improving Scale Inhibitor Retention and Treatment Lifetimes—Use of Ionic Polymers in the Overflush, Society of Petroleum Engineers, 2012, January 1, https://doi.org/10.2118/154935-MS.

[156] O.M. Selle, R.M.S. Wat, O. Vikane, H. Nasvik, P. Chen, T. Hagen, et al., A way beyond scale inhibitors—extending scale inhibitor squeeze life through bridging, in: International Symposium on Oilfield Scale, 29–30 January, Aberdeen, United Kingdom, 2003, https://doi.org/10.2118/80377-MS.

[157] H.C. Lau, Laboratory development and field testing of succinoglycan as fluid-loss-control fluid, in: Paper SPE 26724 Presented at the 1993 Offshore Europe Conference, Aberdeen, September 7-10, 1993.

[158] E.J. Mackay, K.S. Sorbie, Simulation of scale inhibitor treatments in horizontal wells using existing full field models, in: Presented at the 1998 NIF Ninth International Oil Field Chemicals Symposium, Geilo, Norway, March 22-25, 1998.

[159] M.M. Jordan, M.C. Edgerton, J. Cole-Hamilton, K. Mackin, The application of novel wax divertor technology to allow successful scale inhibitor squeeze treatment into a sub sea horizontal well, North Sea Basin, in: Paper Presented at the SPE Annual Technical Conference and Exhibition, New Orleans, Louisiana, September, 1998, https://doi.org/10.2118/49196-MS.

[160] S. Bhaduri, C. Lowe, D. Shen, M. Usie, D.V.S. Gupta, The Application of Solid Scale Inhibitor in Deepwater Completion to Reduce Well Intervention Costs, NACE International, 2018, July 20.

[161] R. Stalker, F. Wahid, G.M. Graham, Chemical placement in heterogeneous and long reach horizontal wells, in: 17th International Oil Field Chemistry Symposium, Geilo (Norway), 19-22 March 2006, 2006.

[162] O. Vazquez, E. Mackay, M. Raga, G. Ross, Effect of squeeze treatment stages mixing during injection on lifetime, in: SPE International Oilfield Scale Conference and Exhibition, 2018, https://doi.org/10.2118/190751-ms.

[163] A.T. Kan, L. Yan, P.B. Bedient, J.E. Oddo, M.B. Tomson, Sorption and fate of phosphonate scale inhibitors in the sandstone reservoir, in: Paper SPE 21714 Presented at the Production Operation Symposium Held in Oklahoma, April 7-9, 1991.

[164] M.M. Jordan, K.S. Sorbie, P. Griffin, S. Hennessey, K.E. Hourston, P. Waterhouse, Scale inhibitor adsorption/desorption vs. precipitation: the potential for extending squeeze life while minimising formation damage, in: SPE European Formation Damage Conference, 1995, https://doi.org/10.2118/30106-ms.

[165] K.S. Sorbie, R.M.S. Wat, A.C. Todd, Interpretation and theoretical modelling of scale inhibitor/tracer core floods, SPE Prod. Eng. 7 (August 1992) 307–312.

[166] K.S. Sorbie, P. Jiang, M.D. Yuan, P. Chen, M.M. Jordan, A.C. Todd, The Effect of pH, Calcium, and Temperature on the Adsorption of Phosphonate Inhibitor Onto Consolidated and Crushed Sandstone, Society of Petroleum Engineers, 1993, January 1, https://doi.org/10.2118/26605-MS.

[167] M. Kahrwad, K.S. Sorbie, L.S. Boak, Coupled adsorption/precipitation of scale inhibitors: experimental results and modeling, SPE Prod. Oper. 24 (2009) 481–491, https://doi.org/10.2118/114108-PA.

[168] H. Zhang, E.J. Mackay, P. Chen, K.S. Sorbie, Non-equilibrium adsorption and precipitation of scale inhibitors: corefloods and mathematical modelling, in: Paper Presented at the International Oil and Gas Conference and Exhibition in China, Beijing, China, November, 2000, https://doi.org/10.2118/64755-MS.

[169] K.S. Sorbie, M.D. Yuan, P. Chen, A.C. Todd, R.M.S. Wat, The Effect of pH on the Adsorption and Transport of Phosphonate Scale Inhibitor Through Porous Media, Society of Petroleum Engineers, 1993, January 1, https://doi.org/10.2118/25165-MS.

[170] J.L. Przybylinski, Adsorption and desorption characteristic of mineral scale inhibitors as related to the design of squeeze treatment, in: SPE 18486 Presented at the 1989 SPE International Symposium on Oil-Field Chemistry, Houston, TX, 8-10 February, 1989.

[171] G.P. Funkhouser, R.D. Gdanski, Improved adsorption-isotherm modelling for phosphonate scale inhibitors, in: Proceedings of the NIF International Oilfield Chemistry Symposium, Geilo, Norway, 1-4 April, 2001.

[172] R.D. Gdanski, G.P. Funkhouser, Mineralogy driven scale inhibitor squeeze designs, in: Paper SPE 94510 Presented at the SPE European Formation Damage Conference Held in Scheveningen, The Netherlands, 25-27 May, 2005.

[173] G.M. Graham, K.S. Sorbie, The effect of molecular weight on the adsorption/desorption characteristics of polymeric scale inhibitors on silica sand and in sandstone cores, in: Corrosion 94: National Association of Corrosion Engineers (NACE) International Annual Conference, Baltimore, MD, United States, 28 February-4 March, 1994.

[174] J.E. Pardue, Results of field tests with a new extended squeeze life scale inhibitor, in: Paper No. 35 Presented at the NACE Annual Conference and Corrosion Show Held in Nashville, Tennessee, April 27-May 1, 1992.

[175] L.S. Boak, G.M. Graham, K.S. Sorbie, The influence of divalent cations on the performance of BaSO4 scale inhibitor species, in: SPE 50771, SPE International Symposium on Oilfield Chemistry, Houston, Texas, 16-19 February, 1999.

[176] A.T. Kan, G. Fu, M.B. Tomson, M. Al-Thubaiti, A.J. Xiao, Factors affecting scale inhibitor retention in carbonate-rich formation during squeeze treatment, SPE J. 9 (2004) 280–289, https://doi.org/10.2118/80230-PA.

[177] N. Fleming, K. Ramstad, A.C. Nelson, S. Kidd, Innovative use of kaolinite in downhole scale management: squeeze life enhancement and water shutoff, in: Paper Presented at the SPE International Oilfield Scale Conference, Aberdeen, UK, May, 2008, https://doi.org/10.2118/113656-MS.

[178] K. Cheremisov, D. Oussoltsev, K.K. Butula, A. Gaifullin, I. Faizullin, D. Senchenko, First Application of Scale Inhibitor During Hydraulic Fracturing Treatments in Western Siberia, Society of Petroleum Engineers, 2008, January 1, https://doi.org/10.2118/114255-MS.

[179] O.V. Levanyuk, A.M. Overin, A. Sadykov, S. Parkhonyuk, B.R. Lungwitz, P. Enkababian, S. Legeza, A 3-year results of application a combined scale inhibition and hydraulic fracturing treatments using a novel hydraulic fracturing fluid, Russia, in: SPE International Conference on Oilfield Scale, 2012, https://doi.org/10.2118/155243-ms.

[180] D.V.S. Gupta, M. Brown, S. Szymczak, Multiyear scale inhibition from a solid inhibitor applied during stimulation, in: SPE Annual Technical Conference and Exhibition, 2008, https://doi.org/10.2118/115655-ms.

[181] A.M. Fitzgerald, L.G. Cowie, A history of frac-pack scale-inhibitor deployment, in: Paper Presented at the SPE International Symposium and Exhibition on Formation Damage Control, Lafayette, Louisiana, USA, February, 2008, https://doi.org/10.2118/112474-MS.

[182] M.M. Jordan, Simultaneous well stimulation and scale squeeze treatments in sandstone and carbonate reservoirs, in: SPE International Production and Operations Conference & Exhibition, 11–14 November, Doha, Qatar, 2012, https://doi.org/10.2118/160703-MS.

[183] D. Patterson, M. Kendrick, W. Williams, M.M. Jordan, "Squimulation"—simultaneous well-stimulation and scale-squeeze treatments in Deepwater West Africa, SPE Prod. Oper. 28 (1) (2013) 55–66, https://doi.org/10.2118/15186 3-PA.

[184] M. Marquez, L.A. Schafer, W.D. Norman, Preemptive scale management: treating with scale inhibitor while frac packing a well, in: SPE Annual Technical Conference and Exhibition, 30 October-2 November, Denver, Colorado, USA, 2011, https://doi.org/10.2118/14700 6-MS.

[185] A. Kamispayev, E. Gazizova, A. Skopich, T. Munbayev, E. Neubauer, J. Clarke, Successful application of scale squeeze inhibitor in Tengiz field, in: Paper Presented at the SPE Annual Caspian Technical Conference, Baku, Azerbaijan, October, 2019, https://doi.org/10.2118/198352-MS.

[186] K.U. Raju, Successful scale mitigation strategies in Saudi Arabian oil fields, in: Paper Presented at the SPE International Symposium on Oilfield Chemistry, The Woodlands. Texas, April, 2009, https://doi.org/10.2118/121679-MS.

[187] M. Norris, D. Perez, H.M. Bourne, S.M. Heath, Maintaining fracture performance through active scale control, in: International Symposium on Oilfield Scale, 2001, https://doi.org/10.2118/68300-ms.

[188] A.M. Fitzgerald, L.G. Cowie, A history of Frac-pack scale-inhibitor deployment, in: SPE International Symposium and Exhibition on Formation Damage Control, 2008, https://doi.org/10.2118/112474-ms.

[189] O.M. Selle, F. Haavind, M.H. Haukland, A. Moen, C.F. Hals, K. Oien, G. Clark, Downhole scale control on Heidrun field using scale inhibitor impregnated gravel, in: SPE International Conference on Oilfield Scale, 2010, https://doi.org/10.2118/130788-ms.

[190] M.M. Jordan, J. Buckman, C.J. Johnston, Direct monitoring of the performance of scale control programs across the produced water life cycle via suspended solids analysis, in: SPE—European Formation Damage Conference, Proceedings, EFDC, Society of Petroleum Engineers (SPE), 2005, pp. 29–47, https://doi.org/10.2118/93892-ms.

[191] A. Alabi, M. Chiesa, C. Garlisi, G. Palmisano, Advances in anti-scale magnetic water treatment, Environ. Sci.: Water Res. Technol. (2015), https://doi.org/10.1039/c5ew00052a.

[192] L. Lin, W. Jiang, X. Xu, et al., A critical review of the application of electromagnetic fields for scaling control in water systems: mechanisms, characterization, and operation, npj Clean Water 3 (2020) 25, https://doi.org/10.1038/s41545-020-0071-9.

[193] F.F. Farshad, J. Linsley, O. Kuznetsov, S. Vargas, The Effects of Magnetic Treatment on Calcium Sulfate Scale Formation, Society of Petroleum Engineers, 2002, January 1, https://doi.org/10.2118/76767-MS.

[194] L. Zhao, K. Ma, Z. Yang, Changes of water hydrogen bond network with different externalities, Int. J. Mol. Sci. (2015), https://doi.org/10.3390/ijms16048454.

[195] E. Chibowski, A. Szcześ, L. Hołysz, Influence of magnetic field on evaporation rate and surface tension of water, Colloids Interfaces 2 (4) (2018) 68, https://doi.org/10.3390/colloids2040068.

[196] N.I. Gamayunov, Coagulation of suspension after magnetic treatment., J. Appl. Chem. U.S.S.R. 56 (1983) 975–982.

[197] J.S. Baker, S.J. Judd, Magnetic amelioration of scale formation, Water Res. 30 (1996) 247–260.

[198] F. Alimi, M.M. Tlili, M.B. Amor, G. Maurin, C. Gabrielli, Effect of magnetic water treatment on calcium carbonate precipitation: influence of the pipe material, Chem. Eng. Process. 48 (2009) 1327–1332.

[199] J.D. Donaldson, S.M. Grimes, Control of Scale in Sea Water Applications by Magnetic Treatment of Fluids, Society of Petroleum Engineers, 1987, January 1, https://doi.org/10.2118/16540-MS.

[200] J.M.D. Coey, Magnetic water treatment-how might it work? Philos. Mag. 92 (2012) 3857–3865, https://doi.org/10.1080/14786435.2012.685968.

[201] A. Szkatula, M. Balanda, M. Kopec, Magnetic treatment of industrial water. Silica activation, Eur. Phys. J. Appl. Phys. 18 (2002) 41–49, https://doi.org/10.1051/epjap:2002025.

[202] S. Kobe, G. Draži, P.J. McGuiness, J. Stražišar, The influence of the magnetic field on the crystallisation form of calcium carbonate and the testing of a magnetic water-treatment device, J. Magn. Magn. Mater. 236 (1–2) (2001) 71–76, https://doi.org/10.1016/S0304-8853(01)00432-2.

[203] J.F. Grutsch, J.W. McClintock, Controlling scale by magnetic treatment, Report submitted to the American Petroleum Institute, Committee on Refinery Environmental Conservation, American Petroleum Institute, Washington, DC, 1982.

[204] J.F. Grutsch, J.W. McClintock, Deposit control in alkaline cooling water using magnetic water treatment at Amoco's largest refinery, in: CORROSION 84, National Association of Corrosion Engineers, Houston, TX, 1984 (Paper 330).

[205] A.L. Martins, B.B. Castro, H.E.P. Schluter, M.V.D. Ferreira, A.R.A. Achy, I.M. Pepe, Offshore field experience with non chemical oilfield scale prevention/remediation strategies in Brazil, in: Paper Presented at the Offshore Technology Conference, Houston, Texas, USA, May, 2020, https://doi.org/10.4043/30552-MS.

[206] A.M. Pritchard, W. Birch, S. Brabon, Assessing the Effect of the Application of Magnetic Treatment for the Reduction of Oilfield Scale, NACE International, 2000, January 1.

[207] K.W. Smothers, C.D. Curtiss, V.F. Hock, Demonstration and Evaluation of Magnetic Descalers, NACE International, 2003, January 1.

[208] J.H. Sun, B.Q. Liu, Application of high pressure static resistance scale technology in steam turbine condenser of power plant, Power Syst. Eng. 19 (5) (2003) 35.

[209] Q.S. Ye, S.L. Hu, Application of ion-stick electrostatic water treating device in industrial water treatment, Fuel Chem. Process. 28 (2) (1997) 101–103.

[210] K.Z. Liu, Application of high voltage electrostatic ionizing bar in circulating water system of metallurgical industry, Energy Conserv. 3 (2010) 25–27.

[211] W. Wang, Y. Wu, Z. Chang, F. Chen, H. Wang, G. Gu, Z.L. Wang, Self-powered intelligent water meter for electrostatic scale preventing, rust protection, and flow sensor in a solar heater system, ACS Appl. Mater. Interfaces 11 (6) (2019) 6396–6403, https://doi.org/10.1021/acsami.8b19683.

[212] F.A. Setta, A. Neville, Efficiency assessment of inhibitors on CaCO3 precipitation kinetics in the bulk and deposition on a stainless steel surface (316L), Desalination 281 (2011) 340–347.

[213] W.C. Cheong, P.H. Gaskell, A. Neville, Substrate effects on surface adhesion/crystallisation of calcium carbonate, J. Cryst. Growth 363 (2012) 7–21.

[214] F.A. Setta, H. Chen, A. Neville, A Surface Kinetic Scaling Model for CaCO3 on a Stainless Steel Surface (316L), NACE, October 2012.

[215] W.C. Cheong, A. Neville, P.H. Gaskell, S. Abott, Using nature to provide solutions to calcareous scale deposition, in: SPE 114082, SPE International Oilfield Scale Conference, Aberdeen, UK, 28-29 May, 2008.

[216] M. Li, R. Zhao, S. Ma, T. Yang, Scale deposition inhibiting composites by HDPE/silicified acrylate polymer/nano-silica for landfill leachate piping, Materials 13 (16) (2020), https://doi.org/10.3390/MA13163497.

[217] S. Baraka-Lokmane, T.V.J. Charpentier, A. Neville, C. Hurtevent, J.R. Ordonez-Varela, F. Møller Nielsen, V. Eroini, J.H. Olsen, J.A. Ellingsen, Bache, Comparison of characteristic of anti-scaling coating for subsurface safety valve for use in oil and gas industry, in: International Petroleum Technology Conference 2014, IPTC 2014—Innovation and Collaboration: Keys to Affordable Energy, 3, Society of Petroleum Engineers, 2014, pp. 2150–2165, https://doi.org/10.2523/iptc-17953-ms.

[218] C. Wang, H. Wang, Y. Hu, Z. Liu, C. Lv, Y. Zhu, N. Bao, Anti-corrosive and scale inhibiting polymer based functional coating with internal and external regulation of TiO_2 whiskers, Coatings 8 (2018) 29.

[219] Jansen, et al., Contribution to the "Jaarboek" of the "Mijnbouwkundige Vereeniging", February 2001.

[220] A. Krag, A.M. Fergusson, M. Grainger, Preventing scale deposition downhole using high frequency electromagnetic AC signals from surface to enhance production offshore Denmark, in: Paper Presented at the SPE Annual Technical Conference and Exhibition, Amsterdam, The Netherlands, October, 2014, https://doi.org/10.2118/170898-MS.

[221] R. Martinez, W.L. Parker, E. Hernandez, C. Misino, I.E. Ramírez, An environmentally friendly method to control halite scale deposition: proven case histories from Latin America, in: Paper Presented at the SPE Latin America and Caribbean Petroleum Engineering Conference, Maracaibo, Venezuela, May, 2014, https://doi.org/10.2118/169415-MS.

[222] J. Macadam, S. Parsons, Calcium carbonate scale control, effect of material and inhibitors, Water Sci. Technol. 49 (2) (2004) 153–159.

[223] J. MacAdam, S.A. Parsons, Calcium carbonate scale formation and control, Rev. Environ. Sci. Biotechnol. 3 (2) (2004) 159–169, https://doi.org/10.1007/s11157-004-3849-1.

[224] S.M. Proctor, Scale dissolver development and testing for HP/HT systems, in: SPE 60221 (Paper Presented at the SPE International Symposium on Oilfield Scale, Aberdeen, UK, 26–27 January, 2000.

[225] A.S. Metcalf, C.R. Parker, J.L. Boles, Acetic acid demonstrates greater carbonate dissolution than typically expected, J. Can. Pet. Technol. 44 (12) (2005) 22–24, https://doi.org/10.2118/05-12-TN.

[226] M. Al-Khaldi, H.A. Nasr-El-Din, S. Metha, A. Aamri, Reaction of citric acid with calcite, Chem. Eng. Sci. 62 (2007) 5880.

[227] T. Huang, P.M. McElfresh, A.D. Gabrysch, Acid removal of scale and fines at high temperatures, in: SPE 74678, Paper Presented at the SPE Oilfield Scale Symposium, Aberdeen, UK, 30–31 January, 2002.

[228] T. Chen, P. Chen, H. Montgomerie, T. Hagen, B. Juliussen, T. Haaland, Development of low corrosive environmental friendly calcium carbonate scale dissolver for HTHP wells, in: NACE—International Corrosion Conference Series, 2010.

[229] O.M. Selle, M.S. Rex, H.N. Wat, A. Mebratu, Gelled organic acid system for improved $CaCO_3$ removal in horizontal openhole wells at the heidrun field, in: Proceedings—SPE Annual Technical Conference and Exhibition, Society of Petroleum Engineers (SPE), 2004, pp. 2335–2350, https://doi.org/10.2523/90359-ms.

[230] A. Mohamed, S. Elkatatny, S. Basfar, One-stage calcium carbonate oil-based filter cake removal using a new biodegradable acid system, in: Paper Presented at the SPE Kuwait Oil & Gas Show and Conference, Mishref, Kuwait, October, 2019, https://doi.org/10.2118/198045-MS.

[231] A. Mohamed, S. Elkatatny, A. Al-Majed, Removal of calcium carbonate water-based filter cake using a green biodegradable acid, Sustainability 12 (3) (2020), https://doi.org/10.3390/su12030994.

[232] M. Rietjens, Sense and non-sense about acid-induced sludge, in: SPE—European Formation Damage Control Conference, Proceedings, Society of Petroleum Engineers (SPE), 1997, pp. 195–210, https://doi.org/10.2118/38163-ms.

[233] C.N. Fredd, H. Scott Fogler, Alternative stimulation fluids and their impact on carbonate acidizing, SPE J. 3 (1) (1998) 34–40, https://doi.org/10.2118/31074-pa.

[234] C.M. Shaughnessy, W.E. Kline, EDTA Removes Formation Damage at Prudhoe Bay, Society of Petroleum Engineers, 1983, https://doi.org/10.2118/11188-PA.

[235] T.N. Tyler, R.R. Metzger, L.R. Twyford, Analysis and treatment of formation damage at Prudhoe Bay, ALASKA, J. Pet. Technol. 37 (1985) 1010–1018, https://doi.org/10.2118/12471-pa.

[236] M.W. Bodine, T.H. Fernalld, EDTA dissolution of gypsum, anhydrite, and Ca-Mg carbonates, J. Sediment. Petrol. 43 (4) (1973) 1152–1156, https://doi.org/10.1306/74D72932-2B21-11D7-8648000102C1865D.

[237] M. Tavakkoli, C.A. Sung, J. Kuang, A. Chen, J. Hu, F.M. Vargas, Effect of carbon steel corrosion on asphaltene deposition, Energy Fuel 33 (5) (2019) 3808–3815, https://doi.org/10.1021/acs.energyfuels.8b03238.

[238] K.D. Demadis, P. Lykoudis, R.G. Raptis, G. Mezei, Phosphonopolycarboxylates as chemical additives for calcites scale dissolution and metallic corrosion inhibition based on a calcium–phosphonotricarboxylate organic–inorganic hybrid, Cryst. Growth Des. 6 (5) (2006) 1064.

[239] R.J. Scholze, M.T. Prakash, K. Rainwater, J. Harwood, Application of carbon dioxide treatment to manage scale in a building potable water system, USA-CERL Technical Report, U.S. Army Corps of Engineers Construction Engineering Research Laboratory, Champaign, IL, 1989.

[240] T.M. Prakask, C.H. Neff, R.H. Scholze, J.T. Bandy, E.D. Smith, Carbon dioxide treatment for scale removal and control in potable water systems, USA-CERL Technical Report n-87/16, U.S. Army Corps of Engineers Construction Engineering Research Laboratory, Champaign, IL, 1987.

[241] Z. Li, G.P. Demopoulos, Solubility of $CaSO_4$ phases in aqueous $HCl + CaCl_2$ solutions from 283 K to 353 K, J. Chem. Eng. Data 50 (6) (2005) 1971–1982, https://doi.org/10.1021/je050217e.

[242] P.J. Lucchesi, E.D. Whitney, Solubility of strontium sulphate in water and aqueous solutions of hydrogen chloride, sodium chloride, sulphuric acid and sodium sulphate by the radiotracer method, J. Appl. Chem. 12 (6) (1962) 277–279, https://doi.org/10.1002/jctb.5010120607.

[243] A. Vishwanath Shende, Dissolution of Barite Scale Using Chelating Agents (Master thesis), Texas A&M University, 2012.

[244] S. Muthaiah, A. Bhatia, M. Kannan, Stability of metal complexes, in: Stability and Applications of Coordination Compounds, IntechOpen, 2020, https://doi.org/10.5772/intechopen.90894.

[245] T. Almubarak, J.H. Ng, H. Nasr-El-Din, Chelating agents in productivity enhancement: a review, in: Society of Petroleum Engineers—SPE Oklahoma City Oil and Gas Symposium 2017, Society of Petroleum Engineers, 2017, pp. 88–112, https://doi.org/10.2118/185097-ms.

[246] A. Hassan, M. Mahmoud, I.A. Hussein, Applications of chelating agents in the upstream oil and gas industry: a review, Energy Fuel (2020), https://doi.org/10.1021/acs.energyfuels.0c03279.

[247] C.F. Smith, T.J. Nolan, P.L. Crenshaw, Removal and inhibition of calcium sulfate scale in waterflood projects, J. Pet. Technol. 20 (11) (1968) 1249–1256, https://doi.org/10.2118/1957-pa.

[248] J. Charleston, Scale removal in the Virden, Manitoba, area, J. Pet. Technol. 22 (1970) 701–704, https://doi.org/10.2118/2160-pa.

[249] M. Cikes, et al., A successful treatment of formation damage caused by high-density brine, SPE Prod. Eng. (May) (1990) 175–179.

[250] R.W. Schutte, Designing Gyp Scale Removal Treatments Based on Chemical Characteristics to Optimize Field Performance, 1987, https://doi.org/10.2523/16888-ms.

[251] S. Ren, H. Xu, F. Shen, The application of using EDTA to dissolve calcium sulphate deposits for plugged ESP wells in Rumaila Oilfield, in: IOP Conference Series: Earth and Environmental Science, Institute of Physics Publishing, 2018, https://doi.org/10.1088/1755-1315/170/2/022081.

[252] G. Garcia-Olvera, O. Galicia-Lopez, E. Santillán-Mirsaydi, A novel solution to remove carbonate and sulfate scale in electric submersible pumps, offshore oil wells in the gulf of Mexico, in: Society of Petroleum Engineers—SPE International Oilfield Scale Conference and Exhibition 2018, Society of Petroleum Engineers, 2018, https://doi.org/10.2118/190725-ms.

[253] H. de Vries, F. Arnaud, The first successful downhole barium sulphate scale dissolving operation, in: Paper SPE 26704-MS Presented at the Offshore Europe, Aberdeen, United Kingdom, 7-10 September, 1993, https://doi.org/10.2118/26704-MS.

[254] J.S. Rhudy, Removal of mineral scale from reservoir core by scale dissolver, in: SPE 25161, Paper Presented at the SPE International Symposium on Oilfield Chemistry, New Orleans, LA, 2–5 March, 1993.

[255] A. Putnis, C.V. Putnis, J.M. Paul, The Efficiency of a DTPA-Based Solvent in the Dissolution of Barium Sulfate Scale Deposits, Society of Petroleum Engineers, 1995, January 1, https://doi.org/10.2118/29094-MS.

[256] C. Putnis, M. Kowacz, A. Putnis, The mechanism and kinetics of DTPA-promoted dissolution of barite, Appl. Geochem. 23 (9) (2008) 2778–2788, https://doi.org/10.1016/j.apgeochem.2008.07.006.

[257] K.S. Wang, Y. Tang, T.F. Yen, Effects of scale dissolvers on barium sulfate deposits: a macroscopic and microscopic study, in: NACE—International Corrosion Conference Series, National Association of Corrosion Engineers International, 2002.

[258] I. Lakatos, J. Lakatos-Szabó, B. Kosztin, Optimization of barite dissolvers by organic acids and pH regulation, in: SPE Oilfield Scale Symposium, Society of Petroleum Engineers (SPE), 2002, pp. 147–156, https://doi.org/10.2118/74667-ms.

[259] I. Lakatos, J. Lakatos-Szabó, B. Kosztin, Comparative study of different barite dissolvers: technical and economic aspects, in: Proceedings—SPE International Symposium on Formation Damage Control, Society of Petroleum Engineers (SPE), 2002, pp. 191–199, https://doi.org/10.2118/73719-ms.

[260] G.A.P. Abib, G.F. Da Cruz, A.S.L. Vaz Junior, Study of barium sulfate dissolution by scale dissolver based on solutions of Dtpa, An. Acad. Bras. Cienc. 90 (3) (2018) 3185–3196, https://doi.org/10.1590/0001-37652018 20170728.

[261] B. Bageri, M. Mahmoud, A.S. Abdulraheem, Evaluation of barium sulfate (barite) solubility using different chelating agents at a high temperature, J. Pet. Sci. Technol. 7 (2017) 42–56.

[262] J.M. Paul, E.R. Fieler, New solvent for oilfield scales, in: Proceedings—SPE Annual Technical Conference and Exhibition, Society of Petroleum Engineers of AIME, 1992, pp. 845–855, https://doi.org/10.2523/24847-ms.

[263] B.S. Bageri, Filter Cake Removal of Barite Water-Based Mud (Ph.D. thesis), King Fahd University of Petroleum and Minerals, Saudi Arabia, 2016.
[264] M.A. Mahmoud, B.S. Ba geri, K. Abdelgawad, M.S. Kamal, I.A. Hussein, S. Elkatatny, et al., Evaluation of the reaction kinetics of diethylenetriaminepentaacetic acid chelating agent and a converter with barium sulfate (barite) using a rotating disk apparatus, Energy Fuel 32 (2018) 9813–982110, https://doi.org/10.1021/acs.energyfuels.8b02332.
[265] I. Lakatos, J. Lakatos-Szabó, Potential of different polyamino carboxylic acids as barium and strontium sulfate dissolvers, in: SPE—European Formation Damage Conference, Proceedings, EFDC, Society of Petroleum Engineers (SPE), 2005, pp. 237–244, https://doi.org/10.2118/94633-ms.
[266] M. Blanco, Y. Tang, P. Shuler, W.A. Goddard, Activated complex theory of barite scale control process. III, Mol. Eng. 7 (1997) 491–514.
[267] D. Coll de Pasquali, E. Horai, M.I. Real, S. Castro, F.B. Dunn, G. Gunawan, H.M. Azam, J.M. Wilson, Chemical dissolution of oilfield strontium sulfate (SrSO4) scale by chelating agents, Appl. Geochem. (2019), https://doi.org/10.1016/j.apgeochem.2019.05.004.
[268] M.H. Al-Khaldi, A.M. Al-Juhani, M.N. Gurmen, New insights into the removal of calcium sulfate scale, in: Society of Petroleum Engineers—9th European Formation Damage Conference 2011, Society of Petroleum Engineers, 2011, pp. 1052–1070, https://doi.org/10.2118/144158-ms.
[269] H.K. Kotlar, O.M. Selle, F. Haavind, A "standardized" method for ranking of scale dissolver efficiency. A case study from the Heidrun field, in: Presented at the International Symposium on Oilfield Scale, Aberdeen, United Kingdom, 30–31 January. SPE-74668-MS, 2002, https://doi.org/10.2118/74668-MS.
[270] B.S. Bageri, M.A. Mahmoud, R.A. Shawabkeh, S.H. Al-Mutairi, A. Abdulraheem, Toward a complete removal of barite (barium sulfate $BaSO_4$) scale using chelating agents and catalysts, Arab. J. Sci. Eng. (2017), https://doi.org/10.1007/s13369-017-2417-2.
[271] Z. Luo, N. Zhang, H. Ji, A chelating agent system for the removal of barium sulfate scale, J. Pet. Explor. Prod. Technol. 10 (2020) 3069–3079, https://doi.org/10.1007/s13202-020-00886-5.
[272] K. Robards, P.R. Haddad, P.E. Jackson, High-performance liquid chromatography—separations, in: Principles and Practice of Modern Chromatographic Methods, Elsevier, 2004, pp. 305–380, https://doi.org/10.1016/b978-0-08-057178-2.50009-1.
[273] A. Van Zon, F. de Jong, G.J. Torny-Schutte, Dissolving Barium Sulfate Scale With Aqueous Solutions of Salts of Phosphomethyl and Amino-Substituted Macrocyclic Polyethers, Shell, United States, 1981.
[274] F. De Jong, D.N. Reinhoudt, G. Torny-Schutte, U.S. Patents 4215000 and 4190462, 1980.
[275] N.A. Thiele, S.N. MacMillan, J.J. Wilson, Rapid dissolution of $BaSO_4$ by macropa, an 18-membered macrocycle with high affinity for Ba^2, J. Am. Chem. Soc. 140 (49) (2018) 17071–17078, https://doi.org/10.1021/jacs.8b08704.
[276] E.H. Riad, Soda ash as gypsum dissolver in 8 inch production line case study Gemsa Field, Gulf of Suez, Egypt, in: SPE 142457, Paper Presented at the SPE European Formation Damage Conference, Noordwijk, Netherlands, 7–10 June, 2011.
[277] S.V. Golubev, P. Bénézeth, A. Castillo, Siderite dissolution kinetics in acidic aqueous solutions from 25 to 100°C and 0 to 50 atm pCO2, Chem. Geol. 265 (2009) 13–19, https://doi.org/10.1016/j.chemgeo.2008.12.031.
[278] V. Fajardo, B. Brown, S. Nešić, Study of the solubility of iron carbonate in the presence of acetic acid using an EQCM, in: NACE—International Corrosion Conference Series, 2013.
[279] M. Shailaja, S.V. Narasimhan, Mechanism of oxide scale removal during dilute chemical decontamination of carbon steel surfaces, J. Nucl. Sci. Technol. 30 (1993) 524–532, https://doi.org/10.1080/18811248.1993.9734514.
[280] R.P. Kreh, W.L. Henry, V.R. Kuhn, The use of chelants and dispersants for prevention and removal of rust scale, in: Mineral Scale Formation and Inhibition, Springer, US, 1995, pp. 145–155, https://doi.org/10.1007/978-1-4899-1400-2_13.

[281] J.A. Gray, K. Heinz, A neutral pH process for on-line cleaning and passivation of hot and chilled water systems, in: Paper No. 446, NACE CORROSION 91, Cincinnati, OH, March, 1991.

[282] J. Yap, M.J. Fuller, L.A. Schafer, S. Kelkar, Removing iron sulfide scale: a novel approach, in: Society of Petroleum Engineers—14th Abu Dhabi International Petroleum Exhibition and Conference 2010, ADIPEC 2010, 3, Society of Petroleum Engineers, 2010, pp. 1921–1930, https://doi.org/10.2118/138520-ms.

[283] M.M. Brezinski, Chelating agents in sour well acidizing: methodology or mythology, in: Paper SPE 54721 Presented at the 1999 SPE European Formation Damage Conference Held in The Hague, The Netherlands, 31 May-01 June, 1999.

[284] K.C. Taylor, H.A. Nasr-El Din, J.A. Saleem, Laboratory evaluation of iron-control chemicals for high temperature sour-gas wells, in: Paper SPE 65010 Presented at the 2001 SPE International Symposium on Oilfield Chemistry held in Houston, Texas, 13–16 February, 2001.

[285] H.A. Nasr El-Din, A.Y. Al-Humaidan, Iron sulfide scale: formation, removal, and prevention, in: Paper SPE 68315 Presented at the 2001 SPE International Symposium on Oilfield Scale Held in Aberdeen, UK, 30–31 January, 2001.

[286] M.L. Walker, W.R. Dill, M.R. Besler, D.G. McFatridge, Iron control in West Texas sour-gas wells provides sustained production increases, J. Pet. Technol. 43 (1991) 603.

[287] J. Leal, et al., A systematic approach to remove iron sulfide scale: a case history, in: Paper SPE 105607 Presented at the 2007 SPE Middle East Oil and Gas Show and Conference Held in Kingdom of Bahrain, 11–14 March, 2007.

[288] Q. Wang, H. Ajwad, J.D. Lynn, Iron sulfide scale dissolvers: how effective are they? in: Society of Petroleum Engineers—SPE Saudi Arabia Section Technical Symposium and Exhibition 2013, Society of Petroleum Engineers, 2013, pp. 75–96, https://doi.org/10.2118/168063-ms.

[289] S. Elkatatny, New formulation for iron sulfide scale removal, in: SPE Middle East Oil and Gas Show and Conference, MEOS, Proceedings, Society of Petroleum Engineers (SPE), 2017, pp. 1809–1821, https://doi.org/10.2118/183914-ms.

[290] W.G.F. Ford, M.L. Walker, M.P. Halterman, D.L. Parker, D.G. Brawley, R.G. Fulton, Removing a typical iron sulfide scale: the scientific approach, in: Paper Presented at the SPE Rocky Mountain Regional Meeting, Casper, Wyoming, May, 1992, https://doi.org/10.2118/24327-MS.

[291] J. Leal, J.R. Solares, G. Izquierdo, A systematic approach to remove iron sulphide scale: a case history, in: SPE Middle East Oil and Gas Show and Conference, MEOS, Proceedings, 2007, pp. 1322–1331, https://doi.org/10.2523/105607-ms.

[292] H.A. Nasr-El-Din, B.A. Fadhel, A.Y. Al-Humaidan, W.W. Frenier, D. Hill, An experimental study of removing Iron sulfide scale from well tubulars, in: Society of Petroleum Engineers—International Symposium on Oilfield Scale 2000, Society of Petroleum Engineers, 2000, https://doi.org/10.2523/60205-ms.

[293] C. Reed, J. Foshee, J.E. Penkala, M. Roberson, Acrolein application to mitigate biogenic sulfides and remediate injection well damage in a gas plant water disposal system, in: SPE 93602, Paper Presented at the SPE International Symposium on Oilfield Chemistry, The Woodlands, TX, 2–4 February, 2005.

[294] J. Penkala, M.D. Law, D.D. Horaska, A.L. Dickinson, Acrolein 2-Propenal: a versatile microbiocide for control of bacteria in oilfield systems, in: Paper 04749, Paper Presented at the NACE CORROSION Conference, 2004.

[295] J.E. Penkala, C.A. Reed, J.D. Foshee, Acrolein application to mitigate biogenic sulfides and remediate injection-well damage in a gas-plant water-disposal system, in: Proceedings—SPE International Symposium on Formation Damage Control, Society of Petroleum Engineers (SPE), 2006, pp. 123–131, https://doi.org/10.2523/98067-ms.

[296] D.D. Horaska, J.E. Penkala, C.A. Reed, S.H. Gaffney, M.M. Srour, A.S. Al-Harthy, Field experiences detailing Acrolein (2-propenal) treatment of a produced water injection system in the Sultanate of Oman, in:

SPE Middle East Oil and Gas Show and Conference, MEOS, Proceedings, 2, 2009, pp. 768–780, https://doi.org/10.2118/120238-ms.

[297] P.D. Gilbert, J.M. Grech, R.E. Talbot, M.A. Veale, K.A. Hernandez, Tetrakishydroxymethylphosphonium sulfate (THPS) for dissolving iron sulfides downhole and topside—a study of the chemistry influencing dissolution, in: NACE—International Corrosion Conference Series. Vol. 2002-April, National Association of Corrosion Engineers International, 2002.

[298] J.J. Wylde, I.G. Winning, Challenges and solutions associated with the development of a Iron scale dissolver chemistry, in: Paper Presented at the CORROSION 2004, New Orleans, Louisiana, March 2004 (Paper no. 04730).

[299] R.E. Talbot, C.R. Jones, J. Grech, Treatment of Iron Sulphide Deposits, 2007. US 20070108127: A1.

[300] M.W. Jawish, R. Suleiman, Q. Wang, T. Chen, S.W. Shen, New formulations for the control of iron sulfide deposits in oil production facilities, in: Society of Petroleum Engineers—SPE Kingdom of Saudi Arabia Annual Technical Symposium and Exhibition 2018, SATS 2018, Society of Petroleum Engineers, 2018, https://doi.org/10.2118/192228-ms.

[301] P.R. Rincon, J.P. McKee, C.E. Tarazon, L.A. Guevara, B. Vinccler, Biocide stimulation in oilwells for downhole corrosion control and increasing production, in: SPE 87562, Paper Presented at the SPE International Symposium on Oilfield Corrosion, Aberdeen, UK, 28 May, 2004.

[302] H.A. Nasr-El-Din, A.M. Al-Mohammad, M.A. Al-Hajri, J.B. Chesson, A new chemical treatment to remove multiple damages in a water supply well, in: SPE 95001, Paper Presented at the SPE European Formation Damage Conference, Scheveningen, Netherlands, 25–27 May, 2005.

[303] J. Romaine, T.J. Strawser, M.L. Knippers, Application of chlorine dioxide as an oilfield-facilities-treatment fluid, SPE Prod. Facil. 11 (1996) 18–21, https://doi.org/10.2118/29017-PA.

[304] J.F. McCafferty, E.W. Tate, D.A. Williams, Field performance in the practical application of chlorine dioxide as a stimulation enhancement fluid, SPE Prod. Facil. 8 (1993) 9–14, https://doi.org/10.2118/20626-PA.

[305] P.A. Vella, J.L. Nickerson, Hazardous material decontamination with potassium permanganate for refinery turnarounds, in: NACE—International Corrosion Conference Series, National Association of Corrosion Engineers International, 1998.

[306] P.A. Vella, Improved cleaning method safely removes pyrophoric iron sulfide, Oil Gas J. 95 (1997) 65–68.

[307] T.R. McCartney, R. Shank, Application of stabilized hydrogen peroxide for the removal of iron polysulfide deposits, in: Proceeding of International Conference on Heat Exchanger Fouling and Cleaning 2013, Budapest, Hungary, 2013.

[308] L.D. Martin, Methods and Compositions for Removing Deposits Containing Iron Sulfide From Surfaces Comprising Basic Aqueous Solutions of Particular Chelating Agents, 1981. US Patent No. 4276185.

[309] M.A. Mahmoud, M. Kamal, B.S. Bageri, I. Hussein, Removal of pyrite and different types of iron sulfide scales in oil and gas wells without H2S generation, in: International Petroleum Technology Conference, 2015, December 6, https://doi.org/10.2523/18279-MS.

[310] C. De Wolf, J.N. LePage, H.A. Nasr-El-Din, et al., Ammonium Salts of Chelating Agents and Their Use in Oil and Gas Field Applications, 2012. International Patent Application WO/2012/080299.

[311] R. Ramanathan, H.A. Nasr-El-Din, A comparative experimental study of alternative iron sulfide scale dissolvers in the presence of oilfield conditions and evaluation of new synergists to aminopolycarboxylic acids, SPE J. 26 (2021) 1–23, https://doi.org/10.2118/205005-pa.

[312] H. Gamal, S. Elkatatny, D. Al Shehri, M. Bahgat, A novel low-temperature non-corrosive sulfate/sulfide scale dissolver, Sustainability 12 (6) (2020), https://doi.org/10.3390/su12062455.

[313] T. Hafiz, M. Hoegerl, A. Al Suwaij, A. Almathami, Synthetic iron sulfide scale and polymeric scale dissolvers, in: SPE Middle East Oil and Gas Show and Conference, MEOS, Proceedings, 2017-March, Society of Petroleum Engineers (SPE), 2017, pp. 2723–2727, https://doi.org/10.2118/183954-ms.

[314] X. Wang, Q. Qu, S. Berry, J. Cutler, B. Hughes, Iron sulfide removal: a nonacidic alternative to hydrochloric acid treatment, in: SPE—European Formation Damage Conference, Proceedings, EFDC, vol. 2, 2013, pp. 1214–1226, https://doi.org/10.2118/165199-ms.

[315] M. Ahmed, I.A. Hussein, A.T. Onawole, M.A. Saad, Development of a new borax-based formulation for the removal of pyrite scales, ACS Omega 5 (24) (2020) 14308–14315, https://doi.org/10.1021/acsomega.0c00556 (American Chemical Society).

[316] Improved black powder removal from gas pipelines using innovative gelled fluids, traditional cleaning methods and careful candidate selection, Pipeline Rehabilitation and Maintenance Conference, 2005.

[317] S.L. Berry, J.L. Boles, A. Singh, B. Hughes, I. Hashim, Enhancing production by removing zinc sulfide scale from an offshore well: a case history, SPE Prod. Oper. 27 (3) (2012) 318–326, https://doi.org/10.2118/140751-PA.

[318] J.G.R. Eylander, D.M. Frigo, F.A. Hartog, G. Jonkers, A novel methodology for in-situ removal of NORM from E & P production facilities, in: SPE 46791 (Paper Presented at the SPE International Conference on Health, Safety and Environment in Oil and Gas Exploration and Production, Caracas, Venezuela, 7–10 June, 1998.

[319] J.J. Wylde, J.L. Slayer, Halite scale formation mechanisms, removal and control: a global overview of mechanical, process and chemical strategies, in: Proceedings—SPE International Symposium on Oilfield Chemistry, vol. 1, Society of Petroleum Engineers (SPE), 2013, pp. 402–410, https://doi.org/10.2118/164081-ms.

[320] A. Sadiku, E. Marston, D. Roberts, E. Marinou, R. Gijbels, A. Twerda, A. Davis, B. Dros, D. Westera, E. Kremer, E. Isaj, E. Hornstra, G. Visser, J. Wood, K. van Zadelhoff, T. Moss, Management of Salt Deposition in Gas Wells Guideline, Oil and Gas UK (OGUK), Churchyard, London, 2020.

[321] S.A. Gharaibeh, T.R. McCartney, R.A. Shank, A novel non-toxic method for the decontamination of silicate scales—a case study, in: NACE-International Corrosion Conference Series, National Association of Corrosion Engineers International, 2017, pp. 3896–3906.

[322] D.L. Gallup, C.J. Hinrichsen, Control of silicate scales in steam flood operations, in: Society of Petroleum Engineers—9th International Conference on Oilfield Scale 2008—"Managing Scale Through the Field Lifetime", 2008, pp. 151–161, https://doi.org/10.2118/114042-ms.

[323] G. Wilken, Application of alkylnaphthalene absorption oil as a sulfur solvent in sour-gas wells, SPE Prod. Eng. 6 (1991) 137–141, https://doi.org/10.2118/19813-PA.

[324] T.E. Sample Jr., Method for Removing Elemental Sulfur in Sour Gas Wells, 1975. US Patent US3909422A.

[325] H. Gamal, S. Elkatatny, S. Al-Afnan, M. Bahgat, Toward Developing Non-Corrosive Acid System for Complex Scales Removal, Society of Petroleum Engineers (SPE), 2021, https://doi.org/10.2523/iptc-21400-ms.

[326] H. Gamal, S. Elkatatny, S. Al-Afnan, M. Bahgat, Development of a unique organic acid solution for removing composite field scales, ACS Omega 6 (2) (2021) 1205–1215, https://doi.org/10.1021/acsomega.0c04335.

[327] P. Skeldon, Green scaling with black magic, Process. Eng. 71 (7) (1990) 57–63.

[328] J. Sohaili, H.S. Shi, S.K. Muniyandi, Removal of scale deposition on pipe walls by using magnetic field treatment and the effects of magnetic strength, J. Clean. Prod. 139 (2016) 1393–1399, https://doi.org/10.1016/j.jclepro.2016.09.028.

[329] V.A. Prisyazhniuk, Why "magnetic treatment" prevents scale, Energy Explor. Exploit. 23 (2005) 277–288,-https://doi.org/10.1260/014459805775219175.

[330] L.D. Tijing, H.Y. Kim, D.H. Lee, C.S. Kim, Y.I. Cho, Physical water treatment using RF electric fields for the mitigation of $CaCO_3$ fouling in cooling water, Int. J. Heat Mass Transf. 53 (2010) 1426–1437.

[331] L.D. Tijing, D.H. Lee, D.W. Kim, Y.I. Cho, C.S. Kim, Effect of high-frequency electric fields on calcium carbonate scaling, Desalination 279 (2011) 47–53.

[332] Z. Xu, H. Chang, B. Wang, J. Wang, Q. Zhao, Characteristics of calcium carbonate fouling on heat transfer surfaces under the action of electric fields, J. Mech. Sci. Technol. 32 (7) (2018) 3445–3451, https://doi.org/10.1007/s12206-018-0648-0.

[333] D. Georgiou, D. Bendos, M. Kalis, C. Koutis, Removal and/or prevention of limescale in plumbing tubes by a radiofrequency alternating electric field inductance device, J. Water Process. Eng. 22 (2018) 34–40,- https://doi.org/10.1016/j.jwpe.2017.12.013.

[334] D. Heath, B. Širok, B. Pecnik, The use of the cavitation effect in the mitigation of CaCO3 deposits, Stroj Vestn-J Mech E 59 (2013) 203–215, https://doi.org/10.5545/sv-jme.2012.732.

[335] B. Kieser, R. Phillion, S. Smith, T. McCartney, The application of industrial scale ultrasonic cleaning to heat exchangers, in: M.R. Malayeri, A.P. Watkinson, H. Müller-Steinhagen (Eds.), Proceedings of International Conference on Heat Exchanger Fouling and Cleaning—2011 (Peer-Reviewed) June 05–10, 2011, Crete Island, Greece, 2011.

[336] S. Fujikawa, T. Akamatsu, Effects of the non-equilibrium condensation of vapour on the pressure wave produced by the collapse of a bubble in a liquid, J. Fluid Mech. 97 (1980) 481–512.

[337] H. Kunanz, S. Wölfel, Scale removal with ultrasonic waves, in: Paper Presented at the SPE International Oilfield Scale Conference and Exhibition, Aberdeen, Scotland, May, 2014, https://doi.org/10.2118/SPE-169770-MS.

[338] J. Taheri-Shakib, H. Naderi, Y. Salimidelshad, S. Hosseini, The effect of ultrasonic wave on the removal of inorganic scales: Nacl and Kcl, in: Offshore Mediterranean Conference and Exhibition 2017, OMC 2017, Offshore Mediterranean Conference, 2017.

[339] J. Taheri-Shakib, H. Naderi, Y. Salimidelshad, E. Kazemzadeh, A. Shekarifard, Application of ultrasonic as a novel technology for removal of inorganic scales (KCl) in hydrocarbon reservoirs: an experimental approach, Ultrason. Sonochem. 40 (January) (2018) 249–259, https://doi.org/10.1016/j.ultsonch.2017.06.019.

[340] C. Pu, D. Shi, S. Zhao, H. Xu, H. Shen, Technology of removing near wellbore inorganic scale damage by high power ultrasonic treatment, Pet. Explor. Dev. 38 (2) (2011) 243–248, https://doi.org/10.1016/S1876-3804(11)60030-X.

CHAPTER

Gas Hydrate Management

16

16.1 Introduction

Gas hydrate deposition is a costly and challenging problem in oil and gas systems. Hydrate deposition causes production loss and equipment failure, in addition to environmental and safety issues. Hydrate blockages in deep subsea fields are costly to remove: in some cases, annual costs to clear hydrate blockage in lines may exceed $100 million, at a rate of nearly $1 million per mile of affected lines [1]. Therefore, once detected, gas hydrates must be treated quickly and effectively to avoid their risks.

Hydrate control during startup, shutdown, and steady-state production operations can be maintained by the following [2]:

- Understanding of hydrate formation mechanisms and the factors affecting them.
- Good documentation of the hydrate risks, monitoring, and control methods.
- Improvement in system design and optimization of operating parameters.
- Knowledge sharing and staff training.

Like mineral scales, gas hydrate management involves preventing gas hydrate deposits from forming and removing the formed deposits and blockages. These management methods are based on operational, chemical, and nonchemical methods. Fig. 16.1 summarizes the different methods of gas hydrate management.

16.2 Gas hydrate prevention

16.2.1 Operational gas hydrate prevention

Gas hydrates are formed by combining hydrocarbons and water at high pressure and low temperature. Hence, controlling system parameters that are affecting hydrate formation can reduce or eliminate their risk. Operational methods, if successfully implemented, can eliminate the use of chemical methods or significantly reduce the chemical consumption and costs.

Essentials of Flow Assurance Solids in Oil and Gas Operations. https://doi.org/10.1016/B978-0-323-99118-6.00007-1

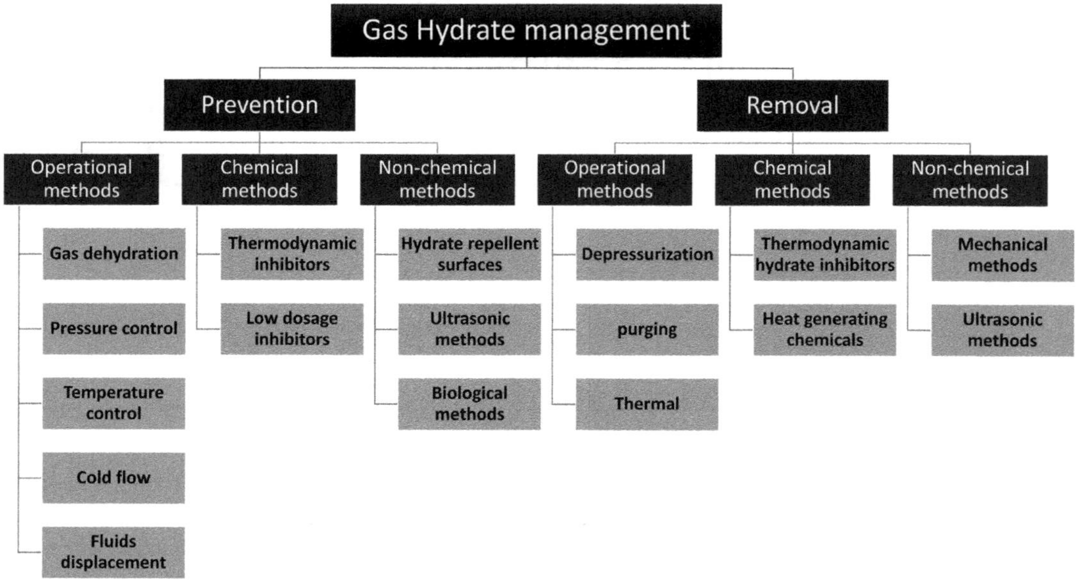

FIG. 16.1

Gas hydrate management methods.

16.2.1.1 Water removal (gas dehydration)

The gas dehydration methods aim to remove water vapor from the gas, thus eliminating a major ingredient from the hydrate recipe and preventing its formation. Water removal is also needed to enhance gas quality, reduce liquid loading in pipelines, reduce acid gases in condensed water, reduce the corrosion risks, and improve the flaring system performance [3,4].

Gas dehydration can be achieved by different techniques, including absorption, adsorption, cooling, membranes, and chemical methods.

Absorption techniques use liquid desiccants (glycols) and are considered the most widely used industrial natural gas dehydration method. Common glycols used in gas dehydration are monoethylene glycol (MEG), diethylene glycol (DEG), and triethylene glycol (TEG), with TEG being the most commonly used in glycol dehydration [5]. A typical gas dehydration unit consists of two sections, namely dehydration (where lean glycol absorbs water from gas) and regeneration (where rich glycol is regenerated) [3,5].

Adsorption techniques use solid desiccants (molecular sieve, on a silica gel or on alumina) to remove water vapor from the gas stream. A minimum of two bed systems are used. Typically one bed dries the gas while the other is being regenerated [4].

Dehydration by cooling/refrigeration is based on condensing the water vapor to turn water molecules into the liquid phase and then remove them from the stream. Cold gas usually carries fewer water molecules than hot warm gas, leading to separating the water from the gas stream.

In the membrane method, gas separation take place by selective permeation [6]. Since water vapor is one of the most permeable gas components as compared to other feed gas components, the method allows membrane units to selectively permeate the water vapor and thus reduce the water content of the resulting product gas [7].

Chemical methods are based on the deliquescing desiccants, which are salts that adsorb water vapor; the water then condenses and dissolves the salt. The water drops down as brine and is removed from the vessel. In the past, the common deliquescing desiccant was calcium chloride ($CaCl_2$) [8].

Subsea water separation and injection represent another solution for hydrate prevention. Subsea separation systems (SSSs) have been used for a long time, with the SUBSIS system in the Troll gas field being the first subsea dehydration and injection in the world. The main processing modules are the horizontal gravity-based separation vessel and the subsea water re-injection pump, with fully automated control system [9]. Other state-of-the-art concepts are British Offshore Engineering Technology (BOET), Kvaerner Booster Station (KBS), Vertical Axial Separation and Pumping System (VASPS), and the more recent developments of PETROBOOST9 and the CoSWaSS JIP [10].

Recently, an integrated system DIPSIS (Deep Integrated Production Separation and Injection System) processing unit was designed to separate the water from the hydrocarbons as close as possible to the subsea production wellheads, with only a maximum of 10% water cut expected downstream of the module. The DIPSIS module has been developed to be part of an antihydrate strategy in deep water at depths of 1500 m or more. The system can be used alone to lower the water percentage in the gas to lower than 10% or to maintain the natural dispersing property of the crude at low water cuts, or it can be used along with chemical additives (KHI or dispersants AA) to mitigate the hydrate problems in subsea, allowing treatment with a low-dosage additive of a reasonable quantity of produced water [11].

Another technique, known as cold flow, is used to eliminate water problems in the gas stream to convert the water into flowable gas hydrates; the technique is used not only to dehydrate the water in the produced gas but also is used in the storage and transport of natural gas [12].

16.2.1.2 Cold flow

Hydrate formed under various flow regimes is described as "slurry-like," "slush-like," or "powder-like," with slurry-like and powder-like hydrates more easily flowable compared to slush-like hydrates, which tend to aggregate. Therefore a potential solution for hydrate deposition is the induction of controlled formation of movable hydrate slurries, which can flow in the pipelines without deposition or causing blockages [13]. The concept is known as cold flow, stabilized flow, or CONWHYP (CONversion of Water to HYdrate Particles). As defined by Straume et al. [14], gas hydrate cold flow is defined as a flow of nonadhesive and noncohesive hydrate particles dispersed in the production fluid flowlines. These hydrates are precipitated in a controlled manner to stay suspended in the solution and to rapidly remove the water from the system to avoid hydrate accumulation [15].

The technique represents a major breakthrough in deepwater hydrate control (although it is not applicable for all oilfield applications) and it reduces or eliminates the use of chemicals, heating, or insulation. This method is also applicable to wax deposits [16,17].

The idea of cold flow is based on the pioneering work of Coberly [18], which showed that the presence of foreign particles decreased the tendency of wax crystals to deposit. Gas hydrates cold flow research started in the late 1990s and continued through the first decade of this century; the Norwegian research institute SINTEF developed and tested their patented cold flow process.

The cold flow approach can be achieved in different ways, including seeding the hydrate particles, or through mixing the cold fluids to produce noncohesive, nonadhesive hydrate particles.

The SINTEF process technology is based on seeding and growing dry hydrate particles and eliminating the availability of free water. SINTEF Petroleum Research developed and patented a concept called SATURN Cold Flow Patented Technology. BP became a partner in it later [15]. Fig. 16.2 illustrates the SINTEF cold flow concept. The hydrate particles are fed into a fast-cooled (shock-chilled)

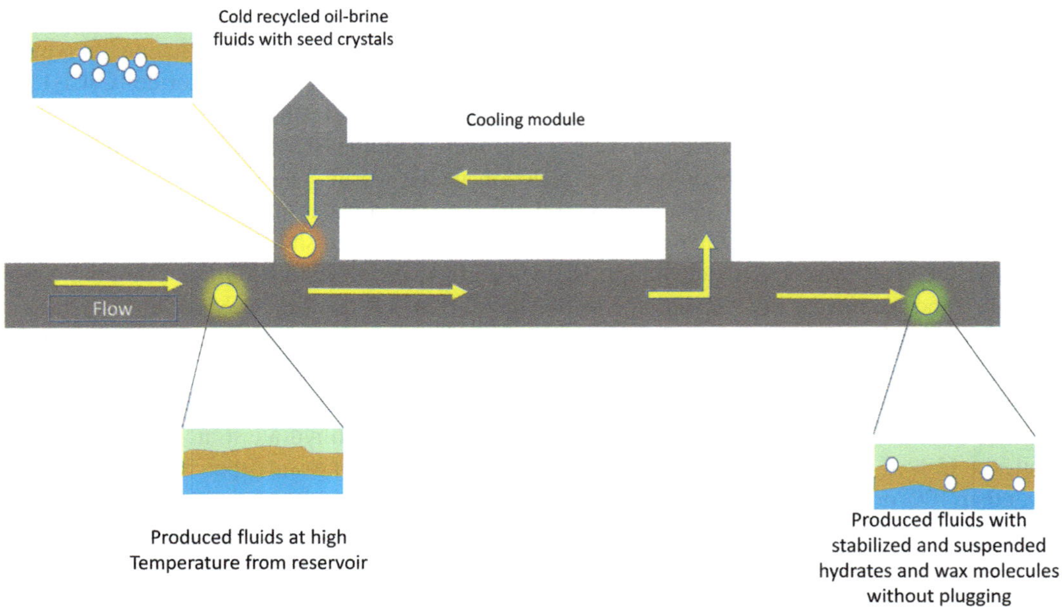

FIG. 16.2

Schematic of the general cold flow concept.

water-containing well stream in which the water is dispersed or emulsified in the liquid hydrocarbons by mixers. The hydrate particles seed further hydrate growth in the water droplets from the inside out and grow quickly and in a controlled manner. The formed hydrate particles are dry, nondepositing, and nonagglomerating, and will eliminate free water from the rest of the transport system [19].

Another concept of the cold flow was developed by the Norwegian University of Science and Technology (NTNU). The NTNU approach considers separation of liquid phases from gas at temperatures above hydrate equilibrium, then cooling the liquid phases, and finally mixing all phases for hydrate formation in a reactor [14]. The hydrate particles will then flow with oil in a three-phase flow together with surplus gas. The main units involved in the NTNU process include wellhead unit, separator unit, heat exchanger unit, and reactor unit [12].

Another concept of cold flow, developed by the Centre for Gas Hydrate Research at Heriot-Watt University, has been patented as HYDRAFLOW. The approach aims to convert all or most of the gas in a hydrocarbon fluid system into hydrates that can be transported in hydrocarbon liquid, water, or a mixture of these two liquid phases, where hydrate control chemicals (antiagglomerants, AAs) may be added to the system to control hydrate crystal size and prevent blockage formation [14]. Water can be added to the system and/or recirculated if this is necessary for conversion of the gas to gas hydrates. HYDRAFLOW is different from the other concepts as it does not recycle hydrates for seeding and controlled hydrate growth, and it is not a "dry" hydrate concept, as hydrate particles are formed in the presence of excess water, and are not agglomerated due to the use of AAs [14,20].

A new approach developed by Empig has two versions, one based on induction heating and the other being a hollow magnetic pig. In both versions, the pipe wall cleaning procedure is planned to operate in a compact cooler module in which warm fluid is cooled to ambient temperatures while hydrates and wax are forming. The cold flow seeding process is implemented by cold fluid with hydrate and wax particles being pumped from the cold outlet and mixed with warm production flow at the inlet of the cooler [14].

Another approach to the cold flow developed by ExxonMobil is based on a once-through method of generating a nonplugging hydrate slurry. This patented method involves the use of static mixers to reduce water droplet diameter in order to facilitate instant conversion of the entire water droplet to hydrates when hydrate formation occurs [14,21].

Since cold flow can convert the water into hydrate particles, which eliminates the free water in the gas stream and can also be used in removing water from gas streams, it is considered as another option for gas dehydration. This approach was patented, as SINTEF has proposed using cold flow as an alternative dehydration process [22], yet it still has not been tested experimentally, but builds on the experimental results from cold flow research with a condensate-like model oil.

Problems with cold flow are [16]:

- Agglomeration of free water (before it reaches the mixers) during unplanned shut-in.
- The technique does not work for high water-cut fields and wet gas fields where hydrate particle flow may not be possible, or it requires additional water or brine to be injected to allow complete gas conversion to hydrates and to make the hydrate slurry transportable in the high water system.
- Possible increased inorganic scale deposition problems.

16.2.1.3 Pressure control

Another way to effectively prevent hydrate blockages is controlling the operating parameters through maintaining the pressure and temperature conditions outside the hydrate formation region [11]. This method requires extensive investigation to completely understand the system design and parameters and fluids composition, and hydrate modeling to generate the hydrate formation/dissociation curve in order to identify the safe operation limits or hydrate-free zones.

Optimizing the system pressure to be lower than the pressure corresponding to the hydrate formation temperature, based on the hydrate formation/dissociation curve, is a common hydrate prevention method. For deep water with an ambient temperature of 4°C, the pressure may need to be 300 psia or less [23]. By using subsea choking and keeping the production flowline at a lower pressure, the difference between hydrate dissociation and operating temperatures (i.e., subcooling) is reduced. This lower subcooling as a result decreases the hydrate formation driving force, which can reduce the hydrate chemical inhibitor consumption [23]. However, operating at low pressures is not an optimum long-term solution, and impractical as it can be uneconomical in many cases and can induce other problems in the system.

16.2.1.4 Heat and temperature control

Maintaining the temperature above the hydrate formation temperature is a common method used to prevent gas hydrate deposits. Additionally, the use of heat and temperature control is also used in preventing and mitigating paraffin deposition, and as a cost-effective method for other flow assurance applications, such as improving flowing fluid viscosity and fluids separation in the topsides [24].

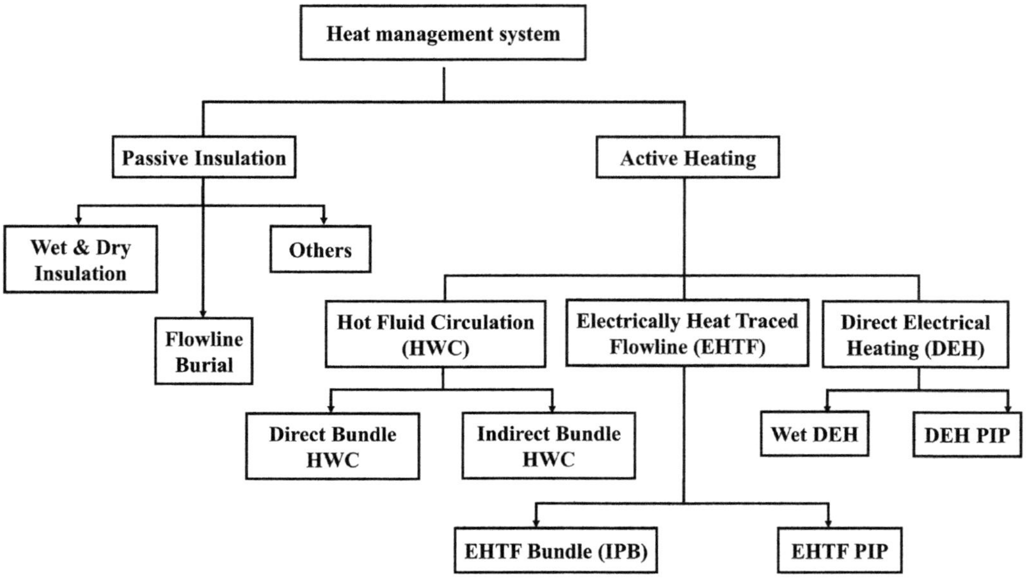

FIG. 16.3

Heat management in subsea flowlines.

From N. Sunday, A. Settar, K. Chetehouna, N. Gascoin, An overview of flow assurance heat management systems in subsea flowlines, Energies 14(2) (2021) 458, https://doi.org/10.3390/en14020458.

Generally speaking, two main methods of temperature control are commonly used in oil and gas fields: passive methods (insulation) and active methods (active heating). Fig. 16.3 summarizes the heat and temperature control methods.

16.2.1.4.1 Thermal insulation

Insulation is used to minimize heat loss from the production system to the surroundings by insulating the flowline with low thermal conductivity material [25,26]. In such a way, temperatures can be maintained above hydrate formation conditions to provide hydrate control.

The low thermal conductivity of insulation material causes an insulated pipeline to lose heat at a slower rate than an uninsulated one; hence, the insulation extends the cooldown time before reaching hydrate formation temperature, which gives operators enough time either to recover from the shutdown and restart a warm system or to prepare and protect the system for a long-term shutdown [23].

Different types of thermal insulation have been applied in oil and gas fields. Some of the useful literature reviews on thermal insulation and heat management were published in Refs. [25–28]. A summary of these methods is given in Table 16.1.

Wet insulation is a common method of heat and temperature control in oil and gas (Fig. 16.4). The materials used for wet insulation are typically polyurethane, syntactic polyurethane, polypropylene, multilayered and syntactic polypropylene, and rubber or glass reinforced plastic. These materials have overall heat transfer coefficients (U-values) of approximately 2 W/m^2K [29]. Some of the wet insulation materials and their properties are summarized in Table 16.2.

Table 16.1 Conventional thermal insulation methods [24–28].

Insulation method	Description	Insulation materials	Advantages	Disadvantages
Wet insulation	Steel pipelines are directly coated with external insulating material and laid on the seabed, generally without external protection or cover The coating can be single layer or multiple layers for different purposes, i.e., thermal insulation and corrosion protection	Polypropylene, syntactic polyurethane, polyurethane, multilayered and syntactic polypropylene, plastic and glass matrices, syntactic phenolic, syntactic epoxy, etc.	– Simple construction with low construction costs – Low life-cycle costs – High inspection reliability	– May not be able to keep temperature above WAT for a long time – Bad contamination if fluids leak
Pipe-in-pipe/dry insulation	A single-insulated inner flowline centrally fitted inside an outer protective pipe; between the protective outer pipe and inner pipe, there is an annular space for insulation to prevent heat losses from the flowing fluid	Different insulation materials occupy the annulus connecting the pipes, such as: aerogel, mineral wool, vacuum or inert gas, polyurethane foam (PUF), thermal cement, fiberglass, phase change materials (PCM), etc.	– Superior performance – No ingress of water to the insulation layer – Better mechanical load – Flexible maintenance and operations – More resistant to harsh environments – Easy detection and containment of leaks	– Higher in cost than wet insulation – High repair costs – Complex design and constructions – Heavier than wet methods and depends on availability of installation equipment
Vacuum systems	Examples are vacuum insulated tubing (VIT) and vacuum insulated panels (VIPs), which are upgrades of the PIP methods. The configuration has a slim space between pipes kept at a vacuum state to decrease the transfer of heat from the hydrocarbon to the surroundings	Vacuum	Very efficient	Expensive
Bundle systems	The configuration has multiple flowlines, hydraulic control and service lines, and electrical umbilicals encased or "integrated" in a single carrier pipe or casing	Foam material	– Allows control and monitoring of system – Cheaper than the PIP system – Easy implementation on flowline – Lower depth of water than PIP	– The module gaps may cause convection current – A huge material quantity is needed – Geometry could have insulation difficulty

Continued

Table 16.1 Conventional thermal insulation methods—cont'd

Insulation method	Description	Insulation materials	Advantages	Disadvantages
Insulation fluids	Chemicals with low thermal conductivity placed in the cold regions surrounding the production tubing, and risers in deep water applications	Steam, base oil, sodium silicate, gelatinous oil-base fluids	– Easy application – Application is common in wellbore area, production tubing, perforation	– Possibility of deposition – Interactions with produced fluids – High viscosity of some of them restricts their placement
Burial	Flowline is buried (fully or partially) or has rocks, grit, or seabed material placed over it below the mudline	Soil, rocks, grit, seabed material	– The period of heat containment is high – The construction and laying are simple – Low costs of construction – Life cycle costs are low	– Employed only for laying pipes in the mudline – Location affects the thermal conductivity of the soil – Unpredictable thermal performance due to changes in the properties of the soil and troubles predicting for measuring them.

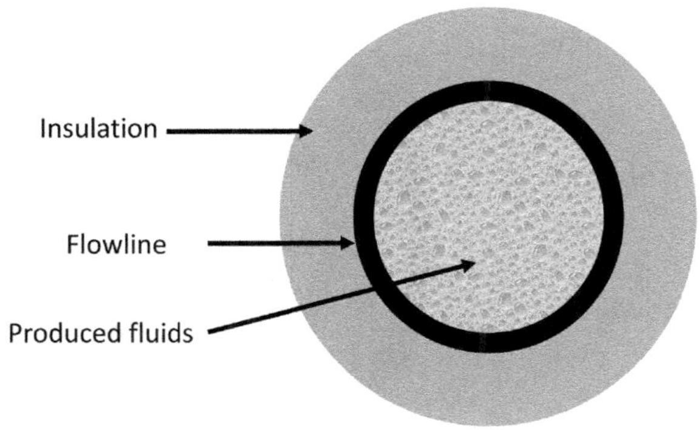

FIG. 16.4
Wet insulation.

Table 16.2 Summary of wet insulation properties [26].

Insulation material	U-value (W/m^2K)	Density (kg/m^3)	Water depth (m)	Service temperature (°C)
Epoxy syntactic	0.10–0.14	590–720	2000–3000	75–100
Syntactic polyurethane (SPU)	0.12–0.15	608.70–848.98	91.40–2743.20	55–115.60
Polypropylene (PP)	0.22	900	3000	145
Glass syntactic polyurethane (GSPU)	0.12–0.17	610–830	2000–3000	55–90

The dry insulations used polyurethane foam and Rockwool, which have a better U-value of approximately 1 W/m^2K [29]. The pipe-in-pipe (PIP) (Fig. 16.5) system was introduced to prevent water ingress to give better insulation performance, and has shown promising flow assurance performance with successful reported cases. Some characteristics of PIP pipeline insulation materials are summarized in Table 16.3.

Pipeline burial is a project-specific selection and depends on the pressure, viscosity, and temperature (PVT) characteristics, soil properties, topography geohazards, and infrastructures [30].

Another new cost-effective technology is the flexible insulated pipe. The flexible pipe insulation takes the form of wrapped syntactic tape layers. This tape material is an extruded polypropylene thermoplastic mixed with hollow glass microspheres. The technology, which is provided by Technip USA, offers simple installations, installation efficiencies, reeling, a lower hang-off angle for a more compact seabed footprint, and can also be recovered for reuse or decommissioning [31].

Other technological advances to address flow assurance problems include: collapsible pipe, phase change insulation, and hybrid thermal insulation with phase change material and aerogel [32].

Thermal insulating fluids are used mostly in deepwater riser and packer applications [16]. For gas systems, insulation is only applicable for high reservoir temperatures and/or short tie-back lengths [23].

A heat bank in which the subsea equipment is covered and the fluid trapped inside is heated during normal production was proposed as an alternative to conventional insulation [33]. The principle of the heat bank is that the hot, flowing hydrocarbons will heat up the steel and seawater inside the cover to

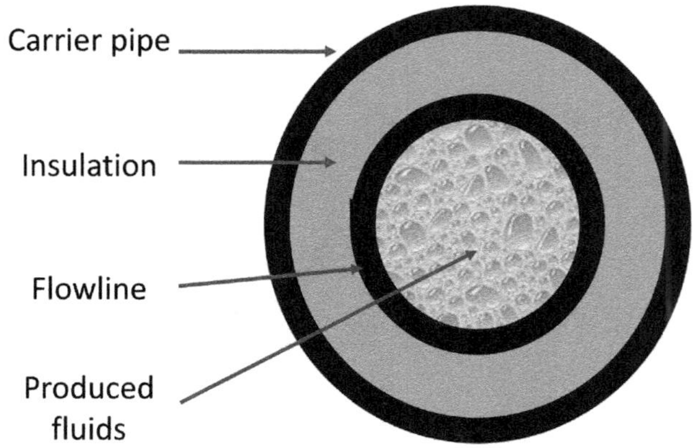

FIG. 16.5

Dry insulation, pipe-in-pipe (PIP) configuration.

Table 16.3 Properties of PIP insulation materials [26].

Insulation material	U-value (W/m^{2}K)	Density (kg/m^3)	Water depth (m)	Service temperature (°C)
Mineral wool	1.60	140	3000	700
Fiberglass	1.36	56.06–88.10	3000	−17.78–454.44
Polyurethane foam	0.76	60	3000	147

form a reservoir of stored thermal energy. In a shutdown situation, the heated fluid inside the cover keeps the production fluid warm through natural convection and conduction. The authors showed how the production fluids were kept warm by water or gel trapped in the heat bank several days after a shutdown [33].

One advantage of an insulated production system is that it can allow higher water production, which would not be economical with continuous inhibitor injection. However, shutdown and restart operations would be more complicated. However, insulation of complex geometries in subsea production equipment is more complex and costly and, in some cases, infeasible [34].

16.2.1.4.2 Active heating

The active heating system involves the external addition of heat to the production flowline generally by an electrical heat source or circulated hot water [26,35,36].

For most fields, the goal is to stay above the region of 15–40°C [37] or to go above the modeled hydrate formation temperature. Therefore for this purpose active heating is an attractive, economical, and effective solution. Compared to passive insulation, active heating can sustain the hydrocarbon fluid temperature beyond the hydrate and wax formation temperatures over long periods and longer distances [26,30]. Using active heating will reduce pigging runs and minimize chemical consumption. In flowlines and risers, active heating must be applied with thermal insulation to minimize power requirements [23].

There are two major types of active heating methods: one is electrical heating where the heat is supplied by an electric cable, and the other is circulation of hot fluid [38], and these have branched out into other types. The different methods of applying active heating are summarized in Table 16.4.

Table 16.4 Summary of active heating methods [24,26,28,30,39].

Method	Description	Advantages	Disadvantages
Hot fluid circulation	– The produced fluid is heated through conduction and convection by the hot fluids circulating around in a little heating pipe (hot fluid PIP) – Circulated hot fluids can be water or oil – Hot water systems operate in a loop where the water is recirculated back to the topsides – The hot water technique uses heat exchangers with an extended secured tube having production fluid on the inner part and the heating medium on the outer part of the tube	– HWC system is used to adequately warm the production fluid during steady-state or restart a line with issues of pour point – HWC-bundle can integrate insulation with good thermal performance with U-value from 0.6–6 W/m^2 K – HWC-bundles have limited thermal expansion and low risk of lateral buckling, as well as good resistance to accidental impacts – Possible combination with hot fluid from more process units on the topside – Can use different hot fluid sources; e.g., hot water can be produced from a heat recovery system using production fluid heat or from fire heaters fed with production gas	– Only used during steady-state production in practice – Demands high power – Conventional HWC-PIP has relatively low wet insulation performance, U-value varies from 3 to 6 W/m^2 K – HWC-bundle construction necessitates a dedicated bundle assembly plant (along with specific equipment) and can only be installed by towing up to 500 m water depth. – A large footprint is required on the topside for heat recovery and water treatment systems – The injected water needs to be treated to avoid corrosion issues inside the circulation path – Material selection of the bundle system to avoid corrosion issues
Hot fluids bundle	– This system composed of a carrier pipe with a combination of individual flowlines and umbilical components are carried and there is hot water/fluids distribution in close vicinity to the production flowline in an assigned line – The bundles can employ direct heating by annulus circulation or indirect using dedicated hot water supply and return lines		
Direct electric heating wet insulated flowline (Wet-DEH)	– The heat is supplied by an electrical cable, where heat is generated by the Joule effect from the current circulating in the flowline wall – The pipe to be heated is an active conductor in a single-phase electric circuit (AC), together with the single core power cable	– It is field proven – Extended heated circuits of approximately 50 km – DEH components can be replaced or retrofitted on already installed wet insulated pipes if anticipated at the design stage – The cord and its components are possibly retrofitted/restored	– High linear power, voltages, and currents (1500 A) – The demand for 3–1 phase adaption – High power requirements – Big cathodic system protection is needed – Piggyback cable requires protection and strapping

Continued

Table 16.4 Summary of active heating methods—cont'd

Method	Description	Advantages	Disadvantages
	(piggyback cable) as the forward conductor, located in parallel with and close to the heated pipe – The return circuit comprises a combination of the pipe wall and the surrounding seawater to allow the flow of electrons	– Small U-value of 2.5 W/m^2 K	– The low heating efficiency of 50%–75% – Not efficient for continuous heating – Nonhomogeneous longitudinal and circumferential heating, depending on steel material electromagnetic properties (pipe sorting required) and coupling with piggyback cable – Low operability – Application limited to deep water (up to 1000 m water depth); also water depth is limited by piggyback cable – A/C corrosion risk, complex open-loop design with anodes – Regular inspection and maintenance are required
Direct electrical heating (DEH)-PIP	– Production line is split into heated PIP segments where each segment is a closed electrical circuit composed of inner and outer pipe, which is terminated at both ends with a steel bulkhead in order to have a closed electrical circuit – Current is supplied at the middle of the flowline and returned through the carrier pipe instead of a piggyback cable	– It is field-proven. – The high heating efficiency of up to 95% – High thermal performance – PIP arrangement mechanical robustness – Suitable for deepwater application	– High linear power and currents (1500 A) – Small flowline lengths – Lengthy power cord needed for mid-flowline electricity provision site – It does not appeal to continual heating – Limited maximum heated length with a single power supply which needs to be located at the mid-point of the line due to risk of electrical arcing between the flowline and carrier pipe – Low electrical efficiency – Low operability

Electrically heat-traced flowline-pipe-in-pipe (EHTF-PIP)	– The inner pipe in the PIP assembly is installed with three-phase insulated heat-traced wires and it terminates in a star end socket, consequently avoiding the desire to return the current back to the topside power generation equipment. The production/inner flowline receives heat by conduction through the Joule resistive effect from the three-phase cord	– A great heating efficiency of about 95% – It is used for continual heating – Accurate fluid temperature monitoring all along the flowline through the optical fibers – Better operability – Low power consumption	– Limited flowline length/little heating loops of approximately 25 km – The flowline inner diameter is restricted to about 12 in. – The EHTF technique needs unique installation equipment – It is not repairable after installation
Electrically heat-traced flowline-bundle or integrated production bundle (IPB)	– An IPB is composed of a core (a standard flexible pipe structure for production fluid transportation) and a bundled assembly of tubes, hoses, cables, and fillers wrapped around the core. These components primarily provide active heating cables, gas lift tubing, and passive insulation material made of syntactic polypropylene foam. Additional umbilical component functionality is available, such as electric and hydraulic hoses and fiber optics	– High operability with temperature monitoring by optical fibers	– IPB internal diameter limited to approximately 11–12 in., depending on insulation requirements, due to present overall diameter manufacturing limitations – Heating efficiency remains around 40% to 60% – Tracing cables cannot be repaired or replaced subsea
Bundle configuration	– The pipeline is combined with hot circulating fluids or electric heating combined and wrapped in insulation and encased in carrier pipe	– Better control and monitoring of the system – Can integrate different methods of heating	– Can be more expensive than other methods

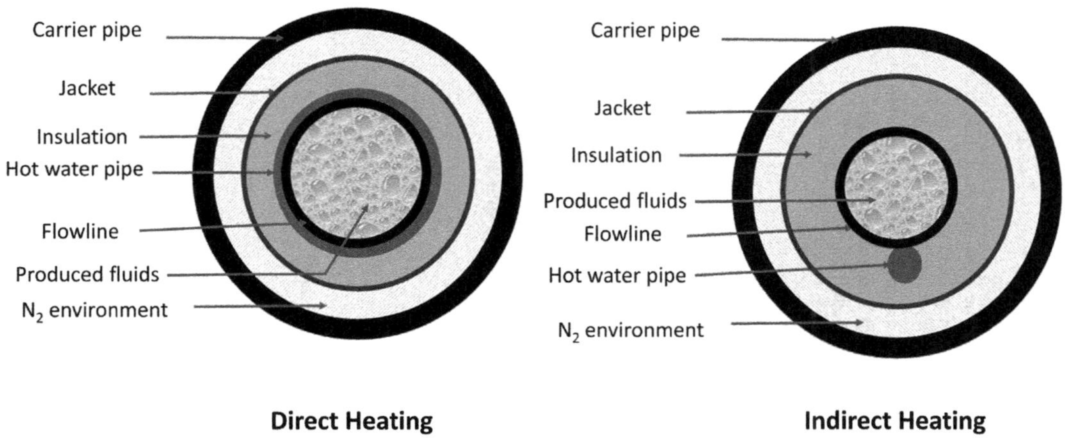

FIG. 16.6

Hot fluids circulation bundle.

Hot water circulation technology was used in Conoco Phillips' Britannia fields, Statoil's Gullfak and Asgard, BP King, and West Africa offshore Angola [24,36]. Hot fluid circulation is illustrated in Fig. 16.6.

Another method is to wrap the pipeline wall with an electrically resistant, heat-tracing cable, or a heat-tracing tube containing a circulating hot fluid, which elevates the temperature of the pipeline wall [16]. A new approach to flowline active heating was introduced [40]. The concept that has been evaluated for the new active-heating system consists of a pipeline with insulation and the active-heating (hot fluid) tube inserted inside of the production conduit. The active-heating system is configured so that the heating fluids exit the tube at a point in the pipeline and commingle with the produced fluids, and return to the host for processing/separation.

Direct electric heating is expensive to use on a continual basis, so it is used only in extended shutdown situations [16]. Electrical heating has been applied in many fields, including Nakika, Serrano, Oregano, and Habanero in the GoM, and Asgard, Huldra, and Sliepner in the North Sea [41]. Figs. 16.7 and 16.8 illustrate the direct electric heating methods.

Bundling allows the integration of flowlines, insulation, chemical lines, power, signal and heating cables, and others inside an outer jacket or carrier pipe. These are efficient systems, with smaller installation campaigns. Fig. 16.9 illustrates a bundle system.

Other approaches include direct electric heating of flexible pipes [42] and through flowline (TFL) [43].

Generally speaking, there is no single technology or architecture that suits all fields; the best solution is based on field composition, needs, costs, and the associated flow assurance problems. Passive insulation systems are not the most suitable, cost-effective techniques for subsea deepwater long-distance transportation of hydrocarbons, while active heating is a game changer for subsea and deepwater production systems, allowing minimizing chemical and mechanical treatments. Combining two

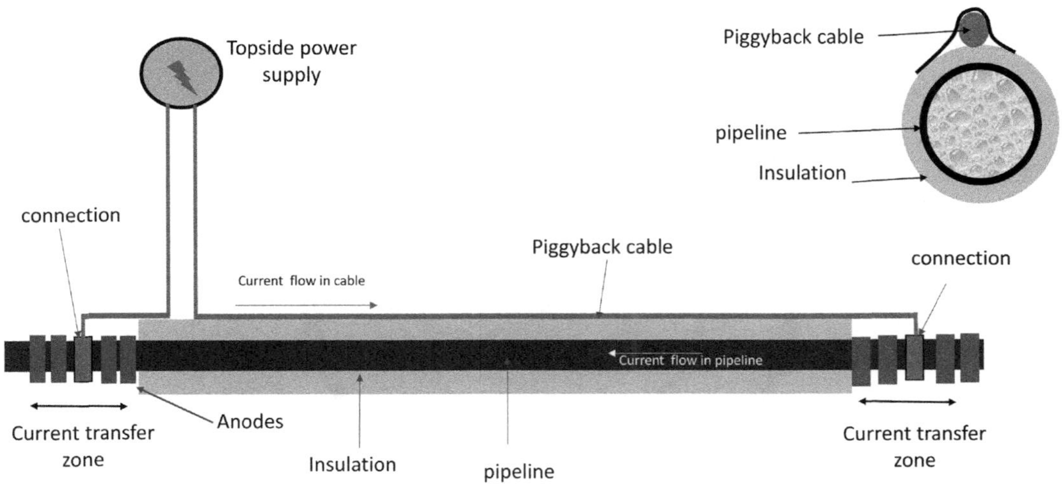

FIG. 16.7

Schematic of the direct electric heating method with cross section view (top) and side view (bottom).

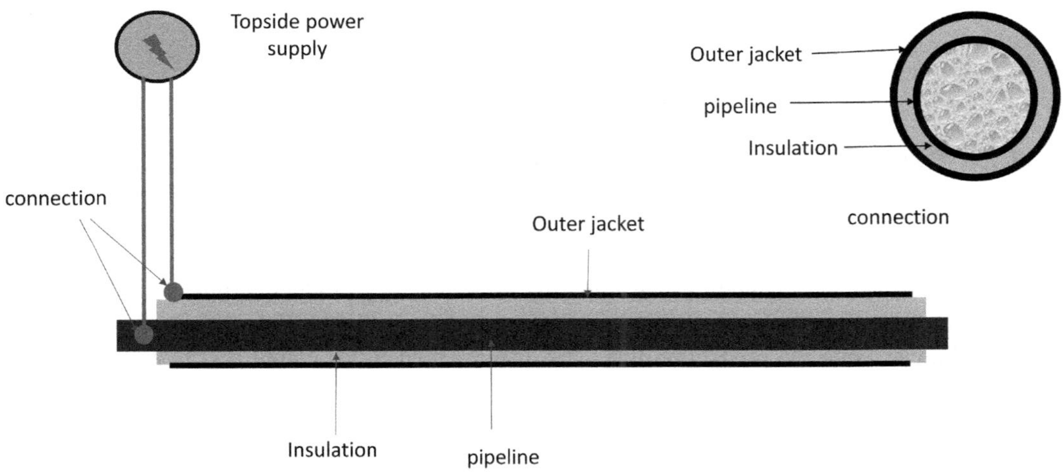

FIG. 16.8

Schematic of the direct electric heating pipe-in-pipe method with cross section view (top) and side view (bottom).

or more heat management technologies is found to be very effective for flow assurance management in the industry. From a cost perspective, active heating systems are the most cost-effective for subsea deepwater fields, because they can lead to a cost reduction in capital and operating expenditures due to the use of a single line architecture instead of a conventional loop, which consequently lowers chemical injection and power requirements [26].

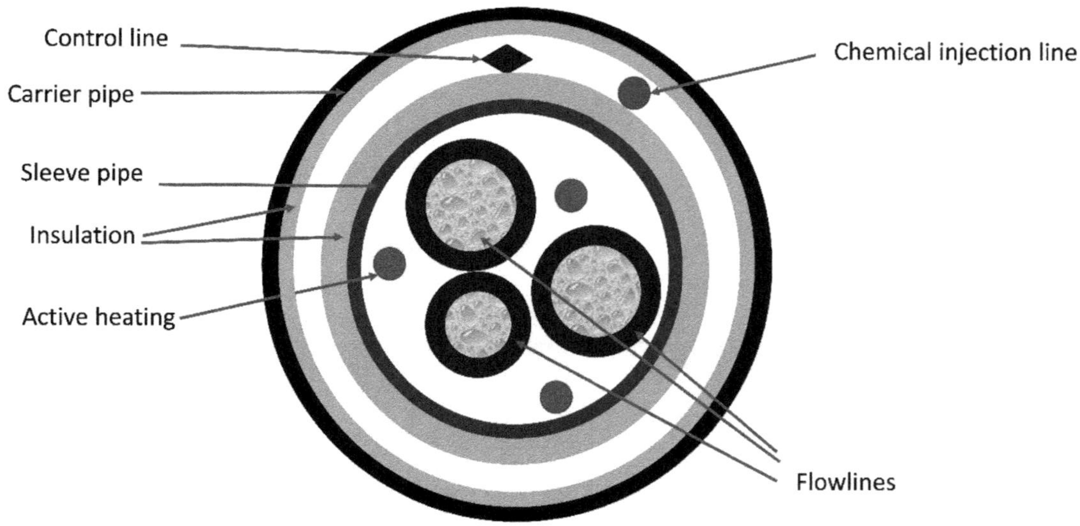

FIG. 16.9

Schematic of bundle configuration.

16.2.1.5 Displacing fluids and removal of the production

This method is based on replacing the fluid in the line with another one that does not form hydrate when cooled, even at system restart. This is normally done by flushing the line, either with gas (N_2 or CO_2) pushing a pig, or by replacing the line content with an inhibited liquid [11].

A fluid displacement method for managing hydrates in subsea production lines has been claimed [44]. The method includes depressurizing displacing production fluids from a service line within the umbilical line and into the production line. The displacement fluids comprise a hydrocarbon-based fluid having a low dosage hydrate inhibitor (LDHI).

One other option is to try to treat the content of the line, mainly the water, with a thermodynamic inhibitor. This requires each flowline to be connected to the topsides with an independent loop, or at least service lines for injection and depressurization [11].

Note that using methanol for the inhibition and displacing fluids of a significant length of nonflowing line after a shutdown is a real challenge. The operator must initiate the chemical injection treatment quickly, before hydrate plugs are formed. Predicting the amount of chemical (to achieve 40% dilution in the line), and the hydrodynamics or flow of methanol (limited velocity of migration of the methanol across the water zone) are challenging issues. The storage, handling, and injection of methanol are other issues. For example, a 10 km 8 in. line half full of liquid and a WC of 80% will require more than 90 m^3 of methanol to be injected. Speaking of time, the injection through a 1 in. line, at a maximum velocity of 2 m/s, would take approximately 30 h [11].

Table 16.5 summarizes the operational methods of gas hydrate control.

Table 16.5 Operational methods for gas hydrate mitigation.

Method	Description
Gas dehydration	Removal of water to avoid hydrate formation
Pressure control	Operate at optimum pressures below the gas hydrate formation pressure
Temperature control	Optimize system temperature to be above gas hydrate formation temperature Achieved using heat management methods; active heating or passive insulation
Cold flow	Controlled precipitation of gas hydrates and allowing safe flow of gas hydrate particle slurry

16.2.2 Chemical prevention of gas hydrates

Chemical methods can be applied individually or in combination with the other operational methods mentioned previously to achieve feasible and economical gas hydrates management.

A common analogy for the chemical methods of controlling gas hydrates are those methods used to combat ice in cold places where snowfall is common. In the winter, salt is often used to remove ice from roads and sidewalks. A glycol-based antifreeze coolant is used for vehicle radiators, and glycol solution is sprayed on airplanes waiting for takeoff, for deicing. Similar techniques are used for combating hydrates [45].

16.2.2.1 Classification and mechanisms of gas hydrate inhibitors

Chemical treatment to prevent hydrate plugging can be accomplished with three different classes of chemical, all of which are now used in the field, as depicted in Fig. 16.10:

- Thermodynamic hydrate inhibitors (THIs)
- Low-dosage hydrate inhibitors (LDHIs); this category can be subdivided into:
  - Kinetic hydrate inhibitors (KHIs)
  - Antiagglomerants (AAs)

The chemical inhibition usually takes place in the aqueous phase [23]. Chemical inhibitors are not normally used continuously for oil systems; instead they are used after shutdown or during re-start-up. Thermodynamic inhibitors are usually used continuously for gas pipelines, because gas pipelines are normally not insulated [46].

THIs like alcohols, glycols, and salts generally work by changing the bulk thermodynamic properties and equilibrium of the fluid system, thereby shifting the equilibrium conditions for gas hydrate formation to lower temperatures or higher pressures, which creates an increased hydrate-free temperature range during operations [16,47], as depicted in Fig. 16.11.

MEOH and MEG reduce water activity and change its chemical potential by the formation of a strong hydrogen bond with the neighboring water molecules; these inhibitor-water bonds compete with the hydrogen bonds between water molecules and water hydrate bonds, preventing them from forming the hydrate cages until much lower temperatures are reached. Fig. 16.12 illustrates the THI hydrogen bonding with the water molecules.

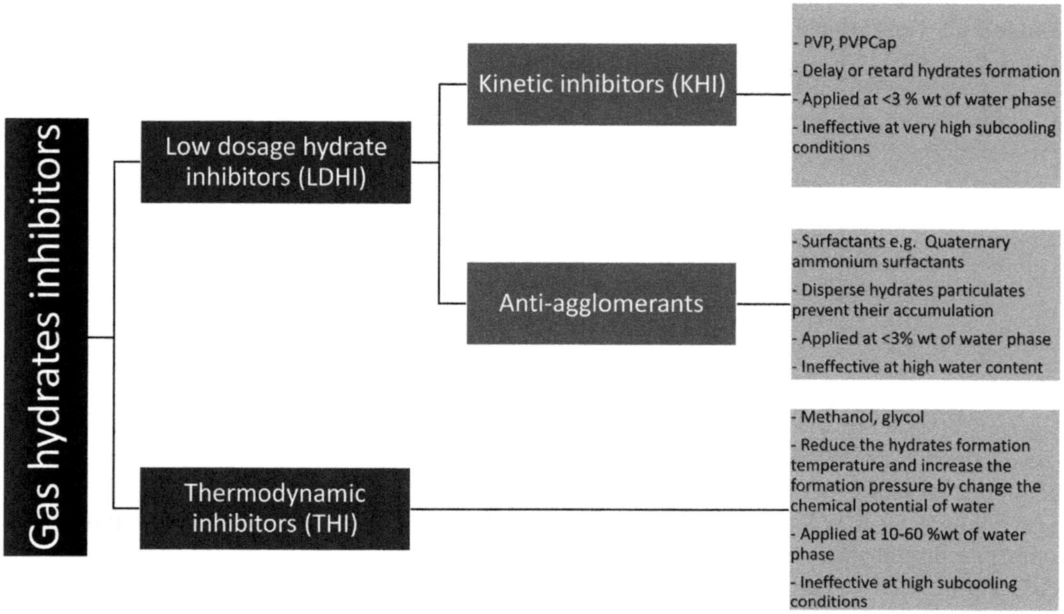

FIG. 16.10

Gas hydrates inhibitors.

MEG molecules were found to prevent the agglomeration of hydrate particles by surrounding the hydrate particles and bonding with free water molecules between hydrates that work as a lubricant to avoid the formation of a capillary bridge between hydrate particles and preventing the particle interactions and agglomeration. This mechanism is depicted in Fig. 16.13 [48].

Low-dosage hydrate inhibitors are classified into kinetic hydrate inhibitors and antiagglomerants. The main mechanism by which these chemicals work is slowing the hydrate crystallization process (by nucleation, crystal growth inhibition, or dispersion), thus delaying hydrate formation and growth such that there is sufficient time to transport the fluids to their destination. However, given sufficient time, hydrates will form even in the presence of the LDHIs [45]. The hold time is the time before hydrates start to form rapidly. The hold time of some kinetic inhibitors can be about 24–48 h [46]. Fig. 16.14 illustrates the general concept of the LDHI inhibition mechanisms.

Various mechanisms have been proposed for KHI. The adsorption-inhibition mechanism (AIM) assumes that polymers adsorb on the nucleus/embryo surface, preventing it from gaining critical size and nucleation cannot continue. Common chemicals that work by this mechanism include PVP, PVCap, and *N*-vinylacetamide (VIMA). The perturbation-inhibition mechanism (PIM) assumes that the chemical is perturbing the water phase, preventing the formation of nuclei therein. Common chemicals that work by this mechanism are amino acids, alkyl-vinylformamide polymers, poly(*N*,

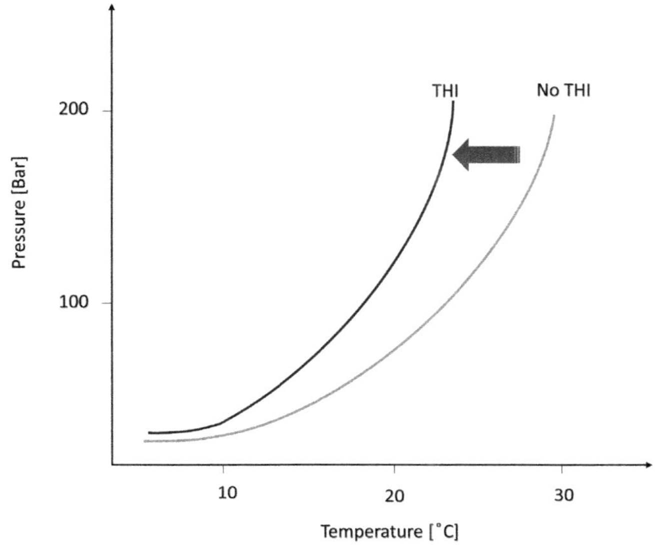

FIG. 16.11

Effect of THI on hydrate formation conditions.

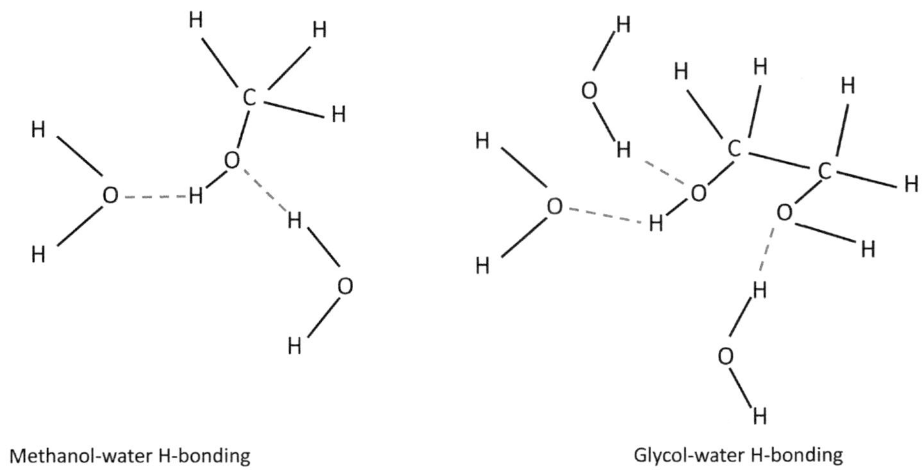

FIG. 16.12

THIs bonding with water molecules.

N-dimethylacrylamide)s (PDMAA), PVP, and PVCap. The crystal growth-inhibition mechanism (CGIM) works by blocking the growing crystals and covering the surfaces to prevent further crystal growth; common chemicals that work by this mechanism are PVCap, PVPC, and VVA. The nucleation site interference-inhibition mechanism (NSIM) is based on covering and interfering with the suitable

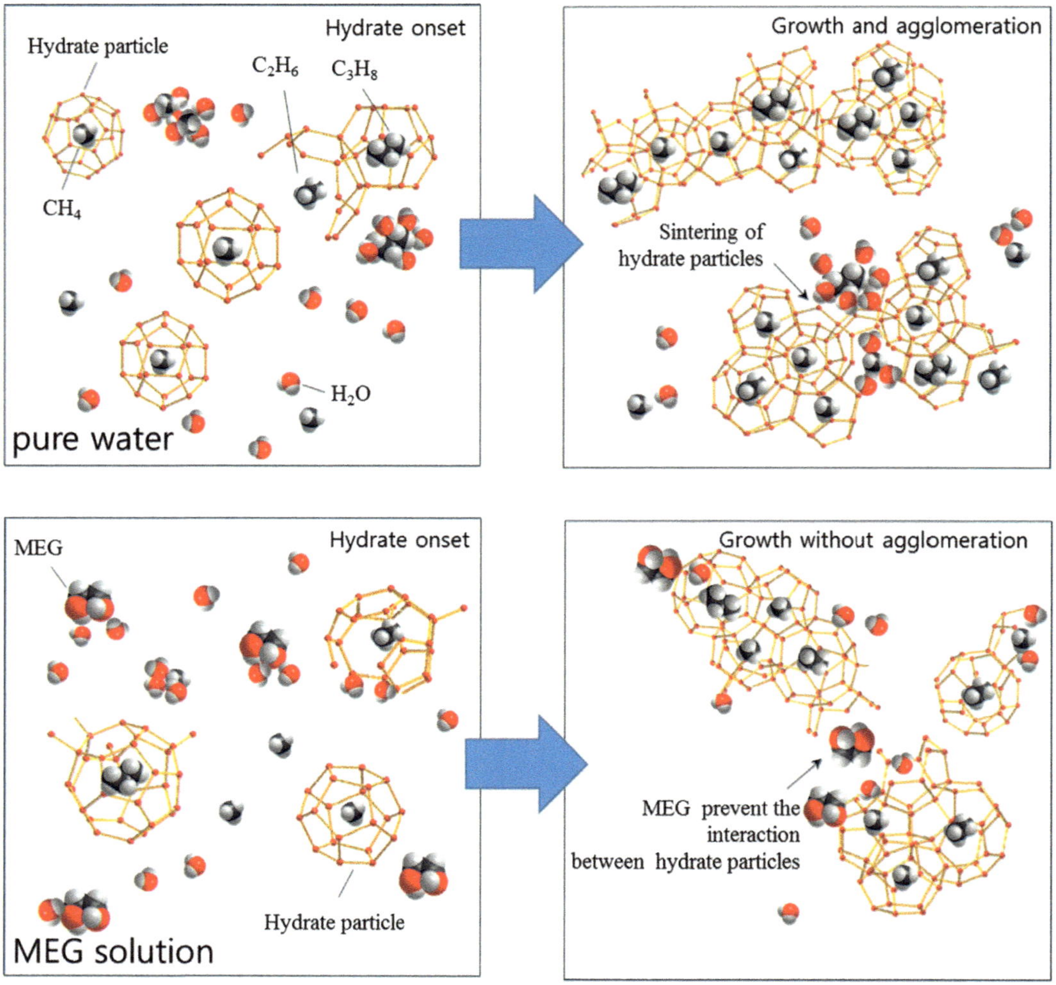

FIG. 16.13

Representation of hydrate formation in pure water (top), and in MEG solution (bottom)

Reproduced with permission from Y. H. Sohn, Y. Seo, Effect of monoethylene glycol and kinetic hydrate inhibitor on hydrate blockage formation during cold restart operation, Chem. Eng. Sci. 168 (2017) 444–455, https://doi.org/10.1016/j.ces.2017.05.010, Copyright (2017), Elsevier.

nucleation sites to prevent nuclei formation from there; common chemicals in this category are chitosan and cationic starches [49]. Fig. 16.15 illustrates some of the KHI proposed mechanisms [50].

AAs incorporate into the hydrate crystals, changing their size and morphology and preventing their agglomeration and forming transportable slush-like hydrates [51,52]. AAs also work by lowering the cohesive forces between hydrate particles [53].

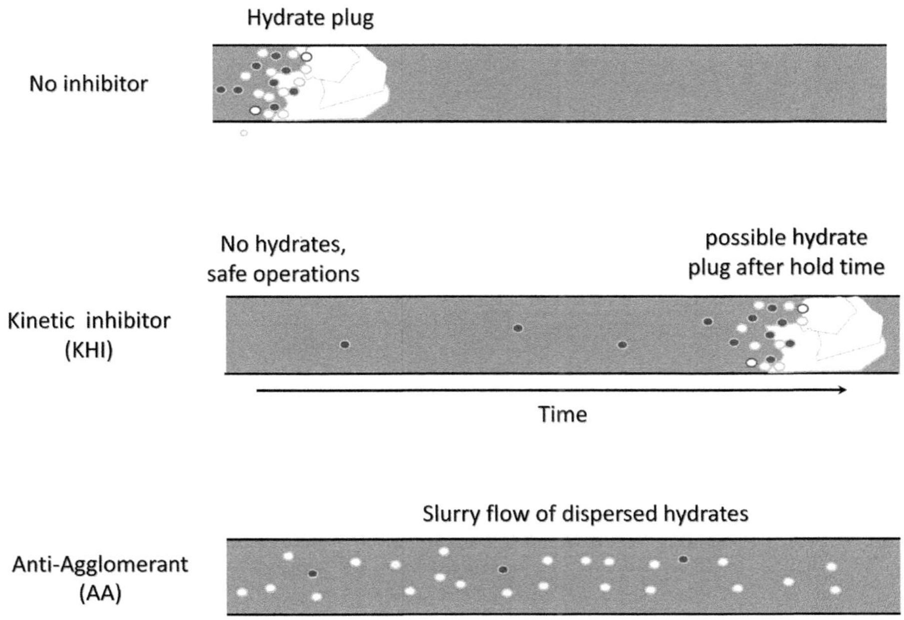

FIG. 16.14

Effect of LDHIs on hydrate formation.

16.2.2.2 Thermodynamic hydrate inhibitors

These are also known as "hydrate antifreeze" or high-dosage hydrate inhibitors. THIs are regarded as the most common chemical class that can be used to prevent hydrate formation and to remove hydrate plugs [16]. The most commonly used classes of THIs are alcohols, glycols, and salts, which are extensively applied in deepwater developments. Fig. 16.16 illustrates the effect of common THIs on hydrate equilibrium phase boundaries [54].

High concentrations of THIs are needed to shift the equilibrium: maybe 20–80 wt% of the water phase needs to be added [55], and in some cases up to two barrels of THI per barrel of water [16] may be required.

16.2.2.2.1 Alcohols

Alcohols like methanol (CH_3OH), ethanol (CH_3CH_2OH), or isopropanol (C_3H_8O) can be used as thermodynamic hydrate inhibitors. Methanol is the most common alcohol that has been proven both effective and economical in use, especially for subsea installations and arctic regions. Methanol basically dissolves in the water phase, yet a large amount of it remains/partitions in the vapor or

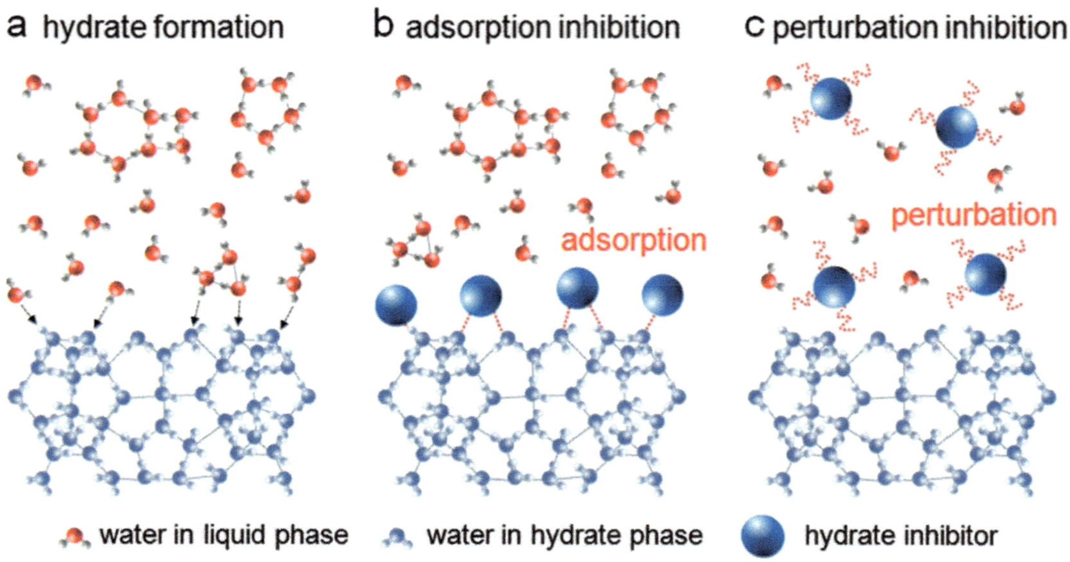

FIG. 16.15

A schematic diagram demonstrating hydrate formation and the inhibition processes by KHI.

From J.H. Sa, G.H. Kwak, B.R. Lee, D.H. Park, K. Han, K.H. Lee, Hydrophobic amino acids as a new class of kinetic inhibitors for gas hydrate formation, Sci. Rep. 3 (2013), https://doi.org/10.1038/srep02428.

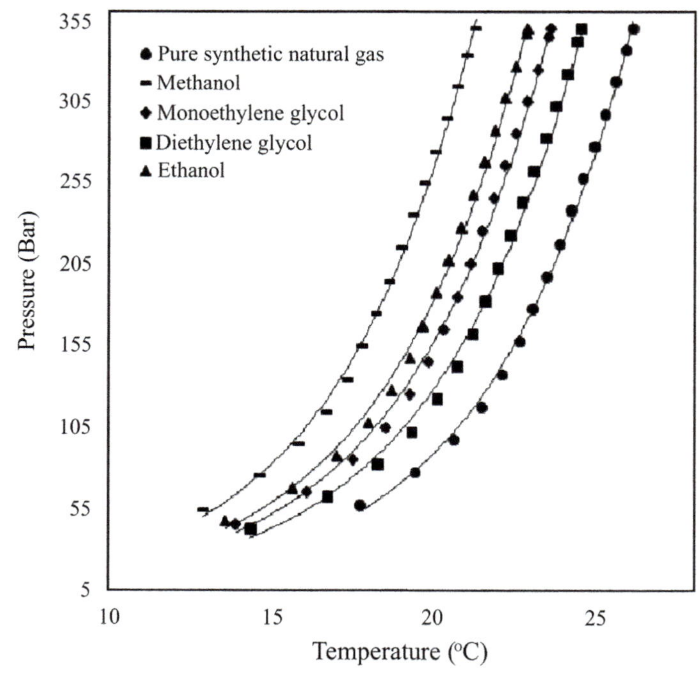

FIG. 16.16

Effects of commonly used inhibitors on hydrate equilibrium phase boundary of natural gas source.

From E. Broni-Bediako, R. Amorin, C.B. Bavoh, Gas hydrate formation phase boundary behaviour of synthetic natural gas system of the Keta basin of Ghana, Open Pet. Eng. J. 10 (2017) 64–72, https://doi.org/10.2174/1874834101701010064.

hydrocarbon phase which is counted as an economic loss and should be accounted for during injection calculations [23].

Ethanol has also been proven to work in Brazil and the North Sea; however, it can be significantly less effective when it forms a binary hydrate with methane above ca. 5.6 mol% [16,56]. Isopropanol (IPA) is less effective and also expensive; however, it was found that IPA is an effective cosurfactant, which reduces the dosage of antiagglomerants used to control gas hydrates [57]. Other alcohols tested as cosurfactants include 1-butanol (1-BtOH), 2-butanol (2-BtOH), tert-butanol (t-BtOH), 1-pentanol, 2-methyl-2-butanol (2-m-2-BtOH), 1-hexanol, 2-ethyl-1-butanol (2-e-1-BtOH); the methyl group next to the hydroxyl group in these molecules was found to bind strongly to the emulsion/hydrate surface, which improves efficiency [57].

16.2.2.2.2 Glycols

Glycols are a class of alcohols that possess two hydroxyl groups (diols). Monoethylene glycol (MEG, $HOCH_2CH_2OH$) is widely used to protect against hydrate formation, especially in deepwater systems. Diethylene glycol (DEG) and triethylene glycol (TEG) have also been reported, although they are less effective. TEG is commonly used in the gas dehydration process, and also can be used for melting hydrate plugs [16].

A comparison between methanol and monoethylene glycol characteristics is furnished in Table 16.6.

Gas producers tend to regenerate THI due to the high volumes used to treat gas hydrates, despite their relatively low cost.

As shown in Fig. 16.16, methanol outperforms ethanol, while monoethylene glycol outperforms diethylene glycol, indicating that the lower the molecular weight of a thermodynamic inhibitor, the better the hydrate inhibition performance.

Table 16.6 Comparison between methanol and glycol [16,58,59].

Methanol	Monoethylene glycol
– Easily vaporized in the gas phase, loss in gas and condensate	– Less soluble in gas and condensate
– Used for flowlines and topsides plugs	– Used for wells and risers plugs
– No salt problems	– Associated fouling and salt precipitation
– Induces severe corrosion problems	– Less corrosive
– Low flash point	– High flash point
– Very hard and costly to recover/regenerate	– Relatively recoverable, easily regenerated
– Rarely used on a continuous basis in oil fields due to the high volumes of water required to be treated	– Glycols in particular are used on a continuous basis in condensate and gas fields
– Low initial costs	– High initial costs, and lower operating costs if regenerated
– Less viscous	– Highly viscous, affects flow and umbilical lines
– Minimal equipment as it requires only a free-water separator and a suitable means for injection and atomizer	– Requires a free-water separator plus a gas-liquid separator and a glycol reconcentration unit at the point of recovery downstream

16.2.2.2.3 Salts

Ionic solids also inhibit the formation of hydrate in much the same way that they inhibit the formation of ice [45]. Like MEOH and MEG, ionic solids disturb the water activity and prevent the formation of hydrate cages. The presence of ions imposes Coulombic effects, which overcome the hydrogen bonding and network formation between water molecules and hinder cages formation and hydrate crystal growth. With increasing the salts concentration, the hydrate formation decelerates, resulting in a hydrate-free zone [60]. Examples of salts are halides (NaCl, $CaCl_2$), salts of organic acids (potassium formate and sodium acetate), salts of nitrates and phosphates (e.g., dipotassium hydrogen phosphate), and others that can be used to suppress hydrate formation [16]. The inhibition effect of cations on hydrate formation is in the order of $Mg^{2+} > Ca^{2+} > Na^{+} > K^{+}$, while the Cl^{-} ion has a much stronger hydrate inhibition effect than SO_4^{2-} ion [61].

Salts like halides and organic acids salts are commonly used in drilling fluids to suppress hydrate formation, sometimes in combination with glycols. Halide salts are less suited for injection into production lines due to the high concentrations needed, produced water incompatibility (for calcium salts), and increased corrosion potential [16].

Other types of THIs include ammonia, which was once suggested as an inhibitor for hydrate formation. It has a relatively low molar mass, 17.03 vs 32.04 g/mol for methanol, so it will be used at smaller concentrations than methanol [45]. Acetone at high concentration can inhibit gas hydrates, while at low concentration it will enhance hydrate formation [62]. Water-soluble solvents such as dimethylformamide, *N*-methyl pyrrolidone, and ethanolamines, and glycerin and other polar compounds from processing the waste stream from biodiesel are also claimed as THIs [16].

While THIs are an effective and economical method of gas hydrate management, they entail numerous problems, including the following:

- High inhibitor consumption, e.g., methanol injection rates can be 0.15–1.5 m^3/day (1–10 bpd) or more. Such high rates can be uneconomical and require storage, transport, injection, etc. [16,45].
- The cost of regeneration units.
- Safety and environmental issues of the THIs regarding toxicity, flammability, and biodegradability [16,45].
- Their effect on downstream operations if they are concentrated in the downstream hydrocarbons (LPG, sales gas) [45].
- THIs can increase the corrosion rates in many ways, especially in sour gas systems as reported by Park et al. [63]:
  - Methanol is capable of dissolving corrosion inhibitors protecting film leading to unexpected corrosion problems.
  - Oxygen ingress in methanol during storage in open areas, leading to accelerated corrosion rates, localized corrosion, elemental sulfur formation, and diminishing corrosion inhibitor efficiency.
  - Methanol increases the risk of sulfide stress cracking (SSC) and stress orientated hydrogen induced cracking (SOHIC) in sour gas conditions.
  - Iron sulfide (FeS) structure can change at certain methanol concentrations, which increases risk of localized corrosion.
  - Methanol can increase the rate of vapor phase corrosion, which directly increases the risk of a top-of-the-line corrosion failure.

- THIs increase the potential for formation of other deposits, including mineral scales (NaCl, carbonates, sulfates) and naphthenates. Methanol is a paraffin wax precipitant. Methanol and ethanol are more detrimental than glycols [16,45].
- THIs require good preparation, especially during shutdowns and restarts. Methanol requires attention during calculating the volumes to account for the loss in the hydrocarbon phase, and its flow in the cold water full lines, while glycols are highly viscous, which requires good preparation especially during injection in narrow, long, cold injection lines [16].

16.2.2.3 Low dosage hydrate inhibitors

Due to the limitations of thermodynamic hydrate inhibitors, the development of alternative, cost-effective, and environmentally acceptable and field applicable hydrate inhibitors is a technological challenge. This led to the development of LDHIs in the 1990s.

Low dosage hydrate inhibitors (LDHIs) were developed because they can be used at comparatively low concentration levels, usually <3 wt% in produced water, compared to the higher concentrations/volumes required of more traditional THIs (10%–60 wt%) [64,65]. Unlike THIs, LDHIs do not significantly change the hydrate equilibrium curve, but rather interfere with the nucleation and crystal growth of hydrate crystal formation. The use of LDHIs reduces the environmental concerns and eliminates the regeneration units requirement, which results in reduction of capital cost.

16.2.2.3.1 Kinetic hydrate inhibitors

Kinetic hydrate inhibitors (KHIs) have been used in the industry for over two decades. Most KHI field applications are based on polymers (water-soluble polymers) with other synergistic smaller organic molecules [16]. Some characteristics of KHIs are [52]:

- More environmentally friendly.
- They work in any water/oil ratio (WOR), and provide a hydrate-free environment.
- KHI chemicals are generally affected by subcooling and fluids residence time, and are limited to a lower degree of subcooling and a shorter shut-in period compared to AA chemicals.
- They have low solubility in hydrocarbons and can be affected by liquid hydrocarbons (positively or negatively) [66].

Some examples of KHIs are listed in Table 16.7.

- **Polyvinyllactam polymers**

A major type of polymeric KHIs, compounds of this category have typical lactam rings. Examples are five-ring polyvinylpyrrolidone (PVP), six-ring polyvinylpiperidone (PVPip), seven-ring polyvinylcaprolactam (PVCap), and eight-ring polyvinylazacyclooctanone (PVACO). Among these, PVP and PVCap (products with five and seven rings) are commercial products, while PVPip and PVACO are not, as they are commercially expensive [49]. Increasing lactam ring size would enhance the KHI performance [49,67]. These chemicals are highly sensitive to pressure-temperature and residence time conditions, for example: VCap polymers, applied in fields up to about 9–10°C subcooling [16]. Fig. 16.17 illustrates different types of poly vinyl lactam polymers.

Vinylpyrrolidone (VP)/and vinyl caprolactam (VCap) are also commercially used, as they have higher cloud points; a mixture of 1:1 VP/VCap is available under the Luvicap series from BASF [68]. The low molecular weight of that 1:1 VP/VCap mixture is considered the most common

Table 16.7 Examples of kinetic hydrate inhibitors.

KHI type	Examples
Vinyllactam polymers	Polyvinylpyrrolidone (PVP) Polyvinylpiperidone (PVPip) Polyvinylcaprolactam (PVCap) Polyvinylazacyclooctanone (PVACO) Vinylpyrrolidone (VP) VinylCaprolactam (VCap) VP/BA (vinyl pyrrolidone/butyl acrylate)
Hyberbranched polymer	Hyperbranched polyesteramide
Other polymers	Polyalkyloxazolines Pyroglutamate polymers Polymaleimides Maleic copolymers Polyaspartamides Polyallylamides
Green inhibitors	Graft polymers Polyester Ionic liquids Natural extracts, e.g., proteins, starch, pectin, glucose

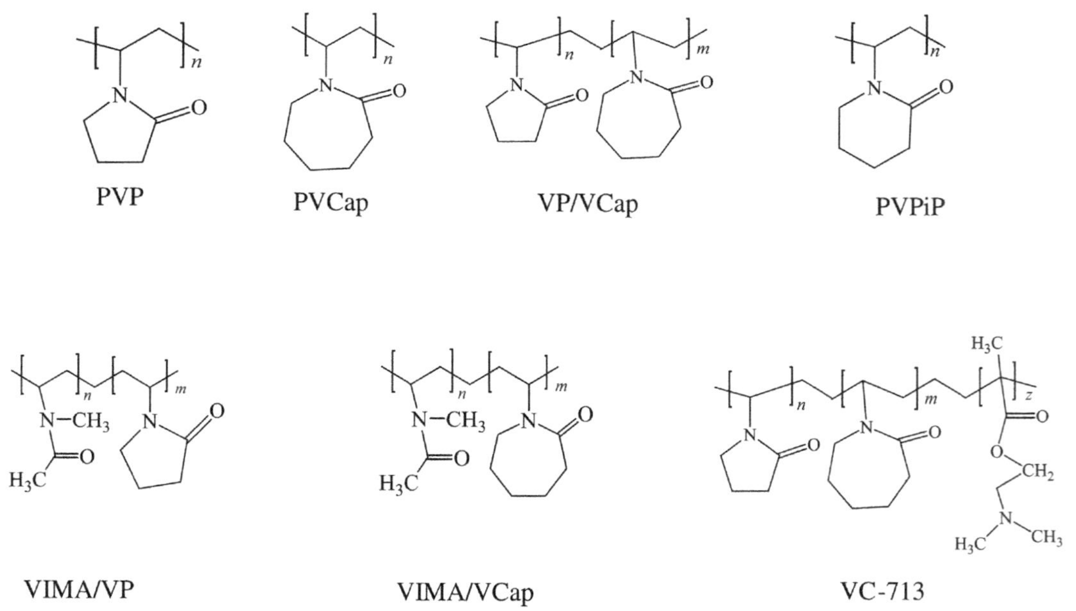

FIG. 16.17

Structures of polyvinyllactam polymers.

Reproduced with permission from Y. Wang, S. Fan, X. Lang, Reviews of gas hydrate inhibitors in gas-dominant pipelines and application of kinetic hydrate inhibitors in China, Chin. J. Chem. Eng. (2019), https://doi.org/10.1016/j.cjche.2019.02.023, Copyright (2019), Elsevier.

high-cloud-point copolymer on the market today, with a performance a little lower than PVCap of the same molecular weight [16].

VP/BA (vinylpyrrolidone/butyl acrylate) is a new copolymer developed by BASF which dominantly prevents hydrate nucleation [69].

Common synergists for VCap include butoxyethanol, quaternary salts, polyamine oxides, amine oxides, and polyethylene oxide [16], while NaCl enhances the performance of PVP [70], and PVCap is enhanced by glycol ether compounds [69].

- **Hyperbranched polyesteramides**

These are claimed to have a better performance than the VCap-based polymers on structure s-I hydrates. However, since the type of hydrate structure usually formed during oil and gas transport is s-II instead of s-I, the actual field performance of hyperbranched polyesteramides remains unknown and may need to be tested on site [49].

Like the VCap polymers, they are also useful for field applications up to about 10°C (18°F) subcooling [16].

- **Other polymeric KHIs**

Other types of polymers include acrylamide polymers, vinylamide polymers, and maleimide polymers. Fig. 16.18 illustrates the structures of these polymers. *Isopropylmethacrylamide* (IPMA) polymers and copolymers have improved their tolerance to saline environments with higher cloud points. Polyalkyloxazolines and pyroglutamates are other polymers, and the latter have been commercialized and used in a few field applications. Also, polymaleimides, maleic copolymers, and alkylamide derivatives, polyaspartamides, polyallylamides, proteins, and modified starch and starch derivatives have been tested as potential KHI candidates [16,49,71,72].

- **Salt (NaCl)**

Studies on sodium chloride, a traditional thermodynamic hydrate inhibitor, have reported that it could act as a kinetic hydrate inhibitor in porous media [73].

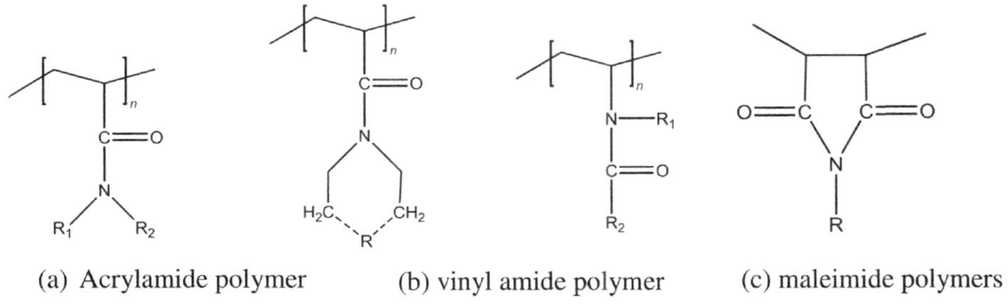

FIG. 16.18

Structures of acrylamide polymers, vinylamide polymers, and maleimide polymers.

Reproduced with permission from Y. Wang, S. Fan, X. Lang, Reviews of gas hydrate inhibitors in gas-dominant pipelines and application of kinetic hydrate inhibitors in China, Chin. J. Chem. Eng. (2019), https://doi.org/10.1016/j.cjche.2019.02.023, copyright (2019), Elsevier.

- **Green hydrate inhibitors**

The green inhibitors were developed to avoid the environmental impacts of the conventional KHIs on the ecosystem in terms of toxicity and biodegradability. Green inhibitors can be synthetic chemicals with enhanced biodegradability, or natural components extracted from natural sources.

The first type involves the classic KHIs with enhanced biodegradability. Some examples were reviewed by Kelland [71]:

- Grafted polymers, with biodegradable backbone polymer and vinylic monomers grafted onto it. Examples are *N*-vinyl lactams grafted onto a biodegradable backbone such as polyalkylene glycol, polyalkyleneimine, polyether, or polyurethane. Currently, it appears that one of the best biodegradable KHIs is a graft polymer with a biodegradable backbone with pendant caprolactam groups [71].
- Other synthetic polyester and polyamide KHI polymers, examples of which are polyesters or polyethyleneimines with pendant pyroglutamate groups that have been developed as high cloud point, biodegradable KHIs. Ester groups are generally biodegrade faster than amide groups, but polymer backbones containing either of these functional groups should be more biodegradable than polyvinyl backbones.
- Ionic liquids, examples of which are quaternary ammonium salts, salts of the cations 1-alkyl-3-methylimidazolium, 1-alkylpyridinium, *N*-methyl-*N*-alkylpyrrolidinium, and imidazolium salts.
- Pseudo-polypeptides, examples of which are polyaspartamide and poly(*N*-alkylglycine)s, were also shown to have good biodegradability.

The other class of green chemicals includes chemicals of natural source. Examples of these include:

- Proteins, such as antifreeze glycoproteins (AFGP) and antifreeze proteins (AFP). These products were inspired by organisms (winter flounder fish) that live in the cold and arctic regions and possess a mixture of proteins (known as thermal hysteresis proteins or ice-structuring proteins (ISPs)) [74]. These proteins feature hydrophilic amino acid residues that interact with the growing surface of an ice crystal through an adsorption-inhibition mechanism. Thus small quantities of AFPs can have large effects on ice-crystal growth [74,75]. However, such AFPs, extracted from fish, insects, plants, fungi, or bacteria, can presumably be made in an enzymatic process, which on an industrial scale is prohibitively expensive [16]. Other non-AFP proteins extracted from wheat, beef, and milk and other natural sources were reported to give comparable performance to some commercial KHIs.
- Amino acids and their effects on gas hydrates show conflicting results, as some studies consider them promoters of gas hydrate formation and others found them to show gas hydrate inhibition.
- Other natural polymers and their derivatives, examples of which are cellulose and its derivatives, xylomannan, which is another polysaccharide extracted from the beetle, chitosan extracted from the outer skeleton of shellfish, starch, pectin, and glucose [71,76]. Fig. 16.19 illustrates the structures of these compounds.

Another way of attenuating the environmental impacts of the KHIs and reducing their discharge into the ecosystem is by recycling them. Adham et al. [77] investigated physical, chemical, and biological methods to remove KHI from produced water, including centrifugation, filtration, and ozonation.

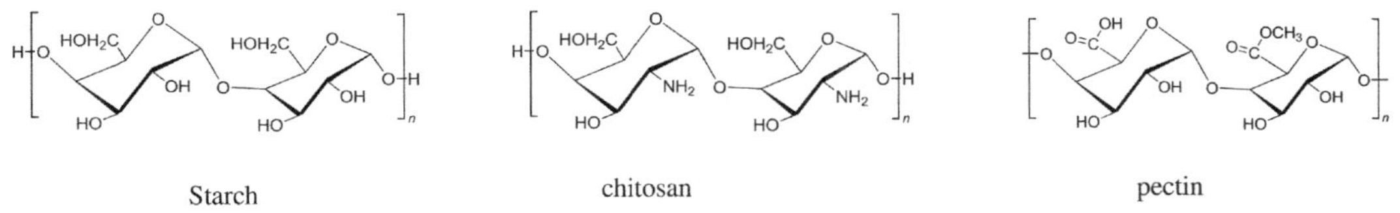

FIG. 16.19

Structures of some of natural KHIs.

Reproduced with permission from Y. Wang, S. Fan, X. Lang, Reviews of gas hydrate inhibitors in gas-dominant pipelines and application of kinetic hydrate inhibitors in China, Chin. J. Chem. Eng. (2019), https://doi.org/10.1016/j.cjche.2019.02.023, Copyright (2019), Elsevier.

Table 16.8 Examples of AAs.

Examples of AAs
Diethanolamides
Dioctylsulfosuccinates
Sorbitans
Ethoxylated polyols
Ethoxylated fatty acids
Ethoxylated amines
Polyetherdiamines
Polyetheramines
Quaternary ammonium surfactants
Quaternary phosphonium surfactans

16.2.2.3.2 Antiagglomerants

Anticoagulant (also called antiagglomeration, AA) inhibitors, are a class of LDHI that works on a different level [45]. Table 16.8 summarizes some examples of AAs.

They are molecules that generally have a hydrophilic head and a hydrophobic tail, the best of which can prevent hydrate accumulation into large masses or a plug (they do not prevent hydrate formation) at higher subcoolings than KHIs. Some AA characteristics include:

- Due to their higher subcooling efficiency, they are applicable for more severe and/or deepwater applications, and in events when the residence time in the flowline is very high (weeks) [16,78].
- Some AAs also show KHI effect.
- AAs require the presence of a carrier liquid to transport the dispersed hydrate particles; the liquid may be liquid hydrocarbons or water.
- AAs are most efficient at water cuts below 50% [79], and a minimum of 50% to 70% oil cut is needed for an AA to work [52].
- AAs are usually applied in a low dosage (1%–3%) [80].

Common AA molecules include diethanolamides, dioctylsulfosuccinates, sorbitans, ethoxylated polyols, ethoxylated fatty acids, and ethoxylated amines. Other polymeric surfactants include those based on polyalkylene glycol derivatives of polyalkenyl succinic anhydrides, or carboxylic acid hydroxycarbylamide, substituted or nonsubstituted, and a carboxylic acid monoethanolamide or diethanolamide containing three to six carbon atoms [16]. These molecules work by forming water-in-oil emulsion, preventing hydrate from forming or agglomerating. They are commonly applied at 0.8–1.0 wt% based on the water phase [16]. Other surfactants include caprolactam and alkylamide surfactants and quaternary ammonium and phosphonium surfactants, which prevent hydrate formation by attaching to the crystal and preventing further growth. These surfactants are optimum for use in pipelines, where WC is below 50% [16].

Examples of AAs that can be used in excess water include polyetherpolyamines, such as polyetherdiamines. Polyetheramines have been used downhole for several years to prevent hydrate blockages in more than 100 gas wells, mostly onshore, but also offshore [16]. New formulations for high water cut and sour gases were suggested and applied in sour gas wells in Alberta, Canada [80].

16.2.2.4 Factors affecting hydrate inhibitor efficiency

Many factors affect the performance of hydrate inhibitors, including the following:

- **System temperature and subcooling**

THIs like methanol, glycols, and salts can inhibit hydrates above and below 0°C for any length of time and can be applied in deepwater installations [81]. KHIs are limited to a maximum of 16–18°F subcooling and thus may not be applicable for most deepwater fields. AAs have better subcooling tolerance than KHIs, as they are effective to around 40°F subcooling, and are suitable for deepwater fields [82]. Another temperature-related term is the compatibility of the chemical itself with the process temperature. Some KHIs exhibit cloud-point temperature (Tcl) [also known as lowest critical solution temperature (LCST)] at which they start to precipitate, often when the aqueous solution of polymers is heated. This is a thermoresponsive behavior in aqueous solutions. Low cloud KHI cloud point is a problem when the temperature of the production fluids at the injection point is hotter than the KHI cloud point, resulting in its deposition and diminishing the efficiency; it is known that some KHIs have Tcls that are not far from the hydrate formation temperature, e.g., PVCap has a Tcl of about 30–40°C [83].

- **System pressure**

System pressure affects the inhibitor performance and its applied dose.

- **Types of hydrates**

KHIs are designed primarily for inhibition of sII hydrates [84]. However, recent reports showed that recently developed KHIs can inhibit sI and sII [49]. In addition, different hydrate types might need different application parameters (P, T) when using KHIs.

- **Fluid residence time**

When designing hydrate mitigation strategies for subsea systems, it is critical to make sure the fluid residence time in the pipeline during steady-state flow is less than the hold time [46]. Hydrates can form in longer fluid residence times in the presence of KHIs, and shorter residence times allow the use of KHIs even in higher subcooling.

- **Flow regime**

KHIs have shown low protection from hydrate formation at the top of the pipeline when stratified flow is the dominating flow regime [85].

- **Acid gases**

H_2S and CO_2 gases considerably impair KHI performance [16]; however, the CO_2 effect depends on the CO_2 concentration and other gas components in the system. CO_2 is found to enhance KHIs at higher pressure, while it is found to reduce KHI performance at lower pressures [84].

- **pH**

Acidic solutions cause coagulation and precipitation of the inhibitor, so they reduce hydrate inhibition performance while encouraging fouling problems [86].

- **Other production chemicals**

Corrosion inhibitors (CIs) and KHIs are not always compatible, and the presence of a CI can significantly decrease the performance of a KHI [87]. Scale inhibitors (SIs) in general are not detrimental to the performance of neutral KHIs and can in fact enhance the performance in some cases [88]. Wax inhibitors (WIs) had a more negative impact on kinetic hydrate inhibitor performance at low dosage, while at high concentrations it could be negligible [89]. Oil-soluble chemicals accumulate at the interface at which KHI acts to inhibit hydrates [52]. THIs and KHIs can also affect each other. Some reports showed improvement in the performance of KHIs from THIs, which cause KHIs to stand higher sub-cooling conditions [90]; combined KHIs and THIs can prolong the induction time and prevent hydrate blockages [91].

- **Chemical properties**

The applied chemical must have acceptable viscosity and good solubility in the produced fluids. THIs with lower molecular weights gave better performance than the higher molecular weight ones. KHIs should have ideal molecular weight (normally around 1500–3000) for enhanced surface/volume ratio and lower viscosity [16].

- **Liquid hydrocarbons**

Kinetic hydrate inhibitors have negligible solubility in hydrocarbon liquids, and the presence of liquid hydrocarbons can have an impact on their performance. In general, light condensate shows less or no impact on KHI performance; crude oil is more prone to varying it. Fu et al. [52] showed KHI was found to be effective in a crude oil system, allowing its economic applications, as in gas-condensate fields [92].

16.2.2.5 Hydrate inhibitor testing and screening

The many ways of testing and screening for hydrate formation range from simple methods to more sophisticated ones.

16.2.2.5.1 Tetrahydrofuran method

One of the simplest methods in experimenting with gas hydrates is use of a mixture of tetrahydrofuran (THF) and water to simulate a wet gas. THF can form hydrate crystals in the same temperature range as a wet gas, but crystal formation takes place at atmospheric pressure. Various concentrations of additives can be added to study their effect on hydrate formation [93].

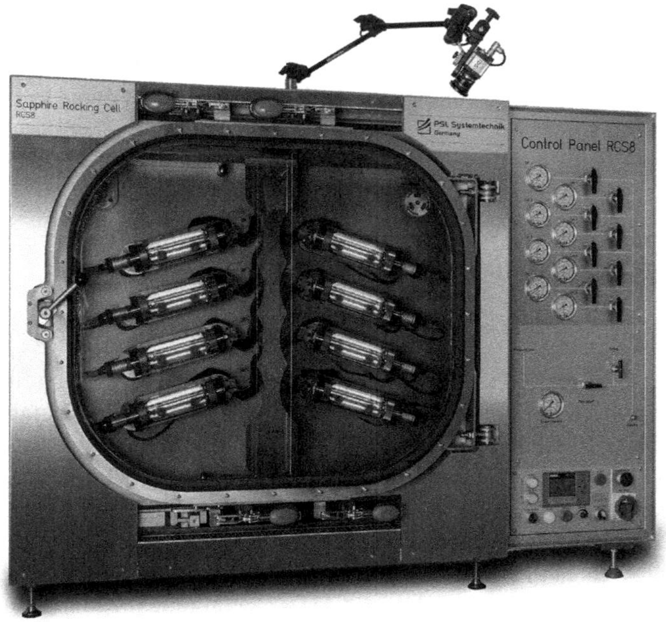

FIG. 16.20

Sapphire rocking cell with camera.

Courtesy of PSL-Systemtechnik, https://psl-systemtechnik.com/, reprinted with permission.

16.2.2.5.2 Rocking cell

The technique of using rocking cells is based mainly on visual observation of a metal ball in motion [79,94]. Fig. 16.20 shows a typical rocking cell unit. The ball moves back and forth in the cell, which rocks at a steady rate in a mixture of oil, water, and gas under pressure and declining temperature, creating enough turbulence to mix the fluids and enhance hydrate formation. When formed, the hydrates impact the movement of the metal ball and can be seen through a sight glass or sapphire window or a whole cell made up of sapphire for closer observation [95]. Fig. 16.21 shows the metal ball motion and the development of gas hydrate deposits. When evaluating AAs, as in the case of hydrates when they are allowed to form, they must be limited in size and spread to allow continuous movement of the ball [94]. Fig. 16.22 illustrates the LDH test results in the rocking cell.

Some advantages of the rocking cell include:

- Ability to simulate different field conditions, as the cell can be rocked (to simulate flow conditions) or held static (to simulate a shut-in condition).
- Multicells allow performing multitests simultaneously and screening many products, which shortens the time of selecting the chemical treatment.
- Small sample volume.
- Ability to monitor rheology changes.

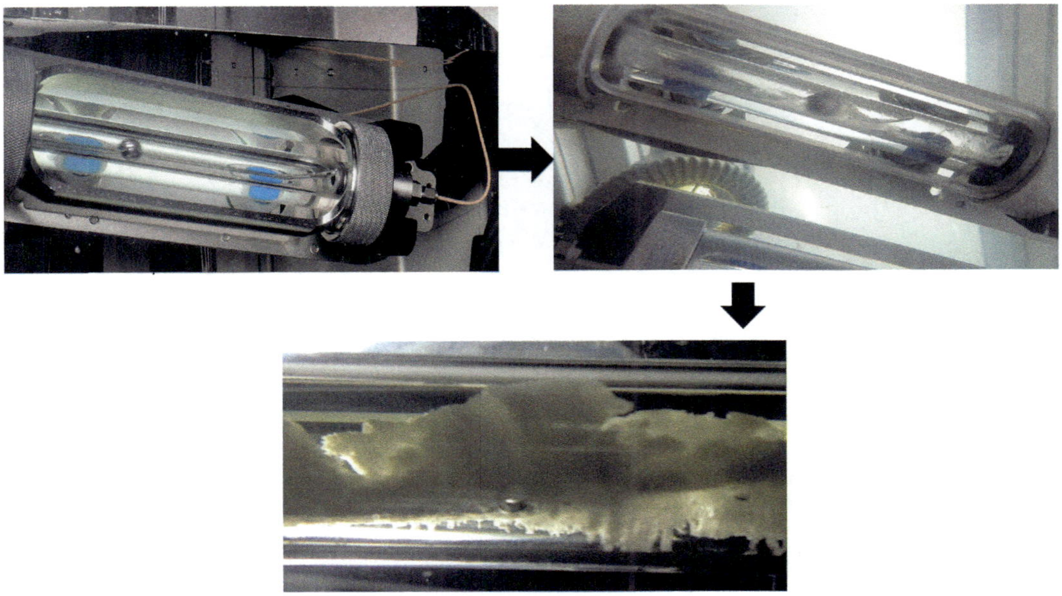

FIG. 16.21

The motion of metal ball and the development of gas hydrates in sapphire rocking cell.

Courtesy of PSL-Systemtechnik, https://psl-systemtechnik.com/, reprinted with permission.

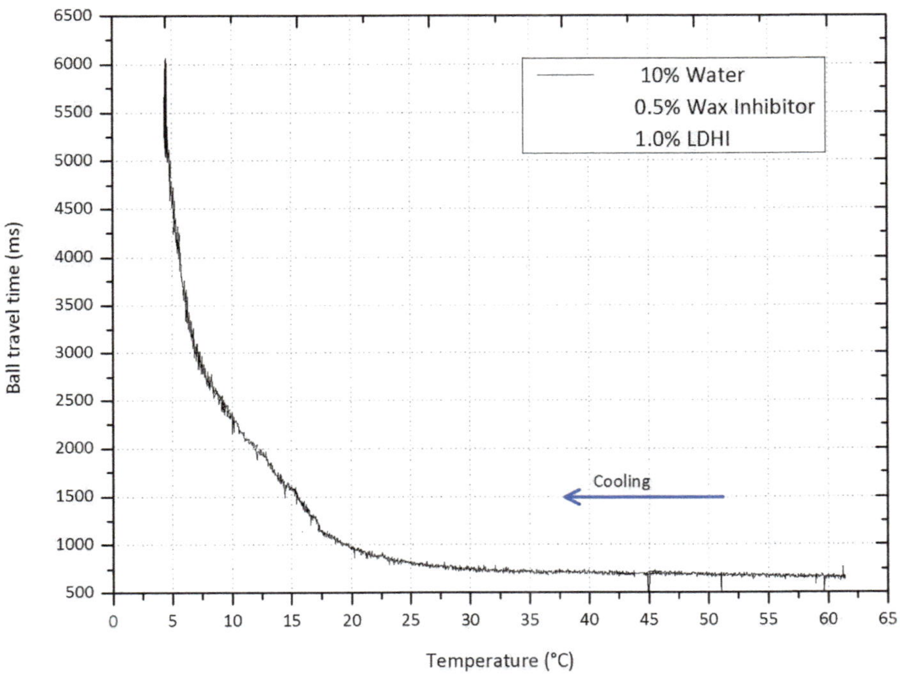

FIG. 16.22

Chemical testing using sapphire rocking cell showing the resultant curve of ball run time vs temperature.

Courtesy of PSL-Systemtechnik, https://psl-systemtechnik.com/, reprinted with permission.

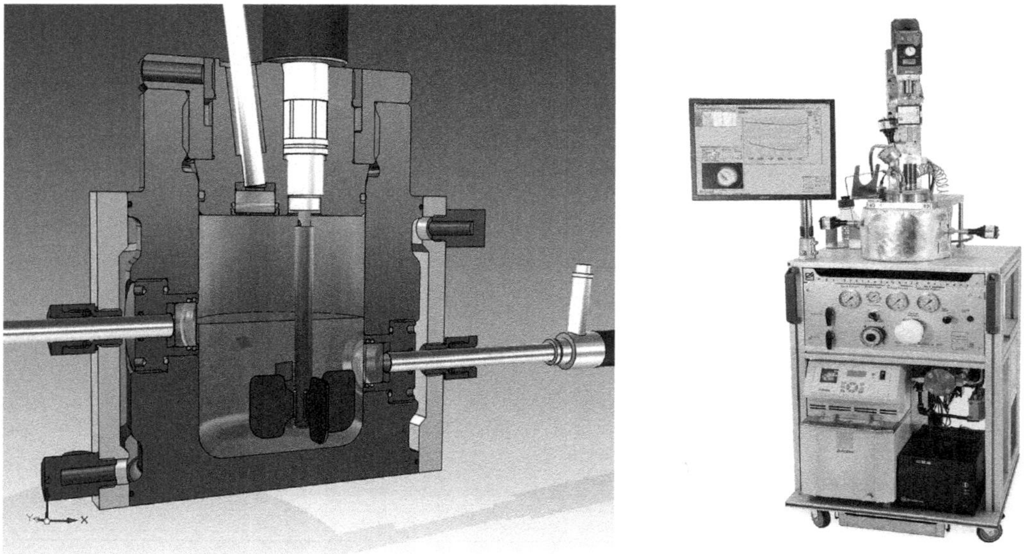

FIG. 16.23

High pressure autoclave with camera for visualization, photo and video recording during testing.

Courtesy of PSL-Systemtechnik, https://psl-systemtechnik.com/, reprinted with permission.

16.2.2.5.3 High pressure autoclave

For years, high-pressure stirred cells or autoclaves have been commonly used as a standard method for chemical inhibitor screening and for fast hydrate inhibitor hold-time determination [96]. The cell is placed in a cooling bath where the pressure, temperature, and sometimes torque exerted on the stirrer are measured. Some cells have windows for visual observations, while others are entirely made of sapphire [79]. Mini-autoclaves have been developed for rapid screening of LDHIs [94,97]. Figs. 16.23 and 16.24 illustrate the gas hydrate autoclave and the gas hydrate crystals at the water surface detected by the water surface camera.

16.2.2.5.4 Wheel loop

A more complex apparatus is the vertically placed pipe wheel or loop wheel. The pipe is usually 1–3 in. in diameter, and can have a window for visual observation [79]. It is pressurized before being rotated in a cooling chamber [79]. Pressure, temperature, and torque exerted on the wheel can be measured [98,99]. Tests in large loops and pipe wheels are generally the last and best step before the field trials or field implementation of the KHI [16].

16.2.2.5.5 Flow loop

Alternatively, a horizontal flow loop can be used. A simple flow loop can be used for tests using THF hydrates at atmospheric pressure; however, this type of loop has limited scope of use [100]. Most loops used today are high-pressure loops using natural gas, condensate, or oil and an aqueous phase [79]. They can range from the mini-loop (e.g., ¼ inches in diameter) to the full-scale pilot loop of larger diameter.

FIG. 16.24

Gas hydrate crystals at the water surface detected by the water surface camera.

Courtesy of PSL-Systemtechnik, https://psl-systemtechnik.com/, reprinted with permission.

16.2.2.5.6 Other methods

Cingotti et al. [101] developed a semi-batch-stirred reactor in which temperature and pressure are maintained at constant values. This reactor is equipped with a turbidimetric sensor used to characterize in situ a methane/water system (particle size distribution of the hydrate suspension or state of dissociation of the undersaturated solution). This method enables quantitative and qualitative determination of the efficiency of kinetic inhibitors, both from the induction time (nucleation stage) and from the quantity of hydrates formed (growth stage) in a very short time. Differential scanning calorimetry (DSC) and high pressure DSC (HP-DSC) have been used to screen KHIs [102,103].

The preceding conventional methods are performed based on the induction/hold time, and they suffer from some problems:

- Lack of repeatability, lack of predictability.
- Suitability questions in shut-in conditions, hydrate formation at the top of the pipeline, time requirements.
- Poor operator confidence: hindering KHI adoption.

Thus other approaches for the selection of hydrate control chemicals include second germination (SG) and crystal growth inhibition (CGI) approaches, which were found to give more repeatable results. These approaches were mentioned in Refs. [104,105].

16.2.2.6 Field application of hydrate inhibitors

THI dosage must be optimized during operations (sufficient concentration and proper distribution), as reports showed that hydrate blockages form more readily in under-inhibited systems than in systems completely without inhibitor, as the underinhibited systems can enhance hydrate formation rates and increase plugging potential [106,107].

Generally speaking, the injected inhibitor may distribute into three possible phases:

1. Vapor hydrocarbon phase.
2. Liquid hydrocarbon phase.
3. Aqueous phase, in which hydrate inhibition occurs, and the inhibitor affects the hydrate-formation condition.

Therefore it is crucial to determine the distribution of the THI between the gas or liquid phases and to calculate the inhibitor concentration in aqueous phase and total lean inhibitor injection rate.

Various commercial prediction and simulation packages are available for predicting the effect of inhibitors on hydrate formation. However, even these models can sometimes give inadequate results, especially at high pressures, high inhibitor concentrations, or when the water also contains significant concentrations of salts [16].

One of the basic and simple correlations to estimate the inhibitor amount is the Hammerschmidt equation (Eq. 16.1), which gives the hydrate depression temperature (ΔT) in °F as a function of the weight percent of the inhibitor (W) in the final water phase and the molecular weight of the inhibitor (M).

$$\Delta T = \frac{KW}{M(100-W)} \tag{16.1}$$

where K is a constant; Hammerschmidt's Fahrenheit-based constant, depends on the nature of the inhibitor (for MEOH = 2335, and for MEG = 2200). Notice that the K factor has different values if represented in °C: for example, if ΔT was measured in °C the factor K will be 1297. Also notice that the ΔT is a temperature difference, not an actual temperature, which means 1°C = 1.8°F and not converted in the regular way from °F to °C, like 40°F = 4.4°C. From this equation, it is also clear that the lower the inhibitor molecular weight, the higher the depression in the hydrate formation temperature.

Example:

Gas hydrates form at 55°F and 12.00 MPa; calculate the amount of MEOH and MEG to suppress the temperature by 20°F. The molecular weights are MEOH = 32.042 g/mol, MEG = 62.07 g/mol.

Answer:

From Eq. (16.1), we find that for MEOH, $20 = \frac{2335 \times w}{32.042(100-W)}$.

Rewriting this equation to obtain W, $w = \frac{100 \times 20 \times 32.042}{2335 + (20 \times 32.041)} = \frac{64,084}{2975.84} = 21.53$ wt%.

Similarly, for glycol we obtain $w = \frac{100 \times 20 \times 62.07}{2200 + (20 \times 62.07)} = \frac{124140}{3441.4} = 36.07$ wt%.

The Hammerschmidt equation is based on experimental determinations of the freezing point depression [108], yet it was proven to be not very accurate and limited to concentrations of about 30 wt% for methanol and EG and only to about 20 wt% for other glycols [45]. For this reason, various modifications and other methods for prediction of freezing point depression have appeared [109], as in Eq. (16.2) from Nielsen and Bucklin:

$$\Delta T = -72 \ln(1-X) \tag{16.2}$$

where ΔT is in °C and X is the mole fraction of methanol. They claim that this equation is accurate up to a mole fraction of 0.8 (about 88 wt%). Although this equation has a wider range of applicability than the Hammerschmidt equation, it has not gained wide acceptance. Most design engineers continue to use the simpler Hammerschmidt equation [45].

Other correlations include the Carroll method [45] and the McCain method [110], which developed a correlation for estimating the effect of brine on the hydrate formation temperature, and the Østergaard et al. [111] method, which presented a correlation applicable to both inorganic salts (such as NaCl) and organic compounds (such as alcohols and glycols). Additionally, graphical methods have been proposed to predict the required THI weight percent and flow rate for a desired depression in hydrate temperature of natural gas mixtures [45,112].

One of the main issues with methanol is its volatility, which causes part of the injected dose to partition into the gas phase, so this amount must be accounted for and more inhibitor must be injected. If liquid hydrocarbons exist in the system, some methanol will partition there too and must be accounted for. Some charts, like those from the GPSA Engineering Data Book, are used to estimate methanol losses. Glycol loss and solubility in the liquid hydrocarbons are very low compared to methanol [45].

Generally speaking, methanol injection rates can swing between 1 and 10 bpd depending on the production capacity, the water cut, fluids composition, and other factors. Injection rates can be at high pressure ~ 1000 psi or higher [45].

Like THIs, KHIs are also expected to be injected at higher doses than the lab tested ones, perhaps at 1–2°C less subcooling (or higher dosage), to be in a safe margin and assure safe operations [16]. KHI subcooling is usually specified by the chemical supplier. The KHI cloud point must be known and its injection should be avoided in locations with high temperature; another way around this is by using water immiscible formulations of KHIs [113]. Case studies of successful application of KHIs are summarized in Ref. [114], with 40% cost savings reported after using KHIs. Following are some notes on LDHIs [52]:

- Both KHIs and AAs are applied in low dosage.
- AAs work at higher subcooling degrees and at longer shut-in periods than KHIs.
- AAs require a minimum of 50 to 70% oil cut to work, while KHIs work at any water-oil ratio.
- KHIs have low hydrocarbon solubility.
- KHIs interfere with corrosion inhibitors and other production chemicals.
- AAs can induce emulsions, and require higher pumping capabilities for the slurry flow.
- KHIs are more environmentally friendly than AAs.
- Both AAs and KHIs have minimal loss to the vapor phase.

16.2.2.7 Recent advances in gas hydrate chemical inhibition

Several advances have been achieved in the realm of hydrate chemical management. KHI performance was improved and optimized by manipulating their structure, using synergists and solvents. The synergistic effect of combining THIs and KHIs was also studied and found to be effective, economical, and avoided the drawbacks of each type. Ionic liquids are considered one of the promising and green inhibitors for gas hydrates with many advantages, including: (1) dual action on gas hydrates (thermodynamic and kinetic inhibitors), (2) less corrosive, and (3) low cost [115]. More expansion of green chemicals has taken place by the use of natural products and extracts. Another emerging advance is the use of nanomaterials, which have been shown to kinetically inhibit hydrates formation and also work as AAs and disperse hydrate particulates [116]. However, some reports have shown that nanomaterials can promote hydrate formation.

16.2.3 Nonchemical gas hydrate prevention

16.2.3.1 Hydrate prevention surfaces

A nonchemical method of preventing gas hydrate accumulation can be achieved by preventing the adhesion of these particles to the internal pipeline walls using an internal low-adhesion surface protective coating. In this way the hydrate particle slurry can be transported safely until they reach a hydrate unstable zone without impeding production.

Gas hydrate particle adhesion to surfaces depends on many factors, including capillary forces, wettability, surface energy, surface roughness, and other factors that are discussed in Chapter 4, "Principles of flow assurance solids formation mechanisms" Gas hydrates formed in the bulk of the solution are less likely to adhere to the pipe wall due to the insufficient capillary forces under the hydrodynamic conditions. However, those formed near or at the pipe wall surface are more likely to adhere to the surface under the flow hydrodynamic conditions [117].

The adhesion of gas hydrates to common pipe surfaces in a water-free environment is sufficiently low to allow debonding by natural flow forces alone [118]. However, in the presence of water and a hydrated surface, the strength of clathrate-surface adhesion is as much as 10 times stronger than clathrate-clathrate adhesion [119].

Different materials were found to reduce hydrate particle adhesion to surfaces. A low surface energy silane layer reduced hydrate adhesion by fourfold [117]. CVD fabricated surfaces of polydivinylbenzene capped by poly(perfluorodecyl acrylate) have been shown to reduce hydrate adhesion [120]. Useful super-hydrophobic, antiicing coatings and an omniphobic and corrosion-resistant coating were effective at reducing the adhesive forces between hydrate particles and steel surfaces, suggesting their use in preventing hydrate plugging at low water cut in corroded pipes [121]. Ultralow adhesion coating by using a textured coating applied to pipe walls to promote the formation of a water-repellent film by attracting liquid hydrocarbons already present in petroleum were found feasible in preventing gas hydrates from sticking [122]. A magnetic slippery surface was introduced to prevent hydrate adhesion at atmospheric pressure. The coating consisted of a hydrocarbon-based magnetic fluid (gel), which was applied to a metal surface to create an interface that lowered the hydrate adhesion strength on the surface [123].

Generally speaking, the successful application of a hydrophobic surface increases the hydrate induction time, reduces hydrate growth, and reduces the hydrate inhibitor chemical consumption [124]. It was seen that 10 to 14 times less KHI is needed in the presence of a hydrophobically coated surface.

16.2.3.2 Ultrasonic gas hydrate prevention methods

Ultrasonic waves have been adopted in monitoring gas hydrate formation [125] and removal of gas hydrate blockages [126].

In fact, there are conflicting results on the effect of ultrasonic waves on gas hydrates. Some reports showed that the application of ultrasonic waves prevented hydrate formation and decreased dissociation time and caused instantaneous removal of formed hydrates in gas pipelines [127,128], while other studies showed that ultrasonic methods can be used to enhance and promote gas hydrate formation [129].

16.2.3.3 Microbial gas hydrate prevention methods

This approach is based on the fact that some microbes are able to survive low temperatures by the production of ice-binding proteins, such as ice-nucleating proteins (INPs) or antifreeze proteins (AFPs), which interfere with hydrate formation and crystal growth [130].

A few research studies have been published concerning the microbial hydrate inhibition approach. One of the earliest attempts was reported by Walker et al. [130], which reported that the bacteria *Chryseobacterium* sp. C14, which shares the ice recrystallization inhibition ability of AFPs, increased induction time of THF hydrate crystallization in isothermal experiments.

Townson [131] reported the use of microbial species for inhibition of methane hydrates in a stirred crystallizer under subcooling of 2.34 K. The results showed that the ice associating *Chryseobacterium* sp. Strain C14, grown in 0.5 wt% tryptic soy broth (TSB) delayed hydrate nucleation, on average, by 30.3 h compared to 37.9 h for the PVP inhibitor solutions. *Escherichia coli* TG2 in 0.5 wt% TSB was used as a nonice-associating bacteria control and surprisingly had the longest induction period of 118.1 h, suggesting that it was three times more effective as a hydrate inhibitor than PVP.

16.3 Gas hydrate blockage removal

There are several operational, chemical, and nonchemical methods for removal of gas hydrate plugs or blockages.

16.3.1 Operational gas hydrate plug removal

This includes changing the operational parameters so that the conditions are favored toward hydrate dissociation. The operational methods are usually economical and depend on the system design complexity.

16.3.1.1 Depressurization

Depressurization, also known as blowdown, is the most common technique used to remediate hydrate blockages in production systems. It is considered the most efficient and economical method of removing gas hydrate blockages. It is also a method of producing methane gas from hydrate-bearing sediments and reservoirs [132,133].

In depressurization, the pipeline pressure is reduced at ambient temperature sufficiently to reverse the equilibrium conditions and to reach dissociation pressure at ambient temperature [134]. The pressure should be reduced evenly on the plug, and the lower the pressure achievable, the loosely it melts [135].

For better depressurization operations, the following factors should be considered [16,23,45,135,136]:

- Hydrate plug circumstances, such as number of plugs, location of plugs, density, shear strength, and size of plugs, should be addressed to optimize the depressurization operations.
- Production system must be studied carefully for suitability of depressurization operations. Depressurization may not be applicable in some system geometries, like fairly high liquid head in the riser, subsea installations, or flowlines, which may prevent sufficient depressurization below hydrate dissociation conditions. Another example is depressurizing a plug in a deepwater systems and in arctic-like systems at subzero temperature conditions; in this case, the hydrate plugs will always convert into ice plugs during depressurization.

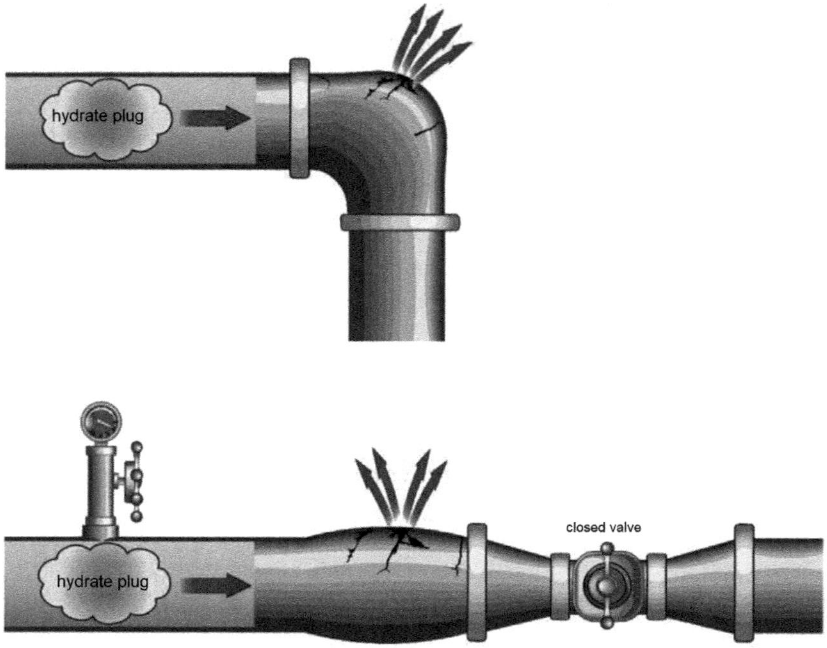

FIG. 16.25

Examples of the damage to process facilities due to hydrate plugs movement.

Reproduced with permission from F. Caputo, F. Cascetta, G. Lamanna, D.G. Rotondo, A. Soprano, Estimation of the damage in a natural gas flow line caused by the motion of methane hydrates, J. Nat. Gas Sci. Eng. 26 (2015) 1222–1231, https://doi.org/10.1016/j.jngse.2015.07.050, copyright (2015), Elsevier.

- Pressure control during the depressurization is crucial. Increasing pressure can convert soft hydrate plugs into dense hard plugs. Rapid depressurization should also be avoided, as it can result in Joule-Thomson cooling, which can worsen the hydrate problem and form ice.
- From both safety and technical standpoints, the preferred method for dissociating hydrates is to depressurize from both sides of the blockage. If only one side of a blockage is depressurized, and once the plug is partially dissolved, the large pressure differential with a high-pressure void behind it will result across the plug, which can potentially create a high-speed projectile. Sudden changes in direction during its travel or a sudden restriction can rupture, distort, or dislocate equipment, resulting in loss of human life or damage to equipment, involving expensive repair and downtime. Fig. 16.25 illustrates some of the damage caused by high-speed hydrate plugs. A possible solution if one-side depressurization is inevitable is to fill the system adjacent to the downstream of the plug with liquids, which can significantly attenuate plug velocity.
- Depressurization is a slow process; it usually takes from days to weeks to remove a plug, and longer time is expected if multiple plugs exist.
- Once partial dissolution of the plug is achieved, establishing a pressure communication through the plug, THI chemical can be injected into the system to speed up the dissociation process and to stabilize the dissolved mixture.

– Depressurization may be followed by flowlines fluid displacement to flush the remains of hydrates and for better results of the cleaning.

Depressurizing simulation tools and methods have been developed for both two-sided and one-sided methods to determine plug movement, dissociation time, complete meltdown time, and what distance would be considered low risk during remediation (e.g., CSMPlug) [137].

Boubenider et al. [138] proposed single-sided depressurization for deepwater flowlines, which can be achieved safely by the use of gas lift depressurization, to avoid the risk associated with one-sided depressurization.

The riser base gas lift (RBGL) method, which is applied for liquid removal/depressurization, may be used to dissociate hydrate. The application of this equipment is to remove liquid from the riser, reducing pressure in the flowline in which there is a hydrate blockage. In this method, pressure reduction will displace the system out of the hydrate region, inducing blockage dissociation. This can be achieved, for example, with gas (or nitrogen) circulation from gas lift to the production flowline. However, the method is not intended to be used as an artificial lift on a steady production basis. The RBGL should be installed at the connection between flowline and riser, usually placed in a horizontal distance of 1000 m from the production platform [139]. A schematic of the RBGL method is depicted in Fig. 16.26.

Kanstad [140] proposed a method of dissociating gas hydrate plugs by depressurization. The method entails fluidically isolating the production station, diverting production flow to a bypass line, and adjusting the pressure in the production station to a level sufficient to melt the hydrate plugs.

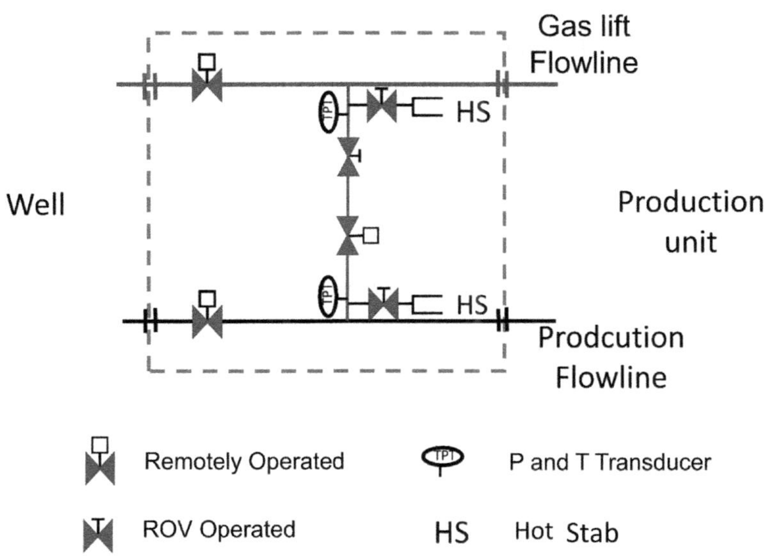

FIG. 16.26

Riser base gas lift schematic.

Reprinted with permission from C.A.B.R. Cardoso, M.A.L. Gonçalves, R.M.T. Camargo, Design Options for Avoiding Hydrates in Deep Offshore Production, J. Chem. Eng. Data 60(2) (2015) 330–335, https://doi.org/10.1021/je500601f.N. Copyright 2015, American Chemical Society.

16.3.1.2 Purging

Purging with nitrogen, air, or some other nonhydrate-forming or hydrate-forming gas that forms hydrates only at a higher pressure than the light hydrocarbon gas-water hydrate is effective in removing the light hydrocarbon gas-water hydrate [141,142]. The method is based on the chemical potential difference for methane in the hydrate and gas phases. Nitrogen is considered to be a thermodynamic inhibitor as it shifts the hydrate stability conditions to higher pressures and lower temperatures [143].

16.3.1.3 Use of heat

Increasing the temperature to melt down the hydrate plugs is well known. Heating can be supplied by many means, as stated in Section 16.2.1.4, including [16,45,144]:

- Line heaters
- Heat tracing
- Hot oiling
- Pipe warm-up
- Hot water circulation through coiled tubing
- Heat-generating chemicals

Heat application must be done properly. Plug location, dimensions, number of plugs, and other parameters must be known. Heat must be applied evenly on the plug on the ends, not in the center point, to avoid pressure build-up by the dissociated gas (knowing that every volume of dissociated hydrates releases 180 volumes of gas [135]), which with the heat application can cause pipeline rupture.

16.3.2 Chemical removal of hydrate plugs

There are two methods for removing a hydrate plug by chemical treatment:

- Use of thermodynamic inhibitors (THIs)
- Use of heat-generating chemicals

16.3.2.1 Removal using thermodynamic inhibitors

THI chemicals are not only useful for inhibiting gas hydrate formation, but can also melt hydrate (or ice) plugs that have formed. Methanol and MEG are the preferred THIs for hydrate plug melting.

The efficiency of this remedy depends on many factors [145,146]:

- The plug properties, (porous, dense, etc.)
- The density and viscosity of the THI
- Distribution of the THI
- The contact area between the THI and the plug
- Turbulence of the system
- The temperature during plug removal

Generally, for very porous plugs, the low viscosity of methanol seems to be beneficial during melting; it is also useful in melting permeable/porous hydrate plugs, or when a high degree of turbulence can be obtained in the liquid system. Otherwise, denser MEG can be more efficient for other types of plugs, as it has the ability to penetrate into a compact plug and cause melting. However, when used at high concentrations, MEG may freeze out as a solid or gel, which can lead to reduced hydrate melting efficiency. At high pressures and in the presence of a hydrocarbon liquid, TEG seems to easily freeze and become inefficient for melting hydrate plugs [146].

These chemicals are usually added by being bullheaded into the plug in the well or in the pipelines. Gounah et al. [147] reported a new, cost-effective, time-saving alternative for removing hydrate plugs offshore, which has been applied three times in the Gulf of Mexico, with 100% success. In this method, instead of deploying the chemicals by bullheading or coil tubing, they instead used 3/8 in. tubing and a capillary tubing unit, to deliver the chemicals directly to the top of the hydrate plug, allowing the plug to be dissolved. The THI chemicals are also used with other depressurization and heat methods to furnish receiving liquids for the loose moving plug and to prevent recrystallization after dissociation.

16.3.2.2 Heat-generating chemicals

Heat-generating chemicals have been used to remove both hydrate and wax deposits [148]; they are also known as nitrogen-generating systems (SGN, or NGS process), nitrogen/heat generating systems (N/HGS), or chemically autogenetic heat. They are based on exothermic chemical reactions that produce heat after being delivered to the location of the plug. The last years have witnessed many advances in such systems. One of the common systems is composed of a mixture of an ammonium salt and sodium nitrite and an acid catalyst (e.g., acetic acid) or precatalyst, which generates acid in situ. The acid catalyzes decomposition of the ammonium nitrite to yield sodium chloride, nitrogen, water, and heat [16].

The method was successfully applied in many cases, e.g., a subsea clogged Christmas tree [149] and oil-based filtercake damage from a sandstone horizontal well [150].

This method must be used with caution, as the chemicals/catalysts must be mixed right before the plug. Wrong design or implementation of the method exerts high stress on the pipes, causing their failure and rupture.

16.3.3 Nonchemical methods of removal of gas hydrate plugs

16.3.3.1 Ultrasonic methods

Miura et al. [126] proposed a method for underwater ultrasonic wave irradiation to promote the dissociation of methane hydrate on the basis of the facilitator effect.

Ultrasonic waves at 20 kHz were found to augment the dissociation of R-134a gas hydrates (a gas that can form hydrates at a relatively low pressure). The results showed that not only does the dissociation rate increase with an increase in the ultrasonic wave power, but also the ultrasonic waves effect superseded the application of an equivalent amount of energy in the form of heat at the same location [128]. Other attempts to study the effect of ultrasonic waves on gas hydrates have been reported [151].

16.3.3.2 Mechanical methods

Mechanical methods such as jetting, milling, pigging, and scraping have been attempted. Better options are by using coiled tubing to free the hydrate plug by removing liquid hydrostatic head at the hydrate face to enhance depressurization, or by jetting inhibitor directly at the hydrate. The performance of coiled tubing can be enhanced by using hot fluids, which combines the effect of different methods.

16.3.4 Tips for efficient gas hydrate blockage removal

- **General considerations**

The removal of gas hydrate plugs entails complex technical operations and is usually associated with high safety concerns such as pipes rupturing, equipment damage, fire hazards, and harm to operators. Therefore there are two mandatory requirements when a hydrate plug is encountered: first, seek help from a gas hydrates specialist to properly assess the risks associated with the removal process, and second, prepare a robust safety risk assessment if any unexpected events arise during the plug removal to assure a safe procedure. The Canadian Association of Petroleum Producers Hydrate Guidelines [152] propose three safety measures:

(1) Always assume multiple plugs.
(2) Understand that an attempt to move ice (hydrates) may rupture pipes and vessels.
(3) Even though heating seldom is an option for buried flowlines, all attempts should be made at the hydrate plug's ends; this way, the gas is released and does not build up a dangerous pressure.

These efficient and safe plug removal procedures are not necessarily standard procedures, as they vary from field to field depending on:

- The production system (design, status).
- Plug size.
- Plug location (well, flowline, equipment, etc.).
- Plug specifications (density, shear strength).
- Available removal technique.
- Available time for plug removal.

Plug location can be estimated using sensors impeded in the system, using predictive models, or can be calculated from system parameters. Solids monitoring methods were reviewed in Chapter 13, "Monitoring of flow assurance solids in oil and gas fields." Locations known for liquid accumulations and hold up are more vulnerable to hydrates formation and should be checked first. Liquid hold-up monitoring devices like densimeters can also give a good indication of plug formation.

Plug size can be estimated using process data, predictive models, and system geometry. Plug size is roughly estimated by assuming that all the water in the line will convert into gas hydrates.

System aspects such as corrosion status, service lifetime, pipe wall thickness, and maximum allowable pressure are important parameters for designing the removal. Natural gas is usually released upon plug dissolution, which increases the system pressure and in severe cases the pipeline can burst.

It is crucial to choose the optimum removal method. Depressurization is a very effective and cheap method that can be used for removal of most plugs; however, it is a slow process that can take weeks to remove a plug. Mechanical methods, however, are fast in removing the plugs but they require access to the system and must be able to reach the plug to interact with it. Chemical methods are also effective, and they are commonly pumped in the system, combined with other methods, to improve the efficiency of the removal, to stabilize the dissolved mixture, and to furnish a plug-receiving matrix/cushion to reduce the impact of projectile-like plugs. Thermal methods are very effective for different applications.

From a technical standpoint, these few tips are recommended:

- A combination of these methods is employed to improve the removal: chemical-mechanical, chemical-thermal-mechanical, etc.
- Increasing the pressure during plug removal is not recommended, since it can worsen the problem by making the plug harder and very difficult to dissolve.
- Always add chemicals to the system to furnish a receiving matrix of the plug to attenuate high-speed plugs, thus reducing their impacts. It also increases dissolution of the plug and prevents re-accumulation of dissolved components.

In the case of wells, the following tips are recommended [135]:

- Plug location in the wells can be identified using production data (P, T, and flow), mechanical intervention jobs (wireline, well gauging, or coil tubing), production logs, downhole nuclear densitometer, DTS temperature monitoring device.
- Simulation models are used to obtain more information about the plugging.
- Make sure the well integrity is in a good case to avoid unexpected failures.
- Thermal methods can be applied by hot fluid circulation in the annulus, injecting hot chemical dissolver or using heating methods equipped in the completion. Thermal methods must be applied properly on the plug or it will create a high-speed plug, causing a safety hazard.
- Mechanical methods can be used and combined with hot fluids circulated in the annulus to improve the efficiency of dissolution. Hot chemical dissolvers can also be added to improve the dissolution efficiency. Coiled tubing was used to circulate hot glycol in a gas condensate well in offshore South America to remove hydrate blockage [153]. Furthermore, heated wirelines can also be applied.
- Hot chemicals should be added properly, meaning methanol should be added below the plug, and glycol added above the plug [this can be achieved through access ports, i.e., gas lift valves (GLVs), or surface controlled subsurface safety valves (SCSSV)], due to their density difference. Chemicals can be added above the plug using capillary tubing for more efficient removal [147]. Chemicals can also be bulheaded above the plug from the Christmas tree, or using coil tubing.
- Depressurization can be achieved one-sided from the Christmas tree and must be achieved slowly to avoid high-speed projectile plugs.
- The system should be flooded with chemical dissolver to improve the dissolution efficiency, to stabilize the dissolved mixture, and to attenuate the impact of high-speed plugs.

In the case of flowlines, the following tips are recommended [135]:

- Locate the plug using process data, using sensors, or using simulation methods.
- Depressurization is a very effective method of removing plugs in flowlines. Several successful case studies were reported, one of the common ones being the use of depressurizing methods to remove hydrate blockage in a $>$20 miles subsea tieback in the Gulf of Mexico [154]. Depressurization can be done one-sided or two-sided. Two-sided is preferred for safety reasons.
- Depressurization may not be applicable for risers and subsea flowlines with fluid head that is forbidding effective depressurization; in this case mechanical methods are the best option. Depressurization cannot be applied if it entails any safety hazards.
- Mechanical methods are also effective in flowlines. Coiled tubing was deployed to remove hydrate blockage in a Statoil offshore gas field in the Barents Sea as well.

- Chemical methods can be applied as hot chemicals and in combination with other methods.
- The system should be flooded with chemical dissolver to improve the dissolution efficiency, to stabilize the dissolved mixture, and to attenuate the impact of high-speed plugs.

The following equipment is recommended:

- Plug formation in equipment is usually monitored by the process parameters and equipment performance. Visual inspection using borescope or similar tools can be used.
- Depressurization is usually applied to remove the plug, which can be supported by chemicals added and mechanical methods to improve the dissolution.
- The applied method will basically depend on the accessibility of the equipment and its location. Inaccessible and subsea equipment will be the hardest to clean. Accessible surface equipment supplied by heating gears will be the easiest to clean.
- Circulating heated chemicals using surface heat exchanger or coil tubing is very effective in cleaning most equipment.

16.4 Strategies for gas hydrates control

16.4.1 Risk assessment

Gas hydrates risk assessment starts with the early stages of field development. The initial step is obtaining as much as possible on the main system specifications, i.e., design, expected operating pressures, temperatures (topsides, subsea, seabed, and seasonal temperatures). Once fluids samples are available, a risk assessment plan should be put into action.

16.4.1.1 Fluids sampling and analysis

The assessment starts with fluids analysis and characterization. Gas, oil, water initial volumes, physical properties, and chemical composition will be the main lab work in this early step.

It is also beneficial to perform gas hydrate deposition studies, using flow loop or rocking cells, especially if the prediction calculations (in the next step) showed gas hydrate formation tendencies or the analogous fields had a history of hydrates formation. Hydrate inhibitors can be tested and screened for application.

16.4.1.2 Modeling and simulation

The next step is to perform modeling and simulations with the data in hand: fluids composition, system design, parameters, hydraulics, etc. are integrated in one or more of the gas hydrate simulation/prediction models in addition to the computationl fluid dynamics modeling (CFD). Gas hydrate prediction is discussed in Chapter 12, "Flow assurance solids prediction and modeling". The outcome from the prediction is a hydrate formation curve that defines the safe zones of operations and the suggested doses of inhibitors. The modeling also indicates the proper design specification to avoid hydrates formation. Furthermore, modeling also indicates the need for thermal methods, e.g., insulation and its proposed properties. Modeling is a crucial step in gas hydrate management.

Besides modeling, the experience from analogous fields can also be studied and combined with the field under investigation to benefit from the previous experience in assessing and managing gas hydrates.

16.4.1.3 Process monitoring

This is based on continuously monitoring the process parameters, to determine the potential of hydrate formation. Locations of low temperature, liquid hold up, and large pressure are more vulnerable to hydrate formation.

Combining the results from the experimental work and the simulation calculations will give a reasonable estimation of the gas hydrates risk and possible locations of formation in the system.

If the field did not start production yet, this risk assessment can be used to modify and optimize the design (adding heat, insulation, chemical injection point) to be less prone to gas hydrates. In systems that are already in the production phase, then the main goal is to optimize the system to minimize the hydrates risk.

16.4.2 Selecting the proper control method

The management strategy will be based on many factors: project size, system design, production fluids volumes, gas hydrate formation rates and sizes, availability of the management technique, and cost of the management method.

Operating conditions management is the cheapest method of management. Operating within the pressure-temperature-free window is a cost-effective method. Although inapplicable in some instances, it is considered a standard approach to operate a system in the pressure-temperature-free zone in deepwater systems [2,143].

Chemical mitigation methods are commonly in use in oil and gas fields. However, they come with a huge cost, especially the THIs when the produced fluids volumes are high. LDHIs are lower in dosage and volumes than the THIs, but they are more expensive than the THIs.

Thermal methods can be used independently or jointly; however, it may not be cost-effective for longer flowlines to carry high GOR (gas/oil ratio) fluids [52].

Large projects with huge production volumes and a high possibility of hydrates formation will consume tremendous amounts of THIs and relatively smaller amounts of LDHIs, which will lead to high OPEX of gas hydrates mitigation and production in general. Therefore in such projects it is more convenient to invest more in operational and thermal methods to lower the use and costs of chemicals. In this case, subsea dehydration, cold flow, internal coating, insulation, and active heating methods will be good CAPEX money to spend to alleviate the bitter effect of gas hydrates.

Gas hydrate mitigation methods are summarized in Table 16.9.

16.4.3 Monitoring and assessment

After applying the mitigation method, continuous monitoring is mandatory to assess the effectiveness of the applied methods and to optimize the operations if necessary.

Fig. 16.27 summarizes the gas hydrates management strategy.

16.5 Recent advances in gas hydrates management methods

Several advances in gas hydrates have been reported. Basically, gas hydrate deposition studies have been improved to understand the various factors that affect gas hydrates formation. These experimental data were then used and implemented in gas hydrates prediction models. Chemical management of gas

Table 16.9 Gas hydrates mitigation methods.

Mitigation method	Description
Operational methods	– Used in gas hydrate prevention and blockage removal – Generally very effective and economical – Depressurization is effective in plugs removal – May require system pre-design or system modification – Cold flow can be applied in severe cases where chemicals and thermal methods are so expensive
Thermal methods	– Used in gas hydrates prevention and removal – Very successful in field applications – Crucial in subsea and deepwater installations with potential wax and hydrate problems – Insulation and electric heating require pre-design or system modification
Chemical methods	– Used in prevention and removal – Wide variety of chemicals for different applications – Generally, do not require system modification – Cost effective, but can be expensive in severe cases that require high volumes of chemicals – Essential in plug removal to prevent reprecipitation, ice formation, and to avoid projectile plugs – Can be combined with heat and mechanical methods to improve performance and remove hard plugs – Mandatory during shutdown and system startup to avoid plugging – Chemicals can have side effects, e.g., methanol can enhance other solids deposition, corrosion. KHIs have compatibility issues with other production chemicals
Mechanical methods	– Used mainly in removal – Routine pigging used to dry pipelines and hence to avoid hydrates formation – Can be used in different cases, e.g., topsides, downhole – Can be used with other methods, e.g., chemical methods, application of heat, to improve its efficiency. Coiled tubing combines heat, chemical, and mechanical effects – System design and accessibility dependent
Other nonchemical methods	– Internal coatings have showed success in field application – Biological, magnetic, ultrasonic, methods are not common in field application – May require system modification for installation – They can be case dependent

hydrates has improved by employing new chemical types and improving the conventional chemicals. Nonchemical methods such as ultrasonic and microbial methods have been reported, and are expected to improve in the future with the current wave of using environmentally friendly methods. Gas hydrate monitoring methods have also improved, e.g., velocity-conductivity techniques and also integration of artificial intelligence methods with them.

16.6 Case studies

Case study: Gas hydrates mitigation during deepwater drilling in Qiongdongnan Basin, South China Sea [155].

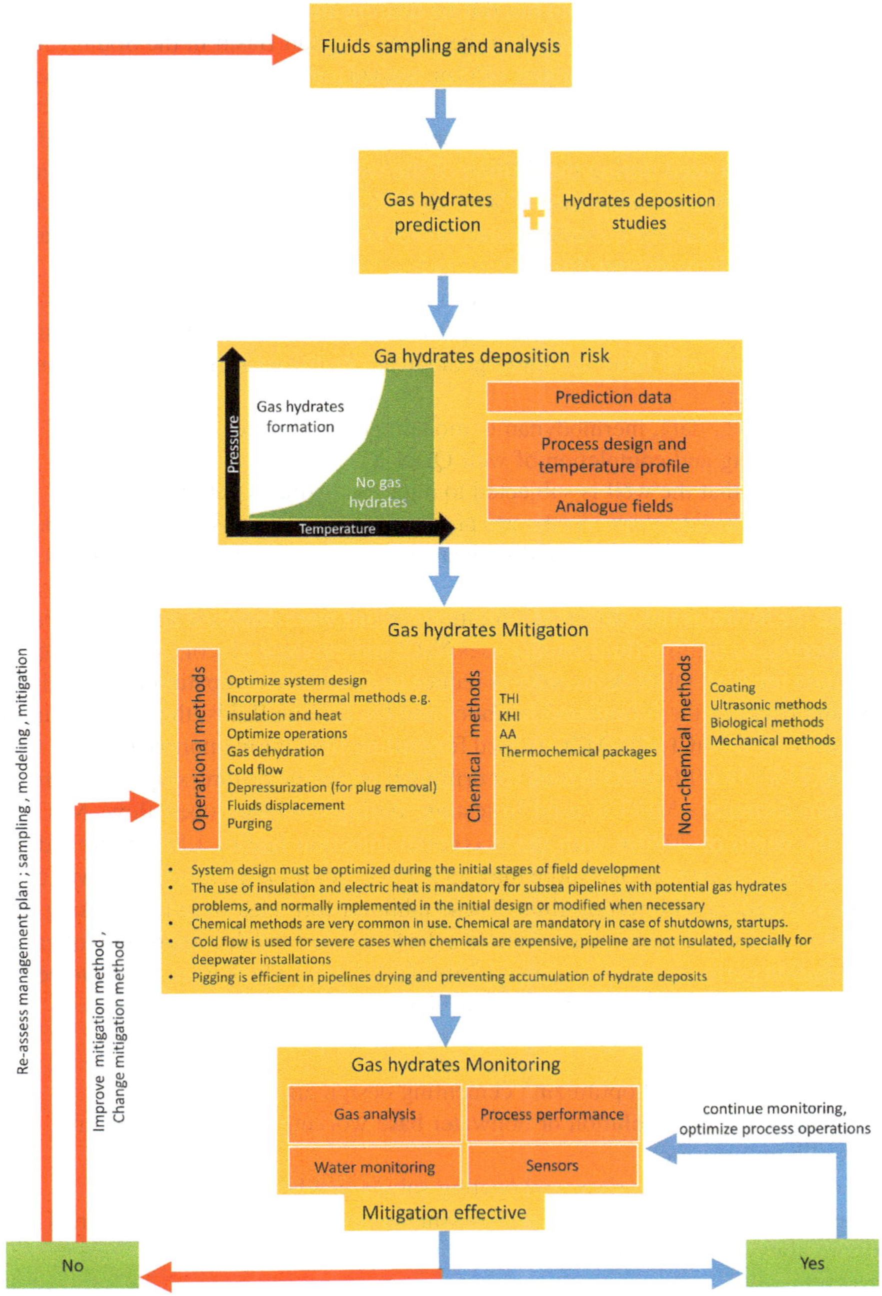

FIG. 16.27

Gas hydrates management strategy.

This case study discusses the hydrate risk during drilling and completion of a deepwater exploration well of natural gas, Well QDN-X, located in the Qiongdongnan Basin in the South China Sea.

The well area has a water depth of about 1455 m, seabed temperature of 3–4°C, geothermal gradient of 4.4°C/100 m, and pressure coefficient of 1.24–1.30. The well, designed with a depth of 3561 m, is expected to have a bottom temperature of 95°C and maximum bottom pressure of 45.32 MPa. While oil-based mud has been used during the drilling of the wells in the Qiongdongnan Basin to avoid gas hydrates formation problems, in the case of Well QDN-X, water-based muds were selected in favor of environmental and cost-saving considerations; however, using water-based muds was expected to cause gas hydrate problems during drilling.

To estimate the risk, a field analogy was used, as the operators used the gas composition and hydrate phase curves from the test data of Well LS22-Y, as both of them were located in the central valley belt of Qiongdongnan Basin, and have the same geological conditions of reservoir forming in the same target natural gas layer.

Using the well LS22 data, thermodynamic modeling was applied, which indicated a high risk of hydrates during drilling and completion of well QDN-X.

The conclusion is that the well is subjected to high gas hydrates risk during normal drilling operations and also when drilling stops. As the gas production rate increases, the well section with hydrate risk becomes smaller and smaller, until it disappears in the whole wellbore when the gas rate surpasses $25 \times 104\,m^3/d$.

To mitigate the hydrates risk, different inhibition options were designed and theoretically predicted. Currently, water-based drilling fluid usually uses NaCl and ethylene glycol (MEG) as the main hydrate inhibitor formula for Well QDN-X, in order to reduce costs and the compatibility issues in the inhibitor formulation and between the formulated inhibitor and the drilling fluids; the drilling fluid with an inhibitor formula of 17% NaCl+2% MEG is used for drilling fluid during normal drilling; the formula of 20% NaCl+10.71%–18.00% MEG is used for drilling fluid during drilling stops; the formula of $CaCl_2$/KFo (potassium formate)+MEG is used in testing; during throttling open flow, MeOH is injected downhole at the depth of 2080 m with the maximum injection rate of 1.86 L/min, and the injection is stopped when the gas rate reaches more than $25 \times 104\,m^3/d$; if the well is shut down for a long time, the test string is prefilled with testing fluid. These inhibitor formulas were tested using high pressure stirred hydrate experimental apparatus to simulate the formation and decomposition of gas hydrate and assess the inhibition performance of different kinds of inhibitors, which were shown to be effective and to meet the requirements of inhibition during drilling and testing.

Other measures that were designed to avoid the risk of hydrates include:

- monitoring gas invasion and optimizing cementing design and operation
- prefilling hydrate inhibitor solution or nonwater base working fluid in the pipelines
- reducing wellbore pressure
- applying insulation to marine riser, and heating subsea wellhead
- monitoring the changes of temperature and pressure at key parts of the wellbore in real time to ensure the wellbore always stays out of hydrate stability zone

In the case of gas hydrate blockages in the wellbore, the treatment measures include:

- pumping heated thermodynamic inhibitor slug through coiled tubing (e.g., MeOH, MEG, NaCl, and $CaCl_2$) into the wellbore to flush and remove solid hydrate

- injecting light density drilling fluid into the marine riser to reduce the pressure in the wellbore to remove solid hydrate
- pulling out BOP. The method most commonly used is injecting hot thermodynamic inhibitors to flush the hydrate plugging

The well was drilled in 43 days at a total depth of 3510 m, water depth of 1447 m, and mud line temperature of 3.60°C. The well was drilled at the drilling fluid pumping rate of 63 L/s in the target, the measured temperatures of returned drilling fluid at wellhead and mud line annulus of 11.90°C and 23.50°C, respectively. The hydrate preventive measures taken during the drilling process include adopting drilling fluid with NaCl + MEG inhibitors, pre-filling inhibitor solution into the control pipelines, and monitoring the gas invasion in the wellbore and the temperatures and pressures at subsea BOP and the surface. Field application shows that no hydrate blockage in the wellbore and control pipelines was observed during the drilling process; moreover, the predicted wellbore temperature only has a small error with the measured temperature (the predicted seabed temperature is 3.04°C; the predicted temperatures of returned drilling fluid at mud line and at surface are 24.05°C and 12.99°C, respectively).

16.7 Summary

In this chapter, gas hydrates mitigation methods were reviewed. Gas hydrates mitigation basically depends on operational, chemical, and nonchemical methods. The operational methods are based on optimizing the operating conditions (P, T) to be outside the hydrate formation zone, eliminating water from the flowing stream by dehydration, using cold flow, and displacing fluids. The chemical methods are based on using thermodynamic hydrate inhibitors, kinetic inhibitors, or the antiagglomerants. The nonchemical methods are based on using thermal methods, hydrate repellent surfaces, and ultrasonic or biological methods.

Gas hydrates removal is based on using operational methods such as thermal methods, or depressurization. Chemical methods are based on using thermodynamic hydrate inhibitors to dissolve the plug. Mechanical and ultrasonic methods can also be used.

Gas hydrate management involves fluid characterization, simulation and modeling, mitigation method selection and application, and then the process is monitored for optimization.

References

[1] R. Sivaraman, Flow Assurance: "Understanding and Controlling Natural Gas Hydrates", Gas Technology Institute, 2002, pp. 18–23.

[2] A. Ballard, G. Shoup, D. Sloan, Industrial operating procedures for hydrate control, in: Natural Gas Hydrates in Flow Assurance, Elsevier, 2011, pp. 145–162, https://doi.org/10.1016/B978-1-85617-945-4.00007-8.

[3] N. NouriSamie, Systems and equipment for offshore platform design, in: Practical Engineering Management of Offshore Oil and Gas Platforms, Gulf Professional Publishing, USA, 2016.

[4] M. Netušil, P. Ditl, Natural gasp dehydration, in: S.B. Gupta (Ed.), Natural Gas Extraction to End Use, IntechOpen, October 31, 2012, https://doi.org/10.5772/45802. Available from: https://www.intechopen.com/books/natural-gas-extraction-to-end-use/natural-gas-dehydration.

[5] M. Stewart, K. Arnold, Dehydration considerations, in: Gas Dehydration Field Manual, Elsevier, 2011, pp. 55–168, https://doi.org/10.1016/b978-1-85617-980-5.00002-1.
[6] F. Fournié, J.P. Agostini, Permeation: a new competitive process for offshore gas dehydration, in: Proceedings of the Annual Offshore Technology Conference, 1984, May, Offshore Technology Conference, 1984, pp. 113–120, https://doi.org/10.4043/4659-ms.
[7] M. Mitariten, S. Mokhatab, Efficient Membrane Systems for Natural Gas Dehydration. Applications of Membrane Technology for Reducing Water Content in Natural Gas to Meet Pipeline Specifications, 2018, Digital Refining Retrieved From: https://www.digitalrefining.com/article/1001543/efficient-membrane-systems-for-natural-gas-dehydration#.YHsC0-hKg2x. (Accessed April 2021).
[8] F.R. Redus, Field operating experience with calcium chloride gas dehydrators, World Oil (1 February) (1966) 63.
[9] R. Stromquist, S. Gustafson, SUBSIS—the World's first subsea separation and injection system, World Pumps 1999 (390) (1999) 33–36, https://doi.org/10.1016/S0262 1762(00)87640-3.
[10] B. Bringedal, T. Ingebretsen, K. Haugen, Subsea separation and reinjection of produced water, in: Offshore Technology Conference, 1999, January 1, https://doi.org/10.4043/10967-MS.
[11] V. Alary, F. Marchais, T. Palermo, Subsea water separation and injection: a solution for hydrates, in: Offshore Technology Conference, 2000, January 1, https://doi.org/10.4043/12017-MS.
[12] J. Gudmundsson, Cold flow hydrate technology, in: 4th International Conference on Gas Hydrates, May 19–23, 2002, Yokohama, 2002.
[13] D. Turner, L. Talley, Hydrate inhibition via cold flow—no chemicals or insulation, in: Proceedings of the 6th International Conference on Gas Hydrates (ICGH 2008), Vancouver, British Columbia, CANADA, July 6-10 2008, 2008.
[14] E.O. Straume, R.E.M. Morales, A.K. Sum, Perspectives on gas hydrates cold flow technology, Energy Fuel 33 (1) (2019) 1–15, https://doi.org/10.1021/acs.energyfuels.8b02816.
[15] T. Vinatovskaja, Cold Flow in the Arctic: A Feasibility Study (Master's thesis), University of Stavanger, 2015.
[16] M.A. Kelland, Production Chemicals for the Oil and Gas Industry, CRC Press, 2014, https://doi.org/10.1201/9781420092974.
[17] M. Ilahi, Cold Flow Concepts Compared, NTNU, Norway, 2005.
[18] C.J. Coberly, Method for Preventing Wax Deposits in Tubing, 1942. US Patent 2303823.
[19] R. Larsen, A. Lund, C.B. Argo, Cold flow—a practical solution, in: 11th International Conference on Multhiphase Flow "Multiphase 03" BHR Group, San Remo, 11–13 June 2003, 2003. https://www.forskningsradet.no/CSStorage/Vedlegg/149604artikkel,0.pdf.
[20] H. Haghighi, R. Azarinezhad, A. Chapoy, R. Anderson, B. Tohidi, Hydraflow: Avoiding Gas Hydrate Problems, Society of Petroleum Engineers, 2007, January 1, https://doi.org/10.2118/107335 MS.
[21] L.D. Talley, D.J. Turner, D.K. Priedeman, Method of Generating a Non-Plugging Hydrate Slurry, International Patent No. WO 2007/095399, September 23, 2007.
[22] A. Lund, R. Larsen, J.H. Kaspersen, E.O. Straume, M. Fossen, K.W. Hjarbo, Treatment of Produced Hydrocarbon Fluid Containing Water, U.S. Patent US2011/0220352 A1, September 15, 2011.
[23] Y. Bai, Q. Bai, Y. Bai, Q. Bai, Hydrates, in: Subsea Engineering Handbook, Elsevier, 2010, pp. 451–481, https://doi.org/10.1016/B978-1-85617-689-7.10015-9 (Chapter 15).
[24] O.C. Aja, M. Ramasamy, Thermal management of flow assurance challenges in offshore fields—a review, ARPN J. Eng. Appl. Sci. 11 (2006) 6415–6422.
[25] R.N. Tucker, P.R. Hays, J.K. Antani, Insulated flowline technology for deep water, in: Offshore Technology Conference, 1996, https://doi.org/10.4043/8247-ms.
[26] N. Sunday, A. Settar, K. Chetehouna, N. Gascoin, An overview of flow assurance heat management systems in subsea flowlines, Energies 14 (2) (2021) 458, https://doi.org/10.3390/en14020458.

[27] X. Wang, Q. Qu, P.H. Javora, R.G. Pearcy, New trend in oilfield flow assurance management: a review of thermal insulating fluids, in: International Oil & Gas Conference and Exhibition in China, 2006, https://doi.org/10.2118/103829-ms.

[28] B. Ansart, A. Marret, T. Parenteau, O. Rageot, Technical and economical comparison of subsea active heating technologies, in: Proceedings of the Annual Offshore Technology Conference, vol. 1, 2014, pp. 332–345, https://doi.org/10.4043/24711-ms.

[29] L. Jaeyoung, Design and Installation of Deepwater Petroleum Pipelines, 2002. Houston, Texas.

[30] M.M. Myo Thant, M.T. Mohd Sallehud-Din, G.F. Hewitt, C.P. Hale, G.L. Quarini, Mitigating flow assurance challenges in deepwater fields using active heating methods, in: SPE Middle East Oil and Gas Show and Conference, MEOS, Proceedings, 1, 2011, pp. 570–579, https://doi.org/10.2118/140997-ms.

[31] R. Hunter, Wet insulation evolves to meet subsea flowline performance demands, as production moves deeper, new technologies emerge, Offshore Magazine (2008).

[32] K. Shukla, Non-Chemical Products Offer Effective Flow Assurance Solutions, Offshore, McDermott Corporation, Texas, 2014.

[33] K. Falk, V. Kaste, S. Håheim, P. Damsleth, Subsea Heat Bank—an alternative thermal insulation method, in: Proceedings—SPE Annual Technical Conference and Exhibition, Society of Petroleum Engineers (SPE), 2002, pp. 2123–2132, https://doi.org/10.2118/77577-ms.

[34] G.J. Zabaras, J. Zhang, Steady-state and transient thermal performance of subsea hardware, in: Paper Presented at the Offshore Technology Conference, Houston, Texas, May, 1997, https://doi.org/10.4043/8544-MS.

[35] Z.M. Aman, M. Di Lorenzo, K. Kozielski, C.A. Koh, P. Warrier, M.L. Johns, E.F. May, Hydrate formation and deposition in a gas-dominant flowloop: initial studies of the effect of velocity and subcooling, J. Nat. Gas Sci. Eng. 35 (2016) 1490–1498.

[36] F. Fleyfel, O. Hernandez, R. Sturgis, W. Meng, Evaluation of pipeline configurations with active heating for export of waxy crude oil, in: Proceedings of the SPE Annual Technical Conference and Exhibition, Society of Petroleum Engineers, Richardson, TX, USA, 2004.

[37] E. Louvet, S. Giraudbit, B. Seguin, R. Sathananthan, Active heated pipe technologies for field development optimisation, in: Offshore Technology Conference Asia 2016, OTCA 2016, 2016, pp. 58–78, https://doi.org/10.4043/26578-ms.

[38] F. Fouad, H. Oris, S. Rusty, M. Weihong, Evaluation of pipeline configurations with active heating for export of waxy crude oil, in: SPE 90054, Houston, Texas, 26-29 September 2004, 2004.

[39] C. Candelier, S. Durica, F. Beys, Subsea pipeline electrical heat trace (EHT)—'active' heating—application for a deep water brown field development, in: Offshore Mediterranean Conference and Exhibition, OMC 2015, 2015.

[40] L. Abney, A new approach to flowline active heating, in: SPE Asia Pacific Oil and Gas Conference and Exhibition, 2010, https://doi.org/10.2118/133370-ms.

[41] F.M. Pattee, F. Kopp, Impact of electrically-heated systems on the operation of deep water subsea oil flowlines, in: OTC11894, Offshore Technology Conference, Houston, Texas, May, 2000.

[42] T. Hoist, T. Larsen, A. Straarup, K. Glejbol, K.S. Olsen, Direct electrical heating of a flexible pipe, in: Proceedings of the Annual Offshore Technology Conference, vol. 1, 2016, pp. 824–832, https://doi.org/10.2118/1116-0076-jpt.

[43] S. Chioetto, G. De Ghetto, Testing program probes capabilities and potentialities of TFL systems, in: Paper Presented at the Offshore Technology Conference, Houston, Texas, May, 1992, https://doi.org/10.4043/6990-MS.

[44] R.F. Stoisits, D.C. Lucas, L.D. Talley, D.P. Shatto, J. Cai, International Patent Application WO/2009/042319, 2009.

[45] J. Carroll, Inhibiting hydrate formation with chemicals, in: Natural Gas Hydrates, Elsevier, 2014, pp. 163–208, https://doi.org/10.1016/b978-0-12-800074-8.00005-3.

[46] B. Guo, S. Song, A. Ghalambor, T. Ran Lin, B. Guo, S. Song, A. Ghalambor, T. Ran Lin, Flow assurance, in: Offshore Pipelines, 2014, pp. 179–231, https://doi.org/10.1016/B978-0-12-397949-0.00015-7 (Chapter 15).

[47] H. Williams, T. Herrmann, M. Jordan, C. McCallum, The impact of thermodynamic hydrate inhibitors (MEG and methanol) on scale dissolver performance, in: Paper presented at the SPE International Oilfield Scale Conference and Exhibition, Aberdeen, Scotland, UK, May, 2016, https://doi.org/10.2118/179864-MS.

[48] Y.H. Sohn, Y. Seo, Effect of monoethylene glycol and kinetic hydrate inhibitor on hydrate blockage formation during cold restart operation, Chem. Eng. Sci. 168 (2017) 444–455, https://doi.org/10.1016/j.ces.2017.05.010.

[49] W. Ke, D. Chen, A short review on natural gas hydrate, kinetic hydrate inhibitors and inhibitor synergists, Chin. J. Chem. Eng. (2018), https://doi.org/10.1016/j.cjche.2018.10.010.

[50] J.H. Sa, G.H. Kwak, B.R. Lee, D.H. Park, K. Han, K.H. Lee, Hydrophobic amino acids as a new class of kinetic inhibitors for gas hydrate formation, Sci. Rep. 3 (2013), https://doi.org/10.1038/srep02428.

[51] T. Knott, Holding Hydrates at Bay, Offshore Engineer, 2001, p. 29.

[52] B. Fu, S. Neff, A. Mathur, K. Bakeev, Application of low-dosage hydrate inhibitors in deepwater operations, SPE Prod. Facil. 17 (03) (2002) 133–137, https://doi.org/10.2118/78823-pa.

[53] S.A. Morrissy, T.O. Kuteyi, M.L. Johns, E.F. May, Z.M. Aman, S.F. McKay, Quantitative ranking and development of hydrate anti agglomerants, in: Offshore Technology Conference Asia, 2018, https://doi.org/10.4043/28578-ms.

[54] E. Broni-Bediako, R. Amorin, C.B. Bavoh, Gas hydrate formation phase boundary behaviour of synthetic natural gas system of the Keta basin of Ghana, Open Pet. Eng. J. 10 (2017) 64–72, https://doi.org/10.2174/1874834101701010064.

[55] M.A. Kelland, Challenges with gas hydrate formation, in: IOP Conference Series: Materials Science and Engineering, vol. 700, IOP Publishing Ltd, 2019, https://doi.org/10.1088/1757-899X/700/1/012057.

[56] R. Anderson, A. Chapoy, J. Tanchawanich, H. Haghighi, J. Lachwa-Langa, B. Tohidi, Binary ethanol-methane clathrate hydrate formation in the system CH4-C2H5OH-H2O: experimental data and thermodynamic modelling, in: Proceedings of the 6th International Conference on Gas Hydrates Vancouver, British Columbia, Canada, 6–10 July, 2008.

[57] M. Sun, Y. Wang, A. Firoozabadi, Effectiveness of alcohol cosurfactants in hydrate antiagglomeration, Energy Fuel 26 (9) (2012) 5626–5632, https://doi.org/10.1021/ef300922h.

[58] Stewart, Hydrate prediction and prevention, in: Surface Production Operations, third ed., Elsevier, 2014.

[59] E.D. Sloan, Hydrate engineering, in: SPE Monograph, vol. 21, Society of Petroleum Engineers, Richardson, Texas, 2000.

[60] M.S. Khan, B. Lal, M.A. Bustam, Gas hydrate inhibitors, in: Green Energy and Technology, Springer Verlag, 2020, pp. 27–46, https://doi.org/10.1007/978-3-030 30750-9_2.

[61] Y. Wang, S. Fan, X. Lang, Reviews of gas hydrate inhibitors in gas-dominant pipelines and application of kinetic hydrate inhibitors in China, Chin. J. Chem. Eng. (2019), https://doi.org/10.1016/j.cjche.2019.02.023.

[62] H.-J. Ng, D.B. Robinson, New developments in the measurement and prediction of hydrate formation for processing needs, Ann. NY Acad. Sci. 715 (1994) 450–462.

[63] N.G. Park, L. Morello, J.E. Wong, S.A. Maksoud, The Effect of Oxygenated Methanol on Corrosion of Carbon Steel in Sour Wet Gas Environments, NACE International, 2007, January 1.

[64] P.F. Pickering, B. Edmonds, R.A.S. Moorwood, R. Szczepanski, M.J. Watson, Evaluating new chemicals and alternatives for mitigating hydrates in oil and gas production, in: A Paper Presented at the IIR Conference, Aberdeen, Scotland, 2001.

[65] B.S. Kiran, P.S.R. Prasad, Inhibition of methane hydrates using biodegradable additives, ACS Omega 6 (12) (2021) 8261–8270, https://doi.org/10.1021/acsomega.0c06328.

[66] B. Fu, S. Neff, A. Mathur, K. Bakeev, Novel low dosage hydrate inhibitors for deepwater operations, in: Proceedings - SPE Annual Technical Conference and Exhibition, Society of Petroleum Engineers (SPE), 2001, pp. 1271–1277, https://doi.org/10.2118/71472-ms.

[67] P.C. Chua, M.A. Kelland, Poly(N-vinyl azacyclooctanone): 1 a more powerful structure II kinetic hydrate inhibitor than poly(N-vinyl caprolactam), Energy Fuel 26 (2012) 4481–4485.

[68] Global Oilfield Solutions Low Dosage Hydrate Inhibitors for the Oilfield Industry: Luvicap, BASF, 2017.

[69] J. Yang, B. Tohidi, Characterization of inhibition mechanisms of kinetic hydrate inhibitors using ultrasonic test technique, Chem. Eng. Sci. 66 (3) (2011) 278–283, https://doi.org/10.1016/j.ces.2010.10.025.

[70] X. Zhao, Z. Qiu, G. Zhou, W. Huang, Synergism of thermodynamic hydrate inhibitors on the performance of poly (vinyl pyrrolidone) in deepwater drilling fluid, J. Nat. Gas Sci. Eng. 23 (2015) 47–54, https://doi.org/10.1016/j.jngse.2014.12.031.

[71] M.A. Kelland, A review of kinetic hydrate inhibitors from an environmental perspective, Energy Fuel (2018), https://doi.org/10.1021/acs.energyfuels.8b03363.

[72] W. Ke, M.A. Kelland, Kinetic hydrate inhibitor studies for gas hydrate systems: a review of experimental equipment and test methods, Energy Fuel (2016), https://doi.org/10.1021/acs.energyfuels.6b02739.

[73] Z.R. Chong, A.H.M. Chan, P. Babu, M. Yang, P. Linga, Effect of NaCl on methane hydrate formation and dissociation in porous media, J. Nat. Gas Sci. Eng. 27 (2015) 178–189, https://doi.org/10.1016/j.jngse.2015.08.055.

[74] A. Perrin, O.M. Musa, J.W. Steed, The chemistry of low dosage clathrate hydrate inhibitors, Chem. Soc. Rev. 42 (5) (2013) 1996, https://doi.org/10.1039/c2cs35340g.

[75] V.K. Walker, H. Zeng, H. Ohno, N. Daraboina, H. Sharifi, S.A. Bagherzadeh, P. Englezos, Antifreeze proteins as gas hydrate inhibitors, Can. J. Chem. 93 (8) (2015) 839–849, https://doi.org/10.1139/cjc-2014-0538.

[76] S. Xu, S. Fan, S. Fang, X. Lang, Y. Wang, J. Chen, Pectin as an extraordinary natural kinetic hydrate inhibitor, Sci. Rep. 6 (March) (2016), https://doi.org/10.1038/srep23220.

[77] S. Adham, S. Gharfeh, A. Hussain, J. Minier-Matar, A. Janson, Kinetic hydrate inhibitor removal by physical, chemical and biological processes, in: Offshore Technology Conference Asia, 2014, https://doi.org/10.4043/24807-ms.

[78] S. Gao, Hydrate risk management at high watercuts with anti-agglomerant hydrate inhibitors, Energy Fuel 23 (4) (2009) 2118–2121, https://doi.org/10.1021/ef8009876.

[79] M.A. Kelland, History of the development of low dosage hydrate inhibitors, Energy Fuel 20 (3) (2006) 825–847.

[80] D.S. Treybig, R. Fowles, D. Stannard, C.S. Smith, Use of anti-agglomerates for sour gases, high water cuts and its performance evaluation, in: SPE International Symposium on Oilfield Chemistry, 2015, https://doi.org/10.2118/173724-ms.

[81] T.Y. Makogon, Flow restrictions and blockages in operations, in: Handbook of Multiphase Flow Assurance, Elsevier, 2019, pp. 95–189, https://doi.org/10.1016/b978-0-12-813062 9.00005-1.

[82] P. Kulkarni, Flow assurance: hydrate inhibitors for deepwater flow assurance, Offshore 63 (3) (2003). https://www.offshore-mag.com/pipelines/article/16755698/flow-assurance-hydrate-inhibitors-for-deepwater-flow-assurance.

[83] E.G. Dirdal, M.A. Kelland, Does the cloud point temperature of a polymer correlate with its kinetic hydrate inhibitor performance? Energy Fuel 33 (8) (2019) 7127–7137, https://doi.org/10.1021/acs.energyfuels.9b01185.

[84] F. Tohidi, R. Anderson, B. Tohidi Kalorazi, Effect of acid gases on kinetic hydrate inhibitors, J. Chem. Thermodyn. (2017), https://doi.org/10.1016/j.jct.2017.09.008.

[85] M. Nazeri, B. Tohidi, A. Chapoy, An Evaluation of Risk of Hydrate Formation at the Top of a Pipeline, Society of Petroleum Engineers, 2012, January 1, https://doi.org/10.2118/160404-MS.

[86] X. Yu, P. Somasundaran, Kinetics of polymer conformational changes and its role in flocculation, J. Colloid Interface Sci. 178 (1996) 770–774.

[87] B. Fu, Sour gas pipelines with co-injection of kinetic hydrate inhibitors, in: NACE 2007, 2007.
[88] R. Masoudi, B. Tohidi, Experimental investigation on the effect of commercial oilfield scale inhibitors on the performance of low dosage hydrate inhibitors (LDHI), in: Proceedings of the 5th International Conference on Gas Hydrates, Trondheim, Norway, 12–16 June, 2006.
[89] X. Lang, P. Lv, S. Xu, B. Li, S. Fan, Y. Wang, Investigation of the performance of low-dosage hydrate inhibitors combined with wax inhibitors for a mixed CH4-CO2 gas hydrate formation, Can. J. Chem. 93 (2015) 1–6, https://doi.org/10.1139/cjc-2014-0595.
[90] H.J. Huo, R.H. Wang, H.J. Ni, Y.L. Liu, An experimental study on the synergetic effects of kinetic and thermodynamic gas hydrate inhibitors, Pet. Sci. Technol. 32 (16) (2014) 1940–1947, https://doi.org/10.1080/10916466.2012.756014.
[91] Y. Gao, Y. Wang, G. Xin, X. Wang, C. Yue, L. Chen, Experimental study on the effect of combination of thermodynamic inhibitors' and kinetic inhibitors' hydrate inhibition, Front. Energy Res. 9 (2021) 718673, https://doi.org/10.3389/fenrg.2021.718673.
[92] G.F. Mitchell, L.D. Talley, Application of kinetic hydrate inhibitor in black-oil flowlines, in: Proceedings—SPE Annual Technical Conference and Exhibition, vol. 2 (PI), Soc. Pet. Eng. (SPE), 1999, pp. 341–348, https://doi.org/10.2523/56770-ms.
[93] J. Fink, Petroleum Engineer's Guide to Oil Field Chemicals and Fluids, Elsevier, 2012, pp. 253–275, https://doi.org/10.1016/C2009-0-61871-7.
[94] H. Oskarsson, I. Uneback, et al., Evaluation of anti-agglomerants as gas-hydrate dispersant using multi-cell technology, in: Proceedings of the Fifth International Conference on Gas Hydrates, vol. 4, 2005, p. 12.
[95] PSL-Systemtechnik, Research of Gas Hydrates With the Sapphire Rocking Cell, 21 February 2011, Available from: http://www.psl-systemtechnik.de/sapphire_rocking_cell0.html?&L=1.
[96] M. Arjmandi, B. Tohidi, et al., Is subcooling the right driving force for testing low dosage hydrate inhibitors? Chem. Eng. Sci. 60 (2005) 8.
[97] A. Lund, D.E. Akporiaye, et al., Patent 6688180, 2004.
[98] D. Lippmann, D. Kessel, et al., Proceedings of the 5th International Offshore and Polar Engineering Conference, The Hauge, The Netherlands, 1995.
[99] A. Lund, O. Urdahl, et al., Proceedings of the 2nd International Conference on Natural Gas Hydrates, Toulouse, France, 1996, pp. 407–414.
[100] M. Pakulski, Proceedings of the SPE International Symposium on Oilfield Chemistry, Houston, Texas, SPE 37285, 1997.
[101] B. Cingotti, A. Sinquin, J.S. Pic, J.M. Herri, M. Cournil, A new method to investigate kinetic hydrate inhibitors, in: SPE International Symposium on Oilfield Chemistry, 1999, https://doi.org/10.2118/50757-ms.
[102] K.P. Mcnamee, Evaluation of kinetic hydrate inhibitor performance by high pressure differential scanning calorimetry, in: Offshore Technology Conference, 2011, January 1, https://doi.org/10.4043/21604-MS.
[103] C. Duchateau, T.-E. Pou, M. Hidalgo, P. Glénat, C. Dicharry, Chem. Eng. Sci. 71 (2012) 220–225.
[104] P. Bourg, P. Glénat, M.-L. Bousque, Selection of commercial kinetic hydrate inhibitors using a new crystal growth inhibition approach highlighting major differences between them, in: SPE Middle East Oil and Gas Show and Conference, 2013, https://doi.org/10.2118/164258-ms.
[105] R. Anderson, H. Mozaffar, B. Tohidi, Development of a crystal growth inhibition based method for the evaluation of kinetic hydrate inhibitors, in: Proceedings of the 7th International Conference on Gas Hydrates, July 17–21, Edinburgh, Scotland, 2011, Domestic Organizing Committee ICGH-7, 2011.
[106] P.V. Hemmingsen, X. Li, K. Kinnari, S. Asa, N. Trondheim, N. Stavanger, Hydrate plugging potential in underinhibited systems, in: Proceedings of the 6th International Conference on Gas Hydrates (ICGH 2008), Vancouver, British Columbia, CANADA, July 6–10, 2008, 2008.
[107] M.H. Yousif, Effect of under-inhibition with methanol and ethylene glycol on the hydrate control process, in: Paper presented at the Offshore Technology Conference, Houston, Texas, May, 1996, https://doi.org/10.4043/8234-MS.

[108] E.D. Sloan, C.A. Koh, Clathrate Hydrates of Natural Gases, third ed., vol. 119 of Chemical Industries Series, CRC Press, Boca Raton, FL, 2008.

[109] R.B. Nielsen, R.W. Bucklin, Why not use methanol for hydrate control? Hydrocarb. Process. 62 (4) (1983) 71–78.

[110] W.D. McCain, The Properties of Petroleum Fluids, second ed., PennWell Publishing, Tulsa, OK, 1990.

[111] K.K. Østergaard, R. Masoudi, B. Tohidi, A. Danesh, A.C. Todd, A general correlation for predicting the suppression of hydrate dissociation temperature in the presence of thermodynamic inhibitors, J. Pet. Sci. Eng. 48 (2005) 70–80.

[112] M. Moshfeghian, R. Taraf, A new shortcut/graphical method to determine the required MEG injection rate for natural gas hydrate inhibition, in: 87th Annual Gas Processors Association Convention March 2-5, 2008, Grapevine, Texas, 2008.

[113] F. Tohidi, R. Anderson, B. Tohidi, Evaluation of a novel water-immiscible kinetic hydrate inhibitor formulation, Energy Fuel 32 (6) (2018) 6518–6523, https://doi.org/10.1021/acs.energyfuels.8b00437.

[114] S.B. Fu, L.M. Cenegy, C.S. Neff, A Summary of Successful Field Applications of a Kinetic Hydrate Inhibitor, Society of Petroleum Engineers (SPE), 2001, https://doi.org/10.2118/65022-ms.

[115] M. Tariq, D. Rooney, E. Othman, S. Aparicio, M. Atilhan, M. Khraisheh, Gas hydrate inhibition: a review of the role of ionic liquids, Ind. Eng. Chem. Res. (2014), https://doi.org/10.1021/ie503559k.

[116] S. Baek, J. Min, Y.-H. Ahn, M. Cha, J.W. Lee, Effect of hydrophobic silica nanoparticles on the kinetics of methane hydrate formation in water-in-oil emulsions, Energy Fuel 33 (2018) 523–530, https://doi.org/10.1021/acs.energyfuels.8b03210.

[117] J.D. Smith, A.J. Meuler, H.L. Bralower, R. Venkatesan, S. Subramanian, R.E. Cohen, K.K. Varanasi, Hydrate-phobic surfaces: fundamental studies in clathrate hydrate adhesion reduction, Phys. Chem. Chem. Phys. 14 (17) (2012) 6013–6020, https://doi.org/10.1039/c2cp40581d.

[118] J.W. Nicholas, L.E. Dieker, E.D. Sloan, C.A. Koh, Assessing the feasibility of hydrate deposition on pipeline walls—adhesion force measurements of clathrate hydrate particles on carbon steel, J. Colloid Interface Sci. 331 (2009) 322–328, https://doi.org/10.1016/j.jcis.2008.11.070.

[119] G. Aspenes, L. Dieker, Z. Aman, S. Høiland, A. Sum, C. Koh, E. Sloan, Adhesion force between cyclopentane hydrates and solid surface materials, J. Colloid Interface Sci. 343 (2010) 529–536, https://doi.org/10.1016/j.jcis.2009.11.071.

[120] H. Sojoudi, G.H. McKinley, K.K. Gleason, Linker-free grafting of fluorinated polymeric cross-linked network bilayers for durable reduction of ice adhesion, Mater. Horiz. 2 (2015) 91–99, https://doi.org/10.1039/C4MH00162A.

[121] E. Brown, S. Hu, S. Wang, J. Wells, M. Nakatsuka, V. Veedu, C. Koh, Low-adhesion coatings as a novel gas hydrate mitigation strategy, in: Paper Presented at the Offshore Technology Conference, Houston, Texas, USA, May, 2017, https://doi.org/10.4043/27874-MS.

[122] A. Das, T.A. Farnham, S.B. Subramanyam, K.K. Varanasi, Designing ultra-low hydrate adhesion surfaces by interfacial spreading of water-immiscible barrier films, ACS Appl. Mater. Interfaces 9 (25) (2017) 21496–21502, https://doi.org/10.1021/acsami.7b00223.

[123] T. Ragunathan, X. Xu, J. Aris Shuhili, C.D. Wood, Preventing hydrate adhesion with magnetic slippery surfaces, ACS Omega 4 (14) (2019) 15789–15797, https://doi.org/10.1021/acsomega.9b01232.

[124] C.M. Perfeldt, H. Sharifi, N. von Solms, P. Englezos, Oil and gas pipelines with hydrophobic surfaces better equipped to deal with gas hydrate flow assurance issues, J. Nat. Gas Sci. Eng. 27 (November) (2015) 852–861, https://doi.org/10.1016/j.jngse.2015.09.044.

[125] X. Li, Y. Liu, J. Zhao, A hydrate blockage detection apparatus for gas pipeline using ultrasonic focused transducer and its application on a flow loop, Energy Sci. Eng. 8 (2020) 1770–1780, https://doi.org/10.1002/ese3.631.

[126] H. Miura, M. Takata, D. Tajima, K. Tsuyuki, Promotion of methane hydrate dissociation by underwater ultrasonic wave, Jpn. J. Appl. Phys. Part 1 45 (5 B) (2006) 4816–4823, https://doi.org/10.1143/JJAP.45.4816.

[127] M. Faryadi, F. Parvizian, M. Rahimi, Prevention and Removal of Gas Hydrates in Pipelines Using Ultrasonic Waves, ICHEC07_618, 2011.

[128] H.J. Hong, C.H. Ko, M.H. Song, S. Lee, K. Seong, Effect of ultrasonic waves on dissociation kinetics of tetrafluoroethane (CH2FCF3) hydrate, J. Ind. Eng. Chem. 41 (September) (2016) 183–189, https://doi.org/10.1016/j.jiec.2016.07.021.

[129] S.S. Park, N.J. Kim, Study on methane hydrate formation using ultrasonic waves, J. Ind. Eng. Chem. 19 (5) (2013) 1668–1672, https://doi.org/10.1016/j.jiec.2013.02.004.

[130] E.I. Huva, R.V. Gordienko, J.A. Ripmeester, H. Zeng, V.K. Walker, The search for'green inhibitors': perturbing hydrate growth with bugs, in: Proceedings of the 6th International Conference on Gas Hydrates (ICGH 2008), Vancouver, British Columbia, Canada, 2008.

[131] I.M. Townson, Microbial Inhibition of Methane Clathrate Hydrates (Electronic Theses and Dissertations (ETDs) 2008+), University of British Columbia, 2012, https://doi.org/10.14288/1.0059103.

[132] A. Demirbas, Processes for methane production from gas hydrates, in: Green Energy and Technology, vol. 34, Springer Verlag, 2010, pp. 161–181, https://doi.org/10.1007/978-1-84882-872-8_5.

[133] Y. Konno, Y. Masuda, K. Akamine, M. Naiki, J. Nagao, Sustainable gas production from methane hydrate reservoirs by the cyclic depressurization method, Energy Convers. Manag. 108 (January) (2016) 439–445, https://doi.org/10.1016/j.enconman.2015.11.030.

[134] E.M. Reyna, S.R. Stewart, Case history of the removal of a hydrate plug formed during deep water well testing, in: SPE/IADC Drilling Conference, 2001.

[135] N. McMullen, How hydrate plugs are remediated, in: Natural Gas Hydrates in Flow Assurance, Elsevier, 2011, pp. 49–86, https://doi.org/10.1016/B978-1-85617-945-4.00004-2.

[136] F. Caputo, F. Cascetta, G. Lamanna, G. Rotondo, A. Soprano, Estimation of the damage in a natural gas flow line caused by the motion of methane hydrates, J. Nat. Gas Sci. Eng. 26 (2015) 1222–1231, https://doi.org/10.1016/j.jngse.2015.07.050.

[137] P. Bollavaram, E.D. Sloan, Hydrate plug dissociation by method of depressurization, in: Paper Presented at the Offshore Technology Conference, Houston, Texas, May, 2003, https://doi.org/10.4043/15257-MS.

[138] R. Boubenider, A. Rebello, L. Alexander, R. Spradling, Deepwater hydrate plug remediation using gas lift—a unique approach & novel application of fundamental physics, in: Offshore Technology Conference, 2016, https://doi.org/10.4043/27146-ms.

[139] C.A.B.R. Cardoso, M.A.L. Gonçalves, R.M.T. Camargo, Design options for avoiding hydrates in deep offshore production, J. Chem. Eng. Data 60 (2) (2015) 330–335, https://doi.org/10.1021/je500601f.

[140] S. Kanstad, Method and Apparatus for Removing Hydrate Plugs, 2016. US Patent US9303488B2.

[141] H. Haneda, Y. Sakamoto, T. Kawamura, T. Komai, Experimental study on dissociation behavior of methane hydrate by air, in: Proceedings of the 5th International Conference on Gas Hydrates; Trondheim, Norway, 2005.

[142] Y. Masuda, U. Konno, T. Hasegawa, H. Haneda, H. Ouchi, M. Kurihara, Prediction of methane hydrate dissociation behavior by nitrogen injection, in: Proceedings of the 6th International Conference on Gas Hydrates, Vancouver, British Columbia, Canada, 2008.

[143] J.L. Panter, A.L. Ballard, A.K. Sum, E.D. Sloan, C.A. Koh, Hydrate plug dissociation via nitrogen purge: experiments and modeling, Energy Fuel 25 (6) (2011) 2572–2578, https://doi.org/10.1021/ef200196z.

[144] M.H. Yousif, V.A. Dunayevsky, A.H. Hale, Hydrate plug remediation: options and applications for deep water drilling operations, in: SPE/IADC Drilling Conference, 1997, https://doi.org/10.2118/37624-ms.

[145] X. Li, L.H. Gjertsen, T. Austvik, Melting hydrate plugs by thermodynamic inhibitors—plug properties and melting efficiencies, in: Proceedings of the 4th International Conference on Gas Hydrates, Yokohama, Japan, 19–23 May, 2002.

[146] X. Li, L.H. Gjertsen, T. Austvik, Thermodynamic inhibitors for hydrate plug melting, Ann. NY Acad. Sci. 912 (1) (2006) 822–831, https://doi.org/10.1111/j.1749-6632.2000.tb06836.x.

[147] T.A. Gounah, K. Renfro, J.L. Bolding, M.R. Embrey, Remediation of hydrate plugs in offshore wells via capillary tubing—a cost effective alternative, in: Paper presented at the SPE Annual Technical Conference and Exhibition, Denver, Colorado, USA, September 2008, https://doi.org/10.2118/116148-MS.

[148] T.I. Mitchell, S.C. Donovan, J.B. Collesi, H.W. McSpadden, Field application of a chemical heat and nitrogen generating system, in: SPE California Regional Meeting, 1984, https://doi.org/10.2118/12776-ms.

[149] L.C.C. Marques, C.A. Pedroso, L.F. Neumann, A new technique to troubleshoot gas hydrates buildup problems in subsea christmas-trees, in: Proceedings—SPE Annual Technical Conference and Exhibition, Society of Petroleum Engineers (SPE), 2002, pp. 2075–2081, https://doi.org/10.2118/77572-ms.

[150] A.A. Al-Taq, H. Al-Haji, J.A. Saleem, First successful filtercake damage removal treatment utilizing in-situ nitrogen/heat generating system for relatively heavy oil wells, in: SPE Annual Technical Conference and Exhibition, 2014, https://doi.org/10.2118/170832-ms.

[151] M. Pohl, M. Prasad, M.L. Batzle, Ultrasonic attenuation of pure tetrahydrofuran hydrate, Geophys. Prospect. 66 (7) (2018) 1349–1357, https://doi.org/10.1111/1365-2478.12534.

[152] R. King, et al., Canadian Association of Petroleum Producers Hydrate Guidelines, Canadian Association of Petroleum Producers, Calgary, Alberta, Canada, 1994.

[153] J. Davalath, J.W. Barker, Hydrate inhibition design for deepwater completions, SPE Drill. Complet. 10 (1995) 115–121, https://doi.org/10.2118/26532-PA.

[154] J. Lee, B. Hampton, R.R. Alapati, E.A. Sanford, S. O'Brien, OTC 20171 innovative technique for flowline plug remediation, in: Otc 4–7, 2009.

[155] L. Zhang, C. Zhang, H. Huang, D. Qi, Y. Zhang, S. Ren, M. Fang, Gas hydrate risks and prevention for deep water drilling and completion: a case study of well QDN-X in Qiongdongnan Basin, South China Sea, Pet. Explor. Dev. 41 (6) (2014) 824–832, https://doi.org/10.1016/S1876-3804(14)60100-2.

CHAPTER

Wax Management

17

17.1 Introduction

Wax deposition is one of the major flow assurance challenges in oil and gas fields. Waxes deposit when the hydrocarbons temperature falls below the wax appearance temperature (WAT). The impact of wax deposition in oil and gas fields was reported to be deleterious on all levels, from formation damage, equipment impairment, pressure drop, loss of production, and major CAPEX and OPEX investments to fight the wax problems, to complete field abandonment and decommissioning [1]. Consequently, rigorous management measures must be attained to contain wax deposits and reduce their impact on production. Such management measures should start with the early stages of field development and be continuously monitored and optimized during the field lifetime.

Similar to scale deposits and gas hydrates, wax control strategy is based on preventing wax crystallization in the early stages and removing the formed wax deposits. Fig. 17.1 summarizes the wax management methods.

17.2 Wax deposit prevention methods

Several methods can be used to prevent wax from depositing. They can be classified into operational, chemical, and nonchemical methods.

17.2.1 Operational methods of wax deposit prevention

Operational methods are based on optimizing the system parameters to operate within the wax-free zone. This accordingly relies on identifying the main factors affecting wax deposition in addition to the extensive fluids characterization, modeling, and simulation work done to map the wax deposition envelope. These operational management methods are conditional, so what can be applicable in one field may not be applicable in another.

17.2.1.1 High rate production

High shear environment can increase or decrease wax formation. In the general sense, high flow rates increase heat transfer, causing the fluids to reach the WAT rapidly, which disrupts the wax nucleation process, hence diminishing their growth and plugging [2]. Furthermore, the shear action associated

Essentials of Flow Assurance Solids in Oil and Gas Operations. https://doi.org/10.1016/B978-0-323-99118-6.00014-9

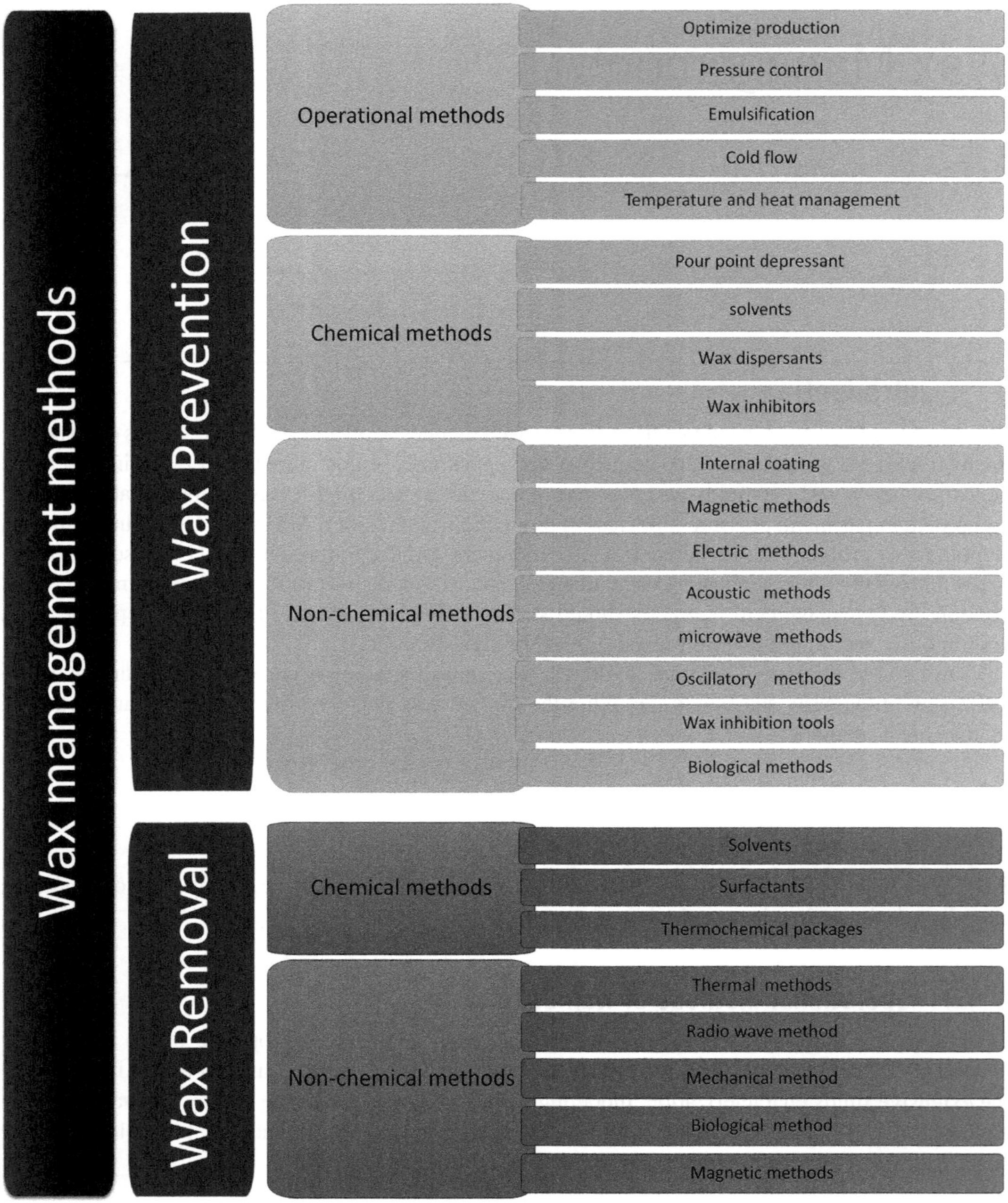

FIG. 17.1

Wax management methods.

with high rates helps prevent wax adhesion and slough the accumulated wax particles. Therefore the high flow rate technique can be adopted to avoid the risk of wax deposition. To practically achieve that, a high production rate is optimized to guarantee minimum heat loss from the reservoir to the surface so that the flow temperature is kept above the cloud point or WAT, so wax deposits will not appear [3,4]. However, the advantages and disadvantages of increasing the production rates should be evaluated in terms of their effect on the other production chemistry problems and fuel consumption. Moreover, some paraffin waxes can deposit even at high flow rates/shear rates, and these are known to be compact and harder wax deposits, which are challenging to remove when they stick to the surfaces [5].

17.2.1.2 Pressure control

Pressure affects wax deposition when pressure drop causes Joule-Thomson cooling of fluids below WAT. This is more common in gas condensate systems. Furthermore, depressurization below the bubble point and liberation of light components increases the WAT and increases wax precipitation. Hence, system pressure optimization is crucial in order to avoid these events.

17.2.1.3 Temperature control

Wax deposition is temperature dependent. The thermal management methods are based on keeping the flowing fluids temperature above WAT. In oil and gas fields, the thermal methods are usually applied by two means: passive insulation or active heating.

Thermal insulation is based on heat retention. Since the produced fluids are usually produced with high temperature, the goal is to make them hold that high temperature and prevent heat loss to the surrounding environment by surrounding the pipeline with materials with low thermal conductivity that offer a high resistance to heat transfer.

Active heating is based on applying heat to the flow lines using external means like electric heating methods, or by circulating hot fluids.

These thermal methods are usually applied to fight gas hydrates, paraffin waxes, and other fluid rheology-related issues in oil and gas fields. A thorough discussion of the thermal methods is furnished in Chapter 16. Literature reviews of the heat management and thermal methods in oil and gas fields are provided by [6–8].

17.2.1.3.1 Passive insulation methods

- **Wet insulation**

A wide range of organic and inorganic materials are available for thermal insulation. Among all the materials, plastic materials are the most adequate; examples are polyurethane, polyurethane-syntactic (SPU), polyurethane-glass syntactic (GSPU), isocyanurate, ethylene tetrafluoroethylene (ETFE) plastic pipe coating, polypropylene (PP), polypropylene solid or foam, polypropylene-reinforced foam combination (RPPF), phenolic syntactic (PhS), epoxy syntactic (SEP), and SEP with minispheres (MSEP) [9,10]. A few are suited for application in wells, considering mechanical and chemical resistance requirements, access, installation and maintenance limitations, and economics [3,11].

- **Thin film insulation**

Extensive research indicates that multiple layers have lower thermal conductivity than a much thicker single layer of the same material. TF insulation is a liquid epoxy coating that has been modified to provide resistance to heat flow. It is normally applied in layers that range from a thickness of

0.010″ to 0.020″ per layer with no limit on the number of layers that can be used. It may be used on either the ID or the OD of pipe [12].

- **Pipe-in-pipe (PIP) systems**

Pipe-in-pipe is also known as dry insulation. In a PIP system, a pipe is inserted inside another pipe, or the pipe is double walled. An insulation material is placed in the created vacant annulus and is protected by the outer pipe from hydrostatic pressure and water [3,13]. Examples of insulating materials used include aerogel, mineral wool, alumina silicate microspheres, thermal cement, polyurethane foam, or microporous silica blanket [10,14,15].

One of the earliest forms of high temperature insulated tubing is using inert "insulating" gas such as argon or nitrogen to fill in the gaps in the annulus instead of insulating material, whereby the gas reduces the convective heat transfer [10,14]. Another method is by creating a vacuum in the annular space (hence the name vacuum-insulated tubing or VIT); this technique is effective, with improved performance as it minimizes the convection; however, it is still costly [3,10,14]. Vacuum insulated tubing PIP design was patented, where the annular space between the tubes is evacuated to vacuum conditions, with reported uses in downhole, subsea completions, and other applications [16].

- **Burial**

Pipeline burial is one of the effective methods of heat management in onshore and shallow water fields [17]. Trenching and backfilling can be an effective method of increasing the amount of insulation, because the heat capacity of soil is significant and acts as a natural heat store. In offshore, pipeline burial depends on the pressure, viscosity, and temperature characteristic of the site, soil properties, topography geohazards, and infrastructures [18]. The effect of burial can typically decrease the U-value and significantly increase the pipeline cooldown period [10].

- **Subsea heat bank**

Falk et al. [19] introduced the subsea heat bank, where the subsea equipment is covered and the fluid trapped inside is heated during normal production (by the effect of hot flowing fluids), which can be an alternative to conventional insulation.

17.2.1.3.2 Active heating methods

When the system temperature is not high enough to support fluids flow and avoid wax deposition, or when the insulation use is not sufficient, the use of direct heating is necessary.

Actively heated systems generally use hot fluid or electricity as a heating medium. The main attraction of active heating is its flexibility, as it can be used in [10,20]:

- Maintaining the oil temperature above the pour point during shut-in period
- Maintaining the oil temperature above the WAT to prevent wax precipitation
- Removing wax deposits by heating above the wax deposition temperature (WDT); it has also been recognized as an effective method in removing hydrate plugs within an estimated time of 3 days, while depressurization methods employed in deepwater developments can take up to several months.
- Direct heating by applying an electrical current in the pipe has been used since the 1970s. Electrically heated systems are diverse, with diverse applications. Ansart et al. [21] reviewed the active heating technologies with their economical approach.

- ***Direct electric heating wet insulation (Wet-DEH)***

In this system, the pipe to be heated is an active conductor together with a power cable (piggyback cable) as the forward conductor, located in parallel with and close to the heated pipe. Heat is generated by the Joule effect from the current circulating in the pipeline wall [21,22]. The supply current is usually generated by a topsides current generator, or from the platform main power source [23]. Pipeline anodes are used to provide cathodic protection of the pipeline and to limit the stray current effects to nearby [10,21,23].

The following information is required when designing DEH [23]:

- Cross-sectional layout drawing with dimensions of the pipeline, including piggyback cable cross section and cable mechanical protection;
- Length of pipeline;
- Electrical conductivity and magnetic permeability of the pipeline;
- Thermal conductivity with corresponding U-value and heat capacity for the pipe insulation, cable protection, and surrounding seabed (including depth of gravel, rock dumping, etc.);
- Thermal properties of the pipe fluids in different operating modes;
- Temperature and electrical conductivity of sea water;
- Temperature requirements of the pipe fluids to prevent hydration;
- Required maximum time for heating from a cold state.

Generally speaking, Wet-DEH is a mature technology used primarily for preservation; however, for long pipelines it may show low efficiency and high power requirements, besides bulky and expensive topside equipment for additional power generation [21].

- ***Direct electric heating–Pipe-in-pipe (DEH-PIP)***

The system has been developed by Shell and is relevant for the PIP technology. DEH-PIP works on the same principle as Wet-DEH except that current is supplied at the middle of the flowline and returned through the carrier pipe instead of a piggyback cable [21]. The production line is split into heated segments (each segment is equivalent to a closed electrical circuit composed of inner and outer pipe). Each heated PIP segment is terminated at both ends with a steel bulkhead in order to have a closed electrical circuit [22].

Generally, DEH-PIP global efficiency is better than Wet-DEH, due to the better thermal insulation provided by the PIP arrangement (heating efficiency: 95%–100%). Electrical efficiency remains relatively low in practice (50%–70%) and long static power umbilicals are required for power supply at the midpoint of the flowline [21].

- ***Electrically trace heated pipe-in-pipe (ETH-PIP***

In this method the flowline is wrapped with trace heating cables, below the insulation layer, providing a very high heating efficiency. Optical fiber cables are also incorporated in the system to permanently monitor the internal fluid and cable temperature all along the flowline through a DTS system (Distributed Temperature Sensing) [21]. ETH-PIP is an efficient method that requires less power, with small impact on topside power generation and control equipment. The application of EHTF-PIP for maximum size flowlines (ID 12 in.) is currently restricted to just a small distance (about 25 km) and low-medium voltage (1–3.6 kV) cord used [22].

- ***Bundling***

In bundles, a combination of multiple flowlines, insulation, chemical lines, power, signal and heating cables, and others is contained inside an outer jacket or carrier pipe.

A bundle configuration, the Integrated Production Bundle (IPB), was developed for heat management within flexible pipes for dynamic riser and static flowline applications. IPB is basically composed of: (1) a core flexible pipe structure, (2) a bundle of components: steel tubes, hoses; electrical and optical cables, and (3) external insulation and protection layers [21]. Such a configuration allowed hot gas lift injection at the riser base, electrical heating, and temperature monitoring using fiber optics DTS.

Hot water and hot oiling are commonly circulated around the production fluids to increase their temperature above wax or hydrates formation temperatures.

- ***Hot water circulation (HWC)***

This heating technology relies on the circulation of hot water in close vicinity to the production flowline, either in the annular space (hot water pipe-in-pipe) or in a dedicated line (hot water bundle) [7]. Hot water circulation (HWC) systems generally operate in a loop, as the water needs to be recirculated back to the topsides (usually heat exchangers with an extended secured tube having production fluid on the inner part and the heating medium on the outer part of the tube). Nonetheless, the water may be injected in the well at the extremity of the system, but creates a constraining coupling between the production and injection systems [7].

The HWC technology is one of the first active heating technologies historically developed and installed offshore. Hot Water Bundles have been installed on several projects, such as Asgard and Gullfaks for Statoil, and on Conoco Britannia [24].

- ***Hot oiling***

Hot oiling is another hot fluids circulation method similar to HWC. The method is frequently used to remove wax, so it will be discussed with wax deposits removal.

Other technologies have been emerging besides the previous methods. Among the new technologies are THOR—In-Pipe Solution (Permanent/Semipermanent), an actively mechanical heat generator in which the heating element is powered by a downhole motor running on the fluid dynamics. The technique is promising to save powering energy and can be predesigned with the completion. The technique is used to remove wax and hydrate plugs as well as the basic application for heavy oil enhanced oil recovery (EOR). Field application of the technique includes two successful case studies [25].

17.2.1.4 Minimum gas production

The loss of light ends and fluids cooling by gas expansion usually lead to paraffin deposition [26]. Therefore producing oil with a minimum of gas and maintaining a steady flow of oil will help prevent paraffin particle precipitation and adherence to the pipe walls [27]. Such conditions can often be regulated by the use of proper size tubing, choke, a back pressure regulator, or a combination of these. In the case of a high gas-oil ratio (GOR), stressing back pressure is beneficial to prevent wax precipitation (assuring that it does not affect well productivity). Keeping the amount and pressure of the gas as low as possible (while assuring good productivity) also minimizes local cooling [26].

Furthermore, the tubing should be full of oil at all times, preferably not allowing the oil level to fall between times of production, if production is intermittent. By maintaining the pressure and fluid level constant, the maximum amount of oil can be produced with a minimum of paraffin [26].

17.2.1.5 Optimizing gas lift operations

Gas-lift operations must be optimized to prevent gas expansion and localized fluid cooling, which causes wax deposition in the tubing. Elhaddad et al. [28] showed that injecting the gas at a pressure of 86.3 bar and a relatively high temperature of 65°C results in a surface casing temperature above WAT and the pour point temperature (52°C). The technique was applied in low-pressure/low-temperature wells (LPLT) in Libya, and the results were found to be promising.

Narvaez and Ferrer [29] showed that the use of a plunger lift device for both gas lift and natural flow wells has proven to be a viable, economical method by which to eliminate paraffin deposition problems by continual scraping of the deposited paraffin deposits off the production tubing. The technique was applied on a Venezuelen oilfield, and the results were promising, as the lost productivity in 17 initial candidate wells of some 200 BPD has been restored.

Moreover, theoretical and experimental studies demonstrated the use of periodic injection of hot associated petroleum gas (APG) into the annulus of an oil-producing well to improve the efficiency of gas-lift wells during the production of high-wax oil [30].

17.2.1.6 Fluids dilution

Waxy crude oils can be mixed with a proper diluent principally to avoid wax problems, reduce oil viscosity, and to improve its mobility. The diluent can be a gas condensate, natural gas liquids, light hydrocarbon naphtha, and some organic solvents [31,32]. Solvents include xylene, toluene, kerosene, diesel, nonaromatic cyclic solvents, alcohol-based solvents, especially pentanol, and terpenes [32,33]. Other effective organic solvents with proven success include benzene, chlorinated hydrocarbons, and carbon disulfide, but these solvents are not environmentally friendly and pose higher risk [34]. Butane was injected downhole in 12 production wells in a 3:1 ratio (60 gal oil:20 gal butane), resulting in a reduction in the cost of paraffin control by as much as 90% [35].

Other methods to enhance the solvent efficiency are [36]: (1) Mixing xylene or toluene and aliphatic solvent to increase the removal of waxes; (2) incorporating surfactant to enhance the performance of the solvent in dispersing the waxes; and (3) heating up the xylene and toluene solvent to increase the removal efficiency of waxes. Some limitations of the dilutions using solvents are low specific gravity and low flash point.

Moreover, light crude with a lower WAT or pour point, or crude oil produced with a very high temperature, can be used, which is considered a cost-effective method [33]. The final mix with any of these diluents should have lower wax content, which reduces its WAT or pour point to lower temperatures [31]. The resulting viscosity of the mixture depends on the dilution rate and the respective viscosities and densities of the crude oil and diluents [32]. Compatibility issues must be evaluated before using the diluent, as light hydrocarbons and solvents can strip off resins from waxy crudes and cause asphaltene deposition problems.

17.2.1.7 Emulsification

Wax and other solids are known to stabilize oil/water emulsions; these emulsions can be promoted using surfactants and controlled enough to prevent wax deposition without impeding phase separations in the next processing stages. Obviously, the presence of water generates the water wet form, preventing wax from deposition [37].

The field application of this method reported that it requires produced water of 35% or more to form O/W stable emulsion by applying a surfactant, production GOR less than 1000 SCF/BBL to keep the O/W emulsion stable, and hot oiling to dissolve the deposited wax and to promote the water, oil, and surfactant interaction [38]. The method was applied in six pumping wells that produce a paraffinic crude, with a 105°F pour point, resulting in reducing wax deposition in wells and flowlines, permitting 44% fewer hot-oil jobs. Average net gain from reduced operating cost and increased production from the test wells was about $1870 per well per year [38].

However, this technique seems to be feasible only in specific cases, like severe wax problems, such as HMW wax crudes with high WAT, when strong paraffin chemicals like EVA and comb copolymers cannot effectively mitigate the problem.

Two commercial emulsion pipeline systems for crude transport are known. One in Indonesia [39] carries 6359.6 m^3/day of a 70 vol% emulsion of a high pour-point waxy crude in a 0.508 m diameter pipeline that is 238 km long. The other, in California [40], carries a 50 vol% emulsion of highly viscous 13.5 API gravity crude for a distance of 20.92 km in a 0.203 m diameter pipeline.

The major application of heavy oil aqueous emulsion is the ORIMULSION process, developed by PDV (Petroléos de Venezuela) in the 1980s. ORIMULSION is an aqueous bitumen emulsion, made of 30% water and 70% natural extra-heavy oil, which is directly used as a feedstock for power generation in thermoelectrical plants [41]. Stability of the emulsion depends on many parameters, including oil composition in terms of surface active molecules, salinity, and pH of the water, oil/water volume ratio, droplet size and polydispersity, temperature, surfactant type, concentration, and mixing energy [41].

17.2.1.8 Cold flow

Cold flow technology, commonly used for gas hydrates and paraffin wax control, is based on the pioneering work of Coberly [42]. Based on the fact that precipitation does not imply deposition, in cold flow the precipitated solids are safely transported as a solid dispersion without deposition or blocking. A discussion of cold flow technology for gas hydrates is provided in Chapter 16. The technique is viewed as a different type of thermal method: instead of heating the fluids to keep the waxes in the solution, the fluids are cooled down below WAT to allow precipitation of waxes in a controlled way so that they are suspended in the bulk solution and flowing like a slurry in the production system, with limited wax particles adhering to the internal pipe walls. It is claimed that, if solid slurry is formed in the first section of the pipe, it will be transported in a stable way without further solid deposition [43]. To apply the technique safely and efficiently, the temperature gradient (heat flux) between hot oil and cold wall should be eliminated so that the wax particles remain suspended in the bulk solution without depositing on the pipe surface [3,44].

Merino-Garcia and Correra [44] reviewed existing patents on the methods for creating a wax-oil suspension mixture for cold flow technology. Fig. 17.2 illustrates some of these methods.

- Cold seeding (NEI, Calgary, Canada, and Marathon Oil, Houston, Texas)

Wax crystallization can be induced by a seeding method using wax solids. Nenniger [45,46] proposed seeding crude with waxy solids to reduce wax deposition in a one-pass system with controlled temperature and shear stress. Argo et al. [47] proposed a seeding method using a cold fluid (below WAT) and different solid particles acting as nucleating sites for wax, asphaltenes, and other solid-forming species from the flow of fluid hydrocarbons.

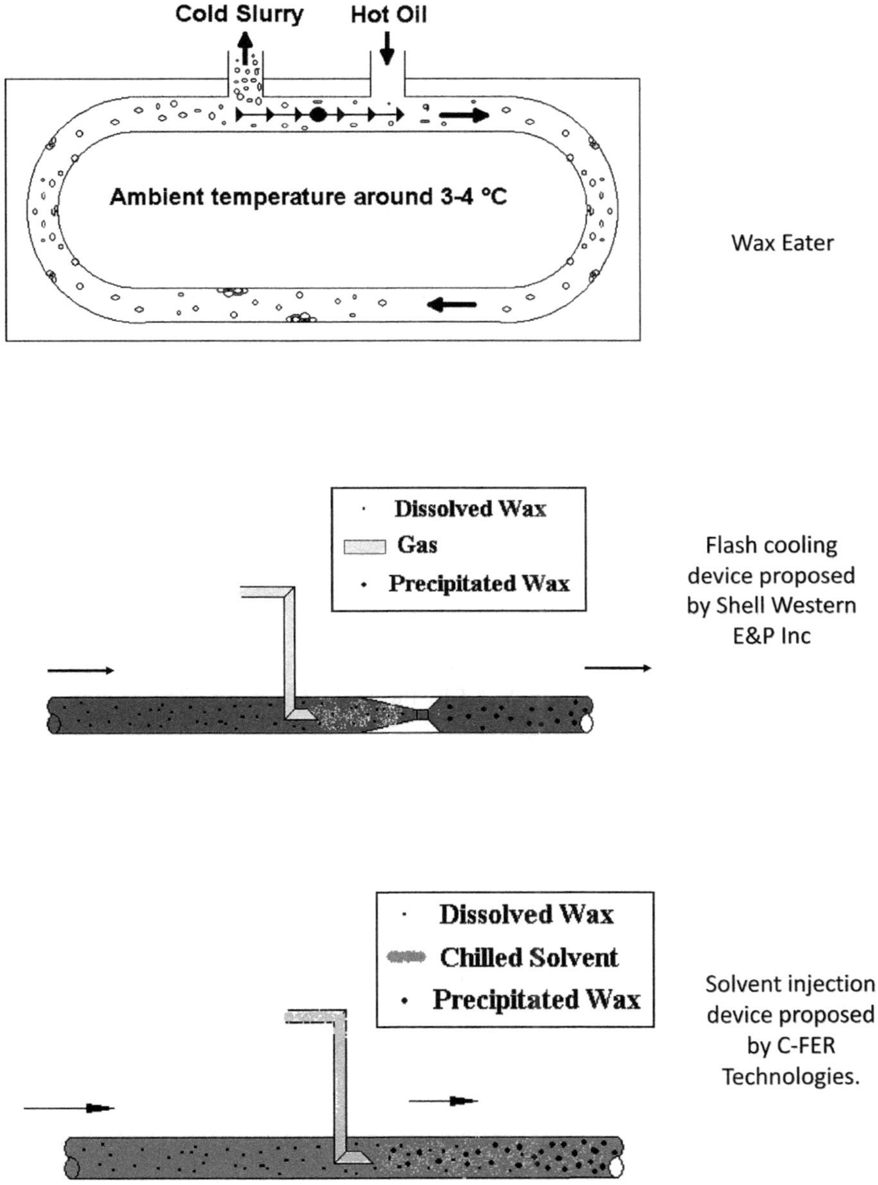

FIG. 17.2

Cold flow methods.

Reproduced from D. Merino-Garcia, S. Correra, Cold flow: a review of a technology to avoid wax deposition, Pet. Sci. Technol. 26 (4) (2008) 446–459. https://doi.org/10.1080/10916460600809741, by permission of Taylor & Francis Ltd., http://www.tandfonline.com.

- The wax eater (the technology is used by KBR, Inc., Halliburton)

Wax solids are induced to precipitate in a loop where the external temperature is below WAT as oil is cooled to the seabed temperature [48,49]; then the formed wax deposits are constantly removed from the loop's walls by using a unique mechanical device (WaxEater), which consists of a train of circular discs continuously circulating, scrubbing the formed deposits [50].

- High-shear heat exchanger (KBR, Inc., Halliburton):

This method is based on a treatment loop where the incoming fluid stream rapidly cools, mixes, or changes the pressure to precipitate wax solids, while an auxiliary pressure-driven device increases the speed of the oil stream through the heat exchanger, which provides continuous cleaning of the formed deposits off the walls and slurry production [49]. In principle, the use of high speed would imply an increase in size of the heat exchanger, as the time to transfer heat is reduced by the high flow rates.

- Pressure surges (KBR, Inc., Halliburton):

This method combines the pig action with induced pressure surges. The discontinuous release of pressure courses at or near sonic conditions is supposed to aid in the release of the deposited solids [49].

- Flash cooling (Shell Western E&P Inc., Houston, Texas):

The oil is mixed with a stream of gas where sudden pressure drop is induced using a choke, causing fluids cooling down to temperatures below WAT by the Joule-Thomson effect, which leads to the precipitation of waxes in the bulk [51].

- Oil or solvent injection (C-FER Technologies, Edmonton, Canada):

C-FER proposes cooling by addition of a recirculating current of cold oil or solvent. Also a supercooled gas can be used to create the slurry.

- Magnetic-fluid conditioning:

Magnetic conditioning is useful toward preventing wax deposition and clogging. In this technology, the fluid is exposed to a magnetic field that alters the growth of the wax crystals and causes changes in solids that are being carried or precipitated from that fluid, preventing their deposition. Several patents have been derived from this technology, However, this method is expected to be expensive compared to the conventional methods.

Table 17.1 summarizes the operational methods used in managing wax deposition.

17.2.2 Chemical prevention of wax deposition

Chemical wax management is the most widely used method of wax deposit management. Chemical methods are characterized by low initial cost and minimum maintenance costs, besides the diverse types of chemicals that are suitable for use with different types of crudes. These chemicals basically work as a kinetic barrier, as they delay the formation and accumulation of hard deposits and allow safe production operations.

Table 17.1 Summary of operational methods for wax deposits prevention.

Method	Description
Optimizing production rates	Using high production rates to strip off the wax deposits layer
Pressure control	Avoid drastic pressure drop that can cause fluid cooling or loss of light ends
Temperature management	Operating above WAT Using thermal insulation Using direct heating methods
Fluids dilution	Diluting the fluids with less viscous, high temperature, or low WAT fluids
Emulsification	Formation of O/W emulsion with low viscosity and good flowable properties
Cold flow	Formation of flowable suspended wax slurry Used when chemical methods are expensive or require high volumes of chemicals in deepwater fields

17.2.2.1 Classification and mechanisms of wax prevention chemicals

A wide variety of chemicals can be used to control wax problems in oilfields. They are generally classified according to application purposes into wax or paraffin inhibitors (WI, PI), pour point dispersants (PPD), and wax/paraffin dispersants (WD, PD). They also can be classified according to chemistry and mechanism of action to polymeric inhibitors (crystal modifiers), surfactants (film formers), and solvents (dissolvers). These groups of chemicals are also called flow improvers, based on their action in enhancing fluids flow and reducing wax problems. Fig. 17.3 illustrates the wax control chemicals.

As previously indicated in Chapter 7 (wax deposition), the wax deposit formation process involves nucleation and growth of particles that then interlock into a 3D network that attaches to the pipe wall where they age and harden. Generally speaking, wax control chemicals interfere with the crystallization process (by distorting the crystal shape), resulting in reducing the WAT or pour point temperatures. Chemicals that reduce the WAT are usually referred to as wax inhibitors or wax crystal modifiers, while chemicals that affect the pour point are referred to as PPDs or flow improvers. These two classes of chemicals have quite similar chemistry and mechanisms of action, so some WIs can work as PPDs. PDs are surfactants that act using a different mechanism, where they prevent particle adherence to the pipe wall [31].

Comprehensive reviews of paraffin wax control chemicals and their mechanism of action have been published [31,36,52,53].

17.2.2.1.1 Wax inhibitors and pour point depressants (WIs and PPDs) mechanisms of action

WIs and PDDs are classified as crystal modifiers as they affect the morphologies of the wax crystals. They basically are incorporated into the wax crystals and alter their surface characteristics (incorporation-perturbation), which inhibits wax deposition or results in formation of soft deposits.

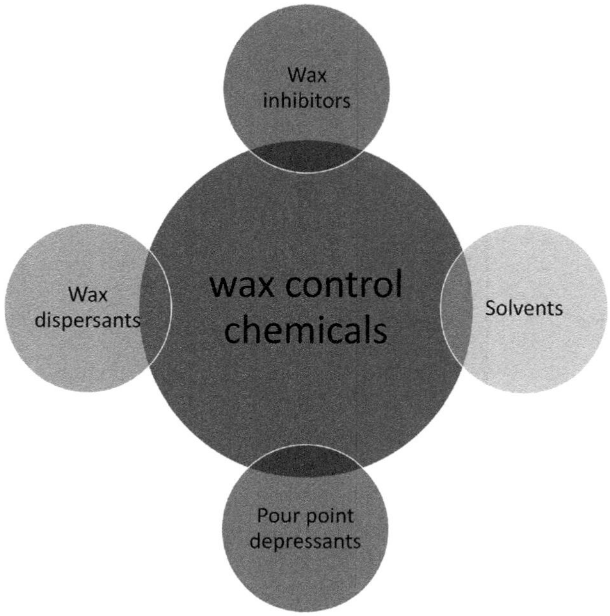

FIG. 17.3

Wax control chemicals.

Despite the extensive studies of these chemicals, the mechanisms of WIs and PPDs are not fully understood [31,36,52,53]. Some of the mechanisms mentioned in the literature are discussed in the following sections.

- ***Nucleation sequestering***

Some wax inhibitors can self-assemble into micelle-like aggregates exhibiting a crystalline core and soluble hairy brushes surrounding the core, even when the bulk temperature is above WAT. These finally form a large number of subcritical nuclei (also called polynucleation), reducing the supersaturation and delaying the nucleation of wax crystals, which consequently reduces crystal growth rates and facilitates the formation of more abundant smaller wax crystals, which are less stable in a flowing solution [53–55]. WI long chain adsorbs on the surface of wax nuclei, distorting the wax surface, which weakens interaction with the surrounding crystals and hinders the crystal growth rate [56]. Fig. 17.4 depicts the nucleation sequestration mechanism.

- ***Incorporation and perturbation***

Wax inhibitors either cocrystallize with wax molecules or incorporate on the growing surface of precipitated wax crystals. WIs are usually composed of hydrophobic ethylene segments (side chains) that are similar to the wax molecules, and a bulky substituent group that is dissimilar to the wax crystals. Due to the favorable van der Waals interaction of the long paraffin chains with the similar long side chains of inhibitors, the inhibitors preferentially bind to paraffin molecules out of the oil phase while the bulky side chain of wax inhibitors places a steric hindrance responsible for preventing and

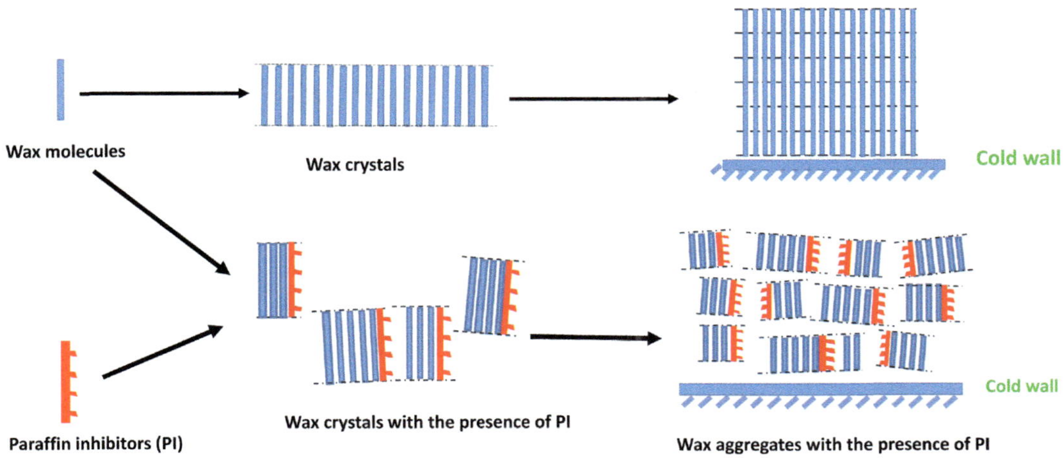

FIG. 17.4

Nucleation sequestering mechanism.

Reprinted with permission from Y. Chi, J. Yang, C. Sarica, N.A. Daraboina, Critical review of controlling paraffin deposition in production lines using chemicals, Energy Fuels 33 (2019) 2797–2809. Copyright {2019} American Chemical Society.

disrupting the crystal growth, changing the wax crystal habits and preventing the formation of interlocking structures. This leads to inhibiting wax deposition, forming abundant small wax crystals, or making the deposit weaker. This allows the flow (drag) forces to slough off the weakened wax deposits [31,57–59]. It is reported that a good match between the WI side chain length and the crude paraffin wax molecule carbon numbers gives better efficiency of the WI. Fig. 17.5 illustrates the incorporation

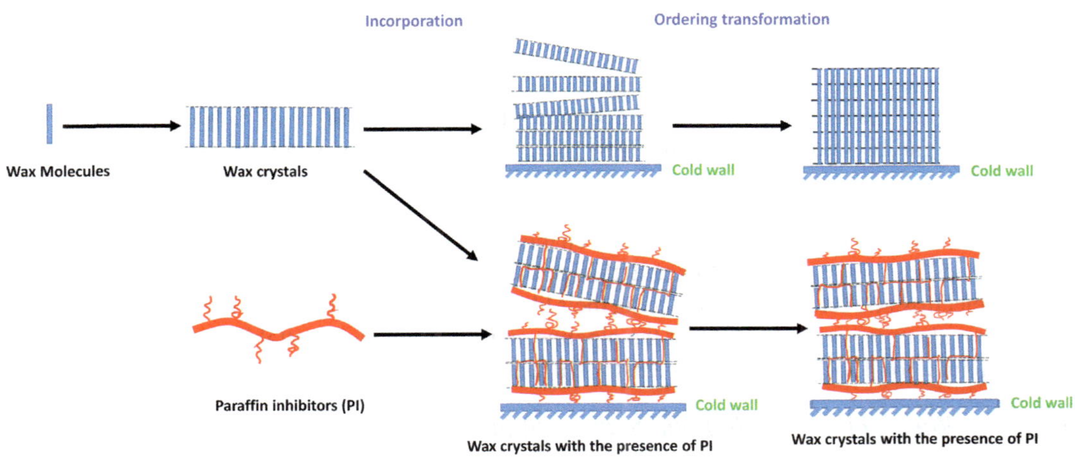

FIG. 17.5

Schematic of incorporation and perturbation mechanism.

Reprinted with permission from Y. Chi, J. Yang, C. Sarica, N.A. Daraboina, Critical review of controlling paraffin deposition in production lines using chemicals, Energy Fuels 33 (2019) 2797–2809. Copyright {2019} American Chemical Society.

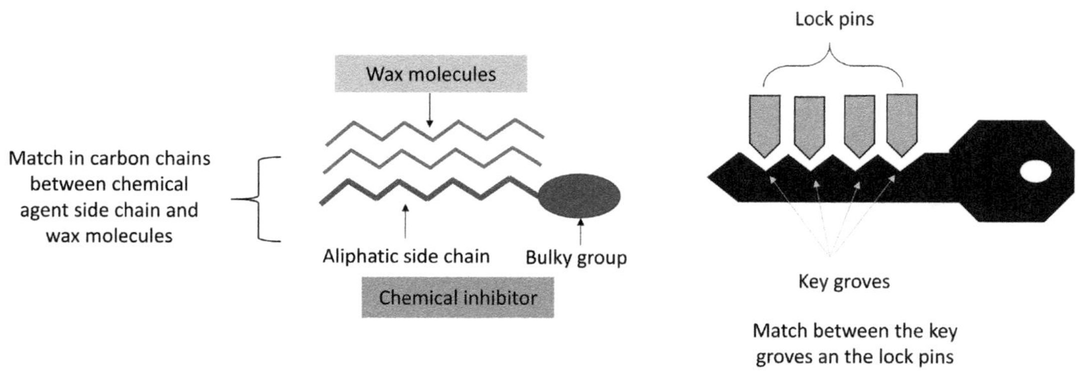

FIG. 17.6

Key-lock analogy with inhibitor-wax molecules to simplify the necessary matching in carbon numbers between the inhibitor and the wax molecules.

and perturbation model. This can be mimicked by the lock and key analogy (Fig. 17.6), as the number of the pins inside the lock cylinder must be matched with the cuts (grooves) in the key. Hence, the WI should contain reasonable amounts of ethylene groups to match the same number in the wax molecules in the crude oil. The interaction between wax molecules and the WI alkyl side chain increases the wax solubility, keeping them in the solution.

Another way was shown to mimic paraffin inhibitors, which was for WI to serve as a "wrapper" that envelops the wax molecules and prevents their growth with reduced crystal-crystal adhesion [60]. Also, the cohesive energy density (driving force for the ordering transformation) was reduced by three times in the presence of an inhibitor, compared to the pure system [61].

- ***Adsorption on pipe wall***

In this mechanism, the inhibitors adsorb on the pipe wall, creating oleophobic surfaces, and prevent adsorption/attachment of paraffin molecules on the wall. This results in the formation of a weak deposit layer, which can be sloughed off by the flow [62]. Fig. 17.7 illustrates this mechanism.

17.2.2.1.2 Wax dispersant mechanism of action

PDs, on the other hand, employ a different mechanism than the crystal modification. These are surfactants that adsorb on the wax crystals and change their surface energy, water-wet them, and disperse them by electrostatic forces, hence reducing their likelihood of depositing. Another way the dispersants work is by adsorbing on the surface of the pipeline and changing their wettability, thus preventing the wax accumulation [36,52]. This mechanism is illustrated in Fig. 17.8.

17.2.2.2 Wax inhibitors (WIs) and pour point depressants (PPDs)

Several wax crystal modifiers (WIs and PPDs) have been evaluated and showed promising results toward wax inhibition. Some examples of these chemicals are given in Table 17.2.

As indicated in the preceding lock and key example, since wax molecules contain several ethylene groups, it is expected that the best WI and PPD structures are those that have a number of ethylene groups in their structure, similar to those in paraffin molecules, which allows the interaction, incorporation, and inhibition action in the wax crystal network [52].

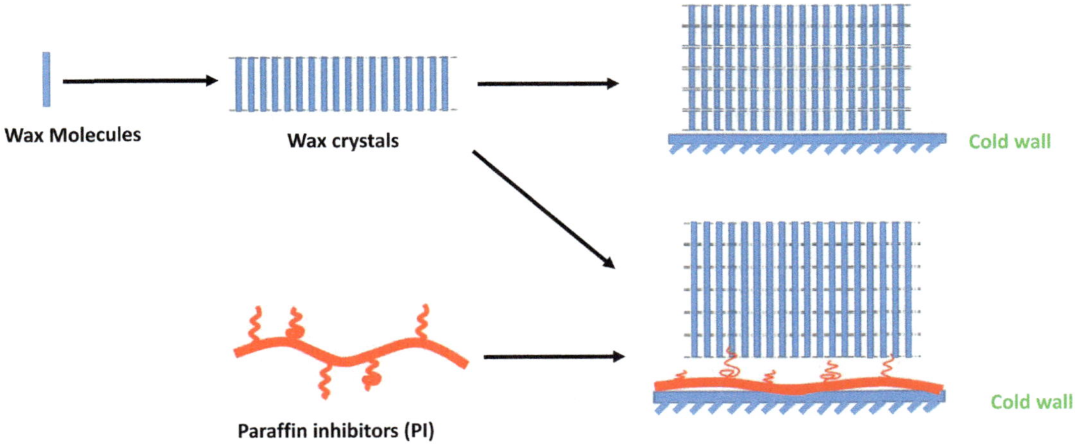

FIG. 17.7

Schematic of adsorption on pipe wall mechanism.

Reprinted with permission from Y. Chi, J. Yang, C. Sarica, N.A. Daraboina, Critical review of controlling paraffin deposition in production lines using chemicals, Energy Fuels 33 (2019) 2797–2809. Copyright {2019} American Chemical Society.

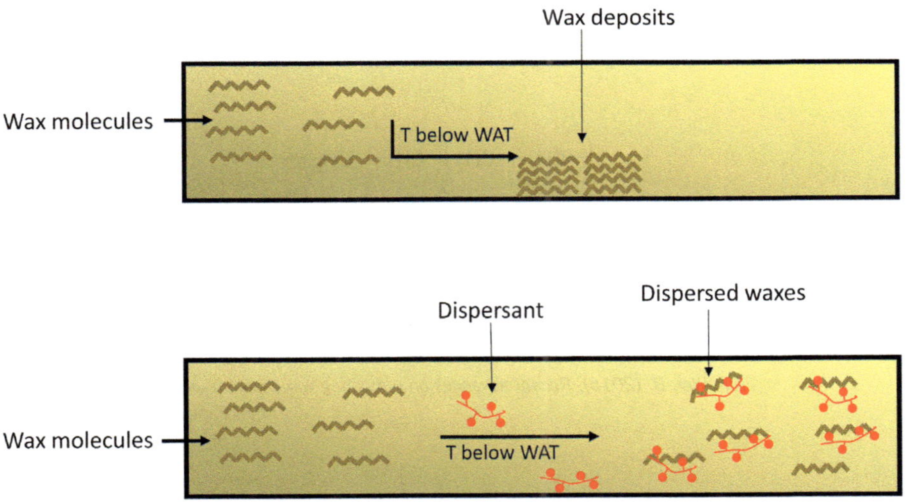

FIG. 17.8

Wax dispersant mechanism.

- **Linear polymers and copolymers**

Common examples of these chemicals include linear copolymers. Amorphous high molecular weight polyethylene (PE) has been used as a wax crystal modifier or PPD in the past; also poly(ethylene butene) (PEB) (Fig. 17.9) and polyethylene-poly(ethylene-*b*-propylene) (PE-PEP) (Fig. 17.10) are examples of crystalline-amorphous diblock copolymers, which contain polyethylene (PE) groups (as

Table 17.2 Wax inhibitors and pour point depressant chemicals.	
Chemical inhibitor category	**Example**
Linear polymer/copolymer	Polyethylene with amorphous structure Poly(ethylene butene) (PEB) Polyethylene poly(ethylene-*b*-propylene) (PE-PEP) Ethylene-vinyl acetate copolymers (EVA) Ethylene/acrylonitrile copolymers
Comb polymers	(Meth)acrylate ester polymers Maleic copolymers
Nanohybrid pour point depressants	Various modified nanomaterials
Surfactants	Ethoxylated aliphatic alcohol Poly-oxyethylene surfactants Carboxylic esters Alkyl benzene sulfonates Alkyl benzene sulfonate Bolaform surfactant Ethanolamine-based surfactant Amphoteric surfactants Gemini surfactants
Ionic liquids	Imidazolium ionic liquids
Green chemicals	Nanohybrid materials Natural extracts: sunflower oil, vegetable oils

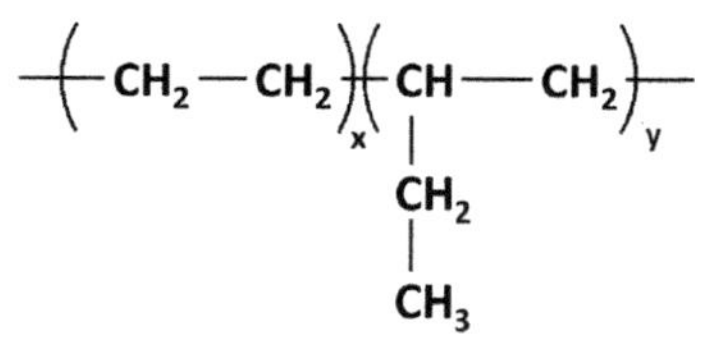

FIG. 17.9

Structure of PEB.

From Wei, B. (2015). Recent advances on mitigating wax problem using polymeric wax crystal modifier. J. Pet. Explor. Prod. Technol., 5(4), 391–401.

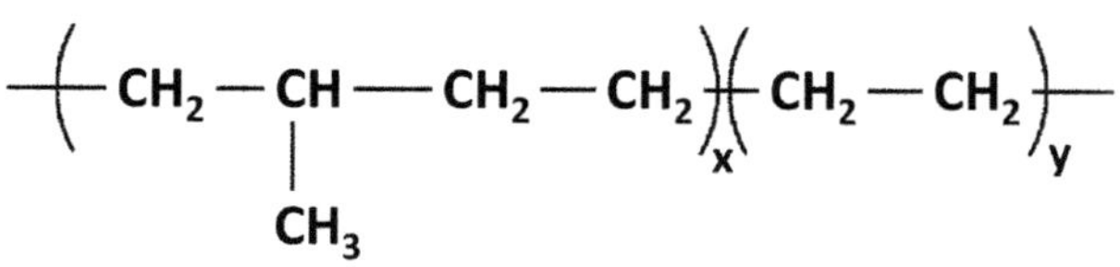

FIG. 17.10

Structure of PE-PEP.

Source: B. Wei, Recent advances on mitigating wax problem using polymeric wax crystal modifier, J. Pet. Explor. Prod. Technol. 5 (4) (2015) 391–401.

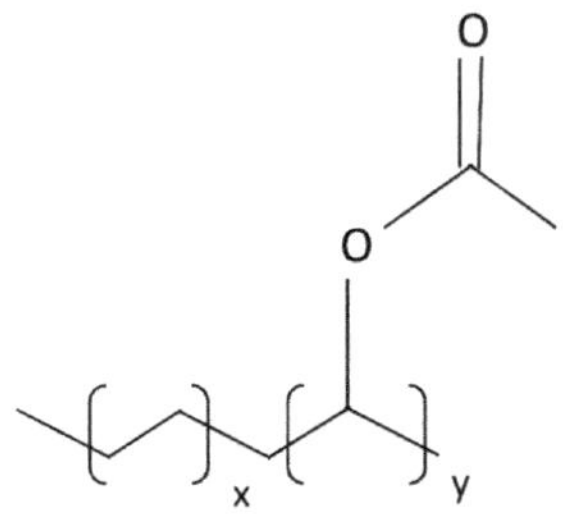

FIG. 17.11

Structure of EVA.

crystalline groups) and polybutene (PB) or polyethylenepropylene (PEP) (as amorphous moieties) which have been investigated as wax deposition inhibitors and flow improvers [31,36,63,64]. These polymers are found to act by nucleation sequestering and cocrystallization mechanisms leading to formation of small wax crystals [55,65]. They also enhance the yield strength of the crude oil, which makes them effective as WIs and PPDs given their efficiency at lower concentrations. However, their applications are not really presented [36].

Ethylene vinyl acetate copolymers (EVAs) are the most commonly used wax inhibitors and flow improvers [52]. The structural formula of EVA is given in Fig. 17.11. The structure displays strong van der Waals interaction and other intermolecular interactions via oxygen and hydrogen atoms, which increase wax solubility and solid wax deposit inhibition probability [36]. The EVA inhibits wax deposition when side chains cocrystallize with paraffin molecules, while the polar moiety causes steric hindrance that interferes with new paraffin molecule alignment, resulting in reducing WAT and improving flow [66,67].

The important structural features of EVA copolymers that affect their properties are the vinyl acetate content, average molecular weight, short chain branching, long chain branching, molecular weight distribution, and distribution of vinyl acetate groups [58]. VA content decreases crystallinity and increases polarity and solubility in solvents, so it is generally advisable to be in the range of 25%–30% for optimum performance [68,69]. The molecular weight decreases the solubility. EVA copolymers are not effective comb polymers as PPDs [31]. Ethylene/acrylonitrile copolymers are cheap inhibitors and PPDs [31].

- **Comb polymers**

Comb-shaped copolymers are generally regarded as the most effective wax inhibitors; they are also accepted as both WIs and PPDs. A comb-shaped copolymer has both a nonpolar group (side alkyl chain) and a polar group (e.g., ethyl vinyl ether, styrene, esters, carboxylic, amide groups [52]). Matching the length of the side chain of comb-shaped copolymers and the paraffin molecules in crude oil is crucial for optimum inhibition. The higher the length of the side chain, the more the pour point depression occurs [70]. It was suggested that the melting point of the PPD should match the melting point of the waxes for limiting the growth of the crystals [71].

Two main classes of the *comb*-type polymers are commonly used, namely, (1) maleic anhydride copolymers (MACs) and (2) polyacrylate or methacrylate (PA or PMA) ester polymers [31,36]. Some examples of MAC structures are given in Fig. 17.12.

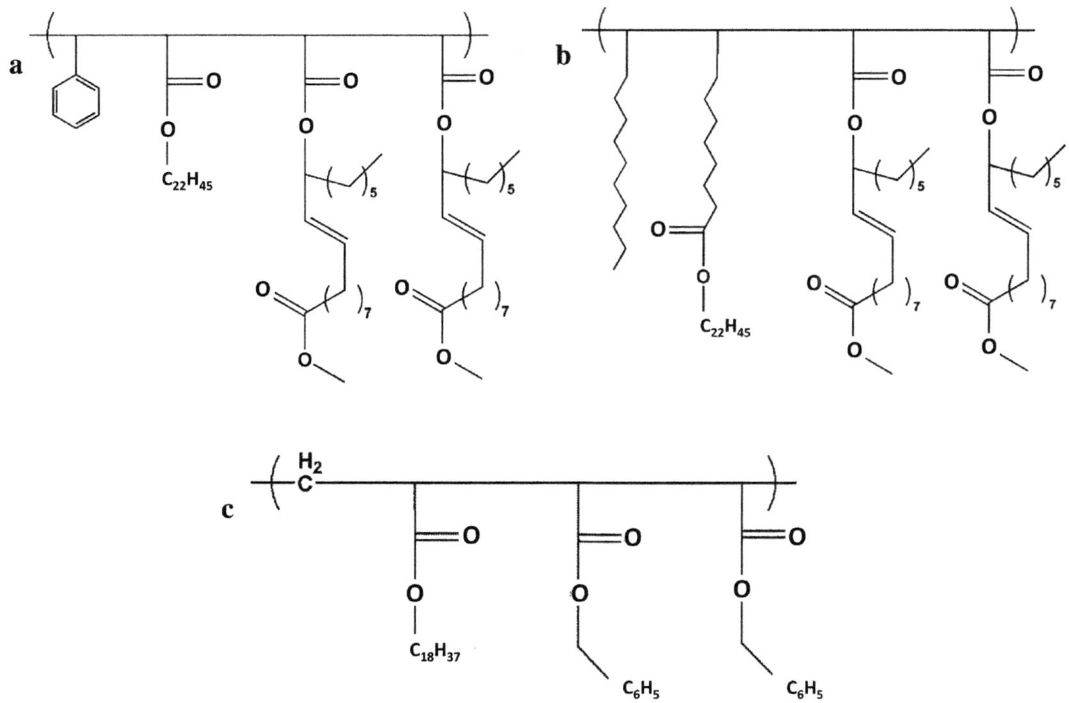

FIG. 17.12

Examples of chemical structures of MAC copolymers.

Source: B. Wei, Recent advances on mitigating wax problem using polymeric wax crystal modifier, J. Pet. Explor. Prod. Technol. 5 (4) (2015) 391–401.

MACs, PA, and PMA are widely used as WIs and PPDs. PMA is considered a better PPD than PA.

Addition of xylene has been shown to improve the performance of commercial polymeric wax inhibitors. For example, poly(behenylacrylate) gave a lower pour point when mixed with xylene or other aromatic solvents used to remove asphaltenes [72,73]. Other unsaturated carboxylic acid monomers besides maleic acid can be used to make wax inhibitors. For example, long-chain alkyl fumarate/vinyl acetate copolymers were shown to be good as flow improvers for high waxy Indian crude oils [74].

MAC comb copolymers have been reported to improve crude oil flowability by interacting with the different components, including paraffin waxes and asphaltenes, as illustrated in Fig. 17.13 [75].

- **Other polymers**

Alkylphenol-formaldehyde resins can also be used as PPDs and asphaltene inhibitors. Branched polymers can also be used as wax inhibitors and flow improvers: for example, hyperbranched polyesters showed improved flow properties of North Sea crude oil [76], and they showed a decline in crude viscosity up to 65% at 50°C [77].

Polybetaine wax inhibitors have been assessed as candidates for squeeze treatments. The presence of the carboxylic acid groups undoubtedly helps their adsorption onto rock compared with more traditional wax inhibitors [78].

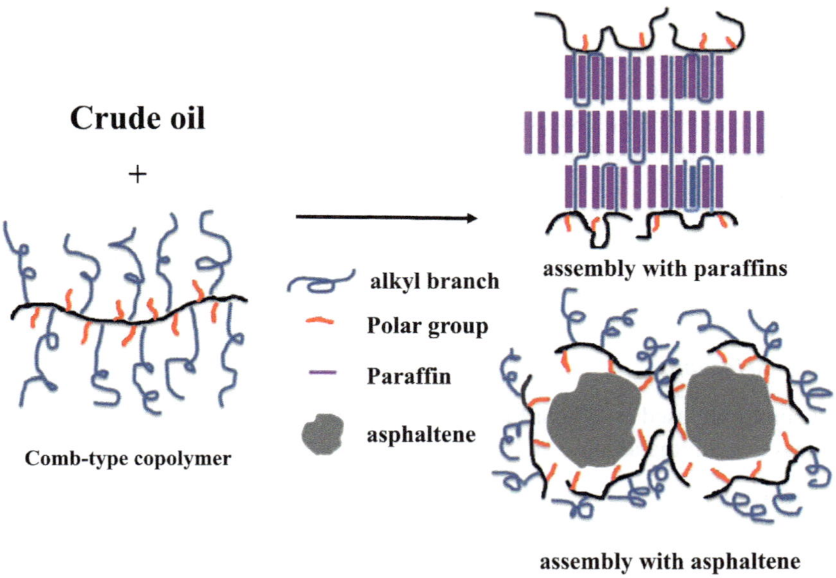

FIG. 17.13

The proposed mechanism by which comb copolymers interact with crude oil components and improve its flowability.

Reprinted from reference J. Xu, S. Xing, H. Qian, S. Chen, X. Wei, R. Zhang, et al., Effect of polar/nonpolar groups in comb-type copolymers on cold flowability and paraffin crystallization of waxy oils, Fuel 103 (2013) 600–605, copyright (2013) with permission from Elsevier.

- **Nanohybrid pour point depressants**

Nanoparticles exhibit potential applications in polymer modification due to unique size, high surface, and quantum tunneling effects. Compared with pure polymers, nanocomposite properties such as mechanical and thermal stability, abrasion resistance, and tenacity are greatly improved due to introduction of dispersed nanoparticles [53].

Nanohybrid PPDs (modified nanomaterials by polymeric PPDs such as polyoctadecyl acrylate, EVA copolymers, and methacrylate) have been demonstrated in many studies in which they showed improved viscosity/pour point reduction, shear resistance, and crude flowability over traditional PPDs [53,54,79,80]. Their improved performance was attributed to their effects on nucleation (the nanoparticles work as dispersed nucleation sites or heterogeneous nucleating agent), cocrystallization (the incorporation into the wax crystals and hindering the growth), adsorption on the surface of wax crystals (which hinders wax crystal growth), and also their dispersion abilities [79,80].

Novel nanocomposite PPD (graphene oxide-based nanocomposite) was developed, leading to 30°C reduction in crude oil pour point, besides reducing oil gelling, aging, and viscosity [81].

- ***Nonionic surfactant***

Some nonionic surfactants are ethoxylated aliphatic alcohol, polyoxyethylene surfactants, carboxylic esters, carboxylic amides, and others are considered good PPDs. They are capable of modifying wax crystals and changing their morphology, hence reducing their tendency of interconnection to form a three-dimensional structure [82].

- ***Anionic surfactants***

Examples, including calcium *o,p*-dioctyl benzene sulfonate, alkyl benzene sulfonates, and calcium hexadecyl benzene sulfonate, have been reported, where *p*-dioctyl calcium benzene sulfonate showed the best performance as a PPD [83,84]. Generally in these compounds, their aliphatic part binds to wax molecules in crude oil while hydrophilic heads retard wax crystallization.

- ***Cationic surfactants***

Examples of this type include triethanolamine and vinyl pyridine-based pour point depressants and wax inhibitors [85]. Khidr and Ahmed [86] synthesized bolaform surfactants, which have exhibited good results as PPDs for waxy crudes.

- ***Amphoteric surfactants***

These are not commonlyused as WIs or PPDs on their own, but they can be generally used to enhance the effectiveness of both the anionic and nonionic surfactants [85].

- ***Gemini surfactants***

Gemini surfactant (GS) is considered to be a novel surfactant for wax inhibition. GS contains more than one ionic or polar hydrophilic head and hydrophobic hydrocarbon tails, which are chemically bonded by a spacer. With that, GS is able to perform more efficiently than the other conventional surfactants [36]. GS is usually used as a wax dispersant in diesel fuel [87], and their use as a crude oil wax inhibitor was reported [88,89].

The mechanism of action depends on their surface properties, especially the interfacial tension (IFT) of the GS and the side chain length. GSs adsorb on the surface of growing wax crystals, impeding the formation of an interlocking network of waxes in the crude oils. Similar to the polymeric WIs, the longer the Gemini long-chain esters, the higher the reduction of PP and surface tension [88]. GSs can act as synergists for styrene/stearyl methacrylate copolymers [87].

- ***Water-dispersible wax inhibitors***

This type of chemical was inspired by wax depositions during fracturing operations due to the low temperature of fracking fluids and low reservoir temperature.

In this approach, wax control chemicals are incorporated into a microdispersion system that is fully dispersible in water, which is applicable for use during fracturing operations. In such micron-sized liquid-in-liquid or solid-in-liquid colloidal dispersed systems, active chemistries comprising polymers from poly(EVA), poly(alkylacrylate), poly(EVA-alkylacrylate), poly(α-olefine-MAA) esters/amides/imides, and selected dispersants and surfactants are brought together to deliver immediate and short-term inhibition for paraffin wax control [90].

- **Emulsified WI and PPD**

The commercial WIs and PPDs are typical polymeric compounds dissolved in an organic solvent; such formulations can suffer from solidification and plugging issues in the injection lines in deepwater and cold arctic regions, which reduce their efficiency. Thus emulsified PPDs and WIs have been developed to improve the PPD adaptability and deliverability in different oilfield climate temperatures. The emulsified chemicals avoid solidification in the injection lines and offer better physical handling properties compared to the solvent-based chemicals [91,92].

Crystal modifiers like EVA can be formulated in emulsion form for deepwater applications, allowing effective wax control at lower dosage concentration and lower dosage injection rate without causing viscosity or plugging issues [93,94]. They can also be formulated with antifreeze compounds to allow wider range of use at the very low temperatures of deepwater and arctic locations [95].

- **Ionic liquids**

Imidazolium ionic liquids were found to reduce the pour point to 12°C, and to improve the rheological behavior of the crude oil by reducing the viscosity by 73% and 87% at 23.5°C and 15°C, respectively [96]. A novel ionic liquid PPD was found to reduce pour point by 18°C [81].

- ***Green wax inhibitors and pour point depressants***

Although some conventional WIs and PPDs are considered relatively safe, like EVA, some other environmentally friendly chemicals have been studied in the last few years to work as alternatives to conventional chemicals.

Nanohybrid PPDs in which inert nanomaterial is used are considered efficient and environmentally friendly chemicals. Ionic liquids are also considered environmentally benign chemicals.

Other types of green WI and PPD include natural products and extracts. Kumar et al. [97] discovered that *Sapindus mukorossi*, as a natural surfactant, significantly reduced the viscosity by 80% when it was used as a PPD on heavy crude oil. Kumar et al. [98] extracted fatty acids present in sunflower (*Helianthus annuus*) oil to prepare oil-in-water emulsions of a heavy crude oil from the western oil field of India, used to overcome flow assurance problems. Akinyemi et al. [99] explored the application of vegetable oils, such as jatropha, rubber, and castor, on the flowability of waxy crude oil, which revealed the capability of these vegetable oils to depress the PP and the viscosity of waxy crude oil appreciably within the dosage of 0.1–0.3 vol%.

17.2.2.3 Wax dispersants

Wax dispersants are generally film-forming surfactants that penetrate wax deposits, adsorb on the particles, and disperse them, preventing their accumulation. They can also lower the PP by breaking the wax network [100].

Typical commercial wax dispersants are based on alkyl sulfonates, alkyl aryl sulfonates, fatty amine ethoxylates, and alkoxylated products [31].

Dispersants can be used alone or, most frequently, blended with polymeric wax inhibitors to enhance their performance. Some case studies have shown limited success of dispersants when used on their own, although some reports showed their sole success [100,101]. Also, it was proved that wax dispersants were successfully applied in field trials where the wax inhibitors failed [102]. Long chain oleic film formers like imidazolines were found to enhance the performance of wax inhibitors [101].

Another way of applying the surfactants is by emulsification of waxy crude oil in the presence of water, which was discussed earlier in the section on operational techniques of wax control. Ahn et al. [37] showed that Triton-X and Tween show lower wax deposition, with an increase in their HLB number (hydrophilic lipophilic balance) and generally lower IFT. Field application of this process is reported using nonionic surfactant (a synthetic tridecanol with 9–10 ethylene oxide groups) [38].

17.2.2.4 Factors affecting the performance of wax control chemicals

- **Molecular structure of the chemical agent:**
  - The bulky group composition and spacing.
  - The alkyl side chain length matching with the crude oil waxes.

- The molecular weight of the chemical.
- The surface properties.
- Melting point matching with the crude waxes.

- **Crude oil composition:**
  - paraffin waxes: heavy *n*-paraffins (*n*-C24+paraffins) were deleterious for the efficiency of a paraffin inhibitor, while cyclo+branched alkanes enhanced the activity of the crystal modifier; in fact, most commercial wax inhibitors reduce deposition of low molecular weight paraffins (C34 and below), while exhibiting little effect on deposition of high molecular weight paraffins (C35–C44) [53,101]. Severely waxy crudes with high WAT value may not be solved, even with highly effective inhibitors like comb polymers over extended periods. This is because the bulk of the wax alkanes are considerably longer than the alkyl chains in the comb polymer [31]. They also may show limited applicability for microcrystalline waxes containing branched and cyclic components.
  - Asphaltenes: There are conflicting accounts in the literature about the effect of asphaltenes on wax; some studies proved the effect of asphaltenes as natural wax inhibitors, and others showed that they are wax deposition promoters, and a third team showed that there is no significant effect of asphaltenes on wax deposition [103]. However, asphaltenes can interact with some PPDs and increase the dose required to perform efficiently [31]. Other authors showed that asphaltenes enhanced the performance of WIs and PPDs [104,105].
  - Besides, the naphthenic acids were found to cause a small increase in pour point [104].
  - Other crude components or chemical additives that might be competitive in adsorption or interfere with the WIs and PPDs.
- **Operating conditions:**
  - Shear: high shear can affect the PPD efficiency, especially below WAT [53].
  - Temperature: increasing cooling rate along the pipeline reduces the PPD efficiency, especially in long subsea pipelines [53].
  - At times some PPDs showed short-lived viscosity improvements [53]. Static long-term stability of treated crudes may not satisfy transport requirements due to sequestration or depletion processes [31].

17.2.2.5 Testing and screening of wax control chemicals

Evaluation of wax control chemicals can be performed either by evaluating crude oil property like WAT, pour point, or carbon numbers, or by evaluating wax deposition rates.

17.2.2.5.1 WAT measurement

A change in crude oil WAT is one of the main goals of the wax control chemicals, especially the crystal modifiers class, i.e., WI and PPD. Consequently, the wax control chemicals can be screened and ranked based on their effect on WAT. Different standard test methods can be used to measure the crude WAT or cloud point, including ASTM D2500, ASTM D5773, ASTM D7397, and ASTM D5772, where ASTM D2500 is considered the most commonly used manual method by the operators, based on a visual inspection technique. Several other techniques have been effectively used to monitor WAT of crude oils before and after adding the chemicals, to evaluate their efficiency. Methods include DSC, HP-DSC, cross-polarization microscopy, IR spectroscopy, sonic testing technique, nuclear magnetic resonance, viscometry, and others. DSC is the most commonly used technique, having high accuracy and applicability. It is preferred for WAT to be tested on live oil rather than stock tank oil (STO). A significant decrease in the WAT value is a good sign of effectiveness of a WI or PPD chemical.

17.2.2.5.2 Pour point measurement

ASTM standard D97 is a method devised for stock tank oil samples and has been used over the years as a standard method to measure pour point. ASTM D5853 has been defined specifically for crude oil pour point measurement. The method involves determination of lower and upper pour point (minimum and maximum pour point), which provides a temperature window where a crude oil, depending on its thermal history, might appear in the liquid as well as the solid state. This test method can be used to supplement other measurements of cold flow behavior. It is especially useful for the screening of the effect of wax interaction modifiers on the flow behavior of crude oils.

The repeatability of the test can be poor, but it allows a ranking of PPD efficiency [85]. The dosage of PPD needed in a laboratory test will not be representative and will almost always be higher than that necessary for field use [31]. Fig. 17.14 shows a typical D97 pour point test jar with oil sample.

17.2.2.5.3 Carbon number distribution

Various research studies showed that the best-performing wax control chemicals are those that are matched in their carbon numbers with the crude paraffins they intend to inhibit [106]. Therefore it has been industry practice recently to determine the carbon distribution of the crude oil as well as the paraffin control chemical used, to measure their degree of matching. The wax carbon number distribution is commonly determined by high-temperature gas chromatography (HTGC) analysis, which is also used to identify what is called the critical carbon number (CCN), which is another term related to

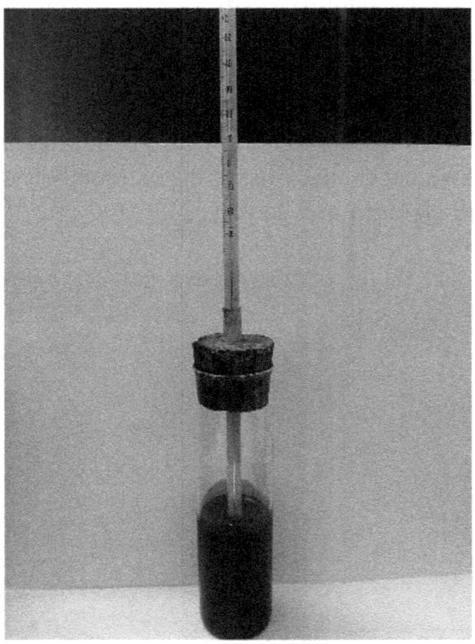

FIG. 17.14

Pour point test jar with a crude oil sample.

the number after which paraffin waxes showed hard-to-remove deposits [31,107]. The ASTM D5442 method is used to quantitatively determine the carbon number distribution of petroleum waxes in the n-C17 to n-C44 range by gas chromatography.

17.2.2.5.4 Boiling point measurement

The boiling method is basically used to assess the hydrocarbons deposits rather than the treatment chemicals. True boiling point (TBP) or TBP distillation technique can provide data on the mass, mole, and volume fractions of distillation cuts with measured molecular weight, specific gravity, and boiling points [108].

Recently, high-temperature simulated distillation (HTSD) analysis (measuring carbon number fractions) can be used to determine wax amounts in deposits collected on the cold finger during chemical inhibitors assessment [109]. The method was found to improve the cold finger measurements, as it eliminates the uncertainties from the bulk weighting procedure (deposits scrubbed from cold finger surface and weighted), due to the occluded oil variations. As a result, this generates an accurate baseline for wax control chemical screening.

17.2.2.5.5 Rheology measurements

WIs and PPDs are flow improvers, hence their effect will be reflected on the crude rheology. Viscosity and yield stress are the most common methods to evaluate rheology. The crude oil is usually tested at the process temperature to evaluate the efficiency of the applied chemical.

Various standard methods can be used to measure crude viscosity, e.g., the ASTM D445 method using a calibrated capillary viscometer under a reproducible driving head and at a closely controlled and known temperature; and the ASTM D7042 method, using a Stabinger viscometer (measuring cells consist of a pair of rotating concentric cylinders and an oscillating U-tube). Other commercial rheometers can be used, considering their reliability [110]. Fig. 17.15 illustrates some of the viscosity measurement methods.

Becker [111] has mentioned that pour point serves as a good indication of chemical effectiveness if the crude shows a high tendency to congeal and little or no tendency to precipitate. However, viscosity measurements should be conducted in combination with the pour point tests to determine the magnitude of a chemically induced physical change, due to the extremely low shear imparted by the method, which can represent very small changes in viscosity.

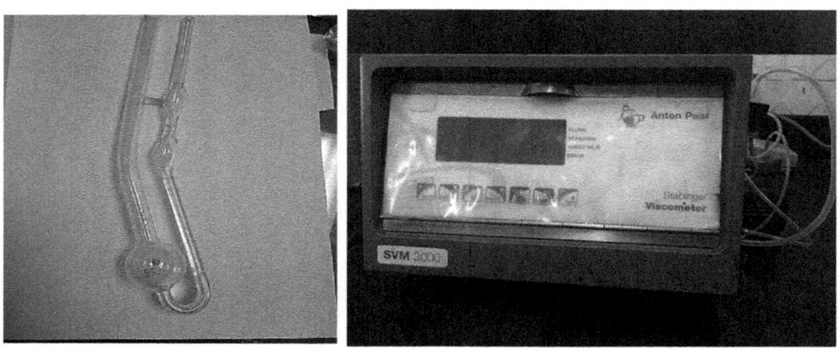

FIG. 17.15

Viscosity measurement methods: capillary viscometer (left), Stabinger viscometer (right).

17.2.2.5.6 Cold finger (CF) test

The most common apparatus used in the laboratory for chemical additive screening is the cold finger device, which has been used for decades in the industry. It is simple, easy to use, requires small sample volumes, and interpretation of results does not require high skills. In a cold finger test, heated oil stirs around a cooled metal finger (its temperature is maintained by an external chiller) that simulates a cold pipeline's inner wall, leaving wax deposits after the finger's temperature falls below the crude WAT. Agitation or circular flow is applied by a magnetic stirrer or rotary outer vessel to impose a shear on the surface of the cold finger probe [52]. Fig. 17.16 illustrates the cold finger test system, and Figs. 17.17 and 17.18 show wax deposition tests from uninhibited crude oil and condensate.

CF can only qualitatively evaluate the chemical performance and rank different chemicals rather than quantitatively predict the performance of PI under field conditions. After ranking and identifying PIs using CF, a field trial is usually done to confirm the applicability of the chemical and to optimize the dosage [52,109].

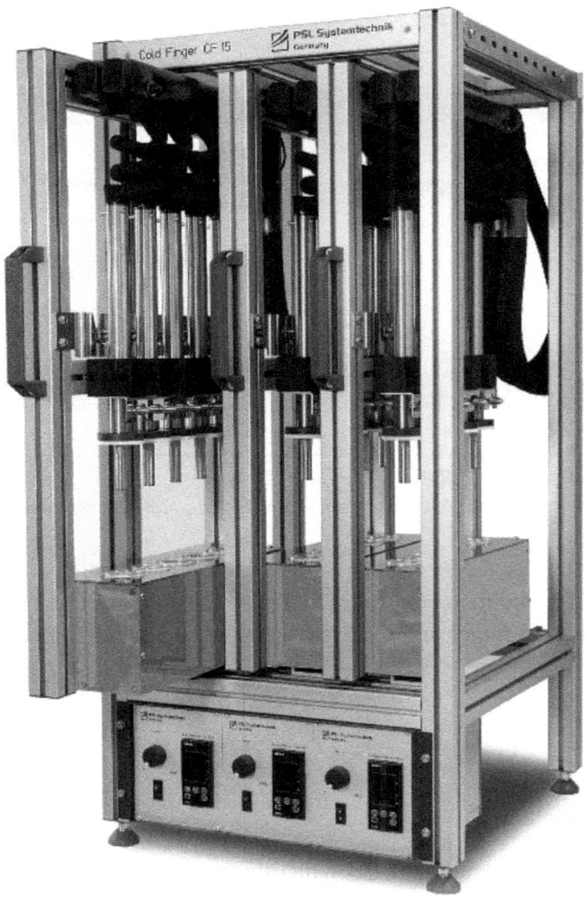

FIG. 17.16

Cold finger test system.

Courtesy of PSL Systemtechnik GmbH (https://psl-systemtechnik.com/), reprinted with permission.

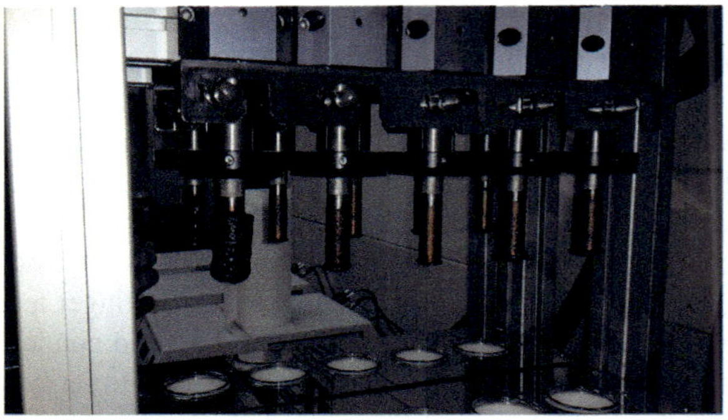

FIG. 17.17

Crude oil wax deposition on cold finger.

Courtesy of PSL Systemtechnik GmbH (https://psl-systemtechnik.com/), reprinted with permission.

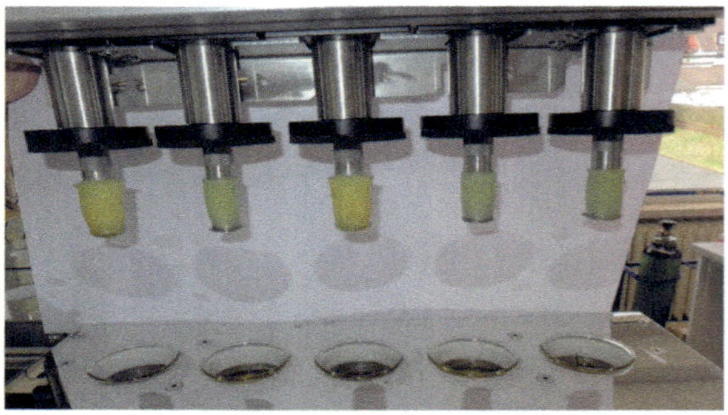

FIG. 17.18

Condensate wax deposition on cold finger.

Courtesy of PSL Systemtechnik GmbH (https://psl-systemtechnik.com/), reprinted with permission.

The cold finger is subjected to uncertainties and may not fully represent the field conditions due to differences between the cold finger and the actual pipeline conditions, including:

- Temperature,
- Flow rates and flow regimes,
- Level of agitation,
- Shear stress,
- Cold finger surface roughness.

These differences lead to the fact that the carbon distribution of untreated deposit on the cold finger test might not be representative of the carbon distribution of the field deposits [112]. Roughened CF can be more stringent when screening WIs, as rougher CFs tend to make certain WIs less effective under low shear conditions. On the other hand, CF with very low roughness (e.g., a polished surface) may cause repeatability issues or give spurious high WI efficiencies [109]. Estimation and characterization of the waxes deposited on the cold finger are crucial. Incorporating techniques like HTSD or HTGC is necessary to determine wax deposit carbon distribution, rather than bulk deposit weighting to reduce experimental uncertainty due to occluded crude oil during the test and to accurately assure WI and PPD efficiency by identifying the specific carbon range that actually deposits in the field and affording the necessary carbon matching between the wax deposits and the WIs/PPDs, rather than the overall protection that is conventionally applied [109,112].

Weispfennig's model, or the Chilton-Colburn analog, has been used by researchers to correlate wax deposition using cold finger with that in pipeline flow [113,114].

Another design of the cold finger, known as the coaxial shear cold finger (CSCF), where the mixing is provided by the rotation of the refrigerated finger itself, allowing different mixing regimes and giving a better representation of the field shear conditions.

17.2.2.5.7 Flow loop

Flow loop or tube blocking rig is more representative of the field pipeline conditions. They can be designed in different sizes, e.g., microbore capillary or larger diameter pipe pilot loops. In flow loops, the change in pressure across the microbore capillary or pipe is measured due to the buildup of wax on the internal walls of the loop [31]. They can be placed in a cold water bath or jacketed where the coolant flows through the annulus in the opposite direction of the tested crude that is flowing internally. The pipe flow simulates to some degree field flow conditions, which imposes a constant shear stress on the surface of the pipe wall or wax deposits [52]. Fig. 17.19 illustrates an automated wax deposition flow loop, and Fig. 17.20 shows typical results during chemical screening.

FIG. 17.19

Automated wax flow loop.

Courtesy of PSL Systemtechnik GmbH (https://psl-systemtechnik.com/), reprinted with permission.

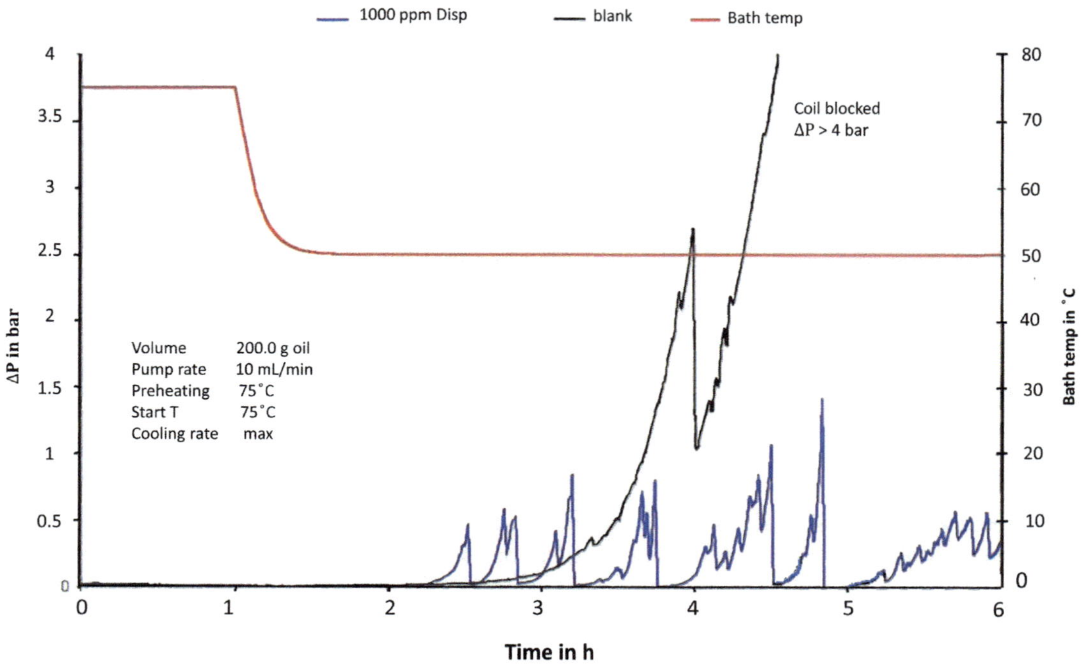

FIG. 17.20

Results from wax flow loop during chemical screening.

Courtesy of PSL Systemtechnik GmbH (https://psl-systemtechnik.com/), reprinted with permission.

Daraboina et al. [115] found that flow loop deposits always have a higher normalized deposit mass and lower wax content compared to the cold finger deposits. They also concluded that the flow loop has a much higher wax mass flux than the cold finger, which is mainly due to the higher concentration gradient, resulting from a thinner mass transfer boundary layer thickness.

17.2.2.5.8 Dynamic paraffin deposition cell (DPDC)

One of the recently patented devices is the DPDC. The method allows testing different chemistries in the presence of brines under field dynamic conditions. An emulsion is created where the water phase could be dispersed into the oil phase, causing wax deposition on a cold finger.

In this method, multiple cells are filled with test fluid and inserted into a shaker cage, along with a cold finger connected to a chiller bath which is loaded onto a horizontal shaker inside of a shaker bath, where horizontal shaking is created. The horizontal shaking should be able to generate high shear inside the cell, making product screening in an oil/water system feasible [116].

Russell et al. [117] showed that, despite the difference in the mixing mode and shear between CF (cold finger), CSCF (coaxial shear cold finger), and DPDC (dynamic paraffin deposition cell), the formed wax deposits were very similar in composition, where the paramount factor in generating laboratory deposits similar to those in the field is the careful selection of temperatures for the experiment, which includes the utilization of a small temperature gradient (5°C) that is close to the WAT.

Other methods are used to test the chemicals to assess their flowability and thermal properties, especially for applications in harsh conditions like deepwater and subsea applications. These include static-freezer stability, rheological profiling and no-flow points, thermal cycling, and cold centrifugation [118].

In addition to these methods, for batch and squeeze treatments, core flooding tests are used to estimate the efficiency of the applied chemicals and their effect on the permeability of the producing formation.

17.2.2.6 Application of wax control chemicals

After selecting the proper wax control chemical and estimating the proper dose, the next step is to also properly deploy them to the optimum location. Wax control chemicals can be directly injected into the system using injection points or umbilical lines, or can be squeezed into the reservoir.

- **Continuous injection of paraffin control chemicals**

Wax control chemicals can be directly injected into the system. The injection point should be located upstream of the wax problem, meaning upstream cold spots in the system where the temperature is expected to fall below WAT or where paraffin usually accumulates at low-speed flow rates. That requires good monitoring and profiling of the system temperature, and precise measurement of crude oil WAT and fluid hydrodynamics.

Wax control chemicals can be injected through injection points that are already installed in the pipelines, or injected at the Christmas tree, offering safety to the whole system. The concentration needed of wax inhibitor or PPD depends on the severity of the wax problem, crude composition, and other factors, while 100–2000 ppm is commonly used to cover different field applications. Wax dispersants can be injected in a range of dosages between 50 and 300 ppm [31].

Some polymeric paraffin inhibitors are supplied solid and need to be dissolved in a proper solvent [aromatic solvent, preferably] to be pumped in liquid form. Their solubility in the solvent determines the polymer amount in the finished product to meet viscosity, flowability, and safety specifications.

Wax chemicals can be injected downhole using a surface chemical injection system, umbilical lines, using the gas lift system, down the annulus to a bottomhole valve, or with no packer, simply down the annulus and around the end of the tubing [119]. Umbilical lines are also mainly used for subsea systems. The application of wax control chemicals in umbilical lines encounters some problems, like their high viscosity due to their polymeric nature, suspended solids, aggregation, and precipitations, particularly for application in deepwater, subsea, and cold climates [120]. To avoid the complication of these chemicals:

- The chemicals are deployed at low-concentration mixtures or solutions of the active ingredient in a solvent, often an aromatic hydrocarbon. In this case the chemicals are applied at high dosing rates [31].
- Some low-temperature stable chemicals were developed for subsea and deepwater applications, like alkyphenol-formaldehyde (APF) polymers [118].
- Some chemicals can be formulated as low-viscosity emulsions or water-dispersed formulation, allowing their delivery without complications.

- **Squeeze treatments**

Like mineral scales, squeeze jobs were also applied to protect the reservoir and downhole regions from the detrimental effects of paraffin wax deposits. Some important aspects for the squeeze treatments include:

- Wettability
- Design for a minimum inhibitor concentration in the produced fluids
- Ability to trace/measure the inhibitor in the produced fluids

Although theoretically thought to be inapplicable, there have been reports of successful wax chemical squeeze treatments, some lasting in excess of 6months [111,121]. Haynes and Lenderman [119] applied an inhibitor by squeeze, followed by squeezing a chemical that precipitated it. The precipitate releases the chemical slowly over a long period, prolonging the life of the treatment.

Wax dispersants can also be applied in batch and squeeze treatments. Newberry and Barker [122] squeezed a dispersant and after a week shut in were able to remove paraffin from formation and restore a well. McClaflin and Whitfill [102] used water-soluble dispersant in wells with high water cut and oil soluble dispersant in high oil cut wells; the dispersants were batch treated in the wells, resulting in reducing the hot oiling frequency in the wells, and a savings of $10,000 per year on three wells.

- **Application with hydraulic fracturing**

To extend the lifetime of the treatment downhole and to reach the inaccessible and far locations in the reservoir, solid slow release paraffin inhibitors were proposed in which the solid inhibitors are added to the fracturing fluids for enhanced chemical delivery. The method can be applied for all production chemistry problems where the solid inhibitors for scale, corrosion, or for biocide activity desorb into the produced water, while the solid inhibitors for paraffin or asphaltene inhibition desorb into the hydrocarbon phase [123].

One of the ways to apply the slow-release paraffin chemicals is during hydraulic fracturing operations, where the solid inhibitor composite is adsorbed/distributed throughout the proppant pack during the hydraulic fracturing treatment. The paraffin inhibitor-proppant is designed to slowly release the adsorbed paraffin inhibitor into the bulk oil as it flows through the propped zone and to inhibit paraffin deposition [124]. Gupta et al. [125] proposed using high-strength bauxite proppant as the substrate for solid inhibitors for high inhibitor load and higher closure pressure, which can extend treatment lifetime, especially for deepwater applications. Generally speaking, the release rate depends on the temperature of the oil and the rate of production. Higher production rate wells correlate with faster consumption of the solid inhibitor [126].

Since 2005, this chemical delivery technology has been applied in over 15,000 wells to prevent various production chemistry issues, e.g., scale, paraffin or asphaltene deposition, where solid paraffin inhibitors alone have been applied in over 2000 wells primarily in the US and Canada. since 2007 [124,127,128] Wornstaff et al. [128] reported increasing cumulative production by approximately 40% and increase in revenue by about 15.8 million USD in the first 350days of applying solid paraffin inhibitors.

Frances, and Bhaduri [129] reported using combination of treatments, whereby a compatible submicron-sized paraffin inhibitor dispersion is added to the fracturing fluid, in addition to a dispersion based paraffin inhibitor was adsorbed onto a proppant-like substrate that can be mixed into the proppant and delivered deep into the formation with fracturing treatments, affording long term wax deposition treatment. In another study, colloidal microdispersion comprised of poly different inhibitor chemistries were added fully dispersed into the fracturing liquids to give long term protection from paraffin deposition [90].

17.2.2.7 Recent advances in chemical inhibition of wax deposits

Recently, the chemical management of wax deposition has witnessed various advances. New chemicals in different formulations were developed for different purposes, such as emulsified, dispersed, solid inhibitors used to extend the treatment life and reach different regions in the reservoir and also to allow applicability in harsh conditions. Nanohybrid chemicals and ionic liquids are other emerging technologies found to improve the efficiency of chemical treatment. In the meantime, major advancements have been made in testing and screening these chemicals. For example, cold finger, which is considered the industry main screening method, has witnessed a number of critiques of the effect of surface roughness, shear, mixing, differential temperature, and whether the method is representing the actual field conditions. Recent research is more inclined to recommend the use of flow loops incorporated with more advanced techniques, e.g., HTGC, HTSD to characterize the lab-created deposits and relate them to the field deposits and to match them with the candidate chemical for better performance. A DPDC device was recently developed to allow chemical screening in the presence of brines and enhanced mixing and shear for better representation of the field conditions.

17.2.3 Nonchemical prevention of wax deposition

These methods include nonchemical means (physical, mechanical, or biological) to prevent wax deposition. Magnetic, ultrasonic, electric, wax repellent surface, and biological means are the most common methods of wax mitigation methods.

17.2.3.1 Coatings and wax repellent surfaces

The chemical and operational methods aim to prevent wax deposition by disturbing the nucleation and crystal growth of wax crystals. Internal coating materials (wax-repellent surfaces) inhibit wax deposition by preventing the precipitated waxes from adhering to and accumulating on the tubing surfaces, causing plugging problems.

Paso et al. [130] comprehensively reviewed the use of surfaces with wax-repellent materials to prevent or reduce wax deposition based on numerous analogies of nonstick and antiadhesive materials from the marine, biomedical, paint, food, and tribological industries. They concluded that self-assembled polymers and nanocomposite materials exhibiting smooth surfaces and low surface free energies may provide appropriate surface technologies for prevention of wax deposition. Common types of effective coatings include fluoro-siloxanes, fluoro-urethanes, oxazolane-based polymers, and DLC-polymer hybrids. In recent years, more attention has been paid to the interfacial wettability between coatings and crude oil. Novel functional coatings have been designed by using bionic concepts, including superhydrophilic micro-nanostructure composite coatings, and self-reinforcing ultralow adhesion surfaces [131].

Some examples of the deposit-repellent surfaces that have been used for different applications, including wax prevention, are:

- *Hydrophilic coatings*: including lacquered compositions, Bakelite, enamels, epoxides, and glass, had been used in the Tatneft fields [132]. Other types are epoxies as well as polymer or copolymer segments containing polyhydroxystyrene as well as polyhydroxyethyl methacrylate, polyallylamine, polyaminostyrene, polyacrylamide, polyacrylic acid, and polyhydroxymethylene, superhydrophilic $TiO_2{=}SiO_2$ composite thin films formed by sol–gel deposition, and sulfur trioxide surface treated tubulars has successfully showed promising results when applied in the field [133,134].
- Plastic coatings showed lower wax deposition compared to steel surfaces [135]. They include phenol-formaldehyde, an epoxy-phenolic, and a polyurethane, with phenol-formaldehyde having superior performance. Phenol-formaldehyde coatings have been proven to control wax deposition when applied in the field [136]. Zhang et al. [137] showed that fluoropolymer and vulcanized silicone rubber coatings with low surface energy had significant effects on drag reduction and paraffin deposition prevention. The silicone rubber coating could achieve an efficiency of up to 75% in paraffin deposition prevention and could reduce drag by 22% in crude oil at 26°C. Liu et al. [138] applied a fluorocarbon coating for anticorrosion and antiwax in alkaline-surfactant-polymer (ASP) flooding production.
- *Glass-lined piping* was introduced to manage wax deposition issues in the Jilin Oilfield in northeastern China. The coating showed complete success in eliminating paraffin blockages over a 10-year test in 34 wells [139].
- Glass-reinforced epoxy resin (GRE) tubing was applied in a real production field located in Oman, [140] and it also was used for corrosion control purposes [141].
- Diamond-like carbon (DLC) coatings that can be applied to prevent mineral scales, wax, asphaltenes, gas hydrates, biofouling, and also aids in corrosion prevention [142].
- Specific surface treatment can also be beneficial in reducing solids deposition in general that includes ion implantation, ion sputtering, carbo-nitriding, oxidizing, electrodeposited films [130,143]. Ion implantation is also applied to reduce the metal surface energy by introducing heteroatoms into the solid surface by energetic bombardment of ions in the keV to MeV range. Ion implantation of F^+, Si^+, and SiF^+ resulted in a reduction in the amount of deposited wax by approximately 90% [130].
- Other polymer and copolymer based coatings: fluorosiloxane polymers, polysiloxane networks, teflon-based films, oxazoline-based polymer films, urethane-based polymer films, acrylate homopolymers, acrylate copolymers, crosslinked and grafted acrylates, hyperbranched polymers, styrenes, and others. Some of these have been used in the oil industry and some not, but they were used in other industries [130]. Rashidi et al. [144] and Mombekov et al. [145] reported polyvinyl chloride (PVC) pipe and ethylene-tetra-fluoro-ethylene (ETFE) with ETFE giving the best performance.
- Bioinspired superhydrophilic coatings that are inspired by nature. In recent years, a large number of researchers have applied the idea of bionics to the study of functional materials and have tried to learn from and imitate the effective structures in nature, such as superhydrophobic structures on the upper surface of leaves [146], and the water-collecting structure of the backs of beetles [147].
- Chemical conversion coatings: Guo et al. [148] developed an excellent nonwax-stick coating prepared based on a Zn coating by chemical conversion treatment. The chemical conversion coating had hydrophilic and superoleophobic properties in a water/oil mixture system, which enabled it to

trap water molecules on the surface and reduce wax deposition. Pyrophosphate-based conversion coating showed a good antiwax performance, reaching about 80% [149].

- Organogel coatings: A state of ultralow adhesion has been developed by reducing the drag between the surface and the interface, to prepare a new surface. Yao et al. [150] prepared a new type of organogel (OG) material by polydimethylsiloxane (PDMS) and showed that it provides a good antiwax surface because the adhesive force between paraffin crystals and the surface of PDMS was very low, 500 times smaller than that of traditional materials.

For antiwax coating to perform properly, some criteria must be met, including [3,130]:

- The coating must be chemically inert, thermally stable, sustaining long-term operability at high flow rates and high shear.
- The coating must have anticorrosion, and antiabrasion properties.
- The coating material should also be suitable for multiphase flow involving water, oil, and gas phases.
- The material should be resistant to other deposits such as asphaltenes, resins, and scale.

17.2.3.2 Magnetic methods

Crude oil rheology and paraffin wax crystallization was reported to be affected by magnetic fields [151,152]. Crystallization and wax deposition was lower when a magnetic field was applied; also rheological changes in crude oil due to magnetic field exposure were reported and found to be dependent on the crude composition [153,154].

Generally, a magnetic field is utilized through the application of a magnetic fluid controller (MFC) or magnetic field conditioner (MFC), where magneto hydrodynamics and Lorentz force governs the magnetic effect on deposits [155]. The magnetic fields interact with the growing nuclei in the fluids, affecting precipitation [156]. Under the effect of magnetic field, paraffin molecules tend to align their poles with the ones of the magnetic field (if the thermal agitation is not excessive), which changes their orbital angular momentum and causes a disturbance in the crystal agglomeration processes [154]. Magnetic orientation and disaggregation of wax crystals are found to be the microscopic mechanism of magnetization [157].

Tao and Xu [158] reported that magnetic field pulse was successfully used to reduce the viscosity of paraffin-base crude oil, while the effect was weak on the asphalt-base crude oil or mixed-base crude oil, the viscosity of which was reduced by an electric field pulse. Also, magnetic field conditioning was found to lower the WAT [156,157], and to reduce oil viscosity by 82% under the optimum magnetization parameters [157]. However, Chow et al. [159] showed that no changes in WAT were detected under magnetic fields and an increase in crude viscosity was observed when experiments were conducted at temperatures close to WAT.

Practically, quite a few companies adopted the technique and are manufacturing magnetic fluid conditioners like (Algae X International Inc.), (MundiMex Inc.), (Mag Tek Inc.), (Fuel Mag International Ltd.). In field application, Magnets have been used to prevent wax deposition, mostly downhole. The technology is commonly used in Asia, but less in the West; for example, by 1995, as many as 14,440 magnets had been installed in China with a claimed good success rate [160].

However, as was reported in the case of mineral scale deposits, the magnetic field conditioning is still controversial and some negative results were reported. The industry currently has less confidence in this method and is reluctant to implement it.

17.2.3.3 Electric field methods

Pulsed electric field was found to reduce the viscosity of paraffin base and asphaltenes base crude oils [158]. Strong local electric field was able to polarize oil-suspended particles in pipelines, forcing aggregate particles into short chains, thus decreasing crude oil viscosity. The method to be effective for several types of crude oil, including asphalt base crude oil and paraffin base crude oil, with instantaneous effect that lasts for more than 11 h [161].

Krasnov et al. [162] applied electrostatic field using cathode polarization which was created by employing electric current throughout pipelines or downhole equipment (either by external source of current or artificial creation of galvanic elements). This method was able to disperse the organic deposits and create localized heat that melted the formed deposits.

17.2.3.4 Acoustic methods

Generally speaking, ultrasonic radiation improves heavy and viscous oil flowability by cavitation, mechanical vibration, and thermal effects to destroy the bonds formed between wax-asphaltene-resin compounds [163,164].

The viscosity of the ultrasound treated oil decreased by 1.7 times, and the pour point was reduced by 32°C compared to untreated oil [165]. Mullakaev et al. [166] reported that radiating systems of downhole tools with ultrasound was proven to be very effective for oil recovery. Ultrasonic treatment can increase oil extraction by 50% or even more in some cases in wells with a permeability of more than 20 mD and a porosity of more than 15% [167]. On the other side, Towler et al. [168] tried to use ultrasonic waves at a frequency of 120 kHz for wax deposition prevention, but the results were not promising.

17.2.3.5 Microwave irradiation

Microwave radiation was reported to improve the viscosity of heavy and high pour point crude oils by the effect of localized heat generated, and due to the microcracking of asphaltenes and other heavy molecules to lighter ones [169]. The technology was adopted and provided by Baker Hughes, and is known as Ecowave [170,171].

The equipment consists of an Ecowave unit, tuner, antenna system, and portable power source. After an appropriate choice of a well is approved by the operator, this is then followed by the deployment of the antenna in an annulus or tubing at the surface and attached to the Ecowave. The system is then powered, and the treatment is commenced, which normally lasted for about 2–4 h. The efficiency of the method has been field-proved with increase in production rate from 20% to more than 120% and treatment durability of more than 60 days [172,173].

17.2.3.6 Oscillatory motion method

Oscillatory motion has been proposed as a method for wax deposition avoidance and removal based on the Avrami theory [174], but its practical application has not been proven yet. The method depends on the wax content: in low wax content crudes, oscillatory motion reduced the wax deposition by 40%–60%, but at higher wax contents, the results were contrary.

17.2.3.7 Wax inhibition tools

Wax inhibiting tools (WIT) are devices made in such way that can change crude oil properties, while it flows through them, thus prevent wax deposition based on the physical and hydrodynamic phenomena.

A Silver-Hawg WIT is an alloy composed of dissimilar metals that is based on the fact that physical characteristics of a flowing liquid are modified by contact with a particular alloy. The crude oil is sucked inside this tool and creates multiple streams or jets through the radial holes in the tool, which bombard the copper nickel walls of the annular chamber and the center insert releases, which combines with molecules of the hydrocarbons and other minerals, thereby altering certain physical characteristics of the produced fluids and preventing solids precipitation, breaking long chain hydrocarbon and keeping paraffin suspended in the solution [175]. The tool has been assessed by Sulaimon et al. ([176] (Fig. 17.21) and field applications have proven the functionality of this tool, especially when (WIT) is installed 100–500 ft. below the tubing depth corresponding to the predicted WAT or at the wax nucleation depth.

Enercat wax elimination tool: This tools is based on the piezoelectric effect (the ability of some materials to generate electric potential in respond to mechanical stress). The Enercat tool looks like a standard production tubing pup joint with a jacket in an aluminum casting. Within this jacket are quartz crystals and semiprecious metals that generate a passive energy wave. Enercat creates a passive vibrational energy to stabilize the original micelle structure to prevent mineral and paraffin deposition and viscosity changes. This allows the solution to move smoothly and cleanly through the pipe without causing deposition problems [177].

Sulaiman et al. [178,179] developed a piezoelectric WIT using semiprecious metals and quartz. Zinc and lead were mixed with quartz and subsequently used to fabricate the tool with an aluminum matrix. Site testing in a flowline was able to provide encouraging results, as the thickness of the assessed deposits was reduced significantly (by over 32%).

Relief flow swirlers: this technique is based on flow passing through the direct-flow swirler and its transformation into a pulsating turbulent flow with pressure fluctuations that occur in the peripheral

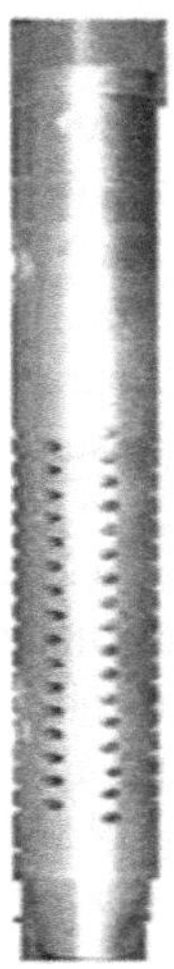

FIG. 17.21

Wax inhibition tool.

Source: A.A. Sulaimon, G.K. Falade, W. Delandro, A proactive approach for predicting and preventing wax deposition in production tubing strings: a Niger delta experience, J. Pet. Gas Eng. 1 (4) (2010) 26–36.

zone, and flow rates are redistributed [180]. This results in flow pressure fluctuations affecting the tubing walls and preventing the formation of deposits, contributing to the increase in flow temperature. It should be noted that this method assumes that it is possible to determine in advance, almost reliably, the locations of deposit formation in the tubing [180].

Another WIT was reported by Anyanwu et al. [181]. In the field application, a Xylene—Diesel chemical solvent was used to dissolve all the wax in the tubing and restore the well to production, while the WIT, which prevents wax deposition, was installed at the XN-Nipple of the tubing. The well was opened up to a potential of 1000 BPD after successfully removing the wax and installing WIT and it produced for about 4 months before getting waxed up again.

17.2.3.8 Biological wax prevention

Although using biological means to mitigate wax deposits is not very common in the oil and gas industry, some of these techniques have been used for decades in microbial enhanced oil recovery (MEOR), by using the bacteria that generate bioproducts such as gases, acids, surfactants, and polymers to change the physicochemical properties of reservoir oil, such as viscosity, flowability, and rock wettability and to stimulate oil-water-rock interactions that improve oil recovery [182].

Bishop and Woodward [183] used three product systems comprising hydrocarbon oxidizing microorganisms for the treatment of paraffin problems in production wells. The technique showed effective treatment of paraffin wax deposits and reduced maintenance cycles in over 500 wells and also showed production increase in some wells. Santamaria and George [184] used bacteria to control paraffin problems in five wells and the results were reduction of the periodicity of heavy maintenance production interruptions, from twice a week or twice a month to once every 6 months, with total savings of $8000/month. However, the technique was limited to wells that produce water, are pumping wells, and have bottomhole temperatures below 210°F. Besides, flourishing of the harmful SRB was observed too. Lazar et al. [185] found that special bacterial consortium (SBC1) facultative anaerobes have high performance in biosurfactant and biosolvent production and in degradation of waste hydrocarbons, which can be suitable for applications to remove solid or semisolid paraffin depositions. Etoumi [186] described the use of *Pseudomonas* bacteria in the reduction of wax precipitation in waxy crude oils through the biodegradation of heavy paraffinic hydrocarbons.

Rana et al. [187] used two microbial consortia, PDS-10 and FIB-19, to control paraffin wax deposition in downhole and surface flowline applications, respectively. The mechanism of action is believed to be that bacteria produces biofilm at the inner surface of the tubing, which hinders the growth, aggregation, and adhesion of wax crystals to the pipe wall. Also, the bacteria in the biofilm degraded the fresh wax crystals and the higher molecules of wax into smaller molecules, which then easily were dispersed into the crude oil. She et al. [188] found that bacterial strains such as *Acinetobacter* sp., *Pseudomonas* sp., and *Bacillus* sp. played important roles in deposit reduction and resulted in EOR. Their microbial action led to degrading the long-chain n-alkanes and the longer branched side chain aromatic hydrocarbons, decreasing polar compounds with N and increasing the O and O_2 species, and decreasing the pour point of the crude oil by 2–3°C. Lei et al. [189] evaluated two paraffin-degrading bacterial strains, namely, CYY0807 and CYY0810, which can degrade the heavy components in waxy crude oil. Wang et al. [190] found that *Pseudomonas* sp. DG2 exhibits good ability in paraffin degradation and wax removal efficiency. The bacteria produces a biosurfactant that emulsifies the oil and reduces the oil-water IFT value. In addition, the flowability improved by reducing the WAT of waxy crude oil by 4.16°C, and it reduced the viscosity by 23%. Liu et al. [191] isolated two bacterial strains *Bacillus cereus* QAU68 and *Bacillus subtilis* XCCX from oil production wells in the Daqing oilfield of China,

Table 17.3 Summary of nonchemical methods of wax management.

Method	Description
Internal coating	Wax-repellent surfaces Have shown success in field applications
Magnetic methods	Magnetic field interacts with wax molecules, preventing their deposition, improving crude oil rheology
Electric methods	Electric field interacts with molecules to prevent their precipitation, improving crude oil viscosity
Ultrasonic methods	Ultrasonic radiation prevents wax deposition and decrease viscosity by cavitation, heat generation and mechanical vibrations
Microwave methods	Microwave irradiation prevents wax deposition and improves crude viscosity by heat generation and heavy molecules cracking
Biological method	The bacteria can degrade the high molecular weight wax molecules and are capable of generating wax-repellent surface besides generating biosurfactants

and they showed the degradation rate of paraffin could reach 64% and the prevention rate of paraffin could reach 55%. Zhang et al. [192] showed that biosurfactant producer strains *Pseudomonas aeruginosa* (N2) and *Bacillus licheniformis* (KB18) showed promising efficiency of 79% of wax removal and also wax prevention action. Yen et al. [193] showed that three groups of bacteria, *Bacillus* sp. (LWH 1), *Bacillus* sp. (LWH 2), and *Pseudomonas* sp. (LWH 3) were able to remove wax deposits and reduce maintenance costs.

Table 17.3 summarizes the different nonchemical methods of controlling wax deposition.

17.3 Wax deposits removal

The other method of managing wax deposits is based on removing the formed wax deposits. The removal methods include chemical, thermal, mechanical, and other physical methods.

17.3.1 Chemical methods

Chemical wax removal is one of the efficient, cost effective, and easily applicable methods of removing wax. Different classes of chemicals can be used to remove wax deposits, including wax solvents, dispersants, surfactants, and heat-producing chemicals. The first three types are the most commonly used in oil and gas fields and depend on the solubility parameter of the solvent to break the bonds between the wax molecules in the hard deposits 3D network. The latter depends on the heat produced by an exothermic chemical reaction to melt the wax deposits.

17.3.1.1 Wax solvents

Using organic solvents is a common practice in the industry to remove wax deposits. Samuelson [194] listed different solvents as alternatives to aromatics to dissolve organic deposits. Xylene is one of the most used solvents in downhole and topsides applications, due to its superior performance [195].

For downhole applications, different batches of xylene might be needed if the problem is severe, or xylene can be left soaked at the affected area for a specific time to gain the advantage of the downhole high temperature as well; also agitation can be used by applying gas from any source at the well head, followed by bleeding the pressure at the production riser. For topside applications, the common practice is flushing the lines or soaking them with xylene by filling the whole line or filling the xylene between two pigs, followed by pigging. One problem with xylene is its low flash point (mixed xylene FP is 77–90°F), therefore usually it is mixed with diesel to avoid the lower FP and to lower the costs of the chemicals, as diesel is a bit cheaper than the xylene. A xylene-to-diesel ratio of 1:1 or 1:3 should be enough to dissolve or remove wax deposits. Furthermore, xylene use entails safety and environmental hazards in long-term and short-term exposure.

Other conventional solvents that can be used are toluene, kerosene, naphtha, gas oil, petroleum distillates with high aromatic content, and compatible hot oil. Also aliphatic solvents can be used, and different mixtures of these solvents can be formulated and used. A mixture of xylene or toluene together with an aliphatic solvent has been shown to increase the wax removal efficiency, and the addition of surfactants can also serve to enhance the performance of the solvent by increasing the dispersion action of the waxes [196]. The advantage of the aromatic solvents is that they can also dissolve asphaltene deposits and they do not disturb the crude oil equilibrium, causing asphaltene deposition like aliphatic solvents. Benzene, chlorinated hydrocarbons, and carbon disulfide have seen a good level of success. However, many of these solvents are not environmentally friendly [26,34]. Chlorinated hydrocarbons are good because of their high specific gravity, which enables them to penetrate the wax deposits [34]. Naphtha/toluene mixtures were claimed to treat a 250 bbl tank with 15-in. residues in the bottom, using low volumes of this mixture [196].

Recently, a new method for wax removal was proposed using CO_2 switchable-hydrophilicity solvents (SHSs), which exhibit an excellent dissolving capacity for paraffin wax, especially *N*, *N*-dimethylcyclohexylamine (DMCHA). All SHSs can effectively dissolve paraffin wax at a dissolution rate >0.03 g/min [197].

Environmentally friendly solvents are terpene, which showed good efficiency in removing wax deposits and can be used alone or in combination with other solvents and surfactants [31,198], and methyl ester, ethyl lactate, and cardanol [52].

As mentioned earlier, it is better to apply these solvents at high temperature, since elevated temperature increases wax solubility and increases the rate of wax removal. The effect of heat may be naturally applicable in downhole treatments and can be achieved using heaters or heat exchangers to circulate hot solvents downhole or in topside flowlines. Using mechanical means during solvent application also improves the efficiency of deposits removal, especially when a thick layer of deposit exists, and when other types of hard deposits exist, like mineral scales and asphaltenes. After enough time for dissolution, hot solvents must be pumped off the well to prevent paraffin resolidification after the liquids cool down.

Surfactants and dispersants can be used to dissolve wax deposits or be added to the solvents packages and formulations to improve the efficiency. Surfactants penetrate, break, and remove the wax deposit layer, besides preventing the dissolved wax from redepositing. Batch application of surfactants is preferred over continuous application for wax deposits removal. Surfactants are also used in emulsification to reduce wax deposition rates. They may also be used with modifiers for enhanced removal of wax deposits.

Dispersants consist of solvent, dispersant, and demulsifier and they disperse paraffin and prevent redeposition/sticking. Their major use is resolving interface and tank bottom problems. They are usually applied with hot water to improve wax deposits removal. Chemical dispersant-solvent packages, combined with the right mechanical application (involving a snubbing unit inside the pipeline that jetted the chemical dispersant and drilled through the wax mass), have been used to remove a wax blockage in a deepwater pipeline, in which earlier remedial treatments had failed and compounded the problem [199].

Hicks [200] recommended the use of hot water with a nonionic surfactant in the annulus while treating surface lines with hot oil containing either a paraffin solvent or a dispersing agent. Periodic application of dispersants down the annulus provided little help.

Solvents in well tubing and casing in shallow and short depths are usually applied by bullheading. However, for deepwater and subsea configurations, more tools and equipment such as coil tubing are needed. A new method of applying solvents/dispersants via capillary tube to remove wax plugs in deepwater wells was reported [201]. In this method, a capillary tubing unit was used to deploy the capillary tubing to the paraffin plug. A paraffin solvent was then pumped down the capillary tubing to remove the plug, and after sufficient soaking time the blockage was cleared and the operator gained communication with the well. The method was applied in the Gulf of Mexico and was able to remove 113 ft. of wax plug, saving operating coiled tubing expenses [201].

A novel chemical treatment that controls wax precipitation was developed. In this method the chemical works when mixed and activated with crude oil by a proprietary nanochemical mechanism and attacks wax differently. Flushing wax with activated crude oil removes the wax by layers and creates a waxophobic (wax repelling) condition for an extended period. The chemistry is also environmentally friendly and offers other advantages like treatment longevity, less labor, and low power consumption on top of wax trouble-free production [202].

Table 17.4 summarizes some of the wax deposits solvents.

Table 17.4 Examples of wax deposits solvents.

Examples of wax deposits solvents
Hexane
Pentane
Heptane
Xylene
Toluene
Kerosene
Diesel
Naphtha
Gas oil
Hot oil
Methyl ester
Ethyl lactate
Cardanol
Dispersants
Surfactants

17.3.1.2 Thermochemical packages

In situ heat generation by a chemical reaction can be used to melt wax deposits, as long as it is well designed and applied and catastrophic side effects been avoided. The method is commonly known as "the nitrogen-generating system" (NGS), or SGN after the original Portuguese, introduced by Petrobras, which is a thermochemical cleaning method based on an exothermic reaction (reaction between ammonium chloride and sodium nitrite catalyzed by acid), producing nitrogen gas and an immense amount of heat. The method has been reported to remove gas hydrates and paraffin wax plugs [10]. The method is also known as a fused chemical reaction (FCR) and has been extensively studied [203–205] (Fig. 17.22).

The recently applied packages combine different factors affecting wax dissolution, including:

- Chemical effect due to the added organic solvents.
- Thermal effect due to the heat generated, which melts the wax and also increases the dissolution efficiency of the organic solvents.
- Mechanical effect by the released gases in the thermochemical reaction providing enough agitation to speed up the dissolution process.

The method has been used to stimulate highly paraffinic producing wells in the Gulf of Mexico where the previous attempts to stimulate these wells with paraffin solvents and acid systems have been unsuccessful [206]. In this method, the in-situ heat produced downhole is generated by the exothermic reaction of sodium nitrite ($NaNO_2$) and ammonium nitrate (NH_4NO_3) or ammonium chloride (NH_4CI) in an aqueous solution, where the reaction is catalyzed by acid [206].

Hassan et al. [207] reported thermochemical fluids were utilized to remove the accumulated wax in the production tubing. The chemicals used were catalyzed by acetic acid as an activating agent, generating a significant amount of heat (up to 500°F) and pressure (up to 2000psi). The reported benefits were: more than 95% of the deposit can be removed, the generated heat is able to liquefy the precipitated wax, and then the induced pressure buildup due to the chemical reaction can flush the wax out of the production tubing, and finally no damage was observed in the production tubing.

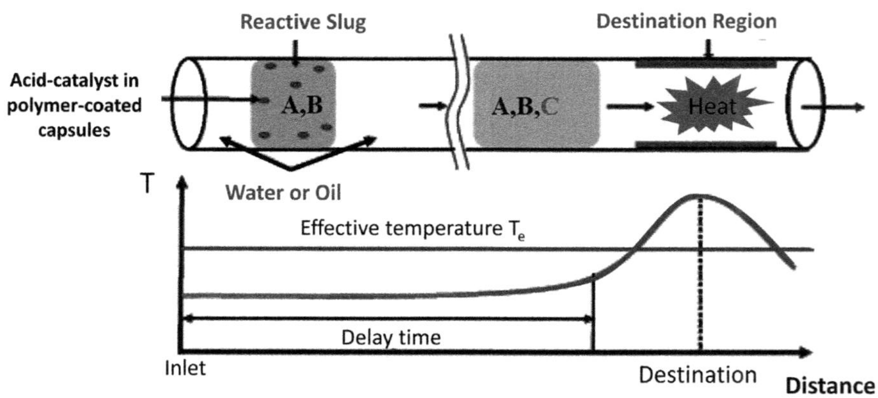

FIG. 17.22

Schematic of FCR system used in wax deposits removal and temperature control.

Reprinted with permission from D.A. Nguyen, F.F. de Moraes, H.S. Fogler, Fused chemical reactions. 3. Controlled release of a catalyst to control the temperature profile in tubular reactors. Ind. Eng. Chem. Res. 43 (18) (2004) 5862–5873. https://doi.org/10.1021/ie049933k. Copyright {2004} American Chemical Society.

For optimum application of the technique, it is necessary to know:

- The thickness profile of the wax deposits
- Wax fraction of the deposits (to estimate the melting point and consequently the amount of heat required)
- Location of the deposition problem.

The reactants should be supplied properly in such a way that the reaction is delayed until it reaches the exact deposits location and produces an amount of heat sufficient to melt them. In case the exothermic reaction takes place before reaching the deposits, e.g., just at the pipeline entrance, the resulting hot fluid will cool prior to reaching the wax deposit because of low subsea temperatures and lack of pipeline insulation [203].

The different ways of deploying the thermochemical packages and assuring reaching the deposits location include [203–205]:

- Physical separation: The two reactants are separated physically (by pumping one through coiled tubing and the other being flowed through the annulus) so that they only meet at the wax deposition region. This technique is effective only for short pipelines (<8 km).
- Dispersion: By injecting pulses of the reactants separated by an inert. However, the large amount of inert used in the technique reduces the amount of heat produced.
- Emulsification. Here an oil-in-water emulsion of an aqueous-based fluid that contains the reactants and an oil-based fluid that consists of aliphatic and/or aromatic hydrocarbons is used to delay the reaction. The method is applicable for short pipelines. Several dewaxing operations were performed with success at the Campos Basin (Brazil).
- Encapsulation. With this technique, either one of the reactants or the catalyst is encapsulated. The delay time is obtained from the controlled release of the encapsulated substance to the bulk solution.

The main drawbacks of the method are that it needs good detailed design, it also uses a lot of chemicals to work, which makes it an expensive method of treatment as it is more expensive than the hot oiling method. This process is limited to noncalciferous formations, since that kind of formation could react with the acid catalyst [31].

17.3.2 Nonchemical methods

Besides the conventional chemical methods, a variety of nonchemical methods can be efficiently used to remove wax deposits.

17.3.2.1 Thermal methods

Thermal methods are based on heating the wax affected to temperatures much higher than the wax melting point or the pour point, to dissolve the formed deposits and to liquefy them to transport them with the flowing stream. Different thermal methods have been applied in oil and gas fields with great success.

- **Direct heating methods**

Direct heating can be applied by downhole heaters, direct electric heating supplied in the well completion or in the pipeline installations. The direct heating methods discussed earlier in the prevention methods can be used in removing the wax deposits.

Daraboina et al. [208] showed that the removal of wax deposits was possible at temperatures lower than the WAT (WAT − 4%); they also showed that the removal temperature decreased from WAT +20% to WAT − 9%; the deposits underwent complete removal, partial removal, and no removal.

- **Hot oiling**

In this method, hot oil is used to remove hydrocarbon deposits from the tubing and casing. This method works by melting the hydrocarbon deposits.

In this method, the crude oil is collected from the battery or shipping tanks and is heated to a temperature as high as 125°C (in a heater treater or hot oiler trucks available on the market) and pumped down the annulus. The annular fluid level gradually increases until the oil begins to circulate back up the tubing [209]. Some additives are added along with the oil during hot oiling, including a demulsifier to facilitate the removal of solids and water; PD and crystal modifiers are added to prevent reaccumulation of the dissolved deposits [31,209].

The quantity of fluid pumped, pumping rates, and temperatures vary from well to well based on the severity of the problem. An efficient hot oiling job should deliver the maximum BTUs in the shortest time using the least fluid. Also the efficiency of the hot oiling depends on the skill of the hot oiling truck operator [210].

Critics of the hot oiling are [31,210]:

- This method can be expensive in some cases.
- It can cause formation damage: if the oil injected into the annulus goes into the formation, the undissolved particulates in stock tank oil used in hot oiling may pass into the formation, causing formation damage.
- It can worsen the wax problem: Hot oiling may strip only the light ends off the wax, thus the wax remains becomes more difficult to dissolve or disperse.
- Hot oiling can cause pump damage if it has enough paraffin content and cooled down reaching the pump, or if the casing happened to have a wax problem so the hot oil will dissolve these and then carry it to the pump, where it cools down and reprecipitates.

Mansure and Barker [210] set guidance for the best practice during hot oiling to overcome these drawbacks:

- Paraffin treatment should be well specific (not all wells should be hot oiled).
- Frequency of treatment should be minimized.
- Good quality fluid should be used (STO with low paraffin, asphaltene particulates).
- Injection should be down the annulus (not in the tubing).
- The tubing should be full and the well producing oil, not just gas.
- BTUs per hour injected should be maximized.
- The volume injected should be limited.
- Thermodynamics alone should not be the deciding factor in choosing between hot oil and hot water.

A mixture of hot oil and nitrogen gas was used so as to reduce the density of the circulating fluid in low bottomhole pressure wells, thereby avoiding loss of injected fluid to the formation [211]. Also, Brock [212] lists several alternatives to hot oiling, such as hot diesel, hot kerosine, or hot xylene. These are more expensive than hot oiling, but have less tendency to deposit paraffin in the formation.

Heating the riser by circulating heated fluids down the annulus is used occasionally. This method can be effective if there is just a thin layer of deposition on the tubing wall. If the entire tubing is plugged, there is often not enough heat transferred into the tubing to dissolve the plug [201].

- **Hot watering**

Hot watering is the same principle as hot oiling, as hot water is used to melt the wax deposits.

Hot water is often useful since it has a greater heat capacity than hot oil, although in some cases water contact changes the wax from a soft to a very heavy, hard material similar to canning wax [26].

Hot water treatments involve surfactant packages [dispersants, crystal modifiers, and surfactants] to aid in the dispersion of wax into the water phase [213]. Also demulsifiers may be added to separate the melted accumulations of organic matter. The amount of heat reaching the bottom of the well may be slightly greater than that obtained by hot oiling due to the higher heat capacity of water; however, the nonexistent solvency possessed by water greatly reduces the wax-carrying capacity of water. The addition of surfactants or dispersants to the water aids in wax removal, yet the reduced wax particle separation distance acts to reduce the added temperature benefits [213].

Other problems caused by using hot watering include [209]:

- Higher fluid loss due to high back pressure on formation.
- Hot watering using produced water causes scaling.
- Hot watering with fresh water causes clay swelling.
- Hot watering can cause emulsions and fluids separation problems.

Becker [213] recommended combining the use of hot oiling and hot watering along with crystal modifiers and dispersants to make of use of the benefits of each process. Green [214] advocated the use of a hot air heater to melt wax.

- **Steam injection**

Steam can be injected into the space between the tubing and the casing to dissolve the paraffinic waxes. Some notes to effectively apply the method are [3]:

- The steam should be applied to a depth below the lowest point of accumulation.
- The steam pressure must be higher than the formation pressure.
- It is necessary to guarantee that the condensate steam will not damage the formation or emulsify with the oil, causing emulsion troubles.

The amount of heat and the duration required will depend on the well conditions and the severity of the problem. It is common to leave the steam to produce its effect for 3–6 h, and then the well should be put back on production as fast as possible right after steaming; otherwise the melted wax can accumulate below the valve and with fluids cooling, the waxes can reprecipitate. The usage of steam is not as economical or practical as using a scraper. However, in some oil fields where the steam is available without extra expense, this method must be used [3].

A downhole electrosteam generator that can provide high steam quality, which increases effectiveness of the heat treatment, was reported [211,215].

17.3.2.2 Magnetic methods

Magnetic fields were reported to affect the crude oil and other production fluids. However the technique is still surrounded by ambiguity and the mechanism of action is not very clear. Yet several case studies have been reported with some success of using magnetic fields in removing wax deposits, reducing oil viscosity, and enhancing heavy oil recovery, as mentioned in Section 17.2.3.2.

A new technique involved the use of a magnetic field to effectively remove wax deposits. The technique involves the use of *Co*—Ni nanoparticles along with an exclusive polymer to be pumped inside the annulus between production tubing and casing at the time of well commissioning. The nanoparticles remain suspended in the polymer filled inside the annulus throughout the life of the well. The application of a magnetic field across the casing causes the nanoparticles to vibrate and be heated. This increase in temperature is transferred to the tubing by convection. The increase in temperature causes the wax to melt and is produced along with the oil stream [216].

17.3.2.3 Radio waves method

A device emits specific radio waves downhole in order to affect the hydrogen bond at a subatomic level. By affecting the proton and electronic spin states of a statistically small sample of molecules with this device, it is possible to interfere with crystallization and this can lead to disruption of hydrogen bonding of wax molecules with each other and water molecules on the reservoir surfaces, which solubilize paraffin and mitigating deposition. Field trials of the method showed promising results by reducing maintenance frequency and costs in addition to increasing production [217].

17.3.2.4 Pipe-pulse method

This method was developed to treat fully blocked flowlines. It is based on sending high-frequency, low-amplitude pressure pulses into the flowline. The pulses work in two different ways: by breaking the bond between the plug and the pipe or by fissuring the plug itself, creating a path for fluid to flow. With the latter, once communication is restored between the two ends of the plug, the pipeline can be flushed with solvent to complete the wax cleaning operations [218].

17.3.2.5 Mechanical methods

Conventional mechanical methods like well gauge cutters, scratchers, wireline and slickline tubing cleaners, coil tubing, and pigging are consistently used as routine measure to fight deposit accumulations in production tubings and flowlines. Full discussion of these conventional mechanical methods is furnished in Chapter 21.

Moreover, the industry adopted other unconventional mechanical methods to improve the efficiency of removing wax in locations deemed inapplicable for the conventional methods, like the improvements added to coiled tubing to improve its efficiency using Flexi-Coil and CoilTAC technologies.

17.3.2.6 Biological methods

The biological methods of controlling wax are biodegrading the long chain paraffin molecules, breaking them apart into smaller molecules, or by producing biosurfactants that change the wettability of the steel surfaces and make them repellent to wax adherence and accumulation.

Several cases studies were mentioned in Section 17.2.3.8, demonstrating successful application of different bacterial strains to prevent and remove wax deposition.

Lazar et al. [185] used occurring bacterial strains isolated from waste hydrocarbon-contaminated sites, with which they were able to remove solid and semisolid paraffin depositions on flow equipment.

Zhang et al. [219] showed that extracellular enzymes from *Aspergillus* spp. can efficiently degrade paraffin wax and improve the solubility of paraffin wax in n-hexane. These enzyme preparations have the potential for use in oil wells with paraffin deposition problems.

Novel thermophilic bacterial strains isolated from seawater and oil-contaminated soil samples were reported to achieve more than 70% biodegradation efficiency of crude oil in a short period (3 days). Notably, most of the strains could completely degrade C_{37}–C_{40} and increase the ratio of C_{14}–C_{18}, especially during the initial 2 days incubation [220].

Dutta and Harayama [221] have observed 28% of crude oil degradation at 20°C in 8 weeks using the microbial consortium. Liu et al. [191] have observed 64% paraffin degradation with two *Bacillus* species. Thavasi et al. [222] have observed 89% of degradation of crude oil in 7 days at 25°C using *P. aeruginosa*, using fertilizer as a nutrient supplement.

The technology has been adopted by some companies like CoreBiologic, which has provided the Paraffin Destroyer™ product with a list of successful case studies in Ohio, Michigan, Pennsylvania, Texas, and Oklahoma [223].

17.3.3 Tips for efficient wax deposit removal

Despite the various methods that can be used to remediate and remove wax deposits from oil and gas fields, the wax removal job must be thoroughly designed and optimized to achieve its main purpose. Otherwise it may not only be ineffective, but it may worsen the problem and aggravate other problems, like those associated with improper hot oiling, asphaltene problems associated with using the wrong solvents, and phase separation problems in general. Furthermore, normal operations like pigging, if not performed properly, can turn the loose wax in the pipeline into a solid hard plug that requires expensive treatment with long downtimes.

Generally speaking, the cleaning/removal job design depends on some factors, including:

- Wax deposit composition
- Wax layer thickness, age, hardness/strength
- Location of wax deposits problem, accessibility
- System design
- Availability of the treatment method
- Economics

Wax deposits must be analyzed properly to determine their detailed composition. Wax deposits encompass a wide range of long chain hydrocarbons, which can be determined using standard method UOP46, and can be characterized in further detail using SARA, modified SARA, and gas chromatography analysis. Modified SARA analysis can specify the wax deposits and distinguish microcrystalline and macrocrystalline waxes, which are known to have different solubilities. Carbon distribution is used during the selection of the chemical inhibitor/dispersant. Besides, the wax deposits usually incorporate mineral scales, sand, fines, asphaltenes, and other solids in their 3D network. Mineral scales, sand, and fines will drop off the solution and accumulate and might cause plugging. The asphaltenes will dissolve in aromatic solvents but will not dissolve in aliphatic dissolvers. Sometimes asphaltenes are misdiagnosed by operators as paraffins, and vice versa, so an accurate analysis of the deposit sample is crucial. Deposit characterization is furnished in Chapter 14.

Deposit physical properties are crucial to properly design the removal job. Deposit thickness or estimated volume in the pipeline is used to estimate the relative amount of chemicals required. Deposit texture or hardness is indicative of the difficulty of dissolving the wax: fresh and loose deposits will be easier to remove than hard and old ones. Wax gel strength is a critical factor used to optimize the pigging operations, in addition to optimizing the restart parameters of clogged/gelled systems. Breakdown of the wax-oil gel occurs if the shear stress exerted on the gel due to the applied pressure exceeds the yield stress of the gel [224].

The location of the deposit determines the accessibility of the removal method, volumes of liquids (solvents, or liquids used in hot oiling/watering), power usage, pumps and tools required, and other requirements. For example, subsea and deepwater systems will be the hardest to deal with, since they have limited accessibility for the mechanical methods and will need huge volumes of solvents to fill the line to reach a wax plug.

Chemical (solvents) and mechanical methods (pigging, coiled tubing, wireline, etc.) are commonly used to remove wax deposits. A combination of removal methods is recommended, like heated solvent (chemical + thermal effects) or heated solvent with the aid of a pigging or scratching tool (chemical + thermal + mechanical effects). Using a heated chemical in coiled tubing combines the mechanical action and chemical dissolution as well as the thermal effect. In addition, chemical formulations can involve solvent, PD, or paraffin inhibitor to improve the dissolution performance. Heat-producing chemicals are used when the conventional methods fail to remove the deposits.

Pigging operations require special attention since it can turn a loose wax into a solid hard plug that is impossible to remove, or the pig can be stuck. Removing a large amount of wax at high pressure using a single pig can turn the loose wax into a hard solid plug several meters in length. Pigging at low pressure can cause the plug to get stuck if it is not able to break the wax deposits. The operators usually run a series of pigs with increasing size, so that each pig will remove a fraction of the deposit without plugging the pipeline. Using specially designed pigs, such as bypass pigs or jet pigs, has been reported as beneficial without causing plugging.

Removal of wax from producing wells:

- Location of the obstruction is determined using well gauging, where a deposit sample can also be collected.
- Determine the chemical composition and the physical properties of the collected deposit sample.
- Chemical methods are usually applied using a diesel/xylene mixture. Dispersants and surfactants can be added to improve their dissolution and to prevent reprecipitation of the dissolved deposits.
- Applying these chemicals at high temperature improves the dissolution efficiency. Hot solvent is preferably injected, especially when the downhole temperature may be high enough to warm up the liquids to improve their efficiency. Another efficient method of warming up the liquids by circulating warm liquids in the annulus.
- Mechanical methods can be used to improve the efficiency of the removal. Cutters, scratchers, and other wireline/slickline tools can be used. However, coiled tubing is capable of handling injected hot chemicals and thus combines the mechanical, thermal, and chemical effects.
- If the deposit showed mineral scales, then depending on the scale composition, a scale dissolver can be applied to prevent their accumulation after dissolving the wax deposits.

- If asphaltenes deposit is present, then aromatic-based solvent is preferred since it can remove both paraffins and asphaltenes. Dispersants or surfactants can be added to it to improve its efficiency.
- Hot oiling is extensively used in downhole applications with a successful track record. Hot oiling should include dispersants or surfactants to improve the efficiency of the treatment.
- After treatments, a paraffin prevention program is designed using inhibitor or dispersant, and system monitoring plans should be developed.

Wax removal in flowlines and equipment:

- Determine the location of the deposit, and obtain a sample.
- If possible, visual inspection of the affected location using borescope helps to realistically assess the situation inside the pipeline.
- Determine the sample chemical composition and physical properties.
- Estimate the volumes of solvents required for treatments.
- Apply the chemical dissolver and let it soak for the proper time. Chemical solvents are formulated with dispersants and surfactants to improve the dissolution efficiency.
- Pigging operations should be optimized to avoid plug formation or stuck pig situations. Prediction and simulation models for wax removal and pigging can be used to evaluate the problem. The use of special pigs for wax removal is recommended.
- Mechanical methods can be used besides chemical methods to improve the removal treatment. Solvents can be applied during pigging operations, in front of pigs or between pigs. Scratchers or coiled tubing can be used.
- Coiled tubing is usually used to remove wax and hydrate plugs from subsea pipelines and employs thermal, chemical, and mechanical features.
- After treating the affected areas, an inhibition program should be designed and a thorough monitoring program must be attained.

Table 17.5 summarizes the different methods of wax removal.

Table 17.5 Summary of wax removal methods.

Method	Description
Chemical methods	- Solvents, surfactants, dispersants, thermochemical packages - Wide variety of chemicals for different applications
Thermal methods	- Using heat to melt the formed wax deposits. Methods include: electric heating, hot oiling, hot watering, steam injection - Very effective method and very common in use - Can be expensive - Hot oiling can cause asphaltene problems
Physical methods (magnetic, radio waves)	- The physical method reduces oil viscosity, degrades the high molecular weight wax molecules, or generates heat that melts the formed deposits
Biological methods	- Bacteria degrades the formed wax deposit molecules, generates biosurfactants that help dissolve the wax deposits
Mechanical methods	- Using pigging, wireline, coiled tubing, etc. - Very effective - Cost effective - Can be combined with chemical methods

17.4 Wax control strategies and philosophies

A wax control strategy should start with the early stages of the field development. Once available, collect a fluids sample and start the journey to assess the risks of wax deposition and develop mitigation methods.

17.4.1 Wax problem risk assessment

17.4.1.1 Fluids analysis

Once available, fluid samples should be collected and sent to a well-equipped lab to characterize the produced fluids. Common tests performed are:

- PVT analysis
- SARA analysis, modified SARA analysis
- Wax content
- Viscosity
- Yield stress/gel strength
- WAT
- Pour point
- Carbon distribution using gas chromatography.

In addition, other experimental work is needed to assess the wax problems, especially if the simulation/modeling calculations (Section 17.4.1.3) show potential wax deposition, including a deposition simulation:

- Cold finger experiments
- Flow loop experiments
- Testing spool
- Dynamic paraffin wax deposition cell

These tests are performed to assess the wax deposition problem and also to evaluate the possible treating chemicals that can be used to inhibit or attenuate the risks of the problem. The next step is to evaluate these data, and to use them in the prediction models.

17.4.1.2 System temperature profile

Wax deposits are directly related to the temperature of the system. Therefore the next step in assessing the wax problems is assessing the system temperature and detecting the cold spots in the system where the temperature falls below WAT. A typical system profile is then established and used as a reference by the engineers and operators. Various models can be used to estimate the system temperature profile.

System pressure is also important to evaluate the wax risks correlated to the system parameters, and to evaluate if system pressure can cause a Joule-Thomson effect at some locations in the system.

17.4.1.3 Modeling and simulation

After enough data are gathered on the produced fluids (chemical composition and physical properties) and the production system operating parameters (P, T, flow rates), these data are used to model paraffin precipitation.

Various types of simulation models can be used to simulate wax deposition in steady state and transient state production. Computational fluid dynamics modeling (CFD) is also recommended. These models give the wax deposition curve of the wax deposition envelope. The modeling is also used to determine and optimize the system design and determine the insulation requirement for pipelines.

Fields analogy is an important method that has been used to assure the risk assessment of the wax deposits is accurate. Fields with similar fluids composition and wax deposition characteristics are usually treated the same.

17.4.2 Choosing the optimum control method

As mentioned before, the earlier the fluids analysis, lab work, modeling, and risk assessment are performed, the better the control of the problem.

Early wax deposits risk assessment helps in:

- Optimizing the system design
- Providing thermal insulation or electric heating
- Optimizing the operating conditions to be within a wax-free zone.
- Optimizing the chemical injection points in the system: for example, operate the well at sufficiently high production rates to avoid deposition in the wellbore and tree.
- Install and optimize monitoring devices.

For full system control, the general steps to be followed are:

- Schedule a routine fluids sampling and analysis to measure basic chemical and physical properties of the crude oil: basically, wax content, pour point, WAT, viscosity yield stress, etc.
- Model and simulate the produced fluids at the system parameters.
- Monitor system parameters for changes in pressure or temperature.
- Systems with potential wax deposition must be fitted with insulation or heating systems.
- Predetermine the maintenance and shutdowns so that optimum actions are applied, e.g., use of PPD, dispersants or paraffin solvents.
- During unplanned shutdowns, the of solvents, and pumps to displace and remove the gelled crudes.

Table 17.6 summarizes the different wax mitigation methods.

17.4.3 Monitoring an assessment

After applying the control method, it is crucial to monitor the operating system to evaluate its effectiveness. This can be achieved by the following:

- Routine system parameters monitoring (P, T, flow rates, etc.).
- Fluids sampling and analysis, e.g., WAT, pour point, wax content, viscosity, etc.
- Using any of the physical devices that give real-time monitoring of the deposition status. Fig. 17.23 summarizes the wax management strategies.

Table 17.6 Wax management methods.

Mitigation method	Description
Operational methods	- Generally used in wax prevention - Requires system predesign, or system modification - Cold flow can be applied in severe cases where chemicals and thermal methods are expensive - Case dependent
Thermal methods	- Used in wax prevention and removal - Very successful in field applications - Crucial in subsea and deepwater installations with potential wax problems - Insulation and electric heating require predesign or system modification
Chemical methods	- Used in prevention and removal - Wide variety of chemicals for different applications - Generally, do not require system modification - Generally cost effective, but can be expensive in severe cases that require high volumes of chemicals - Can be applied in combination with other methods, e.g., mechanical, heat - Very effective for use during shutdowns to protect the system - Chemicals can have side effects, e.g., incompatibility with other chemicals, produced fluids, or can enhance other production chemistry problems
Mechanical methods	- Used mainly in removal - Can be used in different cases, e.g., topsides, downhole - Can be used with other methods, e.g., chemical methods, application of heat, to improve its efficiency - System design and accessibility dependent
Other nonchemical methods	- Internal coatings have shown success in field application - Biological methods have reported with success in field applications - Magnetic, ultrasonic, electric, microwave methods were reported to improve crude oil viscosity - Physical methods may require system modification to be installed - These methods are not generally common in field application compared to chemical methods - They can be case dependent

17.5 Recent advances in wax deposit management methods

Several advances have been reported in wax deposition and its mitigation studies. These include advances in wax deposition studies and prediction. Different devices have been developed to study wax deposition, e.g., modified cold finger devices, flow loops, and organic deposition devices, which helped in understanding the different factors affecting the wax deposition. This was also reflected in the prediction models, which used some of these experimental data to build the robust models. Besides, several advances have been witnessed in chemical management methods and implementation of new types of chemicals such as nanochemicals and ionic liquids. Thermal methods have been expanding to meet the needs for deepwater and subsea installations. Physical methods, although not very common, have been extensively studied so that they can be applicable in the near future as a green nonchemical solution. Biological methods have also reported several successful case studies.

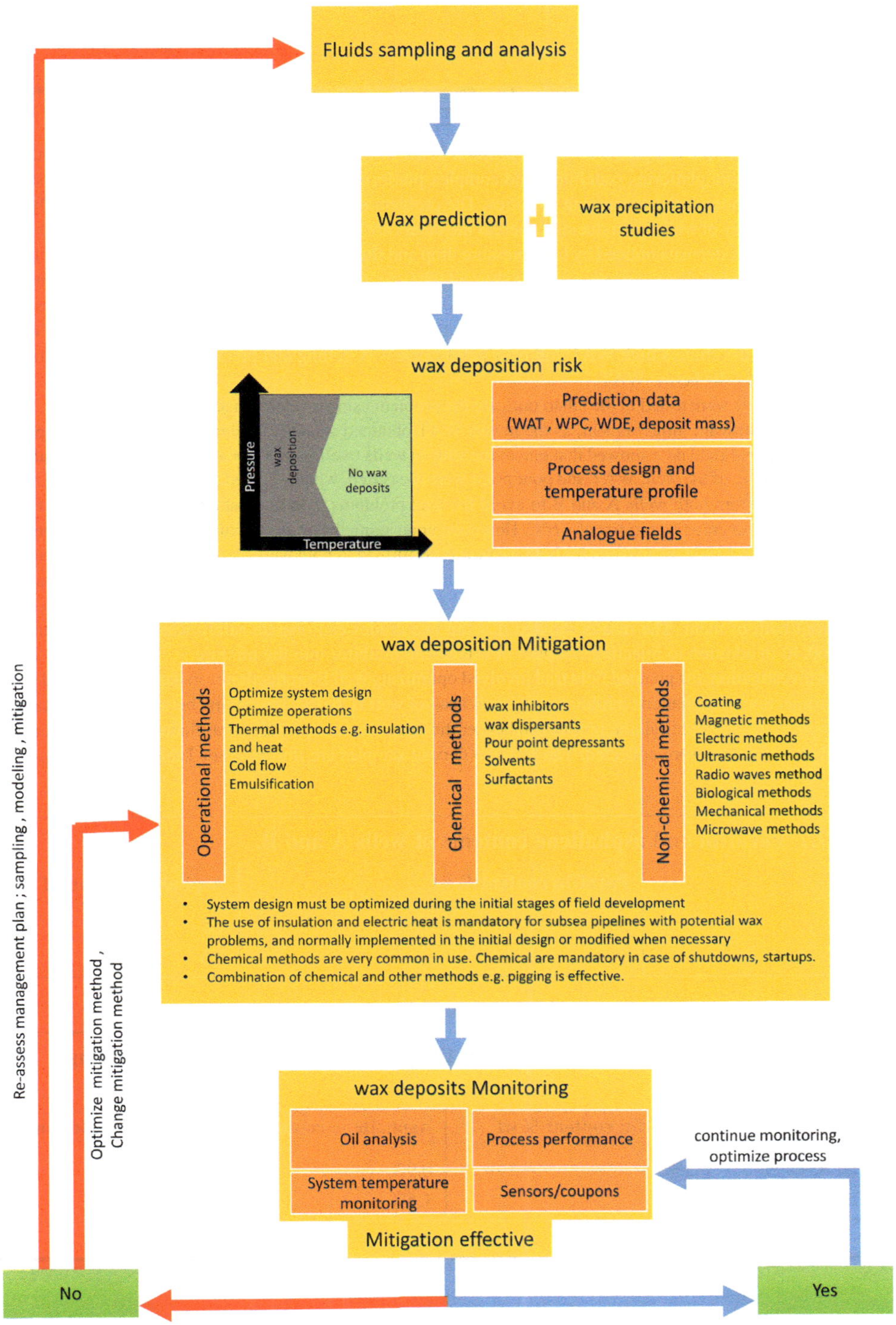

FIG. 17.23

Wax management strategies.

17.6 Case studies

Case study: Solving paraffin problems in Egyptian field

This case study discusses the mitigation of wax deposition in the flowline in an offshore field in eastern Egypt. The field features multiple offshore platforms (satellites) and complex platforms where the production from different satellites is collected and sent onshore to the processing facilities. The problem started in one of the satellites, where only one well (well A) among a battery of wells is producing, whereas the others are shut-in. The main production header in this satellite encountered severe wax deposits noticed by high pressure drop and fluctuations on the topsides flow meters. A lab sample was collected to the lab for characterization, which showed 98% paraffin wax deposits. Furthermore, fluids samples were collected and lab analysis was performed, where the pour point, WAT, and wax content were analyzed to address the roots of the problem. Lab analysis showed that the oil had wax content 5%, WAT of 42°C, pour point is 29°C. The reasonable explanation was that the crude's high WAT and pour point cause wax gelling and precipitation in the production header when the temperature falls below WAT.

Different chemicals were tested in the lab to mitigate the problem using the cold finger technique. However, the results did not show any significant improvement, as the lowest WAT obtained was 20°C, which is not sufficient to prevent the wax deposition. The old field data showed that one of the shut-in wells (well B) has a pour point of −10°C, so, after proper restart and conditioning of the well, WAT and pour point were measured, which showed a pour point of −12°C. Hence the plan was to produce from both wells A and B to avoid the wax problems. Lab work showed that mixing the two crudes from A and B gave an average pour point of −6°C. However, upon aging the crudes mixture, asphaltenes problems showed up in the test container. Crude oil analysis for both wells was performed to both wells to address this mixing problems. The results are shown in Table 17.7, which indicates that well A has high asphaltenes content while well B is low in asphaltenes.

Therefore extensive lab work was conducted to assess the proper ratio and method of mixing the crudes and hence producing from both of them. The results showed that the two crudes can operate safely when mixed at ratios of well B:well A ~60:40 in addition to injecting 200 ppm of asphaltene inhibitor into the mixture.

After this investigation, the planned field trial involved optimizing well B production to make up for the final calculated ratio, besides injection of asphaltene inhibitor at the well head of well B, before the mixing between the wells takes place.

After 3 months of the field trial, the production header did not encounter paraffin or asphaltene deposition, and only a very low volume of debris were collected. The last two deposit samples are illustrated in Table 17.8.

Table 17.7 Paraffin and asphaltene contents of wells A and B.

Well ID	Paraffin content %wt	Asphaltenes content %wt
Well A	5	12
Well B	3	0.5

Table 17.8 Scale sample composition from the production header after applying mitigation method.

Sample	Paraffin content %wt	Asphaltene content	Mineral scales and fines
#1 (after 43 days)	2%	0.9%	97.1%
#2 (after 62 days)	1.5%	1.5%	97%

17.7 Summary

Wax deposit mitigation methods were reviewed in this chapter. Wax mitigation methods are based on operational, chemical, and nonchemical methods. The operational methods are based on optimizing the operating temperature to avoid wax formation, hence thermal methods are commonly used. Other operational methods include optimizing flow rates, cold flow, and emulsification. The chemical methods include using wax inhibitors, dispersants, or pour point depressants to prevent wax deposition. The nonchemical methods involve using physical or biological methods.

Wax removal is based on using chemical methods such as solvents or heat-generating chemicals. Mechanical methods are also commonly used.

Wax management strategy is based on characterizing the fluids, modeling, and simulation to estimate the risk of wax deposition and then mitigating by applying the proper treatment method. Finally, rigorous monitoring is mandatory to assess the applied treatment and optimize the system operations.

References

[1] T. Zhu, J.A. Walker, J. Liang, Evaluation of Wax Deposition and its Control during Production of Alaska North Slope Oils, University of Alaska, Anchorage, AK, USA, 2008.

[2] A. Hammami, M.A. Raines, Paraffin deposition from crude oils: comparison of laboratory results with field data, SPE J. 4 (1999) 9–18, https://doi.org/10.2118/54021-PA.

[3] A.L. Sousa, H.A. Matos, L.P. Guerreiro, Preventing and removing wax deposition inside vertical wells: a review, J. Pet. Explor. Prod. Technol. (2019), https://doi.org/10.1007/s13202-019-0609-x.

[4] Y. Bai, Q. Bai, Y. Bai, Q. Bai, Chapter 16—Wax and asphaltenes, in: Subsea Engineering Handbook, 2010, pp. 483–504, https://doi.org/10.1016/B978-1-85617-689-7.10016-0.

[5] V. Mahto, A. Kumar, Effect of several parameters on wax deposition in the flow line due to Indian waxy crude oil, Int. J. Appl. Eng. Res. Dev. 3 (4) (2013) 1–10.

[6] X. Wang, Q. Qu, P.H. Javora, R.G. Pearcy, New trend in oilfield flow assurance management: a review of thermal insulating fluids, in: International Oil & Gas Conference and Exhibition in China, 2006, https://doi.org/10.2118/103829-ms.

[7] N. Sunday, A. Settar, K. Chetehouna, N. Gascoin, An overview of flow assurance heat management systems in subsea flowlines, Energies 14 (2) (2021) 458, https://doi.org/10.3390/en14020458. MDPI AG.

[8] E. Bell, Y. Lu, N. Daraboina, C. Sarica, Thermal methods in flow assurance: a review, J. Nat. Gas Sci. Eng. (2021), https://doi.org/10.1016/j.jngse.2021.103798. Elsevier B.V.

[9] M. Bagdat, R. Masoud, Control of paraffin deposition in production operation by using ethylene-tetrafluoroethylene (ETFE), in: Proc. Int. Conf. Integr. Pet. Eng. Geosci. ICIPEG 2014, 2015, pp. 13–22, https://doi.org/10.1007/978-981-287-3682 (Internet) http://www.sprin. ger.com/978-981-287-367-5.

[10] Y. Bai, Q. Bai, Heat transfer and thermal insulation, in: Subsea Engineering Handbook, Elsevier, 2010, pp. 401–450, https://doi.org/10.1016/b978-1-85617-689-7.10014-7.

[11] A. Quenelle, M. Gunaltun, Comparison between thermal insulation coatings for underwater pipelines, in: Offshore Technology Conference, 1987, https://doi.org/10.4043/5500-ms.

[12] G. Lively, Flow assurance begins with downhole insulation, in: Offshore Technology Conference, 2002, https://doi.org/10.4043/14118-ms.

[13] K. Shukla, M.V. Labh, Managing Paraffin/Wax Deposition Challenges in Deepwater Hydrocarbon Production Systems, 2019, https://doi.org/10.5772/intechopen.83564.

[14] M. Zatka, Application of vacuum-insulated tubing VIT in thermal oil sand projects, in: SPE Heavy Oil Conference and Exhibition, 2016, https://doi.org/10.2118/184155-ms.

[15] S. Palle, L. Ror, Thermal insulation of flowlines with polyurethane foam, in: Offshore Technology Conference, 1998, https://doi.org/10.4043/8783-ms.
[16] S. Feeney, Vacuum-insulated tubing: project case histories, future applications, Offshore 59 (10) (1999).
[17] J. Lee, Design and installation of deepwater petroleum pipelines, in: Proc. of the World Congress of Korean and Korean Ethnic Scientists and Engineers. Seoul, Korea, 2002.
[18] M.M.M. Thant, Experimental Investigation of Natural Convection Heat Transfer in Bundle Pipeline, Ph.D. Thesis, Imperial College London, 2012.
[19] K. Falk, V. Kaste, S. Håheim, P. Damsleth, Subsea Heat Bank—An Alternative Thermal Insulation Method, Society of Petroleum Engineers, 2002, https://doi.org/10.2118/77577-MS.
[20] P.-S. Kang, D.-G. Lee, J.-S. Lim, Status of Wax Mitigation Technologies in Offshore Oil Production, International Society of Offshore and Polar Engineers, 2014.
[21] B. Ansart, A. Marret, T. Parenteau, O. Rageot, Technical and economical comparison of subsea active heating technologies, in: Offshore Technology Conference-Asia, 2014, https://doi.org/10.4043/24711-ms.
[22] C. Candelier, S. Durica, F. Beys, Subsea pipeline electrical heat trace (EHT)—'Active' heating—application for a deep water brown field development, in: Offshore Mediterranean Conference and Exhibition, OMC 2015. Offshore Mediterranean Conference, 2015.
[23] A. Nysveen, H. Kulbotten, J.K. Lervik, A.H. Børnes, M. Høyer-Hansen, J.J. Bremnes, Direct electrical heating of subsea pipelines—technology development and operating experience, IEEE Trans. Ind. Appl. 43 (1) (2007) 118–129, https://doi.org/10.1109/TIA.2006.886425.
[24] Cavitas Energy Limited, THOR Tool, Cavitas Energy Limited, 2021. Retrieved from: https://www.cavitas.co.uk/enhanced-oil-recovery/. (Last visited November 2021).
[25] L.D. Brown, J. Clapham, Design of Britania's Subsea Heated Bundle for 25 Years Service Life—OTC 11017, 1999.
[26] N. Leo, Treating Paraffin Deposits in Producing Oil Wells, Technical Report. Prepared for, U.S. Department oi Energy, Assistant Secretary for Fossil Energy, Bartlesville, OK, USA, 1992.
[27] C.E. Reistle Jr., O.C. Blade, Paraffin and Congealing-Oil Problems, 1932. BuMines Bulletin 348. 171 pp.
[28] E.E. Elhaddad, A. Bahadori, M.E. AbdelRaouf, S. Elkatatny, A new experimental method to prevent paraffin—wax formation on the crude oil wells: a field case study in Libya, Hem. Ind. 69 (3) (2015) 269–274,-https://doi.org/10.2298/HEMIND130717040E.
[29] C. Narvaez, A.A. Ferrer, Prevention of paraffin well plugging by plunger-lift use, in: SPE Production Operations Symposium, 1991, https://doi.org/10.2118/21640 ms.
[30] M.K. Rogachev, T. Nguyen Van, A.N. Aleksandrov, Technology for preventing the wax deposit formation in gas-lift wells at offshore oil and gas fields in Vietnam, Energies 14 (16) (2021), https://doi.org/10.3390/en14165016.
[31] M.A. Kelland, Production Chemicals for the Oil and Gas Industry, Production Chemicals for the Oil and Gas Industry, CRC Press, 2009, https://doi.org/10.1201/9781420092974.
[32] P. Gateau, I. Hénaut, L. Barré, J.F. Argillier, Heavy oil dilution, Oil Gas Sci. Technol. 59 (5) (2004) 503–509, https://doi.org/10.2516/ogst:2004035. Editions Technip.
[33] J. Bellarby, Production chemistry, in: Developments in Petroleum Science, 2009, https://doi.org/10.1016/S0376-7361(08)00207-0 (Chapter 7).
[34] G.T. Woo, S.J. Garbis, T.C. Gray, Long-term control of paraffin deposition, in: SPE Annual Technical Conference and Exhibition. Houston, Texas, 1984.
[35] C.O. Lukehart, Down-hole injection of butane to control paraffin deposition in pumping wells, J. Petrol. Tech. 12 (01) (1960) 14–16, https://doi.org/10.2118/1315-g. Society of Petroleum Engineers (SPE).
[36] L.Z. Hao, H.S. Al-Salim, N. Ridzuan, A review of the mechanism and role of wax inhibitors in the wax deposition and precipitation, Pertanika J. Sci. Technol. 27 (1) (2019) 499–526.

[37] S. Ahn, K.S. Wang, P.J. Shuler, J.L. Creek, Y. Tang, Paraffin crystal and deposition control by emulsification, in: SPE 93357 (Paper Presented at the SPE International Symposium on Oilfield Chemistry, The Woodlands, TX, 2–4 February), 2005.

[38] A.H. Beyer, D.E. Osborn, Downhole emulsification for improving paraffinic crude production, in: Fall Meeting of the Society of Petroleum Engineers of AIME, 1969, https://doi.org/10.2118/2676-ms.

[39] H.J. Sommer, W.C. Simpson, Pipeline transportation of Waxladen crude oil as water suspension, in: Sixth World Petroleum Congress, Section VII, Paper 13, Frankfurt, 1983.

[40] S.S. Marsden, R. Raghavan, A system for producing and transporting crude oil as oil/water emulsion, J. Inst. Pet. 59 (570) (1973) 273.

[41] D. Langevin, S. Poteau, I. Henaut, J.F. Argillier, Crude oil emulsion properties and their application to heavy oil transportation, Oil Gas Sci. Technol. 59 (5) (2004) 511–521.

[42] C.J. Coberly, Method for Preventing Wax Deposits in Tubing, 1942. U.S. Patent 2303823, December 1, 1942.

[43] M. Al-Yaari, Paraffin wax deposition: mitigation & removal techniques, in: SPE Saudi Arabia Section Young Professionals Technical Symposium, Dhahran, 14–16 March 2011, SPE-155412-MS, 2011.

[44] D. Merino-Garcia, S. Correra, Cold flow: a review of a technology to avoid wax deposition, Pet. Sci. Technol. 26 (4) (2008) 446–459, https://doi.org/10.1080/10916460600809741.

[45] J. Nenniger, Process for Inhibiting Formation of Wax Deposits, CA Patent 1289497, December, 30, 1991.

[46] J. Nenniger, Cold Seeding Lab Experiments, DeepStar Report 4202-b2, 2001.

[47] C.B. Argo, P. Bollavaram, K.W. Hjarbo, et al., Method and System for Transporting Flows of Fluid Hydrocarbons Containing Wax, Asphaltenes, and/or Other Precipitating Solids, 2007. US Patent No. 7,261,810.

[48] R.A. Benson, Method for Making Slurry, U.S. Patent 6070417, March 29, 2000.

[49] G.S. Fung, R.M. Amin, B. Kalpakci, F. Fleyfel, J.F. O'Sullivan, Method for Reducing Solids Build-up in Hydrocarbon Streams Produced from Wells, U.S. Patent 6656366. December 2, 2003.

[50] R. Amin, T. Win, B. Kalpakci, A novel approach for treatment and transport of waxy crude oils in single, bare subsea flowlines, in: SPE Paper 62947. Proceedings SPE Annual Technical Conference and Exhibition, Dallas, TX, October 1–4, 2000.

[51] W.T. Knowles Jr., A Stream of Gas and Waxy Crude is Cooled by Expansion in a Choke and Form a Stable Slurry, U.S. Patent 4697426, October 6, 1986.

[52] Y. Chi, J. Yang, C. Sarica, N. Daraboina, A critical review of controlling paraffin deposition in production lines using chemicals, Energy Fuel (2019), https://doi.org/10.1021/acs.energyfuels.9b00316.

[53] F. Yang, Y. Zhao, J. Sjöblom, C. Li, K.G. Paso, Polymeric wax inhibitors and pour point depressants for waxy crude oils: a critical review, J. Dispers. Sci. Technol. 36 (2) (2015) 213–225, https://doi.org/10.1080/01932691.2014.901917. Taylor and Francis Inc.

[54] F. Yang, K. Paso, J. Norrman, C. Li, H. Oschmann, Sjöblom, J., Hydrophilic nanoparticles facilitate wax inhibition, Energy Fuel 29 (3) (2015) 1368–1374.

[55] W. Leube, M. Monkenbusch, D. Schneiders, D. Richter, D. Adamson, L. Fetters, P. Dounis, R. Lovegrove, Wax-crystal modification for fuel oils by self-aggregating partially crystallizable hydrocarbon block copolymers, Energy Fuel 14 (2) (2000) 419–430.

[56] D.M. Duffy, P.M. Rodger, Modeling the activity of wax inhibitors: a case study of poly(octadecyl acrylate), J. Phys. Chem. B 106 (43) (2002) 11210–11217, https://doi.org/10.1021/jp026501j.

[57] G. Jing, P. Ye, Y. Zhang, The action mechanism of wax inhibitors (WI) on pour point and viscosity of mixed waxy oil, Pet. Chem. 57 (3) (2017) 293–298, https://doi.org/10.1134/s0965544117020153.

[58] G. Gilby, The use of ethylene-vinyl acetate copolymers as flow improvers in waxy crude oil, in: Chemicals in the Oil Industry: The Proceedings of a Symposium Organized by the North West Region of the Industrial Division of the Royal Society of Chemistry, University of Manchester, 1983.

[59] T.J. Behbahani, A.A. Miran Beigi, Z. Taheri, B. Ghanbari, Synthesis of some additives and study effect of gas oil composition on flow properties, J. Mol. Liq. 211 (2015) 308–314.
[60] B. Wei, Recent advances on mitigating wax problem using polymeric wax crystal modifier, J. Pet. Explor. Prod. Technol. 5 (4) (2015) 391–401.
[61] Y.H. Jang, M. Blanco, J. Creek, Y. Tang, W.A. Goddard, Wax inhibition by comb-like polymers: support of the incorporation- perturbation mechanism from molecular dynamics simulations, J. Phys. Chem. B 111 (46) (2007) 13173–13179, https://doi.org/10.1021/jp072792q.
[62] D. Groffe, P. Groffe, S. Takhar, S.I. Andersen, E.H. Stenby, N. Lindeloff, M. Lundgren, A WAX inhibition solution to problematic fields: a chemical remediation process, Pet. Sci. Technol. 19 (1–2) (2001) 205–217,-https://doi.org/10.1081/lft-100001235.
[63] H.S. Ashbaugh, L.J. Fetters, D.H. Adamson, R.K. Prud'homme, Flow improvement of waxy oils mediated by self-aggregating partially crystallizable diblock copolymers, J. Rheol. 46 (4) (2002) 763–776.
[64] H.S. Ashbaugh, A. Radulescu, R.K. Prud'homme, D. Schwahn, D. Richter, L.J. Fetters, Interaction of paraffin wax gels with random crystalline/amorphous hydrocarbon copolymers, Macromolecules 35 (18) (2002) 7044–7053.
[65] F. Yang, L. Cheng, H. Liu, B. Yao, C. Li, G. Sun, Y. Zhao, Comb-like polyoctadecyl acrylate (POA) wax Inhibitor triggers the formation of heterogeneous waxy oil gel deposits in a cylindrical Couette device, Energy Fuel 32 (1) (2018) 373–383.
[66] N. Ridzuan, F. Adam, Z. Yaacob, Effects of shear rate and inhibitors on wax deposition of Malaysian crude oil, Orient. J. Chem. 31 (4) (2015) 1999–2004.
[67] A.L.C. Machado, E.F. Lucas, G. González, Poly(ethylene-co-vinyl acetate) (EVA) as wax inhibitor of a Brazilian crude oil: oil viscosity, pour point and phase behavior of organic solutions, J. Petrol. Sci. Eng. 32 (2–4) (2001) 159–165, https://doi.org/10.1016/S0920-4105(01)00158-9.
[68] L.A. McDougall, A. Rossi, M.J. Wisotsky, Crude Oil Recovery Method Using a Polymeric Wax Inhibitor, US3693720 A, Sep 26, 1972.
[69] B. Yao, C. Li, Z. Mu, X. Zhang, F. Yang, G. Sun, Y. Zhao, Ethylene-vinyl acetate copolymer (EVA) and resin-stabilized asphaltenes synergistically improve the flow behavior of model waxy oils. 3. Effect of vinyl acetate content, Energy Fuel 32 (8) (2018) 8374–8382.
[70] R.A. El-Ghazawy, A.M. Atta, K.I. Kabel, Modified maleic anhydride-co-octadecene copolymers as flow improver for waxy Egyptian crude oil, J. Petrol. Sci. Eng. 122 (2014) 411–419.
[71] H.K. Singhal, G.C. Sahai, G.S. Pundeer, K. Chandra, Designing and selecting wax crystal modifier for optimum field performance based on crude oil composition, in: SPE Annual Technical Conference and Exhibition, 1991, https://doi.org/10.2118/22784-ms.
[72] D. Chanda, A. Sarmah, A. Borthakur, K.V. Rao, B. Subrahmanyam, H.C. Das, Combined effect of asphaltenes and flow improvers on the rheological behaviour of Indian waxy crude oil, Fuel 77 (11) (1998) 1163–1167, https://doi.org/10.1016/s0016-2361(98)00029-5.
[73] O.O. Bello, S.O. Fasesan, C. Teodoriu, K.M. Reinicke, An evaluation of the performance of selected wax inhibitors on paraffin deposition of Nigerian crude oils, Pet. Sci. Technol. 24 (2006) 195.
[74] A. Borthakur, D. Chanda, S.R. Dutta Choudhury, K.V. Rao, B. Subrahmanyam, Alkyl fumarate-vinyl acetate copolymer as flow improver for high waxy Indian crude oils, Energy Fuel 10 (3) (1996) 844.
[75] J. Xu, S. Xing, H. Qian, S. Chen, X. Wei, R. Zhang, et al., Effect of polar/nonpolar groups in comb-type copolymers on cold flowability and paraffin crystallization of waxy oils, Fuel 103 (2013) 600–605 (2013).
[76] M.L. Howarth, Design and Testing of a New Generation of Pour Point Depressants, Introduced for MASTER'S THESIS, University of Stavanger, Faculty of Science and Technology, 2014.
[77] H. Quan, L. Chen, Z. Huang, J. Wang, C. Zheng, The effect of a kind of hyperbranched polyester with different carbon length on flowability for crude oil, Fuel 214 (2018) 356–362, https://doi.org/10.1016/j.fuel.2017.11.010.

[78] N. Halim, S. Ali, M. Nadeem, P. Abdul Hamid, I. Mohd Tan, Synthesis of wax inhibitor and assessment of squeeze technique application for Malaysian waxy crude, in: SPE 142288 (Paper Presented at the SPE Asia and Pacific Oil and Gas Conference, Jakarta, Indonesia, 20–22 September), 2011.
[79] F. Wang, D. Zhang, Y. Ding, L. Zhang, M. Yang, B. Jiang, et al., The effect of nanohybrid materials on the pour-point and viscosity depressing of waxy crude oil, Chin. Sci. Bull. 56 (1) (2011) 14–17.
[80] Z. Sun, G. Jing, Z. Tu, Effect of modified nano-silica/EVA on flow behavior and wax crystallization of model oils with different wax contents, J. Dispers. Sci. Technol. 39 (1) (2018) 71–76.
[81] R. Sharma, B. Deka, V. Mahto, A. Barifcani, H. Vuthaluru, Experimental investigation into the development and evaluation of ionic liquid and its graphene oxide nanocomposite as novel pour point depressants for waxy crude oil, J. Petrol. Sci. Eng. 0920-4105, 208 (Part D) (2022), 109691, https://doi.org/10.1016/j.petrol.2021.109691.
[82] T.T. Khidr, S.A. Mahmoud, Dispersion of waxy gas oil by some nonionic surfactants, J. Dispers. Sci. Technol. 28 (2007) 1309–1315.
[83] T.T. Khidr, E.M.S. Azzam, S.S. Mutawaa, A.M.A. Omar, Study of some anionic surfactants as pour point depressants additives for a waxy gas oil. Ind. Lub. Trib. 59 (2) (2007) 64–68, https://doi.org/10.1108/00368790710731855.
[84] T.T. Khidr, Synthesis and evaluation of copolymer as pour-point depressants, Pet. Sci. Technol. 25 (2007) 671–681.
[85] P. Sivakumar, A. Sircar, D. Barasha, A. Silviya Anumegalai, P. Suresh Moorthi, N. Yasvanthrajan, Flow improvers for assured flow of crude oil in midstream pipeline—a review, J. Petrol. Sci. Eng. (2018), https://doi.org/10.1016/j.petrol.2018.01.022. Elsevier B.V.
[86] T.T. Khidr, S.M. Ahmed, The effect of cationic surfactant additives on flow properties of crude oil, Pet. Sci. Technol. 34 (2016) 1219–1225.
[87] M.N. Maithufi, D.J. Joubert, B. Klumperman, Application of Gemini surfactants as diesel fuel wax dispersants, Energy Fuel 25 (1) (2011) 162–171, https://doi.org/10.1021/ef1006287.
[88] S.M. Ahmed, T.T. Khidr, D.A. Ismail, Effect of Gemini surfactant additives on pour point depressant of crude oil, J. Dispers. Sci. Technol. 39 (8) (2017) 1160–1164.
[89] M. Sahai, R.K. Singh, A. Kukrety, A. Kumar, S.S. Ray, V.K. Chhibber, S. Kumar, Application of triazine based Gemini surfactants as viscosity reducing agents of tar sand derived bituminous crude, Energy Fuel 32 (3) (2018) 3031–3038.
[90] A. Mahmoudkhani, M. Feustel, W. Reimann, M. Krull, Wax and Paraffin Control by Fracturing Fluids: Understanding Mode of Actions and Benefits of Water-Dispersible Wax Inhibitors, Society of Petroleum Engineers, 2017, https://doi.org/10.2118/184594-MS.
[91] J.R. Becker, Winterized paraffin crystal modifiers, in: SPE Annual Technical Conference and Exhibition, Houston, Texas, 1999/1/1, 1999.
[92] M.R.H. Rosdi, M.A.A. Razali, K.M.K. Ishak, A. Ariffin, The impact of ethylene glycol on droplet growth inhibition in ethylene vinyl acetate emulsions, ACS Omega 5 (24) (2020) 14473–14480, https://doi.org/10.1021/acsomega.0c01114. American Chemical Society.
[93] N. Umoruddin, M.H. Rosdi, A. Ariffin, Mixed surfactant enabled EVA emulsion for PPD applications, J. Dispers. Sci. Technol. 40 (2019) 361–368.
[94] D.W. Jennings, M.E. Newberry, Paraffin inhibitor applications in deepwater offshore developments, in: International Petroleum Technology Conference: Kuala Lumpur, Malaysia, 2008.
[95] N. Nasir, M. Rosdi, A. Ariffin, The impact of different antifreeze agents on the thermal properties of ethylene vinyl acetate emulsion, Polym. Adv. Technol. 29 (2018) 708–715.
[96] N. Hazrati, M. Abdouss, A.A. Miran-Beigi, A.A. Pasban, Long chain alkylated ionic liquids as pour point depressant and rheology improver for crude oil, Pet. Chem. 61 (2) (2021) 206–213, https://doi.org/10.1134/S0965544121020079.

[97] R. Kumar, S. Banerjee, A. Mandal, T.K. Naiya, Flow improvement of heavy crude oil through pipelines using surfactant extracted from soapnuts, J. Petrol. Sci. Eng. 152 (2017) 353–360.
[98] S. Kumar, V. Mahto, Emulsification of Indian heavy crude oil using a novel surfactant for pipeline transportation, Pet. Sci. 14 (2) (2017) 372–382.
[99] O.P. Akinyemi, J.D. Udonne, V.E. Efeovbokhan, A.A. Ayoola, A study on the use of plant seed oils, triethanolamine and xylene as flow improvers of Nigerian waxy crude oil, J. Appl. Res. Technol. 14 (3) (2016) 195–205.
[100] L.K. Verma, S.N. Mukhdeo, Effects of surfactants on pour point of crude oil, in: Paper Presented at the 1st Natl. Conv. of Chem. Eng., Calcutta, 21–23 February, 1986.
[101] K.S. Wang, C.H. Wu, J.L. Creek, P.J. Shuler, Y. Tang, Evaluation of effects of selected wax inhibitors on paraffin deposition, Pet. Sci. Technol. 21 (2003) 369–379. Marcel Dekker Inc. https://doi.org/10.1081/LFT-120018526.
[102] G. McClaflin, D. Whitfill, Control of paraffin deposition in production operations, J. Petrol. Tech. 36 (11) (1984) 1,965–1,970.
[103] E. Ariza-León, D.R. Molina-Velasco, A. Chaves-Guerrero, Review of studies on asphaltene—wax interaction and the effect thereof on crystallization, CT&F—Ciencia, Tecnol. Futuro 5 (5) (2014) 39–53.
[104] G.E. Oliveira, C.R.E. Mansur, E.F. Lucas, G. González, W.F. de Souza, The effect of asphaltenes, naphthenic acids, and polymeric inhibitors on the pour point of paraffins solutions, J. Dispers. Sci. Technol. 28 (3) (2007) 349–356, https://doi.org/10.1080/01932690601107526.
[105] L. Li, J. Xu, J. Tinsley, D.H. Adamson, B.A. Pethica, J.S. Huang, R.K. Prud'homme, X. Guo, Improvement of oil flowability by assembly of comb-type copolymers with paraffin and asphaltene, AIChE J. 58 (2012) 2254–2261, https://doi.org/10.1002/aic.12729.
[106] O.A. Adeyanju, L.O. Oyekunle, Influence of long chain acrylate ester polymers as wax inhibitors in crude oil pipelines, in: SPE Nigeria Annual International Conference and Exhibition, 2014, https://doi.org/10.2118/172844-ms.
[107] Q. Quan, J. Gong, W. Wang, G. Gao, Study on the aging and critical carbon number of wax deposition with temperature for crude oils, J. Petrol. Sci. Eng. 130 (2015) 1–5, https://doi.org/10.1016/j.petrol.2015.03.026.
[108] A.N. Nwachukwu, G.A. Chukwu, S.I. Onwukwe, F.I. Chukwu, B. Obah, Synopsis of laboratory methods for characterizing waxy crude oils, ADR J. 2015 (2015) 1–11.
[109] Y. Peng, A. Golchha, G. Broze, Cold finger benchmarking study for paraffin inhibitor selection, in: Offshore Technology Conference, 2018, https://doi.org/10.4043/28714-ms.
[110] T.J. Bruno, M.L. Huber, A. Laesecke, E.W. Lemmon, R.A. Perkins, Thermochemical and Thermophysical Properties of JP-10, Gaithersburg, MD, 2006, https://doi.org/10.6028/NIST.IR.6640.
[111] J.R. Becker, Paraffin-crystal-modifier studies in field and laboratory, in: Paper Presented at the SPE Permian Basin Oil and Gas Recovery Conference, Midland, Texas, May, 2001, https://doi.org/10.2118/70030-MS.
[112] A. Golchha, P. Stead, Study and analysis of cold finger tests for effective selection of paraffin inhibitors, in: NACE Corrosion Conference 2015, 2015, pp. 1–14. No. 5803.
[113] K. Weispfennig, Advancements in paraffin testing methodology, in: SPE International Symposium on Oilfield Chemistry, Houston, Texas, 2001.
[114] Y. Chi, Investigation of Wax Inhibitors on Wax Deposition Based on Flow Loop Testing, M.S. Thesis, The University of Tulsa, Tulsa, OK, 2015.
[115] Y. Chi, N. Daraboina, C. Sarica, Effect of the flow field on the wax deposition and performance of wax inhibitors: cold finger and flow loop testing, Energy Fuel 31 (5) (2017) 4915–4924, https://doi.org/10.1021/acs.energyfuels.7b00253.
[116] S. Gon, Dynamic Wax Deposition Testing Systems and Methods, 2018. United States Patent Application 20180284097.

[117] L. Padula, S. Gon, C. Russell, E. Pohl, I. Littlehales, A. Neilson, Systematic study of temperature and different types of mixing in paraffin deposition test methods, ACS Omega 5 (21) (2020) 12320–12328, https://doi.org/10.1021/acsomega.0c01069.

[118] K. Chichak, C. Khandekar, B. Christofel, J. Doane, A. Vandish, Novel low-temperature stable paraffin inhibitors for subsea application, in: Paper Presented at the Offshore Technology Conference, Houston, Texas, May, 2019, https://doi.org/10.4043/29254-MS.

[119] H.H. Haynes, G.L. Lenderman, Cost-effective paraffin inhibitor squeezes can improve production economics, in: Paper Presented at the SPE Rocky Mountain Regional Meeting, Billings, Montana, May, 1986, https://doi.org/10.2118/15178-MS.

[120] D.W. Jennings, M.E. Newberry, Application of paraffin inhibitor treatment programs in offshore developments, in: Paper Presented at the Offshore Technology Conference, Houston, Texas, USA, May, 2008, https://doi.org/10.4043/19154-MS.

[121] J.B. Dobbs, A unique method of paraffin control in production operations, in: SPE 55647 (Paper Presented at the SPE Rocky Mountain Regional Meeting, Gillette, WY, 15–18 May), 1999.

[122] M.E. Newberry, K.M. Barker, Formation Damage Prevention Through the Control of Paraffin and Asphaltene Deposition, SPE 13796, 1985.

[123] D.V.S. Gupta, S. Szymczak, J.M. Brown, Solid Production Chemicals Added with the Frac for Scale, Paraffin and Asphaltene Inhibition, 2009. SPE 119393.

[124] S. Szymozak, D.V.S. Gupta, W. Steiner, S. Bolton, J. Romano, Well stimulation using a solid, proppant-sized, paraffin inhibitor to reduce costs and increase production for a South Texas, Eagle Ford Shale oil operator, in: Paper Presented at the SPE International Symposium and Exhibition on Formation Damage Control, Lafayette, Louisiana, USA, February, 2014, https://doi.org/10.2118/168169-MS.

[125] D.V.S. Gupta, D. Shen, S.J. Szymczak, A. Chemical Inhibitor-infused, in: High- Strength Proppant Additive for Reservoir Flow Assurance against Scale Deposition in Wells with High Intervention Costs, SPE 165078, 2013.

[126] F. Liang, M. Sayed, G.A. Al-Muntasheri, F.F. Chang, L. Li, A comprehensive review on proppant technologies, Petroleum (2016), https://doi.org/10.1016/j.petlm.2015.11.001. KeAi Communications Co.

[127] T. Smith, S. Szymczak, D.V.S. Gupta, J.M. Brown, Solid Paraffin Inhibitor Pumped in a Hydraulic Fracture Provides Long-term Paraffin Inhibition in Permian Basin Wells, 2009. SPE 124868.

[128] V. Wornstaff, S. Hagen, T. Ignacz, M. Chorney, B. Pedersen, Solid Paraffin Inhibitors Pumped in Hydraulic Fractures Increase Oil Recovery in Viking Wells, 2014. SPE 168147.

[129] F. DeBenedictis, S. Bhaduri, Dispersion-based paraffin inhibitors adsorbed in solid substrates affords long term flow assurance, in: Paper Presented at the SPE International Conference and Exhibition on Formation Damage Control, Lafayette, Louisiana, USA, February, 2018, https://doi.org/10.2118/189503-MS.

[130] K. Paso, T. Kompalla, N. Aske, H.P. Rønningsen, G. Øye, J. Sjöblom, Novel surfaces with applicability for preventing wax deposition: a review, J. Dispers. Sci. Technol. 30 (6) (2009) 757–781, https://doi.org/10.1080/01932690802643220.

[131] J. Bai, X. Jin, J.-T. Wu, Multifunctional anti-wax coatings for paraffin control in oil pipelines, Pet. Sci. (2019), https://doi.org/10.1007/s12182-019-0309-7.

[132] N.G. Ibragimov, Improving methods for protecting the tubing string from asphalt, resin and paraffin deposits in the fields of Tatarstan, Neft. Khoz. 6 (2005) 110–112.

[133] H.D. Gesser, R.T. Lafreniere, A drag-reducing water insoluble hydrophilic marine coating, Prog. Org. Coat. 49 (2004) 372–374.

[134] J.G. Charles, Unique paraffin inhibition technique reduces well maintenance, J. Can. Pet. Technol. 25 (2) (1986) 22, https://doi.org/10.2118/86-04-04.

[135] F.W. Jessen, J.N. Howell, Effect of flow rate on paraffin accumulation in plastic, steel, and coated pipe, Soc. Pet. Eng. J. 213 (4) (1958) 80–84.

[136] R.M. Jorda, Paraffin deposition and prevention in oil wells, J. Petrol. Tech. 18 (1966) 1605–1612.
[137] X.J. Zhang, J. Tian, L.J. Wang, Z.F. Zhou, Wettability effect of coatings on drag reduction and paraffin deposition prevention in oil, J. Petrol. Sci. Eng. 36 (1–2) (2002) 87–95, https://doi.org/10.1016/S0920-4105(02)00267-X.
[138] Y. Liu, X. Zhuge, Z. Wang, B. Huang, X. Le, Case study on fluorocarbons interior coating for anticorrosion and wax-deposition inhibition in ASP flooding production, in: Paper Presented at the SPE Annual Technical Conference and Exhibition, Dallas, Texas, USA, September, 2018, https://doi.org/10.2118/191656-MS.
[139] M.Y. Li, J.G. Su, Z.L. Wu, Y.D. Yang, S.L. Ji, Study of the mechanisms of wax prevention in a pipeline with glass inner layer, Colloids Surf. A 123 (1997) 635–649.
[140] R.M. Archibald, R.A. Bulstra, Use of glass reinforced epoxy resin tubing in Oman, in: Middle East Tech Conf and Exhibition (9–12 March, Bahrain), SPE, 1981, pp. 565–572, https://doi.org/10.2118/9636-MS.
[141] L. Duthie, H. Almarri, H. Saiood, A. Lata, G. Radhakrishnan, Effective corrosion mitigation exploiting glass reinforced epoxy lined tubulars in offshore producing & injection wells, in: Offshore Technology Conference, 2019, https://doi.org/10.4043/29266-MS.
[142] M. Heydrich, S. Choudhary, A. Hammami, J. Ratulowski, O. Bukuaghangin, T. Charpentier, A. Neville, A systematic characterization of a novel diamond-like coating to reduce solids deposition in production systems, in: Offshore Technology Conference Asia 2020, OTCA 2020. Offshore Technology Conference, 2020, https://doi.org/10.4043/30383-ms.
[143] H.U. Zettler, M. Weiß, Q. Zhao, H. Müller-Steinhagen, Influence of surface properties and characteristics on fouling in plate heat exchangers, Heat Transfer Eng. 26 (2) (2005) 3–17, https://doi.org/10.1080/01457630590897024.
[144] M. Rashidi, B. Mombekov, M. Marhamati, A study of a novel inter pipe coating material for paraffin wax deposition control and comparison of the results with current mitigation technique in oil and gas industry, in: Paper Presented at the Offshore Technology Conference Asia, Kuala Lumpur, Malaysia, March, 2016, https://doi.org/10.4043/26695-MS.
[145] B. Mombekov, S.A. Aliyu, I.B. Saaid, Z. Dairov, A study on new copolymer coated material for managing paraffin wax deposition in pipelines production operations, in: Paper Presented at the SPE Annual Caspian Technical Conference, Virtual, October, 2020, https://doi.org/10.2118/202560-MS.
[146] Q.F. Cheng, M.Z. Li, Y.M. Zheng, B. Su, S.T. Wang, L. Jiang, Janus interface materials: superhydrophobic air/solid interface and superoleophobic water/solid interface inspired by a lotus leaf, Soft Matter 7 (13) (2011) 5948–5951, https://doi.org/10.1039/C1SM05452J.
[147] A.R. Parker, C.R. Lawrence, Water capture by a desert beetle, Nature 414 (6859) (2001) 33–34, https://doi.org/10.1038/35102108.
[148] Y.Z. Guo, W.P. Li, L.Q. Zhu, H.C. Liu, An excellent non-wax-stick coating prepared by chemical conversion treatment, Mater. Lett. 72 (4) (2012) 125–127, https://doi.org/10.1016/j.matlet.2011.12.079.
[149] Z.W. Wang, L.Q. Zhu, H.C. Liu, W.P. Li, A conversion coating on carbon steel with good anti-wax performance in crude oil, J. Petrol. Sci. Eng. 112 (3) (2013) 266–272, https://doi.org/10.1016/j.petro l.2013.11.013.
[150] X. Yao, S.W. Wu, L. Chen, J. Ju, Z.D. Gu, M.J. Liu, et al., Self-replenishable anti-waxing organogel materials, Angew. Chem. Int. Ed. 54 (31) (2015) 8975–8979, https://doi.org/10.1002/anie.201503031.
[151] J.L. Evans, Apparatus and Technique for the Evaluation of Magnetic Conditioning as a Means of Retarding Wax Deposition in Petroleum Pipelines, 1997. DeepStar Report 3208-2.
[152] W.W. Zhang, T.T. Wang, X. Li, S.C. Zhang, The effect of magnetic field on the deposition of paraffin wax on the oil pipe, Adv. Mater. Res. 788 (2013) 719–722, https://doi.org/10.4028/www.scientific.net/AMR.788.719.
[153] J.L. Gonçalves, A.J.F. Bombard, D.A.W. Soares, G.B. Alcantara, Reduction of paraffin precipitation and viscosity of Brazilian crude oil exposed to magnetic fields, Energy Fuel 24 (5) (2010) 3144–3149.

[154] N.P. Tung, N. Van Vuong, B.Q.K. Long, N.Q. Vinh, P.V. Hung, V.T. Hue, L.D. Hoe, Studying the mechanism of magnetic field influence on paraffin crude oil viscosity and wax deposition reductions, in: SPE 68749 (Paper Presented at the SPE Asia Pacific Oil and Gas Conference and Exhibition, Jakarta, Indonesia, 17–19 April), 2001.
[155] C.S. Lee, Use of Magnetic Field in Paraffin Wax Deposition Control for Surface Facilities, 2008.
[156] W. Furlow, Magnetic fluid conditioners' success depends on factors, Offshore 58 (9) (1998).
[157] J. Jing, W. Shi, Q. Wang, B. Zhang, Viscosity-reduction mechanism of waxy crude oil in low-intensity magnetic field, Energy Sources A (2019), https://doi.org/10.1080/15567036.2019.1657991. Taylor and Francis Inc.
[158] R. Tao, X. Xu, Reducing the viscosity of crude oil by pulsed electric or magnetic field, Energy Fuel 20 (5) (2006) 2046–2051, https://doi.org/10.1021/ef060072x.
[159] R. Chow, R. Sawatzky, D. Henry, A. Babchin, Y. Wang, L. Cherney, R. Humphreys, Precipitation of wax from crude oil under the influence of a magnetic field, J. Can. Pet. Technol. 39 (2000), https://doi.org/10.2118/00-06-05.
[160] B. Wang, L. Dong, Paraffin characteristics of waxy crude oils in China and the methods of paraffin removal and inhibition, in: SPE 29954 (Paper Presented at the SPE International Meeting on Petroleum Engineering, Beijing, China, 14–17 November), 1995.
[161] R. Tao, H. Tang, Reducing viscosity of paraffin base crude oil with electric field for oil production and transportation, Fuel 118 (February) (2014) 69–72, https://doi.org/10.1016/j.fuel.2013.10.056. Elsevier Ltd.
[162] A.N. Krasnov, E.A. Khoroshavina, M.Y. Prakhova, Preventing paraffination of pumping equipment of oil wells, Adv. Eng. Res. 133 (2017) 370–375. Actual Issues of Mechanical Engineering (AIME 2017).
[163] J. Taheri-Shakib, A. Shekarifard, H. Naderi, S.A. Hosseini, Effect of microwave irradiation on wax and asphaltene content of heavy crude oil, in: 79th EAGE Conference and Exhibition 2017, 2017.
[164] X. Huang, C. Zhou, Q. Suo, L. Zhang, S. Wang, Experimental study on viscosity reduction for residual oil by ultrasonic, Ultrason. Sonochem. 41 (2018) 661–669.
[165] M.S. Mullakaev, D.F. Asylbaev, V.G. Prachkin, G.I. Volkova, Influence of ultrasound and heat treatment on the rheological properties of Ust-Tegusskoe oil, Chem. Pet. Eng. 49 (9–10) (2014) 584–587, https://doi.org/10.1007/s10556-014-9799-8.
[166] M.S. Mullakaev, V.O. Abramov, A.V. Abramova, Development of ultrasonic equipment and technology for well stimulation and enhanced oil recovery, J. Petrol. Sci. Eng. 125 (2015) 201–208, https://doi.org/10.1016/j.petrol.2014.10.024. Elsevier.
[167] V.O. Abramov, M.S. Mullakaev, A.V. Abramova, I.B. Esipov, T.J. Mason, Ultrasonic technology for enhanced oil recovery from failing oil wells and the equipment for its implementation, Ultrason. Sonochem. 20 (2013) 1289–1295.
[168] B.F. Towler, A. Chejara, mokhatab, S., Experimental investigations of ultrasonic waves effects on wax deposition during crude oil production, in: SPE Annual Technical Conference and Exhibition, 2007, https://doi.org/10.2118/109505-ms.
[169] H. Shang, Y. Yue, J. Zhang, J. Wang, Q. Shi, W. Zhang, et al., Effect of microwave irradiation on the viscosity of crude oil: a view at the molecular level, Fuel Process. Technol. 170 (2018) 44–52, https://doi.org/10.1016/j.fuproc.2017.10.021.
[170] A.A. Olajire, Review of wax deposition in subsea oil pipeline systems and mitigation technologies in the petroleum industry, Chem. Eng. J. Adv. 6 (May) (2021) 100104, https://doi.org/10.1016/j.ceja.2021.100104. Elsevier BV.
[171] K. Bimuratkyzy, B. Sagindykov, The review of flow assurance solutions with respect to wax and asphaltene, Braz. J. Pet. Gas 10 (2) (2016) 119–134, https://doi.org/10.5419/bjpg2016-0010.
[172] G. Dean, R. Perkins, Chemical—free stimulation renews mature well production, DEW J. (2010) 21–25.
[173] M.A. Theyab, A review of wax mitigation methods through hydrocarbon production, J. Pet. Environ. Biotechnol. 11 (5) (2020) 412.

[174] L. Ismail, R.E. Westacott, X. Ni, On the effect of wax content on paraffin wax deposition in a batch oscillatory baffled tube apparatus, Chem. Eng. J. 137 (2) (2008) 205–213.
[175] WEAFRI, Catalytic fluid conditioning treatment of paraffin, scale, corrosion, iron sulfide, and hydrogen sulfide, in: Corrosion Inhibitor Systems Manual, 2005.
[176] A.A. Sulaimon, G.K. Falade, W. Delandro, A proactive approach for predicting and preventing wax deposition in production tubing strings: a Niger delta experience, J. Pet. Gas Eng. 1 (4) (2010) 26–36.
[177] Synergetic Oil Tools Inc, Stabilize Production, Synergetic Oil Tools Inc., 2021. Retrieved from: https://www.enercattool.com/about/. (Last visit Dec 2021).
[178] A.D.I. Sulaiman, A.J. Ajienka, I.S. Sunday, Application of piezoelectric energy generated from quartz plus semiprecious metals on wax deposition control, J. Pet. Gas Eng. 2 (May) (2011) 93–98.
[179] A.D. Sulaiman, J.A. Ajienka, Studies on the mechanism of using piezoelectric energy to control wax deposition, J. Emerg. Trends Eng. Appl. Sci. 2 (6) (2011) 914–920 (JETEAS).
[180] K.A. Bashmur, E.A. Petrovsky, Y.A. Geraschenko, V.A. Makolov, Y.N. Shadchina, V.S. Tynchenko, Relief flow swirlers to control wax deposition formation inside downhole equipment, J. Phys. Conf. Ser. 1399 (2019), https://doi.org/10.1088/1742-6596/1399/5/055029. Institute of Physics Publishing.
[181] C. Anyanwu, E. Agbo, B. Olusola, O. Olowoyeye, B. Njoku, Wax removal and prevention in an old well—lessons learnt, in: Paper Presented at the SPE Nigeria Annual International Conference and Exhibition, Virtual, August, 2020, https://doi.org/10.2118/203660-MS.
[182] J. Li, J. Liu, M.G. Trefry, J. Park, K. Liu, B. Haq, et al., Interactions of microbial-enhanced oil recovery processes, Transp. Porous Media 87 (1) (2010) 77–104, https://doi.org/10.1007/s11242-010-9669-6.
[183] M.D. Bishop, D.R. Woodward, Biological Paraffin Control Systems Show Production Increases in Addition to Controlling Paraffin, Petroleum Society of Canada, 1990, https://doi.org/10.2118/90-56.
[184] M.M. Santamaria, R.E. George, Controlling paraffin-deposition-related problems by the use of Bacteria treatments, in: SPE Annual Technical Conference and Exhibition, 1991, https://doi.org/10.2118/22851-ms.
[185] I. Lazar, A. Voicu, C. Nicolescu, D. Mucenica, S. Dobrota, I. Petrisor, et al., The use of naturally occurring selectively isolated bacteria for inhibiting paraffin deposition, J. Petrol. Sci. Eng. 22 (1–3) (1999) 161–169,-https://doi.org/10.1016/s0920-4105(98)00065-5.
[186] A. Etoumi, Microbial treatment of waxy crude oils for mitigation of wax precipitation, J. Petrol. Sci. Eng. 55 (2007) 111–121.
[187] D.P. Rana, S. Bateja, S.K. Biswas, A. Kumar, T.R. Misra, B. Lal, Novel Microbial Process for Mitigating Wax Deposition in Down Hole Tubular and Surface Flow Lines, Society of Petroleum Engineers, 2010, https://doi.org/10.2118/129002-MS.
[188] Y. He, F. Zhang, I.M. Banat, P. Chunsheng, H. Dong, F. Shu, Q. Shi, Y. She, Deposit reduction in a high pour point oil reservoir due to the activity of indigenous bacterial communities, Int. Biodeterior. Biodegrad. 110 (May) (2016) 87–98, https://doi.org/10.1016/j.ibiod.2016.02.009. Elsevier Ltd.
[189] X.Y. Lei, J. You, Y.B. Li, R. Wang, X.K. Shan, X. Zhou, Screening and evaluation of microorganism species in oil recovery in Ji 45 block, IOP Conf. Ser. Earth Environ. Sci. 81 (2017), 012080, https://doi.org/10.1088/1755-1315/81/1/012080.
[190] C. Luo, W. Wang, H. Zhang, X. Yu, G. Wang, Microbial treatment of waxy crude oil for mitigating wax precipitation and improving liquidity, Pet. Sci. Technol. (2019), https://doi.org/10.1080/10916466.2018.1552968.
[191] J.H. Liu, Y.P. Jia, Y.T. Chen, X. Rui Dan, Microbial treatment for prevention and removal of paraffin deposition on the walls of crude pipelines, Indian J. Microbiol. 53 (4) (2013) 482–484, https://doi.org/10.1007/s12088-013-0402-3.
[192] M. Xiao, W.-H. Li, M. Lu, Z.-Z. Zhang, Y.-J. Luo, W. Qiao, et al., Effect of microbial treatment on the prevention and removal of paraffin deposits on stainless steel surfaces, Bioresour. Technol. 124 (227) (2012) 232, https://doi.org/10.1016/j.biortech.2012.07.063.

[193] Z. He, B. Mei, W. Wang, J. Sheng, S. Zhu, L. Wang, T.F. Yen, A pilot test using microbial paraffin-removal technology in Liaohe oilfield, Pet. Sci. Technol. 21 (1–2) (2003) 201–210, https://doi.org/10.1081/lft 120016942.

[194] M.L. Samuelson, Alternatives to aromatics for solvency of organic deposits, in: SPE Formation Damage Control Symposium, 1992, https://doi.org/10.2118/23816 ms.

[195] T.J. Straub, S.W. Autry, G.E. King, An Investigation into Practical Removal of Downhole Paraffin by Thermal Methods and Chemical Solvents, SPE, 1989. SPE 18889.

[196] C.L. Thierheimer Jr., Solvent for Paraffin Removal from Oilfield Equipment, 1990. United States Patent 4925497.

[197] X. Lv, X. Li, Y. Ge, Q. Li, H. Lu, Sustainable separation of paraffin wax by span 80 combined with switchable water, Energy Fuel 35 (9) (2021) 7907–7912, https://doi.org/10.1021/acs.energyfuels.1c00401.

[198] J. Curtis, Environmentally favorable terpene solvents find diverse applications in stimulation, in: Sand Control and Cementing Operations, Society of Petroleum Engineers, 2003, https://doi.org/10.2118/84124-MS.

[199] G.L. Poole, G. Brock, S. Szymczak, G. Casey, Successful pipeline cleanout: lessons learned from cleaning paraffin blockage from a deepwater pipeline, in: Paper Presented at the SPE Annual Technical Conference and Exhibition, Denver, Colorado, USA, September, 2008, https://doi.org/10.2118/115658-MS.

[200] G.A. Hicks, Uniform program reduces paraffin removal costs, Oil Gas J. (1990) 85–90. Pug. 20.

[201] M.R. Embrey, J. Larke, Paraffin-plug remediation in deepwater wells via capillary tubing: a cost effective alternative, in: SPE Annual Technical Conference and Exhibition, 2010, https://doi.org/10.2118/135136-ms.

[202] D. Tukenov, Technology update: nanochemistry drives new method for removal and control of wax, J. Petrol. Tech. 66 (12) (2014) 30–33, https://doi.org/10.2118/1214-0030-jpt. Society of Petroleum Engineers (SPE).

[203] P. Singh, H. Scott Fogler, Fused chemical reactions: the use of dispersion to delay reaction time in tubular reactors, Ind. Eng. Chem. Res. 37 (6) (1998) 2203–2207, https://doi.org/10.1021/ie9706020.

[204] D.A. Nguyen, H.S. Fogler, S. Chavadej, Fused chemical reactions. 2. Encapsulation: application to remediation of paraffin plugged pipelines†, Ind. Eng. Chem. Res. 40 (23) (2001) 5058–5065, https://doi.org/10.1021/ie0009886.

[205] D.A. Nguyen, F.F. de Moraes, H.S. Fogler, Fused chemical reactions. 3. Controlled release of a catalyst to control the temperature profile in tubular reactors, Ind. Eng. Chem. Res. 43 (18) (2004) 5862–5873, https://doi.org/10.1021/ie049933k.

[206] J.P. Ashton, L.J. Kirspel, H.T. Nguyen, D.J. Credeur, In-situ heat system stimulates paraffinic-crude producers in Gulf of Mexico, SPE Prod. Eng. 4 (02) (1989) 157–160, https://doi.org/10.2118/15660-pa.

[207] A. Hassan, O. Alade, M. Mahmoud, A. Al-Majed, A novel technique for removing wax deposition in the production system using thermochemical fluids, in: Paper Presented at the Abu Dhabi International Petroleum Exhibition & Conference, Abu Dhabi, UAE, November, 2019, https://doi.org/10.2118/197323-MS.

[208] E. Bell, Y. Lu, N. Daraboina, C. Sarica, Experimental investigation of active heating in removal of wax deposits, J. Petrol. Sci. Eng. 200 (May) (2021), https://doi.org/10.1016/j.petrol.2021.108346. Elsevier B.V.

[209] J. Nanniger, G. Nanniger, Optimizing hot oiling/watering jobs to minimize formation damage, in: Annual Technical Meeting, 1990, https://doi.org/10.2118/90-57.

[210] A.J. Mansure, K.M. Barker, Insights into Good Hot Oiling Practices, Society of Petroleum Engineers, 1993, https://doi.org/10.2118/25484-MS.

[211] Z.M.W. Haindade, A.M. Mamen, Innovative Hot Oiling Operation in a Low BHP Well Minimizing Formation Damage: A Case Study, Society of Petroleum Engineers, 2014, https://doi.org/10.2118/172195-MS.

[212] R. Brock, An Experimental Paraffin Control Treatment Procedure, Southwestern Petroleum Short Course, 1989, pp. 91–106.

[213] J.R. Becker, Oilfield paraffin treatments: hot oil and hot water compared to crystal modifiers, in: SPE Annual Technical Conference and Exhibition, 2000, https://doi.org/10.2118/63123-ms.
[214] R.S. Green, Transportable System for Providing Heat to Flowing Materials, US 4,720,263 Jan. 19, 1988.
[215] V.O. Zyrin, Electrothermal complex with downhole electrical heating generators for enhanced heavy oil recovery, Int. J. Appl. Eng. Res. 11 (3) (2016) 1859–1866.
[216] Z.M.W. Haindade, A.D. Bihani, S.M. Javeri, C.B. Jere, Enhancing Flow Assurance Using Co-Ni Nanoparticles for Dewaxing of production Tubing, Society of Petroleum Engineers, 2012, https://doi.org/10.2118/157119-MS.
[217] H.L. Becker, J.M. Brown, Quantum Effects Imparted by Radio Frequencies as a Stimulation Method of Oil Production, Society of Petroleum Engineers, 2009, https://doi.org/10.2118/124144-MS.
[218] F. Librino, Unconventional Methods Offer Wax Remediation Options, offshore-mag, 2015.
[219] J. Zhang, Q. Xue, H. Gao, P. Wang, Biodegradation of paraffin wax by crude aspergillus enzyme preparations for potential use in removing paraffin deposits, J. Basic Microbiol. 55 (11) (2015) 1326–1335, https://doi.org/10.1002/jobm.201500290.
[220] N.A. Adlan, S. Sabri, M. Masomian, M.S.M. Ali, R.N.Z.R.A. Rahman, Microbial biodegradation of paraffin wax in Malaysian crude oil mediated by degradative enzymes, Front. Microbiol. 11 (September) (2020), https://doi.org/10.3389/fmicb.2020.565608. Frontiers Media S.A.
[221] T.K. Dutta, S. Harayama, Fate of crude oil by the combination of photooxidation and biodegradation, Environ. Sci. Technol. 34 (2000) 1500–1505.
[222] R. Thavasi, S. Jayalakshmi, I.M. Banat, Effect of biosurfactant and fertilizer on biodegradation of crude oil by marine isolates of *Bacillus megaterium*, *Corynebacterium kutscheri* and *Pseudomonas aeruginosa*, Bioresour. Technol. 102 (2011) 772–778.
[223] https://www.corebiologic.com/paraffin-destroyer-industrial.
[224] R. Venkatesan, N.R. Nagarajan, K. Paso, Y.B. Yi, A.M. Sastry, H.S. Fogler, The strength of paraffin gels formed under static and flow conditions, Chem. Eng. Sci. 60 (2005) 3587–3598.

CHAPTER

Asphaltene Management

18

18.1 Introduction

Asphaltenes are one of the major flow assurance problems, resulting in serious complications during oil production. Asphaltenes aggregate and deposit in the production system due to various thermodynamic and operational factors, including pressure drop, pH change, mixing of incompatible fluids, streaming potential, and other hydrodynamic factors. The nasty outcomes of asphaltene aggregation are clogging production tubulars, damaging production equipment, more costly production, troubled phase separation, and other safety and hazardous impacts. A full discussion of asphaltene aggregation and deposition is given in Chapter 8.

The early detection and management of asphaltene aggregation helps production operators to avoid its drastic impacts and keep production moving in a smooth, steady-state flow.

Management of asphaltenes is preferably to be initiated during the early stages of field development. During production, changes in production parameters and fluid properties are considered early signs of potential asphaltene deposition, and asphaltene management studies should be promptly started.

As with scale deposits, gas hydrates, and wax, asphaltene control strategy is based on either preventing deposition or removal of the deposits formed. Asphaltene deposit prevention methods are based on operational, chemical, and nonchemical methods, while asphaltene removal methods are generally based on chemical and nonchemical methods. Fig. 18.1 illustrates the asphaltene management methods.

18.2 Asphaltene deposition prevention

Asphaltene prevention basically depends on understanding the mechanism and the factors affecting asphaltene aggregation and deposition, and using practical mitigation methods to reduce or eliminate these factors. Asphaltene prevention can be achieved operationally by optimizing the operating parameters to be within the asphaltene stability zone; otherwise system intervention using chemical, physical, mechanical, or biological methods will be required to keep asphaltenes suspended in the flowing fluids.

As stated before, it is recommended to start asphaltene deposition prevention and management at the early stages of field development, to be ready with all the possible management methods and to avoid production interruptions.

Essentials of Flow Assurance Solids in Oil and Gas Operations. https://doi.org/10.1016/B978-0-323-99118-6.00021-6

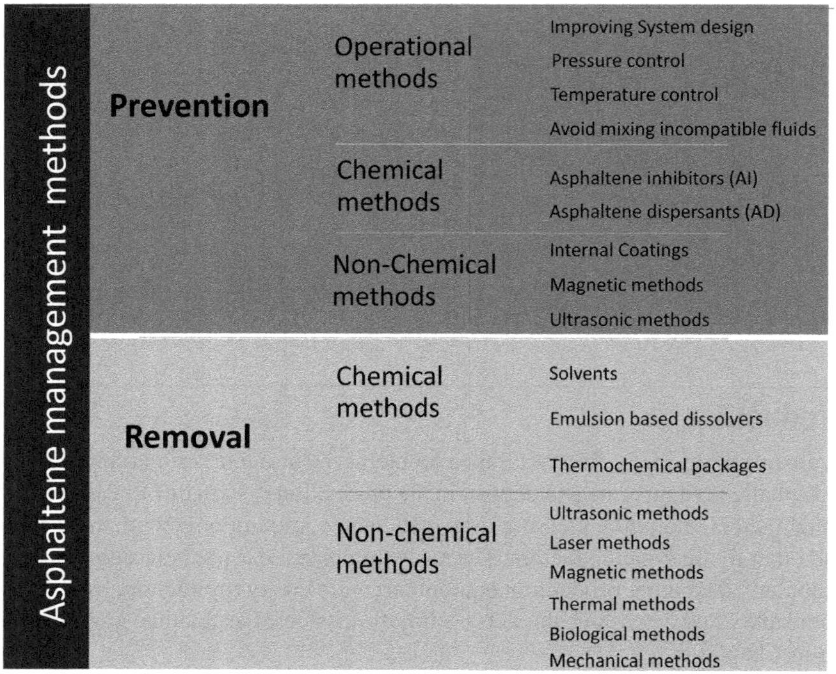

FIG. 18.1

Asphaltene management methods.

18.2.1 Operational methods of asphaltene deposition prevention

These methods depend on optimizing the operating parameters to operate outside the asphaltene thermodynamic instability region or the ADE (asphaltene deposition envelope) without significant loss of production. Since the stability of asphaltenes is a function of pressure, temperature, composition, and flow, by controlling these parameters, it may be possible to minimize the occurrence of conditions that promote asphaltene deposition [1,2].

18.2.1.1 Pressure control

Pressure plays a critical role in asphaltene aggregation and deposition. When pressure drops to the asphaltene onset pressure (AOP), asphaltenes start to precipitate and reach their maximum precipiation at the bubble point (BP). Thus by optimizing the operating pressure and preventing it from reaching the AOP, the chances of asphaltene aggregation and precipitation are reduced. Fig. 18.2 illustrates a typical wax asphaltene deposition envelope indicating the areas of stability and instability of asphaltenes in a crude oil system.

As indicated in this figure, the production system goes from one asphaltene-stable region to another, passing through a region of asphaltene instability, where the maximum amounts of asphaltenes are

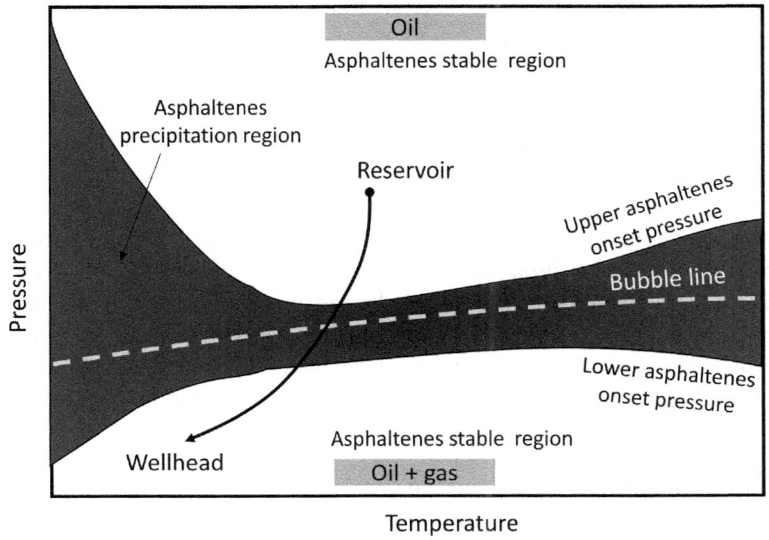

FIG. 18.2

Asphaltene deposition envelope showing asphaltene stability and instability (precipitation) regions.

precipitated around the BP. If pressure can be optimized to be within these stable regions, there will be fewer asphaltene problems [2]. Higher pressure than AOP or safe operating pressure can be achieved in many ways, including:

- The reservoir oil pressure can be maintained above the onset pressure of asphaltene precipitation by early water flooding injection [3]. Also, gas injection can be effective; however, gas dissolution in crude oil can destabilize the asphaltenes and cause their aggregation.
- In undersaturated reservoirs, the pressure is kept at the BP by producing from the upper part of the reservoir [4].
- In retrograde reservoirs, asphaltene drop-out around the wellbore might occur where asphaltene stability is very sensitive to the dissolution of condensate and can be reduced by specifying a constraint, such as producing a lower gas-to-oil ratio (GOR) [5].
- Another method is to optimize the production equipment design and sizes. Ahmadi et al. [6] studied the effect of adjusting choke valve sizes on thermodynamic parameters of fluid flowing and asphaltene deposition in a vertical well. They demonstrated that the optimum choke valve sizes could keep producing vertical wells away from asphaltene precipitation. The results of this study were implemented on a well in Darquin Reservoir that had been experiencing asphaltene precipitation, and after replacing the well's original choke valve with the suggested choke valve, the asphaltene precipitation problem diminished. Dawood et al. [7] modeled a reduction in wellhead pressure and oil flow rate and increase in the tubing size, leading to a decrease in the asphaltene precipitation. However, these changes could not resolve the problem completely.

18.2.1.2 Temperature control

Temperature control can also offer a reasonable asphaltene management method. While the effect of temperature on asphaltenes is still a point of conflict as to whether it increases or decreases the asphaltene deposition potential [8,9], it is believed that for typical reservoir temperature, the higher the temperature in the wellbore, the lower the amount of precipitated asphaltenes [2].

Accordingly, temperature control measures such as thermal insulation, active heating methods, and steam injection can be applied to prevent and remove asphaltene deposits if a temperature reduction was found to trigger asphaltene deposition, taking into consideration that the application of heat sometimes makes the asphaltene deposition problem worse, as in the case of steam injection systems [10]. The applicability of these methods can sometimes depend on whether or not the asphaltene deposits are boosted by or accompany paraffin wax deposits. Asphaltene deposits are usually accompanied by wax deposits at different ratios, which are temperature dependent. Table 18.1 summarizes the percent composition of organic deposits of producing wells in the Egyptian oil field. Thus reducing paraffins deposition may reduce the chances of forming sticking asphaltene deposits.

18.2.1.3 Avoid incompatible fluids mixing

One of the main causes of asphaltene deposition is mixing of incompatible fluids. The blending of incompatible crudes is a common cause of asphaltene precipitation. Light, nonasphaltic crude is a possible precipitant for heavier crudes [1]. Crude oils with high colloidal instability index (CII) (higher than 0.9) have high potential of developing asphaltene problems, and will have compatibility issues when mixed with other crudes [11].

Consequently, stability studies like CII, de Boer plots, toluene equivalence tests, and heptane dilution tests (Wiehe method) must be performed to check the compatibility of the mixed crude streams. When incompatibility is observed, mixing these crudes must be avoided if possible, or diluted and mixed to be within stable asphaltene ratios.

The same incompatibility issues can arise under the effect of gas-lift operations and miscible gas injection on asphaltene stability, which demands a scheme of safe operations or alternative solutions to avoid asphaltene issues.

Table 18.1 Coprecipitation of asphaltenes and waxes from different wells in Egyptian field (inorganic scales and fines are not mentioned).

Well ID	Paraffin deposits %	Asphaltene deposits %
WA1	37.6	60.8
WA4	47.6	50.1
W5A	23.0	30.9
W9A	22.5	57.4
W4A	24	23
W4E	50	13
W2B	8	2
W56	11	11
W44	32	2
W5A	98	2

In the meantime, some crude oil components can be utilized to inhibit or at least reduce asphaltene aggregation and deposition. Recently, Joonaki et al. [12] showed that high molecular weight waxes can decrease the asphaltene deposition rate and shift the AOP due to the higher solubility of asphaltenes in hydrocarbons with higher carbon number compared to those with lower carbon number. Auflem et al. [13] found that crude oil naphthenic acids adsorb to asphaltenes and disperse them, but the performance was lower than a commercial AD based on fatty acids and amines. Tojima et al. [14] presented a method to divide crude oil asphaltenes into two light and heavy fractions in such a way that the light asphaltenes fraction acted as peptizing agents for heavy asphaltenes.

Deasphalted oil has been found to be a strong asphaltene solvent and asphaltene inhibitor (AI), presumably because of its native resin and aromatic contents [15,16]. Khaleel et al. [17] proposed a novel method that reduces the amount of asphaltene precipitation by changing the composition of the system. In this method, maltenes or dead oil are recycled, as good asphaltene solvents, into the wellbore region to alter the composition of the system. Unlike the currently used chemical injection mitigation techniques, this method is considered universal, as the injected solution is tailored for each well: the decision on whether to recycle dead oil or maltenes is based on each well, the saturates, aromatics, resins, and asphaltenes (SARA) analysis of the oil, viscosity, and technoeconomic analysis. Fig. 18.3 illustrates the deasphalted oil injection approach.

Another source of incompatibility is the solvents that are added to improve the oil viscosity and to treat the wax deposit issues, especially the short chain n-alkanes. Therefore special attention should be given during chemical wax removal, crude oil dilution, or viscosity improvement trials to assure compatibility with crude oil asphaltenes.

Compatibility of produced fluids with treating chemicals should be evaluated before application. Acids used in stimulation and cleaning jobs are a potential cause of asphaltene deposition.

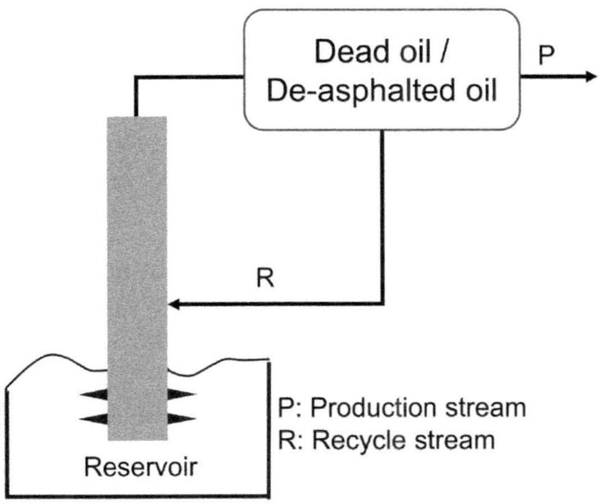

FIG. 18.3

Asphaltene mitigation proposed technique using deasphalted oil.

Reprinted from A.T. Khaleel, M.I.L. Abutaqiya, C.J. Sisco, F.M. Vargas, Mitigation of asphaltene deposition by re-injection of dead oil, Fluid Phase Equilib. 514 (June) (2020). https://doi.org/10.1016/j.fluid.2020.112552 (Elsevier B.V.), copyright 2020, with permission from Elsevier.

18.2.1.4 Other operating parameters control

These include reduction of the factors that are believed to impact asphaltene deposition like shear and neutralization of electrostatics [18]. Ihtsham and Ghosh [19] showed that promising laboratory results were obtained when they used direct current to control asphaltene deposition. However, these methods may not be that effective alone and require a good understanding of the asphaltene characteristics before application.

Salimi et al. [11] showed that decreasing fluid viscosity increases deposit thickness, so fluids viscosity should be high enough to erode and wear down the deposited asphaltenes.

Gas lift operations must be optimized to avoid their drastic impact on asphaltene stability. Similarly, CO_2 injection operations must be optimized to be less disturbing to the thermodynamic stability of asphaltenes. Modeling studies are very beneficial for estimating the risk of using gas lift and in the case of CO_2 injection.

Addressing and mitigating corrosion problems in the system assists in attenuating the asphaltene deposition problem. Corrosion provides high iron content, which destabilizes asphaltenes and furnishes a rough surface for asphaltene particle nucleation and growth.

The operational methods for managing asphaltene problems are summarized in Table 18.2.

18.2.2 Chemical prevention of asphaltene deposition

Although operational methods can be effective in attenuating asphaltene problems, chemical methods are commonly used either alone or in combination with operational methods to give complete control of the problem.

Chemical methods for asphaltene control are based on using specialty chemicals to interfere with asphaltene aggregation and deposition processes. Many studies have reported the successful field application of asphaltene control chemicals [20,21].

18.2.2.1 Classification and mechanisms of asphaltene control chemicals

Two main chemical types are used in asphaltene prevention: AIs and asphaltene dispersants (ADs). AIs provide inhibition by preventing the aggregation of asphaltene molecules, while ADs work by reducing the particle size and keeping them suspended in the solution [1]. Solvents are also conventionally used

Table 18.2 Summary of the operational methods of preventing asphaltene deposition.

Method	Description
Pressure control	- Avoid pressure drop especially near the asphaltene onset pressure and bubble point
Temperature control	- Optimize system temperature to avoid asphaltene destabilization and aggregation - Optimize system temperature to be above WAT to avoid wax deposition, which might enhance asphaltene deposition
Fluids mixing	- Avoid mixing incompatible crudes - Avoid using alkanes as solvents in the system - Optimize gas lift operations to reduce their impact on asphaltenes - Optimize CO_2 injection operations to avoid their impact on asphaltenes
Optimize system design	- Optimize system components to be less impacting on asphaltenes and cause less pressure drop

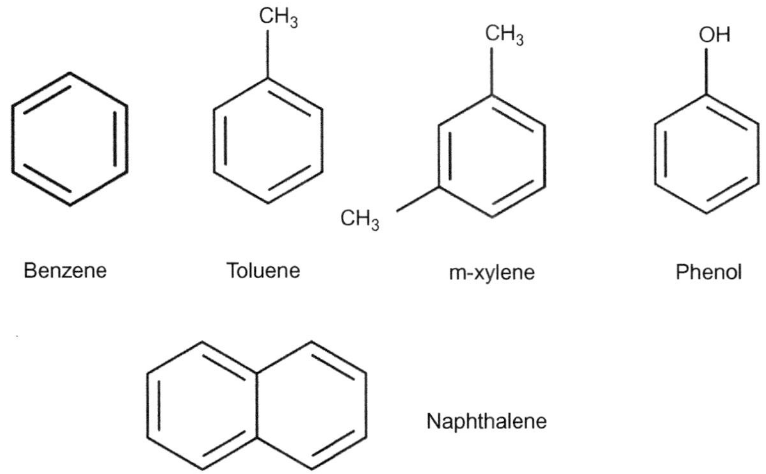

FIG. 18.4

Some examples of organic solvents used in asphaltene mitigation.

either to dissolve the asphaltene deposits, prevent deposition, or act as a carrier for the other asphaltene control chemicals. Examples of asphaltene control chemicals are illustrated in Fig. 18.4.

Generally speaking, AIs and ADs are amphiphiles that have structure similar to the resins that naturally exist in crude oil and that are known to peptize and stabilize asphaltenes. They have polar parts that interact and bridge with asphaltene polar parts and nonpolar (paraffinic) parts that act like a tail that keeps the final compound suspended or dissolved in the bulk hydrocarbon matrix [22,23].

The adsorption interaction between the chemical and asphaltene molecules is a very important parameter for stabilization of asphaltenes in crude oil. The chemical needs several molecular points of interaction with the asphaltene molecules [24]. The possible interactions between asphaltenes and ADs/inhibitors include [1]:

- π-π interactions between their aromatic rings,
- Hydrogen bonds,
- Acid-base interactions,
- Dipole-dipole interactions, and
- Complex metal ions.

Therefore it is expected that the asphaltene control chemical structure will exhibit one or more functional groups such as aromatic rings, heterocyclic rings, hetero atoms, acidic/basic groups, and branched alkyl tails.

Kuang et al. [2] comprehensively summarized how asphaltene control chemicals work in three main mechanisms:

1- Dispersion: here the polar group in the amphiphile interacts and bridges with the polar groups in the asphaltene molecule, while the aliphatic tail suspends the asphaltene in the bulk oil.

2- Aging disruption: in this category the amphiphiles interfere with asphaltene crystallization by preventing the interlocking of asphaltene cores and the aliphatic chains, hence preventing the formation of compact and solid-like asphaltenes.
3- Electrostatic interactions: in this category, the chemicals are based on the fact that the majority of asphaltene deposition is caused by smaller particle sizes, whereas larger aggregates are less susceptible to deposit [25]. Thus traditional dispersants usually work by stabilizing asphaltenes through electrostatically maintaining small particle sizes (or electrostatically maintaining specific average asphaltene size) by manipulation of the surface charge of the asphaltene molecules.

Alberto et al. [26] proposed two different mechanisms for asphaltene control chemicals, namely bulk and surface mechanisms:

- In the bulk mechanism the inhibitor is dissolved in the bulk oil and at a concentration above a given concentration (called critical additive concentration, CAC), it keeps asphaltenes either in solution or as a stable dispersion. This is the mechanism by which most commercial inhibitors generally work.
- In the surface mechanism, the AI acts on the surfaces of rock and tubing by limiting the rate of deposition/adhesion of asphaltene particles.

18.2.2.2 Asphaltene inhibitors (AIs)

AIs can reduce the average asphaltene size, prevent their aggregation, and shift the onset of asphaltene flocculation pressure [1,2]. Some commercial AIs were found to reduce the cumulative particle count relative to an untreated sample, rather than reducing the average particle size [27].

There is no universal AI as such: AIs are crude-oil specific. This is due to the complex asphaltene structures and the various interactions between the inhibitor and the polar parts of the aspaltenes in the crude oil [28]. Some AIs can also be used as demulsifiers or as synergists for the real demulsifiers [1]. Table 18.3 illustrates the different types of AIs.

- **Organic solvents**

Aromatic solvents, such as benzene, toluene, xylene, phenanthrene, and naphthalene, can aid in stabilizing the asphaltene micelles and inhibit precipitation [29–32]. Organic solvents were injected through gas lift valves to prevent asphaltene deposition in production tubing in the Egyptian field. Generally, the benzene ring on these molecules causes asphaltene inhibition [30]. Benzoic acid and phenol yielded better results than toluene, while M-xylene was weaker than toluene, because M-xylene has two methyl groups but toluene has only one methyl group, and this causes a reduction of the aromaticity factor [30]. Fig. 18.4 illustrates some of the common organic solvents.

Although the asphaltene is soluble in toluene, the toluene inhibitory effect on precipitation is weaker than that of common commercial inhibitors. The reason is that toluene contains no polar groups; the effectiveness of any inhibitors is determined by the acid-base interaction, which causes adsorption of the inhibitor on the surface of the asphaltene molecules [31].

However, high volumes of solvents are required to prevent asphaltene deposition, which deems the method economically infeasible. In addition, periodic solvent injections might alter the physiochemical properties of crude oil [33].

- **Organic acids**

Table 18.3 Common examples of asphaltene inhibitors.

Asphaltene inhibitor type	Examples
Organic solvents	Benzene Toluene Xylene Naphthalene Phenol
Organic acids	Benzoic acid Caprylic acid Palmitic acid Linoleic acid
Amphiphiles	Cetyltrimethylammonium bromide (CTAB) Sodium dodecyl sulfate (SDS) Cetylpyridinium chloride (CPC) Dodecyl benzene sulfonic acid (DBSA) Nonylphenol (NP) *n*-Nonylbenzene di-oxyethylene (NBDO) *n*-Nonylbenzene (NB)
Polymeric inhibitors	Alkyl phenol-aldehyde resin oligomers and polymers Polyester Polyamide/imide Polyisobutylene succinimides
Green inhibitors	Green polyalkylphenol-based resins Natural extracts, e.g., coconut essential oil, sweet almond, henna, etc.
Nanoparticles	Metal oxide nanoparticles, e.g., TiO_2, Al_2O_3, CaO
Ionic liquids	Tetrachloride aluminate-based ionic liquids Bromide-based ionic liquids Imidazolium and pyridinium-based ionic liquids

Acid-base interactions are part of asphaltene intermolecular reactions, so it is expected that organic acids can stabilize asphaltenes by participating in these acid-base reactions. Examples are benzoic acid, phthalic acid, salicylic acid [32], linoleic acid, caprylic acid, and palmitic acid [34]. Organic acids usually show good performance if their alkyl chain length is sufficiently large to give steric stabilization around the asphaltene. Fig. 18.5 shows some examples of organic acids.

- **Amphiphiles**

Some commercially available surfactants that have been used for different industrial applications have also been studied for asphaltene inhibition and mitigation. Examples are sodium dodecyl sulfate (SDS), cetyl-pyridinium chloride (CPC), cetyltrimethylammonium bromide (CTAB), and Triton X-100, which are used in biotechnology and personal hygiene products [29,30,35]. Some examples of these amphiphiles are given in Fig. 18.6.

Alkylbenzene-derived amphiphiles include alkylphenols, hydroxyethoxy alkylbenzene, and alkylbenzene sulfonic acid (Fig. 18.7). These amphiphiles are conventionally used as AIs as well as

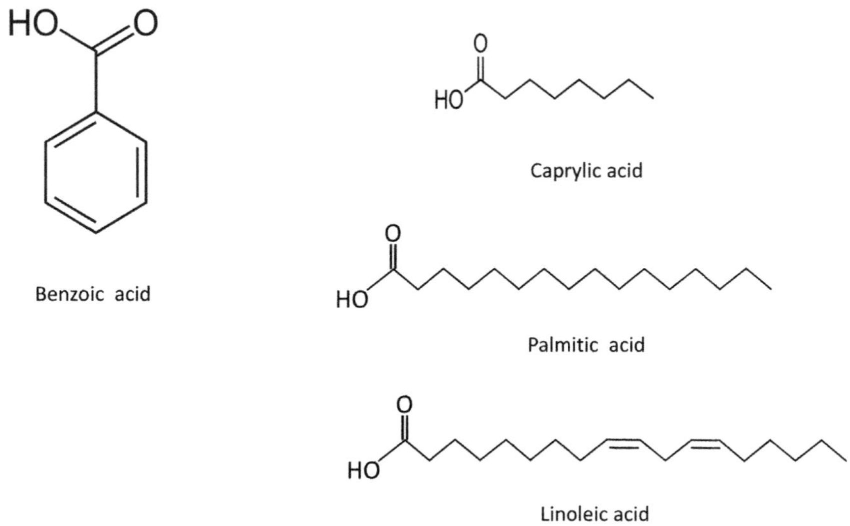

FIG. 18.5

Some examples of organic acids used in asphaltene mitigation.

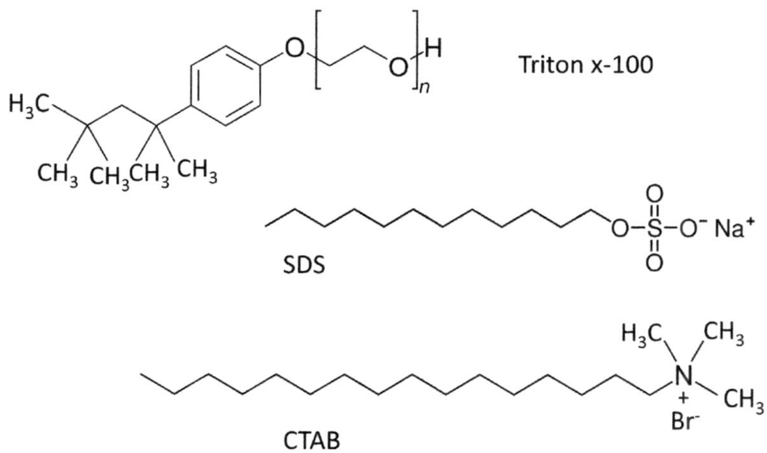

FIG. 18.6

Examples of commercial amphiphiles used in asphaltene mitigation.

dispersants. Alkyl benzene sulfonic acids are stronger than alkylphenols, and alkylphenols are stronger than hydroxyl ethoxy alkyl benzene [36]. Alkylphenols interact with asphaltenes via hydrogen bonding between amphiphile and the interaction sites (—OH and —N) of asphaltenes located at the periphery [37]. Surfactant examples of these categories include nonylphenol (NP), dodecyl benzene sulfonic acid (DBSA), and dodecyl resorcinol (DR), where the polarization of the polar head in their structures has a

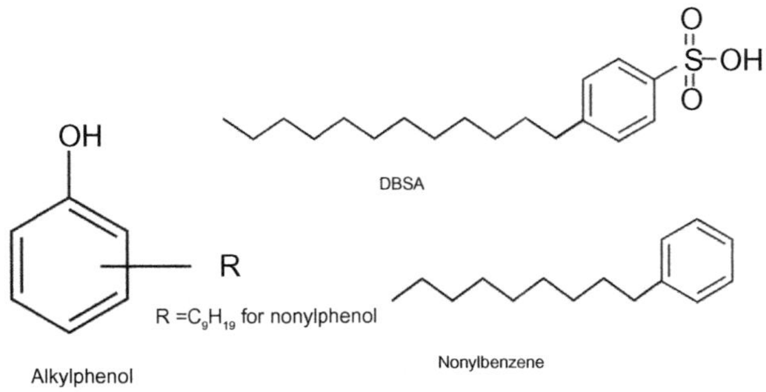

FIG. 18.7

Alkylbenzene derivatives used as asphaltene inhibitors and dispersants.

crucial role in the stability of micelles of asphaltenes [38]. Chang et al. [39] found that the inhibition efficiency is in the following order: DBSA > NP > NBDO > NB. The superior solubilizing ability of DBSA has been attributed to the presence of a larger and more polar sulfonic acid (—SO_3H) group in DBSA [39]. Spiecker et al. [40] ranked the effectiveness of some inhibitors to be in the following order: DR > DBSA > NP > resins (R) > toluene (T) > deasphalted crude oil (DO). Farag et al. [41] showed that DBSA, 4-nonylphenyl-polyethylene glycol, and synthetic cationic gemini surfactant: N2,N3-didodacyl-N2,N2,N3,N3-tetramethylbutane diaminium bromide displayed the highest capacity to inhibit asphaltene deposition, in addition to their nature as corrosion inhibitors. Ethoxylated phenols and ethoxylated alcohols were found to be promising AIs [42].

Generally speaking, DBSA is one of the commonly studied amphiphiles as an AI and dispersant. It is a cheap chemical with reported successful application in the field. Recently, molecular dynamics simulations showed that DBSA weakens the aggregates by breaking the hydrogen bonds between the molecules within the aggregates and forms a protective shell of DBSA molecules around the aggregates, which hinders the flocculation of these aggregates [43]. DBSA nanoemulsions have been introduced recently and have been found more effective, significantly reducing the inhibitor concentration by ~20 times with extended release time [44]. However, they have to be applied in a proper dose under the right conditions, as improper application of DBSA can cause aggregation and precipitation of asphaltenes.

- **Oligomers and polymeric inhibitors**

Alkyl phenol-aldehyde resin oligomers and polymers are regarded as the most investigated product in the oil industry [1]. Their efficiency depends on the temperature, molecular weight, and concentration [45]. Examples of this category include alkyl phenol formaldehyde resin, nonylphenolic formaldehyde resin (NPR) and native resins (NRs), cardanol (phenolic lipid), and polycardanol. Cardanol was reported to be an effective AI [46]; however, polycardanol was ineffective as an inhibitor or even worsened asphaltene flocculation, while it may work as a dispersant at high concentrations [47]. Also cardanol-aldehyde resins were found to be good AIs [6].

Polyester and polyamide/imide have been used as commercial AIs. Examples are (meth)acrylate copolymers, styrene/maleate ester, and alkene/maleate ester copolymers. Other commercially available chemicals include polyalkylene succinimide copolymers, poly(ethylene glycol) esters of maleic anhydride, and α-octadecene copolymers [1].

Polyisobutylene succinimides show good efficiency as asphaltene aggregation inhibitors and dispersants and as viscosity reducers [48]. Sulfonated polystyrene has shown flexible action as an AD/flocculant depending on its hydrophobicity and concentration [49].

- **Environmentally friendly inhibitors**

Various types of environmentally friendly products have been developed recently. Examples are poly(vinyltoluene-*co*-alpha-methylstyrene) [50], green polyalkylphenol-based resins [51], hydrophobic deep eutectic solvent, glycolipid biosurfactant [52], and others.

Other types of green chemicals include the natural products or extracts. Hu and Guo [53] showed that nut oils have good efficiency for inhibition processes. Rocha et al. [34] used vegetable oils including coconut essential oil, sweet almond, andiroba, and sandalwood oil for asphaltene inhibition. In an experimental coreflooding study, henna extract displayed promising asphaltene inhibition performance [54]. Moreira et al. [55] studied the phenolic compounds extracted from cashew-nut shell liquid.

Moreover, nanomaterials and ionic liquids are considered effective and environmentally friendly chemicals.

- **Nanoparticles**

Mohammadi et al. [56] investigated the capacity of nanoparticles TiO_2, ZrO_2, and SiO_2 in organic-based nanofluids for stabilizing asphaltene's particles in the oil. The results showed that TiO_2 nanoparticles in the acidic environment can greatly increase the stability of the asphaltenes. On the other hand, ZrO_2 and SiO_2 nanoparticles have less effect on the stability.

Cortés et al. [57] studied 12 nanoparticles as AIs for asphaltene damage in porous media. They concluded that the injection of nanofluids into porous media showed an inhibition in the agglomeration, precipitation, and deposition of asphaltenes on the rock surfaces. They were also able to restore production and improve the recovery due to their ability to adsorb and stabilize the asphaltene content of the system.

Zabala et al. [58] presented nanoparticles of nanosized metal oxides and high solubility in formation brines, which were used to adsorb and to carry asphaltenes along the produced oil up to the surface, avoiding flocculation and precipitation in the reservoir and the near wellbore. Applying these nanofluids in the reservoir restored the damaged permeability and improved production.

Lu et al. [59] demonstrated the use of Al_2O_3 nanoparticles for inhibition of asphaltene deposition during CO_2 flooding. The coreflood tests showed that the injection of Al_2O_3 nanofluid inhibits the deposition of asphaltenes in porous media (by keeping them in the bulk oil and preventing asphaltenes from accumulation at the CO_2-oil interface), thus avoiding permeability reduction.

Aminshahidy and Ahmadi [60] showed that CaO and SiO_2 nanoparticles decreased asphaltene precipitation in the presence of CO_2, where CaO showed better performance in reducing asphaltene precipitation.

Franco et al. [61] reported that the inclusion of SiO_2/cardanol nanocomposites during coreflooding tests enhanced crude oil mobility by altering the rock surface and reducing interfacial forces in the pores, helping to prevent formation damage.

Recently, single-walled carbon nanotubes (SWCNTs) were found to efficiently reduce the size of asphaltene aggregates and control their growth in live oil, while they reduced the WAT (by 7–9°C) besides significant modification of wax crystals (from dense spherical-like with average size of 756 nm in untreated oil into thin needle-like with an average size of 86 nm in treated oil) [62].

- **Ionic liquids**

This is another category of recently developed effective and green chemicals. Organic salts, ILs, may become strong H-bond donors, so can act as an asphaltene stabilizer with good dispersion properties [63]. Tetrachloride aluminate-based ([BMIM][$AlCl_4$]) and bromide-based (1-propyl boronic acid-3-alkylimidazolium bromides and 1-propenyl-3-alkylimidazolium bromides) ionic liquids have been used to treat heavy oils [64,65]. Hydrophobic ionic liquids, e.g., imidazolium- and pyridinium-based ionic liquids were found to reduce asphaltene aggregation and flocculation [63]. Ismail et al. [66] synthesized imidazolium asphaltene poly (ionic liquid) from asphaltene colloids in the petroleum sludge and used it to disperse and inhibit asphaltene aggregation.

18.2.2.3 Asphaltene dispersants (ADs)

Aromatic organic solvents have been found to be ADs. Examples are benzene, toluene, xylene, quinoline, alkylquinoline, dimethyl formamide, *n*-methyl pyrrolidone (NMP), pyridine, tetrahydrofuran [67–69].

Alkyl benzene-derived amphiphiles include the compounds stated to be AIs: DBSA, noyllphenol, dodecylresocinol, and others, as explained earlier in the AIs section. These amphiphiles can be applied as inhibitor or dispersant at the proper dosage. They are cheap and available commercially.

Another type of chemical includes organometallic polymers such as butyl substituted bis-di-organotin (IV) compound [70], and cupric-grafted polymer [71], which have been reported to be effective in upgrading heavy oil and exhibit multifunctioning.

Polymeric AIs are also common. Goual et al. [72] reviewed examples and mechanisms of polymeric dispersants, including poly(dodecyl phenol formaldehyde), propoxylated dodecyl phenol formaldehyde polymers, polyisobutylene succinic anhydride (PIBSA), polymers or *co*-polymers of maleic anhydride, dendronized polyamidoamine (PAMAM) polymers, terpolymer of styrene, *n*-butylacrylate and vinyl acetate, and polyolefin alkeneamines.

18.2.2.4 Factors affecting the performance of asphaltene control chemicals

Extensive research has been conducted to investigate different variables affecting the performance of the chemical additives. These factors include:

- Asphaltene characteristics: different acid/base groups, aromaticity, heteroatoms, and charges on the asphaltene surface play important roles in inhibitor specificity. Protic amphiphiles react with basic nitrogen species in asphaltenes, while amine amphiphiles react with acidic species in asphaltenes. The negatively charged asphaltenes tend to be dispersed by cationic amphiphiles, whereas the positively charged asphaltenes tend to be dispersed by anionic amphiphiles. Inhibitor effectiveness was also found related to heteroatom content [73–75]. Thus it is crucial to fully characterize the asphaltene molecules to find a good chemical inhibitor/dispersant match.
- Solvent conditions: experimental and theoretical simulation results showed that the inhibitor molecules with more polar head manifest higher adsorption on the asphaltene surface in a polar solvent, while at high concentration they prefer to get together in the bulk rather than adsorb on the surface [76].

- Chemical structures of the additives include:
  - Amphiphile head group: more polar head groups strengthen the attraction with the asphaltenes, which makes them more efficient in inhibiting and stabilizing asphaltenes. Also the head group type is important, for example alkylphenol amphiphiles are more efficient than alkylpyridine amphiphiles, even though alkylpyridine head groups are more polar, that may be attributed to the fact that alkylpyridines only associate with acidic asphaltene functional groups while alkylphenols can react with both acidic and basic groups [36,53,77,78].
  - The tails (alkyl chain group): the amphiphile tail should have minimum length (six carbons for alkylphenols) and increasing the tail length improves its effectiveness in sterically stabilizing asphaltenes; however, this increase may render their affinity towards asphaltenes [36,53,77,78].
- The amount of chemicals adsorbed on asphaltenes: inhibitor efficiency is directly related to the adsorption on the surface of asphaltene or its complexes. It has been found that a self-assembly of inhibitor molecules, induced by relative lyophilic or lyophobic interactions, may be a reason for inhibitor efficacy declining [76,79]. Higher concentrations of DBSA were found to exaggerate asphaltene flocculation.
- Polymer molecular weight. Too high a molecular weight may lead to solubility and coagulation issues and also can lead to flocculation of the asphaltenes.
- Other factors include temperature, chemical thermal stability, aging, and competing components in the produced fluids [79].

18.2.2.5 Testing of asphaltene prevention chemicals

18.2.2.5.1 Asphaltene dispersion test

The asphaltene dispersion test (ADT) is one of the most commonly used techniques to assess the efficacy of AIs and dispersants. The method is a simple, straightforward, and qualitative way of screening and ranking asphaltene control additives [2,79].

In this test, the oil samples are placed in graduated containers, where each sample unit is then treated with a dispersant at different concentration to evaluate the dispersant performance at different concentrations [79]. Then an asphaltene precipitant such as *n*-heptane is added to the well-mixed oil-dispersant sample at a precipitant-to-oil ratio of 40:1. Then the sample is aged for a specific time (usually 24 h) after which the level of asphaltene sedimentation is compared visually (through the cylinder marks). The lower the amount of sedimented asphaltenes, the better the performance of the dispersant [2]. Reversibility of asphaltene flocculation was studied by mixing the precipitant with the oil first, letting the sample settle for some time, and then adding the dispersant [80]. Screening of AIs using ADT is illustrated in Fig. 18.8.

Some drawbacks of the ADT include the following [79,81]:

- ADT is generally regarded as a qualitative technique. No accurate quantitative values can be obtained from measuring the height/level of the sediment, which might cause confusion, especially in aged samples (sediment more compact).
- The method is applied at room temperature and does not represent the high temperature and pressure conditions of the field.
- The high volumes of solvent/precipitant are not representative of the real situation from one side, and from the other side it induces the highest volumes of precipitates or largest precipitate sizes, which may not be ideal.

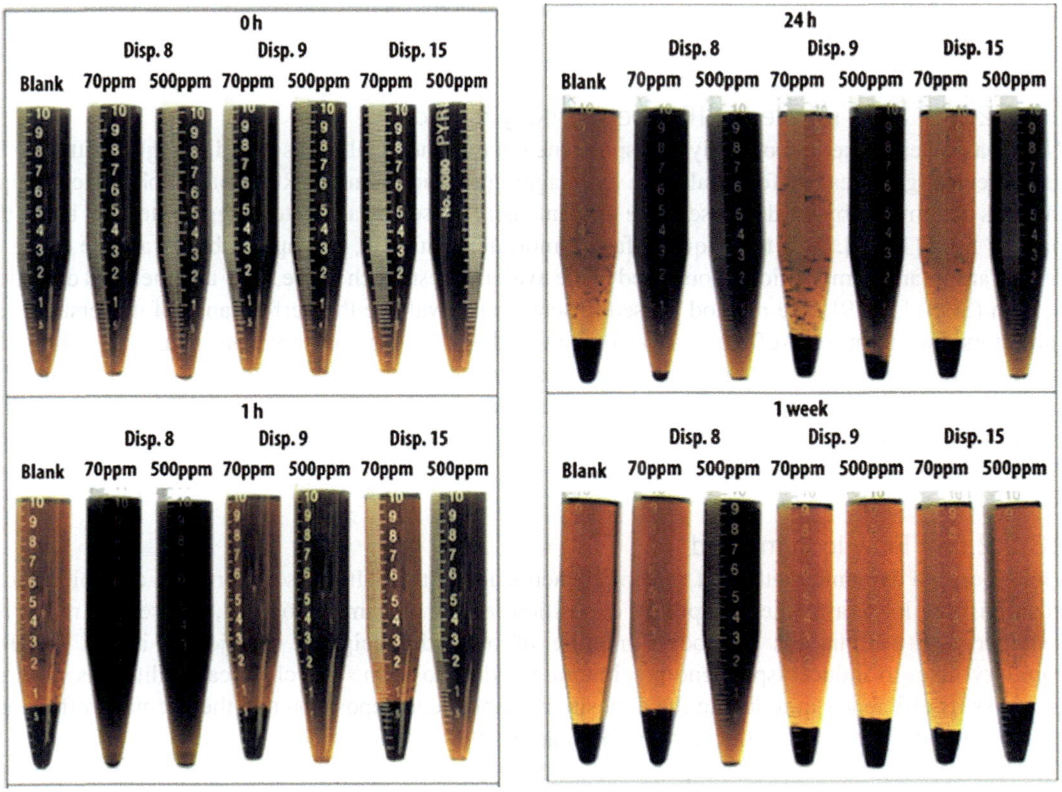

FIG. 18.8

ADT results for a crude oil treated with different doses of different dispersants and aged for different times.

Reprinted from A.A. Melendez-Alvarez, M. Garcia-Bermudes, M. Tavakkoli, R.H. Doherty, S. Meng, D.S. Abdallah, F.M. Vargas, On the evaluation of the performance of Asphaltene dispersants, Fuel 179 (September) (2016) 210–220. https://doi.org/10.1016/j.fuel.2016.03.056 (Elsevier Ltd), copyright (2016), with permission from Elsevier.

- The method screens chemical performance to inhibit precipitation rather than deposition. Real field deposition takes place by multistep diffusion on the metallic surface, while the method ranks the chemicals based on the reduction of sediment amounts/volumes.

Therefore more realistic and field-representative methods should be used before building and applying the treatment strategy.

18.2.2.5.2 Turbidity measurements

The turbidity test is based on the fact that a high turbidity value indicates a significant number of well-dispersed asphaltenes, indicating a good AD performance [82].

This method was used to determine the onset of asphaltene flocculation by titrating the sample with a precipitant [83] and to identify the effectiveness of different inhibitors to disperse precipitated asphaltenes [82–84].

The method is usually performed with a turbidity analyzer such as Turbiscan® or any equivalent and can be performed in accordance with ASTM D7061.

18.2.2.5.3 Near-infrared (NIR) spectroscopy method

NIR is a powerful method of studying asphaltene control chemicals. It is based on light scattering by asphaltene aggregates (preferentially the large aggregates) in the near IR region. Asphaltene precipitation is determined by the decrease in the light intensity caused by asphaltene aggregates that block the path of light [13,79]. This technique is faster, more reproducible, and applicable in a wide range of temperatures and compositions, compared to the available tests such as the ADT and the solid detection system (SDS) [13,79]. The method is used to screen and evaluate the performance of dispersants and inhibitors. Furthermore, Auflem et al. [13] showed that NIR spectroscopy can be used to follow the disintegration of asphaltene aggregates upon addition of chemicals. Melendez-Alvarez et al. [79] reported that, using this technique, they were able to reproduce the effect of inhibitors on the shift of the onset of asphaltene precipitation that is obtained by the SDS at high pressure and temperature, using dead-oil samples at ambient pressure.

18.2.2.5.4 Flow loop method

Flow loop is a dynamic method of screening chemical additives. It is also referred to as capillary deposition flow loop, asphaltene capillary deposition test, or dynamic loop test, as used by many researchers [85–87]. In this method, a mixture of oil and precipitant is injected into a metallic capillary tube, to induce asphaltene precipitation and deposition. The chemical additive is injected to assess its efficiency in reducing the deposition. Asphaltene deposition and the chemical efficiency are monitored by measuring the pressure drop across the capillary tube.

The method can be adjusted to perform the tests at the field conditions (P,T) and affords dynamic representation of the field conditions. Also it has good reproducibility in its results. However, the method is considered time consuming, and requires more sample volume and labor. In addition, the extent of deposition is better inferred from the mass of the deposits rather than the pressure drop. Furthermore, the amount of the inhibitor obtained from the test may be much higher than the amount required in field conditions [86], which leads to the conclusion that the capillary deposition test is not the most appropriate technique to assess the performance of asphaltene control chemicals.

18.2.2.5.5 Packed bed column

The packed bed column is one of the recently developed techniques to study asphaltene deposition [88,89].

The method comprises a glass (or any suitable material, e.g., PTFE) packed bed column packed with stainless steel spheres, where an oil-precipitant (n-heptane) mixture is maintained flowing upwards through the column. Then after the test time, the column is drained and the precipitated material is washed by chloroform and the mass of the precipitate is quantified after evaporating the chloroform and after excluding the occluded oil.

Some aspects of the packed bed column are [2,90]:

- Low sample volume
- Reproducible results

- Ability to determine the impacts of surface properties, e.g., surface roughness, surface area, corrosion, and coating materials, on the deposition of asphaltenes. Also, inhibitor-surface interactions can be elucidated.
- The deposited material can be characterized to understand the asphaltene-inhibitor interactions, which helps in developing and improving future chemical additives.

Vargas et al. [90] used a multisection packed bed column to study the performance of AIs. They demonstrated that the dispersive performance of the inhibitors is not directly related to their ability to prevent asphaltene deposition. In some cases, the chemicals with the highest dispersive efficiency produce the largest amount of asphaltene deposition. They also concluded that the dispersant with longer alkyl chains performed better; however, it is not necessarily the longer the chain, the better the performance, but rather the alkyl chain has to be of optimum length.

18.2.2.5.6 Solid detection system (SDS)

The SDSs technique has the advantage of evaluating live oils at high-pressure and high-temperature conditions to measure the AOP.

The method measures the transmittance of light at a fixed wavelength in the near infrared (NIR) region and provides high-pressure microscopy (HPM) images to detect and quantify AOP at a particular temperature and composition [91]. In evaluating the chemical additives, the SDS system determines the shift in the AOP.

However, the method has been subjected to criticism, saying that the SDS system cannot detect the true AOP because of the lack of sensitivity of the instrument, which can give misleading results on the performance of the chemical additives [79,90].

Fig. 18.9 illustrates AOP measurement using SDS with and without chemical dispersant at 93°C.

18.2.2.5.7 Organic solid deposition and control (OSDC) device

The method was introduced by Zougari et al. [92] to study wax and asphaltene deposition from live oil under turbulent flow conditions. This technique uses a closed cell based on the Taylor-Couette flow principles and takes into account important parameters, including shear, pressure, temperature, and flow time [92,93].

The OSDC cell is efficient in evaluating the chemical effectiveness in the prevention of actual asphaltene deposition; however, it requires a relatively large amount of oil, close to 1 L, especially for oils with low asphaltene content [2].

18.2.2.5.8 The coupon deposition test (CDT)

This method assesses the tendency of asphaltenes to deposit onto metal surfaces in the presence of liquid precipitant, water phase (brines), and other treating chemicals. The resulting deposits are quantified and compared for product performance in the prevention of asphaltene deposition versus chemically untreated samples. With the availability of metal surface deposition, CDT clearly has demonstrated the difference between deposition and precipitation behavior and the impact of the water phase and other treating chemicals on inhibitor performance [94].

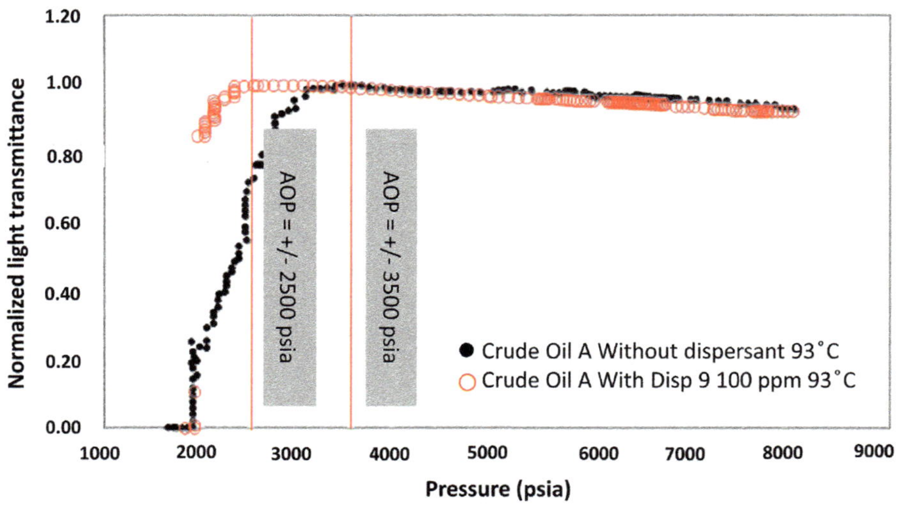

FIG. 18.9

AOP measurement using SDS system in absence and in presence of dispersant (100 ppm).

Reprinted from A.A. Melendez-Alvarez, M. Garcia-Bermudes, M. Tavakkoli, R.H. Doherty, S. Meng, D.S. Abdallah, F.M. Vargas, On the evaluation of the performance of Asphaltene dispersants, Fuel 179 (September) (2016) 210–220. https://doi.org/10.1016/j.fuel.2016.03.056 (Elsevier Ltd), copyright 2016, with permission from Elsevier.

18.2.2.5.9 The high pressure asphaltene rocking cell (ARC)

This test method facilitates the assessment of asphaltene stability at elevated temperatures and pressures, from 20°C to 150°C and up to 20,000 psi. The sample cell contains a magnetic ball that freely moves between the distal ends of the cell, where two magnetic sensors detect the ball motion. Asphaltene precipitation and/or deposition are indicated by a significant decrease in the rate of ball motion [94]. The rocking cell is similar to that used to study hydrate inhibitors. The technique provides the ability to simulate field conditions, at various gas oil ratios, different water cut, and brine compositions.

18.2.2.5.10 Microfluidic device

Microfluidic devices have been employed to simulate the asphaltene deposition phenomenon inside porous media in a small benchtop microscale system [95]. In this technique the asphaltene precipitation process can be visualized and recorded using an optical microscope, after crude oil and precipitant have been injected into the porous media. The precipitate properties are collected by the acquired images. Additionally, the pressure drop across the microchannel is measured by a differential pressure transducer to evaluate overall permeability impairment caused by the deposition [95–97].

Using this method, Lin et al. [97] was able to show that the addition of chemicals increased the initial deposition rate; however, the treatment with additives generated softer deposits that were more easily removed with the shear flow, resulting in a lower overall deposition rate. This finding highlights the limitations of dispersion tests in capturing the effect of chemical additives on the aging process of asphaltene deposits [95].

18.2.2.5.11 Quartz crystal microbalance (QCM)

QCM recently has received more attention concerning the adsorption of asphaltenes on solid surfaces. QCMs are ultrasensitive devices that work according to the piezoelectric effect. The resonance frequency (RF) generated by the quartz crystal is inversely proportional to the mass on top of the quartz crystal surface. Therefore the amount of asphaltene adsorption on the crystal surface can be determined by monitoring the changes in RF [95].

Joonaki et al. [98] used an HPHT QCM device to assess the effect of three different chemical additives on reducing asphaltene adsorption on the quartz crystal surface. Joonaki et al. [12] used an HPHT QCM technique for the first time to investigate the effect of waxes and related chemicals on asphaltene aggregation and deposition phenomena. The results showed that some wax control chemicals can prevent asphaltene deposition and that high wax content (especially the high molecular-weight waxes) can decrease the asphaltene deposition rate and shift the AOP.

18.2.2.5.12 Coreflooding tests

Coreflooding studies are usually performed to investigate the deposition of asphaltenes in porous media, to evaluate the efficiency of inhibitor chemicals for squeeze treatments and their effect on the formation permeability [99]. In those coreflooding tests, crude oil and an asphaltene prcipitant are coinjected into a core plug to induce asphaltene precipitation and deposition inside the porous medium. Differential pressures along the core plug are continuously measured with multiple taps to interpret the deposition inside each segment. These experiments can be conducted under high-pressure and high-temperature conditions [95,100].

Other methods include the use of the Heithaus titration method. The method can be applied according to ASTM D6703. Some automatic devices based on this technology have been made commercially and can be applied for dead and live oil as well. One of these systems is shown in Fig. 18.10, and the results are shown in Fig. 18.11.

- **Best practice for asphaltene chemical testing**

Based on the author's knowledge, there is no standard method to test asphaltene control chemicals in oilfields. Therefore operators have their own workflow and preferences for the deposition tests used for chemical screening, product selection, and field application. The variations between the testing methods and the lack of standardization contribute to confusing results, which are difficult to interpret and to obtain promising solutions for successful field implementation [95].

Generally speaking, the lab evaluation methods discussed earlier can be divided into two main categories: (1) the precipitation and aggregation tests like ADT, NIR, and SDS and (2) the deposition tests, like QCM, flow loop, and rocking cell [95].

Most of the current precipitation and aggregation techniques are based on the dispersion of asphaltenes. However, the efficacy of these methods is not well-corroborated by field success [101]. Furthermore, it was shown that a chemical with good dispersion efficiency does not necessarily reduce asphaltene deposition [86,94]. Therefore the industry is shifting towards the deposition tests, which are more realistic and representative of field conditions.

A few points to keep in mind when testing asphaltene treating chemicals [95] are:

- ADT is commonly conducted in field practice and its results are taken for granted in most cases. The method suffers from various inaccuracies, as mentioned already.

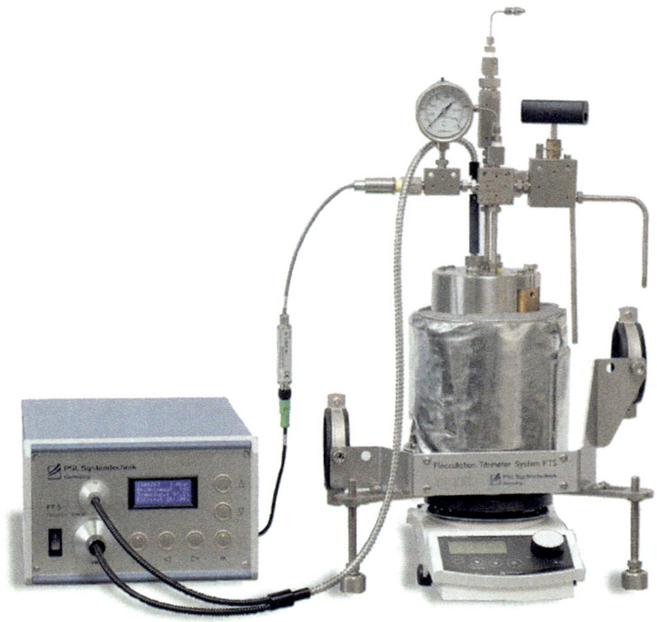

FIG. 18.10

Asphaltene flocculation titrimeter, with high pressure vessel.

Courtesy of PSL Systemtechnik GmbH (https://psl-systemtechnik.com/), reprinted with permission.

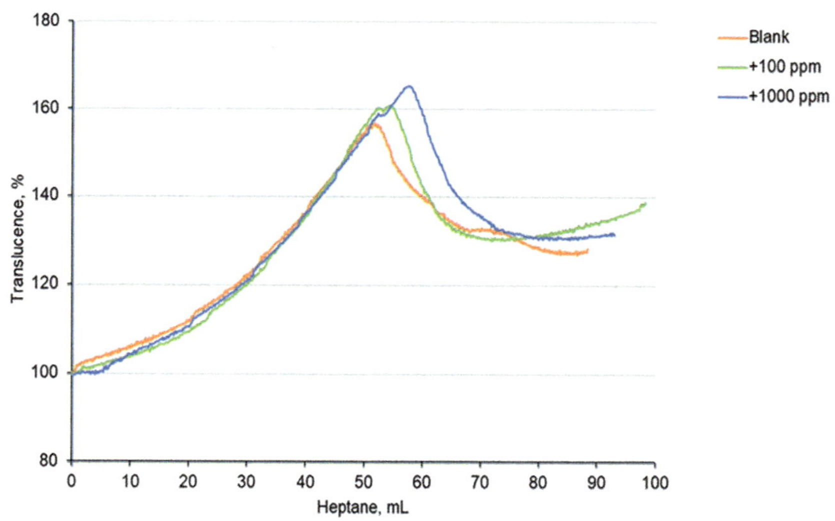

FIG. 18.11

Results from flocculation measurements, showing uninhibited and inhibited crude oil, where translucency increased with inhibition.

Courtesy of PSL Systemtechnik GmbH, (https://psl-systemtechnik.com/) reprinted with permission.

- The amount of inhibitor/dispersant derived from lab tests may vary from the required for lab applications (due to changes in T, P, hydrodynamics). They can be two or three times higher than the amount required for field application. Bouts et al. [20] reported that the chemical dose applied in the field was almost 20 times less than the lab-derived dose at ambient temperature.
- It is believed that there is no single testing method currently available that can find the proper chemistry for the field.
- A combination of different deposition tests is needed to create a degree of certainty in product selection.
- Lab deposition tests are more accurate than the aggregation and precipitation tests.
- Lab tests must be set up in such a way that they are representative of the field conditions.
- Besides efficiency, other features must be tested to assure good application of the chemical, including sludging and compatibility with produced fluids and other injected chemicals, thermal stability, foaming and emulsion tendencies, and corrosivity.

18.2.2.6 Application of asphaltene control chemicals

Asphaltene chemicals can be deployed in different ways to control asphaltene deposition. Continuous injection in the topsides and downhole of the tubing string is usually applied. Squeezing asphaltene chemicals into the producing formation has also been developed and optimized.

- **Continuous injection**

Continuous injection can be applied into the topsides and in the downhole. That requires full investigation of the process parameters, including measuring system BP AOP, streaming potential, and fluid compatibility.

In the topsides, the chemical should be injected upstream of the problem; locations of fluid mixing, locations of pressure drop (upstream of process BP, AOP), and locations of high shear or electrostatic forces. Treatment levels of good commercial AIs are often in the range of 20–100ppm [1].

In the downhole applications, the chemical is injected through a capillary tube to continuously inject AI below obstruction depth or below the measured AOP, so that the inhibitor can mix with the produced fluid up the string before it reaches the depth where asphaltene is supposed to deposit [102].

Chemicals can also be mixed with the proper solvents and injected down the gas lift system. Although some field applications showed success with applying the AIs through the gas lift, others recommended avoiding asphaltene chemical injection through the gas lift, due to the expected effect of gas lifting on asphaltene deposition.

- **Batch treatment (squeeze jobs)**

In this method, inhibitor fluid is squeezed into the formation through a producing well, where the inhibitor fluid is gradually dispersed with the reservoir fluid as the well resumes production [102].

Several tests are performed to assess the suitability of the chemical for a squeeze job and to improve the squeeze treatment performance, including the following [102]:

- The chemical efficiency as dispersant or inhibitor and expected minimum inhibitory dosage/concentration.
- The chemical compatibility with the reservoir fluids and the possibility of reservoir damage.
- The chemical adsorption, retention, release, and dispersion in the reservoir fluids.

Generally speaking, for squeeze purposes, the presence of acid groups in the chemical additive gives it the affinity to bind well to the rock, which increases the amount of adsorbed chemical, which extends the lifetime of the squeeze job [1]. The technical, economical application of squeeze jobs in the field is well established in the literature [20,22,103].

Using nanofluids was found to improve the squeeze job performance, elongate the squeeze lifetime, and effectively restore the reservoir permeability. Zabala et al. [58] reported restoring a damaged reservoir and increasing the production by 118,000 barrels of oil after 4 months of squeeze with the nanofluids.

- **Chemical application with the fracturing fluids and fracturing proppant**

Adding treating chemicals to the fracturing fluids is another effective way to deliver the chemicals to targeted locations in the reservoir where other treatments cannot reach. The method is efficient and cost effective, allowing proper and controlled use of the chemical.

A recent growing technology is adding solid treating chemicals to the fracturing proppant pack, where the solid chemical slowly desorbs from the proppant and is released in the reservoir depending on the temperature of the oil and the rate of production [104]. The applications of this technology have been reported in deepwater, tight gas, and coal bed methane formations [104].

Bestaoui-spurr et al. [105] reported that slow release asphaltene additive has no conductivity losses at increased loadings and can provide a cost-effective, long-term flow assurance solution and can be used in the same range as an intermediate strength proppant (ISP).

18.2.2.7 Recent advances in chemical inhibition of asphaltene deposits

The chemical methods of inhibiting asphaltenes have witnessed several advances. New polymeric products have been developed, after long-term dependence on nonpolymeric amphiphiles. Furthermore, nanomaterials have been introduced to the industry for different applications, and have been proven an efficient and long-lived treatment during squeeze jobs. Moreover, ionic liquids have also been reported. New, environmentally friendly chemicals have been developed using synthesized green polymers, natural extracts, nonchemicals, and ionic liquids.

Testing and evaluating of chemical additives have also witnessed advances, from the introduction of new techniques that simulate the actual field conditions and that help to understand the possible interactions, including AI interactions, asphaltene surface interactions, and inhibitor surface interactions. These techniques have shown that asphaltene deposition is a complex process that involves aggregation, precipitation, and deposition, all in competition, and the best chemical inhibitor is not the one with the best dispersion power, but the one that can participate in this competition and interfere with it, and stabilize the asphaltene particles in the solution.

Finally, more advances have appeared in applications of the asphaltene chemicals, with several reported successful squeeze jobs, and the introduction of solid inhibitors and their application with fracturing fluids.

18.2.3 Nonchemical prevention of asphaltene deposition

These methods use physical means like internal coatings, ultrasound waves, magnetic fields, or electric fields to control asphaltene aggregation and deposition. These methods can be used solely or combined with chemical methods to successfully prevent asphaltene deposition.

18.2.3.1 Coatings and deposit-preventing surfaces

The effectiveness of certain coating systems in the prevention of paraffin and scale deposition has been proven; however, success in asphaltene mitigation has not fully developed and has emerged only recently due to the complexity of the asphaltene deposition problem, which requires using all the possible methods to control it.

As a rule of thumb, the lower or more negative the Hamaker value (surface energy) in conjunction with a lower or more negative sticking tendency, the more effective the system will be in mitigating/ preventing asphaltene deposition [106].

Gonzalez ([107] reported that asphaltenes are least likely to stick on materials made of polytetrafluoroethylene (PTFE) and polyvinylidene fluoride (PVDF) due to their lower surface energy than metals.

Moradi et al. [108] showed that fluoropolymer coatings had the best performance among investigated samples, with the highest antiasphaltene deposition efficiency, followed by superhydrophilic phosphate coatings, and finally hydrophobic polyurethane and epoxy resin coatings showed weak antiasphaltene performance. However, difficulties have been reported in using fluoropolymer coatings to prevent internal deposition in oilfield conditions. Such coatings can be degraded by the abrasive production conditions [106,107].

Marshall et al. [109] reported that omniphobic-based coatings with a water-based polymeric surface that displays repellency to both oil and water phases can be used to prevent asphaltenes, wax gas, hydrates, and biofilms.

Edappillikulangara et al. [110] reported the first known case of its kind worldwide of using fiberglass-lined tubing for preventing aspahltene deposition, after unsuccessful use of conventional internal coatings. The glass-fiber reinforced (GRE) is known to have a smoother surface, lower zeta energy, and thermal insulation. The results over a period of 13–15 months showed improving well flowability and pressure drop by up to 74% reduction in WHP decline rate compared to preinstallation periods, a reduction in cleanout frequency with their associated operational and environmental challenges, with overall annual operational savings of US$519,750 per well, reduction of downtimes, and extended production time by 1 to 1½ months.

A diamond-like coating to reduce solids deposition was reported by Bethke et al. [111] and Heydrich et al. [112]. A commercial product is available on the market under the trademark LotusFlo provided by Shawcor.

StreaMax new fluoropolymer-based coating was installed in the Yagual 12 well in southern Mexico; it performed flawlessly, leading to efficient production with zero cleanups for the 4 years that the well-produced oil through the coated tubing. It also slightly improved production and improved flow, due to a lower friction factor with coated tubing [113].

18.2.3.2 Magnetic methods

Magnetic treatment is often used in certain oil industry applications for alterations in petroleum rheological properties [114]. A magnetic field can be used to mitigate asphaltenes indirectly, by improving the crude oil viscosity, or directly, by reducing asphaltene particles size (microcracking) and affecting the asphaltene aggregation (dis-aggregating the asphaltenes).

Marques et al. [115] demonstrated improved crude oil viscosity and reduced organic deposits upon applying magnetic fields. Tung et al. [116] studied the influence of magnetic field on paraffinic oil viscosity and deposition rate. They found that oils with a high asphaltene-resin content were highly affected due to their polarity. Also, they found that magnetically treated oils behaved as Newtonian fluids and there were reductions in wax crystal size and wax deposition rate.

Since magnetic fields have effects on polar species, it is believed asphaltene molecules present in oils are influenced by magnetic fields due to their asymmetrical charge distribution and polarity [115,116]. Mansoori et al. [114] showed that the onset of asphaltene aggregation in petroleum fluids is affected by the magnetic field and it is a function of its strength. The magnetic fields seems to produce changes in the average number of hydrogen bonds. These changes are due to changes in the orientation and the distances between asphaltene molecules.

18.2.3.3 Ultrasonic methods

Ultrasonic waves have a significant impact on the rheology of crude oil. Ultrasonic waves can play four main roles in crude oil [117,118]:

1. Dissolution of suspended soluble particles in crude oil,
2. Increasing the temperature of crude oil,
3. Disintegration/formation of long chain molecules and asphaltene flocs, and
4. Inversion of reversed micelles and creation of self-induced surfactant molecules.

The resultant of these roles in this study leads to a total increase in crude oil viscosity.

Gunel and Islam [119] showed that in the case of asphaltenic crude oil, ultrasonic waves can change the rheological properties of oil samples, but these alterations are not long lasting. Dunn and Yen [120] concluded that sonication would lead to both dehydrogenation and cracking of asphaltenes in bitumen. Shedid and Attallah [121] showed ultrasonic radiation decreases the size of asphaltene clusters, thus increasing the suspension of asphaltene in the crude oil and reducing its tendency to precipitate. Najafi and Amani [122] concluded that ultrasonic radiation inhibits asphaltene flocculation by particle disintegration and rejoining of free radicals to form new asphaltene particles. Amani et al. [123] concluded that ultrasonic waves have the ability to break down asphaltene conglomerates and, hence, reduce asphaltene content in crude oil samples and that the process is time dependent, as there is an optimum time for sonication at which highest asphaltene particle disintegration is achieved. Mohebbi et al. [124] presented a novel technique for prevention or elimination of asphaltene by spreading out the high energy ultrasound wave within the oil reservoir.

Despite all the research in this area, using ultrasonic radiation in preventing asphaltene deposition is still not common in oil and gas fields.

A commercial ultrasonic device has been developed to combat asphaltene aggregation in topsides, refineries, and petrochemicals by Hielscher Ultrasonics [125].

18.3 Asphaltene deposits removal

Asphaltene deposits removal is a challenging job. Asphaltene deposits usually deposit with paraffins; while many applications can be used to remove paraffins, some of these methods do not apply for asphaltenes, but can worsen the problem.

For example, asphaltene deposits are much harder than paraffin deposits, and paraffin deposits can be melted and dissolved with heat application, while asphaltene deposits will not melt using heat. Also paraffin deposits can redissolve in flowing crude oil, especially if the crude temperature was higher than WAT or pour point, while asphaltene deposits may be irreversible and will not dissolve in the crude oil, yet deasphalted oil can still dissolve asphaltenes. Additionally, paraffin deposits can be dissolved with a wide range of aliphatic and aromatic solvents and amphiphiles, whereas asphaltenes are preferably dissolved in aromatic solvents since they are insoluble in short chain aliphatic solvents, e.g., pentane and heptane, which can exaggerate the problem. Surprisingly, a recent study showed that aromatic-based asphaltene solvents xylene and toluene can also increase the crude oil WAT and affect paraffin deposition [126]. Thus the utmost caution while removing asphaltenes is necessary to avoid exaggerating the problem.

Asphaltene removal methods in the literature are basically chemical, depending on using different solvents and surfactants, and nonchemical, based on using physical, biological, or mechanical means.

18.3.1 Chemical asphaltene deposits removal

Various chemical solvents and surfactants have been used effectively to remove asphaltene deposits. In addition, thermal fusion packages have been reported to serve the same purpose. Table 18.4 summarizes some of the common solvents for asphaltene deposits.

18.3.1.1 Solvents

One of the major applications for the removal of asphaltenes is the use of solvents. Most asphaltene dissolvers are based on aromatic solvents, and sometimes additives can be added to enhance their performance.

Common and simple solvents are benzene, xylene mixtures, toluene, and propylbenzene, with xylene and toluene as the most commonly used solvents. The only problem with them is their low flash points, 28°C for xylene and 5°C for toluene. To avoid this safety risk, these solvents are usually mixed with diesel at different percentages. Additionally, Kuang et al. [127] demonstrated the solvency power of some chemicals and showed that a solvent wash by a toluene/diesel mixture (50/50 by volume) and

Table 18.4 Examples of common asphaltene deposit solvents.

Chemical agents	Examples
Organic solvents	Benzene Toluene Xylene Propylbenzene Carbon disulfide Imidazolines Glycol ethers Alkanolamines Pyridine Tetraline Naphthalene and its derivatives Quinolone Isoquinoline
Surfactants	DBSA Nonyl phenol (NP)

diesel significantly reduces the redissolution efficiency by 31.1% and 74.3% compared to toluene. The authors suggested that screening the chemical solvents according to their solubility parameters may not be practical for asphaltene removal.

Various case studies have reported the use of aromatic solvents, e.g., xylene and toluene with and without other surfactants (dispersants) and mutual solvents to remove asphaltene deposits [21,128–130].

Adding surfactants with polar heads (polymeric or DBSA) was found to enhance the dissolution performance of aromatic solvents like toluene for bulk and adsorbed asphaltene dissolution [131]. The acid surfactant DBSA was shown to completely dissolve asphaltenes by means of acid-base chemistry reactions at heteroatomic sites on the asphaltene molecules. The lab study showed that DBSA was found to effectively remove asphaltene deposits at concentrations roughly 10 times less than that required by toluene [132]. Additives with phenolic, alcohol, or amine functional groups were found to enhance the dissolution power of xylene and increase asphaltene desorption from mineral surfaces of reservoir rock after matching the overall cosolvent polarity with the asphaltene type [133]. Chang et al. [78] demonstrated the dissolution of asphaltenes using a mixture of alkane and alkylebenzene-derived amphiphiles, e.g., NP and DBSA. The factors affecting the rate of dissolution were:

- Type and concentration of the amphiphile
- Type of solvent (higher alkanes were more effective in asphaltene dissolution with amphiphiles)
- Temperature (increasing the temperature increases the dissolution process)
- Flow rate (higher flow rates of dissolver increase the dissolution process)

Al-Hajri- et al. [134] used different combinations to dissolve asphaltenes, including: (1) xylene in diesel used with 5 vol% mutual solvent, (2) n-methylpyrrolidone, diesel, surfactant, and an aromatic solvent, and (3) chemical solvent, mutual solvent, and diesel.

Pyridine was found to be a good asphaltene dissolver, but its use is limited due to toxicity and incompatibility with elastomers [135]. Other solvents that can be used are tetraline and naphthalene (and its derivatives like methyl naphthalene); quinolone and isoquinoline were also reported [131,136]. Bicyclic aromatic solvents are the best aromatic solvent, and monocyclic and bicyclic aromatic solvents are better than tricyclic and polycyclic solvents. The aromatic solvents interact via π-π orbital overlap with the aromatic components of asphaltene aggregates [131,136].

Other nonaromatic solvents are carbon disulfide, imidazolines [137], glycol ethers, and alkanolamines [138]. Samuelson [139] listed different solvents as alternatives to aromatics to dissolve organic deposits.

Hamzaoui et al. [140] investigated the effectiveness of solvent type and additives to dissolve asphaltenes and concluded that heteroatom, surfactants, and hydrogen donor compounds increased the solubility by a moderate degree, whereas aromatics increased the solubility by the highest degree.

18.3.1.2 Emulsion-based dissolvers

The main drawbacks of the organic solvents are their safety and environmental hazards (lower flash points and the presence of benzene, ethyl benzene, toluene, or xylene (BETX) components), and their relatively expensive prices. In addition, using the organic solvents can cause production problems, as the hydrocarbon-based solvents leave the formation in an oil-wet state after asphaltene removal instead of reestablishing the water-wet condition that acts as a barrier to slow down the deposition of the asphaltene on the formation. This makes the asphaltene aggregation and deposition more frequent, along with production decline and expensive treatments [141].

Some emulsion-based asphaltene dissolvers were introduced and found to outperform regular organic solvents [141,142]. The emulsion dissolvers comprise:

- Water
- Solvent (polar or nonpolar)
- Surfactant (to help in forming the emulsion and interacting with the asphaltenes to dissolve them)

The method was applied successfully in the field in Italy, where previous treatments based on hydrocarbon solvent had failed to make significant improvements [141].

An emulsified solvent system comprising n-methylpyrrolidone, fresh water, emulsifier, and an aromatic solvent was introduced [134]. These emulsion systems could also be designed with acid (HCl) as an aqueous phase to remove inorganic minerals that could be found in conjunction with asphaltene in the composite deposits. The mixture also might include an iron-controlling agent for better acid performance [143]. Acid emulsified in xylene was used to remove asphalting deposition and enhance well productivity [144]. An emulsified acid stimulation mixture was used to stimulate producing formations with problematic asphaltene deposits [145]. Quintero et al. [146] demonstrated the use of microemulsion propriety formulations to clean near wellbore asphaltene deposits.

Emulsions of 70/30 (w/o) were able to remove the deposited asphaltene in a porous medium despite having a lower volume fraction of toluene. They also swept the whole area of porous media, whereas toluene left more than half the area of the micromodel unswept [147].

These emulsion systems afford efficient cleaning performance with fewer safety and environmental hazards, as they have higher flashpoint and more enhanced production due to the fact that they leave the formation water wet, which prevents asphaltene redepositing and enhances oil flow.

18.3.1.3 Thermochemical packages

The same method of removing paraffin deposits using fused chemical reactions can be used in removing asphaltene deposits.

Change et al. [78] showed the use of such technology to remove paraffin and asphaltene deposits. The method is not intended to directly prevent asphaltene deposition, but it targets the wax deposition, which can affect the deposition of asphaltenes.

The chemical system consists of two components/pills and a catalyst, which are injected in a controlled manner into the well so that their mixing will be delayed until reaching the affected area. Upon mixing, the acid-catalyzed exothermic reaction will produce heat and reaction products. The heat generated is capable of melting/detaching the organic deposits, while the solvent and reaction product (ester) will further dissolve and disperse the deposits; in the meantime the produced gases afford some mechanical agitation. The technology has been applied in 23 wells in Malaysia, 2 wells in Brunei, and a pipeline in India. From the two case studies, average production increased to more than 100% with cost savings of up to US$100,000/well [148].

18.3.2 Nonchemical asphaltene deposits removal

These methods depend on using physical methods, i.e., ultrasonic or magnetic, biological, or mechanical methods, to remove asphaltenes.

18.3.2.1 Ultrasonic methods

Ultrasonic treatment has been proven to reduce viscosity to facilitate flow in pipelines and to clean asphaltene-clogged sections of oil wells and reservoirs [149,150].

Shedid [151,152] investigated the role of ultrasound waves in asphaltene damage removal in core plugs. The application of the ultrasound irradiation not only removed the deposited asphaltene from the surface, but also created small microcavities at the plug. In addition, he found that increasing the ultrasonic irradiation time interval and/or frequency significantly improved the damaged permeability.

An acoustic method was able to restore the permeability from damaged plug cores. The method is considered effective to treat damaged long sections of pay (horizontal wells), where chemical methods may be too expensive [153].

Paddok and Islam [154] reported a new method that involves the use of ultrasonic treatment coupled with foam treatment to remove deposits from the wellbore region. The technique is believed to work by viscosity reduction and resuspending the asphaltene particles in the solution due to microbubble generation. In situ foam formation is achieved using surfactant.

Other studies that have also showed the promising effect of ultrasonic techniques include [155,156].

18.3.2.2 Laser methods

Several tests were performed to study the use of lasers to clean asphaltene deposition and treat permeability damage [157,158]. The test results showed restored permeability of the tested core samples. It was thought that the laser energy works through different mechanisms:

- Altering the thermodynamics of the system, resulting in a reversible redissolution of the asphaltene deposits.
- The heat generated from the laser irradiation helps in redissolving the deposits.

Higher laser intensity provided more improvement of the rock-damaged permeability. They also reported that there is an optimum exposure time beyond which no additional improvement of the damaged core occurs.

18.3.2.3 Magnetic methods

Magnetic methods have been used to upgrade heavy crudes. Magnetic fields can help remove the asphaltene deposits by:

- Improving the surrounding crude oil viscosity and disaggregating the asphaltenes, which might help to reversibly redissolve these deposits [159].
- The magnetic field might cause degradation (microcracking) of the aged deposits.
- The electromagnetic field can induce localized heat that can be useful in removing the deposits.

Although not many studies have been published on this topic, it is open for future research considering the high prices of solvents and chemicals, and the rising trend of using green methods of treatment and reducing the use of chemicals.

18.3.2.4 Thermal treatments

This category of treating methods includes hot oiling, bottomhole heaters, water or steam, and the use of heat-liberating chemicals. Asphaltene deposits melt at relatively high temperatures, and recent studies have shown that asphaltene deposits are partially reversible under temperature changes, and precipitated asphaltenes should not be expected to redissolve, regardless of the magnitude of the temperature change [160,161]. Therefore thermal methods are intended to dissolve the coprecipitated waxes, which helps in removing asphaltene deposits and is also intended to decrease the surrounding matrix viscosity, and hence improve the solubility of the formed deposits, or at least dislodge or soften them. The good news here is that most asphaltene deposits are accompanied by wax deposits at different ratios, as shown in Table 18.1.

The different methods to apply heat include hot oiling, downhole heaters, direct electric heating, stem injection, and the use of heat-generating chemicals.

18.3.2.5 Biological methods

One of the recently proposed green and cost-effect remedial methods is the application of microorganisms capable of consuming the heavy hydrocarbon chains [162].

The possibility of asphaltene molecule degradation by microorganisms has been investigated by several researchers [163,164]. Asphaltenes were considered to be recalcitrant or exhibiting only very low rates of biodegradation [165]; however, Ferrari et al. [166] showed that microorganisms with higher metabolic capacity were able to consume several types of hydrocarbons with higher biodegradation resistance, including aromatics, resins, and asphaltenes.

Zekri and El-Mehaideb [167] showed that bacteria or steam can be used to remove asphaltene damage in carbonate reservoirs. They used two strains of bacteria, rounded and rod-shape type, both belonging to the Bacillus family, and showed that bacteria treatment resulted in a 41%–129% improvement in the asphaltene damaged core permeability.

Asadollahi et al. [168] used *Bacillus cereus* as a biosurfactant-producing bacterium in the oil-contaminated soil and oily sludge samples around the Kermanshah Oil Refining Company, which could degrade 40% of the asphaltenes after 60 days at 28°C.

Staphylococcus saprophyticus sp. and *B. cereus* were able to biodegrade 46.41% of asphaltenes at 45°C, salinity 160 $g L^{-1}$, pH 6.5, and 25 $g L^{-1}$ initial asphaltene concentration [169].

Biological enzymes, a new type of efficient plug-removal agent, had good application results in such countries as Venezuela and Indonesia, among others. Also, they showed very good results in removing asphaltene formation damage in wells in China [170].

18.3.2.6 Mechanical methods

Various mechanical methods have been applied in the field. Pipeline pigging, coiled tubing, milling, scratching, well gauging, getting, and others have been applied to remove asphaltene deposits. The mechanical method performance can be improved by combining it with solvents at high temperature. The mechanical methods are discussed in detail in Chapter 21.

Other methods were also mentioned in the literature. Honda et al. [171] demonstrated that seismic waves removed plugged asphaltene in the pore space. In Japan, a field consists of two production wells; the first well has been hit by 250 earthquakes since 2003, while the second well has been hit by 205 earthquakes since 2008. The reservoir responded to five earthquakes. Seismic intensity of the earthquake was 3 or higher, based on the Japan Meteorological Agency's 10-point scale.

18.3.3 Tips for efficient asphaltene deposits removal jobs

The removal of asphaltene deposits depends on many factors including:

- System design, operating parameters, and location of the deposit sample.
- Deposit thickness, hardness, age, and chemical composition.

- Availability of treating methods.
- Economics of the removal process.

To remove asphaltic deposits in flowlines, dissolvers are either batch treated or recirculated to the affected zone for a number of hours [1].

In the Deepwater Gulf of Mexico, 12 stimulation jobs were pumped down chemical injection umbilical tubing to remediate suspected asphaltene-related productivity decreases. All 12 jobs were done on dry tree wells but the procedures could be adapted for wet tree applications [172].

Also, the asphaltene removal chemicals can be formulated with the stimulation and fracturing fluids to dissolve asphaltenes in the reservoir, as indicated in the emulsion asphaltene dissolvers.

18.4 Asphaltene deposition control strategies and philosophies

Asphaltene deposition is a complex problem that requires a solid management strategy based on multi-step investigation, modeling, mitigation, and rigorous monitoring.

18.4.1 Risk assessment

This involves various investigation processes to identify the asphaltene risk sources and factors and estimate the risk level.

18.4.1.1 Fluids analysis

Fluids analysis and characterization represent the first step in the risk assessment. The tests that are usually conducted are:

- PVT tests
- SARA analysis
- Asphaltene onset precipitation, and onset pressure AOP
- De Boer plot
- Colloidal instability index (CII)
- Different crudes compatibility (Wiehe method)

In addition to the preceding tests, recent studies showed good characterization of the asphaltenes can be obtained using more sophisticated techniques, e.g., mass spectrometry and chromatography, which can be used to identify the acidic and basic fractions of petroleum asphaltenes, which helps in matching them with the proper inhibitor, i.e., an acidic inhibitor used to treat asphaltenes with dominant basic fractions. Also, alkyl chain matching was found to be crucial.

18.4.1.2 System pressure profile

Asphaltenes are known to be sensitive to pressure changes. Thus while investigating asphaltenes, recent and updated pressure measurement of the production system is mandatory. This updated pressure values will be utilized in the next steps of modeling.

18.4.1.3 Root cause investigation

Asphaltenes are known to be sensitive to many factors, e.g., pressure, temperature, hydrodynamics, fluids mixing, fluids composition, electrostatic and streaming potential effects, corrosion, and others. It is crucial to perform rigorous investigation to study the effect of the process variables and their effects on asphaltene stability. For example, the system parameters might be operating in the asphaltene safe zone, and crude oil is stable enough to be lifted to the surface facilities, while asphaltene deposition can be generated due to severe corrosion problems in the system or fluids incompatibility issues.

18.4.1.4 Modeling

After fluid characterization and system parameter updating, modeling is performed using one or more of the asphaltene prediction software packages.

18.4.1.5 Analogous field cases

Analogous field cases are commonly used when no samples are available in the field and they are also employed to use their experience to proactively optimize the system design and operations, to have fewer asphaltene problems. While it is highly recommended to study each case of asphaltenes independently, due to the different behavior of asphaltenes under different conditions, some case histories from analogous fields might lead to identifying the root cause of the problem, gaining knowledge of using some failed or successful treatments, whether chemical or nonchemical, highlight the importance of specific variables in the design, or highlight the importance of monitoring methods. The results from modeling and analogous fields combined can build a solid risk assessment, which can be used to plan the mitigation method.

18.4.2 Choosing the best management method

The mitigation method depends on the status of the field, e.g., whether it is still in early development stages where the design can be modified, in steady-state production, shutdown, etc.; based on the risk assessment results and the status of the field, the mitigation method can be applied.

Operational methods can be optimized, like modifying the system design by optimizing equipment sizing and operating parameters. Gas lift operations can be optimized, to limit their effect on asphaltenes. Avoiding mixing incompatible crudes and changing the incompatible crude routes are other options. Furthermore, adding heating systems to the system design can also be done during the early design steps. Moreover, the injection points and downhole umbilical lines are also designed if the risk assessment determines that asphaltenes are an imminent problem.

During production, the operating parameters are maintained at lower pressure and temperature loss; addition of asphaltene destabilizing fluids, e.g., acids and low chain aliphatic solvents, is avoided; mixing incompatible crudes is avoided; gas lift operations are optimized; sands and fines production are reduced; and corrosion problems are contained. Also, routine mechanical removal of deposits in the system is feasible. In cases where none of these methods can show any improvement or when the asphaltenes problem is severe, chemical inhibitors and dispersants are added. The chemicals can be used alone or in combination with these operational methods or with physical methods, depending on the severity of the problem and the operator evaluations.

Table 18.5 Comparison between the different methods of asphaltene mitigation.

Mitigation method	Description
Operational methods	- Used in asphaltene prevention - Requires system predesign or system modification - Case dependent
Chemical methods	- Used in prevention and removal - Used in all cases, and in different applications (e.g., downhole and topsides) and for different system designs - Wide variety of chemicals - Generally, do not require system modification - Can be expensive in severe cases or when high volumes of chemicals are required - Can be applied in combination with other methods - Chemicals can have side effects, e.g., incompatibility with other chemicals and produced fluids, or can enhance other production chemistry problems
Mechanical methods	- Used mainly in removal - Can be used in different cases, e.g., topsides, downhole - Can be used with other methods e.g. chemical methods - System design and accessibility dependent
Other nonchemical methods	- Internal coatings have showed success in field application - Biological methods have reported with success in field applications - Magnetic and ultrasonic methods were reported to improve crude oil viscosity - Thermal methods are used to dissolve paraffin and asphaltene deposits - These methods are not generally common in field application compared to chemical methods, except for thermal methods, which are common - They can be case dependent

During shutdown and unsteady state operations, more control methods are applied. To avoid unprecedented events, mechanical cleaning must be utilized on a routine basis, with the use of chemical dispersants. In severe cases, fluids displacement with inhibited fluids is also used to avoid fluids gelling and asphaltene particle flocculation, especially during shutdown.

Table 18.5 compares the different methods of mitigating asphaltenes.

18.4.3 Monitoring and assessment

After the mitigation method is applied, monitoring of the system is mandatory to evaluate the applied method efficiency. The monitoring method can be as simple as fluids analysis for asphaltene content and emulsion problems, routine system pressure and temperature monitoring, and flow rate monitoring, or more sensitive online monitoring devices may be needed that can give real-time measurement and assessment of the problem. The monitoring method is applied based on the severity of the problem and evaluation of the operator. Unsuccessful or insufficient treatments will require reassessing the problem to either optimize the current treatment or change the mitigation plan. Successful treatments will be monitored and optimized if needed.

Fig. 18.12 summarizes the asphaltene deposition management strategy.

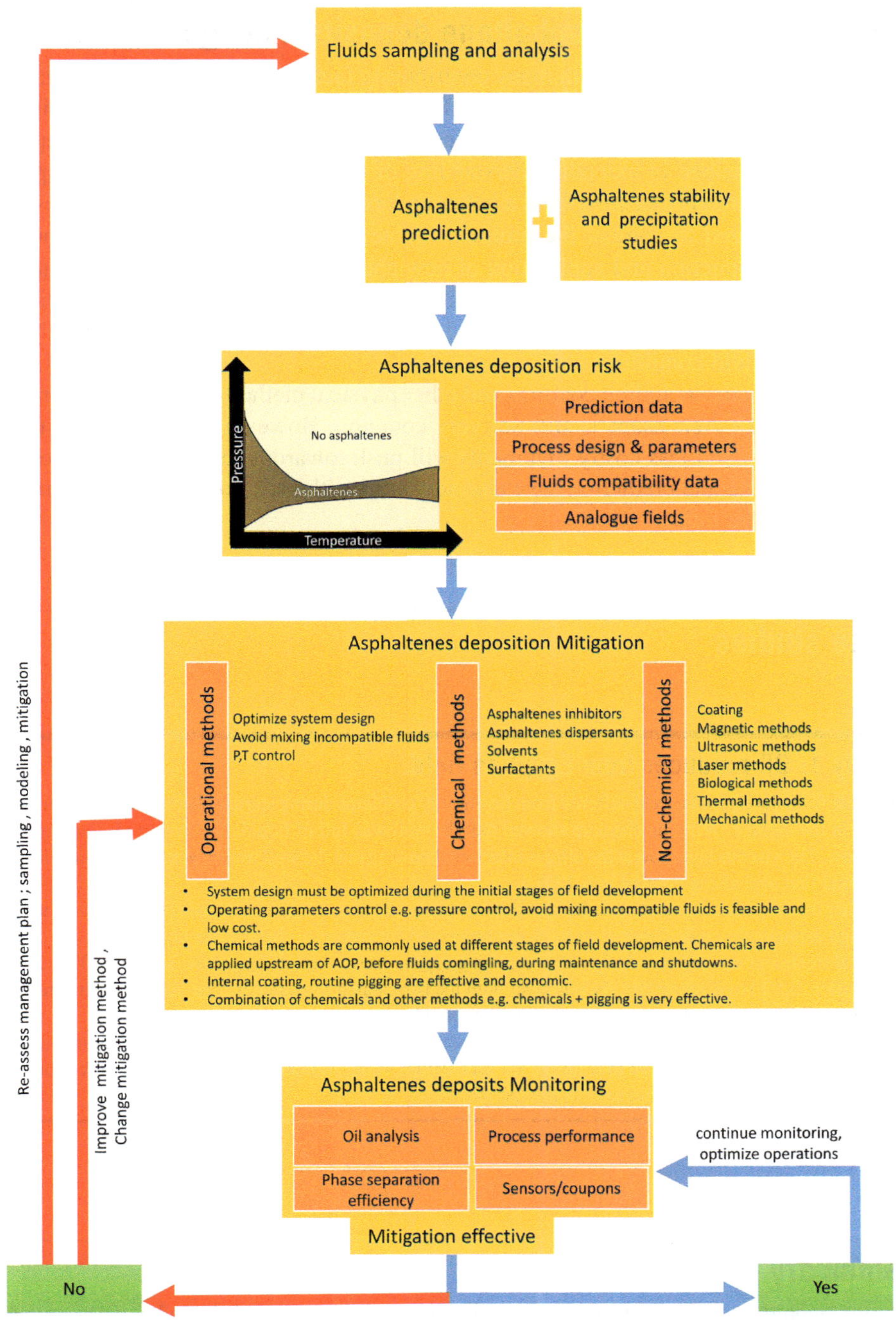

FIG. 18.12

Asphaltenes management strategy.

18.5 Recent advances in asphaltene deposits management methods

Asphaltenes are a complex problem with many contributing factors. This has opened the gate for many recent research advances. Initially, the methods of experimenting with asphaltene precipitation and deposition improved by introducing new techniques that are more sensitive and better represent field conditions, which has helped in indepth understanding of the problem origins, hence improving mitigation. Operational methods that stem from a better understanding of the asphaltene deposition mechanisms have been utilized to combat the problem. Some advances in the mitigation methods have emerged from the introduction and application of new chemicals, including polymeric and nonpolymeric types, nanomaterials, ionic liquids, and environmentally friendly chemicals. In addition, testing and screening these chemicals have been improved by applying deposition experiments that are more representative of the field conditions.

Nonchemical methods, including coatings and other physical methods, have shown some improvements as well, although their usage is still not very common. However, the inclination to use green methods for mitigation and less costly treatments will push towards more usage of these types.

Asphaltene modeling and prediction have also advanced with more robust packages that have been proven applicable in the field.

18.6 Case studies

Case study 1: Oil producer from Egyptian field.

An oil producer from the Egyptian field suffered from flowability problems during startup after shutdown, as no flow liquid was received at the surface. Wireline checkup showed that the well was full of liquid; however, a viscous layer was preventing the flow. The viscous layer was at the oil/gas interface, knowing that the well was on gas lift. By analyzing the crude oil, it was found that asphaltene content was 12%.

With further investigation, it was concluded that the root cause of the problem was the gas lift operations, which destabilized the high asphaltenic crude, leading to its flocculation and formation of the viscous layer.

A chemical treatment method was designed, comprising organic solvent (89.5%), mutual solvent (10%), and conventional demulsifier (0.5%) to disperse the viscous layer with the flocculated asphaltenes. The volume of this treatment was designed to be sufficient to displace the oil in the production tubing.

After this treatment was successfully applied, the well started to flow, and a polymeric AI was continuously injected into the well through the gas lift system to stabilize the asphaltenes. After months of production, the oil viscosity was noticeably improving with significant increase in production.

18.7 Summary

In this chapter, asphaltene aggregation and deposition mitigation methods have been reviewed. Asphaltene mitigation basically depends on the understanding of the problem's root causes and the factors affecting it. Asphaltene mitigation must start from the early stages of field development, as removing and cleaning asphaltene deposits is always higher in cost than preventing their aggregation and deposition. Mitigation of asphaltenes is based on operational, chemical, or nonchemical methods.

Operational methods are based on optimizing the process parameters to operate in asphaltene stabilized conditions, avoiding all detrimental factors that may disturb their thermodynamic stability, like pressure, temperature drop, adding incompatible fluids such as acids or short chain paraffins; also, routine mitigation of corrosion, avoiding high shear, avoiding high electrostatic charges, and optimizing gas lift operations are feasible in avoiding asphaltene instability.

Chemical methods are the most commonly used, and are based on using organic solvents (mainly aromatic solvents), chemical dispersants, or chemical inhibitors to stabilize asphaltenes and keep them in solution. These chemicals are amphiphiles in nature and interact with the asphaltene molecules through intermolecular forces to disturb the aggregation, flocculation, and crystal growth of asphaltene particles. Such chemicals have to be applied upstream of the problem.

Nonchemical methods are based on using internal coating, heating systems (direct electric heat, hot fluids circulation, heat-generating chemicals, insulation, steam), physical methods (magnetic, acoustic, electric), and biological methods to prevent and remove the asphaltene deposits.

A good understanding of the system design and operating parameters as well as routine monitoring of the running processes helps in detecting any early signs of asphaltene problems and proactively mitigating them or optimizing the treatment method already in place.

References

[1] M.A. Kelland, Production Chemicals for the Oil and Gas Industry, CRC Press, 2009, https://doi.org/10.1201/9781420092974.

[2] J. Kuang, A.T. Khaleel, J. Yarbrough, P. Pourreau, M. Tavakkoli, F.M. Vargas, Strategies for mitigation and remediation of asphaltene deposition, in: F. Vargas, M. Tavakkoli (Eds.), Asphaltene Deposition, CRC Press, Boca Raton, 2018, https://doi.org/10.1201/9781315268866 (Chapter 7).

[3] S. Wang, F. Civan, Preventing Asphaltene Deposition in Oil Reservoirs by Early Water Injection, Society of Petroleum Engineers, 2005, https://doi.org/10.2118/94268-MS.

[4] A. Al-Qasim, A. Al-Anazi, A.B. Omar, M. Ghamdi, Asphaltene Precipitation: A Review on Remediation Techniques and Prevention Strategies, Society of Petroleum Engineers, 2018, https://doi.org/10.2118/192784-MS.

[5] S. Afshari, R. Kharrat, M.H. Ghazanfari, Asphaltene Precipitation Study During Natural Depletion at Reservoir Conditions, Society of Petroleum Engineers, 2010, https://doi.org/10.2118/130071-MS.

[6] M.Z. Hasanvand, M. Ali Ahmadi, R.M. Behbahani, Solving asphaltene precipitation issue in vertical wells via redesigning of production facilities, Petroleum 1 (2) (2015) 139–145, https://doi.org/10.1016/j.petlm.2015.07.002 (KeAi Communications Co.).

[7] S.S. Bahram, R. Davood, T. Bahman, J. Mohammad, A novel method for mitigation of asphaltene deposition in the wellstring, Iran. J. Chem. Chem. Eng. 29 (2) (2010) 131–142.

[8] S. Mohammadi, F. Rashidi, S. Ali Mousavi-Dehghani, M. Hossein Ghazanfari, On the effect of temperature on precipitation and aggregation of asphaltenes in light live oils, Can. J. Chem. Eng. 94 (9) (2016) 1820–1829, https://doi.org/10.1002/cjce.22555 (Wiley-Liss Inc.).

[9] S. Verdier, H. Carrier, S.I. Andersen, J.L. Daridon, Study of pressure and temperature effects on asphaltene stability in presence of CO_2, Energy Fuel 20 (4) (2006) 1584–1590, https://doi.org/10.1021/ef050430g.

[10] S. Canbolat, S. Akin, A.R. Kovscek, Asphaltene deposition during steam-assisted gravity drainage: effect of non-condensable gases, Pet. Sci. Technol. 24 (1) (2006) 69–92, https://doi.org/10.1081/LFT-200041109.

[11] F. Salimi, J. Salimi, M. Abdollahifar, Investigation of asphaltene deposition under dynamic flow conditions, Pet. Sci. 13 (2) (2016) 340–346, https://doi.org/10.1007/s12182-016-0080-y (China University of Petroleum Beijing).

[12] E. Joonaki, A. Hassanpouryouzband, R. Burgass, A. Hase, B. Tohidi, Effects of waxes and the related chemicals on asphaltene aggregation and deposition phenomena: experimental and modeling studies, ACS Omega 5 (13) (2020) 7124–7134, https://doi.org/10.1021/acsomega.9b03460 (American Chemical Society).

[13] I.H. Auflem, T. Erik Havre, J. Sjöblom, Near-IR study on the dispersive effects of amphiphiles and naphthenic acids on asphaltenes in model heptane-toluene mixtures, Colloid Polym. Sci. 280 (8) (2002) 695–700, https://doi.org/10.1007/s00396-002-0660-9.

[14] M. Tojima, S. Suhara, M. Imamura, A. Furuta, Effect of heavy asphaltene on stability of residual oil, Catal. Today 43 (1998) 347–351.

[15] A.K.M. Jamaluddin, T.W. Nazarko, S. Sills, B.J. Fuhr, Deasphalted Oil—A Natural Asphaltene Solvent, Society of Petroleum Engineers, 1996, August 1, https://doi.org/10.2118/28994-PA.

[16] S.F. Alkafeef, F. Al-Medhadi, A.D. Al-Shammari, Asphaltene Remedial Technology Using Advanced Deasphalted Oil, Society of Petroleum Engineers, 2003, January 1, https://doi.org/10.2118/81570 MS.

[17] A.T. Khaleel, M.I.L. Abutaqiya, C.J. Sisco, F.M. Vargas, Mitigation of asphaltene deposition by re-injection of dead oil, Fluid Phase Equilib. 514 (June) (2020), https://doi.org/10.1016/j.fluid.2020.112552 (Elsevier B.V.).

[18] G.A. Mansoori, Remediation of asphaltene and other heavy organic deposits in oil wells and in pipelines, in: SOCAR Proceedings, 2010, 2010, pp. 12–23, https://doi.org/10.5510/OGP20100400039.

[19] M. Ihtsham Hashmi, B. Ghosh, Dynamic asphaltene deposition control in pipe flow through the application of DC potential, J. Pet. Explor. Prod. Technol. 5 (1) (2015) 99–108, https://doi.org/10.1007/s13202-014-0113-2 (Springer Verlag).

[20] M.N. Bouts, R.J. Wiersma, H.M. Muijs, A.J. Samuel, An evaluation of new asphaltene inhibitors; laboratory study and field testing, J. Pet. Technol. 47 (9) (1995) 782–787, https://doi.org/10.2118/28991-pa (Society of Petroleum Engineers (SPE).

[21] A. Yen, Y.R. Yin, S. Asomaning, Evaluating asphaltene inhibitors: laboratory tests and field studies, in: Paper Presented at the SPE International Symposium on Oilfield Chemistry, Houston, Texas, February, 2001, https://doi.org/10.2118/65376-MS.

[22] S.J. Allenson, M.A. Walsh, A Novel Way to Treat Asphaltene Deposition Problems Found in Oil Production, Society of Petroleum Engineers, 1997, https://doi.org/10.2118/37286-MS.

[23] E. Rogel, O. León, Study of the adsorption of alkyl-benzene-derived amphiphiles on an asphaltene surface using molecular dynamics simulations, Energy Fuel 15 (5) (2001) 1077–1086, https://doi.org/10.1021/ef000152f (American Chemical Society).

[24] H. Pan, A. Firoozabadi, Thermodynamic micellization model for asphaltene precipitation inhibition, AICHE J. 46 (2000) 416.

[25] F.M. Vargas, J.L. Creek, W.G. Chapman, On the development of an asphaltene deposition simulator, Energy Fuel 24 (4) (2010) 2294–2299, https://doi.org/10.1021/ef900951n.

[26] A.G. Di Lullo, T.P. Lockhart, C. Carniani, M. Tambini, A techno economic feasibility study of asphaltene inhibition squeeze treatments, in: Paper Presented at the European Petroleum Conference, The Hague, Netherlands, October, 1998, https://doi.org/10.2118/50656-MS.

[27] K. Karan, A. Hammami, M. Flannery, B. Artur Stankiewicz, Evaluation of asphaltene instability and a chemical control during production of live oils, Pet. Sci. Technol. 21 (2003) 629–645, https://doi.org/10.1081/LFT-120018543 (Marcel Dekker Inc.).

[28] X. Fu, A. Finley, S. Carpenter, Formation damage problems associated with CO_2 flooding, in: Formation Damage During Improved Oil Recovery, Elsevier, 2018, pp. 305–359, https://doi.org/10.1016/B978-0-12-813782-6.00008-7.

[29] M. Madhi, R. Kharrat, T. Hamoule, Screening of inhibitors for remediation of asphaltene deposits: experimental and modeling study, Petroleum 4 (2) (2018) 168–177, https://doi.org/10.1016/j.petlm.2017.08.001 (KeAi Communications Co.).

[30] A. Naseri, M. Nikazar, S.A.M. Dehghani, B. Dabir, O. Gohari, The role of inhibitors' molecular structure on asphaltene deposition in reservoir conditions, Pet. Sci. Technol. 29 (10) (2011) 988–999.

[31] M.M. Shadman, A.H. Saeedi Dehaghani, M. Vafaie Sefti, M. Dehghanizadeh, The effect of inhibitors on asphaltene precipitation in crude oil using the viscometric method, Energy Sources A 34 (9) (2012) 827–838, https://doi.org/10.1080/15567036.2011.654312.

[32] M.A. Karambeigi, R. Kharrat, S. Mahdavi, Investigation of inhibitors performance on different asphaltenic crude oils, Energy Sources A 37 (16) (2015) 1715–1722, https://doi.org/10.1080/15567036.2011.585380 (Taylor and Francis Inc.).

[33] E.Y. Sheu, D.A. Storm, M.M. De Tar, Asphaltenes in polar solvents, J. Non-Cryst. Solids 131 (133) (1991) 341–347, https://doi.org/10.1016/0022-3093(91)90326-2.

[34] L.C. Rocha, M.S. Ferreira, A.C. da Silva Ramos, Inhibition of asphaltene precipitation in Brazilian crude oils using new oil soluble amphiphiles, J. Pet. Sci. Eng. 51 (1–2) (2006) 26–36, https://doi.org/10.1016/j.petrol.2005.11.006.

[35] E. Hong, P. Watkinson, A study of asphaltene solubility and precipitation, Fuel 83 (2004) 1881–1887.

[36] C.L. Chang, H.S. Fogler, Stabilization of asphaltenes in aliphatic solvents using alkylbezene-derived amphiphiles. 1. Effect of the chemical structure of amphiphiles on asphaltenes stabilization, Langmuir 10 (1994) 1749–1757.

[37] L. Goual, M. Sedghi, X. Wang, Z. Zhu, Asphaltene aggregation and impact of alkylphenols, Langmuir 30 (19) (2014) 5394–5403, https://doi.org/10.1021/la500615k (American Chemical Society).

[38] T.A. Al Sahhaf, M.A. Fahim, A.S. Elkilani, Retardation of asphaltene precipitation by addition of toluene, resins, deasphalted oil and surfactant, Fluid Phase Equilib. 194–197 (2002) 1045–1057.

[39] C.-L. Chang, H. Fogler, Asphaltene stabilization in alkyl solvents using oilsoluble amphiphiles, in: SPE International Symposium on Oilfield Chemistry, Society of Petroleum Engineers, 1993.

[40] P.M. Spiecker, K.L. Gawrys, P.K. Kilpatrick, Aggregation and solubility behavior of asphaltenes and their subfractions, J. Colloid Interface Sci. 267 (2003) 178–193.

[41] R.M. Hegazey, R.K. Farag, M.I. Nessim, A.A. El-Feky, Y.M. Moustafa, A.I. Hashem, Dispersion of asphaltene precipitation in Egyptian crude oil using corrosion inhibitors, Pet. Sci. Technol. 32 (19) (2014) 2337–2344, https://doi.org/10.1080/10916466.2013.806931 (Taylor and Francis Inc.).

[42] A. Carlos da Silva Ramos, L. Haraguchi, F.R. Notrispe, W. Loh, R.S. Mohamed, Interfacial and colloidal behavior of asphaltenes obtained from Brazilian crude oils, J. Pet. Sci. Eng. 32 (2–4) (2001) 201–216, https://doi.org/10.1016/S0920-4105(01)00162-0.

[43] B. Jiang, R. Zhang, N. Yang, L. Zhang, Y. Sun, C. Jian, L. Liu, Z. Xu, Molecular mechanisms of suppressing asphaltene aggregation and flocculation by dodecylbenzenesulfonic acid probed by molecular dynamics simulations, Energy Fuel 33 (6) (2019) 5067–5080, https://doi.org/10.1021/acs.energyfuels.9b00821 (American Chemical Society).

[44] M. Alhreez, D. Wen, Controlled releases of asphaltene inhibitors by nanoemulsions, Fuel 234 (2018) 538–548.

[45] C. Ovalles, E. Rogel, H. Morazan, K. Chen, M.E. Moir, The use of nonylphenol formaldehyde resins for preventing asphaltene precipitation in vacuum residues and hydroprocessed petroleum samples, Pet. Sci. Technol. 34 (4) (2016) 379–385, https://doi.org/10.1080/10916466.2015.1126605 (Taylor and Francis Inc.).

[46] C.A. Guerrero-Martin, E. Montes-Páez, M.C. Khalil de Oliveira, J. Campos, E.F. Lucas, Calculating asphaltenes precipitation onset pressure by using cardanol as precipitation inhibitor: a strategy to increment the oil well production, in: Paper Presented at the SPE Trinidad and Tobago Section Energy Resources Conference, Port of Spain, Trinidad and Tobago, June, 2018, https://doi.org/10.2118/191275-MS.

[47] A.F. Lima, C.R.E. Mansur, E.F. Lucas, G. González, Polycardanol or sulfonated polystyrene as flocculants for asphaltene dispersions, Energy Fuel 24 (2010) 2369–2375, https://doi.org/10.1021/ef901031h.
[48] T.E. Chávez-Miyauchi, L.S. Zamudio-Rivera, V. Barba-López, Aromatic polyisobutylene succinimides as viscosity reducers with asphaltene dispersion capability for heavy and extra-heavy crude oils, Energy Fuel 27 (2013) 1994–2001, https://doi.org/10.1021/ef301748n (American Chemical Society).
[49] C.P.P. Mazzeo, F.A. Stedille, C.R.E. Mansur, A.C.S. Ramos, E.F. Lucas, Flocculation of asphaltenes by polymers: influence of polymer solubility conditions, Energy Fuel 32 (2) (2018) 1087–1095, https://doi.org/10.1021/acs.energyfuels.7b02577 (American Chemical Society).
[50] L. Ali, A novel inhibitor for controlling Iraqi asphaltene problems, in: 2017 International Conference on Environmental Impacts of the Oil and Gas Industries: Kurdistan Region of Iraq as a Case Study. EIOGI 2017, 17–19 April 2017, Koya-Erbil, Iraq. IEEE, 2018, pp. 37–41, ISBN: 9781509030620, https://doi.org/10.1109/EIOGI.2017.8267622.
[51] P.G.P. Fiorio, M.J.O.C. Guimarães, P.R. Seidl, Study on the inhibitory potential of green polyalkylphenol based resins for stabilizing asphaltene molecules, in: Rio Oil and Gas Expo and Conference, 20, 2020, pp. 27–28, https://doi.org/10.48072/2525-7579.rog.2020.027.
[52] A. Sanati, M.R. Malayeri, Hydrophobic deep eutectic solvent and glycolipid biosurfactant as green asphaltene inhibitors: experimental and theoretical studies, Energy Fuel 35 (6) (2021) 4791–4802, https://doi.org/10.1021/acs.energyfuels.0c03922.
[53] Y.-F. Hu, T.-M. Guo, Effect of the structures of ionic liquids and alkylbenzenederived amphiphiles on the inhibition of asphaltene precipitation from CO_2-injected reservoir oils, Langmuir 21 (2005) 8168–8174.
[54] E. Monjezi, A. Rostami, E. Joonaki, S. Reza Shadizadeh, New application of henna extract as an asphaltene inhibitor: an experimental study, Asia Pac. J. Chem. Eng. 11 (6) (2016) 1027–1034, https://doi.org/10.1002/apj.2043 (John Wiley and Sons Ltd).
[55] L.F.B. Moreira, E.F. Lucas, G. González, Stabilization of asphaltenes by phenolic compounds extracted from cashew-nut shell liquid, J. Appl. Polym. Sci. 73 (1) (1999) 29–34.
[56] M. Mohammadi, M. Akbari, Z. Fakhroueian, A. Bahramian, R. Azin, S.H. Arya, Inhibition of asphaltenes precipitation by TiO_2, SiO_2, and ZrO_2 nanofluids, Energy Fuel 25 (2011) 3150–3156.
[57] C.A. Franco, N.N. Nassar, M.A. Ruiz, P. Pereira-Almao, F.B. Cortés, Nanoparticles for inhibition of asphaltenes damage: adsorption study and displacement test on porous media, Energy Fuel 27 (6) (2013) 2899–2907, https://doi.org/10.1021/ef4000825.
[58] R. Zabala, E. Mora, O.F. Botero, C. Cespedes, L. Guarin, C.A. Franco, F.B. Cortes, J.E. Patino, N. Ospina, Nano-technology for asphaltenes inhibition in Cupiagua South Wells, in: Society of Petroleum Engineers—International Petroleum Technology Conference 2014, IPTC 2014: Unlocking Energy Through Innovation, Technology and Capability, vol. 2, Society of Petroleum Engineers, 2014, pp. 1260–1274, https://doi.org/10.3997/2214-4609-pdb.395.iptc-17344-ms.
[59] T. Lu, Z. Li, W. Fan, X. Zhang, Q. Lv, Nanoparticles for inhibition of asphaltenes deposition during CO_2 flooding, Ind. Eng. Chem. Res. 55 (23) (2016) 6723–6733, https://doi.org/10.1021/acs.iecr.5b04893 (American Chemical Society).
[60] Y. Ahmadi, B. Aminshahidy, Effects of hydrophobic CaO and SiO_2 nanoparticles on asphaltene precipitation envelope (APE): an experimental and modeling approach, Oil Gas Sci. Technol. 73 (2018) 56. https://doi.org/10.2516/ogst/2018052ff. ffhal-01930405f. Revue d'IFP Energies nouvelles, Institut Français du Pétrole.
[61] D. López, J.E. Jaramillo, E.F. Lucas, M. Riazi, S.H. Lopera, C.A. Franco, F.B. Cortés, Cardanol/SiO_2 nanocomposites for inhibition of formation damage by asphaltene precipitation/deposition in light crude oil reservoirs. Part II: nanocomposite evaluation and coreflooding test, ACS Omega 5 (43) (2020) 27800–27810, https://doi.org/10.1021/acsomega.0c02722 (American Chemical Society).
[62] S. Mohammadi, F.M. Alemi, Simultaneous control of formation and growth of asphaltene solids and wax crystals using single-walled carbon nanotubes: an experimental study under real oilfield conditions, Energy Fuel 35 (18) (2021) 14709–14724, https://doi.org/10.1021/acs.energyfuels.1c02244.

[63] Z. Rashid, C.D. Wilfredand, T. Murugesan, Effect of hydrophobic ionic liquids on petroleum asphaltene dispersion and determination using UV-visible spectroscopy, in: AIP Conference Proceedings, vol. 1891, American Institute of Physics Inc., 2017, https://doi.org/10.1063/1.5005451.

[64] Z.-X. Fan, T.-F. Wang, Y.-H. He, Upgrading and viscosity reducing of heavy oils by [BMIM][$AlCl_4$] ionic liquid, J. Fuel Chem. Technol. 37 (2009) 690–693 (12).

[65] M. Boukherissa, F. Mutelet, A. Modarressi, A. Dicko, D. Dafri, M. Rogalski, Ionic liquids as dispersants of petroleum asphaltenes, Energy Fuel 23 (2009) 2557–2564.

[66] M.E. El-Hefnawy, A.M. Atta, M. El-Newehy, A.I. Ismail, Synthesis and characterization of imidazolium asphaltenes poly(ionic liquid) and application in asphaltene aggregation inhibition of heavy crude oil, J. Mater. Res. Technol. 9 (6) (2020) 14682–14694, https://doi.org/10.1016/j.jmrt.2020.10.038.

[67] D. Espinat, J.C. Ravey, V. Guille, J. Lambard, T. Zemb, J.P. Cotton, Colloidal macrostructure of crude oil studied by neutron and X-ray small angle scattering techniques, J. Phys. IV 3 (C8) (1993) 181.

[68] J. Marugán, Characterization of the onset of asphaltenes by focus-beamed laser reflectance: a tool for chemical additives screening, in: Paper Presented at the 9th International Conference on Petroleum Phase Behavior and Fouling, Victoria, BC, 15–19 June, 2008.

[69] T. Takanohashi, S. Sato, R. Tanaka, Molecular dynamics simulation of structural relaxation of asphaltene aggregates, Pet. Sci. Technol. 21 (2003) 491.

[70] R. Hernández-Altamirano, V.Y. Mena-Cervantes, T.E. Chávez-Miyauchi, D.A. Nieto-Álvarez, M.A. Domínguez-Aguilar, L.S. Zamudio-Rivera, et al., New bis-di-organotin compounds derived from aminoacid-imine-hexadentate ligands. Multifunctional evaluation as corrosion inhibitors, antibacterials and asphaltenes dispersants/inhibitors, Polyhedron 52 (2013) 301–307, https://doi.org/10.1016/j.poly.2012.09.022.

[71] J.J. Li, X. Wang, X.D. Tang, F. Wang, D.Y. Qing, Effect of transition metal polymers with varying side alkyl chain on viscosity reduction of crude oil and aggregation behavior of asphaltene, Energy Fuel 29 (12) (2015) 7771–7780, https://doi.org/10.1021/acs.energyfuels.5b01013.

[72] E. Lowry, M. Sedghi, L. Goual, Polymers for asphaltene dispersion: interaction mechanisms and molecular design considerations, J. Mol. Liq. 230 (2017) 589–599, https://doi.org/10.1016/j.molliq.2017.01.028.

[73] H.H. Ibrahim, R.O. Idem, Interrelationships between asphaltene precipitation inhibitor effectiveness, asphaltenes characteristics, and precipitation behavior during *n*-heptane (light paraffin hydrocarbon)-induced asphaltene precipitation, Energy Fuel 18 (4) (2004) 1038–1048, https://doi.org/10.1021/ef0340460.

[74] J. Wang, C. Li, L. Zhang, G. Que, Z. Li, The properties of asphaltenes and their interaction with amphiphiles, Energy Fuel 23 (7) (2009) 3625–3631, https://doi.org/10.1021/ef801148y.

[75] P. Juyal, A.T. Yen, R.P. Rodgers, S. Allenson, J. Wang, J. Creek, Compositional variations between precipitated and organic solid deposition control (OSDC) asphaltenes and the effect of inhibitors on deposition by electrospray ionization Fourier transform ion cyclotron resonance (FT-ICR) mass spectrometry, Energy Fuel 24 (4) (2010) 2320–2326, https://doi.org/10.1021/ef900959r.

[76] M. Barcenas, P. Orea, E. Buenrostro-González, L.S. Zamudio-Rivera, Y. Duda, Study of medium effect on asphaltene agglomeration inhibitor efficiency, Energy Fuel 22 (3) (2008) 1917–1922, https://doi.org/10.1021/ef700773m.

[77] C.-L. Chang, H.S. Fogler, Stabilization of asphaltenes in aliphatic solvents using alkylbenzene-derived amphiphiles. 2. Study of the asphaltene-amphiphile interactions and structures using Fourier transform infrared spectroscopy and small-angle X-ray scattering techniques, Langmuir 10 (1994) 1758–1766.

[78] C.L. Chang, V. Nalwaya, P. Singh, H.S. Fogler, Removal of Asphaltene and Paraffin Deposits Using Micellar Solutions and Fused Reactions. Final Report, 1995–1997. United States, 1998, https://doi.org/10.2172/290886. Web.

[79] A.A. Melendez-Alvarez, M. Garcia-Bermudes, M. Tavakkoli, R.H. Doherty, S. Meng, D.S. Abdallah, F.M. Vargas, On the evaluation of the performance of Asphaltene dispersants, Fuel 179 (September) (2016) 210–220, https://doi.org/10.1016/j.fuel.2016.03.056 (Elsevier Ltd).

[80] P. Juyal, V. Ho, A. Yen, S.J. Allenson, Reversibility of asphaltene flocculation with chemicals, Energy Fuel 26 (5) (2012) 2631–2640, https://doi.org/10.1021/ef201389e.
[81] A. Khaleel, M. Abutaqiya, M. Tavakkoli, A.A. Melendez-Alvarez, F.M. Vargas, On the Prediction, Prevention and Remediation of Asphaltene Deposition, Society of Petroleum Engineers, 2015, https://doi.org/10.2118/177941-MS.
[82] E.L. Abrahamsen, Organic Flow Assurance: Asphaltene Dispersant/Inhibitor Formulation Development Through Experimental Design, University of Stavanger, Norway, 2012. Retrieved from: http://brage.bibsys.no/xmlui/handle/11250/182547.
[83] C. Chaogang, J. Guo, N. An, B. Ren, Y. Li, Q. Jiang, Study of asphaltene deposition from Tahe crude oil, Pet. Sci. 10 (1) (2013) 134–138, https://doi.org/10.1007/s12182-013-0260-y.
[84] K. Kraiwattanawong, H.S. Fogler, S.G. Gharfeh, P. Singh, W.H. Thomason, S. Chavadej, Effect of asphaltene dispersants on aggregate size distribution and growth, Energy Fuel 23 (3) (2009) 1575–1582, https://doi.org/10.1021/ef800706c.
[85] R.B. De Boer, K. Leerlooyer, M.R.P. Eigner, A.R.D. Van Bergen, Screening of crude oils for asphalt precipitation: theory practice and the selection of inhibitors, SPE Prod. Facil. 10 (1) (1995) 55–61.
[86] S. Gon, D.M. Fouchard, Modified asphaltene capillary deposition unit: a novel approach to inhibitor screening, Energy Fuel 30 (5) (2016) 3687–3692.
[87] X. Liu, T. Chen, P. Chen, H. Montgomerie, T.H. Hagen, B. Wang, X. Yang, Understanding Mechanisms of Scale Inhibition Using Newly Developed Test Method and Developing Synergistic Combined Scale Inhibitors, Society of Petroleum Engineers, 2012, https://doi.org/10.2118/156008-MS.
[88] C. Vilas Boas Favero, A. Hanpan, P. Phichphimok, K. Binabdullah, H.S. Fogler, Mechanistic investigation of asphaltene deposition, Energy Fuel 30 (11) (2016) 8915–8921, https://doi.org/10.1021/acs.energyfuels.6b01289.
[89] J. Kuang, A. Melendez-Alvarez, J. Yarbrough, M. Garcia-Bermudes, M. Tavakkoli, D.S. Abdallah, F.M. Vargas, Novel Way to Assess the Performance of Asphaltenes Dispersants on the Prevention of Asphaltene Deposition, 2017.
[90] J. Kuang, A.A. Melendez-Alvarez, J. Yarbrough, M. Garcia-Bermudes, M. Tavakkoli, D.S. Abdallah, S. Punnapala, F.M. Vargas, Assessment of the performance of asphaltene inhibitors using a multi-section packed bed column, Fuel (2019) 247–254, https://doi.org/10.1016/j.fuel.2018.11.059.
[91] A.A. Melendez Alvarez, Experimental Study of the Effect of Commercial Dispersants on the Precipitation, Aggregation and Deposition of Asphaltenes, Master Thesis, Rice University, 2015.
[92] M.I. Zougari, J. Ratulowski, A. Hammami, A. Kharrat, Live Oils Novel Organic Solid Deposition and Control Device: Wax and Asphaltene Deposition Validation, Society of Petroleum Engineers, 2005, January 1, https://doi.org/10.2118/93558-MS.
[93] M. Zougari, S. Jacobs, J. Ratulowski, A. Hammami, G. Broze, M. Flannery, A. Stankiewicz, K. Karan, Novel organic solids deposition and control device for live-oils: design and applications, Energy Fuel 20 (4) (2006) 1656–1663, https://doi.org/10.1021/ef050417w.
[94] J. Bae, D. Fouchard, S. Garner, J. Macias, Advantages of applying a multifaceted approach to asphaltene inhibitor selection, in: Offshore Technology Conference, 2016, https://doi.org/10.4043/27171-MS.
[95] S. Enayat, M. Tavakkoli, A. Yen, S. Misra, F.M. Vargas, Review of the current laboratory methods to select asphaltene inhibitors, Energy Fuel 34 (12) (2020) 15488–15501, https://doi.org/10.1021/acs.energyfuels.0c02554.
[96] Y. Zhuang, A. Goharzadeh, Y.J. Lin, Y.F. Yap, J.C. Chai, N. Mathew, F. Vargas, S.L. Biswal, Three dimensional measurements of asphaltene deposition in a transparent micro-channel, J. Pet. Sci. Eng. 145 (2016) 77–82.
[97] Y.-J. Lin, P. He, M. Tavakkoli, N.T. Mathew, Y.Y. Fatt, J.C. Chai, A. Goharzadeh, F.M. Vargas, S.L. Biswal, Characterizing asphaltene deposition in the presence of chemical dispersants in porous media micromodels, Energy Fuel 31 (2017) 11660–11668.

[98] E. Joonaki, R. Burgass, A. Hassanpouryouzband, B. Tohidi, Comparison of experimental techniques for evaluation of chemistries against asphaltene aggregation and deposition: new application of high-pressure and high-temperature quartz crystal microbalance, Energy Fuel 32 (2018) 2712–2721.

[99] M. Tavakkoli, S.R. Panuganti, V. Taghikhani, M.R. Pishvaie, W.G. Chapman, Asphaltene deposition in different depositing environments. Part 2. Real oil, Energy Fuel 28 (2014) 3594–3603.

[100] M. Tavakkoli, P. He, P.-H. Lin, S. Rezaee, M. Puerto, R. Doherty, J. Creek, J. Wang, G. Kusinski, J. Gomes, W. Chapman, S.L. Biswal, F.M. Vargas, Asphaltene deposition and fouling in reservoirs, in: Proceedings of the Offshore Technology Conference, Houston, TX, May 1–4, 2017, https://doi.org/10.4043/27933-MS.

[101] A. Montesi, R.A. Pinnick, S. Subramanian, J. Wang, J.L. Creek, Asphaltene management in GOM DW subsea development, in: Proceedings of the Offshore Technology Conference, Houston, TX, May 2–5, 2011, https://doi.org/10.4043/21587-MS.

[102] A. Albannay, T. Masuzawa, C.A. Dabbouk, Alternative Methods in Preventing Asphaltene Deposition, Society of Petroleum Engineers, 2010, https://doi.org/10.2118/137556-MS.

[103] A. Zouaoui, et al., Curative and preventive treatment of asphaltenes deposits on Hassi Messaoud oilfield, in: Paper Presented at 2nd International Symposium on Colloid Chemistry in Oil Production (ISCOP'97), September 1997, Rio de Janeiro, Brazil, 1997.

[104] D.V.S. Gupta, S. Szymczak, J.M. Brown, Solid production chemicals added with the frac for scale, paraffin and asphaltene inhibition, in: Paper Presented at the SPE Hydraulic Fracturing Technology Conference, The Woodlands, Texas, January 2009, Society of Petroleum Engineers, 2009. SPE 119393.

[105] N. Bestaoui-Spurr, F. DeBenedictis, M. Usie, S. Bhaduri, A new-intermediate strength proppant additive to combat asphaltene deposition for delivering long term flow assurance, in: Paper presented at the SPE International Conference and Exhibition on Formation Damage Control, Lafayette, Louisiana, USA, February, 2020, https://doi.org/10.2118/199283-MS.

[106] R.S. Lauer, The use of high performance polymeric coatings to mitigate corrosion and deposit formation in pipeline applications, in: Corrosion 2007, Tennessee, TN, 2007.

[107] D.L. Gonzalez, Modeling of Asphaltene Precipitation and Deposition Tendency Using the PC-SAFT Equation of State, Rice University, 2008. Retrieved from: https://scholarship.rice.edu/handle/1911/22140.

[108] S. Moradi, S. Amirjahadi, I. Danaee, B. Soltani, Experimental investigation on application of industrial coatings for prevention of asphaltene deposition in the well-string, J. Pet. Sci. Eng. 181 (2019), https://doi.org/10.1016/j.petrol.2019.05.046.

[109] M.A. Pickarts, E. Brown, J. Delgado-Linares, G. Blanchard, V. Veedu, C.A. Koh, Deposition mitigation in flowing systems using coatings, in: Paper Presented at the Offshore Technology Conference, Houston, Texas, May, 2019, https://doi.org/10.4043/29380-MS.

[110] R.E. Chinnappan, M. Tilang, R. Quttainah, G. Radhakrishnan, A. Fernandes, K. Rajendran, First time worldwide application of glass reinforced epoxy lined tubing for prevention of asphaltene deposition on tubing in oil wells—a case study from Kuwait, in: Paper Presented at the International Petroleum Technology Conference, Virtual, March, 2021, https://doi.org/10.2523/IPTC-21441-MS.

[111] G.K. Bethke, B. Snook, G. Herrera, A.E. Kelly, S. Joshi, S. Jain, S. Choudhary, A. Hammami, L. Evans, A novel coating to reduce solids deposition in production systems, in: Paper Presented at the Offshore Technology Conference, Houston, Texas, USA, April, 2018, https://doi.org/10.4043/28902-MS.

[112] M. Heydrich, S. Choudhary, A. Hammami, J. Ratulowski, O. Bukuaghangin, T. Charpentier, A. Neville, A systematic characterization of a novel diamond-like coating to reduce solids deposition in production systems, in: Paper Presented at the Offshore Technology Conference Asia, Kuala Lumpur, Malaysia, November, 2020, https://doi.org/10.4043/30383-MS.

[113] L.W. McKeen, Commercial applications and uses of fluorocoatings, in: Fluorinated Coatings and Finishes Handbook, Elsevier, 2016, pp. 307–333, https://doi.org/10.1016/b978-0-323-37126-1.00017-5.

[114] M.H. Khalaf, G. Ali Mansoori, C.W. Yong, Magnetic treatment of petroleum and its relation with asphaltene aggregation onset (an atomistic investigation), J. Pet. Sci. Eng. (2019) 926–933, https://doi.org/10.1016/j.petrol.2019.01.059 (Elsevier B.V.).

[115] L.C.C. Marques, N.O. Rocha, A.L.C. Machado, G.B.M. Neves, L.C. Vieira, C.H. Dittz, Study of paraffin crystallization process under the influence of magnetic fields and chemicals, in: Latin American and Caribbean Petroleum Engineering Conference, Society of Petroleum Engineers, 1997, pp. 1–6.

[116] N.P. Tung, N. Van Vuong, B.Q.K. Long, N.Q. Vinh, P.V. Hung, V.T. Hue, L.D. Hoe, Studying the mechanism of magnetic field influence on paraffin crude oil viscosity and wax deposition reductions, in: Paper Presented at the SPE Asia Pacific Oil and Gas Conference and Exhibition, Jakarta, Indonesia, April, 2001, https://doi.org/10.2118/68749-MS.

[117] K.S. Suslick, The chemical effects of ultrasound, Sci. Am. 260 (2) (1989) 80–86.

[118] A.S. Lee, X.W. Xu, T.F. Yen, Upgrading of heavy emde at low-temperature and ambient atmosphere, in: Proceedings of Fourth UNITARIUNDP International Conference an Heavy Crude and Tar Sands, vol. 5, 1989, pp. 109–116.

[119] G.O. Gunel, M.R. Islam, Alteration of asphaltic crude rheology with electromagnetic and ultrasonic irradiation, J. Pet. Sci. Eng. 26 (2000) 263–272.

[120] K. Dunn, T.Y. Yen, A plausible reaction pathway of asphaltene under ultrasound, Fuel Process. Technol. 73 (2001) 59–71.

[121] S.A. Shedid, S.R. Attallah, Influences of Ultrasonic Radiation on Asphaltene Behavior With and Without Solvent Effects, Society of Petroleum Engineers, 2004, https://doi.org/10.2118/86473-MS.

[122] I. Najafi, M. Amani, Asphaltene flocculation inhibition with ultrasonic wave radiation: a detailed experimental study of the governing mechanisms, Adv. Pet. Explor. Dev. 2 (2) (2011) 32–36, https://doi.org/10.3968/j.aped.1925543820110202.108.

[123] M. Amani, A. Retnanto, S. AlJuhani, M. Al-Jubouri, S. Shehada, R. Yrac, Investigating the role of ultrasonic wave technology as an asphaltene flocculation inhibitor, an experimental study, in: International Petroleum Technology Conference, 2015, December 6, https://doi.org/10.2523/IPTC-18473-MS.

[124] R. Rezaei Dehshibi, A. Mohebbi, M. Riazi, M. Niakousari, Experimental investigation on the effect of ultrasonic waves on reducing asphaltene deposition and improving oil recovery under temperature control, Ultrason. Sonochem. 45 (July) (2018) 204–212, https://doi.org/10.1016/j.ultsonch.2018.03.023. Elsevier B.V.

[125] Reduced Asphaltene Deposition with Power Ultrasonics, Hielscher Ultrasonics GmbH, 2022. Retrieved from: https://www.hielscher.com/reduced-asphaltene-deposition-with-power-ultrasonics.htm. (Accessed December 2021).

[126] J.I.S. Aguiar, A.A. Pontifes, A. Nerris, J. Rogers, A. Mahmoudkhani, Impact of solvent treatments for asphaltenes on wax deposition and an efficient alternative with green surfactants, in: Paper Presented at the Offshore Technology Conference, Houston, Texas, USA, May, 2020, https://doi.org/10.4043/30695-MS.

[127] J. Kuang, J. Yarbrough, S. Enayat, N. Edward, J. Wang, F.M. Vargas, Evaluation of solvents for in situ asphaltene deposition remediation, Fuel 241 (April) (2019) 1076–1084, https://doi.org/10.1016/j.fuel.2018.12.080 (Elsevier Ltd).

[128] S. Schantz, W. Stephenson, Asphaltene deposition: development and application of polymeric asphaltene dispersants, Soc. Pet. Eng. J. (1991), https://doi.org/10.2118/22783-MS.

[129] D. Thomas, et al., Controlling asphaltene deposition in oil wells, Soc. Pet. Eng. J. (1995), https://doi.org/10.2118/25483 -PA.

[130] S. Fakher, M. Ahdaya, M. Elturki, A. Imqam, Critical review of asphaltene properties and factors impacting its stability in crude oil, J. Pet. Explor. Prod. Technol. (2020), https://doi.org/10.1007/s13202-019-00811-5 (Springer).

[131] L. Barberis Canonico, A. Del Bianco, G. Piro, F. Stroppa, C. Carniani, E.I. Mazzolini, A comprehensive approach for the evaluation of chemicals for asphaltene deposit removal, in: Recent Advances in Oilfield Chemistry, Chemicals in the Oil Industry V, vol. 220, The Royal Society of Chemistry, Ambleside, Cumbria, UK, 1994.

[132] S.M. Hashmi, A. Firoozabadi, Effective removal of asphaltene deposition in metal-capillary tubes, SPE J. 21 (2016) 1747–1754. Society of Petroleum Engineers https://doi.org/10.2118/166404-PA.

[133] L. Minssieux, Removal of asphalt deposits by cosolvent squeeze: mechanisms and screening, SPE J. 6 (1) (2001) 39 (SPE 69672).

[134] N.M. Al-Hajri, A.A. Taq, M.D. AlAjmi, A.A. Al-Ghamdi, Laboratory Analysis of Several Solvent Systems to Dissolve Wellbore Organic Deposits, Society of Petroleum Engineers, 2016, https://doi.org/10.2118/184083-MS.

[135] W. Kleinitz, Asphaltene precipitates in oil production wells, in: Paper Presented at the 8th Oil Field Chemical Symposium, NIF, Geilo, Norway, 1997.

[136] C.W. Benson, R.A.S. Simcox, I.C. Huldal, Tailoring aromatic hydrocarbons for asphaltene removal, in: Chemicals in the Oil Industry IV, vol. 215, Royal Society of Chemistry, 1991.

[137] H.L. Becker, B.W. Wolf, U.S. Patent 5504063, 1996.

[138] D. Fluck, A. Sehgal, S. Trivedi, R.T. Pabalan, C. Aymes, International Patent Application WO/2012/078193, 2012.

[139] M.L. Samuelson, Alternatives to aromatics for solvency of organic deposits, in: SPE Formation Damage Control Symposium, 1992, https://doi.org/10.2118/23816-ms.

[140] B. Hamzaoui, A.M. Moajil, A. Al Ghamdi, Structure-Solubility Correlation for Asphaltene Deposit Removal, Society of Petroleum Engineers, 2018, https://doi.org/10.2118/192277-MS.

[141] S.C. Lightford, E. Pitoni, L. Mauri, F. Armesi, Development and Field Use of a Novel Solvent Water Emulsion for the Removal of Asphaltene Deposits in Fractured Carbonate Formations, Society of Petroleum Engineers, 2006, https://doi.org/10.2118/101022-MS.

[142] K.A. Frost, R.D. Daussin, M.S. Van Domelen, New, Highly Effective Asphaltene Removal System With Favorable HSE Characteristics, Society of Petroleum Engineers, 2008, https://doi.org/10.2118/112420-MS.

[143] L. Salgaonkar, A. Danait, Environmentally Acceptable Emulsion System: An Effective Approach for Removal of Asphaltene Deposits, Society of Petroleum Engineers, 2012, https://doi.org/10.2118/160877-MS.

[144] W.A. Fattah Ahmed, H.A. Nasr-El-Din, Acid Emulsified in Xylene: A Cost-Effective Treatment to Remove Asphalting Deposition and Enhance Well Productivity, Society of Petroleum Engineers, 2008, https://doi.org/10.2118/117251-MS.

[145] D. Appicciutoli, R.W. Maier, P. Strippoli, A. Tiani, L. Mauri, Novel Emulsified Acid Boosts Production in a Major Carbonate Oil Field with Asphaltene Problems, Society of Petroleum Engineers, 2010, https://doi.org/10.2118/135076-MS.

[146] L. Quintero, R. Ponnapati, M.J. Felipe, Cleanup of organic and inorganic wellbore deposits using microemulsion formulations: laboratory development and field applications, in: Offshore Technology Conference, 2017, https://doi.org/10.4043/27653-MS.

[147] N. Shahsavar, M. Riazi, M.R. Malayeri, Removal of asphaltene deposition in porous media using emulsified solvents—a visual study, J. Pet. Sci. Eng. 191 (August) (2020), https://doi.org/10.1016/j.petrol.2020.107207 (Elsevier B.V.).

[148] K. Singh, A. Saidu Mohamed, S.S. Alian, M. Ismail, M. Anwar, W.W. Mohamad, S. Abdul Ghani, Thermo chemical in-situ heat generation technique to remove organic solid deposition: effective tool for production enhancement and flow assurance, in: Paper Presented at the Offshore Technology Conference, Houston, Texas, USA, May 2013, 2013, https://doi.org/10.4043/23933-MS.

[149] M.R. Islam, Potential of ultrasonic generators for use in oil wells and heavy crude oil/bitumen transportation facilities, in: E.Y. Sheu, O.C. Mullins (Eds.), Asphaltenes, Springer, Boston, MA, 1995.

[150] U.K. Gollapudi, S.S. Bang, M.R. Islam, Ultrasonic Treatment for Removal of Asphaltene Deposits during Petroleum Production, Society of Petroleum Engineers, 1994, https://doi.org/10.2118/27377-MS.
[151] S.A. Shedid, A novel technique of asphaltene deposition treatment using ultrasonic irradiation, J. Pet. Sci. Technol. 20 (9–10) (2002) 1097–1118.
[152] S.A. Shedid, An ultrasonic irradiation technique for treatment of asphaltene deposition, J. Pet. Sci. Eng. 42 (1) (2004) 57–70.
[153] P.M. Roberts, V. Adinathan, M.M. Sharma, Ultrasonic removal of organic deposits and polymer induced formation damage, in: Paper presented at the SPE Formation Damage Control Symposium, Lafayette, Louisiana, February 1996, https://doi.org/10.2118/31129-MS.
[154] J. Paddock, M.R. Islam, A new technique for cleaning horizontal wellbores, in: Paper Presented at the Technical Meeting/Petroleum Conference of The South Saskatchewan Section, Regina, October 1999, https://doi.org/10.2118/99-98.
[155] X. Xu, T. Bao, Research on the removal of near-well blockage caused by asphaltene deposition using Sonochemical method, Ultrason. Sonochem. 64 (June) (2020), https://doi.org/10.1016/j.ultsonch.2019.104918. Elsevier B.V.
[156] A.N. Tutuncu, R. Joha, An experimental study for removal of near-wellbore asphaltene deposits using ultrasonics, in: Paper Presented at the 2008 SEG Annual Meeting, Las Vegas, Nevada, November, 2008.
[157] A.Y. Zekri, S.A. Shedid, H. Alkashef, A Novel Technique for Treating Asphaltene Deposition Using Laser Technology, Society of Petroleum Engineers, 2001, https://doi.org/10.2118/70050-MS.
[158] A.Y. Zekri, S.A. Shedid, H. Alkashef, A new technique for treatment of permeability damage due to asphaltene deposition using laser technology, J. Pet. Sci. Eng. 59 (3–4) (2007) 300–308, https://doi.org/10.1016/j.petrol.2007.05.005.
[159] C. Jiang, et al., Probing disaggregation of crude oil in a magnetic field with terahertz time-domain spectroscopy, Energy Fuel 28 (1) (2014) 483–487.
[160] W. Chaisoontornyotin, A.W. Bingham, M.P. Hoepfner, Reversibility of asphaltene precipitation using temperature-induced aggregation, Energy Fuel 31 (4) (2017) 3392–3398, https://doi.org/10.1021/acs.energyfuels.6b02344 (American Chemical Society).
[161] S. Shedid, E.A. Abbas, An Experimental Approach of the Reversibility of Asphaltene Deposition under Dynamic Flow Conditions, Society of Petroleum Engineers (SPE), 2007, https://doi.org/10.2523/92159-ms.
[162] S. Iraji, S. Ayatollahi, Experimental investigation on asphaltene biodegradability using microorganism: cell surface properties' approach, J. Pet. Explor. Prod. Technol. 9 (2) (2019) 1413–1422, https://doi.org/10.1007/s13202-018-0537-1.
[163] M.V. Reddy, M.P. Devi, K. Chandrasekhar, R.K. Goud, S.V. Mohan, Aerobic remediation of petroleum sludge through soil supplementation: microbial community analysis, J. Hazard. Mater. 197 (2011) 80–87.
[164] Y. Wang, N.F.Y. Tam, Microbial remediation of organic pollutants, in: C. Sheppard (Ed.), World Seas: An Environmental Evaluation, second ed., Academic Press, MA, 2019.
[165] J.G. Leahy, R.R. Colwell, Microbial degradation of hydrocarbons in the environment, Microbiol. Rev. 54 (3) (1990) 305–315.
[166] M. Ferrari, C. Albornoz, E. Neirotti, Biodegradability in soil of residual hydrocarbons in petroleum tank bottoms, Rev. Argent. Microbiol. 26 (1993) 157–170.
[167] A.Y. Zekri, R. El-Mehaideb, Steam/bacteria to treatment of asphaltene deposition in carbonate rocks, J. Pet. Sci. Eng. 37 (1–2) (2003) 123–133, https://doi.org/10.1016/s0920-4105(02)00315-7.
[168] L. Asadollahi, H. Salehizadeh, N. Yan, Investigation of biosurfactant activity and asphaltene biodegradation by *Bacillus cereus*, J. Polym. Environ. 24 (2016) 119–128, https://doi.org/10.1007/s10924-016-0754-y.
[169] Y. Shahebrahimi, A. Fazlali, H. Motamedi, S. Kord, A.H. Mohammadi, Effect of various isolated microbial consortiums on the biodegradation process of precipitated asphaltenes from crude oil, ACS Omega 5 (7) (2020) 3131–3143, https://doi.org/10.1021/acsomega.9b02056 (American Chemical Society).

[170] Y. Wang, A. Kantzas, B. Li, Z. Li, Q. Wang, M. Zhao, New agent for formation-damage mitigation in heavy-oil reservoir: mechanism and application, in: Paper Presented at the SPE International Symposium and Exhibition on Formation Damage Control, Lafayette, Louisiana, USA, February, 2008, https://doi.org/10.2118/112355-MS.

[171] T. Uetani, T. Matsuoka, H. Honda, Removal of Plugged Asphaltene in a Reservoir by Earthquakes, Society of Petroleum Engineers, 2014, https://doi.org/10.2118/168137-MS.

[172] T. Peruzzi, T. Coulon, J. Fauria, D. Frey, H. Marechal, R. Sloan, Umbilical Deployed Stimulations for Asphaltene Related Damage in Offshore Oil Fields, Society of Petroleum Engineers, 2012, https://doi.org/10.2118/151813-MS.

CHAPTER

Naphthenate and Carboxylate Soap Management

19

19.1 Introduction

For decades, the industry has tackled, with considerable interest, the major flow assurance issues, such as mineral scales, gas hydrates, waxes, asphaltenes, and biofouling. However, in recent years interest in naphthenate and carboxylate deposits has been sparked by their detrimental effects on production operations, which have been reported from different fields around the globe [1].

Heidrun, a Norwegian oilfield, is a famous example of these detrimental consequences, where naphthenate soaps were the most serious problem in the field, growing in severity and affecting all parts of the liquid systems after 2 years of production and only 2–3 months of water breakthrough [2].

Naphthenic acids are corrosive and have dreadful effects on the production system, storage system, and transport facilities integrity. Naphthenate and carboxylate soaps can form in different locations in the system, stabilizing emulsions and causing phase separation problems, and they can deposit, forming hard scales that cause flow restriction, pressure drop, and production losses with expensive maintenance. These issues are not new, but their severity and frequent occurrence highlight the need for a successful management strategy. Such a strategy should be able to analyze and assess the risks of the naphthenate/carboxylate soaps at the early stages of field development, and then develop a long-term treatment method to avoid their harsh effects on production. It is also recommended that the risk assessment and mitigation plans be developed case by case (or field by field), since an optimum treatment method that is effective in one case may not be effective in another, due to the continuous changes in production conditions (water volumes, water composition, pressure, temperature, hydrodynamics, etc.), and due to the complex nature of physical conditions and chemical reactions governing naphthenate deposition [3].

The management strategy in a general sense involves preventing the formation of soaps by means of operational and chemical methods, as well as removing and mitigating the formed deposits by chemical and nonchemical methods. Fig. 19.1 illustrates these mitigation methods.

19.2 Naphthenate and carboxylate soap prevention

Naphthenate prevention methods are based on using operational and chemical methods to avoid the partitioning of naphthenic acids at the oil/water interface and to avoid interactions between the naphthenic acids and metal ions.

Essentials of Flow Assurance Solids in Oil and Gas Operations. https://doi.org/10.1016/B978-0-323-99118-6.00001-0

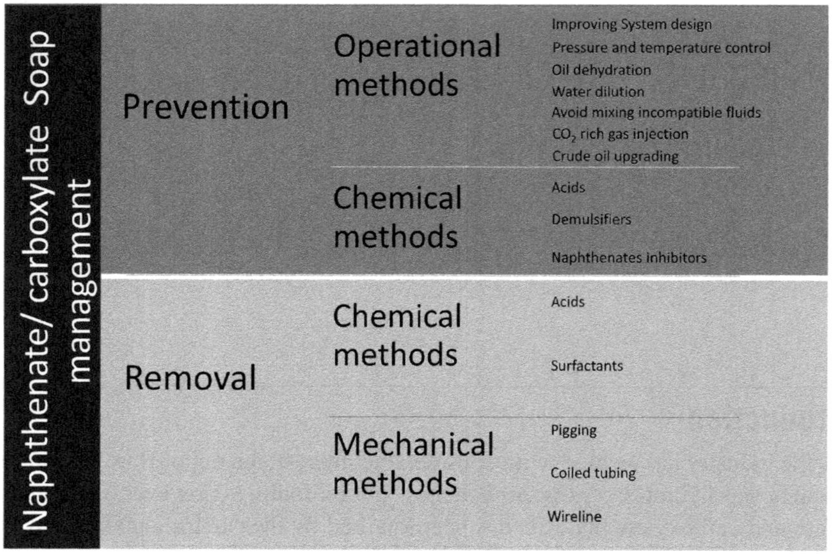

FIG. 19.1

Naphthenate and carboxylate soap management methods.

19.2.1 Operational methods of soap prevention

Naphthenate and carboxylate soap formation is complex and depends on changes in different operating parameters such as pressure, temperature, pH, phase separation, and others. By controlling these variables, the system can be adjusted to minimize their formation.

19.2.1.1 System design

Identifying the potential for soap problems at the early stages of field development is key to process control, as it helps to optimize the system design concept rather than reoptimizing and modifying running process facilities and operating parameters. Early detection of the problem highlights the need for smart or less troubled design concepts, rather than using a conventional design with a high risk of soap problems. The new design should avoid the sources of soap triggers, e.g., high pressure drop and fluids mixing, and enable the inclusion of mitigation methods, e.g., wash tank, chemical injection, etc. Furthermore, the smart design features should also anticipate the effect of a problem when it occurs and attenuate its effect on production [4].

Runham and Smith [1] reviewed various system design and other operational methods that can effectively improve naphthenate management.

One of the effective design features is using the wash tank, which is based on diluting the problem crude with other crude (with low risk) and with water. In this approach, the problem crude is mixed with low-risk crude to dilute the high ARN acid content. The low-risk crudes supposedly contain higher levels of monoacids, which prevent naphthenate mass formation and reduce their deposition. Additionally, water is added to dilute the naphthenic acid concentration, to wash the salts, and to decrease the emulsion stability, which significantly improves oil dehydration [1]. This option does have the advantage of avoiding the use of acid, reducing CAPEX for materials and operational concerns of HSE and logistics.

To apply this approach, some mandatory prerequisites must be managed, including [1]:

- Fully addressing the chemistry of crude oils (the problem crude and the low-risk crude), crudes compatibility produced water, and wash water, in addition to understanding the naphthenate problem chemistry and the factors affecting it.
- While wash water can attenuate the problem, the process is more reliant on the presence of the crude with high monoacids. The wash water can be mixed with a chemical inhibitor for improved performance.
- The best practice of this method involves washing with no pH control to allow monoacid dissociation and intervention with naphthenate deposition and low temperature to reduce the naphthenate deposition rate.
- Future plans of the field such as mixing compatibility with other process fluids and fields that may be tied back to the current system.

A new process for improved crude dehydration and prevention of naphthenate soap formation was claimed. The process separates produced fluids into gas and degassed emulsion, and then the degassed emulsion is separated into water and oil. The process also includes washing of the emulsion at an oil/water interface [5].

Other design features that should be included in the design concept or modified in the existing design to attenuate the effect of naphthenate include: [1]:

- Installing bypass at locations with high potential of naphthenate formation to reroute the produced fluids during cleaning and maintenance operations in these locations.
- Modifying vulnerable equipment and installation parts to be less affected and to also be less affecting of naphthenate. A common example is level controls like floats and bridles, which become heavily fouled; they should be modified with alternative technology. Separator internals like plate packs and demisters are vulnerable to fouling and should be modified in severe cases of naphthenate soaps. The same thing applies for heaters and coalescers, wherein the design should be modified, e.g., to proper sizes and flow rates, to attenuate the effect of naphthenates. Hydrocyclones and media filters are clogged easily with naphthenate/carboxylate soaps and should be replaced with less vulnerable alternative equipment.
- Naphthenate/carboxylate management usually involves pH control and acid addition to inhibit their formation; thus this approach should be considered during system material selection to be more corrosion resistant.
- Sampling point locations should be considered to obtain representative samples where naphthenates are likely to form, e.g., at the interfaces, to provide good monitoring of the system.
- Chemical and injection points should be considered during system design.
- The design should include equipment to reduce slugging.
- The design features of the system must involve operating at high pressure, since it is considered one of the main methods to combat naphthenate deposition.

19.2.1.2 Pressure control

Pressure drop is one of the common causes of naphthenate deposition. Pressure drop causes expulsion of dissolved CO_2 gas, which increases the solution pH and precipitates naphthenate/carboxylate. Thus, operating pressure must be optimized and kept high as long as possible in the process [6]. Since dissolved gases basically separate from the produced fluids once pressure drops below bubble point, keeping production pressure above bubble point should work as a safeguard from naphthenate problems. Another way to determine the safe working pressure is by modeling, using software such as ScaleSoftPitzer or ScaleChem to determine the minimum pressure that can achieve $pH<6$ (since most naphthenates will start depositing around $pH\geq 6$) [1].

This technique can be applied as the primary method of preventing naphthenate/carboxylate emulsions and deposits in oil production systems. The practical application of this method involves addressing the following:

- Determine the fluids chemistry.
- Measure dissolved gases concentration (CO_2, alkalinity, H_2S, etc.).
- Measure the bubble point pressure.
- Lab work must be done to estimate the pH range of relative naphthenate stability.
- Using commercial or academic software (scale prediction software can be used) to determine the pressure required to keep the system pH above the asphaltene instability point.

Field tests have demonstrated increase in degasser/separator pressure was correlated with a reduction in calcium naphthenate formation [4].

In cases like shutdown, where system depressurization occurs, CO_2 is lost, and pH increases, it is important to either displace the fluids, or use chemical additives (acids, inhibitors) to protect the system [1].

19.2.1.3 Temperature control

Temperature is a major factor in soap formation. Increasing temperature increases the potential of naphthenate deposit formation; however, increasing temperature is beneficial to treat carboxylate emulsions [3,7]. Thus the operator must be aware of the different effects of temperature on the two different kinds of soaps.

For fluids with a high tendency of forming naphthenate soaps, the operations are optimized at minimum heat requirements that will achieve the targeted phase separation requirement, or BS&W. This requires identifying the optimum operating window (minimum temperature achieving targeted BS&W). The use of cold treating chemical demulsifiers helps reduce the heat requirements [1,4]. On the other hand, high temperature is applied for fluids with carboxylate emulsion soaps issues. Therefore temperature optimization is a key to control both naphthenate and carboxylate soaps.

To practically apply temperature control, the following must be addressed:

- Fluids chemistry to identify whether the problem is with naphthenate soaps or carboxylate emulsions. Besides, using fluid chemistry the main drivers for forming these issues should be identified, e.g., metal ions, pH, alkalinity, dissolved gases, etc.

- Bottle test to evaluate the effect of temperature and demulsifiers on the naphthenate and carboxylate stability, so that a working temperature window is established, where minimum working temperature is identified.

19.2.1.4 Oil dehydration

Presence of water in the produced fluids introduces the oil-water interface required for interaction between oil and water, and supplies the required cations for the reaction. Increasing the water cut was found to increase the amounts of naphthenate deposits [8].

Accordingly, removal of water early in the process is an effective method to reduce the amount of soap generated (and reduce the amount of chemical required to control the soap) [4]. Also, periodic produced water recycle from the water-handling facility to the separation train front end, to assist oil/water separation and to keep the water droplet size distribution to larger droplet sizes. Since heat is applied during dehydration, fluids chemistry must be addressed beforehand to avoid the detrimental effect of heat on naphthenate deposition, as mentioned in Section 19.2.1.3.

19.2.1.5 Water dilution

Wash water can be added to dilute the soap-promoting components, e.g., cations and bicarbonates. Any suitable water source can be used, including treated potable, seawater, or softened water streams to the separation. A chemical inhibitor, acid, or demulsifier can be added to the wash water to improve the efficiency of the dilution method. If the injection water is used during maintaining reservoir pressure breakthrough, it can gradually reduce the tendency of soap formation [4]. Blending in crude oil having no soap-forming tendency and rich in monoacids can dilute the soap-forming crude oil, as mentioned in Section 19.2.1.1.

19.2.1.6 Avoid mixing incompatible fluids

Mixing incompatible fluids streams, e.g., one stream rich in naphthenic and carboxylic acids with another stream rich in Ca, Na cations, is one of the main causes of soap formation. Thus incompatible fluid mixing should be avoided. To achieve this:

- Produced fluid chemistry must be fully addressed.
- Bypass and rerouting of the fluid streams must be predesigned or modified in the system.
- Chemical injection points must be predesigned to allow chemical injection if needed.

19.2.1.7 CO_2-rich gas injection

CO_2-rich gas injection is an option, as it could depress the water pH outside the soap formation envelope, until most of the water has been separated from the oil [4].

19.2.1.8 Crude oil upgrading

Different methods of removal of naphthenic acids from crude oil have been proposed, which is also known as crude oil upgrading. Detailed methods of removal of naphthenic acids from crude oil have been reported by Gates et al. [9].

There are two main approaches to separate naphthenic acids from crude oil. One is to destroy the carboxylic group through chemical reactions such as esterification, hydrogenation, and thermal decomposition. The other is to separate naphthenic acids for other uses by techniques such as alkali washing, alcohol and ammonia processing, solvent extraction, filtration, and adsorption [10].

19.2.1.9 Other operating condition control methods

These include [1]:

- Minimize the use of process equipment that has high shear, e.g., this reduces the need for ESPs in wells and topsides high shear pumps; use low shear level control valves, optimize pipe sizes, and reduce complex routed system.
- Reduce the use of electrostatic treaters that may concentrate and exacerbate an existing soap problem.
- Anticipate future developments, where tie-backs may introduce incompatible fluids.

Table 19.1 summarizes the operational methods of naphthenate and carboxylate soap control.

19.2.2 Chemical prevention of naphthenate and carboxylate soaps

Chemical methods have demonstrated success in the prevention of naphthenate formation, with reports coming from the North Sea, Asia, West Africa, Venezuela, and Brazil. Different types of chemicals were used, including acids, demulsifiers, surfactants, and other polymeric chemicals. Table 19.2 summarizes the chemical agents used in soaps mitigation.

19.2.2.1 Classification and mechanism of soap prevention chemicals

Different types of chemicals have been successfully used to prevent naphthenate/carboxylate soaps in oilfields. They basically fall into three different categories: acids, demulsifiers, and naphthenate inhibitors. The mechanisms of these chemicals are elaborated in the following sections.

The acid usage is based on the fact that naphthenate and carboxylate formation is fundamentally pH dependent; thus the traditional way to prevent their formation is to lower the pH (to be <6) by adding an acid with a lower pK_a value than naphthenic acids [4,11]. The acid injection reverses the naphthenate formation, thus preventing naphthenic acids dissociation, or solubilizing already formed deposits in the

Table 19.1 Operational methods of preventing naphthenate and carboxylate soaps.

Method	Description
Improving system design	- Improve system components to be less vulnerable to soap deposition - Avoid mixing incompatible streams - Proper materials selection to be corrosion resistant
Pressure and temperature control	- Avoid pressure drop and pH increase - Calculate optimum pressure to keep pH below 6 - Optimize temperature to allow phase separation and to avoid soap formation
Oil dehydration	- Continuous removal of water to avoid naphthenate-metal ions interactions and soaps deposition
Fluids dilution	- Using crudes that are rich in monoacids to dilute crudes with high ARN acids - Using fresh water to dilute streams that are rich in Ca and Na ions

Table 19.2 Chemical agents used in naphenate and carboxylate soap mitigation.

Chemical agent type	Example
Acids	Mineral acids Hydrochloric acid Phosphoric acid Organic acids Acetic acid Sulfonic acids Myristic acid Lauric acid Citric acid Formic acid
Naphthenate inhibitors	Phosphate esters Quaternary ammonium compounds Quaternary phosphonium compounds
Demulsifiers	Alkyl phenol formaldehyde Ethoxylated polymers Alkyl phenols Acid demulsifiers

bulk phase, and forming calcium soluble salt according to Eq. (19.1), where $(RCOO)_2Ca$ is the formed soap and HA(aq) is the added acid [2]:

$$(RCOO)_2Ca(s) + 2HA(aq) \leftrightharpoons 2RCOOH(oil) + CaA_2(aq) \tag{19.1}$$

Demulsifiers and naphthenate inhibitors are similar in their chemistry. They are molecules with amphiphilic characteristics (surface active agents), they are polymeric in nature and work by interfacial crowding, surfactant-reactant interactions, and surfactant-particle interactions. The interfacial crowding is the main mechanism by which these surfactants retard the soap formation. The interface-active inhibitor/demulsifier self-assembles at the oil-water interface and hinders contact between the tetraacids and the metal cation by aligning themselves in the interfacial area between oil and water and dispersing between the naphthenic acids to increase the lateral distance between the dissociated acid monomers, thus blocking the cation-nephthenic/carboxylate interaction and reducing the rate of metal naphthenate/carboxylate formation overall (Fig. 19.2). The inhibitor or its complex is preferably at least as interfacially active as the naphthenic acid, if not more interfacially active than naphthenic acid [6,12–15]. Lucas et al. [16] showed that the mechanism of delaying film formation of calcium naphthenates is more related to the molar mass and structure of the additive than to its ability in reducing interfacial tension.

Other suggested mechanisms include the use of molecules capable of sequestering the ions present in the production water, inhibiting the formation of metallic naphthenate salts [17]. Another chemical approach to naphthenate inhibitors is that these inhibitors also contain compounds that react with the carboxylic groups of the naphthenic acids in the crude, making them no longer available for any reaction with the cations in the produced waters [18].

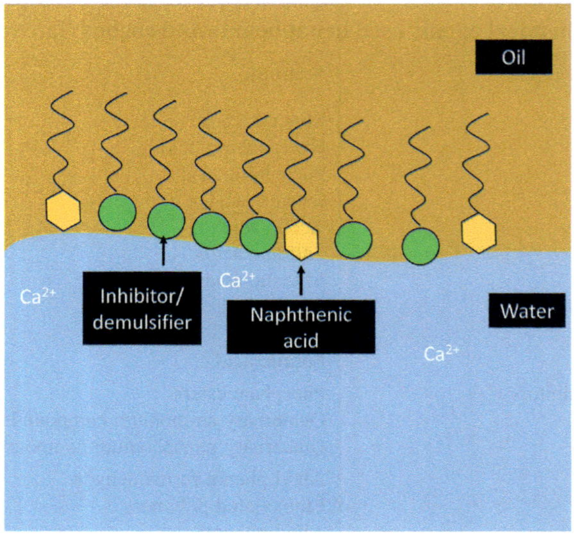

FIG. 19.2

Effect of inhibitors/surfactants on naphthenate formation.

19.2.2.2 Acids

Acid injection is the most common method to control calcium naphthenates [4,12,19–21]. Mineral and organic acids have been used as an effective method of controlling the pH in some fields. Lowering pH to around 6 is sufficient to prevent soap formation, but further lowering the pH might not add any value and can cause corrosion problems.

Acetic acid and acetic acid-based products, including glacial, hydroxyacetic, and glycolic acid, are the most commonly used chemicals in naphthenate soaps and deposits control, with proven lab and field success in removal and inhibiting of naphthenate deposits and emulsions [4,6,22–24]. As acetic acid is increasingly undissociated and it is fairly volatile, it can partition in the gas phase, causing 12 o'clock corrosion in the flowline [6,20,25–27]. So it is convenient to keep the pH at the optimum value of 6 [20,24,28]. Large quantities of acetic acid are often required for successful inhibition because a significant portion of the acid is used to overcome bicarbonate buffering in the brine [29,30]. For instance, a platform was reported to inject, on average, 3000 L of acetic acid per day for control of metal naphthenate formation [31]. Therefore, it is important to consider the cost, logistics, and storage, especially for offshore applications [27]. Another consideration is that low-dose acetic acid injection can exacerbate the corrosion problem, even for inhibited systems [29,32].

Other organic acids include sulfonic acids, myristic acid, and lauric acid. Sulfonic acids (commonly dodecyl benzene sulfonic acid, DDBSA) have been widely used, often in combination with acetic acid, for soap control [4,19,33,34]. Other acids like formic acid and citric acid were used in the lab during extraction procedures.

Hydrochloric acid can also be used. Metal naphthenate can be hydrolyzed by dilute hydrochloric acid in the presence of toluene under reflux [35]. Mineral acids like HCl have been used in the past to temporarily solve naphthenate and scaling problems in the Heidrun field (used with solvents like an alcohol) [19] and in the Gulf of Mexico [1]. However, naphthenate soaps are generally poorly soluble in mineral acids (although the solubility can be improved by addition of a mutual solvent such as isopropyl alcohol [4]), in addition to the corrosion problems that may arise because of HCl.

Phosphoric acid and phosphoric acid-based products are used less commonly than acetic acid, but have been reported. For example, two phosphoric acid-based products (one being alcoholic acid, the other being a phosphoric acid-based demulsifier) were successfully used to treat a sodium carboxylate soap-stabilized emulsion problem in offshore East Kalimantan, Indonesia, instead of acetic acid-based chemicals that did not prove as effective in this case [36]. Phosphoric acid will not partition in the gas phase and may not cause gas-side corrosion; however, overtreatments can cause corrosion (due to its acidity, since it is a stronger acid than acetic acid), and scaling problems when the acid interacts with Ca ion, forming hydroxyapatite and vivianite/polyphosphate scale [1,36].

Some issues with the acid treatment method include [2]:

- Large volumes of acid are required, especially with the increasing water volumes during the well lifetime, and due to the buffering capacity (alkalinity) of the produced water, which makes the method expensive.
- Corrosion problems are associated with acid additions. Corrosion can be close to the injection point and in parts of the process that were not made of acid-proof steel, in the gas side of the steam due to partitioning of the acid in the gas phase, and in systems injected with low-dose acetic acid [37]. This requires good material selections in the design concept.
- Acid injection had a negative effect on the oil dehydration efficiency.
- Acetic acid dissociates to acetate, which can be used as a substrate by some bacteria, e.g., methanogens and some strains of SRB [38], which can promote biocorrosion problems.
- Acid addition can complicate the solids problem, so instead of solving one problem it can create multiple ones. For example, phosphoric acid can trigger phosphate scales, and acids in general are known to destabilize asphaltenes, causing their aggregation and deposition in the system.
- The safety and environmental hazards associated with handling acids.
- The carryover of some of these acids causes serious complications in the refineries.

19.2.2.3 Naphthenate inhibitors

Most of these chemicals are indicated in the literature as proprietary chemicals. Kelland [11] reviewed the patent literature regarding naphthenate ion inhibitors.

Compounds mentioned in the patent literature include monophosphate and diphosphate esters [12]; tris-esters and alkoxylate [14]; phosphonate end-capped water-soluble polymers and polyvinyl polymers, which can also contain sulfonate and/or carboxylate groups; these chemicals work by chelating or scavenging the calcium ion from the oil phase and migrating to the water phase, and they can also work as scale inhibitors [39]; surface-active acrylamido(methyl)propylsulfonic acid (AMPS) copolymers with comonomers such as acrylic acid, acrylate esters, vinyl aromatics, or vinyl ethoxylates [40]; quaternary ammonium or quaternary phosphonium compounds [41]; substituted primary amines like polyalkylene-polyamines or a polyalkylene-polyamine-hydrocarbyl succinimide residue [42]; and rosin amines and quaternary ammonium derivatives [43].

Other chemicals in the literature include different surfactants with dispersing power [44], amphiphilic macromolecules produced by cardanol polyaddition (PCA) and polycondensation (PCC) (cardanol is a naturally occurring alkylphenol obtained from cashew nut shell liquid (CNSL) [16]. Such amphiphiles work by increasing the average intermolecular distance between the naphthenic acid monomers, leading to restructuring them at the interface and reducing their rate of formation [44].

Polymeric fatty acid ester amide compounds, or fatty hydrazide, were tested for their potential as naphthenate inhibitors and demulsifiers on Malaysian oil and showed promising performance [45].

Ethoxylated nonylphenols (NP resins) and glycolesters were studied and found to be efficient in demulsifying and preventing naphthenate soaps [6].

19.2.2.4 Demulsifiers

Since interaction between oil and water is the main source of problems in naphthenates, a good demulsification might help in reducing the impact of the problem. The demulsifier molecules pass between the naphthenates at the interface and diminish the strong interactions between the naphthenate chains, so the gel-like soap structure is inhibited [6].

Convenient demulsifiers used in the industry have been applied with success. Common demulsifiers comprise hydrophobic and hydrophilic parts, which determine the effectiveness. The hydrophilic parts are composed of oxyethylene, hydroxyl, carboxyl, or amine groups, while the hydrophobic parts contain alkyl, alkylphenol, or oxypropylene groups [45].

To improve the demulsifier efficiency, acid can be added to its formulation. An acid demulsifier, comprising acetic acid in an aromatic solvent mixture, and a nonacid demulsifier consisting of ethoxylates and alcohol have been successfully used to mitigate sludge deposition in oil storage tanks and water-handling facilities in the Indonesian oilfield, where the treatment costs were offset by the incremental oil that is recovered after dissolving the sludge [30]. A microemulsion formulation consisting of 30% organic acid and 70% solvent was designed and applied in the field in two producers in Malaysia, which showed no blockage after both cleaning operations and also an increase in the production flow rate [46].

Also, conventional and reverse demulsifiers can be used to combat the surface active properties of naphthenates and facilitate water separation. Kroknes et al. [47] demonstrated the efficiency of reverse emulsion breakers in controlling naphthenate soap.

Different combinations can be used to enhance the performance, such as demulsifier + naphthenate inhibitor, demulsifier + acid, demulsifier + scale inhibitor + naphthenate inhibitor, and others [6]. Such combinations can impart different functionalities during mitigating soaps, including soap inhibition and dissolution, and mineral scale inhibition and dissolution.

Panuganti et al. [48] demonstrated the use of different combinations including "surfactant + chelating agent + acid + solvent" combinations such as: mono alkyl palm ester (MAPE) + acid (HCl) + derivative of linear alkylbenzene sulfonic acid (DLABSA) as surfactant + glycol ether (GE) as co-surfactant combination, MAPE + alkyl sulfonic acid (ASA) + DLABSA + GE combination, and aromatic naphtha (ARN) + chelating agent (EDTA or HEDTA) + DLABSA + primary alcohol (PA) combination.

Rousseau et al. [6] emphasized that to efficiently prevent problems associated with stabilized soap emulsion and deposits, three simultaneous actions are required:

- Demulsification.
- Mineral scale inhibition.
- Naphthenate dispersion.

Recently, ionic liquids have been tested as potential naphthenate inhibitors. Two ionic liquids (ILs), 1-ethyl-3-methylimidazolium chloride (EMIM-Cl) and 1-butyl-3-methylimidazolium chloride (BMIM-Cl), have been tested and showed promising results in inhibiting naphthenate precipitation, with the BMIM-Cl showing better performance than the EMIM-Cl [49].

19.2.2.5 Crude oil natural naphthenic acids as naphthenate inhibitors

Some components in crude oils were found to work as natural naphthenate soap inhibitors. These are the short chain naphthenic/carboxylic acids, also known as the monoacids. Such short-chain acids can ionize at lower pH than the ARN naphthenic acids, which prevents their deposition. While the ARN might be still able to react with the Ca cations, the formation of a polymer-like structure is prevented by reaction with shorter acids [26]. However, if the short-chain acids are less interfacially active than the ARN or they ionize at higher pH, they might not be able to compete with the ARN, and so will be less effective in preventing ARN deposition.

Naphthenic acids are also thought to be a crosslinking termination point at the oil-water interface [50]. Crudes with short-chain acids can be used to mitigate and dilute crudes with naphthenate high risk, as mentioned in Section 19.2.1.1.

Another crude oil component that was found to reduce naphthenate formation are asphaltenes. They either inhibit the formation of ARN/Ca^{2+} interfacial gel or desorb or disperse if already formed. That was explained either by interfacial competition with the ARN molecules or by bulk interaction with the ARN molecules reducing their tendency to interact with Ca cations and deposit, with the latter mechanism more acceptable than the former as asphaltenes are less surface active than ARNs [51].

19.2.2.6 Methods of screening naphthenate/carboxylate control chemicals

19.2.2.6.1 Bottle test/jar test

Screening and ranking naphthenate control chemicals follow some classical test methods, like bottle test and jar test, commonly used to screen oilfield demulsifiers. The interface layer can be separated and characterized, using mass spectroscopy methods for naphthenic acids and the metal ion measurements.

However, bottle/jar tests suffer from enormous limitations relating to their ability to mimic production conditions such as appropriate emulsification, in situ pH control, ability to differentiate between the type of interface deposit formed (e.g., CaN vs. soap scales vs. stabilized tight emulsion) residence times [3,52].

19.2.2.6.2 Naphthenate testing rig

Recently, a small scale pilot plant has been developed by Statoil (now known as Equinor) at the Research and Technology Center in Trondheim, Norway [2].

The test rig has been designed to replicate conditions in low pressure (5–10 bar) separators, in order to test for calcium naphthenate deposition under more "field" representative conditions [3]. The maximum operating temperature is 90°C at 100 bars for the components, piping, and pipe couplings [2].

It consists of feed tanks, the weight of which can be constantly monitored, connected to liquid pumps for oil and water. The design also allows for injecting chemicals after the fluids mixing point. The rig is connected to three gas lines: CO_2, N_2, and CH_4. The gas may mix with the liquid flow upstream of choke or straight into the separator cell. The feeding tanks are at atmospheric pressure, but the components in the flowline, pipe, and pipe coupling are designed for higher pressures than 100 bars [2].

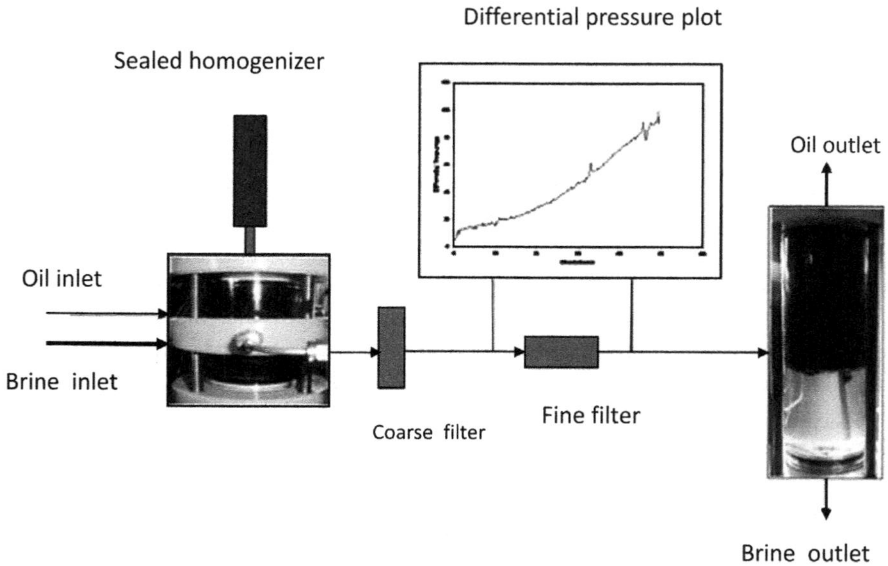

FIG. 19.3

Schematic of Mini loop.

Source: N. Borhan, A. Ramli, Inhibition of naphthenates emulsion soaps chemical EOR using synthesis compound derived from fatty hydrazide, J. Phys. Conf. Ser. 1123 (2018). Institute of Physics Publishing, https://doi.org/10.1088/1742-6596/1123/1/012057.

Another method of dynamic testing of naphthenate deposition and testing the chemicals involves use of a flow loop. Borhan and Ramli [45] reported the use of a multifunctional mini flow loop (MMFL). In this MMFL, crude oil and brine were injected into a mixing cell equipped with a sealed homogenizer operating at 13,500 rpm. The combined fluids then flowed from the mixing cell through the two in-line filters. The first was a coarse filter/baffle and the second was a 7 μm sintered metal filter. The differential pressure (dP) across the fine filter was used to assess the viscosity of the fluids and any arising blockages. The fluids then entered a glass separator where the oil/brine interface and any emulsion formed were observed and photographed. The entire apparatus was housed inside an oven and pressurized to replicate the conditions in a low-pressure separator in the field. Fig. 19.3 illustrates the mini flow loop.

19.2.2.6.3 Autoclave naphthenate tests

These tests were developed to simulate the production of a mixed oil-water system into a separator and the effect of pressure drop and associated pH change. During the test, brine (containing bicarbonate ions and buffered with CO_2) and crude oil (saturated with CO_2) are initially mixed at high pressure (100 bar), and then the mixture is continually mixed overnight at 50°C at a constant pressure of 1000 psi, to ensure formation of a stable emulsion. After this, the pressure is dropped to ambient while maintaining constant stirring (1000 rpm) and temperature (50°C) conditions; then the mixing is stopped, and the contents are collected and analyzed for assessment. When chemical treatments were being tested, they were injected into the vessel after aging, and mixing was continued for a further 2 h prior to pressure release [3].

19.2.2.6.4 Other screening methods

Other tests that were mentioned in the literature include interfacial properties measurements and rheology measurements.

Nordgård et al. [32] reported three new methods of screening naphthenate inhibitors, namely interfacial activity measurement, gravimetric determination of interfacial layer, and UV depletion method.

Interfacial tension measurements are based on the fact that some inhibitors are more interfacially active than the naphthenic acids. In this test, an effective inhibitor significantly lowers the interfacial tension of the mixture, hence suppresses the interfacial activity of naphthenic acids and prevent soap formation (Fig. 19.4).

In the gravimetric methods, the weight of the soap formed at the interface is measured. A blank solution comprising the oil is continuously mixed with water/brine, adjusted at a high enough pH to cause naphthenate formation, and is compared with other solutions with the same composition in addition to different doses of inhibitors. After sufficient mixing, the oil and water phases are removed, and the interfacial layer is weighted. The lower the weight of the deposit, the better the inhibitor.

In the UV depletion method, the depletion (reduction) of naphthenic acid concentration in the oil phase during the two-phase reaction is measured. The lower the depletion, the lower the available naphthenic acids at the interface for the reaction, and hence the better the inhibitor.

19.2.2.7 Application of naphthenate/carboxylate prevention chemicals

Continuous injection is the most common method of delivering the naphthenate mitigation chemicals. Like conventional demulsifiers, the naphthenate control chemicals are injected upstream of the expected problem locations; the wellhead injection, first point in the flowlines, and inlet separators are commonly used. Upstream locations of fluids mixing, pressure drop, and temperature change are also common locations for better control of the problem. Chemicals can be injected downhole using the gas lift valves, after being diluted in the proper solvent.

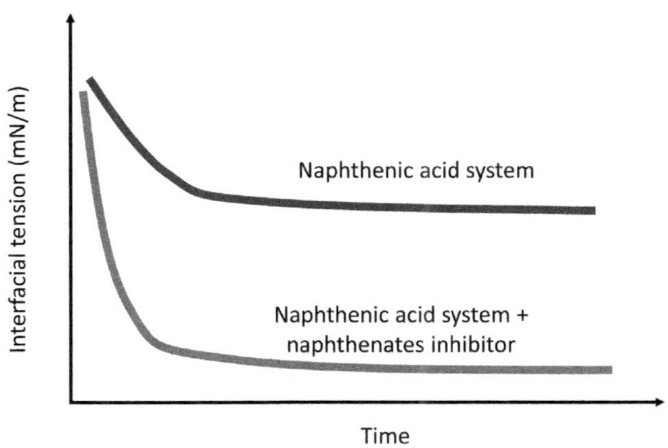

FIG. 19.4

Interfacial method for screening naphthenate control chemicals.

To properly apply the chemical, it is critical to address the following:

- Determine the fluid chemistry to identify whether the problem is naphthenate or carboxylate soap, which allows either use of heat (in the case of carboxylates) to improve the chemical performance, or eliminate the use of heat in the case of naphthenate soap.
- Investigate whether mineral scales are involved, which might worsen the problem, and require the use of a combination of naphthenates inhibitor + scale inhibitor.
- Examine the use of acid in the inhibitor/demulsifier blend to improve the performance and remove the formed deposits.
- Assess the robustness of the system integrity before using acidic material to avoid causing corrosion problems.

Downhole batch treatments of naphthenate chemicals have been applied to remove the formed deposits in the near wellbore area and production tubing. A microemulsion formulation consisting of 30% organic acid and 70% solvent has been pumped and soaked in the near wellbore area in two producers, to remove and inhibit naphthenate deposits [46].

19.2.2.8 Recent advances in chemical inhibition of naphthenate and carboxylate soaps

The early uses of naphthenate control chemicals involved the use of acids (mineral or organic acids) to remove and prevent the formation of naphthenate deposits. Demulsifiers and acid-demulsifiers have also been used.

Recently, polymeric materials have been developed to allow a fast, efficient, less corrosive solution for soap problems. These polymeric materials were developed after the recent advances in characterizing and understanding the main mechanisms by which naphthenates form. Recently, ionic liquids have been tested as naphthenate inhibitors, where they represent effective and environmentally friendly chemicals.

Furthermore, more advances in testing and screening these chemicals have been witnessed, including developing the naphthenate rig, besides other physical properties such as interfacial shear/dilational rheometry, pendant drop tensiometry, gravimetry, Langmuir balance technique, light-scattering near-infrared spectroscopy, and others.

19.3 Napthenate deposits removal

The simplest and most cost-effective method of removing naphthenate and carboxylate soaps is using acids. Several acids, both organic and inorganic acids, have been used.

The efficiency of the acid is improved by adding surfactant or demulsifier to disperse the dissolved soaps and prevent their reprecipitation.

Microemulsions have been applied to remove blockages of naphthenate deposits downhole by producers in oilfields from offshore Malaysia. The formulation comprised 30% acid and 70% organic solvent. The organic solvent helps remove the hydrocarbon residues and improve the dissolution.

Mechanical methods can also be applied to improve the removal efficiency when combined with acids or microemulsions.

Mechanical methods are commonly applied to remove any accumulated deposits in the pipeline or in production tubing. Routine pipeline pigging with proper pig selection can effectively remove accumulated soaps and reduce their risk. Similarly, wireline and coiled tubing methods can be used to clean the wellbore and subsea installations.

19.4 Strategies and philosophies to control naphthenate and carboxylate soaps

19.4.1 Risk assessment

To assess naphthenate problems in a field, sufficient data about the produced fluids, system parameters, and other factors that could possibly affect naphthenates must be obtained.

19.4.1.1 Fluids analysis

Different chemical and physical parameters are measured to characterize the produced fluids, including:

- Crude oil composition, using SARA or modified SARA analysis.
- Crude oil total acid number (TAN): although the test is not fully representative of the naphthenic acids in crude oil, it is still a quick and reliable way to screen the most problematic crudes.
- Detection and quantification of naphthenic acids. A variety of investigative techniques has been used to determine the naphthenic acids in crude oil and in naphthenate deposits, including SEM/EDAX, elemental analysis, nuclear magnetic resonance (NMR) spectroscopy, vapor pressure osmometry (VPO), high temperature gas chromatography [HTGC) and liquid chromatography mass spectrometry (LCMS) [2,3]. FT-IR can also be used to detect carboxylic acid. Other methods include distillation combined with resin column and scanning UV spectroscopy. These methods are used to give a good estimation of the naphthenic/carboxylic acids, and thus estimate their risk.
- Crude oil physical properties: viscosity, density, etc.
- Water chemistry: pH, cations, anions, alkalinity, specific gravity, salinity, dissolved gases, etc. Although pH is usually measured, it is preferred that during the naphthenates risk assessment, the pH is calculated.
- A quick method to understand the partitioning/behavior of the naphthenic acids between oil/water phases is $pH_{initial}/pH_{final}$. The test is performed by shaking small volumes of crude/water mixture in a vial and measuring their pH before and after shaking to observe the change in pH.
- A good knowledge of the reservoir and reservoir fluids properties are needed. Reservoir pressure and H_2S, CO_2 partial pressure are of great importance.

19.4.1.2 Possible root cause analysis

Full system analysis must be done in this step to look for the origins of the problem. Pressure drop, temperature change, incompatible fluids mixing, and other hydrodynamic factors should be investigated. These factors are combined with the rest of the risk assessment factors to build a holistic risk assessment matrix.

19.4.1.3 Modeling and simulation

Simulation models for naphthenate deposits are not very common commercially, yet a few indoor trials have been made by some academic institutions.

Another parameter to model is the solution. This can be done using one of the scale prediction models (ScaleSoftpitzer, ScaleChem, etc.); pH is preferably calculated since the measured pH is subjected to errors, e.g., sampling error, loss of gas, secondary precipitations, contaminations, etc.

Another parameter to model is the minimum working pressure, which can be used to keep the pH <6, which is considered a safe zone for operating without naphthenate problems.

19.4.2 Choose proper control method

The control method is applied depending on the phase of the field development and the severity of the problem. Before production starts, design modification and optimization is possible by reducing the causes of pressure drop, reducing fluids mixing, adding heating equipment, adding washing tanks, adding chemical injection points, adding more sampling points to the design, in addition to proper selection of corrosion resistant materials, and optimizing the design of other production equipment to be less prone to soap formation.

During production, the methods will depend on optimizing the pressure to be above bubble point, and keeping the pH below 6. Applying heat or operating at the lowest temperature depend on the type of soap forming in the system. Further bypassing of incompatible fluids may be necessary, to keep the production running while cleaning the soaps from the system. Chemical methods are efficient and cost effective when they are applied in the proper location at the proper dosage.

Mechanical cleaning is always a feasible method, although it is not applicable where naphthenates are more likely to form, e.g., separators. However, routine pigging can reduce the amount of sludge in the system, which might help in reducing the overall solids buildup.

Shutdowns must be planned properly, as depressurization causes pH change and possible naphthenate deposition. The fluids must be inhibited, using proper chemicals or fluids displacement using inhibited solution.

19.4.3 Monitoring and assessment

After the mitigation method is applied, monitoring of the system is crucial to estimate the effectiveness of the applied methods. This is usually carried out by routine monitoring of the system parameters, and monitoring of phase separation operations, separation equipment, and heating equipment, and by fluids characterization and analysis.

A general strategy of naphthenate management is illustrated in Fig. 19.5.

19.5 Recent advances in naphthenate/carboxylate soap management methods

Several advances have been reported regarding naphthenate and carboxylate soaps. These advances stemmed from the newly reported field cases and experiences, as well as the immense research efforts that have been implemented in this field.

Design optimization and modification have been demonstrated and proven as a primary protection method against soaps formation. Optimizing system components and operating parameters in the earliest stages of the field development is a key in fighting soaps. In addition, development of new chemical inhibitor formulations, microemulsions, and polymeric materials and ionic liquids are considered

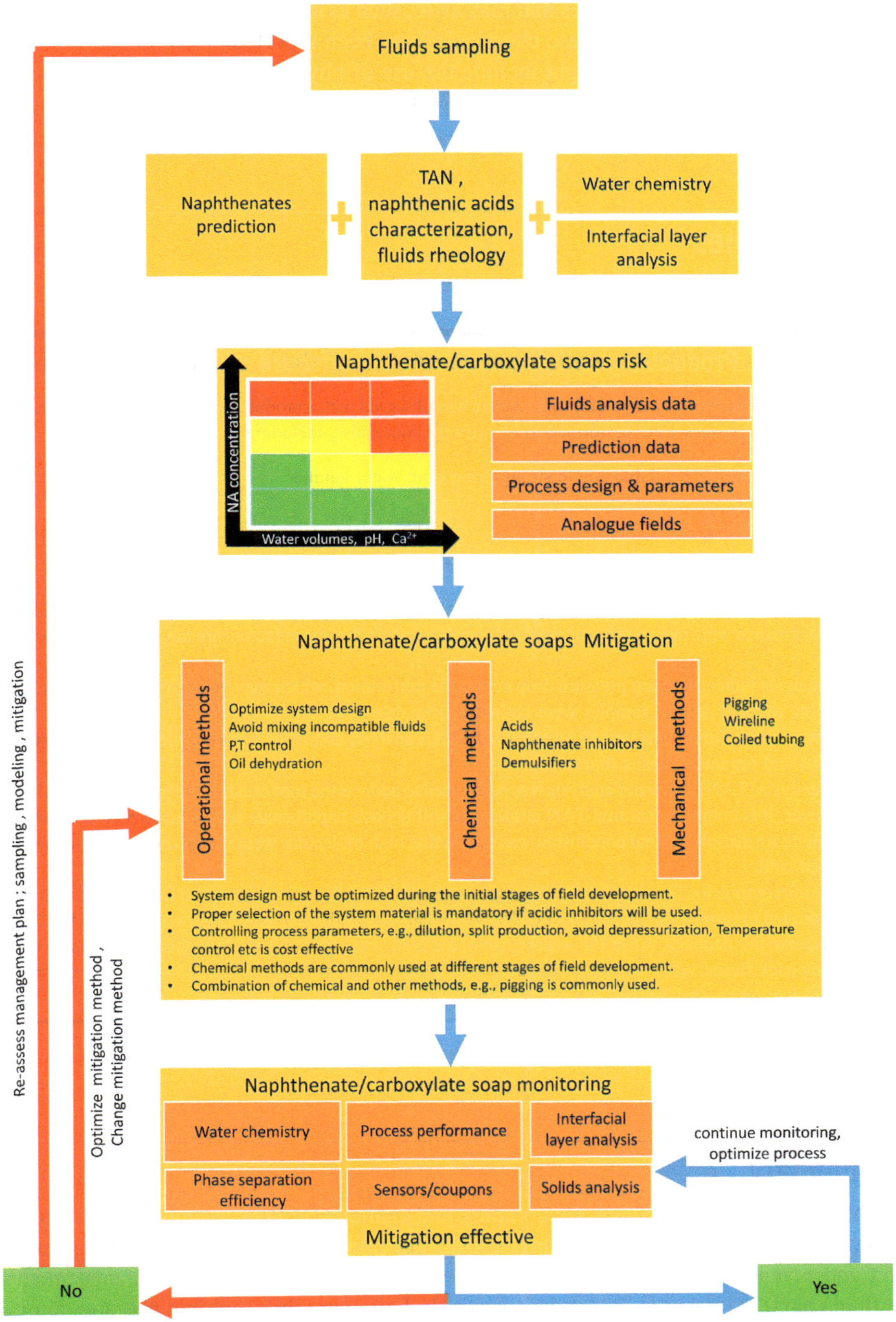

FIG. 19.5

Naphthenate soaps management strategy.

as a start for upcoming future advances that may arise, such as using nanomaterials. Furthermore, more advances in testing and screening these chemicals have been reported.

Additionally, future advances might include the use of physical methods like magnetic methods, acoustic methods, electric fields, and others as green alternatives to the conventional chemical methods.

19.6 Case studies

CASE STUDY 1: Production wells from an oilfield offshore of East Malaysia [46].

A dual string oil producer from offshore of East Malaysia was observed with a production drop from 1200 bpd to 300 bpd in 7 years. Several investigations were performed to identify the root cause, including:

- Reservoir pressure was fairly stable due to strong drive from water aquifer
- Surface choke
- Gas lift performance
- Skin factor

The reservoir pressure was found to be fairly stable due to the strong drive from the water aquifer. The surface choke and gas lift performance were optimized, and the final conclusion was that these factors are not the cause of the production decline.

The skin analysis indicated sharp pressure drop across the perforation, which may be attributed to the sand, fines, and solids deposition, so solid deposit samples were collected from downhole of two wells.

The deposit samples were characterized by modified SARA analysis and XRD analysis, which showed inorganic scale (mainly $CaCO_3$) and organic deposits (naphthenates).

Crude oil analysis (TAN) and water analysis was performed to address the root cause of the problem. The oil TAN value was low; however, it is common that low TAN crudes can still deposit naphthenate soaps. Scale prediction showed high potential of calcite formation. A final conclusion was drawn that high molecular weight naphthenates were the main issue for the specified well.

A specialty microemulsion formulation (30% organic acid and 70% solvent) was designed for the well treatment which, upon application, led to improving the production. The treatment was bullheaded down the two wells and left for 8 h soaking time to disperse the deposits. After the soaking time, the well was put back on production and samples collected to monitor the flowback fluids. After 48 h of production, well gauging was performed. The ultimate result reported was increasing the production rate by 700 bpd of liquids, with significant reduction in water cut.

19.7 Summary

Naphthenates and carboxylate soaps are complex and challenging flow assurance issues in oil and gas fields. They form sticky emulsions and hard deposits that interfere with the major field operations. Early detection and management of such deposits is key to steady operations. The management methods involve fluid characterization to assess the soaps risk. Following this, multiple operational or chemical methods can be applied to attenuate or eliminate their risk based on the severity of the problem.

Operational methods include optimizing and modifying the system design and operating parameters to eliminate the soap triggering factors. This includes optimizing pipelines and equipment sizes to reduce pressure drop, modifying equipment designs to reduce accumulation of solids, adding washing tank, split production, adding heating equipment, adding chemical injection lines, and sampling points.

Chemical methods are based on using acids (organic, inorganic), demulsifiers, and naphthenate inhibitors. Acids are usually used with proven success in different fields, with a drawback of being corrosive. Demulsifiers have also been reported with successful treatments. Combinations of acids + demulsifiers, acid + organic solvents, or acid + demulsifier + surfactant have also been reported. Naphthenate inhibitors are polymeric materials that interfere with the naphthenic acids at the oil-water interface, and most of these are proprietary to chemical suppliers.

Naphthenate removal is achieved basically by acids. A combination of acid and organic solvents as microemulsions was found to be effective. Other combinations include acid + solvent + surfactants, which helps in washing organic residues, dissolving the soaps and keeping them dispersed without reprecipitation.

References

[1] G. Runham, C. Smith, Successful naphthenate scale and soap emulsion management, in: Proceedings—SPE International Symposium on Oilfield Chemistry, vol. 1, Society of Petroleum Engineers (SPE), 2009, pp. 326–337, https://doi.org/10.2118/121522-ms.

[2] H. Mediaas, K. Grande, B.-M. Hustad, K.R. Hovik, H. Kummernes, B. Nergard, J.E. Vindstad, A unique laboratory test rig reduces the need for offshore tests to combat calcium naphthenate deposition in oilfield process equipment, in: Presented at Tekna Oilfield Chemistry Symposium, Geilo, Norway, March 13–16, 2005.

[3] G.M. Graham, K.B. Melvin, A.E. Gabb, F. Haider, H. Williams, S. Dyer, C. Cummine, Simulating Calcium Naphthenate Formation and Mitigation Under Laboratory Conditions, Society of Petroleum Engineers, 2006, https://doi.org/10.2118/100632-MS.

[4] M.S. Turner, P.C. Smith, Controls on soap scale formation, including naphthenate soaps—drivers and mitigation, in: SPE 94339 (Paper Presented at the SPE International Symposium on Oilfield Scale, Aberdeen, UK), 2005.

[5] V. Van-Khoi, Process for the Treatment of Crude Oil, Process for the Separation of a Water-in-Oil Hydrocarbon Emulsion and Apparatus for Implementing the Same, US Patent US20070267325A1, 2005.

[6] A. Goldszal, C. Hurtevent, G. Rousseau, Scale and Naphthenate Inhibition in Deep-Offshore Fields, SPE 74661, 2002.

[7] A. Celsie, J.M. Parnis, D. Mackay, Impact of temperature, pH, and salinity changes on the physico-chemical properties of model naphthenic acids, Chemosphere 146 (2016) 40–50, https://doi.org/10.1016/j.chemosphere.2015.11.122.

[8] A.A. Sulaimon, S. Arumugam, B.J. Adeyemi, Investigating the influence of water cut on naphthenate precipitation in oil production facilities, in: ICIPEG 2014, Springer Singapore, 2015, pp. 145–153, https://doi.org/10.1007/978-981-287-368-2_14.

[9] C. Wu, A. De Visscher, I.D. Gatesa, On naphthenic acids removal from crude oil and oil sands process-affected water, Fuel 253 (2019) 1229–1246.

[10] Y.-z. Wang, H.-l. Duan, C.-M. Song, X.-T. Han, X.-R. Ma, Removal of naphthenic acids from crude oils by catalytic decomposition using Mg-Al hydrotalcite/γ-Al_2O_3 as a catalyst, Fuel 134 (2014) 499–504.

[11] M.A. Kelland, Production Chemicals for the Oil and Gas Industry. Production Chemicals for the Oil and Gas Industry, CRC Press, 2014, https://doi.org/10.1201/9781420092974.
[12] C. Hurtevent, S. Ubbels, Preventing naphthenate stabilised emulsions and naphthenate deposits at ekoundou oil field in cameroon, in: Paper Presented at the SPE International Oilfield Scale Symposium, Aberdeen, UK, 2006, https://doi.org/10.2118/100430-MS.
[13] S. Ubbels, P. Venter, V. Nace, Low Dosage Naphthenate Inhibitors, U.S. Patent WO2006025912A3, 2006.
[14] M. Hellsten, I. Uneback, A Method for Preventing the Formation of Calcium Carboxylate Deposits in the Dewatering Process for Crude Oil/Water Streams, U.S. Patent 8,653,009, 2008.
[15] A.M.D. Hanneseth, Ø. Brandal, J. Sjöblom, Formation, growth, and inhibition of calcium naphthenate particles in oil/water systems as monitored by means of near infrared spectroscopy, J. Dispers. Sci. Technol. 27 (2) (2006) 185–192, https://doi.org/10.1080/01932690500265896.
[16] R.M.M. Dip, M.F.P. Costa, E.F. Lucas, Interfacial films of calcium naphthenate: influence of polycardanol structures, Energy Fuel 31 (9) (2017) 9275–9282, https://doi.org/10.1021/acs.energyfuels.7b01675.
[17] J.C. Raymond, Soap Control Agent, U.S. Patent 20080029438A1, 2008.
[18] S.J. Dyer, G.M. Graham, C. Arnott, Naphthenate Scale Formation—Examination of Molecular Controls in Idealised Systems, Society of Petroleum Engineers, 2003, https://doi.org/10.2118/80395-MS.
[19] J.E. Vindstad, A.S. Bye, K.V. Grande, B.M. Hustad, E. Hustvedt, B. Nergård, Fighting Naphthenate Deposition at the Heidrun Field, Society of Petroleum Engineers, 2003, https://doi.org/10.2118/80375-MS.
[20] G. Rousseau, H. Zhou, C. Hurtevent, Calcium Carbonate and Naphthenate Mixed Scale in Deep Offshore Fields, SPE 68307, 2001.
[21] P.C. Smith, M. Turner, Naphthenate and caboxylate soap scales workshop, in: IQPC Flow Assurance Europe, London, 2003.
[22] S.J. Ubbels, Preventing naphthenate stabilized emulsions and naphthenate deposits during crude oil 5th processing, in: Proceedings From the International Conference on Petroleum Phase Behaviour and Fouling, Banff (June 13–17, 2004), 2004.
[23] P.C. Smith, Soaps (naphthenates) scales management from deepwater flow assurance aspects to oil terminal sludge processing workshop, in: IQPC—Flow Assurance: A Holistic Approach, Kuala Lumpur, 2004.
[24] M.H. Ese, K.P. Kilpatrick, Stabilization of water-in-oil emulsion by naphthenic acids and their salts: model compounds, role of pH and soap/acid ratio, J. Dispers. Sci. Technol. 25 (3) (2004) 253–261.
[25] L.H. Allen, The importance of pH in controlling metal soap deposition, TAPPI J. 71 (1988) 61–64.
[26] B. Brocart, C. Hurtevent, Flow assurance issues and control with naphthenic oils, J. Dispers. Sci. Technol. 29 (10) (2008) 1496–1504, https://doi.org/10.1080/01932690802316827.
[27] K. Melvin, C. Cummine, J. Youles, H.L. Williams, G.M. Graham, S.J. Dyer, Optimising calcium naphthenate control in the Blake field. SPE 114123, in: SPE International Scale Conference, Aberdeen, 2008.
[28] K.S. Sorbie, A. Shepherd, P.C. Smith, M. Turner, R.E. Westacott, Naphthenate formation in oil production: general theories and field observations, in: Proceedings of Chemistry in the Oil Industry IX, RSC/EOSCA, 2005.
[29] N. Bretherton, R. Smith, G. Keilty, S. Ubbels, Naphthenate control: is acetic acid injection the answer? SPE 95115, in: International Symposium on Oilfield Corrosion, Aberdeen, 2005.
[30] D.L. Gallup, J. Star, Soap sludges: aggravating factors and mitigation measures, in: 6th International Symposium on Oilfield Scale, Aberdeen Society of Petroleum Engineers, 2004, https://doi.org/10.2118/87471-MS.
[31] J. Debord, P. Srivastava, Development and field application of a novel non-acid calcium naphthenate inhibitor. SPE 123660, in: SPE Annual Technical Conference and Exhibition, Louisiana, 2009.
[32] E.L. Nordgård, A.D. Hanneseth, J. Sjöblom, Inhibition of calcium naphthenate. Experimental methods to study the effect of commercially available naphthenate inhibitors, J. Dispers. Sci. Technol. 31 (5) (2010) 668–675.
[33] D.L. Gallup, J.A. Curiale, Characterization of sodium emulsion soaps formed from production fluids of Kutel Basin, Indonesia, Energy Fuel 21 (2007) 1741–1759.

[34] D.L. Gallup, P.C. Smith, J. Chipponeri, A. Abuyazid, D. Mulyono, Formation & mitigation of "metallic soap" sludge, Attaka, Idonesia field. SPE 73960, in: SPE International Conference on Health and Safety and Environment in Oil and Gas Exploration and Product, Kuala Lumpur, 2002.

[35] M.A. Mohammed, K.S. Sorbie, Naphthenic acid extraction and characterization from naphthenate field deposits and crude oils using ESMS and APCI-MS, Colloids Surf. A Physicochem. Eng. Asp. 349 (1–3) (2009) 1–18.

[36] D.L. Gallup, P.C. Smith, J.F. Star, S. Hamilton, West Seno Deepwater Development Case History—Production Chemistry, Society of Petroleum Engineers, 2005, https://doi.org/10.2118/92969-MS.

[37] W.I. Eke, V.-O. Cynthia, A. Onyewuchi, Oilfield metal naphthenate formation and mitigation measures: a review, J. Pet. Explor. Prod. Technol. 10 (2) (2020) 805–819, https://doi.org/10.1007/s13202-019-00797-0. Springer.

[38] C.I. Chen, M.A. Reinsal, Characterisation of microbial souring in berea-san porous medium with a north sea oil field inoculum, Biofouling 9 (3) (1996).

[39] C.R. Jones, International Patent Application WO/2005/085392, 2005.

[40] J.P. Vijn, P. In't Veld, International Patent Application WO/2012/154378, 2012.

[41] J.E. Vindstad, K.V. Grande, K.R. Hoevik, H. Kummernes, H. Mediaas, Applying laboratory techniques and test equipment for efficient management of calcium naphthenate deposition at oil fields in different life stages, in: Paper Presented at the 8th International Conference on Petroleum Phase Behavior and Fouling, Pau, France, 10–14 June 2007, 2007.

[42] D. Smith, C. Smith, D. Watson, U.S. Patent Application 20090301936, 2009.

[43] C. Khandekar, S. Gopal, J. Smith, International Patent Application WO/2011/063459, 2011.

[44] O. Brandal, Interfacial (O/W) Properties of Naphthenic Acids and Metal Naphthenates, Naphthenic Acid Characterization and Metal Naphthenate Inhibition, Norwegian University of Science and Technology, 2005. Doctoral Thesis.

[45] N. Borhan, A. Ramli, Inhibition of naphthenates emulsion soaps chemical EOR using synthesis compound derived from fatty hydrazide, J. Phys. Conf. Ser. 1123 (2018), https://doi.org/10.1088/1742-6596/1123/1/012057. Institute of Physics Publishing.

[46] A.S. Mohamed, A. Goyal, M.D. Mohd Ismail, G. Munainni, J.K. Amar Singh, M.Z. Ismail, M. Anwar, Detection and removal of naphthenate deposit in oil well, in: Offshore Technology Conference, 2014, https://doi.org/10.4043/25003-MS.

[47] K. Kroknes, S. Ubbels, L. Bergersen, G.M. Bergås, J.T. Thingvoll, G. Ulfeng, Improved dehydration and discharge water quality by changing from demulsifier to 15th reverse demulsifier, in: Presented at International Oil Field Chemistry Symposium, Geilo, Norway, 2004.

[48] I.K. Salleh, S. Misra, J.M.B.M. Ibrahim, et al., Micro-emulsion-based dissolver for removal of mixed scale deposition, J. Pet. Explor. Prod. Technol. 9 (2019) 2635–2641, https://doi.org/10.1007/s13202-019-0643-8.

[49] A.A. Sulaimon, A.N. Masri, U.A.A.J. Sabri, B.J. Adeyemi, Predicting naphthenate precipitation and evaluating the effect of ionic liquids on its deposition, J. Petrol. Sci. Eng. 209 (2022), 109865, https://doi.org/10.1016/j.petrol.2021.109865.

[50] S. Simon, S. Subramanian, B. Gao, J. Sjöblom, Interfacial shear rheology of gels formed at the oil/water interface by tetrameric acid and calcium ion: influence of tetrameric acid structure and oil composition, Ind. Eng. Chem. Res. 54 (35) (2015) 8713–8722, https://doi.org/10.1021/acs.iecr.5b02165.

[51] S. Simon, B. Gao, S. Tofte, J. Sjöblom, N. Passade-Boupat, T. Palermo, M. Rondon-Gonzalez, Influence of asphaltenes on gelation of tetrameric acid with calcium ion at the oil/water interface under flow-model condition, Energy Fuel 34 (5) (2020) 5846–5855, https://doi.org/10.1021/acs.energyfuels.0c00829.

[52] E.E. Johnsen, H. Kallevik, A. Westvik, The importance of performing water/oil separation tests at 15th elevated pressure and temperature, in: Presented at the International Oil Field Chemistry Symposium, Geilo, Norway, 28–31 March, 2004.

CHAPTER

20 Biofouling Management

20.1 Introduction

Microbial activities in oil and gas fields are a complex problem leading to a variety of challenges, including biocorrosion, biofouling, reservoir souring, degradation of chemicals, and other hazardous effects. Every one of these issues has a serious impact on production operations and requires a credible strategy based on early detection of the microbial activity and quick application of effective mitigation methods before the problems exacerbate.

Detection of microbial activity is basically achieved by fluids sampling and analysis to identify the bacterial types and determine their counts in order to estimate their potential risk. The management strategy for biofouling involves controlling the bacterial growth and/or controlling the outcomes of bacterial growth, e.g., MIC, scale and biomasses, and H_2S. Therefore controlling bacterial activity and biofouling involves a variety of methods and chemicals.

Generally speaking, methods of controlling bacteria and biofouling can be operational, physical, chemical, biological, or mechanical, and of course a combination of two or more of these methods may be necessary, depending on the severity of the problem. Fig. 20.1 summarizes these methods.

20.2 Operational methods

Bacterial activity is highly dependent on a large number of factors, including physical (temperature, pressure, flow rates, system design, and surface roughness), chemical (phases present, salinity, cations, anions, pH, organic sources, dissolved gases, and treatment chemicals), and operational (parameters such as cleaning and pigging operations, system outage and stagnancy, and changes in produced/injected fluids) [1].

Controlling these parameters and optimizing them can lead to better management of the bacterial and biofouling problems.

20.2.1 System design and operating parameters optimization

Several studies have shown that the design philosophy and operating parameters can affect and shape the microbial communities. Locations that are stagnant, have lower temperature ($<37°C$), experience long retention times, and have recycled fluids from stagnant sites, closed drains, and loops are more vulnerable to higher MIC and biofouling risk [2,3].

Essentials of Flow Assurance Solids in Oil and Gas Operations. https://doi.org/10.1016/B978-0-323-99118-6.00005-8

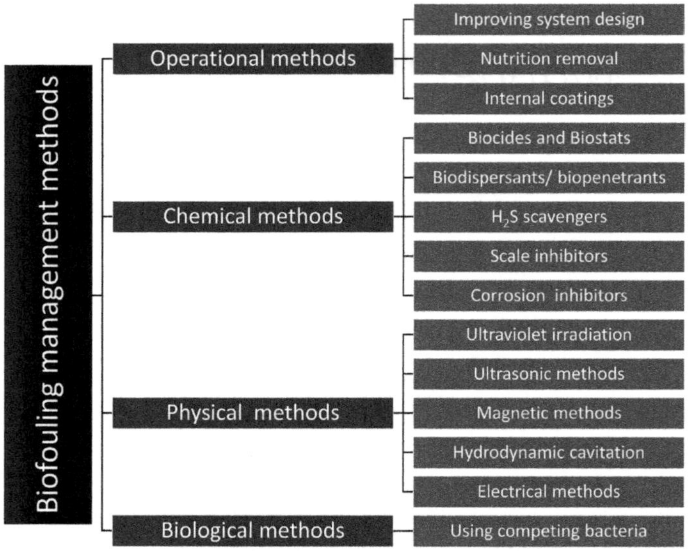

FIG. 20.1

Biofouling management methods.

Therefore understanding the effects of system design and operating parameters helps in modifying and optimizing them to better control biofouling. A few examples of useful practices are:

- Good choice of materials, less corrosive, smooth, coated, etc.
- Operating at higher flow rates. Licina [4] suggested that a flow rate of $1\,m\,s^{-1}$ ($3\,ft.\,s^{-1}$) should be maintained to prevent thick biofilm formation.
- Reduce dead legs, fluids recycling, closed drains, and the number of bends and elbows.
- Having good access to surfaces.
- Decrease the retention times, improving the fluids separation.
- Apply rigorous pigging and maintenance strategies.
- Reduce the shutdown times, avoid transient state production or intermittent production.
- Apply good monitoring of surface biofouling and MIC.
- Good choice of fluids introduced to the formation, e.g., fracturing fluids, waterflood, acidizing fluids, hydrotesting, scale squeeze fluids, etc., that are clean or uncontaminated with bacteria, and have low sulfate and other nutrient concentrations. Important water-quality criteria identified by Lutey [5] include low organic content, minimum soluble iron, minimum microflora and fauna, minimum turbidity and suspended solids, minimum dissolved gases, including H_2S, NH_3, O_2, and CO_2, and essentially no hydrocarbons such as oils and greases.

20.2.2 Nutrition removal

SRB and other bacteria require specific and suitable levels of nutrients, e.g., N, P, SO_4^{2-}, and carbon sources, to support their growth; thus reducing one or all of these nutrients to below limiting levels may provide a means to control bacterial activity [6].

Sulfates are sometimes absent in the reservoir prior to introduction of high sulfate content seawater during waterflood operations. Therefore removing sulfate from the injection water may be a feasible method to control SRB activity in the reservoirs. Sulfate removal can be achieved by nanofiltration systems, as discussed in Chapter 15.

Reducing P and N in the injection water may be of less use in controlling SRB, since SRB only uses limited amounts of these elements and they already exist in produced water (either naturally or by the injected chemicals, e.g., phosphonate scale inhibitors are a main source of P, quats are a main source of N) [6].

20.2.3 Coatings

Coatings have been reported to control different flow assurance issues, e.g., scale, hydrates, wax, asphaltenes, and have also been used in corrosion control. In fact, most of the cost of corrosion prevention (about 89.5% of total costs) is ascribed to the usage of protective coatings [7]. Coatings have been used since the 1980s to control microbial activity and biocorrosion, with several advances to fulfill the needs of the oil and gas operators [8,9]. Yuan et al. [8] reviewed the protective coatings for combating microbial corrosion and biofouling.

Different types of inorganic, organic, and hybrid (inorganic/organic) coating have been reported [10–12]. Inorganic coatings include surface conversion, anodizing, galvanizing, metallic coatings, and others. Organic coatings include epoxy resins, fluorinated compounds, and other polymeric coatings.

Some problems associated with the coatings include low bonding with the metal surface, occurrence of defects, cracks, crevices, and biodegradation (especially for polymeric material, as bacteria use it as nutrition). Good coating materials must fulfill the following [11]:

- Corrosion resistant
- Resistant to bacterial attack and degradation
- Solids and biofilm repellent

Silicon-based coatings have been shown to have good performance [13], while galvanized steel was reported to be corroded by SRB [14]. The inorganic zinc coatings were also found to suffer from biocorrosion by SRB [15]. Titanium oxide coatings were susceptible to bacterial adhesion and biofilm formation [16]. Electrodeposited nanoporous tungstite films on steel were presented as a biofouling-resistant metallic coating [17]. Polyamide-epoxy coating outperformed other polymeric materials and showed the lowest degradation and bacteria [18]. Coal-tar and epoxy resin coatings showed good performance, whereas PVC-based coatings showed poor performance [10]. Fiberglass resin and epoxy resin coatings are generally effective but they are still costly [9]. Cement linings may reduce microbial fouling, although they may be attacked by sulfur-oxidizing bacteria such as *Thiobacillus* [19]. Polytetrafluoroethylene (PTFE) is known for its nonstick properties, poor abrasion resistance, thermal conductivity, and adhesion to the metal surface, making it incompatible for commercial use. To improve its performance, an electroless nickel phosphorous, PTFE (Ni-P-PTFE), composite coating was developed [20].

Polymeric coatings incorporated with biocides can give optimum protection against microbial activity. Different types of polymers with biocides are available, including:

(i) traditional polymers incorporated with biocides, examples being polyurethane, fluorinated compounds, epoxy resins, polyimides, silicone, coal-tar epoxy, and polyvinyl chloride; (ii) antibacterial polymers containing quaternary ammonium compounds; and (iii) conductive polymers, examples being polymer polypyrrole (PPy), polyaniline (PANI), and polythiophene (PBT) [8].

Other types of antifouling coating that have been used in the marine industry include controlled depletion polymer (CDP), self-polishing copolymer (SPC), and foul release (FR) [21].

20.2.4 Pigging

Pigging operations are crucial in mitigating biofouling and MIC. Pigging removes the pipeline debris and deposits that form due to bacterial activities. Routine pigging assures cleanliness of the pipe, reducing the chances of biofilm accumulations. Pigging combined with biocides exposes the biofilm and its bacteria to the biocides, improving the biocides killing efficiency and sterilizing the system in general.

20.3 Chemical biofouling control

As indicated above, biofouling is a complex problem that involves diverse processes, e.g., biofilm formation, biocorrosion, biomineralization, and H_2S secretion. Thus different types of chemicals are usually used to control the microbial growth, such as biocides, biostats, cleaners (biodispersants and biopenetrants), or to control the problem resulting from microbial activity, e.g., H_2S scavengers, scale inhibitors, and corrosion inhibitors. Combining these chemicals sometimes is recommended in severe cases.

20.3.1 Biocides and biostats

Using biocides is the most common method of combating biofouling and microbial activity in general. A review of biocides was reported by [9–11,22–25]. Biocides, literally meaning "killer of living organisms," can be generally divided into inorganic, such as chlorine, ozone, bromine, etc., or organic, including isothiazolones, quarternary ammonium compounds, and aldehydes. However, most common biocides are classified into two large categories [11]:

- Oxidizing and
- Nonoxidizing biocides.

The major feature of a biocide, whether oxidizing or nonoxidizing, is being "broad spectrum," which means being able to kill as many diverse types of microorganisms as possible [9]. Thus the action of a biocide used to disinfect any system is expected to be bactericidal, fungicidal, and algicidal [11].

20.3.1.1 Oxidizing biocides

Oxidizing biocides basically cause irreversible cell damage by hydrolysis/oxidation of the proteins in the microorganism cell membrane and in the polysaccharides that bind the microorganisms to surfaces, causing loss of normal enzyme activity, disrupting nutrient passage through cell walls, and finally cell death [22,26].

The most common oxidizing biocides are chlorine- and bromine-based compounds, ozone, hydrogen peroxide, and others, as shown in Table 20.1.

Table 20.1 Some examples of oxidizing biocides.

Oxidizing biocide type	Example
Chlorine-based biocides	Chlorine gas Hypochlorous acid Hypochlorite Chlorine dioxide Chloramines
Bromine-based biocides	Hypobromous acid Hypobromite Bromo-chloro-dimethyl-hydantoin (BCDMH) Bromo-chloro ethylmethyl-hydantoin (BCEMH) Bromine chloride
Peracids	Peracetic acid performic acid
Other oxidizing agents	Ozone Hydrogen peroxide Permanganate

- ***Chlorine and its compounds***

Chlorine and its compounds are commonly deployed in the oil-producing industry, primarily for water injection systems. Chlorine is usually generated in oilfields by means of electrolytic oxidation of the chloride ions in seawater. Chlorine is an excellent algicidal and bactericidal agent.

Chlorine gas reacts with water to produce hypochlorous acid (HOCl) and hydrochloric acids:

$$Cl_2 + H_2O \rightarrow HOCl + HCl \tag{20.1}$$

And also hypochlorite ions (OCl^-) are produced, which are biocidal:

$$OH^- + HOCl \rightarrow OCl^- + H_2O \tag{20.2}$$

During injection water treatment, chlorination is used as a primary bactericidal control and is usually applied upstream of the deaerators (or preferably at the seawater lift pumps where the seawater is pre-treated before a first filtration takes), and nonoxidizing biocides are then applied downstream of the deaerator. Continuous treatments using 0.1–0.2 mg/L of chlorine are common, while for intermittent treatments the dose can be between 0.5 and 1 mg/L [11].

The presence of chlorine, hypochlorous, or hypochlorite is pH dependent [27]. Fig. 20.2 shows the pH profile of the reactive chlorine species.

HOCl is considered the most effective form of the chlorine species. Its ideal pH range is from pH 6.5 to 7.5 [11]. Fig. 20.3 illustrates the proposed mechanism of action of HOCl on the bacterial cell.

Other sources of chlorine are salts of hypochloric acids such as sodium hypochlorite (NaOCl) or calcium hypochlorite ($Ca(OCl)_2$); these compounds can be supplied in solid or liquid form, and their action is similar to that of gaseous chlorine. Chloroisocyanurates are another source of chlorine that are easily handled powdered compounds, which hydrolyze in water to slowly release chlorine and cyanuric acid [11,22].

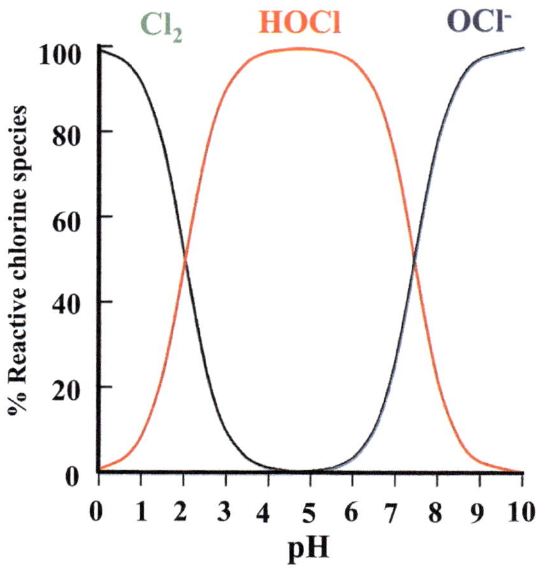

FIG. 20.2

pH profile of chlorine species in water at 140mM chloride.

Reprinted from A.J. Kettle, A.M. Albrett, A.L. Chapman, N. Dickerhof, L.V. Forbes, I. Khalilova, R. Turner, Measuring chlorine bleach in biology and medicine, Biochim. Biophys. Acta 1840 (2014) 781–793 copyright (2014) with permission from Elsevier.

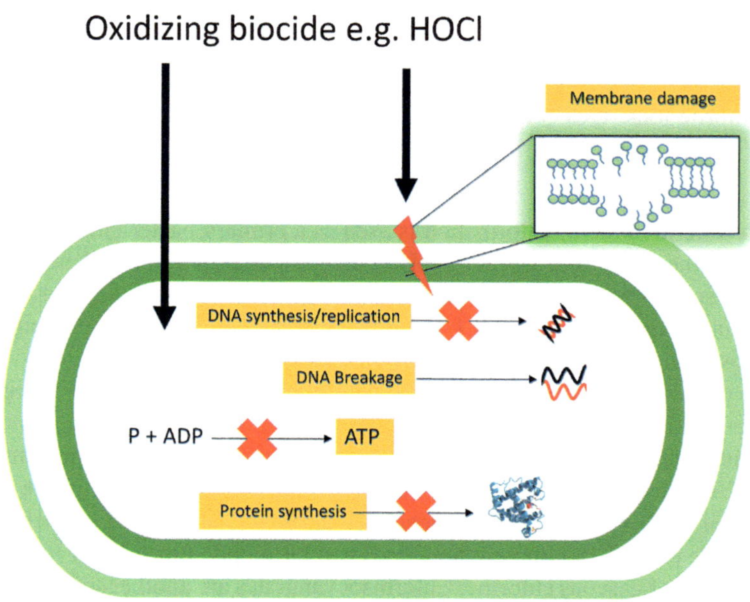

FIG. 20.3

Mechanism of action of HOCl on the bacterial cell.

Chlorine dioxide (ClO_2) is also used as a substitute of gaseous chlorine and was found to be effective in remediation of bacterial and sulfide contamination in oilfields with excellent biofilm-removing characteristics [11,22]. ClO_2 was reported to clean cooling systems that were heavily fouled with biofilm and calcium carbonate [28].

Common drawbacks of chlorine and its products are that chlorine gas is hazardous and difficult to handle, and not sufficiently effective against biofilm, and is pH sensitive and corrosive to metals [29,30]. Hypochlorites are more expensive than chlorine, and can cause scaling issues [9]. ClO_2 is expensive and considered hazardous.

- ***Chloramines***

Chloramines are used in papermaking and other industries as superior biocides to the hypochlorites. They have good biofilm activity and they were found to be more effective than chlorine and hypochlorite on biofilm because they could penetrate the biofilm [31,32]. They also have low corrosivity, but they are difficult to handle and a bit costly [9]. An example of chloramine is given in Fig. 20.4.

- ***Bromine and its compounds***

Carpenter and Nalepa [33] reviewed bromine-based biocides. The compounds of bromine form hypobromous acid (HOBr) with contact with water. The biocidal action of the bromine compounds is more effective at a wider pH range than HOCl [11], and the same for the hypobromite ion, which is a much more powerful oxidizing biocide than hypochlorite [22].

A common method for applying bromine involves using solid brominated hydantoins using BCDMH (Fig. 20.5) and BCEMH, which also can be a source of chlorine, as they react slowly with water-releasing HOCl and HOBr [11,22]. These compounds are more active than sodium hypochlorite, at the same dosage, on biofilms and planktonic bacteria and are less corrosive [34,35]. The biocidal and biofilm removal activity of BCDMH was superior to BCEMH [11].

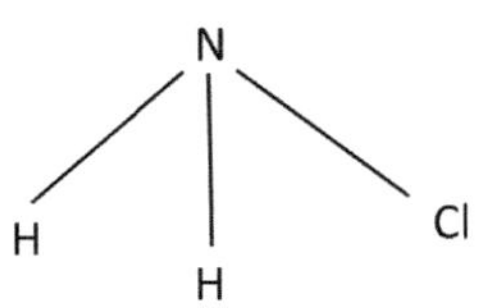

FIG. 20.4

Monochloramine.

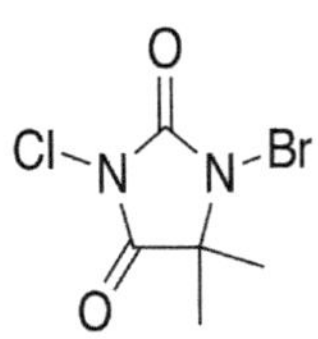

FIG. 20.5

Structure of BCDMH.

Source: https://en.wikipedia.org/wiki/BCDMH#/media/File:BCDMH.png, photo credit Edgar181.

Bromine-based biocides have been shown to penetrate and inactivate biofilm better than chlorine, but it is less effective for control of algae than chlorine, and it is a bit costly [22,35]. Bromine-based biocides were found to cause partial oxidative decomposition of AMP scale inhibitor and affect its efficiency [36].

- ***Bromine chloride***

This biocide is generally used for the treatment of industrial waters. It reacts with water, producing HOBr at near neutral pH [33]. The activity of the formulation is comparable to that of a fresh bleach, while still delivering the benefits of a stabilized bromine system [23]. It is an effective biofouling control agent at low doses (0.5 ppm or less) [33].

- ***Ozone***

Ozone has been widely used for drinking water and food disinfection [37]. Ozone is a wide spectrum biocide with high oxidative power against bacterial cells and biofilms. Ozone is less corrosive and can prevent scale formation, as ozonated water showed less potential for scale deposition, which is explained as due to chelating the scale-forming cations and dispersing the suspended solids by the oxidized end products of the biofilms, including biopolymers, carboxylic acids, and other organic compounds [38,39].

In general, ozone is used in the range of 0.01–0.05 mg/L to prevent biofilm formation, while higher concentrations (between 0.2 and 1.0 mg/L) are required if surfaces are covered with biological deposits, to remove biofouling [11].

Major problems with ozone are that its use is limited to small systems or specific sites within larger systems, and it is corrosive at high concentrations. Also, ozone must be generated on site, requiring investment for installation and running the equipment [9].

- ***Hydrogen peroxide***

It is stable and very safe (frequently used in pharmaceutical products). In practice, it can work as an effective biocide and H_2S scavenger. Peroxide can be used at concentrations between 50 and 100 ppm to inhibit microbial growth when water is in contact with metals for prolonged periods of time [9,11]. Hydrogen peroxide has also been reported to be used in stimulation operations, drilling, and other applications in oil and gas fields.

- ***Peracids***

Peracids are regarded as clean, effective biocides for oil and gas applications. A common example of peracids is peracetic acid (PAA), which is a strong oxidizer (Fig. 20.6), and a highly effective biocide and chlorine alternative that has gained in popularity over the past several years for different

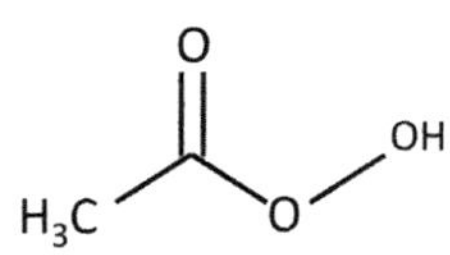

FIG. 20.6

Structure of peracetic acid.

applications in the petroleum industry [40]. PAA is effective at low concentrations against a broad spectrum of microorganisms, including gram-positive and gram-negative bacteria, yeasts, molds, and algae under a wide variety of conditions. It is also effective against anaerobic and spore-forming bacteria. Most importantly, PAA can penetrate the biofilm and effectively kill biofilm microorganisms at low concentrations and short contact times. PAA is not sensitive to pH changes and is not affected by water hardness, organic matter, or other contaminants [41]. An improved, compatible, and environmentally friendly formulation of PAA for fracturing fluids was introduced [42]. The only problems related to PAA are that it needs to be generated onsite, it can be corrosive, and still contains residuals of hydrogen peroxide.

Performic acid was also reported to be effective in removing biofilm, inhibiting bacterial growth (both sessile and planktonic), controlling H_2S, and removing iron sulfide scales. Treatment using performic acid, at 0.5 ppm, provides a microbial reduction equivalent to 3 logs of bacteria/mL in a planktonic kill study [43].

- ***Permanganate***

This compound is used as a disinfectant in the health care industry. It was also used as an iron sulfide deposits dissolver in refineries [44].

20.3.1.2 Nonoxidizing biocides

Nonoxidizing biocides are commonly used in different applications in oil and gas fields, downhole, injection water, flowlines, vessels, tanks, etc. Common chemistries of nonoxidizing biocides include aldehydes, quaternary ammonium salts, and quaternary phosphonium salts. Nonoxidizing biocides can supersede oxidizing biocides owing to their overall control of bacteria, algae, and fungi, their greater persistence as many of them are pH independent, and their applicability in different field conditions [11]. Table 20.2 summarizes the different types of nonoxidizing biocides.

- ***Aldehydes***

Common compounds used in oilfields are formaldehyde, glutaraldehyde, and acroline. Aldehydes are electrophilic biocides that interact and cross link with amino and nucleic acids, leading to cell wall damage and cytoplasmic coagulation [24].

- *Glutaraldehyde*

Glutaraldehyde, or glut for short, is a common oil and gas biocide used to control fungi, algae, and bacteria (including sulfate-reducing bacteria in bacterial biofilms) [11,45]. Glutaraldehyde is a fast-acting biocide (3–6 h), compatible with chlorine and production fluids, nonionic, nonfoaming, effective over a wide temperature and pH range, typically at pH 6.5–9.0 (7–10) as hydroxyl, carbonyl, and amino groups are formed at the bacterial cell surface at this higher pH range, inducing a faster bactericidal effect [25,46]. Glutaraldehyde is one of the main biocides, being used in offshore hydraulic fracturing operations, besides its conventional use in waterflood operations [47]. Typical use concentration is 40–125 mg/L.

Glut is insensitive to sulfides, and displays only low to moderate toxicity toward aqueous organisms [11,22]. Nonetheless, studies have shown limited efficiency of glutaraldehyde to penetrate and control organisms in biofilms [48]. Glutaraldehyde is also incompatible with strong acids, ammonia, amines, and bisulfite oxygen scavengers. Fig. 20.7 illustrates glutaraldehyde structure.

Table 20.2 Common types of nonoxidizing biocides.

Nonoxidizing biocide type	Examples
Aldehydes	Formaldehyde Glutaraldehyde Acrolein
Quaternary phosphonium compounds	Tetrakis(hydroxymethyl)phosphonium sulfate (THPS) Tributyltetradecylphosphonium chloride (TTPC)
Quaternary ammonium salts (QUATS)	Alkyldimethylbenzylammonium chloride Hexadecyltrimethylammonium bromide Chlorhexidine Benzalkonium chloride
Guanides	Dodecylguanidine hydrochloride Polyhexamethylene biguanide hydrochloride Tetradecylguanidine
Organic bromides	Bronopol DBNPA BHAP
Imidazole	Metronidazole
Isothiazolones	5-Chloro-2-methyl-4-isothiazoline-3-one (CMI) 2-Methyl-4-isothiozolin-3-one (MIT)
Thiones	Tetrahydro-3,5-dimethyl-1,3,5-thiadiazine-2-thione (Dazomet)

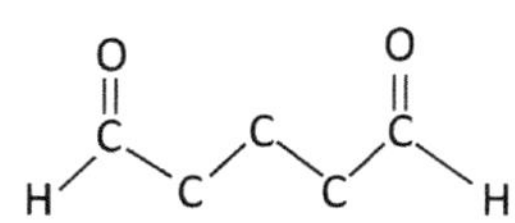

FIG. 20.7

Glutaraldehyde structure.

- *Formaldehyde*

Formaldehyde is another low-cost aldehyde that has been reported as being effective in killing sessile microorganisms in established biofilms in water treatment systems [49] and in coreflood experiments [50]. Formaldehyde was successfully used in eight wells in West Texas fields as an inhibitor for corrosion promoted by biogenic hydrogen sulfide, with important net savings in operating costs [51]. Formaldehyde can also be used as an H_2S scavenger. However, formaldehyde is classified as a probable human carcinogen by the US Environmental Protection Agency (EPA), which limits its use in the fields. Fig. 20.8 illustrates the structure of formaldehyde.

- *Acrolein*

Acrolein is considered a cost-effective, multifunctioning agent that has been used to solve different production chemistry problems, including scavenging H_2S, scavenging mercaptans R-SH, odor control, and acting as a potent biocide and a dissolver for iron sulfide scales [52–54]. Its dual solubility

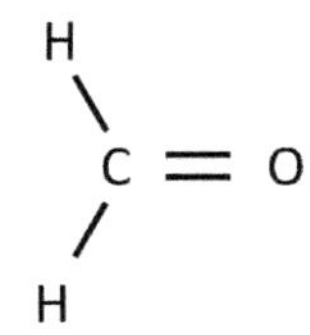

FIG. 20.8

The structure of formaldehyde.

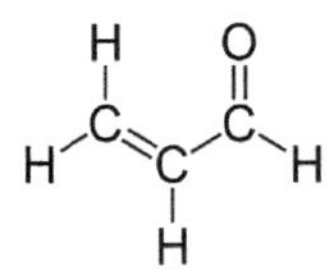

FIG. 20.9

Structure of acrolein.

in oil and water enables it to penetrate and remove oil-coated FeS solids and sessile bacteria from production systems, which helps in controlling microbial growth and improves phase separation [55]. Additionally, acroline-based polymers were found to be useful as scale inhibitors [56]. However, although biodegradable, it should be noted that acrolein has high acute toxicity and is a suspected carcinogen, so it needs to be handled carefully. Acrolein also has limited penetration into the biofilm [57]. Other drawbacks of acrolein are that it is potentially flammable and reactive with polymers, scavengers, and violently reactive with strong acid and alkalis [9]. Fig. 20.9 illustrates the structure of acrolein.

- ***Quaternary phosphonium compounds***
  - *THPS*

Tetrakis(hydroxymethyl)phosphonium sulfate (THPS) (Fig. 20.10) is considered an effective oilfield biocide for different surface and downhole applications. It is considered nonfoaming and has wide applications as a biocide that appears to kill all anaerobic strains of SRB at pH 3–10 [22]. Under proper conditions, THPS releases the THP ion, which cleaves sulfur-sulfur bonds in the disulfide amino acids of the microbial cell wall, causing cell wall damage [24].THPS is also considered an effective iron sulfides dissolver, with enhanced performance when formulated with ammonium chloride [58,59], and the mixture was also found to be an effective zinc and lead sulfides dissolver [60]. In one field, using THPS with a surfactant killed the SRB, removed iron sulfide scale, and increased well production

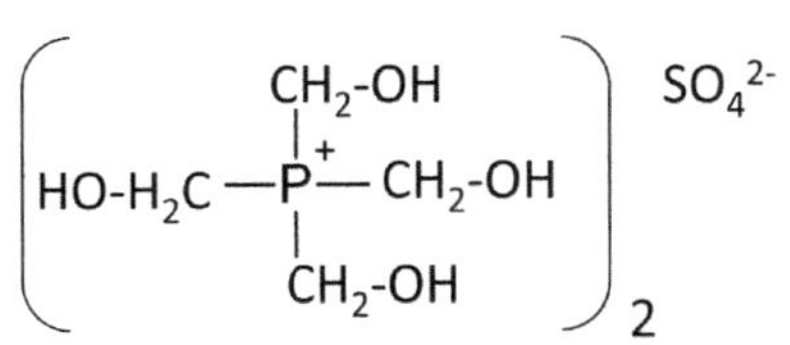

FIG. 20.10

Structure of THPS.

in several wells by as much as 300% [61]. Downhole injection of THPS retards the growth rate of SRB and the rate of H_2S production, which results in savings on H_2S scavenger consumption [58].

THPS is usually mixed with surfactants to increase its biofilm penetration efficiency: for example, surfactants such as benzalkonium chloride (BAC) were found to be very effective in improving the penetration of THPS into bacterial slimes [62].

THPS is generally injected in the range of 10–90ppm (as active THPS) [11] or 50–100ppm for planktonic treatments; for biofilm treatments, the concentrations might reach 250ppm.

The main drawbacks of THPS are that it is incompatible with commonly used oxygen scavengers, sulfite-based scavengers, and erythorbic acid, [22,63] leading to inefficient deareation, or deposit formation, in addition to molecular oxygen formation [63]. Besides, it can be a source of sulfate ion in water and cause scale formation. THPS is also considered a corrosive compound and care should be taken when it is needed to be injected at higher temperatures.

- *TTPC*

Tributyltetradecylphosphonium chloride (TTPC) (Fig. 20.11) is a broad spectrum microbiocide, with good algicide properties [45]. TTPC is effective at low concentrations, is fast acting, and is effective against both acid-producing and sulfate-reducing bacteria. It is a lytic biocide that binds to the cell membrane and disrupts it, causing cell lysis. It has outperformed both glutaraldehyde and THPS in comparative biocidal tests. TTPC is compatible with other production chemicals such as oxidizing biocides, hydrogen sulfide, and oxygen scavengers, and has good thermal stability and corrosion inhibition properties. However, TTPC can cause some low foaming and will not dissolve iron sulfide scale [22].

The typically used concentration is 20–40mg/L as product (at 50% active material), for periodic cleanup programs, either alone or in combination with other nonoxidizers, such as bronopol, guanadines, DBNPA, and isothiazolines [45].

- ***Quaternary ammonium salts (QUATS)***

Quaternary ammonium compounds (quats or QAC) form a class of cationic amphiphiles that have been used to enhance the performance of other biocides or used as a biocide themselves in addition to being a highly effective corrosion inhibitor [11,22]. As biocides, they are lytic biocides that bind to anionic functional groups on the cell membrane, disrupting its structure and function, and causing eventual lysis of the cells [24]. Quats can also be used as biopenetrants/biodispersants with biocides to increase biocide efficiency in removing biofilms [64].

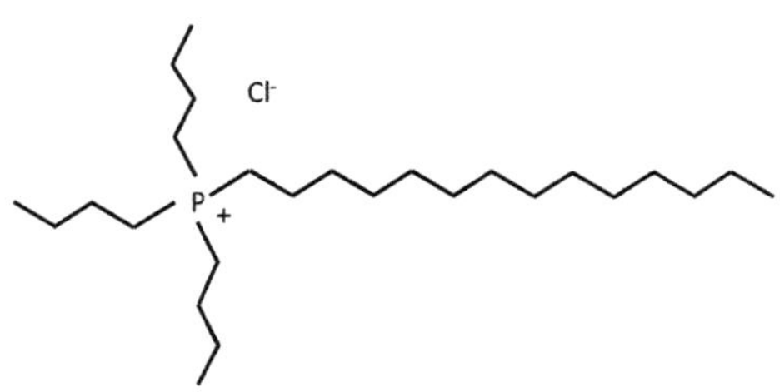

FIG. 20.11

Structure of TTPC.

Quats are effective algicides and reasonably good bactericides, and can be applied over a wide range of pH (optimum pH 6.5–8.5) [45]. Low concentrations (<250ppm) of a common quat such as alkyldimethylbenzylammonium surfactant have been shown to reduce the growth of numerous bacterial strains, including SRB [22].

Quats are deactivated by high hardness (typically over 500mg/L), chlorides, peroxides, chromates, permanganates, oil, dirt, silt and debris. Quats have poor compatibility with polyanionic polymers [11,45].

- **Guanides**

These biocides are cationic surfactants with good dispersing properties and are readily biodegradable. They act as toxicants due to their high surface activity, which disrupts the microorganism's cell wall and cytoplasm, causing eventual cell lysis [45]. Examples are dodecylguanidine hydrochloride and acetate, also polyhexamethylene biguanide hydrochloride and tetradecylguanidine.

- ***Organic bromides***
  - *Bronopol*

Bronopol (2-bromo-2-nitropropane-1,3-diol) (Fig. 20.12), also known as BNPD, is a general-purpose biocide in oilfield operations [65]. Bronopol is commonly used in the tight and shale gas industry due to its rapid degradation in slightly alkaline environments [66]. It is also suitable for cooling waters, operating at lower pH levels [45]. Furthermore, it is also used during fracturing and stimulation jobs.

Its biocidal efficacy is based on the reaction with sulfur-containing nucleophiles such as glutathione or cysteine, thereby disrupting key cell components and biological functions [24]. It is also believed that the biocidal efficacy of bronopol is different under oxic and anoxic conditions [66]. Bronopol is compatible with most other biocides, such as quats, methylene bis(thiocyanate), and isothiazolinones, and all common cooling water inhibitors [45].

Disadvantages of bronopol include its extreme sensitivity toward sulfides, and its instability in heated environments above 40°C [66].

- *2,2-Dibromo-3-nitrilopropionamide (DBNPA)*

DBNPA, or 2,2-dibromo-3-nitrilopropionamide (Fig. 20.13), is an effective, general-purpose biocide that is commonly applied in various decontamination operations, such as hydrotesting or for cleaning topside production facilities, as it is suitable for bacterial slime cleanup [45,66]. DBNPA is also approved for use in seawater sulfate removal units (SRU) to avoid biofilm formation on the nanofiltration membranes, since the deaerator is upstream of the SRU [66].

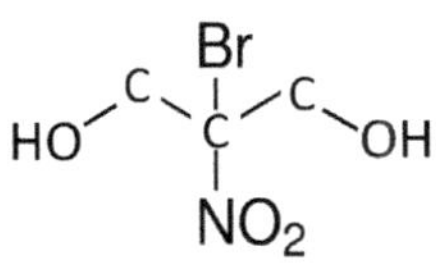

FIG. 20.12

Structure of bronopol.

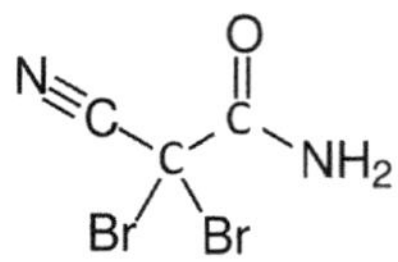

FIG. 20.13

Structure of DBNPA.

Similar to bronopol, the mechanism of action of DBNPA is by interfering with thiol and amine groups within cellular components, and thereby disrupting normal cellular components and biological functions [66].

DBNPA has rapid mode of action (contact time often less than 1 h), dosage levels are generally low (<50 ppm), and it can be suitable for once-through cooling systems [45,66]. However, DBNPA is easily hydrolyzed at alkaline pH, is temperature and light sensitive, is incompatible with sulfites and sulfides, is not effective against fungi and algae, is foaming, and is also expensive [9,45].

- *2-Bromo-4'-hydroxyacetophenone (BHAP)*

BHAP or 2-bromo-4'-hydroxyacetophenone (Fig. 20.14) is a good biocide for bacterial slimes, and biofilms due to its good penetration properties. It is pH independent, effective at high pH levels, and useful for applications at low dose level, such as once-through cooling systems, where the dose rate is only 1–3 mg/L [45].

- ***Imidazoles***

Imidazoles such as metronidazole show biocidal and corrosion inhibition performance. The biocide efficacy is thought to be from disrupting the DNA helical structure, and inhibiting nucleic acid synthesis. It has been shown to be effective in controlling biogenic sulfide production in oilfield water injectors [67].

- ***Isothiazolones***

Isothiazolones are broad-spectrum antimicrobials that are effective against sessile organisms (biofilms). [9]. Examples of isothiazolones are CMI and MIT (Fig. 20.15).

Their biocidal efficacy utilizes a two-step mechanism involving rapid inhibition (minutes) of growth and metabolism, followed by irreversible cell damage resulting in loss of viability (hours) [68].

Isothiazolones are effective over a broad pH range, are compatible with other water treatment chemicals and other biocides, and have good biodegradability [11]. However, isothiazolone use in the oil and gas industry is limited due to their extreme H_2S sensitivity [11,66]. However, they have been

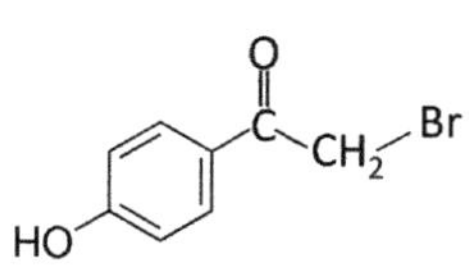

FIG. 20.14

Structure of BHAP.

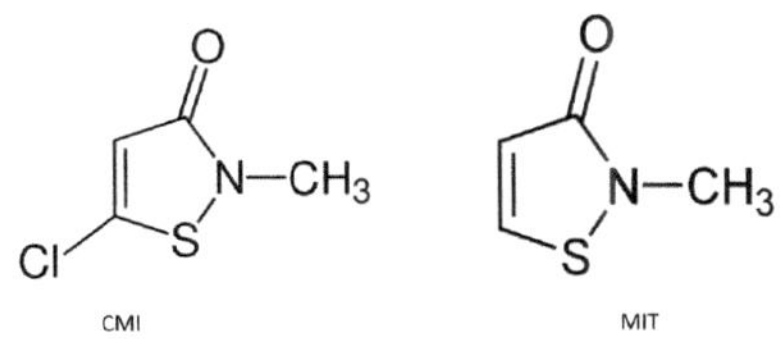

FIG. 20.15

Structure of CMI and MIT.

Source: Wikipedia.

reported to mitigate SRB activity, thus preventing H_2S production [69]. They are also used in the treatment of injection and cooling water, paper mills, and metal working fluid systems [70].

- ***Thiones***

A good example is tetrahydro-3,5-dimethyl-1,3,5-thiadiazine-2-thione (Dazomet) (Fig. 20.16), which been successfully used in the Barnett, Marcellus, Eagle Ford, and Haynesville gas shale basins in the United States. This biocide is effective for long-term maintenance against most aerobic and anaerobic bacteria (very effective against SRB) encountered in water sourced for fracturing applications [71].

- ***Thiocyanates***

A common example is methylene bis(thiocyanate) or MBT, which is a relatively quick-kill biocide, with a 2–4h contact time, that is very effective against SRBs. Their kill mechanism is to affect and poison essential enzymatic reactions. They are pH sensitive, hydrolyse quickly at pH greater than 7.5, and deactivate at high levels of dissolved ferric iron [45].

Other common biocide types include dithiocarbamates, which are general-purpose biocides, with high efficiency against fermentation-producing organisms with optimum performance at pH between 7 and 8.5; DTEA (2-(decylthio)ethanamine hydrochloride), which is a multifunctioning corrosion inhibitor, biodispersant, and wide spectrum biocide with good biofilm-removing properties. Other examples of multifunctioning compounds include amines like alkylamines, diamines, and triamines, which are also corrosion inhibitors, phenolic biocides, triazine derivatives, which are also used as H_2S scavengers, oxazolidines, which are also H_2S scavengers [22], sulfones like Bis(trichloromethyl) sulfone and Bis(tributyltin) oxide TBTO [45].

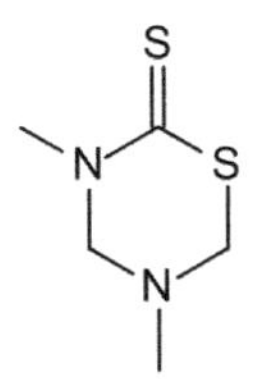

FIG. 20.16

Structure of Dazomet.

Source: https://en.wikipedia.org/wiki/Dazomet#/media/File:Dazomet.svg, photo credit: Epop.

20.3.1.3 Biostats

Biostats basically interfere with the main biological and metabolic processes in bacterial cells, and hence control their growth or indirectly cause their death [22]. Common examples of biostats include:

- Anthraquinone
- Rotenone
- Azide ions
- Nitrite and nitrate ions
- Molybdate or tungstate ions
- Selenate ions

Biostats are not frequently used like biocides, but mixing biostat and biocide can enhance the biocide efficiency, reduce the kill time, and decrease the required dose to kill the bacteria. Batch treatment of anthraquinone together with THPS biocide has also been shown to be effective in stopping biogenic sulfide production in produced water tanks in slop-handling systems [72,73].

20.3.1.4 Metals

Some metals in extremely small doses act like powerful germicides and this action is termed oligodynamic action. A common example is silver. Silver is found to inhibit enzymatic activities, disrupt cell membranes, besides binding to DNA molecules clumping or separating the double helix, which eventually cause cell damage and death [74]. A common application bactericidal effect of silver is the use of Katadyn water filter.

Antimicrobial effect is shown by metals such as mercury, silver, copper, lead, zinc, gold, aluminum, and other metals, and the concentration of the metal needed for this antimicrobial effect is extremely small [75].

20.3.1.5 Biopenetrants, biodispersants and enhancers

Biopenetrants and biodispersants are chemicals that can penetrate and break the complex matrix of biofilms to allow biocides to reach the organisms for more effective kill and control. They are surfactants that are able to also emulsify oils and remove mineral scales and foulants and disperse biopolymers to improve biocide treatments for biofilm/biofouling removal [76].

Some biopenetrants or biodispersants include:

- EDTA
- DMAD (dimethyl amide of fatty acid)
- Dodecylamine acetate
- EDDS
- DOSS (Docusate or dioctyl sulfosuccinate)
- Polyoxyalkylenes
- Enzymes

Biopenetrants were found to enhance the performance of THPS and increase the rate of removing the biofilm [62]. Oxidizing biocide and a wetting/penetrating agent, along with pH adjustments, were used to control biocides in nuclear power plants [77].

Different polymers and copolymers, e.g., polyacrylates, polymaleates, and copolymers of partially hydrolyzed polyacrylamides, are commonly used as dispersants to remove biofouling deposits [11].

EDTA is an effective "antibiofilm" solution, causing biofilm dispersal [78]. EDTA is also known to disrupt the outer membrane of Gram-negative bacteria [79], without exerting any direct actions on Gram-positive bacteria. Divalent cations, such as Ca^{+2} and Mg^{+2}, are found in the cell wall of bacteria and in the EPS of biofilm to give them structural stability; EDTA basically interacts with and chelates these cations, which destabilizes and disrupts the cell membrane and biofilm structure, allowing biofilm penetration and detachment [80]. Thus EDTA has been used as a biocide performance enhancer.

Another example of a chelating agent that is used to remove biofilm and enhance biocide efficiency is ethylenediamine disuccinate (EDDS). Addition of EDDS to glutaraldehyde and THPS showed enhanced performance and a smaller dose of these chemicals to be injected [81].

Other commercial surfactants were used as biofilm removers/dispersants, like Triton X-100, CTAB, TSP, sodium tripolyphosphate, and SDS [82]. Surfactants are mixed with a biocide to enhance the biocide efficiency and to deliver the biocide where it is really needed (e.g., a pipe surface instead of the bulk fluid) [83].

D-amino acids have been found to trigger biofilm disassembly, presumably because they are signal molecules or because they interfere with protein synthesis. Many studies have reported that D-amino acids enhanced THPS against the SRB biofilm [84,85].

20.3.1.6 Testing and screening biocides

Biocide screening and selection are usually known as kill studies, kill tests, or time kill tests. A broth bottle or nutrition media that represents the produced fluid salinity is inoculated with the bacteria culture under study to determine the efficiency of the biocide to achieve the highest killing of bacteria (99% killing). The test is performed as described in NACE Standard TM0194.

The main problems with the time kill test include:

- It only represents planktonic bacteria, not the sessile and biofilm-forming bacteria.
- It measures the bacteria using a most probable number (MPN) technique, which underestimates the problem, e.g., measures between 1% and 10% of the actual bacteria count.
- It is time consuming, as the incubation period takes up to 28 days.

Therefore technically the dosage determined from a kill study is supposedly the minimum dosage that should be present at the end of the system, where the biocide is no longer required [48].

For studying sessile bacteria and the actual field conditions, a more representative test setup in addition to a faster and more accurate measurement technique are needed, so the following are recommended:

- Use more dynamic methods, e.g., reactors, Robbins device, coupons, bioprobe, and other methods that involve biofilm formation.
- Use a fast and sensitive method, such as the molecular microbiological methods (MMM), qPCR, DGGE.

20.3.1.7 Field application of biocides

Biocide applications depend on many factors, including the following:

- Type of system, e.g., water injection, oil system, cooling water system
- Location of the problem, e.g., reservoir, downhole, topsides, flowlines, etc.
- The type of biocide and its properties, e.g., oxidizing, nonoxidizing, fast acting, pH dependent, etc.
- Operating conditions

Generally speaking, the methods by which biocides are applied include:

- Continuous injection
- Slug dosing
- Squeeze treatments
- Added with fracturing fluids

If continuous injection to be used, a higher dose is recommended to achieve biofilm penetration and bacteria killing. However, continuous injection of biocides is not common in production systems, due to the following reasons [48]:

- It is costly, especially with high volumes of production fluids.
- At the low doses, the biocide acts as a biostat rather than biocide.
- Bacteria are able to adapt and be more resistant to biocides with continuous injection at low doses.
- In oil-water systems, biocide loss is possible by partitioning in the oil phase.

Most commonly, biocides are introduced to a system via a slug dose once or twice per week, depending on the application, for a predetermined contact time (2–8h). The high concentration over a short time achieves biofilm removal and bacterial killing more effectively than treating with lower doses for longer periods of time [86]. It is recommended to use a different biocide in each slug treatment by alternating between two or more products (e.g., THPS, glutaraldehyde), to avoid the bacterial adaptation and resistance to biocides.

It is a common practice to apply slug doses of nonoxidizing biocides to sterilize flowlines and equipment during pigging operations, as shown in Fig. 20.17.

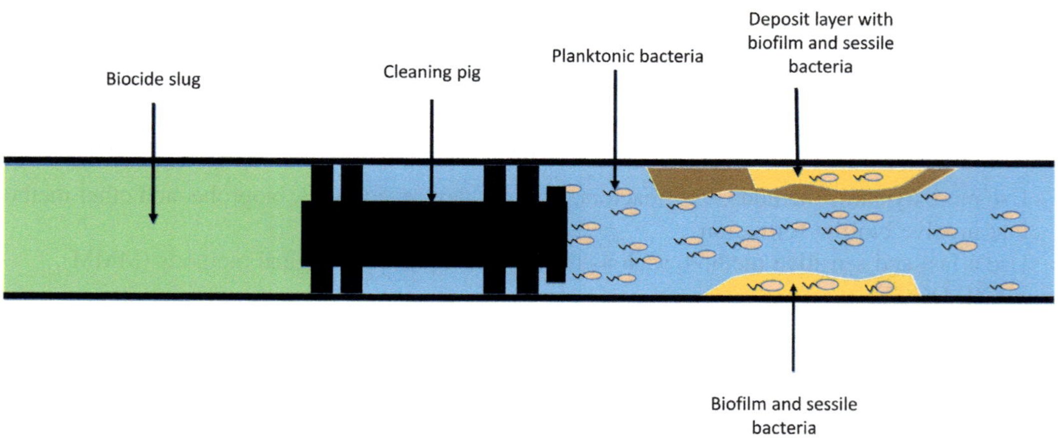

FIG. 20.17

Biocide application during pigging operations. The mechanical cleaning effect of the pig removes the deposits and the biofilms, hence exposes the bacteria to the subsequent biocide to effectively kill the bacteria.

Other applications of biocides include system mothballing, hydrotesting, and sterilizing after shutdowns. Biocides used for these applications are acroline, glutaraldehyde, THPS, and quaternary ammonium compounds.

For fracturing operations, oxidizing biocides like persulfate salts, chlorine, ClO_2, and chlorite, are often used successfully [87,88]. However, the use of oxidizing biocides in fracturing fluids is limited to the pretreatment, as they are significantly affected by the downhole conditions, and hence they are combined with a nonoxidizing biocide for effective microbial control [24,89]. The biocides that are commonly used to treat fracturing fluids include DBNPA and bronopol as solid products, and THPS in smaller liquid quantities [25].

Nonoxidizing biocides can also be applied downhole as squeeze treatments to control bacterial activity in producing reservoirs, or in water supply wells for water treatment at the source [61,90]. Nonoxidizing biocides like THPS and glutaraldehyde are the best choice for this application.

The most commonly used biocides for the preservation of drilling muds are glutaraldehyde and isothiazolones. With drilling muds having a pH of 8 or lower, biocides such as DBNPA or bronopol may be effective. Quats, glutaraldehyde, and isothiazolone are usually used for the preservation of water-based packer fluids. Glutaraldehyde is also preferred for use in completion and workover fluids [25].

For effective application of the biocide, the following points should be considered:

- Accurate identification and count of the bacteria.
- Proper selection of the biocide: the selected biocide must [91]:
  - Be effective against the targeted bacteria
  - Have an effective time/kill profile, e.g., fast-acting biocides are required for injection and production systems
  - Be able to penetrate the biofilm, and achieve the bacteria killing
  - Be stable and effective under operating conditions, e.g., P, T, pH, salinity, etc.
  - Be compatible with other chemical and production fluids, and not cause other production problems, e.g., foaming, emulsions
  - Have the ability to be monitored and traced
  - Be cost effective
  - Be ecofriendly
- Biocide doses are based on the time kill test, which are preferred to be based on sessile bacteria and biofilm studies rather than planktonic bacteria.
- Biocides are applied where the problem originates or preferably upstream of the problem. If the problem is in the reservoir, squeeze is necessary, or well head injection can be used if squeeze is not possible. Injection should be targeted in the water phase, to avoid loss in the oil phase [48].
- A combination of biocides can be used in severe cases, one having high biofilm penetration features and properties and another with high killing action.
- In the case of heavily fouled systems, cleaning and killing are necessary, and this is achieved by:
  - Using scale/deposit dissolvers to remove the deposit layers
  - Using biodispersants and biopenetrants to enhance biofilm penetration
  - Using combination of biocides
  - Applying biocides with pigging. Biocide application during pigging is very effective in removing the deposit layer and biofilm, allowing effective killing by the biocides.

- After biocide application, system monitoring is mandatory, which is achieved by the following:
  - Monitoring of biocide concentration
  - Frequent bacteria detection and count
  - MIC and corrosion in general monitoring

20.3.2 H_2S scavengers

H_2S is the main product of the bacterial activity of SRBs. H_2S reacts with the metallic components of the production system and causes severe corrosion and corrosion scales, and it can react with formation waters that are rich in Zn, Pb, Fe and cause severe sulfide deposition. H_2S is a hazardous gas that causes health issues and even death in very high concentrations. H_2S removal or scavenging can be achieved using a variety of chemicals, including those described in the following sections.

20.3.2.1 Amines

These are most commonly used to remove H_2S from produced natural gas; the process is known as amine scrubbing, gas sweetening, or acid gas removal. Aqueous alkanolamines such as monoethanolamine (MEA), diethanolamine (DEA), or *N*-methyldiethanolamine (MDEA) are usually used in these units to remove H_2S and CO_2. These solutions can also be regenerated and reused [22].

20.3.2.2 Oxidizing chemicals

Chemicals such as chlorites ($NaClO_2$), bromates/iodates ($NaBrO_3$), nitrites (NaNO2), or persalts can be used and can work as biocides and H_2S scavengers as well [22].

20.3.2.3 Aldehydes

Examples are formaldehyde, acrolein, glutaraldehyde, and glyoxal. These compounds work as biocides, H_2S scavengers, and sulfide scale dissolvers [22].

20.3.2.4 Triazine

Triazine scavengers are a common class of H_2S scavengers. The application of triazines requires high temperature and longer contact time. A major drawback of this class is that they can increase the pH of the produced water, which can exacerbate carbonate scaling.

20.3.2.5 Metal carboxylates and chelates

Metal ions such as (II) or iron(III), and chelating agents such as nitrilotrisacetic acid, EDTA, polyaminodisuccinic acid, or gluconate [22] have been used as H_2S scavengers.

20.3.2.6 Metal-based scavengers

Metal oxide beds like zinc oxide and iron oxide are used to control sour gas facilities. These are not recommended in upstream facilities, as their main reaction product is sulfide deposits.

20.3.2.7 Testing and screening of H_2S scavengers

Different methods have been reported for screening and evaluating the H_2S scavengers. In the so-called "gas breakthrough test" H_2S is bubbled through the chemical scavenger solution in a lab-designed glass tower until H_2S breakthrough in the headspace is measured, after which sampling analysis of the gas phase is conducted. The Parr reactor evaluation method is an autoclave method in which H_2S is monitored after the addition of a scavenger to a liquid phase, which is saturated with H_2S [92]. Other new designs have been reported [92,93].

20.3.2.8 Field application of H_2S scavengers

Field application of H_2S scavengers is very complicated and requires a good design. One method reported by Petrobras involved injecting the H_2S through the gas lift lines [93]. They also injected flowlines and other locations.

For injection in flowlines, the scavenger requires spray nozzles or atomizers rather than using open end quills, so that the scavenger will be atomized into the gas phase, increasing the surface area for the reaction. The reaction also might need enough retention time and high temperature. Four H_2S scavenger field trials have failed in the Egyptian field, due to using an improper injection nozzle (open end quill instead of atomizer), low temperature, and short contact time. Injection should be avoided near some specific design elements, e.g., bends, since the scavenger aerosol will condensate forming droplets due to its density. Water-based H_2S scavengers require that the gas be water saturated to prevent solid reaction products [22].

20.3.3 Scale inhibitors and dispersants

A main result of biofouling is the formation of different mineral scale deposits, including the sulfides of Fe, Zn, Pb, iron oxides, and Ca and Fe carbonates. Using scale inhibitors, dissolvers, and dispersants can ease the severity of the biofouling problem in oilfields. These chemicals can be applied in the reservoir to mitigate formation damage, in the near wellbore area to mitigate tubing scaling, or on topsides like flowlines and separators.

Chemical dissolvers are also used to dissolve these mineral scales, including mineral acids, organic acids, and chelating agents. Scale control chemicals and their application are discussed in detail in Chapter 15.

20.3.4 Corrosion inhibitors

A variety of corrosion inhibitors arer used to control corrosion problems in oil and gas operations. These inhibitors work by preventing the anodic reaction, the cathodic reaction, or both, or by forming a protective hydrophobic film on the metal surface, which provides a barrier to the metal dissolution [94]. So these inhibitors do not interrupt the biological activities directly, but rather they try to extend the lifetime of the metallic parts. However, some corrosion inhibitors showed antimicrobial performance against some bacteria, like quats and imidazolines.

20.4 Physical methods of controlling biofouling

20.4.1 Irradiation with UV

Ultraviolet (UV) radiation is a common physical method of controlling microbial activity. The UV radiation kills the microbial cells by destructive effects of the UV on the bacterial DNA [95]. Ultraviolet light interacts with and is absorbed by pyrimidine bases (e.g., thymine and cytosine in DNA), causing their dimerization. This ultimately makes the bacteria unable to produce proteins or reproduce, which eventually leads to the death of the organisms [96]. Generally, UV rays can be in a wide range of wavelengths; for the best result in microbial control, it is recommended to use UV radiation with a wavelength of 254 nm, which is in the germicidal UV-C spectrum [97].

Ultraviolet light irradiation can be as effective as the common biocides, but is less expensive [98]. UV irradiation was tenfold more effective than chemical biocides in controlling metatrophic bacteria and iron bacteria, with only one-fifth of the cost [99]. In addition, Gloe et al. [100] reported that the use of UV irradiation to disinfect water on-the-fly for stimulation treatments greatly reduced the required biocide volumes and hence their cost.

The efficiency of the UV irradiation process depends on different factors, including the turbidity of the water/fluid, the flow rate, suspended and colloidal solids, initial bacterial density, and other physical and chemical parameters [101].

20.4.2 Ultrasonic treatment

The ultrasonic (UT) treatment, or acoustic cavitation (AC), is based on the cavitation mechanism, i.e., bubble generation and collapse, where the collapsed bubble generates a localized high temperature (thousands of degrees) and high pressure (hundreds of atmospheres), which then kills the bacteria by destroying the cell components or by the formation of chemical species with biocidal effects, e.g., hydrogen peroxide [9,102].

The techniques have been reported in different applications in drinking water and wastewater. [103,104]. Some studies reported the use of UT alone or in combination with chemical biocides. The combination of UT and 5 ppm gluteraldehyde decreased the bacterial concentration to undetectable levels within 4 h after sonication [105]. UT has also been reported with good efficiency against soft biofilms [106].

20.4.3 Hydrodynamic cavitation treatment

Hydrodynamic cavitation (HC) is generated by fluctuations in fluids pressure. Similar to ultrasonic cavitation, HC is based on bubble formation and collapse, causing localized high pressure and temperature. The method has been proposed as a possible method of disinfection for potable water, wastewater, and industrial process water [107].

The technique can also be used to decrease heavy crudes viscosity and improve their transportation.

20.4.4 Magnetic methods

The application of magnetic field was found to mitigate mineral scales, paraffins, and asphaltenes. However, not much work has been done on bacteria control by magnetic field. Alkhazan and Sadddiq [108] found that the application of a magnetic field decreased the bacterial count in sewage water, Liu et al. [109] showed that using a low-frequency square wave pulsed magnetic field can inhibit heterotrophic bacteria activities in circulating cooling water.

20.4.5 Electrical methods

20.4.5.1 Cathodic protection

Cathodic protection (CP) is a common technique that has been widely used in mitigating corrosion in oil and gas fields. In this method, a continuous stream of electrons is supplied to the targeted metal to replace the corroding metal electrons and hence prevent its dissolution. That electron source could be a sacrificial anode or electric current.

CP is thought to mitigate MIC and biofouling by making the metal surfaces more repellent to the bacteria, preventing their attachment, or by creating alkaline conditions that are not tolerated by the bacteria [9,110]. However, CP may not be able to prevent biofilm formation and once biofilm is formed CP cannot prevent MIC unless excessively high potential is applied, which is not economical [111].

Commonly reported CP criteria to protect against SRB-induced MIC are −0.95 V (vs. Cu-$CuSO_4$ reference electrode) [9,112,113]. However, recent investigations of CP effects on pure iron surfaces in the presence of SRB have showed that cathodic polarization of −1070 mV (vs. Cu-$CuSO_4$ reference electrode) was not sufficient to combat SRB MIC [9,114].

20.4.5.2 Pulsed-power and electrostatic treatment

Pulsed-power treatment, also known as pulsed electric field (PEF) treatment [115] or electropulse treatment [116], involves the bombardment of the metal surface with electromagnetic energy pulses, which may inactivate the microorganisms present in the treated liquid. The method has been evaluated for use in the food processing industry [117], as well as for cooling tower process water treatment [118,119].

20.4.5.3 Electrochemical production of biocides

These methods produce the biocides electrochemically. The most common examples are production of chlorine in waterflood plants and cooling water systems, and silver production by the Electro-Katadyn method.

20.5 Biological methods of controlling MIC and biofouling

In these methods, good (less destructive) bacteria are used to limit the growth of the bad (highly destructive) bacteria. An example is the nitrate-reducing and sulfide-oxidizing bacteria (hNRB and NR-SOB), which reduce nitrate ions to nitrite, nitrous oxide, and other end products that can inhibit SRB growth [120]. Therefore some techniques have used nitrate-reducing bacteria (NRB) against SRB [9,22]. Common methods of implementing this technique include [121] (a) adding the essential nutrients for the good bacteria that exist in the system in low counts (biocompetitive exclusion), e.g., adding nitrate, the NRB already in the system will outnumber SRB, and (b) adding both the essential nutrients and bacteria for a system that does not have any indigenous good bacteria (bioaugmentation), e.g., addition of ex situ grown NRB and nitrate into a system that may have no "indigenous nitrate-reducing bacteria" to outnumber SRB [9]. These methods are illustrated in Fig. 20.18.

This method of using NRB and nitrate has shown effective mitigation of souring problems in over 1000 assets [122].

Such treatments have been improved by adding molybdenum compound (molybdate) as it is a known enzymatic inhibitor of the hydrogenase enzyme found in SRB [123], also a sulfate analog, which inhibits dissimilatory sulfate reduction and enhances mitigation, was reported [122]. A study

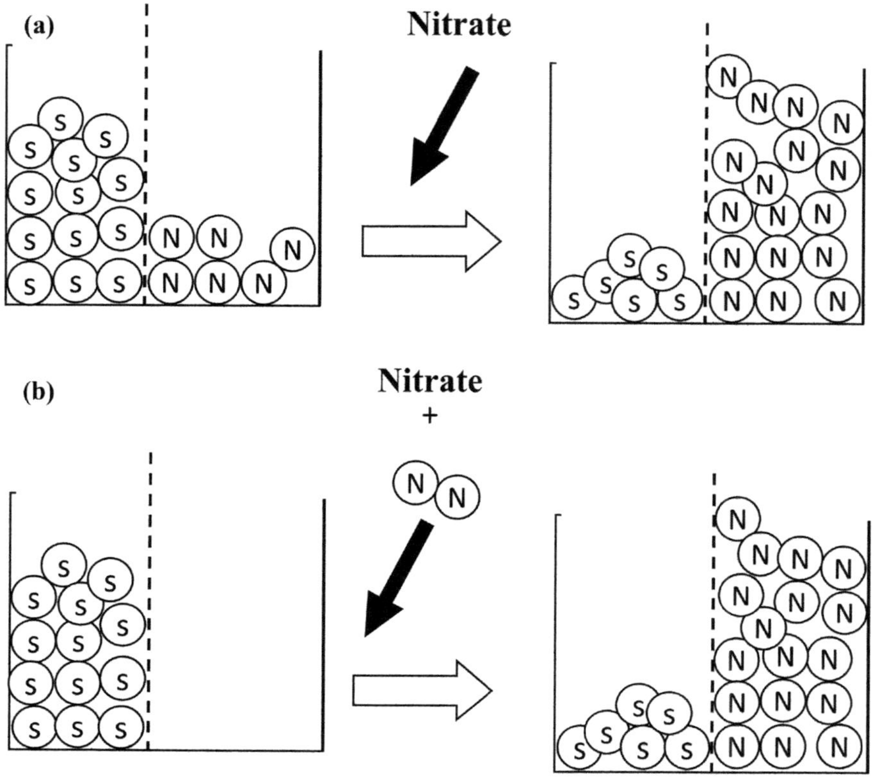

FIG. 20.18

Biological methods of controlling SRB growth using NRB bacteria: (a) biocompetitive exclusion, (b) bioaugmentation.

Reprinted by permission from Springer Nature: (Springer, Microbiologically Influenced Corrosion: An Engineering Insight, by Reza Javaherdashti, copyright (2017).

showed that nitrite treatment alone may be preferable in reservoirs with only SRB present [124]. A novel class of bacteria that oxidize sulfide as well as oil organics with nitrate has been reported [125].

Nitrite (not nitrate) can be added to production systems to inhibit SRB growth and to scavenge the already formed H_2S [22].

20.6 Removing biofouling deposits

The removal of biofouling deposits is a complex process that requires thorough investigation and good selection of the treatment options. Effective removal of biofouling depends on

- Biofouling deposit composition and properties, e.g., texture, hardness
- Location of deposit

- System design and accessibility
- Economics

Generally, if the deposit comprises organic (paraffin, asphaltenes) or mineral scales, a proper solvent/dissolver must be applied to remove the deposits layer, before using the biocide to remove the biofilm and then kill the bacteria. In pipelines, pigging followed by biocide applications is an effective method to remove the deposits and expose the bacteria to the biocide solution to achieve the killing. The removal of organic deposits is discussed in Chapters 17–19. The removal of mineral scales is discussed in Chapter 15. Special attention should be given to locations with relative stagnancy, e.g., low velocity spots, separators, dead legs, drain lines, etc., as these locations usually include volumes of solids and biomass and should be treated carefully.

Most of the biofouling removal problems arise from the assumption that killing is equal to cleaning or that biocides application will do the job, which is a vague and misleading assumption, due to the following: first, killing the planktonic bacteria will not clean the system and will not solve the problem of the destructive sessile bacteria, and second, this denies the fact that biofouling always entails a complex mixture of organic deposits, scale, biomass, bacterial cells, and fines, which can be arranged in layers that hinder effective treatment. Besides, partial cleaning will leave a portion of the bacteria population available to recover after the biocide is rinsed off and use the biomass to reproduce.

Thus, an effective biofouling mitigation depends on these few tips:

- Fully characterize the biofouling deposits and determine their physical properties.
- Understand the difference between killing and cleaning.

Based on the deposit composition, full cleaning of the deposits is mandatory for effective bacteria killing. Proper dissolvers are designed to remove the deposit layers, and biocide is designed (as mentioned in Sections 20.3.1.6 and 20.3.1.7) to remove the biofilm and fully mitigate the bacteria. Some chemicals can work as scale dissolver and biocide, e.g., THPS and acrolein,

- Use a combination of biocides.

Combining two biocides in treatment has the advantage of biofilm penetration of one biocide and the effective killing of the other. It also helps in offsetting the physiological resistance of bacteria against biocides. Combinations of both oxidizing and nonoxidizing biocides in one treatment affords a wide spectrum of biocidal efficiency because of their dual mechanism of action.

- Use biodispersants, biopenetrants and surfactants.

These compounds help in penetrating the biofilm, reducing the attachment of the bacteria to the surface.

- Use pigging frequently.

The use of pigging is very efficient in removing biofilms and biomass from the production system. It is highly recommended to make the pigging operations more frequent for highly infected areas. Biocides are often used in pigging operations, since this is a very efficient way of treating the freshly exposed surfaces of the pipe with a fairly high concentration of biocide [126]. Biocides used in pigging operations must be fast acting and not be deactivated by hydrogen sulfide [25].

20.7 Strategies and philosophies of controlling biofouling

20.7.1 Fluids analysis

20.7.1.1 Water chemistry

Water chemistry includes measuring the concentration of cations, anions, pH, salinity, TSS, alkalinity, dissolved gases, and others. High Fe, Ni, Mn can be indicative of a corrosion problem. High sulfide, CO_2, and low SO_4 can be indicative of bacterial activity. Furthermore, the water chemistry is used in corrosion and MIC modeling.

20.7.1.2 Microbiological analysis

This includes collecting samples from the system from locations that are thought to be vulnerable to bacterial activity, and performing microbiological analysis to identify the type of bacteria and count them. It is recommended to use highly sensitive techniques such as MMM. The different methods of monitoring and performing microbiological analysis are reviewed in Chapter 13.

20.7.2 Pig returns and surface analysis

Microbiological analysis is performed on the deposits from pig returns and surfaces to identify the bacteria and count them. The use of MMM methods is highly recommended. Chemical analysis of pig returns and surface analysis is also aimed at identifying the deposit composition as an indicator of the microbial activity. Microscopic methods are also effective in surface analysis for corrosion investigations. Scale coupons, corrosion coupons, and bioprobes are effective methods for monitoring MIC and biofouling; they are removed from the system periodically and investigated using microbiological methods, chemical analysis, and surface analysis.

20.7.3 Modeling and simulation

Some companies have developed their indoor MIC modeling and simulation. Corrosion modeling is also generally useful.

20.7.4 Choosing the proper method of control

Based on the lab analysis and modeling methods, the optimum mitigation method is designed. Biocides are commonly used, and commonly applied by slug/batch treatments. A combination of different biocides is preferred. The use of biodispersants and biopenetrants is recommended in severe cases. In the case of severe fouling with organic and inorganic deposits, deposits dissolution should be planned. Pigging is an effective method of removing debris and exposing bacteria for biocide. Table 20.3 summarizes the different methods of biofouling mitigation.

20.7.5 Monitoring and assessment

After applying the treatment method, system monitoring should be maintained. This is achieved by:

- Monitoring system parameters (P, T, etc.),
- Performing microbiological analysis frequently on the water and on the pig returns,

Table 20.3 Summary of biofouling mitigation methods.

Mitigation method	Description
Operational methods	- Generally used in MIC and biofouling prevention - Requires system proper predesign, or system modification - Coatings are effective in protecting from MIC and corrosion in general - Pigging can be used in combination with chemical methods
Chemical methods	- Used in prevention and removal - Wide variety of chemicals for different applications, depending on the case. Biocides are commonly used. H_2S scavenger are used to mitigate reservoir souring and sour systems. Corrosion inhibitors are used to protect the production facilities - Generally, do not require system modification - Two or more different biocides are required to avoid bacteria adaptation with biocide - Generally cost effective, but can be expensive in severe cases that require high volumes of chemicals - Can be applied combined with mechanical methods for efficient treatment - Very effective for use during shut downs to protect the system - Chemicals can have side effects, e.g., incompatibility with other chemicals, produced fluids, or enhance other production chemistry problems
Other nonchemical methods	- UV irradiation has been successfully used and reduced chemical consumption - Magnetic and ultrasonic methods have been reported - Electrical methods protect the system from MIC - Biological methods have been reported in the field - They can be case dependent

- Monitoring corrosion rates, e.g., UT inspection, LPR,
- Use online methods, coupons, bioprobes, etc.,
- Measurement of biocide concentrations in the produced fluids,
- Frequent analysis of water chemistry can be indicative of microbiological activities.

A general scheme for MIC and biofouling management is illustrated in Fig. 20.19.

20.8 Recent advances in biofouling management

The realm of MIC and biofouling have witnessed several advances related to the factors affecting the problems and the methods of mitigation. New types of biocides have been developed, including nanomaterials as biocides and ecofriendly biocides. Further improvements in internal coatings have been seen, including the applications of nanomaterials. Coatings provide long-term protection from MIC and biofouling, but they can be susceptible to mechanical damage. Further research in the physical methods such as electrical, ultrasonic, magnetic, and electromagnetic is expected to be driven by the movement toward ecofriendly treatment methods. Generally, a holistic approach combining different methods is required to combat MIC and biofouling, rather than using simple methods.

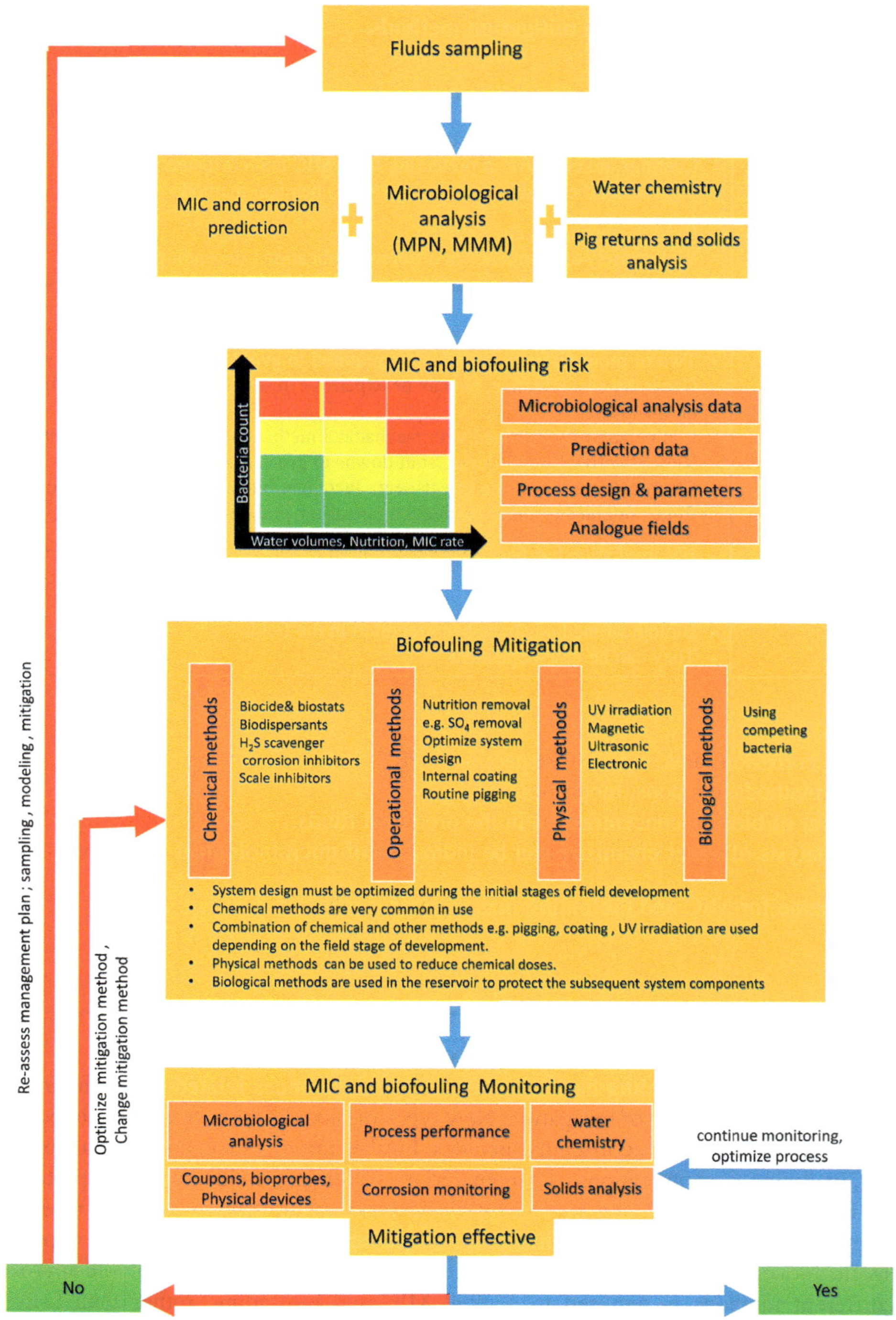

FIG. 20.19

Biofouling management strategy.

20.9 Case studies

20.9.1 MIC risk assessment in oil field [127]

The Halfdan is a chalk field located in the Danish Sector of the North Sea. In the platform under investigation (Halfdan Platform A), the initial investigation of the field showed that MIC and biofouling are not a threat due to the relatively low water cut (BS&W < 1%–2%). Water injection was used (currently at rate of 200,000 BPD) and treatment of water injection was applied, using nitrate as well as THPS biocide on the water injection supply platform in order to minimize biofilm build-up in the near wellbore area of the Halfdan water injectors due to nitrate injection.

Three spools were removed from the system for investigation of system integrity. Initial results indicated ongoing corrosion with deposits of biofoulant accumulation (2–3 cm thick layer of mixtures of scales and corrosion products) and underdeposit corrosion. Microbiological analysis of the inner and outer deposit layers indicated the presence of bacteria and archaea (SRB, SRP, and methanogens), with an active biofilm formation. Deposits analysis showed that the composition was mainly FeS, iron oxides, biofilm, with other components $CaCO_3$, $BaSO_4$, and NaCl.

A generic risk assessment of the MIC problem was performed to calculate the MIC risk factor, to estimate the time it takes to initiate MIC below a biofilm, especially for the SRP, SRB, and methanogens.

After addressing the problem and estimating the risks, extra measures were put in place to monitor and control it. These included the following:

- Installation of new chemical injection points
- Actively applying new MMM methods for detecting microorganisms
- Ongoing evaluation of scale and MIC control strategies (new chemistry to be introduced)
- Flush corrosion coupons installed for better warning of scale and MIC issues (and to monitor the effect of applied chemistry).

20.10 Summary

In this chapter, the methods of mitigating MIC and biofouling were reviewed. These include operational, chemical, and nonchemical methods. The operational methods involve adjusting the system design, the removal of essential nutrition required for the microorganisms, and the use of coatings. The chemical methods are based on different chemicals to control the microorganism activity or to control the harmful end products they produce. Thus different types of biocides, H_2S scavengers, corrosion inhibitors, and scale inhibitors are used. Biocides are the common chemical method for combating MIC and biofouling, and these chemicals should be screened and applied properly to achieve effective biofilm penetration and killing of the harmful sessile bacteria. This is achieved by accurately identifying the bacteria. The nonchemical methods involve the use of UV irradiation, ultrasonic cavitation, HC, and electrical methods, such as CP.

MIC and biofouling management involves fluids sampling and analysis to determine the water chemistry and microbiological species and their counts. Pig returns, coupons, and bioprobes are also analyzed to determine the deposits composition and microbiological speciation. MIC modeling and simulation can also be performed if possible. This will be combined with system analysis to determine the spots with high MIC and biofouling potential. After assessing the risk of MIC and biofouling, a mitigation method is attained, using biocides, H_2S scavengers, corrosion inhibitors, or scale inhibitors. A combination of these chemicals can be used to fully mitigate the problem. Pigging and biocide application is an effective method to control microbial activities. After applying the mitigation method, microbiological analysis, water analysis, and biocide tracing are commonly used to monitor the system.

References

[1] T.L. Skovhus, R.B. Eckert, Practical aspects of MIC detection, monitoring and management in the oil and gas industry, in: Paper #3920, Corrosion 2014, San Antonio TX, USA, March 9–13, 2014.

[2] K.E. Duncan, I.A. Davidova, H.S. Nunn, B.W. Stamps, B.S. Stevenson, P.J. Souquet, J.M. Suflita, Design features of offshore oil production platforms influence their susceptibility to biocorrosion, Appl. Microbiol. Biotechnol. 101 (16) (2017) 6517–6529, https://doi.org/10.1007/s00253-017-8356-8. Springer Verlag.

[3] M.P. Shaw, H. Hoffmann, M. Home, Case study: comparison of microbial monitoring techniques used in the field and how their complementarity can be harnessed to build a full picture of the microbial life in the field, in: SPE International Oilfield Corrosion Conference and Exhibition, Aberdeen, Scotland, UK, 9–10 May, 2016. SPE-179936-MS.

[4] G.T. Licina, Sourcebook for Microbiologically Influenced Corrosion, Electric Power Institute, Palo Alto, CA, 1988, pp. 3–5. ERRI NP-6815–D.

[5] R.W. Lutey, Treatment for the mitigation of MIC, in: J. Stoecker (Ed.), A Practical Manual on Microbiologically Influenced Corrosion, NACE International, Houston, TX, 2001, pp. 9.1–9.30.

[6] G.G. Geesey, H.-E. Flemming, Biofouling and Biocorrosion in Industrial Water Systems, CRC Press, 1994.

[7] G.H. Koch, M.P.H. Brongers, N.G. Thompson, Y.P. Virmani, J.H. Payer, Corrosion Costs and Preventive Strategies in the United States. Report FHWA-RD-01-156, NACE International, Houston, TX, 2002.

[8] J. Guo, S. Yuan, W. Jiang, L. Lv, B. Liang, S.O. Pehkonen, Polymers for combating biocorrosion, Front. Mater. 5 (2018) 10, https://doi.org/10.3389/fmats.2018.00010.

[9] R. Javaherdashti, Microbiologically Influenced Corrosion: An Engineering Insight, Springer International Publishing, 2017.

[10] H.A. Videla, Manual of Biocorrosion, CRC Lewis Publishers, Boca Raton, FL, 1996.

[11] H.A. Videla, Prevention and control of biocorrosion, Int. Biodeterior. Biodegradation 49 (2002) 259–270,-https://doi.org/10.1016/S0964-8305(02)00053-7.

[12] S. Holberg, C. Bischoff, Application of a repellent urea-siloxane hybrid coating in the oil industry, Prog. Org. Coat. 77 (10) (2014) 1591–1595.

[13] D. Wiebe, J. Connor, G. Dolderer, R. Riha, B. Dyas, Protection of concrete structures in immersion service from biological fouling with silicone-based coatings, Mater. Perform. 36 (5) (1997) 26–31.

[14] E. Ilhan-Sungur, N. Cansever, A. Cotuk, Microbial corrosion of galvanized steel by a freshwater strain of sulphate reducing bacteria (*Desulfovibrio* sp.), Corros. Sci. 49 (2007) 1097–1109, https://doi.org/10.1016/j.corsci.2006.05.050.

[15] Z. Muntasser, M. Al-Darbi, M. Tango, M.R. Islam, Prevention of microbiologically influenced corrosion using coatings, in: Corrosion 2002, NACE International, Houston, TX, 2002. Paper 02029.

[16] S.J. Yuan, F.J. Xu, S.O. Pehkonen, Y.P. Ting, E.T. Kang, K.G. Neoh, Biocorrosion behavior of titanium oxide/butoxide-coated stainless steel, J. Electrochem. Soc. 155 (2008) C196–C210, https://doi.org/10.1149/1.2885073.

[17] A.B. Tesler, P. Kim, S. Kolle, C. Howell, O. Ahanotu, J. Aizenberg, Extremely durable biofouling-resistant metallic surfaces based on electrodeposited nanoporous tungstite films on steel, Nat. Commun. 6 (2015), https://doi.org/10.1038/ncomms9649. Nature Publishing Group.

[18] B.J. Little, R.I. Ray, P.A. Wagner, J. Jones-Meehan, C.C. Lee, F.B. Mansfeld, Spatial relationships between marine bacteria and localized corrosion on polymer coated steel, Biofouling 13 (1999) 301–321, https://doi.org/10.1080/08927019909378387.
[19] S.W. Borenstein, Microbiologically Influenced Corrosion Handbook, Industrial Press Inc., New York, 1994.
[20] Q. Zhao, Y. Liu, H. Müller-Steinhagen, G. Liu, Graded Ni-P-PTFE coatings and their potential applications, Surf. Coat. Technol. 155 (2–3) (2002) 279–284.
[21] S.B. Ulaeto, J.K. Pancrecious, T.P.D. Rajan, B.C. Pai, Smart coatings, in: S. Mohapatra, T.A. Nguyen, P. - Nguyen-Tri (Eds.), Noble Metal-Metal Oxide Hybrid Nanoparticles Fundamentals and Applications Micro and Nano Technologies, Elsevier, Amsterdam, 2019, pp. 341–372, https://doi.org/10.1016/B978-0-12 814134-2.00017-6.
[22] M.A. Kelland, Production Chemicals for the Oil and Gas Industry, CRC Press, 2014, pp. 1–439, https://doi.org/10.1201/9781420092974.
[23] J. Fink, Petroleum Engineer's Guide to Oil Field Chemicals and Fluids, Elsevier Inc., 2012, https://doi.org/10.1016/C2009-0-61871-7.
[24] G.A. Kahrilas, J. Blotevogel, P.S. Stewart, T. Borch, Biocides in hydraulic fracturing fluids: a critical review of their usage, mobility, degradation, and toxicity, Environ. Sci. Technol. (2015), https://doi.org/10.1021/es503724k. American Chemical Society.
[25] D.B. McIlwaine, Oilfield application for biocides, in: Directory of Microbicides for the Protection of Materials, Springer, Netherlands, 2004, pp. 157–175, https://doi.org/10.1007/1-4020-2818-0_10.
[26] A.S.H. Makhlouf, M.A. Botello, Failure of the metallic structures due to microbiologically induced corrosion and the techniques for protection, in: Handbook of Materials Failure Analysis, Elsevier, 2018, pp. 1–18, https://doi.org/10.1016/b978-0-08-101928-3.00001-x.
[27] A.J. Kettle, A.M. Albrett, A.L. Chapman, N. Dickerhof, L.V. Forbes, I. Khalilova, R. Turner, Measuring chlorine bleach in biology and medicine, Biochim. Biophys. Acta 1840 (2014) 781–793.
[28] W.J. Ward, J.W. Lee, S.G. Freymark, Advantages of chlorine dioxide as a biocide, Ammonia Plant Saf. 20 (1978) 64.
[29] W.F. McCoy, E.J. Allain, S. Yang, A.W. Dallmier, Strategies Used in Nature for Microbial Fouling Control: Applications for Industrial Water Treatment, NACE International, 1998. Paper No. 98520.
[30] D. de Beer, R. Srinivasan, P. Stewart, Direct measurement of chlorine penetration into biofilms during disinfection, Appl. Environ. Microbiol. 60 (12) (1994) 4339.
[31] M.W. LeChevallier, C.D. Cawthon, R.G. Lee, Inactivation of biofilm bacteria, Appl. Environ. Microbiol. 54 (10) (1988) 2492–2499.
[32] E.T. Ben-Ari, Not just slime: beneath the slippery exterior of a microbial bio film lies a remarkably organized community of organisms, Bioscience 49 (9) (1999) 689–695, https://doi.org/10.2307/1313592.
[33] J.F. Carpenter, C.J. Nalepa, Bromine-Based Biocides for Effective Microbiological Control in the Oil Field, Society of Petroleum Engineers, 2005, https://doi.org/10.2118/92702-MS.
[34] C.J. Nalepa, H. Ceri, C.A. Stremick, A novel technique for evaluating the activity of biocides against biofilm bacteria, in: Paper Presented at the NACE CORROSION Conference, 2000. Paper 00347.
[35] J.F. Kramer, Biofilm control with bromo-chloro-dimethylhydantoin, in: Corrosion 2001, NACE International, Houston, TX, 2001 (Paper no. 1277).
[36] K.D. Demadis, A. Ketsetzi, Degradation of phosphonate-based scale inhibitor additives in the presence of oxidizing biocides: 'collateral damages' in industrial water systems, Sep. Sci. Technol. 42 (2007) 1639–1649, https://doi.org/10.1080/01496390701290532.
[37] W.H. Glaze, J.W. Kang, D.H. Chapin, The chemistry of water treatment processes involving ozone, hydrogen peroxide and W radiation, Ozone Sci. Eng. 9 (4) (1987) 335–352.
[38] P.D. McNicholas, Economical Calculations for the Use of Ozone in Cooling Towers, VS/PMcN/Cooling Towers/10-10-02, Duebendorf, Switzerland, 2002.

[39] N. Nombuso Gina, Effective and Efficient Ozone Use on Cooling Water Systems (Master Thesis), School of Chemical and Metallurgical Engineering, Faculty of Engineering and the Built Environment, University of the Witwatersrand, Johannesburg, South Africa, 2014.

[40] M. Kitis, Disinfection of wastewater with peracetic acid: a review, Environ. Int. 30 (2004) 47–54.

[41] J.F. Kramer, Peracetic Acid: A New Biocide for Industrial Water Applications, NACE International, 1997.

[42] R.M. De Paula, V. Keasler, J. Li, D. McSherry, R. Staub, Development of Peracetic Acid (PAA) as an Environmentally Safe Biocide for Water Treatment during Hydraulic Fracturing Applications, Society of Petroleum Engineers, 2013, https://doi.org/10.2118/164088-MS.

[43] R. Balasubramanian, J. Li, I. Yunus, T.J. Tidwell, V. Keasler, R. Staub, A Next Generation Peracid for Onshore and Offshore Water Treatment Applications, Society of Petroleum Engineers, 2017, https://doi.org/10.2118/184560-MS.

[44] P.A. Vella, Improved cleaning method safely removes pyrophoric iron sulfide, Oil Gas J 95 (8) (1997) 65–68.

[45] C. Frayne, The Selection and Application of Nonoxidizing Biocides for Cooling Water Systems, Metro Group, New York, NY, 2011.

[46] W.A. Rutala, D.J. Weber, Disinfection and sterilization in health care facilities: what clinicians need to know, Clin. Infect. Dis. 39 (2004) 702–709.

[47] H. McGinley, M.V. Enzien, G.E. Jenneman, J.B. Harris, Studies on the Chemical Stability of Glutaraldehyde in Produced Water, Society of Petroleum Engineers, 2011, https://doi.org/10.2118/141449-MS.

[48] V. Keasler, R.M. De Paula, G. Nilsen, L. Grunwald, T.J. Tidwell, Biocides overview and applications in petroleum microbiology, in: Trends in Oil and Gas Corrosion Research and Technologies: Production and Transmission, Elsevier Inc., 2017, pp. 539–562, https://doi.org/10.1016/B978-0-08-101105-8.00023-1.

[49] W.G. McLelland, Results of using formaldehyde in a large north slope water treatment system, in: SPE 35675, Paper Presented at the SPE Computer Applications, April 1997, p. 55.

[50] B.G. Kriel, A.B. Crews, E.D. Burger, E. Vanderwende, D.O. Hitzman, The Efficacy of Formaldehyde for the Control of Biogenic Sulfide Production in Porous Media, Society of Petroleum Engineers, 1993, https://doi.org/10.2118/25196-MS.

[51] P.L. Menaul, T.H. Dunn, Formaldehyde as an Inhibitor of Corrosion Caused by Hydrogen Sulphide, Society of Petroleum Engineers, 1946, https://doi.org/10.2118/946026-G.

[52] R.M. Jorda, Aqualin biocide in injection waters, in: SPE 280, Paper Presented at the SPE Production Research Symposium, Tulsa, OK, 1962.

[53] C. Reed, J. Foshee, J.E. Penkala, M. Roberson, Acrolein application to mitigate biogenic sulfides and remediate injection well damage in a gas plant water disposal system, in: SPE 93602, Paper Presented at the SPE International Symposium on Oilfield Chemistry, The Woodlands, TX, 2005.

[54] J.J. Howell, M.B. Ward, The Use of Acrolein as a Hydrogen Sulfide Scavenger in Multiphase Production, Society of Petroleum Engineers, 1991, https://doi.org/10.2118/21712-MS.

[55] D.D. Horaska, C.M. San Juan, A.L. Dickinson, S.L. Lear, A. Colquhoun, Acrolein Provides Benefits and Solutions to Offshore Oilfield Production Problems, Society of Petroleum Engineers, 2011. https://doi.org/10.2118/146080-MS.

[56] R. Siegmeier, M. Kirschey, M. Voges, Acrolein Based Polymers as Scale Inhibitors, NACE International, 1998.

[57] J.M. Walsh, Water management for hydraulic fracturing in unconventional resources—part 5, J. Pet. Technol. SPE V5 (1) (2014).

[58] J. Larsen, P. Sanders, R.E. Talbot, Experience with the use of tetrakishydroxymethylphosphonium sulfate (THPS) for the control of downhole hydrogen sulfide, in: Corrosion 2000, NACE International, Houston, TX, 2000. Paper No. 00123.

[59] F.F. Chang, Q. Wang, S. Shen, H. Badairy, T. Shafai, Y. Jeshi, T. Chen, Laboratory assessment of tetrakis(hydroxymethyl)phosphonium sulfate as dissolver for scales formed in sour gas wells, Int. J. Corros. Scale Inhib. 4 (3) (2015) 235–254, https://doi.org/10.17675/2305-6894-2015-4-3-235-254.

[60] A.E. Hussein, H.S. Mohamed, Studying the use of Tetrakis(hydroxymethyl)phosphonium sulfate (THPS) as zinc sulfide and lead sulfide scales dissolver and the factors influencing the dissolution, Int. J. Corros. Scale Inhib. 6 (3) (2017) 349–358, https://doi.org/10.17675/2305-6894-2017-6-3-9. Russian Association of Corrosion Engineers.

[61] P.R. Rincon, J.P. McKee, C.E. Tarazon, L.A. Guevara, B. Vinccler, Biocide stimulation in oilwells for downhole corrosion control and increasing production, in: SPE 87562, Paper Presented at the SPE International Symposium on Oilfield Corrosion, Aberdeen, United Kingdom, 2004.

[62] C.R. Jones, R.E. Talbot, B.L. Downward, S.D. Fidoe, K. Hernandez, R. Diaz, Keeping Pace with the Need for Advanced, High Performance Biocide Formulations of Oil Production Applications, NACE International, 2006.

[63] J. Moore, V. Keasler, B. Bennett, Compatibility of Tetrakis(Hydroxymethyl) Phosphonium Sulfate (Thps) and Ammonium Bisulfite (Abs), NACE International, 2010.

[64] L.A. Grab, A.B. Theis, Comparative biocidal efficacy vs. sulfate-reducing bacteria, in: Corrosion 92, NACE International, Houston, TX, 1992. Paper No. 184.

[65] L.G. Kleina, M.H. Czechowski, J.S. Clavin, W.K. Whitekettle, C.R. Ascolese, Performance and monitoring of a new non-oxidizing biocide—the study of BNPD/ISO and ATP, in: Paper 97403, Paper Presented at the NACE CORROSION Conference, 1997.

[66] B.E.L. Morris, G.M. van der Kraan, Application of biocides and chemical treatments to both combat microorganisms and reduce (bio)corrosion, in: Microbiologically Influenced Corrosion in the Upstream Oil and Gas Industry, CRC Press, 2017, pp. 229–253, https://doi.org/10.1201/9781315157818.

[67] E.S. Littmann, T.L. McLean, Chemical control of biogenic H_2S in producing formations, in: SPE 16218, Paper Presented at the SPE Production Operations Symposium, Olahoma City, OK, 1987.

[68] T.M. Williams, The mechanism of action of isothiazolone biocides, in: NACE—International Corrosion Conference Series, 2006. 060901-17.

[69] T.W. Williams, R. Levy, B. Hegarty, Control of SRB biofouling and MIC by chloromethylmethylisothiazolone, in: Corrosion 2001, NACE International, Houston, TX, 2001. Paper No. 01273.

[70] R. Prasad, Assessment and control of MIC in the oil industry in the 20th century, in: Corrosion 2000, NACE International, Houston, TX, 2000. Paper No. 00390.

[71] J.C. Dawson, M. Wood, A New Approach to Biocide Application Offers Improved Efficiency in Fracturing Fluids, Society of Petroleum Engineers, 2012, January 1, https://doi.org/10.2118/152051-MS.

[72] A. Dickinson, G. Peck, B. Arnold, Effective chemistries to control SRB, H_2S, and FeS problems, in: SPE 93007 Paper Presented at the SPE Western Regional Meeting, Irvine, CA, 30 March–1 April, 2005.

[73] E.D. Burger, C.A. Andrade, M. Rebelo, R. Ribeiro, Flexible treatment program for controlling H_2S in FPSO produced-water tanks, in: SPE 106106, Paper Presented at the SPE International Symposium on Oilfield Chemistry, Houston, TX, 28 February–2 March, 2007.

[74] R.B. Thurman, C.P. Gerba, Molecular mechanisms of viral inactivation by water disinfectants, Adv. Appl. Microbiol. 33 (C) (1988) 75–105, https://doi.org/10.1016/S0065-2164(08)70205-3.

[75] R.K. Jain, Copper-resistant microorganisms and their role in the environment, World J. Microbiol. Biotechnol. 6 (4) (1990) 356–365.

[76] I. de Almeida Simões, Biofilm Removal – The Effects of Biodispersant and Chlorine Dioxide, Dissertation for Master in Bioengineering, Faculty of Engineering, University of Porto, 2018.

[77] J.P. Terry, G. Schultz, S. Ewens, Control and Reversal of Biofilm Episode in a Nuclear Power Plant Main Condenser System, NACE International, 1996.

[78] J. de Almeida, M. Hoogenkamp, W.T. Felippe, W. Crielaard, S.V. van der Waal, Effectiveness of EDTA and modified salt solution to detach and kill cells from *Enterococcus faecalis* biofilm, J. Endod. 42 (2016) 320–323.

[79] L. Leive, Release of lipopolysaccharide by EDTA treatment of *E. coli*, Biochem. Biophys. Res. Commun. 21 (1965) 290–296.
[80] S.V. van der Waal, L.W.M. van der Sluis, Potential of calcium to scaffold an endodontic biofilm, thus protecting the micro-organisms from disinfection, Med. Hypotheses 79 (2012) 1–4.
[81] D. Xu, J. Wen, T. Gu, I. Raad, Biocide cocktail consisting of glutaraldehyde, ethylene diamine disuccinate (EDDS), and methanol for the mitigation of souring and biocorrosion, Corrosion 68 (11) (2012) 994–1002,- https://doi.org/10.5006/0605.
[82] C. Whittaker, H. Ridgway, B.H. Olson, Evaluation of cleaning strategies for removal of biofilms from reverse-osmosis membranes, Appl. Environ. Microbiol. 48 (2) (1984) 395–403.
[83] D.S. Xu, Y. Li, A.L. Lindenberger, H. Liu, H. Liu, Green Chemicals for Enhanced Biofilm Mitigation, 2013.
[84] D. Xu, Y. Li, T. Gu, A synergistic D-tyrosine and tetrakis hydroxymethyl phosphonium sulfate biocide combination for the mitigation of an SRB biofilm, World J. Microbiol. Biotechnol. 28 (2012) 3067–3074.
[85] R. Jia, D. Yang, H. Bin Abd Rahman, T. Gu, Laboratory testing of enhanced biocide mitigation of an oilfield biofilm and its microbiologically influenced corrosion of carbon steel in the presence of oilfield chemicals, Int. Biodeterior. Biodegrad. 125 (2017) 116–124, https://doi.org/10.1016/j.ibiod.2017.09.006.
[86] K.J. Grobe, P.S. Stewart, Characterization of glutaraldehyde efficacy against bacteria biofilm, in: Corrosion 2000, NACE 2000, Orlando, FL, 2000. Paper No. 124.
[87] A.C. Malcolm, S.P. Dempster, B. Chandler, S. Setiawan, Evaluation of chlorine dioxide as an oxidative biocide for hydraulic fracturing in Western Canada, in: SPE/CSUR Unconventional Resources Conference, Society of Petroleum Engineers, 2015.
[88] R. Barati, J.T. Liang, A review of fracturing fluid systems used for hydraulic fracturing of oil and gas wells, J. Appl. Polym. Sci. 131 (16) (2014), https://doi.org/10.1002/app.40735.
[89] D. Flynn, The Nalco Water Handbook, third ed., McGraw-Hill, New York, 2009.
[90] H.R. Rosser, H.A. Nasr-El-Din, N.A. Al-Shammari, A.M. Al-Dhafeeri, Injection Water Treatment at the Source: Biocide-Enhanced Corrosion Inhibitor Squeeze Treatments of Water Supply Wells in a Central Arabia Oilfield, Society of Petroleum Engineers, 1999, January 1, https://doi.org/10.2118/50740-MS.
[91] C.R. Jones, B.L. Downward, K. Hernandez, T. Curtis, F. Smith, Extending Performance Boundaries With Third Generation THPS Formulations, NACE International, 2010.
[92] A. Rafferty, C. Stewart-Liddon, C. Simpson, P. Hammonds, G.M. Graham, P. Maskell, Methodology to evaluate the performance and stability of hydrogen sulphide scavengers, in: Proceedings—SPE International Symposium on Oilfield Chemistry, vol. 2019, Society of Petroleum Engineers (SPE), 2019, https://doi.org/10.2118/193571-ms.
[93] C. Mendez, A.A.O. Magalhaes, P. Altoe, E. Acosta, H2S Scavengers Injection: A Novel Evaluation Protocol to Enhance the Integrity of Production Lines, Society of Petroleum Engineers (SPE), 2009, https://doi.org/10.4043/20168-ms.
[94] C.G. Dariva, A.F. Galio, in: M. Aliofkhazraei (Ed.), Corrosion Inhibitors—Principles, Mechanisms and Applications, Developments in Corrosion Protection, IntechOpen, 2014, https://doi.org/10.5772/57255.
[95] J.A. Haggstrom, J.D. Weaver, L.M. Gloe, J. Holtsclaw, UV Light Treatment Methods and System, United States Patent Application Publication, United States, 2011.
[96] H. Brumner, H. Klein, Ultraviolet disinfection of injection seawater for secondary oil recovery, in: Paper SPE 14007 Presented for the SPE International Symposium on Oilfield Corrosion, Aberdeen, UK, 10–13 May, 1985, https://doi.org/10.2118/14007-MS.
[97] C.C.E. Meulemans, The basic principles of UV-sterilization of water, in: Ozone and Ultraviolet Water Treatment, International Ozone Association, Aquatec Amsterdam, Paris, 1986, pp. B.1.1–B.1.13.
[98] J.B. Clark, J.C. Luppens, P.T. Tucker, Using Ultraviolet Radiation for Controlling Sulfate-Reducing Bacteria in Injection Water, Society of Petroleum Engineers, 1984, https://doi.org/10.2118/13245-MS.

[99] J. Wang, F. Yang, X. Yuan, B. Liu, H. Wu, X. Sui, Successfully Sterilizing the Sulfate Reducing Bacteria With Ultraviolet Radiation in Produced-Water Treatment in Daqing Oilfield, Society of Petroleum Engineers, 2005, https://doi.org/10.2118/93148-MS.

[100] L.M. Gloe, G. Neal, K. Kleinwolterink, UV Light Treatment Reduces the Amount of Biocide Required to Disinfect Water for Fracturing Fluids, Society of Petroleum Engineers, 2010, January 1, https://doi.org/10.2118/126851-MS.

[101] G. Neal, K. Kleinwolterink, L. Abney, L.M. Gloe, Nonchemical Bacteria-Control Process, Society of Petroleum Engineers, 2010, https://doi.org/10.2118/133368-MS.

[102] B.G. Pound, Y. Gorfu, P. Schattner, K.E. Mortelmans, Ultrasonic mitigation of microbiologically influenced corrosion, Corrosion 61 (5) (2005) 452–463.

[103] M. Furuta, M. Yamaguchi, T. Tsukamoto, B. Yim, C.E. Stavarche, K. Hasiba, Y. Maeda, Inactivation of *Escherichia coli* by ultrasonic irradiation, Ultrason. Sonochem. 11 (2004) 57–60.

[104] I. Hua, J. Thompson, Inactivation of *Escherichia coli* by sonication at discrete ultrasonic frequencies, Water Res. 34 (2000) 3888–3893.

[105] B.G. Pound, Y. Gorfu, P. Schattner, K.E. Mortelmans, Ultrasonic Mitigation of Microbiologically Influenced Corrosion, NACE International, 2005, May 1.

[106] H.-C. Flemming, Biofouling in water treatment, in: H.-C. Flemming, G.G. Geesey (Eds.), Biofouling and Biodeterioration in Industrial Water Systems. Proceedings of the International Workshop on Industrial Biofouling and Biocorrosion, Stuttgart, 13–14 September 1990, Springer, Berlin, Heidelberg, 1991.

[107] S. Duda, J.E. Stout, R. Vidic, Biological control in cooling water systems using nonchemical treatment devices, HVAC&R Res. 17 (5) (2011) 872–890, https://doi.org/10.1080/10789669.2011.587588.

[108] M.M. Khazan Alkhazan, A.A. Nasser Saddiq, The effect of magnetic field on the physical, chemical and microbiological properties of the lake water in Saudi Arabia, J. Evol. Biol. Res. 2 (1) (2010) 7–14.

[109] Z. Liu, X. Gao, J. Zhao, Y. Xiang, The sterilization effect of solenoid magnetic field direction on heterotrophic bacteria in circulating cooling water, Procedia Eng. 174 (2017) 1296–1302. Elsevier Ltd. https://doi.org/10.1016/j.proeng.2017.01.274.

[110] G.G. Geesey, Biofilm formation, in: G. Kobrin (Ed.), A Practical Manual on Microbiologically-Influenced Corrosion, NACE, Houston, TX, USA, 1993.

[111] B.J. Little, D.J. Blackwood, J. Hinks, F.M. Lauro, E. Marsili, A. Okamoto, et al., Microbially influenced corrosion—any progress? Corros. Sci. (2020), https://doi.org/10.1016/j.corsci.2020.108641 (Elsevier Ltd.).

[112] F. Kajiyama, K. Okamura, Evaluating cathodic protection reliability on steel pipes in microbially active soils, Corrosion 55 (1) (1999) 74–80.

[113] K.P. Fischer, Cathodic protection criteria for saline mud containing SRB at ambient and higher temperatures, in: Corrosion 81, NACE International, USA, 1981. Paper No. 110.

[114] M.F. de Romero, J. Parra, R. Ruiz, L. Ocando, M. Bracho, O.T. de Rincón, G. Romero, A. Quintero, Cathodic polarisation effects on sessile SRB growth and iron protection, in: Corrosion, NACE International, USA, 2006. Paper No. 06526.

[115] A. Rieder, T. Schwartz, K. Schön-Hölz, S.M. Marten, J. Süß, C. Gusbeth, W. Kohnen, W. Swoboda, U. Obst, W. Frey, Molecular monitoring of inactivation efficiencies of bacteria during pulsed electric field treatment of clinical wastewater, J. Appl. Microbiol. 105 (6) (2008) 2035–2045, https://doi.org/10.1111/j.1365-2672.2008.03972.x.

[116] N.B. Danilenko, G.G. Savel'ev, N.A. Yavorovskii, M.B. Khaskel'berg, T.A. Yurmazova, V.V. Shamanskii, Water purification to remove As(V) by electropulse treatment of an active metallic charge, Russ. J. Appl. Chem. 78 (10) (2005) 1631–1635, https://doi.org/10.1007/s11167-005-0575-6.

[117] C. Feng, K. Suzuki, S. Zhao, N. Sugiura, S. Shimada, T. Maekawa, Water disinfection by electrochemical treatment, Bioresour. Technol. 94 (1) (2004) 21–25, https://doi.org/10.1016/j.biortech.2003.11.021.

[118] D. Opheim, Biological control in cooling towers treated with pulsed-power systems, in: Paper Presented at IWC-01-54. Presented at the International Water Conference, October 2001, 2001.
[119] K.A. Kitzman, E.F. Maziarz, B. Padgett, C.D. Blumenschein, Alan, Chemical vs non-chemical cooling water treatments a side-by-side comparison IWC-03-22, in: Presented at the International Water Conference, October, 2003.
[120] J.F.D. Scott, Modern concepts of chemical treatment for the control of microbially-induced corrosion in oilfield water systems, in: Proceedings of Chemistry in the Oil Industry IX Symposium, 31 October–2 November, Royal Society of Chemistry Publications, Manchester, UK, 2005.
[121] B. Little, J. Lee, R. Ray, New development in mitigation of microbiologically influenced corrosion, in: MIC "An International Perspective" Symposium, Extrin Corrosion Consultants-Curtin University, Perth-Australia, 14–15 February 2007, 2007.
[122] K. Gawas, D. Safarin, H. Riley, J. Ogle, Improved Efficiency of a Nitrate Reducing Bacteria/Nitrate Treatment by the Incorporation of a Sulfate Analog, Society of Petroleum Engineers, 2018, https://doi.org/10.2118/191773-MS.
[123] D.M. Dennis, D.O. Hitzman, Advanced nitrate-based technology for sulfide control and improved oil recovery, in: SPE 106154. Paper Presented at the SPE International Symposium on Oilfield Chemistry, Houston, TX, 28 February–2 March, 2007.
[124] G. Voordouw, B. Buziak, S. Lin, A. Grigoryan, K.M. Kaster, G.E. Jenneman, J.J. Arensdorf, Use of nitrate or nitrite for the management of the sulfur cycle in oil and gas fields, in: SPE 106288. Paper presented at the SPE International Symposium on Oilfield Chemistry, Houston, TX, 28 February–2 March, 2007.
[125] C. Hubert, G. Voordouw, J. Arensdorf, G.E. Jenneman, Control of souring through a novel class of bacteria that oxidize sulfide as well as oil organics with nitrate, in: Paper Presented at the NACE CORROSION Conference, 2006. Paper No. 06669.
[126] C. Patton, Applied Water Technology, second ed., Campbell Petroleum Series, Norman, OK, 1995.
[127] T.L. Skovhus, L. Holmkvist, K. Andersen, H. Pedersen, J. Larsen, MIC risk assessment of the halfdan oil export spool, in: Society of Petroleum Engineers—SPE International Conference and Exhibition on Oilfield Corrosion 2012, 2012, pp. 139–151, https://doi.org/10.2118/155080-ms.

CHAPTER

Mechanical Removal of Flow Assurance Solids 21

21.1 Introduction

One of the most common techniques for mitigating solids deposition in oil and gas systems is the use of mechanical methods. They are efficient, easy to operate, relatively cheap, and, most importantly, they are affordable in most fields.

Mechanical methods have numerous benefits and contribute in solid deposits management in many different ways, including the following:

- Principally, mechanical methods are used to remove the accumulated solids in production tubulars. They have been applied with highly successful rates to remove mineral scales, gas hydrates, wax, asphaltenes, and biofouling deposits. Frequent and scheduled mechanical cleaning keeps the tubulars clean and reduces the chances of pipe blockages and plug formation, and maintains pipe integrity.
- Solid samples collected from the mechanical cleaning returns are used to characterize and determine the deposit composition, which helps in understanding the root cause of the problem and designing optimal mitigation methods, e.g., chemical, physical, or operational.
- Mechanical methods are flexible, as they can be used in combination with other methods such as chemicals and heat to improve solid removal efficiency.
- Mechanical methods can be used to monitor solids formation and assess the applied treatment methods, as the data from pigging operations or tubing gauging during and after treatment are indicative of increasing or decreasing solids accumulation rates.
- Mechanical methods can be used for chemical delivery to the required locations.
- Mechanical method applications have been extended through smart applications. For example, besides cleaning applications, smart pigs are now widely used to detect, locate, and identify the corrosion in pipelines and deliver the necessary chemicals for it.

One of the earliest scale-removal methods was the use of explosives outgrowth and string shots to rattle pipe and break off brittle scale. However, explosives can damage the production tubulars [1]. The more convenient pigging, wirelines, slicklines, mills, jets, and coil tubing methods are the most common mechanical methods applied in oil and gas fields.

Essentials of Flow Assurance Solids in Oil and Gas Operations. https://doi.org/10.1016/B978-0-323-99118-6.00022-8

21.2 Pigging

21.2.1 Introduction

A pig is a device that travels along the inside of the pipeline while performing different tasks, ranging from visualizing the inside of the pipeline to cleaning and inspections [2,3]. The name itself is an acronym of "pipeline inspection gauge" (or it could also have originated from the sound created by these devices) [2,4].

Generally, pigs come in different shapes and sizes, allowing them to be suitable for the pipeline size and the intended purpose, but they should be flexible around bends and small areas of the pipes. They have been used for multiple purposes in oil and gas fields, as discussed in the following section.

21.2.2 Purpose of pipeline pigging

Pipeline pigging has gained considerable attention in recent years, since these devices are very useful in managing the oil and gas pipeline integrity, traveling over long pipeline distances and overcoming the different obstacles that conventional inspection methods can face [5]. Pigs are basically used to maintain the integrity and optimum efficiency of the pipeline, in addition to the major cost savings from early warnings. Stopping operations is not required during pigging, which is an additional advantage from an economics point of view.

As stated, pigs can be used for different purposes in oil and gas operations, including the following [6,7] (Fig. 21.1):

- Cleaning flowlines of debris after construction and installation, and removing solids, sludge, and foulants during operations.
- Improving flow efficiency.
- Flowlines drying and dewatering after hydrostatic testing, and drying wet gas pipelines during operations.
- Inspecting pipeline internals: surveying the ID, detecting corrosion and defects, leak detection, line cover and spanning (subsea), and other developing problems.
- Liquids displacement: displacing fluids during shutdowns, condensate removal.
- Chemicals deployment: deploying inhibitors to affected locations using smart pigs, applying the chemicals in front of pigs or between two pigs.
- Product separation (batching).

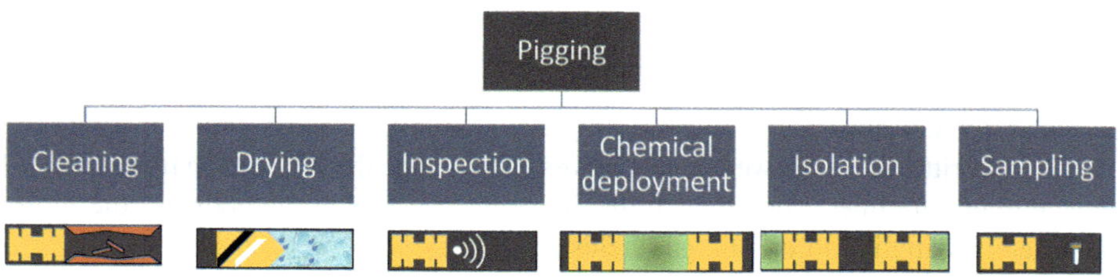

FIG. 21.1

Purposes of pigging.

- Sampling.
- Isolation specialty pigs are used to plug the line or isolate certain areas of the line for maintenance work.

With respect to scale deposit and corrosion problems, the pig offers two main uses: cleaning and inline inspection by intelligent or smart pigs [8]. Routine gas line drying using pigs reduces the potential of gas hydrate formation. Routine pigging in general reduces the chances of pipeline blockages and plug formation, which require intense safety and maintenance work.

21.2.3 Types of pigs

Pigs can be generally classified into three main categories: utility pigs (for cleaning and sealing), inline inspection tools (for wall thickness/loss), and gel pigs (for purging, cleaning, drying, inhibiting, and commissioning) [9–12]. The main pig types within each category are summarized in Fig. 21.2.

21.2.3.1 Utility pigs

Utility pigs can be cleaning pigs, to remove solids and debris, or they can be sealing pigs, to remove accumulated liquids, separate dissimilar fluids, etc. They are generally classified into four main types:

- Mandrel pigs: they consist of a steel mandrel body with a series of replaceable cups and/or discs (sealing elements). Mandrel pigs can contain special brushes that remove debris from the internal sections of pipelines. Figs. 21.3 and 21.4 illustrate different types of mandrel pigs.

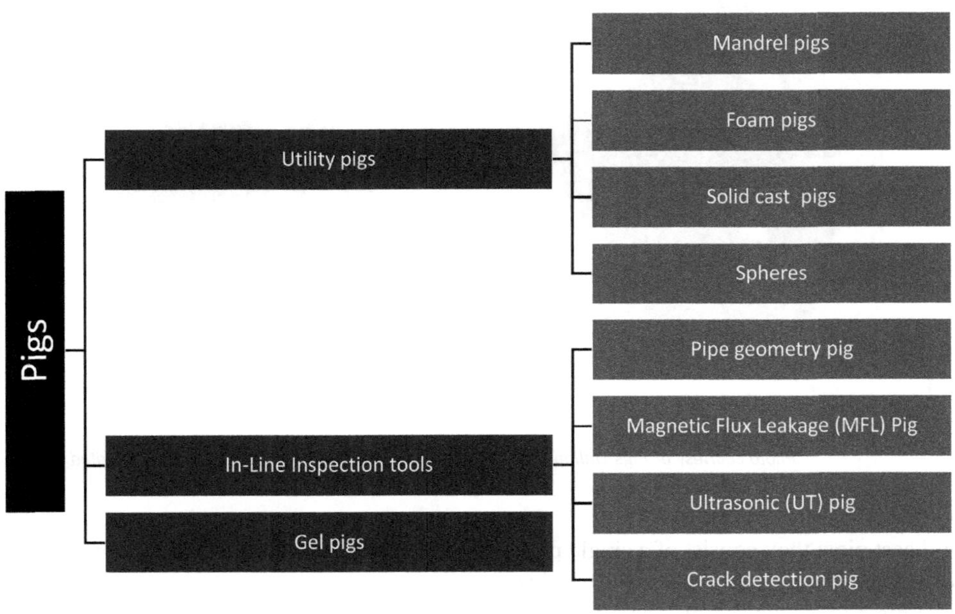

FIG. 21.2

Types of pipeline pigs.

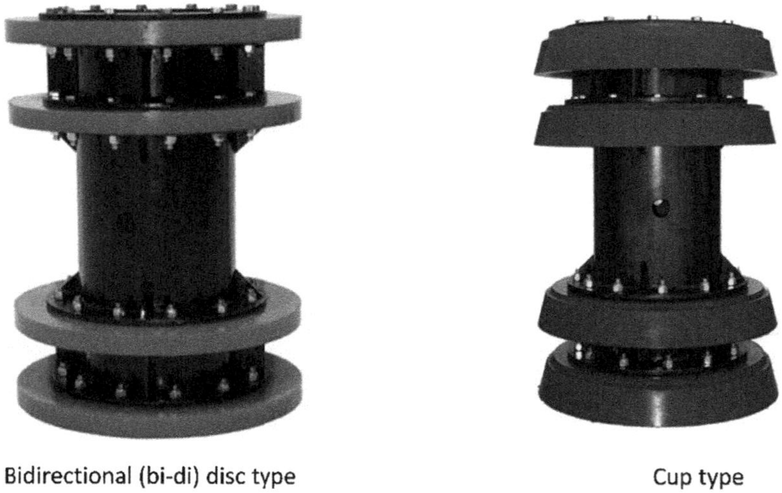

FIG. 21.3

Disc type and cup type mandrel pigs.

Photo courtesy of Pigs Unlimited International LLC, https://pigsunlimited.com/, reprinted with permission.

FIG. 21.4

Mandrel pigs with brushes.

Photo courtesy of Pigs Unlimited International LLC, https://pigsunlimited.com/, reprinted with permission.

- Solid cast pigs: they consist of a single piece molded from polyurethane polymer or steel. Fig. 21.5 illustrates different types of solid cast pigs.
- Foam pigs: these consist of a combination of polymer foam (polyethylene, polyurethane, or polypropylene rubber) embedded with solid abrasive materials, e.g., wire bristles, studs, etc. Figs. 21.6 and 21.7 illustrate types of foam pigs.

FIG. 21.5

Solid cast pigs with brushes.

Photo courtesy of Pigs Unlimited International LLC, https://pigsunlimited.com/, reprinted with permission.

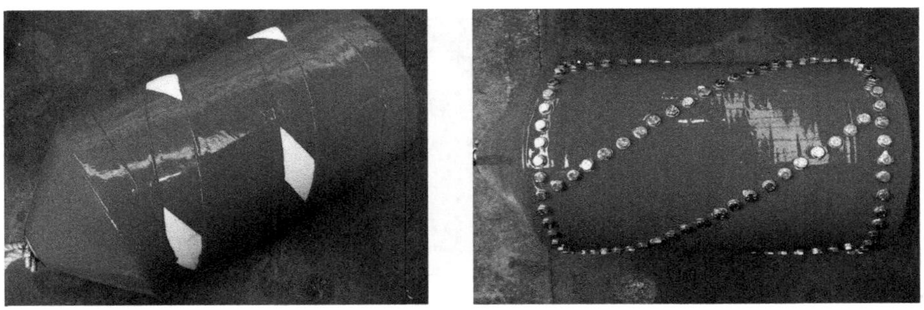

FIG. 21.6

Foam pigs: conventional foam pig (left), studded foam pig (right).

Photo courtesy of Propipe Limited (https://www.propipe.co.uk/), reprinted with permission.

- Spheres are hollow balls composed of polyurethane or neoprene; they can be inflated to an optimum size and can be filled with a solution like water or glycol. Fig. 21.8 illustrates spheres.

Pigs equipped with different brushes, scrapers, and aggressive cleaning accessories such as carbide pins or blades are more efficient in cleaning hard and aged deposits. The addition of magnets to these pigs can aid in removing ferrous debris. Pinwheel pigs are also effective in removing hard scale. They are similar to a standard mandrel pig but the polyurethane discs are fitted with pins into which tungsten

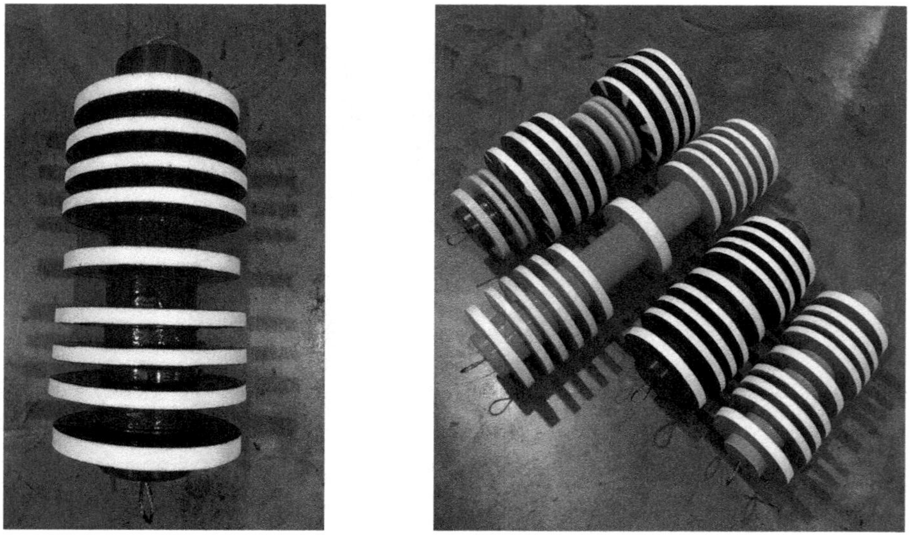

FIG. 21.7

Foam disc pigs.

Photo courtesy of Propipe Limited (https://www.propipe.co.uk/), reprinted with permission.

FIG. 21.8

Pipeline sphere.

Photo courtesy of Pigs Unlimited International LLC, https://pigsunlimited.com/, reprinted with permission.

carbide or ceramic balls are mounted, which protrude from the edge of the disc. Brushwheel pigs can be used to clean corrosion pits and spots that cannot be cleaned by the conventional pigs.

21.2.3.2 In-line inspection tools and intelligent pigs

These are pigs equipped with sensors, electronic instrumentation, and data processing that allow them to trace any defects and map the inside of the pipe with great precision, while locating and recording any damage, corrosion, or scale deposits, such as wax or lime scale. The main types of in-line inspection (ILI) tools or intelligent pigs that could be used to monitor different structural performance in

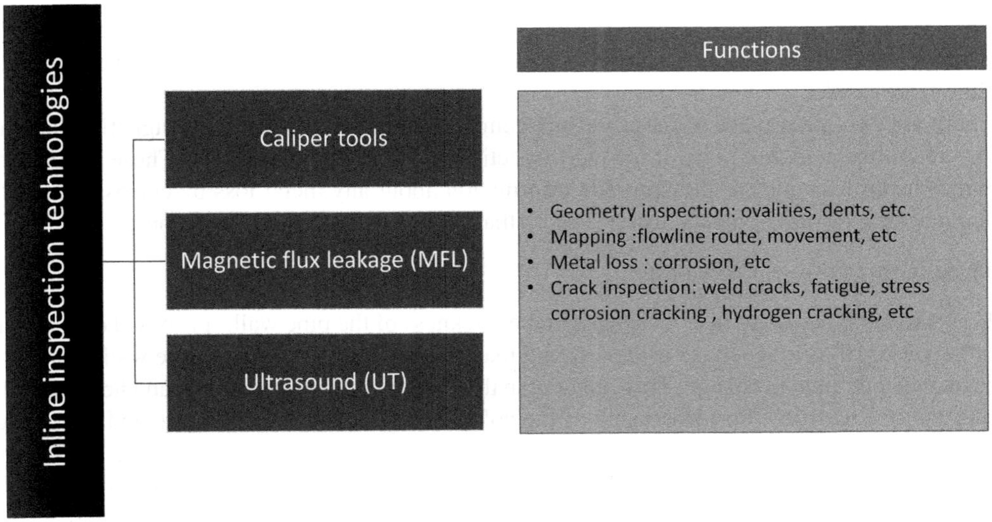

FIG. 21.9

Technologies and functions of ILI pigs.

pipelines include: pipe geometry pig, metal loss pig, and crack detection pig [8,13,14]. Fig. 21.9 summarizes the functions covered by ILI tools.

NACE SP0102-2010 provides Standard Practice In-Line Inspection of Pipelines, and API STD 1163 provides In-Line Inspection Systems Qualifications; this standard is an umbrella document covering all aspects of ILI systems, including procedures, personnel, equipment, and associated software. It is performance-based, but it does not define how to meet qualification requirements [6,15].

- **Pipe geometry pig**

This type of inspection tool is the least complex in terms of sensors and data capture capabilities. The most common examples are [5]:

- Caliper pigs: used to measure pipeline internal bore.
- Internal navigation pigs: used to accurately map a pipeline route.
- Linear/scale thickness pigs: used to measure the thickness of any material on the inside wall of an iron/steel pipeline.
- Shock logger pigs: used to identify the location of blockages or restrictions in a pipeline.

Caliper pigs allow a data acquisition system (DAS), an inertial measurement unit (IMU), a large number of sensing fingers, and an odometer to swim through a pipeline in contact with the pipeline walls, to collect data on pipe wall deformations and other issues [16]. They can be more sophisticated and equipped with more sensory parts for other applications, e.g., GPS, friction, pressure, temperature, etc. Also, the scale thickness pig allows measurement of the thickness of deposits and their distribution inside the pipeline.

- **Metal loss pigs**

These are devices or tools that commonly use a nondestructive testing (NDT) technique to inspect the pipeline from inside.

There are two main types of these sensors commonly applied in industry, magnetic flux leakage (MFL) and ultrasound, but most of the pig inspection relies on MFL [8,13,14]. These are commonly known as metal loss pigs, as they provide information about any metal loss by corrosion and stress-related cracking or other defects, using sensors that can withstand harsh conditions.

- Magnetic flux leakage (MFL) pigs

MFL pigs use magnetism to detect changes in the thickness of the pipe wall. It is based on the fact that when the pipe wall is magnetized using a strong magnet, any metal loss in the pipe wall will cause the induced magnetic field to change. This change can then be sensed using a detector and hence metal-loss defects can be identified. Postprocessing of the collected data can be used to size and in some cases characterize (i.e., manufacturing, corrosion, etc.) the features identified in the pipeline. MFL pigs can be run in both liquid or gas pipelines and are capable of measuring metal losses, even in thin-walled pipes. The inspection provides a relative (i.e., percentage wall thickness) measurement of defect depth [17,18].

- Ultrasonic pigs

An ultrasonic inspection pig uses ultrasound to detect changes in the thickness of the pipe wall. An ultrasonic transducer (UT) transmits signals perpendicular to the pipe wall surface and receives signals back from both the internal and the external surface of the pipe wall. The time taken to receive the signals can then be used to determine the wall thickness of the pipeline. Ultrasonic pigs must be run in a liquid, since they require a couplant around the pig. They give an absolute measurement of defect depth and have the advantage that they can measure metal losses in much thicker pipes than is possible with MFL technology [13,14,17,18]. However, the UT pigs use a piezoelectric transducer that requires a liquid medium; this has been improved with new electromagnetic transducers (EMATs).

- **Crack detection pigs**

Crack detection can be achieved by means of ultrasonic crack detection, transverse MFL, or elastic wave tools [8,13].

Leakage detection pigs are used to pinpoint leaks in water pipelines; the system known as Sahara uses a microphone, which travels along the pipeline with the water flow listening for leaks, and signals are transmitted to the surface via an umbilical, where they are recorded. Leaks as low as 1–2 L/h have been detected. The distance of the leak from the point of insertion can be estimated from the length of the umbilical that has been played out [19].

Also, some intelligent pigs are optical pigs equipped with a camera or laser that is able to collect enough data to produce a 3D reconstruction of the inside of the pipeline within 1 mm resolution. This allows detection of corrosion, erosion, and any geometric deformation with a very good accuracy inside the pipeline. This tool can be deployed in pipelines, on land, and underwater [20].

Normally, smart pigs are preceded by another pig type, such as a gauge pig, to ensure the safety of the intelligent pigs by using a thin metal plate to check the minimum diameter of the pipe, so the pig will not stuck inside the pipe.

21.2.3.3 Gel pigs

Gel pigs are another type of pig that can remove debris and reduce wear of the pig, as a good alternative to mechanical pigs. They are elastic and flexible pigs chemically constructed with a rigid or semirigid structure. They have the advantage of easy extrusion. The higher the load of debris, the higher viscosity and yield point required in the type of gel used to keep the debris in suspension until removed. Successful application of gel pigs was recorded in different countries, such as Saudi Arabia, Nigeria, and the North Sea and the United States [21]. Besides their use in debris removal, gel pigs can be used as separators, for corrosion inhibitor applications, for removing stuck pigs, for hydrotesting, and for drying purposes.

21.2.3.4 Other types

Pig manufacturers have developed other types of pigs made for specific purposes that the conventional ones cannot perform. These pigs' design can be similar to the standard ones mentioned earlier with minor modifications to fulfill their purpose. Some of these are shown in Figs. 21.10–21.14. These include multidiameter pigs for multidiameter pipes. Also, dual mode pigs can be used for the same purpose. Articulated pigs can be used for bends and pipelines with wye pieces. Bypass pigs are used for wax deposit pigging. Filming pigs are another type, used to deliver chemical inhibitors to the affected areas. Ice pigging is another technology that can be applied especially in unpiggable pipelines. Other types of specialty pigs are manufactured for specific maintenance or in-line inspection purposes.

21.2.4 Piggable and unpiggable pipelines

21.2.4.1 Pipeline design and parameters for pigging

Piggable pipelines must have certain features related to their design, components, and operating conditions, so that the pigging operations will be effective without causing any damage to the pig or to the pipeline. Some of these specifications are discussed in the following sections.

FIG. 21.10

Dual diameter pig.

Photo courtesy of Pigs Unlimited International LLC, https://pigsunlimited.com/, reprinted with permission.

FIG. 21.11

Multidiameter pigs.

Photo courtesy of Propipe Limited (https://www.propipe.co.uk/), reprinted with permission.

FIG. 21.12

Articulated pigs.

Photo courtesy of Propipe Limited (https://www.propipe.co.uk/), reprinted with permission.

- **Pipeline dimensions and materials**

An ideal pipeline for pigging would be perfectly round with constant diameter and straight from end to end, with no or minimal valves or other components that could hinder the motion of the pig [6].

Some pigs have been designed to transverse pipes that change in size. Dual diameter pigs and multidiameter pigs have been developed for dual diameter and multidiameter pipelines, allowing negotiation of the various pipeline diameter changes.

- **Pipeline components**

For pigging purposes, the pipeline bend usually follows the general rule of minimum bend radius (MBR), which depends on diameter and wall thickness, but is generally three to five times the pipe diameter. Miter bends should be avoided if the pipeline is to be pigged.

FIG. 21.13

Jetting bypass pig.

Photo courtesy of Propipe Limited (https://www.propipe.co.uk/), reprinted with permission.

FIG. 21.14

Wax cleaning pig.

Photo courtesy of Propipe Limited (https://www.propipe.co.uk/), reprinted with permission.

Barred tees are recommended for pigging operations, especially with tees outlets greater than 50% of the main diameter [6]. Other components like wyes and laterals must be taken into consideration during pigging. Dual mode and articulated pigs can be used for better pigging in pipelines with tight bends and wye components.

Tees and bends must be spaced properly, as pigs can stall between two tees, two bends, or between tee and bend. It is recommended that a minimum straight length of pipe, equivalent to *three* pipe diameters, be installed between any two components or features [6].

Valves are known to interfere the most with pigging operations. A general rule for valves is to have smooth bore with a diameter approximately equal to the bore of the pipeline; guaranteed 100% opening. Butterfly valves, choke valves, globe valves, and most plug type valves are not piggable and cannot be used in a pipeline to be pigged unless a special pigging method or pig is used [22]. Spheres are not suited for dependable passage through check valves, as they tend to sit in the bowl and be held there by the weight of the clapper [6].

Other pipeline features that must be optimized during a design for good pigging operations is the pipeline elevations, laterals, wyes, and others.

- **Operating conditions**

The pipeline product should be compatible with the pig polymeric or metallic material used. Some pigs like foam pigs are known to absorb materials, so they should be avoided in pigging flammable hydrocarbons from the pipeline, or they should be properly cleaned after operations.

Proper operating pressure should be maintained so that it will not damage the pig's body, and will be effective in giving the pig sufficient power to scrub the formed deposits. Low pigging pressure causes the pig to stop at minor obstructions and turns pigging inefficient. Similarly, temperatures during pigging should not cause damage to the pig material.

Sufficient flow rates/velocities are usually maintained during on-stream pigging. The pigging speed depends on the pipeline status, e.g., 1–5 mph during new construction pipelines, and the pipeline product, e.g., 5–15 mph for on-stream gas pipelines, and 2–10 mph for on-stream liquid pipelines [6].

21.2.4.2 Pigging unpiggable pipelines

Unpiggable pipelines are those that cannot be pigged or inspected internally. This can be due to the pipeline dimensions and component designs that do not fit with the specifications mentioned earlier, such as no access, no pigging stations, buried pipelines, small diameter with tight bends, dual/multi-diameters, unbarred tees, containing unpiggable valve, low or no flow, very high temperature, old, worn or damaged pipelines, and other similar pigging impediments.

Thus to perform pigging and keep such pipelines maintained, a variety of conventional and unconventional techniques can be used, including the following:

- Using temporary or portable pigging stations for pipelines with no pigging stations.
- Using small diameter pigs for small diameter pipelines.
- Using gel pigs.
- Using ice pigging, in which high-solids ice slurry is pumped under pressure to deliver a wall shear stress on the pipe, cleaning it through physical abrasion. It can be used for different diameters and lengths of pipelines, and can pass through various types of process equipment, such as valves and pumps, and can be used in different environments [23].
- For buried pipelines, bell holes are usually made, making pipeline cut points and pigged using bi-di pigs.

21.2.5 Pig stations

21.2.5.1 General design specifications

A pig station consists of a device called a pig "trap" for launching or receiving a pig, while the pipeline product passes through or around the pig trap without interruption of flow. Pig stations include the traps and their associated piping and instrumentation, as well as all the valves needed to isolate the trap from the operating pipeline.

The main objectives considered during the design of pig stations include [24]:

- The purpose of the pigging operations: cleaning, sealing, etc.
- The pipeline dimensions: the pig traps have to be compatible with the pipeline being pigged.
- Pig dimensions: utility pig dimensions and specifications for launcher and receiver are different from those of ILI pigs and spheres.
- Product type: Gas launchers and receivers differ from liquid launchers and receivers.

Pigging stations for natural gas and liquid can differ, as mentioned already, but they still share common features. Generally, the main design components of the pig traps include the following (Figs. 21.15 and 21.16):

- Major barrel: a section of pipe that is larger than the nominal pipeline size where the pig is introduced.

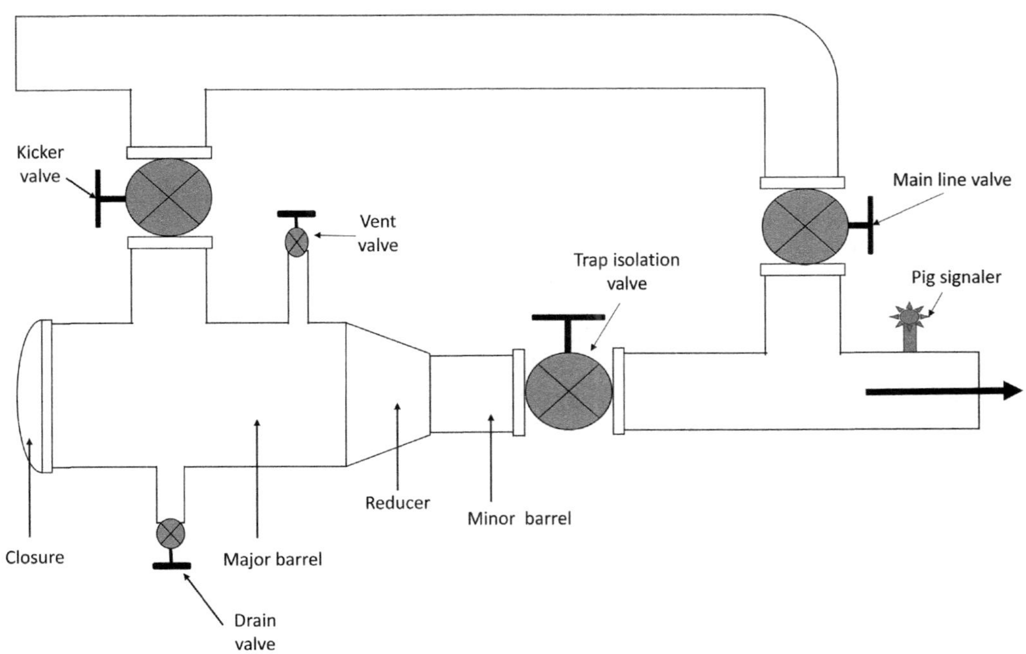

FIG. 21.15

Schematic of pig launcher.

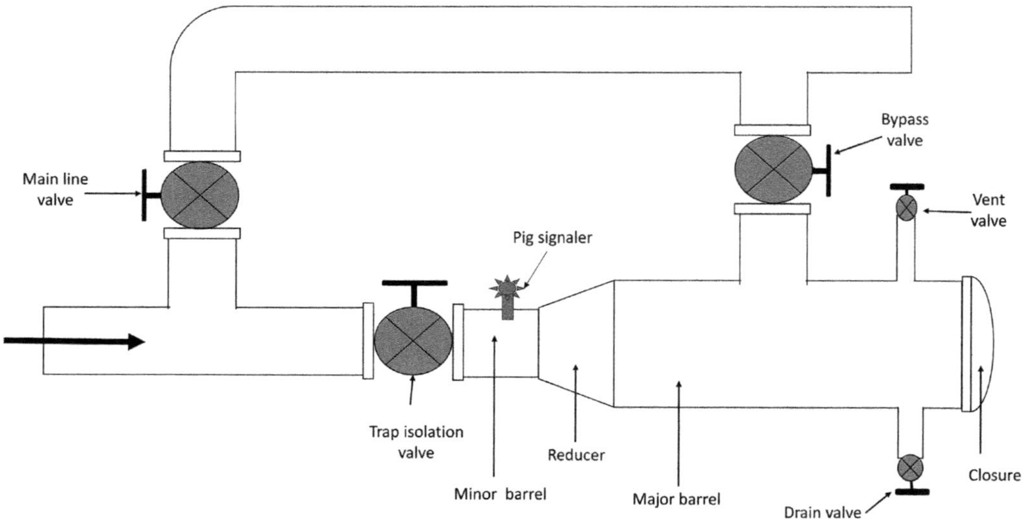

FIG. 21.16

Schematic of pig receiver.

- Minor barrel: a smaller section of pipe that matches with the pipeline size, also referred to as the "nominal bore section."
- A reducer barrel connecting the major and minor barrels.
- A closure: a quick opening door on the major barrel end of the pig trap.
- Kicker line: this is used in the launcher to pressurize the upstream side of the pig so it can be kicked inside the pipeline or launched. In a receiver, it is known as the bypass and it is used as an outlet for fluids arriving in the pig trap.
- Drain and vent lines.
- Valves, pressure gages, pig signalers, and other utility connections.

The design of pig launchers, pig traps, and their related equipment is achieved in accordance with standards developed by several organizations (ASME B31.4 for Liquid Pipelines, ASME B31.8 for Gas Pipelines, ASME B31.3 for Process Piping). The launcher is generally shorter than the receiver. The launcher and receiver have the same number of nozzles, but differ in location for a variety of reasons [24].

The length of pipeline between pig traps depends on the pipeline length, system design and available footprints, and service product type, and is also determined by factors other than pigging, such as the location of pump stations or compressor stations. Generally, the following parameters are applied for the distance between pig launcher and receiver [6]:

- 50–100 miles apart for gas pipelines,
- 100–150 miles apart for refined products pipelines,
- 150–200 miles apart for crude oil pipelines.

Different types of pig traps are available depending on the applications, including automated pig traps, sloped automated pig traps, vertical automated pigs traps, batch automated systems, and temporary pig traps. Also, subsea pig traps have been used; these are useful in the case of offshore sites where space is limited, and vertical units are more suitable. For wet gas gathering systems and systems that require pigging several times a day or week, automated systems are beneficial. Some companies attempt to design a combination pig launcher and receiver unit to perform both functions [24].

21.2.5.2 Pig launcher

A pig launcher is a device or pressurized vessel used to shoot the pig inside the pipeline for a specific purpose. Launchers must have specific designs to be able to fit the pig and operate properly.

The barrel diameter should be 2″ over the pipeline size for pipelines in sizes up to 10″. For pipelines from 12″ up to 26″ the barrel diameter should be 4″ larger than the pipeline and for pipelines 28″ and above, the barrel should be 6″ oversize [6].

The launcher barrel specification are 1.5 times the length of the longest pig or 2.5D, where D is the nominal pipeline diameter if the pig length was not known during trap design [6]. ILI pigs need extra length due to the different modules, so the barrel should be at least 1D longer than the overall length of the longest pig. For sphere pigs, the launcher barrel depends on the number of spheres loaded (usually 6–12) and the barrel has to be inclined at a 10–15-degree angle; one sphere can be released at a time with the use of two launch pins (hydraulic, pneumatic, or electric actuated) or special valve [25].

For launchers, the nominal bore section need only be long enough to prevent the nose of the pig touching the valve ball or gate of the trap isolation valve. A length of 1.5D should be adequate. [6].

The kicker line in the launcher usually has the size of one-third (D/3) in the case of a liquid pipeline and one-half (D/2) in the case of a gas pipeline. For example, for 12″ pipeline, a kicker line will usually be 6″ for a gas pipeline and 4″ for a liquid pipeline. The kicker line should be near the closure, allowing flow to be directed behind the pig for a successful launch [24].

On the launcher, there should be a length of at least 2D between the kicker and the reducer.

The reducer is preferred to be eccentric with flat side on the bottom, allowing easy launching of the pig. Concentric reducers may cause the pig nose to be stuck into the edge of the reducer, requiring a sliding carriage to slide it into the pipeline [6,24].

The barrel must be fitted with a closure, which is basically a door to gain access to the launcher. Different types of closures are available [26]:

- Weld cap (flat, ellipsoidal, spherical, conical)
- Bolted blind flange
- Quick opening closure

Quick closure allows quick, easy, and safe access to the contents of the launcher or receiver than would be expected with a bolted flange. Pig launcher and receiver closures are normally designed to pipeline design codes ASME B31.4 for liquid pipelines or ASME B31.8 for gas pipelines [26].

The different designs of the pig trap closures include:

- Threaded type
- Door latch type
- Clamp ring type
- Bandlock closure

Closures have to be safe to operate due to the possibility of pressure accumulation in the launcher, which can be very dangerous. Besides, closure must be equipped with pressure indicators not only to give the operator an indication of internal pressure, but also to prevent the operator from opening the door-latching mechanism as long as internal pressure exists. They also have to be well maintained and serviced.

21.2.5.3 Pig receiver

The pig receiver is longer than the launcher, as mentioned above. For utility pigging the length of the receiver barrel is 2.5 times the length of the longest pig, or long enough to decelerate the pig or to receive two pigs in the case of unexpected events. If the pig length is not known during the trap design, a 4.5D barrel length is a general rule to follow. For ILI tools, the receiver barrel should be 2.5D longer than the longest pig to allow for deceleration. The nominal bore should be longer than the pig as well. For spheres, the design is similar to that of the launcher [6].

In the receiver, the kicker line is called the "bypass line." It follows the same size as the launcher. This line should be located near the reducer entering the oversize barrel [6]. The distance between the bypass and the reducer should be the minimum possible.

Drain line is necessary for pig traps and it has to be in the lowest point of the trap. Some trap designs can involve two drain lines (one in front of and one behind the pig). The two drains prevent the drain line from being blocked by the pig, allowing the entire barrel to be drained [6].

21.2.6 Pigging operations

As mentioned earlier, pigging is applied during pipeline construction, precommissioning, and after installation and during regular operations, for many different purposes. In the following section the general pigging procedure is discussed briefly.

21.2.6.1 General pigging procedure

When precommissioning the pipeline, pigging is used in different activities including filling, cleaning and gauging, hydrotesting, dewatering, and drying.

After pipeline installation and during regular operations, pigging is used for different purposes including cleaning, batching, displacement, and inspection. For cleaning purposes, routine pigging can keep the pipelines running at over 90%–95% capacity, reduce plugging and corrosion problems, and improve flow rates [6].

The pigging procedures will vary from one pipeline company to another. The pig launching and receiving procedures given here are just generic guidelines and it is recommended to consult professional personnel for developing operating procedures in conjunction with company-specified safety and technical protocols.

Pigging generally involves safety and technical aspects to assure safe and effective operations. Some of the safety risks include:

- Exposure to high-pressure hydrocarbon gas or condensate.
- Potential hydrocarbons leakage, and potential fire hazard.
- Wrong operations of launcher or receiver can create an explosive environment.
- Physical exertion and bodily harm during operation of manual valves and handling the pigs.

Thus the first thing is to maintain safety precautions before starting the pigging operations. This includes:

- Personnel have to wear proper PPE.
- Assuring pig traps are depressurized before draining, venting, opening closures, or any action that can lead to personal exposure to high pressure.
- Follow manual handling procedures during opening manual valves, or handling the pigs.
- Trolleys and trays can be used to insert or remove the pigs.
- Areas around pig traps have to be cleared and have good housekeeping to prevent unexpected events.

Pigs must be routinely checked and properly maintained for any defects and replaceable parts if necessary. A final check on the pig integrity and operability is necessary before starting pigging operations. The proper pig must be selected for operations based on the type of deposit and expected hardness and volume. Pig selection is case dependent and should be made by a specialist.

For pig launching, a general procedure can be schemed as follows [27]:

1. First of all, the status of the pig traps must be checked from the last pigging operation before proceeding. The pig traps must be checked as to whether they were left on stream pressurized or isolated, drained, and vented from last use before commencing any part of pig launching procedure.
2. Assure that the isolation valve and the kicker valve are closed.
3. Assure the launcher is completely drained and vented.
4. Once completely drained and vented, open the closure and install the pig, with the pig nose in contact with the reducer.
5. After installing the pig, assure the closure seal is clean, and then close and secure the closure.
6. Close the drain valve, open the kicker line valve, and fill the trap.
7. When filling is complete, close the vent valve to allow pressure to equalize across the isolation valve.
8. Open the isolation valve. The pig is ready for launching.
9. The main valve is kept partially closed, to increase the flow through the kicker valve and behind the pig. Continue to close the main line valve until the pig leaves the trap into the main line, as indicated by the pig signaler.
10. After the pig leaves the trap and enters the main line, fully open the main line valve. Close the isolation valve and the kicker valve.
11. The pig launching is complete.

The pig is propelled using different propelling media. The propelling medium depends on many factors:

- Product fluid properties; gas, liquid, viscous, etc.
- Availability and cost of propellant.
- Pipeline length, pigging runs, and frequency.
- Environmental and safety concerns.

Common propelling media include compressed air, nitrogen, CO_2, or liquids. Compressed air is cheap and commonly utilized; however, it can become uneconomical with larger sizes or in long pigging runs

[6]. The pressure required for pigging is different depending on the pipeline dimensions and purpose of pigging. General values can be found in the production engineering textbooks.

A general procedure for receiving the pig can be schemed as follows [27]:

1. Assure that the receiver is pressurized.
2. Fully open the bypass valve, and then fully open the isolation valve and partially close the main line valve.
3. Monitor the pig signaler for pig arrival.
4. Close the isolation valve and bypass valve.
5. Drain and vent the trap and assure it is completely depressurized.
6. Open the trap closure and remove the pig from the receiver, and then close and secure the closure after checking the seal.
7. Return the receiver to the original condition.

Some of the problems associated with pigging operations include:

- Safety concerns due to incorrect pigging operations.
- Damage to the pipeline or its components.
- A pig can become stuck in the pipeline.
- Insufficient pigging operations.
- Inability to navigate varying diameters or sharp bends.

To improve pigging operations, mathematical models of pigging operations for different applications have been developed, which can optimize the pigging operations and determine the required pigging runs or pigging frequency. Pigging frequency generally depends on the type of the deposit, its thickness and expected volumes, efficiency and type of pig, pipeline length, and economics of the pigging runs. Fig. 21.17 shows different pigs being received at pig stations with different amounts of solids.

21.2.6.2 Pigging efficiency

The main pigging operational goal is to keep the pipeline clean and running at its highest throughput. The first pig run usually shows the greatest improvement; however, sometimes many pig runs are required to achieve optimum efficiency [6].

FIG. 21.17

Pigs being received with varying volumes of deposits.

Photo courtesy of Propipe Limited (https://www.propipe.co.uk/), reprinted with permission.

Simple methods to determine the pigging efficiency are by monitoring the improvement of flow rate and throughput after pigging. This can be represented by Eq. (21.1):

$$\eta_{pig} = \left(\frac{\left(\frac{Q}{\Delta p}\right)_{ap} - \left(\frac{Q}{\Delta p}\right)_{bp}}{\left(\frac{Q}{\Delta p}\right)_{bp}} \right) \times 100 \tag{21.1}$$

where η_{pig} is the pigging efficiency, Q is the flow rate, Δp is the differential pressure, ap is after pigging, and bp is before pigging.

Also, records are usually made for such data, where an increase or a decrease is indicative of the pigging efficiency.

Another method of determining the efficiency of pigging is by monitoring the amount of solids and debris material that accumulates in the receiving pig trap after each run. This is usually used to assess the applied treatments in the system. This can also be used to determine the number of runs required to get the best performance.

Discoloration of the liquid received at the pig receiver is another simple method of estimating pigging efficiency.

21.3 Wellbore and production tubing cleaning tools

During production operations, a number of well intervention operations are performed for data collection, diagnostics, workover maintenance, evaluation, and management. Cabling technologies such as wireline and slickline are commonly used methods for well intervention.

Some applications of the cabling technology include but are not limited to the following:

- Formation of geophysical data and downhole engineering data collection using logging tools.
- Tubing gauging, solid deposits tagging (locating) and removal (cutting), bailer runs.
- Liquid tagging.
- Equipment maintenance: sleeve shifting, gas lift valve replacement and maintenance.
- Setting/pulling plugs and chokes.
- Spinner surveys (detecting locations at which fluids are escaping from the wellbore into a porous formation and also to determine point of formation fluid entry).
- Fishing operations.

21.3.1 Wireline and slickline tools

Many types of wireline tools have been used to remove solid deposition. These mainly include scratchers, brushes, gauge cutters, bailers, snippers, and other tools specialized for removing solids. Some of these tools are shown in Figs. 21.18–21.24.

For efficient operations of the wireline and slickline tools in removing deposits, they must:

- Have effective diameter to be able to reach the deposit layer to loosen or remove it.
- Be compatible and not affected by the produced fluids, e.g., sour conditions.

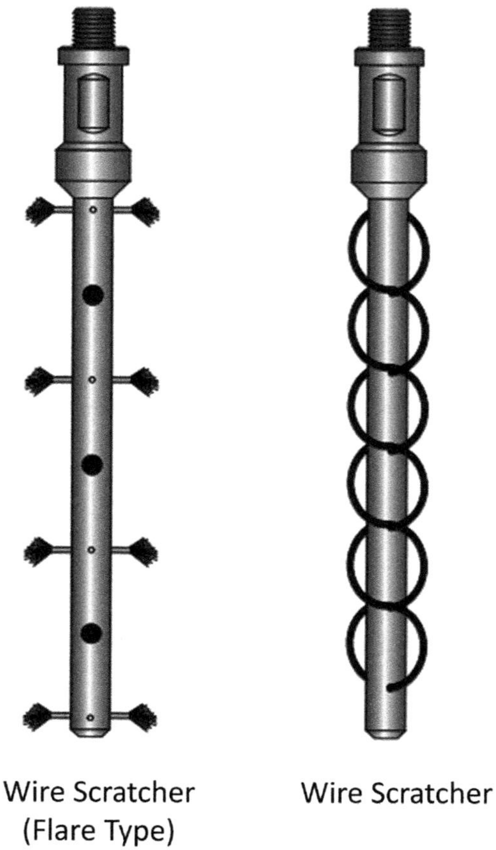

FIG. 21.18

Different types of tubing scratchers.

Photo courtesy of Friedrich Leutert GmbH & Co. KG, (https://www.leutert.com/) reprinted with permission.

- Be compatible with and not affected by any chemical dissolvers if they were chosen to be used with the mechanical removal methods.

21.3.2 Other downhole methods

Oil and gas service companies have developed different types of tools that have been successfully applied to remove different types of downhole deposits.

For example, SageRider has patented a tool that uses heat as well as cutting capabilities to remove wax deposits. The tool is battery powered and can be conveyed on all forms of wireline [28]. Peakwell Systems has developed a downhole hard scale removal tool with enhanced performance and less operating time than regular wireline tools [29].

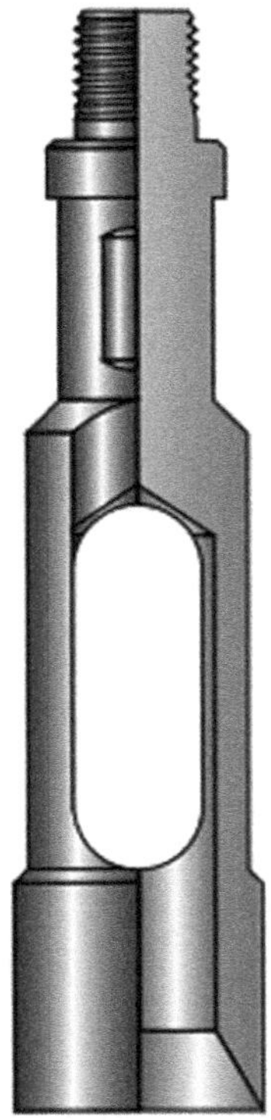

FIG. 21.19

Tubing gauge or cutter, commonly used to remove paraffins and loose mineral scales.

Photo courtesy of Friedrich Leutert GmbH & Co. KG, (https://www.leutert.com/) reprinted with permission.

Schwanitz et al. [30] reported successful use of a combination of well tractor technology and a new vacuum/bailer system (with different cleaning heads for different environments), all conveyed on electric wireline to mitigate sand and scale remediation in high-angle wells.

21.4 Jets

Jetting is a very common method to clean various downhole and surface tubulars and equipment. It can also be used to remove scale from the perforations. Jetting is also known as hydro-milling or hydro-jetting, where a nozzle with multiple orifices shoots fluids at high pressure to erode the deposits from the metal surface. Jets remove deposits from surfaces by different mechanisms, including erosion, abrasion, stress cycling, and cavitation. Jetting can be used with chemical washes to attack hard sticking deposits.

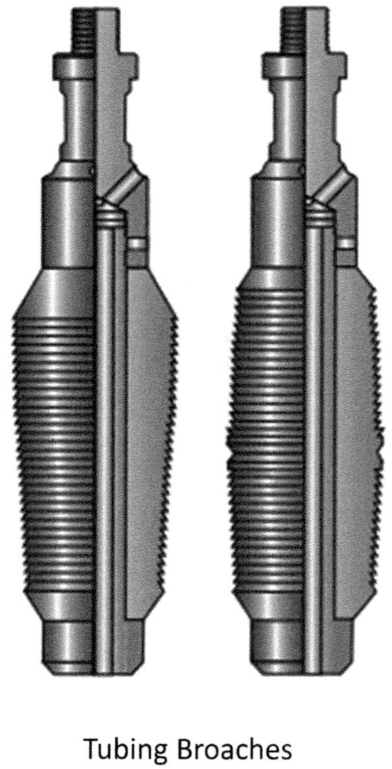

FIG. 21.20

Tubing broach.

Photo courtesy of Friedrich Leutert GmbH & Co. KG, (https://www.leutert.com/) reprinted with permission.

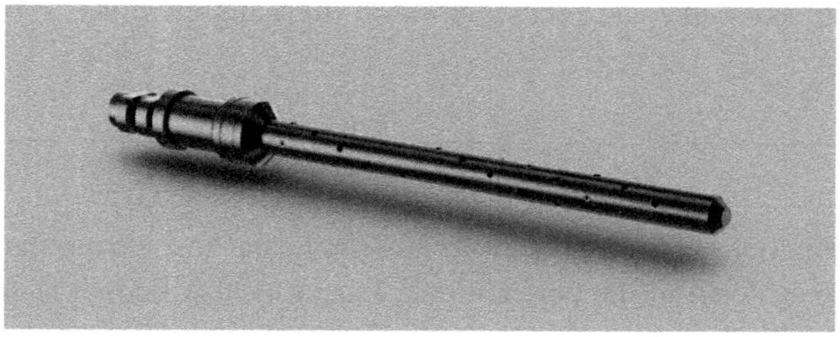

FIG. 21.21

Wire scratcher, used to remove sand, wax, and mineral scale buildup in the tubing.

Photo courtesy of LiMAR Well Intervention Products, (https://www.limaroiltools.com/) reprinted with permission.

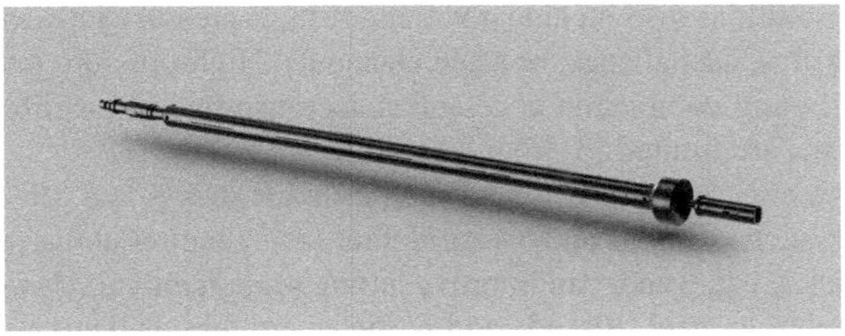

FIG. 21.22

Junk basket tool, used to drift the tubing to confirm it is free from any restrictions and to trap the debris for retrieval back to surface.

Photo courtesy of LiMAR Well Intervention Products, (https://www.limaroiltools.com/) reprinted with permission.

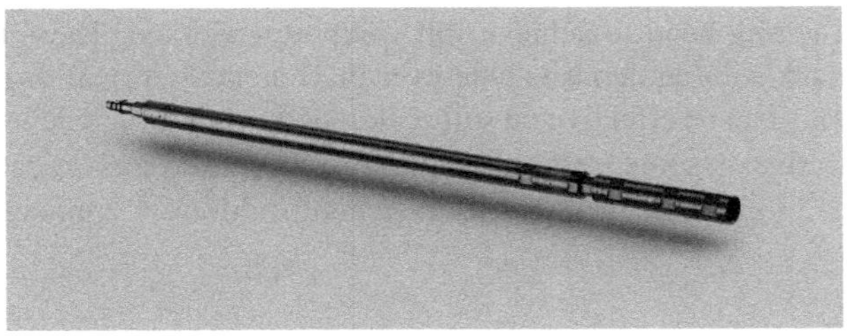

FIG. 21.23

Bailer tool, used to remove and recover sand and debris from the wellbore.

Photo courtesy of LiMAR Well Intervention Products, (https://www.limaroiltools.com/) reprinted with permission.

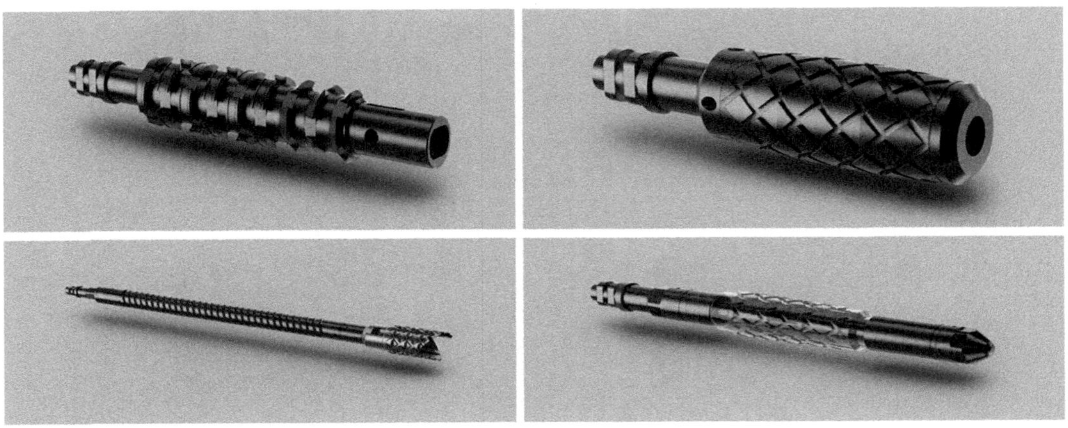

FIG. 21.24

Broach tools, very effective tools in removing hard scales and tubing restrictions. Different designs from LiMAR oil tools allowing effective removal of hard scales.

Photo courtesy of LiMAR Well Intervention Products, (https://www.limaroiltools.com/) reprinted with permission.

Wellbore cleaning solutions also include a variety of high-pressure jet washing tools. These tools are used for sand, paraffins, asphaltenes, or scale cleanouts. High-pressure jet washing tools are also highly effective for cleaning completion accessories such as nipples, gas lift mandrels, and screens where mechanical means are limited [31].

Water jetting can be effective on soft scale, such as halite, but is less effective on some forms of medium-to-hard scales, such as calcite and barite. Harder scale requires abrasive slurries during jetting. Adding 1% solid particles, e.g., sand, can improve jetting scale removal. However, sharp-edged sand particles can be damaging to steel. Round sand particles are less damaging than sharp-edged ones. "Sterling beads" is an alternative abrasive material for barite hard scale removal by jetting. This material matches the erosive performance of sand on hard, brittle scales, while being 20 times less erosive of steel [1].

For optimum jetting operations, good selection of the abrasive material is necessary. It is mandatory that the abrasive material on the steel avoids any damage to the tubulars.

Many downhole fluid-jetting systems are available on the market. The jetting tools use multiple jet orifices or an indexed jetting head to achieve full wellbore coverage. These tools can be used with chemical washes to attack soluble deposits wherever placement is critical to prevent bullheading reagent losses. Water jetting can be effective on soft scale, such as halite, and debris or fill, but experience shows that it is less effective on some forms of medium-to-hard scale such as calcite and barium sulfate [1]. Figs. 21.25 and 21.26 illustrate different jet systems for different applications.

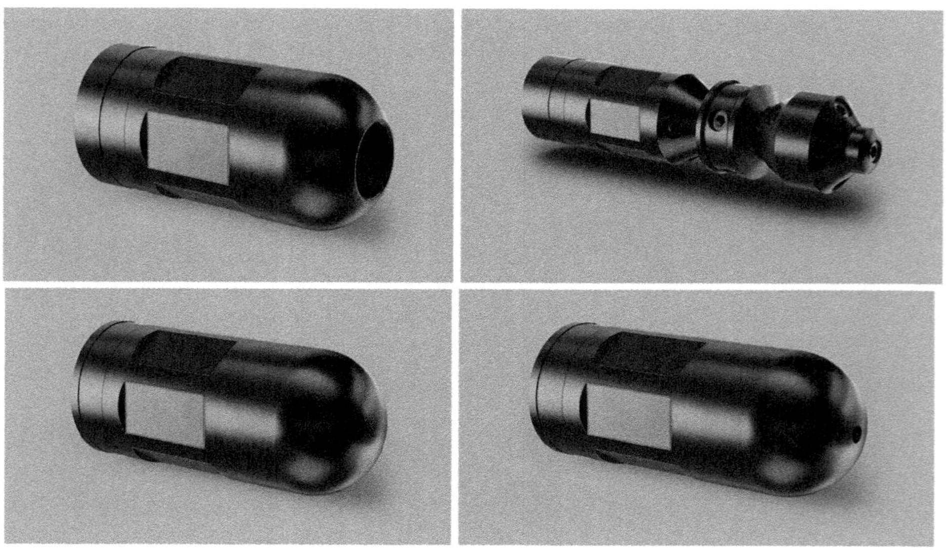

FIG. 21.25

Different jet nozzles.

Photo courtesy of LiMAR Well Intervention Products, (https://www.limaroiltools.com/) reprinted with permission.

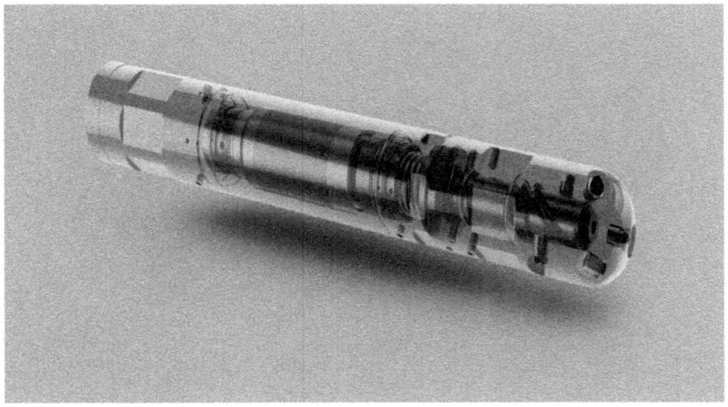

FIG. 21.26

Abrasive rotary jetting tool, with high impact 360° jetting coverage to target and dislodge debris, paraffins, waxes, and scales from the completion tubulars.

Photo courtesy of LiMAR Well Intervention Products, (https://www.limaroiltools.com/) reprinted with permission.

21.5 Mills

Milling and drilling operations are conventional and necessary tools for scale and other deposit removal, especially in the case of thick, hard, and aged deposits that are highly insoluble in most chemical dissolvers. Impact drilling can provide a viable solution. However, milling is the most commonly used method.

Impact-drill systems are an attractive and efficient method of removing scale, because they perform reliably in adverse conditions [32]. Impact drilling is also used in different applications including cement milling, sand removal, and gravel removal. Impact drilling can operate under high temperature (above 600°F) and can work in hostile conditions like H_2S and acidic media.

A cut-button bit is the optimum impact bit to use on cement, rock, or hard scales. It is also effective in removing sand and other types of fill. Impact-drilling penetration rates can vary greatly depending on the amount and hardness of the scale or cement. In extreme cases, rates can be as slow as 3 ft./h (usually through the upsets). An acceptable penetration rate in badly scaled wells is 80–100 ft./h [32].

The mill bit runs by a powerful workover motor that supplies constant torque, which gives it advantages and makes it more efficient [32].

The typical milling assembly includes standard running tools, a nonrotating stabilizer, a workover motor, and a mill bit. Many types of mill designs can be used to mill scale. Cutting matrixes including diamonds, polycrystalline diamond compact (PDC), and tungsten carbide (Fig. 21.27), have been successfully used [32]. The most used matrix is tungsten carbide in the shape of a dome.

The designed head presents a small contact area to the scale, which reduces torque during milling. Although this mill has an aggressive face, it does not damage tubing walls. Specially designed stabilizing pads directly behind the milling matrix on the outside diameter (OD) of the mill body provide the required protection [32]. Impact tools like motors and mills usually need fullbore access and seldom clean scale completely to the steel walls.

The under-reamer is also used to clean out scale and hard fill that cannot be removed from liners by jet-washing tools [32]. An under-reamer is designed to pass through a downhole restriction, open, clean

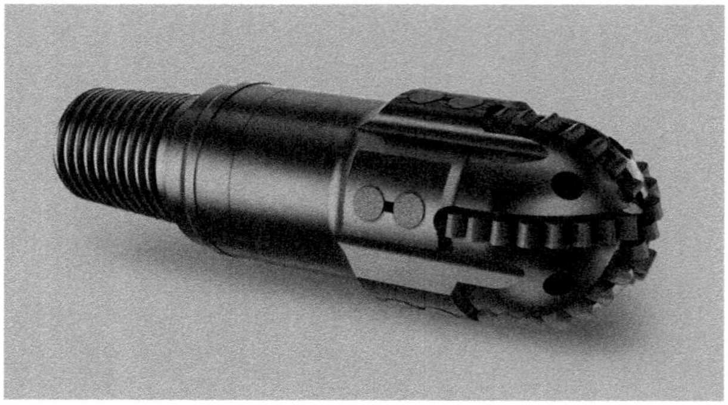

FIG. 21.27

Tungsten carbide mill.

Photo courtesy of LiMAR Well Intervention Products, (https://www.limaroiltools.com/) reprinted with permission.

the hole to full gauge, and close for retrieval back through the restriction [31]. Figs. 21.28 and 21.29 show the under-reamer tools.

Under-reamers provide a solution to complete debris removal in large diameter bores where a smaller ID restriction exists above. Typically by increasing flow rate, cutting blades expand from the body of the tool, which effectively creates a larger diameter mill than could be deployed through that ID restriction. Under-reamers are extremely effective when there's a need to remove scales and cement from the production tubular walls in order to minimize flow restrictions and maximize production [33]. The under-reamer is also effective in partial access situations, as it can increase the effective diameter by moving the milling blades outward in response to pump pressure and rate. However, under-reaming mills can remove scale only at about half the rate of a typical mill [1].

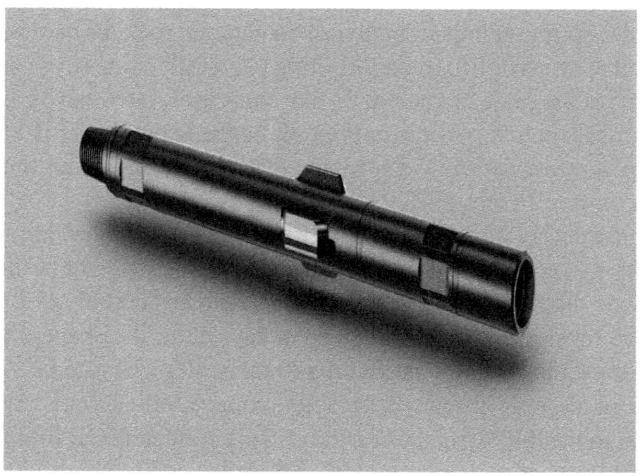

FIG. 21.28

Under-reamer tool.

Photo courtesy of LiMAR Well Intervention Products, (https://www.limaroiltools.com/) reprinted with permission.

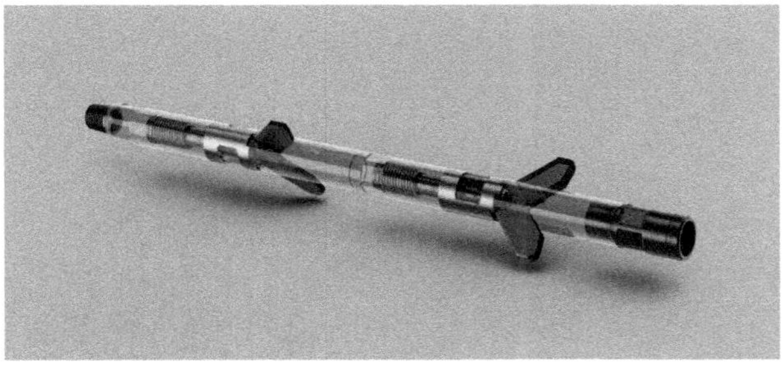

FIG. 21.29

High extension under-reamer to remove compacted debris in wellbores with a large difference in internal diameter between the tubing and the casing or liner.

Photo courtesy of LiMAR Well Intervention Products, (https://www.limaroiltools.com/) reprinted with permission.

Hipp-Tripper tools from Baker Hughes are good impact tools that can work in acid and high H_2S environments. They can be used for scale milling.

Milling operations are not only restricted to downhole tubing deposits removal. They can also be applied to surface pipelines and equipment. Pipeline deposits milling usually takes place in pipes that are taken from service during the process of rehabilitation of the pipelines.

21.6 Coiled tubing

Coiled tubing (CT) refers to a continuous length of metal or composite tubing with no joints, which is very long and flexible, to be supplied spooled on a large reel. The CT has many applications in the petroleum industry, which include:

- Drilling (CT drilling),
- Cementing,
- Wellbore cleanout,
- Logging and perforating,
- Production: CT is used as a production pipe in gas wells,
- Chemical circulation,
- Acidizing,
- Hydraulic fracturing,
- Downhole installation, i.e., electrical submersible pump installation, placing bridge plugs and packers for zonal isolation.

It is also used in other well intervention applications. The main benefits over wireline are the ability to pump chemicals through the coil and the ability to push it into the hole rather than relying on gravity. Therefore they can efficiently be used to deliver the chemicals to specific spots. Their use is not limited to the downhole applications; they can also be used in long flowlines, subsea, and deepwater flowlines as well [34]. Fig. 21.30 shows a typical CT unit.

FIG. 21.30

Coiled tubing unit.

Source: https://commons.wikimedia.org/wiki/File:%D0%9A%D0%BE%D0%BB%D1%82%D1%8E%D0%B1%D1%96%D0%BD%D0%B3%D0%BE%D0%B2%D0%B0_%D1%83%D1%81%D1%82%D0%B0%D0%BD%D0%BE%D0%B2%D0%BA%D0%B0.jpg.

More than half of all coiled-tubing jobs are wellbore cleanouts. A significant number of these cleanouts are performed using workover motors and milling assemblies. While the value of a milling assembly is well documented, downhole conditions such as high temperatures (above 400°F) and hostile fluids can significantly reduce the life of a workover motor and limit milling operations [32].

The milling and jetting methods previously described are commonly applied through the CT units. CT affords the ability of using chemicals to aid in removing the deposits. Circulating hot fluids can also be employed to enhance the dissolution. A number of field cases have reported the use of CT with hot/warm solvent to remove deposits.

21.7 Other methods

Different methods have been created to fill industry needs where conventional methods seem inefficient or uneconomical.

Electrohydraulic pulsing (EHP) is a unique and new technology for cleaning scale in the upper completion with a form of high-pulsed power (HPP). In EHP, a small amount of electrical energy is amplified, stored, and then released in an extremely short time, creating a shock wave and a pressure pulse, which is sufficient to dislodge the scale and deposits in the tubing [35].

The impulse deposit removal system is another newly developed method for removal of deposits formed on heat transfer surfaces of boilers. The design of the unit is based on combined combustion technology that involves several fully controllable combustion methods, such as deflagrative, pulse, and detonative combustion.

Librino [36] reported some unconventional methods of wax removal. CoilTAC technology, originally developed by Superior Energy Services Co., significantly extends the reach of CT. It was successful in removing wax blockages at 20,000 ft. Flexi-Coil, which is similar to CoilTAC but with smaller components than conventional CT, makes it easier to install the equipment in limited space production facilities. Another technology is the pipe pulse, where pressure pulses are used to clean blocked pipelines.

Dynamic underbalance is another technology that has been successfully used to remove downhole hard scales [37].

21.8 Field application of mechanical methods for solids removal

Mechanical methods have been used successfully to clean flow assurance solids in oil and gas systems. The selection of the method and field application depends on many factors related to the deposits and the cleaning method. Some of these factors include the following:

- Type of deposits
- Age, thickness, and hardness of deposit layer
- Presence of restrictions or blockages
- Expected volume of deposits
- Location and access to the deposit
- Efficiency and abrasiveness of the cleaning method
- Status of the pipe or tubing, e.g., design, duration in service, corrosion, etc., to determine if the mechanical method can cause any damage to it
- Availability of using chemical dissolvers with mechanical methods
- Compatibility of mechanical methods with the produced fluids
- Compatibility of the used mechanical methods with the chemical dissolver if allowed for use

Knowledge of the preceding information is critical to design and optimize the cleaning job. It is very common to find composite scales (mixture of organic and inorganic deposits), so organic solvent washes are usually applied to help in removing organic deposits or loosening them, to aid in removing the mineral scales. The thickness of the deposits can be obtained using inline solids monitoring sensors, mentioned in Chapter 13. Also, it can be obtained from the pressure pulse method or calculated from pressure drop readings.

The status of the deposits inside the pipeline or the tubing can be visualized for better assessment of the case using a borescope, which can give a good indication whether the deposits are evenly forming on the pipe wall or if there exist unequal distributions of deposits, to avoid events of mechanical tools being stuck, and also to reduce the chances of damaging the cleaning tool or the pipe itself.

Routine pipeline pigging prevents accumulation of high amounts of solid and plug formation. Pigging has been used to clean mineral scales. Foam pigs are usually used to remove halite and other soft mineral scales. For harder scales such as barite, pigs with abrasive parts or with metal brushes are commonly used.

Wax deposit pigging is a confusing problem. The wax removal depends on wax breaking, plug formation, accumulation, and production phases. The pigging force during wax removal can be divided into three parts: baseline force, wax-breaking force, and wax transport force. Wax-breaking force increases when wax thickness rises; also, wax plug transportation-force gradient is independent of wax plug length. In addition, it was found that the geometrical shape of the pig has a strong impact on wax removal efficiency [38,39].

Several models have been developed to predict wax removal by pigging. However, wax removal and pigging are still confusing due to the following associated problems [38,39]:

- Wax deposit hardness is a critical factor in pigging. Wax deposit hardness is not known and is hard to predict, although some factors are known to be contributing, such as crude oil composition, wax content, carbon distribution, and others.
- Besides, the pig driving forces usually squeeze the oil from the wax deposits network, leading to increasing the wax deposit hardness and finally forming a wax plug several meters long.
- Overestimation of the wax hardness causes unnecessary pigging operations.
- If the wax-breaking force exceeds the driving force behind the pig, the pig can become stuck and cause a problem.
- The geometrical shape of the pig has a strong impact on wax removal efficiency; however, the rules of selecting a pig for wax removal still use the rule of thumb, which may not be optimum in certain situations. However, some pigs designed for wax removal have been developed to avoid problems associated with wax plugging.

Hence it is very important to properly design pigging for waxy systems. Some methods of efficiently removing wax include:

- By adding solvent between two pigs, or in front of a pig, benefiting from the combined action of the pig and the solvent.
- By performing multiple runs, starting with smaller diameter pigs and gradually increasing the pig diameter in each run to avoid compacting the wax and forming a plug.
- Using bypass pigs, which are similar to normal pigs but have bypass ports to allow for a certain amount of fluid flowthrough, can also be efficient in removing wax deposits.
- Using jetting pigs that shoot crude oil or liquid to clear the front pathway of the pig and prevent wax plug formation. Fig 21.31 illustrates wax pigging with conventional pigs and with bypass pigs.

Pigging can also be used to control gas hydrate deposits. Pigging operations are used to dry pipelines and to reduce the chances of hydrate formation, especially for long cold subsea pipelines. Routine pigging of pipelines dries them out and prevents accumulation of the gas hydrates, hence preventing gas hydrate plug formation, as these are aggressive and require intensive safety and workover operations.

CT methods have been used to remove deposits in topsides, downhole, and subsea applications. They employ erosive mechanical effects in addition to the ability to flush solvents at high temperature. Barium sulfate is commonly removed from downhole using CT, especially when the chemical treatment is ineffective or highly expensive. CT was deemed the safest and most economical method of removing iron sulfide scales in the Ghawar field [40]. Wax deposits and gas hydrates have been reported to be successfully removed using CT. CT has the benefit of circulating hot solvent, which is superior in removing wax and gas hydrates deposits. The benefits of CT in removing gas hydrate plugs include [41]:

- Small footprint with lower manpower requirements
- Flexible, not rig-dependent
- Lower risk of pipeline damage or equipment getting stuck

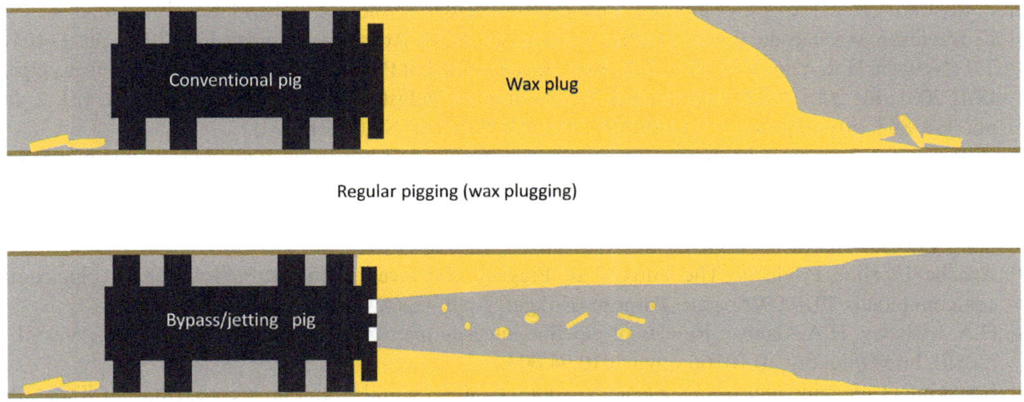

FIG. 21.31

Wax pigging using conventional pigs and bypass/jetting pigs.

- Returns pipelines to transportation faster
- Cost-effective
- HSE advantages

Downhole mechanical cleaning is very common in oil and gas fields. Fullbore access to deposits can be partially blocked by restrictions such as decreasing tubing diameter or downhole equipment; in this case, tools that can change diameter are used to remove scale below the restriction, or if such equipment is not available, then a small hole—less than full tubing size—can usually be drilled through the scale below the restriction to allow increased flow [1]. Downhole mills, jets, scratchers, and brushes can be used alone or with the aid of chemical dissolvers.

21.9 Summary

In this chapter, the different mechanical methods of removing flow assurance deposits have been reviewed. Mechanical deposit removal is one of the most common and successful methods in the oil and gas industry. Pigging, CT, and wireline tools are the most common methods.

The use of these methods depends basically on the scale properties, the mechanical method efficiency, and the pipeline/tubing condition.

The mechanical methods can be combined with other methods such as chemical and thermal methods to enhance the performance and improve deposit removal efficiency.

References

[1] M. Crabtree, D. Eslinger, P. Fletcher, A. Johnson, G. King, Fighting Scale—Removal and Prevention, Oilfield Review, 1999.

[2] P. Rostron, Critical review of pipeline scale measurement technologies, Indian J. Sci. Technol. 8 (1) (2018).

[3] R. Davidson, An introduction to pipeline pigging, in: PPSA Aberdeen Seminar, UK, 2002, pp. 1–104.

[4] T.T. Nguyen, H.R. Yoo, Y.W. Rho, S.B. Kim, Speed control of PIG using bypass flow in natural gas pipeline. ISIE 2001, in: 2001 IEEE International Symposium on Industrial Electronics Proceedings, vol. 2, 2001, pp. 863–868, https://doi.org/10.1109/ISIE.2001.931581 (Cat. No.01TH8570).

[5] C. Reed, A.J. Robinson, D. Smart, Techniques for Monitoring Structural Behaviour of Pipeline Systems, American Water Works Association, USA, 2004, pp. 1–246.

[6] J. Cordell, H. Vanzant, The Pipeline Pigging Handbook, third ed., Clarion Technical Publishers and Scientific Surveys Ltd, 2003.

[7] Apache Pipeline Products, The Purpose of Pigging, Retrieved from: https://www.apachepipe.com/wp-content/uploads/2017/09/Apache_Purpose-of-Pigging.pdf. (Last visit January 2022).

[8] H.A. Kishawy, H.A. Gabbar, Review of pipeline integrity management practices, Int. J. Press. Vessel. Pip. (2010), https://doi.org/10.1016/j.ijpvp.2010.04.003.

[9] Bob winters, cleaning pigs design and applications, NACE Central Area Conference, Houston, Texas, April, 2014.

[10] O. Uzu, R. Napier, K. Ngwuobia, Gel pig technology applications in pipeline servicing, SPE-66068-MS, in: Nigerian Annual International Conference and Exhibition, 7–9 August, Abuja, Nigeria, Society of Petroleum Engineers, 2000, https://doi.org/10.2118/66068-MS.

[11] Rigzone, How Does Pipeline Pigging Work? Rigzone Training, 2022. https://www.rigzone.com/training/insight.asp?insight_id=310. (Last visited November 2021).

[12] Enggcyclopedia, Pipe Pigs, Enggcyclopedia, 2022. http://www.enggcyclopedia.com/2011/07/pipe-pigs/. (Last visited December 2021).

[13] A.A. Carvalho, J.M.A. Rebello, L.V.S. Sagrilo, C.S. Camerini, I.V.J. Miranda, MFL signals and artificial neural networks applied to detection and classification of pipe weld defects, NDT&E Int. 39 (8) (2006) 661–667, https://doi.org/10.1016/j.ndteint.2006.04.003.

[14] A. Barbian, M. Beller, S. Hartmann, U. Schneider, High resolution ultrasonic in-line inspection: added value and special applications, in: 6th Pipleine Technology Conference 2011, vol. 1, 2011.

[15] NACE International Standard Practice, In-line inspection of pipelines, NACE SP0102-2010 (Formerly RP0102), NACE International.

[16] J. Yu, J.G. Lee, C.G. Park, H.S. Han, An off-line navigation of a geometry PIG using a modified nonlinear fixed-interval smoothing filter, Control. Eng. Pract. 13 (11) (2005) 1403–1411, https://doi.org/10.1016/j.conengprac.2004.12.016.

[17] A. Pople, Magnetic flux leakage pigs or ultrasonic pigs? The case for combined intelligent pig inspections, in: 6th International Conference, Pipeline Rehabilitation and Maintenance, vol. 53, 2003, pp. 1–14, https://doi.org/10.1017/CBO9781107415324.004 (9).

[18] L. Pirtle, In-Line Inspection & More, T.D. Williamson, Inc. in the United States and in foreign countries, 2013. © Copyright 2013.

[19] H. Chapman, Development of a successful internal leak detection and pipeline condition assessment technology for large diameter pipes, in: 6th Annual WIOA NSW Water Industry Engineers & Operators Conference Tamworth Regional Entertainment & Conference Centre, 27–29 March, 2012.

[20] U. Ekpemu, S. Aloba, Pipeline laser imaging and metrology using laser smart pig, SPE-140641-MS, in: Nigeria Annual International Conference and Exhibition, 31 July–7 August, Tinapa—Calabar, Nigeria, Society of Petroleum Engineers, 2010, https://doi.org/10.2118/140641-MS.

[21] O. Uzu, R. Napier, K. Ngwuobia, Gel pig technology applications in pipeline servicing, SPE-66068-MS, in: Nigerian Annual International Conference and Exhibition, 7–9 August, Abuja, Nigeria, Society of Petroleum Engineers, 2000, January 1, https://doi.org/10.2118/66068-MS.

[22] How It Works: Pipeline Pigging, Valves that Enable Utility, Inspection, Specialty, and Gel Pigs, SLB.com (Last visit Jan 2022).

[23] S. Wheeler, D. Rhys, Wax Removal from Large Diameter Pipes Using Ice Slurries, Pigging Products & Services Association, 2016.

[24] L.D. Payne, J.D. Lollis, What everyone needs to know about building pig launchers, receivers, Pipeline Gas J. 246 (2) (2019).

[25] L. Payne, The re-emergence of spherical pigs, Pipeline Gas J. 245 (1) (2018).

[26] L.D. Payne, More to quick-opening closures than meets the eye, Pipeline Gas J. (2017). Oildom Publishing Co. of Texas Inc.

[27] Girard Industries, Pig Launching & Receiving Procedures, Girard industries, girardindustries.com, 2021 (Last visited November 2021).

[28] Seagerider Paraffin Melting Tool, Retrieved from: https://www.sageriderinc.com/products/sagerider-paraffin-melting-tool. (Last visited November 2021).

[29] Peak Well System, Scale Removal in Less than One Day with Peak's Unique Torque Action Debris Breaker, 2021, Retrieved from: www.peakwellsystems.com. (Last visited November 2021).

[30] B.J. Schwanitz, A.S. McKay, An E-line solution to sand and scale remediation in high-angle wells, in: Paper Presented at the SPE/ICoTA Coiled Tubing and Well Intervention Conference and Exhibition, The Woodlands, Texas, USA, March 2007, https://doi.org/10.2118/107243-MS.

[31] Baker Hughes, Baker Hughes Coiled Tubing Handbook, Solve Downhole Problems with Reliable, Cost-Effective Technology, Baker Hughes, 2003.

[32] J. DeGeare, The Guide to Oilwell Fishing Operations Tools, Techniques, and Rules of Thumb, Elsevier, 2015.

[33] Limaroiltools, What are CT Milling & Under Reaming Tools?, LiMAR Oiltools (UK) Ltd, 2021. https://www.limaroiltools.com/. (Last visited September 2021).

[34] Z. Huang, G. Li, S. Tian, X. Song, M. Sheng, S. Shah, Flow behavior and friction characteristics of fluid flow in coiled tubing, in: Abrasive Water Jet Perforation and Multi-Stage Fracturing, Elsevier, 2018, pp. 153–190, https://doi.org/10.1016/b978-0-12-812807-7.00005-4.

[35] Mike Perri Cleaning Scale from Upper Completion Equipment with EHP, Blue Spark Energy Inc.

[36] F. Librino, Unconventional Methods Offer Wax Remediation Options, Offshore, PennWell Corporation, 2015.

[37] E. Vegliante, A. Al Busaidy, Dynamic underbalance for selective perforation cleaning from hard scale deposits, successful and fully controlled case histories, in: Paper Presented at the SPE EUROPEC/EAGE Annual Conference and Exhibition, Vienna, Austria, May 2011, https://doi.org/10.2118/140689-MS.

[38] W. Wang, Q. Huang, Y. Liu, K. Sepehrnoori, Experimental study on mechanisms of wax removal during pipeline pigging, in: Paper Presented at the SPE Annual Technical Conference and Exhibition, Houston, Texas, USA, September 2015, https://doi.org/10.2118/174827-MS.

[39] W. Li, Q. Huang, W. Wang, Y. Ren, X. Dong, Q. Zhao, L. Hou, Study on wax removal during pipeline-pigging operations, SPE Prod. Oper. 34 (2019) 216–231, https://doi.org/10.2118/194010-PA.

[40] M. Buali, N. Ginest, J. Leal, O. Sambo, A. Chacon, J. Vielma, Extreme challenges in FeS scale cleanout operation overcome using temporary isolation, high pressure coiled tubing, and tailored fluid systems, in: Paper Presented at the SPE International Oilfield Scale Conference and Exhibition, Aberdeen, Scotland, May 2014, https://doi.org/10.2118/SPE-169759-MS.

[41] M. Nepomiluev, V. Streletskaya, Subsea Gas Pipeline Coiled Tubing Intervention for Hydrate Plug Removal, Paper Presented at the SPE Russian Oil and Gas Exploration & Production Technical Conference and Exhibition, Moscow, Russia, October 2014, https://doi.org/10.2118/171322-MS.

CHAPTER

Chemical Injection Systems 22

22.1 Introduction

The use of chemical agents is common in oil and gas fields. Chemicals are used from the earliest stages of field development until field decommissioning and abandonment. During production stages, treatment chemicals are used extensively to mitigate flow assurance problems, as discussed in Chapters 15–20. These chemicals are typically injected in predetermined amounts to mitigate processing problems and protect the equipment. For injection purposes, the proper chemical injection system must be installed to assure accurate and proper delivery of the chemicals in the different locations of production facilities.

22.2 Production treatment chemicals

Numerous types of specialty chemicals are normally used during the various phases of field development. The chemicals principally work by imparting some functionality, using specific function groups in their composition to change the physical or chemical behavior of the production fluids, mitigating the problem by attenuating its impact or preventing its occurrence.

As mentioned, specialty chemicals are used in all the stages of field development: drilling, cementing, completion, production, EOR, and abandonment. During the production stage, chemicals such as scale inhibitors, wax inhibitors/dispersants, low-dosage hydrate inhibitors, asphaltene inhibitors/dispersants, biocides, H_2S and O_2 scavengers, and others are used to mitigate the flow assurance problems. The selection of the suitable chemical for the operation depends on some factors, including [1]:

- Chemical performance
- Economics
- Health and safety in handling and storage
- Environmental restrictions
- Stability during field applications and storage
- Compatibility with the production fluids, other chemicals, and system materials.

Essentials of Flow Assurance Solids in Oil and Gas Operations. https://doi.org/10.1016/B978-0-323-99118-6.00019-8

Depending on the application, the treating chemicals are deployed in the field using different methods, including:

- Continuous injection in topsides or downhole.
- Slug dosing or batch injection: injection of an overdose of the chemical to achieve a specific purpose, e.g., slug dosing of biocides.
- Squeeze treatments in which the chemicals are pumped down into the reservoir to mitigate a problem in the reservoir and to give protection to the wellbore, tubing, and the subsequent installations in the system; the chemical delivery in squeeze treatments is implemented by methods such as bullheading or coiled tubing.
- Added to the fracturing fluids, where the treatment chemicals are added to the fracturing chemicals to assure that they reach the targeted spots in the reservoir.
- Slow release methods, for example solid-scale inhibitor chemicals can be introduced down into the rathole of a well, where they slowly are released to assure protection against scale formation. Chemicals are also added to the fracturing proppant, where they can slowly release in the reservoir.

The optimum location and conditions for these chemicals to be injected are [1]:

- Upstream of the location of the problem, in order to hinder any possible interactions that might be responsible for generating problems/damage and to assure protection of the subsequent installations. Therefore wax inhibitors are deployed before the wax appearance temperature; asphaltene inhibitors before asphaltene onset precipitation/flocculation; hydrate inhibitors are injected before the hydrate formation temperature; and scale inhibitors before incompatible waters mixing.
- Provide enough mixing of the chemicals with fluids, so that the chemical is uniformly distributed and has interacted with the fluids.
- Have enough residence time. For example, H_2S scavengers require sufficient residence time to effectively work; also enough residence time is required for demulsification.

For any chemical injection system, some particular specifications must be considered for optimum selection and operations. These include the following:

- Aim and frequency of injection.
- The injected chemical properties: compatibility with other chemicals, compatibility with material of construction of injection system, viscosity, density, etc.
- Location of the facility and location where the injection system will be installed in the facility. The injection system should be installed in a safe and accessible location, where power sources are available.
- System parameters, such as discharge pressure, flow rate, and other system design specifications.

The amount of chemical to be delivered depends on the type of chemical as well as the production volumes of oil, gas, and/or water phases. Some chemicals require knowledge of only one of these phases, whereas some require two, or all three. The injected dosage is specified by the manufacturer or by the screening tests performed in the labs, and is then correlated to the production volumes to determine the actual dose rates. The chemical injection system is then set accordingly to achieve the targeted dosage.

The main injection system components include (Fig. 22.1):

- Metering pump
- Tank for storing and blending the chemicals

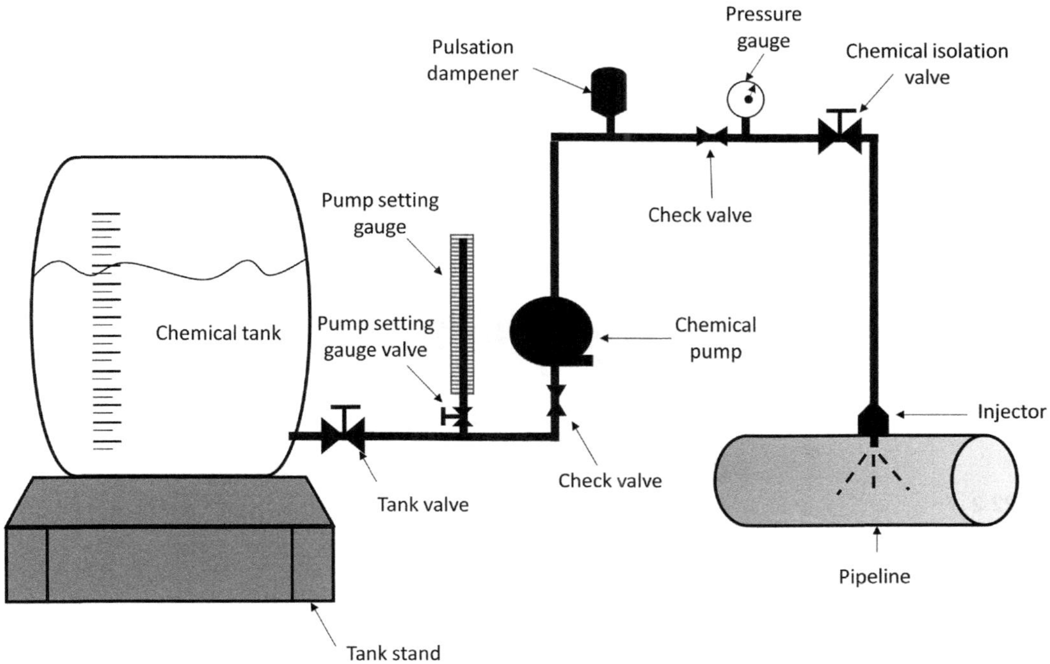

FIG. 22.1

Schematic of chemical injection system.

- Injection lines
- Control system
- Other components: level measurement devices, check valves, gauges, temperature transmitters, other instrumentation.

22.3 Chemical pumps

Pump types typically used for chemical injection are positive displacement pumps, reciprocating units designed to inject precise amounts of chemical into a process stream [2]. The pumps have to be in compliance with API 675, which is a comprehensive standard for specifications of the positive displacement pumps in oil and gas industry. Two types of pumps are commonly used in chemical injection in oil and gas fields: the packed plunger pump and diaphragm pump. These pumps operate in the same manner, but with different reciprocating mechanisms. The packed plunger pumps work by a reciprocating piston or plunger moving back and forth, and directly pumping the chemical. In the diaphragm design, the reciprocating action of a piston or plunger is used to transmit pressure to a flexible diaphragm, which isolates and displaces the chemical fluids. Other types of pumps with suitable size can also be used, e.g., rotary gear and progressive cavity pumps. Figs. 22.2 and 22.3 show the packed plunger pump and the diaphragm pump.

FIG. 22.2

Packed plunger metering pump.

FIG. 22.3

Diaphragm metering pump.

The motive force of chemical pumps is used to supply force to the plunger or diaphragm system, which then injects the chemical at a measured rate. The motive force can be:

- Pneumatic: compressed air or gas. Natural gas from processing facilities is commonly used.
- Electrical: using municipal electricity.
- Solar: solar panels are used to generate electricity to power the motor.

Some parameters should be considered during the selection of the metering pumps used in chemical injection, including:

- Injected chemical properties: viscosity, volatility, composition.
- Pump material compatibility with the injected chemicals.
- Pump compatibility with the system design, suction side static head, net positive suction head (NPSH), system flow rates, and discharge pressure.
- Pump accuracy, or turndown, which is the ratio of the maximum to minimum output of the pump.
- Power source availability.

22.4 Chemical tanks

Different tank designs and capacities are used in oil and gas fields. Essential specifications for chemical tanks include:

- Enough capacity
- Easy to handle
- Cost effective
- Chemical resistant
- Corrosion resistant
- Thermally stable
- Proper specific gravity rating
- Proper tank wall thickness
- Impact and stress cracking resistant
- UV stable
- Suitable for outdoors
- Durable one-piece seamless construction
- Proper venting
- Appropriately graduated
- Supplied with level sensors/indicators

Different tank designs, including vertical and horizontal configurations and capacities: for example, commonly 65, 125, 225, 335 gal or higher are commercially available, depending on the need. High-density polyethylene (HDPE) tanks are commonly used. However, they can have limited life if misused. The standard specifications of polyethylene upright tanks are given in ASTM D1998-15. Crosslinked polyethylene (XLPE) tanks have more advantages than HPDE, as they are known to be more resistant to heat and chemicals (even for hypochlorite and other harsh chemicals), corrosion, stress cracking, and have improved impact strength. The tanks are supplied with a leg tank, steel stand, tie-down bands, and stainless steel connections. Steel tanks can be also used and they have to be

FIG. 22.4

Different types of chemical tanks: steel tank (left) and HPDE tank (right).

corrosion resistant. Other materials are also used, such as fiberglass or fiber-reinforced polymer (FRP has also been used). Fig. 22.4 illustrates different types of chemical tanks.

The tank design specifications depend on its application, chemical types and volumes, work environment, and other specifications. Tank wall thickness depends on the chemical type, temperature, and tank dimensions (wall thicknesses between 0.2 and 2.0 in. are commercially available). Specific gravity rating is a measure of the tank ability to hold materials. Higher specific gravity ratings indicate a greater ability to withstand hydrostatic stresses caused by a stored liquid (specific gravity ratings between 1.0 and 1.9 are commercially available).

Accurate determination of the liquid level in the chemical tank is critical for an optimum chemical management program, for the following reasons:

- It enables workers to estimate the need for tank refill before it empties out and causes production problems.
- It is used to determine the actual dosing rates.
- Accurate level management is crucial for good inventory management.

Therefore the liquid level in chemical tanks can be automated by using proper level sensors. Different types of sensors can be used, including switches, hydrostatic pressure, ultrasonic, guided wave radar, capacitance, and other methods.

The accuracy of level detection depends on the sensor technology and other factors, including:

- Temperature variations.
- Tank venting.
- Work environment vibration, pump stroking and other signal interfering noise.
- Tank leveling (tilted/and leaning tanks).

- Compatibility between chemical and sensor material.
- Chemical properties, such as specific gravity and foaming tendency.
- Tank fills.
- Fluid expansion/contraction.
- Tank expansion/contraction.

22.5 Chemical transfer and injection lines

The chemicals are pumped through chemical transfer of injection lines to the injection point, where they are introduced to the production fluids. The injection lines are usually ¼-inch seamless stainless steel 316 lines in accordance with ASTM A789, and NACE Standard MR0175 for corrosion-resistant alloys.

The chemical is introduced to the production system through the injection assembly by different types of nozzles, including the following:

- Quill: an open-ended tube cut at a 45-degree angle with a slot.
- Open tube: a straight open tube.
- NPT: similar to open tube, with threaded ending allowing the addition of other nozzles that can be perpendicular to the flow (straight nozzle), or parallel (right-angle nozzle).
- Atomizers: mostly used with scavenger to atomize the chemical to the gas phase for high surface area for reaction.
- Mixing probe: these are used for thick liquids and batch injections.

Fig. 22.5 illustrates some of the injectors used for chemical injections.

Injection is usually maintained in the middle of the pipeline and in the same direction of the flow, where the highest speed of the flow is, to take advantage of the turbulence for effective mixing and distribution of the chemical in the flowing stream. If the pipeline is piggable, the injection point may be flush with the pipe wall.

Capillary tubing is also used for chemical injection downhole in the wells. The capillary tubing can be installed internally (inside the production tubing) or externally (outside the production tubing). Fig. 22.6 illustrates the methods of using the capillary tubing for downhole chemical injection.

- Capillary string

The capillary tubing used is usually ¼ inch (or 3/8 in. in some cases) and is most commonly manufactured super duplex steel 2205, or other optimum alloys (825 or 625) for better resistance to the harsh downhole conditions.

- Capillary hanger

This hanger is mounted on the Christmas tree to hang the capillary tube and hold the well pressure, in a similar way to coiled tubing.

- Chemical injection manifold

For pressure and connection monitoring for the injected chemical from the chemical pump.

- Downhole chemical injection valves (DHCIVs), or chemical injection metering valves (CIMVs)

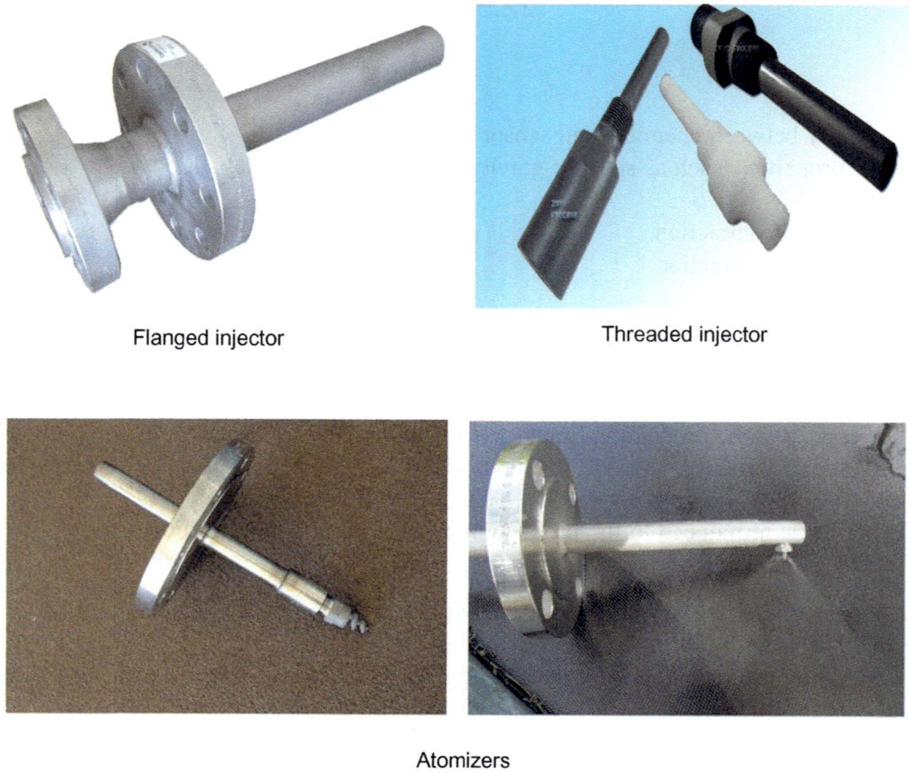

FIG. 22.5

Different types of nozzles used for chemical injection.

Photo courtesy Inyo Process, https://inyoprocess.com/, reprinted with permission.

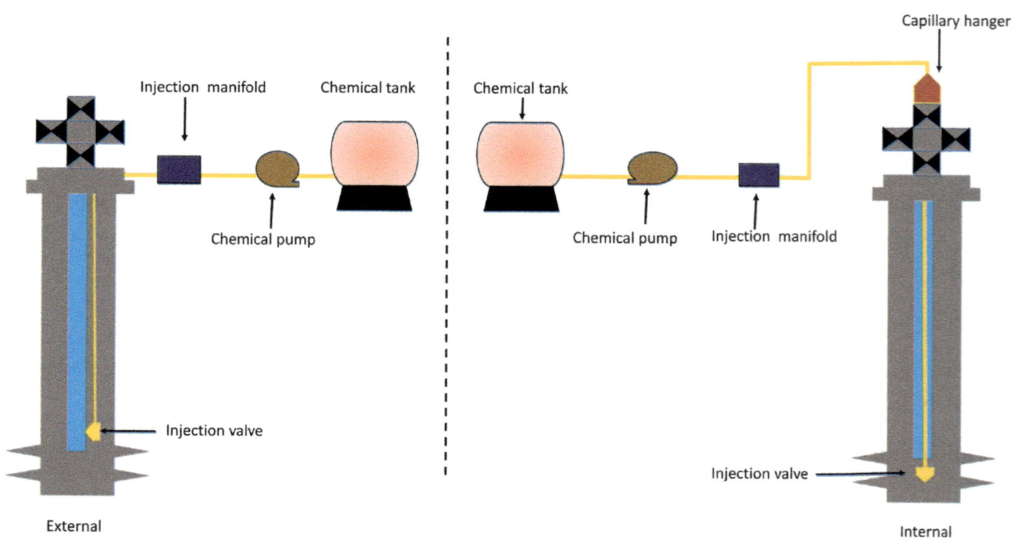

FIG. 22.6

Schematic of methods of downhole chemical injection using capillary tubing.

These have to be corrosion resistant and be able to precisely deliver the correct amounts of chemicals.

Further details of downhole chemical injection systems are furnished by Refs. [3–5].

Umbilicals are very commonly used in subsea and deepwater systems. They are used to link surface and seafloor oil and gas equipment for controls, power, or heat. They provide electric and fiber-optic signals, electrical power and hydraulic and chemical injection fluids to the subsea unit. Usually an injection tubing or capillary (1/4 in.) delivers the chemical to the targeted point. The umbilical lines cover long vertical and horizontal distances [6,7]. Bundles are another version of pipelines, with different tubes all together in one set, wrapped in insulation. The bundle contains the chemical injection, heat, and cables for controllers and monitoring.

The optimum design and operations of DHCI depend on the following:

- The completion and well designs
- Well fluids composition and properties: viscosity, density, compressibility
- Well/system temperature: temperature change causes changes in the injected fluid properties, which affects the pumping/injection rates
- Well/system pressure: pressure fluctuations in the system cause vibrations, which affect the injection/pumping rates
- Well/system flow rates and flow regimes
- Material selection.

Some of the problems encountered in DHCI include the following:

- Corrosion
- Particulate formation or chemical gunking result in line plugging or blockage of injection valves and filters
- Column separation
- Changes in properties due to changes in temperature and pressure, thickening, vaporization.

Therefore it is of utmost importance that the selected chemical products have the appropriate physical and chemical characteristics essential for secure delivery through a downhole/subsea capillary tube/umbilical line for use at the intended injection point.

An API Technical Report (17TR5) addresses the problems encountered in subsea production control and chemical injection systems and provides the necessary measures to prevent blockages in subsea chemical injection systems. In addition, Liddon et al. [8] suggested a series of test methods to assist in the selection of chemicals suitable for application in DHCI systems, including filter test, distillation test, visual assessment, and product stability under high temperature/high pressure conditions.

22.6 Other components

In addition to the main components discussed previously, other components are also essential, including pulsating dampener, pressure gauges, transmitters, level sensors, etc.

22.7 Control systems

The control system is the method by which the chemical dose into the system is adjusted and controlled. Chemical control systems are important, as they affect chemical consumption, optimum delivery of the chemical, manpower requirements, and other aspects of chemical injection.

Manual control involves adjusting the pump to continuously injecting the chemical at a predetermined constant rate, and any adjustments have to be made by operators. Manual control is usually used when chemical control is not critical, established control ranges are wide, and system retention time is long.

Simple automated methods include an on-off mode where the pump is adjusted at on-off cycles by the control signal. Control systems such as Supervisory Control and Data Acquisition (SCADA) systems have been used extensively in oil and gas operations. Such systems not only control and report system measurements such as flow rates and discharge pressures, but they are also able to predict upcoming problems and diagnose them.

22.8 Chemical skids

Usually in oil and gas field applications, the previously mentioned pumps, tanks, and instrumentation are skid mounted for easy installation and usage. The whole set is known as a chemical injection skid (CIS) or chemical injection package (CIP).

The features of these skids are different from one supplier to another based on the application, cost, capacity, work environment, and space constraints.

A standard CIS will have a chemical pump, tank, and instrumentation to inject one treatment chemical to the pipeline or well head. Some of these simple installations can be installed mobilely on a specially designed vehicle, allowing its use at different points in the same site or on different sites.

A multipoint CIS is used when the chemical is injected at different points in a platform when there are space constraints, eliminating the need for multiple injection point skids, which is considered a cost-effective solution.

A multicompartmental chemical injection system is a skid that has multiple compartmental chemical tanks allowing the storage and injection of different chemicals. Each compartment in such skids has a separate set of pumps, valves, and instruments all integrated in one control system. In 2015, LEWA completed its largest package to date, including 40 diaphragm pumps, 120 chemical injection points, and 18 chemical storage tanks [9].

A multihead pump CIS is used so that two or more chemicals can be injected through one pump in the skid (diaphragm or plunger) instead of using different pumps for different chemicals. In this case, the pump employs a single drive mechanism coupled with multiple pumping chambers, each capable of handling different types of chemicals.

Chemical injection skids have some advantages:

- Cost-effective
- Space efficient
- Greater safety and ability to be used in hazardous areas
- Better quality assurance

- Easy identification in asset management
- Can be fabricated and installed quickly
- Ensures reduced downtime at the worksite

22.9 Recent advances in chemical injection systems

One of the common problems associated with chemical injection is overdosing of the chemical or underdosing. That is more common in big fields with a high number of producers, especially with the process and process fluids changing parameters, such as:

- Varying production phases (oil, gas, or water) volumes
- Varying process operating parameters (P, T, flow rates)

Thus, instead of the conventional method of chemical delivery, an effective, or more specifically, a smart chemical injection system may help to cope with the field needs. Potential savings from using such smart systems were estimated to be $30,000 on average per well per year [10]. Several fully automated, or smart, controlled chemical injection systems have been developed and applied in the fields.

Such smart systems can guarantee the reliability of the chemical treatment, as its fully automatized injection features allow adjustments of the injection rate in real time and remotely, along with visualizing the chemical consumption through level control monitoring.

22.10 Summary

Chemical agents are widely used in production operations in oil and gas fields. To mitigate flow assurance problems, different types of inhibitors, dispersants, biocides, and scavengers are routinely injected in production systems. These need specially designed chemical injection systems or skids to precisely inject a predetermined amount of the chemical to a specific location in the production system. These injection systems include chemical pumps, tanks, injection lines, and other instrumentation and control systems. The aspects of these components were briefly reviewed in this chapter in light of the recent advances.

References

[1] M.A. Kelland, Introduction and environmental issues, in: Production Chemicals for the Oil and Gas Industry, CRC Press, 2014, pp. 19–38, https://doi.org/10.1201/9781420092974-4.

[2] United states environmental protection agency (EPA), Methane emissions from the L Gas Industry Volume 13—Chemical Injection Pumps, 1996. GRI-94 I 0?573C, EPA 600lR9G OIlOm.

[3] D.J. Peng, A. Low, Quantitative risk assessment of downhole continuous chemical injection for subsea multiphase production system, in: Paper Presented at the 17th International Conference on Multiphase Production Technology, Cannes, France, 2015.

[4] M. Shamlooh, S. Abdelrady, M. Buasali, K. Farouque, Well integrity protection using corrosion inhibitor treatment system for sour gas producers in Bahrain field, in: SPE Middle East Oil and Gas Show and

Conference, MEOS, Proceedings Society of Petroleum Engineers (SPE), vol. 2019, 2019, https://doi.org/10.2118/194893-ms.

[5] B.M. Hustad, O.G. Svela, J.H. Olsen, K. Ramstad, T. Tjomsland, Downhole Chemical Injection Lines—Why Do They Fail? Experiences, Challenges and Application of New Test Methods, Society of Petroleum Engineers, 2012, https://doi.org/10.2118/154967-MS.

[6] L.A. Guerra, G.H. Oliveira, Designing Downhole Chemical Injection Systems, Society of Petroleum Engineers, 2014, https://doi.org/10.2118/170772-MS.

[7] Q. Lu, Z. Yang, J. Yan, Q. Yue, Design of cross-sectional layout of steel tube umbilical, J. Offshore Mech. Arct. Eng. 136 (4) (2014), 041401, https://doi.org/10.1115/1.4027800.

[8] C. Stewart-Liddon, N.J. Goodwin, G.M. Graham, T. Tjomsland, B.M. Hustad, O.G. Svela, et al., Qualification of Chemicals/Chemical Injection Systems for Downhole Continuous Chemical Injection, Society of Petroleum Engineers, 2014, https://doi.org/10.2118/169782-MS.

[9] Largest LEWA Chemical Injection Package, 2021. Retrieved From: https://www.lewa.com/en/applications/lewas-biggest-chemical-injection-package#position-facts, last visited November 2021.

[10] Detechtion Technologie, Chemical Injection Monitoring and Control, 2018. Retrieved From: https://www.detechtion.com/hubfs/Chemical-Injection-Monitoring-and-Control-2.pdf, last visited November 2021.

CHAPTER

23 Environmental Impacts of Flow Assurance Solids and Their Management

23.1 Introduction

Oil and gas exploration and production operations have several impacts on the environment. These impacts depend on the project size and complexity, stage of project development, nature of the affected surrounding environment, and effectiveness of treatment plans. Such impacts can be classified into [1]:

- Human, socioeconomic, and cultural impacts: such as the changes in land patterns after using the land to create access to oilfields, new transportation systems, population changes due to immigration for jobs, economic changes due to employment and new jobs income, and others.
- Atmospheric impacts, which basically include the gas emission produced from the different sources in oil and gas fields, including flaring, venting, and purging gases, combustion processes, and airborne particulates from different operations.
- Aquatic impacts due to the discharge of different production fluids and chemicals into the ocean. These include produced water, drilling fluids, process drainage, oil spills, cooling water, and treating chemicals. For example, discharges of produced water in the North Sea are in the range of 2400–40,000 m^3/day for oil installations and 2–30 m^3/day for gas installations.
- Terrestrial impacts such as ground disturbances, oil spills, etc.
- Accidental emergencies such as oil spills, fires, well blowouts, etc.

The potential impacts of oil and gas operations on the environment are listed in Table 23.1 [1–3].

The formation of flow assurance solids is a major problem in oil and gas operations. Not only can they interrupt production operations and cause expensive maintenance, but they also have several environmental impacts, as discussed in the following sections.

23.2 Environmental laws and legislations

Due to the previously mentioned impacts of the oil and gas industry on the environment, several laws and guidelines have been developed to serve as a standard framework that should be followed in order to reduce such environmental impacts. Various international, regional, and national environmental regulations have been issued. Some of the common pieces of legislation that apply to oil operations are [1]:

Essentials of Flow Assurance Solids in Oil and Gas Operations. https://doi.org/10.1016/B978-0-323-99118-6.00011-3

Table 23.1 The environmental impacts of the oil and gas industry [1–3].

Activity/concern	Potential impact
Exploration	- Aerial and seismic noise - Emissions - Fluid discharges
Drilling	- Vegetation cleared - Change of surface hydrology - Soil contamination - Machinery noise - Emissions - Discharges (drilling cuts, drilling chemicals, cementing materials, hydrocarbons)
Development and production	- Machinery noise - Emissions (flare, combustions, etc.) - Discharge (produced water, other waste) - Land disturbance and usage for processing equipment and pipeline extensions - Radiation use in production - Radioactive materials production and exposure - Oil spill - Impacts on aquatic systems
Transport	- Oil spill - Land use for pipelines and stations

- Petroleum laws
- Environmental Protection Act
- Environmental impact assessments
- Clean Air Act
- Marine pollution
- Standards for noise, radiation, chemical exposure
- Wastes discharge and management regulations
- Land contamination or land disturbance
- Water Act
- Safety and fire regulations
- Control of major hazards
- Storage and usage of chemicals
- Public and worker health and safety
- National Park or Protected Area laws
- Forest protection laws
- Protection of indigenous and cultural heritage
- Fishery protection, marine navigation, and safety

Some of the international environmental conventions include [1]:

- Montreal Protocol of the Vienna Convention (1985, 1987)
- Basel Convention (1989)

- Convention on Migratory Species of Wild Animals (1979)
- The United Nations Framework Convention on Climate Change (1992)
- Biodiversity Convention (1992)
- United Nations Convention on the Law of the Sea (UNCLOS) (1982)
- Regional Sea Conventions (Barcelona, OSPAR, Kuwait, etc.)

The application of such legislation of regulations is implemented by the local/national authority, and their application depends on the use of the appropriate laws with clearly defined responsibilities, appropriate monitoring and consulting, and adequate performance reporting.

23.3 Environmental risks associated with flow assurance solids

Solid deposits pose serious environmental and safety challenges. When solids form inside certain equipment, they increase the equipment fuel consumption, which increases the greenhouse gas emissions. In addition, solids form in huge amounts that require removal, handling, and disposal. Some of these deposits are pyrophoric or radioactive in nature. Fig. 23.1 summarizes the different environmental issues associated with solids deposition.

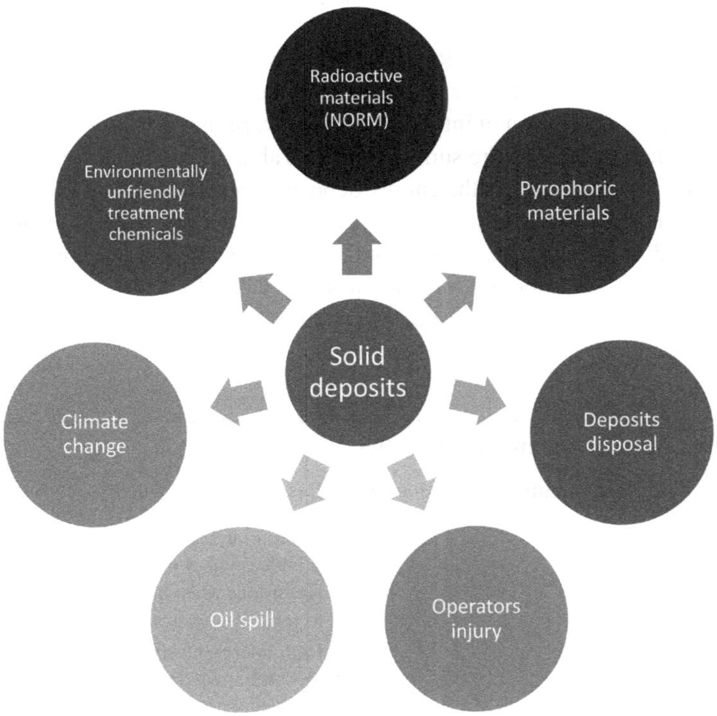

FIG. 23.1

The environmental issues associated with flow assurance solids deposition.

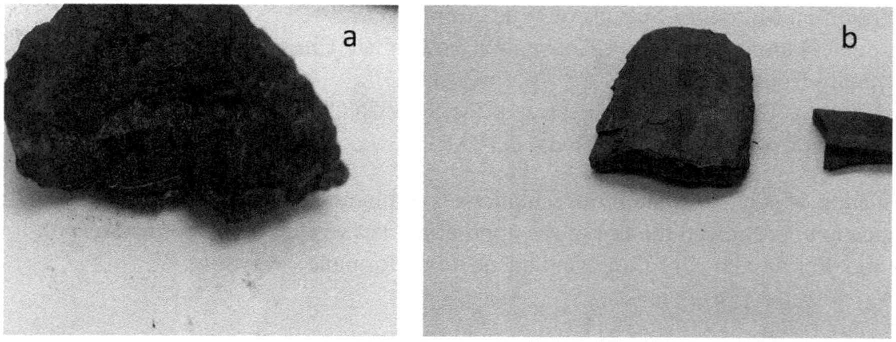

FIG. 23.2

NORMs accumulated with sulfate scale (a), sulfate and sulfide scale (b).

23.3.1 Naturally occurring radioactive materials

Naturally occurring radioactive material (NORM) is a type of waste from oil and gas production that can accumulate at different locations along the production system, such as wellheads, separation vessels, pumps, and other processing equipment. NORMs can accumulate in the form of scale, sludge, scrapings, and other waste media, causing flow restrictions and exposing the operators to potential radiological health hazards [4].

The mechanism of formation of NORMs and their impacts on flow assurance are discussed in Chapter 5. Fig. 23.2 shows mineral scales with accumulated NORM.

There are three types of radiation of interest emitted in various amounts by NORMs: alpha (α), beta (β), and gamma (γ); their properties are summarized in Table 23.2. The SI unit for absorbed (energy) dose is the gray (Gy), which quantifies the energy deposited per unit mass: 1 Gy equals 1 J/kg. Because not all types of radiation produce the same biological effect, and the dose equivalent is often used instead of the absorbed dose. The dose equivalent or the effective dose is the product of the absorbed dose and a radiation weighting factor and is expressed in sieverts (Sv) [5]. Sieverts are frequently expressed per unit time such as per hour (Sv/h) or per year (Sv/year). The relationship between the absorbed dose, equivalent dose, and effective dose is illustrated in Fig. 23.3.

Table 23.2 Alpha, beta, and gamma properties.

Radiation	Mass (amu)	Charge	Description
Alpha (α)	4	+2	Helium nucleus, high ionizing power, can easily be shielded with a piece of paper, coverall, or gloves. Hazardous when inhaled or ingested and causes damage to sensitive tissues
Beta (β)	1/1850	−1	High-energy electrons, moderate ionizing power, moderate penetration, harmful to living cells, can be shielded with a thin piece of aluminum, steel, or plastic
Gamma (γ)	0	0	High energy photons, low ionization power, very high penetration power, very harmful to all bodies

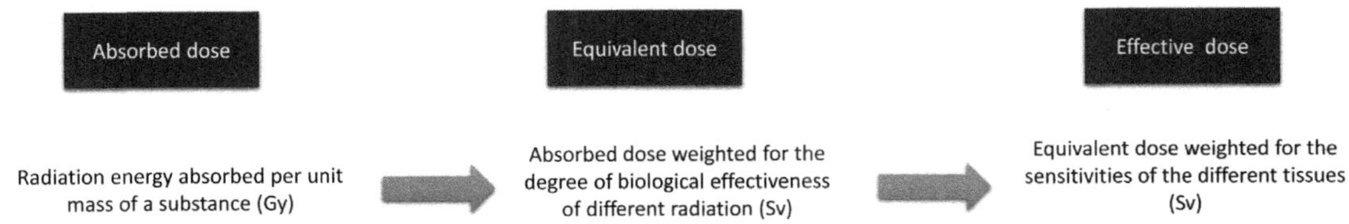

FIG. 23.3

Relationship between absorbed dose, equivalent dose, and effective dose.

Workers are exposed to NORM radiation in two main ways: external or internal. External exposure occurs during daily routine work by the materials emitting radiations when the systems are closed, whereas internal exposure takes place by inhalation or ingestion from open systems during shutdowns and maintenance, if proper precautions have not been taken. Alpha particles are the most harmful internal hazard, while gamma photos are the most harmful external hazard. Fig. 23.4 illustrates these methods of exposure.

Irradiation from external sources occurs in the proximity of the source and decreases with distance, dropping to effectively zero at a sufficient distance. In the case of NORM, this is usually a few meters from contaminated equipment and follows the inverse square law, whereby intensity is inversely proportional to the square of the distance from the source. Absolute exposure at 1 m will be four times greater than exposure at 2 m [6]. The dose rates in some production facilities are summarized in Table 23.3.

The health effects of NORM are a function of the energy transmitted to the body, as the α, β, or γ radiation dissipates excess energy into living cells, which may result in cellular damage and genetic mutation. The severity of the radiation effect increases with increasing dose and dose rate. The effects on humans can be classified into [6]:

- *Deterministic effects* in which harmful tissue reactions takes place and in some instances results in organ damage following high doses. Deterministic effects occur at doses exceeding 0.5–1 Gy and are usually associated with incidents in the nuclear industry and not encountered in oil and gas fields.
- *Stochastic effects* are usually associated with long-term, low-level exposure, such as cancer and heritable effects in offspring due to mutations of reproductive (germ) cells. NORM radiation protection deals basically with the stochastic effects.

23.3.2 Pyrophoric deposits

Pyrophoric materials are substances that ignite instantly upon exposure to air (oxygen), or moisture (water). The interaction of pyrophoric materials with oxygen or water produces a tremendous amount of heat that can ignite the materials and cause hazardous fires.

A common pyrophoric material that is commonly found in oil and gas fields is iron sulfide scale. Iron sulfide is principally formed due to the interaction of the hazardous H_2S gas, which is produced by microbiological activity of sulfate-reducing bacteria (SRB) with the iron and steel in process facilities. The mechanisms of formation and polymorphs of iron sulfides are detailed in Chapter 5.

Pyrophoric iron sulfide scale can form in the downhole, topsides flowlines, process equipment, and tank bottoms. Once exposed to oxygen, it reacts rapidly, and the exothermic reaction releases large

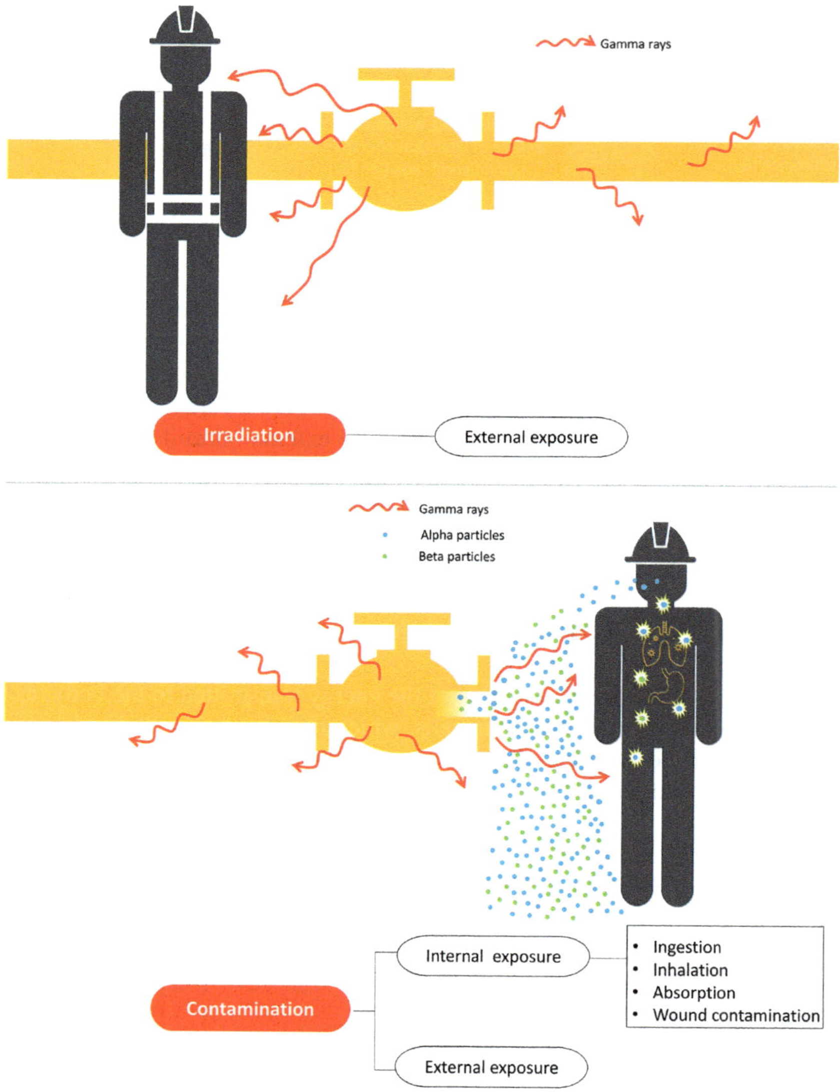

FIG. 23.4

Workers exposure to NORM from internal and external sources.

amounts of heat, which is enough to make the deposits glow red and it serves as an ignition source for any hydrocarbon (oil or gas) that is present.

Fires and explosions caused by pyrophoric materials can occur during shutdowns when tanks or vessels are emptied, or equipment and piping are opened for inspections and maintenance [7].

For example, in three reported incidents of tank explosions during maintenance, it was found that iron sulfide scale was the main culprit. In these incidents, oxygen came in contact with pyrophoric iron

Table 23.3 Dose rates in production facilities.

Item type	Location	Dose rate (μSv/h)
Tubulars	Downhole tubing, safety valves (internal)	Up to 300
	Wellheads, production manifold	Up to 20
	Flowlines (external)	Up to 8
Equipment	Separator (external)	Up to 30
	Pumps (external)	Up to 20

sulfide scale, which autoignites in the presence of the flammable mixture of hydrocarbon vapors and oxygen, resulting in catastrophic explosions [7]. Tanks that have sat empty for an extended time can still be hazardous. Pyrophoric iron sulfide can remain on tank walls for long periods of time before finally igniting when exposed to oxygen.

23.3.3 Deposits disposal

Different types of solid deposits are formed in oil and gas fields. These deposits can form in thin films or in thick layers that can eventually block the whole diameter of the flowlines. Considering the size and layout of the production systems, massive amounts of solids can accumulate in the production system. A millimeter of wax deposited over 20 km of a 20″ pipeline equates to a volume of 31 m^3 of wax (a solid plug of 157 m or 515 ft) [8]. Halite can precipitate at the rate of 20 lbm for each barrel of water produced, leading to many tons of scale every day in a single well producing water at a rate of 1000 B/D [159 m^3/d] [9]. A commonly reported case that represents the size of the accumulated deposits is the case in the Nigerian field when only a few milligrams per liter of naphthenate deposits over 6 weeks led to an accumulation of 30 tons of deposits, costing $6M USD to remove the deposits [10].

These solids can be treated with chemicals to dissolve and remove them. However, chemicals are quite expensive, time consuming, and can be ineffective in some instances, such as the cases of hard thick layers of scale. Therefore mechanical methods like pigging, wireline, slickline, coiled tubing, mills, and jets are commonly used to remove deposits. For example, pigging is conventionally used on a routine basis to clean and remove deposits from flowlines (Fig. 23.5). Pigging runs usually retrieve different kinds of scale, organic deposits, and sludge, with volumes ranging from a few kilograms to metric tons of debris, depending on the severity of the problem. This poses an environmental challenge due to the high volumes of these deposits and their nature, since they contain insoluble hazardous materials (hydrocarbon sludge, wax, NORMS, pyrophoric scale), which contaminate the environment and must be disposed of properly.

23.3.4 Climate change

Climate change is one of the biggest challenges faced globally. Global warming is an alarming indicator of climate change that has been gaining attention since the mid-20th century. Global warming is happening at a faster rate than ever before, and scientists are projecting this to continue in the decades to

FIG. 23.5

Various amounts of deposits, from a few kilograms to tons, can be removed during pigging.

come. Global warming causes extreme weather, melting ice, rising sea levels, floods, drought, environmental pollution, and health issues for individuals.

The impact of solid depositions on climate change is not minuscule. They participate indirectly in increasing CO_2 emissions when they deposit on the internals of equipment, serving as an insulation layer, reducing efficiency, and increasing fuel consumption, which when combusted, leads to liberation of extra CO_2. For example, in studying the effects of heat-exchanger fouling on power plant condenser efficiency and on the environment, the authors found that, as the thickness of the biofilm increases, the carbon dioxide emissions increase in proportion to the thickness of the biofilm. They estimated that, for the thickest biofilm studied (10^3 μm), the sum of the additional carbon dioxide produced amounts to 6.2 t/h. Although this is a relatively large discharge of carbon dioxide, it represents only a small percentage of the total carbon dioxide produced by the power station when operating on full load [11].

Data from refineries suggest that crude oil fouling accounts for around 10% of their total CO_2 footprint. A conservative estimate of the approximate total amount of emitted CO_2 due to fouling is approximately 88 million tons per annum. This accounts for about 2.5% of the total anthropogenic emission of carbon dioxide [12]. In a study of the effect of fouling in the Brazilian field, the authors estimated that for a 200,000 bpd refinery the crude preheat train fouling would be responsible for US $6 million/year extra energy costs and 62,500 metric tons/year of CO_2 emissions. Considering an estimate of the total emissions equivalent to about 0.16 t CO_2/t oil processed, this corresponds to 3.9% of the total carbon footprint of the refinery [13].

Similar to the increase in CO_2 emissions in the membrane and desalination industry, some estimations can be made for the effects of fouling in oil and gas field operations.

In the meantime, global warming directly affects gas hydrate sediments in the marine and permafrost areas, leading to destabilizing them. Methane and methane-derived carbon that escapes from sediments and soils and reaches the atmosphere could exacerbate greenhouse warming. The synergy between the warming climate and gas hydrate dissociation feeds a popular perception that global warming could drive catastrophic methane releases from the contemporary gas hydrate reservoir [14].

23.3.5 Treating chemicals with environmental risks

Solid deposits have another environmental impact when they are treated using chemicals. The treating chemicals used in solid deposit mitigation pose environmental impacts in two different ways:

- The reaction between the treating chemical and the deposits releases environmentally unfriendly products.
- The treating chemical itself or its decomposition products are environmentally unfriendly due to their toxicity, low biodegradability, bioaccumulation, etc.

The first way includes chemical dissolvers that interact with the solid deposits, aiming to dissociate them into soluble form. This dissociation reaction can lead to generation of hazardous materials. One of the common examples is H_2S generated from dissolution of sulfide scales when they are treated with acids. H_2S is highly flammable and toxic, even at low concentrations, which can cause severe complications for the personnel at high concentrations. THPS is also used in dissolving sulfide scales, which usually does not generate H_2S. However, when mixed with NH_4Cl (as a catalyst) and used at high temperatures, HCl is produced, which consequently could produce H_2S when dissolving sulfide scales. THPS also decomposes, producing formaldehyde, which is a strong-smelling, flammable chemical and probable carcinogen. Dissolution of sulfate hard scales, especially barium and strontium scales, leads to the release of radioactive radionuclides into the final spent solution, which is usually discharged and treated with the produced water. In a similar way, removing the organic deposits increases the load of high molecular weight compounds in the spent solvent, which with the effect of turbulence can be emulsified as oil in water emulsions requiring excessive treatments in the ballast process to prevent their discharge with the produced water.

The second way the treating chemicals are environmentally affecting the operations occurs when these treating chemicals are carried over with the produced streams and partition between hydrocarbons and produced water, ending up mostly in the discharged water, which affects the aquatic systems, soil, and the ecosystem in general. Fig. 23.6 illustrates some of the components of environmental interest in produced waters from Norwegian offshore production platforms [15]. High volumes of some chemicals can also decrease the quality of the produced hydrocarbons, such as methanol, which lowers the economic value of natural gas.

Chemicals discharged into the marine environment have the potential for acute or long-term effects on aquatic organisms. These effects depend on a number of factors, such as the inherent toxicity of the material, the quantities discharged, and the accumulated concentrations in the water column, the exposure time of biota to that chemical, and the sensitivity of the organisms to particular chemicals [16]. Furthermore, the magnitude of the impact of the chemical discharge depends on environmental and hydrogeological factors characteristic of the sea, including bathymetry, waves, currents, and depth of the water column. These factors determine the extent of mixing and, therefore, the geographical range of the impact [16]. The chemistry of these chemicals is largely based on amine salts, phosphonates and sulfonates, alkyl phenols, and other polymers, in which the formulations also have proportions of surfactants. The polymers used are frequently based on organic acids, which exhibit low to moderate biodegradation combined with low potential for bioconcentration [17]. They also can include catalysts or organic solvents that can increase their toxicity and bioaccumulations risk.

The discharge of chemicals has a strong potential to impact both physicochemical and ecological attributes of receiving marine environments [18]. Generally speaking, an ideal benign chemical inhibitor/dissolver has to have [18]:

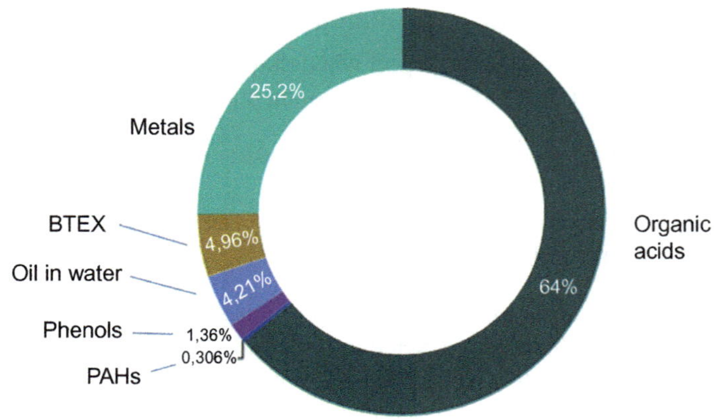

FIG. 23.6

Relative emissions of naturally occurring substances in PW discharges from Norwegian offshore production platforms in 2019.

Source: J. Beyer, J., A. Goksøyr, D.Ø. Hjermann, J. Klungsøyr, Environmental effects of offshore produced water discharges: a review focused on the Norwegian continental shelf, Mar. Environ. Res. 162 (2020) 105155, https://doi.org/10.1016/j.marenvres.2020.105155.

(1) Excellent functioning (inhibition, dissolution, etc.);
(2) Low aquatic and human toxicity;
(3) High biodegradability;
(4) A low water hazard class;
(5) No phosphorus, nitrogen, or heavy metals;
(6) Cost effectiveness.

A biodegradable polymer criteria is different from one country to another. For example, according to the Convention for the Protection of the Marine Environment of the North-East Atlantic (known as the OSPAR Convention), a polymer is considered readily biodegradable if it satisfies the following criteria [18]:

(1) Over 60% biodegradation within 28 days (based on ASTM and/or OECD biodegradation tests), whereas for less than 20% biodegradation, the chemical should be replaced;
(2) Toxicity of LC50 (lethal concentration for 50% of the sample population) or EC50 (half-maximal effective concentration) >1 mg/L for inorganic species and LC50 or EC50 >10 mg/L for organic species; and
(3) Bioaccumulation of log-(POW) <3 (POW = partition coefficient in octanol/water).

A chemical that fulfills two of the three requirements and has biodegradability greater than 20% in 28 days is eligible to be listed on the PLONOR list (poses little or no risk).

Taking some scale inhibitors as an example, their biodegradability was reported as: 17% in 28 days for PBTC, 23% in 28 days for ATMP, 33% in 28 days for HEDP, while for PPCA it is 0% in 35 days [19]. The toxicity and biodegradation data of some phosphonate scale inhibitors are published elsewhere [20].

The use of organic solvents like xylene, toluene, or ethylbenzene to remove organic deposits, especially asphaltenes, is very common in field operations. These compounds are known as BTEX, and it is one of the produced water pollutants that have various toxic effects on aquatic microorganisms and require specialized water treatments.

Residual oil in separated produced water can contain toxic carcinogens such as polyaromatic hydrocarbons (PAHs); therefore there are regulations on its discharge level. For example, in the North Sea, this level is currently 30 ppm; however, for some regions, it can be as low as 5–10 ppm, which may be difficult to achieve.

23.3.6 Other environmental impacts

Deposits accumulation in equipment and pipes represents a major safety hazard during their maintenance. Full blockage of pipes and equipment can lead to serious incidents. Some of these deposits can build up pressure behind them and, once the system is opened, the depressurization breaks the solid blockage, or parts of it, which can be like projectiles that can be fatal. In addition, fully blocked pipelines during operations can cause pipeline rupturing (or cause corrosion that leads to pipeline rupturing), leading to hydrocarbon spills into the surrounding environment.

One of the famous incidents is the North Sea Piper Alpha Disaster, which is related to gas hydrates buildup. The incident killed 165 out of 226 crew members and two crew from the standby vessel Sandhaven, which makes it the world's deadliest offshore disaster. In addition, the property damage from the Piper Alpha explosion was estimated to be $1.4 billion dollars.

Recently, and surprisingly, new analysis of the Deepwater Horizon disaster indicated that gas hydrate buildup might have contributed to the problem [21] (Fig. 23.7). The Deepwater Horizon is considered the worst oil spill incident in the United States, after releasing over 130 million gallons of crude oil into the Gulf of Mexico (Fig. 23.8).

23.4 Environmental risks management

23.4.1 NORM scales management

The basic NORM management strategy depends on following some key steps, as illustrated in Fig. 23.9 [4,22,23]:

- NORM survey and monitoring
- Worker protection
- Control of NORM scale formation
- Decontamination of NORM-contaminated equipment
- Disposal of NORM scale
- Knowledge sharing and training

The radioactivity surveys are crucial to determine the level of radiation and baseline spread of NORM in the facility and to prepare a safe work environment for the workers, and to determine their optimal personal protective equipment (PPE). Full scale deposits analysis and characterization represent another crucial step. Common instruments used in radiation monitoring include [6]:

FIG. 23.7

Deepwater Horizon after the explosion. *Source: https://commons.wikimedia.org/wiki/File:Deepwater_Horizon_offshore_drilling_unit_on_fire_2010.jpg.*

FIG. 23.8

Oil from the Deepwater Horizon oil spill approaches the coast of Mobile, Alabama. *Source: https://commons.wikimedia.org/wiki/File:Defense.gov_photo_essay_100506-N-6436W-023.jpg.*

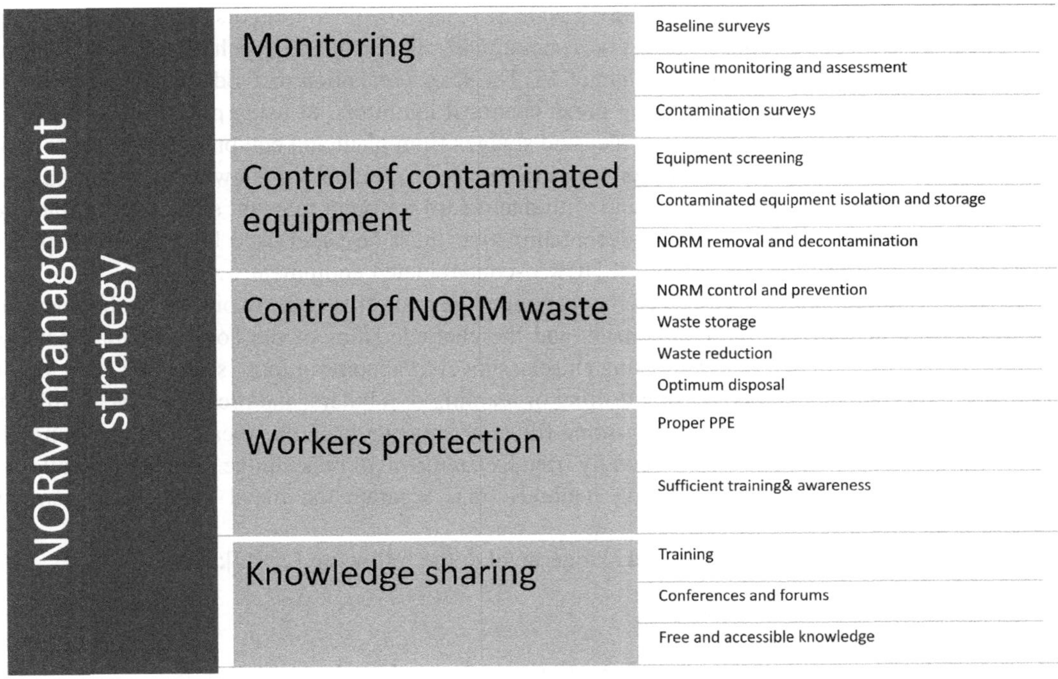

FIG. 23.9

NORM management strategy.

- Dose rate meters
- Dosimeters
- Surface contamination meters
- Airborne contamination meters and gas monitors.

Some practical measures are usually maintained to protect against the radiation and to keep radiation doses as low as reasonably achievable.

Some of the basic measures for the external exposure include [4,22,23]:

- Minimize the duration of external exposure;
- Maintain optimum distances between any accumulation of NORM (installation part) and exposed personnel;
- Maintain shielding material or equipment between the NORM and exposed personnel.

For internal exposure, some basic measures are [22]:

- Wear proper PPE and protective clothing during work, and apply industry hygiene after work.
- Refrain from any activities that may lead to inhaling or ingesting NORMs, e.g., eating, drinking, smoking, chewing gum, etc.
- Keep the NORM scale confined, in a dedicated place.

One of the most effective methods of fighting NORM scale is by preventing its formation. Scale prevention can be achieved by many means, i.e., operational, chemical, and nonchemical. A full discussion on scale inhibition is furnished in Chapter 15. The scale prevention method must be efficient and environmentally friendly itself, i.e., using green chemical inhibitors, or using physical methods.

The removal of NORM-containing scales and sludges from plant and equipment, whether for production and safety reasons or during decommissioning, needs to be carried out with adequate radiation protection measures having been taken and with due regard for other relevant safety, waste management, and environmental aspects. The decontamination must be supervised by a qualified person [6,22]. Fig. 23.10 illustrates the control of NORM-contaminated equipment.

Various decontamination methods are being applied on and off site, the choice of method depending on the type and size of the components and the characteristics of the contaminating material. Methods range from manual removal (using chisels, shovels, vacuum for loose scale, and other simple machinery), mechanical methods (using drilling or reaming of tubulars, and preferably wet drilling to lower the generation of dusts or applying some filters or abrasive jets), chemical cleaning (using different dissolvers that must be environmentally friendly), removal of bulk sludge from vessels followed by rinsing with water, to using melting methods (that separate the metal component from the NORM) [22].

The disposal of the NORM scale and sludge involve the following [22,23]:

- Disposal by shallow land burial
- Deep underground disposal
- Injection into hydraulically fractured formations, or disposal in abandoned wells
- Land spreading
- Salt cavern disposal
- Offshore discharge
- Wetlands disposal
- Recycling by melting of contaminated scrap metal

Before disposal, NORM scales are temporarily stored in a dedicated place, where it is plainly indicated that NORM scale exists, using proper storage vessels or containers.

Sharing knowledge between national and international oil companies, oil and gas service companies, and the other agencies that are concerned with NORM and radioactivity management is another key point in NORM management. Shared experience of a common problem can provide benefits to all concerned, along with providing training in light of the shared experience and the latest advances [4,22,23].

23.4.2 Deposits disposal management

As mentioned earlier, large amounts of deposits usually form during production operations. These solids must be disposed of properly to lower their impact on the environment.

First, these deposits must be identified and characterized. Then it needs to be sorted (if possible) and stored, waiting for the proper method of disposal.

Mineral scale has to be sorted into NORM scale, pyrophoric scale, and other types of scale before disposal, with specific measures applied for each type. NORM scale must be managed as mentioned in Section 23.4.1. Pyrophoric scale must be quenched before disposal. Mineral scale can be dissolved

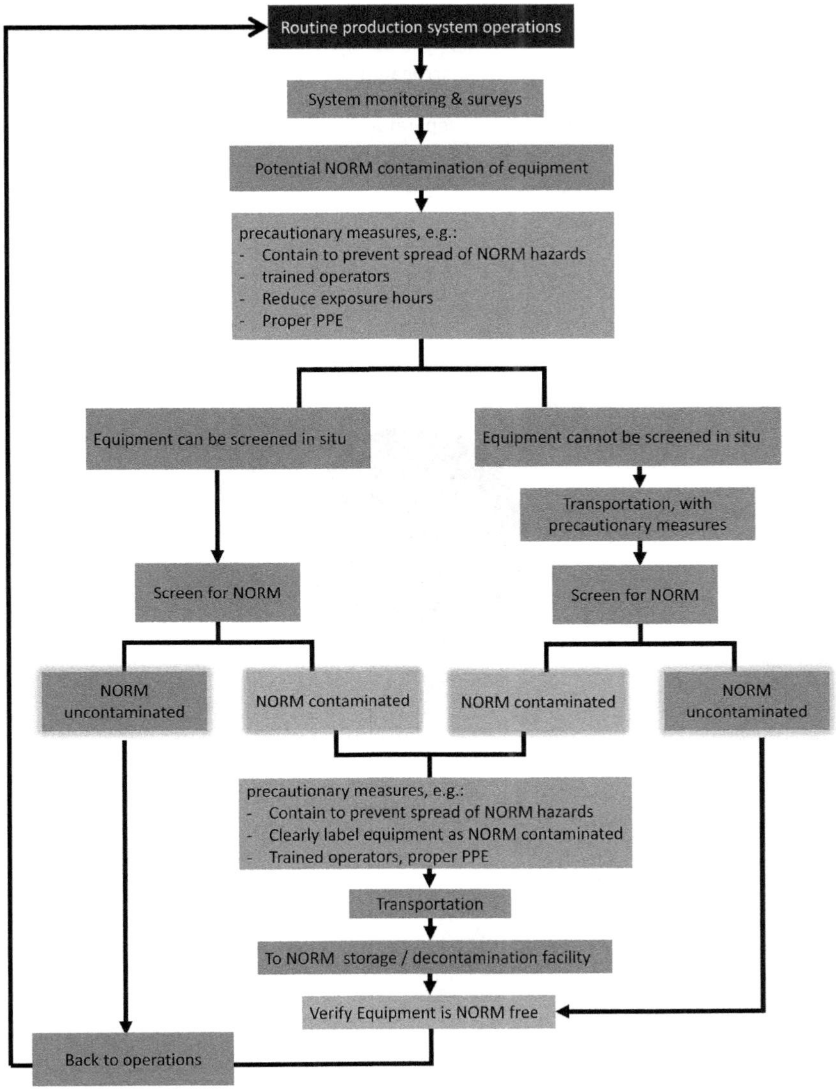

FIG. 23.10

Control of NORM-contaminated equipment.

using environmentally friendly chemicals. Another method of disposing them is by land dispersal or shallow or deep land burial.

Organic deposits and sludges can also be dissolved using proper environmentally friendly dissolvers added to the hydrocarbons, considering the compatibility of the dissolved deposits with the hydrocarbon streams. They also can be used for other applications like reclamation and road paving. Finally, like mineral scales, they can be disposed of on land.

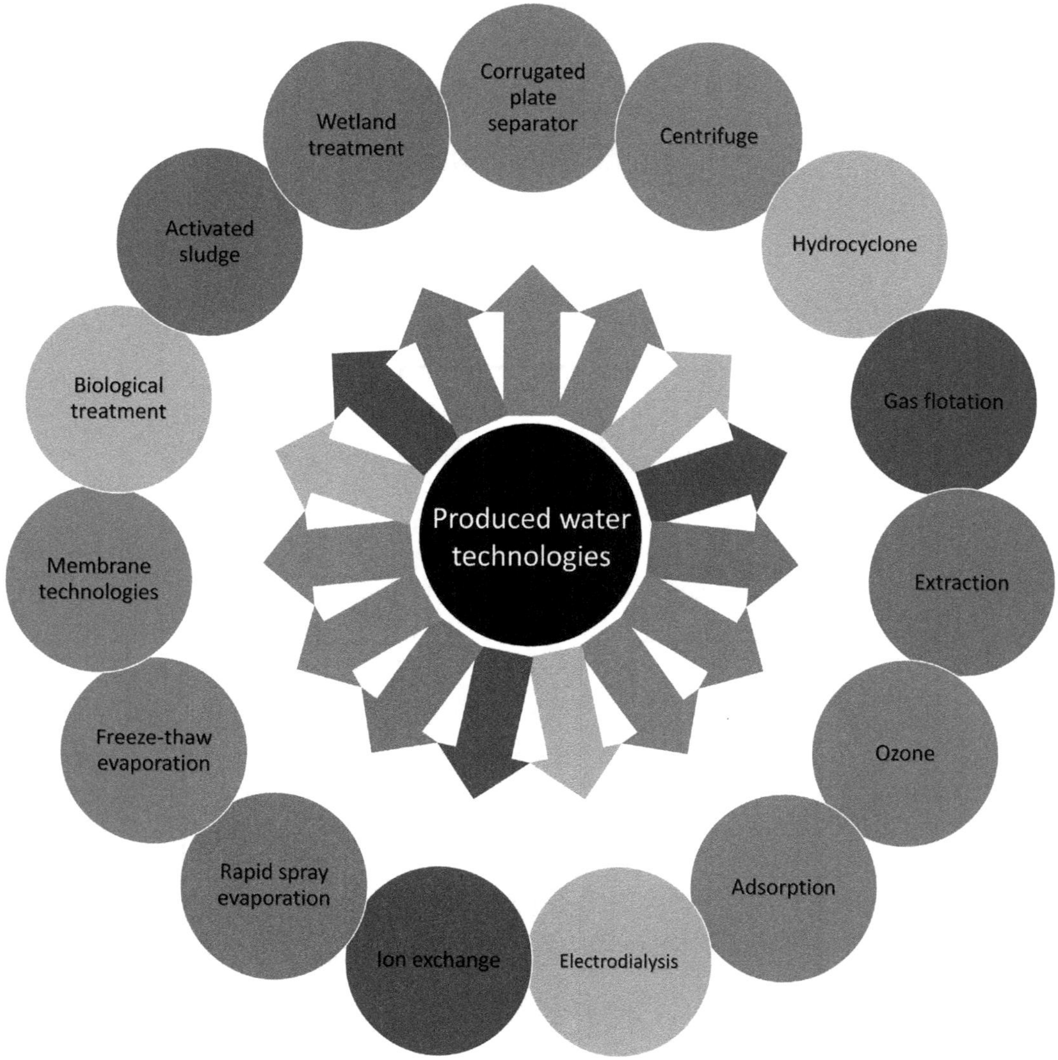

FIG. 23.11

Produced water treatment methods.

23.4.3 Produced water treatment and reuse

Produced water that is contaminated with different types of hazardous material must be treated before discharged into the ocean. Fig. 23.11 illustrates the different technologies in produced water treatment.

The main goal of produced water treatment technologies is to reduce the levels of aliphatic hydrocarbons, benzene, heavy aromatic compounds (PAHs), alkylated phenols, and some production chemicals. For example, phosphorous-based chemicals can be removed by chemical precipitation, biological methods, adsorbed methods, or filtration methods.

One of the methods that is considered economic and effective in safely disposing of the produced water is produced water reinjection (PWRI), where produced water (instead of sea water) is injected into the formation to maintain the reservoir pressure. This method minimizes the expensive wastewater treatments, works as a sustainable water source, and prevents discharging contaminated water into the environment. Produced water has to be treated prior to injection to remove the high suspended solids, the high oil content, and to check the compatibilities with the reservoir fluids.

23.4.4 Environmentally friendly treating chemicals (green chemicals)

Production chemicals pose a threat to the ecosystem once they are partitioned in the produced water and discharged to the ocean. New types of chemicals have been emerging, considered environmentally friendly or green chemicals. Such chemicals are considered less toxic or nontoxic, biodegradable, and safer to the ecosystem, besides being technically efficient in controlling the production problem. Table 23.4 summarizes some of the conventional treatment chemicals and their environmentally friendly counterparts.

Additionally, operational methods that optimize the conditions to work in the deposit-free zones are another low-cost environmentally friendly option. Furthermore, other nonchemical methods like internal coatings, deposit repellent surfaces, electromagnetic, acoustic, and biological methods are efficient methods.

Table 23.4 Green chemicals as alternative to conventional chemicals.

Deposit type	Conventional chemicals	Green chemicals
Mineral scales	- Phosphates - Phosphonates - Polycarboxylates	- Polyaspartic acid (PASP) - Carboxymethylinulin (CMI) - Polyepoxysuccinic acid (PESA) - Natural scale inhibitors/potato extract, aloe vera, cornstalks, proteins
Gas hydrates	- Methanol - Glycol - Vinyllactam polymers - hyper-branched polyesteramide	- Proteins and amino acids - Polyester and polyamide with high biodegradability - Pseudo-polypeptides - Naturally extracted polymers - Ionic liquids
Wax	- Ethylene vinyl acetate copolymers - Comb-shaped copolymers - Maleic copolymers	- Alkylene oxide, modified alcohol surfactant and aminodicarboxylic acid diester with improved biodegradability - Using bacteria-degrading wax - Biobased surfactants
Asphaltenes	- Aromatic organic solvents - Alkyle phenols - Surfactants SDS, CTAB, DBSA - Polyester and polyamide with high biodegradability	- Methyl ester oleate - Natural product and extracts; vegetable oils, cashew nut shell liquid (CNSL) Pepper Jamaican essential oil, Jamaican essential oil, Brazil nut oil, coconut essential oil, andiroba oil, grapeseed oil, pinewood essential oil, sweet almond oil

23.5 Summary

Flow assurance solids are major challenges in the oil and gas industry due to their detrimental impact on the production operations, causing production loss and reduction in revenue. In addition, such solid deposits have another global impact due to their effects on the environment. Deposit formation causes extra fuel consumption, leading to extra CO_2 emissions, which is a main contributor to climate change. Some of these deposits are hazardous materials like NORM scales, which have radioactive components that lead to health issues when workers are exposed to them externally or internally. Pyrophoric scale, such as iron sulfides that autoignite in the air, are another threat. Additionally, such deposits are formed in huge amounts that must be disposed of properly. Furthermore, some of the chemical agents used to manage such deposits are unsafe for the ecosystem.

The management strategy for deposits is based on preventing the formation of such deposits. The deposit prevention methods must be environmentally friendly as well, such as using green chemicals, or using other physical and operational methods that are efficient in preventing the deposits.

References

[1] E&P Forum/UNEP, Environmental Management in Oil and Gas Exploration and Production. Industry and Environment, 1997, pp. 1–67.

[2] G. Amoasah, The Potential Impacts of Oil and Gas Exploration and Production on the Coastal Zone of Ghana. An Ecosystem Services Approach, University Netherlands, Wageningen, 2010. Master's Thesis.

[3] S. Oppong, Common health, safety and environmental concerns in upstream oil and gas sector: implications for HSE management in Ghana, Academicus Int. Sci. J. 9 (2015) 93–106, https://doi.org/10.7336/academicus.2014.09.07.

[4] M. Cowie, K. Mously, O. Fageeha, R. Nassar, NORM management in the oil and gas industry, Ann. ICRP 41 (3–4) (2012) 318–331, https://doi.org/10.1016/j.icrp.2012.06.008.

[5] J.I. Gallin, F.P. Ognibene, L.L. Johnson, Principles and Practice of Clinical Research. Principles and Practice of Clinical Research, Elsevier, 2017, pp. 1–806, https://doi.org/10.1097/ccm.0b013e3181a5c47f.

[6] Managing Naturally Occurring Radioactive Material (NORM) in the Oil and Gas Industry, International Association of Oil and Gas Producers (IOGP), 2016. Report 412.

[7] Worksafe, Pyrophoric Materials Create Fire and Explosion Hazard, 2016. WS 2016-13.

[8] A. O'Donoghue, Pigging as a flow assurance solution: estimating pigging frequency for dewaxing, Pipeline World (2004) 13–17.

[9] M. Crabtree, D. Eslinger, P. Fletcher, M. Miller, A. Johnson, G. King, Fighting scale: removal and prevention, Oilfield Rev. 11 (03) (1999) 30–45.

[10] C. Igwebueze, L. Oduola, O. Smith, P. Vijn, A.G. Shepherd, Calcium naphthenate solid deposit identification and control in offshore Nigerian fields, in: Proceedings—SPE International Symposium on Oilfield Chemistry, vol. 1, 2013, pp. 156–168, https://doi.org/10.2118/164055-ms.

[11] T. Casanueva-Robles, T.R. Bott, The environmental effect of heat exchanger fouling: a case study, in: Proceedings of the 6th International Conference on Heat Exchanger Fouling and Cleaning, Challenges and Opportunities, vol. RP2, 2005, pp. 5–10.

[12] H. Müller-Steinhagen, M.R. Malayeri, A.P. Watkinson, Heat exchanger fouling: environmental impacts, Heat Transfer Eng. (2009), https://doi.org/10.1080/01457630902744119.

[13] A.L.H. Costa, V.B.G. Tavares, E.M. Queiroz, F.L.P. Pessoa, F.S. Liporace, S.G. Oliveira, J.L. Borges, Analysis of the environmental and economic impact of fouling in crude preheat trains for petroleum distillation, in: International Conference on Heat Exchanger Fouling and Cleaning, 2011, 2007, pp. 47–49.

[14] C.D. Ruppel, J.D. Kessler, The interaction of climate change and methane hydrates, Rev. Geophys. (2017), https://doi.org/10.1002/2016RG000534. Blackwell Publishing Ltd.

[15] J. Beyer, A. Goksøyr, D.Ø. Hjermann, J. Klungsøyr, Environmental effects of offshore produced water discharges: a review focused on the Norwegian continental shelf, Mar. Environ. Res. 162 (2020), 105155, https://doi.org/10.1016/j.marenvres.2020.105155.

[16] Oil & Gas UK, Knowledge Center: Chemical Discharges, Oil & Gas UK, 2009. Retrieved from: http://www.oilandgasuk.co.uk/knowledgecentre/chemical_discharges.cfm (Accessed December 2010) last visited October 2021.

[17] D. Sheahan, R. Rycroft, Y. Allen, A. Kenny, R. Irish, Contaminant Status of the North Sea. Assessment, 2001, p. 101. Retrieved from: http://www.offshoresea.org.uk/site/scripts/downloads.php?categoryID=9, last visited October 2021.

[18] D. Hasson, H. Shemer, A. Sher, State of the art of friendly 'green' scale control inhibitors: a review article, Ind. Eng. Chem. Res. 50 (12) (2011) 7601–7607, https://doi.org/10.1021/ie200370v.

[19] S. Jenkins, K. Harris, Biodegradation and testing of scale inhibitors, Chem. Eng. 118 (4) (2011).

[20] Human & Environmental Risk Assessment (HERA) on Ingredients of Household Cleaning Products: Phosphonates, HERA, 2004, pp. 1–114.

[21] D. Vavik, Implications of Gas Hydrates in Drilling and Completion, A Scientific Root Cause Analysis of the Deepwater Horizon Disaster, Doctoral Theses at NTNU, 2020.

[22] IAEA, Radiation protection and the management of radioactive waste in the oil and gas industry, Environ. Health Perspect. 34 (34) (2003) 130. Retrieved from: https://www.pub.iaea.org/MTCD/publications/PDF/TCS-40_web.pdf.

[23] M. Cowie, K. Mously, O. Fageeha, R. Nassar, NORM management in the oil & gas industry, in: AIP Conference Proceedings 1034, 2008, pp. 319–321, https://doi.org/10.1063/1.2991237.

Index

Note: Page numbers followed by *f* indicate figures, *t* indicate tables, and *b* indicate boxes.

B

C

D

E

F

G

O

P

X

Y

Z

9780323991186